Student Solutions Manual to accompany Physical Chemistry: Quanta, Matter, and Change

SECOND EDITION

Charles Trapp

Marshall Cady

Carmen Giunta

OXFORD
UNIVERSITY PRESS

W. H. Freeman and Company

OXFORD
UNIVERSITY PRESS

Great Clarendon Street, Oxford, OX2 6DP,
United Kingdom

Oxford University Press is a department of the University of Oxford. It furthers the University's objective of excellence in research, scholarship, and education by publishing worldwide. Oxford is a registered trade mark of Oxford University Press in the UK and in certain other countries

First edition 2009

British Library Cataloguing in Publication Data
Data available
Library of Congress Control Number: 2013950759

ISBN 978–0–19–870128–6

Printed in Great Britain by
CPI Group (UK) Ltd, Croydon, CR0 4YY

Published in the United States and Canada by
W.H. Freeman and Company
41 Madison Avenue
New York, NY 10010
www.whfreeman.com

ISBN-10: 1-4641-2442-6
ISBN-13: 978-1-4641-2442-6

Contents

Topic 1 Matter

Discussion questions

1.1 The **nuclear atomic model** consists of atomic number Z protons concentrated along with all atomic neutrons within the nucleus, an extremely small central region of the atom. Z electrons occupy **atomic orbitals**, which are voluminous regions of the atom that describe where electrons are likely to be found with no more than two electrons in any orbital. The electrostatic attraction binds the negatively charged electrons to the positively charged nucleus and the so-called strong interaction binds the protons and neutrons within the nucleus.

The atomic orbitals are arranged in shells around the nucleus, each shell being characterized by a **principal quantum number,** $n = 1, 2, 3, 4 \ldots$. A shell consists of n^2 individual orbitals, which are grouped together into n subshells. The **subshells**, and the orbitals they contain, are denoted s, p, d, and f. For all neutral atoms other than hydrogen, the subshells of a given shell have slightly different energies.

The **atomic number, Z,** is the number of protons in an atom. Theses protons are located within the nucleus. The **nucleon number, A,** which is also commonly called the **mass number**, is the total number of protons and neutrons in an atom. Theses nucleons are located within the nucleus.

1.3 A single bond is a shared pair of electrons between adjacent atoms within a molecule while a multiple bond involves the sharing of either two pairs of electrons (a double bond) or three pairs of electrons (a triple bond).

1.5 (a) The solid phase of matter has a shape that is independent of the container it occupies. It has a density compatible with the close proximity of either its constituent elemental atoms or its constituent molecules and, consequently, it has low compressibility. Constituent atoms, or molecules, are held firmly at specific lattice sites by relatively strong, net forces of attraction between neighboring constituents. Solids may be characterized by terms such as brittle, ductile, tensile strength, toughness, and hardness.

A liquid adopts the shape of the part of the container that it occupies; it can flow under the influence of gravitational attraction to occupy any shape. Like a sold, it has a density caused by the close proximity of either its constituent elemental atoms or its constituent molecules and it has low compressibility. Liquids can flow because the constituent atoms or molecules have enough average kinetic energy to overcome the attractive forces between neighboring constituents, thereby, making it possible for them to slip past immediate neighbors. This causes constituents to be placed randomly in contrast to the orderly array in crystals. Liquids are characterized by terms such as surface tension, viscosity, and capillary action. Liquids within a vertical, narrow tube exhibit a meniscus that is either concave-up or concave-down depending upon the nature of the attractive or repulsive forces between the liquid and the material of the tube.

Gases have no fixed shape or volume. They expand to fill the container volume. The constituent molecules move freely and randomly. A perfect gas has a total molecular volume that is negligibly small compared to the container volume and, because of the relatively large average distance between molecules, intermolecular attractive forces are negligibly small. Gases are compressible.

(b) Condensed forms of matter (liquids and solids) have relatively high densities because of the close proximity of constituent elemental atoms or constituent molecules; compressibility is low and attractive forces are strong between neighbors. Perfect gases have low densities and they are highly compressible; intermolecular forces of attraction are negligibly small.

Exercises

1.1(a)

	Example	Element	Ground-state Electronic Configuration
(a)	Group 2	Ca, calcium	$[Ar]4s^2$
(b)	Group 7	Mn, manganese	$[Ar]3d^5 4s^2$
(c)	Group 15	As, arsenic	$[Ar]3d^{10}4s^2 4p^3$

1.2(a)

(a) chemical formula and name: $MgCl_2$, magnesium chloride
ions: Mg^{2+} and Cl^-
oxidation numbers of the elements: magnesium, +2; chlorine, −1

(b) chemical formula and name: FeO, iron(II) oxide
ions: Fe^{2+} and O^{2-}
oxidation numbers of the elements: iron, +2; oxygen, −2

(c) chemical formula and name: Hg_2Cl_2, mercury(I) chloride
ions: Cl^- and Hg_2^{2+} (a polyatomic ion)
oxidation numbers of the elements: mercury, +1; chlorine, −1

1.3(a)

(a) Methylamine, a gas $(H_3C){-}NH_2$
(b) Acetoxime, a solid $(H_3C)_2C{=}NOH$
(c) Acetonitrile, a liquid $H_3CC{\equiv}N$

1.4(a)

(a) Sulfite anion, SO_3^{2-}

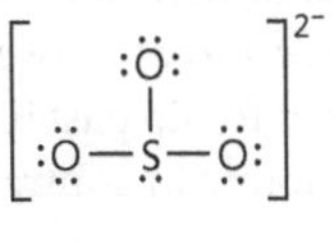

Alternatively, resonance structures may be drawn and, if desired, formal charges (shown in circles below) may be indicated.

$$\left[\overset{:\ddot{O}:^{\ominus}}{\underset{:\ddot{O}-\overset{\ominus}{S}-\ddot{O}:^{\ominus}}{\ominus}} \right]^{2-} \longleftrightarrow \left[\overset{\ddot{O}:}{\underset{:\ddot{O}^{\ominus}-S-\ddot{O}:^{\ominus}}{\parallel}} \right]^{2-} \longleftrightarrow \left[\overset{:\ddot{O}:^{\ominus}}{\underset{:O=S-\ddot{O}:^{\ominus}}{\mid}} \right]^{2-} \longleftrightarrow \left[\overset{:\ddot{O}:^{\ominus}}{\underset{:\ddot{O}^{\ominus}-S=O:}{\mid}} \right]^{2-}$$

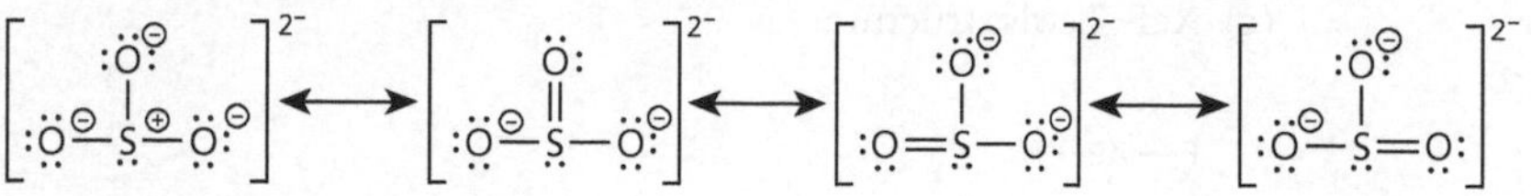

(b) Xenon tetrafluoride, XeF_4

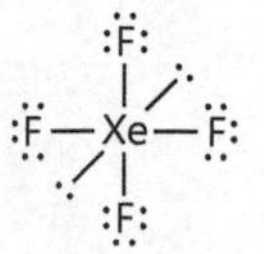

(c) White phosphorus, P_4

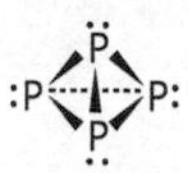

1.5(a)

Beryllium dichloride	$:\ddot{C}l—Be—\ddot{C}l:$
Chlorine dioxide	$:\ddot{O}—\dot{C}l—\ddot{O}:$
Nitrogen monoxide	$\dot{N}=\ddot{O}:$

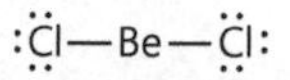

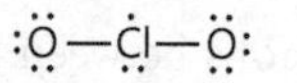

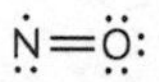

1.6(a)   Molecular and polyatomic ion shape are predicted by drawing the Lewis structure and applying the concepts of VSEPR theory.

(a) PCl_3 Lewis structure:

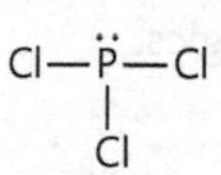

Orientations caused by repulsions between one lone pair and three bonding pair:

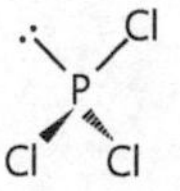

Molecular shape: trigonal pyramidal and bond angles somewhat smaller than 109.5°

(b) PCl_5 Lewis structure:

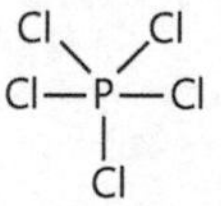

Orientations caused by repulsions between five bonding pair (no lone pair):

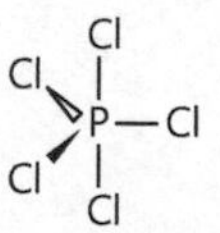

Molecular shape: trigonal bipyramidal with equatorial bond angles of 120° and axial bond angles of 90°

(c) XeF_2 Lewis structure:

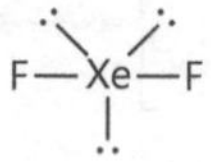

Orientations caused by repulsions between three lone pair and two bonding pair:

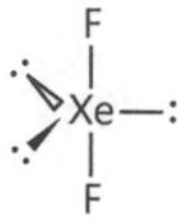

Molecular shape: linear with a 180° bond angle.

(d) XeF_4 Lewis structure:

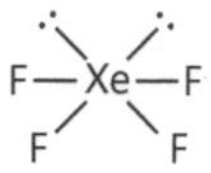

Orientations caused by repulsions between two lone pair and four bonding pair:

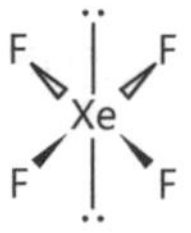

Molecular shape: square planar with a 90° bond angles.

1.7(a)

(a) $\overset{\delta^+}{C}$—$\overset{\delta^-}{Cl}$ (b) P—H Nonpolar or weakly polar.

(c) $\overset{\delta^+}{N}$—$\overset{\delta^-}{O}$

1.8(a)

(a) CO_2 is a linear, nonpolar molecule.

(b) SO_2 is a bent, polar molecule.

(c) N_2O is linear, polar molecule.

(d) SF_4 has a seesaw molecule and it is a polar molecule.

1.9(a) In the order of increasing dipole moment: CO_2, N_2O, SF_4, SO_2

1.10(a)

(a) Mass is an extensive property.

(b) Mass density is an intensive property.

(c) Temperature is an intensive property.

(d) Number density is an intensive property.

1.11(a)

(a) $n = \dfrac{m}{M} = 25.0\ \text{g}\left(\dfrac{1\ \text{mol}}{46.069\ \text{g}}\right) = \boxed{0.543\ \text{mol}}$ [1.3]

(b) $N = nN_{\text{A}} = 0.543\ \text{mol}\left(\dfrac{6.0221\times10^{23}\ \text{molecules}}{\text{mol}}\right) = \boxed{3.27\times10^{23}\ \text{molecules}}$

1.12(a)

(a) $m = n\,M = 10.0\ \text{mol}\left(\dfrac{18.015\ \text{g}}{\text{mol}}\right) = \boxed{180.\ \text{g}}$ [1.3]

(b) $\text{weight} = F_{\text{gravity on Earth}} = m\,g_{\text{Earth}}$

$$= (180.\ \text{g})\times(9.81\ \text{m s}^{-2})\times\left(\frac{1\ \text{kg}}{1\,000\ \text{g}}\right) = 1.77\ \text{kg m s}^{-2} = \boxed{1.77\ \text{N}}$$

1.13(a) $p = \dfrac{F}{A} = \dfrac{mg}{A}$

$$= \frac{(65\ \text{kg})\times(9.81\ \text{m s}^{-2})}{150\ \text{cm}^2}\left(\frac{1\ \text{cm}^2}{10^{-4}\ \text{m}^2}\right) = 4.3\times10^4\ \text{N m}^{-2} = 4.3\times10^4\ \text{Pa}\left(\frac{1\ \text{bar}}{10^5\ \text{Pa}}\right)$$

$$= \boxed{0.43\ \text{bar}}$$

1.14(a) $0.43\ \text{bar}\left(\dfrac{1\ \text{atm}}{1.01325\ \text{bar}}\right) = \boxed{0.42\ \text{atm}}$

1.15(a)

(a) $1.45\ \text{atm}\left(\dfrac{1.01325\times10^5\ \text{Pa}}{1\ \text{atm}}\right) = \boxed{1.47\times10^5\ \text{Pa}}$

(b) $1.45\ \text{atm}\left(\dfrac{1.01325\ \text{bar}}{1\ \text{atm}}\right) = \boxed{1.47\ \text{bar}}$

1.16(a) $T/\text{K} = \theta/{}^\circ\text{C} + 273.15 = 37.0 + 273.15 = 310.2$ [1.4]

$\boxed{T = 310.2\ \text{K}}$

1.17(a) To devise an equation relating the Fahrenheit and Celsius scales requires that consideration be given to both the degree size and a common reference point in each scale. Between the normal freezing point of water and its normal boiling point there is a degree scaling of 100 °C and 180 °F. Thus, the scaling ratio is 5 °C per 9 °F. A convenient reference point is provided by the normal freezing point of water, which is 0 °C and 32 °F. So to calculate the Celsius temperature from a given Fahrenheit temperature, 32 must be subtracted from the Fahrenheit temperature (θ_{F}) followed by scaling to the Celsius temperature (θ) with the ratio 5 °C/9 °F.

$$\boxed{\theta/{}^\circ\text{C} = \tfrac{5}{9}\times(\theta_{\text{F}}/{}^\circ\text{F} - 32) \quad \text{or} \quad \theta_{\text{F}}/{}^\circ\text{F} = \tfrac{9}{5}\times\theta/{}^\circ\text{C} + 32}$$

$\theta_{\text{F}}/{}^\circ\text{F} = \tfrac{9}{5}\times\theta/{}^\circ\text{C} + 32 = \tfrac{9}{5}\times78.5 + 32 = 173$

$\boxed{\theta_{\text{F}} = 173\ {}^\circ\text{F}}$

$$\theta_R/°R = \tfrac{9}{5}\times(T/K) = \tfrac{9}{5}\times(273.15) = 491.67$$

$$\theta_R = \boxed{491.67\ °R}$$

1.18(a) $$110\ \text{kPa}\times\left(\frac{(7.0+273.15)\ \text{K}}{(20.0+273.15)\ \text{K}}\right) = \boxed{105\ \text{kPa}}$$

1.19(a) $$pV = nRT\ [1.5] = \frac{mRT}{M}$$

$$M = \frac{mRT}{pV} = \frac{\rho RT}{p} \quad \text{where } \rho \text{ is the mass density [1.2]}$$

$$= \frac{(3.710\ \text{kg m}^{-3})(8.314\ \text{J K}^{-1}\ \text{mol}^{-1})(773.15\ \text{K})}{93.2\times10^3\ \text{Pa}} = 0.256\ \text{kg mol}^{-1} = 256\ \text{g mol}^{-1}$$

The molecular mass is eight times as large as the atomic mass of sulfur (32.06 g mol^{-1}) so the molecular formula is S_8.

1.20(a) $$n = 22\ \text{g}\times\left(\frac{1\ \text{mol}}{30.07\ \text{g}}\right) = 0.73\ \text{mol [1.3]}$$

$$p = \frac{nRT}{V}\ [1.5] = \frac{(0.73\ \text{mol})(8.314\ \text{J K}^{-1}\ \text{mol}^{-1})(298.15\ \text{K})}{1000.\ \text{cm}^3}\left(\frac{\text{cm}^3}{10^{-6}\ \text{m}^3}\right)$$

$$= 1.8\times10^6\ \text{Pa} = \boxed{1.8\ \text{MPa}}$$

1.21(a) $$n_{N_2} = 1.0\ \text{mole} \quad \text{and} \quad n_{H_2} = 2.0\ \text{mole}$$

$$p_{N_2} = \frac{n_{N_2}RT}{V}\ [1.5] = \frac{(1.0\ \text{mol})(8.314\ \text{J K}^{-1}\ \text{mol}^{-1})(278.15\ \text{K})}{10.0\ \text{dm}^3}\left(\frac{\text{dm}^3}{10^{-3}\ \text{m}^3}\right)$$

$$= \boxed{2.3\times10^5\ \text{Pa}}$$

Since there are twice as many moles of hydrogen as nitrogen, the hydrogen partial pressure must be twice as large.

$$p_{H_2} = \boxed{4.6\times10^5\ \text{Pa}} \qquad p = p_{N_2} + p_{H_2}\ [1.6] = \boxed{6.9\times10^5\ \text{Pa}}$$

Topic 2 Energy

Discussion questions

2.1 In the classical physics of Newton the force $\boldsymbol{F}$ acting on a body so as to cause acceleration $\boldsymbol{a}$ equals the mass m of the object multiplied by its acceleration: $\boldsymbol{F}=m\boldsymbol{a}$ where the bold faced symbols represent vector quantities. Coulomb's law describes the particularly important electrostatic force between two point charges Q_1 and Q_2 separated by the distance r:

$$F=\frac{Q_1Q_2}{4\pi\varepsilon_0 r^2} \text{ in a vacuum } (\varepsilon_0 \text{ is the vacuum permittivity})$$

and

$$F=\frac{Q_1Q_2}{4\pi\varepsilon_r\varepsilon_0 r^2} \text{ in a medium that has the relative permittivity } \varepsilon_r \text{ (formerly, dielectric constant)}$$

Convention assigns a negative value to the Coulomb force when it is attractive and a positive value when it is repulsive. The SI unit of force is the newton (N) and $1\,\text{N}=1\,\text{kg m s}^{-2}$.

An infinitesimal amount of **work**, dw, done on a body when it experiences an infinitesimal displacement, $d\boldsymbol{s}$, is defined by the scalar product (sometimes called the dot-product): $dw=-\boldsymbol{F}\cdot d\boldsymbol{s}$ where $\boldsymbol{F}$ is now the force that opposes the displacement. When the opposing force and displacement lie along the same direction, the z direction for example, the infinitesimal work done on the body simplifies to: $dw=-Fdz$ where F is the magnitude of the opposing force. The SI unit of work is the joule (J) and $1\,\text{J}=1\,\text{N m}=1\,\text{Pa m}^3=1\,\text{kg m}^2\,\text{s}^{-2}$.

Continuing the example of displacement along the z direction alone, integration of the infinitesimal work between the initial position "i" and the final position "f" gives the total work, w, done on the object: $w=-\int_{\text{i}}^{\text{f}} F\,dz$. Provided that the opposing force is constant over the displacement, the integral simplifies to give: $w=-F\times(z_{\text{f}}-z_{\text{i}})=-F\Delta z$ where $\Delta z=z_{\text{f}}-z_{\text{i}}$.

Energy is the capacity to do work. It is a property and the SI unit of energy, like the unit of work, is the joule. The **law of conservation of energy** states that the total energy E of an isolated system is conserved; that is, the total energy of an isolated system is a constant. Even in an isolated system, however, energy can be transferred from one location to another and transformed from one form to another. The transfers and transformations involve heat, work, gravitational potential energy, Coulomb potential energy, and electromagnetic radiation.

2.3 The second law of thermodynamics states that any spontaneous (that is, natural) change in an isolated system is accompanied by an increase in the entropy of the system. This tendency is commonly expressed by saying that the natural direction of change is accompanied by dispersal of energy from a localized region to a less organized form.

Yes, the entropy of a non-isolated system can decrease during a spontaneous process but for this to happen the total entropy ($S_{\text{system}} + S_{\text{environment}}$) must increase.

2.5 Kinetic molecular theory, a model for a perfect gas, assumes that the molecules, imagined as particles of negligible size, are in ceaseless, random motion and do not interact except during their brief collisions.

Exercises

2.1(a) $a = \mathrm{d}\upsilon/\mathrm{d}t = g$ so $\mathrm{d}\upsilon = g\,\mathrm{d}t$. The acceleration of free fall is constant near the surface of the Earth.

$$\int_{\upsilon=0}^{\upsilon(t)} \mathrm{d}\upsilon = \int_{t=0}^{t=t} g\,\mathrm{d}t$$
$$\upsilon(t) = gt$$

(a) $\upsilon(1.0\text{ s}) = (9.81\text{ m s}^{-2}) \times (1.0\text{ s}) = \boxed{9.81\text{ m s}^{-1}}$

$$E_{\mathrm{k}} = \tfrac{1}{2}m\upsilon^2 = \tfrac{1}{2}(0.0010\text{ kg}) \times (9.81\text{ m s}^{-1})^2 = \boxed{48\text{ mJ}}$$

(b) $\upsilon(3.0\text{ s}) = (9.81\text{ m s}^{-2}) \times (3.0\text{ s}) = \boxed{29.4\text{ m s}^{-1}}$

$$E_{\mathrm{k}} = \tfrac{1}{2}m\upsilon^2 = \tfrac{1}{2}(0.0010\text{ kg}) \times (29.4\text{ m s}^{-1})^2 = \boxed{0.43\text{ J}}$$

2.2(a) The terminal velocity occurs when there is a balance between the force exerted by the electric field and the force of frictional drag. It will be in the direction of the field and have the magnitude s_{terminal}.

$$ze\mathcal{E} = 6\pi\eta R s_{\text{terminal}}$$

$$\boxed{s_{\text{terminal}} = \frac{ze\mathcal{E}}{6\pi\eta R}}$$

2.3(a) $x(t) = A\sin(\omega t) + B\cos(\omega t)$

$$\begin{aligned}\frac{\mathrm{d}^2x}{\mathrm{d}t^2} &= A\omega\frac{\mathrm{d}\cos(\omega t)}{\mathrm{d}t} - B\omega\frac{\mathrm{d}\sin(\omega t)}{\mathrm{d}t}\\ &= A\omega \times (-\omega\sin(\omega t)) - B\omega \times (\omega\cos(\omega t))\\ &= -\omega^2(A\sin(\omega t) + B\cos(\omega t)) = -\omega^2 x \quad \text{where } \omega = (k/m)^{1/2}\end{aligned}$$

$$m\frac{\mathrm{d}^2x}{\mathrm{d}t^2} = -m\omega^2 x = -m \times \left(\frac{k}{m}\right)x = -kx$$

This confirms that $x(t)$ satisfies the harmonic oscillator equation of motion.

2.4(a) The electron acceleration is caused by the centripetal electrostatic force of attraction of the positively charged nucleus for the negatively charged electron. The force, and its resultant acceleration, always points from the electron to the nucleus. A path of constant radius r is only possible when the electron speed v creates a centrifugal force, mv^2/r, that exactly balances the centripetal force. The magnitude of the centripetal acceleration is found by equating the force of Newton's second law of motion and the centrifugal force.

$$F = ma = \frac{mv^2}{r}$$

$$a = \frac{v^2}{r} = \frac{(2\,188\times10^3\ \text{m s}^{-1})^2}{53\times10^{-12}\ \text{m}} = 9.0\times10^{22}\ \text{m s}^{-2}$$

2.5(a) The relationship between angular velocity ω and the speed of an electron in a stable circular path (i.e., an "orbit" of radius r) is found by recognizing that, when the electron travels through one orbit, it traverses 2π radians while traveling the distance $2\pi r$. Thus, $\omega = v \times (2\pi/2\pi r) = v/r$.

$$J = I\omega = (mr^2)\times\left(\frac{v}{r}\right) = mrv \quad [2.3]$$

$$= (9.10938\times10^{-31}\ \text{kg})\times(53\times10^{-12}\ \text{m})\times(2\,188\times10^3\ \text{m s}^{-1})$$

$$= 1.1\times10^{-34}\ \text{kg m}^2\ \text{s}^{-1} = 1.1\times10^{-34}\ \text{J s}$$

$$= (1.1\times10^{-34}\ \text{J s})\times\left(\frac{2\pi}{h}\right)\hbar \quad \text{where } \hbar = h/2\pi$$

$$= (1.1\times10^{-34}\ \text{J s})\times\left(\frac{2\pi}{6.63\times10^{-34}\ \text{J s}}\right)\hbar = \boxed{1.0\,\hbar}$$

2.6(a) $w = \tfrac{1}{2}kx^2$ where $x = R - R_e$ is the displacement from equilibrium

(a) $w = \tfrac{1}{2}(450\ \text{N m}^{-1})\times(10\times10^{-12}\ \text{m})^2 = 2.25\times10^{-20}\ \text{N m} = \boxed{2.25\times10^{-20}\ \text{J}}$

(b) $w = \tfrac{1}{2}(450\ \text{N m}^{-1})\times(20\times10^{-12}\ \text{m})^2 = 9.00\times10^{-20}\ \text{N m} = \boxed{9.00\times10^{-20}\ \text{J}}$

2.7(a) $E_k = e\Delta\phi$

$$\tfrac{1}{2}mv^2 = e\Delta\phi \quad \text{or} \quad v = \left(\frac{2e\Delta\phi}{m}\right)^{1/2}$$

$$v = \left(\frac{2(1.6022\times10^{-19}\ \text{C})\times(100\times10^3\ \text{V})}{9.10938\times10^{-31}\ \text{kg}}\right)^{1/2} = 1.88\times10^8\left(\frac{\text{C V}}{\text{kg}}\right)^{1/2} = 1.88\times10^8\left(\frac{\text{J}}{\text{kg}}\right)^{1/2}$$

$$= 1.88\times10^8\left(\frac{\text{kg m}^2\ \text{s}^{-2}}{\text{kg}}\right)^{1/2} = \boxed{1.88\times10^8\ \text{m s}^{-1}}$$

$$E = E_k = e\Delta\phi = e\times(100\ \text{kV}) = \boxed{100\ \text{keV}}$$

2.8(a) The work needed to separate two ions to infinity is identical to the Coulomb potential drop that occurs when the two ions are brought from an infinite separation, where the interaction potential equals zero, to a separation of r.

In a vacuum:

$$w=-V=-\left(\frac{Q_1Q_2}{4\pi\varepsilon_0 r}\right)[2.14]=-\left(\frac{(e)\times(-e)}{4\pi\varepsilon_0 r}\right)=\frac{e^2}{4\pi\varepsilon_0 r}$$

$$=\frac{(1.6022\times10^{-19}\ \mathrm{C})^2}{4\pi(8.85419\times10^{-12}\ \mathrm{J^{-1}\ C^2\ m^{-1}})\times(200\times10^{-12}\ \mathrm{m})}=\boxed{1.15\times10^{-18}\ \mathrm{J}}$$

In water:

$$w=-V=-\left(\frac{Q_1Q_2}{4\pi\varepsilon r}\right)=-\left(\frac{(e)\times(-e)}{4\pi\varepsilon r}\right)=\frac{e^2}{4\pi\varepsilon r}=\frac{e^2}{4\pi\varepsilon_r\varepsilon_0 r}\ [2.15]\ \text{where } \varepsilon_r=78 \text{ for water at 25°C}$$

$$=\frac{(1.6022\times10^{-19}\ \mathrm{C})^2}{4\pi(78)\times(8.85419\times10^{-12}\ \mathrm{J^{-1}\ C^2\ m^{-1}})\times(200\times10^{-12}\ \mathrm{m})}=\boxed{1.48\times10^{-20}\ \mathrm{J}}$$

2.9(a) We will model a solution by assuming that the LiH pair consists of the two point charge ions Li^+ and H^-. The electric potential will be calculated along the line of the ions.

$$\phi=\phi_{\mathrm{Li^+}}+\phi_{\mathrm{H^-}}\ [2.17]=\frac{e}{4\pi\varepsilon_0 r_{\mathrm{Li^+}}}+\frac{(-e)}{4\pi\varepsilon_0 r_{\mathrm{H^-}}}\ [2.16]=\frac{e}{4\pi\varepsilon_0}\left(\frac{1}{r_{\mathrm{Li^+}}}-\frac{1}{r_{\mathrm{H^-}}}\right)$$

$$\phi=\frac{1.6022\times10^{-19}\ \mathrm{C}}{4\pi(8.85419\times10^{-12}\ \mathrm{J^{-1}\ C^2\ m^{-1}})}\left(\frac{1}{200\times10^{-12}\ \mathrm{m}}-\frac{1}{150\times10^{-12}\ \mathrm{m}}\right)$$

$$=-2.40\ \mathrm{J\ C^{-1}}=-2.40\ \mathrm{C\ V\ C^{-1}}=\boxed{-2.40\ \mathrm{V}}$$

2.10(a) We will assume that the electric circuit has a negligibly small heat capacity so that all of the electrically generated heat energy is received by the water.

$$\Delta U_{\mathrm{H_2O}}=\text{energy dissipated by the electric circuit}$$

$$=I\Delta\phi\,\Delta t\ [2.20]$$

$$=(2.23\ \mathrm{A})\times(15.0\ \mathrm{V})\times(720\ \mathrm{s})=24.1\times10^3\ \mathrm{C\ s^{-1}\ V\ s}=\boxed{24.1\ \mathrm{kJ}}$$

$$\Delta U_{\mathrm{H_2O}}=(nC_{\mathrm{m}}\Delta T)_{\mathrm{H_2O}}\ [2.21]$$

$$\Delta T=\frac{\Delta U_{\mathrm{H_2O}}}{(nC_{\mathrm{m}})_{\mathrm{H_2O}}}=\frac{\Delta U_{\mathrm{H_2O}}}{(mC_{\mathrm{m}}/M)_{\mathrm{H_2O}}}=\frac{24.1\times10^3\ \mathrm{J}}{(200\ \mathrm{g})\times(75.3\ \mathrm{J\ K^{-1}\ mol^{-1}})/(18.02\ \mathrm{g\ mol^{-1}})}$$

$$=28.8\ \mathrm{K}=\boxed{28.8\ \mathrm{°C}}$$

2.11(a) $$\Delta T=\frac{\Delta U}{C}[2.21]=\frac{100.\ \mathrm{J}}{3.67\ \mathrm{J\ K^{-1}}}=\boxed{27.2\ \mathrm{K\ or\ 27.2\ °C}}$$

2.12(a) $$n=100.\ \mathrm{g}\times\left(\frac{1\ \mathrm{mol}}{207.2\ \mathrm{g}}\right)=0.483\ \mathrm{mol}$$

$$\Delta U=C\Delta T\ [2.21]=nC_{\mathrm{m}}\Delta T$$

$$=(0.483\ \mathrm{mol})(26.44\ \mathrm{J\ K^{-1}\ mol^{-1}})(10.0\ \mathrm{K})=\boxed{128\ \mathrm{J}}$$

2.13(a) $C_s = C_m / M = (111.46\ \text{J K}^{-1}\ \text{mol}^{-1}) \times \left(\frac{1\ \text{mol}}{46.069\ \text{g}}\right) = \boxed{2.4194\ \text{J K}^{-1}\ \text{g}^{-1}}$

2.14(a) $C_m = C_s M = (4.18\ \text{J K}^{-1}\ \text{g}^{-1}) \times \left(\frac{18.02\ \text{g}}{\text{mol}}\right) = \boxed{75.3\ \text{J K}^{-1}\ \text{mol}^{-1}}$

2.15(a) Dividing Eq. 2.23 by the number of moles n and using the molar properties (intensive) H_m, U_m, and V_m, yields the equation:

$$H_m - U_m = pV_m$$

Substitution of the perfect gas equation of state $pV_m = RT$ yields:

$$H_m - U_m = RT = (8.3145\ \text{J mol}^{-1}\ \text{K}^{-1}) \times (1000\ \text{K}) = \boxed{8.3145\ \text{kJ mol}^{-1}}$$

2.16(a) $S_{H_2O(g)} > S_{H_2O(l)}$

2.17(a) $S_{Fe(3000\ K)} > S_{Fe(300\ K)}$

2.18(a) In a state of dynamic equilibrium, which is the character of all chemical equilibria, the forward and reverse reactions are occurring at the same rate and there is no net tendency to change in either direction. Examples:

$2\ SO_2(g) + O_2(g) \rightleftharpoons 2\ SO_3(g)$ Addition of oxygen shifts the equilibrium to the right. An increase in pressure also shifts it to the right so as to reduce the number of moles of gas.

$CaCO_3(s) \rightleftharpoons CaO(s) + CO_2(g)$ Addition of carbon dioxide shifts the equilibrium to the left. An increase in pressure also shifts it of the left so as to reduce the number of moles of gas.

$CO(g) + H_2O(g) \rightleftharpoons CO_2(g) + H_2(g)$ Addition of carbon monoxide shifts the equilibrium to the right. An increase in pressure has no effect on the equilibrium because there are equal numbers of moles of gas on left and right.

2.19(a) $$\frac{N_i}{N_j} = e^{-(E_i - E_j)/kT} = e^{-\Delta E_{ij}/kT} \quad [2.25a]$$

(a) $$\frac{N_2}{N_1} = e^{-(1.0\ \text{eV}) \times (1.602\times10^{-19}\ \text{J eV}^{-1}) \times (1.381\times10^{-23}\ \text{J K}^{-1})^{-1} \times (300\ \text{K})^{-1}} = 1.6\times10^{-17}$$

(b) $$\frac{N_2}{N_1} = e^{-(1.0\ \text{eV}) \times (1.602\times10^{-19}\ \text{J eV}^{-1}) \times (1.381\times10^{-23}\ \text{J K}^{-1})^{-1} \times (3\,000\ \text{K})^{-1}} = 0.021$$

2.20(a) $$\lim_{T\to0}\left(\frac{N_{\text{upper}}}{N_{\text{lower}}}\right) = \lim_{T\to0}\left(e^{-\Delta E/kT}\right)\ [2.25a] = e^{-\infty} = 0$$

In the limit of the absolute zero of temperature all particles occupy the lower state. The upper state is empty.

2.21(a) $$\Delta E = E_{\text{upper}} - E_{\text{lower}} = \tilde{\nu}hc = (2500\ \text{cm}^{-1})(6.626\times10^{-34}\ \text{J s})(3.000\times10^{10}\ \text{cm s}^{-1})$$

$$= 4.970\times10^{-20}\ \text{J}$$

$$\frac{N_{\text{upper}}}{N_{\text{lower}}} = e^{-\Delta E/kT} \ [\text{F.9}] = e^{-(4.970\times10^{-20}\ \text{J})/\{(1.381\times10^{-23}\ \text{J K}^{-1})\times(293\ \text{K})\}} = \boxed{4.631\times10^{-6}}$$

The ratio $N_{\text{upper}}/N_{\text{lower}}$ is so small that the population of the upper level is approximately zero.

2.22(a) Molecules can survive for long periods without undergoing chemical reaction at low temperatures when few molecules have the requisite speed and corresponding kinetic energy to promote excitation and bond breakage during collisions.

2.23(a) $\upsilon_{\text{mean}} \propto (T/M)^{1/2}$ [2.26]

$$\frac{\upsilon_{\text{mean}}(T_2)}{\upsilon_{\text{mean}}(T_1)} = \frac{(T_2/M)^{1/2}}{(T_1/M)^{1/2}} = \left(\frac{T_2}{T_1}\right)^{1/2}$$

$$\frac{\upsilon_{\text{mean}}(313\ \text{K})}{\upsilon_{\text{mean}}(273\ \text{K})} = \left(\frac{313\ \text{K}}{273\ \text{K}}\right)^{1/2} = \boxed{1.07}$$

2.24(a) $\upsilon_{\text{mean}} \propto (T/M)^{1/2}$ [2.26]

$$\frac{\upsilon_{\text{mean}}(M_2)}{\upsilon_{\text{mean}}(M_1)} = \frac{(T/M_2)^{1/2}}{(T/M_1)^{1/2}} = \left(\frac{M_1}{M_2}\right)^{1/2}$$

$$\frac{\upsilon_{\text{mean}}(\text{N}_2)}{\upsilon_{\text{mean}}(\text{CO}_2)} = \left(\frac{44.0\ \text{g mol}^{-1}}{28.0\ \text{g mol}^{-1}}\right)^{1/2} = \boxed{1.25}$$

2.25(a) A gaseous argon atom has three translational degrees of freedom (the components of motion in the *x*, *y*, and *z* directions). Consequently, the equipartition theorem assigns a mean energy of $\tfrac{3}{2}kT$ to each atom. The molar internal energy is

$$U_{\text{m}} = \tfrac{3}{2}N_{\text{A}}kT = \tfrac{3}{2}RT = \tfrac{3}{2}(8.3145\ \text{J mol}^{-1}\ \text{K}^{-1})(298\ \text{K}) = 3.72\ \text{kJ mol}^{-1}$$

$$U = nU_{\text{m}} = mM^{-1}U_{\text{m}} = (5.0\ \text{g})\left(\frac{1\ \text{mol}}{39.95\ \text{g}}\right)\left(\frac{3.72\ \text{kJ}}{\text{mol}}\right) = \boxed{0.47\ \text{kJ}}$$

2.26(a)

(a) A gaseous, linear, carbon dioxide molecule has three quadratic translational degrees of freedom (the components of motion in the *x*, *y*, and *z* directions) but it has only two rotational quadratic degrees of freedom because there is no rotation along the internuclear line. There is a total of five quadratic degrees of freedom for the molecule. Consequently, the equipartition theorem assigns a mean energy of $\tfrac{5}{2}kT$ to each molecule. The molar internal energy is

$$U_{\text{m}} = \tfrac{5}{2}N_{\text{A}}kT = \tfrac{5}{2}RT = \tfrac{5}{2}(8.3145\ \text{J mol}^{-1}\ \text{K}^{-1})(293\ \text{K}) = 6.09\ \text{kJ mol}^{-1}$$

$$U = nU_{\text{m}} = mM^{-1}U_{\text{m}} = (10.0\ \text{g})\left(\frac{1\ \text{mol}}{44.0\ \text{g}}\right)\left(\frac{6.09\ \text{kJ}}{\text{mol}}\right) = \boxed{1.38\ \text{kJ}}$$

(b) A gaseous, nonlinear, methane molecule has three quadratic translational degrees of freedom (the components of motion in the *x*, *y*, and *z* directions) and three quadratic rotational degrees of freedom. Consequently, the equipartition theorem assigns a mean energy of $^6/_2\, kT$ to each molecule. The molar internal energy is

$$U_m = \tfrac{6}{2} N_A kT = 3RT = 3(8.3145\ \text{J mol}^{-1}\ \text{K}^{-1})(293\ \text{K}) = 7.31\ \text{kJ mol}^{-1}$$

$$U = nU_m = mM^{-1}U_m = (10.0\ \text{g})\left(\frac{1\ \text{mol}}{16.04\ \text{g}}\right)\left(\frac{7.31\ \text{kJ}}{\text{mol}}\right) = \boxed{4.56\ \text{kJ}}$$

2.27(a) See exercise 2.25(a) for the description of the molar internal energy of argon.

$$C_{V,m} = \frac{\partial U_m}{\partial T} = \frac{\partial(\frac{3}{2}RT)}{\partial T} = \tfrac{3}{2}R = \tfrac{3}{2}(8.3145\ \text{J mol}^{-1}\ \text{K}^{-1}) = \boxed{12.47\ \text{J mol}^{-1}\ \text{K}^{-1}}$$

2.28(a) See exercise 2.26(a) for the description of the molar internal energies of carbon dioxide and methane.

(a) $U_m = \tfrac{5}{2}RT$ for carbon dioxide

$$C_{V,m} = \frac{\partial U_m}{\partial T} = \frac{\partial(\frac{5}{2}RT)}{\partial T} = \tfrac{5}{2}R = \tfrac{5}{2}(8.3145\ \text{J mol}^{-1}\ \text{K}^{-1}) = \boxed{20.79\ \text{J mol}^{-1}\ \text{K}^{-1}}$$

(b) $U_m = 3RT$ for methane

$$C_{V,m} = \frac{\partial U_m}{\partial T} = \frac{\partial(3RT)}{\partial T} = 3R = 3(8.3145\ \text{J mol}^{-1}\ \text{K}^{-1}) = \boxed{24.94\ \text{J mol}^{-1}\ \text{K}^{-1}}$$

Topic 3 **Waves**

Discussion questions

3.1 The study of wave motion includes the examination of water waves, sound waves, electromagnetic waves (microwave, infrared, visible, ultraviolet, etc.), radio waves, seismic waves, and de Broglie waves. Mechanical waves, generated by a source that causes particulate displacements within an elastic medium, are **transverse** when the particles experience displacements perpendicular to the direction of propagation while they are **longitudinal** when the particles are displaced parallel to the direction of wave propagation. Water molecules are displaced in ocean waves. Gas molecules are displaced in sound waves. Differential equations are used to describe the wave motion in space and time.

Exercises

3.1(a) $$c_{H_2O} = \frac{c}{n_r}\ [3.4] = \frac{3.00\times10^{8}\ \text{m s}^{-1}}{1.33} = \boxed{2.26\times10^{8}\ \text{m s}^{-1}}$$

3.2(a) $$\lambda = \frac{1}{\tilde{\nu}}\ [3.5] = \frac{1}{2\,500\ \text{cm}^{-1}}\left(\frac{10^{6}\ \mu\text{m}}{10^{2}\ \text{cm}}\right) = \boxed{4.00\ \mu\text{m}}$$

$$\nu = \frac{c}{\lambda}\ [3.1] = \frac{3.00\times10^{8}\ \text{m s}^{-1}}{4.00\times10^{-6}\ \text{m}} = 7.50\times10^{13}\ \text{s}^{-1} = \boxed{7.50\times10^{13}\ \text{Hz}}$$

Focus 1: Integrated activities

F.1

(a) Plots of the Maxwell–Boltzmann distribution at $T = 200$ K, 500 K, 1 000 K, and 2 000 K with $M = 100$ g mol^{-1} are calculated in the following Mathcad Prime 2 worksheet. The worksheet looks much the same in all versions of Mathcad. How would you change the worksheet so as to plot the distribution at the fix temperature $T = 300$ K for a range of molar masses such as 20 g mol^{-1}, 50 g mol^{-1}, and 100 g mol^{-1}?

$M := 100 \cdot gm \cdot mol^{-1}$ $\quad R := 8.3145 \cdot J \cdot K^{-1} \cdot mol^{-1}$

$f(\upsilon, T) := 4 \cdot \pi \cdot \left(\frac{M}{2 \cdot \pi \cdot R \cdot T} \right)^{\frac{3}{2}} \cdot \upsilon^2 \cdot e^{\frac{-M \cdot \upsilon^2}{2 \cdot R \cdot T}}$ $\quad i := 0..1000$ $\quad \upsilon_i := i \cdot m \cdot s^{-1}$

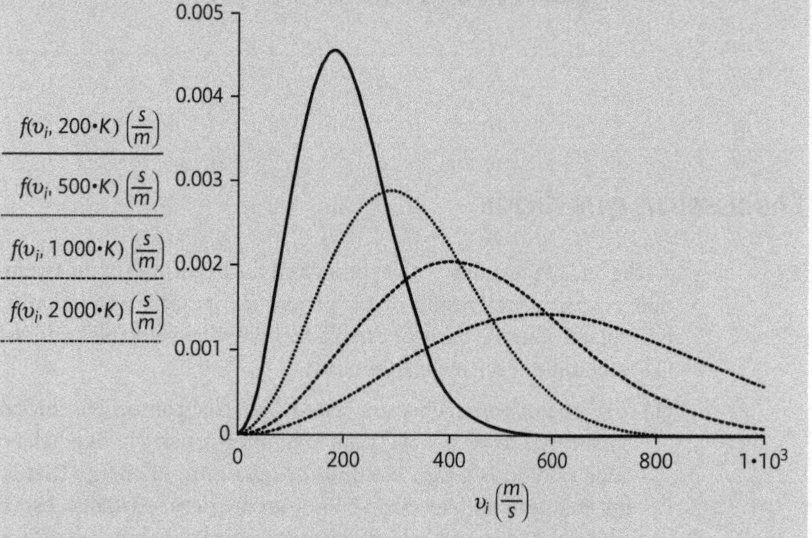

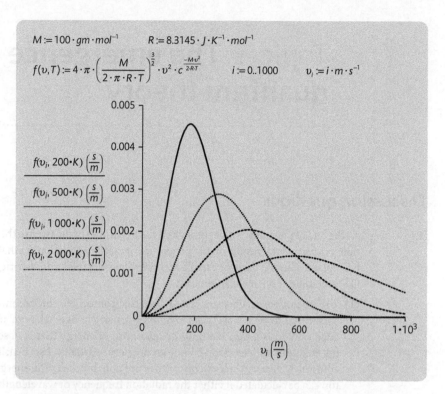

(b) The fraction of molecules with speeds in the range 100 m s^{-1} to 200 m s^{-1} at 300 K and 1 000 K is calculated with a numerical integration of the Maxwell–Boltzmann distribution over the desired range of speeds at fixed temperature. The following integration calculations are a continuation of the above Mathcad worksheet. Note that the integration over all possible speeds (0-∞ m s^{-1}) equals 1 because the sum of all fractions equals 1.

$fraction(T) := \int_{100 \cdot m \cdot s^{-1}}^{200 \cdot m \cdot s^{-1}} f(\upsilon, T) \, d\upsilon$

$fraction(300 \cdot K) = 0.281$ $\quad fraction(1000 \cdot K) = 0.066$

Topic 4 **The emergence of quantum theory**

Discussion questions

4.1 At the end of the nineteenth century and the beginning of the twentieth, there were many experimental results on the properties of matter and radiation that could not be explained on the basis of established physical principles and theories. Here we list only some of the most significant.

(1) The photoelectric effect revealed that electromagnetic radiation, classically considered to be a wave, also exhibits the particle-like behavior of photons. Each photon is a discrete unit, or quantum, of energy that is absorbed during collisions with electrons. Photons are never partially absorbed. They either completely give up their energy or they are not absorbed. The energy of a photon can be calculated if either the radiation frequency or wavelength is known: $\boldsymbol{E_{\text{photon}} = h\nu = hc/\lambda}$.

(2) Absorption and emission spectra indicated that atoms and molecules can only absorb or emit discrete packets of energy (i.e., photons). This means that an atom or molecule has specific, allowed energy levels and we say that their energies are quantized. During a spectroscopic transition the atom or molecule gains or loses the energy ΔE by either absorption of a photon or emission of a photon, respectively. Thus, spectral lines must satisfy the **Bohr frequency condition:** $\boldsymbol{\Delta E = h\nu}$.

(3) Neutron and electron diffraction studies indicated that these particles also possess wave-like properties of constructive and destructive interference. The joint particle and wave character of matter and radiation is called **wave-particle duality**. The **de Broglie relation**, $\lambda_{\text{de Broglie}} = h/p$, connects the wave character of a particle ($\lambda_{\text{de Broglie}}$) with its particulate momentum (p).

Evidence that resulted in the development of quantum theory also included:

(4) The energy density distribution of blackbody radiation as a function of wavelength.

(5) The heat capacities of monatomic solids such as copper metal.

Exercises

4.1(a) $\Delta E = h\nu = h/T$ where the period T equals $1/\nu$ $(T = 1/\nu)$

(a) $\Delta E = \dfrac{6.626\times10^{-34}\,\text{J s}}{20\times10^{-15}\,\text{s}} = 3.3\times10^{-20}\,\text{J} = \boxed{33\,\text{zJ}}$

This corresponds to $N_A \times (3.3\times10^{-20}\,\text{J}) = \boxed{20.\,\text{kJ mol}^{-1}}$

(b) $\Delta E = \dfrac{6.626\times10^{-34}\,\text{J s}}{2.0\,\text{s}} = \boxed{3.3\times10^{-34}\,\text{J}}$, which corresponds to

$N_A \times (3.3\times10^{-34}\,\text{J}) = \boxed{0.20\,\text{nJ mol}^{-1}}$

This is much too small to be measurable, thereby, demonstrating that for practical purposes the energy of a macroscopic object is a non-quantized, continuous variable.

4.2(a) $E_k = \frac{1}{2}m_e\upsilon^2 = h\nu - \Phi = \dfrac{hc}{\lambda} - \Phi$ [4.5] and $\upsilon = \left\{\dfrac{2}{m_e}\times E_k\right\}^{1/2}$

$\Phi = 2.14\,\text{eV} = (2.14)\times(1.602\times10^{-19}\,\text{J}) = 3.43\times10^{-19}\,\text{J}$

(a) $\dfrac{hc}{\lambda} = \dfrac{(6.626\times10^{-34}\,\text{J s})\times(2.998\times10^{8}\,\text{m s}^{-1})}{580\times10^{-9}\,\text{m}} = 3.42\times10^{-19}\,\text{J}$

The photon energy is very nearly equal to the value of the work function. Consequently, if photoejection occurs, the electrons will have $\boxed{\text{no kinetic energy and zero speed}}$.

(b) $\dfrac{hc}{\lambda} = \dfrac{(6.626\times10^{-34}\,\text{J s})\times(2.998\times10^{8}\,\text{m s}^{-1})}{250\times10^{-9}\,\text{m}} = 7.95\times10^{-19}\,\text{J}$

$E_k = \frac{1}{2}m\upsilon^2 = (7.95-3.43)\times10^{-19}\,\text{J} = 4.52\times10^{-19}\,\text{J} = \boxed{0.452\,\text{aJ}}$

$\upsilon = \left(\dfrac{2E_k}{m}\right)^{1/2} = \left(\dfrac{(2)\times(4.52\times10^{-19}\,\text{J})}{9.109\times10^{-31}\,\text{kg}}\right)^{1/2} = \boxed{996\,\text{km s}^{-1}}$

4.3(a) $E_k = \frac{1}{2}m_e\upsilon^2 = h\nu - \Phi = \dfrac{hc}{\lambda} - \Phi$ [4.5] or $\Phi = \dfrac{hc}{\lambda} - E_k$

$\Phi = \dfrac{hc}{\lambda} - E_k = \dfrac{(6.626\times10^{-34}\,\text{J s})\times(2.998\times10^{8}\,\text{m s}^{-1})}{465\times10^{-9}\,\text{m}} - (2.11\,\text{eV})\times\left(\dfrac{1.602\times10^{-19}\,\text{J}}{\text{eV}}\right)$

$= 8.92\times10^{-20}\,\text{J}$

The maximum wavelength needed for photoejection leaves the electron with zero kinetic energy. If the wavelength is longer, the radiation has insufficient energy for photoejection. If the absorbed wavelength is shorter, the electron will have a non-zero kinetic energy after photoejection. In the former case,

$$\lambda_{\max} = \frac{hc}{\Phi} = \frac{(6.626\times10^{-34}\ \mathrm{J\ s})\times(2.998\times10^{8}\ \mathrm{m\ s^{-1}})}{8.92\times10^{-20}\ \mathrm{J}} = 2.23\times10^{-6}\ \mathrm{m} = \boxed{2.23\ \mu\mathrm{m}}$$

4.4(a) $$E_{\text{binding}} = E_{\text{photon}} - E_{\text{k}} = h\nu - \tfrac{1}{2}m_{\text{e}}\upsilon^2 = \frac{hc}{\lambda} - \tfrac{1}{2}m_{\text{e}}\upsilon^2$$

$$\begin{aligned} E_{\text{binding}} &= \frac{hc}{\lambda} - \tfrac{1}{2}m_{\text{e}}\upsilon^2 \\ &= \frac{(6.626\times10^{-34}\ \mathrm{J\ s})\times(2.998\times10^{8}\ \mathrm{m\ s^{-1}})}{150\times10^{-12}\ \mathrm{m}} - \tfrac{1}{2}(9.109\times10^{-31}\ \mathrm{kg})\times(2.14\times10^{7}\ \mathrm{m\ s^{-1}})^2 \\ &= (1.12\times10^{-15}\ \mathrm{J})\times\left(\frac{1\ \mathrm{eV}}{1.602\times10^{-19}\ \mathrm{J}}\right) = \boxed{6.96\ \mathrm{keV}}\ \text{without a relativist mass correction} \end{aligned}$$

Note: The photoelectron is moving at 7.1% of the speed of light. So, in order to calculate a more accurate value of the binding energy, it would be necessary to use the relativistic mass in place of the rest mass.

$$m = \frac{m_{\text{e}}}{\left(1-(\upsilon/c)^2\right)^{1/2}} = \frac{9.109\times10^{-31}\ \mathrm{kg}}{\left(1-\left(2.14\times10^{7}\ \mathrm{m\ s^{-1}}/2.998\times10^{8}\ \mathrm{m\ s^{-1}}\right)^2\right)^{1/2}} = 9.13\times10^{-31}\ \mathrm{kg}$$

$$\begin{aligned} E_{\text{binding}} &= \frac{hc}{\lambda} - \tfrac{1}{2}m_{\text{e}}\upsilon^2 \\ &= \frac{(6.626\times10^{-34}\ \mathrm{J\ s})\times(2.998\times10^{8}\ \mathrm{m\ s^{-1}})}{150\times10^{-12}\ \mathrm{m}} - \frac{1}{2}(9.13\times10^{-31}\ \mathrm{kg})\times(2.14\times10^{7}\ \mathrm{m\ s^{-1}})^2 \\ &= (1.12\times10^{-15}\ \mathrm{J})\times\left(\frac{1\ \mathrm{eV}}{1.602\times10^{-19}\ \mathrm{J}}\right) = \boxed{6.96\ \mathrm{keV}}\ \text{with the relativistic mass correction.} \end{aligned}$$

The relativistic mass correction did not make a difference in this exercise.

4.5(a) $$E = h\nu = \frac{hc}{\lambda}, \qquad E(\text{per mole}) = N_{\text{A}}E = \frac{N_{\text{A}}hc}{\lambda}$$

$$hc = (6.62608\times10^{-34}\ \mathrm{J\,s})\times(2.99792\times10^{8}\ \mathrm{m\,s^{-1}}) = 1.986\times10^{-25}\ \mathrm{J\,m}$$

$$N_{\text{A}}hc = (6.02214\times10^{23}\ \mathrm{mol^{-1}})\times(1.986\times10^{-25}\ \mathrm{J\,m}) = 0.1196\ \mathrm{J\ m\ mol^{-1}}$$

$$\text{Thus, } E = \frac{1.986\times10^{-25}\ \mathrm{J\,m}}{\lambda}; \qquad E(\text{per mole}) = \frac{0.1196\ \mathrm{J\ m\ mol^{-1}}}{\lambda}$$

We can therefore draw up the following table.

	λ/nm	E/J	E/(kJ mol^{-1})
(a)	620	3.20×10^{-19}	193
(b)	570	3.49×10^{-19}	210
(c)	380	5.23×10^{-19}	315

4.6(a) Power is energy per unit time; hence

$$\frac{N}{\Delta t}=\frac{P}{h\nu}[P=\text{power in watts, } 1\text{ W}=1\text{ J s}^{-1}]=\frac{P\lambda}{hc}$$

$$=\frac{P\lambda}{(6.626\times10^{-34}\text{ Js})\times(2.998\times10^{8}\text{ ms}^{-1})}=\frac{(P/\text{W})\times(\lambda/\text{nm})\text{s}^{-1}}{1.99\times10^{-16}}$$

$$=5.03\times10^{15}(P/\text{W})\times(\lambda/\text{nm})\text{s}^{-1}$$

(a) $N/\Delta t=(5.03\times10^{15})\times(10)\times(590)\text{ s}^{-1}=\boxed{3.0\times10^{19}\text{ s}^{-1}}$

(b) $N/\Delta t=(5.03\times10^{15})\times(250)\times(590)\text{ s}^{-1}=\boxed{7.4\times10^{20}\text{ s}^{-1}}$

4.7(a) $\lambda=\frac{h}{p}=\frac{h}{m\upsilon}$ [4.6]

(a) $\lambda=\frac{6.626\times10^{-34}\text{ Js}}{(2.0\times10^{-3}\text{ kg})\times(1.0\times10^{-2}\text{ ms}^{-1})}=\boxed{3.3\times10^{-29}\text{ m}}$

For a macroscopic object (e.g., 2 g) the de Broglie wavelength is much too small to be observed. Its particulate character predominates and energy levels are extraordinarily close, thereby, producing an apparent continuum of possible energies.

(b) $\lambda=\frac{6.626\times10^{-34}\text{ Js}}{(2.0\times10^{-3}\text{ kg})\times(2.5\times10^{5}\text{ m s}^{-1})}=\boxed{1.3\times10^{-36}\text{ m}}$

(c) $\lambda=\frac{6.626\times10^{-34}\text{ Js}}{(4.003)\times(1.6605\times10^{-27}\text{ kg})\times(1\,000\text{ m s}^{-1})}=\boxed{99.7\text{ pm}}$

Comment. The wavelengths in (a) and (b) are smaller than the dimensions of any known particle, whereas that in (c) is comparable to atomic dimensions.

Question. For stationary particles, $\upsilon=0$, corresponding to an infinite wavelength. What meaning can be ascribed to this result?

4.8(a) $\lambda=\frac{h}{p}=\frac{h}{m_e\upsilon}$ [4.6] so $\upsilon=\frac{h}{m_e\lambda}=\frac{6.626\times10^{-34}\text{ J s}}{(9.109\times10^{-31}\text{ kg})\times(100\times10^{-12}\text{ m})}$

$=\boxed{7.27\times10^{6}\text{ m s}^{-1}}$

When an electron is accelerated from rest through a Coulomb potential difference $\Delta\phi$, $E_k=e\Delta\phi$ Thus,

$$\Delta\phi=\frac{E_k}{e}=\frac{\tfrac{1}{2}m_e\upsilon^2}{e}$$

$$=\frac{\tfrac{1}{2}(9.109\times10^{-31}\text{ kg})\times(7.27\times10^{6}\text{ m s}^{-1})^2}{1.602\times10^{-19}\text{ C}}\left(\frac{1\text{ C V}}{\text{J}}\right)=\boxed{150\text{ V}}$$

Low energy electron diffraction (LEED) uses a beam of electrons of a well-defined low energy (typically in the range 20-200 eV) incident to a single crystal with a well-ordered surface structure in order to generate a back-scattered electron diffraction pattern.

4.9(a) $$E_k = \tfrac{1}{2}m_e\upsilon^2 = h\nu = \frac{hc}{\lambda} \quad \text{or} \quad \upsilon = \left(\frac{2hc}{m_e\lambda}\right)^{1/2}$$

$$\upsilon = \left(\frac{2(6.626\times10^{-34}\ \text{J s})\times(2.998\times10^{8}\ \text{m s}^{-1})}{(9.109\times10^{-31}\ \text{kg})\times(150\times10^{-9}\ \text{m})}\right)^{1/2} = \boxed{1.71\times10^{6}\ \text{m s}^{-1}}$$

Problems

4.1 (a) A plot of Planck's spectral energy density at 298 K is shown in Figure 4.1.

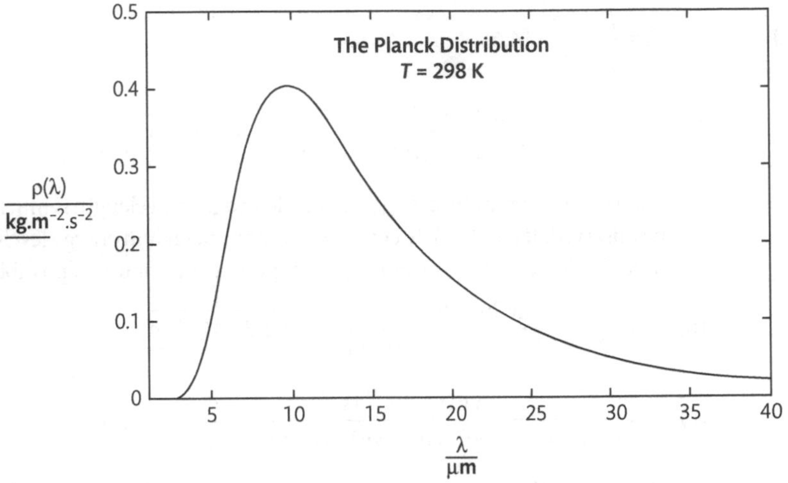

Figure 4.1

(b) $$\lim_{\lambda\to0}\rho(\lambda) = \lim_{\lambda\to0}\left\{\frac{8\pi hc}{\lambda^5}\left(\frac{1}{e^{hc/\lambda kT}-1}\right)\right\} = \lim_{\lambda\to0}\left\{\frac{8\pi hc}{\lambda^5}\left(\frac{1}{e^{hc/\lambda kT}}\right)\right\}$$

In the limit as $\lambda \to 0$, $e^{hc/\lambda kT} \to \infty$ more rapidly than $\lambda^5 \to 0$. Hence, $\rho \to 0$ as $\lambda \to 0$. To see this in a series of mathematical manipulations, let $a = hc/kT$ and examine the denominator of the above expression in the limit as $\lambda \to 0$.

$$\lim_{\lambda\to0}\rho(\lambda) = 8\pi hc\lim_{\lambda\to0}\left\{\frac{1}{\lambda^5 e^{a/\lambda}}\right\} = 8\pi hc\lim_{\lambda\to0}\left\{\frac{1}{e^{\ln\lambda^5}e^{a/\lambda}}\right\} = 8\pi hc\lim_{\lambda\to0}\left\{\frac{1}{e^{\lambda\ln\lambda^5}e^{a}}\right\}^{1/\lambda}$$

$$= 8\pi hc\lim_{\lambda\to0}\left\{\frac{1}{e^{\ln\lambda^{5\lambda}}e^{a}}\right\}^{1/\lambda} = 8\pi hc\lim_{\lambda\to0}\left\{\frac{1}{e^{\lim_{\lambda\to0}(\ln\lambda^{5\lambda})}e^{a}}\right\}^{1/\lambda} = 8\pi hc\lim_{\lambda\to0}\left\{\frac{1}{e^{\lim_{\lambda\to0}(\ln\lambda^{0})}e^{a}}\right\}^{1/\lambda}$$

$$= 8\pi hc\lim_{\lambda\to0}\left\{\frac{1}{e^{\ln1}e^{a}}\right\}^{1/\lambda} = 8\pi hc\lim_{\lambda\to0}\left\{\frac{1}{e^{0}e^{a}}\right\}^{1/\lambda} = 8\pi hc\lim_{\lambda\to0}\left\{\frac{1}{e^{a}}\right\}^{1/\lambda} = 8\pi hc\lim_{\lambda\to0}\left\{\frac{1}{e^{a/\lambda}}\right\}$$

$$= \lim_{\lambda\to0}\left\{\frac{8\pi hc}{e^{hc/\lambda kT}}\right\} = 0$$

(c) As λ increases, $hc/\lambda kT$ decreases, and at very long wavelength $hc/\lambda kT \ll 1$. Hence we can expand the exponential in a power series. Let $x = hc/\lambda kT$, then

$$e^x = 1 + x + \frac{1}{2!}x^2 + \frac{1}{3!}x^3 + \cdots$$

$$\rho = \frac{8\pi hc}{\lambda^5}\left[\frac{1}{1 + x + \frac{1}{2!}x^2 + \frac{1}{3!}x^3 + \cdots - 1}\right]$$

$$\lim_{\lambda\to\infty} \rho = \frac{8\pi hc}{\lambda^5}\left[\frac{1}{1 + x - 1}\right] = \frac{8\pi hc}{\lambda^5}\left(\frac{1}{hc/\lambda kT}\right)$$

$$= \frac{8\pi kT}{\lambda^4}$$

This is the Rayleigh–Jeans law.

4.3‡

$$\lambda_{\max} = \frac{hc}{5kT}\,[\text{P4.2}] = \frac{(6.626\times10^{-34}\ \text{J s})\times(2.998\times10^{8}\ \text{m s}^{-1})}{5\times(1.381\times10^{-23}\ \text{J K}^{-1})\times(5\,800\ \text{K})}$$

$$= 4.96\times10^{-7}\ \text{m} = \boxed{49\overline{6}\ \text{nm, blue-green}}$$ (See text Fig. 3.4.)

Topic 5 The wavefunction

Discussion questions

5.1 With the scientifically observed failures of classical physics it became apparent during about the first-quarter of the twentieth century that the Newtonian concept of a deterministic particle path over which the precise position and momentum could be specified at each instant is wrong at the atomic/molecular level. The Schrödinger equation, discovered in 1926, resolved the theoretical difficulties of the particulate-wave duality of nature at the microscopic scale by providing a quantum method for calculating atomic and molecular energy levels along with other observable properties. The method involved the calculation and use of the wavefunction ψ, called the **probability amplitude**. The probability amplitude is a property that is often imaginary and, consequently, non-observable, so the question arose 'Is the wavefunction an artifact that provides a method for calculating observables or is there a useful physical interpretation for it?' Max Born suggested the interpretation that came to be accepted, and found to be very useful, by the scientific community. It is: The probability of finding a particle in a small region of space of volume δV is proportional to $|\psi|^2\,\delta V$, where ψ is the value of the wavefunction in the region. (The modulus square of the wavefunction, $|\psi|^2 = \psi^* \times \psi$, is always real.) This is the probabilistic **Born interpretation** of the wavefunction and $|\psi|^2$ is called the **probability density** as its SI unit is m^{-3} for one particle quantum systems in three dimensions (like the probability density for an electron in a hydrogen atom). In addition to being an integral part of the quantum methodology, the probability density and Born interpretation are regularly used to provide a rationale for quick understanding of the quantum method for calculating expectation values of observables.

The analogy between the Born interpretation of the probability density and the square of the amplitude of an electromagnetic wave is very supportive of the Born interpretation. In classical electromagnetic theory the square of the amplitude is the radiation intensity and therefore proportional to the number of photons present.

5.3 A wavefunction that is normalized to 1 satisfies the condition:

$$\int_{\text{all space}} \psi^* \psi \, d\tau = 1 \quad [5.4].$$

The simplicity of normalization to 1 means that the probability P of find a particle in a region of space is given by an integration over that region:

$P = \int_{\text{region}} \psi^*\psi \mathrm{d}\tau$. Also, the expectation value of observable Ω of the quantum operator $\hat{\Omega}$ is given simply as $\langle\Omega\rangle = \int \psi^*\hat{\Omega}\psi \mathrm{d}\tau$.

Exercises

5.1(a) The time-independent wavefunction in three-dimensional space is a function of x, y, and z so we write $\psi(x,y,z)$ or $\psi(\boldsymbol{r})$. The infinitesimal space element is $\mathrm{d}\tau = \mathrm{d}x\mathrm{d}y\mathrm{d}z$ with each variable ranging from $-\infty$ to $+\infty$. The time-independent wavefunction is said to be a **stationary state**.

It is reasonable to expect that in some special cases the probability densities in each of the three independent directions should be mutually independent. This implies that the probability density for the time-independent wavefunction $\psi(\boldsymbol{r})$ should be the product of three probability densities, one for each coordinate: $|\psi(\boldsymbol{r})|^2 \propto |X(x)|^2 \times |Y(y)|^2 \times |Z(z)|^2$. Subsequently, we see that the wavefunction is the product of three independent wavefunctions in such a special case and we write $\psi(\boldsymbol{r}) \propto X(x) \times Y(y) \times Z(z)$. Such a wavefunction is said to exhibit the **separation of variables.**

For the special case of a particle free to move in a cube of volume L^3, we may generalize the wavefunction provided in Example 1.4 of the text, $X(x) \propto \sin(\pi x/L)$, to $Y(y) \propto \sin(\pi y/L)$ and $Z(z) \propto \sin(\pi z/L)$ and conclude that $\psi(\boldsymbol{r}) \propto \sin(\pi x/L) \times \sin(\pi y/L) \times \sin(\pi z/L)$. Alternatively, for a particle free to move in a rectangular parallelepiped of sides L_x, L_y, and L_z: $\psi(\boldsymbol{r}) \propto \sin(\pi x/L_x) \times \sin(\pi_y/L_y) \times \sin(\pi z/L_z)$

Remarkably, when the potential energy term of the hamiltonian is either zero or a constant value throughout space, the time-independent wavefunction does not depend upon the particle mass! Mass does appear in both the kinetic energy operator and eigenvalues of operators that contain the kinetic energy operator. Similarly, electrical charge does not appear in the time-independent wavefunction in this particular case.

5.2(a) An isolated, freely moving hydrogen atom is expected to have a translational, time-independent wavefunction that is a function of the center-of-mass coordinates x_{cm}, y_{cm}, and z_{cm} so we write $\psi_{\text{cm}}(x_{\text{cm}},y_{\text{cm}},z_{\text{cm}})$ or $\psi_{\text{cm}}(\boldsymbol{r}_{\text{cm}})$ with each variable ranging from $-\infty$ to $+\infty$. The infinitesimal space element for the center of mass variables is $\mathrm{d}\tau_{\text{cm}} = \mathrm{d}x_{\text{cm}}\mathrm{d}y_{\text{cm}}\mathrm{d}z_{\text{cm}}$. We expect that there are special cases for which the translational wavefunction exhibits the separation of variables: $\psi_{\text{cm}}(\boldsymbol{r}_{\text{cm}}) \propto X_{\text{cm}}(x_{\text{cm}}) \times Y(y_{\text{cm}}) \times Z_{\text{cm}}(z_{\text{cm}})$.

In general we expect that the total wavefunction of the isolated hydrogen atom is the product of the center-of-mass wavefunction and an electronic wavefunction, $\psi_{\text{el}}(\boldsymbol{r})$ or $\psi_{\text{el}}(x,y,z)$, that originates from the interaction between the electron and proton (see E1.12a):

$$\Psi_{\text{total}} = \psi_{\text{cm}}(\boldsymbol{r}_{\text{cm}}) \times \psi_{\text{el}}(\boldsymbol{r}) = \psi_{\text{cm}}(\boldsymbol{r}_{\text{cm}}) \times \psi_{\text{el}}(x,y,z)$$

with the origin of x, y, and z at the center-of-mass, variables ranging from $-\infty$ to $+\infty$, and the infinitesimal space element being $\mathrm{d}\tau = \mathrm{d}x\,\mathrm{d}y\,\mathrm{d}z$. The electronic wavefunction does not exhibit the separation of Cartesian variables because the electrostatic potential between the electron and nucleus is proportional to $1/r$, which cannot be written as a sum of separate terms in the variables x, y, and z. But the hydrogen atom, as viewed from the center-of-mass, has spherical symmetry which suggests that the electronic wavefunction variables may be separated when using the spherical polar coordinates r, θ, ϕ where $0 \le r \le \infty$, $0 \le \theta \le \pi$, $0 \le \phi \le 2\pi$, and $d\tau = r^2 \sin\theta \, dr \, d\theta \, d\phi$. The separated functions are symbolized as $R(r)$, $\Theta(\theta)$, and $\Phi(\phi)$ and the electronic wavefunction becomes: $\psi_{el} = R(r) \times \Theta(\theta) \times \Phi(\phi)$.

5.3(a) The normalized wavefunction has the form $\psi(\phi) = N e^{i\phi}$ where N is the normalization constant.

$$\int_0^{2\pi} \psi^* \psi \, d\phi = 1 \ [5.4]$$

$$N^2 \int_0^{2\pi} e^{-i\phi} e^{i\phi} \, d\phi = N^2 \int_0^{2\pi} d\phi = 2\pi N^2 = 1$$

$$\boxed{N = \left(\frac{1}{2\pi}\right)^{1/2}}$$

5.4(a) $\psi(\phi) = (1/2\pi)^{1/2} e^{i\phi}$ so $|\psi(\phi)|^2 = (1/2\pi) e^{-i\phi} e^{i\phi} = 1/2\pi$. Thus, the probability of finding the atom in an infinitesimal volume element at any angle is $\boxed{(1/2\pi)\, d\phi}$.

5.5(a) The normalized wavefunction is $\psi = \left(\frac{1}{2\pi}\right)^{1/2} e^{i\phi}$.

$$\text{Probability that } \pi/2 \le \phi \le 3\pi/2 = \int_{\pi/2}^{3\pi/2} \psi^* \psi \, d\phi \quad \text{[Postulate II]}$$

$$= \left(\frac{1}{2\pi}\right) \int_{\pi/2}^{3\pi/2} e^{-i\phi} e^{i\phi} \, d\phi = \left(\frac{1}{2\pi}\right) \int_{\pi/2}^{3\pi/2} d\phi = \left(\frac{1}{2\pi}\right) \phi \Big|_{\phi=\pi/2}^{\phi=3\pi/2} = \left(\frac{1}{2\pi}\right) \times \left(\frac{3\pi}{2} - \frac{\pi}{2}\right)$$

$$= \boxed{\frac{1}{2}}$$

Problems

5.1 $$\psi = \left(\frac{2}{L}\right)^{1/2} \sin\frac{\pi x}{L} \quad \text{and} \quad \psi^2 = \frac{2}{L} \sin^2 \frac{\pi x}{L}$$

The probability P that the particle will be found in the region between a and b is the integral summation of all the probabilities of finding the particle within infinitesimally small volume elements within the region ($\psi^2 dx$ according to Postulate II).

$$P(a,b) = \int_a^b \psi^2 \, dx \quad \text{[Postulate II]}$$

$$= \frac{2}{L} \int_a^b \sin^2 \frac{\pi x}{L} dx = \frac{2}{L} \times \left(\frac{x}{2} - \frac{L}{4\pi} \sin\frac{2\pi x}{L}\right)\Bigg|_{x=a}^{x=b} = \left(\frac{x}{L} - \frac{1}{2\pi} \sin\frac{2\pi x}{L}\right)\Bigg|_{x=a}^{x=b}$$

$$= \frac{b-a}{L} - \frac{1}{2\pi}\left(\sin\frac{2\pi b}{L} - \sin\frac{2\pi a}{L}\right)$$

$L = 10.0$ nm

(a) $P(4.95\text{ nm}, 5.05\text{ nm}) = \dfrac{0.10}{10.0} - \dfrac{1}{2\pi}\left(\sin\dfrac{(2\pi)\times(5.05)}{10.0} - \sin\dfrac{(2\pi)\times(4.95)}{10.0}\right)$

$= 0.010 + 0.010 = \boxed{0.020}$

(b) $P(7.95\text{ nm}, 9.05\text{ nm}) = \dfrac{1.10}{10.0} - \dfrac{1}{2\pi}\left(\sin\dfrac{(2\pi)\times(9.05)}{10.0} - \sin\dfrac{(2\pi)\times(7.95)}{10.0}\right)$

$= 0.110 - 0.063 = \boxed{0.047}$

(c) $P(9.90\text{ nm}, 10.0\text{ nm}) = \dfrac{0.10}{10.0} - \dfrac{1}{2\pi}\left(\sin\dfrac{(2\pi)\times(10.0)}{10.0} - \sin\dfrac{(2\pi)\times(9.90)}{10.0}\right)$

$= 0.010 - 0.009993 = \boxed{7\times10^{-6}}$

(d) $P(0\text{ nm}, 5.0\text{ nm}) = P(5.0\text{ nm}, 10.0\text{ nm}) = \boxed{0.5}$ [by symmetry]

(e) $P\left(\frac{1}{3}L, \frac{2}{3}L\right) = \frac{1}{3} - \frac{1}{2\pi}\left(\sin\frac{4\pi}{3} - \sin\frac{2\pi}{3}\right) = \boxed{0.61}$

5.3 The normalization constant for this wavefunction is

$$N = \left(\frac{1}{\int_0^{\pi} x^2 \times x^2 \, dx}\right)^{1/2} [5.2] = \left(\frac{5}{\pi^5}\right)^{1/2}$$

so the normalized wavefunction and desired probability are given by

$$\psi = \left(\frac{5}{\pi^5}\right)^{1/2} x^2 \text{ and } P = \int_0^a \psi^2 \, dx = \left(\frac{5}{\pi^5}\right)\int_0^a x^4 \, dx = \left(\frac{a}{\pi}\right)^5$$

Consequently, $P = ½$ when $\boxed{a = \pi/2^{1/5}}$.

5.5 The normalization constant for this wavefunction is

$$N = \left(\frac{1}{\int_0^{2\pi} e^{im\phi} \times e^{-im\phi} \, d\phi}\right)^{1/2} [5.2] = \left(\frac{1}{\int_0^{2\pi} d\phi}\right)^{1/2} = \left(\frac{1}{2\pi}\right)^{1/2}$$

so the normalized wavefunction is $\psi = (2\pi)^{-1/2} e^{-im\phi}$.

$$\langle\phi\rangle = \int_0^{2\pi} \psi^* \phi \psi \, d\phi = (2\pi)^{-1}\int_0^{2\pi} \phi e^{im\phi} e^{-im\phi} \, d\phi = (2\pi)^{-1}\int_0^{2\pi} \phi \, d\phi = (2\pi)^{-1}\left[\frac{\phi^2}{2}\right]_{\phi=0}^{\phi=2\pi}$$

$$= \left(\frac{1}{2\pi}\right) \times \left(\frac{4\pi^2}{2}\right) = \boxed{\pi}$$

5.7 The most probable location occurs when the probability density, $|\psi|^2$, is a maximum. Thus, we wish to find the value $x = x_{\max}$ such that $\mathrm{d}|\psi|^2/\mathrm{d}x = 0$.

$$\psi(x) = Nx\mathrm{e}^{-x^2/2a^2}$$
$$|\psi^2| = N^2x^2\mathrm{e}^{-x^2/a^2}$$
$$\mathrm{d}|\psi^2|/\mathrm{d}x = N^2\left\{2x\mathrm{e}^{-x^2/a^2} - \left(2x^3/a^2\right)\mathrm{e}^{-x^2/a^2}\right\} = 2N^2x\left\{1 - x^2/a^2\right\}\mathrm{e}^{-x^2/a^2}$$

The above derivative equals zero when the factor $1 - x^2/a^2$ equals zero so we conclude that $\boxed{x_{\max} = \pm a}$.

Topic 6 Extracting information from the wavefunction

Discussion questions

6.1 See Figs. 4.12, 5.2, 5.3, 6.2, 8.2, and 8.3 of the text. The key points of these figures include the statements:

A particle with high momentum has a wavefunction with a short wavelength, and vice versa.

Both negative and positive regions of a wavefunction may correspond to a high probability of finding a particle in a region.

A wavefunction must be continuous, have a continuous slope, be single-valued, and be square-integrable.

A sharply curved wavefunction is associated with a high kinetic energy, and one with a low curvature is associated with a low kinetic energy.

If a particle is at a definite location, its wavefunction must be large there and zero everywhere else.

The superposition of a few harmonic functions gives a wavefunction that spreads over a range of locations.

Exercises

6.1(a) The classical expression for the potential energy of a harmonic oscillator is given by $E_p = {}^1/_2 k_f x^2$ [Topic 2.3(b)] where k_f is the force constant. To construct the potential energy operator, $\hat{V}$, we replace the position x in the classical expression with the position operator identity $\hat{x} = x\times$, which is presented in Postulate III (Topic 6.1). This rather simple procedure yields: $\boxed{\hat{V} = \tfrac{1}{2} k_f x^2}$.

6.2(a) Let a and b be any real or complex number and let $f(x)$ be defined as: $f(x) = ae^{2ix} + be^{-2ix}$.

Then,

$$\frac{df}{dx} = 2iae^{2ix} - 2ibe^{-2ix}$$

$$\frac{d^2 f}{dx^2} = (2i)^2 ae^{2ix} + (2i)^2 be^{-2ix} = -4\left(ae^{2ix} + be^{-2ix}\right)$$

$$= -4f(x)$$

$f(x)$ is an eigenfunction of the operator d^2/dx^2. The eigenvalue is $\boxed{-4}$.

6.3(a)

(a) $\dfrac{\mathrm{d}e^{ikx}}{\mathrm{d}x} = (ik)e^{ikx}$ Thus, e^{ikx} is an eigenfunction of d/dx. The eigenvalue is $\boxed{ik}$.

(b) $\dfrac{\mathrm{d}e^{ax^2}}{\mathrm{d}x} = (2ax)e^{ax^2}$ Thus, e^{ax^2} is $\boxed{\text{not an eigenfunction of d/d}x}$.

(c) $\dfrac{\mathrm{d}x}{\mathrm{d}x} = 1$ Thus, x is $\boxed{\text{not an eigenfunction of d/d}x}$.

(d) $\dfrac{\mathrm{d}x^2}{\mathrm{d}x} = 2x$ Thus, x^2 is $\boxed{\text{not an eigenfunction of d/d}x}$.

(e) $\dfrac{\mathrm{d}(ax+b)}{\mathrm{d}x} = a$ Thus, $ax + b$ is $\boxed{\text{not an eigenfunction of d/d}x}$.

(f) $\dfrac{\mathrm{d}\sin(x+3a)}{\mathrm{d}x} = \cos(x+3a)$ Thus, $\sin(x+3a)$ is $\boxed{\text{not an eigenfunction of d/d}x}$.

6.4(a) Let f and g be functions of x and examine the integral $\int_{-\infty}^{\infty} f^*\left(-\dfrac{\hbar^2}{2m}\dfrac{\mathrm{d}^2}{\mathrm{d}x^2}\right)g\,\mathrm{d}x$. Integrate successively by parts (see Justification 6.1) and use the fact that these functions must be well behaved at the boundaries (i.e., they equal zero at infinity in either direction).

$$\int_{-\infty}^{\infty} f^*\left(-\frac{\hbar^2}{2m}\frac{\mathrm{d}^2}{\mathrm{d}x^2}\right)g\,\mathrm{d}x = \left(-\frac{\hbar^2}{2m}\right)\int_{-\infty}^{\infty} f^*\left(\frac{\mathrm{d}^2}{\mathrm{d}x^2}\right)g\,\mathrm{d}x = \left(-\frac{\hbar^2}{2m}\right)\int_{-\infty}^{\infty} f^*\left(\frac{\mathrm{d}}{\mathrm{d}x}\right)\frac{\mathrm{d}g}{\mathrm{d}x}\,\mathrm{d}x$$
$$= \left(-\frac{\hbar^2}{2m}\right)\times\left\{\left[f^*\frac{\mathrm{d}g}{\mathrm{d}x}\right]_{-\infty}^{\infty} - \int_{-\infty}^{\infty}\frac{\mathrm{d}f^*}{\mathrm{d}x}\times\frac{\mathrm{d}g}{\mathrm{d}x}\,\mathrm{d}x\right\} = -\left(-\frac{\hbar^2}{2m}\right)\times\left\{\int_{-\infty}^{\infty}\frac{\mathrm{d}f^*}{\mathrm{d}x}\times\frac{\mathrm{d}g}{\mathrm{d}x}\,\mathrm{d}x\right\}$$
$$= -\left(-\frac{\hbar^2}{2m}\right)\times\left\{\left[g\frac{\mathrm{d}f^*}{\mathrm{d}x}\right]_{-\infty}^{\infty} - \int_{-\infty}^{\infty} g\times\frac{\mathrm{d}^2 f^*}{\mathrm{d}x^2}\,\mathrm{d}x\right\} = \left(-\frac{\hbar^2}{2m}\right)\times\left\{\int_{-\infty}^{\infty} g\times\frac{\mathrm{d}^2 f^*}{\mathrm{d}x^2}\,\mathrm{d}x\right\}$$
$$= \int_{-\infty}^{\infty} g\left(-\frac{\hbar^2}{2m}\frac{\mathrm{d}^2}{\mathrm{d}x^2}\right)f^*\,\mathrm{d}x = \left\{\int_{-\infty}^{\infty} g^*\left(-\frac{\hbar^2}{2m}\frac{\mathrm{d}^2}{\mathrm{d}x^2}\right)f\,\mathrm{d}x\right\}^*$$

Thus, $\displaystyle\int_{-\infty}^{\infty} f^*\left(-\frac{\hbar^2}{2m}\frac{\mathrm{d}^2}{\mathrm{d}x^2}\right)g\,\mathrm{d}x = \left\{\int_{-\infty}^{\infty} g^*\left(-\frac{\hbar^2}{2m}\frac{\mathrm{d}^2}{\mathrm{d}x^2}\right)f\,\mathrm{d}x\right\}^*$

This is exactly the criteria that a hermitian operator must satisfy [6.7] so we conclude that the kinetic energy operator is hermitian.

6.5(a) A quick examination of its expectation value reveals that it is complex because the expectation values of both position and momentum are real. All observables must be real.

$$\int\psi^*(\hat{x}+\mathrm{i}a\hat{p}_x)\psi\,\mathrm{d}\tau = \int\psi^*(\hat{x})\psi\,\mathrm{d}\tau + \mathrm{i}a\int\psi^*(\hat{p}_x)\psi\,\mathrm{d}\tau = \langle x\rangle + \mathrm{i}a\langle p_x\rangle$$

Also, x and p_x are complementary observables so they cannot be simultaneously known.

The hermiticity of the operator may also be checked using the fact the both the position and linear momentum operators are hermitian.

$$\int g(\hat{x}+i\hat{p}_x) f\,d\tau = \int g(\hat{x}) f\,d\tau + i\int g(\hat{p}_x) f\,d\tau$$
$$= \left\{\int f(\hat{x}) g\,d\tau\right\}^* + i\left\{\int g(\hat{p}_x) f\,d\tau\right\}^*$$
$$= \left\{\int f(\hat{x}) g\,d\tau\right\}^* - i^*\left\{\int g(\hat{p}_x) f\,d\tau\right\}^*$$
$$= \left\{\int f(\hat{x}-i\hat{p}_x) g\,d\tau\right\}^*$$

This is not the relationship required by eqn 6.7 so we conclude that the operator is not hermitian and cannot correspond to an observable.

Problems

6.1 Time independent Schrödinger equation of a single particle:

$$\left(-\frac{\hbar^2}{2m}\frac{d^2}{dx^2}+V\right)\psi = E\psi \quad [6.1]$$

(a) $\left(-\frac{\hbar^2}{2m_e}\frac{d^2}{dx^2}-\frac{e^2}{4\pi\varepsilon_0 x}\right)\psi = E\psi$

(b) $\left(-\frac{\hbar^2}{2m}\frac{d^2}{dx^2}\right)\psi = E\psi$

(c) $F=c$ (a constant) implies that $V=-cx$ because $F=-dV/dx$.

$$\left(-\frac{\hbar^2}{2m}\frac{d^2}{dx^2}-cx\right)\psi = E\psi$$

6.3

(a) $\hat{i}(x^3-kx)=(-x)^3-k(-x)=-x^3+kx=-1\times(x^3-kx)$

x^3-kx is an eigenfunction of $\hat{i}$. Its eigenvalue is $\boxed{-1}$.

(b) $\hat{i}(\cos(kx))=\cos(-kx)=\cos(kx)$

$\cos(kx)$ is an eigenfunction of $\hat{i}$. Its eigenvalue is $\boxed{+1}$.

(c) $\hat{i}(x^2+3x-1)=(-x)^2+3(-x)-1=x^2-3x-1$

x^2+3x-1 is $\boxed{\text{not an eigenfunction of } \hat{i}}$.

Topic 7 **Predicting the outcome of experiments**

Discussion question

7.1 In quantum mechanics all dynamical properties of a physical system have associated with them a corresponding operator. The system itself is described by a wavefunction. The observable properties of the system can be obtained in one of two ways from the wavefunction depending upon whether or not the wavefunction is an eigenfunction of the operator.

When the function representing the state of the system is an eigenfunction of the operator $\hat{\Omega}$, we solve the eigenvalue equation (eqn 6.6) $\hat{\Omega}\psi = \omega\psi$ in order to obtain the observable values, ω, of the dynamical property.

When the function is not an eigenfunction of $\hat{\Omega}$, we can only find the average or expectation value of dynamical property by performing the integration shown in eqn 7.2: $\langle\Omega\rangle = \int \psi^*\hat{\Omega}\psi\,d\tau$.

Exercises

7.1(a) ψ_i and ψ_j are orthogonal if $\int \psi_i^*\psi_j\,d\tau = 0$ [7.3a]. Where $n \neq m$ and both n and m are integers,

$$\int_0^L \sin(n\pi x/L)\times\sin(m\pi x/L)\,dx = \left[\frac{\sin(\pi(n-m)x/L)}{2\pi(n-m)/L} - \frac{\sin(\pi(n+m)x/L)}{2\pi(n+m)/L}\right]_{x=0}^{x=L}.$$

$$= \frac{\sin(\pi(n-m))}{2\pi(n-m)/L} - \frac{\sin(\pi(n+m))}{2\pi(n+m)/L} - \left\{\frac{\sin(0)}{2\pi(n-m)/L} - \frac{\sin(0)}{2\pi(n+m)/L}\right\}$$

$= 0$ because the sine of an integer multiple of π equals zero.

Thus, the functions $\sin(n\pi x/L)$ and $\sin(m\pi x/L)$ are orthogonal in the region $0 \leq x \leq L$.

7.2(a) ψ_i and ψ_j are orthogonal if $\int \psi_i^*\psi_j\,d\tau = 0$ [7.3a].

$$\int_0^{2\pi} \left(e^{i\phi}\right)^* \times e^{2i\phi}\,d\varphi = \int_0^{2\pi} e^{-i\phi}\times e^{2i\phi}\,d\varphi = \int_0^{2\pi} e^{i\phi}\,d\phi = \frac{1}{i}e^{i\phi}\Big|_{\phi=0}^{\phi=2\pi}$$

$$= \frac{1}{i}\left(e^{2\pi i} - e^0\right) = \frac{1}{i}\left(e^{2\pi i} - 1\right) = \frac{1}{i}\left(\cos(2\pi) - i\sin(2\pi) - 1\right) = \frac{1}{i}(1-0-1) = 0$$

(The Euler identity $e^{ai} = \cos(a) - i\sin(a)$ has been used in the math manipulations.)

Thus, the functions $e^{i\phi}$ and $e^{2i\phi}$ are orthogonal in the region $0 \le \phi \le 2\pi$

7.3(a) The normalized form of this wavefunction is:

$$\psi(x) = \left(\frac{2}{L}\right)^{1/2} \sin(2\pi x/L) \qquad \text{(See Self-test 5.3)}$$

The expectation value of the electron position is:

$$\langle x\rangle = \int_0^L \psi^* x \psi \,\mathrm{d}x \;[7.2] = \left(\frac{2}{L}\right)\int_0^L x \sin^2\left(\frac{2\pi x}{L}\right)\mathrm{d}x$$

$$= \left(\frac{2}{L}\right)\times\left[\frac{x^2}{4} - \frac{x\sin\left(\frac{4\pi x}{L}\right)}{8\pi/L} - \frac{\cos\left(\frac{4\pi x}{L}\right)}{8(2\pi/L)^2}\right]_{x=0}^{x=L}$$

$$= \left(\frac{2}{L}\right)\times\left[\frac{L^2}{4} - \frac{L\sin(4\pi)}{8\pi/L} - \frac{\cos(4\pi)}{8(2\pi/L)^2} - \left\{\frac{0^2}{4} - \frac{0\times\sin(0)}{8\pi/L} - \frac{\cos(0)}{8(2\pi/L)^2}\right\}\right]$$

$$= \left(\frac{2}{L}\right)\times\left(\frac{L^2}{4}\right)$$

$$= \boxed{\frac{L}{2}}$$

7.4(a) The normalized form of this wavefunction is:

$$\psi(x) = \left(\frac{2}{L}\right)^{1/2} \sin(\pi x/L) \qquad \text{(See Example 5.2)}$$

$$\frac{\mathrm{d}\psi}{\mathrm{d}x} = \left(\frac{2}{L}\right)^{1/2}\left(\frac{\pi}{L}\right)\cos(\pi x/L)$$

The expectation value of the electron momentum is:

$$\langle p_x\rangle = \int_0^L \psi^* \hat{p}_x \psi\,\mathrm{d}x\;[7.2] = \int_0^L \psi^*\left(\frac{\hbar}{\mathrm{i}}\frac{\mathrm{d}}{\mathrm{d}x}\right)\psi\,\mathrm{d}x = \left(\frac{\hbar}{\mathrm{i}}\right)\int_0^L \psi^*\left(\frac{\mathrm{d}\psi}{\mathrm{d}x}\right)\mathrm{d}x$$

$$= \left(\frac{\hbar}{\mathrm{i}}\right)\left(\frac{2}{L}\right)^{1/2}\left(\frac{\pi}{L}\right)\int_0^L \psi^* \cos\left(\frac{\pi x}{L}\right)\mathrm{d}x = \left(\frac{\hbar}{\mathrm{i}}\right)\left(\frac{2}{L}\right)\left(\frac{\pi}{L}\right)\int_0^L \sin\left(\frac{\pi x}{L}\right)\cos\left(\frac{\pi x}{L}\right)\mathrm{d}x$$

$$= \left(\frac{h}{\mathrm{i}L^2}\right)\int_0^L \sin\left(\frac{\pi x}{L}\right)\cos\left(\frac{\pi x}{L}\right)\mathrm{d}x$$

$$= \left(\frac{h}{\mathrm{i}L^2}\right)\times\left[\frac{\sin^2\left(\frac{\pi x}{L}\right)}{2\pi/L}\right]_{x=0}^{x=L} = \left(\frac{h}{\mathrm{i}L^2}\right)\times\left(\frac{\sin^2(\pi)}{2\pi/L} - \frac{\sin^2(0)}{2\pi/L}\right) = \left(\frac{h}{\mathrm{i}L^2}\right)\times(0+0) = \boxed{0}$$

Problems

7.1 $\psi = a^{1/2}\mathrm{e}^{-ax/2}$

$$\langle x\rangle = \int_0^\infty \psi^* x\psi\,\mathrm{d}x = a\int_0^\infty x\mathrm{e}^{-ax}\,\mathrm{d}x \quad \text{[Integrate by parts]}$$

$$= a\left\{\left[x\times\left(-\frac{1}{a}\mathrm{e}^{-ax}\right)\right]_0^\infty - \int_0^\infty\left(-\frac{1}{a}\mathrm{e}^{-ax}\right)\mathrm{d}x\right\} = \int_0^\infty \mathrm{e}^{-ax}\,\mathrm{d}x$$

$$= \left[-\frac{1}{a}\mathrm{e}^{-ax}\right]_0^\infty = -\frac{1}{a}\left[\mathrm{e}^{-\infty} - \mathrm{e}^{-0}\right] = -\frac{1}{a}[0-1] = \boxed{\frac{1}{a}}$$

7.3 $$\hat{E}_\mathrm{k} = \frac{\hat{p}_x^2}{2m_\mathrm{e}} = \frac{1}{2m_\mathrm{e}}\left(\frac{\hbar}{\mathrm{i}}\frac{\mathrm{d}}{\mathrm{d}x}\right)\left(\frac{\hbar}{\mathrm{i}}\frac{\mathrm{d}}{\mathrm{d}x}\right) = -\frac{\hbar^2}{2m_\mathrm{e}}\frac{\mathrm{d}^2}{\mathrm{d}x^2}$$

$$\psi = (\cos\chi)\mathrm{e}^{ikx} + (\sin\chi)\mathrm{e}^{-ikx} = c_1\mathrm{e}^{ikx} + c_2\,\mathrm{e}^{-ikx}$$

$$\langle\hat{E}_\mathrm{k}\rangle = N^2\int\psi^*\left(\frac{\hat{p}_x^2}{2m_\mathrm{e}}\right)\psi\,\mathrm{d}\tau = \int\psi^*\left(\frac{\hat{p}_x^2}{2m_\mathrm{e}}\right)\psi\,\mathrm{d}\tau\Big/\int\psi^*\psi\,\mathrm{d}\tau \quad \left[5.2,\ N^2 = 1/\int\psi^*\psi\,\mathrm{d}\tau\right]$$

$$= \frac{\frac{-\hbar^2}{2m_\mathrm{e}}\int\psi^*\frac{\mathrm{d}^2}{\mathrm{d}x^2}\left(\mathrm{e}^{ikx}\cos\chi + \mathrm{e}^{-ikx}\sin\chi\right)\mathrm{d}\tau}{\int\psi^*\psi\,\mathrm{d}\tau}$$

$$= \frac{\frac{-\hbar^2}{2m_\mathrm{e}}\int\psi^*(-k^2)\times(\mathrm{e}^{ikx}\cos\chi + \mathrm{e}^{-ikx}\sin\chi)\mathrm{d}\tau}{\int\psi^*\psi\,\mathrm{d}\tau} = \frac{\hbar^2k^2\int\psi^*\psi\,\mathrm{d}\tau}{2m_\mathrm{e}\int\psi^*\psi\,\mathrm{d}\tau} = \boxed{\frac{\hbar^2k^2}{2m_\mathrm{e}}}$$

7.5 In each case the normalization constant $N = \left(1/\int\psi^*\psi\,\mathrm{d}\tau\right)^{1/2}$ [5.2] must be evaluated by analytically determining the integral over the whole space of the wavefunction.

(a) (i) The unnormalized wavefunction is $\psi = \left(2 - \frac{r}{a_0}\right)\mathrm{e}^{-r/a_0}$ and

$$\psi^2 = \left(2 - \frac{r}{a_0}\right)^2\mathrm{e}^{-2r/a_0}$$

$$\int\psi^2\,\mathrm{d}\tau = \int_{r=0}^\infty\int_{\theta=0}^\pi\int_{\varphi=0}^{2\pi}\left(4 - \frac{4r}{a_0} + \frac{r^2}{a_0^2}\right)\mathrm{e}^{-2r/a_0}\,r^2\sin\theta\,\mathrm{d}r\,\mathrm{d}\theta\,\mathrm{d}\phi$$

$$= \int_0^\infty\left(4r^2 - \frac{4r^3}{a_0} + \frac{r^4}{a_0^2}\right)\mathrm{e}^{-2r/a_0}\,\mathrm{d}r\int_0^\pi\sin\theta\,\mathrm{d}\theta\int_0^{2\pi}\mathrm{d}\phi$$

$$= 4\pi\int_0^\infty\left(4r^2 - \frac{4r^3}{a_0} + \frac{r^4}{a_0^2}\right)\mathrm{e}^{-2r/a_0}\,\mathrm{d}r \quad \left[\int_0^\pi\sin\theta\,\mathrm{d}\theta\int_0^{2\pi}\mathrm{d}\phi = 4\pi\right]$$

Using the standard integral $\int_0^\infty x^n\mathrm{e}^{-ax}\,\mathrm{d}x = \frac{n!}{a^{n+1}}$ where $n = 0,1,2\ldots$, the working equation becomes

$$\int \psi^2 d\tau = 4\pi \int_0^\infty \left(4r^2 - \frac{4r^3}{a_0} + \frac{r^4}{a_0^2}\right) e^{-2r/a_0}\, dr$$

$$= 4\pi \left\{ \frac{4\times 2!}{(2/a_0)^3} - \frac{4\times 3!}{a_0(2/a_0)^4} + \frac{4!}{a_0^2(2/a_0)^5} \right\} = \pi a_0^3$$

Hence $\boxed{N = \left(\frac{1}{\pi a_0^3}\right)^{1/2}}$ and the normalized wavefunction is

$$\psi = \left(\frac{1}{\pi a_0^3}\right)^{1/2} \left(2 - \frac{r}{a_0}\right) e^{-r/a_0}$$

(ii) The unnormalized wavefunction is

$$\psi = r \sin\theta \cos\varphi e^{-r/2a_0} \quad \text{and} \quad \psi^2 = r^2 \sin^2\theta \cos^2\varphi e^{-r/a_0}$$

$$\int \psi^2 d\tau = \int_{r=0}^\infty \int_{\theta=0}^\pi \int_{\phi=0}^{2\pi} (r^2 \sin^2\theta \cos^2\varphi e^{-r/a_0}) r^2 \sin\theta\, dr\, d\theta\, d\phi$$

$$= \int_{r=0}^\infty r^4 e^{-r/a_0}\, dr \int_{\theta=0}^\pi \sin^2\theta \sin\theta\, d\theta \int_{\varphi=0}^{2\pi} \cos^2\phi\, d\phi$$

$$= 4! a_0^5 \int_{-1}^1 (1 - \cos^2\theta) d\cos\theta \times \pi$$

$$= 4! a_0^5 \left(2 - \frac{2}{3}\right)\pi = 32\pi a_0^5$$

Hence, $\boxed{N = \left(\frac{1}{32\pi a_0^5}\right)^{1/2}}$ and the normalized wavefunction is

$$\psi = \left(\frac{1}{32\pi a_0^5}\right)^{1/2} r \sin\theta \cos\phi e^{-r/2a_0}$$

In the above manipulations, we have used $\int_0^\pi \cos^n\theta \sin\theta\, d\theta = -\int_1^{-1} \cos^n\theta\, d\cos\theta = \int_{-1}^1 x^n\, dx$.

(b) The functions will be orthogonal if the following integral, which uses the unnormalized functions, proves to equal zero.

$$\int \psi_1 \psi_2 d\tau = \int \left\{ \left(2 - \frac{r}{a_0}\right) e^{\frac{r}{a_0}} \right\} \left\{ r \sin\theta \cos\phi\, e^{-\frac{r}{2a_0}} \right\} d\tau$$

$$= \int_{r=0}^\infty \int_{\theta=0}^\pi \int_{\phi=0}^{2\pi} \left\{ \left(2 - \frac{r}{a_0}\right) e^{-\frac{r}{a_0}} \right\} \left\{ r \sin\theta \cos\phi\, e^{-\frac{r}{2a_0}} \right\} r^2 \sin\theta\, dr\, d\theta\, d\phi$$

$$= \int_{r=0}^\infty \left\{ \left(2r^3 - \frac{r^4}{a_0}\right) e^{-\frac{3r}{2a_0}} \right\} dr \int_{\theta=0}^\pi \sin^2\theta\, d\theta \int_{\varphi=0}^{2\pi} \cos\phi\, d\phi$$

The integral on the far right equals zero: $\int_{\phi=0}^{2\pi}\cos\phi\,\mathrm{d}\phi = \sin\phi\,|_0^{2\pi} = \sin(2\pi) - \sin(0) = 0 - 0 = 0$

Consequently, the functions are orthogonal.

7.7 The normalized wavefunction (see Problem 5.2) is $\psi = \left(\dfrac{1}{\pi a_0^3}\right)^{1/2} \mathrm{e}^{-r/a_0}$

(a) $\langle V\rangle = \int \psi^* \hat{V}\psi\,\mathrm{d}\tau \quad \left[\hat{V} = -\dfrac{e^2}{4\pi\varepsilon_0 r}\right]$

$$= \int \psi^*\left(\frac{-e^2}{4\pi\varepsilon_0}\cdot\frac{1}{r}\right)\psi\,\mathrm{d}\tau = \frac{1}{\pi a_0^3}\left(\frac{-e^2}{4\pi\varepsilon_0}\right)\int_0^\infty r\mathrm{e}^{-2r/a_0}\,\mathrm{d}r \times 4\pi$$

$$= \frac{1}{\pi a_0^3}\left(\frac{-e^2}{4\pi\varepsilon_0}\right)\times\left(\frac{a_0}{2}\right)^2 \times 4\pi = \boxed{\frac{-e^2}{4\pi\varepsilon_0 a_0}}$$

(b) In one-dimension: $\hat{E}_\mathrm{k} = \dfrac{\hat{p}_x^2}{2m} = \dfrac{1}{2m}\left(\dfrac{\hbar}{\mathrm{i}}\dfrac{\mathrm{d}}{\mathrm{d}x}\right)\left(\dfrac{\hbar}{\mathrm{i}}\dfrac{\mathrm{d}}{\mathrm{d}x}\right) = \boxed{-\dfrac{\hbar^2}{2m}\dfrac{\mathrm{d}^2}{\mathrm{d}x^2}}$ [6.5]. For three-dimensional systems such as the hydrogen atom the kinetic energy operator is

$$\hat{E}_\mathrm{k} = \frac{\hat{p}_x^2}{2m}+\frac{\hat{p}_y^2}{2m}+\frac{\hat{p}_z^2}{2m} = -\frac{\hbar^2}{2m}\left\{\frac{\partial^2}{\partial x^2}+\frac{\partial^2}{\partial y^2}+\frac{\partial^2}{\partial z^2}\right\} = \boxed{-\frac{\hbar^2}{2m}\nabla^2}$$

where $\dfrac{\partial^2}{\partial x^2} = \left(\dfrac{\partial^2}{\partial x^2}\right)_{y,z}$, $\dfrac{\partial^2}{\partial y^2} = \left(\dfrac{\partial^2}{\partial y^2}\right)_{x,z}$, $\dfrac{\partial^2}{\partial z^2} = \left(\dfrac{\partial^2}{\partial z^2}\right)_{x,y}$,

and $\nabla^2 = \dfrac{\partial^2}{\partial x^2}+\dfrac{\partial^2}{\partial y^2}+\dfrac{\partial^2}{\partial z^2}$

The ∇^2 operator, called the laplacian operator or the del-squared operator, is advantageously written in spherical coordinates because the wavefunction has its simplest form in spherical coordinates. Mathematical handbooks report that

$\nabla^2 = \dfrac{\partial^2}{\partial x^2}+\dfrac{\partial^2}{\partial y^2}+\dfrac{\partial^2}{\partial z^2} = \dfrac{\partial^2}{\partial r^2}+\dfrac{2}{r}\dfrac{\partial}{\partial r}+\dfrac{1}{r^2}\Lambda^2$ where the Λ^2 operator, called the legendrian operator, is an operator of angle variables only.

$$\Lambda^2 = \frac{1}{\sin^2\theta}\frac{\partial^2}{\partial\phi^2}+\frac{1}{\sin\theta}\frac{\partial}{\partial\theta}\sin\theta\frac{\partial}{\partial\theta}$$

Since our wavefunction has no angular dependence, $\Lambda^2\psi = 0$ and the laplacian simplifies to

$$\nabla^2 = \frac{\partial^2}{\partial r^2}+\frac{2}{r}\frac{\partial}{\partial r}$$

$$\nabla^2\psi = \left\{\frac{\partial^2}{\partial r^2}+\frac{2}{r}\frac{\partial}{\partial r}\right\}\left(\left(\frac{1}{\pi a_0^3}\right)^{1/2} e^{-r/a_0}\right) = \left(\frac{1}{\pi a_0^3}\right)^{1/2}\left\{\frac{1}{a_0^2}-\frac{2}{ra_0}\right\}e^{-r/a_0}$$

$$\langle \hat{E}_k\rangle = -\frac{\hbar^2}{2m_e}\int \psi^*\nabla^2\psi\,d\tau = -\frac{\hbar^2}{2m_e}\int_{r=0}^{\infty}\int_{\theta=0}^{\pi}\int_{\varphi=0}^{2\pi}\psi^*\nabla^2\psi\, r^2\sin\theta\, dr\, d\theta\, d\phi$$

$$= -\frac{\hbar^2}{2m_e}\int_{r=0}^{\infty} r^2\psi^*\nabla^2\psi\, dr\int_{\theta=0}^{\pi}\sin\theta\, d\theta\int_{\varphi=0}^{2\pi}d\phi = -\frac{\hbar^2}{2m_e}\int_{r=0}^{\infty} r^2\psi^*\nabla^2\psi\, dr\times 4\pi$$

$$= -\frac{4\pi\hbar^2}{2m_e}\left(\frac{1}{\pi a_0^3}\right)\int_{r=0}^{\infty}\left\{\frac{r^2}{a_0^2}-\frac{2r}{a_0}\right\}e^{-2r/a_0}\,dr = -\frac{4\pi\hbar^2}{2m_e}\left(\frac{1}{\pi a_0^3}\right)\times a_0\left(\frac{1}{4}-\frac{1}{2}\right)$$

$$= \boxed{\frac{\hbar^2}{2m_e a_0^2}} \qquad \left[\text{Note that } \int_0^{\infty} x^n e^{-ax}\,dx = \frac{n!}{a^{n+1}}\right]$$

Topic 8 The uncertainty principle

Discussion questions

8.1 If the wavefunction describing the linear momentum of a particle is precisely known, the particle has a definite state of linear momentum; but then according to the uncertainty principle (eqn 8.2), the position of the particle is completely unknown. Conversely, if the position of a particle is precisely known, its linear momentum cannot be described by a single wavefunction. Rather, the wavefunction is a superposition of many wavefunctions, each corresponding to a different value for the linear momentum. All knowledge of the linear momentum of the particle is lost when its position is specified exactly. In the limit of an infinite number of superposed wavefunctions, the wavepacket illustrated in text Fig. 8.3 turns into the sharply spiked packet shown in text Fig. 8.2. But the requirement of the superposition of an infinite number of momentum wavefunctions in order to locate the particle means a complete lack of knowledge of the momentum.

Exercises

8.1(a) $\Delta p \approx 0.0100\,\text{per cent of } p_0 = p_0 \times (1.00\times10^{-4}) = m_p v \times (1.00\times10^{-4}) \quad (p_0 = m_p v)$

$$\Delta q \approx \frac{\hbar}{2\Delta p}[8.2] \approx \frac{1.055\times10^{-34}\,\text{J s}}{2\times(1.673\times10^{-27}\,\text{kg})\times(6.1\times10^{6}\,\text{m s}^{-1})\times(1.00\times10^{-4})}$$
$$\approx 5.2\times10^{-11}\,\text{m} \approx \boxed{52\,\text{pm}}$$

8.2(a) The minimum uncertainty in position and momentum is given by the uncertainty principle in the form $\Delta p\Delta q \geq \tfrac{1}{2}\hbar$ [8.2] with the choice of the equality. The uncertainty in momentum is $\Delta p = m\Delta v$ so:

(a) $$\Delta v_{\text{min}} = \frac{\hbar}{2m\Delta q} = \frac{1.055\times10^{-34}\,\text{J s}}{2\times(0.500\,\text{kg})\times(1.0\times10^{-6}\,\text{m})} = \boxed{1.1\times10^{-28}\,\text{m s}^{-1}}$$

(b) $$\Delta q_{\text{min}} = \frac{\hbar}{2m\Delta v} = \frac{1.055\times10^{-34}\,\text{J s}}{2\times(5.0\times10^{-3}\,\text{kg})\times(1\times10^{-5}\,\text{m s}^{-1})} = \boxed{1\times10^{-27}\,\text{m}}$$

Comment. These uncertainties are extremely small; thus, the ball and bullet are effectively classical particles.

Question. If the ball were stationary (no uncertainty in position) the uncertainty in speed would be infinite. Thus, the ball could have a very high speed, contradicting the fact that it is stationary. What is the resolution of this apparent paradox?

8.3(a)

(a) $[\hat{x},\hat{y}] = xy - yx = xy - xy = \boxed{0}$

(b) $[\hat{p}_x,\hat{p}_y] = \left(\frac{\hbar}{\mathrm{i}}\frac{\partial}{\partial x}\right)\left(\frac{\hbar}{\mathrm{i}}\frac{\partial}{\partial y}\right) - \left(\frac{\hbar}{\mathrm{i}}\frac{\partial}{\partial y}\right)\left(\frac{\hbar}{\mathrm{i}}\frac{\partial}{\partial x}\right) = \left(\frac{\hbar}{\mathrm{i}}\right)^2\left(\frac{\partial^2}{\partial x\partial y} - \frac{\partial^2}{\partial y\partial x}\right)$

$= \boxed{0}$ because $\frac{\partial^2}{\partial x\partial y} = \frac{\partial^2}{\partial y\partial x}$ for all well-behaved functions.

(c) $[\hat{x},\hat{p}_x] = x\left(\frac{\hbar}{\mathrm{i}}\frac{\partial}{\partial x}\right) - \left(\frac{\hbar}{\mathrm{i}}\frac{\partial}{\partial x}\right)x = x\left(\frac{\hbar}{\mathrm{i}}\frac{\partial}{\partial x}\right) - \left(\frac{\hbar}{\mathrm{i}}\right) - \left(\frac{\hbar}{\mathrm{i}}x\frac{\partial}{\partial x}\right) = \boxed{\mathrm{i}\hbar}$

(d) $[\hat{x}^2,\hat{p}_x] = x^2\left(\frac{\hbar}{\mathrm{i}}\frac{\partial}{\partial x}\right) - \left(\frac{\hbar}{\mathrm{i}}\frac{\partial}{\partial x}\right)x^2 = x^2\left(\frac{\hbar}{\mathrm{i}}\frac{\partial}{\partial x}\right) - \frac{2x\hbar}{\mathrm{i}} - x^2\left(\frac{\hbar}{\mathrm{i}}\frac{\partial}{\partial x}\right) = \boxed{2\mathrm{i}x\hbar}$

(e) $[\hat{x}^n,\hat{p}_x] = x^n\left(\frac{\hbar}{\mathrm{i}}\frac{\partial}{\partial x}\right) - \left(\frac{\hbar}{\mathrm{i}}\frac{\partial}{\partial x}\right)x^n = x^n\left(\frac{\hbar}{\mathrm{i}}\frac{\partial}{\partial x}\right) - \frac{nx^{n-1}\hbar}{\mathrm{i}} - x^n\left(\frac{\hbar}{\mathrm{i}}\frac{\partial}{\partial x}\right)$

$= \boxed{n\mathrm{i}x^{n-1}\hbar}$

Problems

8.1 According to the uncertainty principles, $\Delta p\Delta q \geq \frac{1}{2}\hbar$, where Δq and Δp are root-mean-square deviations: $\Delta q = (\langle x^2\rangle - \langle x\rangle^2)^{1/2}$ and $\Delta p = (\langle p^2\rangle - \langle p\rangle^2)^{1/2}$. To verify whether the relationship holds for the particle in a state whose normalized wavefunction is $\psi = (2a/\pi)^{1/4}\mathrm{e}^{-ax^2}$, we need the quantum-mechanical averages $\langle x\rangle, \langle x^2\rangle, \langle p\rangle$, and $\langle p^2\rangle$.

$$\langle x\rangle = \int \psi^* x\psi \mathrm{d}\tau = \int_{-\infty}^{\infty}\left(\frac{2a}{\pi}\right)^{1/4}\mathrm{e}^{-ax^2}x\left(\frac{2a}{\pi}\right)^{1/4}\mathrm{e}^{-ax^2}\mathrm{d}x = \left(\frac{2a}{\pi}\right)^{1/2}\int_{-\infty}^{\infty} x\mathrm{e}^{-2ax^2}\mathrm{d}x = 0$$

$$\langle x^2\rangle = \int_{-\infty}^{\infty}\left(\frac{2a}{\pi}\right)^{1/4}\mathrm{e}^{-ax^2}x^2\left(\frac{2a}{\pi}\right)^{1/4}\mathrm{e}^{-ax^2}\mathrm{d}x = \left(\frac{2a}{\pi}\right)^{1/2}\int_{-\infty}^{\infty} x^2\mathrm{e}^{-2ax^2}\mathrm{d}x = \left(\frac{2a}{\pi}\right)^{1/2}\frac{\pi^{1/2}}{2(2a)^{3/2}} = \frac{1}{4a}$$

So, $\Delta q = \left(\langle x^2\rangle - \langle x\rangle^2\right)^{1/2} = \left(\frac{1}{4a} - 0\right)^{1/2} = \frac{1}{2a^{1/2}}$

$$\langle p\rangle = \int_{-\infty}^{\infty}\psi^*\left(\frac{\hbar}{\mathrm{i}}\frac{\mathrm{d}\psi}{\mathrm{d}x}\right)\mathrm{d}x \text{ and } \langle p^2\rangle = \int_{-\infty}^{\infty}\psi^*\left(-\hbar^2\frac{\mathrm{d}^2\psi}{\mathrm{d}x^2}\right)\mathrm{d}x.$$

We need to evaluate the derivatives within the integrand expressions:

$$\frac{\mathrm{d}\psi}{\mathrm{d}x} = \left(\frac{2a}{\pi}\right)^{1/4}(-2ax)\mathrm{e}^{-ax^2}$$

$$\frac{d^2\psi}{dx^2}=\left(\frac{2a}{\pi}\right)^{1/4}[(-2ax)^2 e^{-ax^2}+(-2a)e^{-ax^2}]=\left(\frac{2a}{\pi}\right)^{1/4}(4a^2x^2-2a)e^{-ax^2}$$

$$\text{So, } \langle p\rangle=\int_{-\infty}^{\infty}\left(\frac{2a}{\pi}\right)^{1/4} e^{-ax^2}\left(\frac{\hbar}{i}\right)\left(\frac{2a}{\pi}\right)^{1/4}(-2ax)e^{-ax^2}dx$$

$$=-\frac{2a\hbar}{i}\left(\frac{2a}{\pi}\right)^{1/2}\int_{-\infty}^{\infty} xe^{-2ax^2}\,dx=0;$$

$$\langle p^2\rangle=\int_{-\infty}^{\infty}\left(\frac{2a}{\pi}\right)^{1/4} e^{-ax^2}(-\hbar^2)\left(\frac{2a}{\pi}\right)^{1/4}(4a^2x^2-2a)e^{-ax^2}dx,$$

$$\langle p^2\rangle=(-2a\hbar^2)\left(\frac{2a}{\pi}\right)^{1/2}\int_{-\infty}^{\infty}(2ax^2-1)e^{-2ax^2}\,dx,$$

$$\langle p^2\rangle=(-2a\hbar^2)\left(\frac{2a}{\pi}\right)^{1/2}\left(2a\frac{\pi^{1/2}}{2(2a)^{3/2}}-\frac{\pi^{1/2}}{(2a)^{1/2}}\right)=a\hbar^2;$$

$$\text{So, } \Delta p=\left(\langle p^2\rangle-\langle p\rangle^2\right)^{1/2}=\left(a\hbar^2-0\right)^{1/2}=a^{1/2}\hbar.$$

Finally, $\Delta q\Delta p=\dfrac{1}{2a^{1/2}}\times a^{1/2}\hbar=\tfrac{1}{2}\hbar$, which is the minimum product consistent with the uncertainty principle.

8.3

(a) (i) $V(x)=V$

$$[\hat{H},\hat{p}_x]=\left[\frac{\hat{p}_x^2}{2m}+V,\hat{p}_x\right]=\frac{\hat{p}_x^3}{2m}+V\hat{p}_x-\left\{\frac{\hat{p}_x^3}{2m}+V\hat{p}_x\right\}=\boxed{0}$$

(ii) $V(x)=\tfrac{1}{2}kx^2$

$$[\hat{H},\hat{p}_x]=\left[\frac{\hat{p}_x^2}{2m}+\tfrac{1}{2}kx^2,\hat{p}_x\right]=\frac{\hat{p}_x^3}{2m}+\tfrac{1}{2}kx^2\hat{p}_x-\left\{\frac{\hat{p}_x^3}{2m}+\hat{p}_x\left(\tfrac{1}{2}kx^2\right)\right\}$$

$$=\tfrac{1}{2}kx^2\hat{p}_x-\left(\frac{\hbar}{i}\right)\left(\frac{d}{dx}\right)\left(\tfrac{1}{2}kx^2\right)=\tfrac{1}{2}kx^2\hat{p}_x-\left(\frac{\hbar}{i}\right)\left\{kx+\left(\tfrac{1}{2}kx^2\right)\left(\frac{d}{dx}\right)\right\}$$

$$=\tfrac{1}{2}kx^2\hat{p}_x-\left(\frac{\hbar}{i}\right)kx-\left(\tfrac{1}{2}kx^2\right)\left(\frac{\hbar}{i}\right)\left(\frac{d}{dx}\right)=-\left(\frac{\hbar}{i}\right)kx=\boxed{i\hbar kx}$$

(b) (i) $V(x)=V$

$$[\hat{H},\hat{x}]=\left[\frac{\hat{p}_x^2}{2m}+V,x\right]=\frac{-\hbar^2}{2m}\left(\frac{d}{dx}\right)\left(\frac{d}{dx}\right)x+Vx-\left\{\frac{x\hat{p}_x^2}{2m}+Vx\right\}=\boxed{-\frac{x\hat{p}_x^2}{2m}}$$

(ii) $V(x) = ½\,kx^2$

$$\left[\hat{H},\hat{x}\right]=\left[\frac{\hat{p}_x^2}{2m}+½kx^2,x\right]=\frac{\hat{p}_x^2 x}{2m}+½kx^3-\left\{\frac{x\hat{p}_x^2}{2m}+½kx^3\right\}$$

$$=\frac{-\hbar^2}{2m}\left(\frac{\mathrm{d}}{\mathrm{d}x}\right)\left(\frac{\mathrm{d}}{\mathrm{d}x}\right)x-\frac{x\hat{p}_x^2}{2m}=\frac{-\hbar^2}{2m}\left(\frac{\mathrm{d}}{\mathrm{d}x}\right)\left(\frac{\mathrm{d}x}{\mathrm{d}x}+x\frac{\mathrm{d}}{\mathrm{d}x}\right)-\frac{x\hat{p}_x^2}{2m}$$

$$=\frac{-\hbar^2}{2m}\left(\frac{\mathrm{d}}{\mathrm{d}x}\right)\left(1+x\frac{\mathrm{d}}{\mathrm{d}x}\right)-\frac{x\hat{p}_x^2}{2m}$$

$$=\frac{-\hbar^2}{2m}\left(\frac{\mathrm{d}}{\mathrm{d}x}\right)\left(x\frac{\mathrm{d}}{\mathrm{d}x}\right)-\frac{x\hat{p}_x^2}{2m}=\frac{-\hbar^2}{2m}\left(\frac{\mathrm{d}}{\mathrm{d}x}+\cancel{x\frac{\mathrm{d}^2}{\mathrm{d}x^2}}\right)-\cancel{\frac{x\hat{p}_x^2}{2m}}$$

$$=\frac{-\hbar^2}{2m}\frac{\mathrm{d}}{\mathrm{d}x}=\boxed{-\frac{i\hbar\hat{p}_x}{2m}}$$

Focus 2: Integrated activities

F2.1

(a) The eigenvalue equation $\hat{\Omega}\psi=\omega\psi$ provides the relation between the operator for an observable $(\hat{\Omega})$ and the value of the observable (ω). Whenever the system is described by a wavefunction ψ, which is an eigenfunction of $\hat{\Omega}$, the outcome of a measurement of the observable Ω will be the eigenvalue ω. This is Postulate IV of the text.

(b) Should the system wavefunction be a superposition of eigenfunctions $(\psi=\sum_k c_k\psi_k)$ the probability of observing the eigenvalue ω_k is proportional to $|c_k|^2$. This is Postulate V of the text.

F2.3

(a) In the momentum representation $\hat{p}_x = p_x\times$, consequently

$$[\hat{x},\hat{p}_x]\phi=[\hat{x},p_x\times]\phi=\hat{x}p_x\times\phi-p_x\times\hat{x}\phi=i\hbar\phi \quad 8.6$$

Suppose that the position operator has the form $\hat{x}=a\frac{\mathrm{d}}{\mathrm{d}p_x}$ where a is a complex number. Then,

$$a\frac{\mathrm{d}}{\mathrm{d}p_x}(p_x\times\phi)-p_x\times\left(a\frac{\mathrm{d}}{\mathrm{d}p_x}\phi\right)=i\hbar\phi$$

$$\frac{\mathrm{d}}{\mathrm{d}p_x}(p_x\times\phi)-p_x\times\left(\frac{\mathrm{d}}{\mathrm{d}p_x}\phi\right)=\frac{i\hbar}{a}\phi$$

$$\frac{\mathrm{d}p_x}{\mathrm{d}p_x}\phi+p_x\frac{\mathrm{d}\phi}{\mathrm{d}p_x}-p_x\frac{\mathrm{d}\phi}{\mathrm{d}p_x}=\frac{i\hbar}{a}\phi \quad \text{[Rule for differentiation of } f(x)g(x).]$$

$$\phi=\frac{i\hbar}{a}\phi$$

This is true when $a = \mathrm{i}\hbar$. We conclude that $\boxed{\hat{x} = \mathrm{i}\hbar\frac{\mathrm{d}}{\mathrm{d}p_x}}$ in the momentum representation.

(b) The fact that integration is the inverse of differentiation suggests the guess that in the momentum representation

$$\hat{x}^{-1}\phi = \left(\mathrm{i}\hbar\frac{\mathrm{d}}{\mathrm{d}p_x}\right)^{-1}\phi = \left(\frac{1}{\mathrm{i}\hbar}\int_{-\infty}^{p_x}\mathrm{d}p_x\right)\phi = \frac{1}{\mathrm{i}\hbar}\int_{-\infty}^{p_x}\phi\,\mathrm{d}p_x$$

where the symbol $\int_{-\infty}^{p_x}\mathrm{d}p_x$ is understood to be an integration operator which uses any function on its right side as an integrand. To validate the guess that $\boxed{\hat{x}^{-1} = \frac{1}{\mathrm{i}\hbar}\int_{-\infty}^{p_x}\mathrm{d}p_x}$ we need to confirm the operator relationship $\hat{x}^{-1}\hat{x} = \hat{x}\hat{x}^{-1} = \hat{1}$.

Using the Leibnitz's rule for differentiation of integrals:

$$\hat{x}\hat{x}^{-1}\phi = \left(\mathrm{i}\hbar\frac{\mathrm{d}}{\mathrm{d}p_x}\right)\left(\frac{1}{\mathrm{i}\hbar}\int_{-\infty}^{p_x}\mathrm{d}p_x\right)\phi = \left(\frac{\mathrm{d}}{\mathrm{d}p_x}\right)\left(\int_{-\infty}^{p_x}\phi\,\mathrm{d}p_x\right)$$

$$= \left(\int_{-\infty}^{p_x}\frac{\mathrm{d}\phi}{\mathrm{d}p_x}\,\mathrm{d}p_x\right) + \phi(p_x)\lim_{c\to-\infty}\frac{\mathrm{d}c}{\mathrm{d}p_x} - \phi(-\infty)\frac{\mathrm{d}p_x}{\mathrm{d}p_x} = \left(\int_{-\infty}^{p_x}\frac{\mathrm{d}\phi}{\mathrm{d}p_x}\,\mathrm{d}p_x\right) - \phi(-\infty)$$

Since $\phi(-\infty)$ must equal zero, we find that

$$\hat{x}\hat{x}^{-1}\phi = \int_{-\infty}^{p_x}\frac{\mathrm{d}\phi}{\mathrm{d}p_x}\,\mathrm{d}p_x = \phi$$

from which we conclude that $\hat{x}\hat{x}^{-1} = \hat{1}$.

$$\hat{x}^{-1}\hat{x}\phi = \left(\frac{1}{\mathrm{i}\hbar}\int_{-\infty}^{p_x}\mathrm{d}p_x\right)\left(\mathrm{i}\hbar\frac{\mathrm{d}}{\mathrm{d}p_x}\right)\phi = \int_{-\infty}^{p_x}\mathrm{d}\phi = \phi(p_x) - \phi(-\infty) = \phi(p_x) = \phi$$

Topic 9 The emergence of quantum theory

Discussion question

D9.1 In quantum mechanics, particles are said to have wave characteristics. The fact of the existence of the particle then requires that the wavelengths of the waves representing it be such that the wave does not experience destructive interference upon reflection by a barrier or in its motion around a closed loop. This requirement restricts the wavelength to values $\lambda = 2/n \times L$, where L is the length of the path and n is a positive integer. Then using the relations $\lambda = h/p$ and $E = p^2/2m$, the energy is quantized at $E = n^2h^2/8mL^2$. This derivation applies specifically to the particle in a box.

Exercises

E9.1(a) If the wavefunction is an eigenfunction of an operator, the corresponding eigenvalue is the value of corresponding observable (Topic 6.2, Postulate IV). Applying the linear momentum operator $\hat{p} = \frac{\hbar}{\mathrm{i}}\frac{\mathrm{d}}{\mathrm{d}x}$ (eqn 6.4b) to the wavefunction yields

$$\hat{p}\psi = \frac{\hbar}{\mathrm{i}}\frac{\mathrm{d}}{\mathrm{d}x}\psi = \frac{\hbar}{\mathrm{i}}\frac{\mathrm{d}}{\mathrm{d}x}\mathrm{e}^{\mathrm{i}kx} = \hbar k\mathrm{e}^{\mathrm{i}kx}$$

so the wavefunction is an eigenfunction of the linear momentum; thus, the value of the linear momentum is the eigenvalue

$$\hbar k = 1.0546\times10^{-34}\ \mathrm{J\,s}\times 3\ \mathrm{m}^{-1} = \boxed{3\times10^{-34}\ \mathrm{kg\,m\,s}^{-1}}$$

Similarly, applying the kinetic energy operator $\hat{E}_\mathrm{k} = -\frac{\hbar^2}{2m}\frac{\mathrm{d}^2}{\mathrm{d}x^2}$ (eqn 6.5) to the wavefunction yields

$$\hat{E}_\mathrm{k}\psi = -\frac{\hbar^2}{2m}\frac{\mathrm{d}^2}{\mathrm{d}x^2}\psi = -\frac{\hbar^2}{2m}\frac{\mathrm{d}^2}{\mathrm{d}x^2}\mathrm{e}^{\mathrm{i}kx} = \frac{\hbar^2k^2}{2m}\mathrm{e}^{\mathrm{i}kx}$$

so the wavefunction is an eigenfunction of this operator as well; thus, its value is the eigenvalue

$$\frac{\hbar^2 k^2}{2m} = \frac{(1.0546\times10^{-34}\ \text{J s}\times 3\ \text{m}^{-1})^2}{2\times 9.11\times10^{-31}\ \text{kg}} = \boxed{5\times10^{-38}\ \text{J}}$$

E9.2(a) The wavefunction for the particle is (eqn 9.3 with $B=0$ because the particle is moving toward positive x)

$$\psi_k = \boxed{Ae^{ikx}}$$

The index k is given by the relationship

$$E_k = \frac{\hbar^2 k^2}{2m} = 20\ \text{J}$$

so $$k = \frac{\sqrt{2mE_k}}{\hbar} = \frac{\left(2(2.0\times10^{-3}\ \text{kg})(20\ \text{J})\right)^{1/2}}{1.0546\times10^{-34}\ \text{J s}} = \boxed{2.7\times10^{33}\ \text{m}^{-1}}$$

E9.3(a) $$E_n = \frac{n^2h^2}{8m_eL^2} \quad [9.8b]$$

$$\frac{h^2}{8m_eL^2} = \frac{(6.626\times10^{-34}\ \text{J s})^2}{8(9.11\times10^{-31}\ \text{kg})\times(1.0\times10^{-9}\ \text{m})^2} = 6.0\bar{2}\times10^{-20}\ \text{J}$$

The conversion factors required are

$1\ \text{kJ mol}^{-1} = 6.022\times10^{20}\ \text{J}$ and $1\ \text{eV} = 1.602\times10^{-19}\ \text{J}$.

Strictly speaking, wavenumbers are not an energy unit, but we can still consider a "conversion factor" between joules and cm^{-1} because of the relationship

$$\Delta E = hc\tilde{\nu}$$

Thus, the energy that corresponds to $1\ \text{cm}^{-1}$ is $hc\times 1\ \text{cm}^{-1} = 1.986\times10^{-23}\ \text{J}$

(a) $$E_2 - E_1 = (4-1)\frac{h^2}{8m_eL^2} = \frac{3h^2}{8m_eL^2} = (3)\times(6.0\bar{2}\times10^{-20}\ \text{J})$$

$$= \boxed{1.8\times10^{-19}\ \text{J}},\ \boxed{1.1\times10^{2}\ \text{kJ mol}^{-1}},\ \boxed{1.1\ \text{eV}},\ \boxed{9.1\times10^{3}\ \text{cm}^{-1}}$$

(b) $$E_6 - E_5 = (36-25)\frac{h^2}{8m_eL^2} = \frac{11h^2}{8m_eL^2} = (11)\times(6.0\bar{2}\times10^{-20}\ \text{J})$$

$$= \boxed{6.6\times10^{-19}\ \text{J}},\ \boxed{4.0\times10^{2}\ \text{kJ mol}^{-1}},\ \boxed{4.1\ \text{eV}},\ \boxed{3.3\times10^{4}\ \text{cm}^{-1}}$$

Comment. The energy level separations increase as n increases.

E9.4(a) The wavefunctions are

$$\psi_n = \left(\frac{2}{L}\right)^{1/2} \sin\left(\frac{n\pi x}{L}\right) \quad [9.8a]$$

The required probabilities are

$$P_n = \int_{0.49L}^{0.51L} \psi_n^2 \, dx \approx \psi_n^2 \Delta x$$

where $\Delta x = 0.02L$ and the function is evaluated at the midpoint of the interval, at $x = L/2$.

(a) $P_1 = \left(\frac{2}{L}\right)\sin^2\left(\frac{\pi}{2}\right) \times 0.02L = \boxed{0.04}$

(b) $P_2 = \left(\frac{2}{L}\right)\sin^2\left(\frac{2\pi}{2}\right) \times 0.02L = \left(\frac{2}{L}\right)\sin^2 \pi \times 0.02L = \boxed{0}$

E9.5(a) The wavefunction for a particle in the state $n = 1$ in a square-well potential is

$$\psi_1 = \left(\frac{2}{L}\right)^{1/2} \sin\left(\frac{\pi x}{L}\right) \quad [9.8a]$$

The expectation value of the momentum operator, $\hat{p} = \frac{\hbar}{\mathrm{i}} \frac{\mathrm{d}}{\mathrm{d}x}$, is

$$\langle p \rangle = \int_0^L \psi_1 * \hat{p}\psi_1 \, \mathrm{d}x = \frac{2\hbar}{\mathrm{i}L} \int_0^L \sin\left(\frac{\pi x}{L}\right)\frac{\mathrm{d}}{\mathrm{d}x}\sin\left(\frac{\pi x}{L}\right)\mathrm{d}x$$
$$= \frac{2\pi\hbar}{\mathrm{i}L^2}\int_0^L \sin\left(\frac{\pi x}{L}\right)\cos\left(\frac{\pi x}{L}\right)\mathrm{d}x = \boxed{0}$$

The expectation value of the momentum squared operator, $\hat{p}^2 = -\hbar^2 \frac{\mathrm{d}^2}{\mathrm{d}x^2}$, is

$$\langle p^2 \rangle = -\frac{2\hbar^2}{L}\int_0^L \sin\left(\frac{\pi x}{L}\right)\frac{\mathrm{d}^2}{\mathrm{d}x^2}\sin\left(\frac{\pi x}{L}\right)\mathrm{d}x = \left(\frac{2\hbar^2}{L}\right)\times\left(\frac{\pi}{L}\right)^2 \int_0^L \sin^2\left(\frac{\pi x}{L}\right)\mathrm{d}x$$

Use $\int (\sin^2 ax)\mathrm{d}x = \frac{x}{2} - \frac{1}{4a}\sin 2ax$ with $a = \pi/L$. Thus

$$\langle p^2 \rangle = \left(\frac{2\hbar^2}{L}\right)\times\left(\frac{\pi}{L}\right)^2\left(\frac{x}{2} - \frac{L}{4\pi}\sin\left(\frac{2\pi x}{L}\right)\right)\Bigg|_0^L = \left(\frac{2\hbar^2}{L}\right)\times\left(\frac{\pi}{L}\right)^2\times\left(\frac{L}{2}\right) = \boxed{\frac{h^2}{4L^2}}$$

Comment. The expectation value of $\hat{p}$ is zero because on average the particle moves to the left as often as the right.

E9.6(a) The zero-point energy is the ground-state energy, that is, with $n = 1$:

$$E = \frac{n^2h^2}{8m_\mathrm{e}L^2} \, [9.8b] = \frac{h^2}{8m_\mathrm{e}L^2}$$

Set this equal to the rest energy $m_e c^2$ and solve for L:

$$m_e c^2 = \frac{h^2}{8m_e L^2} \quad \text{so} \quad L = \frac{h}{8^{1/2} m_e c}$$

In absolute units, the length is

$$L = \frac{6.63 \times 10^{-34}\ \text{J s}}{8^{1/2} \times (9.11 \times 10^{-31}\ \text{kg}) \times (3.00 \times 10^{8}\ \text{m s}^{-1})} = 8.58 \times 10^{-13}\ \text{m} = 0.858\ \text{pm}$$

In terms of the Compton wavelength of an electron, $\lambda_C = \dfrac{h}{m_e c}$, $L = \boxed{\dfrac{\lambda_C}{8^{1/2}}}$

E9.7(a) $$\psi_5 = \left(\frac{2}{L}\right)^{1/2} \sin\left(\frac{5\pi x}{L}\right) \quad [9.8a]$$

$$P(x) \propto \psi_5^2 \propto \sin^2\left(\frac{5\pi x}{L}\right)$$

The maxima and minima in $P(x)$ correspond to $\dfrac{dP(x)}{dx} = 0$.

$$\frac{dP(x)}{dx} \propto \frac{d\psi^2}{dx} \propto \sin\left(\frac{5\pi x}{L}\right)\cos\left(\frac{5\pi x}{L}\right) \propto \sin\left(\frac{10\pi x}{L}\right) \quad [2\sin\alpha\cos\alpha = \sin 2\alpha]$$

$\sin\theta = 0$ when θ equals an integer (call it n') times π:

$$\frac{10\pi x}{L} = n'\pi \ \text{ for } n' = 0, 1, 2, \ldots \text{ which corresponds to } x = \frac{n'L}{10},\ n' \le 10.$$

$n' = 0, 2, 4, 6, 8$, and 10 correspond to minima in ψ^2, leaving $n' = 1, 3, 5, 7, 9$ for the maxima, that is

$$\boxed{x = \frac{L}{10}, \frac{3L}{10}, \frac{L}{2}, \frac{7L}{10}, \text{ and } \frac{9L}{10}}$$

Comment. Maxima in ψ^2 correspond to maxima *and* minima in ψ itself, so one can also solve this exercise by finding all points where $\dfrac{d\psi}{dx} = 0$.

Problems

P9.1 $$E = \frac{n^2h^2}{8mL^2} \ [9.8b], \quad E_2 - E_1 = \frac{3h^2}{8mL^2}$$

We take $m(O_2) = (32.000) \times (1.6605 \times 10^{-27}\ \text{kg})$, and find

$$E_2 - E_1 = \frac{(3)\times(6.626\times10^{-34}\ \mathrm{J\ s})^2}{(8)\times(32.00)\times(1.6605\times10^{-27}\ \mathrm{kg})\times(5.0\times10^{-2}\ \mathrm{m})^2} = \boxed{1.24\times10^{-39}\ \mathrm{J}}$$

We set $E = \dfrac{n^2h^2}{8mL^2} = \dfrac{1}{2}kT$ and solve for n.

From above $\dfrac{h^2}{8mL^2} = \dfrac{E_2 - E_1}{3} = 4.13\times10^{-40}$ J; then

$$n^2 \times (4.13\times10^{-40}\ \mathrm{J}) = \left(\tfrac{1}{2}\right)\times(1.381\times10^{-23}\ \mathrm{J\,K^{-1}})\times(300\ \mathrm{K}) = 2.07\times10^{-21}\ \mathrm{J}$$

We find $n = \left(\dfrac{2.07\times10^{-21}\ \mathrm{J}}{4.13\times10^{-40}\ \mathrm{J}}\right)^{1/2} = \boxed{2.2\times10^{9}}$

At this level,

$$E_n - E_{n-1} = \{n^2 - (n-1)^2\}\times\frac{h^2}{8mL^2} = (2n-1)\times\frac{h^2}{8mL^2} \approx (2n)\times\frac{h^2}{8mL^2}$$

$$= (4.4\times10^{9})\times(4.13\times10^{-40}\ \mathrm{J}) \approx \boxed{1.8\times10^{-30}\ \mathrm{J}}\ \ [\text{or } 1.1\ \mu\mathrm{J\ mol^{-1}}]$$

P9.3 The probability is given by

$$P = \int |\psi^2|\,\mathrm{d}x$$

where $|\psi^2| = \psi^2 = \dfrac{2}{L}\sin^2\left(\dfrac{\pi x}{L}\right)$ [9.9].

Use the integral

$$\int \sin^2 ax\ \mathrm{d}x = \frac{x}{2} - \frac{\sin 2ax}{4a}$$

(a) $P = \dfrac{2}{L}\displaystyle\int_0^{L/2} \sin^2\left(\dfrac{\pi x}{L}\right)\mathrm{d}x = \dfrac{2}{L}\left(\dfrac{x}{2} - \dfrac{L}{4\pi}\sin\left(\dfrac{2\pi x}{L}\right)\right)\Bigg|_0^{L/2} = \boxed{\dfrac{1}{2}}$

(b) $P = \dfrac{2}{L}\left\{\dfrac{x}{2} - \dfrac{L}{4\pi}\sin\left(\dfrac{2\pi x}{L}\right)\right\}\Bigg|_0^{L/4} = \dfrac{2}{L}\left\{\dfrac{L}{8} - \dfrac{L}{4\pi}\sin\left(\dfrac{\pi}{2}\right)\right\} = \boxed{\dfrac{1}{4} - \dfrac{1}{2\pi} = 0.0908}$

(c) Here the interval of integration is small, so the integral is equal to the integrand evaluated at the midpoint of the interval times the size of the interval:

$$P = \frac{2}{L}\int_{\frac{1}{2}L-\delta x}^{\frac{1}{2}L+\delta x} \sin^2\left(\frac{\pi x}{L}\right)\mathrm{d}x = \frac{2}{L}\sin^2\left(\frac{\pi}{2}\right)\times 2\delta x = \boxed{\frac{4\delta x}{L}}$$

P9.5 The normalization integral is

$$1=\int_0^L \psi^*\psi\,\mathrm{d}x=\int_0^L (N\mathrm{e}^{-\mathrm{i}kx})(N\mathrm{e}^{\mathrm{i}kx})\,\mathrm{d}x=N^2\int_0^L \mathrm{d}x=N^2L$$

So the normalization constant is

$$N=\boxed{\left(\frac{1}{L}\right)^{1/2}}$$

P9.7 Taking the hint, we compute Δx for the classical limit, *i.e.*, for a uniform distribution. The average position is

$$\langle x\rangle=\int_0^L\left(\frac{1}{L}\right)^{1/2}x\left(\frac{1}{L}\right)^{1/2}\mathrm{d}x=\frac{1}{L}\frac{x^2}{2}\bigg|_0^L=\frac{L}{2}$$ (which we would also expect by symmetry).

$$\langle x^2\rangle=\int_0^L\left(\frac{1}{L}\right)^{1/2}x^2\left(\frac{1}{L}\right)^{1/2}\mathrm{d}x=\frac{1}{L}\frac{x^3}{3}\bigg|_0^L=\frac{L^2}{3}$$

So $\Delta x=\left(\langle x^2\rangle-\langle x\rangle^2\right)^{1/2}\ [8.3]=\left(\frac{L^2}{3}-\frac{L^2}{4}\right)^{1/2}=L\left(\frac{1}{12}\right)^{1/2}$

in the classical limit. Now use particle-in-a-box wavefunctions:

$$\langle x\rangle=\int_0^L\left(\frac{2}{L}\right)^{1/2}\sin\left(\frac{n\pi x}{L}\right)x\left(\frac{2}{L}\right)^{1/2}\sin\left(\frac{n\pi x}{L}\right)\mathrm{d}x$$

$$\langle x\rangle=\left(\frac{2}{L}\right)\int_0^L x\sin^2 ax\ \mathrm{d}x\quad\left[a=\frac{n\pi}{L}\right]$$

$$=\left(\frac{2}{L}\right)\times\left(\frac{x^2}{4}-\frac{x\sin 2ax}{4a}-\frac{\cos 2ax}{8a^2}\right)\bigg|_0^L$$ [Trig functions vanish or cancel at limits.]

$$=\left(\frac{2}{L}\right)\times\left(\frac{L^2}{4}\right)=\frac{L}{2}$$ [by symmetry also].

$$\langle x^2\rangle=\frac{2}{L}\int_0^L x^2\sin^2 ax\,\mathrm{d}x=\left(\frac{2}{L}\right)\times\left[\frac{x^3}{6}-\left(\frac{x^2}{4a}-\frac{1}{8a^3}\right)\sin 2ax-\frac{x\cos 2ax}{4a^2}\right]\bigg|_0^L$$

$$=\left(\frac{2}{L}\right)\times\left(\frac{L^3}{6}-\frac{L^3}{4n^2\pi^2}\right)=L^2\left(\frac{1}{3}-\frac{1}{2n^2\pi^2}\right).$$

$$\Delta x=\left\{L^2\left(\frac{1}{3}-\frac{1}{2n^2\pi^2}\right)-\frac{L^2}{4}\right\}^{1/2}=L\left(\frac{1}{12}-\frac{1}{2\pi^2n^2}\right)^{1/2}$$

Take the limit of this expression as n approaches infinity: the second term in parentheses vanishes, so the particle-in-a-box expression approaches the classical limit.

Topic 10 Tunnelling

Discussion question

D10.1 The physical origin of tunnelling is related to the probability density of the particle, which according to the Born interpretation is the square of the wavefunction that represents the particle. This interpretation requires that the wavefunction of the system be everywhere continuous, even at barriers. Therefore, if the wavefunction is non-zero on one side of a barrier it must be non-zero on the other side of the barrier and this implies that the particle can tunnel through the barrier (*i.e.*, that it has a non-zero probability of tunnelling through). The transmission probability depends upon the mass of the particle (specifically on $m^{1/2}$, through eqns 10.1 and 10.4): the greater the mass the smaller the probability of tunnelling. Electrons and protons have small masses, and molecular groups have large masses; therefore, tunnelling effects are more observable in processes involving electrons and protons. Thus tunnelling can provide important pathways for electron-transfer reactions, from such simple examples as the formation and bonding of monatomic ions to complex biochemical processes such as photosynthesis. Tunnelling can also be significant in proton-transfer reactions (acid-base reactions). It is unimportant in reactions involving the migration of larger groups of atoms, even relatively small ones like methyl groups.

Exercise

E10.1(a) The transmission probability (eqn 10.6) depends on the energy of the tunneling particle relative to the barrier height ($\varepsilon = E/V = 1.5\ \text{eV}/(2.0\ \text{eV}) = 0.75$), on the width of the barrier ($L = 100$ pm), and on the decay parameter of the wavefunction inside the barrier (κ), where (eqn 10.3)

$$\kappa = \frac{\{2m(V-E)\}^{1/2}}{\hbar} = \frac{\{2\times 9.11\times 10^{-31}\ \text{kg}\times(2.0-1.5)\ \text{eV}\times 1.602\times 10^{-19}\ \text{J eV}^{-1}\}^{1/2}}{1.0546\times 10^{-34}\ \text{J s}}$$

$$= 3.\overline{6}\times 10^{9}\ \text{m}^{-1}$$

We note that $\kappa L = 3.\overline{6}\times 10^{9}\ \text{m}^{-1}\times 100\times 10^{-12}\ \text{m} = 0.3\overline{6}$ is not large compared to 1, so we must use eqn 10.6 for the transmission probability.

$$T=\left\{1+\frac{(e^{\kappa L}-e^{-\kappa L})^2}{16\varepsilon(1-\varepsilon)}\right\}^{-1}=\left\{1+\frac{(e^{0.3\overline{6}}-e^{-0.3\overline{6}})^2}{16\times 0.75\times(1-0.75)}\right\}^{-1}=\boxed{0.8}$$

Problems

P10.1 The text defines the transmission probability and expresses it as the ratio of $|A'|^2/|A|^2$, where the coefficients A and A' are introduced in eqns 10.1 and 10.4. Eqn 10.5 lists four equations for the six unknown coefficients of the full wavefunction. Once we realize that we can set B' to zero, these equations in five unknowns are:

(a) $A+B=C+D$

(b) $Ce^{\kappa L}+De^{-\kappa L}=A'e^{ikL}$

(c) $ikA-ikB=\kappa C-\kappa D$

(d) $\kappa Ce^{\kappa L}-\kappa De^{-\kappa L}=ikA'e^{ikL}$

We need A' in terms of A alone, which means we must eliminate B, C, and D. Notice that B appears only in eqns (a) and (c). Solving these equations for B and setting the results equal to each other yields:

$$B=C+D-A=A-\frac{\kappa C}{ik}+\frac{\kappa D}{ik}.$$

Solve this equation for C:

$$C=\frac{2A+D\left(\dfrac{\kappa}{ik}-1\right)}{\dfrac{\kappa}{ik}+1}=\frac{2Aik+D(\kappa-ik)}{\kappa+ik}.$$

Now note that the desired A' appears only in (b) and (d). Solve these for A' and set them equal:

$$A'=e^{-ikL}(Ce^{\kappa L}+De^{-\kappa L})=\frac{\kappa e^{-ikL}}{ik}(Ce^{\kappa L}-De^{-\kappa L})$$

Solve the resulting equation for C, and set it equal to the previously obtained expression for C:

$$C=\frac{\left(\dfrac{\kappa}{ik}+1\right)De^{-2\kappa L}}{\dfrac{\kappa}{ik}-1}=\frac{(\kappa+ik)De^{-2\kappa L}}{\kappa-ik}=\frac{2Aik+D(\kappa-ik)}{\kappa+ik}.$$

Solve this resulting equation for D in terms of A:

$$\frac{(\kappa+ik)^2 e^{-2\kappa L}-(\kappa-ik)^2}{(\kappa-ik)(\kappa+ik)}D=\frac{2Aik}{\kappa+ik},$$

so $$D=\frac{2Aik(\kappa-ik)}{(\kappa+ik)^2 e^{-2\kappa L}-(\kappa-ik)^2}.$$

Substituting this expression back into an expression for C yields:

$$C=\frac{2Aik(\kappa+ik)e^{-2\kappa L}}{(\kappa+ik)^2 e^{-2\kappa L}-(\kappa-ik)^2}.$$

Substituting for C and D in the expression for A' yields:

$$A'=e^{-ikL}(Ce^{\kappa L}+De^{-\kappa L})=\frac{2Aike^{-ikL}}{(\kappa+ik)^2 e^{-2\kappa L}-(\kappa-ik)^2}[(\kappa+ik)e^{-\kappa L}+(\kappa-ik)e^{-\kappa L}],$$

so $$\frac{A'}{A}=\frac{4ik\kappa e^{-\kappa L}e^{-ikL}}{(\kappa+ik)^2 e^{-2\kappa L}-(\kappa-ik)^2}=\frac{4ik\kappa e^{-ikL}}{(\kappa+ik)^2 e^{-\kappa L}-(\kappa-ik)^2 e^{\kappa L}}.$$

The transmission coefficient is:

$$T=\frac{|A'|^2}{|A|^2}=\left(\frac{4ik\kappa e^{-ikL}}{(\kappa+ik)^2 e^{-\kappa L}-(\kappa-ik)^2 e^{\kappa L}}\right)\left(\frac{-4ik\kappa e^{ikL}}{(\kappa-ik)^2 e^{-\kappa L}-(\kappa+ik)^2 e^{\kappa L}}\right).$$

The denominator is worth expanding separately in several steps. It is:

$$\begin{aligned}&(\kappa+ik)^2(\kappa-ik)^2 e^{-2\kappa L}-(\kappa-ik)^4-(\kappa+ik)^4+(\kappa-ik)^2(\kappa+ik)^2 e^{2\kappa L}\\&=(\kappa^2+k^2)^2(e^{2\kappa L}+e^{-2\kappa L})-(\kappa^2-2i\kappa k-k^2)^2-(\kappa^2+2i\kappa k-k^2)^2\\&=(\kappa^4+2\kappa^2k^2+k^4)(e^{2\kappa L}+e^{-2\kappa L})-(2\kappa^4-12\kappa^2k^2+2k^2).\end{aligned}$$

If the $12\kappa^2k^2$ term were $-4\kappa^2k^2$ instead, we could collect terms still further (completing the square), but of course we must also account for the difference between those quantities, making the denominator:

$$(\kappa^4+2\kappa^2k^2+k^4)(e^{2\kappa L}-2+e^{-2\kappa L})+16\kappa^2k^2=(\kappa^2+k^2)^2(e^{\kappa L}-e^{-\kappa L})^2+16\kappa^2k^2$$

So the coefficient is:

$$T=\frac{16k^2\kappa^2}{(\kappa^2+k^2)^2(e^{\kappa L}-e^{-\kappa L})^2+16\kappa^2k^2}.$$

We are almost there. To get to eqn 10.6, we invert the expression:

$$T=\left(\frac{(\kappa^2+k^2)^2(e^{\kappa L}-e^{-\kappa L})^2+16\kappa^2k^2}{16k^2\kappa^2}\right)^{-1}=\left(\frac{(\kappa^2+k^2)^2(e^{\kappa L}-e^{-\kappa L})^2}{16k^2\kappa^2}+1\right)^{-1}$$

Finally, we try to express $\dfrac{(\kappa^2+k^2)^2}{k^2\kappa^2}$ in terms of a ratio of energies, $\varepsilon = E/V$. Eqns 10.3 and 10.4 define κ and k. The factors involving 2, $\hbar$, and the mass cancel leaving $\kappa \propto (V-E)^{1/2}$ and $k \propto E^{1/2}$, so:

$$\frac{(\kappa^2+k^2)^2}{k^2\kappa^2} = \frac{[E+(V-E)]^2}{E(V-E)} = \frac{V^2}{E(V-E)} = \frac{1}{\varepsilon(1-\varepsilon)},$$

which makes the transmission coefficient:

$$T = \boxed{\left(\frac{(e^{\kappa L}-e^{-\kappa L})^2}{16\varepsilon(1-\varepsilon)}+1\right)^{-1}} \quad [10.6,\ \text{QED}]$$

If $\kappa L >> 1$, then the negative exponential is negligible compared to the positive, and the 1 inside the parentheses is negligible compared to the exponential:

$$T \approx \left(\frac{e^{2\kappa L}}{16\varepsilon(1-\varepsilon)}\right)^{-1} = \frac{16\varepsilon(1-\varepsilon)}{e^{2\kappa L}} = \boxed{16\varepsilon(1-\varepsilon)e^{-2\kappa L}} \quad [10.7,\ \text{QED}]$$

P10.3 (a) The wavefunctions in each region (see Fig. 10.1) are (eqns 10.1, 10.3, and 10.4)

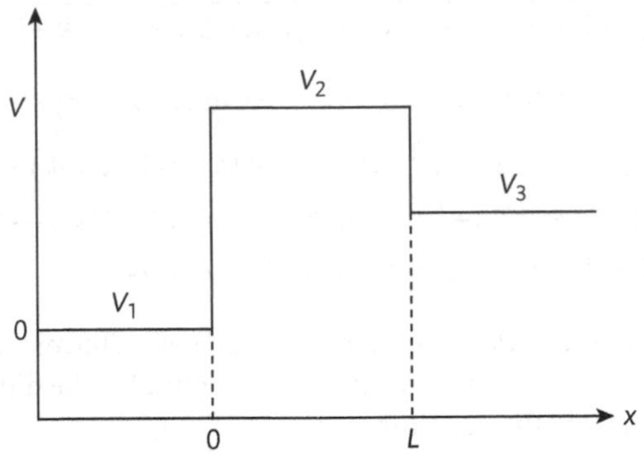

Fig 10.1

$$\psi_1(x) = e^{ik_1x} + B_1e^{-ik_1x}$$

$$\psi_2(x) = A_2e^{k_2x} + B_2e^{-k_2x}$$

$$\psi_3(x) = A_3e^{ik_3x}$$

With the above choice of $A_1 = 1$ the transmission probability is simply $T = |A_3|^2$. The wavefunction coefficients are determined by the criteria that both the wavefunctions and their first derivatives with respect to x be continuous at potential boundaries

$$\psi_1(0)=\psi_2(0);\quad \psi_2(L)=\psi_3(L)$$

$$\frac{d\psi_1(0)}{dx}=\frac{d\psi_2(0)}{dx};\quad \frac{d\psi_2(L)}{dx}=\frac{d\psi_3(L)}{dx}$$

These criteria establish the algebraic relationships:

$$1+B_1-A_2-B_2=0$$
$$ik_1-ik_1B_1-k_2A_2+k_2B_2=0$$
$$A_2e^{k_2L}+B_2e^{-k_2L}-A_3e^{ik_3L}=0$$
$$A_2k_2e^{k_2L}-B_2k_2e^{-k_2L}-iA_3k_3e^{ik_3L}=0$$

Solving the simultaneous equations for A_3 gives

$$A_3=\frac{4k_1k_2\,e^{-ik_3L}}{(ia+b)e^{k_2L}-(ia-b)e^{-k_2L}}$$

where $a=k_2^2-k_1k_3$ and $b=k_1k_2+k_2k_3$.
Since $\sinh(z)=(e^z-e^{-z})/2$ or $e^z=2\sinh(z)+e^{-z}$, substitute $e^{k_2L}=2\sinh(k_2L)+e^{-k_2L}$ giving

$$A_3=\frac{2k_1k_2e^{-ik_3L}}{(ia+b)\sinh(k_2L)+b\,e^{-k_2L}}$$

$$T=|A_3|^2=A_3\times A_3^*=\frac{4k_1^2k_2^2}{(a^2+b^2)\sinh^2(k_2L)+b^2}$$

where $a^2+b^2=(k_1^2+k_2^2)(k_2^2+k_3^2)$ and $b^2=k_2^2(k_1+k_3)^2$

(b) In the special case for which $V_3=0$, eqns 10.1 and 10.4 require that $k_1=k_3$. Additionally,

$$\left(\frac{k_1}{k_2}\right)^2=\frac{E}{V_2-E}=\frac{\varepsilon}{1-\varepsilon}\quad\text{where }\varepsilon=E/V_2.$$

$$a^2+b^2=(k_1^2+k_2^2)^2=k_2^4\left\{1+\left(\frac{k_1}{k_2}\right)^2\right\}^2$$

$$b^2=4k_1^2k_2^2$$

$$\frac{a^2+b^2}{b^2}=\frac{k_2^2\left\{1+\left(\frac{k_1}{k_2}\right)^2\right\}^2}{4k_1^2}=\frac{1}{4\varepsilon(1-\varepsilon)}$$

$$T = \frac{b^2}{b^2 + (a^2 + b^2)\sinh^2(k_2 L)} = \frac{1}{1 + \left(\dfrac{a^2 + b^2}{b^2}\right)\sinh^2(k_2 L)}$$

$$T = \left\{1 + \frac{\sinh^2(k_2 L)}{4\varepsilon(1-\varepsilon)}\right\}^{-1} = \left\{1 + \frac{(e^{k_2 L} - e^{-k_2 L})^2}{16\varepsilon(1-\varepsilon)}\right\}^{-1}$$

This is eqn 10.6. In the high, wide barrier limit $k_2L >> 1$. This implies both that e^{-k_2L} is negligibly small compared to e^{k_2L} and that 1 is negligibly small compared to $e^{2k_2L}/\{16\varepsilon(1-\varepsilon)\}$. The previous equation simplifies to eqn 10.7.

$$T = 16\,\varepsilon(1-\varepsilon)e^{-2k_2L}$$

P10.5 The rate of tunnelling is proportional to the transmission probability, so a ratio of tunnelling rates is equal to the corresponding ratio of transmission probabilities (given in eqn 10.6). The desired factor is T_1/T_2, where the subscripts denote the tunnelling distances in nanometers:

$$\frac{T_1}{T_2} = \frac{1 + \dfrac{(e^{\kappa L_2} - e^{-\kappa L_2})^2}{16\varepsilon(1-\varepsilon)}}{1 + \dfrac{(e^{\kappa L_1} - e^{-\kappa L_1})^2}{16\varepsilon(1-\varepsilon)}}$$

If $\kappa L>>1$, then each transmission coefficient is approximately $T \approx 16\varepsilon(1-\varepsilon)e^{-2\kappa L}$ [10.7], and $\frac{T_1}{T_2} \approx \frac{(e^{\kappa L_2} - e^{-\kappa L_2})^2}{(e^{\kappa L_1} - e^{-\kappa L_1})^2} \approx e^{2\kappa(L_2 - L_1)} = e^{2(7/\text{nm})(2.0-1.0)\text{nm}} = \boxed{1.2\times10^6}$.

That is, the tunneling rate increases about a million-fold. Note: κL need not be terribly large for the first approximation to hold; see Exercise 10.1(b).

Topic 11 **Translational motion in several dimensions**

Discussion question

D11.1 Because translational motion in two or three dimensions can be separated into independent one-dimensional motions, many features of the one-dimensional solutions carry over. For example, quantization occurs in each dimension just as in the one-dimensional case; the energy associated with motion in each dimension has the same form as in the one-dimensional case; one-dimensional wavefunctions are factors in the multi-dimensional wavefunction. The concept of degeneracy—of more than one distinct wavefunction having the same energy—did not arise for a particle in a one-dimensional box, but it can arise in the multi-dimensional case (depending on the proportions of the box). For example, in a square two-dimensional box, the state with $n_1 = 1$ and $n_2 = 2$ has the same energy as the state $n_1 = 2$ and $n_2 = 1$.

Exercises

E11.1(a) $$E_{n_1,n_2} = \frac{(n_1^2+n_2^2)h^2}{8m_eL^2} \quad [11.4\text{b, with } L_1 = L_2 = L]$$

$$\frac{h^2}{8m_eL^2} = \frac{(6.626\times10^{-34}\text{ J s})^2}{8(9.11\times10^{-31}\text{ kg})\times(1.0\times10^{-9}\text{ m})^2} = 6.0\bar{2}\times10^{-20}\text{ J}$$

The conversion factors required are

$$1\text{ kJ mol}^{-1} = 6.022\times10^{20}\text{ J} \quad\text{and}\quad 1\text{ eV} = 1.602\times10^{-19}\text{ J}.$$

Strictly speaking, wavenumbers are not an energy unit, but we can still consider a "conversion factor" between joules and cm^{-1} because of the relationship

$$\Delta E = hc\tilde{\nu}$$

Thus, the energy that corresponds to 1 cm^{-1} is $hc \times 1\text{ cm}^{-1} = 1.986\times10^{-23}\text{ J}$

$$\text{(a)}\ E_{2,2}-E_{1,1}=\frac{(2^2+2^2-1^2-1^2)h^2}{8m_eL^2}=\frac{6h^2}{8m_eL^2}=6\times6.0\bar{2}\times10^{-20}\ \text{J}$$

$$=\boxed{3.6\times10^{-19}\ \text{J}}=\boxed{2.2\times10^{2}\ \text{kJ mol}^{-1}}=\boxed{2.3\ \text{eV}}=\boxed{1.8\times10^{4}\ \text{cm}^{-1}}$$

$$\text{(b)}\ E_{6,6}-E_{5,5}=\frac{(6^2+6^2-5^2-5^2)h^2}{8m_eL^2}=\frac{22h^2}{8m_eL^2}=22\times6.0\bar{2}\times10^{-20}\ \text{J}$$

$$=\boxed{1.3\times10^{-18}\ \text{J}}=\boxed{8.0\times10^{2}\ \text{kJ mol}^{-1}}=\boxed{8.3\ \text{eV}}=\boxed{6.7\times10^{4}\ \text{cm}^{-1}}$$

E11.2(a) (a) Radiation stimulates a transition if the energy of the photon is equal to the difference between energy levels. The photon energy is $h\nu$, so

$$\nu=\frac{\Delta E}{h}=\frac{3.6\times10^{-19}\ \text{J}}{6.626\times10^{-34}\ \text{J s}}=\boxed{5.5\times10^{14}\ \text{s}^{-1}}$$

The wavelength is the reciprocal of the wavenumber, so

$$\lambda=\frac{1}{1.8\times10^{4}\ \text{cm}^{-1}}=5.5\times10^{-5}\ \text{cm}=\boxed{5.5\times10^{-7}\ \text{m}}=\boxed{550\ \text{nm}}$$

$$\text{(b)}\ \nu=\frac{\Delta E}{h}=\frac{1.3\times10^{-18}\ \text{J}}{6.626\times10^{-34}\ \text{J s}}=\boxed{2.0\times10^{15}\ \text{s}^{-1}}$$

$$\text{and}\ \lambda=\frac{1}{6.7\times10^{4}\ \text{cm}^{-1}}=1.5\times10^{-5}\ \text{cm}=\boxed{1.5\times10^{-7}\ \text{m}}=\boxed{150\ \text{nm}}$$

E11.3(a) (a) The energy is

$$E_{n_1,n_2}=\left(\frac{n_1^2}{L_1^2}+\frac{n_2^2}{L_2^2}\right)\frac{h^2}{8m_e}\qquad[11.4\text{b}],$$

so the zero-point energy is the ground-state energy

$$E_{1,1}=\left(\frac{1^2}{(1.0\times10^{-9}\,\text{m})^2}+\frac{1^2}{(2.0\times10^{-9}\,\text{m})^2}\right)\frac{(6.626\times10^{-34}\ \text{J s})^2}{8\times9.11\times10^{-31}\,\text{kg}}=7.5\times10^{-20}\ \text{J}$$

The temperature at which this energy equals the typical thermal energy kT is

$$T=\frac{E_{1,1}}{k}=\frac{7.5\times10^{-20}\ \text{J}}{1.381\times10^{-23}\ \text{J K}^{-1}}=\boxed{5.5\times10^{3}\ \text{K}}$$

(b) The first excited energy level (minimal excitation in the longer dimension) is

$$E_{1,2}=\left(\frac{1^2}{(1.0\times10^{-9}\ \text{m})^2}+\frac{2^2}{(2.0\times10^{-9}\ \text{m})^2}\right)\frac{(6.626\times10^{-34}\ \text{J s})^2}{8\times9.11\times10^{-31}\ \text{kg}}=1.2\bar{0}\times10^{-19}\ \text{J},$$

so the first excitation energy is

$$E_{1,2}-E_{1,1}=(1.2\bar{0}\times10^{-19}-7.5\times10^{-20})\ \text{J}=4.\bar{5}\times10^{-20}\ \text{J}$$

The temperature at which this energy equals the typical thermal energy kT is

$$T=\frac{4.\overline{5}\times10^{-20}\ \text{J}}{1.381\times10^{-23}\ \text{J K}^{-1}}=\boxed{3.\overline{3}\times10^{3}\ \text{K}}$$

Comment. The temperature computed in part (b) is meaningful in the sense that a significant fraction of nanostructures at that temperature would be in the first excited state because there is sufficient thermal energy to excite them there. The temperature computed in part (a) is not similarly meaningful, though, because the zero-point energy is, by definition, the energy of the lowest-energy state, and that state is occupied even at zero temperature. The computed temperature in part (a) corresponds to the difference in energy between the zero-point energy and a hypothetical motionless state (which is precluded according to the uncertainty principle.)

E11.4(a) The wavefunctions are

$$\psi_{n_1,n_2}=\frac{2}{L}\sin\left(\frac{n_1\pi x}{L}\right)\sin\left(\frac{n_2\pi y}{L}\right)\ [11.4\text{a, with } L_1=L_2=L]$$

The required probability is

$$P_{n_1,n_2}=\int_{0.49L}^{0.51L}\int_{0.49L}^{0.51L}\psi_{n_1,n_2}^2\,\mathrm{d}x\,\mathrm{d}y\approx\psi_{n_1,n_2}^2\,\Delta x\Delta y$$

where $\Delta x=\Delta y=0.02L$, and the function is evaluated at $x=y=L/2$.

(a) $P_{1,1}=\left(\frac{2}{L}\right)^2\sin^2\left(\frac{\pi}{2}\right)\sin^2\left(\frac{\pi}{2}\right)\times0.02L\times0.02L=\boxed{0.0016}$

(b) $P_{2,2}=\left(\frac{2}{L}\right)^2\sin^2\left(\frac{2\pi}{2}\right)\sin^2\left(\frac{2\pi}{2}\right)\times0.02L\times0.02L=\boxed{0}$

Comment. Check whether the results make sense. The region is a small square in the center of the two-dimensional square box comprising a fraction of only 0.0004 of the area of the box. In part (a), both factors of the wavefunction have maxima in the region, so we would expect the electron to be **more likely** to be found there than if the probability were uniform. Indeed, the computed probability of 0.0016, though small compared to 1, is in fact larger than 0.0004, which is what the probability would be for a uniform distribution. In part (b), both factors of the wavefunction have nodes in the region, so we expect a low probability, and once again the calculation is consistent with this expectation.

E11.5(a) See the comment at the end of the solution to Exercise 9.7(a): Maxima in the probability distribution, which is equal to ψ^2, correspond to maxima *and* minima in ψ itself, so we look for points where $\frac{\partial\psi}{\partial x}=\frac{\partial\psi}{\partial y}=0$.

$$\psi_{2,3}=\frac{2}{L}\sin\left(\frac{4\pi x}{L}\right)\sin\left(\frac{5\pi y}{L}\right)\ [11.4\text{a, with } L_1=L_2]$$

$$\frac{\partial\psi_{2,3}}{\partial x}=\frac{8\pi}{L^2}\cos\left(\frac{4\pi x}{L}\right)\sin\left(\frac{5\pi y}{L}\right)=0$$

$\cos\theta = 0$ when θ equals an odd integer times $\pi/2$:

$$\frac{4\pi x}{L} = \frac{2n'+1}{2}\pi \text{ for } n' = 0, 1, 2, \ldots \text{ which corresponds to } x = \frac{L}{8}, \frac{3L}{8}, \frac{5L}{8}, \text{ and } \frac{7L}{8}.$$

(Note: we disregard positions where the sine factor vanishes because they correspond to positions where the wavefunction vanishes as well.)

$$\frac{\partial\psi_{2,3}}{\partial y} = \frac{10\pi}{L^2}\sin\left(\frac{4\pi x}{L}\right)\cos\left(\frac{5\pi y}{L}\right) = 0$$

$$\frac{5\pi y}{L} = \frac{2n''+1}{2}\pi \text{ for } n'' = 0, 1, 2, \ldots \text{ which corresponds to}$$

$$y = \frac{L}{10}, \frac{3L}{10}, \frac{5L}{10} = \frac{L}{2}, \frac{7L}{10}, \text{ and } \frac{9L}{10}.$$

Thus, the positions of maximum probability, expressed as ordered pairs $(x/L, y/L)$ are

(1/8, 1/10),	(1/8, 3/10),	(1/8, 1/2),	(1/8, 7/10),	(1/8, 9/10)
(3/8, 1/10),	(3/8, 3/10),	(3/8, 1/2),	(3/8, 7/10),	(3/8, 9/10)
(5/8, 1/10),	(5/8, 3/10),	(5/8, 1/2),	(5/8, 7/10),	(5/8, 9/10)
(7/8, 1/10),	(7/8, 3/10),	(7/8, 1/2),	(7/8, 7/10),	(7/8, 9/10)

E11.6(a) Nodes of the wavefunction are sets of points where the wavefunction vanishes.

$$\psi_{2,3} = \frac{2}{L}\sin\left(\frac{2\pi x}{L}\right)\sin\left(\frac{3\pi y}{L}\right) \quad [11.4a]$$

$\sin\theta = 0$ when θ equals an integer times π:

$$\frac{2\pi x}{L} = n'\pi \text{ for } n' = 0, 1, 2, \ldots \text{ which corresponds to } \boxed{\frac{x}{L} = 0, \frac{1}{2}, 1}$$

$$\frac{3\pi y}{L} = n'\pi \text{ for } n' = 0, 1, 2, \ldots \text{ which corresponds to } \boxed{\frac{y}{L} = 0, \frac{1}{3}, \frac{2}{3}, 1}$$

Of course, if one factor of the wavefunction vanishes, the whole wavefunction vanishes. Thus, the wavefunction vanishes not at single points, but on the vertical and horizontal lines defined above.

Comment. Some of these lines correspond to edges of the square box. The term *node* is often used to describe places other than the boundaries where the wavefunction vanishes. These are $x/L = 1/2$ and $y/L = 1/3, 2/3$.

E11.7(a) See Exercise 9.5(a) for the one-dimensional analog to this question. The wavefunction is

$$\psi_{1,1} = \frac{2}{L}\sin\left(\frac{\pi x}{L}\right)\sin\left(\frac{\pi y}{L}\right) \quad [11.4a]$$

The expectation value of the x component of the momentum operator, $\hat{p}_x = \dfrac{\hbar}{\mathrm{i}}\dfrac{\partial}{\partial x}$, is

$$\langle p_x \rangle = \int_0^L \psi_{1,1} * \hat{p}_x \psi_{1,1}\, \mathrm{d}x\mathrm{d}y = \frac{4\hbar}{\mathrm{i}L^2}\int_0^L\int_0^L \sin\left(\frac{\pi x}{L}\right)\sin\left(\frac{\pi y}{L}\right)\frac{\partial}{\partial x}\sin\left(\frac{\pi x}{L}\right)\sin\left(\frac{\pi y}{L}\right)\mathrm{d}x\mathrm{d}y$$

$$= \frac{4\pi\hbar}{\mathrm{i}L^3}\int_0^L \sin\left(\frac{\pi x}{L}\right)\cos\left(\frac{\pi x}{L}\right)\mathrm{d}x\int_0^L \sin^2\left(\frac{\pi y}{L}\right)\mathrm{d}y = \boxed{0}$$

(The $\mathrm{d}x$ integral vanishes.) To find the expectation value of the momentum squared operator, $\hat{p}^2 = -\hbar^2\left(\dfrac{\partial^2}{\partial x^2}+\dfrac{\partial^2}{\partial y^2}\right)$, we need

$$\hat{p}^2\psi_{1,1} = -\frac{2\hbar^2}{L}\left(\frac{\partial^2}{\partial x^2}+\frac{\partial^2}{\partial y^2}\right)\sin\left(\frac{\pi x}{L}\right)\sin\left(\frac{\pi y}{L}\right)$$

$$= \frac{2\pi^2\hbar^2}{L^3}\sin\left(\frac{\pi x}{L}\right)\sin\left(\frac{\pi y}{L}\right) + \frac{2\pi^2\hbar^2}{L^3}\sin\left(\frac{\pi x}{L}\right)\sin\left(\frac{\pi y}{L}\right)$$

$$= \frac{2\pi^2\hbar^2}{L^2}\psi_{1,1} = \frac{h^2}{2L^2}\psi_{1,1}$$

$$\langle p^2 \rangle = \int_0^L\int_0^L \psi_{1,1}\hat{p}^2\psi_{1,1}\,\mathrm{d}x\mathrm{d}y = \frac{h^2}{2L^2}\int_0^L\int_0^L \psi_{1,1}^2\mathrm{d}x\mathrm{d}y = \boxed{\frac{h^2}{2L^2}}$$

E11.8(a) The zero-point energy is the ground-state energy, that is, with $n_1 = n_2 = 1$:

$$E_{1,1} = \frac{(n_1^2+n_2^2)h^2}{8m_\mathrm{e}L^2} = \frac{(1^2+1^2)h^2}{8m_\mathrm{e}L^2} = \frac{h^2}{4m_\mathrm{e}L^2} \quad [11.4\mathrm{b}]$$

Set this equal to the rest energy $m_\mathrm{e}c^2$ and solve for L:

$$m_\mathrm{e}c^2 = \frac{h^2}{4m_\mathrm{e}L^2} \quad \text{so} \quad L = \boxed{\frac{h}{2\,m_\mathrm{e}c}}$$

In absolute units, the length is

$$L = \frac{6.63\times10^{-34}\ \mathrm{J\,s}}{2\times(9.11\times10^{-31}\ \mathrm{kg})\times(3.00\times10^{8}\ \mathrm{m\,s^{-1}})} = 1.21\times10^{-12}\ \mathrm{m} = 1.21\ \mathrm{pm}$$

In terms of the Compton wavelength of an electron, $\lambda_\mathrm{C} = \dfrac{h}{m_\mathrm{e}c}$, $L = \boxed{\dfrac{\lambda_\mathrm{C}}{2}}$

E11.9(a) The energy levels are given by

$$E_{n_1,n_2} = \left(\frac{n_1^2}{L_1^2}+\frac{n_2^2}{L_2^2}\right)\frac{h^2}{8m}\,[11.4\mathrm{b}] = \left(\frac{n_1^2}{1}+\frac{n_2^2}{4}\right)\frac{h^2}{8mL^2}$$

$$E_{2,2} = \left(\frac{2^2}{1}+\frac{2^2}{4}\right)\frac{h^2}{8mL^2} = \frac{5h^2}{8mL^2}$$

We are looking for another state that has the same energy. By inspection we note that the first term in parentheses in $E_{2,2}$ works out to be 4 and the second 1; we can arrange for those values to be reversed:

$$E_{1,4} = \left(\frac{1^2}{1}+\frac{4^2}{4}\right)\frac{h^2}{8mL^2} = \frac{5h^2}{8mL^2}$$

So in this box, the state $\boxed{n_1 = 1, n_2 = 4}$ is degenerate to the state $n_1 = 2, n_2 = 2$. The question notes that degeneracy frequently accompanies symmetry, and suggests that one might be surprised to find degeneracy in a box with unequal lengths. Symmetry is a matter of degree. This box is less symmetric than a square box, but it is more symmetric than boxes whose sides have a non-integer or irrational ratio. Every state of a square box except those with $n_1 = n_2$ is degenerate (with the state that has n_1 and n_2 reversed). Only a few states in this box are degenerate. In this system, a state (n_1, n_2) is degenerate with a state $(n_2/2, 2n_1)$ as long as the latter state (a) exists (*i.e.*, $n_2/2$ must be an integer) and (b) is distinct from (n_1, n_2). (See Discussion question 11.2.) A box with incommensurable sides, say, L and $2^{1/2}L$, would have no degenerate levels.

E11.10(a) $E_{n_1,n_2,n_3} = \dfrac{(n_1^2+n_2^2+n_3^2)h^2}{8m_eL^2}$ [11.6b]

$$E_{1,1,1} = \frac{3h^2}{8mL^2}, \qquad 3E_{1,1,1} = \frac{9h^2}{8mL^2}$$

Hence, we require the values of n_1, n_2 and n_3 that make

$$n_1^2+n_2^2+n_3^2 = 9$$

Therefore, $(n_1, n_2, n_3) = (1, 2, 2)$, $(2, 1, 2)$, and $(2, 2, 1)$ and the degeneracy is $\boxed{3}$.

E11.11(a) $E_{n_1,n_2,n_3} = \dfrac{(n_1^2+n_2^2+n_3^2)h^2}{8m_eL^2}$ [11.6b]

$$\frac{\Delta E}{E} = \frac{\dfrac{K}{(0.9L)^2} - \dfrac{K}{L^2}}{\dfrac{K}{L^2}} = \frac{1}{0.81} - 1 = \boxed{0.23} \text{ or } \boxed{23\%}$$

E11.12(a) Set the particle in a cubic box energy equal to $3kT/2$

$$E = \frac{(n_1^2+n_2^2+n_3^2)h^2}{8mL^2} \text{ [11.6b]} = \frac{n^2h^2}{8mL^2} = \tfrac{3}{2}kT$$
$$= (\tfrac{3}{2}) \times (1.381\times10^{-23}\ \mathrm{J\,K^{-1}}) \times (300\ \mathrm{K}) = 6.21\bar{4}\times10^{-21}\ \mathrm{J}$$

So $n^2 = \dfrac{8mL^2}{h^2}E$

If $L^3 = 2.00\ \mathrm{m^3}$, then $L^2 = (2.00\ \mathrm{m^3})^{2/3} = 1.59\ \mathrm{m^2}$,

$$\frac{h^2}{8mL^2} = \frac{(6.626\times10^{-34}\ \mathrm{J\,s})^2}{(8)\times\left(\dfrac{0.03200\ \mathrm{kg\,mol^{-1}}}{6.022\times10^{23}\ \mathrm{mol^{-1}}}\right)\times1.59\ \mathrm{m^2}} = 6.51\times10^{-43}\ \mathrm{J},$$

and $n^2 = \dfrac{6.21\bar{4}\times10^{-21}\ \mathrm{J}}{6.51\times10^{-43}\ \mathrm{J}} = 9.55\times10^{21}$; $n = \boxed{9.77\times10^{10}}$

$$\Delta E = E_{n+1} - E_n = \frac{(n+1)^2h^2}{8mL^2} - \frac{n^2h^2}{8mL^2} = (2n+1)\times\left(\frac{h^2}{8mL^2}\right),$$

$$\Delta E = E_{9.77\times10^{10}+1} - E_{9.77\times10^{10}} = [(2)\times(9.77\times10^{10})+1]\times\left(\frac{h^2}{8mL^2}\right) = \frac{19.5\overline{5}\times10^{10}h^2}{8mL^2}$$

$$= (19.5\overline{5}\times10^{10})\times(6.51\times10^{-43}\ \text{J}) = \boxed{1.27\times10^{-31}\ \text{J}}$$

The de Broglie wavelength is obtained from

$$\lambda = \frac{h}{p} = \frac{h}{mv} \quad [4.6]$$

The velocity is obtained from

$$E_K = \tfrac{1}{2}mv^2 = \tfrac{3}{2}kT = 6.21\overline{4}\times10^{-21}\ \text{J}$$

$$\text{so} \quad v^2 = \frac{6.21\overline{4}\times10^{-21}\ \text{J}}{\left(\frac{1}{2}\right)\times\left(\frac{0.03200\ \text{kg mol}^{-1}}{6.022\times10^{23}\ \text{mol}^{-1}}\right)} = 2.34\times10^5\ \text{m}^2\ \text{s}^{-2}; \quad v = 484\ \text{m s}^{-1}$$

$$\lambda = \frac{6.626\times10^{-34}\ \text{J s}}{(5.31\times10^{-26}\ \text{kg})\times(484\ \text{m s}^{-1})} = 2.58\times10^{-11}\ \text{m} = \boxed{25.8\ \text{pm}}$$

The conclusion to be drawn from all of these calculations is that the translational motion of the oxygen molecule can be described classically. The energy of the molecule is essentially continuous,

$$\frac{\Delta E}{E} <<< 1$$

Problems

P11.1 The energy of the particle in a cubic box is

$$E = \frac{(n_1^2 + n_2^2 + n_3^2)h^2}{8mL^2} \quad [11.6]$$

so the difference between the two lowest energy levels is

$$\Delta E = \frac{(1^2+1^2+2^2)h^2}{8mL^2} - \frac{(1^2+1^2+1^2)h^2}{8mL^2} = \frac{3h^2}{8mL^2}$$

The mass is $m(O_2) = (32.00)\times(1.6605\times10^{-27}\ \text{kg}) = 5.31\times10^{-26}\ \text{kg}$, so

$$\Delta E = \frac{3h^2}{8mL^2} = \frac{3\times(6.626\times10^{-34}\ \text{J s})^2}{8\times5.31\times10^{-26}\ \text{kg}\times(5.0\times10^{-2}\ \text{m})^2} = \boxed{8.86\times10^{-39}\ \text{J}}$$

Set the energy equal to $3kT/2$:

$$E = \frac{3n^2h^2}{8mL^2} = \frac{3kT}{2}$$

$$n = \left(\frac{4kTmL^2}{h^2}\right)^{1/2}$$
$$= \left(\frac{4\times1.381\times10^{-23}\ \text{J K}^{-1}\times300\ \text{K}\times5.31\times10^{-26}\ \text{kg}\times(5.0\times10^{-2}\ \text{m})^2}{(6.626\times10^{-34}\ \text{J s})^2}\right)^{1/2}$$
$$= 2.2\underline{4}\times10^9$$

The degenerate energy level immediately below this level has one of the quantum numbers equal to $n-1$ and the other two still equal to n.

$$\Delta E = E_{n,n,n} - E_{n,n,n-1} = \frac{3n^2h^2}{8mL^2} - \frac{\{2n^2+(n-1)^2\}h^2}{8mL^2} = \frac{(2n-1)h^2}{8mL^2}$$
$$= \frac{(2\times2.2\underline{4}\times10^9-1)\times(6.626\times10^{-34}\ \text{J s})^2}{8\times5.31\times10^{-26}\ \text{kg}\times(5.0\times10^{-2}\ \text{m})^2} = \boxed{1.11\times10^{-30}\ \text{J}}$$

Comment. The problem is similar to Exercise 11.12(a). The separation in energy levels is much smaller than the energy of the occupied level. The energy of the molecule is essentially continuous.

P11.3 In C_{60}, each carbon atom is bound to three others, so each makes three sigma bonds. That leaves one p orbital and one electron per carbon atom to go into the π system. Thus, C_{60} has 60 π electrons, which would doubly occupy the lowest 30 particle-in-a-cube "orbitals." The energy levels are

$$E_{n_1,n_2,n_3} = \frac{(n_1^2+n_2^2+n_3^2)h^2}{8m_eL^2} \quad [11.6b]$$

Levels in which all three quantum numbers are the same are non-degenerate. The degeneracy of levels that have two quantum numbers in common is three, for the different quantum number can be in any of three places. The degeneracy of levels that have three different quantum numbers is six, the number of possible different ordered triplets. The energy levels are shown in Fig. 11.1, labeled by one corresponding set of quantum numbers.

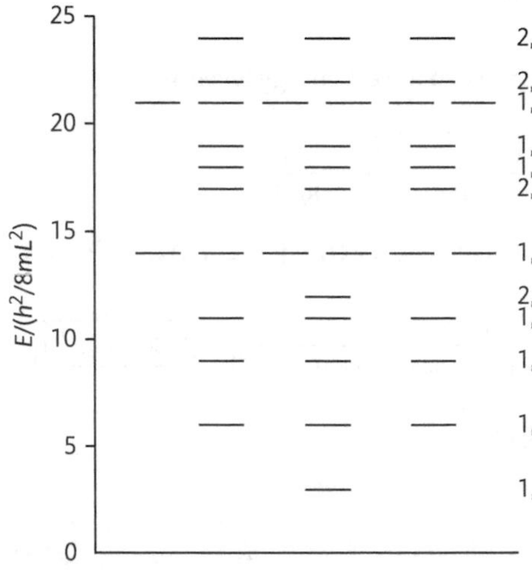

Fig 11.1

Filling the orbitals from the bottom makes the level marked (1,2,4) the highest occupied level. Its energy is $21h^2/8mL^2$. The next level has an energy of $22h^2/8mL^2$ (the level marked (2,3,3)). So the excitation energy is the difference between these levels, namely

$$\Delta E = \frac{h^2}{8m_eL^2} = \frac{(6.626\times10^{-34}\ \mathrm{J\ s})^2}{8\times9.11\times10^{-31}\ \mathrm{kg}\times(0.7\times10^{-9}\ \mathrm{m})^2} = 1.2\times10^{-19}\ \mathrm{J}$$

$$\Delta E = E_{\mathrm{photon}} = \frac{hc}{\lambda},$$

$$\text{so } \lambda = \frac{hc}{\Delta E} = \frac{6.626\times10^{-34}\ \mathrm{J\ s}\times2.998\times10^{8}\ \mathrm{m\ s^{-1}}}{1.2\times10^{-19}\ \mathrm{J}} = \boxed{1.6\times10^{-6}\ \mathrm{m}} = \boxed{1.6\ \mu\mathrm{m}}$$

P11.5 The excitation energy is the difference between the two lowest energy levels, and that energy difference is equal to the energy of the photon.

$$\Delta E = \frac{(F_{1,1}^2 - F_{1,0}^2)h^2}{8m_ea^2} = E_{\mathrm{photon}} = \frac{hc}{\lambda}$$

$$\text{so } a = \left(\frac{(F_{1,1}^2 - F_{1,0}^2)h\lambda}{8m_ec}\right)^{1/2} = \left(\frac{(1.430^2 - 1^2)\times6.626\times10^{-34}\ \mathrm{J\ s}\times1500\times10^{-9}\ \mathrm{m}}{8\times9.11\times10^{-31}\ \mathrm{kg}\times2.998\times10^{8}\ \mathrm{m\ s^{-1}}}\right)^{1/2}$$

$$= \boxed{6.9\times10^{-10}\ \mathrm{m}} = \boxed{0.69\ \mathrm{nm}}$$

P11.7 (a) The circle of radius n is the set of all ordered pairs of *real numbers* (n_1,n_2) such that $n_1^2 + n_2^2 = n^2$. The circumference of the circle, $2\pi n$, is proportional to the number of points *on* the circle. We are not interested in all of the points, though; rather, we want to count the points in which n_1 and n_2 are integers. We can do so by moving around the circle in integer steps. The number of such steps *is* the circumference; because the dimensions in this two-dimensional space are pure numbers, the "unit" in this space is simply 1. Finally, we are interested only in the first quadrant, so the degeneracy is 1/4 the circumference: $\boxed{g = n\pi/2}$.

(b) In three dimensions, the corresponding measure of the number of points that satisfies $n_1^2 + n_2^2 + n_3^2 = n^2$ is the surface area of a sphere of radius n. Once again, we restrict ourselves to the first quadrant, which is 1/8 of the whole:

$$g = \frac{1}{8}\times\frac{4\pi n^2}{3} = \boxed{\frac{\pi n^2}{6}}$$

P11.9 The Schrödinger equation is

$$-\frac{\hbar^2}{2m}\left(\frac{\partial^2\psi}{\partial x^2} + \frac{\partial^2\psi}{\partial y^2}\right) = E\psi \quad [11.1]$$

To verify that

$$\psi(x,y) = \frac{2}{(L_1L_2)^{1/2}}\sin\frac{n_1\pi x}{L_1}\sin\frac{n_2\pi y}{L_2} \quad [11.4\mathrm{a}]$$

is actually a solution, we take the required partial derivatives.

$$\frac{\partial\psi}{\partial x}=\frac{2}{(L_1L_2)^{1/2}}\times\frac{n_1\pi}{L_1}\cos\frac{n_1\pi x}{L_1}\sin\frac{n_2\pi y}{L_2},$$

$$\text{so } \frac{\partial^2\psi}{\partial x^2}=-\frac{2}{(L_1L_2)^{1/2}}\times\left(\frac{n_1\pi}{L_1}\right)^2\sin\frac{n_1\pi x}{L_1}\sin\frac{n_2\pi y}{L_2}=-\left(\frac{n_1\pi}{L_1}\right)^2\psi$$

$$\text{Similarly } \frac{\partial^2\psi}{\partial y^2}=-\frac{2}{(L_1L_2)^{1/2}}\times\left(\frac{n_2\pi}{L_2}\right)^2\sin\frac{n_1\pi x}{L_1}\sin\frac{n_2\pi y}{L_2}=-\left(\frac{n_2\pi}{L_2}\right)^2\psi$$

Putting these derivatives into the Schrödinger equation yields

$$\frac{\hbar^2}{2m}\left\{\left(\frac{n_1\pi}{L_1}\right)^2\psi+\left(\frac{n_2\pi}{L_2}\right)^2\psi\right\}=E\psi$$

This expression is a true equation if

$$\frac{\hbar^2}{2m}\left\{\left(\frac{n_1\pi}{L_1}\right)^2+\left(\frac{n_2\pi}{L_2}\right)^2\right\}=E$$

With a little rearrangement, we arrive at eqn 11.4b for the energy:

$$\frac{(h/2\pi)^2}{2m}\times\pi^2\left(\frac{n_1^2}{L_1^2}+\frac{n_2^2}{L_2^2}\right)=\boxed{\frac{h^2}{8m}\left(\frac{n_1^2}{L_1^2}+\frac{n_2^2}{L_2^2}\right)=E}$$

P11.11 (a) In the box, the Schrödinger equation is

$$-\frac{\hbar^2}{2m}\left(\frac{\partial^2}{\partial x^2}+\frac{\partial^2}{\partial y^2}+\frac{\partial^2}{\partial z^2}\right)\psi=E\psi$$

Assume that the solution is a product of three functions of a single variable; that is, let

$$\psi(x,y,z)=X(x)Y(y)Z(z)$$

Substituting into the Schrödinger equation gives

$$-\frac{\hbar^2}{2m}\left(YZ\frac{\partial^2 X}{\partial x^2}+XZ\frac{\partial^2 Y}{\partial y^2}+XY\frac{\partial^2 Z}{\partial z^2}\right)=EXYZ$$

Divide both sides by XYZ:

$$-\frac{\hbar^2}{2m}\left(\frac{1}{X}\frac{\partial^2 X}{\partial x^2}+\frac{1}{Y}\frac{\partial^2 Y}{\partial y^2}+\frac{1}{Z}\frac{\partial^2 Z}{\partial z^2}\right)=E$$

For the purposes of illustration, isolate the terms that depend on x on the left side of the equation:

$$-\frac{\hbar^2}{2m}\left(\frac{1}{X}\frac{\partial^2 X}{\partial x^2}\right)=E+\frac{\hbar^2}{2m}\left(\frac{1}{Z}\frac{\partial^2 Z}{\partial z^2}+\frac{1}{Y}\frac{\partial^2 Y}{\partial y^2}\right)$$

Note that the left side depends only on one variable, x, while the right side depends on two different and independent variables, y and z. The only way that the two sides can be equal to each other for all x, y, and z is if they are both equal to a constant. Call that constant E_x, and we have, from the left side of the equation:

$$-\frac{\hbar^2}{2m}\left(\frac{1}{X}\frac{\partial^2 X}{\partial x^2}\right)=E_x \quad \text{so} \quad -\frac{\hbar^2}{2m}\frac{\partial^2 X}{\partial x^2}=E_x X$$

Note that this is just the Schrödinger equation for a particle in a one-dimensional box. Note also that we could just as easily have isolated y terms or z terms, leading to similar equations.

$$-\frac{\hbar^2}{2m}\frac{\partial^2 Y}{\partial y^2}=E_y Y \quad \text{and} \quad -\frac{\hbar^2}{2m}\frac{\partial^2 Z}{\partial z^2}=E_z Z$$

The assumption that the wavefunction can be written as a product of single-variable functions is a valid one, for we can find ordinary differential equations for the assumed factors. That is what it means for a partial differential equation to be separable.

(b) Because X, Y, and Z are particle-in-a-box wavefunctions of independent variables x, y, and z respectively, each of them has its own quantum number. The three-dimensional wavefunction is a product of the three one-dimensional wavefunctions, and therefore depends on all three quantum numbers:

$$\psi(x,y,z)=X(x)Y(y)Z(z)=\left(\frac{2}{L_1}\right)^{1/2}\sin\frac{n_x\pi x}{L_1}\times\left(\frac{2}{L_2}\right)^{1/2}\sin\frac{n_y\pi y}{L_2}\times\left(\frac{2}{L_3}\right)^{1/2}\sin\frac{n_z\pi z}{L_3}$$

Each constant of separation (E_x, E_y, and E_z) depends on its own quantum number. The three constants of separation add up to the total energy, which therefore depends on all three quantum numbers:

$$E=E_x+E_y+E_z=\frac{h^2}{8m}\left(\frac{n_x^2}{L_1^2}+\frac{n_y^2}{L_2^2}+\frac{n_z^2}{L_3^2}\right)$$

(c) For a cubic box, $L_1=L_2=L_3=L$, so

$$E=\frac{h^2(n_1^2+n_2^2+n_3^2)}{8mL^2}=\frac{(6.626\times10^{-34}\text{ J s})^2}{8\times9.11\times10^{-31}\text{ kg}\times(5\times10^{-9}\text{ m})^2}\times(n_1^2+n_2^2+n_3^2)$$
$$=2.4\times10^{-21}\text{ J}\times(n_1^2+n_2^2+n_3^2)$$

The energy levels are shown in Fig. 11.2 in units of $h^2/8mL^2$.

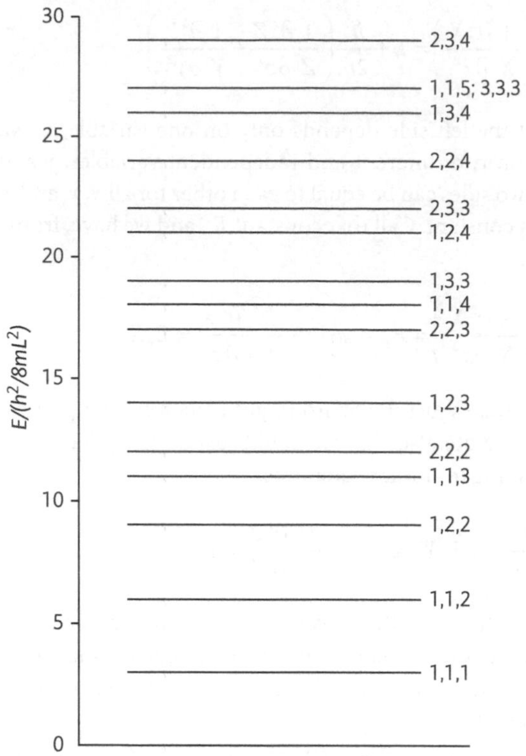

Fig 11.2

(d) Compare this energy-level diagram to Fig. 9.3 of the textbook. The energy levels here are much more closely spaced. In a one-dimensional box, the 15th energy level is not reached until $\dfrac{E}{h^2/8mL^2} = 225$, and the previous level is 29 units below that. In the three-dimensional box, the first 15 energy levels all fit within the range of 29 units. The energy levels in a one-dimensional box are sparse compared to those in a three-dimensional box.

Topic 12 **Vibrational motion**

Discussion questions

D12.1 The separation of vibrational energy levels of the harmonic oscillator is uniform, $\hbar\omega$ (eqn 12.5), where $\omega = (k_f/m)^{1/2}$ (eqn 12.4), regardless of the level. Thus the energy separation varies in direct proportion to the square root of the force constant and in inverse proportion to the square root of the mass.

D12.3 The harmonic oscillator serves as the first approximation for describing molecular vibrations. The excitation energies for stretching and bending bonds are manifested in infrared and Raman spectroscopy.

Exercises

E12.1(a) The zero-point energy of a harmonic oscillator is (eqn 12.6)

$$E_0 = \frac{1}{2}\hbar\omega = \frac{\hbar}{2}\left(\frac{k_f}{m}\right)^{1/2} \quad [12.4] = \frac{1.0546\times10^{-34}\ \mathrm{J\,s}}{2}\times\left(\frac{155\ \mathrm{N\,m^{-1}}}{1.67\times10^{-27}\ \mathrm{kg}}\right)^{1/2}$$

$$= \boxed{1.61\times10^{-20}\ \mathrm{J}}$$

E12.2(a) The difference in adjacent energy levels is

$$\Delta E = E_{\upsilon+1} - E_{\upsilon} = \hbar\omega\ [12.5] = \hbar\left(\frac{k_f}{m}\right)^{1/2} \quad [12.4]$$

Hence

$$k_f = m\left(\frac{\Delta E}{\hbar}\right)^2 = (1.33\times10^{-25}\ \mathrm{kg})\times\left(\frac{4.82\times10^{-21}\ \mathrm{J}}{1.055\times10^{-34}\ \mathrm{J\,s}}\right)^2 = 278\ \mathrm{kg\,s^{-2}} = \boxed{278\ \mathrm{N\,m^{-1}}}$$

E12.3(a) The requirement for a transition to occur is that $\Delta E(\text{system}) = E(\text{photon})$, where $\Delta E(\text{system}) = \hbar\omega \quad [12.5]$

and $E(\text{photon}) = h\nu = \dfrac{hc}{\lambda}$

Therefore, $\dfrac{hc}{\lambda}=\dfrac{h\omega}{2\pi}=\left(\dfrac{h}{2\pi}\right)\times\left(\dfrac{k}{m}\right)^{1/2}$

$$\lambda=2\pi c\left(\frac{m}{k}\right)^{1/2}=(2\pi)\times(2.998\times10^{8}\ \mathrm{m\ s^{-1}})\times\left(\frac{1.673\times10^{-27}\ \mathrm{kg}}{855\ \mathrm{N\ m^{-1}}}\right)^{1/2}$$

$$=2.63\times10^{-6}\,\mathrm{m}=\boxed{2.63\ \mu\mathrm{m}}$$

E12.4(a) $\lambda=2\pi c\left(\dfrac{m}{k}\right)^{1/2}$ [E12.3a]

Since $\lambda\propto m^{1/2}$, $\lambda_{\mathrm{new}}=2^{1/2}\lambda_{\mathrm{old}}=(2^{1/2})\times(2.63\ \mu\mathrm{m})=\boxed{3.72\ \mu\mathrm{m}}$

E12.5(a) The harmonic oscillator wavefunctions have the form

$$\psi_v(x)=N_vH_v(y)\exp\left(-\tfrac{1}{2}y^2\right)\ \text{with}\ y=\frac{x}{\alpha}\ \text{and}\ \alpha=\left(\frac{\hbar^2}{mk}\right)^{1/4}\qquad[12.8]$$

The exponential function approaches zero only as x approaches $\pm\infty$, so the nodes of the wavefunction are the nodes of the Hermite polynomials.

$H_4(y)=16y^4-48y^2+12=0$ [Table 12.1]

Dividing through by 4 and letting $z=y^2$, we have a quadratic equation $4z^2-12z+3=0$.

$$\text{So}\ z=\frac{-b\pm\sqrt{b^2-4ac}}{2a}=\frac{12\pm\sqrt{12^2-4\times4\times3}}{2\times4}=\frac{3\pm\sqrt{6}}{2}$$

$$\text{and}\ y=\pm\sqrt{\frac{3\pm\sqrt{6}}{2}}\ \text{and}\ x=\alpha\left(\pm\sqrt{\frac{3\pm\sqrt{6}}{2}}\right)$$

Evaluating the result numerically yields $z=0.275$ or 2.72, so $y=\pm0.525$ or ±1.65. Therefore, $x=\pm0.525\alpha$ or $\pm1.65\alpha$.

Comment. Numerical values could also be obtained graphically by plotting $H_4(y)$.

E12.6(a) The harmonic oscillator wavefunctions have the form

$$\psi_v(x)=N_vH_v(y)\exp\left(-\tfrac{1}{2}y^2\right)\ \text{with}\ y=\frac{x}{\alpha}\ \text{and}\ \alpha=\left(\frac{\hbar^2}{mk}\right)^{1/4}\qquad[12.8]$$

so the wavefunction in question is

$$\psi_2(x)=N_2(4y^2-2)\mathrm{e}^{-y^2/2}.$$

Normalization requires

$$1=\int_{-\infty}^{\infty}\psi_2{}^*\psi_2\,\mathrm{d}x=\int_{-\infty}^{\infty}\psi_2^2\alpha\,\mathrm{d}y=\int_{-\infty}^{\infty}\left(N_2(4y^2-2)\mathrm{e}^{-y^2/2}\right)^2\alpha\,\mathrm{d}y$$
$$=4N_2^2\alpha\int_{-\infty}^{\infty}(4y^4-4y^2+1)\mathrm{e}^{-y^2}\mathrm{d}y$$

To evaluate the integral, note that the integrand is even, so the integral is twice the integral from zero to infinity of the same function. Use

$$\int_0^{\infty}y^{2n}\mathrm{e}^{-ay^2}\,\mathrm{d}y=\frac{1\cdot3\cdots(2n-1)}{2^{n+1}a^n}\sqrt{\frac{\pi}{a}}\quad\text{and}\quad\int_0^{\infty}\mathrm{e}^{-ay^2}\,\mathrm{d}y=\frac{1}{2}\sqrt{\frac{\pi}{a}}$$

Thus $1=8N_2^2\alpha\left(4\times\frac{3\sqrt{\pi}}{2^3}-4\times\frac{\sqrt{\pi}}{2^2}+\frac{\sqrt{\pi}}{2}\right)=8N_2^2\alpha\sqrt{\pi}$

and $N_2=\frac{1}{(8\alpha)^{1/2}\pi^{1/4}}$

Confirming orthogonality amounts to demonstrating that

$$\int_{-\infty}^{\infty}\psi_4{}^*\psi_2\,\mathrm{d}x=0\qquad\text{where}\quad\psi_4(x)=N_4(16y^4-48y^2+12)\mathrm{e}^{-y^2/2}$$

The integral in question is

$$\int_{-\infty}^{\infty}N_4(16y^4-48y^2+12)\mathrm{e}^{-y^2/2}N_2(4y^2-2)\mathrm{e}^{-y^2/2}\alpha\,\mathrm{d}y$$
$$=8N_4N_2\alpha\int_{-\infty}^{\infty}(4y^4-12y^2+3)(2y^2-1)\mathrm{e}^{-y^2}\,\mathrm{d}y$$
$$=8N_4N_2\alpha\int_{-\infty}^{\infty}(8y^6-28y^4+18y^2-3)\mathrm{e}^{-y^2}\,\mathrm{d}y$$
$$=8N_4N_2\alpha\left(8\times\frac{15}{2^4}-28\times\frac{3}{2^3}+18\times\frac{1}{2^2}-3\times\frac{1}{2}\right)\sqrt{\pi}$$

The terms in parentheses, $\frac{15}{2}-\frac{21}{2}+\frac{9}{2}-\frac{3}{2}$, do indeed add up to zero, making the integral vanish, as required.

E12.7(a) The effective mass is

$$\mu=\frac{m_{\mathrm{A}}m_{\mathrm{B}}}{m_{\mathrm{A}}+m_{\mathrm{B}}}=\frac{m_{\mathrm{Cl}}m_{\mathrm{Cl}}}{m_{\mathrm{Cl}}+m_{\mathrm{Cl}}}=\frac{m_{\mathrm{Cl}}}{2}=\frac{34.9688\,m_{\mathrm{u}}\times1.66054\times10^{-27}\ \mathrm{kg}\ m_{\mathrm{u}}^{-1}}{2}$$
$$=2.90335\times10^{-26}\ \mathrm{kg}$$

The difference in adjacent energy levels, which is equal to the energy of the photon, is

$$\Delta E=\hbar\omega\ [12.5]=h\nu$$

so, in terms of wavenumbers

$$\hbar\left(\frac{k}{m}\right)^{1/2} [12.4] = hc\tilde{\nu} \quad \text{and} \quad \tilde{\nu} = \frac{\hbar}{hc}\left(\frac{k}{m}\right)^{1/2} = \frac{1}{2\pi c}\left(\frac{k}{m}\right)^{1/2}$$

$$\tilde{\nu} = \frac{1}{2\pi \times 2.998 \times 10^{10}\ \text{cm s}^{-1}}\left(\frac{329\ \text{N m}^{-1}}{2.90335 \times 10^{-26}\ \text{kg}}\right)^{1/2} = \boxed{565\ \text{cm}^{-1}}$$

E12.8(a) Example 12.5 analyzes the classical turning points of the harmonic oscillator. In terms of the dimensionless variable y, the turning points are $y_{\text{tp}} = \pm(2v+1)^{1/2}$. The probability of extension beyond the classical turning point is

$$P = \int_{x_{\text{tp}}}^{\infty} \psi_v^2\, \text{d}x = \alpha N_v^2 \int_{y_{\text{tp}}}^{\infty} \{H_v(y)\}^2 \text{e}^{-y^2}\, \text{d}y$$

For $v = 1$, $H_1(y) = 2y$ and $N_1 = \left(\dfrac{1}{2\alpha\pi^{1/2}}\right)^{1/2}$

$$P = 4\alpha N_1^2 \int_{3^{1/2}}^{\infty} y^2 \text{e}^{-y^2}\, \text{d}y$$

Use integration by parts:

$$\int u\,\text{d}v = uv - \int v\,\text{d}u$$

where $u = y$, $\text{d}v = y\text{e}^{-y^2}\,\text{d}y$

so $\text{d}u = \text{d}y$, $v = -\tfrac{1}{2}\text{e}^{-y^2}$

$$\text{and } P = -2\alpha N_1^2\left(y\text{e}^{-y^2}\Big|_{3^{1/2}}^{\infty} - \int_{3^{1/2}}^{\infty} \text{e}^{-y^2}\, \text{d}y\right)$$

$$= \pi^{-1/2}\left(3^{1/2}\text{e}^{-3} + \int_{3^{1/2}}^{\infty} \text{e}^{-y^2}\, \text{d}y\right)$$

The remaining integral can be expressed in terms of the error function.

$$\text{erf}\, z = 1 - \frac{2}{\pi^{1/2}}\int_z^{\infty} \text{e}^{-y^2}\, \text{d}y$$

$$\text{so } \int_{3^{1/2}}^{\infty} \text{e}^{-y^2}\, \text{d}y = \frac{\pi^{1/2}(1 - \text{erf}\, 3^{1/2})}{2}$$

Finally, using $\text{erf}\, 3^{1/2} = 0.986$,

$$P = \pi^{-1/2}\left(3^{1/2}\text{e}^{-3} + \frac{\pi^{1/2}(1 - \text{erf}\, 3^{1/2})}{2}\right) = \boxed{0.056}$$

Comment. This is the probability of an extension greater than the positive classical turning point. There is an equal probability of a compression smaller than the negative classical turning point, so the total probability of finding the oscillator in a classically forbidden region is $\boxed{0.112}$.

Comment. Note that molecular parameters such as m and k_{f} do not enter into the calculation.

E12.9(a) Mean kinetic and potential energies are related by the virial theorem (eqn 12.14)

$2\langle E_k\rangle = b\langle V\rangle$, where $V = ax^b$.

So for $V \propto x^3$, we have $2\langle E_k\rangle = 3\langle V\rangle$, or $\boxed{\langle E_k\rangle = 3\langle V\rangle/2}$

Problems

P12.1 The harmonic oscillator wavefunctions have the form

$$\psi_v(x) = N_v H_v(y)\exp\left(-\tfrac{1}{2}y^2\right) \text{ with } y = \frac{x}{\alpha} \text{ and } \alpha = \left(\frac{\hbar^2}{mk}\right)^{1/4} \quad [12.8]$$

The exponential factor is even, since it is a function of x^2, and the normalization constant is even (as is any constant factor) because it is invariant upon changing x to $-x$. Thus

$$\psi_v(x) = \text{even} \times H_v(y) \times \text{even}$$

so the symmetry of the entire wavefunction is the same as that of its Hermite polynomial (because the product of a function with an even factor does not change the symmetry of the original function). Looking at Table 12.1, we see that each Hermite polynomial contains only even or only odd powers of y (and therefore also of x), making them even or odd functions respectively:

$$\boxed{\psi_0(x) \text{ is even;} \quad \psi_1(x) \text{ is odd;} \quad \psi_2(x) \text{ is even;} \quad \psi_3(x) \text{ is odd.}}$$

The quantum number v is equal to the degree of the Hermite polynomial. Thus,

$$\boxed{\psi_v(x) \text{ is even if } v \text{ is even,} \quad \text{odd if } v \text{ is odd.}}$$

P12.3 The Schrödinger equation is $-\dfrac{\hbar^2}{2m}\dfrac{d^2\psi}{dx^2} + \tfrac{1}{2}k_f x^2\psi = E\psi$

We write the trial solution $\psi = e^{-\kappa x^2}$, so $\dfrac{d\psi}{dx} = -2\kappa x e^{-\kappa x^2}$,

and $\dfrac{d^2\psi}{dx^2} = -2\kappa e^{-\kappa x^2} + 4\kappa^2 x^2 e^{-\kappa x^2} = -2\kappa\psi + 4\kappa^2 x^2\psi$

Insert the trial solution into the Schrödinger equation.

$$\left(\frac{\hbar^2\kappa}{m}\right)\psi - \left(\frac{2\hbar^2\kappa^2}{m}\right)x^2\psi + \tfrac{1}{2}k_f x^2\psi = E\psi,$$

$$\text{so} \quad \left[\left(\frac{\hbar^2\kappa}{m}\right) - E\right]\psi + \left(\tfrac{1}{2}k_f - \frac{2\hbar^2\kappa^2}{m}\right)x^2\psi = 0$$

This equation is satisfied if

$$E=\frac{\hbar^2\kappa}{m} \quad \text{and} \quad 2\hbar^2\kappa^2=\tfrac{1}{2}mk_f,\ \text{so}\ \kappa=\boxed{\frac{1}{2}\left(\frac{mk_f}{\hbar^2}\right)^{1/2}}$$

The corresponding energy is

$$E=\tfrac{1}{2}\hbar\left(\frac{k_f}{m}\right)^{1/2}=\tfrac{1}{2}\hbar\omega \quad \text{with}\ \omega=\left(\frac{k_f}{m}\right)^{1/2}$$

P12.5 $\langle x^n\rangle=\alpha^n\langle y^n\rangle=\alpha^n\int_{-\infty}^{+\infty}\psi y^n\psi\,dx=\alpha^{n+1}\int_{-\infty}^{+\infty}\psi^2y^n\,dy\quad [x=\alpha y]$

$$\langle x^3\rangle\propto\int_{-\infty}^{+\infty}\psi^2y^3\,dy=\boxed{0}\ \text{by symmetry}\quad [y^3\ \text{is an odd function of}\ y]$$

$$\langle x^4\rangle=\alpha^5\int_{-\infty}^{+\infty}\psi y^4\psi\,dy$$

$$y^4\psi=y^4NH_v e^{-y^2/2}$$

Now use the relations in Table 12.1:

$$\begin{aligned}
y^4H_v &= y^3\left(\tfrac{1}{2}H_{v+1}+vH_{v-1}\right)=y^2\left[\tfrac{1}{2}\left(\tfrac{1}{2}H_{v+2}+(v+1)H_v\right)+v\left(\tfrac{1}{2}H_v+(v-1)H_{v-2}\right)\right]\\
&= y^2\left[\tfrac{1}{4}H_{v+2}+\left(v+\tfrac{1}{2}\right)H_v+v(v-1)H_{v-2}\right]\\
&= y\left[\tfrac{1}{4}\left(\tfrac{1}{2}H_{v+3}+(v+2)H_{v+1}\right)+\left(v+\tfrac{1}{2}\right)\times\left(\tfrac{1}{2}H_{v+1}+vH_{v-1}\right)\right.\\
&\quad \left.+v(v-1)\times\left(\tfrac{1}{2}H_{v-1}+(v-2)H_{v-3}\right)\right]\\
&= y\left(\tfrac{1}{8}H_{v+3}+\tfrac{3}{4}(v+1)H_{v+1}+\tfrac{3}{2}v^2H_{v-1}+v(v-1)\times(v-2)H_{v-3}\right)
\end{aligned}$$

Only yH_{v+1} and yH_{v-1} lead to H_v and contribute to the expectation value (since H_v is orthogonal to all except H_v) [Table 12.1]; hence, using … to denote all the terms that produce Hermite polynomials other than H_v,

$$\begin{aligned}
y^4H_v &= \tfrac{3}{4}y\{(v+1)H_{v+1}+2v^2H_{v-1}\}+\ldots\\
&= \tfrac{3}{4}\left[(v+1)\left(\tfrac{1}{2}H_{v+2}+(v+1)H_v\right)+2v^2\left(\tfrac{1}{2}H_v+(v-1)H_{v-2}\right)\right]+\ldots\\
&= \tfrac{3}{4}\{(v+1)^2H_v+v^2H_v\}+\ldots\\
&= \tfrac{3}{4}(2v^2+2v+1)H_v+\ldots
\end{aligned}$$

Therefore

$$\int_{-\infty}^{+\infty}\psi y^4\psi\,dy=\tfrac{3}{4}(2v^2+2v+1)N^2\int_{-\infty}^{+\infty}H_v^2e^{-y^2}\,dy=\frac{3}{4\alpha}(2v^2+2v+1)$$

and so

$$\langle x^4\rangle=(\alpha^5)\times\left(\frac{3}{4\alpha}\right)\times(2v^2+2v+1)=\boxed{\frac{3}{4}(2v^2+2v+1)\alpha^4}$$

P12.7 (a) The problem requires us to consider small displacements from the minimum-energy or equilibrium value of ϕ. Diagram 3 in the problem depicts a staggered conformation with $\phi = \pi/3$; this is an equilibrium conformation. Use the first few terms of the Taylor series expansion of cosine about this angle:

$$V = V_0 \cos 3\phi \approx V_0 \left\{ \cos\left(3\times\frac{\pi}{3}\right) - 3\sin\left(3\times\frac{\pi}{3}\right)\left(\phi-\frac{\pi}{3}\right) - \frac{9}{2}\cos\left(3\times\frac{\pi}{3}\right)\left(\phi-\frac{\pi}{3}\right)^2 \right\}$$

$$\approx -V_0 + \frac{9V_0}{2}\left(\phi-\frac{\pi}{3}\right)^2.$$

This potential energy function is harmonic: the difference in energy from the minimum energy, namely $V-(-V_0)$, is proportional to the square of the displacement from the equilibrium position. Can we put the Schrödinger equation for this system into the form of a harmonic oscillator? We need the kinetic energy operator, which, we might suspect, involves a second derivative with respect to ϕ. In fact, the kinetic energy operator is $-\frac{\hbar^2}{2I}\frac{\partial^2}{\partial\phi^2}$. (In rotational motion, the moment of inertia, I, plays a role analogous to mass in translational motion. (Look ahead to eqn 13.11a.) Now in order to put the Schrödinger equation into the form of a harmonic oscillator, let us choose the displacement, $\phi - \pi/3$, to be our independent variable, rather than ϕ itself; that is, let $x = \phi - \pi/3$, so that

$$V \approx -V_0 + \frac{9V_0}{2}x^2$$

Because $\partial/\partial\phi = \partial/\partial x$, the Schrödinger equation is

$$-\frac{\hbar^2}{2I}\frac{\partial^2\psi}{\partial x^2} + \left(\frac{9V_0}{2}x^2 - V_0\right)\psi = E\psi$$

Because V_0 is a constant, we may combine it with the energy eigenvalue E to turn this differential equation into an eigenvalue equation with the same form as the Schrödinger equation for the harmonic oscillator (eqn 12.3):

$$-\frac{\hbar^2}{2I}\frac{\partial^2\psi}{\partial x^2} + \frac{9V_0}{2}x^2\psi = (E+V_0)\psi = E'\psi$$

The energy eigenvalues are:

$$E'_v = (v+\tfrac{1}{2})\hbar\omega = E_v + V_0 \text{ where } \omega = \left(\frac{9V_0}{I}\right)^{1/2} \quad \text{[adapting 12.4]}$$

So $E_v = (v+\tfrac{1}{2})\hbar\omega - V_0$

and the difference in adjacent energy levels is

$$E_1 - E_0 = \hbar\omega \quad [12.5]$$

The molar excitation energy is

$$N_A\hbar\omega = N_A\hbar\left(\frac{9V_0}{I}\right)^{1/2}$$

(b) Harmonic motion has a potential energy of $k_f x^2/2$ where x is the displacement and k_f the force constant. So the force constant is two times whatever constant multiplies x^2 in the potential, namely $\boxed{9V_0}$. Note: this torsional "force constant" has different dimensions than a force constant for a linear displacement, because the angular displacement has different dimensions than linear displacement.

(c) The rotational barrier for ethane is relatively small compared to chemical bond energies, about 3 kcal mol^{-1} = 12 kJ mol^{-1}. (See most organic chemistry texts or Sidney Benson, *Thermochemical Kinetics*.) This corresponds to $N_A V_0$. Compare this energy to $RT = N_A kT$. At room temperature, $RT = (8.3145\ \text{J mol}^{-1}\ \text{K}^{-1})(300\ \text{K}) = 2.5\ \text{kJ mol}^{-1}$. At room temperature, then, the oscillations are not excited in most molecules at any given time.

Comment. Even though the oscillations are not excited in a majority of the molecules at any given time, the faction of excited molecules is not negligible (as can be estimated from the Boltzmann distribution; see Topic 2.3a). A chemical reaction that proceeds through excitation of this oscillation can still be rapid at room temperature.

(d) If the displacements are sufficiently large, the potential energy does not rise as rapidly with the angle as would a harmonic potential, because the next term in the expansion of the potential energy detracts from the quadratic term. Therefore, the magnitude of the wavefunction will not fall off so sharply with increasing displacement as a harmonic-oscillator wavefunction. Each successive energy level would become lower than that of a harmonic oscillator, so the energy levels will become progressively closer together.

Question. The next non-vanishing term in the Taylor series for the potential energy is $-\dfrac{27V_0}{8}x^4$. Treat this as a perturbation to the harmonic oscillator wavefunction and compute the first-order correction to the energy.

P12.9 Assuming that one can identify the CO peak in the infrared spectrum of the CO-myoglobin complex, taking infrared spectra of each of the isotopic variants of CO-myoglobin complexes can show which atom binds to the haem group and determine the C≡O force constant. Compare isotopic variants to $^{12}C^{16}O$ as the standard; when an isotope changes but the vibrational frequency does not, then the atom whose isotope was varied is the atom that binds to the haem. See the table below, which includes predictions of the wavenumber of all isotopic variants compared to that of $\tilde{\nu}(^{12}C^{16}O)$. (As usual, the better the experimental results agree with the whole set of predictions, the more confidence one would have with the conclusion.)

Wavenumber for isotopic variant	If O binds	If C binds
$\tilde{\nu}(^{12}C^{18}O) =$	$\tilde{\nu}(^{12}C^{16}O)^*$	$(16/18)^{1/2}\tilde{\nu}(^{12}C^{16}O)$
$\tilde{\nu}(^{13}C^{16}O) =$	$(12/13)^{1/2}\tilde{\nu}(^{12}C^{16}O)$	$\tilde{\nu}(^{12}C^{16}O)^*$
$\tilde{\nu}(^{13}C^{18}O) =$	$(12/13)^{1/2}\tilde{\nu}(^{12}C^{16}O)$	$(16/18)^{1/2}\tilde{\nu}(^{12}C^{16}O)$

*That is, no change compared to the standard.

The wavenumber is related to the force constant as follows:

$$\omega = 2\pi c\tilde{\nu} = \left(\frac{k_f}{m}\right)^{1/2} \quad \text{so} \quad k_f = m(2\pi c\tilde{\nu})^2$$

$$k_f = (1.66\times10^{-27}\ \text{kg}\ m_u^{-1})[(2\pi)(2.998\times10^{10}\ \text{cm s}^{-1})\tilde{\nu}(^{12}\text{C}^{16}\text{O})]^2$$

$$\text{and}\ k_f/(\text{kg s}^{-1}) = (5.89\times10^{-5})(m/\text{u})[\tilde{\nu}(^{12}\text{C}^{16}\text{O})/\text{cm}^{-1}]^2$$

Here m is the mass of the atom that is not bound, *i.e.*, 12 m_u if O is bound and 16 m_u if C is bound. (Of course, one can compute k_f from any of the isotopic variants, and take k_f to be a mean derived from all the relevant data.)

P12.11 (a) Normalization requires

$$\int_{-\infty}^{\infty}|\psi_\alpha|^2\,dx = 1 = \int_{-\infty}^{\infty}\left(N\sum_{v=0}^{\infty}\frac{\alpha^v}{(v!)^{1/2}}\psi_v\right)^2 dx$$

The fact that the harmonic oscillator wavefunctions are orthogonal means that all of the "cross" terms in the square of the sum vanish; the fact that the functions are also normalized means that all of the remaining integrals are 1:

$$1 = N^2\sum_{v=0}^{\infty}\frac{\alpha^{2v}}{v!}\int_{-\infty}^{\infty}\psi_v^2\,dx = N^2\sum_{v=0}^{\infty}\frac{\alpha^{2v}}{v!}$$

The sum is a power-series expansion of the exponential function:

$$e^x = \sum_{n=0}^{\infty}\frac{x^n}{n!} \quad \text{so} \quad \sum_{v=0}^{\infty}\frac{\alpha^{2v}}{v!} = e^{\alpha^2}$$

This leads to

$$1 = N^2 e^{\alpha^2} \quad \text{and} \quad N = \boxed{e^{-\alpha^2/2}}$$

The above tacitly assumed that α is real. If α is complex, then factors of α^v multiply factors of $(\alpha^*)^v$, yielding $|\alpha|^{2v}$ and $N = \boxed{e^{-|\alpha|^2/2}}$.

(b) Orthogonality would require

$$\int_{-\infty}^{\infty} \psi_\beta {}^* \psi_\alpha \, dx = 0$$

so we evaluate the integral

$$\int_{-\infty}^{\infty} \psi_\beta {}^* \psi_\alpha \, dx = \int_{-\infty}^{\infty} \left(N_\beta \sum_{v'=0}^{\infty} \frac{(\beta^*)^{v'}}{(v'!)^{1/2}} \psi_{v'} \right) \left(N_\alpha \sum_{v=0}^{\infty} \frac{\alpha^v}{(v!)^{1/2}} \psi_v \right) dx$$

The harmonic oscillator eigenfunctions *are* orthogonal, so terms in which $v' \neq v$, vanish, but terms in which $v' = v$ do not:

$$\int_{-\infty}^{\infty} \psi_\beta {}^* \psi_\alpha \, dx = N_\alpha N_\beta \sum_{v=0}^{\infty} \frac{(\beta^*)^v \alpha^v}{v!} \int_{-\infty}^{\infty} \psi_v \psi_v \, dx = N_\alpha N_\beta \sum_{v=0}^{\infty} \frac{(\beta^* \alpha)^v}{v!} = N_\alpha N_\beta e^{\beta^* \alpha} \neq 0$$

(c) $\Delta x = \left(\langle x^2 \rangle - \langle x \rangle^2 \right)^{1/2}$ and $\Delta p = \left(\langle p^2 \rangle - \langle p \rangle^2 \right)^{1/2}$

$$\langle x \rangle = \int_{-\infty}^{\infty} \psi_\alpha {}^* x \psi_\alpha \, dx = \int_{-\infty}^{\infty} \left(N_\alpha \sum_{v'=0}^{\infty} \frac{(\alpha^*)^{v'}}{(v'!)^{1/2}} \psi_{v'} \right) x \left(N_\alpha \sum_{v=0}^{\infty} \frac{\alpha^v}{(v!)^{1/2}} \psi_v \right) dx$$

Most of the terms in the double sum vanish. To see which ones remain, recall that the harmonic oscillator eigenfunctions are

$$\psi_v = N_v H_v(y) e^{-y^2/2} \quad [12.8]$$

We need some properties of Hermite polynomials from Table 12.1. In particular,

$$H_{v+1} - 2yH_v + 2vH_{v-1} = 0 \quad \text{so} \quad 2yH_v = H_{v+1} + 2vH_{v-1}$$

Recall that the displacement x is related to y by

$$y = \frac{x}{a} \text{ where } a = \left(\frac{\hbar^2}{mk_f} \right)^{1/4}$$ [12.8, with a because this problem already has an α].

Therefore

$$xH_v = \frac{a}{2} \times 2yH_v = \frac{a}{2}(H_{v+1} + 2vH_{v-1})$$

and $$x\psi_v = \frac{aN_v}{2}(H_{v+1} + 2vH_{v-1}) e^{-y^2/2} = \frac{aN_v}{2N_{v+1}} \psi_{v+1} + \frac{avN_v}{N_{v-1}} \psi_{v-1}$$

The harmonic oscillator normalization constants are

$$N_v = \left(\frac{1}{a\pi^{1/2} 2^v v!} \right)^{1/2} \quad \text{[Example 12.2]}$$

so $$\frac{N_v}{N_{v+1}} = \{2(v+1)\}^{1/2} \quad \text{and} \quad \frac{N_v}{N_{v-1}} = \left(\frac{1}{2v} \right)^{1/2}$$

making $$x\psi_v = a\left(\frac{v+1}{2} \right)^{1/2} \psi_{v+1} + a\left(\frac{v}{2} \right)^{1/2} \psi_{v-1}$$

Now back to the expectation values:

$$\langle x\rangle = N_\alpha^2\sum_{v=0}^{\infty}\frac{\alpha^v}{(v!)^{1/2}}\sum_{v'=0}^{\infty}\frac{(\alpha^*)^{v'}}{(v'!)^{1/2}}\int_{-\infty}^{\infty}\psi_{v'}x\psi_v\,\mathrm{d}x$$

$$= aN_\alpha^2\sum_{v=0}^{\infty}\frac{\alpha^v}{(v!)^{1/2}}\sum_{v'=0}^{\infty}\frac{(\alpha^*)^{v'}}{(v'!)^{1/2}}\int_{-\infty}^{\infty}\psi_{v'}\left\{\left(\frac{v+1}{2}\right)^{1/2}\psi_{v+1}+\left(\frac{v}{2}\right)^{1/2}\psi_{v-1}\right\}\mathrm{d}x$$

All of the terms in the inner sum vanish except when $v' = v+1$ or $v-1$.

$$\langle x\rangle = aN_\alpha^2\sum_{v=0}^{\infty}\frac{\alpha^v}{(v!)^{1/2}}\frac{(\alpha^*)^{v+1}}{\{(v+1)!\}^{1/2}}\left(\frac{v+1}{2}\right)^{1/2}+aN_\alpha^2\sum_{v=1}^{\infty}\frac{\alpha^v}{(v!)^{1/2}}\frac{(\alpha^*)^{v-1}}{\{(v-1)!\}^{1/2}}\left(\frac{v}{2}\right)^{1/2}$$

Notice the numbering of the second sum. In the expression for $x\psi_v$ above, the second term vanishes for $v = 0$, so starting the second sum at $v = 1$ does not change the value of the sum. We finish evaluating this expectation value by cleaning up the sums:

$$\langle x\rangle = aN_\alpha^2\sum_{v=0}^{\infty}\frac{|\alpha|^{2v}}{v!}\frac{\alpha^*}{2^{1/2}}+aN_\alpha^2\sum_{v=1}^{\infty}\frac{|\alpha|^{2(v-1)}}{(v-1)!}\frac{\alpha}{2^{1/2}}$$

$$= aN_\alpha^2\sum_{v=0}^{\infty}\frac{|\alpha|^{2v}}{v!}\frac{\alpha^*}{2^{1/2}}+aN_\alpha^2\sum_{u=0}^{\infty}\frac{|\alpha|^{2u}}{u!}\frac{\alpha}{2^{1/2}}=\frac{a(\alpha^*+\alpha)}{2^{1/2}}$$

The next to last step renumbers the second sum, introducing a counter $u = v-1$, where u starts at zero. The last step uses the result of part (a) of this problem.

Next $$\langle x^2\rangle = \int_{-\infty}^{\infty}\psi_\alpha{}^*x^2\psi_\alpha\,\mathrm{d}x = \int_{-\infty}^{\infty}\left(N_\alpha\sum_{v'=0}^{\infty}\frac{(\alpha^*)^{v'}}{(v'!)^{1/2}}\psi_{v'}\right)x^2\left(N_\alpha\sum_{v=0}^{\infty}\frac{\alpha^v}{(v!)^{1/2}}\psi_v\right)\mathrm{d}x$$

Distribute one power of x to each of the wavefunctions so that

$$\langle x^2\rangle = a^2N_\alpha^2\sum_{v=0}^{\infty}\frac{\alpha^v}{(v!)^{1/2}}\sum_{v'=0}^{\infty}\frac{(\alpha^*)^{v'}}{(v'!)^{1/2}}$$

$$\times\int_{-\infty}^{\infty}\left\{\left(\frac{v'+1}{2}\right)^{1/2}\psi_{v'+1}+\left(\frac{v'}{2}\right)^{1/2}\psi_{v'-1}\right\}\left\{\left(\frac{v+1}{2}\right)^{1/2}\psi_{v+1}+\left(\frac{v}{2}\right)^{1/2}\psi_{v-1}\right\}\mathrm{d}x$$

The only terms that survive the inner sum have $v' = v$, $v+2$, or $v-2$:

$$\langle x^2\rangle = a^2N_\alpha^2\sum_{v=0}^{\infty}\frac{\alpha^v}{(v!)^{1/2}}\left\{\frac{(\alpha^*)^v}{(v!)^{1/2}}\left(\frac{v+1}{2}+\frac{v}{2}\right)\right.$$

$$\left.+\frac{(\alpha^*)^{v+2}}{\{(v+2)!\}^{1/2}}\left(\frac{v+2}{2}\right)^{1/2}\left(\frac{v+1}{2}\right)^{1/2}+\frac{(\alpha^*)^{v-2}}{\{(v-2)!\}^{1/2}}\left(\frac{v-1}{2}\right)^{1/2}\left(\frac{v}{2}\right)^{1/2}\right\},$$

$$\langle x^2\rangle = a^2N_\alpha^2\left\{\sum_{v=0}^{\infty}\frac{|\alpha|^{2v}}{v!}\left(v+\frac{1}{2}\right)+\sum_{v=0}^{\infty}\frac{|\alpha|^{2v}}{v!}\frac{(\alpha^*)^2}{2}+\sum_{v=2}^{\infty}\frac{|\alpha|^{2(v-2)}}{\{(v-2)!\}^{1/2}}\frac{\alpha^2}{2}\right\}$$

The last sum can be renumbered as was done above. The presence of terms in α^2 and $(\alpha^*)^2$ suggests that we might be able to complete a square, $(\alpha + \alpha^*)^2$. To do so, look at the first summation term:

$$\sum_{v=0}^{\infty}\frac{|\alpha|^{2v}}{v!}\left(v+\frac{1}{2}\right)=\frac{1}{2}\sum_{v=0}^{\infty}\frac{|\alpha|^{2v}}{v!}+\sum_{v=0}^{\infty}\frac{|\alpha|^{2v}\,v}{v!}=\frac{1}{2}\sum_{v=0}^{\infty}\frac{|\alpha|^{2v}}{v!}+\sum_{v=1}^{\infty}\frac{|\alpha|^{2(v-1)}|\alpha|^2}{(v-1)!}$$
$$=\frac{1}{2}\sum_{v=0}^{\infty}\frac{|\alpha|^{2v}}{v!}+\sum_{v=0}^{\infty}\frac{|\alpha|^{2v}|\alpha|^2}{v!}$$

Finally, we can put all the sums back together to yield

$$\langle x^2\rangle=a^2N_\alpha^2\left(\frac{1}{2}+\frac{(\alpha+\alpha^*)^2}{2}\right)\sum_{v=0}^{\infty}\frac{|\alpha|^{2v}}{v!}=\frac{a^2}{2}+\frac{a^2(\alpha+\alpha^*)^2}{2}$$

The uncertainty in position is

$$\Delta x=\left(\langle x^2\rangle-\langle x\rangle^2\right)^{1/2}=\left\{\frac{a^2}{2}+\frac{a^2(\alpha+\alpha^*)^2}{2}-\left(\frac{a(\alpha^*+\alpha)}{2^{1/2}}\right)^2\right\}^{1/2}=\frac{a}{2^{1/2}}$$

The momentum operator is $\hat{p}=\frac{\hbar}{\mathrm{i}}\frac{\mathrm{d}}{\mathrm{d}x}$, so we need

$$\hat{p}\psi_\alpha=\frac{\hbar}{\mathrm{i}}\frac{\mathrm{d}}{\mathrm{d}x}N_\alpha\sum_{v=0}^{\infty}\frac{\alpha^v}{(v!)^{1/2}}\psi_v=\frac{N_\alpha\hbar}{\mathrm{i}}\sum_{v=0}^{\infty}\frac{\alpha^v}{(v!)^{1/2}}\frac{\mathrm{d}\psi_v}{\mathrm{d}x}$$

where $\frac{\mathrm{d}\psi_v}{\mathrm{d}x}=N_v\frac{\mathrm{d}y}{\mathrm{d}x}\frac{\mathrm{d}}{\mathrm{d}y}H_v(y)\mathrm{e}^{-y^2/2}=\frac{N_v}{a}\left(\frac{\mathrm{d}H_v}{\mathrm{d}y}\mathrm{e}^{-y^2/2}-yH_v\mathrm{e}^{-y^2/2}\right)$

We need a recursion relationship between Hermite polynomials and their first derivatives, available in Table 12.1 or from a mathematical handbook,

$$\frac{\mathrm{d}H_v}{\mathrm{d}y}=2vH_{v-1}$$

and the expression derived above for $x\psi_v$ to express $\mathrm{d}\psi_v/\mathrm{d}x$:

$$\frac{\mathrm{d}\psi_v}{\mathrm{d}x}=\frac{N_v}{a}(2vH_{v-1}-yH_v)\mathrm{e}^{-y^2/2}=\frac{vN_v}{aN_{v-1}}\psi_{v-1}-\frac{N_v}{2aN_{v+1}}\psi_{v+1}$$
$$=\frac{1}{a}\left(\frac{v}{2}\right)^{1/2}\psi_{v-1}-\frac{1}{a}\left(\frac{v+1}{2}\right)^{1/2}\psi_{v+1}$$

Now back to the expectation values:

$$\langle p\rangle=\frac{N_\alpha^2\hbar}{\mathrm{i}a}\sum_{v=0}^{\infty}\frac{\alpha^v}{(v!)^{1/2}}\sum_{v'=0}^{\infty}\frac{(\alpha^*)^{v'}}{(v'!)^{1/2}}\int_{-\infty}^{\infty}\psi_{v'}\left\{\left(\frac{v}{2}\right)^{1/2}\psi_{v-1}-\left(\frac{v+1}{2}\right)^{1/2}\psi_{v+1}\right\}\mathrm{d}x$$

Steps analogous to those applied to the expectation value of x lead to

$$\langle p\rangle = \frac{\hbar}{ia}\frac{(\alpha-\alpha^*)}{2^{1/2}} \quad \text{so} \quad \langle p\rangle^2 = \frac{\hbar^2}{-a^2}\frac{(\alpha-\alpha^*)^2}{2} = -\frac{\hbar m\omega}{2}(\alpha-\alpha^*)^2$$

We can evaluate $\langle p^2\rangle$ from other expectation values. Recall that the harmonic-oscillator hamiltonian can be written as

$$\hat{H} = \frac{\hat{p}^2}{2m} + \frac{1}{2}k_f x^2 \quad \text{so} \quad \hat{p}^2 = 2m\left(\hat{H} - \frac{1}{2}k_f x^2\right) = 2m\hat{H} - mk_f x^2$$

and $\langle p^2\rangle = 2m\langle H\rangle - mk_f\langle x^2\rangle$

We need

$$\langle H\rangle = \int_{-\infty}^{\infty}\psi_\alpha{}^*\hat{H}\psi_\alpha\,dx = \int_{-\infty}^{\infty}\left(N_\alpha\sum_{v'=0}^{\infty}\frac{(\alpha^*)^{v'}}{(v'!)^{1/2}}\psi_{v'}\right)\hat{H}\left(N_\alpha\sum_{v=0}^{\infty}\frac{\alpha^v}{(v!)^{1/2}}\psi_v\right)dx$$

$$= \hbar\omega N_\alpha^2\sum_{v'=0}^{\infty}\frac{(\alpha^*)^{v'}}{(v'!)^{1/2}}\sum_{v=0}^{\infty}\frac{\alpha^v}{(v!)^{1/2}}\left(v+\frac{1}{2}\right)\int_{-\infty}^{\infty}\psi_{v'}\psi_v\,dx$$

Only the terms where $v' = v$ are non-zero, so

$$\langle H\rangle = \hbar\omega N_\alpha^2\sum_{v=0}^{\infty}\frac{|\alpha|^{2v}}{v!}\left(v+\frac{1}{2}\right) = \hbar\omega\left(|\alpha|^2 + \frac{1}{2}\right)$$

(For the last step, notice that we have seen the sum containing $v + 1/2$ above.) Therefore,

$$\langle p^2\rangle = 2m\langle H\rangle - mk_f\langle x^2\rangle = 2m\hbar\omega\left(|\alpha|^2 + \frac{1}{2}\right) - \frac{mk_f a^2}{2}\{1+(\alpha+\alpha^*)^2\}$$

Using the definitions of a and ω to note that $mk_f a^2 = m\hbar\omega$ and collecting terms,

$$\langle p^2\rangle = \frac{m\hbar\omega}{2}\{1-(\alpha-\alpha^*)^2\}$$

$$\Delta p = \left(\langle p^2\rangle - \langle p\rangle^2\right)^{1/2} = \left(\frac{m\hbar\omega}{2}\{1-(\alpha-\alpha^*)^2 + (\alpha-\alpha^*)^2\}\right)^{1/2} = \left(\frac{m\hbar\omega}{2}\right)^{1/2}$$

Finally $\Delta x\Delta p = \frac{a}{2^{1/2}}\left(\frac{m\hbar\omega}{2}\right)^{1/2} = \left(\frac{\hbar}{2m\omega}\right)^{1/2}\left(\frac{m\hbar\omega}{2}\right)^{1/2} = \boxed{\frac{\hbar}{2}}$

Topic 13 Rotational motion in two dimensions

Discussion question

D13.1 In quantum mechanics, particles are said to have characteristics of waves. The fact that the particle exists then requires that the waves representing it not experience destructive interference upon reflection by a barrier or in its motion around a closed loop. This requirement restricts the wavelength of a particle on a ring to wavelengths such that a whole number of wavelengths must fit on the circumference of the ring. That means

$$n\lambda = 2\pi r, \text{ where } n = 0, 1, 2, \ldots$$

Exercise

E13.1(a) Orthogonality requires that, if $m \neq n$, then

$$\int \psi_m^* \psi_n \, \mathrm{d}\tau = 0$$

Performing the integration

$$\int \psi_m^* \psi_n \, \mathrm{d}\tau = \int_0^{2\pi} N\mathrm{e}^{-\mathrm{i}m\phi} N\mathrm{e}^{\mathrm{i}n\phi} \, \mathrm{d}\phi = N^2 \int_0^{2\pi} \mathrm{e}^{\mathrm{i}(n-m)\phi} \, \mathrm{d}\phi$$

If $m \neq n$, then

$$\int \psi_m^* \psi_n \, \mathrm{d}\tau = \frac{N^2}{\mathrm{i}(n-m)} \mathrm{e}^{\mathrm{i}(n-m)\phi} \Big|_0^{2\pi} = \frac{N^2}{\mathrm{i}(n-m)}(1-1) = 0$$

Therefore, they are orthogonal.

Problems

P13.1 (a) The normalization integral is

$$1=\int_0^{2\pi} N(\psi_{-1}{}^{*}-3^{1/2}\mathrm{i}\psi_{+1}{}^{*})N(\psi_{-1}+3^{1/2}\mathrm{i}\psi_{+1})\mathrm{d}\varphi$$
$$=N^2\int_0^{2\pi}(|\psi_{-1}|^2+3|\psi_{+1}|^2+3^{1/2}\mathrm{i}\psi_{-1}{}^{*}\psi_{+1}-3^{1/2}\mathrm{i}\psi_{+1}{}^{*}\psi_{-1})\mathrm{d}\varphi=4N^2$$

In the last step, we used the fact that the particle-on-a-ring eigenfunctions are themselves normalized and orthogonal. The overall normalization constant, then, is $\boxed{N=1/2}$, and the normalized wavefunction can be written as a superposition of particle-on-a-ring eigenfunctions

$$\psi=\frac{\psi_{-1}+3^{1/2}\mathrm{i}\psi_{+1}}{2}$$

(b) The particle-on-a-ring wavefunctions are eigenfunctions of the angular momentum, so any measurement of angular momentum will result in a corresponding eigenvalue (Topic 7.1, Postulate V). The measurements will be $m_l\hbar$ [13.16]. In particular

$\boxed{-\hbar}$, the eigenvalue of ψ_{-1}, with probability $|(1/2)|^2=\boxed{1/4}$,

and $\boxed{+\hbar}$, the eigenvalue of ψ_{+1}, with probability $|(3^{1/2}\mathrm{i}/2)|^2=\boxed{3/4}$.

(c) The particle-on-a-ring wavefunctions are also eigenfunctions of the energy, so any energy measurement will result in a corresponding eigenvalue. However, both of the energy eigenstates that comprise the superposition have the same energy eigenvalue, so any measurement will yield that value, namely

$$E_{m_l}=\frac{m_l^2\hbar^2}{2I}=\boxed{\frac{\hbar^2}{2m_\mathrm{p}r^2}}\quad[13.7\mathrm{b}]$$

(d) Expectation values are weighted averages of eigenvalues; the weights are the squares of the magnitude of the coefficients in the superposition wavefunction:

$$\langle\Omega\rangle=\sum_k|c_k|^2\omega_k\quad[7.4]$$

So the expectation value of angular momentum is

$$\langle l_z\rangle=\left|\frac{1}{2}\right|^2\times(-\hbar)+\left|\frac{3^{1/2}\mathrm{i}}{2}\right|^2\times\hbar=\boxed{\frac{\hbar}{2}}$$

(Note that none of the angular momentum measurements will yield the expectation value.) The expectation value of the energy can be computed in the same way. We have already noted, though, that *every* energy measurement will yield the same value; therefore, the expectation value is also equal to that value.

$$\langle H\rangle = \boxed{\frac{\hbar^2}{2m_p r^2}}$$

P13.3 (a) $E_{m_l} = \frac{m_l^2\hbar^2}{2I}[13.7\text{b}] = \frac{m_l^2\hbar^2}{2m_e r^2}$

The ground-state energy is 0, when $m_l = 0$. The first excited state energy is

$$E_1 = \frac{\hbar^2}{2mr^2} = \frac{(1.055\times10^{-34}\ \text{J s})^2}{(2)\times(1.008)\times(1.6605\times10^{-27}\ \text{kg})\times(160\times10^{-12}\ \text{m})^2} = 1.30\times10^{-22}\ \text{J}$$

This is also the excitation energy (the difference between the first two levels) and the energy of the photon that can excite rotation. Thus

$$\Delta E = E_{\text{photon}} = \frac{hc}{\lambda}$$

so $$\lambda = \frac{hc}{\Delta E} = \frac{6.626\times10^{-34}\ \text{J s}\times2.998\times10^{8}\ \text{m s}^{-1}}{1.30\times10^{-22}\ \text{J}} = \boxed{1.53\times10^{-3}\ \text{m}} = \boxed{1.53\ \text{mm}}$$

(b) The minimum angular momentum is $\boxed{\pm\hbar}$

P13.5 The average angular position is

$$\langle\varphi\rangle = \int_0^{2\pi} \frac{e^{-im_l\varphi}}{(2\pi)^{1/2}}\varphi\frac{e^{im_l\varphi}}{(2\pi)^{1/2}}\text{d}\varphi = \frac{1}{2\pi}\int_0^{2\pi}\varphi\,\text{d}\varphi = \frac{1}{2\pi}\times\frac{\varphi^2}{2}\Big|_0^{2\pi} = \boxed{\pi}$$

At least that is the result if the range of φ is defined conventionally. The word conventional is important here, because the angle is a cyclical variable. One cycle of its range can be specified with respect to any arbitrary angle, φ_0. In that case, the expectation value would be

$$\langle\varphi\rangle = \frac{1}{2\pi}\int_{\varphi_0}^{\varphi_0+2\pi}\varphi\,\text{d}\varphi = \frac{1}{2\pi}\times\frac{\varphi^2}{2}\Big|_{\varphi_0}^{\varphi_0+2\pi} = \frac{(\varphi_0+2\pi)^2-\varphi_0^2}{4\pi} = \boxed{\varphi_0+\pi}$$

That is, the expectation value is the midpoint of the range of angle—however it is specified. The endpoints of the range, however specified, are not really boundaries. The expectation value, really, is arbitrary.

P13.7 The elliptical ring to which the particle is confined is defined by the set of all points that obey a certain equation. In Cartesian coordinates, that equation is

$$\frac{x^2}{a^2} + \frac{y^2}{b^2} = 1$$

as you may remember from analytical geometry. An ellipse is similar to a circle, and an appropriate change of variable can transform the ellipse of this problem into a circle. That change of variable is most conveniently described in terms of new Cartesian coordinates (X,Y) where

$$X = x \quad \text{and} \quad Y = ay/b$$

In this new coordinate system, the equation for the ellipse becomes:

$$\frac{x^2}{a^2}+\frac{y^2}{b^2}=1 \quad \Rightarrow \quad \frac{X^2}{a^2}+\frac{Y^2}{a^2}=1 \Rightarrow \quad X^2+Y^2=a^2,$$

which we recognize as the equation of a circle of radius a centered at the origin of our (X,Y) system. The text found the eigenfunctions and eigenvalues for a particle on a circular ring by transforming from Cartesian coordinates to plane polar coordinates. Consider plane polar coordinates (R,Φ) related in the usual way to (X,Y):

$$X = R\cos\Phi \quad \text{and} \quad Y = R\sin\Phi$$

In this coordinate system, we can simply quote the results obtained in the text. The energy levels are

$$E=\frac{m_l^2\hbar^2}{2I} \quad [13.7\text{b}]$$

where the moment of inertia is the mass of the particle times the radius of the circular ring

$$I = ma^2$$

The eigenfunctions are

$$\psi=\frac{e^{im_l\Phi}}{(2\pi)^{1/2}} \quad [13.7\text{a}]$$

It is customary to express results in terms of the original coordinate system, so express Φ in terms first of X and Y, and then substitute the original coordinates:

$$\frac{Y}{X}=\tan\Phi \quad \text{so} \quad \Phi=\tan^{-1}\frac{Y}{X}=\tan^{-1}\frac{ay}{bx}$$

P13.9 (a) By hypothesis, the angular momentum is $+\hbar$, so the corresponding wavefunction is

$$\psi_{+1}(\phi)=\frac{e^{+i\phi}}{(2\pi)^{1/2}}$$

There is $\boxed{\text{no uncertainty in } \Delta l_z}$ because the wavefunction is an eigenfunction of $\hat{l}_z$. Therefore, the uncertainty product $\Delta l_z \, \Delta\sin\phi$ must also be zero. Because the question asks for all of the pieces in the uncertainty expression, we evaluate $\Delta\sin\phi$:

$$\langle \sin\phi \rangle = \int_0^{2\pi} \frac{e^{-i\phi}}{(2\pi)^{1/2}} \sin\phi \frac{e^{+i\phi}}{(2\pi)^{1/2}} d\phi = \frac{1}{2\pi} \int_0^{2\pi} \sin\phi d\phi = \frac{1}{2\pi} \times \left. \frac{-\cos\phi}{2} \right|_0^{2\pi} = 0$$

$$\langle \sin^2\phi \rangle = \int_0^{2\pi} \frac{e^{-i\phi}}{(2\pi)^{1/2}} \sin^2\phi \frac{e^{+i\phi}}{(2\pi)^{1/2}} d\phi = \frac{1}{2\pi} \int_0^{2\pi} \sin^2\phi d\phi$$

$$= \frac{1}{2\pi} \times \left. \left(\frac{\phi}{2} - \frac{\sin 2\phi}{4} \right) \right|_0^{2\pi} = \frac{1}{2}$$

So $\Delta \sin\phi = \left(\langle \sin^2\phi \rangle - \langle \sin\phi \rangle^2 \right)^{1/2} = \left(\frac{1}{2} - 0 \right)^{1/2} = \boxed{\frac{1}{2^{1/2}}}$

The only way the uncertainty relationship can hold is if the expectation value of cos φ also vanishes:

$$\langle \cos\phi \rangle = \int_0^{2\pi} \frac{e^{-i\phi}}{(2\pi)^{1/2}} \cos\phi \frac{e^{+i\phi}}{(2\pi)^{1/2}} d\phi = \frac{1}{2\pi} \int_0^{2\pi} \cos\phi d\phi = \frac{1}{2\pi} \times \left. \frac{\sin\phi}{2} \right|_0^{2\pi} = 0,$$

so $\Delta l_z \Delta \sin\varphi \geq \frac{1}{2}\hbar |\langle \cos\varphi \rangle|$ because $0 \geq 0$.

(b) A wavefunction proportional to $\cos\varphi$ is a mixture of the two particle-on-a-ring wavefunctions that have $m_l = +1$ and $m_l = -1$, because $e^{-i\varphi} + e^{+i\varphi} = 2\cos\varphi$:

$$\psi(\varphi) = N \frac{e^{+i\varphi} + e^{-i\varphi}}{(2\pi)^{1/2}}$$

In order to compute expectation values, we need the normalization constant:

$$1 = \int_0^{2\pi} N \frac{e^{+i\phi} + e^{-i\phi}}{(2\pi)^{1/2}} N \frac{e^{-i\phi} + e^{+i\phi}}{(2\pi)^{1/2}} d\phi = \frac{N^2}{2\pi} \int_0^{2\pi} (2 + e^{-2i\phi} + e^{+2i\phi}) d\phi$$

$$= \frac{N^2}{2\pi} \left. \left(2\phi - \frac{e^{-2i\phi}}{2i} + \frac{e^{+2i\phi}}{2i} \right) \right|_0^{2\pi} = 2N^2$$

So $N = 1/2^{1/2} = c_{-1} = c_{+1}$, where the c are coefficients in the superposition. Because particle-on-a-ring wavefunctions are eigenfunctions of angular momentum, we can use

$$\langle \Omega \rangle = \sum_k |c_k|^2 \omega_k \quad [7.4]$$

So the expectation value of angular momentum is

$$\langle l_z \rangle = \frac{1}{2} \times (-\hbar) + \frac{1}{2} \times \hbar = 0$$

Both particle-on-a-ring wavefunctions have the same eigenvalue of angular momentum squared, namely $\hbar^2$, so this is the expectation value of that operator.

$$\Delta l_z = \left(\langle l_z^2\rangle - \langle l_z\rangle^2\right)^{1/2} = \left(\hbar^2 - 0\right)^{1/2} = \boxed{\hbar}$$

Before we look at the expectation values of trigonometric functions, note that the wavefunction can be written as

$$\psi(\phi) = \frac{1}{2^{1/2}} \times \frac{e^{+i\phi} + e^{-i\phi}}{(2\pi)^{1/2}} = \frac{\cos\phi}{\pi^{1/2}}$$

So $\quad \langle \sin\phi\rangle = \int_0^{2\pi} \frac{\cos\phi}{\pi^{1/2}} \sin\phi \frac{\cos\phi}{\pi^{1/2}} d\phi = \frac{1}{\pi}\int_0^{2\pi} \cos^2\phi \sin\phi d\phi = \frac{1}{\pi} \times \left.\frac{-\cos^3\phi}{3}\right|_0^{2\pi} = 0,$

$$\langle \sin^2\phi\rangle = \int_0^{2\pi} \frac{\cos\phi}{\pi^{1/2}} \sin^2\phi \frac{\cos\phi}{\pi^{1/2}} d\phi = \frac{1}{\pi}\int_0^{2\pi} \cos^2\phi \sin^2\phi d\phi$$

$$= \frac{1}{\pi} \times \left.\left(\frac{\phi}{8} - \frac{\sin 4\phi}{32}\right)\right|_0^{2\pi} = \frac{1}{4}$$

and $\quad \Delta \sin\phi = \left(\langle \sin^2\phi\rangle - \langle \sin\phi\rangle^2\right)^{1/2} = \left(\frac{1}{4} - 0\right)^{1/2} = \boxed{\frac{1}{2}}$

$$\langle \cos\phi\rangle = \int_0^{2\pi} \frac{\cos\phi}{\pi^{1/2}} \cos\phi \frac{\cos\phi}{\pi^{1/2}} d\phi = \frac{1}{\pi}\int_0^{2\pi} \cos^3\phi d\phi$$

$$= \frac{1}{\pi} \times \left.\left(\sin\phi - \frac{\sin^3\phi}{3}\right)\right|_0^{2\pi} = 0$$

Finally, $\quad \Delta l_z \Delta \sin\varphi \geq \tfrac{1}{2}\hbar \left|\langle \cos\phi\rangle\right| \quad$ because $\tfrac{1}{2}\hbar \geq \tfrac{1}{2}\hbar \times 0 = 0.$

The uncertainty principle is trivially satisfied in both cases, because the minimum uncertainty is zero. The cases are somewhat different, though, in that there is no uncertainty in the angular momentum in the first case, and there is uncertainty in the second case.

Topic 14 **Rotational motion in three dimensions**

Discussion question

D14.1 Rotational motion on a ring and on a sphere share features such as the form of the energy (square of the angular momentum over twice the moment of inertia) and the lack of zero-point energy because the ground state does not restrict the angular position of the particle. Degeneracy is possible in both cases. In the case of the ring, the axis of rotation is specified, but not in the case of the sphere. This distinction gives rise to the fact that angular momenta about different perpendicular axes **cannot** be simultaneously specified: $\hat{l}_x$, $\hat{l}_y$, and $\hat{l}_z$ are complementary in the sense described in Topic 8.3.

Exercises

E14.1(a) Magnitude of angular momentum $=\{l(l+1)\}^{1/2}\hbar$ [14.10]

Projection on arbitrary axis $=m_l\hbar$ [14.11]

Thus,

Magnitude $=(2^{1/2})\times\hbar=\boxed{1.49\times10^{-34}\ \mathrm{J\,s}}$

Possible projections 0 or $\pm\hbar$

E14.2(a) The diagrams are drawn by forming a vector of length $\{l(l+1)\}^{1/2}$ with a projection m_l on the z-axis. (See Fig. 14.1.) Each vector represents the edge of a cone around the z-axis (that for $m_j=0$ represents the side view of a disk perpendicular to z).

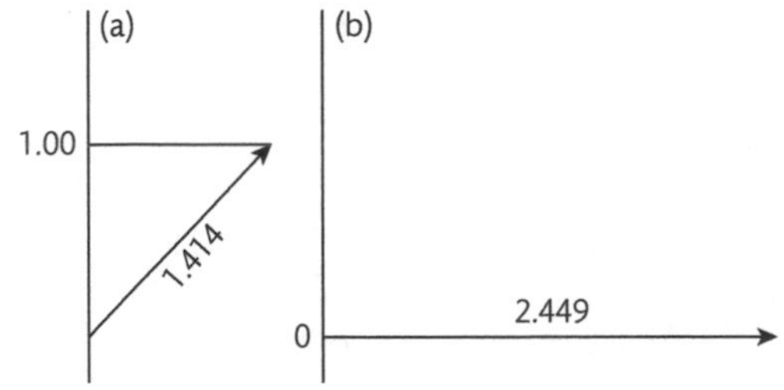

Fig 14.1

Let θ be the angle of the vector with the positive z-axis. The cosine of that angle is the ratio of $m_l/\{l(l+1)\}^{1/2}$

(a) $\cos\theta = \dfrac{1}{2^{1/2}}$ so $\theta = \boxed{\pi/4 \text{ radians or } 45°}$

(b) $\cos\theta = 0$ so $\theta = \boxed{\pi/2 \text{ radians or } 90°}$

E14.3(a) The rotational energy depends only on the quantum number l (eqn 14.5), but there are distinct states for every allowed value of m_l, which can range from $-l$ to l in integer steps. For $l = 3$, possible values of $m_l = -3, -2, -1, 0, 1, 2, 3$. There are 7 such values, so the degeneracy is $\boxed{7}$.

Problems

P14.1 $$E_l = \frac{l(l+1)\hbar^2}{2I}\,[14.5] = \frac{l(l+1)\hbar^2}{2\mu R^2}$$

The reduced mass is

$$\mu = \frac{1.008\, m_u \times 126.90\, m_u}{1.008\, m_u + 126.90\, m_u} = 1.000\, m_u$$

so $$E_l = \left(\frac{l(l+1)\times(1.055\times10^{-34}\ \mathrm{J\,s})^2}{(2)\times(1.000\times1.6605\times10^{-27}\ \mathrm{kg})\times(160\times10^{-12}\ \mathrm{m})^2}\right)$$
$$= l(l+1)\times(1.31\times10^{-22}\ \mathrm{J}).$$

The degeneracy of level l is $2l+1$, because each l has states characterized by different m_l, which can range from $-l$ to l in integer steps. The energies and degeneracies (g) are

l	0	1	2	3
$10^{22}\,E$/J	0	2.62	7.86	15.72
g	1	3	5	7

The excitation energy is equal to the energy of the photon that excites the rotation. Thus

$$\Delta E = E_{\text{photon}} = \frac{hc}{\lambda}$$

so $$\lambda = \frac{hc}{\Delta E} = \frac{6.626\times10^{-34}\ \mathrm{J\,s}\times2.998\times10^{8}\ \mathrm{m\,s^{-1}}}{2.62\times10^{-22}\ \mathrm{J}} = \boxed{7.59\times10^{-4}\ \mathrm{m}} = \boxed{0.759\ \mathrm{mm}}.$$

This electromagnetic radiation is a millimeter wave, near the fuzzy boundary between microwaves and the far infrared.

P14.3 (a) The normalization integral is

$$1 = \int_0^{2\pi}\int_0^{\pi} N(2^{1/2}Y_{2,+1}{}^* - 3iY_{2,+2}{}^* + Y_{1,+1}{}^*)N(2^{1/2}Y_{2,+1} + 3iY_{2,+2} + Y_{1,+1})\sin\theta\,d\theta\,d\phi$$

The spherical harmonics are themselves normalized and orthogonal, so none of the cross terms will contribute to the integral.

$$1 = N^2 \int_0^{2\pi}\int_0^{\pi}(2|Y_{2,+1}|^2 + 9|Y_{2,+2}|^2 + |Y_{1,+1}|^2)\sin\theta\,d\theta\,d\varphi = 12N^2$$

The overall normalization constant, then, is $\boxed{N = 1/(12)^{1/2}}$, and the normalized wavefunction can be written as a superposition of spherical harmonic states:

$$\psi = \frac{2^{1/2}Y_{2,+1} + 3iY_{2,+2} + Y_{1,+1}}{12^{1/2}} = \frac{Y_{2,+1}}{6^{1/2}} + \frac{3^{1/2}iY_{2,+2}}{2} + \frac{Y_{1,+1}}{12^{1/2}}$$

(b) The spherical harmonics are eigenfunctions of the angular momentum operator—or rather the angular momentum squared operator, so any measurement of the magnitude of angular momentum will result in a corresponding eigenvalue (Topic 7.1, Postulate V). The measurements will be $\{l(l+1)\}^{1/2}\hbar$ [14.10]. In particular

$6^{1/2}\hbar$, the eigenvalue of $Y_{2,+1}$ and of $Y_{2,+2}$, with probability 11/12,

$2^{1/2}\hbar$, the eigenvalue of $Y_{1,+1}$, with probability 1/12.

(The squares of the magnitudes of the coefficients of the spherical harmonics in the normalized wavefunction are the probabilities of finding the system in a state described by the corresponding spherical harmonic. The probability of finding a given angular momentum eigenvalue is equal to the sum of probabilities of the corresponding states.)

(c) The spherical harmonics are eigenfunctions of the z-component of the angular momentum operator, so any measurement of that component will result in a corresponding eigenvalue. The measurements will be $m_l\hbar$ [14.11]. In particular

$+\hbar$, the eigenvalue of $Y_{2,+1}$ and of $Y_{1,+1}$, with probability $3/12 = 1/4$,

$+2\hbar$, the eigenvalue of $Y_{2,+2}$, with probability $9/12 = 3/4$.

(d) The spherical harmonics are also eigenfunctions of the energy operator, so any energy measurement will result in a corresponding eigenvalue. The measurements will be

$$E_l = \frac{l(l+1)\hbar^2}{2I} \quad [14.5]$$

In particular, the energy measurements, in units of $\hbar^2/2I$, will be

6, the eigenvalue of $Y_{2,+1}$ and of $Y_{2,+2}$, with probability 11/12,

2, the eigenvalue of $Y_{1,+1}$, with probability 1/12.

(e) Expectation values are weighted averages of eigenvalues; the weights are the squares of the magnitude of the coefficients in the superposition wavefunction:

$$\langle\Omega\rangle = \sum_k |c_k|^2\,\omega_k \quad [7.4]$$

Take the z-component of angular momentum first. Its expectation value is

$$\langle l_z\rangle = \left|\frac{1}{6^{1/2}}\right|^2 \times\hbar + \left|\frac{3^{1/2}i}{2}\right|^2 \times 2\hbar + \left|\frac{1}{12^{1/2}}\right|^2 \times\hbar = \boxed{\frac{7\hbar}{4}}$$

The expectation value of the energy is

$$\langle H\rangle = \left|\frac{1}{6^{1/2}}\right|^2 \times\frac{6\hbar^2}{2I} + \left|\frac{3^{1/2}\mathrm{i}}{2}\right|^2 \times\frac{6\hbar^2}{2I} + \left|\frac{1}{12^{1/2}}\right|^2 \times\frac{2\hbar^2}{2I}$$

$$= \frac{\hbar^2}{2I} + \frac{9\hbar^2}{4I} + \frac{\hbar^2}{12I} = \boxed{\frac{17\hbar^2}{6I}}$$

To obtain a numerical result, we need the moment of inertia,
$I = m_{\mathrm{He}}r^2 = (4.003\times 1.6605\times10^{-27}\ \mathrm{kg})\times(0.35\times10^{-9}\ \mathrm{m})^2 = 8.1\times10^{-46}\ \mathrm{kg\ m^2}$,

$$\text{so}\quad \langle H\rangle = \frac{17\hbar^2}{6I} = \frac{17\times(1.0546\times10^{-34}\ \mathrm{J\ s})^2}{6\times8.1\times10^{-46}\ \mathrm{kg\ m^2}} = \boxed{3.9\times10^{-23}\ \mathrm{J}}$$

The angular momentum operator is related to the energy operator by

$$\hat{H} = \frac{\hat{l}^2}{2I} \quad\text{so}\quad \langle l^2\rangle^{1/2} = (2I\langle H\rangle)^{1/2} = \left(2I\times\frac{17\hbar^2}{6I}\right)^{1/2} = \boxed{\left(\frac{17}{3}\right)^{1/2}\hbar}$$

P14.5 Mathematical software can animate the real part or the imaginary part of $\Psi(\phi,t)$, or you may wish to have it display $|\Psi(\phi,t)|^2$. Try a "pure" state, that is, let $c = 1$ for one value of m_l and 0 for all others. This "packet" does not spread, but only circulates. Also try making all the coefficients in the sum equal (all 1, for example). Whatever your choice of coefficients, the pattern will repeat with a period T such that all the time-dependent factors equal the exponential of ($2\pi\mathrm{i}\times$ an integer), making the exponent $\dfrac{\mathrm{i}E_{m_l}t}{\hbar}$ equal to $2\pi\mathrm{i}m_l^2$ when $t = T$ and at intervals of T thereafter. That is,

$$\frac{\mathrm{i}T}{\hbar}\times\frac{m_l^2\hbar^2}{2I} = 2\pi\mathrm{i}m_l^2 \quad\text{so}\quad T = \frac{4\pi I}{\hbar} = \frac{8\pi^2 I}{h}$$

An example of this approach using Mathcad is illustrated below:

Wavepacket on a Ring as a MathCad Document. Let $\tau = \dfrac{h\cdot t}{4\cdot\pi\cdot I}$ and let each function in the superposition of $m + 1$ functions contribute with equal probability. The normalized angular functions are

$$\psi(m,\phi) := \left(\frac{1}{2\cdot\pi}\right)^{\frac{1}{2}}\cdot e^{i m\phi}$$ ([13.7a] where m is an integer)

The normalized superposition is

$$\psi(m_{max},\phi,\tau) := \left(\frac{1}{m+1}\right)^{\frac{1}{2}}\cdot\sum_{m=0}^{m_{max}}\psi(m,\phi)\cdot e^{-i m^2\cdot\tau}$$

$N := 500 \quad j := 0..N \quad \phi_j := \dfrac{2\cdot\pi\cdot j}{N} \quad m_{max} := 8 \quad \Delta\tau := .03$

The probability density of the superposition is

$P(\phi,\tau) := \psi(m_{max},\phi,\tau)\cdot\overline{\psi(m_{max},\phi,\tau)}$.

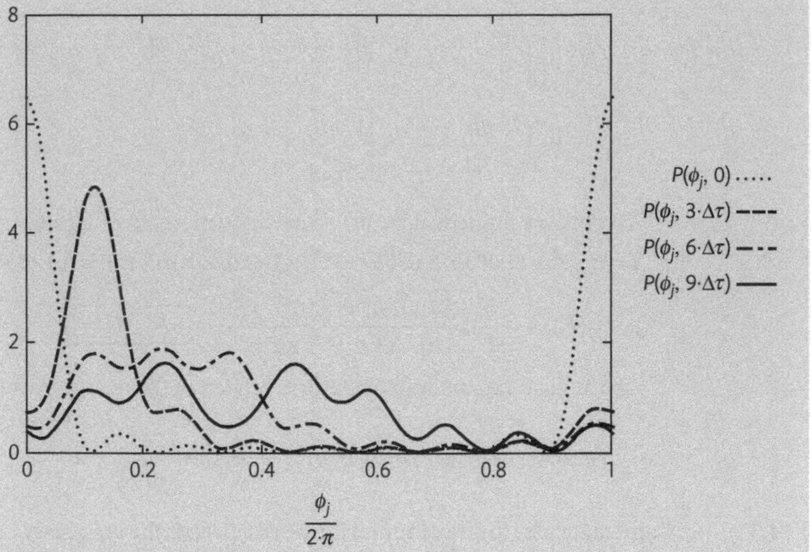

Fig 14.2

The above plots (Fig. 14.2) show that as the initially localized wave propagates around the ring it spreads with time and the uncertainty in knowing particle position increases. The effect of increasing or decreasing the energies accessible to the particle may be explored by increasing or decreasing the value of m_{max} in the MathCad document.

P14.7 The required integral is over a surface of a sphere. In spherical polar coordinates, that corresponds to "all space" that has a given (constant) value of r. Thus, we need only the angular portions of the volume element from Chemist's Toolkit 14.1. Substitute the appropriate spherical harmonics from Table 14.1, and call the integral I; we will ignore multiplicative constants (since we are to prove that I vanishes).

$$I \propto \int_0^{2\pi}\int_0^{\pi} Y_{1,+1}{}^* Y_{2,0} \sin\theta \,\mathrm{d}\theta\,\mathrm{d}\phi \propto \int_0^{2\pi}\int_0^{\pi} (\sin\theta \mathrm{e}^{+\mathrm{i}\phi})^* (3\cos^2\theta - 1)\sin\theta\,\mathrm{d}\theta\,\mathrm{d}\phi \,.$$

Collect terms:

$$I \propto \int_0^{2\pi} \mathrm{e}^{-\mathrm{i}\phi}\,\mathrm{d}\phi \int_0^{\pi} \sin^2\theta(3\cos^2\theta - 1)\,\mathrm{d}\theta = \left(\frac{\mathrm{e}^{-\mathrm{i}\phi}}{-\mathrm{i}}\right)\Bigg|_0^{2\pi} \int_0^{\pi} \sin^2\theta(3\cos^2\theta - 1)\,\mathrm{d}\theta = \boxed{0}$$

The ϕ integral has the same value at both limits of integration, so it vanishes, making the entire integral vanish.

P14.9 $\hat{\mathbf{l}} = \hat{\mathbf{r}} \times \hat{\mathbf{p}} = \begin{vmatrix} \mathbf{i} & \mathbf{j} & \mathbf{k} \\ \hat{x} & \hat{y} & \hat{z} \\ \hat{p}_x & \hat{p}_y & \hat{p}_z \end{vmatrix}$ [see any book treating the vector product of vectors]

$$= \mathbf{i}(\hat{y}\hat{p}_z - \hat{z}\hat{p}_y) + \mathbf{j}(\hat{z}\hat{p}_x - \hat{x}\hat{p}_z) + \mathbf{k}(\hat{x}\hat{p}_y - \hat{y}\hat{p}_x)$$

Therefore,

$$\hat{l}_x = (\hat{y}\hat{p}_z - \hat{z}\hat{p}_y) = \boxed{\frac{\hbar}{\mathrm{i}}\left(y\frac{\partial}{\partial z} - z\frac{\partial}{\partial y}\right)},$$

$$\hat{l}_y = (\hat{z}\hat{p}_x - \hat{x}\hat{p}_z) = \boxed{\frac{\hbar}{\mathrm{i}}\left(z\frac{\partial}{\partial x} - x\frac{\partial}{\partial z}\right)},$$

and $\hat{l}_z = (\hat{x}\hat{p}_y - \hat{y}\hat{p}_x) = \boxed{\frac{\hbar}{\mathrm{i}}\left(x\frac{\partial}{\partial y} - y\frac{\partial}{\partial x}\right)}$

We have used $\hat{p}_x = \frac{\hbar}{\mathrm{i}}\frac{\partial}{\partial x}$, etc. The commutator of $\hat{l}_x$ and $\hat{l}_y$ is $[\hat{l}_x, \hat{l}_y] = (\hat{l}_x\hat{l}_y - \hat{l}_y\hat{l}_x)$. Recall that the operators always imply operation on a function. We form

$$\hat{l}_x\hat{l}_y f = -\hbar^2\left(y\frac{\partial}{\partial z} - z\frac{\partial}{\partial y}\right)\left(z\frac{\partial}{\partial x} - x\frac{\partial}{\partial z}\right)f$$

$$= -\hbar^2\left(yz\frac{\partial^2 f}{\partial z\partial x} + y\frac{\partial f}{\partial x} - yx\frac{\partial^2 f}{\partial z^2} - z^2\frac{\partial^2 f}{\partial y\partial x} + zx\frac{\partial^2 f}{\partial z\partial y}\right)$$

and $\hat{l}_y\hat{l}_x f = -\hbar^2\left(z\frac{\partial}{\partial x} - x\frac{\partial}{\partial z}\right)\left(y\frac{\partial}{\partial z} - z\frac{\partial}{\partial y}\right)f$

$$= -\hbar^2\left(zy\frac{\partial^2 f}{\partial x\partial z} - z^2\frac{\partial^2 f}{\partial x\partial y} - xy\frac{\partial^2 f}{\partial z^2} + xz\frac{\partial^2 f}{\partial z\partial y} + x\frac{\partial f}{\partial y}\right)$$

Since multiplication and differentiation is each commutative, the results of the operation $\hat{l}_x\hat{l}_y$ and $\hat{l}_y\hat{l}_x$ differ only in one term. For $\hat{l}_y\hat{l}_x f$, $x\frac{\partial f}{\partial y}$ replaces $y\frac{\partial f}{\partial x}$. Hence, the commutator of the operations, $(\hat{l}_x\hat{l}_y - \hat{l}_y\hat{l}_x)$ is $-\hbar^2\left(y\frac{\partial}{\partial x} - x\frac{\partial}{\partial y}\right)$ or $\boxed{-\frac{\hbar}{\mathrm{i}}\hat{l}_z}$

Comment. We also would find

$$(\hat{l}_y\hat{l}_z - \hat{l}_z\hat{l}_y) = -\frac{\hbar}{\mathrm{i}}\hat{l}_x \quad \text{and} \quad (\hat{l}_z\hat{l}_x - \hat{l}_x\hat{l}_z) = -\frac{\hbar}{\mathrm{i}}\hat{l}_y$$

P14.11 We are to show that $[\hat{l}^2, \hat{l}_z] = 0$. Start by expanding the first operator:

$$[\hat{l}^2, \hat{l}_z] = [\hat{l}_x^2 + \hat{l}_y^2 + \hat{l}_z^2, \hat{l}_z] = [\hat{l}_x^2, \hat{l}_z] + [\hat{l}_y^2, \hat{l}_z] + [\hat{l}_z^2, \hat{l}_z]$$

The three commutators are:

$$[\hat{l}_z^2,\hat{l}_z]=\hat{l}_z^2\hat{l}_z-\hat{l}_z\hat{l}_z^2=\hat{l}_z^3-\hat{l}_z^3=0$$

$$\begin{aligned}[\hat{l}_x^2,\hat{l}_z]&=\hat{l}_x^2\hat{l}_z-\hat{l}_z\hat{l}_x^2=\hat{l}_x^2\hat{l}_z-\hat{l}_x\hat{l}_z\hat{l}_x+\hat{l}_x\hat{l}_z\hat{l}_x-\hat{l}_z\hat{l}_x^2\\&=\hat{l}_x(\hat{l}_x\hat{l}_z-\hat{l}_z\hat{l}_x)+(\hat{l}_x\hat{l}_z-\hat{l}_z\hat{l}_x)\hat{l}_x=\hat{l}_x[\hat{l}_x,\hat{l}_z]+[\hat{l}_x,\hat{l}_z]\hat{l}_x\\&=\hat{l}_x(-\mathrm{i}\hbar\hat{l}_y)+(-\mathrm{i}\hbar\hat{l}_y)\hat{l}_x=-\mathrm{i}\hbar(\hat{l}_x\hat{l}_y+\hat{l}_y\hat{l}_x)\ [14.7]\end{aligned}$$

$$\begin{aligned}[\hat{l}_y^2,\hat{l}_z]&=\hat{l}_y^2\hat{l}_z-\hat{l}_z\hat{l}_y^2=\hat{l}_y^2\hat{l}_z-\hat{l}_y\hat{l}_z\hat{l}_y+\hat{l}_y\hat{l}_z\hat{l}_y-\hat{l}_z\hat{l}_y^2\\&=\hat{l}_y(\hat{l}_y\hat{l}_z-\hat{l}_z\hat{l}_y)+(\hat{l}_y\hat{l}_z-\hat{l}_z\hat{l}_y)\hat{l}_y=\hat{l}_y[\hat{l}_y,\hat{l}_z]+[\hat{l}_y,\hat{l}_z]\hat{l}_y\\&=\hat{l}_y(\mathrm{i}\hbar\hat{l}_x)+(\mathrm{i}\hbar\hat{l}_x)\hat{l}_y=\mathrm{i}\hbar(\hat{l}_y\hat{l}_x+\hat{l}_x\hat{l}_y)\ [14.7]\end{aligned}$$

Therefore, $\quad [\widehat{l^2},\hat{l}_z]=-\mathrm{i}\hbar(\hat{l}_x\hat{l}_y+\hat{l}_y\hat{l}_x)+\mathrm{i}\hbar(\hat{l}_x\hat{l}_y+\hat{l}_y\hat{l}_x)+0=0$

We may also conclude that $[\widehat{l^2},\hat{l}_x]=0$ and $[\widehat{l^2},\hat{l}_y]=0$ because $\hat{l}_x$, $\hat{l}_y$, and $\hat{l}_z$ occur symmetrically in $\widehat{l^2}$.

Focus 3: Integrated activities

F3.1 The correspondence principle says that classical mechanics is a limiting case of quantum mechanics in the limit of sufficiently large quantum numbers. One example is the translational motion of molecules in macroscopic containers at ambient temperatures, as explored in Exercises 11.12. In those exercises, the quantum numbers of such molecules, treated as particles in a three-dimensional box, were high indeed (on the order of 10^{11}) and the separation of energy levels tiny compared to the energy levels themselves. Another example is the increasing uniformity of the probability distribution of a particle in a box of any dimension with increasing quantum number, as illustrated in Example 9.1. In this case, a uniform probability distribution corresponds to classical behavior. A third example is that the probability of finding a quantum harmonic oscillator in a classically forbidden region diminishes as the vibrational quantum number increases.

F3.3 The particle in a box can serve a model for many kinds of bound particles. Perhaps most relevant to chemistry, it is a model for π-electrons in linear systems of conjugated double bonds. See Example 9.2, in which it is the basis for estimating the absorption wavelength for an electronic transition in the linear polyene β-carotene. Electrons in nanostructures such as quantum dots can be modeled as particles in a three-dimensional box or sphere. More crudely yet more fundamentally, the particle in a box model can provide order-of-magnitude excitation energies for bound particles such as an electron confined to an atom-sized box around a nucleus or even for a nucleon confined to a nucleus-sized box. The harmonic oscillator serves as the first approximation for describing molecular vibrations. The excitation energies for stretching and bending bonds are manifested in infrared and Raman spectroscopy.

F3.5 (a) $\langle x\rangle = \int_0^L \left(\frac{2}{L}\right)^{1/2} \sin\left(\frac{\pi x}{L}\right) x \left(\frac{2}{L}\right)^{1/2} \sin\left(\frac{\pi x}{L}\right) dx$

$$\langle x\rangle = \left(\frac{2}{L}\right)\int_0^L x \sin^2 ax \; dx \quad \left[a = \frac{\pi}{L}\right]$$

$$= \left(\frac{2}{L}\right)\times\left(\frac{x^2}{4} - \frac{x\sin 2ax}{4a} - \frac{\cos 2ax}{8a^2}\right)\Bigg|_0^L \quad \text{[Trig functions vanish or cancel at limits.]}$$

$$= \left(\frac{2}{L}\right)\times\left(\frac{L^2}{4}\right) = \frac{L}{2} \quad \text{[as could be anticipated by symmetry]}$$

$$\langle x^2\rangle = \frac{2}{L}\int_0^L x^2 \sin^2 ax\, dx = \left(\frac{2}{L}\right)\times\left\{\frac{x^3}{6} - \left(\frac{x^2}{4a} - \frac{1}{8a^3}\right)\sin 2ax - \frac{x\cos 2ax}{4a^2}\right\}\Bigg|_0^L$$

$$= \left(\frac{2}{L}\right)\times\left(\frac{L^3}{6} - \frac{L^3}{4\pi^2}\right) = L^2\left(\frac{1}{3} - \frac{1}{2\pi^2}\right).$$

So $\Delta x = \left[L^2\left(\frac{1}{3}-\frac{1}{2\pi^2}\right)-\frac{L^2}{4}\right]^{1/2} = \boxed{L\left(\frac{1}{12}-\frac{1}{2\pi^2}\right)^{1/2}}$

$\langle p\rangle = 0$ [by symmetry, also see Exercise 9.5(a)]

and $\langle p^2\rangle = \frac{h^2}{4L^2}$ [from $E=\frac{p^2}{2m}$, also Exercise 9.5(a)],

so $\Delta p = \left(\frac{h^2}{4L^2}\right)^{1/2} = \boxed{\frac{h}{2L}}$

$$\Delta p\Delta x = \frac{h}{2L}\times L\left(\frac{1}{12}-\frac{1}{2\pi^2}\right)^{1/2} = \frac{h}{2}\left(\frac{2\pi^2-12}{24\pi^2}\right)^{1/2} = \boxed{\frac{\hbar}{2}\left(\frac{\pi^2-6}{3}\right)^{1/2}} > \frac{\hbar}{2}$$

(b) $\langle x\rangle = \alpha^2\int_{-\infty}^{+\infty}\psi^2 y\,\mathrm{d}y[x=\alpha y] = 0$ [by symmetry, y is an odd function]

$$\langle x^2\rangle = \frac{2}{k}\left\langle\frac{1}{2}kx^2\right\rangle = \frac{2}{k}\langle V\rangle$$

But $\langle V\rangle = \langle E_{\mathrm{k}}\rangle = \frac{E}{2} = \frac{\hbar\omega}{4}$ [12.13]

so $\langle x^2\rangle = \frac{2}{k}\times\frac{\hbar\omega}{4} = \frac{\hbar}{2\omega m}$

and $\Delta x = \boxed{\left(\frac{\hbar}{2\omega m}\right)^{1/2}}$

$\langle p\rangle = 0$ [by symmetry, or by noting that the integrand is an odd function of x]

$$\langle p^2\rangle = 2m\langle E_{\mathrm{k}}\rangle = (2m)\times\left(\frac{1}{2}\right)\times\left(v+\frac{1}{2}\right)\times\hbar\omega = \frac{m\hbar\omega}{2} \quad [12.13c]$$

$$\Delta p = \boxed{\left(\frac{\hbar\omega m}{2}\right)^{1/2}}$$

$$\Delta p\Delta x = \boxed{\left(\frac{\hbar\omega m}{2}\right)^{1/2}\left(\frac{\hbar}{2\omega m}\right)^{1/2}} = \frac{\hbar}{2}$$

Both results are consistent with the uncertainty principle, $\Delta p\Delta x \geq \frac{\hbar}{2}$. Note that for the ground-state harmonic oscillator, the equality holds.

F3.7 (a) Notice that regions of high probability and low probability alternate on an increasingly smaller scale as n increases. The distribution of the particle's probability becomes more uniform if one ignores the fluctuations or (what amounts to the same thing) if one looks for the probability of finding the particle in a region small compared to the box as a whole but large compared to the very short-wavelength $n = 50$ state. For example, compare the probability of finding the particle in the central 10% of the box to the first 10% of the box. The difference in probabilities is huge for the ground state and minuscule for the $n = 50$ state. One more way to think of the distribution as becoming more uniform is that both regions of high probability and regions of low probability become more and more widely and evenly distributed.

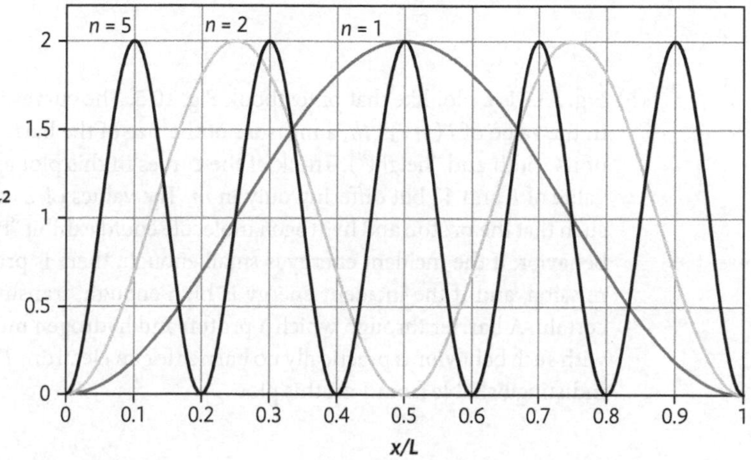

Fig F3.1a

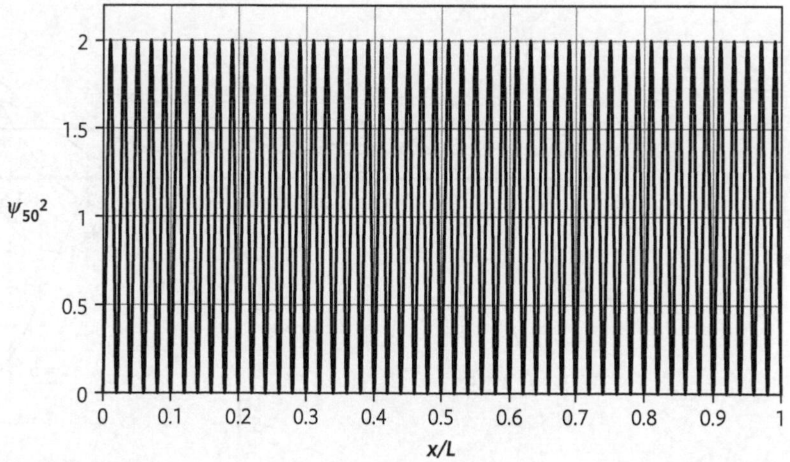

Fig F3.1b

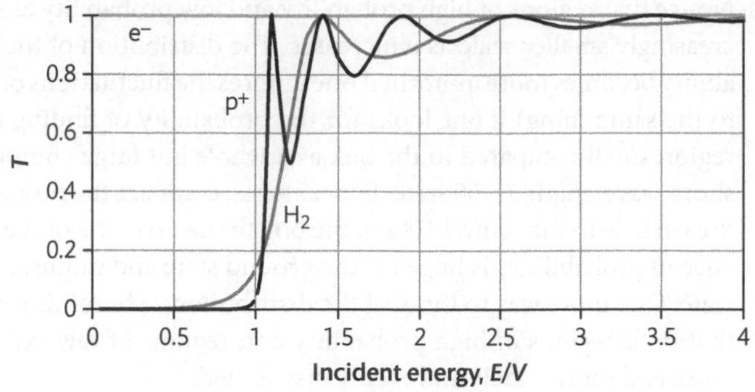

Fig F3.2

(b) Fig. F3.2 is a plot like that of textbook Fig. 10.5. The curves in the figure differ in the value of $L(mV)^{1/2}/\hbar$, a measure of the size of the barrier (a combination of its width and "height"). Think of the curves in this plot as having the same value of L and V, but differing only in m. The values of L and V were chosen such that the proton and hydrogen molecule could exhibit "typical" tunnelling behavior: if the incident energy is small enough, there is practically no transmission, and if the incident energy is high enough, transmission is virtually certain. A barrier through which a proton and hydrogen molecule can tunnel with such behavior is practically no barrier for an electron: T for the electron is indistinguishable from 1 on this plot.

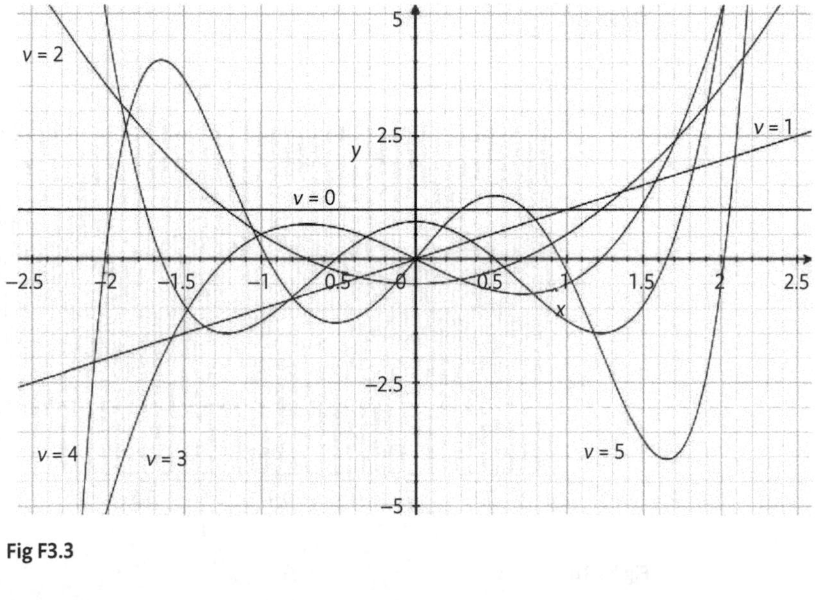

Fig F3.3

(c) In Fig. F3.3, $y = H_v(x)$ is plotted against x. The quantum number v is equal to the number of nodes. Notice that the number of nodes increases as v increases and that the position of those nodes spreads out.

(d) The number of nodal lines (excluding those that fall on the boundaries) is $n_1 + n_2 - 2$. The number of nodal lines for motion along a given axis is one less than n for that motion.

Topic 15 **Time-independent perturbation theory**

Discussion questions

D15.1 Perturbation theory in general is useful in quantum mechanics because the Schrödinger equation can be solved exactly only for a small number of fairly simple physical systems. Quantum mechanical treatments of more complex systems and even more realistic treatments of simple systems require approximation methods, and perturbation theory is a powerful and widely applicable approximation method. In general, a system can be treated by perturbation theory if its hamiltonian can be written as a sum of two parts, one of which yields an exactly soluble Schrödinger equation and the remainder of which is considered a perturbation.

The two forms of perturbation theory are time-independent (where the perturbation **does not** depend on time) and time-dependent (where the perturbation **does** depend on time). The former, treated in the current Topic, is useful in describing the static properties of quantum mechanical systems (for example electrical and magnetic properties). The latter, treated in Topic 16, is useful in describing how quantum mechanical systems interact with electromagnetic radiation (i.e., an oscillating electromagnetic field), collisions, and other time-dependent disturbances.

D15.3 Perturbation theory considers first-order corrections to the ground-state wavefunction to be a linear combination of zero-order wavefunctions of different states (eqn 15.5):

$$\psi_0^{(1)} = \sum_{n\neq 0} c_n \psi_n^{(0)} \quad \text{where} \quad c_n = \frac{\int \psi_n^{(0)*} \hat{H}^{(1)} \psi_n^{(0)} \, d\tau}{E_n^{(0)} - E_n^{(0)}}$$

Thus the first-order correction to the wavefunction can be thought of as introducing some character of excited states of the zero-order (unperturbed) system into the ground state of the perturbed system. The coefficients c_n (or more strictly the squares of their magnitudes, $|c_n|^2$) can be thought of as weights: the greater $|c_n|^2$, the more character of the corresponding unperturbed state in the first-order wavefunction. In general, the c_n are inversely proportional to the difference in (unperturbed) energy levels between the ground state and the nth state, so all other things being equal, low-lying zero-order states contribute more to the perturbed ground-state wavefunction than do highly excited states. Even low-lying unperturbed states may not contribute to the first-order wavefunction, though, if

$$\int \psi_n^{(0)} * \hat{H}^{(1)} \psi_n^{(0)} \, d\tau = 0$$

which may occur because of the symmetry of the zero-order wavefunctions and the perturbation hamiltonian. (There are several examples of this sort of vanishing integral in the Exercises.)

Exercises

E15.1(a) The perturbation is

$$\hat{H}^{(1)} = V(x) = -\varepsilon \sin\left(\frac{2\pi x}{L}\right)$$

The first-order correction to the ground-state energy, E_1, is [15.4]

$$E_1^{(1)} = \int_0^L \psi_1^{(0)*}\hat{H}^{(1)}\psi_1^{(0)}\,dx = -\int_0^L \left(\frac{2}{L}\right)^{1/2} \sin\left(\frac{\pi x}{L}\right)\varepsilon \sin\left(\frac{2\pi x}{L}\right)\left(\frac{2}{L}\right)^{1/2} \sin\left(\frac{\pi x}{L}\right)dx$$

$$E_1^{(1)} = -\frac{2\varepsilon}{L}\int_0^L \sin^2\left(\frac{\pi x}{L}\right)\sin\left(\frac{2\pi x}{L}\right)dx$$

Notice that the $\sin^2$ term is symmetric about the midpoint of the region of integration (Fig. 15.1) while the sin term is antisymmetric. Therefore, the integral vanishes, and $E_1^{(1)} = \boxed{0}$.

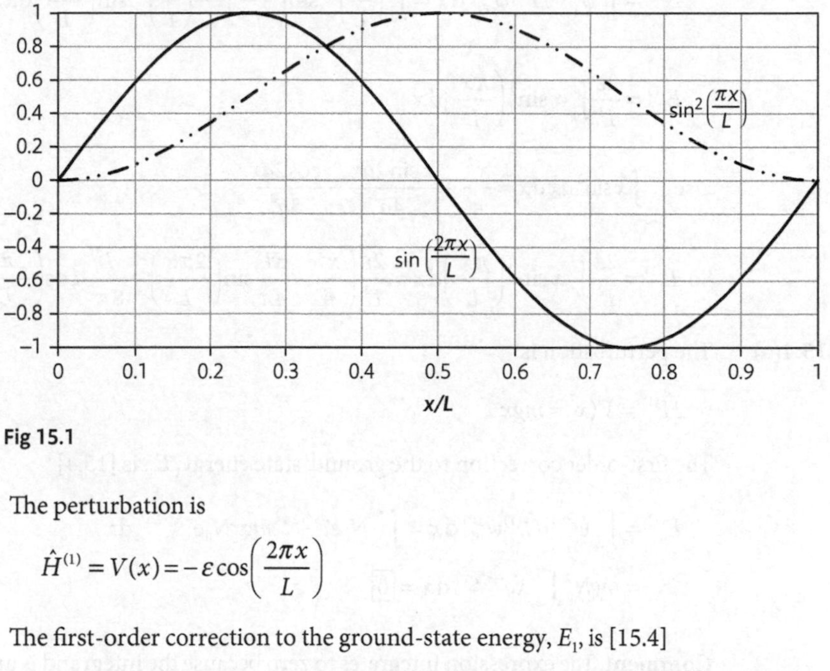

Fig 15.1

E15.2(a) The perturbation is

$$\hat{H}^{(1)} = V(x) = -\varepsilon \cos\left(\frac{2\pi x}{L}\right)$$

The first-order correction to the ground-state energy, E_1, is [15.4]

$$E_1^{(1)} = \int_0^L \psi_1^{(0)*}\hat{H}^{(1)}\psi_1^{(0)}\,dx = -\int_0^L \left(\frac{2}{L}\right)^{1/2} \sin\left(\frac{\pi x}{L}\right)\varepsilon \cos\left(\frac{2\pi x}{L}\right)\left(\frac{2}{L}\right)^{1/2} \sin\left(\frac{\pi x}{L}\right)dx$$

$$E_1^{(1)} = -\frac{2\varepsilon}{L}\int_0^L \sin^2\left(\frac{\pi x}{L}\right)\cos\left(\frac{2\pi x}{L}\right)dx$$

Use the trigonometric identity

$$\sin^2\left(\frac{\pi x}{L}\right)=\frac{1}{2}-\frac{1}{2}\cos\left(\frac{2\pi x}{L}\right),$$

so $$E_1^{(1)}=-\frac{\varepsilon}{L}\left\{\int_0^L \cos\left(\frac{2\pi x}{L}\right)\mathrm{d}x-\int_0^L \cos^2\left(\frac{2\pi x}{L}\right)\mathrm{d}x\right\}$$

Use $\int \cos^2 ax\,\mathrm{d}x=\frac{x}{2}+\frac{\sin 2ax}{4a}$

for the second integral, so

$$E_1^{(1)}=-\frac{\varepsilon}{L}\left\{\frac{L}{2\pi}\sin\left(\frac{2\pi x}{L}\right)-\frac{x}{2}-\frac{L}{8\pi}\sin\left(\frac{4\pi x}{L}\right)\right\}\Bigg|_0^L=\boxed{\frac{\varepsilon}{2}}$$

E15.3(a) The functional form of a linear slope that rises from 0 at $x=0$ to ε at $x=L$ is

$$\hat{H}^{(1)}=V(x)=\frac{\varepsilon x}{L}$$

The first-order correction to the ground-state energy, E_1, is [15.4]

$$E_1^{(1)}=\int_0^L \psi_1^{(0)*}H^{(1)}\psi_1^{(0)}\,\mathrm{d}x=\int_0^L\left(\frac{2}{L}\right)^{1/2}\sin\left(\frac{\pi x}{L}\right)\frac{\varepsilon x}{L}\left(\frac{2}{L}\right)^{1/2}\sin\left(\frac{\pi x}{L}\right)\mathrm{d}x$$

$$E_1^{(1)}=\frac{2\varepsilon}{L^2}\int_0^L x\sin^2\left(\frac{\pi x}{L}\right)\mathrm{d}x$$

Use $\int x\sin^2 ax\,\mathrm{d}x=\frac{x^2}{4}-\frac{x\sin 2ax}{4a}-\frac{\cos 2ax}{8a^2}$

So $$E_1^{(1)}=\frac{2\varepsilon}{L^2}\int_0^L x\sin^2\left(\frac{\pi x}{L}\right)\mathrm{d}x=\frac{2\varepsilon}{L^2}\left(\frac{x^2}{4}-\frac{xL}{4\pi}\sin\left(\frac{2\pi x}{L}\right)-\frac{L^2}{8\pi^2}\cos\left(\frac{2\pi x}{L}\right)\right)\Bigg|_0^L=\boxed{\frac{\varepsilon}{2}}$$

E15.4(a) The perturbation is

$$\hat{H}^{(1)}=V(x)=mgx$$

The first-order correction to the ground-state energy, E_0, is [15.4]

$$E_0^{(1)}=\int_{-\infty}^{\infty}\psi_0^{(0)*}\hat{H}^{(1)}\psi_0^{(0)}\,\mathrm{d}x=\int_{-\infty}^{\infty}N_0\mathrm{e}^{-x^2/2\alpha^2}mgxN_0\mathrm{e}^{-x^2/2\alpha^2}\,\mathrm{d}x$$

$$=mgN_0^2\int_{-\infty}^{\infty}x\mathrm{e}^{-x^2/\alpha^2}\,\mathrm{d}x=\boxed{0}$$

Comment. The expression integrates to zero because the integrand is antisymmetric about the midpoint of the region of integration (*i.e.*, an odd function of x). The result should not be terribly surprising, as the perturbation affects stretching and compressing the bond to an equal but opposite degree.

E15.5(a) The second-order correction to the ground-state energy, E_0, is [15.6]

$$E_0^{(2)} = \sum_{v=1}^{\infty} \frac{\left|\int_{-\infty}^{\infty} \psi_v^{(0)*} \hat{H}^{(1)} \psi_0^{(0)}\, dx\right|^2}{E_0^{(0)} - E_v^{(0)}}$$

where $\hat{H}^{(1)} = mgx, \psi_v^{(0)} = N_v H_v(y) e^{-y^2/2}$ and $E_v = (v+\frac{1}{2})\hbar\omega$

The denominator in the sum is:

$$E_0^{(0)} - E_v^{(0)} = \tfrac{1}{2}\hbar\omega - (v+\tfrac{1}{2})\hbar\omega = -v\hbar\omega.$$

The integral in the sum is:

$$\int_{-\infty}^{\infty} \psi_v^{(0)*} \hat{H}^{(1)} \psi_0^{(0)}\, dx = \int_{-\infty}^{\infty} N_v H_v(y) e^{-y^2/2} mgx N_0 e^{-y^2/2}\, dx$$
$$= mgN_v N_0 \alpha \int_{-\infty}^{\infty} H_v(y) e^{-y^2/2} y e^{-y^2/2}\, dx,$$

where we have used $H_0(y) = 1$ and $y = x/\alpha$. Now notice that $H_1(y) = 2y$, so the integral can be expressed as

$$\int_{-\infty}^{\infty} \psi_v^{(0)*} \hat{H}^{(1)} \psi_0^{(0)}\, dx = \frac{mgN_0\alpha}{2N_1} \int_{-\infty}^{\infty} N_v H_v(y) e^{-y^2/2} N_1 H_1(y) e^{-y^2/2}\, dx$$
$$= \frac{mgN_0\alpha}{2N_1} \int_{-\infty}^{\infty} \psi_v^{(0)*} \psi_1^{(0)}\, dx$$

Because the harmonic-oscillator wavefunctions are orthogonal, this integral vanishes unless $v = 1$. In that case, the remaining integral is the normalization integral.

$$E_0^{(2)} = \sum_{v=1}^{\infty} \frac{\left|\frac{mgN_0\alpha}{2N_1}\int_{-\infty}^{\infty} \psi_v^{(0)*}\, \psi_1^{(0)}\, dx\right|^2}{E_0^{(0)} - E_v^{(0)}} = \frac{\left|\frac{mgN_0\alpha}{2N_1}\int_{-\infty}^{\infty} \psi_1^{(0)*}\, \psi_1^{(0)}\, dx\right|^2}{E_0^{(0)} - E_1^{(0)}} = \frac{\left(\frac{mgN_0\alpha}{2N_1}\right)^2}{-\hbar\omega}$$

Using $N_0 = \left(\frac{1}{\alpha\pi^{1/2}}\right)^{1/2}$, $N_1 = \left(\frac{1}{2\alpha\pi^{1/2}}\right)^{1/2}$, $\alpha = \left(\frac{\hbar^2}{mk_f}\right)^{1/}$, and $k_f = m\omega^2$,

this result simplifies to

$$E_0^{(2)} = -\frac{m^2 g^2 N_0^2 \alpha^2}{4N_1^2 \hbar\omega} = \boxed{-\frac{mg^2}{2\omega^2}}$$

Problems

P15.1 (a) Treat the small step in the potential energy function as a perturbation in the energy operator:

$$H^{(1)} = \begin{cases} 0 & \text{for } 0 \le x \le (1/2)(L-a) \text{ and } (1/2)(L+a) \le x \le L \\ \varepsilon & \text{for } (1/2)(L-a) \le x \le (1/2)(L+a) \end{cases}$$

The first-order correction to the ground-state energy, E_1, is [15.4]

$$E_1^{(1)} = \int_0^L \psi_1^{(0)*} H^{(1)} \psi_1^{(0)}\, dx = \int_{(L-a)/2}^{(L+a)/2} \left(\frac{2}{L}\right)^{1/2} \sin\left(\frac{\pi x}{L}\right) \varepsilon \left(\frac{2}{L}\right)^{1/2} \sin\left(\frac{\pi x}{L}\right) dx$$

$$E_1^{(1)} = \frac{2\varepsilon}{L}\int_{(L-a)/2}^{(L+a)/2} \sin^2\left(\frac{\pi x}{L}\right) dx = \frac{\varepsilon}{L\pi}\left(\pi x - L\cos\left(\frac{\pi x}{L}\right)\sin\left(\frac{\pi x}{L}\right)\right)\Big|_{(L-a)/2}^{(L+a)/2}$$

$$E_1^{(1)} = \frac{\varepsilon a}{L} - \frac{\varepsilon}{\pi}\cos\left(\frac{\pi(L+a)}{2L}\right)\sin\left(\frac{\pi(L+a)}{2L}\right) + \frac{\varepsilon}{\pi}\cos\left(\frac{\pi(L-a)}{2L}\right)\sin\left(\frac{\pi(L-a)}{2L}\right)$$

This expression can be simplified considerably with a few trigonometric identities. The product of sine and cosine is related to the sine of twice the angle:

$$\cos\left(\frac{\pi(L\pm a)}{2L}\right)\sin\left(\frac{\pi(L\pm a)}{2L}\right) = \frac{1}{2}\sin\left(\frac{\pi(L\pm a)}{L}\right) = \frac{1}{2}\sin\left(\pi \pm \frac{\pi a}{L}\right)$$

and the sine of a sum simplifies considerably since one of the terms in the sum is π:

$$\sin\left(\pi \pm \frac{\pi a}{L}\right) = \sin\pi\cos\left(\frac{\pi a}{L}\right) \pm \cos\pi\sin\left(\frac{\pi a}{L}\right) = \pm\sin\left(\frac{\pi a}{L}\right)$$

Thus $E_1^{(1)} = \boxed{\dfrac{\varepsilon a}{L} + \dfrac{\varepsilon}{\pi}\sin\left(\dfrac{\pi a}{L}\right)}$

(b) If $a = L/10$, the first-order correction to the ground-state energy is:

$$E_1^{(1)} = \boxed{\frac{\varepsilon}{10} + \frac{\varepsilon}{\pi}\sin\left(\frac{\pi}{10}\right)} = \boxed{0.1984\varepsilon}$$

P15.3 The second-order correction to the ground-state energy, E_1, is [15.6]

$$E_1^{(2)} = \sum_{n=2}^{\infty} \frac{\left|\int_0^L \psi_n^{(0)*} H^{(1)} \psi_1^{(0)}\, dx\right|^2}{E_1^{(0)} - E_n^{(0)}}$$

where $\hat{H}^{(1)} = mgx, \psi_n^{(0)} = \left(\frac{2}{L}\right)^{1/2} \sin\frac{n\pi x}{L}$, and $E_n = \frac{n^2h^2}{8mL^2}$

The denominator in the sum is

$$E_1^{(0)} - E_n^{(0)} = \frac{h^2}{8mL^2} - \frac{n^2h^2}{8mL^2} = \frac{(1-n^2)h^2}{8mL^2}$$

The integral in the sum is

$$\int_0^L \psi_n^{(0)*} H^{(1)} \psi_1^{(0)}\, dx = \int_0^L \left(\frac{2}{L}\right)^{1/2} \sin\left(\frac{n\pi x}{L}\right) mgx \left(\frac{2}{L}\right)^{1/2} \sin\left(\frac{\pi x}{L}\right) dx$$

$$\int_0^L \psi_n^{(0)*} H^{(1)} \psi_1^{(0)} \, dx = \frac{2mg}{L} \int_0^L x \sin ax \sin bx \, dx$$

where $a = n\pi/L$ and $b = \pi/L$.

Using the integral from the *Resource section*

$$\int x \sin ax \sin bx \; dx = -\frac{d}{da} \int \cos ax \sin bx \; dx \quad [\text{T.9}]$$

allows this integral to be expressed as

$$-\frac{2mg}{L}\frac{d}{da}\int_0^L \cos ax \sin bx \, dx = -mg\frac{d}{da}\left(\frac{\cos(a-b)x}{2(a-b)} - \frac{\cos(a+b)x}{2(a+b)}\right)\Bigg|_0^L$$

$$= -\frac{2mg}{L}\left(\frac{-x\sin(a-b)x}{2(a-b)} - \frac{\cos(a-b)x}{2(a-b)^2} + \frac{x\sin(a+b)x}{2(a+b)} + \frac{\cos(a+b)x}{2(a+b)^2}\right)\Bigg|_0^L$$

where we have also used eqn T.8 from the *Resource section.* The arguments of the trigonometric functions at the upper limit are:

$$(a-b)L = (n-1)\pi \quad \text{and} \quad (a+b)L = (n+1)\pi.$$

Therefore, the sine terms vanish. Similarly, the cosines are ± 1 depending on whether the argument is an even or odd multiple of π; they simplify to $(-1)^{n+1}$. At the lower limit, the sines are still zero, and the cosines are all 1. The integral, evaluated at its limits with π / L factors pulled out from the as and bs in its denominator, becomes

$$\frac{mgL}{\pi^2}\left(\frac{(-1)^{n+1}-1}{(n-1)^2} - \frac{(-1)^{n+1}-1}{(n+1)^2}\right) = \frac{mgL[(-1)^{n+1}-1]}{\pi^2}\left(\frac{(n+1)^2-(n-1)^2}{(n-1)^2(n+1)^2}\right)$$

$$= \frac{4mgL[(-1)^{n+1}-1]n}{\pi^2(n^2-1)^2}$$

The second-order correction, then, is

$$E_1^{(2)} = \sum_{n=2}^{\infty} \frac{\left(\dfrac{4mgL[(-1)^{n+1}-1]n}{\pi^2(n^2-1)^2}\right)^2}{\dfrac{(1-n^2)h^2}{8mL^2}} = -\frac{128m^3g^2L^4}{\pi^4h^2}\sum_{n=2}^{\infty}\frac{[(-1)^n+1]^2n^2}{(n^2-1)^5}$$

Note that the terms with odd n vanish. Therefore, the sum can be rewritten, changing n to $2k$, as:

$$E_1^{(2)} = -\frac{2048m^3g^2L^4}{\pi^4h^2}\sum_{k=1}^{\infty}\frac{k^2}{(4k^2-1)^5}.$$

The sum converges rapidly to 4.121×10^{-3}, as can easily be verified numerically; in fact, to three significant figures, terms after the first do not affect the sum. So the second-order correction is:

$$E_1^{(2)} = \boxed{-\frac{0.08664m^3g^2L^4}{h^2}}$$

The first-order correction to the ground-state wavefunction is also a sum:

$$\psi_0^{(1)} = \sum_n c_n\psi_n^{(0)}$$

$$\text{where} \quad c_n = -\frac{\int_0^L \psi_n^{(0)*}H^1\psi_1^{(0)}\mathrm{d}x}{E_n^{(0)} - E_1^{(0)}} \quad [15.5]$$

$$= -\frac{\dfrac{4mgL[(-1)^{n+1}-1]n}{\pi^2(n^2-1)^2}}{\dfrac{(n^2-1)h^2}{8mL^2}} = \boxed{\frac{32m^2gL^3[(-1)^n+1]n}{\pi^2h^2(n^2-1)^3}}$$

Once again, the odd n terms vanish.

How does the first-order correction alter the wavefunction? Recall that the perturbation raises the potential energy near the top of the box (near L) much more than near the bottom (near $x = 0$); therefore, we expect the probability of finding the particle near the bottom to be enhanced compared with that of finding it near the top. Because the zero-order ground-state wavefunction is positive throughout the interior of the box, we thus expect the wavefunction itself to be raised near the bottom of the box and lowered near the top. In fact, the correction terms do just this. First, note that the basis wavefunctions with odd n are symmetric with respect to the center of the box; therefore, they would have the same effect near the top of the box as near the bottom. The coefficients of these terms are zero: they do not contribute to the correction. The even-n basis functions all start positive near $x = 0$ and end negative near $x = L$; therefore, such terms must be multiplied by positive coefficients (as the result provides) to enhance the wavefunction near the bottom and diminish it near the top.

Topic 16 **Transitions**

Discussion questions

D16.1 A suddenly applied perturbation is likely to excite many states of the unperturbed system, just as hitting a bell with a hammer would excite many of its vibrational modes. A slowly applied perturbation produces a smooth transition from the initial unperturbed state to a state that would arise from a time-independent perturbation.

D16.3 Perhaps the most systematically studied kind of time-dependent perturbation of molecules is the interaction of molecules with electromagnetic radiation (that is, with a periodically oscillating electromagnetic field). Molecular spectroscopy involving electronic, vibrational, and rotational transitions is the subject of Focus 9. Another common kind of time-dependent perturbation arises from collisions between molecules.

Exercise

E16.1(a) The ratio of coefficients A/B is

$$\frac{A}{B}=\frac{8\pi h\nu^3}{c^3}\quad [16.18]$$

The frequency is

$$\nu=\frac{c}{\lambda}\quad \text{so}\quad \frac{A}{B}=\frac{8\pi h}{\lambda^3}$$

$$\text{(a)}\quad \frac{A}{B}=\frac{8\pi(6.626\times10^{-34}\ \text{J s})}{(70.8\times10^{-12}\ \text{m})^3}=\boxed{0.0469\ \text{J m}^{-3}\ \text{s}}$$

$$\text{(b)}\quad \frac{A}{B}=\frac{8\pi(6.626\times10^{-34}\ \text{J s})}{(500\times10^{-9}\ \text{m})^3}=\boxed{1.33\times10^{-13}\ \text{J m}^{-3}\ \text{s}}$$

$$\text{(c)}\quad \frac{A}{B}=\frac{8\pi h}{\lambda^3}=8\pi h\tilde{\nu}^3=8\pi(6.626\times10^{-34}\ \text{J s})\times\left(3\,000\ \text{cm}^{-1}\times\frac{1\ \text{cm}}{10^{-2}\ \text{m}}\right)^3$$

$$=\boxed{4.50\times10^{-16}\ \text{J m}^{-3}\ \text{s}}$$

Comment. Note that the ratio A/B has units. Comparison of these ratios shows that the relative importance of spontaneous transitions decreases as the frequency decreases.

Problem

P16.1 To address this time-dependent problem, we need a time-dependent wavefunction, made up from solutions of the time-dependent Schrödinger equation

$$\hat{H}\Psi(x,t)=\mathrm{i}\hbar\frac{\partial\Psi(x,t)}{\partial t} \quad [16.2]$$

If $\psi(x)$ is an eigenfunction of the energy operator with energy eigenvalue E, then

$$\Psi(x,t)=\psi(x)\mathrm{e}^{-\mathrm{i}Et/\hbar}$$

is a stationary-state solution (eqn 16.1) of the time-dependent Schrödinger equation, provided the energy operator is not itself time dependent. (Note, by the way, that this problem does not have a time-dependent hamiltonian, unlike much of the material treated in Topic 16.) A more general time-dependent solution is a wavepacket, an arbitrary superposition of time-evolving harmonic oscillator wavefunctions,

$$\Psi(x,t)=\sum_{v=0} c_v\psi_v(x)\mathrm{e}^{-\mathrm{i}E_vt/\hbar}$$

where $\psi_v(x)$ are time-independent harmonic-oscillator wavefunctions and

$$E_v=(v+\tfrac{1}{2})\hbar\omega \quad [12.4]$$

Hence, the wavepacket is

$$\Psi(x,t)=\mathrm{e}^{-\mathrm{i}\omega t/2}\sum_{v=0} c_v\psi_v(x)\mathrm{e}^{-\mathrm{i}v\omega t}$$

The angular frequency ω is related to the period T by $T=2\pi/\omega$, so we can evaluate the wavepacket at any whole number of periods after t, that is at a time $t+nT$, where n is any integer (*not* a quantum number.) Note that

$$t+nT=t+2\pi n/\omega,$$

so

$$\Psi(x,t+nT)=\mathrm{e}^{-\mathrm{i}\omega t/2}\mathrm{e}^{-\mathrm{i}\omega nT/2}\sum_{v=0} c_v\psi_v(x)\mathrm{e}^{-\mathrm{i}v\omega t}\mathrm{e}^{-\mathrm{i}v\omega nT}=\mathrm{e}^{-\mathrm{i}\omega t/2}\mathrm{e}^{-\mathrm{i}\pi n}\sum_{v=0} c_v\psi_v(x)\mathrm{e}^{-\mathrm{i}v\omega t}\mathrm{e}^{-2\pi\mathrm{i}vn}$$

Noting that the exponential of ($2\pi\mathrm{i}\times$ any integer) = 1, we note that the last factor inside the sum is 1 for every state. Also, since $\mathrm{e}^{-\mathrm{i}n\pi}=(-1)^n$, we have

$$\Psi(x,t+nT)=(-1)^n\Psi(x,t)$$

At any whole number of periods after time t, the wavefunction is either the same as at time t or -1 times its value at time t. In any event, $|\Psi|^2$ returns to its original value each period, so the wavepacket returns to the same spatial distribution each period.

Topic 17 Hydrogenic atoms

Discussion questions

D17.1 The Schrödinger equation for the hydrogen atom is a six-dimensional partial differential equation, three dimensions for each particle in the atom. One cannot directly solve a multidimensional differential equation; it must be broken down into one-dimensional equations. This is the separation of variables procedure. The choice of coordinates is critical in this process. The separation of the Schrödinger equation can be accomplished in a set of coordinates that are natural to the system but not in others. These natural coordinates are those directly related to the description of the motion of the atom. The atom as a whole (center of mass) can move from point to point in three-dimensional space. The natural coordinates for this kind of motion are the Cartesian coordinates of a point in space. The internal motion of the electron with respect to the proton is most naturally described with spherical polar coordinates. So the six-dimensional Schrödinger equation is first separated into two three-dimensional equations, one for the motion of the center of mass, the other for the internal motion. The separation of the center of mass equation and its solution is fully discussed in Section 17.1. The equation for the internal motion is separable into three one-dimensional equations, one in the angle ϕ, another in the angle θ, and a third in the distance r. The solutions of these three one-dimensional equations can be obtained by standard techniques and were already well known long before the advent of Quantum Mechanics. Another choice of coordinates would not have resulted in the separation of the Schrödinger equation just described. For the details of the separation procedure, see Section 17.1 as well as *Justification* 17.1.

Exercises

E17.1(a) The ionization energy for one electron atoms and ions (A) is given by the negative of eqn 17.7 with $n=1$.

$$I_A = \frac{Z^2\mu_A e^4}{32\pi^2\varepsilon_0^2\hbar^2 n^2}$$

This may be rewritten in terms of the Rydberg constant as

$$I_A = Z^2hc\tilde{R}_A = Z^2hc\frac{\mu_A}{m_e}\tilde{R}_\infty = Z^2\frac{\mu_A}{\mu_H}I_H$$

$I_H = 2.1788$ aJ. Reduced masses are calculated from eqns 17.3, and the ionization energies from the above equation.

μ_H	μ_{He^+}	μ_{He^+}/μ_H	I_{He^+}
$5.4828\times10^{-4}\,m_u$	$5.4850\times10^{-4}\,m_u$	1.00040	8.7188 aJ

E17.2(a) This is essentially the photoelectric effect with the ionization energy of the ejected electron being the work function Φ.

$$h\nu = \frac{1}{2}m_e v^2 + I$$

$$I = h\nu - \frac{1}{2}m_e v^2 = (6.626\times10^{-34}\text{ J Hz}^{-1})\times\left(\frac{2.998\times10^8\text{ m s}^{-1}}{58.4\times10^{-9}\text{ m}}\right)$$

$$-\left(\frac{1}{2}\right)\times(9.109\times10^{-31}\text{ kg})\times(1.59\times10^6\text{ m s}^{-1})^2$$

$$= 2.25\times10^{-18}\text{ J, corresponding to }\boxed{14.0\text{eV}}$$

E17.3(a) $R_{1,0} = Ne^{-r/a_0}$

$$\int_0^\infty R^2r^2\text{d}r = 1 = \int_0^\infty N^2r^2e^{-2r/a_0}\text{d}r = N^2\times\frac{2!}{\left(\frac{2}{a_0}\right)^3} = 1 \quad \left[\int_0^\infty x^n e^{-ax}\,\text{d}x = \frac{n!}{a^{n+1}}\right]$$

$$N^2 = \frac{4}{a_0^3},\quad \boxed{N = \frac{2}{a_0^{3/2}}}$$

Thus,

$$R_{1,0} = 2\left(\frac{1}{a_0}\right)^{3/2}\text{e}^{-r/a_0},$$

which agrees with Table 17.1.

E17.4(a) $R_{2,0} \propto \left(2-\frac{\rho}{2}\right)\text{e}^{-\rho/4}$ with $\rho = \frac{2r}{a_0}$ [Table 17.1]

$$\frac{\text{d}R}{\text{d}r} = \frac{2}{a_0}\frac{\text{d}R}{\text{d}\rho} = \frac{2}{a_0}\left(-\frac{1}{2}-\frac{1}{2}+\frac{1}{8}\rho\right)\text{e}^{-\rho/4} = 0 \quad \text{when } \rho = 8$$

Hence, the wavefunction has an extremum at $r = \boxed{4a_0}$. Because $2-\frac{\rho}{2}<0$, $\psi<0$ and the extremum is a minimum (more formally: $\frac{\text{d}^2\psi}{\text{d}r^2}>0$ at $\rho = 8$).

The second extremum is at $\boxed{r=0.}$ It is not a minimum and in fact is a physical maximum, though not one that can be obtained by differentiation. To see that it is maximum substitute $\rho=0$ into $R_{2,0}$.

E17.5(a) The 3p radial function is proportional to $(4-\rho)\rho e^{-\rho/2}$. This factor goes to zero when $\rho=4$. Since $\rho=(2Z/na)r$, $r=na\rho/2Z$. If we assume $a=a_0$ then when $\rho=4$, $\boxed{r=6a_0=3.175\times10^{-10}\text{ m.}}$

Problems

P17.1 All lines in the hydrogen spectrum fit the Rydberg formula

$$\frac{1}{\lambda}=\tilde{R}_{\mathrm{H}}\left(\frac{1}{n_1^2}-\frac{1}{n_2^2}\right)\quad\left[\text{Example 17.2, with }\tilde{\nu}=\frac{1}{\lambda}\right]\quad\tilde{R}_{\mathrm{H}}=109\,677\text{ cm}^{-1}$$

Find n_1 from the value of $\lambda_{\max}$, which arises from the transition $n_1+1\to n_1$

$$\frac{1}{\lambda_{\max}\tilde{R}_{\mathrm{H}}}=\frac{1}{n_1^2}-\frac{1}{(n_1+1)^2}=\frac{2n_1+1}{n_1^2(n_1+1)^2}$$

$$\lambda_{\max}\tilde{R}_{\mathrm{H}}=\frac{n_1^2(n_1+1)^2}{2n_1+1}=(12\,368\times10^{-9}\text{ m})\times(109\,677\times10^2\text{ m}^{-1})=135.65$$

Since $n_1=1,2,3$, and 4 have already been accounted for, try $n_1=5,6,\ldots$. With $n_1=6$ we get $\frac{n_1^2(n_1+1)^2}{2n_1+1}=136$. Hence, the Humphreys series is $\boxed{n_2\to6}$ and the transitions are given by

$$\frac{1}{\lambda}=(109\,677\text{cm}^{-1})\times\left(\frac{1}{36}-\frac{1}{n_2^2}\right),\quad n_2=7,8,\cdots$$

and occur at 12 372 nm, 7 503 nm, 5 908 nm, 5 129 nm, . . ., 3 908 nm (at $n_2=15$), converging to 3 282 nm as $n_2\to\infty$, in agreement with the quoted experimental result.

P17.3 Refer to problem 21.6 and its solution.

$$\mu_{\mathrm{H}}=\frac{m_{\mathrm{e}}m_{\mathrm{p}}}{m_{\mathrm{e}}+m_{\mathrm{p}}}\approx m_{\mathrm{e}}\quad[m_{\mathrm{p}}=\text{mass of proton}]$$

$$\mu_{\mathrm{Ps}}=\frac{m_{\mathrm{e}}m_{\mathrm{pos}}}{m_{\mathrm{e}}+m_{\mathrm{pos}}}=\frac{m_{\mathrm{e}}}{2}\quad[m_{\mathrm{pos}}=\text{mass of proton}=m_{\mathrm{e}}]$$

$$a_0=r(n=1)=\frac{4\pi\hbar^2\varepsilon_0}{e^2m_{\mathrm{e}}}[17.8\text{ and }17.9]$$

To obtain a_{Ps} the radius of the first Bohr orbit of positronium, we replace m_{e} with $\mu_{\mathrm{Ps}}=\frac{m_e}{2}$; and,

$$\boxed{a_{\mathrm{Ps}}=2a_0}=\frac{8\pi\hbar^2\varepsilon_0}{e^2m_{\mathrm{e}}}$$

The energy of the first Bohr orbit of positronium is

$$E_{1,\mathrm{Ps}} = -hc\tilde{R}_{\mathrm{Ps}} = -\frac{hc}{2}\tilde{R}_{\infty} \text{ [Problem 21.6]}$$

Thus, $\boxed{E_{1,\mathrm{Ps}} = \tfrac{1}{2}E_{1,\mathrm{H}}}$

Question. What modifications are required in these relations when the finite mass of the hydrogen nucleus is recognized?

P17.5 A stellar surface temperature of 3 000 K–4 000 K (a "red star") does not have the energetic particles and photons that are required for either the collisional or radiative excitation of a neutral hydrogen atom. Atomic hydrogen affects neither the absorption nor the emission lines of red stars in the absence of excitation. "Blue stars" have surface temperature of 15 000 K–20 000 K. Both the kinetic energy and the blackbody emissions display energies great enough to completely ionize hydrogen. Lacking an electron, the remaining proton cannot affect absorption and emission lines either.

In contrast, a star with a surface temperature of 8 000 K–10 000 K has a temperature low enough to avoid complete hydrogen ionization but high enough for blackbody radiation to cause electronic transitions of atomic hydrogen. Hydrogen spectral lines are intense for these stars.

Simple kinetic energy and radiation calculations confirm these assertions. For example, a plot of blackbody radiation against the radio photon energy and the ionization energy, I, is shown below in Fig. 17.1. It is clearly seen that at 25 000 K a large fraction of the radiation is able to ionize the hydrogen ($h\nu / I$). It is likely that at such high surface temperatures all hydrogen is ionized and, consequently, unable to affect spectra.

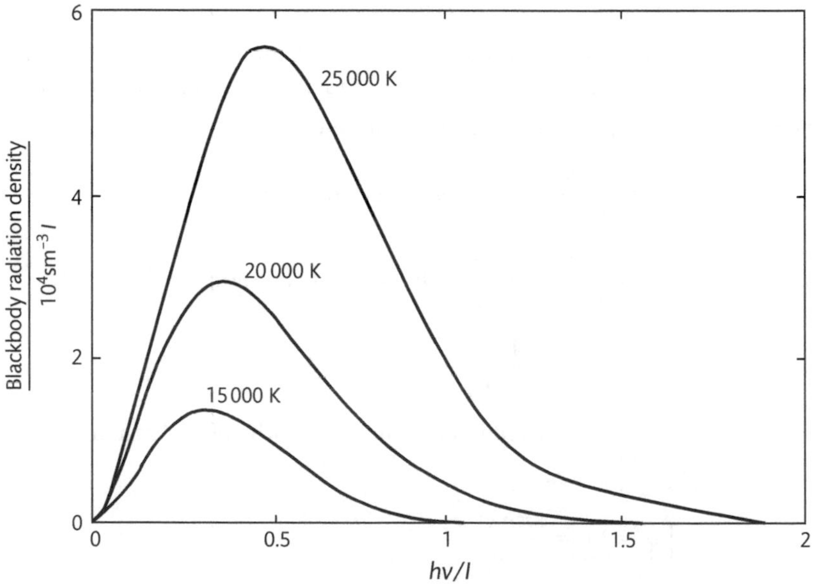

Fig 17.1

Alternatively, consider the equilibrium between hydrogen atoms and their component charged particles:

$$H = H^{+} + e^{-}$$

The equilibrium constant is:

$$K = \frac{p_{+}p_{-}}{p_{H}p^{\ominus}} = \exp\left(\frac{-\Delta G^{\ominus}}{RT}\right) = \exp\left(\frac{-\Delta H^{\ominus}}{RT}\right) \times \exp\left(\frac{-\Delta S^{\ominus}}{R}\right).$$

Clearly $\Delta S^{\ominus}$ is positive for ionization, which makes two particles out of one, and $\Delta H^{\ominus}$, which is close to the ionization energy, is also positive. At a sufficiently high temperature, ions will outnumber neutral molecules. Using concepts developed in statistical mechanics later in this text, one can compute the equilibrium constant; it turns out to be 60. Hence, there are relatively few undissociated H atoms in the equilibrium mixture that is consistent with the weak spectrum of neutral hydrogen observed.

The details of the calculation of the equilibrium constant based on the methods of statistical mechanics (Focuses 11, 12, and 13) follows. Consider the equilibrium between hydrogen atoms and their component charged particles:

$$H = H^{+} + e^{-}.$$

The equilibrium constant is:

$$K = \frac{p_{+}p_{-}}{p_{H}p^{\ominus}} = \exp\left(\frac{-\Delta G^{\ominus}}{RT}\right)$$

Jump ahead to the topics on statistical mechanics (Focuses 11, 12, and 13) to use the statistical thermodynamic analysis of a dissociation equilibrium:

$$K = \frac{q_{+}^{\ominus}q_{-}^{\ominus}}{q_{H}^{\ominus}N_{A}}e^{-\Delta_{r}E_{0}/RT}$$

where $q^{\ominus} = \frac{gRT}{p^{\ominus}\Lambda^{3}}$ and $\Lambda = \left(\frac{h^{2}}{2\pi kTm}\right)^{1/2}$

and where g is the degeneracy of the species. Note that $g_{+} = 2, g_{-} = 2$, and $g_{H} = 4$. Consequently, these factors cancel in the expression for K.

So $K = \frac{RT}{p^{\ominus}N_{A}}\left(\frac{2\pi kT}{h^{2}}\right)^{3/2}\left(\frac{m_{-}m_{+}}{m_{H}}\right)^{3/2} e^{-\Delta_{r}E_{0}/RT}$

Note that the Boltzmann, Avogadro, and perfect gas constants are related ($R = N_{A}k$), and collect powers of kT; note also that the product of masses is the reduced mass, which is approximately equal to the mass of the electron; note finally that the molar energy $\Delta_r E_0$ divided by R is the same as the atomic ionization energy 2.179×10^{-18} J divided by k:

$$K=\frac{(kT)^{5/2}(2\pi m_e)^{3/2}}{p^{\ominus}h^3}e^{-E/kT}$$

$$K=\frac{\left[(1.381\times10^{-23}\ \mathrm{J\,K^{-1}})(25\,000\ \mathrm{K})\right]^{5/2}\left[2\pi(9.11\times10^{-31}\ \mathrm{kg})\right]^{3/2}}{(10^5\ \mathrm{Pa})(6.626\times10^{-34}\ \mathrm{J\,s})^3}$$

$$\times\exp\left(\frac{-2.179\times10^{-18}\ \mathrm{J}}{(1.381\times10^{-23}\ \mathrm{J\,K^{-1}})(25\,000\ \mathrm{K})}\right)$$

$$\boxed{K=60.}$$

Thus, the equilibrium favors the ionized species, even though the ionization energy is greater than kT.

Topic 18 **Hydrogenic atomic orbitals**

Discussion questions

D18.1 Atomic orbitals (wavefunctions) are products of a radial and angular function. The angular functions are the eigenfunctions of the angular momentum operator, the eigenvalues of which are the angular momenta of the orbitals. The angular functions (spherical harmonics) are characterized by the quantum numbers l and m_l from which the angular momenta can be calculated through eqn 14.10. The quantum number l is related to the shape of the orbital and m_l to its orientation in space. Conversely, we might say that angular momentum determines the angular portion of the orbital, hence its shape.

Exercises

E18.1(a) Identify l and use angular momentum $=\{l(l+1)\}^{1/2}\hbar$.

(a) $l=0$, so angular momentum $=0$

(b) $l=1$, so angular momentum $=\sqrt{2}\hbar$

(c) $l=3$, so angular momentum $=2\sqrt{3}\hbar$

The total number of nodes is equal to $n-1$ and the number of angular nodes is equal to l; hence the number of radial nodes is equal to $n-l-1$. We can draw up the following table

	2s	**3p**	**5f**
n, *l*	2,0	3,1	5,3
Angular nodes	0	1	3
Radial nodes	1	1	1

E18.2(a) The L shell corresponds to $n=2$. For $n=2, l=1$ (p subshell) and 0 (s subshell). The degeneracies of these subshells are $2l+1$. Hence the total degeneracy of the L shell is $\boxed{4}$.

E18.3(a) The energies are $E=-\dfrac{hc\tilde{R}_{\rm H}}{n^2}$ [17.7 with 17.12], and the orbital degeneracy g of an energy level of principal quantum number n is

$$g=\sum_{l=0}^{n-1}(2l-1)=1+3+5+\cdots+2n-1=\frac{(1+2n-1)n}{2}=n^2$$

(a) $E=-hc\tilde{R}_{\rm H}$ implies that $n=1$, so $\boxed{g=1}$ [the 1s orbital].

(b) $E=-\dfrac{hc\tilde{R}_{\rm H}}{4}$ implies that $n=2$, so $\boxed{g=4}$ (2s orbital and three 2p orbitals).

(c) $E=-\dfrac{hc\tilde{R}_{\rm H}}{16}$ implies that $n=4$, so $\boxed{g=16}$ (the 4s orbital, the three 4p orbitals, the five 4d orbitals, and the seven 4*f* orbitals).

E18.4(a) We will solve this exercise by straightforward integration.

$$\psi_{1,0,0}=R_{1,0}Y_{0,0}=\left(\frac{Z^3}{\pi a_0^3}\right)^{1/2}{\rm e}^{-r/a_0}\quad \text{[Tables 14.1 and 17.1]}$$

The potential energy operator is

$$V=-\frac{Ze^2}{4\pi\varepsilon_0}\times\left(\frac{1}{r}\right)=-k\left(\frac{1}{r}\right)$$

$$\langle V\rangle=-k\left\langle\frac{1}{r}\right\rangle\left[k=\frac{e^2}{4\pi\varepsilon_0}\right]=-k\int_0^\infty\int_0^\pi\int_0^{2\pi}\left(\frac{1}{\pi a_0^3}\right){\rm e}^{-r/a_0}\left(\frac{1}{r}\right){\rm e}^{-r/a_0}r^2{\rm d}r\sin\theta\,{\rm d}\theta\,{\rm d}\phi$$

$$=-k\times(4\pi)\times\left(\frac{1}{\pi a_0^3}\right)\int_0^\infty r{\rm e}^{-2r/a_0}\,dr=-k\times\left(\frac{4}{a_0^3}\right)\times\left(\frac{a_0^2}{4}\right)=-k\left(\frac{1}{a_0}\right)$$

$$\left[\text{We have used}\int_0^\pi\sin\theta{\rm d}\theta=2,\ \int_0^{2\pi}{\rm d}\theta=2\pi,\text{ and }\int_0^\infty x^n{\rm e}^{-ax}\,{\rm d}x=\frac{n!}{a^{n+1}}\right]$$

Hence,

$$\langle V\rangle=-\frac{Ze^2}{4\pi\varepsilon_0 a_0}=\boxed{2E_{1s}}$$

The kinetic energy operator is $-\dfrac{\hbar^2}{2\mu}\nabla^2$ [17.3 and *Justification* 17.1]; hence

$$\langle E_{\rm K}\rangle\equiv\langle T\rangle=\int\psi_{1s}^*\left(-\frac{\hbar^2}{2\mu}\right)\nabla^2\psi_{1s}\,d\tau$$

$$\nabla^2\psi_{1s} = \frac{1}{r}\frac{\partial^2(r\psi_{1s})}{\partial r^2} + \frac{1}{r^2}\Lambda^2\psi_{1s} \text{ [Problem 14.6]}$$

$$= \left(\frac{1}{\pi a_0^3}\right)^{1/2} \times \left(\frac{1}{r}\right) \times \left(\frac{d^2}{dr^2}\right) re^{-r/a_0}$$

$[\Lambda^2\psi_{1s} = 0, \psi_{1s}$ contains no angular variables]

$$= \left(\frac{Z^3}{\pi a_0^3}\right)^{1/2} \left[-\left(\frac{2}{a_0 r}\right) + \left(\frac{1}{a_0^2}\right)\right] e^{-r/a_0}$$

$$\langle T\rangle = -\left(\frac{\hbar^2}{2\mu}\right) \times \left(\frac{1}{\pi a_0^3}\right) \int_0^\infty \left[-\left(\frac{2}{a_0 r}\right) + \left(\frac{1}{a_0^2}\right)\right] e^{-2r/a_0} r^2\, dr \times \int_0^\pi \sin\theta\, d\theta \int_0^{2\pi} d\phi$$

$$= -\left(\frac{2\hbar^2}{\mu a_0^3}\right) \int_0^\infty \left[-\left(\frac{2r}{a_0}\right) + \left(\frac{r^2}{a_0^2}\right)\right] e^{-2r/a_0}\, dr = -\left(\frac{2\hbar^2}{\mu a_0^3}\right) \times \left(-\frac{a_0}{4}\right) = \frac{\hbar^2}{2\mu a_0^2}$$

$$= \boxed{-E_{1s}}$$

Hence, $\langle T\rangle + \langle V\rangle = -E_{1s} + 2E_{1s} = E_{1s}$

We determined the ionization energy of the He^+ ion in the solution to Exercise 17.1(a) with the result $I(He^+) = 8.7188$ aJ. Using the relationships demonstrated above we obtain for the potential and kinetic energies of the He^+ ion

$$V(He^+) = 2)E_{1s}(He^+) = -2I(He^+) = -2 \times 8.7188 \text{ aJ} = \boxed{-17.4376 \text{ aJ}}$$

$$T(He^+) = -E_{1s}(He^+) = I(He^+) = \boxed{8.7188 \text{ aJ}}$$

Comment. E_{1s} may also be written as

$$E_{1s} = -\frac{\mu e^4}{32\pi^2\varepsilon_0^2\hbar^2}$$

Question. Are the three different expressions for E_{1s} given in this exercise all equivalent?

E18.5(a) See Example 18.1. The mean radius is given by $\langle r\rangle = \int_0^\infty r^3 R_{n,l}^2 dr$. After substituting eqn 17.10 for $R_{n,l}$ and integrating, the general expression for the mean radius of a hydrogenic orbital with quantum numbers l and n is obtained (see eqn 10.19 of Atkins and DePaula, *Physical Chemistry*, 8th edition).

$$\langle r_{n,l}\rangle = n^2\left\{1 + \frac{1}{2}\left(1 - \frac{l(l+1)}{n^2}\right)\right\}\frac{a_0}{Z}$$

For the 2s orbital $n = 2$ and $l = 0$, hence $\boxed{\langle r_{2s}\rangle = \frac{6a_0}{Z}}$.

See Example 18.3, which illustrates the calculation of the most probable radius. It is obtained by solving $dP(r)/dr = 0$ with $P(r) = r^2R(r)^2$. For the 2s orbital

$P(r) \propto r^2\left(2-\frac{Z}{a}r\right)^2 e^{-\frac{Z}{a}r}$. Differentiating this expression with respect to r and solving for those values of r that make the derivative equal to zero is most readily performed with mathematical software. Here we use MathCad®.

$$r^2\cdot\left(2-Z\cdot\frac{r}{a}\right)^2\cdot e^{-Z\frac{r}{a}}$$

$$2\cdot r^2\cdot\left(2-Z\cdot\frac{r}{a}\right)^2\cdot\exp\left(-Z\cdot\frac{r}{a}\right)-2\cdot r^2\cdot\left(2-Z\cdot\frac{r}{a}\right)\cdot\exp\left(-Z\cdot\frac{r}{a}\right)\cdot\frac{Z}{a}-r^2\cdot\left(2-Z\cdot\frac{r}{a}\right)^2\cdot$$

$$\frac{Z}{a}\cdot\exp\left(-Z\cdot\frac{r}{a}\right)=0 \text{ solve},r\rightarrow\begin{bmatrix}0\\ 2\cdot\frac{a}{Z}\\ 2\cdot\left(\frac{3}{2}+\frac{1}{2}\cdot\sqrt{5}\right)\cdot\frac{a}{Z}\\ 2\cdot\left(\frac{3}{2}-\frac{1}{2}\cdot\sqrt{5}\right)\cdot\frac{a}{Z}\end{bmatrix}$$

The first and second of these roots correspond to minima in the probability. The third and fourth roots correspond to maxima, with the third being the largest and, hence, it is the most probable value of r. This may be rewritten as $\boxed{r^* = (3+\sqrt{5})\frac{a}{Z}}$

E18.6(a) The radial distribution function is defined as

$$P = 4\pi r^2\psi^2 \quad \text{so} \quad P_{3s} = 4\pi r^2(Y_{0,0}R_{3,0})^2,$$

$$\boxed{P_{3s}=4\pi r^2\left(\frac{1}{4\pi}\right)\times\left(\frac{1}{243}\right)\times\left(\frac{Z}{\alpha_0}\right)^3\times(6-6\rho+\rho^2)^2e^{-\rho}}$$

where $\rho \equiv \frac{2Zr}{na_0} = \frac{2Zr}{3a_0}$ here. But we want to find the most likely radius, so it would help to simplify the function by expressing it in terms either of r or ρ, but not both. To find the most likely radius, we could set the derivative of P_{3s} equal to zero; therefore, we can collect all multiplicative constants together (including the factors of a_0/Z needed to turn the initial r^2 into ρ^2) because they will eventually be divided into zero

$$P_{3s} = C^2\rho^2(6-6\rho+\rho^2)^2e^{-\rho}$$

Note that not all the extrema of P are maxima; some are minima. But all the extrema of $(P_{3s})^{1/2}$ correspond to maxima of P_{3s}. So let us find the extrema of $(P_{3s})^{1/2}$

$$\frac{\mathrm{d}(P_{3s})^{1/2}}{\mathrm{d}\rho} = 0 = \frac{\mathrm{d}}{\mathrm{d}\rho} C\rho(6-6\rho+\rho^2)\mathrm{e}^{-\rho/2}$$

$$= C[\rho(6-6\rho+\rho^2)\times(-\tfrac{1}{2})+(6-12\rho+3\rho^2)]\mathrm{e}^{-\rho/2}$$

$$0 = C(6-15\rho+6\rho^2-\tfrac{1}{2}\rho^3)\mathrm{e}^{-\rho/2} \quad \text{so} \quad 12-30\rho+12\rho^2-\rho^3=0$$

Numerical solution of this cubic equation yields

$$\rho = 0.49,\ 2.79,\ \text{and } 8.72$$

corresponding to

$$r = \boxed{0.74a_0/Z,\ 4.19a_0/Z,\ \text{and } 13.08a_0/Z}$$

Comment. If numerical methods are to be used to find the roots of the equation which locates the extrema, then graphical/numerical methods might as well be used to locate the maxima directly. That is, the student may simply have a spreadsheet compute P_{3s} and examine or manipulate the spreadsheet to locate the maxima.

E18.7(a) See Fig. 18.6 of the text. The number of angular nodes is the value of the quantum number l which for p orbitals is 1. Hence, each of the three p orbitals has one angular node. To locate the angular nodes look for the value of θ that makes the wavefunction zero.

p_z orbital: see Table 18.1, Eqn 18.2, and Fig. 18.6. The nodal plane is the $\boxed{xy}$ plane and $\boxed{\theta=\pi/2}$ is an angular node.

p_x orbital: see Table 18.1, Eqn 18.4, and Fig. 18.6. The nodal plane is the $\boxed{yz}$ plane and $\boxed{\theta=0}$ is an angular node.

p_y orbital: see Table 18.1, Eqn 18.4 and Fig. 18.6. The nodal plane is the $\boxed{xz}$ plane and $\boxed{\theta=0}$ is an angular node.

Problems

P18.1 If we assume that the innermost electron is a hydrogen-like 1s orbital we may write

$$r^* = \frac{a_0}{Z}[\text{Example 18.3}] = \frac{52.92\,\text{pm}}{126} = \boxed{0.420\,\text{pm}}$$

P18.3 (a) Consider $\psi_{2p_z} = \psi_{2,1,0}$ which extends along the z-axis. The most probable point along the z -axis is where the radial function has its maximum value (for ψ^2 is also a maximum at that point). From Table 17.1 we know that

$$R_{21} \propto \rho e^{-\rho/2}$$

and so $\frac{dR}{d\rho} = (1 - \tfrac{1}{2}\rho)e^{-\rho/2} = 0$ when $\rho = 2$.

Therefore, $r^* = \frac{2a_0}{Z}$, and the point of maximum probability lies at $z = \pm\frac{2a_0}{Z} = \boxed{\pm 106\,\text{pm}}$

(b) The most probable radius occurs when the radial distribution function is a maximum. At this point the derivative of the function w/r/t either r or ρ equals zero.

$$\left(\frac{dR_{31}}{d\rho}\right)_{max} = 0 = \left(\frac{d((4-\rho)\rho e^{-\rho/2})}{d\rho}\right)_{max} \text{[Table 17.1]} = \left(4 - 4\rho + \frac{\rho^2}{2}\right)e^{-\rho/2}$$

The function is a maximum when the polynomial equals zero. The quadratic equation gives the roots $\rho = 4 + 2\sqrt{2} = 6.89$ and $\rho = 4 - 2\sqrt{2} = 1.17$. Since $\rho = (2Z/na_0)r$ and n = 3, these correspond to $r = 10.3 \times a_0/Z$ and $r = 1.76 \times a_0/Z$. However,

$\left|\frac{R_{3,1}(\rho_1)}{R_{3,1}(\rho_2)}\right| = \left|\frac{R_{3,1}(1.17)}{R_{3,1}(10.3)}\right| = 4.90$. So, we conclude that the function is a maximum at $\rho = \pm 1.17$ which corresponds to $\boxed{r = \pm 1.76\, a_0/Z}$.

See the solutions to Exercises 18.5(b) and 18.6(b) for the most probable radii of $2p$ and $3p$ electrons. The results are $4a_0/Z$ and $12a_0/Z$ respectively, both much larger than the points of maximum probability.

Comment. Since the radial portion of any 2p or any 3p function is the same, the same result would have been obtained for all of them. The direction of the most probable point would, however, be different.

P18.5 (a) We must show that $\int |\psi_{3_{Px}}|^2 d\tau = 1$. The integrations are most easily performed in spherical coordinates (Fig. 18.2).

$$\int |\psi_{3_{Px}}|^2 d\tau = \int_0^{2\pi}\int_0^{\pi}\int_0^{\infty} |\psi_{3p_x}|^2 r^2 \sin(\theta) dr\, d\theta\, d\phi$$

$$= \int_0^{2\pi}\int_0^{\pi}\int_0^{\infty} \left| R_{3,1}(\rho)\left\{\frac{Y_{1,-1} - Y_{1,1}}{\sqrt{2}}\right\}\right|^2 r^2 \sin(\theta) dr\, d\theta\, d\phi \text{ [Table 17.1, eqns 18.2 and 18.4]}$$

where $\rho = 2r/a_0$, $r = \rho a_0/2$, $dr = (a_0/2)d\rho$.

$$= \frac{1}{2}\int_0^{2\pi}\int_0^{\pi}\int_0^{\infty}\left(\frac{a_0}{2}\right)^3 \left[\left(\frac{1}{27(6)^{1/2}}\right)\left(\frac{1}{a_0}\right)^{3/2}\left(4-\frac{1}{3}\rho\right)\rho e^{-\rho/6}\right]$$

$$\times\left[\left(\frac{3}{8\pi}\right)^{1/2} 2\sin(\theta)\cos(\phi)\right]^2 \rho^2 \sin(\theta)d\rho d\theta d\phi$$

$$= \frac{1}{46\,656\pi}\int_0^{2\pi}\int_0^{\pi}\int_0^{\infty}\left|\left(4-\frac{1}{3}\rho\right)\rho e^{-\rho/6}\sin(\theta)\cos(\phi)\right|^2 \rho^2 \sin(\theta)d\rho d\theta d\phi$$

$$= \frac{1}{46\,656\pi}\underbrace{\int_0^{2\pi}\cos^2(\phi)d\phi}_{\pi}\underbrace{\int_0^{\pi}\sin^3(\theta)d\theta}_{4/3}\underbrace{\int_0^{\infty}\left(4-\frac{1}{3}\rho\right)^2\rho^4 e^{-\rho/3}d\rho}_{34992}$$

$=1$ Thus, $\psi_{3_{Px}}$ is normalized to 1.

We must also show that $\int \psi_{3_{Px}}\psi_{3d_{xy}}\,d\tau = 0$

Using Tables 17.1 and 18.1, we find that

$$\psi_{3_{Px}} = \frac{1}{54(2\pi)^{1/2}}\left(\frac{1}{a_0}\right)^{3/2}\left(4-\frac{1}{3}\rho\right)\rho e^{-\rho/6}\sin(\theta)\cos(\phi)$$

$$\psi_{3d_{xy}} = R_{32}\left\{\frac{Y_{2,2}-Y_{2,-2}}{\sqrt{2}i}\right\}$$

$$= \frac{1}{32(2\pi)^{1/2}}\left(\frac{1}{a_0}\right)^{3/2}\rho^2 e^{-\rho/6}\sin^2(\theta)\sin(2\phi)$$

where $\rho = 2r/a_0$, $r = \rho a_0/2$, $dr = (a_0/2)d\rho$.

$$\int\psi_{3p_x}\psi_{3d_{xy}}\,d\tau = \text{constant}\times\int_0^{\infty}\rho^5 e^{-\rho/3}d\rho\underbrace{\int_0^{2\pi}\cos(\phi)\sin(2\phi)d\phi}_{0}\int_0^{\pi}\sin^4(\theta)d\theta$$

Since the integral equals zero, ψ_{3p_x} and $\psi_{3d_{xy}}$ are orthogonal.

(b) Radial nodes are determined by finding the ρ values ($\rho = 2r/a_0$) for which the radial wavefunction equals zero. These values are the roots of the polynomial portion of the wavefunction. For the 3s orbital, $6-6\rho+\rho^2=0$, when $\boxed{\rho_{\text{node}} = 3+\sqrt{3} \text{ and } \rho_{\text{node}} = 3-\sqrt{3}}$.

The 3s orbital has these two spherically symmetrical nodes. There is no node at $\rho = 0$ so we conclude that there is a finite probability of finding a 3s electron at the nucleus.

For the 3_{Px} orbital, $(4-\rho)(\rho)=0$, when $\boxed{\rho_{\text{node}} = 0 \text{ and } \rho_{\text{node}} = 4}$. There is a zero probability of finding a $3p_x$ electron at the nucleus.

For the $3d_{xy}$ orbital $\boxed{\rho_{\text{node}} = 0}$ is the only radial node.

(c) $$\langle r\rangle_{3s} = \int |R_{1,0}Y_{0,0}|^2 r\,d\tau = \int |R_{1,0}Y_{0,0}|^2 r^3 \sin(\theta)dr\,d\theta\,d\phi$$

$$= \int_0^\infty R_{1,0}^2 r^3\,dr \underbrace{\int_0^{2\pi}\int_0^{\pi} |Y_{0,0}|^2 \sin(\theta)d\theta\,d\phi}_{1}$$

$$= \frac{a_0}{3\,888}\underbrace{\int_0^\infty (6-2\rho+\rho^2/9)^2\rho^3 e^{-\rho/3}\,d\rho}_{52488}$$

$$\boxed{\langle r\rangle_{3s} = \frac{27a_0}{2}}$$

(d) The plot, Fig. 18.1, shows that the 3*s* orbital has larger values of the radial distribution function for $r < a_0$. This penetration of inner core electrons of multi-electron atoms means that a 3s electron experiences a larger effective nuclear charge and, consequently, has a lower energy than either a 3p or $3d_{xy}$ electron. This reasoning also leads us to conclude that a $3p_x$ electron has less energy than a $3d_{xy}$ electron.

$$E_{3s} < E_{3p_x} < E_{3d_{xy}}.$$

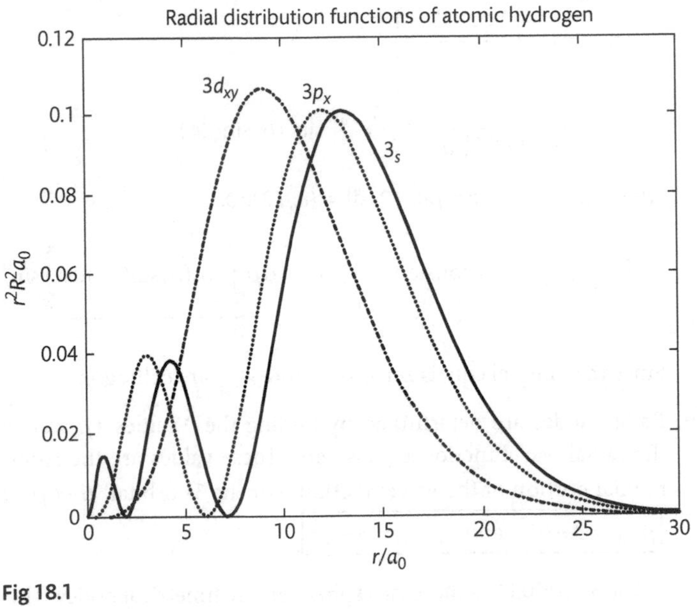

Fig 18.1

(e) Polar plots with $\theta = 90°$, Fig. 18.2

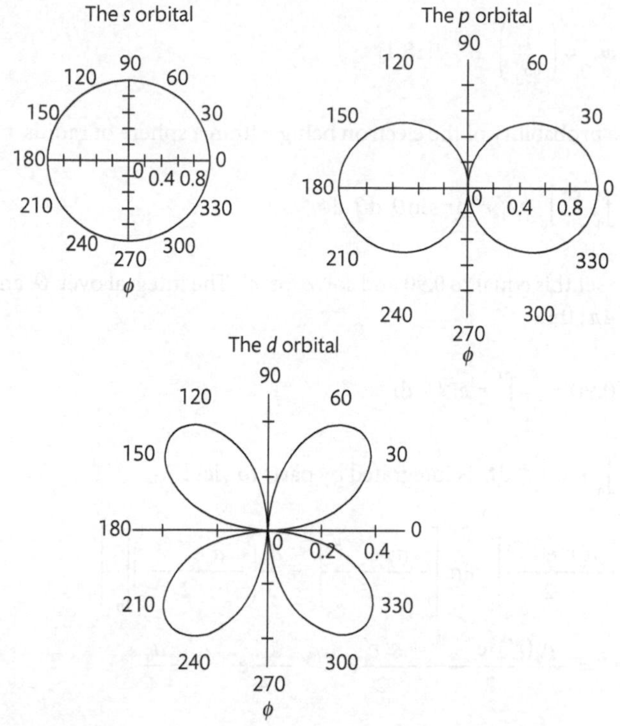

Fig 18.2

Boundary surface plots, Fig. 18.3

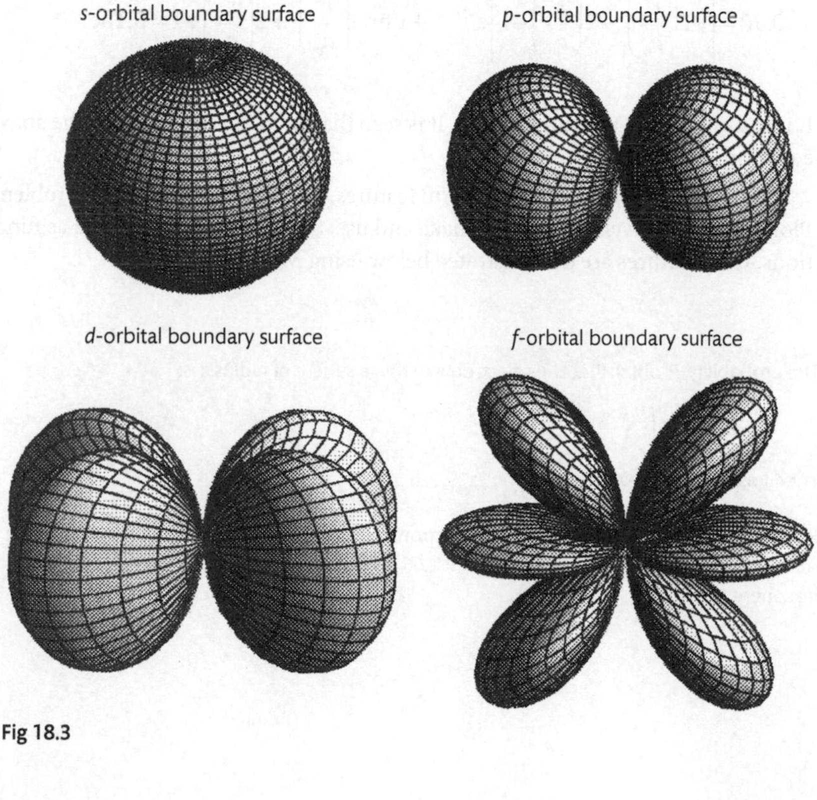

Fig 18.3

P18.7 $$\psi_{1s} = \left(\frac{1}{\pi a_0^3}\right)^{1/2} e^{-r/a_0} \quad [18.1]$$

The probability of the electron being within a sphere of radius r' is

$$\int_0^{r'}\int_0^{\pi}\int_0^{2\pi} \psi_{1s}^2 r^2 \,dr \sin\theta \,d\theta \,d\phi$$

We set this equal to 0.90 and solve for r'. The integral over θ and ϕ gives a factor of 4π; thus

$$0.90 = \frac{4}{a_0^3}\int_0^{r'} r^2 e^{-2r/a_0} \,dr$$

$\int_0^{r'} r^2 e^{-2r/a_0} \,dr$ is integrated by parts to yield

$$-\frac{a_0 r^2 e^{-2r/a_0}}{2}\Bigg|_0^{r'} + a_0\left[-\frac{a_0 r e^{-2r/a_0}}{2}\Bigg|_0^{r'} + \frac{a_0}{2}\left(-\frac{a_0 e^{-2r/a_0}}{2}\right)\Bigg|_0^{r'}\right]$$

$$= -\frac{a_0 (r')^2 e^{-2r'/a_0}}{2} - \frac{a_0^2 r'}{2} e^{-2r'/a_0} - \frac{a_0^3}{4} e^{-2r'/a_0} + \frac{a_0^3}{4}$$

Multiplying by $\frac{4}{a_0^3}$ and factoring e^{-2r'/a_0}

$$0.90 = \left[-2\left(\frac{r'}{a_0}\right)^2 - 2\left(\frac{r'}{a_0}\right) - 1\right] e^{-2r'/a_0} + 1 \text{ or } 2\left(\frac{r'}{a_0}\right)^2 + 2\left(\frac{r'}{a_0}\right) + 1 = 0.10 e^{2r'/a_0}$$

It is easiest to solve this numerically. It is seen that $\boxed{r' = 2.66 a_0}$ satisfies the above equation.

Mathematical software has powerful features for handling this type of problem. Plots are very convenient to both make and use. Solve blocks can be used as functions. Both features are demonstrated below using Mathcad®.

Let $z = r/a_0$. The probability, Prob(z), that is 1 s electron is within a sphere of radius z is:

$$\text{Prob}(z) := 4 \cdot \int_0^z x^2 \cdot e^{-2 \cdot x} \,dx$$

Variables needed for plot: $N := 800$ $\quad i := 0..N$ $\quad z_{max} := 5$ $\quad z_i := \frac{z_{max} \cdot i}{N}$

The plot indicates that the probability of finding the electron in a sphere of radius z is sigmoidal. The trace feature of Mathcad is used to find that with z = 2.66 (r = 2.66 a_0) there is a 90.0% probability of finding the electron in the sphere.

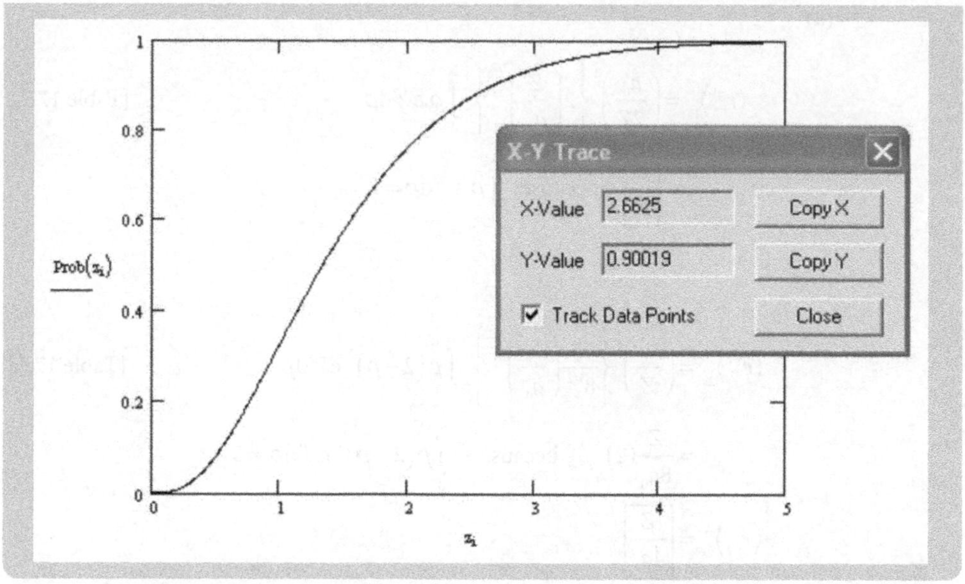

The following Mathcad® document develops a function for calculating the radius for any desired probability. The probability is presented to the function as an argument

z := 2   Estimate of z needed for computation within following Given/Find solve block for the function z(Probability).

Given

$$\text{Probability} := 4 \cdot \int_0^z x^2 \cdot e^{-2 \cdot x}\, dx$$

z(Probability) := Find(z)

z(9) = 2.661

P18.9

$$\langle r^m \rangle_{n,l} = \int r^m \left|\psi_{n,l}\right|^2 \mathrm{d}\tau = \int_0^\infty \int_0^{2\pi} \int_0^{\pi} r^{m+2} \left|R_{n,l} Y_{l,0}\right|^2 \sin(\theta)\mathrm{d}\theta\, \mathrm{d}\phi\, \mathrm{d}r$$

$$= \int_0^\infty r^{m+2} \left|R_{n,l}\right|^2 \mathrm{d}r \int_0^{2\pi} \int_0^{\pi} \left|Y_{l,0}\right|^2 \sin(\theta)\mathrm{d}\theta\, \mathrm{d}\phi = \int_0^\infty r^{m+2} \left|R_{n,l}\right|^2 \mathrm{d}r$$

With $r = (na_0/2Z)\rho$ and $m = -1$ the expectation value is

$$\langle r^{-1} \rangle_{n,l} = \left(\frac{na_0}{2Z}\right)^2 \int_0^\infty \rho \left|R_{n,l}\right|^2 \mathrm{d}\rho$$

(a)

$$\langle r^{-1}\rangle_{1s} = \left(\frac{a_0}{2Z}\right)^2 \left\{2\left(\frac{Z}{a_0}\right)^{3/2}\right\}^2 \int_0^\infty \rho\, e^{-\rho} d\rho \qquad \text{[Table 17.1]}$$

$$= \boxed{\frac{Z}{a_0}} \quad \text{because} \quad \int_0^\infty \rho\, e^{-\rho} d\rho = 1$$

(b)

$$\langle r^{-1}\rangle_{2s} = \left(\frac{a_0}{Z}\right)^2 \left\{\frac{1}{8^{1/2}}\left(\frac{Z}{a_0}\right)^{3/2}\right\}^2 \int_0^\infty \rho(2-\rho)^2\, e^{-\rho} d\rho \qquad \text{[Table 17.1]}$$

$$= \frac{Z}{8a_0}(2) \qquad \text{because} \quad \int_0^\infty \rho(2-\rho)^2\, e^{-\rho} d\rho = 2$$

$$\langle r^{-1}\rangle_{2s} = \boxed{\frac{Z}{4a_0}}$$

(c)

$$\langle r^{-1}\rangle_{2p} = \left(\frac{a_0}{Z}\right)^2 \left\{\frac{1}{24^{1/2}}\left(\frac{Z}{a_0}\right)^{3/2}\right\}^2 \int_0^\infty \rho^3\, e^{-\rho} d\rho \qquad \text{[Table 17.1]}$$

$$= \frac{Z}{24a_0}(6) \qquad \text{because} \quad \int_0^\infty \rho^3\, e^{-\rho} d\rho = 6$$

$$\langle r^{-1}\rangle_{2p} = \boxed{\frac{Z}{4a_0}}$$

The general formula for a hydrogenic orbital is $\langle r^{-1}\rangle_{n,l} = \dfrac{Z}{n^2 a_0}$

Topic 19 Many-electron atoms

Discussion questions

D19.1 Fermions are particles with half-integral spin, 1/2, 3/2, 5/2, . . . whereas bosons have integral spin, 0, 1, 2, ... All fundamental particles that make up matter have spin 1/2 and are fermions, but composite particles can be either fermions or bosons.
Fermions: electrons, protons, neutrons, ^{3}He, ...
Bosons: photons, deuterons

Comment. In the standard model of particle physics, non-composite bosons are carriers of force. Photons are an example: they are not composite particles, but they are not constituents of matter. In the standard model, protons and neutrons are composite particles, but they are still fermions by virtue of their half-integral spin.

D19.3 Spin angular momentum and orbital angular momentum have in common the fact that the magnitude of angular momentum and the z component of angular momentum are both quantized. Both spin and orbital angular momentum give rise to magnetic effects. The spin angular momentum is an intrinsic angular momentum rather than one due to rotational motion in the classical sense.

Exercises

E19.1(a) We can find the speed of a point on the equator if we know the rotational frequency, ω: the speed is $r\omega$, where r is the radius. Classically, angular momentum is equal to moment of inertia, I, times ω. For a sphere of mass m_e, the moment of inertia is $I = 2m_e r^2/5$ [Section 14.1 or any general physics text]. An electron has spin angular momentum of $\sqrt{(1/2)(1+1/2)}\hbar$. Putting it all together,

$$\sqrt{(1/2)(1+1/2)}\hbar = 0.866\hbar = \frac{2m_e r^2}{5}\omega \quad \text{so} \quad \omega = \frac{5\times 0.866\hbar}{2m_e r^2}.$$

The speed is

$$r\omega = \frac{5\times 0.866\hbar}{2m_e r} = \frac{5\times 0.866\times 1.0546\times 10^{-34}\ \text{J s}}{2\times 9.11\times 10^{-31}\ \text{kg}\times 2.82\times 10^{-15}\ \text{m}} = \boxed{8.89\times 10^{10}\ \text{m s}^{-1}}$$

Comment. This speed is about 300 times the speed of light! Clearly this quasi-classical, non-relativistic model does not capture this aspect of the electron.

Problems

P19.1 From the diagram in Fig. 19.1, $\cos\theta = \dfrac{m_l}{\{l(l+1)\}^{1/2}}$ and hence $\boxed{\theta = \cos^{-1}\dfrac{m_l}{\{l(l+1)\}^{1/2}}}$

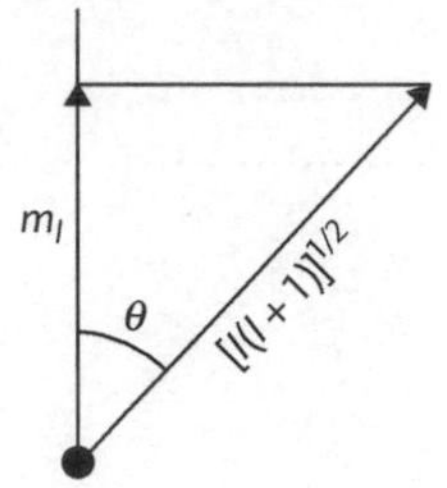

Figure 19.1

For an α electron, $m_s = +\frac{1}{2}, s = \frac{1}{2}$ and (with $m_l \rightarrow m_s, l \rightarrow s$)

$$\theta = \cos^{-1}\frac{\frac{1}{2}}{\left(\frac{3}{4}\right)^{1/2}} = \cos^{-1}\frac{1}{\sqrt{3}} = \boxed{54°44'}.$$

The minimum angle occurs for $m_l = l$;

$$\lim_{l\to\infty}\theta_{\min} = \lim_{l\to\infty}\cos^{-1}\left(\frac{l}{\{l(l+1)\}^{1/2}}\right) = \lim_{l\to\infty}\cos^{-1}\frac{l}{l} = \cos^{-1} 1 = \boxed{0}.$$

Topic 20 Periodicity

Discussion questions

D20.1 See Fig. 18.2 of the text. The periodic table shows the manner in which the chemical and physical properties of an element, composed of atoms of one kind only, varies as a function of the atomic number of the element. This variation is periodic in the atomic numbers of the elements. The configuration of the electrons in an atom is what determines the chemical and physical properties of the atom (element); hence the periodicity of the properties is necessarily a result of the electron configurations.

The order of occupation of atomic orbitals, forming the electron configuration of the atom, mirrors the structure of the periodic table.

The s and p blocks form the main groups of the periodic table. The similar electron configurations for the elements in a main group are the reason for the similar properties of these elements. Each new period corresponds to the occupation of a shell with a higher principal quantum number. This correspondence explains the different lengths of the periods. Period 1 consists of only two elements, H and He, in which the single 1s-orbital of the $n = 1$ shell is being filled with its two electrons. Period 2 consists of the eight elements Li through Ne, in which the one 2s- and three 2p-orbitals are being filled with eight more electrons. In Period 3 (Na through Ar), the 3s- and 3p-orbitals are being occupied by eight additional electrons. In Period 4, not only are the electrons of the 4s- and 4p-orbitals being added but so are the 10 electrons of the 3d-orbitals. Hence there are 18 elements in Period 4. Period 5 elements add another 18 electrons as the 5s-, 4d-, and 5p-orbitals are filled. In Period 6, a total of 32 electrons are added, because 14 electrons are also being added to the seven 4f-orbitals.

Exercises

E20.1(a) The ground state configuration of the carbon atom is $1s^22s^22p^2$. The normal valence of carbon is 4 and each of the electrons with principal quantum number 2 can participate in bonding. In the free carbon atom the three 2p-orbitals are degenerate, hence we have no knowledge in which two of the three 2p-orbitals the electrons actually reside. We do know that the two 2p-electrons are unpaired and that the total spin is 1, but whether they are both $m_s = +1/2$ or both $m_s = -1/2$ is unknown. The table below summarizes one possible set of quantum numbers.

electron	n	l	m_l	s	m_s
1	2	0	0	1/2	+1/2
2	2	0	0	1/2	−1/2
3	2	1	1	1/2	+1/2
4	2	1	0	1/2	+1/2

E20.2(a)
Sc: $[Ar]4s^23d^1$
Ti: $[Ar]4s^23d^2$
V: $[Ar]4s^23d^3$
Cr: $[Ar]4s^23d^4$ or $[Ar]4s^13d^5$ (most probable)
Mn: $[Ar]4s^23d^5$
Fe: $[Ar]4s^23d^6$
Co: $[Ar]4s^23d^7$
Ni: $[Ar]4s^23d^8$
Cu: $[Ar]4s^23d^9$ or $[Ar]4s^13d^{10}$ (most probable)
Zn: $[Ar]4s^23d^{10}$

Problems

20.1 Electronic configurations of neutral, 4th period transition atoms in the ground state are summarized in the following table along with observed, positive oxidation states. The most common positive oxidation states are indicated with bright boxing.

Group	**3**	**4**	**5**	**6**	**7**	**8**	**9**	**10**	**11**	**12**
Oxidation State	Sc	Ti	V	Cr	Mn	Fe	Co	Ni	Cu	Zn
0	$3d4s^2$	$3d^24s^2$	$3d^34s^2$	$3d^54s$	$3d^54s^2$	$3d^64s^2$	$3d^74s^2$	$3d^84s^2$	$3d^{10}4s$	$3d^{10}4s^2$
+1			☺	☺	☺		☺	☺	☺	
+2		☺	☺	☺	☺	☺	☺	☺	☺	☺
+3	☺	☺	☺	☺	☺	☺	☺	☺		
+4		☺	☺	☺	☺	☺	☺	☺		
+5			☺	☺	☺	☺				
+6				☺	☺	☺				
+7					☺					

Toward the middle of the first transition series (Cr, Mn, and Fe) elements exhibit the widest ranges of oxidation states. This phenomenon is related to the availability of both electrons and orbitals favorable for bonding. Elements to the left (Sc and Ti) of the series have few electrons, and a relatively low effective nuclear charge leaves d orbitals at high energies that are relatively unsuitable for bonding. To the far right (Cu and Zn) effective nuclear charge may be higher but there are few, if any, orbitals available for bonding. Consequently, it is more difficult to produce a range of compounds that promote a wide range of oxidation states for elements at either end of the series. At the middle and right of the series the +2 oxidation state is very commonly observed because normal reactions can provide the requisite ionization energies for the removal of 4s electrons. The readily available +2 and +3 oxidation states of Mn, Fe, and the +1 and +2 oxidation states of Cu make these cations useful in electron transfer processes occurring in chains of specialized proteins within biological cells. The special size and charge of the Zn^{2+} cation makes it useful for the function of some enzymes. The tendency of Fe^{2+} and Cu^{+} to bind oxygen proves very useful in hemoglobin and electron transport (respiratory) chain, respectively.

Topic 21 Atomic spectroscopy

Discussion questions

D21.1 When hydrogen atoms are excited into higher energy levels they emit energy at a series of discrete wavenumbers which show up as a series of lines in the spectrum. The lines in the Lyman series, for which the emission of energy corresponds to the final state $n_1 = 1$, are in the ultraviolet region of the electromagnetic spectrum. The Balmer series, for which $n_1 = 2$, is in the visible region. The Paschen series, $n_1 = 3$, is in the infrared. The Brackett series, $n_1 = 4$, is also in the infrared. The Pfund series, $n_1 = 5$, is in the far infrared. The Humphrey series, $n_1 = 6$, is in the very far infrared.

D21.3 An electron has a magnetic moment and magnetic field due to its orbital angular momentum. It also has a magnetic moment and magnetic field due to its spin angular momentum. There is an interaction energy between magnetic moments and magnetic fields. That between the spin magnetic moment and the magnetic field generated by the orbital motion is called spin–orbit coupling. The energy of interaction is proportional to the scalar product of the two vectors representing the spin and orbital angular momenta and hence depends upon the orientation of the two vectors. See Fig. 21.4. The total angular momentum of an electron in an atom is the vector sum of the orbital and spin angular momenta as expressed in eqn 21.5. The spin–orbit coupling results in a splitting of the energy levels associated with atomic terms as shown on the top of Fig. 21.5 This splitting shows up in atomic spectra as a fine structure as illustrated on the bottom of Fig. 21.5.

Exercises

E21.1(a) Eqn 21.1 implies that the shortest wavelength corresponds to $n_2 = \infty$, and the longest to $n_2 = 2$. Solve eqn 21.1 for λ.

$$\lambda = \frac{(1/n_1^2 - 1/n_2^2)^{-1}}{\tilde{R}_H}$$

Shortest: $$\lambda = \frac{(1/1^2 - 1/\infty^2)^{-1}}{109\,677\,\text{cm}^{-1}} = \boxed{9.118 \times 10^{-6}\ \text{cm}}$$

Longest: $$\lambda = \frac{(1/1^2 - 1/2^2)^{-1}}{109\,677\,\text{cm}^{-1}} = \boxed{1.216 \times 10^{-5}\ \text{cm}}$$

E21.2(a) For atoms A, eqn 17.7 may be rewritten in terms of the Rydberg constant $\tilde{R}$ as

$$E_n = \frac{Z^2\mu_A hc\tilde{R}}{m_e n^2} \approx \frac{Z^2 hc\tilde{R}}{n^2}$$

where to within 0.01% the ratio μ_A/m_e is unity. Eqn 21.1 can then be rewritten as

$$\tilde{\nu} = Z^2\tilde{R}\left(\frac{1}{n_1^2}-\frac{1}{n_2^2}\right) \qquad \lambda = \frac{1}{\tilde{\nu}} \qquad \nu = \frac{c}{\lambda}$$

$$\tilde{\nu} = 4\times 109\,737\ \text{cm}^{-1}\left(\frac{1}{1^2}-\frac{1}{2^2}\right) = \boxed{3.292\times 10^5\ \text{cm}^{-1}} \qquad \lambda = \boxed{3.038\times 10^{-6}\ \text{cm}}$$

$$\nu = \frac{2.9978\times 10^{10}\ \text{cm s}^{-1}}{3.0376\times 10^{-6}\ \text{cm}} = \boxed{9.869\times 10^{15}\ \text{s}^{-1}}$$

E21.3(a) The selection rules for a many-electron atom are given by the set [21.8]. For a single-electron transition these amount to Δn = any integer; $\Delta l = \pm 1$ Hence

(a) $3s \rightarrow 1s$; $\Delta l = 0$, $\boxed{\text{forbidden}}$

(b) $3p \rightarrow 2s$; $\Delta l = -1$, $\boxed{\text{allowed}}$

(c) $5d \rightarrow 2p$; $\Delta l = -1$, $\boxed{\text{allowed}}$

E21.4(a) (a) Pd^{2+}: $\boxed{[\text{Kr}]4d^8}$

(b) All subshells except 4d are filled and hence have no net spin. Applying Hund's rule to $4d^8$ shows that there are two unpaired spins. The paired spins do not contribute to the net spin, hence we consider only $s_1 = \frac{1}{2}$ and $s_2 = \frac{1}{2}$ The Clebsch–Gordan series [21.4] produces

$$S = s_1 + s_2, \ldots, |s_1 - s_2|, \text{ hence } \boxed{S = 1, 0}$$

$$M_S = -S, -S+1, \ldots, S$$

For $S = 1$, $\boxed{M_S = -1, 0, +1}$

$S = 0$, $\boxed{M_S = 0}$

E21.5(a) We use the Clebsch–Gordan series [21.5] in the form

$j = l+s, l+s-1, \ldots, |l-s|$ [lower-case for a single electron]

(a) $l = 2$, $s = \frac{1}{2}$; so $j = \boxed{\frac{5}{2}, \frac{3}{2}}$

(b) $l = 3$, $s = \frac{1}{2}$; so $j = \boxed{\frac{7}{2}, \frac{5}{2}}$

E21.6(a) The Clebsch–Gordan series for $\boxed{l = 3 \text{ or } 2}$ and $s = \frac{1}{2}$ leads to $j = \frac{5}{2}$

The Clebsch–Gordan series for $\boxed{l = 1 \text{ or } 0}$ and $s = \frac{1}{2}$ leads to $j = \frac{1}{2}$

E21.7(a) Use the Clebsch–Gordan series in the form

$$J = j_1 + j_2, j_1 + j_2 - 1, \ldots, |j_1 - j_2|$$

Then, with $j_1 = 1$ and $j_2 = 2$

$J = \boxed{3, 2, 1}$

E21.8(a) The letter P indicates that $L=1$, the superscript 3 is the value of $2S+1$ so $S=1$ and the subscript 2 is the value of J. Hence, $\boxed{L=1,\ S=1,\ J=2}$

E.9(a) Use the Clebsch–Gordan series in the form

$$S'=s_1+s_2,\ s_1+s_2-1,\ldots,|s_1-s_2|$$

and

$$S=S'+s_1,\ S'+s_1-1,\ldots,|S'-s_1|$$

in succession. The multiplicity is $2S+1$

(a) $S=\frac{1}{2}+\frac{1}{2},\ \frac{1}{2}-\frac{1}{2}=\boxed{1,0}$ with multiplicities $\boxed{3,1}$ respectively

(b) $S'=1,0$ then $S=\boxed{\frac{3}{2},\frac{1}{2}}$ [from 1], $\boxed{\text{and}\ \frac{1}{2}}$ [from 0], with multiplicities $\boxed{4,2,2}$

E21.10(a) These electrons are not equivalent (different subshells), hence all the terms that arise from the vector model and the Clebsch–Gordan series are allowed (Example 21.2).

$$L=l_1+l_2,\ldots,|l_1-l_2|\ [21.3]=2\ \text{only}$$

$$S=s_1+s_2,\ldots,|s_1-s_2|=1,0$$

The allowed terms are then $^3\mathrm{D}$ and $^1\mathrm{D}$. The possible values of J are given by

$$J=L+S,\ldots,|L-S|\,[21.5]=3,\ 2,\ 1\ \text{for}\ ^3\mathrm{D}\ \text{and}\ 2\ \text{for}\ ^1\mathrm{D}$$

The allowed complete term symbols are then

$$\boxed{^3\mathrm{D}_3,\ ^3\mathrm{D}_2,\ ^3\mathrm{D}_1,\ ^1\mathrm{D}_2}$$

The $\boxed{^3\mathrm{D}\ \text{set of terms are the lower in energy}}$ [Hund's rule]

Comment. Hund's rule in the form given in the text does not allow the energies of the triplet terms to be distinguished. Experimental evidence indicates that $^3\mathrm{D}_1$ is lowest.

E21.11(a) Use the Clebsch–Gordan series in the form

$$J=L+S,\ L+S-1,\ldots,|L-S|$$

The number of states (M_J values) is $2J+1$ in each case.

(a) $L=0, S=1$; hence $\boxed{J=1}$ $(^3\mathrm{S}_1)$ and there are $\boxed{J=1}$ $(^3\mathrm{S}_1)$

(b) $L=2, S=\frac{1}{2}$ hence $\boxed{J=5/2, 3/2\ \text{respectively}}$ $(^2\mathrm{D}_{5/2},\ ^2\mathrm{D}_{3/2})$ with $\boxed{J=5/2, 3/2\ \text{respectively}}$ $(^2\mathrm{D}_{5/2},\ ^2\mathrm{D}_{3/2})$

(c) $L=1, S=0$; hence $\boxed{J=1}$ $(^1\mathrm{P}_1)$ with $\boxed{3\ \text{states}}$.

E21.12(a) Closed shells and subshells do not contribute to either L or S and thus are ignored in what follows.

(a) Na: $[\mathrm{Ne}]3\mathrm{s}^1$: $S=\frac{1}{2}$, $L=0$; $J=\frac{1}{2}$ so the only term is $\boxed{^2\mathrm{S}_{1/2}}$

(b) K: $[\mathrm{Ar}]3\mathrm{d}^1$: $S=\frac{1}{2}$, $L=2$; $J=\frac{5}{2},\frac{3}{2}$, so the terms are $\boxed{^2\mathrm{D}_{5/2}\ \text{and}\ ^2\mathrm{D}_{3/2}}$

Problems

P21.1 A Lyman series corresponds to $n_1 = 1$; hence

$$\tilde{\nu} = \tilde{R}_{\mathrm{Li}^{2+}}\left(1-\frac{1}{n^2}\right), \quad n = 2,3,\cdots \quad \left[\tilde{\nu} = \frac{1}{\lambda}\right]$$

Therefore, if the formula is appropriate, we expect to find that $\tilde{\nu}\left(1-\frac{1}{n^2}\right)^{-1}$ is a constant $\tilde{R}_{\mathrm{Li}^{2+}}$.

We therefore draw up the following table.

n	2	3	4
$\tilde{\nu}/\mathrm{cm}^{-1}$	740 747	877 924	925 933
$\tilde{\nu}\left(1-\frac{1}{n^2}\right)^{-1}/\mathrm{cm}^{-1}$	987 663	987 665	987 662

Hence, the formula does describe the transitions, and $\boxed{\tilde{R}_{\mathrm{Li}^{2+}} = 987\,663\ \mathrm{cm}^{-1}}$. The Balmer transitions lie at

$$\tilde{\nu} = \tilde{R}_{\mathrm{Li}^{2+}}\left(\frac{1}{4}-\frac{1}{n^2}\right) \quad n = 3, 4, \ldots$$

$$= (987663\,\mathrm{cm}^{-1})\times\left(\frac{1}{4}-\frac{1}{n^2}\right) = \boxed{137\,175\ \mathrm{cm}^{-1}}, \boxed{185\,187\ \mathrm{cm}^{-1}}, \ldots$$

The ionization energy of the ground-state ion is given by

$$\tilde{\nu} = \tilde{R}_{\mathrm{Li}^{2+}}\left(1-\frac{1}{n^2}\right), \quad n \to \infty$$

and hence corresponds to

$$\tilde{\nu} = 987\,663\ \mathrm{cm}^{-1}, \ \text{or} \ \boxed{122.5\ \mathrm{eV}}$$

P21.3 The 7*p* configuration has just one electron outside a closed subshell. That electron has $l = 1$ $s = 1/2$ and $j = 1/2$ or $3/2$ so the atom has $L = 1$ $S = 1/2$, and $J = 1/2$ or $3/2$. The term symbols are $\boxed{{}^2\mathrm{P}_{1/2} \text{ and } {}^2\mathrm{P}_{3/2}}$, of which the former has the lower energy. The 6*d* configuration also has just one electron outside a closed subshell; that electron has $l = 2$, $s = 1/2$, and $J = 3/2$ or $5/2$, so the atom has $L = 2$, $S = 1/2$, and $J = 3/2$ or $5/2$. The term symbols are $\boxed{{}^2\mathrm{D}_{3/2} \text{ and } {}^2\mathrm{D}_{5/2}}$, of which the former has the lower energy. According to the simple treatment of spin–orbit coupling, the energy is given by

$$E_{l,s,j} = \tfrac{1}{2}\lambda\hbar^2[j(j+1) - l(l+1) - s(s+1)]$$

where A is the spin–orbit coupling constant. So

$$E(^2P_{1/2}) = \tfrac{1}{2}\lambda\hbar^2[\tfrac{1}{2}(1/2+1) - 1(1+1) - \tfrac{1}{2}(1/2+1)] = -\lambda\hbar^2$$

and $E(^2D_{3/2}) = \tfrac{1}{2}\lambda\hbar^2[\tfrac{3}{2}(3/2+1) - 2(2+1) - \tfrac{1}{2}(1/2+1)] = -\tfrac{3}{2}\lambda\hbar^2$

This approach would predict the ground state to be $\boxed{^2D_{3/2}}$

Comment. The computational study cited finds the $^2P_{1/2}$ level to be lowest, but the authors caution that the error of similar calculations on Y and Lu is comparable to the computed difference between levels.

P21.5 $\tilde{R}_H = k\mu_H$, $\tilde{R}_D = k\mu_D$, $\tilde{R} = k\mu$ [17.12, 17.13]

where $\tilde{R}$ corresponds to an infinitely heavy nucleus, with $\mu = m_e$.

Since $\mu = \dfrac{m_e m_N}{m_e + m_N}$ [N = p or d]

$$\tilde{R}_H = k\mu_H = \frac{km_e}{1+\frac{m_e}{m_p}} = \frac{\tilde{R}}{1+\frac{m_e}{m_p}}$$

Likewise, $\tilde{R}_D = \dfrac{\tilde{R}}{1+\frac{m_e}{m_d}}$ where m_p is the mass of the proton and m_d the mass of the deuteron. The two lines in question lie at

$$\frac{1}{\lambda_H} = \tilde{R}_H\left(1-\frac{1}{4}\right) = \tfrac{3}{4}\tilde{R}_H \quad \frac{1}{\lambda_D} = \tilde{R}_D(1-\tfrac{1}{4}) = \tfrac{3}{4}\tilde{R}_D$$

and hence

$$\frac{\tilde{R}_H}{\tilde{R}_D} = \frac{\lambda_D}{\lambda_H} = \frac{\tilde{\nu}_H}{\tilde{\nu}_D}$$

Then, since

$$\frac{\tilde{R}_H}{\tilde{R}_D} = \frac{1+\frac{m_e}{m_d}}{1+\frac{m_e}{m_p}}, \quad m_d = \frac{m_e}{\left(1+\frac{m_e}{m_p}\right)\frac{\tilde{R}_H}{\tilde{R}_D} - 1}$$

we can calculate m_d from

$$m_d = \frac{m_e}{\left(1+\frac{m_e}{m_p}\right)\frac{\lambda_D}{\lambda_H} - 1} = \frac{m_e}{\left(1+\frac{m_e}{m_p}\right)\frac{\tilde{\nu}_H}{\tilde{\nu}_D} - 1}$$

$$= \frac{9.10\,939\times10^{-31}\ \text{kg}}{\left(1+\frac{9.1039\times10^{-31}\ \text{kg}}{1.67\,262\times10^{-27}\ \text{kg}}\right)\times\left(\frac{82\,259.098\ \text{cm}^{-1}}{82\,281.476\ \text{cm}^{-1}}\right)^{-1}} = \boxed{3.3\,429\times10^{-27}\ \text{kg}}$$

Since $I = \tilde{R}hc$,

$$\frac{I_D}{I_H} = \frac{\tilde{R}_D}{\tilde{R}_H} = \frac{\tilde{\nu}_D}{\tilde{\nu}_H} = \frac{82\,281.476\ \text{cm}^{-1}}{82\,259.098\ \text{cm}^{-1}} = \boxed{1.000\,272}$$

P21.7 *Justification* 21.1 noted that the transition dipole moment, $\mu_{\rm fi}$ had to be non-zero for a transition to be allowed. The Justification examined conditions that allowed the z component of this quantity to be non-zero; now examine the x and y components.

$$\mu_{x,\rm fi} = -e\int \psi_{\rm f}^* x\psi_i \,\mathrm{d}\tau \quad \text{and} \quad \mu_{y,\rm fi} = -e\int \psi_{\rm f}^* y\psi_i \,\mathrm{d}\tau$$

As in the Justification, express the relevant Cartesian variables in terms of the spherical harmonics, $Y_{l,m}$. Start by expressing them in spherical polar coordinates:

$$x = r\sin\theta\cos\phi \quad \text{and} \quad y = r\sin\theta\sin\phi$$

Note that $Y_{1,1}$ and $Y_{1,-1}$ have factors of $\sin\theta$. They also contain complex exponentials that can be related to the sine and cosine of ϕ through the identities

$$\cos\phi = 1/2(\mathrm{e}^{\mathrm{i}\phi} + \mathrm{e}^{-\mathrm{i}\phi}) \quad \text{and} \quad \sin\phi = 1/2i(\mathrm{e}^{\mathrm{i}\phi} - \mathrm{e}^{-\mathrm{i}\phi}).$$

These relations motivate us to try linear combinations $Y_{1,1} - Y_{1,-1}$ and $Y_{1,1} - Y_{1,-1}$ (from Table 3.2; note c here corresponds to the normalization constant in the table):

$$Y_{1,1} + Y_{1,-1} = -c\sin\theta(\mathrm{e}^{\mathrm{i}\phi} + \mathrm{e}^{-\mathrm{i}\phi}) = -2c\sin\theta\cos\phi = -2cx/r,$$

so $x = -(Y_{1,1} + Y_{1,-1})r/2c;$

$$Y_{1,1} - Y_{1,-1} = c\sin\theta(\mathrm{e}^{\mathrm{i}\phi} - \mathrm{e}^{-\mathrm{i}\phi}) = 2ic\sin\theta\sin\phi = 2icy/r,$$

so $y = (Y_{1,1} - Y_{1,-1})r/2ic$

Now we can express the integrals in terms of radial wavefunctions $R_{n,l}$ and spherical harmonics Y_{l,m_l}

$$\mu_{x,\rm fi} = \frac{e}{2c}\int_0^\infty R_{n_{\rm f},l_{\rm f}} r R_{n_{\rm i},l_{\rm i}} r^2\,dr \int_0^\pi\int_0^{2\pi} Y^*{}_{l_{\rm f},m_{l_{\rm f}}}(Y_{1,1} + Y_{1,-1})Y_{l_{\rm i},m_{l_{\rm i}}} \sin\theta\,\mathrm{d}\theta\,\mathrm{d}\phi$$

The angular integral can be broken into two, one of which contains $Y_{1,1}$ and the other $Y_{1,-1}$. A "triple integral" over spherical harmonics of the form

$$\int_0^\pi\int_0^{2\pi} Y^*_{l_f,m_{l_f}} Y_{1,1} Y_{l_{\rm i},m_{l_{\rm i}}} \sin\theta\,\mathrm{d}\theta\,\mathrm{d}\phi$$

vanishes unless $l_{\rm f} = l_{\rm i} \pm 1$ and $m_{l_{\rm f}} = m_{l_{\rm i}} \pm 1$. The integral that contains $Y_{1,-1}$ introduces no further constraints; it vanishes unless $l_{\rm f} = l_{\rm i} \pm 1$ and $m_{l_{\rm f}} = m_{l_{\rm i}} \pm 1$. Similarly, the y component introduces no further constraints, for it involves the same spherical harmonics as the x component. The whole set of selection rules, then, is that transitions are allowed only if

$$\boxed{\Delta l = \pm 1 \text{ and } \Delta m_l = 0 \text{ or } \pm 1}$$

Focus 5: Integrated activities

F5.1 (a) The splitting of adjacent energy levels is related to the difference in wavenumber of the spectral lines as follows:

$$hc\Delta\tilde{\nu} = \Delta E = \mu_B \mathcal{B}\,, \quad \text{so} \quad \Delta\tilde{\nu} = \frac{\mu_B \mathcal{B}}{hc} = \frac{(9.274\times10^{-24}\ \text{J T}^{-1})(2\ \text{T})}{(6.626\times10^{-34}\ \text{J s})(2.998\times10^{10}\ \text{cm s}^{-1})}$$

$$\Delta\tilde{\nu} = \boxed{0.9\ \text{cm}^{-1}}$$

(b) Transitions induced by absorbing visible light have wavenumbers in the tens of thousands of reciprocal centimeters, so normal Zeeman splitting is **small** compared to the difference in energy of the states involved in the transition. Take a wavenumber from the middle of the visible spectrum as typical:

$$\tilde{\nu} = \frac{1}{\lambda} = \frac{1}{600\ \text{nm}}\left(\frac{10^9\ \text{nm m}^{-1}}{10^2\ \text{cm m}^{-1}}\right) = 1.7\times10^4\ \text{cm}^{-1}$$

Or take the Balmer series as an example, as suggested in the problem; the Balmer wavenumbers are (eqn 21.1):

$$\tilde{\nu} = \tilde{R}_H\left(\frac{1}{2^2} - \frac{1}{n^3}\right)$$

The smallest Balmer wavenumber is

$$\tilde{\nu} = (109\,677\ \text{cm}^{-1})\times(1/4 - 1/9) = 15\,233\ \text{cm}^{-1}$$

and the upper limit is

$$\tilde{\nu} = (109\,677\ \text{cm}^{-1})\times(1/4 - 0) = 27\,419\ \text{cm}^{-1}$$

F5.3 (a) The orthonormal hydrogenic atomic orbitals have the form:

$$\psi_{n,l,m_l} = R_{n,l}Y_{l,m_l}\ [17.4,\ 17.11,\ 17.10,\ \text{Table 14.1}]$$

and the energies $E_n = -hc\tilde{R}_{\text{Li}^{2+}}\left(\frac{Z^2}{n^2}\right) = \frac{-9hc\tilde{R}_{\text{Li}^{2+}}}{n^2}$ [17.7, 17.12]

where $\tilde{R}_{\text{Li}^{2+}} = \frac{\mu_{\text{Li}^{2+}}}{m_e}\tilde{R}_\infty$ [17.13] $\approx 109{,}729\ \text{cm}^{-1}$.

Thus, the wavefunction in question is the linear superposition of the states

$\psi_{4,2,-1}$, $\psi_{3,2,1}$, and $\psi_{1,0,0}$ with the coefficients $c_{4,2,-1} = -(1/3)^{1/2}$, $c_{3,2,1} = \frac{2}{3}i$, and $c_{1,0,0} = -(2/9)^{1/2}$, respectively.

Because $|c_{4,2,-1}|^2 + |c_{3,2,1}|^2 + |c_{1,0,0}|^2 = {}^1/_3 + {}^4/_9 + {}^2/_9 = 1$, the superposition is normalized. If the energy of this state is measured, only energies with the principle quantum numbers corresponding to $n = 4$, 3, or 1 will be observed. These are the energies

$$\boxed{\frac{-9hc\tilde{R}_{\mathrm{Li}^{2+}}}{16},\ -hc\tilde{R}_{\mathrm{Li}^{2+}},\ \text{and}\ -9hc\tilde{R}_{\mathrm{Li}^{2+}}}$$

According to Postulate V, Focus 1 the modulus square of each eigenfunction coefficient gives the probability of observing the eigenvalue of the eigenfunction. Thus, the above energies will be observed with the probabilities $\boxed{{}^1/_3,\ {}^4/_9,\ \text{and}\ {}^2/_9}$, respectively. The average energy of many observations will be the weighted average:

$$\text{average observed energy} = \frac{1}{3}\left(\frac{-9hc\tilde{R}_{\mathrm{Li}^{2+}}}{16}\right) + \frac{4}{9}\left(-hc\tilde{R}_{\mathrm{Li}^{2+}}\right) + \frac{2}{9}\left(-9hc\tilde{R}_{\mathrm{Li}^{2+}}\right) = \boxed{-2.632\ hc\tilde{R}_{\mathrm{Li}^{2+}}}$$

(b) The dipole transition integral specifies the selection rule $\Delta l = \pm 1$ [21.2]. Two of the eigenfunctions have $l = 2$ so for them to exhibit an emission to the ground state ($l = 0$) would require the forbidden transition $\Delta l = 2$. The third eigenfunction is the ground state. We conclude that this superposition will not exhibit a spectral emission to the ground state.

Topic 22 **Valence-bond theory**

Discussion questions

D22.1 The Born-Oppenheimer approximation treats the nuclei of the multi-particle system of electrons and nuclei as if they were fixed. The dependence of energy on nuclear positions is then obtained by solving the Schrödinger equation at many different (fixed) nuclear geometries. Molecular potential energy curves and surfaces are the plots of molecular energy (computed under the Born-Oppenheimer approximation) vs. nuclear coordinates.

D22.3 See Topic 22.2(b) for details on carbon-atom hybridization. The carbon atoms in alkanes are sp^3-hybridized. This explains the nearly tetrahedral bond angles about the carbon atoms in alkanes. The double-bonded carbon atoms in alkenes are sp^2-hybridized. This explains the bond angles of approximately 120° about these atoms. The simultaneous overlap of sp^2-hybridized orbitals and unhybridized p orbitals in C=C double bonds explains the resistance of such bonds to torsion and the co-planarity of the atoms attached to those atoms. The triple-bonded carbon atoms in alkynes are sp-hybridized. This explains the 180° bond angles about these atoms. The arrangement of sp-hybridized orbitals and unhybridized p orbitals in C≡C triple bonds explains the resistance of such bonds to torsion as well. The central carbon atom in allene is also sp-hybridized. Each of its C=C double bonds involves one of its sp-hybrids and one unhybridized p orbital. Its two unhybridized p orbitals are oriented along perpendicular axes, which accounts for the two CH_2 groups to which it is attached lying in different planes.

D22.5 Resonance refers to the superposition of the wavefunctions representing different electron distributions in the same nuclear framework. The wavefunction resulting from the superposition is called a resonance hybrid. Resonance allows us to understand the true electron distribution (or rather a good approximation to it) in terms of contributions of various limiting structures readily pictured and understood in the context of chemical concepts. For example, we can understand polar covalent bonds in terms of greater or lesser amounts of covalent and ionic character in a resonance hybrid wavefunction, or we can understand the properties of a structure that does not correspond to a single Lewis structure as a hybrid of component wavefunctions, each of which *does* correspond to a single Lewis structure.

Exercises

E22.1(a) Let s represent the H1s atomic orbital and p_z an F2p orbital centered on the F nucleus. Then the spatial portion of wavefunction (for the bonding electrons only) would be

$$s(1)p_z(2)+s(2)p_z(1)$$

The spatial factor is symmetric with respect to interchange of the electrons' labels, so the spin factor must be antisymmetric:

$$\alpha(1)\beta(2)-\alpha(2)\beta(1).$$

So the total (unnormalized) wavefunction for this bond would be

$$\psi=\boxed{\{s(1)p_z(2)+s(2)p_z(1)\}\times\{\alpha(1)\beta(2)-\alpha(2)\beta(1)\}}$$

Comment. This simple VB function embodies a perfectly covalent H–F bond. A slightly better wavefunction, particularly for the distribution of the non-bonding valence electrons (*i.e.*, lone pairs of fluorine) would have an sp^3 hybrid rather than a p_z orbital.

E22.2(a) Let s represent the H1s atomic orbital and p_z an F2p orbital centered on the F nucleus. (As in E22.1(a), an sp^3 hybrid would be even better than the p_z.) Call the wavefunction that describes a perfectly covalent H–F bond ψ_{VB}:

$$\psi_{VB}=\{s(1)p_z(2)+s(2)p_z(1)\}\times\sigma_-$$

where σ_- is the spin wavefunction that corresponds to paired spins

$$\sigma_-=\alpha(1)\beta(2)-\alpha(2)\beta(1)$$

The wavefunction that corresponds to H^+F^- is

$$\psi_{H^+F^-}=p_z(1)p_z(2)\times\sigma_-$$

and the wavefunction that corresponds to H^-F^+ is

$$\psi_{H^+F^-}=s(1)s(2)\times\sigma_-$$

The wavefunction corresponding to the resonance hybrid is

$$\boxed{\psi=a\psi_{VB}+b\psi_{H^-F^+}+c\psi_{H^+F^-}}$$

where a, b, and c are coefficients determined by the variation principle and normalization.

Question. Based on your knowledge of the properties of H and F, which coefficient would you expect to have the larger magnitude, b or c?

E22.3(a) The valence bond description of P_2 is similar to that of N_2, a triple bond. The three bonds are a σ from the overlap of sp hybrid orbitals and two π bonds from the overlap of unhybridized 3p orbitals.

In the tetrahedral P_4 molecule there are six single P–P bonds of roughly 200 kJ mol^{-1} bond enthalpy each. So the total bonding enthalpy is roughly 1 200 kJ mol^{-1}. In the transformation

$$P_4\rightarrow 2\,P_2$$

there is a loss of about 800 kJ mol^{-1} in σ-bond enthalpy. This loss is not likely to be made up by the formation of 4 P–P π bonds. Period 3 atoms, such as P, are too large to get close enough to each other to form strong π bonds.

E22.4(a) All of the carbon atoms are sp^2 hybridized. The C–H σ bonds are formed by the overlap of Csp^2 orbitals with H1s orbitals. The C–C σ bonds are formed by the overlap of Csp^3 orbitals. The C–C π bonds are formed by the overlap of C2p orbitals. This description predicts double bonds between carbon atoms 1 & 2 and 3 & 4, but (unlike a simple molecular-orbital description of the bonding) no double-bond character between carbon atoms 2 & 3.

E22.5(a) $h_1 = \mathrm{s} + \mathrm{p}_x + \mathrm{p}_y + \mathrm{p}_z \qquad h_2 = \mathrm{s} - \mathrm{p}_x - \mathrm{p}_y + \mathrm{p}_z$

We need to evaluate

$$\int h_1 h_2 \, \mathrm{d}\tau = \int (\mathrm{s} + \mathrm{p}_x + \mathrm{p}_y + \mathrm{p}_z)(\mathrm{s} - \mathrm{p}_x - \mathrm{p}_y + \mathrm{p}_z)\mathrm{d}\tau$$

We assume that the atomic orbitals are normalized atomic orbitals which are mutually orthogonal. We expand the integrand, noting that all cross terms integrate to zero because of orthogonality. The remaining terms integrate to one, because of normalization, yielding

$$\int h_1 h_2 \, \mathrm{d}\tau = \int \mathrm{s}^2 \mathrm{d}\tau - \int \mathrm{p}_x^2 \mathrm{d}\tau - \int \mathrm{p}_y^2 \mathrm{d}\tau + \int \mathrm{p}_z^2 \mathrm{d}\tau = 1 - 1 - 1 + 1 = 0$$

E22.6(a) Normalization requires

$$\int \psi^* \psi \, \mathrm{d}\tau = 1 \quad \text{where} \quad \psi = Nh = N(\mathrm{s} + 2^{1/2}\mathrm{p})$$

Solve for the normalization constant N:

$$1 = N^2 \int (\mathrm{s} + 2^{1/2}\mathrm{p})^* (\mathrm{s} + 2^{1/2}\mathrm{p}) \mathrm{d}\tau = N^2 \int (|\mathrm{s}|^2 + 2^{1/2}\mathrm{p}^*\mathrm{s} + 2^{1/2}\mathrm{s}^*\mathrm{p} + 2|\mathrm{p}|^2)\mathrm{d}\tau = 3N^2$$

In the last step, we used the fact that the s and p orbitals are each normalized to see that the squared terms integrate to 1 and the fact that they are orthogonal to each other to see that the cross terms integrate to 0. Thus

$\boxed{N = 3^{-1/2}}$ and $\boxed{\psi = 3^{-1/2}(\mathrm{s} + 2^{1/2}\mathrm{p})}$

Problems

P22.1 Substitute explicit forms of the orbitals into the expressions, using 2s and 2p orbitals. First use eqns 18.4a and 18.4b to write the p$_x$ and p$_y$ orbitals in terms of the complex hydrogenic orbitals defined in eqn 17.11:

$$\psi = \frac{1}{3^{1/2}}\left(\mathrm{s} + \frac{1}{2^{1/2}} \times \frac{\mathrm{p}_{+1} - \mathrm{p}_{-1}}{2^{1/2}} + \frac{3^{1/2}\mathrm{i}}{2^{1/2}} \times \frac{\mathrm{p}_{+1} + \mathrm{p}_{-1}}{2^{1/2}} \right)$$

Then substitute the radial wavefunctions from Table 17.1 and spherical harmonics from Table 14.1:

$$\psi = \frac{Z^{3/2}\mathrm{e}^{-\rho/2}}{3^{1/2}a^{3/2}}\left(\frac{1}{8^{1/2}} \times \frac{2-\rho}{2\pi^{1/2}} - \frac{\rho}{2 \times 24^{1/2}} \times \frac{3^{1/2}\sin\theta}{(8\pi)^{1/2}} \times \left\{ (\mathrm{e}^{+\mathrm{i}\phi} + \mathrm{e}^{-\mathrm{i}\phi}) + 3^{1/2}\mathrm{i}(\mathrm{e}^{+\mathrm{i}\phi} - \mathrm{e}^{-\mathrm{i}\phi}) \right\} \right)$$

Pull out some common factors and combine the imaginary exponentials into sine and cosine:

$$\boxed{\psi = \frac{Z^{3/2}e^{-\rho/2}}{(24\pi)^{1/2}a^{3/2}}\left(\frac{2-\rho}{2}+\frac{\rho\sin\theta}{8^{1/2}}\times(-\cos\phi+3^{1/2}\sin\phi)\right)}$$

ρ is defined in Table 17.1, and it is proportional to r. The direction of maximum amplitude is the direction, defined by the angles θ and ϕ, in which ψ is a maximum. To find maxima, take derivatives with respect to angles and set them equal to zero. Actually, the θ dependence can be seen by inspection or by setting $\partial\psi/\partial\theta = 0$. Either way, we note that the maximum is at $\theta = \pi/2$, or in the xy plane.

$$\frac{\partial\psi}{\partial\phi} = 0 = \frac{Z^{3/2}e^{-\rho/2}}{(24\pi)^{1/2}a^{3/2}}\left(\frac{\rho\sin\theta}{8^{1/2}}\times(\sin\phi+3^{1/2}\cos\phi)\right)$$

which requires

$$\sin\phi = -3^{1/2}\cos\phi \quad \text{or} \quad \phi = \tan^{-1}(-3^{1/2}) = \boxed{120°} \text{ or } -60°$$

P22.3 One approach is to construct the explicit forms of the hybrid orbitals and find the values of ϕ that maximize the squares of their magnitudes. We need the component unhybridized orbitals, put together from the radial and angular functions listed in Tables 17.1 and 14.1. This approach is perhaps best applied to sp^2 hybrids, because the form of these hybrids has different numerical coefficients in front of the unhybridized p orbitals, not just different signs.

$$2s = R_{20}Y_{00} = \frac{1}{8^{1/2}}\left(\frac{Z}{a}\right)^{3/2}\times(2-\rho)e^{-\rho/2}\times\left(\frac{1}{4\pi}\right)^{1/2} = \left(\frac{1}{32\pi}\right)^{1/2}\left(\frac{Z}{a}\right)^{3/2}(2-\rho)e^{-\rho/2}$$

$$2p_x = -\frac{1}{2^{1/2}}R_{21}(Y_{1,+1}-Y_{1,-1}) \quad [18.4a]$$

$$= -\frac{1}{2^{1/2}}\times\frac{1}{24^{1/2}}\left(\frac{Z}{a}\right)^{3/2}\rho e^{-\rho/2}\times\left(\frac{3}{8\pi}\right)^{1/2}\sin\theta(-e^{+i\phi}-e^{-i\phi})$$

$$= \left(\frac{1}{32\pi}\right)^{1/2}\left(\frac{Z}{a}\right)^{3/2}\rho e^{-\rho/2}\sin\theta\cos\phi$$

$$2p_y = \frac{i}{2^{1/2}}R_{21}(Y_{1,+1}+Y_{1,-1}) \quad [18.4b]$$

$$= \frac{i}{2^{1/2}}\times\frac{1}{24^{1/2}}\left(\frac{Z}{a}\right)^{3/2}\rho e^{-\rho/2}\times\left(\frac{3}{8\pi}\right)^{1/2}\sin\theta(-e^{+i\phi}+e^{-i\phi})$$

$$= \left(\frac{1}{32\pi}\right)^{1/2}\left(\frac{Z}{a}\right)^{3/2}\rho e^{-\rho/2}\sin\theta\sin\phi$$

where $\rho = \frac{2Zr}{2a} = \frac{Zr}{a}$

In forming each hybrid, we neglect the factor $\left(\frac{1}{32\pi}\right)^{1/2}\left(\frac{Z}{a}\right)^{3/2} e^{-\rho/2}$ common to each component; an angle-independent multiplicative term cannot influence the angle at which the hybrid is maximal.

Next, form the hybrids, using eqn 22.7:

$$h_1 = s + 2^{1/2}p_y = (2-\rho) + 2^{1/2}(\rho \sin\theta \sin\phi) = 2 + \rho(2^{1/2}\sin\theta\sin\phi - 1)$$

$$h_2 = s + \left(\frac{3}{2}\right)^{1/2} p_x - \left(\frac{1}{2}\right)^{1/2} p_y = (2-\rho) + \left(\frac{3}{2}\right)^{1/2} \rho\sin\theta\cos\phi - \left(\frac{1}{2}\right)^{1/2} \rho\sin\theta\sin\phi$$

$$= 2 + \rho\left(\frac{3^{1/2}\cos\phi - \sin\phi}{2^{1/2}}\sin\theta - 1\right)$$

and $$h_3 = s - \left(\frac{3}{2}\right)^{1/2} p_x - \left(\frac{1}{2}\right)^{1/2} p_y = (2-\rho) - \left(\frac{3}{2}\right)^{1/2} \rho\sin\theta\cos\phi - \left(\frac{1}{2}\right)^{1/2} \rho\sin\theta\sin\phi$$

$$= 2 - \rho\left(\frac{3^{1/2}\cos\phi + \sin\phi}{2^{1/2}}\sin\theta + 1\right)$$

To find the angle ϕ at which the hybrids have maximum probability, differentiate with respect to ϕ, and set the derivative equal to zero, because positive maxima and negative minima in the hybrid orbitals correspond to maxima in the probability. Differentiation would work even for h_1, but it is unnecessary there. One can see by inspection that the function is maximized when $\sin\phi$ is maximal, namely at $\phi = \pi/2$ (90°). It should come as no surprise that this orbital points along the positive y-axis.

$$\frac{\partial h_2}{\partial\phi} = 0 = \left(\frac{\rho\sin\theta}{2^{1/2}}\right)\left(-3^{1/2}\sin\phi - \cos\phi\right)$$

so $$\frac{\sin\phi}{\cos\phi} = -\frac{1}{3^{1/2}} = \tan\phi \quad \text{or} \quad \phi = 5\pi/6 \text{ or } 11\pi/6 \text{ (150° or 330°)}.$$

Of these, $11\pi/6$ (330°) is a maximum and $5\pi/6$ a minimum. The larger amplitude is at the maximum ($11\pi/6$ or 330°). Finally

$$\frac{\partial h_3}{\partial\phi} = 0 = -\left(\frac{\rho\sin\theta}{2^{1/2}}\right)\left(-3^{1/2}\sin\phi + \cos\phi\right)$$

so $$\frac{\sin\phi}{\cos\phi} = \frac{1}{3^{1/2}} = \tan\phi \quad \text{or} \quad \phi = \pi/6 \text{ or } 7\pi/6 \text{ (30° or 210°)}.$$

Of these, $7\pi/6$ (210°) is a maximum and corresponds to the larger amplitude. The angles 90°, 210°, and 330° are 120° apart.

So much for the case $\lambda = 2$. The cases $\lambda = 1$ and 3 can be treated from the functional forms of the hybrids and the facts that the unhybridized p orbitals are maximally positive along positive Cartesian axes and maximally negative along the negative Cartesian axes. Since s orbitals have constant positive values for all angles, we can figure out the direction of the hybrid orbitals by figuring out the direction in which the component p orbitals reinforce the s contribution.

The sp hybrids are

$$h_1 = s + p_z \quad \text{and} \quad h_2 = s - p_z \quad [22.8]$$

Therefore, h_1 points in the same direction as the positive lobe of the p_z orbital (i.e., the positive z axis) and h_2 points in the same direction as the negative lobe of the p_z orbital (i.e., the negative z axis, where we have positive s – negative p_z). So equivalent sp orbitals point in directions that differ by 180°.

Finally, consider the sp^3 orbitals (eqn 22.5)

$$h_1 = s + p_x + p_y + p_z \qquad h_2 = s - p_x - p_y + p_z$$
$$h_3 = s - p_x + p_y - p_z \qquad h_4 = s + p_x - p_y - p_z$$

As explained in *Justification 22.2*, we know that h_1 points to the Cartesian point (+1, +1, +1), because the three p orbitals have equal weight and it is their positive lobes that reinforce the positive s contribution. By similar reasoning, h_2 points to (−1, −1, +1), h_3 to (−1, +1, −1), and h_4 to (+1, −1, −1). As illustrated in Fig. 22.1, these points fall on the vertices of a regular tetrahedron.

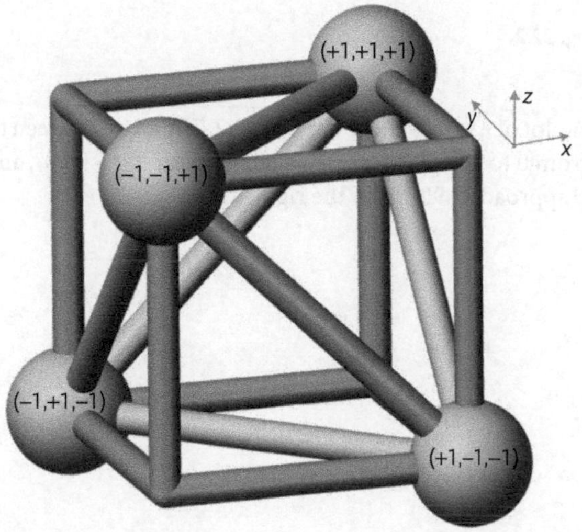

As noted in the text, the angle between the corresponding hybrid orbitals is, therefore, the tetrahedral angle, arccos(−1/3) = 109.47°. Rearranging this equation, we see that

$$-1/3 = \cos 109.47° \qquad \text{or} \qquad -1/(\cos 109.47°) = 3$$

Similarly

$$-1/\cos 120° = 2 \qquad \text{and} \qquad -1/\cos 180° = 1$$

so for these sp^λ hybrids, equivalent orbitals are separated by an angle θ such that

$$\lambda = -1/\cos\theta$$

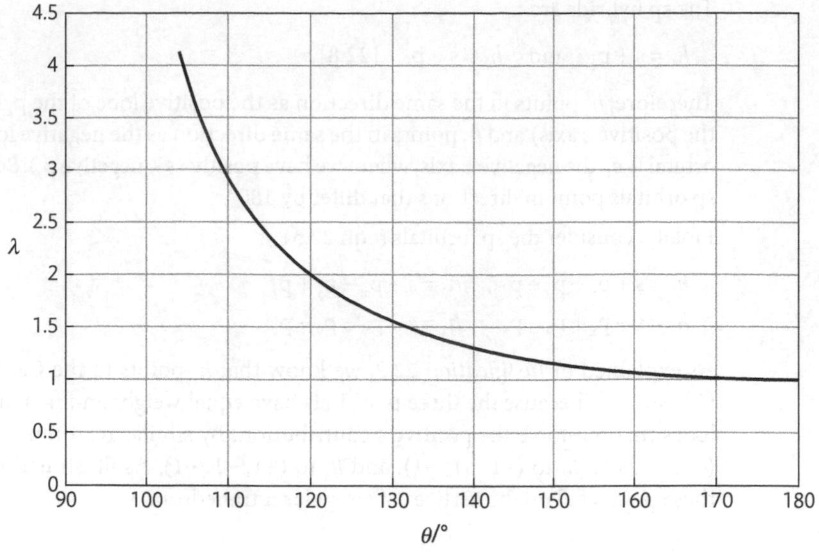

Fig. 22.2

A plot of λ vs. θ is shown in Fig. 22.2. The angle between two orbitals can only range from 0 to 180°; however, λ is negative for $0 < \theta < 90°$, and λ approaches infinity as θ approaches 90° from the right.

Topic 23 **The Principles of molecular orbital theory**

Discussion question

D23.1 As stated in Topic 23.1(b), the cause of reduced energy of bonding molecular orbitals compared to atomic orbitals is not entirely clear, although what is clear is that bonding character correlates strongly with molecular orbitals that have an accumulation of electron density between nuclei due to overlap and constructive interference of their component atomic orbitals. A simple and plausible explanation of this correlation is that enhanced electron probability between nuclei lowers potential energy by putting electrons in a position where they can be attracted to two nuclei at the same time; however, the source of the reduced energy appears to be more complicated.

Exercises

E23.1(a)

$$\int \psi^2 \,\mathrm{d}\tau = N^2 \int (\psi_\mathrm{A} + \lambda\psi_\mathrm{B})^2 \,\mathrm{d}\tau = N^2 \int (\psi_\mathrm{A}^2 + \lambda^2\psi_\mathrm{B}^2 + 2\lambda\psi_\mathrm{A}\psi_\mathrm{B})\mathrm{d}\tau = 1$$

$$= N^2(1+\lambda^2+2\lambda S) \quad \left[\int \psi_\mathrm{A}\psi_\mathrm{B}\,\mathrm{d}\tau = S\right]$$

Hence $\boxed{N = \left(\dfrac{1}{1+2\lambda S+\lambda^2}\right)^{1/2}}$

E23.2(a) Let $\psi_1 = N(0.145A + 0.844B)$ and $\psi_2 = aA + bB$

First, let us normalize ψ_1:

$$\int \psi_1{}^*\psi_1 \,\mathrm{d}\tau = 1 = N^2 \int (0.145A + 0.844B)^*(0.145A + 0.844B)\mathrm{d}\tau,$$

$$0.0210\int |A|^2 \,\mathrm{d}\tau + 0.712\int |B|^2 \,\mathrm{d}\tau + 0.122\int B^* A \,\mathrm{d}\tau + 0.122\int A^* B \,\mathrm{d}\tau = \frac{1}{N^2}$$

The first two integrals are 1 due to normalization and the latter two are the overlap integral, $S = 0.250$. So

$$0.210 + 0.712 + 2\times0.122\times0.250 = \frac{1}{N^2} = 0.795,$$

so $\boxed{N = 1.12}$ which makes $\boxed{\psi_1 = 0.163A + 0.947B}$

Orthogonality of the two molecular orbitals requires

$$\int \psi_1{}^* \psi_2 \, d\tau = 0 = N \int (0.145A + 0.844B)^* (aA + bB) d\tau$$

Dividing by N yields

$$0.145a \int |A|^2 \, d\tau + 0.844b \int |B|^2 \, d\tau + 0.844a \int B^* A d\tau + 0.145b \int A^* B d\tau = 0,$$

$$0.145a + 0.844b + 0.844a \times 0.250 + 0.145b \times 0.250 = 0 = 0.356a + 0.880b$$

Normalization of ψ_2 requires

$$\int \psi_2{}^* \psi_2 \, d\tau = 1 = \int (aA + bB)^* (aA + bB) d\tau \,,$$

$$a^2 \int |A|^2 \, d\tau + b^2 \int |B|^2 \, d\tau + ab \int B^* A d\tau + ab \int A^* B d\tau = 1$$

So $a^2 + b^2 + 2ab \times 0.250 = 1 = a^2 + b^2 + 0.500ab$

We have two equations in the two unknown coefficients a and b. Solve the first equation for a in terms of b:

$$a = -0.880b/0.356 = -2.47b$$

Substitute this result into the second (quadratic) equation:

$$1 = (-2.47b)^2 + b^2 + 0.500(-2.47b)b = 5.88b^2$$

So $\boxed{b = 0.412}$, $\boxed{a = -1.02}$, and $\boxed{\psi_2 = -1.02A + 0.412B}$

E23.3(a) $E_H = E_1 = -hcR_H$ [17.12(a)]

Draw up the following table using the data in the question and using

$$\frac{e^2}{4\pi\varepsilon_0 R} = \frac{e^2}{4\pi\varepsilon_0 a_0} \times \frac{a_0}{R} = \frac{e^2}{4\pi\varepsilon_0 \times (4\pi\varepsilon_0 \hbar^2 / m_e e^2)} \times \frac{a_0}{R}$$

$$= \frac{m_e e^4}{16\pi^2 \varepsilon_0^2 \hbar^2} \times \frac{a_0}{R} = E_h \times \frac{a_0}{R} \quad \left[E_h \equiv \frac{m_e e^4}{16\pi^2 \varepsilon_0^2 \hbar^2} = 2hcR_H \right]$$

so that $\dfrac{e^2}{4\pi\varepsilon_0 R} \times \dfrac{1}{E_h} = \dfrac{a_0}{R}$

R/a_0	0	1	2	3	4	∞
$\frac{e^2}{4\pi\varepsilon_0 R}\times\frac{1}{E_h}$	∞	1	0.500	0.333	0.250	0
$(V_1+V_2)/E_h$	2.000	1.465	0.843	0.529	0.342	0
$(E-E_H)/E_h$	∞	0.212	−0.031	−0.059	−0.038	0

The points are plotted in Fig. 23.1.

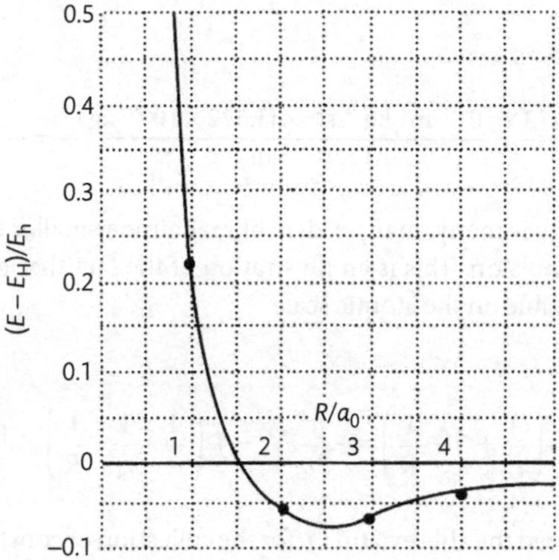

Fig 23.1

The minimum occurs at $R = 2.5a_0$, so $R = 130$ pm. At that bond length

$$E - E_H = -0.07E_h = -1.91\,\text{eV}$$

Hence, the dissociation energy is predicted to be about $\boxed{1.9\ \text{eV}}$ and the equilibrium bond length about $\boxed{130\ \text{pm}}$.

E23.4(a) Fig. 23.2 shows sketches of the orbitals in question. By inspection, one can see that the bonding orbital is odd, $\boxed{u}$, with respect to inversion, and the antibonding orbital is even, $\boxed{g}$.

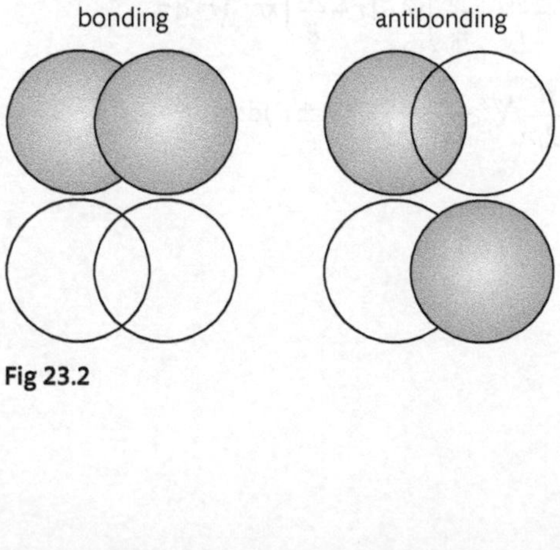

Fig 23.2

Problems

P23.1 The energy of repulsion is given by Coulomb's law:

$$E=\frac{e^2}{4\pi\varepsilon_0 R}=\frac{(1.602\times10^{-19}\ \mathrm{C})^2}{4\pi\times8.854\times10^{-12}\ \mathrm{J^{-1}\ C^2\ m^{-1}}\times74.1\times10^{-12}\ \mathrm{m}}=3.11\times10^{-18}\ \mathrm{J}$$

The repulsion energy of a mole of H_2 is

$$E_{\text{molar}}=N_A E=6.022\times10^{23}\ \mathrm{mol^{-1}}\times3.11\times10^{-18}\ \mathrm{J}=\boxed{1.87\times10^{6}\ \mathrm{J\ mol^{-1}}=1.87\ \mathrm{MJ\ mol^{-1}}}$$

The gravitational energy is

$$E_{\text{grav}}=\frac{-Gm_1m_2}{R}=\frac{-6.673\times10^{-11}\ \mathrm{m^3\ kg^{-1}\ s^{-2}}\times(1.672\times10^{-27}\ \mathrm{kg})^{-2}}{74.1\times10^{-12}\ \mathrm{m}}=-2.52\times10^{-54}\ \mathrm{J}$$

The gravitational energy is many, many orders of magnitude smaller than the energy of electrostatic repulsion. This is an illustration of the fact that gravitational effects tend to be negligible on the atomic scale.

P23.3 The hamiltonian is

$$\hat{H}=-\frac{\hbar^2}{2m}\nabla^2-\frac{e^2}{4\pi\varepsilon_0}\left(\frac{1}{r_A}+\frac{1}{r_B}-\frac{1}{R}\right)=-\frac{\hbar^2}{2m}\nabla^2-j_0\left(\frac{1}{r_A}+\frac{1}{r_B}-\frac{1}{R}\right)\quad[23.1]$$

where we have introduced the abbreviation j_0 for the collection of constants $e^2/4\pi\varepsilon_0$. The expectation value of this operator is

$$\langle H\rangle=\int\psi^*\hat{H}\psi\,\mathrm{d}\tau,$$

where $\psi_\pm=N_\pm(A\pm B)$ [23.2] and $N_\pm=\dfrac{1}{\{2(1+S)\}^{1/2}}$ [Example 23.1]

Thus, $$\langle H\rangle=\int\psi_\pm{}^*\left\{-\frac{\hbar^2}{2m}\nabla^2-j_0\left(\frac{1}{r_A}+\frac{1}{r_B}-\frac{1}{R}\right)\right\}\psi_\pm\,\mathrm{d}\tau$$

Pull out the nuclear repulsion term

$$\langle H\rangle=\int\psi_\pm{}^*\left(-\frac{\hbar^2}{2m}\nabla^2-\frac{j_0}{r_A}-\frac{j_0}{r_B}\right)\psi_\pm\,\mathrm{d}\tau+\frac{j_0}{R}\int\psi_\pm{}^*\psi_\pm\,\mathrm{d}\tau$$

$$=N_\pm^2\int(A\pm B)\left(-\frac{\hbar^2}{2m}\nabla^2-\frac{j_0}{r_A}-\frac{j_0}{r_B}\right)(A\pm B)\mathrm{d}\tau+\frac{j_0}{R}$$

Expand the remaining integral into four integrals, letting $\hat{H}_{\rm H}$ represent the hamiltonian operator of a hydrogen atom. Note that

$$\hat{H}_{\rm H}A=-\frac{\hbar^2}{2m}\nabla^2 A-\frac{j_0}{r_{\rm A}}A=E_{\rm H1s}A \quad \text{and} \quad \hat{H}_{\rm H}B=-\frac{\hbar^2}{2m}\nabla^2 B-\frac{j_0}{r_{\rm B}}B=E_{\rm H1s}B$$

So
$$\langle H\rangle=N_\pm^2\int A\left(\hat{H}_{\rm H}-\frac{j_0}{r_{\rm B}}\right)A\,{\rm d}\tau\pm N_\pm^2\int A\left(\hat{H}_{\rm H}-\frac{j_0}{r_{\rm A}}\right)B\,{\rm d}\tau$$

$$\pm N_\pm^2\int B\left(\hat{H}_{\rm H}-\frac{j_0}{r_{\rm B}}\right)A\,{\rm d}\tau+N_\pm^2\int B\left(\hat{H}_{\rm H}-\frac{j_0}{r_{\rm A}}\right)B\,{\rm d}\tau+\frac{j_0}{R}$$

and
$$\langle H\rangle=N_\pm^2\int A\left(E_{\rm H1s}-\frac{j_0}{r_{\rm B}}\right)A\,{\rm d}\tau\pm N_\pm^2\int A\left(E_{\rm H1s}-\frac{j_0}{r_{\rm A}}\right)B\,{\rm d}\tau$$

$$\pm N_\pm^2\int B\left(E_{\rm H1s}-\frac{j_0}{r_{\rm B}}\right)A\,{\rm d}\tau+N_\pm^2\int B\left(E_{\rm H1s}-\frac{j_0}{r_{\rm A}}\right)B\,{\rm d}\tau+\frac{j_0}{R}$$

Pull out the terms involving $E_{\rm H1s}$:

$$\langle H\rangle=E_{\rm H1s}N_\pm^2\int(A^2\pm 2AB+B^2)\,{\rm d}\tau$$
$$-j_0N_\pm^2\int\frac{A^2}{r_{\rm B}}{\rm d}\tau\mp j_0N_\pm^2\int\frac{AB}{r_{\rm A}}{\rm d}\tau\mp j_0N_\pm^2\int\frac{AB}{r_{\rm B}}{\rm d}\tau-j_0N_\pm^2\int\frac{B^2}{r_{\rm A}}{\rm d}\tau+\frac{j_0}{R}$$

What multiplies $E_{\rm H1s}$ is the normalization integral, which equals 1. Now examine the remaining four integrals. The first and fourth of these are equal to each other; likewise for the second and third. (This can be demonstrated by interchanging the labels A and B, which are arbitrary.) Thus

$$\langle H\rangle=E_{\rm H1s}-2N_\pm^2\left(j_0\int\frac{A^2}{r_{\rm B}}{\rm d}\tau\pm j_0\int\frac{AB}{r_{\rm A}}{\rm d}\tau\right)+\frac{j_0}{R}$$
$$=E_{\rm H1s}-2N_\pm^2(j\pm k)+\frac{j_0}{R}=E_{\rm H1s}-\frac{j\pm k}{1\pm S}+\frac{j_0}{R}$$

where we have used the definitions of j and k in eqn 23.5. The expectation values work out to eqns 23.4 and 23.7:

$$\langle H\rangle_+=\boxed{E_{\rm H1s}-\frac{j+k}{1+S}+\frac{j_0}{R}}=E_{1\sigma} \quad \text{and} \quad \langle H\rangle_-=\boxed{E_{\rm H1s}-\frac{j-k}{1-S}+\frac{j_0}{R}}=E_{2\sigma}$$

P23.5 (a) Start from eqn 23.4 (whose derivation is covered in Problem 23.3).

$$E_{1\sigma}=E_{\rm H1s}-\frac{j+k}{1+S}+\frac{j_0}{R}$$

where we define the Coulomb integral j and the resonance integral k as in eqns 23.5 and use the relationship of the normalization constant N to the overlap integral S as defined in Example 23.1. We must write the integrals S, j, and k explicitly to evaluate them. Use

$$A=\frac{e^{-r_A/a_0}}{(\pi a_0^3)^{1/2}} \quad \text{and} \quad B=\frac{e^{-r_B/a_0}}{(\pi a_0^3)^{1/2}} \quad [\textit{Brief illustration 23.1}]$$

Define a coordinate system centered on nucleus A, so that r_A is a coordinate and

$$r_B=\{x^2+y^2+(z-R)^2\}^{1/2}\ [\textit{Brief illustration 23.1}]=(x^2+y^2+z^2+R^2-2zR)^2$$

$$r_B=(r_A{}^2+R^2-2zR)^{1/2}=(r_A{}^2+R^2-2r_AR\cos\theta)^{1/2}$$

The volume element is

$$d\tau=r_A{}^2\sin\theta\, dr_A\, d\theta\, d\phi$$

Thus, the overlap integral is

$$S=\int AB\,d\tau=\int_0^{2\pi}\int_0^{\pi}\int_0^{\infty}\frac{e^{-r_A/a_0}}{(\pi a_0^3)^{1/2}}\frac{e^{-r_B/a_0}}{(\pi a_0^3)^{1/2}}r_A^2\sin\theta\, dr_A\, d\theta\, d\phi$$

Integration over $d\phi$ yields a factor of 2π. Numerical integration over $d\theta$ and dr_A can be done by approximating the differentials by small but finite $\delta\theta$ and δr_A and summing over a series of θ_j and $r_{A,i}$:

$$S=\frac{2}{a_0^3}\int_0^{\pi}\int_0^{\infty}e^{-r_A/a_0}e^{-r_B/a_0}r_A^2\sin\theta\, dr_A\, d\theta\approx 2\delta\theta\frac{\delta r_A}{a_0}\sum_j\sum_i e^{-r_{A,i}/a_0}e^{-r_{B,ij}/a_0}\left(\frac{r_{A,i}}{a_0}\right)^2\sin\theta_j$$

Similar treatment of the other integrals yields

$$j\approx 2\delta\theta\frac{\delta r_A}{a_0}\times\frac{j_0}{a_0}\sum_j\sum_i\frac{e^{-2r_{A,i}/a_0}}{(r_{B,ij}/a_0)}\left(\frac{r_{A,i}}{a_0}\right)^2\sin\theta_j$$

and $$k\approx 2\delta\theta\frac{\delta r_A}{a_0}\times\frac{j_0}{a_0}\sum_j\sum_i\frac{e^{-r_{A,i}/a_0}e^{-r_{B,ij}/a_0}}{(r_{B,ij}/a_0)}\left(\frac{r_{A,i}}{a_0}\right)^2\sin\theta_j$$

To evaluate the integrals, express the distances in units of a_0. S is dimensionless. j and k have dimensions of energy; they have units of j_0/a_0. In the following table and plots, the numerical values of the integrals were computed using θ_j from 0 to π in increments that correspond to 10° and $r_{A,i}$ from 0 to $5a_0$ in increments of $0.25a_0$.

A plot of S vs. R is shown in Fig. 23.3(a) and a plot of j and k vs. R is shown in Fig. 23.3(b). Note that the numerical results at this level of approximation agree

R/a_0	S_{num}	$\frac{j}{j_0/a_0}$	$\frac{k}{j_0/a_0}$	S_{anal}	$\frac{j_{anal}}{j_0/a_0}$	$\frac{k_{anal}}{j_0/a_0}$
1.00	0.8524	0.7227	0.7220	0.8584	0.7293	0.7358
1.25	0.7875	0.6460	0.6287	0.7939	0.6522	0.6446
1.50	0.7183	0.5781	0.5408	0.7252	0.5837	0.5578
1.75	0.6480	0.5192	0.4606	0.6553	0.5240	0.4779
2.00	0.5789	0.4684	0.3890	0.5865	0.4725	0.4060
2.25	0.5125	0.4249	0.3262	0.5204	0.4284	0.3425
2.50	0.4501	0.3876	0.2718	0.4583	0.3906	0.2873
2.75	0.3924	0.3555	0.2253	0.4009	0.3581	0.2397
3.00	0.3397	0.3279	0.1858	0.3485	0.3300	0.1991
3.25	0.2922	0.3038	0.1525	0.3013	0.3057	0.1648
3.50	0.2498	0.2829	0.1245	0.2592	0.2845	0.1359
3.75	0.2122	0.2644	0.1012	0.2219	0.2660	0.1117
4.00	0.1791	0.2482	0.0817	0.1893	0.2496	0.0916

well but not identically with the analytical ones. The disagreement becomes more noticeable at larger *R*, which is not surprising given that all sums were truncated at $r_A = 5a_0$. The disagreement is more pronounced in *k* than in *S* or *j*. The numerical results can be made arbitrarily close to the analytical by using smaller increments and extending the range of r_A.

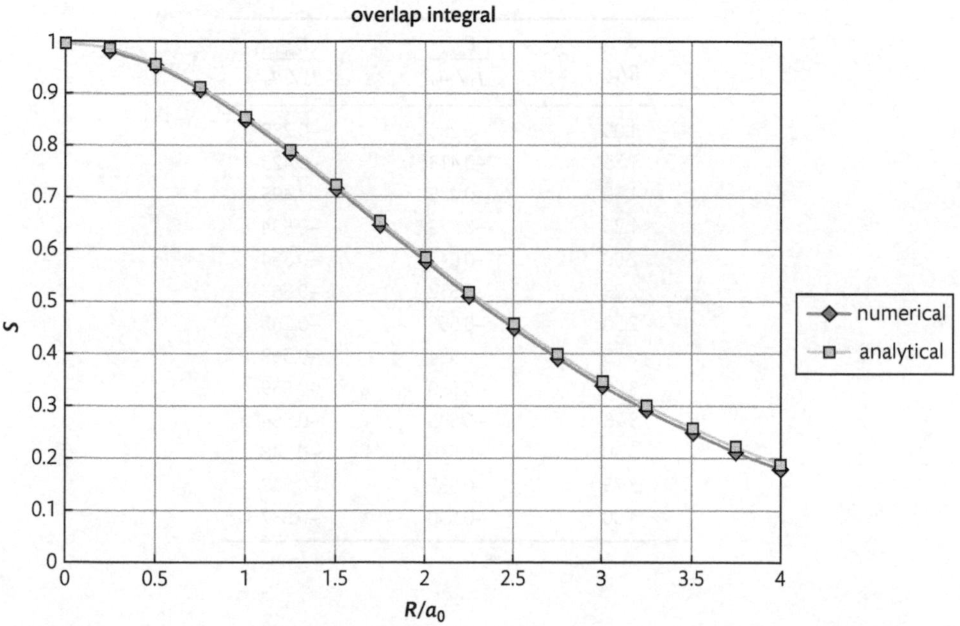

Fig 23.3(a)

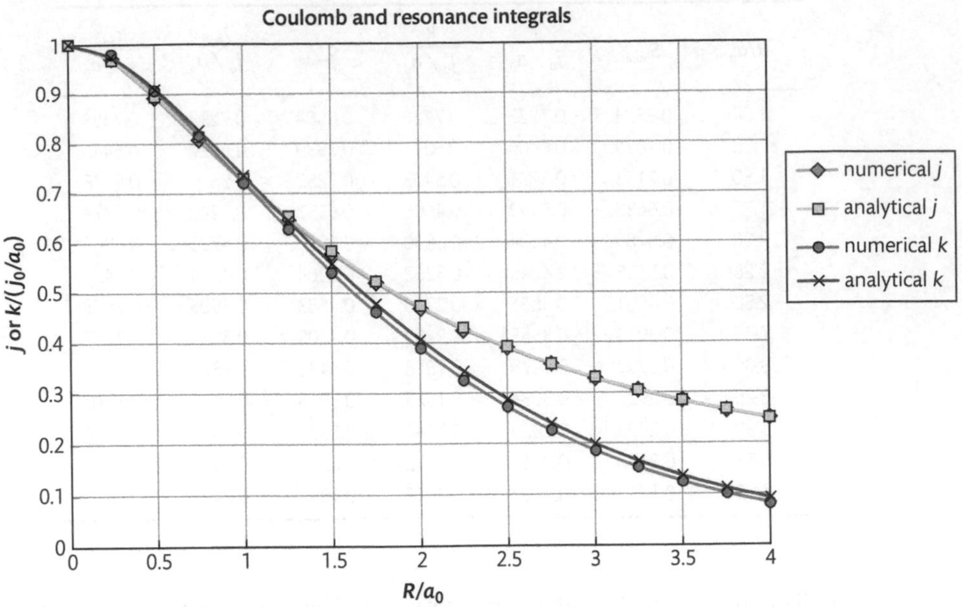

Fig 23.3(b)

(b) Use eqn 23.4 to compute the total energy, whether from numerical or analytical values of the integrals. Results are shown in the table below and in Fig. 23.3(c).

R/a_0	$\dfrac{E_{\text{num}}}{j_0/a_0}$	$\dfrac{E_{\text{anal}}}{j_0/a_0}$
1.00	–0.280	–0.288
1.25	–0.413	–0.423
1.50	–0.485	–0.495
1.75	–0.523	–0.534
2.00	–0.543	–0.554
2.25	–0.552	–0.563
2.50	–0.555	–0.565
2.75	–0.554	–0.563
3.00	–0.550	–0.559
3.25	–0.545	–0.554
3.50	–0.540	–0.548
3.75	–0.535	–0.542
4.00	–0.530	–0.537

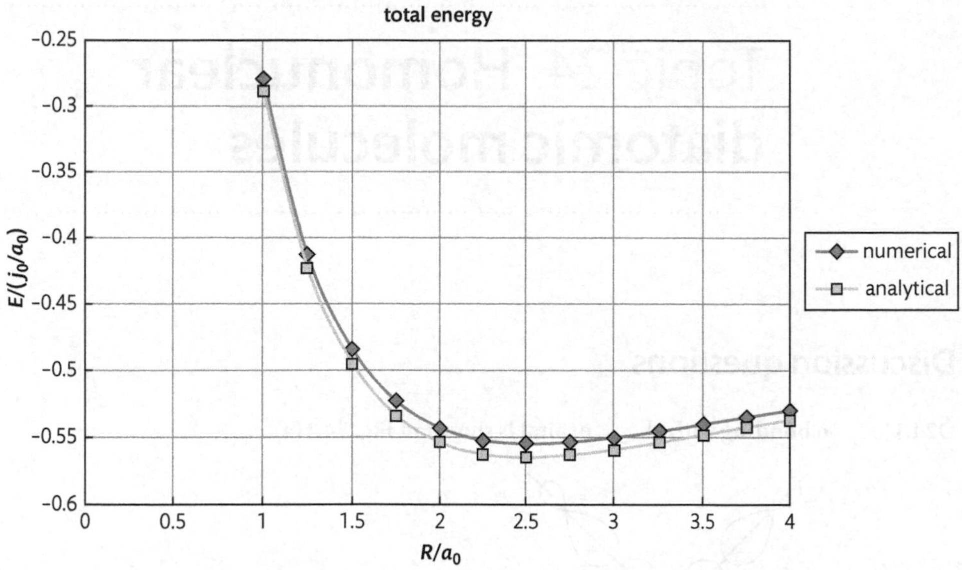

Fig 23.3(c)

The minimum in both total energy curves lies at $R = \boxed{2.5a_0 = 1.3 \times 10^{-10}\ \text{m}}$. The minimum total energy is $E_{\text{min}} = \boxed{-0.555j_0/a_0 = -15.1\ \text{eV}}$ (numerical) and $\boxed{-0.565j_0/a_0 = -15.4\ \text{eV}}$ (analytical). The bond dissociation energy is

$$D_e = E_H - E_{\text{min}} = (-0.500 + 0.555)j_0/a_0 = \boxed{0.055j_0/a_0 = 1.5\ \text{eV}}\ \text{(numerical)}$$

$$D_e = (-0.500 + 0.565)j_0/a_0 = \boxed{0.065j_0/a_0 = 1.8\ \text{eV}}\ \text{(analytical)}$$

Comment. Dissociation energies are typically (relatively) small differences between larger quantities (electron binding energies). Even small errors in the latter lead to relatively large errors in the former. In this case, the numerical binding energy differed by about 2% from the analytical value; the difference in D_e was closer to 20%.

Topic 24 **Homonuclear diatomic molecules**

Discussion questions

D24.1 **σ bonding with $d_{x^2-y^2}$ orbital is shown in Fig. 24.1(a).**

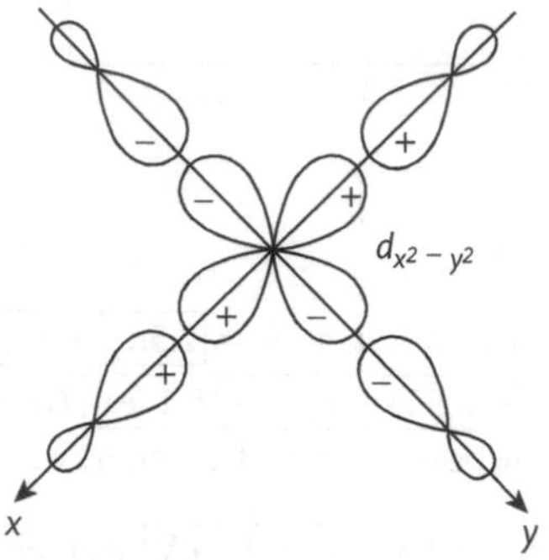

Fig 24.1(a)

The σ antibonding orbital looks the same but with the p orbital lobes pointed in the opposite direction.

***σ* bonding with the d_{z^2} orbital is shown in Figure 24.1(b).**

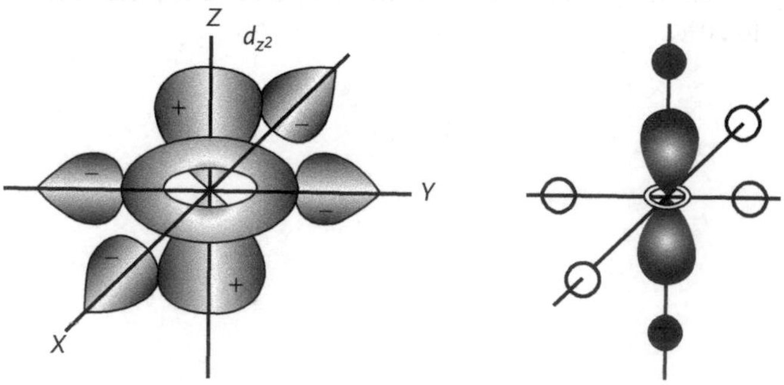

Fig 24.1(b)

In this figure, only one of the p-orbital lobes is shown in each p orbital. p orbitals with positive lobes may also approach this orbital along the + and – z-direction as indicated in the smaller diagram on the right. The antibonding diagrams are similar, but with the signs of the p orbital lobes reversed.

***π* bonding (See Figure 5.1(c).)**

Only the d_{xy}, d_{yz}, and d_{xz} orbitals undergo π bonding with p orbitals on neighboring atoms. The bonding arrangement is pictured below with the d_{xy} orbital. The diagrams for the d_{yz} and d_{xz} orbitals are similar. The antibonding diagrams have the signs of the p orbital lobes reversed.

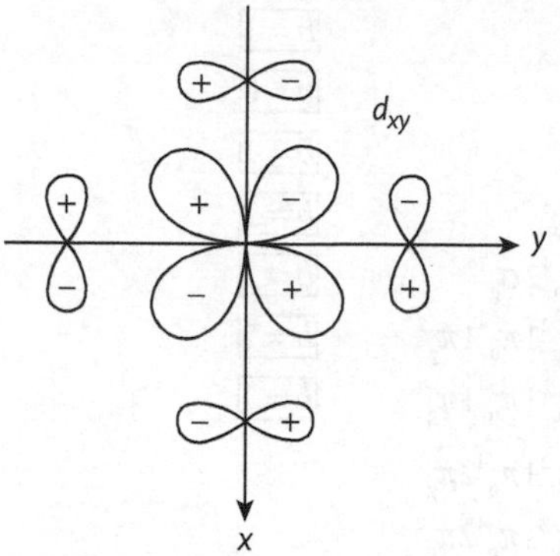

Fig 24.1(c)

D24.3 The Born-Oppenheimer approximation treats the nuclei of the multi-particle system of electrons and nuclei as if they were fixed. The dependence of energy on nuclear positions is then obtained by solving the Schrödinger equation at many different (fixed) nuclear geometries. In molecular orbital theory, specifically, one forms molecular orbitals from linear combinations of atomic orbitals within a fixed framework of nuclei, and then at a slightly different fixed framework, and so on until one finds the framework of nuclei that yields the lowest-energy wavefunction.

D24.5 All other things being equal, bond strength and bond length are inversely related: the shorter the bond, the stronger the bond. (For example, think of carbon-carbon single, double, and triple bonds and their bond dissociation enthalpies. Or consider the lengths and strengths of the single bond in Cl_2, Br_2, and I_2.) Atomic orbital overlap is similarly related to bond length (or internuclear distance): the smaller the internuclear distance, the greater the overlap.

Exercises

E24.1(a) Refer to Fig. 24.10 of the text. Place two of the valence electrons in each orbital starting with the lowest-energy orbital, until all valence electrons are used up. Apply Hund's rule to the filling of degenerate orbitals.

(a) Li_2 (2 electrons) $\boxed{1\sigma_g^2, b=1}$

(b) Be_2 (4 electrons) $\boxed{1\sigma_g^2 1\sigma_u^2, b=0}$

(c) C_2 (8 electrons) $\boxed{1\sigma_g^2 1\sigma_u^2 1\pi_u^4, b=2}$

E24.2(a) B_2 (6 electrons): $1\sigma_g^2 1\sigma_u^2 1\pi_u^2$ $b=1$

C_2 (8 electrons): $1\sigma_g^2 1\sigma_u^2 1\pi_u^4$ $b=2$

The bond orders of B_2 and C_2 are respectively 1 and 2; so $\boxed{C_2}$ should have the greater bond dissociation enthalpy. The experimental values are approximately 4eV and 6eV respectively.

E24.3(a) F_2^+ (13 electrons) $1\sigma_g^2 1\sigma_u^2 2\sigma_g^2 1\pi_u^4 1\pi_g^3$, $b = 1.5$

F_2 (14 electrons) $1\sigma_g^2 1\sigma_u^2 2\sigma_g^2 1\pi_u^4 1\pi_g^4$, $b = 1$

The bond orders of F_2^+ and F_2 are respectively 1.5 and 1; so $\boxed{F_2^+}$ should have the greater bond dissociation energy.

E24.4(a)

Li_2 (2 electrons):	$1\sigma_g^2$	$\boxed{b=1}$
Be_2 (4 electrons):	$1\sigma_g^2 1\sigma_u^2$	$\boxed{b=0}$
B_2 (6 electrons):	$1\sigma_g^2 1\sigma_u^2 1\pi_u^2$	$\boxed{b=1}$
C_2 (8 electrons):	$1\sigma_g^2 1\sigma_u^2 1\pi_u^4$	$\boxed{b=2}$
N_2 (10 electrons):	$1\sigma_g^2 1\sigma_u^2 1\pi_u^4 2\sigma_g^2$	$\boxed{b=3}$
O_2 (12 electrons):	$1\sigma_g^2 1\sigma_u^2 2\sigma_g^2 1\pi_u^4 1\pi_g^2$	$\boxed{b=2}$
F_2 (14 electrons):	$1\sigma_g^2 1\sigma_u^2 2\sigma_g^2 1\pi_u^4 1\pi_g^4$	$\boxed{b=1}$

E24.5(a)

O_2^+ (11 electrons)	$1\sigma_g^2 1\sigma_u^2 2\sigma_g^2 1\pi_u^4 2\pi_g^1$
O_2 (12 electrons)	$1\sigma_g^2 1\sigma_u^2 2\sigma_g^2 1\pi_u^4 2\pi_g^2$
O_2^- (13 electrons)	$1\sigma_g^2 1\sigma_u^2 2\sigma_g^2 1\pi_u^4 2\pi_g^3$
O_2^{2-} (14 electrons)	$1\sigma_g^2 1\sigma_u^2 2\sigma_g^2 1\pi_u^4 2\pi_g^4$

In each case, the HOMO is an antibonding $\boxed{2\pi_g}$ orbital.

E24.6(a) Energy is conserved, so when the photon is absorbed, its energy is transferred to the electron. Part of that energy overcomes the binding energy (ionization energy) and the remainder is manifest as the now freed electron's kinetic energy.

$$E_{\text{photon}} = I + E_{\text{kinetic}}$$

$$\text{so } E_{\text{kinetic}} = E_{\text{photon}} - I = \frac{hc}{\lambda} - I = \frac{(6.626\times10^{-34}\text{ J s})\times(2.998\times10^{8}\text{ m s}^{-1})}{(100\times10^{-9}\text{ m})\times(1.602\times10^{-19}\text{ J eV}^{-1})} - 12.0\text{ eV}$$
$$= 0.4\bar{0}\text{ eV} = 6.\bar{4}\times10^{-20}\text{ J}.$$

The speed is obtained from the kinetic energy:

$$E_{\text{kinetic}} = \frac{mv^2}{2} \quad \text{so} \quad v = \sqrt{\frac{2E_{\text{kinetic}}}{m}} = \sqrt{\frac{2\times6.\bar{4}\times10^{-20}\text{ J}}{9.11\times10^{-31}\text{ kg}}} = \boxed{4\times10^{5}\text{ m s}^{-1}}$$

E24.7(a) The expression in eqn 24.4 cannot be solved analytically, but solving it numerically or graphically is fairly easy. In a spreadsheet, plot S as a function of ZR/a_0, and look for the value of ZR/a_0 where $S = 0.20$. One can change the scale of the plot and the spacing of grid points to find the graphical solution to arbitrary precision.

By inspection of the plot (Fig. 24.2), $S = 0.20$ at approximately $ZR/a_0 = 3.9$. Or one could inspect the spreadsheet in the neighborhood of those points:

ZR/a_0	3.80	3.85	3.90	3.95	4.00
S	0.215	0.208	0.202	0.195	0.189

We can see that the value of S crosses $S = 0.20$ between the values of $ZR/a_0 = 3.90$ and 3.95. So to two significant figures, $ZR/a_0 = 3.9$, agreeing with the graphical estimate.

(a) For H_2, $Z = 1$, so $R = 3.9a_0/Z = 3.9 \times 5.29 \times 10^{-11}$ m/1 = $\boxed{2.1 \times 10^{-10}\text{ m} = 0.21\text{ nm}}$

(b) For He_2, $Z = 2$, so $R = 3.9a_0/Z = 3.9 \times 5.29 \times 10^{-11}$ m/2 = $\boxed{1.0 \times 10^{-10}\text{ m} = 0.10\text{ nm}}$

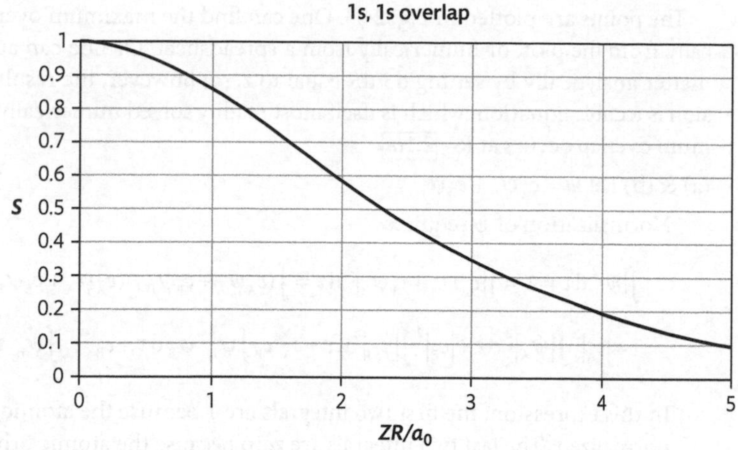

Fig 24.2

Problems

P24.1 Think of the p orbital starting far from the s orbital and approaching it. At long distances, there is no overlap, and as the distance decreases, the positive overlap increases—but only up to a point. Eventually, some of the "other side" of the p orbital (*i.e.*, the region of negative amplitude) begins to overlap the s orbital. When the centres of the orbitals coincide, the region of positive overlap cancels the negative region. Thus, the overlap goes to zero at very large **and** very small separations. Draw up the following table:

R/a_0	0	1	2	3	4	5	6	7	8	9	10
S	0	0.429	0.588	0.523	0.379	0.241	0.141	0.078	0.041	0.021	0.01

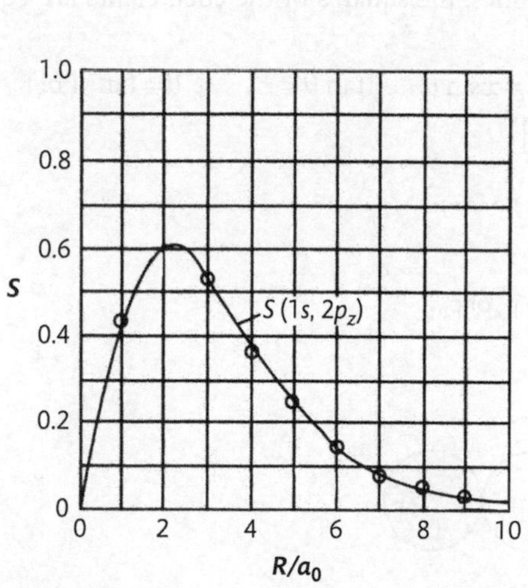

Fig 24.3

The points are plotted in Fig. 24.3. One can find the maximum overlap graphically from the plot, or numerically from a spreadsheet. Or one can approach the matter analytically by setting dS/dR equal to zero; however, the resulting expression is a cubic equation, which is itself most readily solved numerically. The maximum overlap occurs at $R = \boxed{2.1a_0}$.

P24.3 (a) & (b) Let $\psi = c_A\psi_A + c_B\psi_B$

Normalization of ψ requires

$$\int|\psi|^2 d\tau = 1 = \int|c_A\psi_A + c_B\psi_B|^2 d\tau = \int(c_A\psi_A + c_B\psi_B)^*(c_A\psi_A + c_B\psi_B)d\tau$$

$$1 = |c_A|^2 \int|\psi_A|^2 d\tau + |c_B|^2 \int|\psi_B|^2 d\tau + c_A{}^*c_B \int\psi_A{}^*\psi_B d\tau + c_B{}^*c_A \int\psi_B{}^*\psi_A d\tau$$

In this expression, the first two integrals are 1 because the atomic orbitals are normalized. The last two integrals are zero because the atomic orbitals are orthogonal. (Remember, overlap is neglected by hypothesis.) Thus, normalization of ψ requires

$$1 = |c_A|^2 + |c_B|^2$$

If we choose the two coefficients to be cosine and sine of the same parameter θ (note: not the spherical polar coordinate θ), then the squares of the coefficients sum to one as required. Strictly speaking, the parameter θ can range from 0 to π if the wavefunction is defined as above (with a plus sign), because we must allow for the possibility that the coefficients c_A and c_B have different signs; alternatively, if we allow the wavefunction to be

$$\psi = c_A\psi_A \pm c_B\psi_B$$

then we can restrict θ to 0 to $\pi/2$. (Note: There is no loss of generality in forcing one of the coefficients to be positive, as long as we allow the coefficients to have different signs.)

(c) In homonuclear diatomics, the squares of the coefficients are equal by symmetry, so

$\cos^2\theta = \sin^2\theta \qquad \cos\theta = \pm\sin\theta \qquad \tan\theta = \pm 1 \qquad \theta = \tan^{-1}(\pm 1)$

or $\quad \theta = \boxed{\pi/4 \text{ or } 3\pi/4}$

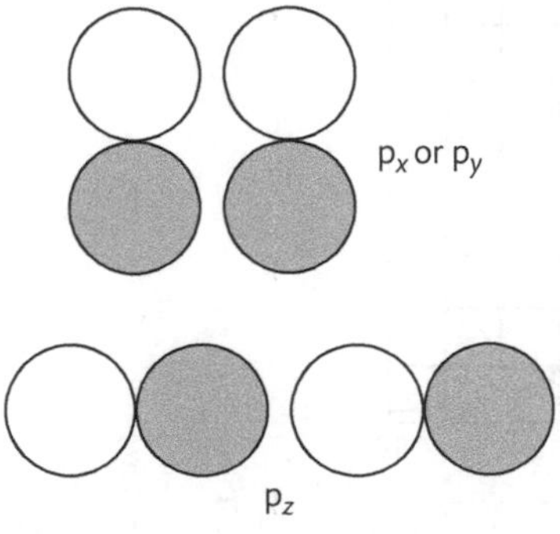

Fig 24.4

Which combination is bonding and which is antibonding depends on what the atomic orbitals are, as illustrated in Fig. 24.4. Constructive interference between the atomic orbitals is characteristic of bonding MOs, and that is sometimes achieved by adding the atomic orbital functions (as in s orbitals or p_x or p_y orbitals) and sometimes by subtracting one atomic orbital function from the other (as in p_z orbitals). Adding the atomic orbitals corresponds to equal coefficients: $\theta = \pi/4$; subtraction corresponds to equal and opposite coefficients: $\theta = 3\pi/4$.

P24.5 The overlap integral is defined by eqn 24.3:

$$S = \int 1s_A 1s_B \, d\tau = \int \left(\frac{Z^3}{\pi a_0^3}\right)^{1/2} e^{-Zr_A/a_0} \left(\frac{Z^3}{\pi a_0^3}\right)^{1/2} e^{-Zr_B/a_0} \, d\tau = \frac{Z^3}{\pi a_0^3} \int e^{-Z(r_A+r_B)/a_0} \, d\tau$$

where we have used the hydrogenic 1s wavefunctions shown in *Brief illustration* 23.1 generalized for a one-electron atom with a nucleus of charge number Z, where r_A and r_B are the distances of the electron from the two nuclei. Unfortunately, r_A and r_B cannot both be coordinates in a spherical polar coordinate system, because such a system has only one origin. (One could choose to place the origin at one of the nuclei, as done in *Brief illustration* 23.1; in that case, r_A is the same as the spherical polar coordinate r.) If one puts the origin at nucleus A, then one can write the integral in spherical polar coordinates using the law of cosines to express r_B in those coordinates.

$r_B{}^2 = r^2 + R^2 - rR \cos \theta$

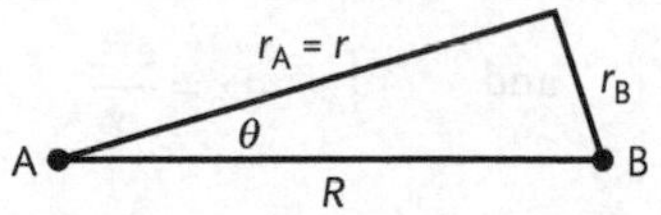

Fig 24.5

$$S = \frac{Z^3}{\pi a_0^3} \int \exp\left\{\frac{-Z\left(r + (r^2 + R^2 - rR\cos\theta)^{1/2}\right)}{a_0}\right\} r^2 \, dr \sin\theta \, d\theta \, d\phi$$

It is one thing to express the integral in this coordinate system, but quite another thing to integrate it! Cartesian coordinates are no better. Following the lead of *Brief illustration* 23.1, the integral can be written in Cartesian coordinates as

$$S = \frac{Z^3}{\pi a_0^3} \int \exp\left\{\frac{-Z\left((x^2+y^2+z^2)^{1/2} + \{x^2+y^2+(z-R)^2\}^{1/2}\right)}{a_0}\right\} dx\,dy\,dz$$

But there is a coordinate system where the integral is relatively straightforward, namely elliptical coordinates, in which the coordinates are

$\lambda = \frac{r_A + r_B}{R}$ $\quad \mu = \frac{r_A - r_B}{R}$ $\quad$ and ϕ (same as in spherical polar coordinates), the volume element is

$$d\tau = \frac{R^3}{8} d\phi(\lambda^2 - \mu^2) d\mu d\lambda$$

(See Carl W. David's Chemistry Education Materials on elliptical coordinates and overlap integrals at the University of Connecticut's DigitalCommons@UConn.) The overlap integral then becomes

$$S = \frac{Z^3 R^3}{8\pi a_0^3} \int_0^{2\pi} d\phi \int_{-1}^{+1} \int_1^{\infty} e^{-Z\lambda R/a_0} (\lambda^2 - \mu^2) d\mu d\lambda$$

Integrate over ϕ first:

$$S = \frac{Z^3 R^3}{4a_0^3} \int_{-1}^{+1} \int_1^{\infty} e^{-Z\lambda R/a_0} (\lambda^2 - \mu^2) d\mu d\lambda$$

Then integrate over μ:

$$S = \frac{Z^3 R^3}{4a_0^3} \int_1^{\infty} e^{-Z\lambda R/a_0} \left(\lambda^2 \mu - \frac{\mu^3}{3} \right)\Bigg|_{-1}^{+1} d\lambda = \frac{Z^3 R^3}{2a_0^3} \int_1^{\infty} e^{-Z\lambda R/a_0} \left(\lambda^2 - \frac{1}{3} \right) d\lambda$$

In the last step, evaluation at the limits of the μ integral yields a factor of 2. Finally, integrate over λ, using the integrals

$$\int e^{\alpha x} x^2 dx = \frac{e^{\alpha x}}{\alpha} \left(x^2 - \frac{2x}{\alpha} + \frac{2}{\alpha^2} \right) \quad \text{and} \quad \int e^{\alpha x} dx = \frac{e^{\alpha x}}{\alpha}$$

so $$S = \frac{Z^3 R^3}{2a_0^3} \frac{e^{-Z\lambda R/a_0}}{-ZR/a_0} \left(\lambda^2 - \frac{2\lambda}{-ZR/a_0} + \frac{2}{(-ZR/a_0)^2} - \frac{1}{3} \right)\Bigg|_1^{\infty}$$

All of the terms vanish at the upper limit. That leaves

$$S = \frac{Z^3 R^3}{2a_0^3} \frac{e^{-ZR/a_0}}{ZR/a_0} \left(1 + \frac{2}{ZR/a_0} + \frac{2}{(ZR/a_0)^2} - \frac{1}{3} \right)$$

$$= e^{-ZR/a_0} \left(\frac{Z^2 R^2}{3a_0^2} + \frac{ZR}{a_0} + 1 \right)$$

Topic 25 **Heteronuclear diatomic molecules**

Discussion questions

D25.1 Both the Pauling and Mulliken methods for measuring the attracting power of atoms for electrons seem to make good chemical sense. If we look at eqn 25.2 (the Pauling scale), we see that if $D(\text{A–B})$ were equal to $\frac{1}{2}[D(\text{A–A}) + D(\text{B–B})]$ the calculated electronegativity difference would be zero, as expected for completely non-polar bonds. Hence, any increased strength of the A–B bond over the average of the A–A and B–B bonds can reasonably be thought of as being due to the polarity of the A–B bond, which in turn is due to the difference in electronegativity of the atoms involved. Therefore, this difference in bond strengths can be used as a measure of electronegativity difference. To obtain numerical values for individual atoms, a reference state (atom) for electronegativity must be established. The value for fluorine is arbitrarily set at 4.0.

The Mulliken scale may be more intuitive than the Pauling scale because we are used to thinking of ionization energies and electron affinities as measures of the electron attracting powers of atoms. The choice of factor $\frac{1}{2}$, however, is arbitrary, though reasonable, and no more arbitrary than the specific form of eqn 25.2 that defines the Pauling scale.

D25.3 It can be proven that if an arbitrary wavefunction is used to calculate the energy of a system, the value calculated is never less than the true energy. This is the variation principle. This principle allows an enormous amount of latitude in constructing wavefunctions. We can continue modifying the wavefunctions in nearly any arbitrary manner (consistent with the minimal constraints of "well-behaved" functions mentioned in Topic 5.2(c)) until we find a set that we feel provides an energy close to the true minimum in energy. Thus we can construct wavefunctions containing many parameters and then minimize the energy with respect to those parameters. These parameters may be interpretable in terms of chemically or physically significant quantities—but they need not be. Examples of the mathematical steps involved are illustrated in Topic 25.2.

D25.5 Carbon is an essential building block of complex biological structures. It can form covalent bonds with many other elements, such as hydrogen, nitrogen, oxygen, sulfur, and, more importantly, other carbon atoms. As a consequence, such networks as long carbon–carbon chains (as in lipids) and chains of peptide links can form readily. Furthermore, carbon atoms can form chains and rings containing single, double, or triple C–C bonds. Such a variety of bonding options leads to the intricate molecular architectures of proteins, nucleic acids, and cell membranes. But the balance of bond strengths is critical to biology: bonds need to be sufficiently strong to maintain the structure of the cell and yet need to be susceptible to dissociation and rearrangement during biochemical reactions.

Intermediate values of ionization energy and electron affinity make carbon's bonding primarily covalent rather than ionic. A low ionization energy would favor cation formation (particularly in interactions with elements whose electron affinity is high) and a high electron affinity would favor anion formation (particularly in interactions with elements whose ionization energy is low). Intermediate values of these quantities also means that the energy of the valence atomic orbitals in carbon is not terribly different from the valence orbital energies of other non-metals; this in turn implies that the molecular orbitals carbon can form with them are significantly different in energy from the component atomic orbitals. That is, the atomic orbitals of carbon can interact with those of both moderately electronegative and electropositive elements to produce molecular orbitals with considerable bonding character—leading to strong covalent bonds. The fact that carbon has four valence electrons is intimately connected to its ability to form four bonds—a connection that has been rationalized by pre-quantum notions such as the octet rule as well as by quantum chemical theories as valence-bond and molecular orbital theory. The ability of carbon to make four bonds is in turn directly related to the incredible variety and complexity of organic chemistry and biochemistry. That carbon can make four bonds allows it to form long chains and rings (requiring two bonds for all but the terminal carbons for the C–C σ skeleton) and still have valences left over for a variety of functional groups. Finally, the strength of C–C bonds relative to C–N bonds and C–O bonds permits large organic molecules to persist in an atmosphere of oxygen and nitrogen.

Exercises

E25.1(a) Refer to Fig. 24.11 of the main text. Note that CO and CN^- are isoelectronic with N_2; note, however, that the σ and π orbitals no longer have u or g symmetry, so they are simply labeled consecutively.

(a) CO (10 electrons) $\boxed{1\sigma^2 2\sigma^2 1\pi^4 3\sigma^2}$

(b) NO (11 electrons) $\boxed{1\sigma^2 2\sigma^2 3\sigma^2 1\pi^4 2\pi^1}$

(c) CN^- (10 electrons) $\boxed{1\sigma^2 2\sigma^2 1\pi^4 3\sigma^2}$

E25.2(a) We can use a version of Fig. 24.12 of the text, but with the energy levels of F lower than those of Xe as in Fig. 25.1.

For XeF we insert 15 valence electrons. The bond order is increased when XeF^+ is formed from XeF, because an electron is removed from an antibonding orbital. Therefore, we predict that XeF^+ has a shorter bond length than XeF.

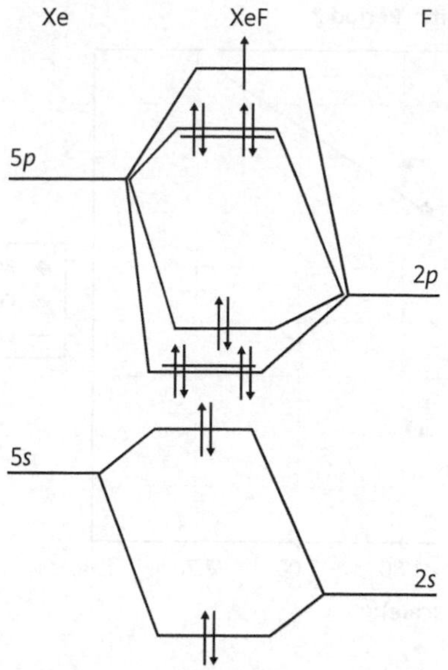

Fig 25.1

E25.3(a) Form the electron configurations and find the bond order. Note that NO^+ is isoelectronic with N_2, so for it we expect 1π to lie below 3σ, and vice versa for NO^-, which is isoelectronic with O_2. See Fig. 24.11 of the main text.

NO^+ (10 electrons) $\quad 1\sigma^2 2\sigma^2 1\pi^4 3\sigma^2 \quad b=3$

NO^- (12 electrons) $\quad 1\sigma^2 2\sigma^2 3\sigma^2 1\pi^4 2\pi^2 \quad b=2$

Based on the electron configurations, we would expect $\boxed{NO^+}$ to have the stronger and therefore the shorter bond.

E25.4(a) Draw up the following table using data from Table 25.1:

element	Li	Be	B	C	N	O	F	Ne
χ_M	1.28	1.99	1.83	2.67	3.08	3.22	4.43	4.60
$(\chi_M)^{1/2}$	1.13	1.41	1.35	1.63	1.75	1.79	2.10	2.14
χ_P (from table)	0.98	1.57	2.04	2.55	3.04	3.44	3.98	
χ_P (from formula)	0.16	0.53	0.46	0.84	1.00	1.05	1.47	1.53

A plot (Fig. 25.2(a) of the Pauling electronegativities (actual and from the formula) vs. the square root of the Mulliken electronegativities shows that the formula does a poor job. The formula consistently underestimates the Pauling electronegativity, and it underestimates the rise in electronegativity across the period.

The problem is that the Mulliken electronegativities given in the table have already been scaled to the range of Pauling electronegativities ***using this very conversion***. That is, the Mulliken electronegativities given in Table 25.1 are not defined by eqn 25.3, but rather by eqn 25.3 and then eqn 25.4. Using Mulliken electronegativities based on eqn 25.3 alone (the "intrinsic Mulliken electronegativities" also known as "Mulliken *a* parameters" from Steven G. Bratsch, "Revised Mulliken

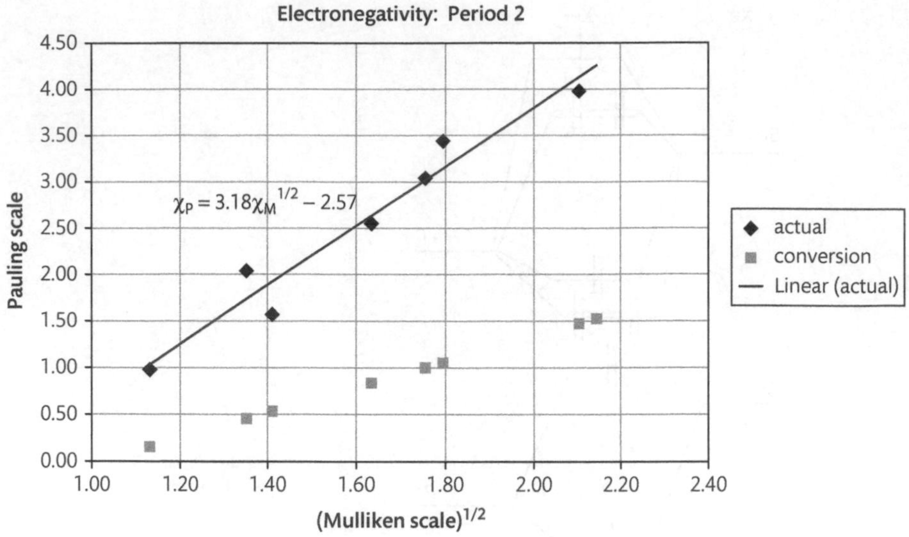

Fig 25.2(a)

electronegativities: I. Calculation and conversion to Pauling units," *J. Chem. Educ.*, **1988**, *65*, 34-41).

element	Li	Be	B	C	N	O	F	Ne
χ_M (Bratch *a*)	3.01	4.65	6.37	8.15	10.00	12.55	15.30	15.71
$(\chi_M)^{1/2}$	1.73	2.16	2.52	2.85	3.16	3.54	3.91	3.96
χ_P (from table)	0.98	1.57	2.04	2.55	3.04	3.44	3.98	
χ_P (from formula)	0.97	1.54	2.04	2.48	2.90	3.41	3.91	

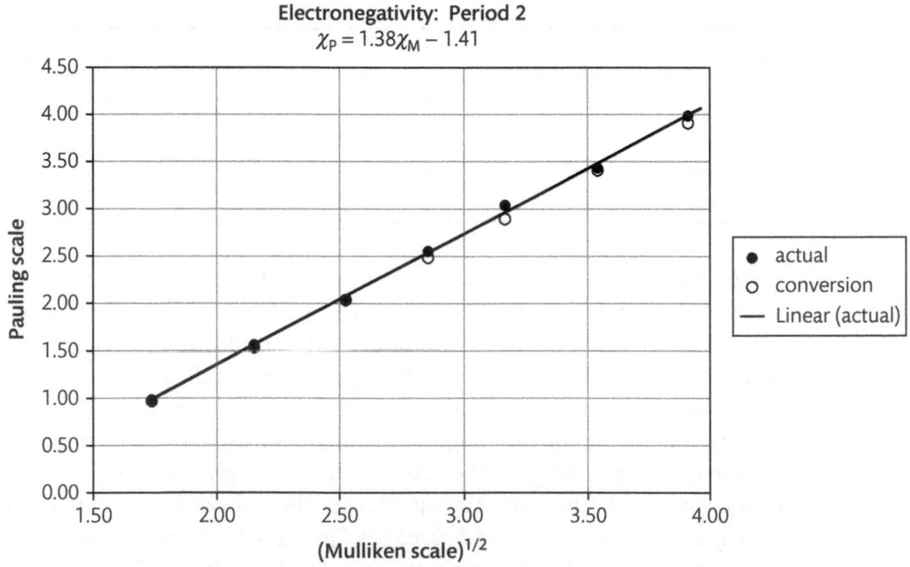

Fig 25.2(b)

A plot (Fig. 25.2(b)) of actual Pauling electronegativities (from Table 25.1; filled circles) vs. the square root of the intrinsic Mulliken electronegativities (from Table 5 of Bratch's article) yields an excellent regression line with an equation very close to that of eqn 25.4. The open circles on the same plot are the result of applying eqn 25.4 to the intrinsic Mulliken electronegativities from Bratch; these are essentially the "scaled" Mulliken electronegativities. As you can see, they fall very close to the best-fit line of the actual Pauling electronegativity values.

E25.5(a) Following *Brief Illustration* 25.1, we draw up a table of ionization energies and electron affinities from Tables 20.2 and 20.3. The mean of those two energies is the estimated orbital energy for each atom.

	I/eV	E_{ea}/eV	$\frac{1}{2}(I+E_{ea})$/eV
H	13.6	0.75	$\boxed{7.2}$
Cl	13.0	3.6	$\boxed{8.3}$

E25.6(a) In the zero overlap approximation, the molecular orbital energies are given by eqn 25.8c:

$$E_{\pm}=\tfrac{1}{2}(\alpha_A+\alpha_B)\pm\tfrac{1}{2}(\alpha_A-\alpha_B)\left\{1+\left(\frac{2\beta}{\alpha_A-\alpha_B}\right)^2\right\}^{1/2}$$

Taking $\beta=-1.0$ eV (a typical value), we have

$$E_{\pm}/\text{eV}=\tfrac{1}{2}(-7.2-8.3)\pm\tfrac{1}{2}(-7.2+8.3)\left\{1+\left(\frac{2(-1)}{-7.2+8.3}\right)^2\right\}^{1/2}=\boxed{-6.6\text{ or }-8.9}$$

E25.7(a) If overlap cannot be neglected, then the molecular orbital energies are given by eqn 25.8a:

$$E_{\pm}=\frac{\alpha_A+\alpha_B-2\beta S\pm\{(\alpha_A+\alpha_B-2\beta S)^2-4(1-S^2)(\alpha_A\alpha_B-\beta^2)\}^{1/2}}{2(1-S^2)}$$

Taking $\beta=-1.0$ eV (a typical value), we have

$$\frac{E_{\pm}}{\text{eV}}=\frac{-7.2-8.3+2(1.0)(0.20)\pm\{(-7.2-8.3+2(1.0)(0.20))^2-4(1-0.20^2)(7.2\times8.3-1.0^2)\}^{1/2}}{2(1-0.20^2)}$$

so $E_{\pm}=\boxed{-5.0\text{ or }-10.7\text{ eV}}$

Problems

P25.1 If overlap is not neglected, the determinant is

$$\begin{vmatrix}\alpha_A-E & \beta-SE\\ \beta-SE & \alpha_B-E\end{vmatrix}$$

and expanding it yields eqn 25.7:

$$0=(1-S^2)E^2+(2\beta S-\alpha_A-\alpha_B)E+\alpha_A\alpha_B-\beta^2$$

This is a quadratic equation in E where $a = 1 - S^2$, $b = 2\beta S - \alpha_A - \alpha_B$, and $c = \alpha_A\alpha_B - \beta^2$. The solution is

$$E_{\pm}=\frac{-b\pm\sqrt{b^2-4ac}}{2a}$$

Let us evaluate $b^2–4ac$:

$$\begin{aligned}b^2-4ac&=4\beta^2S^2-4\alpha_A\beta S-4\alpha_B\beta S+\alpha_A^2+2\alpha_A\alpha_B+\alpha_B^2\\&\quad-4\alpha_A\alpha_B+4\beta^2+4\alpha_A\alpha_BS^2-4\beta^2S^2\\&=-4\alpha_A\beta S-4\alpha_B\beta S+\alpha_A^2+\alpha_B^2-2\alpha_A\alpha_B+4\beta^2+4\alpha_A\alpha_BS^2.\end{aligned}$$

The terms can be gathered into the following factors:

$$b^2-4ac=(\alpha_A-\alpha_B)^2+4(\beta+\alpha_AS)(\beta+\alpha_BS)$$

So $$E_{\pm}=\frac{\alpha_A+\alpha_B-2\beta S}{2(1-S^2)}\pm\frac{\{(\alpha_A-\alpha_B)^2+4(\beta+\alpha_AS)(\beta+\alpha_BS)\}^{1/2}}{2(1-S^2)}$$

$$=\boxed{\frac{\alpha_A+\alpha_B-2\beta S}{2(1-S^2)}\pm\frac{\alpha_A-\alpha_B}{2(1-S^2)}\left(1+\frac{4(\beta+\alpha_AS)(\beta+\alpha_BS)}{(\alpha_A-\alpha_B)^2}\right)^{1/2}}$$

Comment. The effect of overlap is to mix the Coulomb and resonance contributions, producing terms such as $\beta+S\alpha$ and $\alpha-S\beta$ in place of "pure" α or β terms.

Even if overlap is not negligible, one can consider the situation if the atomic orbitals are energetically far apart. In this case, the condition is that

$$(\alpha_A-\alpha_B)^2>>4(\beta+\alpha_AS)(\beta+\alpha_BS)$$

If so, the square root term can be expanded

$$E_{\pm}=\frac{\alpha_A+\alpha_B-2\beta S}{2(1-S^2)}\pm\frac{\alpha_A-\alpha_B}{2(1-S^2)}\left(1+\frac{4(\beta+\alpha_AS)(\beta+\alpha_BS)}{2(\alpha_A-\alpha_B)^2}-\dots\right)$$

so $$E_+\approx\boxed{\frac{\alpha_A-\beta S}{1-S^2}+\frac{(\beta+\alpha_AS)(\beta+\alpha_BS)}{(\alpha_A-\alpha_B)(1-S^2)}}$$

and $$E_-\approx\boxed{\frac{\alpha_B-\beta S}{1-S^2}-\frac{(\beta+\alpha_AS)(\beta+\alpha_BS)}{(\alpha_A-\alpha_B)(1-S^2)}}$$

P25.3 At first blush, we simply have three terms rather than two. But the fact that two of the atomic orbitals belong to the same nucleus alters matters. Overlap, for instance, is zero for atomic orbitals on the same nucleus, because the orbitals are orthogonal. Also the resonance integral for different atomic orbitals on the same nucleus (which we would denote by β_{BC}) vanishes, as it must for orbitals that have zero overlap.

Thus $\quad (\alpha_A - E)c_A + (\beta_{AB} - ES_{AB})c_B + (\beta_{AC} - ES_{AC})c_C = 0$

$\quad (\beta_{AB} - ES_{AB})c_A + (\alpha_B - E)c_B = 0$

and $\quad (\beta_{AC} - ES_{AC})c_A + (\alpha_C - E)c_C = 0$

(i) First consider the case $S = 0$ (without, however, making the resonance integrals also vanish). The secular determinant—call it $f(E)$—is

$$f(E) = \begin{vmatrix} \alpha_A - E & \beta_{AB} & \beta_{AC} \\ \beta_{AB} & \alpha_B - E & 0 \\ \beta_{AC} & 0 & \alpha_C - E \end{vmatrix}$$

Solve for the energies by expanding the secular determinant and setting it equal to zero. Let us expand using the second row:

$$f(E) = \beta_{AB}\{\beta_{AB}(\alpha_C - E)\} - (\alpha_B - E)\{(\alpha_A - E)(\alpha_C - E) - \beta_{AC}^2\} = 0$$

$$0 = E^3 - (\alpha_A + \alpha_B + \alpha_C)E^2 + (\alpha_A\alpha_B + \alpha_A\alpha_C + \alpha_B\alpha_C - \beta_{AB}^2 - \beta_{AC}^2)E + \alpha_B\beta_{AC}^2 + \alpha_C\beta_{AB}^2 - \alpha_A\alpha_B\alpha_C$$

Substituting in the parameters, we have

$$0 = E^3 + (26.0\text{ eV})E^2 + (221\text{ eV}^2)E - 614\text{ eV}^3$$

This cubic equation can be solved numerically and/or graphically. Fig. 25.3(a) displays a plot of $f(E)$ vs. E/eV. The roots are $\boxed{E/\text{eV} = -10.7, -8.7, \text{ and } -6.6}$.

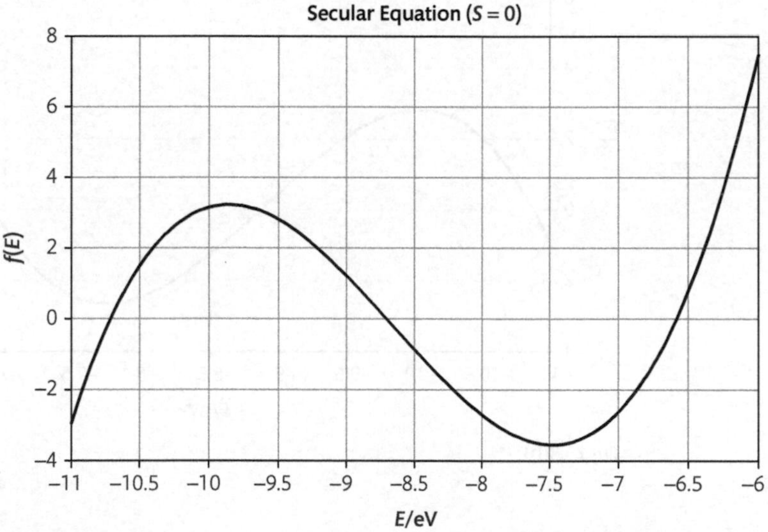

Fig 25.3(a)

(ii) Now consider $S=0.2$. The secular determinant—call it $g(E)$ is

$$g(E)=\begin{vmatrix}\alpha_A-E & \beta_{AB}-ES & \beta_{AC}-ES\\ \beta_{AB}-ES & \alpha_B-E & 0\\ \beta_{AC}-ES & 0 & \alpha_C-E\end{vmatrix}$$

Again, we expand using the second row:

$$g(E)=(\beta_{AB}-ES)\{(\beta_{AB}-ES)(\alpha_C-E)\}-(\alpha_B-E)\{(\alpha_A-E)(\alpha_C-E)-(\beta_{AC}-ES)^2\}=0$$

$$0=E^3(1-2S^2)+\{S^2(\alpha_B+\alpha_C)+2S(\beta_{AB}+\beta_{AC})-\alpha_A-\alpha_B-\alpha_C\}E^2$$
$$+\{\alpha_A\alpha_B+\alpha_A\alpha_C+\alpha_B\alpha_C-\beta_{AB}{}^2-\beta_{AC}{}^2-2S(\alpha_B\beta_{AC}+\alpha_C\beta_{AB})\}E$$
$$+\alpha_B\beta_{AC}{}^2+\alpha_C\beta_{AB}{}^2-\alpha_A\alpha_B\alpha_C$$

Substituting in the parameters, we have

$$0=0.92E^3+(24.5\text{ eV})E^2+(214\text{ eV}^2)E-614\text{ eV}^3$$

This cubic equation can also be solved numerically and/or graphically. Before we do so, however, we note that the coefficients are not markedly different from those considered in part (i) above; therefore, we do not anticipate very different roots of this secular equation. Fig. 25.3(b) displays a plot of $g(E)$ vs. E/eV. The roots are $\boxed{E/\text{eV}=-10.8, -8.9, \text{ and } -6.9}$.

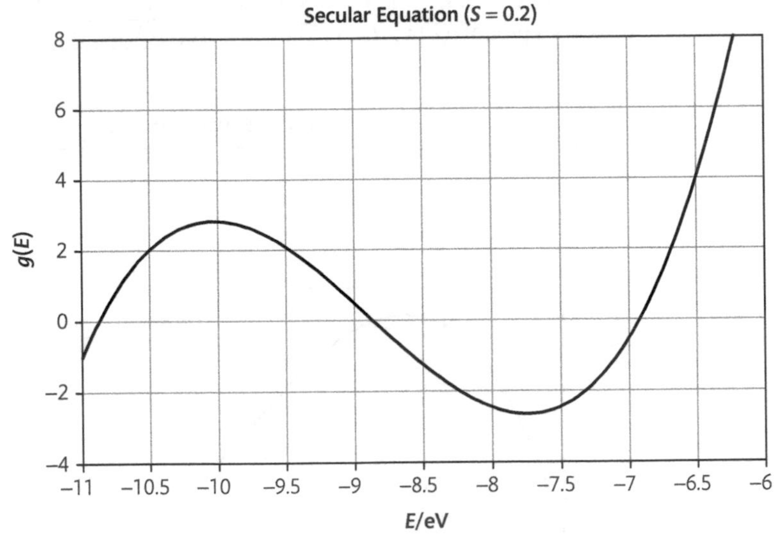

Fig 25.3(b)

Topic 26 Polyatomic molecules

Discussion question

D26.1 The Hückel method parameterizes, rather than calculates, the energy integrals that arise in molecular orbital theory. In the simplest version of the method, the overlap integral is also parameterized. The energy integrals, α and β, are always considered to be adjustable parameters; their numerical values emerge only at the end of the calculation by comparison to experimental energies. The simple form of the method has three other rather drastic approximations, listed in Topic 26.1 of the text, which eliminate many terms from the secular determinant and make it easier to solve. Ease of solution was important in the early days of quantum chemistry before the advent of computers. Without the use of these approximations, calculations on polyatomic molecules would have been difficult to accomplish.

The simple Hückel method is usually applied only to the calculation of π-electron energies in conjugated organic systems. It is based on the assumption of the separability of the σ- and π-electron systems in the molecule. This is a very crude approximation and works best when the energy level pattern is determined largely by the symmetry of the molecule. (See Focus 7.)

Exercises

E26.1(a) In setting up the secular determinant we use the Hückel approximations outlined in Topic 26.1:

$$\text{(a)}\ \begin{vmatrix} \alpha-E & \beta & 0 \\ \beta & \alpha-E & \beta \\ 0 & \beta & \alpha-E \end{vmatrix} \qquad \text{(b)}\ \begin{vmatrix} \alpha-E & \beta & \beta \\ \beta & \alpha-E & \beta \\ \beta & \beta & \alpha-E \end{vmatrix}$$

The atomic orbital basis is $1s_A$, $1s_B$, $1s_C$ in each case; in linear H_3 we ignore A, C overlap because A and C are not neighboring atoms; in cyclic H_3 we include it because they are.

E26.2(a) We use the molecular orbital energy level diagram in Fig. 26.4 of the main text. As usual, we fill the orbitals starting with the lowest-energy orbital, obeying the Pauli principle and Hund's rule. We then write

(a) $C_6H_6^-$ (7 electrons): $\boxed{a_{2u}^2 e_{1g}^4 e_{2u}^1}$

$$E_\pi = 2(\alpha+2\beta)+4(\alpha+\beta)+(\alpha-\beta) = \boxed{7\alpha+7\beta}$$

(b) $C_6H_6^+$ (5 electrons): $\boxed{a_{2u}^2 e_{1g}^3}$

$$E_\pi = 2(\alpha+2\beta)+3(\alpha+\beta) = \boxed{5\alpha+7\beta}$$

E26.3(a) The π-bond formation energy is the difference between the π-electron binding energy and the Coulomb energies α:

$$E_{bf} = E_\pi - N_C\alpha \quad [26.12]$$

The delocalization energy is the difference between the energy of the delocalized π system and the energy of isolated π bonds:

$$E_{delocal} = E_\pi - N_C(\alpha+\beta)$$

(a) $E_\pi = 7\alpha+7\beta$ [E26.2(a)]

so $E_{bf} = 7\alpha+7\beta-7\alpha = \boxed{7\beta}$

and $E_{delocal} = 7\alpha+7\beta-7(\alpha+\beta) = \boxed{0}$

Comment. With an odd number of π electrons, we do not have a whole number of π bonds in the formula for delocalization energy. In effect, we compare the π-electron binding energy to the energy of 3.5 isolated π bonds—whatever that means. The result is that the benzene anion has none of the "extra" stabilization we associate with aromaticity.

(b) $E_\pi = 5\alpha+7\beta$ [E26.2(a)]

so $E_{bf} = 5\alpha+7\beta-5\alpha = \boxed{7\beta}$

and $E_{delocal} = 5\alpha+7\beta-5(\alpha+\beta) = \boxed{2\beta}$

E26.4(a) The structures are numbered to match the row and column numbers shown in the determinants:

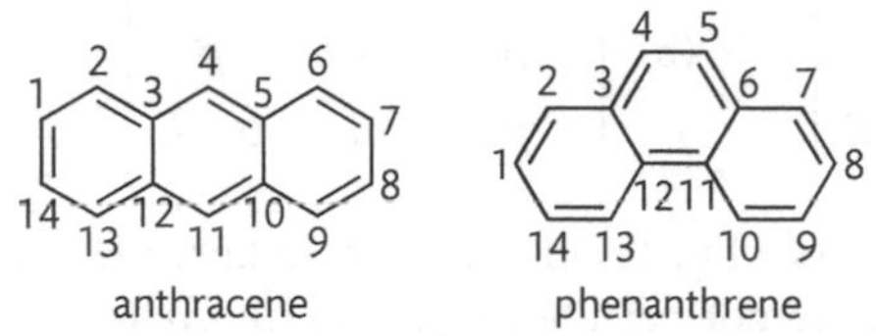

(a) The secular determinant of anthracene in the Hückel approximation is:

	1	2	3	4	5	6	7	8	9	10	11	12	13	14
1	$\alpha-E$	β	0	0	0	0	0	0	0	0	0	0	0	β
2	β	$\alpha-E$	β	0	0	0	0	0	0	0	0	0	0	0
3	0	β	$\alpha-E$	β	0	0	0	0	0	0	0	β	0	0
4	0	0	β	$\alpha-E$	β	0	0	0	0	0	0	0	0	0
5	0	0	0	β	$\alpha-E$	β	0	0	0	β	0	0	0	0
6	0	0	0	0	β	$\alpha-E$	β	0	0	0	0	0	0	0
7	0	0	0	0	0	β	$\alpha-E$	β	0	0	0	0	0	0
8	0	0	0	0	0	0	β	$\alpha-E$	β	0	0	0	0	0
9	0	0	0	0	0	0	0	β	$\alpha-E$	β	0	0	0	0
10	0	0	0	0	β	0	0	0	β	$\alpha-E$	β	0	0	0
11	0	0	0	0	0	0	0	0	0	β	$\alpha-E$	β	0	0
12	0	0	β	0	0	0	0	0	0	0	β	$\alpha-E$	β	0
13	0	0	0	0	0	0	0	0	0	0	0	β	$\alpha-E$	β
14	β	0	0	0	0	0	0	0	0	0	0	0	β	$\alpha-E$

(b) The secular determinant of phenanthrene in the Hückel approximation is:

	1	2	3	4	5	6	7	8	9	10	11	12	13	14
1	$\alpha-E$	β	0	0	0	0	0	0	0	0	0	0	0	β
2	β	$\alpha-E$	β	0	0	0	0	0	0	0	0	0	0	0
3	0	β	$\alpha-E$	β	0	0	0	0	0	0	0	β	0	0
4	0	0	β	$\alpha-E$	β	0	0	0	0	0	0	0	0	0
5	0	0	0	β	$\alpha-E$	β	0	0	0	0	0	0	0	0
6	0	0	0	0	β	$\alpha-E$	β	0	0	0	β	0	0	0
7	0	0	0	0	0	β	$\alpha-E$	β	0	0	0	0	0	0
8	0	0	0	0	0	0	β	$\alpha-E$	β	0	0	0	0	0
9	0	0	0	0	0	0	0	β	$\alpha-E$	β	0	0	0	0
10	0	0	0	0	0	0	0	0	β	$\alpha-E$	β	0	0	0
11	0	0	0	0	0	β	0	0	0	β	$\alpha-E$	β	0	0
12	0	0	β	0	0	0	0	0	0	0	β	$\alpha-E$	β	0
13	0	0	0	0	0	0	0	0	0	0	0	β	$\alpha-E$	β
14	β	0	0	0	0	0	0	0	0	0	0	0	β	$\alpha-E$

E26.5(a) The secular determinants from E26.4(a) can be diagonalized with the assistance of general-purpose mathematical software. Alternatively, programs specifically designed for Hückel calculations (such as the Simple Huckel Molecular Orbital Theory Calculator at the University of Calgary, http://www.chem.ucalgary.ca/SHMO/ or Hückel software in *Explorations in Physical Chemistry*, 2nd Ed. by Julio de Paula, Valerie Walters, and Peter Atkins, http://ebooks.bfwpub.com/explorations.php) can be used. In both molecules, 14 π electrons fill seven orbitals.

(a) In anthracene, the energies of the filled orbitals are $\alpha + 2.414\beta$, $\alpha + 2\beta$, $\alpha + 1.414\beta$ (doubly degenerate), $\alpha + \beta$ (doubly degenerate), and $\alpha + 0.414\beta$, so the total π-electron binding energy is $\boxed{14\alpha + 19.314\beta}$.

(b) For phenanthrene, the energies of the filled orbitals are $\alpha + 2.435\beta$, $\alpha + 1.951\beta$, $\alpha + 1.516\beta$, $\alpha + 1.3060\beta$, $\alpha + 1.142\beta$, $\alpha + 0.769\beta$, and $\alpha + 0.605\beta$, so the total π-electron binding energy is $\boxed{14\alpha + 19.448\beta}$.

Problems

P26.1 In the simple Hückel approximation

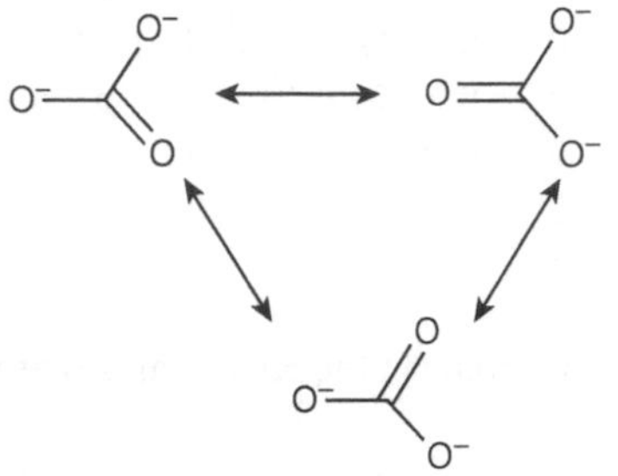

$$\begin{vmatrix} \alpha_O - E & 0 & 0 & \beta \\ 0 & \alpha_O - E & 0 & \beta \\ 0 & 0 & \alpha_O - E & \beta \\ \beta & \beta & \beta & \alpha_C - E \end{vmatrix} = 0$$

$$(E-\alpha_O)^2 \times \{(E-\alpha_O)\times(E-\alpha_C)-3\beta^2\} = 0$$

Therefore, the factors yield

$E - \alpha_O = 0$ (twice) so $\boxed{E = \alpha_O}$,

and $(E-\alpha_O) \times (E-\alpha_C) - 3\beta^2 = 0 = E^2 - E(\alpha_O + \alpha_C) + \alpha_O\alpha_C - 3\beta^2$

so
$$E_\pm = \frac{\alpha_O + \alpha_C \pm \sqrt{(\alpha_O + \alpha_C)^2 - 4(\alpha_O\alpha_C - 3\beta^2)}}{2}$$
$$= \frac{\alpha_O + \alpha_C \pm \sqrt{(\alpha_O - \alpha_C)^2 + 12\beta^2}}{2}$$
$$= \boxed{\frac{1}{2}\left(\alpha_O + \alpha_C \pm (\alpha_O - \alpha_C)\sqrt{1 + \frac{12\beta^2}{(\alpha_O - \alpha_C)^2}}\right)}.$$

Because Coulomb integrals approximate ionization energies in magnitude, α_O is more negative than α_C. Therefore, E_+ corresponds to the lowest-energy orbital and E_- to the highest-energy orbital; the degenerate orbitals at $E = \alpha_O$ fall in between.

The π energies in the absence of resonance are derived for just one of the three structures, *i.e.*, for a structure containing a single localized π bond.

$$\begin{vmatrix} \alpha_O - E & \beta \\ \beta & \alpha_C - E \end{vmatrix} = 0$$

Expanding the determinant and solving for E gives

$$(E-\alpha_O)\times(E-\alpha_C)-\beta^2 = 0 = E^2 - E(\alpha_O+\alpha_C)+\alpha_O\alpha_C-\beta^2,$$

so $$E_{\text{local}\pm} = \frac{1}{2}\left(\alpha_O+\alpha_C \pm (\alpha_O-\alpha_C)\sqrt{1+\frac{4\beta^2}{(\alpha_O-\alpha_C)^2}}\right)$$

There are two π electrons in the system, so the delocalization energy is

$$\boxed{2E_+ - 2E_{\text{local}+} = (\alpha_O-\alpha_C)\left(\sqrt{1+\frac{12\beta^2}{(\alpha_O-\alpha_C)^2}}-\sqrt{1+\frac{4\beta^2}{(\alpha_O-\alpha_C)^2}}\right)}$$

If $12\beta^2 << (\alpha_O-\alpha_C)^2$, we can use $(1+x)^{1/2}\approx 1+x/2$, so the delocalization energy is

$$\approx (\alpha_O-\alpha_C)\left(1+\frac{12\beta^2}{2(\alpha_O-\alpha_C)^2}-1-\frac{4\beta^2}{2(\alpha_O-\alpha_C)^2}\right) = \boxed{\frac{4\beta^2}{\alpha_O-\alpha_C}}$$

P26.3 (a) In the absence of numerical values for α and β, we express orbital energies as $(E_k-\alpha)/\beta$ for the purpose of comparison. Recall that β is negative (as is α for that matter), so the orbital with the greatest value of $(E_k-\alpha)/\beta$ has the lowest energy. Draw up the following table, evaluating

$$\frac{E_k-\alpha}{\beta} = 2\cos\frac{2k\pi}{N}$$

	energy $(E_k-\alpha)/\beta$	
orbital, k	**C_6H_6**	**C_8H_8**
±4		−2.000
±3	−2.000	−1.414
±2	−1.000	0
±1	1.000	1.414
0	2.000	2.000

In each case, the lowest and highest energy levels are non-degenerate, while the other energy levels are doubly degenerate. The degeneracy is clear for all energy levels except, perhaps, the highest: each value of the quantum number k corresponds to a separate MO, and positive and negative values of k therefore give rise to a pair of MOs of the same energy. This is not the case for the highest energy level, though, because there are only as many MOs as there were AOs input to the calculation, which is the same as the number of carbon atoms. Having a doubly-degenerate top energy level would yield one extra MO. (See also P26.5.)

(b) The total energy of the π electron system is the sum of the energies of occupied orbitals weighted by the number of electrons that occupy them. In C_6H_6, each of the three lowest-energy orbitals is doubly occupied, but the second level ($k=\pm 1$) is doubly degenerate, so

$$E_\pi = 2E_0 + 2\times 2E_1 = 2(\alpha + 2\beta\cos 0) + 4\left(\alpha + 2\beta\cos\frac{2\pi}{6}\right) = 6\alpha + 8\beta$$

The delocalization energy is the difference between this quantity and that of three isolated double bonds:

$$E_{\text{deloc}} = E_\pi - 6(\alpha+\beta) = 6\alpha + 8\beta - 6(\alpha+\beta) = \boxed{2\beta}$$

For linear hexatriene, the first three orbitals are also doubly occupied:

$$E_\pi = 2\left(\alpha + 2\beta\cos\frac{\pi}{7}\right) + 2\left(\alpha + 2\beta\cos\frac{2\pi}{7}\right) + 2\left(\alpha + 2\beta\cos\frac{3\pi}{7}\right) = 6\alpha + 6.988\beta$$

so $E_{\text{deloc}} = \boxed{0.988\beta}$. Thus benzene has considerably more delocalization energy (assuming that β is similar in the two molecules). This extra stabilization is an example of the special stability of $\boxed{\text{aromatic}}$ compounds.

(c) In C_8H_8, each of the first three orbitals is doubly occupied, but the second level ($k=\pm 1$) is doubly degenerate. The next level is also doubly degenerate, with a single electron occupying each orbital. So the energy is

$$\begin{aligned} E_\pi &= 2E_0 + 2\times 2E_1 + 2\times 1E_2 \\ &= 2(\alpha + 2\beta\cos 0) + 4\left(\alpha + 2\beta\cos\frac{2\pi}{8}\right) + 2\left(\alpha + 2\beta\cos\frac{4\pi}{8}\right) \\ &= 8\alpha + 9.657\beta. \end{aligned}$$

The delocalization energy is the difference between this quantity and that of four isolated double bonds:

$$E_{\text{deloc}} = E_1 - 8(\alpha+\beta) = 8\alpha + 9.657\beta - 8(\alpha+\beta) = \boxed{1.657\beta}.$$

In linear octatetraene, the first four levels are doubly occupied:

$$\begin{aligned} E_\pi &= 2\left(\alpha + 2\beta\cos\frac{\pi}{9}\right) + 2\left(\alpha + 2\beta\cos\frac{2\pi}{9}\right) + 2\left(\alpha + 2\beta\cos\frac{3\pi}{9}\right) + 2\left(\alpha + 2\beta\cos\frac{4\pi}{9}\right) \\ &= 8\alpha + 9.518\beta, \end{aligned}$$

so $E_{\text{deloc}} = \boxed{1.518\beta}$. Thus cycloocatetraene does not have much additional stabilitzation over the linear structure. Once again, though, we do see that the delocalization energy stabilizes the π orbitals of the closed ring conjugated system to a greater extent than what is observed in the open chain conjugated system. However, the benzene/hexatriene comparison shows a much greater stabilization than does the cyclooctatetraene/octatetraene system. This is an example of the Hückel $4n+2$ rule, which states that any planar, cyclic, conjugated system exhibits unusual aromatic stabilization if it contains $4n+2$ π electrons where n is an integer. Benzene with its 6 π electrons has this aromatic stabilization whereas cycloctatetraene with 8 π electrons doesn't have this unusual stabilization. We can say that it is $\boxed{\text{not aromatic}}$, consistent with indicators of aromaticity such as the Hückel $4n+2$ rule.

P26.5 We use the Hückel approximation, neglecting overlap integrals.

The secular determinant of cyclobutadiene is $\begin{vmatrix} \alpha-E & \beta & 0 & \beta \\ \beta & \alpha-E & \beta & 0 \\ 0 & \beta & \alpha-E & \beta \\ \beta & 0 & \beta & \alpha-E \end{vmatrix}$

Mathematical software (such as the Simple Hückel Molecular Orbital Theory Calculator at the University of Calgary, http://www.chem.ucalgary.ca/SHMO/) diagonalizes the hamiltonian matrix to

$$\boldsymbol{E}=\begin{pmatrix} \alpha+2\beta & 0 & 0 & 0 \\ 0 & \alpha & 0 & 0 \\ 0 & 0 & \alpha & 0 \\ 0 & 0 & 0 & \alpha-2\beta \end{pmatrix}$$

The secular determinant of benzene is

$$\begin{vmatrix} \alpha-E & \beta & 0 & 0 & 0 & \beta \\ \beta & \alpha-E & \beta & 0 & 0 & 0 \\ 0 & \beta & \alpha-E & \beta & 0 & 0 \\ 0 & 0 & \beta & \alpha-E & \beta & 0 \\ 0 & 0 & 0 & \beta & \alpha-E & \beta \\ \beta & 0 & 0 & 0 & \beta & \alpha-E \end{vmatrix}$$

The hamiltonian matrix is diagonalized to

$$\boldsymbol{E}=\begin{pmatrix} \alpha+2\beta & 0 & 0 & 0 & 0 & 0 \\ 0 & \alpha+\beta & 0 & 0 & 0 & 0 \\ 0 & 0 & \alpha+\beta & 0 & 0 & 0 \\ 0 & 0 & 0 & \alpha-\beta & 0 & 0 \\ 0 & 0 & 0 & 0 & \alpha-\beta & 0 \\ 0 & 0 & 0 & 0 & 0 & \alpha-2\beta \end{pmatrix}$$

The secular determinant of cyclooctatetraene is

$$\begin{vmatrix} \alpha-E & \beta & 0 & 0 & 0 & 0 & 0 & \beta \\ \beta & \alpha-E & \beta & 0 & 0 & 0 & 0 & 0 \\ 0 & \beta & \alpha-E & \beta & 0 & 0 & 0 & 0 \\ 0 & 0 & \beta & \alpha-E & \beta & 0 & 0 & 0 \\ 0 & 0 & 0 & \beta & \alpha-E & \beta & 0 & 0 \\ 0 & 0 & 0 & 0 & \beta & \alpha-E & \beta & 0 \\ 0 & 0 & 0 & 0 & 0 & \beta & \alpha-E & \beta \\ \beta & 0 & 0 & 0 & 0 & 0 & \beta & \alpha-E \end{vmatrix}$$

The hamiltonian matrix is diagonalized to

$$E = \begin{pmatrix} \alpha+2\beta & 0 & 0 & 0 & 0 & 0 & 0 & 0 \\ 0 & \alpha+\sqrt{2}\beta & 0 & 0 & 0 & 0 & 0 & 0 \\ 0 & 0 & \alpha+\sqrt{2}\beta & 0 & 0 & 0 & 0 & 0 \\ 0 & 0 & 0 & \alpha & 0 & 0 & 0 & 0 \\ 0 & 0 & 0 & 0 & \alpha & 0 & 0 & 0 \\ 0 & 0 & 0 & 0 & 0 & \alpha-\sqrt{2}\beta & 0 & 0 \\ 0 & 0 & 0 & 0 & 0 & 0 & \alpha-\sqrt{2}\beta & 0 \\ 0 & 0 & 0 & 0 & 0 & 0 & 0 & \alpha-2\beta \end{pmatrix}$$

Recall that β is negative, so energies increase from the upper left to the lower right of the diagonalized matrices. In each of these examples, one can see by inspection that the first and last energy levels are non-degenerate (for there is no other energy value in the diagonalized $\boldsymbol{E}$ matrix equal to them) and that the other levels are two-fold degenerate (for those energy values occur in pairs).

P26.7 The secular determinant for an N-carbon linear polyene (call the determinant P_N) has the form

$$\begin{array}{c} \underline{1\ \ 2\ \ 3\ \ \ldots\ \ \ldots\ \ \ldots\ \ N-1\ \ N} \\ \begin{vmatrix} x & 1 & 0 & 0 & 0 & \ldots & 0 & 0 \\ 1 & x & 1 & 0 & 0 & \ldots & 0 & 0 \\ 0 & 1 & x & 1 & 0 & \ldots & 0 & 0 \\ 0 & 0 & 1 & x & 1 & \ldots & 0 & 0 \\ \vdots & \vdots & \vdots & \vdots & \vdots & \vdots & \vdots & \vdots \\ 0 & 0 & 0 & 0 & 0 & \ldots & x & 1 \\ 0 & 0 & 0 & 0 & 0 & \ldots & 1 & x \end{vmatrix} \end{array} = P_N$$

where $x = \dfrac{\alpha - E}{\beta}$. The determinant can be expanded by cofactors; use the elements of the first row:

$$P_N = M_{11}c_{11} - M_{12}c_{12} + M_{13}c_{13} - \ldots + (-1)^{N+1}M_{1N}c_{1N}$$

In this notation, M_{1n} is the element in the first row, n^{th} column of the determinant and c_{1n} is the cofactor of that element. The cofactor, c_{1n}, is a determinant whose elements are the elements of the original determinant with row 1 and column n removed. Note that the elements of the first row after M_{11} and M_{12} are zero, so $P_N = M_{11}c_{11} - M_{12}c_{12}$, $M_{11} = x$ and c_{11} is the secular determinant of an $(N-1)$-carbon polyene; that is, the c_{11} cofactor in P_N is P_{N-1}. $M_{12} = 1$, so we are almost there:

$$P_N = xP_{N-1} - c_{12}$$

We need to examine c_{12}. As shown below, c_{12} is the determinant left after crossing out the elements in the original P_N determinant that are in the first row or the second column:

$$c_{12} = \begin{vmatrix} \cancel{x} & \boxed{\cancel{1}} & \cancel{0} & \cancel{0} & \cancel{0} & \cancel{\ldots} & \cancel{0} & \cancel{0} \\ 1 & \cancel{x} & 1 & 0 & 0 & \ldots & 0 & 0 \\ 0 & \cancel{1} & x & 1 & 0 & \ldots & 0 & 0 \\ 0 & \cancel{0} & 1 & x & 1 & \ldots & 0 & 0 \\ \vdots & \cancel{\vdots} & \vdots & \vdots & \vdots & \vdots & \vdots & \vdots \\ 0 & \cancel{0} & 0 & 0 & 0 & \ldots & x & 1 \\ 0 & \cancel{0} & 0 & 0 & 0 & \ldots & 1 & x \end{vmatrix}.$$

Now c_{12} is not itself a secular determinant of a polyene; its first row has ones as its first two elements and no x elements. Its first *column*, though, has only one element, a one. Expanding this determinant by the elements and cofactors of its first *column* yields a one-term sum whose cofactor is illustrated here. We cross out elements in the original P_N determinant, not only those in the first row or second column, but also those in the same row or column as the single term in the first column of c_{12}:

$$c_{12} = 1\times \begin{vmatrix} \cancel{x} & \cancel{1} & \cancel{0} & \cancel{0} & \cancel{0} & \cancel{\ldots} & \cancel{0} & \cancel{0} \\ \boxed{\cancel{1}} & \cancel{x} & \cancel{1} & \cancel{0} & \cancel{0} & \cancel{\ldots} & \cancel{0} & \cancel{0} \\ \cancel{0} & \cancel{1} & x & 1 & 0 & \ldots & 0 & 0 \\ \cancel{0} & \cancel{0} & 1 & x & 1 & \ldots & 0 & 0 \\ \cancel{\vdots} & \cancel{\vdots} & \vdots & \vdots & \vdots & \vdots & \vdots & \vdots \\ \cancel{0} & \cancel{0} & 0 & 0 & 0 & \ldots & x & 1 \\ \cancel{0} & \cancel{0} & 0 & 0 & 0 & \ldots & 1 & x \end{vmatrix} = 1\times P_{N-2} = P_{N-2}$$

That is, the cofactor involved in evaluating c_{12} is the secular determinant of an $(N-2)$-carbon polyene, *i.e.*, P_{N-2}. So $\boxed{P_N = xP_{N-1} - P_{N-2}}$.

P26.9 (a) $$\begin{vmatrix} \alpha-E & \beta & \beta \\ \beta & \alpha-E & \beta \\ \beta & \beta & \alpha-E \end{vmatrix} = 0$$

$$(\alpha-E)\begin{vmatrix} \alpha-E & \beta \\ \beta & \alpha-E \end{vmatrix} - \beta\begin{vmatrix} \beta & \beta \\ \beta & \alpha-E \end{vmatrix} + \beta\begin{vmatrix} \beta & \alpha-E \\ \beta & \beta \end{vmatrix} = 0$$

$$(\alpha-E)\times\{(\alpha-E)^2-\beta^2\}-\beta\{\beta(\alpha-E)-\beta^2\}+\beta\{\beta^2-(\alpha-E)\beta\}=0$$

$$(\alpha-E)\times\{(\alpha-E)^2-\beta^2\}-2\beta^2\{\alpha-E-\beta\}=0$$

$$(\alpha-E)\times(\alpha-E+\beta)\times(\alpha-E-\beta)-2\beta^2(\alpha-E-\beta)=0$$

$$(\alpha-E-\beta)\times\{(\alpha-E)\times(\alpha-E+\beta)-2\beta^2\}=0$$

$$(\alpha-E-\beta)\times\{(\alpha-E)\times(\alpha-E+2\beta)-\beta(\alpha-E)-2\beta^2\}=0$$

$$(\alpha-E-\beta)\times\{(\alpha-E)\times(\alpha-E+2\beta)-\beta(\alpha-E+2\beta)\}=0$$

$$(\alpha-E-\beta)\times(\alpha-E+2\beta)\times(\alpha-E-\beta)=0$$

Therefore, the desired roots are $E = \boxed{\alpha-\beta,\ \alpha-\beta,\ \text{and}\ \alpha+2\beta}$. The energy level diagram is shown in Fig. 26.1.

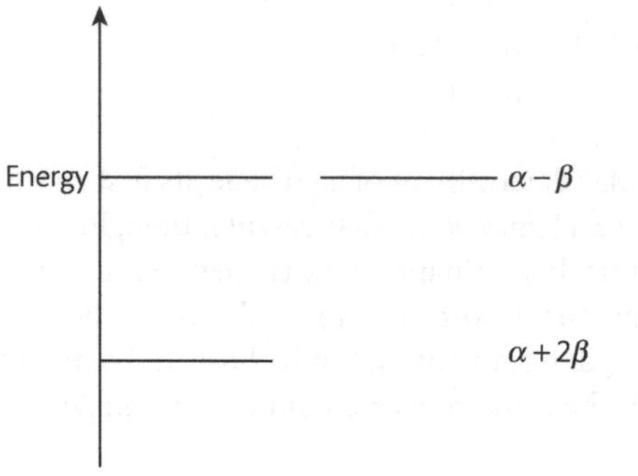

Fig 26.1

The binding energies are shown in the following table:

Species	Number of e⁻	Binding energy
H_3^+	2	$2(\alpha+2\beta)=2\alpha+4\beta$
H_3	3	$2(\alpha+2\beta)+(\alpha-\beta)=3\alpha+3\beta$
H_3^-	4	$2(\alpha+2\beta)+2(\alpha-\beta)=4\alpha+2\beta$

(b)

$H_3^+(g) \rightarrow 2\,H(g) + H^+(g)$	$\Delta H_1 = 849\ \text{kJ mol}^{-1}$
$H^+(g) + H_2(g) \rightarrow H_3^+(g)$	$\Delta H_2 = ?$
$H_2(g) \rightarrow 2\,H(g)$	$\Delta H_3 = \{2(217.97) - 0\}\ \text{kJ mol}^{-1}$

$\Delta H_2 = \Delta H_3 - \Delta H_1 = \{2(217.97) - 849\}\ \text{kJ mol}^{-1}$,

$\Delta H_2 = \boxed{-413\ \text{kJ mol}^{-1}}$

This is only slightly less than the binding energy of H_2 (435.94 kJ mol^{-1})

(c) $2\alpha + 4\beta = -\Delta H_1 = -849\ \text{kJ mol}^{-1}$,

so $\quad \beta = \dfrac{-\Delta H_1 - 2\alpha}{4}\quad$ where $\quad \Delta H_1 = 849\ \text{kJ mol}^{-1}$

Species	Binding energy
H_3^+	$2\alpha + 4\beta = -\Delta H_1 = \boxed{-849\ \text{kJ mol}^{-1}}$
H_3	$3\alpha + 3\beta = 3\left(\alpha - \dfrac{\Delta H_1 + 2\alpha}{4}\right) = 3\left(\dfrac{1}{2}\alpha - \dfrac{\Delta H_1}{4}\right) = \boxed{3(\alpha/2) - 212\ \text{kJ mol}^{-1}}$
H_3^-	$4\alpha + 2\beta = 4\alpha - \dfrac{\Delta H_1 + 2\alpha}{2} = 3\alpha - \dfrac{\Delta H_1}{2} = \boxed{3\alpha - 425\ \text{kJ mol}^{-1}}$

As α is a negative quantity, all three of these species are expected to be stable.

Topic 27 **Self-consistent fields**

Discussion questions

27.1 The Fock operator, $f_1 = h_1 + V_{\text{Coulomb}} + V_{\text{Exchange}}$ [27.3], is the sum of three terms. The **core hamiltonian** for electron 1, h_1 [*Brief illustration* 27.1], is a one-electron hamiltonian that accounts for the electron kinetic energy and the potential energy of attraction between the electron and all nuclei:

$$h_1 = -\frac{\hbar^2}{2m_e}\nabla_1^2 - j_0\sum_I^{N_n}\frac{Z_I}{r_{I1}} \quad \text{where} \quad j_0 = \frac{e^2}{4\pi\varepsilon_0}$$

The **Coulomb repulsion** term, V_{Coulomb}, accounts for the average potential energy due to repulsions between electron 1 and all other electrons. The **exchange potential** term provides an average correction to the Coulomb potential for the phenomena of **electron spin correlation** in which electrons of the same spin tend to avoid each other. Spin correlation reduces the energy description provided by the Coulomb repulsion and, consequently, the exchange potential correction has a negative value.

27.3 The Roothaan equations arise within the context of the Hartree-Fock method by writing the wavefunction as a linear combination of atomic orbitals (LCAOs, 26.1, 27.9). Substitution of the LCAOs into the HF equations yields the Roothaan equations for the simultaneous determination of the LCAO coefficients and the corresponding value for the molecular orbital energy. The atomic orbitals of the LCAO are called **basis functions** and the entire set of basis functions is the **basis set**. The assumption that the wavefunction is an LCAO of a chosen basis set adds another layer of approximation to the HF-SCF methodology.

27.5

$$\psi = \frac{1}{(N_e!)^{1/2}}\begin{vmatrix} \psi_a^{\alpha}(1) & \psi_a^{\beta}(1) & \cdots & \psi_z^{\beta}(1) \\ \psi_a^{\alpha}(2) & \psi_a^{\beta}(2) & \cdots & \psi_z^{\beta}(2) \\ \vdots & \vdots & \vdots & \vdots \\ \psi_a^{\alpha}(N_e) & \psi_a^{\beta}(N_e) & \cdots & \psi_z^{\beta}(N_e) \end{vmatrix} \quad \text{[Slater determinant, 27.2a]}$$

Representing the total wavefunction Ψ as a Slater determinant of spin-orbital functions ψ_a is very important because this mathematical form automatically satisfies the principle that the total wavefunction must be antisymmetric (ie., change sign) under the interchange of any pair of electrons. Additionally, the Slater determinant satisfies Pauli exclusion principle that no two electrons can be described by the same spin-orbital, a principle that is synonymous with the antisymmetry principle. The approximate nature of the Slater determinant wavefunction stems from the selection of trial spin-orbitals. Careful selection may result in a better wavefunction and a lower, closer to the exact, energy but the approximation is not flexible enough to fully account for electron correlation. However, since it is a trial wavefunction, the variation principle guarantees that when the principle is used to calculate the energy, the value calculated will not be less than the true energy. The Hartree-Fock method improves upon the orbital approximation by establishing variationally-based equations for the orbitals that minimize energy.

Exercises

27.1(a)

$$\hat{V} = -3j_0\left\{\frac{1}{r_{\text{Li1}}}+\frac{1}{r_{\text{Li2}}}+\frac{1}{r_{\text{Li3}}}+\frac{1}{r_{\text{Li4}}}\right\} - j_0\left\{\frac{1}{r_{\text{H1}}}+\frac{1}{r_{\text{H2}}}+\frac{1}{r_{\text{H3}}}+\frac{1}{r_{\text{H4}}}\right\} + j_0\left\{\frac{1}{r_{12}}+\frac{1}{r_{13}}+\frac{1}{r_{14}}+\frac{1}{r_{23}}+\frac{1}{r_{24}}+\frac{1}{r_{34}}\right\}$$

27.2(a) $N_e = 2$ for HeH^+.

$$\hat{H} = -\frac{\hbar^2}{2m_e}\{\nabla_1^2+\nabla_2^2\} - 2j_0\left\{\frac{1}{r_{\text{He1}}}+\frac{1}{r_{\text{He2}}}\right\} - j_0\left\{\frac{1}{r_{\text{H1}}}+\frac{1}{r_{\text{H2}}}\right\} + j_0\left\{\frac{1}{r_{12}}\right\}$$

27.3(a) $N_e = 2$ for HeH^+. $\Psi = \frac{1}{\sqrt{2}}\begin{vmatrix} \psi_a^\alpha(1) & \psi_a^\beta(1) \\ \psi_a^\alpha(2) & \psi_a^\beta(2) \end{vmatrix}$

27.4(a) $N_e = 2$ for HeH^+.

$$f_1\psi_a(1) = h_1\psi_a(1) + 2J_a(1)\psi_a(1) - K_a(1)\psi_a(1) = \varepsilon_a\psi_a(1) \quad \text{[27.8] and [27.3]}$$

where

$$h_1 = -\frac{\hbar^2}{2m_e}\nabla_1^2 - j_0\left\{\frac{2}{r_{\text{He1}}}+\frac{1}{r_{\text{H1}}}\right\} \quad \textit{[Brief illustration 27.1]}$$

$$J_a(1)\psi_a(1) = j_0\int \psi_a(1)\frac{1}{r_{12}}\psi_a^*(2)\psi_a(2)\mathrm{d}\tau_2 = K_a(1)\psi_a(1) \quad \text{[27.4] and [27.6]}$$

Consequently,

$$\boxed{f_1\psi_a(1) = h_1\psi_a(1) + J_a(1)\psi_a(1) = \varepsilon_a\psi_a(1)}$$

27.5(a) The complete set of equations for the $N_e = 2$ of HeH^+ is:

(i) $\psi_a = c_{\text{He}a}\chi_{\text{He}} + c_{\text{H}a}\chi_{\text{H}}$ and $\psi_b = c_{\text{He}b}\chi_{\text{He}} + c_{\text{H}b}\chi_{\text{H}}$ [6.9]

(ii) $f_1\chi_{\text{He}}(1) = h_1\chi_{\text{He}}(1) + 2J_a(1)\chi_{\text{He}}(1) - K_a(1)\chi_{\text{He}}(1)$ [27.8]

$f_1\chi_{\text{H}}(1) = h_1\chi_{\text{H}}(1) + 2J_a(1)\chi_{\text{H}}(1) - K_a(1)\chi_{\text{H}}(1)$ [27.8] where

$$h_1 = -\frac{\hbar^2}{2m_e}\nabla_1^2 - j_0\left\{\frac{2}{r_{\text{He1}}} + \frac{1}{r_{\text{H1}}}\right\} \quad [\textit{Brief illustration } 27.1]$$

$$J_a(1)\chi_{\text{He}}(1) = j_0\int \chi_{\text{He}}(1)\frac{1}{r_{12}}\psi_a^*(2)\psi_a(2)\mathrm{d}\tau_2 \quad [27.4]$$

$$K_a(1)\chi_{\text{He}}(1) = j_0\int \psi_a(1)\frac{1}{r_{12}}\psi_a^*(2)\chi_{\text{He}}(2)\mathrm{d}\tau_2 \quad [27.6]$$

$$J_a(1)\chi_{\text{H}}(1) = j_0\int \chi_{\text{H}}(1)\frac{1}{r_{12}}\psi_a^*(2)\psi_a(2)\mathrm{d}\tau_2 \quad [27.4]$$

$$K_a(1)\chi_{\text{H}}(1) = j_0\int \psi_a(1)\frac{1}{r_{12}}\psi_a^*(2)\chi_{\text{H}}(2)\mathrm{d}\tau_2 \quad [27.6]$$

(iii) $\boldsymbol{Fc} = \boldsymbol{Sc\varepsilon}$ [27.10] and [27.11]

where

$$\boldsymbol{F} = \begin{pmatrix} \int \chi_{\text{He}}(1)f_1\chi_{\text{He}}(1)\mathrm{d}\tau_1 & \int \chi_{\text{He}}(1)f_1\chi_{\text{H}}(1)\mathrm{d}\tau_1 \\ \int \chi_{\text{H}}(1)f_1\chi_{\text{He}}(1)\mathrm{d}\tau_1 & \int \chi_{\text{H}}(1)f_1\chi_{\text{H}}(1)\mathrm{d}\tau_1 \end{pmatrix} \quad (F_{12} = F_{21})$$

$$\boldsymbol{c} = \begin{pmatrix} c_{\text{He}a} & c_{\text{He}b} \\ c_{\text{H}a} & c_{\text{H}b} \end{pmatrix}$$

$$\boldsymbol{S} = \begin{pmatrix} \int \chi_{\text{He}}(1)\chi_{\text{He}}(1)\mathrm{d}\tau_1 & \int \chi_{\text{He}}(1)\chi_{\text{H}}(1)\mathrm{d}\tau_1 \\ \int \chi_{\text{H}}(1)\chi_{\text{He}}(1)\mathrm{d}\tau_1 & \int \chi_{\text{H}}(1)\chi_{\text{H}}(1)\mathrm{d}\tau_1 \end{pmatrix} = \begin{pmatrix} 1 & S \\ S & 1 \end{pmatrix}$$

Note: $\int \chi_{\text{He}}(1)\chi_{\text{H}}(1)\mathrm{d}\tau_1 = \int \chi_{\text{H}}(1)\chi_{\text{He}}(1)\mathrm{d}\tau_1 = S$

$$\boldsymbol{\varepsilon} = \begin{pmatrix} \varepsilon_a & 0 \\ 0 & \varepsilon_b \end{pmatrix}$$

27.6(a) For the $N_e = 2$ of HeH^+:

$$\begin{aligned}
F_{\text{AA}} &= \int \chi_{\text{He}}(1)f_1\chi_{\text{He}}(1)\mathrm{d}\tau_1 \\
&= \int \chi_{\text{He}}(1)\{h_1\chi_{\text{He}}(1) + 2J_a(1)\chi_{\text{He}}(1) - K_a(1)\chi_{\text{He}}(1)\}\mathrm{d}\tau_1 \quad [27.11] \text{ and } [27.8] \\
&= \int \chi_{\text{He}}(1)h_1\chi_{\text{He}}(1)\mathrm{d}\tau_1 + 2\int \chi_{\text{He}}(1)\{J_a(1)\chi_{\text{He}}(1)\}\mathrm{d}\tau_1 - \int \chi_{\text{He}}(1)\{K_a(1)\chi_{\text{He}}(1)\}\mathrm{d}\tau_1 \\
&= \int \chi_{\text{He}}(1)h_1\chi_{\text{He}}(1)\mathrm{d}\tau_1 + 2\int \chi_{\text{He}}(1)\left\{j_0\int \chi_{\text{He}}(1)\frac{1}{r_{12}}\psi_a^*(2)\psi_a(2)\mathrm{d}\tau_2\right\}\mathrm{d}\tau_1 \\
&\quad - \int \chi_{\text{He}}(1)\left\{j_0\int \psi_a(1)\frac{1}{r_{12}}\psi_a^*(2)\chi_{\text{He}}(2)\mathrm{d}\tau_2\right\}\mathrm{d}\tau_1
\end{aligned}$$

$$=\int \chi_{\rm He}(1)h_1\chi_{\rm He}(1)\mathrm{d}\tau_1+2j_0\int\int \chi_{\rm He}(1)\chi_{\rm He}(1)\frac{1}{r_{12}}\psi_a^*(2)\psi_a(2)\mathrm{d}\tau_1\mathrm{d}\tau_2$$

$$-j_0\int\int \chi_{\rm He}(1)\psi_a(1)\frac{1}{r_{12}}\psi_a^*(2)\chi_{\rm He}(2)\mathrm{d}\tau_1\mathrm{d}\tau_2$$

$$=E_{\rm He}+2j_0\int\int \chi_{\rm He}(1)\chi_{\rm He}(1)\frac{1}{r_{12}}\{c_{\rm Hea}\chi_{\rm He}(2)+c_{\rm Ha}\chi_{\rm H}(2)\}^2\,\mathrm{d}\tau_1\mathrm{d}\tau_2$$

$$-j_0\int\int \chi_{\rm He}(1)\{c_{\rm Hea}\chi_{\rm He}(1)+c_{\rm Ha}\chi_{\rm H}(1)\}\frac{1}{r_{12}}\{c_{\rm Hea}\chi_{\rm He}(2)+c_{\rm Ha}\chi_{\rm H}(2)\}\chi_{\rm He}(2)\mathrm{d}\tau_1\mathrm{d}\tau_2$$

$$=E_{\rm He}+2j_0\int\int \chi_{\rm He}(1)\chi_{\rm He}(1)\frac{1}{r_{12}}\{c_{\rm Hea}{}^2\chi_{\rm He}{}^2(2)+2c_{\rm Hea}c_{\rm Ha}\chi_{\rm He}(2)\chi_{\rm H}(2)+c_{\rm Ha}{}^2\chi_{\rm H}{}^2(2)\}\,\mathrm{d}\tau_1\mathrm{d}\tau_2$$

$$-j_0\int\int \{c_{\rm Hea}\chi_{\rm He}(1)\chi_{\rm He}(1)+c_{\rm Ha}\chi_{\rm He}(1)\chi_{\rm H}(1)\}\frac{1}{r_{12}}\{c_{\rm Hea}\chi_{\rm He}(2)\chi_{\rm He}(2)+c_{\rm Ha}\chi_{\rm He}(2)\chi_{\rm H}(2)\}\,\mathrm{d}\tau_1\mathrm{d}\tau_2$$

Multiplication of the factors and substituting the symbolism

$$(\mathrm{AB|CD})=j_0\int\int \chi_{\rm A}(1)\chi_{\rm B}(1)\frac{1}{r_{12}}\chi_{\rm C}(2)\chi_{\rm D}(2)\,\mathrm{d}\tau_1\mathrm{d}\tau_2 \text{ gives:}$$

$$F_{\rm AA}=E_{\rm He}+\left\{2c_{\rm Hea}{}^2(\mathrm{HeHe|HeHe})+4c_{\rm Hea}c_{\rm Ha}(\mathrm{HeHe|HeH})+2c_{\rm Ha}{}^2(\mathrm{HeHe|HH})\right\}$$

$$-\left\{c_{\rm Hea}{}^2(\mathrm{HeHe|HeHe})+c_{\rm Hea}c_{\rm Ha}(\mathrm{HeHe|HeH})+c_{\rm Hea}c_{\rm Ha}(\mathrm{HeH|HeHe})+c_{\rm Ha}{}^2(\mathrm{HeH|HeH})\right\}$$

$$\boxed{F_{\rm AA}=E_{\rm He}+c_{\rm Hea}{}^2(\mathrm{HeHe|HeHe})+2c_{\rm Hea}c_{\rm Ha}(\mathrm{HeHe|HeH})+c_{\rm Ha}{}^2\{2(\mathrm{HeHe|HH})-(\mathrm{HeH|HeH})\}}$$

$$F_{\rm AB}=\int \chi_{\rm He}(1)f_1\chi_{\rm H}(1)\mathrm{d}\tau_1=\int \chi_{\rm He}(1)\{h_1\chi_{\rm H}(1)+2J_a(1)\chi_{\rm H}(1)-K_a(1)\chi_{\rm H}(1)\}\mathrm{d}\tau_1 \quad \text{[27.11] and [27.8]}$$

$$=\int \chi_{\rm He}(1)h_1\chi_{\rm H}(1)\mathrm{d}\tau_1+2\int \chi_{\rm He}(1)\{J_a(1)\chi_{\rm H}(1)\}\mathrm{d}\tau_1-\int \chi_{\rm He}(1)\{K_a(1)\chi_{\rm H}(1)\}\mathrm{d}\tau_1$$

$$=\int \chi_{\rm He}(1)h_1\chi_{\rm H}(1)\mathrm{d}\tau_1+2\int \chi_{\rm He}(1)\left\{j_0\int \chi_{\rm H}(1)\frac{1}{r_{12}}\psi_a^*(2)\psi_a(2)\mathrm{d}\tau_2\right\}\mathrm{d}\tau_1$$

$$-\int \chi_{\rm He}(1)\left\{j_0\int \psi_a(1)\frac{1}{r_{12}}\psi_a^*(2)\chi_{\rm H}(2)\mathrm{d}\tau_2\right\}\mathrm{d}\tau_1$$

$$=\int \chi_{\rm He}(1)h_1\chi_{\rm H}(1)\mathrm{d}\tau_1+2j_0\int\int \chi_{\rm He}(1)\chi_{\rm H}(1)\frac{1}{r_{12}}\psi_a^*(2)\psi_a(2)\mathrm{d}\tau_1\mathrm{d}\tau_2$$

$$-j_0\int\int \chi_{\rm He}(1)\psi_a(1)\frac{1}{r_{12}}\psi_a^*(2)\chi_{\rm H}(2)\mathrm{d}\tau_1\mathrm{d}\tau_2$$

$$=\int \chi_{\rm He}(1)h_1\chi_{\rm H}(1)\mathrm{d}\tau_1+2j_0\int\int \chi_{\rm He}(1)\chi_{\rm H}(1)\frac{1}{r_{12}}\{c_{\rm Hea}\chi_{\rm He}(2)+c_{\rm Ha}\chi_{\rm H}(2)\}^2\,\mathrm{d}\tau_1\mathrm{d}\tau_2$$

$$-j_0\int\int \chi_{\rm He}(1)\{c_{\rm Hea}\chi_{\rm He}(1)+c_{\rm Ha}\chi_{\rm H}(1)\}\frac{1}{r_{12}}\{c_{\rm Hea}\chi_{\rm He}(2)+c_{\rm Ha}\chi_{\rm H}(2)\}\chi_{\rm H}(2)\mathrm{d}\tau_1\mathrm{d}\tau_2$$

$$=\int \chi_{\rm He}(1)h_1\chi_{\rm H}(1)\mathrm{d}\tau_1$$

$$+2j_0\int\int \chi_{\rm He}(1)\chi_{\rm H}(1)\frac{1}{r_{12}}\{c_{\rm Hea}{}^2\chi_{\rm He}{}^2(2)+2c_{\rm Hea}c_{\rm Ha}\chi_{\rm He}(2)\chi_{\rm H}(2)+c_{\rm Ha}{}^2\chi_{\rm H}{}^2(2)\}\,\mathrm{d}\tau_1\mathrm{d}\tau_2$$

$$-j_0\int\int \{c_{\rm Hea}\chi_{\rm He}(1)\chi_{\rm He}(1)+c_{\rm Ha}\chi_{\rm He}(1)\chi_{\rm H}(1)\}\frac{1}{r_{12}}\{c_{\rm Hea}\chi_{\rm He}(2)\chi_{\rm H}(2)+c_{\rm Ha}\chi_{\rm H}{}^2(2)\}\mathrm{d}\tau_1\mathrm{d}\tau_2$$

Multiplication of the factors and substituting the symbolism $(\mathrm{AB}|\mathrm{CD}) = j_0\int\int \chi_{\mathrm{A}}(1)\chi_{\mathrm{B}}(1)\frac{1}{r_{12}}\chi_{\mathrm{C}}(2)\chi_{\mathrm{D}}(2)\,\mathrm{d}\tau_1\,\mathrm{d}\tau_2$ gives:

$$F_{\mathrm{AB}} = \int \chi_{\mathrm{He}}(1)h_1\chi_{\mathrm{H}}(1)\mathrm{d}\tau_1 + \left\{2c_{\mathrm{Hea}}{}^2(\mathrm{HeH}|\mathrm{HeHe}) + 4c_{\mathrm{Hea}}c_{\mathrm{Ha}}(\mathrm{HeH}|\mathrm{HeH}) + 2c_{\mathrm{Ha}}{}^2(\mathrm{HeH}|\mathrm{HH})\right\}$$
$$-\left\{c_{\mathrm{Hea}}{}^2(\mathrm{HeHe}|\mathrm{HeH}) + c_{\mathrm{Hea}}c_{\mathrm{Ha}}(\mathrm{HeHe}|\mathrm{HH}) + c_{\mathrm{Hea}}c_{\mathrm{Ha}}(\mathrm{HeH}|\mathrm{HeH}) + c_{\mathrm{Ha}}{}^2(\mathrm{HeH}|\mathrm{HH})\right\}$$

$$\boxed{F_{\mathrm{AB}} = \int \chi_{\mathrm{He}}(1)h_1\chi_{\mathrm{H}}(1)\mathrm{d}\tau_1 + c_{\mathrm{Hea}}{}^2(\mathrm{HeH}|\mathrm{HeHe}) + c_{\mathrm{Hea}}c_{\mathrm{Ha}}\{3(\mathrm{HeH}|\mathrm{HeH}) - (\mathrm{HeHe}|\mathrm{HH})\} + c_{\mathrm{Ha}}{}^2(\mathrm{HeH}|\mathrm{HH})}$$

Notes:

(i) $E_{\mathrm{He}} = \int \chi_{\mathrm{He}}(1)h_1\chi_{\mathrm{He}}(1)\mathrm{d}\tau_1$ [Brief illustration 27.4] is the energy of an electron in orbital χ_{He}, taking into account its interaction with both the He and H nuclei.

(ii) In the above manipulations we have used the relationship [see E27.7(a) and E27.7(b)]:

$$(\mathrm{AB}|\mathrm{AA}) = (\mathrm{AA}|\mathrm{AB}).$$

(iii) We have explicitly shown the integral signs for both the integration over all τ_2 values and the integration over all τ_1 values. For example, we have written the four-centre, two electron integral as

$$(\mathrm{AB}|\mathrm{CD}) = j_0\int\int \chi_{\mathrm{A}}(1)\chi_{\mathrm{B}}(1)\frac{1}{r_{12}}\chi_{\mathrm{C}}(2)\chi_{\mathrm{D}}(2)\,\mathrm{d}\tau_1\,\mathrm{d}\tau_2.$$

This has the same meaning as the equation that shows a single integral sign only:

$$(\mathrm{AB}|\mathrm{CD}) = j_0\int \chi_{\mathrm{A}}(1)\chi_{\mathrm{B}}(1)\frac{1}{r_{12}}\chi_{\mathrm{C}}(2)\chi_{\mathrm{D}}(2)\mathrm{d}\tau_1\mathrm{d}\tau_2 \text{ [27.14]}.$$

We explicitly show the double integration sign to help with the visual recognition of the requisite mathematics.

27.7(a) $$(\mathrm{AA}|\mathrm{AB}) = j_0\int \chi_{\mathrm{A}}(1)\chi_{\mathrm{A}}(1)\frac{1}{r_{12}}\chi_{\mathrm{A}}(2)\chi_{\mathrm{B}}(2)\mathrm{d}\tau_1\mathrm{d}\tau_2 \text{ [6.14]}$$

$$= j_0\int\int \chi_{\mathrm{A}}(1)\chi_{\mathrm{A}}(1)\frac{1}{r_{12}}\chi_{\mathrm{A}}(2)\chi_{\mathrm{B}}(2)\,\mathrm{d}\tau_1\,\mathrm{d}\tau_2$$

This integral is symmetric with the interchange of electron 1 and electron 2. Consequently, (AA|AB) = (AB|AA). Furthermore, the functions A and B, like all functions, commute so the integral may also be written in the forms (AA|BA) and (BA|AA).

27.8(a) CH_3Cl

(a) Minimal basis set: one basis function for the 1s orbital of each of the three hydrogen atoms; one basis function for each of the 1s, 2s, and three 2p orbitals for the carbon atom; one basis function for each of the 1s, 2s, three 2p, 3s, and three 3p orbitals for the chlorine atom. Thus, the minimal basis set consists of $\boxed{\text{17 basis functions}}$.

(b) Split-valence basis set: two basis functions for the 1s orbital of each of the three hydrogen atoms; one basis function for 1s orbital of carbon and two basis functions for each of the 2s and three 2p orbitals of the carbon atom; one basis function for the 1s, 2s, three 2p orbitals of the chlorine atom and two basis functions for each of the 3s and three 3p orbitals of the chlorine atom. Thus, the split-valence basis set consists of $\boxed{28 \text{ basis functions}}$.

(c) Double-zeta basis set: each basis function in the minimal basis is replaced by two functions. Thus, the double-zeta basis set consists of $\boxed{34 \text{ basis functions}}$.

27.9(a) p-type Gaussian: $\chi_p = Nx^i y^j z^k e^{-\alpha r^2}$ where $i+j+k=1$ [27.16]. Example: $\chi_{p_x} = Nxe^{-\alpha r^2}$

27.10(a) The s-type Gaussian, $\chi_s = Ne^{-\alpha r^2}$ [27.16], can be written as the product of these three one-dimensional Gaussian functions: $N_x e^{-\alpha x^2}$, $N_y e^{-\alpha y^2}$, $N_z e^{-\alpha z^2}$.

Proof: $\left(N_x e^{-\alpha x^2}\right)\times\left(N_y e^{-\alpha y^2}\right)\times\left(N_z e^{-\alpha z^2}\right) = N_x N_y N_z e^{-\alpha x^2-\alpha y^2-\alpha z^2} = Ne^{-\alpha(x^2+y^2+z^2)} =$

$Ne^{-\alpha r^2}$ where $N = N_x N_y N_z$

27.11(a) Analysis of the s-type Gaussians for HeH^+ is facilitated by placing the He nucleus at the origin of the Cartesian coordinate system and placing the H nucleus at the coordinate $(x,y,z) = (d,0,0)$. In general an s-type Gaussian centered on the point $(x,y,z) = (x_c, y_c, z_c)$ has the form $Ne^{-\alpha\{(x-x_c)^2+(y-y_c)^2+(z-z_c)^2\}}$ [27.16]. So with our choice for the coordinate system the s-type Gaussians on He and H are:

$$\chi_{He} = N_{He}e^{-\alpha_{He}r^2} = N_{He}e^{-\alpha_{He}\{x^2+y^2+z^2\}} \quad \text{and} \quad \chi_H = N_H e^{-\alpha_H\{(x-d)^2+y^2+z^2\}}$$

Their product is

$$\begin{aligned}\chi_{He}\chi_H &= N_{He}N_H e^{-\alpha_{He}\{x^2+y^2+z^2\}}e^{-\alpha_H\{(x-d)^2+y^2+z^2\}}\\ &= N_{He}N_H e^{-[\alpha_{He}\{x^2+y^2+z^2\}+\alpha_H\{(x-d)^2+y^2+z^2\}]}\end{aligned}$$

and it will have the s-type Gaussian form and be in an intermediate position provided that the exponent equals $-\alpha\{(x-e)^2+y^2+z^2\}-\beta$ where α, e, and β are constants. If we can find these constants, which are independent of coordinates, we will have demonstrated that the product of two s-type Gaussians is a Gaussian.

$$\alpha_{He}\{x^2+y^2+z^2\}+\alpha_H\{(x-d)^2+y^2+z^2\} = \alpha\{(x-e)^2+y^2+z^2\}+\beta$$
$$\alpha_{He}\{x^2+y^2+z^2\}+\alpha_H\{x^2-2dx+d^2+y^2+z^2\} = \alpha\{x^2-2ex+e^2+y^2+z^2\}+\beta$$
$$(\alpha_{He}+\alpha_H-\alpha)(x^2+y^2+z^2)-2(\alpha_H d-\alpha e)x+\alpha_H d^2-\alpha e^2-\beta = 0$$

This last expression can equal zero for any chosen value of the coordinate point (x,y,z) provided that the coefficients of the polynomial expression each equal zero. Thus, from the quadratic term we find that $\alpha = \alpha_{He}+\alpha_H$. The first order term yields

$$e = \frac{\alpha_H d}{\alpha} = \frac{\alpha_H d}{\alpha_{He}+\alpha_H}$$

and the constant term yields

$$\beta=\alpha_{\rm H}d^2-\alpha e^2=\alpha_{\rm H}d^2-(\alpha_{\rm He}+\alpha_{\rm H})\left(\frac{\alpha_{\rm H}d}{\alpha_{\rm He}+\alpha_{\rm H}}\right)^2=\alpha_{\rm H}d^2-\frac{(\alpha_{\rm H}d)^2}{\alpha_{\rm He}+\alpha_{\rm H}}=\frac{\alpha_{\rm He}\alpha_{\rm H}d^2}{\alpha_{\rm He}+\alpha_{\rm H}}.$$

Thus,

$$\chi_{\rm He}\chi_{\rm H}=N_{\rm He}N_{\rm H}{\rm e}^{-\alpha\{(x-e)^2+y^2+z^2\}-\beta}=\left(N_{\rm He}N_{\rm H}{\rm e}^{-\beta}\right)\times{\rm e}^{-\alpha\{(x-e)^2+y^2+z^2\}}$$

$$\text{where } \alpha=\alpha_{\rm He}+\alpha_{\rm H},\ e=\frac{\alpha_{\rm H}d}{\alpha_{\rm He}+\alpha_{\rm H}},\ \text{and } \beta=\frac{\alpha_{\rm He}\alpha_{\rm H}d^2}{\alpha_{\rm He}+\alpha_{\rm H}}$$

This demonstrates that the product is a Gaussian centered at $(x,y,z) = (e,0,0)$, which is positioned at a point intermediate between the two nuclei because $e=\dfrac{\alpha_{\rm H}d}{\alpha_{\rm He}+\alpha_{\rm H}}<d$.

Problems

Many of the following problems call on the use of commercially available software. Use versions that are available with this text or the software recommended by your instructor.

27.1 For H_2 the 6-31G* basis set is equivalent to the 6-31G basis set because the star indicates that the basis set adds d-type polarization functions for each atom other than hydrogen. Consequently, we choose the basis sets **(a)** 6-31G* and **(b)** 6-311+G**. 1 au = 27.2114 eV. Since the calculated energy is with respect to the energy of widely separated stationary electrons and nucli, the experimental ground electronic energy of dihydrogen is calculated as $D_e + 2I$.

Features Calculated with HF-SCF Method*
Bond length (R) in pm and ground-state energy (E_0) in eV

H_2	(a) 6-31G*	(b) 6-311 + G**	exp
R	73.0	73.5	74.1
E_0	–30.6626	–30.8167	–32.06
F_2	(a) 3-31G*	(b) 6-311 + G**	exp
R	134.5	132.9	141.8
E_0	–5406.30	–5407.92	

*Spartan '10TM

Both computational basis sets give satisfactory bond length agreement with the experimental value for H_2. However the 6-31G* basis set is not as accurate as the larger basis set as illustrated by consideration of both its higher ground-state energy and the variation principle that the energy of a trial wavefunction is never less than the true energy. That is, the energy provided by the 6-311 + G** basis set is closer to the true energy. Figure 27.1 shows the variation of the dihyrogen ground-state energy with the internuclear distance.

It is surprising that the 6-311 + G** basis set gives a significantly shorter bond length for F_2. This might be an indication that the method should be used with caution when fluorine is present in a molecule.

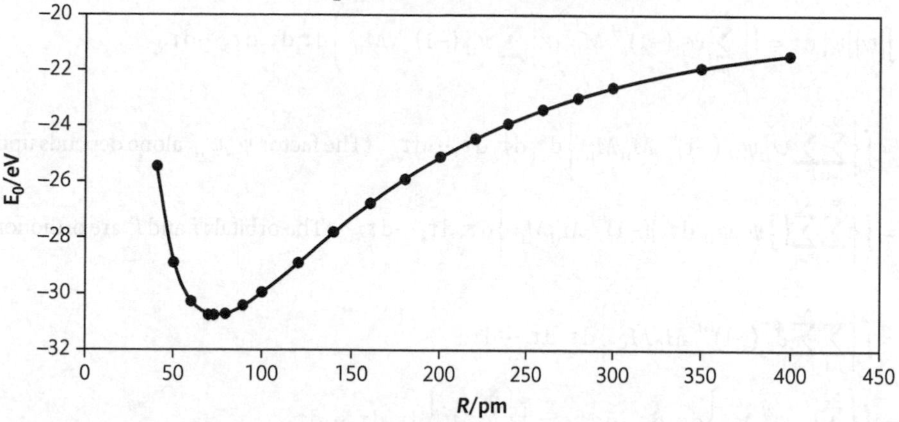

Figure 27.1

27.3 To facilitate the algebraic manipulation of the determinant in equation 27.2a we will use a subscript representation of the determinant elements and expand in terms of the 1st row (the orbitals being occupied by electron 1):

$$|\psi| = \begin{vmatrix} \psi_{11} & \psi_{12} & \cdots & \psi_{1N_e} \\ \psi_{21} & \psi_{22} & \cdots & \vdots \\ \vdots & \vdots & \cdots & \vdots \\ \psi_{N_e1} & \psi_{N_e2} & \cdots & \psi_{N_eN_e} \end{vmatrix} = \sum_{l=1}^{N_e} \psi_{1l}(-1)^{1+l} M_{1l}$$

ψ_{ij} represents electron i in the j'th orthonormal single-electron spin-orbital for which $\int \psi_{kl}\psi_{km}d\tau_k = \delta_{lm}$ (i.e., 1 if $l = m$ and 0 otherwise). M_{kl} is the (determinant) minor of element ψ_{kl} while $M_{kl;mn}$ is the minor of the ψ_{mn} element within the M_{kl} minor. $M_{kl;mn;op}$ is the minor of the ψ_{op} element within the $M_{kl;mn}$ minor. $|\psi|$ has $N_e \times N_e$ elements. M_{kl} has $(N_e - 1)\times(N_e - 1)$ elements. $M_{kl;mn}$ has $(N_e - 2) \times (N_e - 2)$ elements. $M_{kl;mn;op}$ has $(N_e - 3) \times (N_e - 3)$ elements, etc. For example,

$$M_{11} = \begin{vmatrix} \psi_{22} & \psi_{23} & \cdots & \psi_{2N_e} \\ \psi_{32} & \psi_{33} & \cdots & \vdots \\ \vdots & \vdots & \cdots & \vdots \\ \psi_{N_e2} & \psi_{N_e3} & \cdots & \psi_{N_eN_e} \end{vmatrix} \text{ and } M_{11;22} = \begin{vmatrix} \psi_{33} & \psi_{34} & \cdots & \psi_{3N_e} \\ \psi_{43} & \psi_{44} & \cdots & \vdots \\ \vdots & \vdots & \cdots & \vdots \\ \psi_{N_e3} & \psi_{N_e4} & \cdots & \psi_{N_eN_e} \end{vmatrix}$$

We have chosen the determinant expansion $|\psi| = \sum_{l=1}^{N_e} \psi_{1l}(-1)^{1+l} M_{1l}$ because only the ψ_{1l} factors depend upon the coordinates of electron 1 (the minors M_{1l} do not). This will prove to be useful when integrations are performed. The normalization constant of the Slater determinant wavefunction is given by $N = (1/\int |\psi|\,|\psi|\,\mathrm{d}\tau)^{1/2}$ so we find N by evaluating the integral $\int |\psi|\,|\psi|\,\mathrm{d}\tau$.

$$\int |\psi||\psi|\,\mathrm{d}\tau = \int \left(\sum_{l=1}^{N_e} \psi_{1l}(-1)^{1+l} M_{1l} \right) \times \left(\sum_{l'=1}^{N_e} \psi_{1l'}(-1)^{1+l'} M_{1l'} \right) \mathrm{d}\tau_1 \mathrm{d}\tau_2 \mathrm{d}\tau_3 \cdots \mathrm{d}\tau_{N_e}$$

$$= \int \left\{ \sum_{l=1}^{N_e} \sum_{l'=1}^{N_e} \psi_{1l}\psi_{1l'}(-1)^{l+l'} M_{1l} M_{1l'} \right\} \mathrm{d}\tau_1\, \mathrm{d}\tau_2\, \mathrm{d}\tau_3 \cdots \mathrm{d}\tau_{N_e} \quad \text{(The factor } \psi_{1l}\psi_{1l'} \text{ alone depends upon } \tau_1\text{.)}$$

$$= \int \left\{ \sum_{l=1}^{N_e} \sum_{l'=1}^{N_e} \left(\int \psi_{1l}\psi_{1l'}\, \mathrm{d}\tau_1 \right) (-1)^{l+l'} M_{1l} M_{1l'} \right\} \mathrm{d}\tau_2\, \mathrm{d}\tau_3 \cdots \mathrm{d}\tau_{N_e} \quad \text{(The orbitals } l \text{ and } l' \text{ are orthonormal.)}$$

$$= \int \left\{ \sum_{l=1}^{N_e} \sum_{l'=1}^{N_e} \delta_{ll'} (-1)^{l+l'} M_{1l} M_{1l'} \right\} \mathrm{d}\tau_2\, \mathrm{d}\tau_3 \cdots \mathrm{d}\tau_{N_e}$$

$$= \int \left\{ \sum_{l=1}^{N_e} (-1)^{2l} M_{1l}^2 \right\} \mathrm{d}\tau_2\, \mathrm{d}\tau_3 \cdots \mathrm{d}\tau_{N_e} = \int \left\{ \sum_{l=1}^{N_e} M_{1l}^2 \right\} \mathrm{d}\tau_2\, \mathrm{d}\tau_3 \cdots \mathrm{d}\tau_{N_e}$$

The above manipulations depend upon the two facts that $\psi_{kl}\psi_{kl'}$ depend upon τ_k alone and the orbitals are orthonormal. There was no reference to any other property of the orbitals. Since we are going to repeat this procedure, we may immediately recognize that each of the N_e terms in the above integrand summation will behave identically to the $M_{11}{}^2$ term. Thus, for the purpose of finding the normalization constant, we may write

$$\int |\psi||\psi|\,\mathrm{d}\tau = N_e \times \int \{ M_{11}{}^2 \} \mathrm{d}\tau_2\, \mathrm{d}\tau_3 \cdots \mathrm{d}\tau_{N_e}$$

and expand the minor M_{11} in factors of ψ_{2l}, which depend upon the coordinates of electron 2 (the minors $M_{2l;mn}$ do not).

$$\int |\psi||\psi|\,\mathrm{d}\tau = N_e \times \int \left(\sum_{l=2}^{N_e} \psi_{2l}(-1)^{l} M_{2l;mn} \right) \times \left(\sum_{l'=2}^{N_e} \psi_{2l'}(-1)^{l'} M_{2l';mn} \right) \mathrm{d}\tau_2 \mathrm{d}\tau_3 \cdots \mathrm{d}\tau_{N_e}$$

$$= N_e \times \int \sum_{l=2}^{N_e} \sum_{l'=2}^{N_e} \left(\int \psi_{2l}\psi_{2l'}\, \mathrm{d}\tau_2 \right) (-1)^{l+l'} M_{2l;mn} M_{2\ell';mn}\, \mathrm{d}\tau_3 \cdots \mathrm{d}\tau_{N_e} \quad \text{(The orbitals } l \text{ and } l' \text{ are orthonormal.)}$$

$$= N_e \times \int \sum_{l=2}^{N_e} \sum_{l'=2}^{N_e} \delta_{ll'} (-1)^{l+l'} M_{2l;mn} M_{2\ell';mn}\, \mathrm{d}\tau_3 \cdots \mathrm{d}\tau_{N_e}$$

$$= N_e \times \int \sum_{l=2}^{N_e} (-1)^{2l} M_{2l;mn}{}^2\, \mathrm{d}\tau_3 \cdots \mathrm{d}\tau_{N_e} = N_e \times \int \left\{ \sum_{l=2}^{N_e} M_{2l;mn}{}^2 \right\} \mathrm{d}\tau_3 \cdots \mathrm{d}\tau_{N_e}$$

Again, each of the minors in the above integrand will behave like the minor $M_{22;mn}$. The summation involves N_e-1 terms so for our purposes we can write:

$$\int|\psi||\psi|\mathrm{d}\tau = N_e\times(N_e-1)\times\int M_{22;mn}{}^2\mathrm{d}\tau_3\cdots\mathrm{d}\tau_{N_e}$$

By repeating this process we must successively consider minors that have a succession of 1 fewer terms in the determinant expansion and 1 fewer electron coordinate in the integration. So we inductively recognize that: $\int|\psi||\psi|\mathrm{d}\tau = N_e\times(N_e-1)\times(N_e-2)\times(N_e-3)\times\cdots\times 1 = N_e!$
We conclude that $N = (1/\int|\psi|\,|\psi|\,\mathrm{d}\tau)^{1/2} = (1/N_e!)^{1/2}$.

27.5 We begin by choosing a coordinate system centered on the N atom of ammonia, which is shown in Figure 27.2 and is compatible with the one used in E27.11a. d is the bond length. θ is the bond angle and the xz plane bisects the H_3NH_2 bond angle.

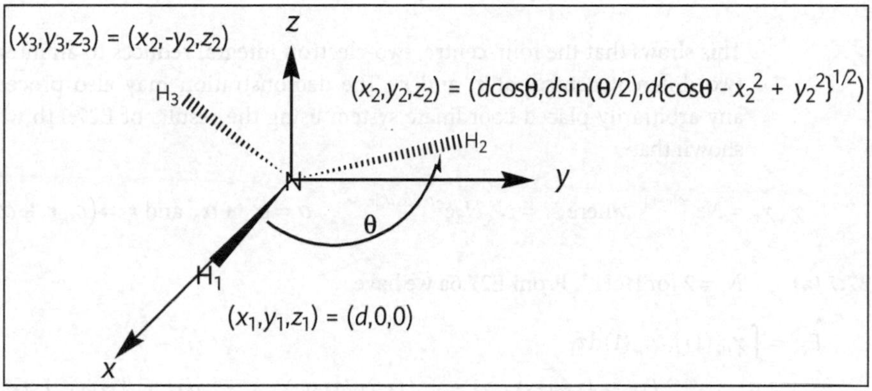

Figure 27.2

The s-type Gaussians on the nitrogen atom and the hydrogen atoms of ammonia are

$$\chi_N = N_N e^{-\alpha_N|r|^2},\ \chi_{H_1} = N_{H_1}e^{\alpha_H|r-r_{H_1}|^2},\ \chi_{H_2} = N_{H_2}e^{-\alpha_H|r-r_{H_2}|^2},\ \chi_{H_3} = N_{H_3}e^{-\alpha_H|r-r_{H_3}|^2}$$

where $\boldsymbol{r}_{H_1}$, $\boldsymbol{r}_{H_2}$, and $\boldsymbol{r}_{H_3}$ are position vectors of the atoms as shown in the figure.

In E27.11a it is shown that

$$\chi_N\chi_{H_1} = \left(N_N N_{H_1} e^{-\beta_1}\right)\times e^{-\alpha_1|\boldsymbol{r}-\boldsymbol{r}_1|^2}$$

where $\alpha_1 = \alpha_N + \alpha_H$, $\boldsymbol{r}_1 = \left(\dfrac{\alpha_H d}{\alpha_N+\alpha_H},0,0\right)$, and $\beta_1 = \dfrac{\alpha_N\alpha_H d^2}{\alpha_N+\alpha_H}$

Thus, the two-centre $\chi_i\chi_j$ s-type Gaussian can be replaced with a one-center Gaussian at the position $\boldsymbol{r}_{NH_1} = (e_1,0,0)$ between N and H_1 atoms. This result also allows us to conclude that the product $\chi_{H_2}\chi_{H_3}$ can be replaced with a one-centre Gaussian that is halfway between H_2 and H_3 because the hydrogen atoms have identical α values.

$$\chi_{H_2}\chi_{H_3} = \left(N_H^2 e^{-\beta_H}\right)\times e^{-\alpha|r-r_c|^2}$$

where $\alpha = 2\alpha_H$, $\beta = \dfrac{d_{HH}}{2} = 2y_2^2$, and $r_c = (x_2, 0, z_2)$.

These products can be substituted in the four-centre, two-electron integral (NH|HH).

$$\begin{aligned}(NH_1|H_2H_3) &= j_0\int \chi_N(1)\chi_{H_1}(1)\frac{1}{r_{12}}\chi_{H_2}(2)\chi_{H_3}(2)d\tau \\ &= j_0\int \chi_N(1)\chi_{H_1}(1)\frac{1}{r_{12}}\chi_{H_2}(2)\chi_{H_3}(2)d\tau_1 d\tau_2 \\ &= N\int e^{-\alpha_1|r(1)-r_1|^2}\frac{1}{r_{12}}e^{-\alpha|r(2)-r_c|^2}d\tau_1 d\tau_2 \quad \text{where} \quad N = j_0 N_N N_H^3 e^{-\beta_1-\beta_H}\end{aligned}$$

This shows that the four-centre, two-electron integral reduces to an integral over two different centres at r_1 and r_c. The demonstration may also proceed using any arbitrarily placed coordinate system using the results of E27.11b where it is shown that

$$\chi_A\chi_B = Ne^{-\alpha|r-r_c|^2} \text{ where } N = N_A N_B e^{-\alpha^{-1}\alpha_A\alpha_B|r_A-r_B|^2},\ \alpha = \alpha_A + \alpha_B \text{ and } r_c = (\alpha_A r_A + \alpha_B r_B)/\alpha.$$

27.7 (a) $N_e = 2$ for HeH^+. From E27.6a we have

$$\begin{aligned}F_{AA} &= \int \chi_{He}(1)f_1\chi_{He}(1)d\tau_1 \\ &= E_{He} + c_{Hea}{}^2(\text{HeHe}|\text{HeHe}) + 2c_{Hea}c_{Ha}(\text{HeHe}|\text{HeH}) + c_{Ha}{}^2\{2(\text{HeHe}|\text{HH}) - (\text{HeH}|\text{HeH})\}\end{aligned}$$

$$\begin{aligned}F_{AB} &= \int \chi_{He}(1)f_1\chi_H(1)d\tau_1 \\ &= \int \chi_{He}(1)h_1\chi_H(1)d\tau_1 + c_{Hea}{}^2(\text{HeH}|\text{HeHe}) + c_{Hea}c_{Ha}\{3(\text{HeH}|\text{HeH}) - (\text{HeHe}|\text{HH})\} \\ &\quad + c_{Ha}{}^2(\text{HeH}|\text{HH})\end{aligned}$$

The above equations arbitrarily utilized the assignments He for "A" an H for "B." To derive equations for F_{BB} and F_{BA}, we need only switch the assignments.

$$\begin{aligned}F_{BB} &= \int \chi_H(1)f_1\chi_H(1)d\tau_1 \\ &= E_H + c_{Ha}{}^2(\text{HH}|\text{HH}) + 2c_{Hea}c_{Ha}(\text{HH}|\text{HeH}) + c_{Hea}{}^2\{2(\text{HeHe}|\text{HH}) - (\text{HeH}|\text{HeH})\}\end{aligned}$$

$$\begin{aligned}F_{BA} &= \int \chi_H(1)f_1\chi_{He}(1)d\tau_1 \\ &= \int \chi_H(1)h_1\chi_{He}(1)d\tau_1 + c_{Ha}{}^2(\text{HeH}|\text{HH}) + c_{Hea}c_{Ha}\{3(\text{HeH}|\text{HeH}) - (\text{HeHe}|\text{HH})\} \\ &\quad + c_{Hea}{}^2(\text{HeH}|\text{HeHe})\end{aligned}$$

(b) $N_e = 2$ for LiH^{2+}. From E27.6b we have

$$F_{AA} = \int \chi_{Li}(1) f_1 \chi_{Li}(1) d\tau_1$$
$$= E_{Li} + c_{Lia}{}^2 (LiLi|LiLi) + 2c_{Lia}c_{Ha}(LiLi|LiH) + c_{Ha}{}^2 \{2(LiLi|HH) - (LiH|LiH)\}$$

$$F_{AB} = \int \chi_{Li}(1) f_1 \chi_{H}(1) d\tau_1$$
$$= \int \chi_{Li}(1) h_1 \chi_{H}(1) d\tau_1 + c_{Lia}{}^2 (LiH|LiLi) + c_{Lia}c_{Ha}\{3(LiH|LiH) - (LiLi|HH)\} + c_{Ha}{}^2 (LiH|HH)$$

The above equations arbitrarily utilized the assignments Li for "A" an H for "B." To derive equations for F_{BB} and F_{BA}, we need only switch the assignments.

$$F_{BB} = \int \chi_{H}(1) f_1 \chi_{H}(1) d\tau_1$$
$$= E_{H} + c_{Ha}{}^2 (HH|HH) + 2c_{Lia}c_{Ha}(HH|LiH) + c_{Lia}{}^2 \{2(LiLi|HH) - (LiH|LiH)\}$$

$$F_{BA} = \int \chi_{H}(1) f_1 \chi_{Li}(1) d\tau_1$$
$$= \int \chi_{H}(1) h_1 \chi_{Li}(1) d\tau_1 + c_{Ha}{}^2 (LiH|HH) + c_{Lia}c_{Ha}\{3(LiH|LiH) - (LiLi|HH)\} + c_{Lia}{}^2 (LiH|LiLi)$$

In the above manipulations we have used the relationships [see E27.7(a) and E27.7(b)]:

$$(AB|AA) = (AA|AB) = (AA|BA)$$

Topic 28 **Semi-empirical methods**

Discussion questions

28.1 In **semi-empirical methods**, many of the integrals that occur in an electronic structure calculation are estimated by appealing to spectroscopic data or physical properties such as ionization energies, or by using a series of rules to set certain integrals equal to zero. The Hückel molecular orbital method is a primitive example of a semi-empirical method: all the properties of the π-system are expressed in terms of the two parameters α and β and all overlap integrals are set equal to zero.

To illustrate how a semi-empirical method estimates the value of a parameter, we summarize the Hückel MO analysis of the resonance integral β using the $\pi^* \leftarrow \pi$ ultraviolet absorptions of the homologous polyene series: ethene (61 500 cm^{-1}), butadiene (46 080 cm^{-1}), hexatriene (39 750 cm^{-1}), and octatetraene (32 900 cm^{-1}). The U.V. absorptions of this series should equal the energy of the LUMO minus the energy of the HOMO. Simple Hückel theory yields the following expression for this difference:

$$\tilde{\nu} = \tilde{E}_{\text{LUMO}} - \tilde{E}_{\text{HOMO}} = (\alpha - x_{\text{LUMO}}\beta) - (\alpha - x_{\text{HOMO}}\beta) = (x_{\text{HOMO}} - x_{\text{LUMO}})\beta$$

with the $x_{\text{HOMO}} - x_{\text{LUMO}}$ values for each member of the series summarized in the following table. Thus, the slope of a plot of $\tilde{\nu}_{\text{observed}}$ against $x_{\text{HOMO}} - x_{\text{LUMO}}$ gives an estimate of β. The plot and linear regression fit are shown in Figure 28.1 where the value of the slope gives the estimate $\beta \sim -21{,}140$ cm^{-1}. Simple Hückel MO theory predicts that the intercept of the plot should equal zero and it is unable to explain the non-zero value.

Substance	Number of π electrons	$x_{\text{HOMO}} - x_{\text{LUMO}}$	$\tilde{\nu}_{\text{observed}}$ / cm^{-1}
Ethene	2	−2.0000	61,500
Butadiene	4	−1.2361	46,080
Hexatriene	6	−0.8981	39,750
Octatetraene	8	−0.6946	32,900

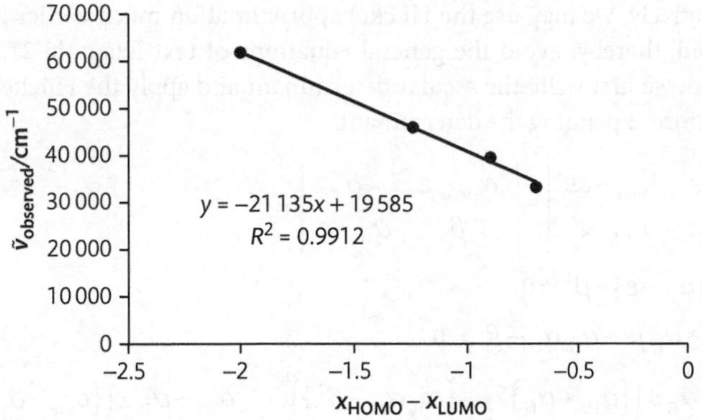

Fig 28.1

Exercises

28.1(a) Very general equations for setting up the Roothaan equations are used in the first and second text Example 27.2. By replacing the index A with He and the index B with H we find that for a linear combination of two basis functions, such as $\chi_{\rm He}$ and $\chi_{\rm H}$, the coefficients of the molecular orbitals in $\psi_a = c_{{\rm He}a}\chi_{\rm He} + c_{{\rm H}a}\chi_{\rm H}$ and $\psi_b = c_{{\rm He}b}\chi_{\rm He} + c_{{\rm H}b}\chi_{\rm H}$ are given by the equations

$$\varepsilon_a = \frac{F_{\rm HeHe}+F_{\rm HH}-S(F_{\rm HeH}-F_{\rm HHe})-\left(\{F_{\rm HeHe}+F_{\rm HH}-S(F_{\rm HeH}-F_{\rm HHe})\}^2-4\{1-S^2\}\{F_{\rm HeHe}F_{\rm HH}-F_{\rm HeH}F_{\rm HHe}\}\right)^{1/2}}{2\{1-S^2\}}$$

and

$$c_{{\rm He}a} = -\frac{F_{\rm HeH}-S\varepsilon_a}{F_{\rm HeHe}-\varepsilon_a}c_{{\rm H}a} \quad \text{and} \quad S=\int \chi_{\rm He}\chi_{\rm H}\,{\rm d}\tau$$

in conjunction with the normalization condition $c_{{\rm He}a}{}^2 + c_{{\rm H}a}{}^2 + 2c_{{\rm He}a}c_{{\rm H}a}S = 1$.

Applying the Hückel approximation $S = 0$ and using the symbols $\alpha_{\rm He} = F_{\rm HeHe}$, $\alpha_{\rm H} = F_{\rm HH}$, and $F_{\rm HeH} = F_{\rm HHe} = \beta$ simplifies the equations to

$$\varepsilon_a = \frac{\alpha_{\rm He}+\alpha_{\rm H}-\left(\{\alpha_{\rm He}+\alpha_{\rm H}\}^2-4\{\alpha_{\rm He}\alpha_{\rm H}-\beta^2\}\right)^{1/2}}{2} = \frac{\alpha_{\rm He}+\alpha_{\rm H}-\left(\{\alpha_{\rm He}-\alpha_{\rm H}\}^2+4\beta^2\right)^{1/2}}{2}$$

and

$$c_{{\rm He}a} = -\frac{\beta}{\alpha_{\rm He}-\varepsilon_a}c_{{\rm H}a}$$

in conjunction with the normalization condition $c_{{\rm He}a}{}^2 + c_{{\rm H}a}{}^2 = 1$. Thus,

$$\left(\frac{\beta}{\alpha_{\rm He}-\varepsilon_a}\right)^2 c_{{\rm H}a}^2 + c_{{\rm H}a}^2 = 1 \text{ or } \boxed{c_{{\rm H}a} = \left(\frac{1}{1+\left(\dfrac{\beta}{\alpha_{\rm He}-\varepsilon_a}\right)^2}\right)^{1/2}}$$

Alternatively, we may use the Hückel approximation much earlier in the derivation and, thereby, avoid the general equations of text Example 27.2. Using this method, we first write the secular determinant and apply the Hückel approximation before expanding the determinant.

$$\begin{vmatrix} F_{\mathrm{HeHe}}-\varepsilon & F_{\mathrm{HeH}}-S\varepsilon \\ F_{\mathrm{HHe}}-S\varepsilon & F_{\mathrm{HH}}-\varepsilon \end{vmatrix} = \begin{vmatrix} \alpha_{\mathrm{He}}-\varepsilon & \beta \\ \beta & \alpha_{\mathrm{H}}-\varepsilon \end{vmatrix} = 0$$

$$(\alpha_{\mathrm{He}}-\varepsilon)(\alpha_{\mathrm{H}}-\varepsilon)-\beta^2=0$$

$$\varepsilon^2-(\alpha_{\mathrm{He}}+\alpha_{\mathrm{H}})\varepsilon+\alpha_{\mathrm{He}}\alpha_{\mathrm{H}}-\beta^2=0$$

$$\varepsilon=\frac{\alpha_{\mathrm{He}}+\alpha_{\mathrm{H}}\pm\left(\{\alpha_{\mathrm{He}}+\alpha_{\mathrm{H}}\}^2-4\{\alpha_{\mathrm{He}}\alpha_{\mathrm{H}}-\beta^2\}\right)^{1/2}}{2}=\frac{\alpha_{\mathrm{He}}+\alpha_{\mathrm{H}}\pm\left(\{\alpha_{\mathrm{He}}-\alpha_{\mathrm{H}}\}^2+4\beta^2\right)^{1/2}}{2}$$

Thus, $\boxed{\varepsilon_a=\frac{\alpha_{\mathrm{He}}+\alpha_{\mathrm{H}}-\left(\{\alpha_{\mathrm{He}}-\alpha_{\mathrm{H}}\}^2+4\beta^2\right)^{1/2}}{2}}$ and

$$\varepsilon_b=\frac{\alpha_{\mathrm{He}}+\alpha_{\mathrm{H}}+\left(\{\alpha_{\mathrm{He}}-\alpha_{\mathrm{H}}\}^2+4\beta^2\right)^{1/2}}{2}.$$

The matrix form of the Roothaan equation can be used to find the wavefunction coefficients.

$$\begin{pmatrix} F_{\mathrm{HeHe}} & F_{\mathrm{HeH}} \\ F_{\mathrm{HHe}} & F_{\mathrm{HH}} \end{pmatrix}\begin{pmatrix} c_{\mathrm{He}a} & c_{\mathrm{He}b} \\ c_{\mathrm{H}a} & c_{\mathrm{H}b} \end{pmatrix}=\begin{pmatrix} S_{\mathrm{HeHe}} & S_{\mathrm{HeH}} \\ S_{\mathrm{HHe}} & S_{\mathrm{HH}} \end{pmatrix}\begin{pmatrix} c_{\mathrm{He}a} & c_{\mathrm{He}b} \\ c_{\mathrm{H}a} & c_{\mathrm{H}b} \end{pmatrix}\begin{pmatrix} \varepsilon_a & 0 \\ 0 & \varepsilon_b \end{pmatrix}$$

$$\begin{pmatrix} \alpha_{\mathrm{He}} & \beta \\ \beta & \alpha_{\mathrm{H}} \end{pmatrix}\begin{pmatrix} c_{\mathrm{He}a} & c_{\mathrm{He}b} \\ c_{\mathrm{H}a} & c_{\mathrm{H}b} \end{pmatrix}=\begin{pmatrix} 1 & 0 \\ 0 & 1 \end{pmatrix}\begin{pmatrix} \varepsilon_a c_{\mathrm{He}a} & \varepsilon_b c_{\mathrm{He}b} \\ \varepsilon_a c_{\mathrm{H}a} & \varepsilon_b c_{\mathrm{H}b} \end{pmatrix}=\begin{pmatrix} \varepsilon_a c_{\mathrm{He}a} & \varepsilon_b c_{\mathrm{He}b} \\ \varepsilon_a c_{\mathrm{H}a} & \varepsilon_b c_{\mathrm{H}b} \end{pmatrix}$$

$$\begin{pmatrix} \alpha_{\mathrm{He}}c_{\mathrm{He}a}+\beta c_{\mathrm{H}a} & \alpha_{\mathrm{He}}c_{\mathrm{He}b}+\beta c_{\mathrm{H}b} \\ \beta c_{\mathrm{He}a}+\alpha_{\mathrm{H}}c_{\mathrm{H}a} & \beta c_{\mathrm{He}b}+\alpha_{\mathrm{H}}c_{\mathrm{H}b} \end{pmatrix}=\begin{pmatrix} \varepsilon_a c_{\mathrm{He}a} & \varepsilon_b c_{\mathrm{He}b} \\ \varepsilon_a c_{\mathrm{H}a} & \varepsilon_b c_{\mathrm{H}b} \end{pmatrix}$$

For example, by equating the (row 1, column 1) elements of the above equation and substituting the normalization condition $c_{\mathrm{He}a}{}^2+c_{\mathrm{H}a}{}^2=1$, we find the coefficient $c_{\mathrm{H}a}$.

$$\alpha_{\mathrm{He}}c_{\mathrm{He}a}+\beta c_{\mathrm{H}a}=\varepsilon_a c_{\mathrm{He}a}$$

$$(\alpha_{\mathrm{He}}-\varepsilon_a)c_{\mathrm{He}a}=-\beta c_{\mathrm{H}a}$$

$$(\alpha_{\mathrm{He}}-\varepsilon_a)(1-c_{\mathrm{H}a}{}^2)^{1/2}=-\beta c_{\mathrm{H}a}$$

$$(\alpha_{\mathrm{He}}-\varepsilon_a)^2(1-c_{\mathrm{H}a}{}^2)=\beta^2 c_{\mathrm{H}a}^2$$

$$\boxed{c_{\mathrm{H}a}=\left(\frac{1}{1+\left(\dfrac{\beta}{\alpha_{\mathrm{He}}-\varepsilon_a}\right)^2}\right)^{1/2}}$$

28.2(a) (a) The semi-empirical **CNDO method (complete neglect of differential overlap)** sets all two-electron integrals of the type (AB|CD) equal to zero unless $\chi_A = \chi_B$ and $\chi_C = \chi_D$. Only integrals of the type (AA|CC) survive.

(b) The semi-empirical **INDO method (intermediate neglect of differential overlap)** sets all two-electron integrals of the type (AB|CD) equal to zero unless $\chi_A = \chi_B$ and $\chi_C = \chi_D$ except for the type (AB|AB) when the basis functions χ_A and χ_B are both centred on the same nucleus.

Problems

Many of the use of commercially available software. Use versions that are available with this text or the software recommended by your instructor.

28.1 (a) Ethanol

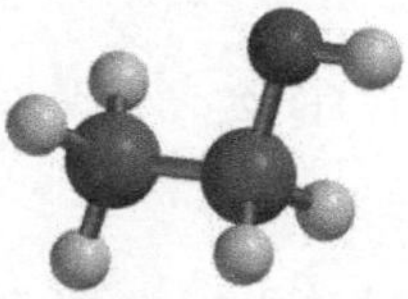

Ethanol	AM1*	PM3*	exp
C_1–C_2/pm	151.2	151.8	153.0
C_1–O/pm	142.0	141.0	142.5
C_1–H/pm	112.4	110.8	110
C_2–H/pm	111.6	109.7	109
O–H/pm	96.4	94.7	97.1
C_2–C_1–O/°	107.34	107.81	107.8
C_1–O–H/°	106.65	106.74	
H–C_1–H/°	108.63	106.98	
H–C_2–H/°	108.24	107.24	
$\Delta_f H^\ominus$(g)/kJ mol^{-1}	−262.18	−237.87	−235.10
Dipole/D	1.55	1.45	1.69

*Spartan '10™

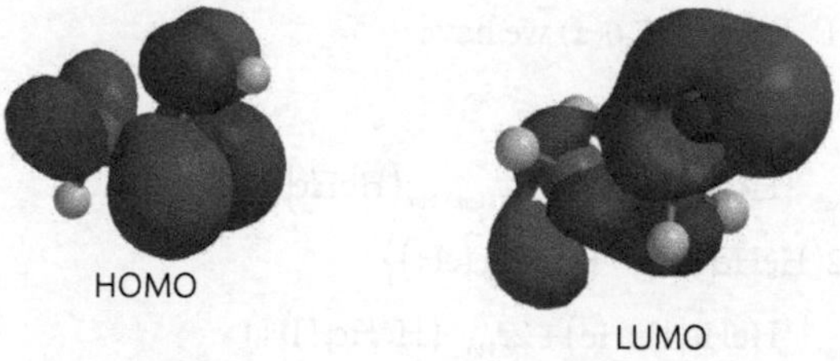

Both methods give fair agreement with experiment but PM3 gives a better estimate of formation enthalpy.

(b) 1,4-dichlorobenzene, D_{2h}

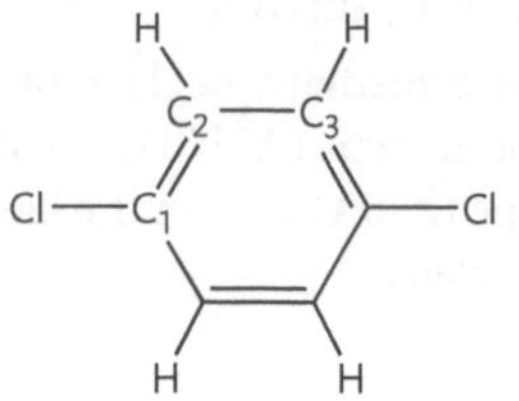

1,4-dichlorobenzene	AM1*	PM3*	exp
C_1–C_2/pm	139.9	139.4	138.8
C_2–C_3/pm	139.3	138.9	138.8
C_1–Cl/pm	169.9	168.5	173.9
C_2–H/pm	110.1	109.6	
Cl–C_1–C_2/°	119.71	119.42	
C_1–C_2–H/°	120.40	120.06	
Cl–C_1–C_2–H/°	0.00	0.00	
$\Delta_f H^\ominus(g)$/kJ mol^{-1}	33.36	42.31	24.6
dipole/D	0.00	0.00	0.0

*Spartan '10™

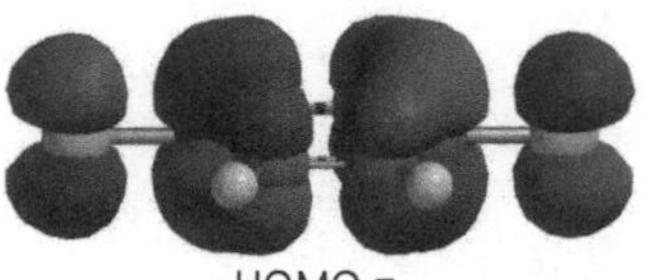

HOMO π_g

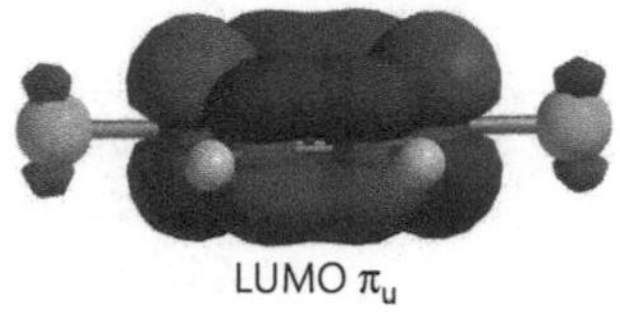

LUMO π_u

Both methods give poor agreement with the experimental formation enthalpy but both methods are in broad agreement with experimental structural values.

28.3 The semi-empirical **CNDO semi-empirical method (complete neglect of differential overlap)** sets all two-electron integrals of the type (AB|CD) equal to zero unless $\chi_A = \chi_B$ and $\chi_C = \chi_D$. Only integrals of the type (AA|CC) survive.

(a) $N_e = 2$ for HeH^+. From E27.6(a) we have

$$\begin{aligned}F_{AA} &= \int \chi_{He}(1) f_1 \chi_{He}(1)\mathrm{d}\tau_1 \\ &= E_{He} + c_{Hea}{}^2(\mathrm{HeHe}|\mathrm{HeHe}) + 2c_{Hea}c_{Ha}(\mathrm{HeHe}|\mathrm{HeH}) \\ &\quad + c_{Ha}{}^2\{2(\mathrm{HeHe}|\mathrm{HH}) - (\mathrm{HeH}|\mathrm{HeH})\} \\ &\simeq E_{He} + c_{Hea}{}^2(\mathrm{HeHe}|\mathrm{HeHe}) + 2c_{Ha}{}^2(\mathrm{HeHe}|\mathrm{HH})\end{aligned}$$

$$\begin{aligned}F_{AB} &= \int \chi_{He}(1) f_1 \chi_H(1)\mathrm{d}\tau_1 \\ &= \int \chi_{He}(1) h_1 \chi_H(1)\mathrm{d}\tau_1 + c_{Hea}{}^2(\mathrm{HeH|HeHe}) + c_{Hea}c_{Ha}\{3(\mathrm{HeH|HeH}) - (\mathrm{HeHe|HH})\} \\ &\quad + c_{Ha}{}^2(\mathrm{HeH|HH}) \\ &\simeq \int \chi_{He}(1) h_1 \chi_H(1)\mathrm{d}\tau_1 - c_{Hea}c_{Ha}(\mathrm{HeHe|HH})\end{aligned}$$

$$\begin{aligned}F_{BB} &= \int \chi_H(1) f_1 \chi_H(1)\mathrm{d}\tau_1 \\ &= E_H + c_{Ha}{}^2(\mathrm{HH|HH}) + 2c_{Hea}c_{Ha}(\mathrm{HH|HeH}) + c_{Hea}{}^2\{2(\mathrm{HeHe|HH}) - (\mathrm{HeH|HeH})\} \\ &\simeq E_H + c_{Ha}{}^2(\mathrm{HH|HH}) + 2c_{Hea}{}^2(\mathrm{HeHe|HH})\end{aligned}$$

$$\begin{aligned}F_{BA} &= \int \chi_H(1) f_1 \chi_{He}(1)\mathrm{d}\tau_1 \\ &= \int \chi_H(1) h_1 \chi_{He}(1)\mathrm{d}\tau_1 + c_{Ha}{}^2(\mathrm{HeH|HH}) + c_{Hea}c_{Ha}\{3(\mathrm{HeH|HeH}) - (\mathrm{HeHe|HH})\} \\ &\quad + c_{Hea}{}^2(\mathrm{HeH|HeHe}) \\ &\simeq \int \chi_H(1) h_1 \chi_{He}(1)\mathrm{d}\tau_1 - c_{Hea}c_{Ha}(\mathrm{HeHe|HH})\end{aligned}$$

(b) $N_e = 2$ for LiH^{2+}. From E27.6(b) we have

$$\begin{aligned}F_{AA} &= \int \chi_{Li}(1) f_1 \chi_{Li}(1)\mathrm{d}\tau_1 \\ &= E_{Li} + c_{Lia}{}^2(\mathrm{LiLi|LiLi}) + 2c_{Lia}c_{Ha}(\mathrm{LiLi|LiH}) + c_{Ha}{}^2\{2(\mathrm{LiLi|HH}) - (\mathrm{LiH|LiH})\} \\ &\simeq E_{Li} + c_{Lia}{}^2(\mathrm{LiLi|LiLi}) + 2c_{Ha}{}^2(\mathrm{LiLi|HH})\end{aligned}$$

$$\begin{aligned}F_{AB} &= \int \chi_{Li}(1) f_1 \chi_H(1)\mathrm{d}\tau_1 \\ &= \int \chi_{Li}(1) h_1 \chi_H(1)\mathrm{d}\tau_1 + c_{Lia}{}^2(\mathrm{LiH|LiLi}) + c_{Lia}c_{Ha}\{3(\mathrm{LiH|LiH}) - (\mathrm{LiLi|HH})\} + c_{Ha}{}^2(\mathrm{LiH|HH}) \\ &\simeq \int \chi_{Li}(1) h_1 \chi_H(1)\mathrm{d}\tau_1 - c_{Lia}c_{Ha}(\mathrm{LiLi|HH})\end{aligned}$$

$$\begin{aligned}F_{BB} &= \int \chi_H(1) f_1 \chi_H(1)\mathrm{d}\tau_1 \\ &= E_H + c_{Ha}{}^2(\mathrm{HH|HH}) + 2c_{Lia}c_{Ha}(\mathrm{HH|LiH}) + c_{Lia}{}^2\{2(\mathrm{LiLi|HH}) - (\mathrm{LiH|LiH})\} \\ &\simeq E_H + c_{Ha}{}^2(\mathrm{HH|HH}) + 2c_{Lia}{}^2(\mathrm{LiLi|HH})\end{aligned}$$

$$\begin{aligned}F_{BA} &= \int \chi_H(1) f_1 \chi_{Li}(1)\mathrm{d}\tau_1 \\ &= \int \chi_H(1) h_1 \chi_{Li}(1)\mathrm{d}\tau_1 + c_{Ha}{}^2(\mathrm{LiH|HH}) + c_{Lia}c_{Ha}\{3(\mathrm{LiH|LiH}) - (\mathrm{LiLi|HH})\} + c_{Lia}{}^2(\mathrm{LiH|LiLi}) \\ &\simeq \int \chi_H(1) h_1 \chi_{Li}(1)\mathrm{d}\tau_1 - c_{Lia}c_{Ha}(\mathrm{LiLi|HH})\end{aligned}$$

Topic 29 *Ab initio* methods

Discussion questions

29.1 A **virtual orbital** is an orbital from a set of MOs obtained as solutions of the SCF equations whose energies are higher than those of doubly occupied MOs producing the single determinant wavefunction of lowest energy for a given system. The virtual orbitals obtained from SCF calculations are not variationally correct approximations to the excited state orbitals. Their energies are not related to electron affinities of the molecular system.

The ground state of an N_e-electron closed-shell species is

$$\Psi_0 = (1/N_e!)^{½}|\psi_a^{\alpha}(1)\psi_a^{\beta}(2)\psi_b^{\alpha}(3)\psi_b^{\beta}(4)..\psi_u^{\beta}(N_e)|$$

where ψ_u is the highest occupied molecular orbital (the HOMO). We can envisage transferring an electron from an occupied orbital to a virtual orbital ψ_v, and forming the corresponding **singly excited determinant**, such as

$$\Psi_1 = (1/N_e!)^{1/2}|\psi_a^{\alpha}(1)\psi_a^{\beta}(2)\psi_b^{\alpha}(3)\psi_v^{\beta}(4)..\psi_u^{\beta}(N_e)|$$

Here a β electron, 'electron 4', has been promoted from ψ_b into ψ_v, but there are many other possible choices. A doubly excited determinant has two electrons transferred to virtual orbitals, and so on. Each of the Slater determinants constructed in this way is called a **configuration state function** (CSF). Ψ_1(⁝) and Ψ_2(⁝) are symbols for singly and doubly excited determinants, respectively.

29.3 In **Møller–Plesset perturbation theory** (MPPT) the hamiltonian is expressed as a sum of a simple, 'model' hamiltonian, $\hat{H}^{(0)}$ and a perturbation $\hat{H}^{(1)}$: $\hat{H} = \hat{H}^{(0)} + \hat{H}^{(1)}$ [15.1] where, in an effort to find the correlation energy, the model hamiltonian is chosen to be the sum of Fock operators $\hat{f}_i$ of the HF-SCF method.

$$\hat{f}_i = \hat{h}_i + \sum_m \left\{2\hat{J}_m(i) - \hat{K}_m(i)\right\} \quad [29.2]$$

The perturbation is the difference between this sum and the true many-electron hamiltonian. That is,

$$\hat{H}^{(1)} = \hat{H} - \hat{H}^{(0)} \text{ with } \hat{H}^{(0)} = \sum_{i=1}^{N_e} \hat{f}_i$$

with

$$\hat{H} = \overbrace{-\frac{\hbar^2}{2m_e}\sum_{i=1}^{N_e}\nabla_i^2 - j_0\sum_{i=1}^{N_e}\sum_{I=i}^{N_n}\frac{Z_I}{r_{Ii}}}^{\sum_{i=1}^{N_e}\hat{h}_i} + \frac{1}{2}j_0\sum_{i=j}^{N_e}\frac{1}{r_{ij}} \quad [29.3b]$$

Because the core hamiltonian (the sum of the $\hat{h}_i$ in the Fock operator in eqn 29.2) cancels the one-electron terms in the full hamiltonian, the perturbation is the difference between the instantaneous interaction between the electrons (the last term in eqn 29.3) and the average interaction (as represented by the operators J and K in the Fock operator). Thus, for electron 1

$$\hat{H}^{(1)}(1) = \sum_i \frac{j_0}{r_{1i}} - \sum_m \{2J_m(1) - K_m(1)\} \quad [29.4]$$

where the first sum (the true interaction) is over all the electrons other than electron 1 itself and the second sum (the average interaction) is over all the occupied orbitals. The true wavefunction is written as a sum of the eigenfunction of the model hamiltonian and higher-order correction terms.

Exercises

29.1(a) Twenty basis set functions generate twenty atomic orbitals. The silicon atom has 14 electrons, which are paired in the 7 lowest AOs to form the ground state. This leaves 13 unoccupied, virtual orbitals.

29.2(a) $N_e = 2$ for H_2. There are a total of four singly excited Slater determinants. Each may be expanded to give the sum of AO spatial wavefunctions multiplied by spin state wavefunctions. Definitions include:

$\psi_a = (1/2)^{1/2}\{\chi_A + \chi_B\}$ and $\psi_b = (1/2)^{1/2}\{\chi_A - \chi_B\}$

"A" represents H_1 while "B" represents H_2 in the molecule $H_1 - H_2$

$\sigma_-(1,2) = (1/2)^{1/2}\{\alpha(1)\beta(2) - \beta(1)\alpha(2)\}$ and $\sigma_+(1,2) = (1/2)^{1/2}\{\alpha(1)\beta(2) + \beta(1)\alpha(2)\}$

$$\boxed{\Psi_1(1\text{ of }4) = (1/2)^{1/2}\begin{vmatrix}\psi_b^\alpha(1) & \psi_a^\beta(1) \\ \psi_b^\alpha(2) & \psi_a^\beta(2)\end{vmatrix}}$$

$$\Psi_1(1\text{ of }4) = (1/2)^{1/2}\{\psi_b(1)\psi_a(2)\alpha(1)\beta(2) - \psi_a(1)\psi_b(2)\beta(1)\alpha(2)\}$$

$$\begin{aligned}2^{3/2}\Psi_1(1\text{ of }4) &= \{\chi_A(1) - \chi_B(1)\}\times\{\chi_A(2) + \chi_B(2)\}\times\alpha(1)\beta(2) \\ &\quad - \{\chi_A(1) + \chi_B(1)\}\times\{\chi_A(2) - \chi_B(2)\}\times\beta(1)\alpha(2) \\ &= \{\chi_A(1)\chi_A(2) - \chi_B(1)\chi_A(2) + \chi_A(1)\chi_B(2) - \chi_B(1)\chi_B(2)\}\times\alpha(1)\beta(2) \\ &\quad - \{\chi_A(1)\chi_A(2) + \chi_B(1)\chi_A(2) - \chi_A(1)\chi_B(2) - \chi_B(1)\chi_B(2)\}\times\beta(1)\alpha(2)\end{aligned}$$

$$= \chi_A(1)\chi_A(2)\{\alpha(1)\beta(2)-\beta(1)\alpha(2)\} - \chi_B(1)\chi_A(2)\{\alpha(1)\beta(2)+\beta(1)\alpha(2)\}$$
$$+\chi_A(1)\chi_B(2)\{\alpha(1)\beta(2)+\beta(1)\alpha(2)\} - \chi_B(1)\chi_B(2)\{\alpha(1)\beta(2)-\beta(1)\alpha(2)\}$$
$$= \{\chi_A(1)\chi_A(2)-\chi_B(1)\chi_B(2)\}\{\alpha(1)\beta(2)-\beta(1)\alpha(2)\}$$
$$+\{\chi_A(1)\chi_B(2)-\chi_B(1)\chi_A(2)\}\{\alpha(1)\beta(2)+\beta(1)\alpha(2)\}$$

$$\Psi_1(1\text{ of }4) = (1/2)\begin{Bmatrix}\{\chi_A(1)\chi_A(2)-\chi_B(1)\chi_B(2)\}\sigma_-(1,2) \\ +\{\chi_A(1)\chi_B(2)-\chi_B(1)\chi_A(2)\}\sigma_+(1,2)\end{Bmatrix}$$

Likewise,

$$\boxed{\Psi_1(2\text{ of }4) = (1/2)^{1/2}\begin{Vmatrix}\psi_a^\alpha(1) & \psi_b^\beta(1) \\ \psi_a^\alpha(2) & \psi_b^\beta(2)\end{Vmatrix}} = (1/2)\begin{Bmatrix}\{\chi_A(1)\chi_A(2)-\chi_B(1)\chi_B(2)\}\sigma_-(1,2) \\ -\{\chi_A(1)\chi_B(2)-\chi_B(1)\chi_A(2)\}\sigma_+(1,2)\end{Bmatrix}$$

$$\boxed{\Psi_1(3\text{ of }4) = (1/2)^{1/2}\begin{Vmatrix}\psi_a^\alpha(1) & \psi_b^\alpha(1) \\ \psi_a^\alpha(2) & \psi_b^\alpha(2)\end{Vmatrix}} = (1/2)^{1/2}\{\chi_B(1)\chi_A(2)-\chi_A(1)\chi_B(2)\}\alpha(1)\alpha(2)$$

$$\boxed{\Psi_1(4\text{ of }4) = (1/2)^{1/2}\begin{Vmatrix}\psi_a^\beta(1) & \psi_b^\beta(1) \\ \psi_a^\beta(2) & \psi_b^\beta(2)\end{Vmatrix}}$$

$$= (1/2)^{1/2}\{\psi_a^\beta(1)\psi_b^\beta(2)-\psi_b^\beta(1)\psi_a^\beta(2)\} = (1/2)^{1/2}\{\psi_a(1)\psi_b(2)-\psi_b(1)\psi_a(2)\}\beta(1)\beta(2)$$
$$= (1/2)^{3/2}\{\{\chi_A(1)+\chi_B(1)\}\{\chi_A(2)-\chi_B(2)\}-\{\chi_A(1)-\chi_B(1)\}\{\chi_A(2)+\chi_B(2)\}\}\beta(1)\beta(2)$$
$$= (1/2)^{3/2}\begin{Bmatrix}\chi_A(1)\chi_A(2)+\chi_B(1)\chi_A(2)-\{\chi_A(1)\chi_B(2)+\chi_B(1)\chi_B(2)\} \\ -\{\chi_A(1)\chi_A(2)-\chi_B(1)\chi_A(2)+\chi_A(1)\chi_B(2)-\chi_B(1)\chi_B(2)\}\end{Bmatrix}\beta(1)\beta(2)$$
$$= (1/2)^{1/2}\{\chi_B(1)\chi_A(2)-\chi_A(1)\chi_B(2)\}\beta(1)\beta(2)$$

Notice that the combinations $\Psi_1(1\text{ of }4)-\Psi_1(2\text{ of }4)$ and $\Psi_1(1\text{ of }4)+\Psi_1(2\text{ of }4)$ have the spin functions $\sigma_+(1,2)$ and $\sigma_-(1,2)$, respectively.

$$\Psi_1(1\text{ of }4)-\Psi_1(2\text{ of }4) = (1/2)\{\chi_A(1)\chi_B(2)-\chi_B(1)\chi_A(2)\}\sigma_+(1,2)$$

$$\Psi_1(1\text{ of }4)+\Psi_1(2\text{ of }4) = (1/2)\{\chi_A(1)\chi_A(2)-\chi_B(1)\chi_B(2)\}\sigma_-(1,2)$$

Obviously, $\Psi_1(3\text{ of }4)$ and $\Psi_1(4\text{ of }4)$ are the $S_z = 1$ and $S_z = -1$ components of a triplet state, respectively, while the $\Psi_1(1\text{ of }4)-\Psi_1(2\text{ of }4)$ combination is the $S_z = 0$ component of the triplet state. The $\Psi_1(1\text{ of }4)+\Psi_1(2\text{ of }4)$ combination has overall Σ_u symmetry, which contrasts with the ${}^1\Sigma_g$ ground state symmetry of dihydrogen. Spin and symmetry considerations reveal that none of these singly excited determinants contributes to the CI description of the ground state of dihydrogen so the eqn 29.1 CI consists of the HF ground state plus a doubly excited determinant only:

$$\boxed{\Psi = C_0\Psi_0 + C_2\Psi_2 = C_0\begin{Vmatrix}\psi_a^\alpha(1) & \psi_a^\beta(1) \\ \psi_a^\alpha(2) & \psi_a^\beta(2)\end{Vmatrix} + C_2\begin{Vmatrix}\psi_b^\alpha(1) & \psi_b^\beta(1) \\ \psi_b^\alpha(2) & \psi_b^\beta(2)\end{Vmatrix}} \quad [29.1]$$

29.3(a) $N_e = 2$ for HeH$^+$. $\psi_a = c_{\text{Hea}}\chi_{\text{He}} + c_{\text{Ha}}\chi_{\text{H}}$ and $\psi_b = c_{\text{Heb}}\chi_{\text{He}} + c_{\text{Hb}}\chi_{\text{H}}$ where the coefficients are determined with the Roothaan equations. Useful spin functions include:

$$\sigma_-(1,2) = (1/2)^{1/2}\{\alpha(1)\beta(2) - \beta(1)\alpha(2)\} \text{ and } \sigma_+(1,2) = (1/2)^{1/2}\{\alpha(1)\beta(2) + \beta(1)\alpha(2)\}$$

We begin by expanding the HF ground state Ψ_0 into a product of spatial and spin functions in order to find the requisite spin state of a CI calculation.

$$\Psi_0 = \frac{1}{2^{1/2}}\begin{vmatrix} \psi_a^\alpha(1) & \psi_a^\beta(1) \\ \psi_a^\alpha(2) & \psi_a^\beta(2) \end{vmatrix} = \frac{1}{2^{1/2}}\psi_a(1)\psi_a(2)\{\alpha(1)\beta(2) - \beta(1)\alpha(2)\} = \frac{1}{2^{1/2}}\psi_a(1)\psi_a(2)\sigma_-(1,2)$$

$$= \{c_{\text{Hea}}\chi_{\text{He}}(1) + c_{\text{Ha}}\chi_{\text{H}}(1)\}\{c_{\text{Hea}}\chi_{\text{He}}(2) + c_{\text{Ha}}\chi_{\text{H}}(2)\}\sigma_-(1,2)$$

$$= \{c_{\text{Hea}}^2\chi_{\text{He}}(1)\chi_{\text{He}}(2) + c_{\text{Hea}}c_{\text{Ha}}\chi_{\text{H}}(1)\chi_{\text{He}}(2) + c_{\text{Hea}}c_{\text{Ha}}\chi_{\text{He}}(1)\chi_{\text{H}}(2) + c_{\text{Ha}}^2\chi_{\text{H}}(1)\chi_{\text{H}}(2)\}\sigma_-(1,2)$$

Thus, the excited determinants must have the $\sigma_-(1,2)$ spin function. We examine two singly excited determinants that have both the $S_z = 0$ property and are expected to be singlets. The normalization constants are absorbed into the variation constants.

$$\Psi_1(1\text{ of }2) = (1/2)^{1/2}\begin{vmatrix} \psi_b^\alpha(1) & \psi_a^\beta(1) \\ \psi_b^\alpha(2) & \psi_a^\beta(2) \end{vmatrix} = (1/2)^{1/2}\{\psi_b^\alpha(1)\psi_a^\beta(2) - \psi_a^\beta(1)\psi_b^\alpha(2)\}$$

$$= (1/2)^{1/2}\{\psi_b(1)\psi_a(2)\alpha(1)\beta(2) - \psi_a(1)\psi_b(2)\beta(1)\alpha(2)\}$$

$$= \{c_{\text{Heb}}\chi_{\text{He}}(1) + c_{\text{Hb}}\chi_{\text{H}}(1)\} \times \{c_{\text{Hea}}\chi_{\text{He}}(2) + c_{\text{Ha}}\chi_{\text{H}}(2)\} \times \alpha(1)\beta(2)$$
$$- \{c_{\text{Hea}}\chi_{\text{He}}(1) + c_{\text{Ha}}\chi_{\text{H}}(1)\} \times \{c_{\text{Heb}}\chi_{\text{He}}(2) + c_{\text{Hb}}\chi_{\text{H}}(2)\} \times \beta(1)\alpha(2)$$

$$= \{c_{\text{Hea}}c_{\text{Heb}}\chi_{\text{He}}(1)\chi_{\text{He}}(2) + c_{\text{Hea}}c_{\text{Hb}}\chi_{\text{H}}(1)\chi_{\text{He}}(2) + c_{\text{Heb}}c_{\text{Ha}}\chi_{\text{He}}(1)\chi_{\text{H}}(2)$$
$$+ c_{\text{Ha}}c_{\text{Hb}}\chi_{\text{H}}(1)\chi_{\text{H}}(2)\}\alpha(1)\beta(2) - \{c_{\text{Hea}}c_{\text{Heb}}\chi_{\text{He}}(1)\chi_{\text{He}}(2) + c_{\text{Heb}}c_{\text{Ha}}\chi_{\text{H}}(1)\chi_{\text{He}}(2)$$
$$+ c_{\text{Hea}}c_{\text{Hb}}\chi_{\text{He}}(1)\chi_{\text{H}}(2) + c_{\text{Ha}}c_{\text{Hb}}\chi_{\text{H}}(1)\chi_{\text{H}}(2)\}\beta(1)\alpha(2)$$

$$= \{c_{\text{Hea}}c_{\text{Heb}}\chi_{\text{He}}(1)\chi_{\text{He}}(2) + c_{\text{Ha}}c_{\text{Hb}}\chi_{\text{H}}(1)\chi_{\text{H}}(2)\}\{\alpha(1)\beta(2) - \beta(1)\alpha(2)\}$$
$$+ \{c_{\text{Hea}}c_{\text{Hb}}\chi_{\text{H}}(1)\chi_{\text{He}}(2) + c_{\text{Heb}}c_{\text{Ha}}\chi_{\text{He}}(1)\chi_{\text{H}}(2)\}\alpha(1)\beta(2)$$
$$- \{c_{\text{Heb}}c_{\text{Ha}}\chi_{\text{H}}(1)\chi_{\text{He}}(2) + c_{\text{Hea}}c_{\text{Hb}}\chi_{\text{He}}(1)\chi_{\text{H}}(2)\}\beta(1)\alpha(2)$$

$$\Psi_1(2\text{ of }2) = (1/2)^{1/2}\begin{vmatrix} \psi_a^\alpha(1) & \psi_b^\beta(1) \\ \psi_a^\alpha(2) & \psi_b^\beta(2) \end{vmatrix} = (1/2)^{1/2}\{\psi_a^\alpha(1)\psi_b^\beta(2) - \psi_b^\beta(1)\psi_a^\alpha(2)\}$$

$$= (1/2)^{1/2}\{\psi_a(1)\psi_b(2)\alpha(1)\beta(2) - \psi_b(1)\psi_a(2)\beta(1)\alpha(2)\}$$

$$= \{c_{\text{Hea}}\chi_{\text{He}}(1) + c_{\text{Ha}}\chi_{\text{H}}(1)\} \times \{c_{\text{Heb}}\chi_{\text{He}}(2) + c_{\text{Hb}}\chi_{\text{H}}(2)\} \times \alpha(1)\beta(2)$$
$$- \{c_{\text{Heb}}\chi_{\text{He}}(1) + c_{\text{Hb}}\chi_{\text{H}}(1)\} \times \{c_{\text{Hea}}\chi_{\text{He}}(2) + c_{\text{Ha}}\chi_{\text{H}}(2)\} \times \beta(1)\alpha(2)$$

$$= \{c_{\text{Hea}}c_{\text{Heb}}\chi_{\text{He}}(1)\chi_{\text{He}}(2) + c_{\text{Heb}}c_{\text{Ha}}\chi_{\text{H}}(1)\chi_{\text{He}}(2) + c_{\text{Hea}}c_{\text{Hb}}\chi_{\text{He}}(1)\chi_{\text{H}}(2)$$
$$+ c_{\text{Ha}}c_{\text{Hb}}\chi_{\text{H}}(1)\chi_{\text{H}}(2)\}\alpha(1)\beta(2) - \{c_{\text{Hea}}c_{\text{Heb}}\chi_{\text{He}}(1)\chi_{\text{He}}(2) + c_{\text{Hea}}c_{\text{Hb}}\chi_{\text{H}}(1)\chi_{\text{He}}(2)$$
$$+ c_{\text{Heb}}c_{\text{Ha}}\chi_{\text{He}}(1)\chi_{\text{H}}(2) + c_{\text{Ha}}c_{\text{Hb}}\chi_{\text{H}}(1)\chi_{\text{H}}(2)\}\beta(1)\alpha(2)$$

$$
\begin{aligned}
&=\{c_{\text{Hea}}c_{\text{Heb}}\chi_{\text{He}}(1)\chi_{\text{He}}(2)+c_{\text{Ha}}c_{\text{Hb}}\chi_{\text{H}}(1)\chi_{\text{H}}(2)\}\{\alpha(1)\beta(2)-\beta(1)\alpha(2)\}\\
&\quad+\{c_{\text{Heb}}c_{\text{Ha}}\chi_{\text{H}}(1)\chi_{\text{He}}(2)+c_{\text{Hea}}c_{\text{Hb}}\chi_{\text{He}}(1)\chi_{\text{H}}(2)\}\alpha(1)\beta(2)\\
&\quad-\{c_{\text{Hea}}c_{\text{Hb}}\chi_{\text{H}}(1)\chi_{\text{He}}(2)+c_{\text{Heb}}c_{\text{Ha}}\chi_{\text{He}}(1)\chi_{\text{H}}(2)\}\beta(1)\alpha(2)
\end{aligned}
$$

Although they are singlets, neither $\Psi_1(1\text{ of }2)$ nor $\Psi_1(2\text{ of }2)$ consist of the pure $\sigma_-(1,2)$ spin state. However, the plus combination does have the desired properties.

$$
\begin{aligned}
\Psi_1(1\text{ of }2)+\Psi_1(2\text{ of }2)&=\left\{\begin{matrix}c_{\text{Hea}}c_{\text{Heb}}\chi_{\text{He}}(1)\chi_{\text{He}}(2)+c_{\text{Ha}}c_{\text{Hb}}\chi_{\text{H}}(1)\chi_{\text{H}}(2)\\
+c_{\text{Heb}}c_{\text{Ha}}\chi_{\text{H}}(1)\chi_{\text{He}}(2)+c_{\text{Hea}}c_{\text{Hb}}\chi_{\text{He}}(1)\chi_{\text{H}}(2)\\
+c_{\text{Hea}}c_{\text{Hb}}\chi_{\text{H}}(1)\chi_{\text{He}}(2)+c_{\text{Heb}}c_{\text{Ha}}\chi_{\text{He}}(1)\chi_{\text{H}}(2)\end{matrix}\right\}\{\alpha(1)\beta(2)-\beta(1)\alpha(2)\}\\
&=\left\{\begin{matrix}c_{\text{Hea}}c_{\text{Heb}}\chi_{\text{He}}(1)\chi_{\text{He}}(2)+c_{\text{Ha}}c_{\text{Hb}}\chi_{\text{H}}(1)\chi_{\text{H}}(2)\\
+c_{\text{Heb}}c_{\text{Ha}}\chi_{\text{H}}(1)\chi_{\text{He}}(2)+c_{\text{Hea}}c_{\text{Hb}}\chi_{\text{He}}(1)\chi_{\text{H}}(2)\\
+c_{\text{Hea}}c_{\text{Hb}}\chi_{\text{H}}(1)\chi_{\text{He}}(2)+c_{\text{Heb}}c_{\text{Ha}}\chi_{\text{He}}(1)\chi_{\text{H}}(2)\end{matrix}\right\}\sigma_-(1,2)
\end{aligned}
$$

So we conclude that the singly excited determinants of an eqn 29.1 CI ground state are combined into the plus combination $\Psi_1(1\text{ of }2)+\Psi_1(2\text{ of }2)$. Absorbing the normalization constant into the coefficients of the CI linear combination of configuration state functions (CSFs) gives the CI ground state wavefunction:

$$
\boxed{\begin{aligned}\Psi=C_0\Psi_0+C_1\{\Psi_1(1\text{ of }2)+\Psi_1(2\text{ of }2)\}=C_0\begin{vmatrix}\psi_a^\alpha(1)&\psi_a^\beta(1)\\ \psi_a^\alpha(2)&\psi_a^\beta(2)\end{vmatrix}\\
+C_1\left\{\begin{vmatrix}\psi_b^\alpha(1)&\psi_a^\beta(1)\\ \psi_b^\alpha(2)&\psi_a^\beta(2)\end{vmatrix}+\begin{vmatrix}\psi_a^\alpha(1)&\psi_b^\beta(1)\\ \psi_a^\alpha(2)&\psi_b^\beta(2)\end{vmatrix}\right\}\end{aligned}}\quad[29.1]
$$

The coefficients of the CI ground state wavefunction are determined with the variation principle.

29.4(a) $N_e = 2$ for HeH^+. The molecular orbitals are $\psi_a=c_{\text{Hea}}\chi_{\text{He}}+c_{\text{Ha}}\chi_{\text{H}}$ and $\psi_b=c_{\text{Heb}}\chi_{\text{He}}+c_{\text{Hb}}\chi_{\text{H}}$ [27.9] while the ground and doubly excited configuration state functions are $\Psi_0=\psi_a(1)\psi_a(2)\sigma_-(1,2)$ and $\Psi_2=\psi_b(1)\psi_b(2)\sigma_-(1,2)$. The first order perturbation correction [29.4] to the hamiltonian is

$$\hat{H}^{(1)}(1)=\frac{j_0}{r_{12}}-2J_a(1)+K_a(1).$$

The second order perturbation correction to the energy [29.5] is

$$E_0^{(2)}=\frac{\left|\int\Psi_2\hat{H}^{(1)}\Psi_0\,d\tau\right|^2}{E_0^{(0)}-E_2^{(0)}}$$

It is our intent to evaluate the integral in the numerator of the $E_0^{(2)}$ expression. Since the spin functions factor out of all integrations, they will not be written in the algebraic manipulations.

$$\hat{H}^{(1)}\Psi_0 = \left\{\frac{j_0}{r_{12}} - 2J_a(1) + K_a(1)\right\}\{\psi_a(1)\psi_a(2)\}$$

$$= \frac{j_0\psi_a(1)\psi_a(2)}{r_{12}} - 2J_a(1)\psi_a(1)\psi_a(2) + K_a(1)\psi_a(1)\psi_a(2)$$

$$= \frac{j_0\psi_a(1)\psi_a(2)}{r_{12}} - 2\psi_a(2)\left(j_0\int\psi_a(1)\frac{1}{r_{12}}\psi_a^*(2)\psi_a(2)\mathrm{d}\tau_2 \;[6.7\mathrm{a}]\right)$$

$$+\psi_a(2)\left(j_0\int\psi_a(1)\frac{1}{r_{12}}\psi_a^*(2)\psi_a(2)\mathrm{d}\tau_2 \;[6.7\mathrm{b}]\right)$$

$$\int\Psi_2\hat{H}^{(1)}\Psi_0\,\mathrm{d}\tau = \int\int\psi_b(1)\psi_b(2)\hat{H}^{(1)}\Psi_0\,\mathrm{d}\tau_1\mathrm{d}\tau_2$$

$$= \int\int\frac{j_0\psi_b(1)\psi_b(2)\psi_a(1)\psi_a(2)}{r_{12}}\mathrm{d}\tau_1\mathrm{d}\tau_2$$

$$-2\int\psi_b(2)\psi_a(2)\mathrm{d}\tau_2\left(j_0\int\psi_b(1)\psi_a(1)\frac{1}{r_{12}}\psi_a^*(2)\psi_a(2)\mathrm{d}\tau_2\mathrm{d}\tau_1\right)$$

$$+\int\psi_b(2)\psi_a(2)\mathrm{d}\tau_2\left(j_0\int\psi_b(1)\psi_a(1)\frac{1}{r_{12}}\psi_a^*(2)\psi_a(2)\mathrm{d}\tau_2\mathrm{d}\tau_1\right)$$

The last two terms vanish because ψ_a and ψ_b are orthogonal. Thus,

$$\int\Psi_2\hat{H}^{(1)}\Psi_0\,\mathrm{d}\tau = \int\int\frac{j_0\psi_b(1)\psi_b(2)\psi_a(1)\psi_a(2)}{r_{12}}\mathrm{d}\tau_1\mathrm{d}\tau_2 = j_0\int\int\psi_a(1)\psi_b(1)\frac{1}{r_{12}}\psi_a(2)\psi_b(2)\mathrm{d}\tau_1\mathrm{d}\tau_2$$

$$= j_0\int\int\frac{1}{r_{12}}\left(c_{\mathrm{He}a}\chi_{\mathrm{He}}(1)+c_{\mathrm{H}a}\chi_{\mathrm{H}}(1)\right)\left(c_{\mathrm{He}b}\chi_{\mathrm{He}}(1)+c_{\mathrm{H}b}\chi_{\mathrm{H}}(1)\right)\left(c_{\mathrm{He}a}\chi_{\mathrm{He}}(2)+c_{\mathrm{H}a}\chi_{\mathrm{H}}(2)\right)$$
$$\times\left(c_{\mathrm{He}b}\chi_{\mathrm{He}}(2)+c_{\mathrm{H}b}\chi_{\mathrm{H}}(2)\right)\mathrm{d}\tau_1\mathrm{d}\tau_2$$

$$= j_0\int\int\frac{1}{r_{12}}\left(c_{\mathrm{He}a}c_{\mathrm{He}b}\chi_{\mathrm{He}}(1)^2 + c_{\mathrm{He}a}c_{\mathrm{H}b}\chi_{\mathrm{He}}(1)\chi_{\mathrm{H}}(1) + c_{\mathrm{H}a}c_{\mathrm{He}b}\chi_{\mathrm{He}}(1)\chi_{\mathrm{H}}(1) + c_{\mathrm{H}a}c_{\mathrm{H}b}\chi_{\mathrm{H}}(1)^2\right)$$
$$\times\left(c_{\mathrm{He}a}c_{\mathrm{He}b}\chi_{\mathrm{He}}(2)^2 + c_{\mathrm{He}a}c_{\mathrm{H}b}\chi_{\mathrm{He}}(2)\chi_{\mathrm{H}}(2) + c_{\mathrm{H}a}c_{\mathrm{He}b}\chi_{\mathrm{He}}(2)\chi_{\mathrm{H}}(2) + c_{\mathrm{H}a}c_{\mathrm{H}b}\chi_{\mathrm{H}}(2)^2\right)\mathrm{d}\tau_1\mathrm{d}\tau_2$$

Multipling the terms of the integrand and using the definition

$$(\mathrm{AB}|\mathrm{CD}) = j_0\int\int\chi_{\mathrm{A}}(1)\chi_{\mathrm{B}}(1)\frac{1}{r_{12}}\chi_{\mathrm{C}}(2)\chi_{\mathrm{D}}(2)\mathrm{d}\tau_1\mathrm{d}\tau_2 \;[28.1]$$

yields the result

$$\boxed{\begin{aligned}\int\Psi_2\hat{H}^{(1)}\Psi_0\,\mathrm{d}\tau &= c_{\mathrm{He}a}{}^2c_{\mathrm{He}b}{}^2(\mathrm{HeHe}|\mathrm{HeHe}) + 2\left(c_{\mathrm{He}a}{}^2c_{\mathrm{He}b}c_{\mathrm{H}b} + c_{\mathrm{He}a}c_{\mathrm{H}a}c_{\mathrm{He}b}{}^2\right)(\mathrm{HeH}|\mathrm{HeHe})\\ &+\left(c_{\mathrm{He}a}{}^2c_{\mathrm{H}b}{}^2 + 2c_{\mathrm{He}a}c_{\mathrm{H}a}c_{\mathrm{He}b}c_{\mathrm{H}b} + c_{\mathrm{H}a}{}^2c_{\mathrm{He}b}{}^2\right)(\mathrm{HeH}|\mathrm{HeH}) + 2c_{\mathrm{He}a}c_{\mathrm{H}a}c_{\mathrm{He}b}c_{\mathrm{H}b}(\mathrm{HH}|\mathrm{HeHe})\\ &+2\left(c_{\mathrm{H}a}{}^2c_{\mathrm{He}b}c_{\mathrm{H}b} + c_{\mathrm{He}a}c_{\mathrm{H}a}c_{\mathrm{H}b}{}^2\right)(\mathrm{HH}|\mathrm{HHe}) + c_{\mathrm{H}a}{}^2c_{\mathrm{H}b}{}^2(\mathrm{HH}|\mathrm{HH})\end{aligned}}$$

where terms have been combined using the relationships (AB|CD) = (CD|AB) and (AB|CD) = (AB|DC) = (BA|CD). See E27.7(a) and E27.7(b).

Problems

29.1‡ Dihelium calculations are performed with Spartan '10™ running in Windows using the "Energy at the Ground State" method "Starting at the Current Geometry." The internuclear distance is set before each run and the minimum energy and equilibrium bond length are found by examination of the energy over a small range of internuclear distances. The energy at infinite separation (~2000 pm) corresponds to the energy of 2 separate He atoms. The depth of the potential well equals the energy at the minimum of the energy against internuclear distance curve minus the energy of the separate atoms: $D_e = E_0 - 2E_{He}$. The results for Advanced Correlated/CCSD(T) and RI–MP2 calculations are reported in the following table. CCSD(T) is a coupled cluster method with single and double determinants and approximate treatment of triple excitations. It is popular computational method and can provide highly accurate results (within about 4 kJ mol^{-1} for thermochemical reactions).

Features Calculated with Full CI, CCSD(T) and RI–MP2 Methods Bond length (*R*) in pm and energies in zJ

He_2	†Full CI	*CCSD(T)/6-311++G(2df,2p)	*RI-MP2/6-311++G(2df,2p)
R	297	321	301
$E_0 - 2E_{He}$	−0.039	−0.00741	−0.165

† T. van Mourik and J.H. van Lenthe, *J. Chem. Phys.*, 102(19), 7479(15 May 1995).
* Spartan '10™

These methods find an extremely weak bond at about 300 pm in agreement with experimentation at temperatures near absolute zero, a remarkable computational achievement. The $E_0 - 2E_{He}$ against *R* variation for each calculation is shown in Fig. 29.1.

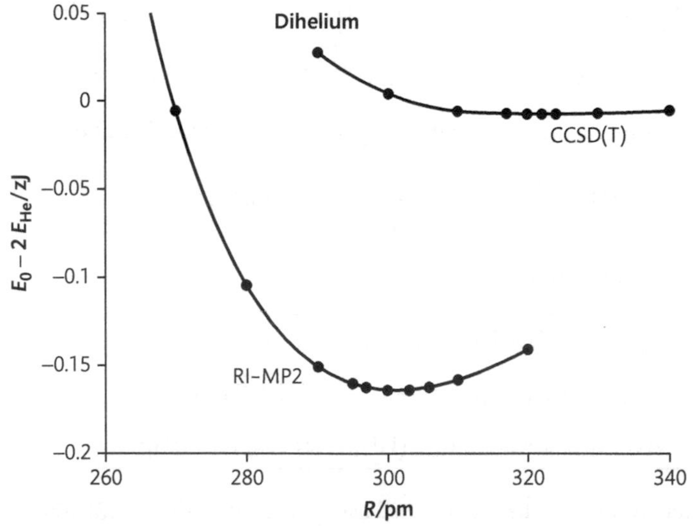

Fig 29.1

29.3 Non-degenerate, time-independent perturbation theory is mathematically developed in Topic 15. We outline the results for the ground-state wavefunction, ψ_0, corrected to first order.

After expanding the hamiltonian, the ground-state wavefunction, and the ground-state energy in successive powers of the perturbation parameter λ, the expansions are substituted into the stationary-state Schrödinger equation and powers of λ are collected. Because λ is an independent parameter, the successive powers in the resultant polynomial must equal zero. The first order term gives:

$$\left(H^{(0)}-E_0^{(0)}\right)\psi_0^{(1)}=\left(E_0^{(1)}-H^{(1)}\right)\psi_0^{(0)} \quad \text{(i)}$$

where $H^{(0)}$, $E_0{}^{(0)}$, and $\psi_0{}^{(0)}$ are unperturbed hamiltonian, ground-state energy, and ground-state wavefunction. $H^{(1)}$, $E_0{}^{(1)}$, and $\psi_0{}^{(1)}$ are the first order perturbations to the hamiltonian, ground-state energy, and ground-state wavefunction. The eigenfunctions of the unperturbed hamiltonian, $\psi_n{}^{(0)}$, form a complete orthonormal basis set of functions such that $H^{(0)}\psi_n^{(0)}=E_n^{(0)}\psi_n^{(0)}$. Let $\psi_0^{(1)}=\sum_n a_n\psi_n^{(0)}$. Substitution of the expanded first order wavefunction correction into equation (i) gives

$$\sum_n a_n\left(E_n^0-E_0^{(0)}\right)\psi_n^{(0)}=\left(E_0^{(1)}-H^{(1)}\right)\psi_0^{(0)}$$

Left multiply by $\psi_k{}^{(0)}$ with $k\neq 0$ and integrate over all space.

$$\sum_n a_n\left(E_n^0-E_0^{(0)}\right)\int\psi_{k(\neq0)}^{(0)}\psi_n^{(0)}\mathrm{d}\tau=E_0^{(1)}\int\psi_{k(\neq0)}^{(0)}\psi_0^{(0)}\mathrm{d}\tau-\int\psi_{k(\neq0)}^{(0)}H^{(1)}\psi_0^{(0)}\mathrm{d}\tau$$

$$\sum_n a_n\left(E_n^0-E_0^{(0)}\right)\delta_{k(\neq0)n}=-\int\psi_{k(\neq0)}^{(0)}H^{(1)}\psi_0^{(0)}\mathrm{d}\tau$$

$$a_{k(\neq0)}\left(E_{k(\neq0)}^0-E_0^{(0)}\right)=-\int\psi_{k(\neq0)}^{(0)}H^{(1)}\psi_0^{(0)}\mathrm{d}\tau$$

$$a_{k(\neq0)}=\frac{\int\psi_{k(\neq0)}^{(0)}H^{(1)}\psi_0^{(0)}\mathrm{d}\tau}{E_0^{(0)}-E_{k(\neq0)}^0}$$

Thus, the ground-state wavefunction corrected to first order in the perturbation is

$$\boxed{\psi_0=\psi_0^{(0)}+\psi_0^{(1)}=\psi_0^{(0)}+\sum_{n(\neq0)}\left\{\frac{\int\psi_n^{(0)}H^{(1)}\psi_0^{(0)}\mathrm{d}\tau}{E_0^{(0)}-E_n^0}\right\}\psi_n^{(0)}}$$

where the first order correction to the hamiltonian is given by equation 29.4.

29.5 $N_e=2$ for H_2 and from E29.2a and E29.2b:

$$\psi_a=(1/2)^{1/2}\{\chi_A+\chi_B\} \text{ and } \psi_b=(1/2)^{1/2}\{\chi_A-\chi_B\} \text{ and}$$
$$\sigma_-(1,2)=(1/2)^{1/2}\{\alpha(1)\beta(2)-\beta(1)\alpha(2)\}$$

$$\Psi_0=\psi_a(1)\psi_a(2)\sigma_-(1,2) \text{ and } \Psi_1=\psi_a(1)\psi_b(2)\sigma_-(1,2) \text{ and}$$
$$\Psi_2=\psi_b(1)\psi_b(2)\sigma_-(1,2)$$

$$H_{22}=\int\Psi_2\left(h_1+h_2+\frac{j_0}{r_{12}}\right)\Psi_2\mathrm{d}\tau$$

$$=\int\psi_b(1)\psi_b(2)\left(h_1+h_2+\frac{j_0}{r_{12}}\right)\psi_b(1)\psi_b(2)\mathrm{d}\tau \quad \text{(The spin states are normalized.)}$$

$$=\int\psi_b(1)\psi_b(2)h_1\psi_b(1)\psi_b(2)\mathrm{d}\tau+\int\psi_b(1)\psi_b(2)h_2\psi_b(1)\psi_b(2)\mathrm{d}\tau$$

$$+\int\psi_b(1)\psi_b(2)\frac{j_0}{r_{12}}\psi_b(1)\psi_b(2)\mathrm{d}\tau \quad \text{(i)}$$

Evaluating the first two integrals to the right, we find:

$$\int\psi_b(1)\psi_b(2)h_1\psi_b(1)\psi_b(2)\mathrm{d}\tau=\int\psi_b(1)h_1\psi_b(1)\mathrm{d}\tau_1\int\psi_b(2)\psi_b(2)\mathrm{d}\tau_2=\int\psi_b(1)h_1\psi_b(1)\mathrm{d}\tau_1 \quad \text{(ii)}$$

Similarly, $\int\psi_b(1)\psi_b(2)h_2\psi_b(1)\psi_b(2)\mathrm{d}\tau=\int\psi_b(2)h_2\psi_b(2)\mathrm{d}\tau_2$, which is identical to integral (ii).

Evaluation of the third integral to the right of equation (i) proceeds as follows.

$$4\int\psi_b(1)\psi_b(2)\frac{j_0}{r_{12}}\psi_b(1)\psi_b(2)\mathrm{d}\tau=4\int\psi_b(1)^2\frac{j_0}{r_{12}}\psi_b(2)^2\mathrm{d}\tau$$

$$=\int\{\chi_\mathrm{A}(1)-\chi_\mathrm{B}(1)\}^2\frac{j_0}{r_{12}}\{\chi_\mathrm{A}(2)-\chi_\mathrm{B}(2)\}^2\mathrm{d}\tau$$

$$=\int\left\{\chi_\mathrm{A}(1)^2-2\chi_\mathrm{A}(1)\chi_\mathrm{B}(1)+\chi_\mathrm{B}(1)^2\right\}\frac{j_0}{r_{12}}\left\{\chi_\mathrm{A}(2)^2-2\chi_\mathrm{A}(2)\chi_\mathrm{B}(2)+\chi_\mathrm{B}(2)^2\right\}\mathrm{d}\tau$$

$$=(\mathrm{AA}|\mathrm{AA})-2(\mathrm{AA}|\mathrm{AB})+(\mathrm{AA}|\mathrm{BB})-2(\mathrm{AB}|\mathrm{AA})+4(\mathrm{AB}|\mathrm{AB})-2(\mathrm{AB}|\mathrm{BB})$$
$$+(\mathrm{BB}|\mathrm{AA})-2(\mathrm{BB}|\mathrm{AB})+(\mathrm{BB}|\mathrm{BB})$$

$$=2(\mathrm{AA}|\mathrm{AA})-8(\mathrm{AA}|\mathrm{AB})+2(\mathrm{AA}|\mathrm{BB})+4(\mathrm{AB}|\mathrm{AB}) \quad \text{(iii)}$$

where we have used $(\mathrm{BB}|\mathrm{BB})=(\mathrm{AA}|\mathrm{AA})$, $(\mathrm{AB}|\mathrm{AA})=(\mathrm{AA}|\mathrm{AB})=(\mathrm{AB}|\mathrm{BB})=(\mathrm{BB}|\mathrm{AB})$, and $(\mathrm{BB}|\mathrm{AA})=(\mathrm{AA}|\mathrm{BB})$.

With (i), (ii) and (iii) we find that

$$\boxed{H_{22}=2\int\psi_b(1)h_1\psi_b(1)\mathrm{d}\tau_1+\tfrac{1}{2}\{(\mathrm{AA}|\mathrm{AA})-4(\mathrm{AA}|\mathrm{AB})+(\mathrm{AA}|\mathrm{BB})+2(\mathrm{AB}|\mathrm{AB})\}}$$

Likewise,

$$\boxed{H_{00}=2E_\mathrm{H}+\tfrac{1}{2}\{(\mathrm{AA}|\mathrm{AA})+4(\mathrm{AA}|\mathrm{AB})+(\mathrm{AA}|\mathrm{BB})+2(\mathrm{AB}|\mathrm{AB})\}} \quad \text{where } E_\mathrm{H}=\int\psi_a(1)h_1\psi_a(1)\mathrm{d}\tau_1$$

$$H_{02}=\int\Psi_0\left(h_1+h_2+\frac{j_0}{r_{12}}\right)\Psi_2\mathrm{d}\tau_1\mathrm{d}\tau_2$$

$$=\int\psi_a(1)\psi_a(2)\left(h_1+h_2+\frac{j_0}{r_{12}}\right)\psi_b(1)\psi_b(2)\mathrm{d}\tau \quad \text{(The spin states are normalized.)}$$

$$=\int\psi_a(1)\psi_a(2)h_1\psi_b(1)\psi_b(2)\mathrm{d}\tau+\int\psi_a(1)\psi_a(2)h_2\psi_b(1)\psi_b(2)\mathrm{d}\tau$$

$$+\int\psi_a(1)\psi_a(2)\frac{j_0}{r_{12}}\psi_b(1)\psi_b(2)\mathrm{d}\tau \quad \text{(iv)}$$

$$\int \psi_a(1)\psi_a(2)h_1\psi_b(1)\psi_b(2)\mathrm{d}\tau = \int \psi_a(1)h_1\psi_b(1)\mathrm{d}\tau_1 \int \psi_a(2)\psi_b(2)\mathrm{d}\tau_2 = 0 \quad \text{[orthogonality]} \quad \text{(v)}$$

Likewise,

$$\int \psi_a(1)\psi_a(2)h_2\psi_b(1)\psi_b(2)\mathrm{d}\tau = 0 \quad \text{(vi)}$$

$$\begin{aligned}
4\int \psi_a(1)\psi_a(2)\frac{j_0}{r_{12}}\psi_b(1)\psi_b(2)\mathrm{d}\tau &= 4\int \psi_a(1)\psi_b(1)\frac{j_0}{r_{12}}\psi_a(2)\psi_b(2)\mathrm{d}\tau \\
&= \int \{\chi_A(1)+\chi_B(1)\}\{\chi_A(1)-\chi_B(1)\}\frac{j_0}{r_{12}}\{\chi_A(2)+\chi_B(2)\}\{\chi_A(2)-\chi_B(2)\}\mathrm{d}\tau \\
&= \int \{\chi_A(1)^2-\chi_B(1)^2\}\frac{j_0}{r_{12}}\{\chi_A(2)^2-\chi_B(2)^2\}\mathrm{d}\tau \\
&= (\mathrm{AA}|\mathrm{AA})-(\mathrm{AA}|\mathrm{BB})-(\mathrm{BB}|\mathrm{AA})+(\mathrm{BB}|\mathrm{BB}) \\
&= 2(\mathrm{AA}|\mathrm{AA})-2(\mathrm{AA}|\mathrm{BB}) \quad \text{(vii)}
\end{aligned}$$

where we have used $(\mathrm{BB}|\mathrm{BB})=(\mathrm{AA}|\mathrm{AA})$ and $(\mathrm{BB}|\mathrm{AA})=(\mathrm{AA}|\mathrm{BB})$.

With (iv)–(vii) we find that

$$\boxed{H_{02} = \tfrac{1}{2}\{(\mathrm{AA}|\mathrm{AA})-(\mathrm{AA}|\mathrm{BB})\} = H_{20}} \quad \text{(by symmetry)}$$

29.7 We are to prove Brillouin's theorem that $\int \Psi_0 H \Psi_a^p \mathrm{d}\tau = 0$ where Ψ_0 is the HF ground-state wavefunction and Ψ_a^p is a singly excited determinant in which the virtual spin orbital ψ_p replaces the occupied spin orbital ψ_a of the ground-state. To simplify expressions, we will represent the spin orbitals as a (1), b (2), . . ., etc., where numerals refer to one of the indistinguishable N_e electrons. The HF ground-state and the singly excited determinant are:

$$\Psi_0 = \left(\frac{1}{N_e!}\right)^{1/2} \begin{vmatrix} a(1) & b(1) & \cdots & z(1) \\ a(2) & b(2) & \cdots & \vdots \\ \vdots & \vdots & \cdots & \vdots \\ a(N_e) & b(N_e) & \cdots & z(N_e) \end{vmatrix} \quad \text{and} \quad \Psi_a^p = \left(\frac{1}{N_e!}\right)^{1/2} \begin{vmatrix} p(1) & b(1) & \cdots & z(1) \\ p(2) & b(2) & \cdots & \vdots \\ \vdots & \vdots & \cdots & \vdots \\ p(N_e) & b(N_e) & \cdots & z(N_e) \end{vmatrix}$$

Eqs i–iv are useful integrals that will facilitate the proof. (A more detailed proof is found as Problem 9.17 in P.W. Atkins and R.S. Friedman, *Solutions Manual for Molecular Quantum Mechanics*, 4th ed., Oxford University Press, 2007.)

(i) $\int a(1)f_1 p(1)\mathrm{d}\tau_1 = \varepsilon_p \int a(1)p(1)\mathrm{d}\tau_1 = 0$ where f_1 is the one-electron Fock operator

This is true because a and p are orthonormal eigenfunctions of the Fock operator.

(ii) $\int a(1)f_1 p(1)\mathrm{d}\tau_1 = \int a(1)h_1 p(1)\mathrm{d}\tau_1 + \sum_m \{(ap|mm)-(am|mp)\}$ where the sum is over spatial wavefunctions

Proof:

$$\int a(1)f_1p(1)\mathrm{d}\tau_1=\int a(1)\left\{h_1p(1)+\sum_m\{J_m(1)-K_m(1)\}p(1)\right\}\mathrm{d}\tau_1 \quad \begin{bmatrix}\text{27.8 written as a sum over spin}\\ \text{wavefunctions}\end{bmatrix}$$

$$=\int a(1)h_1p(1)\mathrm{d}\tau_1+j_0\sum_m\int a(1)\left\{\int p(1)\frac{1}{r_{12}}m(2)m(2)\mathrm{d}\tau_2\right\}\mathrm{d}\tau_1$$

$$-j_0\sum_m\int a(1)\left\{\int m(1)\frac{1}{r_{12}}m(2)p(2)\mathrm{d}\tau_2\right\}\mathrm{d}\tau_1 \quad [27.4 \text{ and } 27.6]$$

$$=\int a(1)h_1p(1)\mathrm{d}\tau_1+\sum_m\{(ap|mm)-(am|mp)\}$$

where $(ab|cd)=j_0\int a(1)b(1)\left(\frac{1}{r_{12}}\right)c(2)d(2)\mathrm{d}\tau_1\mathrm{d}\tau_2$ by definition 28.1

(iii) $\sum_m^{N_e}\int \Psi_0 h_m \Psi_a^p\,\mathrm{d}\tau = N_e\int \Psi_0 h_1 \Psi_a^p\,\mathrm{d}\tau$

Proof:

h_m is a one-electron operator and electrons in the determinant wavefunction are indistinguishable so $\int \Psi_0 h_1\Psi_a^p\,\mathrm{d}\tau=\int\Psi_0h_2\Psi_a^p\,\mathrm{d}\tau=\int\Psi_0h_3\Psi_a^p\,\mathrm{d}\tau=\cdots=\int\Psi_0h_{N_e}\Psi_a^p\,\mathrm{d}\tau$. Thus,

$$\sum_m^{N_e}\int\Psi_0h_m\Psi_a^p\,\mathrm{d}\tau=\int\Psi_0h_1\Psi_a^p\,\mathrm{d}\tau+\int\Psi_0h_2\Psi_a^p\,\mathrm{d}\tau+\int\Psi_0h_3\Psi_a^p\,\mathrm{d}\tau+\cdots+\int\Psi_0h_{N_e}\Psi_a^p\,\mathrm{d}\tau=N_e\int\Psi_0h_1\Psi_a^p\,\mathrm{d}\tau$$

(iv) $\int\Psi_0h_1\Psi_a^p\mathrm{d}\tau=\frac{1}{N_e}\int a(1)h_1p(1)\mathrm{d}\tau$

Proof:

To find a simplifying relationship for $\int\Psi_0h_1\Psi_a^p\,\mathrm{d}\tau$, we must find the number of possible orbital permutations given by the Slater determinants. With a occupied by electron 1 in Ψ_0 and p occupied by electron 1 in Ψ_a^p, there are $(N_e-1)!$ ways of permuting electrons 2, 3, . . ., N_e among the other N_e-1 spin orbitals in each of the determinants Ψ_0 and Ψ_a^p. Each of the $(N_e-1)!$ permutations in Ψ_0 makes a nonzero contribution to the determinant expansion if it matches identically one of the $(N_e-1)!$ permutations in Ψ_a^p. Thus, there are a total of $(N_e-1)!$ terms that contribute to the summation and since the spin orbitals are normalized, each of the $(N_e-1)!$ terms gives an equal contribution of $\int\psi_a(1)h_1\psi_p(1)\mathrm{d}\tau$.

$$\int\Psi_0h_1\Psi_a^p\mathrm{d}\tau=\left(\frac{1}{N_e!}\right)\int\begin{vmatrix}a(1)&b(1)&\cdots&z(1)\\a(2)&b(2)&\cdots&\vdots\\\vdots&\vdots&\cdots&\vdots\\a(N_e)&b(N_e)&\cdots&z(N_e)\end{vmatrix}h_1\begin{vmatrix}p(1)&b(1)&\cdots&z(1)\\p(2)&b(2)&\cdots&\vdots\\\vdots&\vdots&\cdots&\vdots\\p(N_e)&b(N_e)&\cdots&z(N_e)\end{vmatrix}\mathrm{d}\tau$$

$$=\frac{(N_e-1)!}{N_e!}\int\psi_a(1)h_1\psi_p(1)\mathrm{d}\tau=\frac{1}{N_e}\int\psi_a(1)h_1\psi_p(1)\mathrm{d}\tau$$

(v) $$\tfrac{1}{2}\sum_{\substack{i,j\\i\neq j}}^{N_e}\int \Psi_0 \frac{j_0}{r_{ij}}\Psi_a^p \mathrm{d}\tau = \sum_m \left\{(\psi_a\psi_p|\psi_m\psi_m)-(\psi_a\psi_m|\psi_m\psi_p)\right\}$$

Proof: To demonstrate the relationship, first expand the left sum.

$$\tfrac{1}{2}\sum_{\substack{i,j\\i\neq j}}^{N_e}\int \Psi_0 \frac{j_0}{r_{ij}}\Psi_a^p \mathrm{d}\tau = \tfrac{1}{2}j_0\int \Psi_0\left\{r_{12}^{-1}+r_{21}^{-1}+r_{13}^{-1}+r_{31}^{-1}\cdots\right\}\Psi_a^p \mathrm{d}\tau$$

Since $r_{12}=r_{21}$, $r_{13}=r_{31}$, etc.,

$$\tfrac{1}{2}\sum_{\substack{i,j\\i\neq j}}^{N_e}\int \Psi_0 \frac{j_0}{r_{ij}}\Psi_a^p \mathrm{d}\tau = j_0\int \Psi_0\left\{r_{12}^{-1}+r_{13}^{-1}\cdots+r_{23}^{-1}+r_{24}^{-1}+\cdots+r_{(N_e-1)N_e}^{-1}\right\}\Psi_a^p \mathrm{d}\tau$$

The electrons in the determinants are indistinguishable so each term in the summation gives an identical result. There are $\tfrac{1}{2}N_e(N_e-1)$ equal terms so the previous equation simplifies to

$$\tfrac{1}{2}\sum_{\substack{i,j\\i\neq j}}^{N_e}\int \Psi_0 \frac{j_0}{r_{ij}}\Psi_a^p \mathrm{d}\tau = \tfrac{1}{2}N_e(N_e-1)j_0\int \Psi_0 r_{12}^{-1}\Psi_a^p \mathrm{d}\tau$$

The operator r_{12}^{-1} within the above integral is a two electron operator leaving N_e-2 electrons that can be permutated in $(N_e-2)!$ ways in the determinant wavefunctions with each term in the resultant sum being equal.

$$\tfrac{1}{2}\sum_{\substack{i,j\\i\neq j}}^{N_e}\int \Psi_0 \frac{j_0}{r_{ij}}\Psi_a^p \mathrm{d}\tau = \frac{\tfrac{1}{2}N_e(N_e-1)}{N_e!}j_0\int \begin{vmatrix} a(1) & b(1) & \cdots & z(1)\\ a(2) & b(2) & \cdots & \vdots\\ \vdots & \vdots & \cdots & \vdots\\ a(N_e) & b(N_e) & \cdots & z(N_e)\end{vmatrix}$$

$$\times r_{12}^{-1}\begin{vmatrix} p(1) & b(1) & \cdots & z(1)\\ p(2) & b(2) & \cdots & \vdots\\ \vdots & \vdots & \cdots & \vdots\\ p(N_e) & b(N_e) & \cdots & z(N_e)\end{vmatrix}\mathrm{d}\tau$$

$$=\frac{\tfrac{1}{2}N_e(N_e-1)(N_e-2)!}{N_e!}j_0\sum_{\substack{m\\m\neq a,p}}^{N_e}\left\{\begin{matrix}\int a(1)m(2)r_{12}^{-1}p(1)m(2)\mathrm{d}\tau-\int a(1)m(2)r_{12}^{-1}p(2)m(1)\mathrm{d}\tau\\ +\int m(1)a(2)r_{12}^{-1}m(1)p(2)\mathrm{d}\tau-\int m(1)a(2)r_{12}^{-1}m(2)p(1)\mathrm{d}\tau\end{matrix}\right\}$$

The above minus signs appear because permutations in Ψ_a^p differ by a sign if two electrons are interchanged. Because electrons 1 and 2 are indistinguishable, the first and third terms in the sum are equal as are the second and fourth terms. Furthermore, the restriction that $m\neq a,p$ can be removed because, if included in the sum, the subtractions will cause their elimination.

We now use Eqns (i)–(v) to prove Brillouin's theorem.

$$\int \Psi_0 H \Psi_a^p \, d\tau = \int \Psi_0 \left\{ \sum_i^{N_e} h_i + \tfrac{1}{2} \sum_{\substack{i,j \\ i \neq j}}^{N_e} \frac{j_0}{r_{ij}} \right\} \Psi_a^p \, d\tau = \sum_i^{N_e} \int \Psi_0 h_i \Psi_a^p \, d\tau + \tfrac{1}{2} \sum_{\substack{i,j \\ i \neq j}}^{N_e} \int \Psi_0 \left(\frac{j_0}{r_{ij}} \right) \Psi_a^p \, d\tau$$

$$= N_e \int \Psi_0 h_1 \Psi_a^p \, d\tau + \sum_m^{N_e} \{ (ap|mm) - (am|mp) \} \qquad \text{by (iii) and (v)}$$

$$= \int a(1) h_1 p(1) \, d\tau + \sum_m^{N_e} \{ (ap|mm) - (am|mp) \} \qquad \text{by (iv)}$$

$$= \int a(1) f_1 p(1) \, d\tau_1 = 0 \qquad \text{by (ii) and (i)}$$

This proves the Brillouin theorem: $\int \Psi_0 H \Psi_a^p \, d\tau = 0$.

Topic 30 Density functional theory

Discussion questions

30.1 The advantages of the density functional method (DFT) over Hartree–Fock based calculations include less demanding computational effort, less computer time, and—in some cases, particularly for d-metal complexes—better agreement with experimental values.

30.3 The energy functional $E[\rho]$ is written as the sum of two terms:

$$E[\rho] = E_{\text{Classical}}[\rho] + E_{\text{XC}}[\rho] \text{[Energy functional, 30.2]}$$

where $E_{\text{Classical}}[\rho]$ is the sum of the contributions of kinetic energy, electron–nucleus interactions, and the classical electron–electron potential energy, and $E_{\text{XC}}[\rho]$ is the **exchange–correlation energy**. This term takes into account all the non-classical electron–electron effects due to spin and applies small corrections to the kinetic energy part of $E_{\text{Classical}}$ that arise from electron–electron interactions. The **Hohenberg–Kohn theorem** guarantees the existence of $E_{\text{XC}}[\rho]$.

Exercises

30.1(a) The **function** $f(x)$ is a mathematical procedure that assigns a number for a specified value of the independent variable x. The **functional** $G[f]$ assigns a number for a specified function $f(x)$ over a specified range of the variable x. A functional is a function of a function.

Functions: (a) $f(x) = \mathrm{d}(x^3)/\mathrm{d}x = 3x^2$

(c) $f(x) = \int x^3\mathrm{d}x = x^4/4 + \text{constant}$

Functionals: (b) $G[f] = \mathrm{d}(x^3)/\mathrm{d}x|_{x=1} = 3$ because the number 3 is associated with the function $f(x) = x^3$. Basically, we are considering the statement to be

$$G[f] = \mathrm{d}(f)/\mathrm{d}x|_{x=1}$$

such that $G[f{=}x^3] = \mathrm{d}(x^3)/\mathrm{d}x|_{x=1} = 3$ while $G[f{=}\ln(x)] = \mathrm{d}(\ln x)/\mathrm{d}x|_{x=1} = x^{-1}|_{x=1} = 1$. $G[f]$ depends upon the functional form of $f(x)$ over the entire range of x.

(d) $G[f] = \int_1^3 x^3\mathrm{d}x = 20$ because the number 20 is associated with a function $f(x) = x^3$.

30.2(a) $N_e = 4$ for LiH. $\rho(\boldsymbol{r}) = \sum_{i=1}^{N_e=4} |\psi_i(\boldsymbol{r})|^2$ [30.3] $= \boxed{2\{|\psi_a(\boldsymbol{r})|^2 + |\psi_b(\boldsymbol{r})|^2\}}$

30.3(a) $N_e = 2$ for HeH$^+$.

$$\rho(\mathbf{r}) = \sum_{i=1}^{N_e=2} |\psi_i(\boldsymbol{r})|^2 \ [30.3] = 2|\psi_1(\boldsymbol{r}_1)|^2$$

$$\left\{h_1 + j_0\int \frac{\rho(\boldsymbol{r}_2)}{r_{12}}\mathrm{d}\mathbf{r}_2 - \tfrac{3}{2}(3/\pi)^{1/3}\, j_0\rho(\boldsymbol{r}_1)^{1/3}\right\}\psi_1(\boldsymbol{r}_1) = \varepsilon_1\psi_1(\boldsymbol{r}_1) \quad \text{[30.4 and 30.7]}$$

$$\text{where } h_1 = -\frac{\hbar^2}{2m_e}\nabla_1^2 - j_0\left\{\frac{1}{r_{\mathrm{He}1}} + \frac{1}{r_{\mathrm{H}1}}\right\} \quad \text{[Brief illustration 27.1]}$$

Problem

30.1 $E_{\mathrm{XC}}[\rho] = \int C\rho^{5/3}\,\mathrm{d}\boldsymbol{r}$

When the density changes form $\rho[r]$ to $\rho[r] + \delta\,\rho[r]$ at each point, the functional changes form $E_{\mathrm{XC}}[\rho]$ to $E_{\mathrm{XC}}[\rho + \delta\rho]$:

$$E_{\mathrm{XC}}[\rho+\delta\rho] = \int C(\rho+\delta\rho)^{5/3}\,\mathrm{d}\boldsymbol{r}$$

The integrand can be expanded in a Taylor series:

$$(\rho+\delta\rho)^{5/3} = \rho^{5/3} + \tfrac{5}{3}\rho^{2/3}\delta\rho + \tfrac{1}{2}\left(\tfrac{5}{3}\right)\left(\tfrac{2}{3}\right)\rho^{-1/3}\delta\rho^2 + \cdots$$

Discarding terms of order $\delta\rho^2$ and higher we obtain:

$$E_{\mathrm{XC}}[\rho+\delta\rho] = \int C\left(\rho^{5/3} + \tfrac{5}{3}\rho^{2/3}\delta\rho\right)\mathrm{d}\boldsymbol{r} = E_{\mathrm{XC}}[\rho] + \int \tfrac{5}{3}C\rho^{2/3}\,\delta\rho\,\mathrm{d}\boldsymbol{r}$$

Therefore, the differential δE_{XC} of the functional (the difference $E_{\mathrm{XC}}[\rho + \delta\rho] - E_{\mathrm{XC}}[\rho]$ that depends linearly on $\delta\rho$) is

$$\delta E_{\mathrm{XC}}[\rho] = \int \tfrac{5}{3}C\rho^{2/3}\,\delta\rho\,\mathrm{d}\boldsymbol{r}$$

Comparison with $\delta E_{\mathrm{XC}}[\rho] = \int V_{\mathrm{XC}}(\boldsymbol{r})\delta\rho\mathrm{d}\boldsymbol{r}$ [30.6] yields

$$\boxed{V_{\mathrm{XC}}(\boldsymbol{r}) = \tfrac{5}{3}C\rho(\boldsymbol{r})^{2/3}}$$

Focus 6: Integrated activities

F6.1 Our comparison of the two theories will focus on the manner of construction of the trial wavefunctions for the hydrogen molecule in the simplest versions of both theories. In the valence bond method, the trial function is a linear combination of two simple product wavefunctions, in which one electron resides totally in an atomic orbital (AO) on atom A, and the other totally in an orbital on atom B. There is no contribution to the wavefunction from products in which both electrons reside on either atom A or B.

$$\psi_{A-B}(1,2)=\psi_A(1)\psi_B(2)+\psi_A(2)\psi_B(1) \qquad [22.2]$$

So the valence bond approach undervalues, by totally neglecting, any ionic contribution to the trial function. It is a totally covalent function.

The modern one-electron molecular orbital (MO) extends throughout the molecule and is written as a linear combination of atomic orbitals (LCAO).

$$\psi_{MO}(1)=c_A\psi_A(1)+c_B\psi_B(1) \qquad [25.1]$$

The squares of the coefficients give the relative proportions of the AO contributing to the MO.

The two-electron molecular orbital function for the hydrogen molecule is a product of two one-electron MOs. That is

$$\begin{aligned}\psi &= [c_A\psi_A(1)+c_B\psi_B(1)]\times[c_A\psi_A(2)+c_B\psi_B(2)]\\ &= c_A^2\psi_A(1)\psi_A(2)+c_B^2\psi_B(1)\psi_B(2)+c_Ac_B\psi_A(1)\psi_B(2)+c_Ac_B\psi_A(2)\psi_B(1)\end{aligned}$$

The first two terms are ionic forms for which both electrons on either on atom A or on atom B. The molecular orbital approach greatly overvalues the ionic contributions. At these crude levels of approximation, the valence bond method gives dissociation energies closer to the experimental values. However, more sophisticated versions of the molecular orbital approach are the method of choice for obtaining quantitative results on both diatomic and polyatomic molecules.

F6.3 The Hartree-Fock formalism does not explicitly take into account the instantaneous coupling of electron motions that is called **electron correlation**. It considers average electron-electron interaction only. In fact, the correlation energy is defined as the difference in energy between the Hartree-Fock energy and the experimental value.

Configuration interaction accounts for electron correlation by considering the wavefunction to be the summation of a term in which electrons are in occupied molecular orbitals and terms in which electrons have been promoted (excited) into unoccupied molecular orbitals [29.1]. Each excited Slater determinant in the sum is called a **configuration state function** (CSF) and inclusion of an infinite number of CSFs, although impractical, would yield the exact wavefunction and energy.

Many-body perturbation theory accounts for electron correlation by adding perturbation terms to the Hartree-Fock energy to produce a hamiltonion [29.2a] that accounts for the instantaneous interaction between electrons. The perturbation is the difference between the instantaneous interaction between electrons and the average interaction that is utilized in the Fock operator.

F6.5 **Density functional theory** (DFT) is not a semiempirical method. In semi-empirical methods, many of the integrals are expressed in terms of spectroscopic data or physical properties or selectively set equal to zero so as to simplify calculations and simultaneously give reasonable agreement with select experimental data. DFT attempts to evaluate the integrals numerically, using as input only the values of fundamental constants and atomic numbers of the atoms present in the molecule.

In fact, if the **exchange-correlation energy functional** [30.2] were known, DFT would be an exact method.

The exchange-correlation energy functional is modeled in DFT methods. One widely used approximation is the uniform electron gas model for which the exchange-correlation energy is found to be the sum of an exchange contribution and a correlation contribution. Ignoring the complicated correlation contribution for simplicity, the uniform gas approximation for the exchange-correlation energy (see text Example 30.1) is

$$E_{XC}[\rho] = A\int \rho^{4/3}\mathrm{d}r \quad \text{with } A = -(9/8)(3/\pi)^{1/2} j_0$$

An iteration procedure similar to the HF-SCF methods uses the density functional to calculate an improved exchange-correlation energy functional which is then used to find an improved density functional, etc.

F6.7 The six π MOs of benzene are represented in Figure F6.1 as LCAOs of carbon p_z orbitals. By examination of inversion through the centre of symmetry we see that the lowest energy level has ungerade(u) symmetry. The next level is doubly degenerate with gerade(g) symmetry, which is followed by another doubly degenerate level of ungerade symmetry. The highest level is seen to have gerade symmetry.

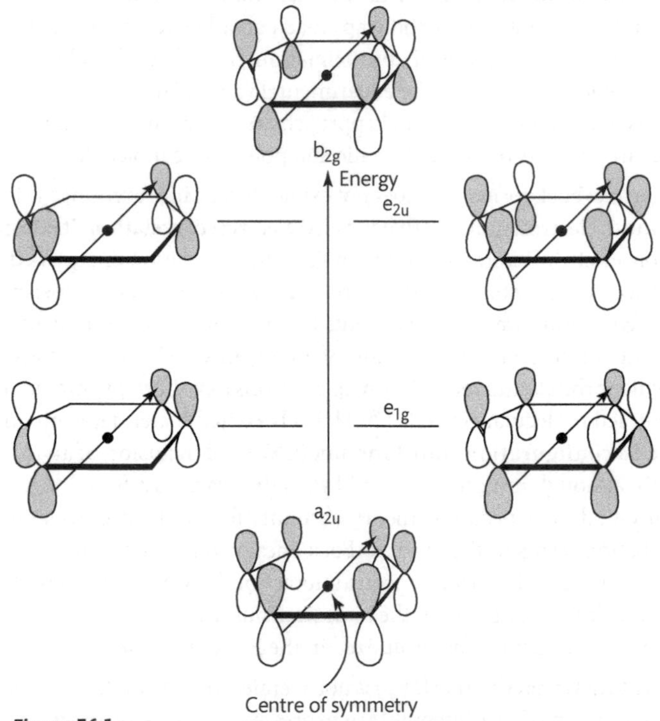

Figure F6.1

F6.9 (a) The table displays PM3 calculations of the LUMO, HOMO, and the predicted $\pi^* \leftarrow \pi$ (ΔE) energies for the all trans alternate double bond polyene series:

ethene, butadiene, hexatriene, octatetraene, and decapentaene. Figure F6.2 plots the transition energies against the number of alternate double bonds for HF/6-31G*, PM3, and DF/B3LYP/6-31G* calculations with Spartan '10™. It is seen that the DF/B3LYP/6-31G* calculation are in very good agreement with experimentation while the HF/6-31G* are in very poor agreement.

PM3 calculations* for trans polyene series

Species	E_{LUMO} / eV*	E_{HOMO} / eV*	ΔE / eV	ΔE / cm^{-1}	$\tilde{\nu}_{obs}$ / cm^{-1}
C_2H_4	1.2282	−10.6411	11.8693	95730	61500
C_4H_6	0.2634	−9.4671	9.7305	78480	46080
C_6H_8	−0.2494	−8.8993	8.6499	69770	39750
C_8H_{10}	−0.5568	−8.5767	8.0199	64680	32900
$C_{10}H_{12}$	−0.7556	−8.3755	7.6199	61460	–

*Spartan '10™

(b) Figure F6.2 shows that the PM3 transition calculations can be "calibrated" to agree with spectroscopic measurements by the subtraction of about 32,000 cm^{-1}. This however has no practical application; having the experimental absorptions for the first four members of the homologous series, it is far better to extrapolate the experimental data to the fifth member rather than to extrapolate a "calibrated" set of computations. In fact, extrapolations are extremely risky and only small extrapolations are generally considered. Figure F6.2 is the demonstration that DF/B3LYP/6-31G* calculations are in far better agreement with spectroscopic measurement than are PM3 and HF computations. So we expect that the direct DF calculation of the decapentaene transition to be very reliable. Also a study of transitions for molecules that cannot be easily examined experimentally might better be undertaken with the DF method. A plot of calculated minus observed wavenumbers against observed, shown in Figure F6.3 leads to identical conclusions. The PM3

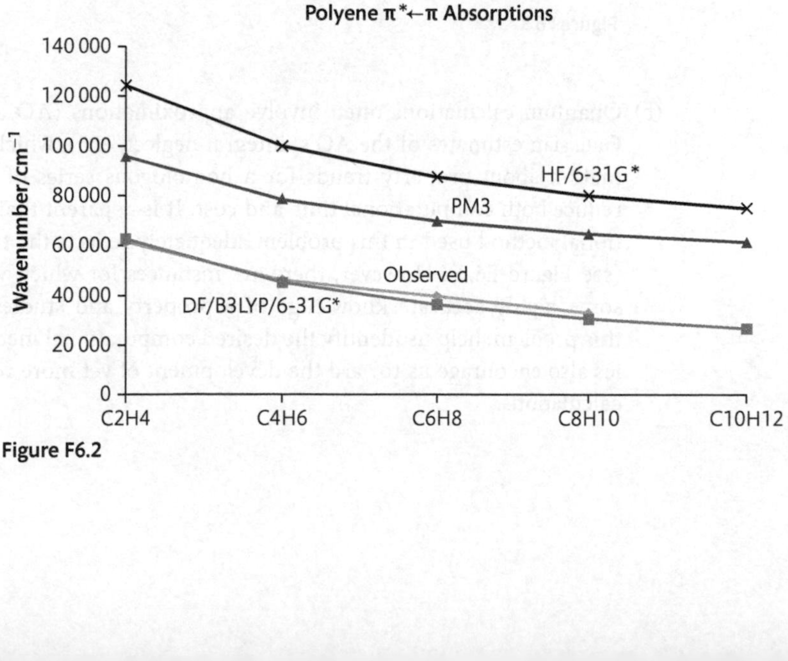

Figure F6.2

calculations are about 32,000 cm^{-1} higher than both the DF calculations and the observed values and the linear regression fits of the Figure F6.3 plots show low correlation coefficients. Accepting the linearity, we can use the regression fits to extrapolate the decapentaene transition. The PM3 transition calculation for decapentaene is 61,460 cm^{-1} while the DF/B3LYP/6-31G* calculation is $\boxed{26\,780\ \text{cm}^{-1}}$. Using PM3 decapentaene calculation and solving the regression fit for the extrapolated observation gives:

$$\text{predicted transition of PM3 regression fit} = \frac{61\,460 - 27\,046}{1.1123} \times \text{cm}^{-1} = 30\,940\ \text{cm}^{-1}$$

Doing this for the DF calculation gives:

$$\text{predicted transition of DF regression fit} = \frac{26\,780 + 7\,911}{1.1452} \times \text{cm}^{-1} = 30\,290\ \text{cm}^{-1}$$

These are reasonable predictions for the decapentaene transition but the direct DF/B3LYP/6-31G* calculation of 26 780 cm^{-1} is expected to be better.

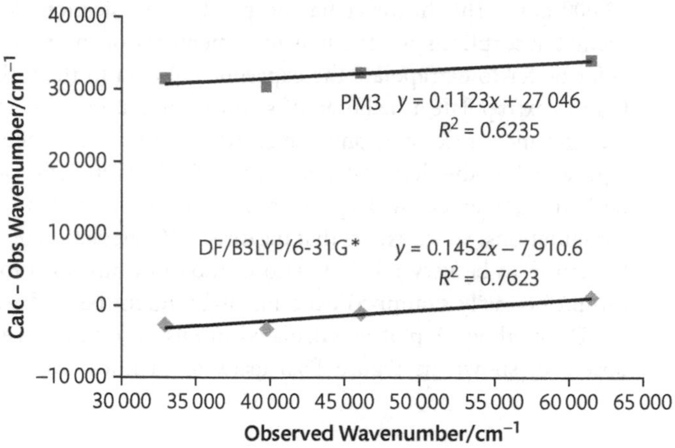

Figure F6.3

(c) Quantum calculations often involve approximations (AO approximations, Gaussian estimates of the AO's, integral neglect, etc.), which provide information about property trends for a homologous series of substances and reduce both computational time and cost. It is apparent that each computational method used in this problem adequately reflects the transition trends (see Figure F6.3). However, there are instances for which we wish to have some highly accurate knowledge of a property and studies, illustrated by this problem, help us identify the desired computational method. Such studies also encourage us toward the development of yet more reliable quantum calculations.

F6.11 (a) The table below defines six 1,4-benzoquinone molecules and displays reduction potentials ($E_{\text{ɩ}}$) and computed LUMO energies calculated with both the PM3 and DF/B3LYP/6-31G* methods.

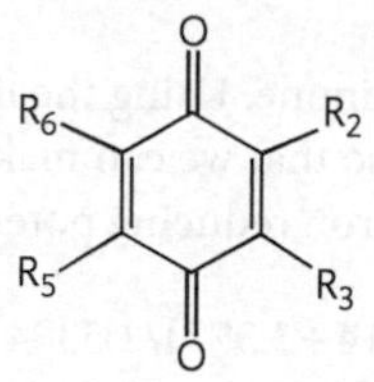

Species	R_2	R_3	R_5	R_6	$E_{\text{ɩ}}$/V	E_{LUMO}/eV*	E_{LUMO}/eV§
1	H	H	H	H	0.078	−1.7063	−3.5377
2	CH_3	H	H	H	0.023	−1.6512	−3.3897
3	CH_3	H	CH_3	H	−0.067	−1.5826	−3.2417
4	CH_3	CH_3	CH_3	H	−0.165	−1.3581	−3.1114
5	CH_3	CH_3	CH_3	CH_3	−0.260	−1.2149	−3.0063
6	CH_3	CH_3	CH_3O	CH_3O		−1.43414	−2.9922

* Spartan '10™ PM3
§ Spartan '10™ DF/B3LYP/6-31G*

The calculated E_{LUMO} values for species 1-5 are plotted against $E_{\text{ɩ}}$ measurements in Figure F6.4. E_{LUMO} values calculated with both the PM3 and DF/B3LYP/6-31G* methods are seen to be linear in $E_{\text{ɩ}}$ and the regression fits are displayed within the figure. Absolute values differ between the two computational methods but it is clear that the two methods produce identical series trends: as $E_{\text{ɩ}}$ increases, E_{LUMO} values decrease at the approximate rate of 1.5 eV per V.

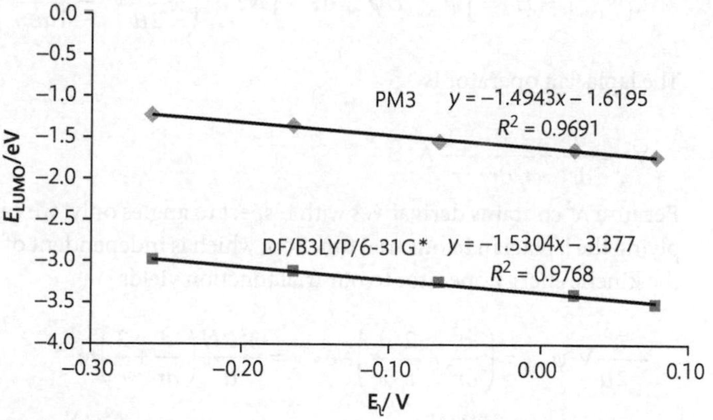

Figure F6.4

(b) We estimate the value of E_1 for species 6, our model for ubiquinone, using the linear regression fit of the DF/B3LYP/6-31G* computations.

$$E_1(\text{species 6}) = -\{(E_{\text{LUMO}}/\text{eV}+3.377)/1.5304\}\text{V} = -\{(-2.9922+3.377)/1.5304\}\text{V} = \boxed{-0.251\text{ V}}$$

(c) Species 4 is the working model for plastoquinone. Using the linear regression fit of the DF/B3LYP/6-31G* computations so that we can make a comparison with the calculation of part (b), its one-electron reducing potential is:

$$E_1(\text{species 4}) = -\{(E_{\text{LUMO}}/\text{eV}+3.377)/1.5304\}\text{V} = -\{(-3.1114+3.377)/1.5304\}\text{V} = \boxed{-0.174\text{ V}}$$

The species 6 analog of ubiquinone has a more negative reduction potential than the species 4 analog of plastoquinone so we conclude that ubiquinone is a better oxidizing agent than plastoquinone or, conversely, plastoquinone is a better reducing agent than ubiquinone.

(d) Coenzyme Q (i.e., ubiquinone) acts as an oxidizing agent when oxidizing NADH and $FADH_2$ in respiration, the overall oxidation of glucose by oxygen. Plastoquinone, on the other hand, acts as a reducing agent by reducing oxidized plastocyanin in photosynthesis. Respiration involves the oxidation of glucose by oxygen, while photosynthesis involves a reduction to glucose and oxygen. It stands to reason that the better oxidizing agent, ubiquinone, is employed in oxidizing glucose in respiration, while the better reducing agent is used to meet the reduction demands in photosynthesis. (Note, however, that both species are recycled to their original forms: reduced ubiquinone is oxidized by iron (III) and oxidized plastoquinone is reduced by water.)

F6.13 $\psi_{\text{trial}} = N\text{e}^{-\alpha r^2}$

We must find the expectation value of the hydrogenic hamiltonian:

$$E[\psi_{\text{trial}}] = \langle H\rangle = \int \psi_{\text{trial}}{}^*\hat{H}\psi_{\text{trial}}\,\text{d}\tau = \int N\text{e}^{-\alpha r^2}\left(-\frac{\hbar^2}{2\mu}\nabla^2 - \frac{e^2}{4\pi\varepsilon_0 r}\right)N\text{e}^{-\alpha r^2}\,\text{d}\tau$$

The laplacian operator is

$$\nabla^2 = \frac{\partial^2}{\partial r^2} + \frac{2}{r}\frac{\partial}{\partial r} + \frac{1}{r^2}\Lambda^2.$$

Because Λ^2 contains derivatives with respect to angles only, we can ignore it in applying the laplacian to our trial function, which is independent of angles. Applying the kinetic energy operator to our trial function yields

$$-\frac{\hbar^2}{2\mu}\nabla^2\psi_{\text{trial}} = \left(\frac{\partial^2}{\partial r^2} + \frac{2}{r}\frac{\partial}{\partial r}\right)N\text{e}^{-\alpha r^2} = \frac{\hbar^2\alpha N}{\mu}\left(\frac{\partial}{\partial r} + \frac{2}{r}\right)r\text{e}^{-\alpha r^2}$$

$$= \frac{\hbar^2\alpha N}{\mu}\left(\text{e}^{-\alpha r^2} - 2\alpha r^2\text{e}^{-\alpha r^2} + 2\text{e}^{-\alpha r^2}\right) = \frac{\hbar^2\alpha N}{\mu}(3 - 2\alpha r^2)\text{e}^{-\alpha r^2}.$$

Inserting this into the energy expectation yields:

$$E = \int N\mathrm{e}^{-\alpha r^2}\left(\frac{\hbar^2\alpha N}{\mu}(3-2\alpha r^2)\mathrm{e}^{-\alpha r^2} - \frac{e^2 N\mathrm{e}^{-\alpha r^2}}{4\pi\varepsilon_0 r}\right)\mathrm{d}\tau$$

To actually evaluate the integral, we must write out $\mathrm{d}\tau$ and the limits of integration explicitly. Here $\mathrm{d}\tau = r^2 \sin\theta\,\mathrm{d}r\mathrm{d}\theta\mathrm{d}\phi$. Other than in $\mathrm{d}\tau$, there is no angular dependence in the integrand, so integrating over the angles yields 4π. Thus the integral becomes

$$E = 4\pi N^2 \int_0^\infty \mathrm{e}^{-2\alpha r^2}\left(\frac{3\hbar^2\alpha r^2}{\mu} - \frac{2\hbar^2\alpha^2 r^4}{\mu} - \frac{e^2 r}{4\pi\varepsilon_0}\right)\mathrm{d}r\,.$$

Consult a table of integrals or symbolic mathematical software to find

$$\int_0^\infty x^n \mathrm{e}^{-kx^2}\,\mathrm{d}x = \frac{\Gamma\{(n+1)/2\}}{2k^{(n+1)/2}}\,.$$

Use this on each term in the integral to obtain

$$E = 4\pi N^2\left(\frac{3\hbar^2\alpha}{\mu}\times\frac{\Gamma(3/2)}{2(2\alpha)^{3/2}} - \frac{2\hbar^2\alpha^2}{\mu}\times\frac{\Gamma(5/2)}{2(2\alpha)^{5/2}} - \frac{e^2}{4\pi\varepsilon_0}\times\frac{\Gamma(1)}{2(2\alpha)}\right).$$

The gamma function with integer or half-integer arguments are readily evaluated. Again, consult a mathematical handbook or software:

$$\Gamma(n+1) = n! \quad \text{so} \quad \Gamma(1) = 0! = 1;$$

$$\Gamma(m+\tfrac{1}{2}) = \frac{1\cdot 3\cdot 5\cdots(2m-1)}{2^m}\pi^{1/2}, \text{ so } \Gamma(3/2) = \frac{\pi^{1/2}}{2} \text{ and } \Gamma(5/2) = \frac{3\pi^{1/2}}{4}\,.$$

$$E = 4\pi N^2\left(\frac{3\hbar^2\pi^{1/2}}{2^{9/2}\alpha^{1/2}\mu} - \frac{e^2}{16\pi\varepsilon_0\alpha}\right)$$

We must now evaluate $N(\alpha)$. Normalization requires

$$\int \psi^*\psi\,\mathrm{d}\tau = 1 = N^2\int \mathrm{e}^{-2\alpha r^2}\,\mathrm{d}\tau = 4\pi N^2\int_0^\infty r^2\mathrm{e}^{-2\alpha r^2}\,\mathrm{d}r$$

$$1 = 4\pi N^2\times\frac{\Gamma(3/2)}{2(2\alpha)^{3/2}} = 4\pi N^2\times\frac{\pi^{1/2}}{2^{7/2}\alpha^{3/2}} \quad \text{or} \quad 4\pi N^2 = \frac{2^{7/2}\alpha^{3/2}}{\pi^{1/2}}$$

$$\text{Thus,}\quad E[\psi_{\text{trial}}] = \frac{2^{7/2}\alpha^{3/2}}{\pi^{1/2}}\left(\frac{3\hbar^2\pi^{1/2}}{2^{9/2}\alpha^{1/2}\mu} - \frac{e^2}{16\pi\varepsilon_0\alpha}\right) = \boxed{\frac{3\hbar^2\alpha}{2\mu} - \frac{e^2\alpha^{1/2}}{2^{1/2}\pi^{3/2}\varepsilon_0}}\,.$$

The variation principle says that the minimum energy is obtained by taking the derivative of the trial energy with respect to adjustable parameters, setting it equal to zero, and solving for the parameters:

$$\frac{\mathrm{d}E}{\mathrm{d}\alpha} = \frac{3\hbar^2}{2\mu} - \frac{e^2}{2^{3/2}\pi^{3/2}\varepsilon_0\alpha^{1/2}} = 0$$

Solving for α yields

$$\frac{3\hbar^2}{2\mu}=\frac{e^2}{2^{3/2}\pi^{3/2}\varepsilon_0\alpha^{1/2}} \quad \text{so} \quad \alpha=\left(\frac{\mu e^2}{3\hbar^2\varepsilon_0}\right)^2\left(\frac{1}{2\pi^3}\right)=\frac{\mu^2 e^4}{18\pi^3\hbar^4\varepsilon_0^2}$$

Substituting this back into the energy expression yields the minimum energy for this trial wavefunction:

$$E[\psi_{\text{trial}}]=\frac{3\hbar^2}{2\mu}\left(\frac{\mu^2e^4}{18\pi^3\hbar^4\varepsilon_0^2}\right)-\frac{e^2}{2^{1/2}\pi^{3/2}\varepsilon_0}\left(\frac{\mu^2e^4}{18\pi^3\hbar^4\varepsilon_0^2}\right)^{1/2}=\frac{\mu e^4}{12\pi^3\varepsilon_0^2\hbar^2}-\frac{\mu e^4}{6\pi^3\varepsilon_0^2\hbar^2}=\boxed{\frac{-\mu e^4}{12\pi^3\varepsilon_0^2\hbar^2}}$$

Notice that the above expression indicates that $\langle V\rangle=-2\langle E_{\text{k}}\rangle$ in accord to the virial theorem for a potential that goes as r^{-1}. Also, compare the above result to the actual hydrogenic energy:

$$E_{\text{H}}=\frac{-\mu e^4}{32\pi^2\varepsilon_0^2\hbar^2}$$

$E[\psi_{\text{trial}}]$ has 12π in the denominator where the true energy has 32. Thus, the trial energy is greater than (not as negative as) the true energy, consistent with the variation principle.

F6.15 (a) The table displays both the experimental and calculated* (HF-SCF/6-311G*) ^{13}C chemical shifts and computed* atomic charges on the carbon atom *para* to a number of substituents in substituted benzenes. Three sets of charges are shown, one derived by fitting the electrostatic potential, another by Mulliken population analysis, and the other by the method of "natural" charges.

Substituent	CH_3	H	CF_3	CN	NO_2
δ_{exp}	128.4	128.5	128.9	129.1	129.4
δ_{calc}*	134.4	133.5	132.1	138.8	141.8
Electrostatic charge*/*e*	-0.240	-0.135	-0.138	-0.102	-0.116
Mulliken charge*/*e*	-0.231	-0.217	-0.205	-0.199	-0.182
Natural charge*/*e*	-0.199	-0.183	-0.158	-0.148	-0.135

Spartan '06™; HF-SCF/6-311G

The idea of atomic charges within molecules is very useful on a conceptual level but this concept is not a measureable property. Nuclei definitely define the positions of atoms but the wave character of electrons causes them to be distributed throughout the molecule in molecular orbitals. Some of the electrons (e.g., core electrons) may be largely localized around a particular nucleus while others are more widely distributed in bonding and anti-bonding MOs. The case for which a bonding MO is localized around two adjacent nuclei illustrates the impossibility of defining an absolute scale of measurable atomic charges as there is no way to define which fraction of the MO should be assigned to one, or the other, nucleus. Consequently, there are many ways of defining atomic charges in a molecule. The most popular three (electrostatic charge, Mulliken charge, and natural charge) have been computed for this molecular series.

The three methods for calculating atomic charge agree that in going across the substituent series $-CH_3$, $-H$, $-CF_3$, $-CN$, $-NO_2$ the magnitude of the negative charge on the para-carbon decreases. The electrostatic charge shows the greatest change across the series; the natural charge shows the least change. The decrease in the magnitude of negative charge across the series is compatible with the elementary concepts of electronegativity effects at a bond and the effects of substituent releasing or withdrawing electrons to or from an aromatic benzene ring. The atomic charges show that an alkyl substituent (e.g., $-CH_3$) is electron-releasing with respect to $-H$ while $-CF_3$, $-CN$, and $-NO_2$ are progressively stronger electron-withdrawing substituents. See Figure F6.5.

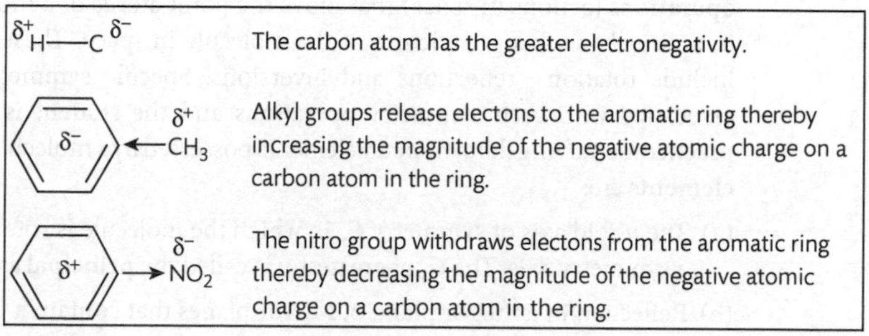

Figure F6.5

(b) The very small increase across the series in the experimental chemical shift of the para-^{13}C strongly correlates with decrease in the magnitude of the negative atomic charge across the series. Furthermore, it is evident that the uncertainty in the computed chemical shifts exceeds the actual variation in the chemical shifts across the series. The calculated shifts do indicate that the cyano and nitro groups are strong electron-withdrawing groups.

Topic 31 The analysis of molecular shape

Discussion questions

31.1 The point group to which a molecule belongs is determined by the **symmetry operations** (actions in space) that move the point atoms of a molecule but result in a visually unchanged picture of the molecule in space. These operations may include rotations, reflections and inversions. Specific symmetry elements are associated with each symmetry operations and the student is well advised to practice identifying the symmetry elements possessed by a molecule. The **symmetry elements** are:

(a) The n-fold axis of symmetry C_n in which the molecule is rotated $360°/n$ about a symmetry axis. The C_n of greatest n is called the **principal axis.**

(b) Reflection in a mirror plane σ. Mirror planes that contain a principal axis are denoted σ_v 'vertical' mirror planes. A vertical mirror plane that bisects two C_2 axes is denoted a σ_d 'dihedral' plane. A mirror plane that is perpendicular to a principal axis is denoted a σ_h 'horizontal' mirror plane.

(c) Inversion i through a centre of symmetry.

(d) The n-fold improper rotation S_n about an n-fold improper rotation axis. This is a rotation through $360°/n$ followed by reflection through a plane that is perpendicular to the axis of rotation.

(e) The identity E in which there is no action in space. Every molecule possesses the E symmetry element.

To assign a molecule to a point group, first identify all the symmetry elements possessed by the molecule. For example, inspection of the NH_3 molecular shape (ignore the lone pair, consider only the positions of the point atoms) reveals that it possesses the symmetry elements E, C_3, and three vertical mirror planes ($3\sigma_v$). This is the collection of elements for the C_{3v} point group so NH_3 belongs to the group C_{3v}.

So, identification of a molecule's point group requires the examination of a molecular model, which may be a plastic model, a drawing, or a mental picture, for its symmetry elements. The point group assignment corresponds to the identified collection of symmetry elements. The flow diagram in Figure 31.7 of the text often simplifies the assignment process.

31.3 The permanent dipole moment is a fixed property of a molecule and as a result it must remain unchanged through any symmetry operation of the molecule. Recall that the dipole moment is a vector quantity; therefore both its magnitude and direction must be unaffected by the operation. That can only be the case if the dipole moment is coincident with *all* of the symmetry elements of the molecule. Hence molecules belonging to point groups containing symmetry elements that do not satisfy this criterion can be eliminated. Molecules with a center of symmetry cannot possess a dipole moment because any vector is changed through inversion. Molecules with more than one C_n axis cannot be polar since a vector cannot be coincident with more than one axis simultaneously. If the molecule has a plane of symmetry, the dipole moment must lie in the plane; if it has more than one plane of symmetry, the dipole moment must lie in the axis of intersection of these planes. A molecule can also be polar if it has one plane of symmetry and no C_n. Examination of the character tables at the end of the data section shows that the only point groups that satisfy these restrictions are C_s, C_n, and C_{nv}.

Exercises

31.1(a) Chloromethane belongs to the point group C_{3v}. The elements, other than the identity E, are a C_3 axis and three vertical mirror planes σ_v. The symmetry axis passes through the C–Cl nuclei. The mirror planes are defined by the three ClCH planes. The C_3 principal axis and one of the σ_v mirror planes are shown in Fig. 31.1.

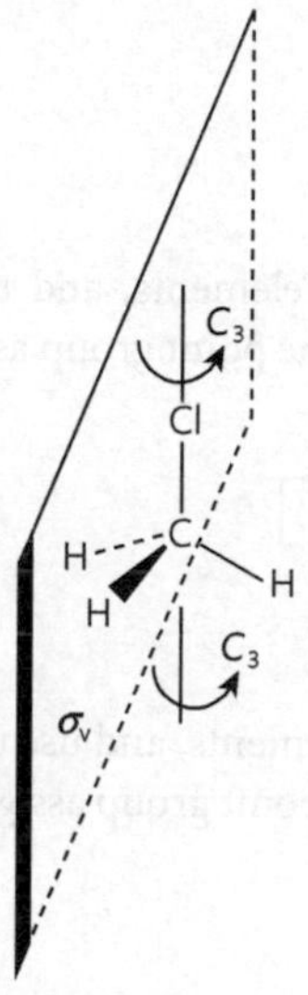

Figure 31.1

31.2(a) Naphthalene belongs to the point group D_{2h} and it has the symmetry elements shown in Fig. 31.2. There are $3C_2$ axes, a centre of inversion, and $3\sigma_h$ mirror planes.

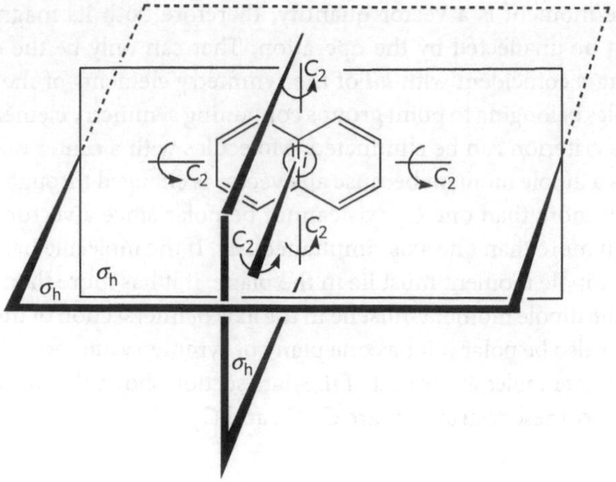

Figure 31.2

31.3(a) Sketch a figure of the object, identify symmetry elements, and use the flow diagram in Figure 31.7 of the text when it simplifies the group assignment.

(a) Sphere: an infinite number of symmetry axes; therefore R_3

(b) Isosceles triangles: E, C_2, σ_v, and σ_v' ; therefore C_{2v}

(c) Equilateral triangle: $\underbrace{\underbrace{E,\ C_3,\ C_2}_{D_3},\ \sigma_h}_{\boxed{D_{3h}}}$

(d) Cylinder: E, C_∞, C_2, σ_h; therefore $\boxed{D_{\infty h}}$

31.4(a) Make a sketch of the molecule, identify symmetry elements, and use the flow diagram in Figure 31.7 of the text when it simplifies the point group assignment.

(a) NO_2: E, C_2, σ_v, σ_v'; C_{2v}

(b) N_2O (linear N–N–O molecule): E, C_∞, C_2, σ_v; $\boxed{C_{\infty v}}$

(c) $CHCl_3$: E, C_3, $3\sigma_v$; $\boxed{C_{3v}}$

(d) $CH_2=CH_2$: E, C_2, $2C_2'$, i, σ_h; $\boxed{D_{2h}}$

31.5(a) Make a sketch of the molecule, identify symmetry elements, and use the flow diagram in Figure 31.7 of the text when it simplifies the point group assignment.

(a) *cis*-CHCl=CHCl: E, C_2, σ_v, σ_v'; $\boxed{C_{2v}}$

(b) *trans*-CHCl=CHCl; E, C_2, σ_h, i; $\boxed{C_{2h}}$

31.6(a) Only molecules belonging to the groups C_n, C_{nv}, and C_s may be polar [Topic 31.3(a)]; hence of the molecules listed only (a) $\boxed{\text{pyridine}}$ (C_{2v}) and (b) $\boxed{\text{nitroethane}}$ (C_s) are polar.

31.7(a) The parent of the dichloronaphthalene isomers is shown to the right. Care must be taken when determining possible isomers because naphthalene is a flat molecule that belongs to the point group D_{2h} as discussed in exercise 31.2(a). It has an inversion centre, mirror planes, and rotational axes that cause superficially distinct visual images to actually be the same molecule viewed from different angles. For example, Fig. 31.3 structures are all 1,3-dichloronaphthalene. By drawing figures that avoid the redundancy caused by the symmetry elements you will find a total of ten dichloronaphthalene isomers.

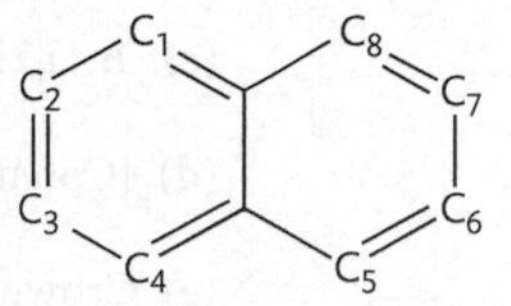

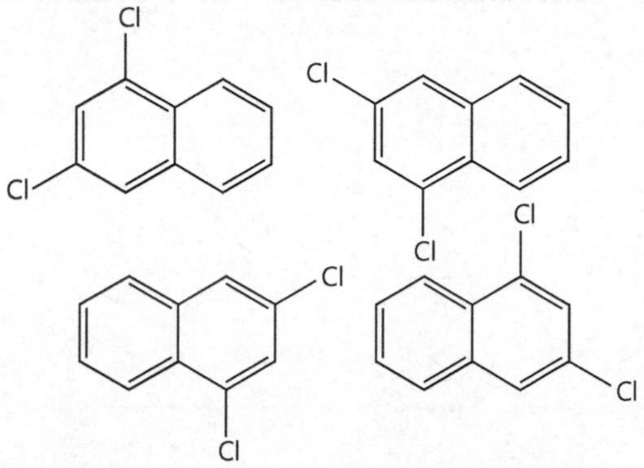

Figure 31.3

The names and point groups of the ten isomers are summarized in the following table.

Isomers and Point Groups of *m*, *n*-Dichloronaphthalene

m,n	1,2	1,3	1,4	1,5	1,6	1,7	1,8	2,3	2,6	2,7
Point Group	C_s	C_s	C_{2v}	C_{2h}	C_s	C_s	C_{2v}	C_{2v}	C_{2h}	C_{2v}

31.8(a) In each molecule we must look for an improper rotation axis, perhaps in the disguised form $S_1 = \sigma$ or $S_2 = i$ (Topic 31.3(b)). If present, the molecule cannot be chiral. D_{2h} contains $\boxed{i}$ and C_{3h} contains $\boxed{\sigma_h}$; therefore, molecules belonging to these point groups cannot be chiral and cannot be optically active.

Problems

31.1

(a) Staggered CH_3CH_3: E, C_3, C_2, i, $2S_6$, $3\sigma_d$; $\boxed{D_{3d}}$

(b) Chair C_6H_{12}: E, C_3, C_2, i, $2S_6$, $3\sigma_d$; $\boxed{D_{3d}}$

Boat C_6H_{12}: E, C_2, σ_v, σ'_v; $\boxed{C_{2v}}$

(c) B_2H_6: E, C_2, $2C_2'$, i, σ_h, $2\sigma_v'$; $\boxed{D_{2h}}$

(d) $[Co(en)_3]^{3+}$: E, $2C_3$, $3C_2$; $\boxed{D_3}$

(e) Crown S_8: E, C_4, C_2, $4C_2'$, $4\sigma_d$, $2S_8$; $\boxed{D_{4d}}$

Only boat C_6H_{12} may be polar, since all the others are D point groups. Only $[Co(en)_3]^{3+}$ belongs to a group without an improper rotation axis ($S_1 = \sigma$), and hence is chiral.

31.2‡ The most distinctive symmetry operation is the $\boxed{S_4}$ axis through the central atom and aromatic nitrogens on both ligands. That axis is also a $\boxed{C_2}$ axis. The group is $\boxed{S_4}$.

Topic 32 Group theory

Discussion questions

32.1 Within the context of quantum theory and molecular symmetry a **group** is a collection of transformations (R, S, T, etc.) that satisfy these criteria:

1. One of the transformations is the identity (E).
2. For every transformation R, the inverse transformation R^{-1} is included in the collection so that the combination RR^{-1} is equivalent to the identity: $RR^{-1} = E$.
3. The product RS is equivalent to a single member of the collection of transformations.
4. Multiple transformations obey the associative rule: $R(ST) = (RS)T$.

32.3 Figure 32.1 uses the C_{3v} point group to illustrate structural features of a character table. The central columns are labeled with the symmetry operations, or **symmetry classes**, of the group. Each operation is prefixed with the number of operations in the class, called the **degeneracy of the class**. For example, the degeneracy of the C_3 class is 2, which corresponds to clockwise and counter-clockwise rotations of the principal axis, shown in Figure 32.2. There are 3 vertical reflection planes in the C_{3v} group; these are also shown in Figure 32.2. The **group order**, h, is the sum of class degeneracies. It is the total number of symmetry operations of the group. The first column of Figure 32.1 shows that the C_{3v} group contains three **symmetry species**. They are the **Mulliken symbols** used to label the irreducible representations of the group. The symmetry species A and B are one-dimensional representations, E is the symbol for a two-dimensional representation, and T is the symbol of a three-dimensional representation. It is useful to remember that E species levels are doubly degenerate while T species levels are triply degenerate.

Number of symmetry species = number of classes [32.5]

Also, the sum of the squares of the dimension d_i of all symmetry species i equals the group order.

$$\sum_i d_i^{\ 2} = h \quad [32.6]$$

For the C_{3v} group the dimensions are 1, 1, and 2 for the A_1, A_2, and E irreducible representations, respectively. Adding the squares gives $1^2 + 1^2 + 2^2 = 6$ and we see that the sum of the squares equals h.

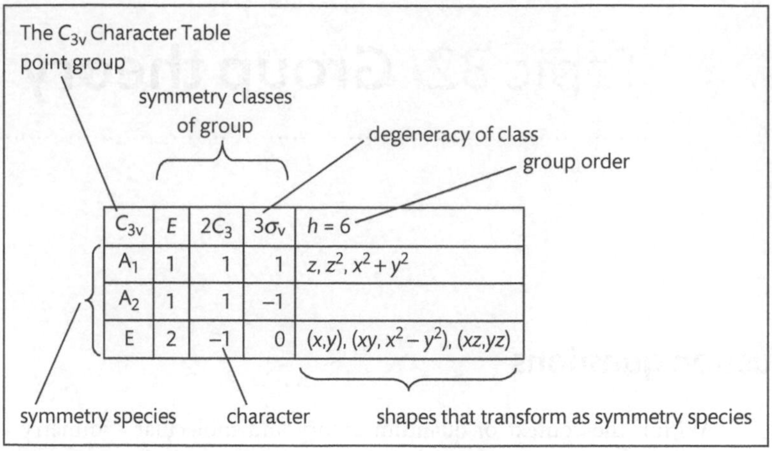

C_{3v}	E	$2C_3$	$3\sigma_v$	$h = 6$
A_1	1	1	1	z, z^2, x^2+y^2
A_2	1	1	−1	
E	2	−1	0	$(x,y), (xy, x^2-y^2), (xz,yz)$

Figure 32.1

The **characters** below a symmetry class of a point group are the traces (the sum of diagonal elements) of the matrix representations of the class operation(s) for a symmetry species. When the character under the principal axis is +1 for a one-dimensional symmetry species, the Mulliken symbol of the species is A; if the character is −1, the Mulliken symbol of the species is B. All characters of the symmetry species A_1 are +1. Characters of +1 are said to be **symmetric** w/r/t the specific operation of the character; characters of −1 are said to be **antisymmetric** w/r/t the operation. The subscript 1 or 2 is added to the Mulliken symbol when a species is either symmetric or antisymmetric under a σ_v operation, respectively. The subscripts g and u (gerade and ungerade) indicate that a species is either symmetric or antisymmetric under the i operation. Single primes or double primes are added to the species symbol to indicate symmetry or antisymmetry under the σ_h operation.

The last column of the character table catalogues the symmetry species to which useful functions belong. Any function that is proportional to one of the summarized functions belongs to the same symmetry species. For example, a quick examination of the C_{3v} group reveals that the basic functions z, z^2, and $x^2 + y^2$ belong to the A_1 **irrep** (irreducible representation) so we immediately recognize that the p_z, d_{z^2} and $d_{x^2+y^2}$ atomic orbitals all belong to the A_1 species. The d_{xy} and $d_{x^2-y^2}$ orbitals **jointly** belong to the E symmetry species of the C_{3v} group; they are said to **span** the representation and to be **basis functions** for the representation.

We take a moment to show that x and y jointly span the E irreducible representation of the C_{3v} group. The operations indicated in Figure 32.2 are useful to this effort.

By inspection we deduce that the matrix representation of σ_v is

$$D(\sigma_v) = \begin{pmatrix} -1 & 0 \\ 0 & 1 \end{pmatrix}$$

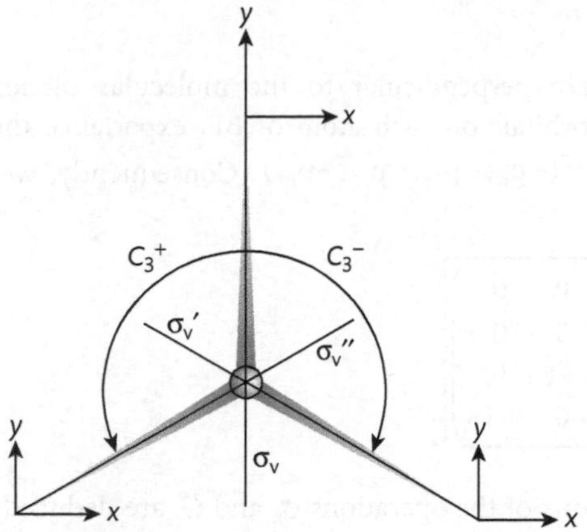

Figure 32.2

because this matrix transforms (x,y) to $(-x,y)$ in accord with the requirements of the σ_v reflection shown in Figure 32.2. The trace of this matrix is $-1+1=0$, which is the character of the σ_v operation of the E symmetry species. The character of the $C_3{}^+$ operation is more difficult to deduce. The general form of the two dimensional rotation of angle ϕ is

$$\boldsymbol{D}(C_\phi)=\begin{pmatrix}\cos\phi & \sin\phi\\ -\sin\phi & \cos\phi\end{pmatrix}$$

so the matrix for a 120° (2π/3 radian) rotation of the principal axis, which is the central z axis in Figure 32.2, is

$$\boldsymbol{D}(C_3^+)=\begin{pmatrix}\cos(2\pi/3) & \sin(2\pi/3)\\ -\sin(2\pi/3) & \cos(2\pi/3)\end{pmatrix}=\begin{pmatrix}-\tfrac{1}{2} & \sqrt{3}/2\\ -\sqrt{3}/2 & -\tfrac{1}{2}\end{pmatrix}$$

The trace of this matrix is $-\tfrac{1}{2}-\tfrac{1}{2}=-1$, which is the character of the C_3 operation of the E symmetry species. Thus, x and y jointly span E.

32.5 The letters and subscripts of a symmetry species provide information about the symmetry behavior of the species. An A or a B is used to denote a one-dimensional representation; A is used if the character under the principal rotation is +1 (symmetric behavior), and B is used if the character is −1 (antisymmetric behavior). Subscripts are used to distinguish the irreducible representations if there is more than one of the same type: A_1 is reserved for the representation with the character 1 under all operations. E denotes a two-dimensional irreducible representation and T a three-dimensional irreducible representation. These labels are called **Mulliken symbols.** For groups with an inversion centre, a subscript g (gerade) indicates symmetric behavior under the inversion operation; a subscript u (ungerade) indicates antisymmetric behavior. A horizontal mirror is assigned ′ or ″ superscripts if the behavior is symmetric or antisymmetric, respectively, under the σ_h operation.

Exercises

32.1(a) Since the p_z orbitals are perpendicular to the molecular plane, we recognize that the set of p_z orbitals on each atom of BF_3 experience the σ_h change $(p_B, p_{F1}, p_{F2}, p_{F3})\boldsymbol{D}(\sigma_h) = (-p_B, -p_{F1}, -p_{F2}, -p_{F3})$. Consequently, we find by inspection that

$$\boxed{\boldsymbol{D}(\sigma_h) = \begin{pmatrix} -1 & 0 & 0 & 0 \\ 0 & -1 & 0 & 0 \\ 0 & 0 & -1 & 0 \\ 0 & 0 & 0 & -1 \end{pmatrix}}$$

32.2(a) The matrix representations of the operations σ_h and C_3 are deduced in exercises 32.1(a) and 32.1(b). According to the precepts of group theory, the successive application of these operations yields another member of the D_{3h} group to which BF_3 belongs and, in fact, by definition the operation $\sigma_h C_3$ should yield the S_3 symmetry operation. The matrix representation of S_3 can be found by matrix multiplication of the component operations.

$$\boldsymbol{D}(\sigma_h)\boldsymbol{D}(C_3) = \begin{pmatrix} -1 & 0 & 0 & 0 \\ 0 & -1 & 0 & 0 \\ 0 & 0 & -1 & 0 \\ 0 & 0 & 0 & -1 \end{pmatrix}\begin{pmatrix} 1 & 0 & 0 & 0 \\ 0 & 0 & 1 & 0 \\ 0 & 0 & 0 & 1 \\ 0 & 1 & 0 & 0 \end{pmatrix} = \boxed{\begin{pmatrix} -1 & 0 & 0 & 0 \\ 0 & 0 & -1 & 0 \\ 0 & 0 & 0 & -1 \\ 0 & -1 & 0 & 0 \end{pmatrix} = \boldsymbol{D}(S_3)}$$

The result may be checked by matrix operation on the pz orbital vector where the effort should yield $(p_B, p_{F1}, p_{F2}, p_{F3})\boldsymbol{D}(S_3) = (-p_B, -p_{F3}, -p_{F1}, -p_{F2})$ and, as expected:

$$(p_B, p_{F1}, p_{F2}, p_{F3})\begin{pmatrix} -1 & 0 & 0 & 0 \\ 0 & 0 & -1 & 0 \\ 0 & 0 & 0 & -1 \\ 0 & -1 & 0 & 0 \end{pmatrix} = (-p_B, -p_{F3}, -p_{F1}, -p_{F2})$$

32.3(a) Consider the equilateral triangle $P_1P_2P_3$, which belongs to the D_{3h} point group (text Fig. 31.8). The three C_2 axes and the three σ_v mirror planes of this triangle are shown in Fig. 32.3.

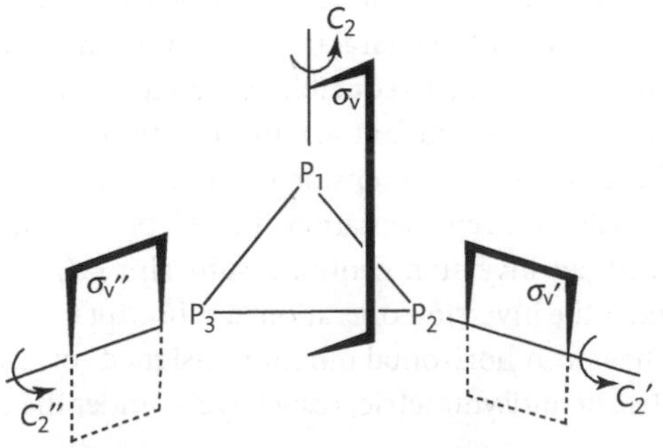

Figure 32.3

The C_2 and C_2' axes belong to the same class if there is a member S of the group such that $C_2' = S^{-1}C_2S$ [7.5] where S^{-1} is the inverse of S. We will work with $S = \sigma_v''$, an operator for which $S^{-1} = \sigma_v''$. By comparison of the action of $S^{-1}C_2S$ upon the vector (P_1,P_2,P_3) with the action of C_2' upon the same vector we can determine whether or not the equality of eqn 32.1 holds. If it does, C_2 and C_2' axes belong to the same class.

$$\begin{aligned} S^{-1}C_2S(P_1,P_2,P_3) &= \sigma_v''C_2\sigma_v''(P_1,P_2,P_3) \\ &= \sigma_v''C_2(P_2,P_1,P_3) \\ &= \sigma_v''(P_2,P_3,P_1) \\ &= (P_3,P_2,P_1) \qquad \text{(i)} \\ C_2'(P_1,P_2,P_3) &= (P_3,P_2,P_1) \qquad \text{(ii)} \end{aligned}$$

Eqs. (i) and (ii) indicate that $C_2' = S^{-1}C_2S$ where $S = \sigma_v''$ and we conclude that C_2 and C_2' belong to the same class. By either using the same argument or seeing the necessities of symmetry we find that C_2 and C_2'' also belong to the same class. Consequently, C_2, C_2' and C_2'' all belong to the same class.

32.4(a) Because the largest character is 3 in the column headed E in the O_h character table, we know that the maximum orbital degeneracy is 3. (See Topic 32.3(a)).

32.5(a) Benzene belongs to the D_{6h} point group. Because the largest character is 2 in the column headed E in the D_{6h} character table, we know that the maximum orbital degeneracy is 2. (See Topic 32.3(a)).

Problems

32.1 The operations are illustrated in Fig. 32.4. Note that $R^2 = E$ for all the operations of the groups, that $ER = RE = R$ always, and that $RR' = R'R$ for this group. Since $C_2\sigma_h = i$, $\sigma_h i = C_2$, and $iC_2 = \sigma_h$, we can draw up the following group multiplication table

$R\downarrow R'\rightarrow$	E	C_2	σ_h	i
E	E	C_2	σ_h	i
C_2	C_2	E	i	σ_h
σ_h	σ_h	i	E	C_2
i	i	σ_h	C_2	E

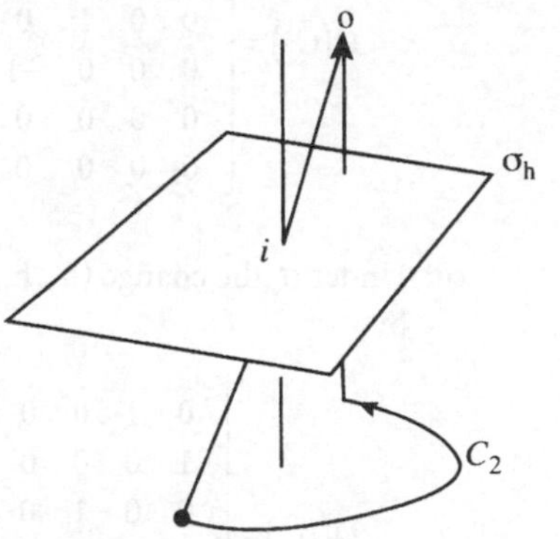

Figure 32.4

The $\boxed{\textit{trans}\text{-CHCl=CHCl}}$ molecule belongs to the group C_{2h}.

32.3 Refer to Fig. 31.3 of the text. Place 1s orbitals h_1 and h_2 on the H atoms and s, p_x, p_y, and p_z on the O atom. The *z*-axis is the C_2 axis; *x* lies perpendicular to σ_v', *y* lies perpendicular to σ_v. Then draw up the following table of the effect of the operations on the basis

	E	C_2	σ_v	σ_v'
h_1	h_1	h_2	h_2	h_1
h_2	h_2	h_1	h_1	h_2
s	s	s	s	s
p_x	p_x	$-p_x$	p_x	$-p_x$
p_y	p_y	$-p_y$	$-p_y$	p_y
p_z	p_z	p_z	p_z	p_z

Each operation column is the changed basis that results from the application of the operation R upon the basis (h_1, h_2, s, p_x, p_y, p_z). We want the 6×6 matrix representative of each R such that the changes are expressed in the form: the columns headed by each operations R in the form (changed basis) = (original basis)$\boldsymbol{D}(R)$. We use the methods set out in Topic 32.2.

(i) Under E the change $(h_1, h_2, s, p_x, p_y, p_z) \leftarrow (h_1, h_2, s, p_x, p_y, p_z)$ is reproduced by:

$$\boldsymbol{D}(E)=\begin{bmatrix} 1 & 0 & 0 & 0 & 0 & 0 \\ 0 & 1 & 0 & 0 & 0 & 0 \\ 0 & 0 & 1 & 0 & 0 & 0 \\ 0 & 0 & 0 & 1 & 0 & 0 \\ 0 & 0 & 0 & 0 & 1 & 0 \\ 0 & 0 & 0 & 0 & 0 & 1 \end{bmatrix}$$

(ii) Under C_2 the change $(h_2, h_1, s, -p_x, -p_y, p_z) \leftarrow (h_1, h_2, s, p_x, p_y, p_z)$ is reproduced by

$$\boldsymbol{D}(C_2)=\begin{bmatrix} 0 & 1 & 0 & 0 & 0 & 0 \\ 1 & 0 & 0 & 0 & 0 & 0 \\ 0 & 0 & 1 & 0 & 0 & 0 \\ 0 & 0 & 0 & -1 & 0 & 0 \\ 0 & 0 & 0 & 0 & -1 & 0 \\ 0 & 0 & 0 & 0 & 0 & 1 \end{bmatrix}$$

(iii) Under σ_v the change $(h_2, h_1, s, p_x, -p_y, p_z) \leftarrow (h_1, h_2, s, p_x, p_y, p_z)$ is reproduced by

$$\boldsymbol{D}(\sigma_v)=\begin{bmatrix} 0 & 1 & 0 & 0 & 0 & 0 \\ 1 & 0 & 0 & 0 & 0 & 0 \\ 0 & 0 & 1 & 0 & 0 & 0 \\ 0 & 0 & 0 & 1 & 0 & 0 \\ 0 & 0 & 0 & 0 & -1 & 0 \\ 0 & 0 & 0 & 0 & 0 & 1 \end{bmatrix}$$

(iv) Under σ_v' the change $(h_1, h_2, s, -p_x, p_y, p_z) \leftarrow (h_1, h_2, s, p_x, p_y, p_z)$ is reproduced by

$$D(\sigma_v') = \begin{bmatrix} 1 & 0 & 0 & 0 & 0 & 0 \\ 0 & 1 & 0 & 0 & 0 & 0 \\ 0 & 0 & 1 & 0 & 0 & 0 \\ 0 & 0 & 0 & -1 & 0 & 0 \\ 0 & 0 & 0 & 0 & 1 & 0 \\ 0 & 0 & 0 & 0 & 0 & 1 \end{bmatrix}$$

(a) Confirm that $C_2\sigma_v = \sigma_v'$ by multiplication of their matrix representations.

$$D(C_2)D(\sigma_v) = \begin{bmatrix} 0 & 1 & 0 & 0 & 0 & 0 \\ 1 & 0 & 0 & 0 & 0 & 0 \\ 0 & 0 & 1 & 0 & 0 & 0 \\ 0 & 0 & 0 & -1 & 0 & 0 \\ 0 & 0 & 0 & 0 & -1 & 0 \\ 0 & 0 & 0 & 0 & 0 & 1 \end{bmatrix} \begin{bmatrix} 0 & 1 & 0 & 0 & 0 & 0 \\ 1 & 0 & 0 & 0 & 0 & 0 \\ 0 & 0 & 1 & 0 & 0 & 0 \\ 0 & 0 & 0 & 1 & 0 & 0 \\ 0 & 0 & 0 & 0 & -1 & 0 \\ 0 & 0 & 0 & 0 & 0 & 1 \end{bmatrix}$$

$$= \begin{bmatrix} 1 & 0 & 0 & 0 & 0 & 0 \\ 0 & 1 & 0 & 0 & 0 & 0 \\ 0 & 0 & 1 & 0 & 0 & 0 \\ 0 & 0 & 0 & -1 & 0 & 0 \\ 0 & 0 & 0 & 0 & 1 & 0 \\ 0 & 0 & 0 & 0 & 0 & 1 \end{bmatrix} = D(\sigma_v')$$

(b) Confirm that $\sigma_v\sigma_v' = C_2$ by multiplication of their matrix representations.

$$D(\sigma_v)D(\sigma_v') = \begin{bmatrix} 0 & 1 & 0 & 0 & 0 & 0 \\ 1 & 0 & 0 & 0 & 0 & 0 \\ 0 & 0 & 1 & 0 & 0 & 0 \\ 0 & 0 & 0 & 1 & 0 & 0 \\ 0 & 0 & 0 & 0 & -1 & 0 \\ 0 & 0 & 0 & 0 & 0 & 1 \end{bmatrix} \begin{bmatrix} 1 & 0 & 0 & 0 & 0 & 0 \\ 0 & 1 & 0 & 0 & 0 & 0 \\ 0 & 0 & 1 & 0 & 0 & 0 \\ 0 & 0 & 0 & -1 & 0 & 0 \\ 0 & 0 & 0 & 0 & 1 & 0 \\ 0 & 0 & 0 & 0 & 0 & 1 \end{bmatrix}$$

$$= \begin{bmatrix} 0 & 1 & 0 & 0 & 0 & 0 \\ 1 & 0 & 0 & 0 & 0 & 0 \\ 0 & 0 & 1 & 0 & 0 & 0 \\ 0 & 0 & 0 & -1 & 0 & 0 \\ 0 & 0 & 0 & 0 & -1 & 0 \\ 0 & 0 & 0 & 0 & 0 & 1 \end{bmatrix} = D(C_2)$$

(a) The **character**, χ (chi), of an operation in a particular matrix representation is the sum of the diagonal elements of the representative of that operation. For the matrix representations the characters of the representation Γ are:

R	E	C_2	σ_v	σ'_v
Γ	6	0	2	4

(b) Compare the set of characters found in part (a) with the set of characters for each of the four irreps in the C_{2v} character table (shown below).

C_{2v}	E	C_2	σ_v	σ'_v
A_1	1	1	1	1
A_2	1	1	−1	−1
B_1	1	−1	1	−1
B_2	1	−1	−1	1

We quickly see that there is no match with the irreps A_1, A_2, B_1, and B_2. Consequently, the representation of this basis is reducible to a combination of these irreps.

(c) Using the inspection method of trial and error, we find in short order that 3× the characters of A_1 plus the characters of B_1 plus 2× the characters of B_2 yields the character set of part (a). The sums are detailed in the following table. Therefore, the representation of this basis set is reduced to $\boxed{\Gamma = 3A_1 + B_1 + 2B_2}$.

	E	C_2	σ_v	σ'_v
$3A_1$	3	3	3	3
B_1	1	−1	1	−1
$2B_2$	2	−2	−2	2
$3A_1 + B_1 + 2B_2$	6	0	2	4

32.5 The basis set consists of four 1s hydrogen orbitals, which is written as $\boldsymbol{f}$= (A,B,C,D) when positioned as shown in Fig. 32.5. When reflected in the mirror plane σ_{dAB} of the figure, the basis vector becomes (B,A,C,D). The order of the T_d group is 24 so there are 24 matrices to find. In addition to E there are $8C_3$ operations; clockwise and counterclockwise C_3 axes along the C—H_A bond, shown in the figure, are labeled C_{3A}^+ and C_{3A}^-. Likewise, the clockwise and counterclockwise C_3 axes along the C—H_B bond, shown in the figure, are labeled C_{3B}^+ and C_{3B}^-. There are $3C_2$ operations; the one that bisects the H_A—C—H_B angle is labeled C_{2AB}. There are 6 S_4. The clockwise and counterclockwise S_4 axes that bisect the H_A—C—H_B angle are labeled S_{4AB}^{+} and S_{4AB}^{-}. Finally, there are 6 σ_d mirror planes; the one, shown in the figure, that bisects the H_A—C—H_B angle is labeled σ_{dAB}.

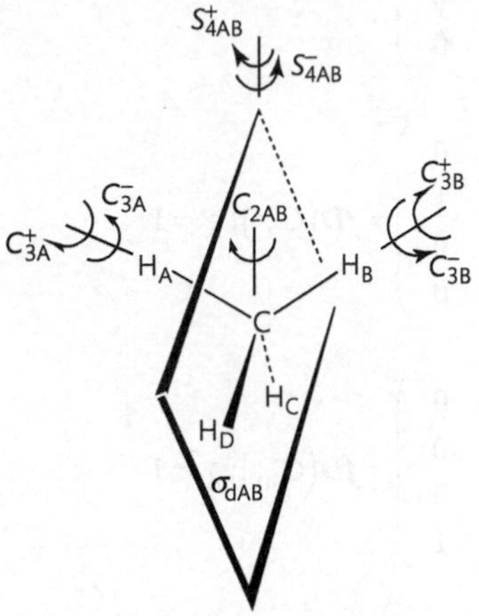

Figure 32.5

$$Ef = (A,B,C,D) = f\begin{pmatrix} 1 & 0 & 0 & 0 \\ 0 & 1 & 0 & 0 \\ 0 & 0 & 1 & 0 \\ 0 & 0 & 0 & 1 \end{pmatrix} = fD(E);\ \chi = 4$$

$$C_{3A}^{+}f = (A,C,D,B) = f\begin{pmatrix} 1 & 0 & 0 & 0 \\ 0 & 0 & 0 & 1 \\ 0 & 1 & 0 & 0 \\ 0 & 0 & 1 & 0 \end{pmatrix} = fD(C_{3A}^{+});\ \chi = 1$$

$$C_{3A}^{-}f = (A,D,B,C) = f\begin{pmatrix} 1 & 0 & 0 & 0 \\ 0 & 0 & 1 & 0 \\ 0 & 0 & 0 & 1 \\ 0 & 1 & 0 & 0 \end{pmatrix} = fD(C_{3A}^{-});\ \chi = 1$$

$$C_{3B}^{+}f = (D,B,A,C) = f\begin{pmatrix} 0 & 0 & 1 & 0 \\ 0 & 1 & 0 & 0 \\ 0 & 0 & 0 & 1 \\ 1 & 0 & 0 & 0 \end{pmatrix} = fD(C_{3B}^{+});\ \chi = 1$$

$$C_{3B}^{-}f = (C,B,D,A) = f\begin{pmatrix} 0 & 0 & 0 & 1 \\ 0 & 1 & 0 & 0 \\ 1 & 0 & 0 & 0 \\ 0 & 0 & 1 & 0 \end{pmatrix} = fD(C_{3B}^{-});\ \chi = 1$$

$$C_{3C}^{+}f = (B,D,C,A) = f\begin{pmatrix} 0 & 0 & 0 & 1 \\ 1 & 0 & 0 & 0 \\ 0 & 0 & 1 & 0 \\ 0 & 1 & 0 & 0 \end{pmatrix} = fD(C_{3C}^{+});\ \chi = 1$$

$$C_{3C}^{-}f = (D,A,C,B) = f\begin{pmatrix} 0 & 1 & 0 & 0 \\ 0 & 0 & 0 & 1 \\ 0 & 0 & 1 & 0 \\ 1 & 0 & 0 & 0 \end{pmatrix} = fD(C_{3C}^{-});\ \chi = 1$$

$$C_{3D}^{+}f = (C,A,B,D) = f\begin{pmatrix} 0 & 1 & 0 & 0 \\ 0 & 0 & 1 & 0 \\ 1 & 0 & 0 & 0 \\ 0 & 0 & 0 & 1 \end{pmatrix} = fD(C_{3D}^{+});\ \chi = 1$$

$$C_{3D}^{-}f = (B,C,A,D) = f\begin{pmatrix} 0 & 0 & 1 & 0 \\ 1 & 0 & 0 & 0 \\ 0 & 1 & 0 & 0 \\ 0 & 0 & 0 & 1 \end{pmatrix} = fD(C_{3D}^{-});\ \chi = 1$$

$$C_{2AB}f = (B,A,D,C) = f\begin{pmatrix} 0 & 1 & 0 & 0 \\ 1 & 0 & 0 & 0 \\ 0 & 0 & 0 & 1 \\ 0 & 0 & 1 & 0 \end{pmatrix} = fD(C_{2AB});\ \chi = 0$$

$$C_{2BC}f = (D,C,B,A) = f\begin{pmatrix} 0 & 0 & 0 & 1 \\ 0 & 0 & 1 & 0 \\ 0 & 1 & 0 & 0 \\ 1 & 0 & 0 & 0 \end{pmatrix} = fD(C_{2BC});\ \chi = 0$$

$$C_{2AC}f = (C,D,A,B) = f\begin{pmatrix} 0 & 0 & 1 & 0 \\ 0 & 0 & 0 & 1 \\ 1 & 0 & 0 & 0 \\ 0 & 1 & 0 & 0 \end{pmatrix} = fD(C_{2AC});\ \chi = 0$$

$$S_{4AB}^{+}f = (D,C,A,B) = f\begin{pmatrix} 0 & 0 & 1 & 0 \\ 0 & 0 & 0 & 1 \\ 0 & 1 & 0 & 0 \\ 1 & 0 & 0 & 0 \end{pmatrix} = fD(S_{4AB}^{+});\ \chi = 0$$

$$S_{4AB}^{-}\boldsymbol{f}=(C,D,B,A)=\boldsymbol{f}\begin{pmatrix}0&0&0&1\\0&0&1&0\\1&0&0&0\\0&1&0&0\end{pmatrix}=\boldsymbol{f}\boldsymbol{D}(S_{4AB}^{-});\ \chi=0$$

$$S_{4BC}^{+}\boldsymbol{f}=(B,D,A,C)=\boldsymbol{f}\begin{pmatrix}0&0&1&0\\1&0&0&0\\0&0&0&1\\0&1&0&0\end{pmatrix}=\boldsymbol{f}\boldsymbol{D}(S_{4BC}^{+});\ \chi=0$$

$$S_{4BC}^{-}\boldsymbol{f}=(C,A,D,B)=\boldsymbol{f}\begin{pmatrix}0&1&0&0\\0&0&0&1\\1&0&0&0\\0&0&1&0\end{pmatrix}=\boldsymbol{f}\boldsymbol{D}(S_{4BC}^{-});\ \chi=0$$

$$S_{4AC}^{+}\boldsymbol{f}=(B,C,D,A)=\boldsymbol{f}\begin{pmatrix}0&0&0&1\\1&0&0&0\\0&1&0&0\\0&0&1&0\end{pmatrix}=\boldsymbol{f}\boldsymbol{D}(S_{4AC}^{+});\ \chi=0$$

$$S_{4AC}^{-}\boldsymbol{f}=(D,A,B,C)=\boldsymbol{f}\begin{pmatrix}0&1&0&0\\0&0&1&0\\0&0&0&1\\1&0&0&0\end{pmatrix}=\boldsymbol{f}\boldsymbol{D}(S_{4AC}^{-});\ \chi=0$$

$$\sigma_{dAB}\boldsymbol{f}=(B,A,C,D)=\boldsymbol{f}\begin{pmatrix}0&1&0&0\\1&0&0&0\\0&0&1&0\\0&0&0&1\end{pmatrix}=\boldsymbol{f}\boldsymbol{D}(\sigma_{dAB});\ \chi=2$$

$$\sigma_{dAC}\boldsymbol{f}=(C,B,A,D)=\boldsymbol{f}\begin{pmatrix}0&0&1&0\\0&1&0&0\\1&0&0&0\\0&0&0&1\end{pmatrix}=\boldsymbol{f}\boldsymbol{D}(\sigma_{dAC});\ \chi=2$$

$$\sigma_{dAD}\boldsymbol{f}=(D,B,C,A)=\boldsymbol{f}\begin{pmatrix}0&0&0&1\\0&1&0&0\\0&0&1&0\\1&0&0&0\end{pmatrix}=\boldsymbol{f}\boldsymbol{D}(\sigma_{dAD});\ \chi=2$$

$$\sigma_{\text{dBC}} f = (\text{A,C,B,D}) = f\begin{pmatrix} 1 & 0 & 0 & 0 \\ 0 & 0 & 1 & 0 \\ 0 & 1 & 0 & 0 \\ 0 & 0 & 0 & 1 \end{pmatrix} = fD(\sigma_{\text{dBC}}); \chi = 2$$

$$\sigma_{\text{dBD}} f = (\text{A,D,C,B}) = f\begin{pmatrix} 1 & 0 & 0 & 0 \\ 0 & 0 & 0 & 1 \\ 0 & 0 & 1 & 0 \\ 0 & 1 & 0 & 0 \end{pmatrix} = fD(\sigma_{\text{dBD}}); \chi = 2$$

$$\sigma_{\text{dCD}} f = (\text{A,B,D,C}) = f\begin{pmatrix} 1 & 0 & 0 & 0 \\ 0 & 1 & 0 & 0 \\ 0 & 0 & 0 & 1 \\ 0 & 0 & 1 & 0 \end{pmatrix} = fD(\sigma_{\text{dCD}}); \chi = 2$$

Exercise: Find the representations for the 1*s* orbital basis at the corners of a regular trigonal bipyramid.

32.7 Representation 1: $\boldsymbol{D}(C_3)\boldsymbol{D}(C_2) = 1 \times 1 = 1 = \boldsymbol{D}(C_6)$

and from the C_{6v} character table this is either A_1 or A_2. Hence, either $\boldsymbol{D}(\sigma_v) = \boldsymbol{D}(\sigma_d)$ $= \boxed{+1 \text{ or } -1}$, respectively.

Representation 2: $\boldsymbol{D}(C_3)\boldsymbol{D}(C_2) = 1 \times (-1) = -1 = \boldsymbol{D}(C_6)$

and from the C_{6v} character table this is either B_1 or B_2. Hence, either

$\boldsymbol{D}(\sigma_v) = -\boldsymbol{D}(\sigma_d) = \boxed{+1}$ or $\boldsymbol{D}(\sigma_v) = -\boldsymbol{D}(\sigma_d) = \boxed{-1}$, respectively.

32.9 f orbitals: $f_{z(5z^2-3r^2)}$, $f_{y(5y^2-3r^2)}$, $f_{x(5x^2-3r^2)}$, $f_{z(x^2-y^2)}$, $f_{y(x^2-z^2)}$, $f_{x(z^2-y^2)}$, f_{xyz}

(a) C_{2v}. The functions x^2, y^2, and z^2 belong to the irreducible species A_1 so they are invariant under all symmetry operations of the group. Thus, $f_{z(5z^2-3r^2)}$ transforms as $z(A_1)$, $f_{y(5y^2-3r^2)}$ as $y(B_2)$, $f_{x(5x^2-3r^2)}$ as $x(B_1)$, and likewise for $f_{z(x^2-y^2)}$, $f_{y(x^2-z^2)}$, and $f_{x(z^2-y^2)}$. The function f_{xyz} transforms as $B_1 \times B_2 \times A_1 = A_2$. Therefore, in group C_{2v}, $f \rightarrow \boxed{2A_1 + A_2 + 2B_1 + 2B_2}$.

(b) C_{3v}. In C_{3v}, both z and z^2 transform as A_1; r^2 is also invariant, and hence $f_{z(5z^2-3r^2)}$ transforms as $z^3(A_1)$. ($f_{z(x^2-y^2)}$, f_{xyz}) transform as (x^2-y^2, xy) so this pair is a joint basis for E (see point group character table). The set ($f_{x(5x^2-3r^2)}$, $f_{y(5y^2-3r^2)}$, $f_{x(z^2-y^2)}$, $f_{y(x^2-z^2)}$) remain to be accounted for in the C_{3v} environment. We will now examine each of these under the class operations: E, $2C_3$, and $3\sigma_v$. Only one representation of each multi-member class need be examined (C_3^+ with C_3 along the z axis, and σ_v with σ_v defined by the yz plane). The irreps spanned by these four is found by the decomposition of the characters $\chi(E)$, $2\chi(C_3^+)$,

$3\chi(\sigma_v)$. $\boldsymbol{D}(E)$ is a 4×4 unit matrix so $\chi(E)=4$ and we turn to the task of finding $\chi(C_3^+)$ and $\chi(\sigma_v)$.

The z-axis is invariant under both C_3^+ and σ_v and it is important to recognize that (see Discussion question 32.1):

$$C_3^+(x,y)\to\left(-\tfrac{1}{2}x+\tfrac{1}{2}\sqrt{3}\,y,-\tfrac{1}{2}\sqrt{3}\,x-\tfrac{1}{2}y\right)\text{ and }\sigma_v(x,y)\to(-x,y).$$

Thus,

$$C_3^+\left(x\left(5x^2-3r^2\right)\right)\to C_3^+(x)\times\left\{5\left(-\tfrac{1}{2}x+\tfrac{1}{2}\sqrt{3}\,y\right)^2-3r^2\right\}=C_3^+(x)\times\left\{5\left(\tfrac{1}{4}x^2-\tfrac{1}{2}\sqrt{3}\,xy+\tfrac{3}{4}y^2\right)-3r^2\right\}$$

$$C_3^+\left(x\left(z^2-y^2\right)\right)\to C_3^+(x)\times\left\{z^2-\left(-\tfrac{1}{2}\sqrt{3}x-\tfrac{1}{2}y\right)^2\right\}=-C_3^+(x)\times\left\{\tfrac{3}{4}x^2+\tfrac{1}{2}\sqrt{3}\,xy+\tfrac{1}{4}y^2-z^2\right\}$$

The xy terms within the above expressions are removed by forming the combination orbital $f_1=f_{x(5x^2-3r^2)}-5\,f_{x(z^2-y^2)}$.

$$C_3^+(f_1)\to C_3^+(x)\times\left\{5\left(x^2+y^2\right)-3r^2-5z^2\right\}$$

All terms in the factor $\left\{5\left(x^2+y^2\right)-3r^2-5z^2\right\}$ belong to A_1 so the character depends upon $C_3^+(x)$ alone.

$$C_3^+(f_1)\to C_3^+(x)=-\tfrac{1}{2}x+\tfrac{1}{2}\sqrt{3}\,y$$

In the same way we define the combination orbital

$$f_2=f_{y(5y^2-3r^2)}+5\,f_{y(x^2-z^2)}$$

and find that

$$C_3^+(f_2)\to C_3^+(y)\times\left\{5\left(x^2+y^2\right)-3r^2-5z^2\right\}\to C_3^+(y)=-\tfrac{1}{2}\sqrt{3}x-\tfrac{1}{2}y.$$

Thus,

$$C_3^+(f_1,f_2)\to\left(-\tfrac{1}{2}x+\tfrac{1}{2}\sqrt{3}\,y,-\tfrac{1}{2}\sqrt{3}x-\tfrac{1}{2}y\right)$$

which has the matrix form

$$(f_1,f_2)\boldsymbol{D}(C_3^+)\to(x,y)\begin{pmatrix}-\tfrac{1}{2} & -\tfrac{1}{2}\sqrt{3}\\ \tfrac{1}{2}\sqrt{3} & -\tfrac{1}{2}\end{pmatrix}$$ and we conclude that $\chi(C_3^+)=-\dfrac{1}{2}-\dfrac{1}{2}=-1$.

Now, consider the effect of σ_v on each of the four orbitals.

$$\sigma_v\left(f_{x(5x^2-3r^2)}\right)\to -f_{x(5x^2-3r^2)}\quad\text{because }\sigma_v(x)\to -x$$

$$\sigma_v\left(f_{y(5y^2-3r^2)}\right) \rightarrow f_{y(5y^2-3r^2)} \quad \text{because } \sigma_v(y) \rightarrow y$$

$$\sigma_v\left(f_{x(z^2-y^2)}\right) \rightarrow -f_{x(z^2-y^2)}$$

$$\sigma_v\left(f_{y(x^2-z^2)}\right) \rightarrow f_{y(x^2-z^2)}$$

The matrix form of these transformations is

$$\left(f_{x(5x^2-3r^2)}, f_{y(5y^2-3r^2)}, f_{x(z^2-y^2)}, f_{y(x^2-z^2)}\right) D(\sigma_v)$$

$$= \left(f_{x(5x^2-3r^2)}, f_{y(5y^2-3r^2)}, f_{x(z^2-y^2)}, f_{y(x^2-z^2)}\right) \begin{pmatrix} -1 & 0 & 0 & 0 \\ 0 & 1 & 0 & 0 \\ 0 & 0 & -1 & 0 \\ 0 & 0 & 0 & 1 \end{pmatrix}$$

and we conclude that $\chi(\sigma_v) = -1+1-1+1 = 0$. In summary, we have found that these four orbitals have the characters $\chi(E) = 4$, $\chi(C_3^+) = -1$, and $\chi(\sigma_v) = 0$. The species of the four is indicated by the total membership of each symmetry operation $\{\chi(E), 2\chi(C_3^+), 3\chi(\sigma_v)\} = \{4,-2,0\}$ for which inspection of the character table immediately yields the decomposition 2E. Finally, we conclude that the set of seven f orbitals in C_{3v} spans $\boxed{A_1 + 3E}$.

(c) T_d. Make the inspired guess that the f orbitals are a basis of dimension 3 + 3 + 1, suggesting the decomposition T + T + A. Is the A representation A_1 or A_2? We see from the character table that the effect of S_4 discriminates between A_1 and A_2. Under S_4, $x \rightarrow y$, $y \rightarrow -x$, $z \rightarrow -z$, and so $S_4\{xyz\} \rightarrow xyz$. The character is $\chi = 1$, and so f_{xyz} spans A_1. Likewise, $(x^3, y^3, z^3) \rightarrow (y^3, -x^3, -z^3)$ and $\chi = 0+0-1 = -1$. Hence, the trio

$$(f_{x(5x^2-3r^2)}, f_{y(5y^2-3r^2)}, f_{z(5z^2-3r^2)})$$

spans T_2. Finally,

$$S_4\{x(z^2-y^2), y(x^2-z^2), z(x^2-y^2)\} \rightarrow \{y(z^2-x^2), -x(z^2-y^2), -z(y^2-x^2)\}$$

resulting in $\chi = 1$, indicating that $(f_{x(z^2-y^2)}, f_{y(x^2-z^2)}, f_{z(x^2-y^2)})$ belongs to T_1. Therefore, in T_d, $f \rightarrow \boxed{A_1 + T_1 + T_2}$.

(d) O_h. Anticipate an A + T + T decomposition as in the other cubic group. Since x, y, and z all have odd parity, all the irreducible representatives will be ungerade (u). Under S_4, $xyz \rightarrow xyz$ (as in (c)), and so the representation is $\chi = 1$, which causes f_{xyz} to belong to A_{2u} (see the character table). Under S_4, $(x^3, y^3, z^3) \rightarrow (y^3, -x^3, -z^3)$, as before, and $\chi = -1$, indicating that ($f_{x(5x^2-3r^2)}$,

$f_{y(5y^2-3r^2)}$, $f_{z(5z^2-3r^2)}$) belongs to T_{1u}. In the same way, the remaining three functions, ($f_{x(z^2-y^2)}$, $f_{y(x^2-z^2)}$, $f_{z(x^2-y^2)}$), span T_{2u}:

$$S_4\{x(z^2-y^2),\, y(z^2-x^2),\, z(x^2-y^2)\} \to \{y(z^2-x^2), -x(z^2-y^2), -z(y^2-x^2)\} \quad \chi = +1$$

$$C_4\{x(z^2-y^2),\, y(z^2-x^2),\, z(x^2-y^2)\} \to \{-y(z^2-x^2), x(z^2-y^2), -z(y^2-x^2)\} \quad \chi = -1$$

Hence, in O_h, $f \to \boxed{A_{2u}+T_{1u}+T_{2u}}$.

The *f* orbitals will cluster into sets according to their irreducible representations. Thus, (a) $f \to A_1+T_1+T_2$ in T_d symmetry, and there is one nondegenerate orbital and two sets of triply degenerate orbitals. (b) In O_h symmetry, $f \to A_{2u}+T_{1u}+T_{2u}$, and the pattern of splitting (but not the order of energies) is the same as that in T_d.

32.11‡ The shape of this molecule is shown in Fig. 32.6.

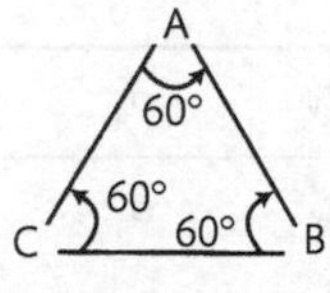

Fig 32.6

(a) Symmetry elements $\boxed{E,\, 2C_3,\, 3C_2,\, \sigma_h,\, 2S_3,\, 3\sigma_v}$, Point group $\boxed{D_{3h}}$

(b) $$\mathbf{D}(C_3)=\begin{pmatrix}0&0&1\\1&0&0\\0&1&0\end{pmatrix}, \quad \mathbf{D}(C_3')=\mathbf{D}^2(C_3)=\begin{pmatrix}0&1&0\\0&0&1\\1&0&0\end{pmatrix}$$

$$\mathbf{D}(S_3)=\mathbf{D}(C_3), \qquad \mathbf{D}(S_3')=\mathbf{D}^2(S_3)=\mathbf{D}(C_3')$$

C_3' and S_3' are counter-clockwise rotations.

σ_v is through A and perpendicular to B–C.

σ_v' is through B and perpendicular to A–C.

σ_v'' is through C and perpendicular to A–B.

$$\mathbf{D}(\sigma_v)=\begin{pmatrix}1&0&0\\0&0&1\\0&1&0\end{pmatrix}, \ \mathbf{D}(\sigma_v')=\begin{pmatrix}0&0&1\\0&1&0\\1&0&0\end{pmatrix}, \ \mathbf{D}(\sigma_v'')=\begin{pmatrix}0&1&0\\1&0&0\\0&0&1\end{pmatrix}$$

$$\mathbf{D}(C_2)=\mathbf{D}(\sigma_v), \qquad \mathbf{D}(C_v')=\mathbf{D}(\sigma_v'), \qquad \mathbf{D}(C_2'')=\mathbf{D}(\sigma_2'')$$

(c) Example of elements of group multiplication table

$$\mathbf{D}(C_3)\mathbf{D}(C_2)=\begin{pmatrix}0&0&1\\1&0&0\\0&1&0\end{pmatrix}\begin{pmatrix}1&0&0\\0&0&1\\0&1&0\end{pmatrix}$$

$$=\begin{pmatrix}0&1&0\\1&0&0\\0&0&1\end{pmatrix}=\mathbf{D}(\sigma_v'')$$

$$\mathbf{D}(\sigma_v')\mathbf{D}(\sigma_v)=\begin{pmatrix}0&0&1\\0&1&0\\1&0&0\end{pmatrix}\begin{pmatrix}1&0&0\\0&0&1\\0&1&0\end{pmatrix}$$

$$=\begin{pmatrix}0&1&0\\0&0&1\\1&0&0\end{pmatrix}=\mathbf{D}(C_v')$$

D_{3h}	E	C_3	C_2	σ_v	σ_v'	σ_h	...
E	E	C_3	C_2	σ_v	σ_v'	σ_h	...
C_3	C_3	C_3'	σ_v''	σ_v''	σ_v	C_3	...
C_2	C_2	σ_v'	E	E	C_3	C_2	...
σ_v	σ_v	σ_v'	E	E	C_3	σ_v	...
σ_v'	σ_v'	σ_v''	C_3	C_3'	E	σ_v'	...
σ_h	σ_h	C_3	C_2	σ_v	σ_v'	E	...
⋮	⋮	⋮	⋮	⋮	⋮	⋮	⋱

Comment: The multiplication table is not strictly speaking the group multiplication; it is instead the multiplication table for the matrix representations of the group in the basis under consideration.

(d) First, determine the number of s-orbitals (the basis has three s orbitals) that have unchanged positions after application of each symmetry species of the D_{3h} point group.

D_{3h}	E	$2C_3$	$3C_2$	σ_h	$2S_3$	$3\sigma_v$
Unchanged basis members	3	0	1	3	0	1

This is not one of the irreducible representations reported in the D_{3h} character table but inspection shows that it is identical to $A_1'+E'$. This allows us to conclude that the three s- orbitals span $\boxed{A_1'+E'}$.

Topic 33 Applications of symmetry

Discussion questions

33.1 Character tables provide a way to: (a) assign symmetry symbols for orbitals, (b) know whether overlap integrals are nonzero, (c) determine what atomic orbitals can contribute to a LCAO-MO, (d) determine the maximum orbital degeneracy of a molecule, and (e) determine whether a transition is allowed.

Exercises

33.1(a) The p*x* orbital spans B_1 of the C_{2v} point group while *z* and p*z* span A_1. Following the Topic 33.1(a) procedure for deducing the symmetry species spanned by the product f_1f_2 and hence to see whether it does indeed span A_1, we write a table of the characters of each function and multiply the rows.

C_{2v}	E	C_2	σ_v	σ_v'
p_x	1	−1	1	−1
z	1	1	1	1
p_z	1	1	1	1
p_xzp_z	1	−1	1	−1

The characters of the product p_xzp_z are those of B_1 alone, so the integrand does not span A_1. It follows that the integral must be zero.

33.2(a) For a C_{3v} molecule, *x* and *y* span E while *z* spans A_1. Thus, the *x* and *y* components of the dipole moment [33.6] have transition integrands that span $A_2 \times E \times A_1$ for the $A_1 \rightarrow A_2$ transition. By inspection of the C_{3v} character table we find the **decomposition of the direct product** to be: $A_2 \times E \times A_1 = E$. The integrand spans E alone. Since it does not span A_1, the *x* and *y* components of the transition integral must be zero. The transition integrand for the *z* component spans $A_2 \times A_1 \times A_1 = A_2$ for the $A_1 \rightarrow A_2$ transition. Consequently, the *z* component of the transition integral must also equal zero and we conclude that the transition is forbidden.

Should these considerations prove confusing, write a table with columns headed by the three components of the electric dipole moment operator, μ.

Component of μ:	x	y	z
A_1	1 1 1	1 1 1	1 1 1
$\Gamma(\mu)$	2 -1 0	2 -1 0	1 1 1
A_2	1 1 -1	1 -1 -1	1 1 -1
$A_1\,\Gamma(\mu)\,A_2$	2 -1 0	2 -1 0	1 1 -1
	E	E	A_2

Since A_1 is not present in any product, the transition dipole moment must be zero.

33.3(a) We first determine how x and y individually transform under the operations of the C_{4v} group. Using these results we determine how the products xy transforms. The transform of xy is the product of the transforms of x and y. Under each operation the functions transform as follows.

	E	C_2	C_4	σ_v	σ_d
x	x	$-x$	y	x	$-y$
y	y	$-y$	$-x$	$-y$	$-x$
xy	xy	xy	$-xy$	$-xy$	xy
χ	1	1	-1	-1	1

From the C_{4v} character table, we see that this set of characters belongs of B_2.

33.4(a) Recall that $p_x \propto x$, $p_y \propto y$, $p_z \propto z$, $d_{xy} \propto xy$, $d_{xz} \propto xz$, $d_{yz} \propto yz$, $d_{z^2} \propto z^2$, $d_{x^2-y^2} \propto x^2 - y^2$ (Topic 18.1 (b) and (c)). Additionally, when the functions f_1 and f_2 of an overlap integral are bases for irreducible representations of a group, the integral must vanish if they are different symmetry species; if they are the same symmetry species, then the integral may be nonzero (Topic 33.1). Since the combination $p_x(A) - p_x(B)$ of the two O atoms (with x perpendicular to the plane) spans A_2, the orbital on N must span A_2 for a nonzero overlap.

Now refer to the C_{2v} character table. The s orbital spans A_1 and the p orbitals of the central N atom span $A_1(p_z)$, $B_1(p_x)$, and $B_2(p_y)$. Therefore, **no N orbitals** span A_2, and hence $p_x(A)-p_x(B)$ is a nonbonding combination. If d orbitals are available, as they are in S of the SO_2 molecule, we could form a molecular orbital with $\boxed{d_{xy}}$, which is a basis for A_2.

33.5(a) The electric dipole moment operator transforms as $x(B_1)$, $y(B_2)$, and $z(A_1)$ of the C_{2v} character table. Transitions are allowed if $\int \psi_f^* \mu \psi_i \, d\tau$ is nonzero (Example 33.5 of text), and hence are forbidden unless $\Gamma_f \times \Gamma(\mu) \times \Gamma_i$ contains A_1. Since $\Gamma_i = A_1$, this requires $\Gamma_f \times \Gamma(\mu) = A_1$. Since $B_1 \times B_1 = A_1$ and $B_2 \times B_2 = A_1$ and $A_1 \times A_1 = A_1$, x-polarized light may cause a transition to a B_1 term, y-polarized light to a B_2 term, and z-polarized light to an A_1 term.

33.6(a)

C_{4v}, $h=8$	E	C_2	$2C_4$	$2\sigma_v$	$2\sigma_d$
A_1	1	1	1	1	1
A_2	1	1	1	−1	−1
B_1	1	1	−1	1	−1
B_2	1	1	−1	−1	1
E	2	−2	0	0	0

$$n(\Gamma)=\frac{1}{h}\sum_R \chi^{(\Gamma)}(R)\chi(R) \quad [33.2] \quad \text{where } \chi(R)=(5,1,1,3,1)$$

$$n(A_1)=\tfrac{1}{8}\{1(1\times5)+1(1\times1)+2(1\times1)+2(1\times3)+2(1\times1)\}=2$$

$$n(A_2)=\tfrac{1}{8}\{1(1\times5)+1(1\times1)+2(1\times1)+2(-1\times3)+2(-1\times1)\}=0$$

$$n(B_1)=\tfrac{1}{8}\{1(1\times5)+1(1\times1)+2(-1\times1)+2(1\times3)+2(-1\times1)\}=1$$

$$n(B_2)=\tfrac{1}{8}\{1(1\times5)+1(1\times1)+2(-1\times1)+2(-1\times3)+2(1\times1)\}=0$$

$$n(E)=\tfrac{1}{8}\{1(2\times5)+1(-2\times1)+2(0\times1)+2(0\times3)+2(0\times1)\}=1$$

Thus, this set of basis functions spans $\boxed{2A_1+B_1+E}$.

33.7(a)

(a) The point group of benzene is D_{6h}. In D_{6h} μ spans $E_{1u}(x, y)$ and $A_{2u}(z)$, and the ground term is A_{1g}. Then, using $A_{2u}\times A_{1g}=A_{2u}$, $E_{1u}\times A_{1g}=E_{1u}$, $A_{2u}\times A_{2u}=A_{1g}$, and $E_{1u}\times E_{1u}=A_{1g}+A_{2g}+E_{2g}$, we conclude that the upper term is $\boxed{\text{either } E_{1u} \text{ or } A_{2u}}$.

(b) Naphthalene belongs to D_{2h}. In D_{2h} itself, the components of μ span $B_{3u}(x)$, $B_{2u}(y)$, and $B_{1u}(z)$ and the ground term is A_g. Hence, since $A_g\times\Gamma=\Gamma$ in this group, the upper terms are $\boxed{B_{3u}(x\text{-polarized})}$, $\boxed{B_{2u}(y\text{-polarized})}$, and $\boxed{B_{1u}(z-\text{polarized})}$.

33.8(a) See Fig. 33.3, which shows that we assume that θ is the angle to the σ_h plane of group C_s and, therefore, it is a dimension that can exhibit a symmetrical integration interval from one side of the plane to the other side in a C_s object like methanol, bromochloromethane, and O=N–Cl. We consider the integral

$$I=\int_{-a}^{a} f_1 f_2\,\mathrm{d}\theta=\int_{-a}^{a}\sin\theta\,\cos\theta\,\mathrm{d}\theta$$

and hence draw up the following table for the effect of operations in the group C_s.

	E	σ_h
$f_1 = \sin\theta$	$\sin\theta$	$-\sin\theta$
$f_2 = \cos\theta$	$\cos\theta$	$\cos\theta$

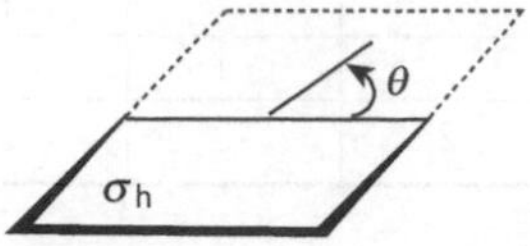

Fig 33.3

In terms of characters:

	E	σ_h	Symmetry Species
f_1	1	−1	A″
f_2	1	1	A′
f_1f_2	1	−1	A″

Since the product does not span the totally symmetric species A′, the integral is necessarily $\boxed{\text{zero}}$.

Problems

33.1 A quick rule for determining the character without first having to set up the matrix representation is to count 1 each time a basis function is left unchanged by the operation, because only these functions give a nonzero entry on the diagonal of the matrix representative. In some cases there is a sign change, $(\ldots -f \ldots) \leftarrow (\ldots f \ldots)$; then −1 occurs on the diagonal, and so count−1. The character of the identity is always equal to the dimension of the basis since each function contributes 1 to the trace.

E: all four orbitals are left unchanged; hence $\chi = 4$

C_3: One orbital is left unchanged; hence $\chi = 1$

C_2: No orbitals are left unchanged; hence $\chi = 0$

σ_d: Two orbitals are left unchanged; hence $\chi = 2$

S_4: No orbitals are left unchanged; hence $\chi = 0$

The character set 4, 1, 0, 2, 0 spans $\boxed{A_1 + T_2}$. Inspection of the character table of the group T_d shows that an s orbital spans A_1 and that the three p orbitals on the C atom span T_2. Hence, the $\boxed{\text{s and p}}$ orbitals of the C atom may form molecular orbitals with the four H1s orbitals. In T_d, the d orbitals of the central atom span E + T_2 (character table, final column), and so only the T_2 set $\boxed{(d_{xy}, d_{yz}, d_{zx})}$ may contribute to molecular orbital formation with the H orbitals.

33.3 Consider the integral $\int_{-a}^{a}(3x^2-1)\mathrm{d}x = 3\int_{-a}^{a}x^2\,\mathrm{d}x - \int_{-a}^{a}1\mathrm{d}x = 3I_2 - I_1$. Integral I_1 has the unit integrand, which spans the totally symmetric irreducible symmetric species in all point groups and is, consequently, nonzero. The integrand of I_2 is the product $x \times x$, which also spans the totally symmetric irreducible representation in all point groups because

$$n(\mathrm{A}_1) = \frac{1}{h}\sum_R \chi^{(\mathrm{A}_1)}(R)\chi(R) = \frac{1}{h}\sum_R \chi(R) = 1 \quad [33.2]$$

whenever $\chi(R)$ are the characters of the square of an irreducible representation (e.g., $\Gamma_i \times \Gamma_i$). Thus, the integral $\int_{-a}^{a}(3x^2-1)\mathrm{d}x$ is not necessarily zero because neither I_1 nor I_2 are necessarily zero. The integral will be accidently zero when the integration interval causes $3I_2$ to equal I_1.

Since the above conclusion is valid for all point groups, in the following parts we simply use eqn 32.2 to show that I_2 is not necessarily zero in each point group because the integrand spans A_1.

(a) A cube, $\mathrm{O_h}$. x spans $\mathrm{T_{1u}}$ and x^2 spans $\mathrm{T_{1u}} \times \mathrm{T_{1u}}$ so the number of times that the integrand of I_2 spans $\mathrm{A_{1g}}$ is:

$$n(\mathrm{A_{1g}}) = \frac{1}{48}\{1\times9+8\times0+6\times1+6\times1+3\times1+1\times9+6\times1+8\times0+3\times1+6\times1\} = 1$$

(b) A tetrahedron, $\mathrm{T_d}$. x spans $\mathrm{T_2}$ and x^2 spans $\mathrm{T_2} \times \mathrm{T_2}$ so the number of times that the integrand of I_2 spans $\mathrm{A_1}$ is:

$$n(\mathrm{A_1}) = \frac{1}{24}\{1\times9+8\times0+3\times1+6\times1+6\times1\} = 1$$

(c) A hexagonal prism, $\mathrm{D_{6h}}$. x spans $\mathrm{E_{1u}}$ and x^2 spans $\mathrm{E_{1u}} \times \mathrm{E_{1u}}$ so the number of times that the integrand of I_2 spans $\mathrm{A_{1g}}$ is:

$$n(\mathrm{A_{1g}}) = \frac{1}{24}\left\{\begin{matrix}1\times4+2\times1+2\times1+1\times4+3\times0+3\times0 \\ +1\times4+2\times1+2\times1+1\times4+3\times0+3\times0\end{matrix}\right\} = 1$$

33.5 The p_z orbitals of the fluorine atoms in $\mathrm{XeF_4}$ are shown in Fig. 33.2. (The orbital has a positive wavefunction sign in shaded lobes and a negative wavefunction sign in unshaded lobes.) These orbitals may form π molecular orbitals with the xenon atom provided that a nonzero overlap integral exists.

To find the symmetry species spanned by the fluorine p_z orbitals, we use a quick rule for determining the character of the basis set under each symmetry operation of the group: count 1 each time a basis function is left unchanged by the operation,

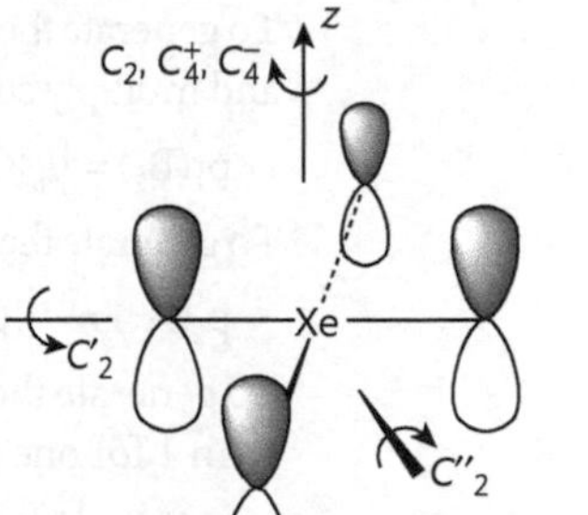

Figure 33.2

because only these functions give a nonzero entry on the diagonal of the matrix representative. In some cases there is a sign change, $(\ldots -f \ldots) \leftarrow (\ldots f \ldots)$; then -1 occurs on the diagonal, and so count -1. The character of the identity is always equal to the dimension of the basis since each function contributes 1 to the trace. Although XeF_4 belongs to the D_{4h} group, we will use the D_4 subgroup for convenience. Here is a summary of the characters exhibited by the fluorine orbital under each symmetry operation:

D_4	E	C_2	$2C_4$	$2C_2'$	$2C_2''$
fluorine p_z orbitals	4	0	0	−2	0

Inspection of the D_4 character table shows that the fluorine p_z orbitals span $A_2 + B_2 + E$. Further inspection of the D_4 character table reveals that p_z belongs to A_2, d_{xy} belongs to B_2, and both the (p_x, p_y) set and the (d_{xz}, d_{yz}) set belong to E. Consequently, only these orbitals of the central atom may possibly have nonzero overlap with the fluorine p_z orbitals. We must now use the procedure of Section 7.5(c) to find the four **symmetry-adapted linear combinations (SALC)** of the fluorine p_z orbitals. Using clockwise labeling of the Fp_z orbitals, we write a table that summarizes the effect of each operation on each orbital. (The symbol A is used to represent the orbital $p_z(A)$ at fluorine atom A, etc.)

D_4	A	B	C	D
E	A	B	C	D
C_2	C	D	A	B
C_4^+	D	A	B	C
C_4^-	B	C	D	A
C_2'(A-C)	−A	−D	−C	−B
C_2'(B-D)	−C	−B	−A	−D
C_2''(A-Xe-B)	−B	−A	−D	−C
C_2''(B-Xe-C)	−D	−C	−B	−A

To generate the B_2 combination, we take the characters for B_2 $(1, 1, -1, -1, -1, -1, 1, 1)$ and multiply column 3 and sum the terms. Then, divide by the order of the group (8):

$$p_1(B_2) = {}^1/_8\{C + A - B - D + C + A - D - B\} = {}^1/_4\{p_z(A) - p_z(B) + p_z(C) - p_z(D)\}$$

To generate the A_2 combination, we take the characters for A_2 and multiply column 1:

$$p_2(A_2) = {}^1/_4\{p_z(A) + p_z(A) + p_z(C) + p_z(D)\}$$

To generate the two E combinations, we take the characters for E and multiply column 1 for one of them and multiply column 2 for the other:

$$p_3(E) = {}^1/_4\{p_z(A) - p_z(C)\}$$

$$p_4(E) = {}^1/_4\{p_z(B) - p_z(D)\}$$

Other column multiplications yield these same combinations or zero.

Fig. 33.3(a) shows that the $p_z(A_2)$ orbital of the central atom does have nonzero overlap with p_2. Fig. 33.3(b) shows that $d_{xz}(E)$ has a nonzero overlap with p_3. However, Fig. 33.3(c) shows that, because of the balance between constructive and destructive interference, there is zero overlap between $d_{xy}(B_2)$ and p_1. Thus, there is no orbital of the central atom that forms a nonzero overlap with the p_1 combination so p_1 is nonbonding.

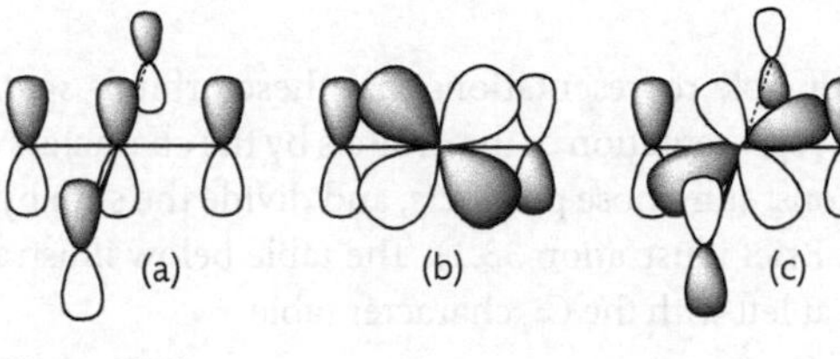

Figure 33.3

33.7 We begin by drawing up the following table for the C_{2v} group:

	N2s	N2p_x	N2p_y	N2p_z	O2p_x	O2p_y	O2p_z	O′2p_x	O′2p_y	O′2p_z	χ
E	N2s	N2p_x	N2p_y	N2p_z	O2p_x	O2p_y	O2p_z	O′2p_x	O′2p_y	O′2p_z	10
C_2	N2s	–N2p_x	–N2p_y	N2p_z	–O′2p_x	–O′2p_y	O′2p_z	–O2p_x	–O2p_y	O2p_z	0
σ_v	N2s	N2p_x	–N2p_y	N2p_z	O′2p_x	–O′2p_y	O′2p_z	O2p_x	–O2p_y	O2p_z	2
σ_v'	N2s	–N2p_x	N2p_y	N2p_z	–O2p_x	O2p_y	O2p_z	–O′2p_x	O′2p_y	O′2p_z	4

The character set (10, 0, 2, 4) decomposes into $\boxed{4A_1+2B_1+3B_2+A_2}$. We then form symmetry-adapted linear combinations as described in Topic 33.2(b).

$\psi(A_1)=N2s$ (column 1)
$\psi(A_1)=N2p_z$ (column 4)
$\psi(A_1)=O2p_z+O'2p_z$ (column 7)
$\psi(A_1)=-O2p_y+O'2p_y$ (column 9)
$\psi(B_1)=N2p_x$ (column 2)

$\psi(B_1)=O2p_x+O'2p_x$ (column 5)
$\psi(B_2)=N2p_y$ (column 3)
$\psi(B_2)=O2p_y+O'2p_y$ (column 6)
$\psi(B_2)=O2p_z-O'2p_z$ (column 7)
$\psi(A_2)=O2p_x-O'2p_x$ (column 5)

The other columns yield the same combinations.

33.9 Consider phenanthrene with carbon atoms as labeled in the structure below.

(a) The 2p orbitals involved in the π system are the basis we are interested in. To find the irreducible representations spanned by this basis, consider how each basis is transformed under the symmetry operations of the C_{2v} group. To find the character of an operation in this basis, sum the coefficients of the basis terms that are unchanged by the operation.

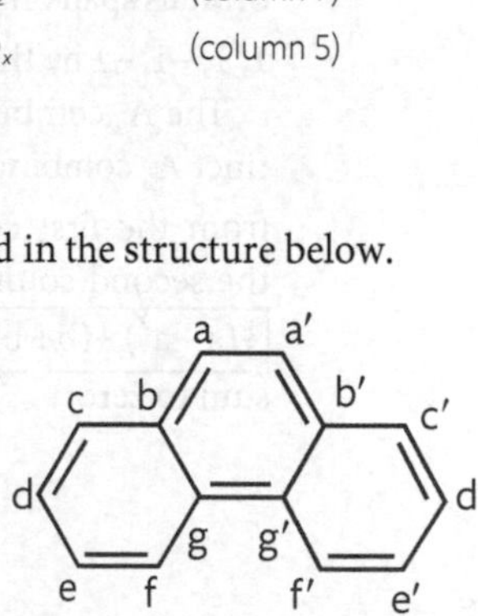

	a	a′	b	b′	c	c′	d	d′	e	e′	f	f′	g	g′	χ
E	a	a′	b	b′	c	c′	d	d′	e	e′	f	f′	g	g′	14
C_2	–a′	–a	–b′	–b	–c′	–c	–d′	–d	–e′	–e	–f′	–f	–g′	–g	0
σ_v	a′	a	b′	b	c′	c	d′	d	e′	e	f′	f	g′	g	0
σ'_4	–a	–a′	–b	–b′	–c	–c′	–d	–d′	–e	–e′	–f	–f′	–g	–g′	–14

To find the irreducible representations that these orbitals span, multiply the characters in the representation of the orbitals by the characters of the irreducible representations, sum those products, and divide the sum by the order h of the group (as in Brief illustration 33.1). The table below illustrates the procedure, beginning at left with the C_{2v} character table.

	E	C_2	σ_v	σ'_v	product	E	C_2	σ_v	σ'_v	sum/h
A_1	1	1	1	1		14	0	0	–14	0
A_2	1	1	–1	–1		14	0	0	14	7
B_1	1	–1	1	–1		14	0	0	14	7
B_2	1	–1	–1	1		14	0	0	–14	0

The orbitals span $\boxed{7A_2+7B_1}$.

To find symmetry-adapted linear combinations (SALCs), follow the procedure described in Topic 33.2(b). Refer to the table above that displays the transformations of the original basis orbitals. To find SALCs of a given symmetry species, take a column of the table, multiply each entry by the character of the species' irreducible representation, sum the terms in the column, and divide by the order of the group. For example, the characters of species A_1 are 1, 1, 1, 1, so the columns to be summed are identical to the columns in the table above. Each column sums to zero, so we conclude that there are no SALCs of A_1 symmetry. (No surprise here: the orbitals span only A_2 and B_1.) An A_2 SALC is obtained by multiplying the characters 1, 1, −1, −1 by the first column: $\frac{1}{4}(a-a'-a'+a)=\frac{1}{2}(a-a')$.

The A_2 combination from the second column is the same. There are seven distinct A_2 combinations in all: $\boxed{\frac{1}{2}(a-a'),\ \frac{1}{2}(b-b'),\ldots,\frac{1}{2}(g-g')}$. The B_1 combination from the first column is: $\frac{1}{4}(a+a'+a'+a)=\frac{1}{2}(a+a')$. The B_1 combination from the second column is the same. There are seven distinct B_1 combinations in all: $\boxed{\frac{1}{2}(a+a'),\frac{1}{2}(b+b'),\ldots,\frac{1}{2}(g+g')}$. There are no B_2 combinations, as the columns sum to zero.

(b) The Hückel secular determinant (Topic 26.1) of phenanthrene is:

	a	b	c	d	e	f	g	g′	f′	e′	d′	c′	b′	a′
a	$\alpha-E$	β	0	0	0	0	0	0	0	0	0	0	0	β
b	β	$\alpha-E$	β	0	0	0	β	0	0	0	0	0	0	0
c	0	β	$\alpha-E$	β	0	0	0	0	0	0	0	0	0	0
d	0	0	β	$\alpha-E$	β	0	0	0	0	0	0	0	0	0
e	0	0	0	β	$\alpha-E$	β	0	0	0	0	0	0	0	0
f	0	0	0	0	β	$\alpha-E$	β	0	0	0	0	0	0	0
g	0	β	0	0	0	β	$\alpha-E$	β	0	0	0	0	0	0
g′	0	0	0	0	0	0	β	$\alpha-E$	β	0	0	0	β	0
f′	0	0	0	0	0	0	0	β	$\alpha-E$	β	0	0	0	0
e′	0	0	0	0	0	0	0	0	β	$\alpha-E$	β	0	0	0
d′	0	0	0	0	0	0	0	0	0	β	$\alpha-E$	β	0	0
c′	0	0	0	0	0	0	0	0	0	0	β	$\alpha-E$	β	0
b′	0	0	0	0	0	0	0	β	0	0	0	β	$\alpha-E$	β
a′	β	0	0	0	0	0	0	0	0	0	0	0	β	$\alpha-E$

We simplify the secular determinant by factoring β out of the determinant and substituting the definition $x = (\alpha - E)/\beta$. The determinant is set equal to zero and solved for x after which the molecular orbital energies are written in the form $E = \alpha - x\beta$. Since the resonance integral β is negative, the lowest energy orbitals have the most negative values of x. The following Mathcad Prime 2.0 worksheet shows how to calculate x values. The columns and rows of the above secular matrix are labeled with the numerals 1, . . .,14.

ORIGIN:= 1

$i := 1..14$ $\quad j := 1..14$ $\quad I_{i,j} := 0$ $\quad I_{i,i} := 1$ $\quad diagonal(x) := x \cdot I$

$off\_diagonal_{i,j} := 0$ $\quad off\_diagonal_{1,14} := 1$ $\quad off\_diagonal_{14,1} := 1$

$off\_diagonal_{2,7} := 1$ $\quad off\_diagonal_{7,2} := 1$

$off\_diagonal_{13,8} := 1$ $\quad off\_diagonal_{8,13} := 1$

$k := 1..13$ $\quad off\_diagonal_{k,k+1} := 1$ $\quad off\_diagonal_{k+1,k} := 1$

$secular\_matrix(x) := off\_diagonal + diagonal(x)$

$$x\_values := |secular\_matrix(x)| \xrightarrow[\text{float, 4}]{\text{solve}} \begin{bmatrix} 1.951 \\ 0.6052 \\ 1.306 \\ -1.142 \\ -0.7691 \\ 2.435 \\ 1.516 \\ -2.435 \\ 1.142 \\ -0.6052 \\ 0.7691 \\ -1.306 \\ -1.516 \\ -1.951 \end{bmatrix} \qquad \text{reverse}(\text{sort}(x\_values)) = \begin{bmatrix} 2.435 \\ 1.951 \\ 1.516 \\ 1.306 \\ 1.142 \\ 0.769 \\ 0.605 \\ -0.605 \\ -0.769 \\ -1.142 \\ -1.306 \\ -1.516 \\ -1.951 \\ -2.435 \end{bmatrix}$$

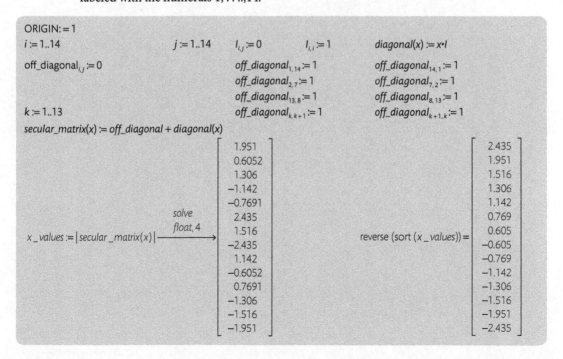

The far-right column of x values indicates the $E_{\text{lowest}} = \alpha + 2.435\beta$. Also, there are 14 π-electrons so the first 7 levels are filled and $E_{\text{HOMO}} = \alpha + 0.605\beta$.

(c) The ground state of the molecule has A_1 symmetry by virtue of the fact that its wavefunction is the product of doubly occupied orbitals, and the product of any two orbitals of the same symmetry has A_1 character. If a transition is to be allowed, the transition dipole must be non-zero, which in turn can only happen if the representation of the product $\Psi_f^* \mu \Psi$ includes the totally symmetric species A_1. Consider first transitions to another A_1 wavefunction, in which case we need the product $A_1 \times \mu \times A_1$. Now $A_1 \times A_1 = A_1$, and the only character that returns A_1 when multiplied by A_1 is A_1 itself. The z component of the dipole operator belongs to species A_1, so z-polarized $A_1 \leftarrow A_1$ transitions are allowed. (Note: transitions from the A_1 ground state to an A_1 excited state are transitions from an orbital occupied in the ground state to an excited-state orbital of the same symmetry.) The other possibility is a transition from an orbital of one symmetry (A_2 or B_1) to the other; in that case, the excited-state wavefunction will have symmetry of $A_2B_1 = B_2$ from the two singly occupied orbitals in the excited state. The symmetry of the transition dipole, then, is $A_1 \times \mu \times B_2 = \mu B_2$, and the only species that yields A_1 when multiplied by B_2 is B_2 itself. Now the y component of the dipole operator belongs to species B_2, so these transitions are also allowed (y-polarized).

Topic 34 Electric properties of molecules

Discussion questions

34.1 A molecule with a permanent separation of electric charge has a **permanent dipole moment** and is said to be **polar molecule.** In molecules containing atoms of differing electronegativity, the bonding electrons may be displaced in such a way as to produce a net separation of charge. Separation of charge may also arise from a difference in atomic radii of the bonded atoms. The separation of charges in the bonds is usually, though not always, in the direction of the more electronegative atom but depends on the precise bonding and nuclear arrangement in the molecule as described in Topic 34.1. A heteronuclear diatomic molecule necessarily has a dipole moment if there is a difference in electronegativity between the atoms, but the situation in polyatomic molecules is more complex. A polyatomic molecule has a permanent dipole moment only if it belongs to the symmetry point group C_n, C_{nv}, or C_s as discussed in Topic 31.3(a) of the text.

An external electric field can distort the electron density in both polar and non-polar molecules. This results in an **induced dipole moment** that is proportional to the field and the constant of proportionality is called the **polarizability**. Molecules with small HOMO–LUMO gaps typically have large polarizabilities.

Exercises

34.1(a) A molecule with a centre of symmetry may not be polar but molecules belonging to the groups C_n and C_{nv} may be polar (Topic 31.3(a)). The C atom of CIF_3 is approximately sp^3 hybridized, which causes the molecule to belong to the C_{3v} point group. The highly electronegative F atoms cause the C–F bonds to be very polar and the average dipole of the three very polar C–F bonds is unbalanced by the less polar C–I bond. Therefore, $\boxed{CIF_3}$ is polar.

Ozone is a bent molecule that belongs to the C_{2v} point group. Lewis resonance structures for the molecule show a central atom with a double bond to an end oxygen atom and a single bond to oxygen at the other end. The central oxygen has a formal charge of +1 while the doubled bonded end oxygen has a formal charge of zero. The single bonded end oxygen has a −1 formal charge. The average position of the −1 formal charge of the two resonance structure predicts a small negative charge that lies half-way between the extremities and a fraction of a bond length away from the central oxygen, which is expected to have a small positive charge. Consequently, $\boxed{O_3}$ is polar.

Hydrogen peroxide belongs to the C_2 point group with each carbon atom being approximately sp^3 hybridized. The H–O–O bond angles are 96.87°. The H–O–O plane of one hydrogen atom is at a 93.85° angle with the O–O–H plane of the other hydrogen atom. Consequently, the dipole moments of the two polar O–H bonds do not cancel and the $\boxed{H_2O_2}$ molecule is polar.

34.2(a)

$$\mu = (\mu_1^2 + \mu_2^2 + 2\mu_1\mu_2\cos\theta)^{1/2} \quad [34.3a]$$

$$= [(1.0)^2 + (2.0)^2 + (2)\times(1.0)\times(2.0)\times(\cos 45^\circ)]^{1/2}\,\mathrm{D} = \boxed{2.8\ \mathrm{D}}$$

34.3(a) The charges and positions of the three particles are summarized in the following table.

Particle	Q/e	x/nm	y/nm
i	3	0	0
ii	-1	0.32	0
iii	-2	0.23 cos(20°) = 0.216	0.23 sin(20°) = 0.0787

Now use equation 34.4a to calculate μ_x and an analogous equation to calculate μ_y. The conversion to Debye units is $1\ \mathrm{D} = 3.33564\times10^{-30}\ \mathrm{C\,m} = 0.020819\ e\ \mathrm{nm}$.

$$\mu_x = 3e\times(0\ \mathrm{nm}) - e\times(0.32\ \mathrm{nm}) - 2e\times(0.216\ \mathrm{nm}) = -0.752e\ \mathrm{nm} = \boxed{-36.1\ \mathrm{D}}$$

$$\mu_y = 3e\times(0\ \mathrm{nm}) - e\times(0\ \mathrm{nm}) - 2e\times(0.0787\ \mathrm{nm}) = -0.157e\ \mathrm{nm} = \boxed{-7.54\ \mathrm{D}}$$

Having the x and y direction components of the dipole moment, the magnitude of the dipole moment is calculated with equation 34.4b.

$$\mu = (\mu_x^2 + \mu_y^2)^{1/2}[34.4b] = ((-36.1\ \mathrm{D})^2 + (-7.54\ \mathrm{D})^2)^{1/2} = \boxed{36.9\ \mathrm{D}}$$

To find the angle θ between the x axis and the dipole moment, to recognize that $\mu_x = \mu\ \cos(\theta)$. Consequently, $\theta = \cos^{-1}(\mu_x/\mu) = \cos^{-1}(-36.1\ \mathrm{D}/36.9\ \mathrm{D}) = \boxed{168^\circ}$ counter-clockwise from the positive x-axis.

34.4(a) The induced dipole moment is

$$\mu^*_{\mathrm{HCl}} = \alpha\mathcal{E}\ [34.5a] = 4\pi\varepsilon_0\alpha'_{\mathrm{HCl}}\mathcal{E}\ [34.6]$$

$$= 4\pi(8.854\times10^{-12}\ \mathrm{J^{-1}\,C^2\,m^{-1}})\times(2.63\times10^{-30}\ \mathrm{m^3})\times(7.5\times10^3\ \mathrm{V\,m^{-1}})$$

$$= \boxed{2.19\times10^{-36}\ \mathrm{C\,m}}$$

which corresponds to $\boxed{0.658\ \mu\mathrm{D}}$.

Problems

34.1 Refer to Fig. 34.1 of the text, and add moments vectorially. Use $\mu = 2\mu_1 \cos \tfrac{1}{2}\theta$ [34.3b] where μ_1 is the dipole moment of toluene and θ is the angle between the C–CH_3 bonds of xylene.

$$o\text{-xylene: } \mu = (2)\times(0.4\ \text{D})\times\cos\left(\tfrac{1}{2}\times 60°\right) = \boxed{0.7\ \text{D}}$$

$$m\text{-xylene } \mu = (2)\times(0.4\ \text{D})\times\cos\left(\tfrac{1}{2}\times 120°\right) = \boxed{0.4\ \text{D}}$$

p-xylene: the resultant is zero (i.e., $\cos(\tfrac{1}{2}\times 180°) = 0$), so $\mu = \boxed{0}$

The p-xylene molecule to the group D_{2h}, and so it is necessarily nonpolar.

34.3 Individual acetic acid molecules have a non-zero dipole moment, but the two dipoles of the dimer are exactly opposed and cancel. At low temperature a significant fraction of molecules are a part of a dimer, which means that their individual dipole moments will not be observed. However, as temperature is increased, hydrogen bonds of the dimer are broken, thereby, releasing individual molecules and the apparent dipole moment increases.

34.5 Let the partial charge on the carbon atom equal δe and the N-to-C distance equal l. Then,

$$\mu = \delta e l \ [34.4a] \quad \text{or} \quad \delta = \frac{\mu}{el}$$

$$\delta = \frac{(1.77\ \text{D})\times(3.3356\times10^{-30}\ \text{C m D}^{-1})}{(1.602\times10^{-19}\ \text{C})\times(299\times10^{-12}\ \text{m})} = \boxed{0.123}$$

34.7 The relative permittivity and density both vary with temperature so it is advantageous to write the Debye equation with these on the left of the equality:

$$f(T) = \frac{1}{\rho}\left(\frac{\varepsilon_r - 1}{\varepsilon_r + 2}\right) = \frac{N_A\alpha}{3M\varepsilon_0} + \left(\frac{N_A\mu^2}{9Mk\varepsilon_0}\right)\frac{1}{T}$$

This equation is a linear relationship between $f(T) = \dfrac{1}{\rho}\left(\dfrac{\varepsilon_r - 1}{\varepsilon_r + 2}\right)$ and $\dfrac{1}{T}$ with intercept $\dfrac{N_A\alpha}{3M\varepsilon_0}$ and slope $\dfrac{N_A\mu^2}{9Mk\varepsilon_0}$. A linear regression fit of $f(T)$ against $1/T$ will give the regression intercept and slope from which α and μ are calculated. We draw a table for calculation of $f(T)$ and $1/T$, prepare a plot, mathematically find the linear regression fit, and calculate the polarizability and dipole moment. Figure 34.1 displays the plot and regression fit, which yields an intercept equal to 0.5487 $\text{cm}^3\ \text{g}^{-1}$ and a slope equal to 69.3 K $\text{cm}^3\ \text{g}^{-1}$.

θ/°C	ρ/g cm^{-3}	ε_r	1000 × K/T	$f(T)$/cm^3 g^{-1}
0	0.99	12.5	3.661	0.8011
20	0.99	11.4	3.411	0.7840
40	0.99	10.8	3.193	0.7734
60	0.99	10	3.002	0.7576
80	0.99	9.5	2.832	0.7466
100	0.99	8.9	2.680	0.7321
120	0.97	8.1	2.544	0.7247
140	0.96	7.6	2.420	0.7161
160	0.95	7.11	2.309	0.7060
200	0.91	6.21	2.113	0.6974

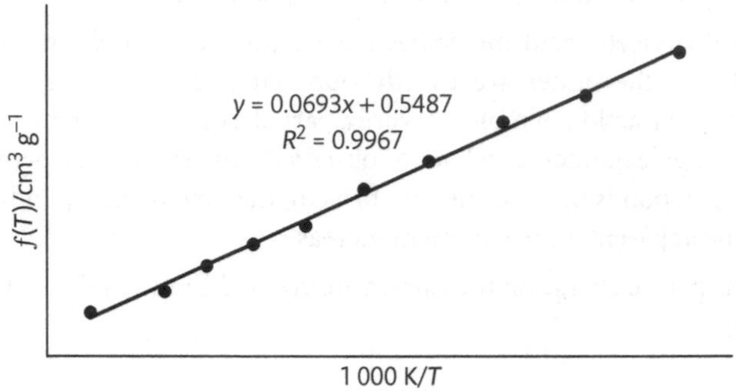

Fig 34.1

$$\alpha = \frac{3M\varepsilon_0 \times (\text{intercept})}{N_A}$$
$$= \frac{3\times(152.24\ \text{g mol}^{-1})\times(8.854\times10^{-12}\ \text{J}^{-1}\ \text{C}^2\ \text{m}^{-1})\times(0.5487\times10^{-6}\ \text{m}^3\ \text{g}^{-1})}{6.022\times10^{23}\ \text{mol}^{-1}}$$
$$= \boxed{3.68\times10^{-39}\ \text{J}^{-1}\ \text{C}^2\ \text{m}^2}$$

$$\alpha' = \frac{\alpha}{4\pi\varepsilon_0}\ [34.6] = \frac{3.68\times10^{-39}\ \text{J}^{-1}\ \text{C}^2\ \text{m}^2}{4\pi\times(8.854\times10^{-12}\ \text{J}^{-1}\ \text{C}^2\ \text{m}^{-1})} = \boxed{3.31\times10^{-29}\ \text{m}^3}$$

$$\mu = \left(\frac{9Mk\varepsilon_0\times(\text{slope})}{N_A}\right)^{1/2}$$
$$= \left(\frac{9\times(152.24\ \text{g mol}^{-1})\times(1.381\times10^{-23}\ \text{J K}^{-1})\times(8.854\times10^{-12}\ \text{J}^{-1}\ \text{C}^2\ \text{m}^{-1})\times(69.3\times10^{-6}\ \text{K m}^3\ \text{g}^{-1})}{6.022\times10^{23}\ \text{mol}^{-1}}\right)^{1/2}$$
$$= 4.39\times10^{-30}\ \text{C m} = \boxed{1.32\ \text{D}}$$

Topic 35 Interactions between molecules

Discussion questions

35.1

(a) $V=-\dfrac{Q_2\mu_1}{4\pi\varepsilon_0 r^2}$ [35.2]

V is the potential energy of interaction between a point dipole μ_1 and the point charge Q_2 at the separation r. The point charge lies on the axis of the dipole, and the separation r is much larger than the separation of charge within the dipole. Therefore, the partial charges of the dipole seem to merge and cancel to create the so-called **point dipole.**

(b) $V=-\dfrac{Q_2\mu_1\cos\theta}{4\pi\varepsilon_0 r^2}$

V is the potential energy of interaction between a point dipole μ_1 and the point charge Q_2 at the separation r. The point charge lies at an angle θ to the axis of the dipole, and the separation r is much larger than the separation of charge within the dipole. Therefore, the partial charges of the dipole seem to merge and cancel.

(c) $V=\dfrac{\mu_1\mu_2 f(\theta)}{4\pi\varepsilon_0 r^3}$ *where* $f(\theta)=1-3\cos^2\theta$ [35.4]

V is the potential energy of interaction between the two point dipoles μ_1 and μ_2 at the separation r. The dipoles are parallel, and the separation distance is at angle θ to the dipoles. The separation r is much larger than the separation of charge within the dipoles so that the partial charges of the dipoles seem to merge and cancel.

35.3 A **hydrogen bond** (···) is an attractive interaction between two species that arises from a link of the form A−H···B, where A and B are highly electronegative elements (usually nitrogen, oxygen, or fluorine) and B possesses a lone pair of electrons. It is a contact-like attraction that requires AH to touch B. Experimental evidences supports a linear or near-linear structural arrangement and a bond strength of about 20 kJ mol^{-1}. The hydrogen bond strength is considerably weaker than a covalent bond but it is larger than, and dominates, other intermolecular attractions such as dipole-dipole attractions. Its formation can be understood in terms of either the (a) electrostatic interaction model or with (b) molecular orbital calculations (Focus 6).

(a) A and B, being highly electronegative, are viewed as having partial negative charges (δ^-) in the electrostatic interaction model of the hydrogen bond. Hydrogen, being less electronegative than A, is viewed as having a partial positive (δ^+). The linear structure maximizes the electrostatic attraction between H and B:

$$\overset{\delta^-}{\mathrm{A}}\text{——}\overset{\delta^+}{\mathrm{H}}\cdots\cdots\cdots:\overset{\delta^-}{\mathrm{B}}$$

This model is conceptually very useful. However, it is impossible to exactly calculate the interaction strength with this model because the partial atomic charges cannot be precisely defined. There is no way to define which fraction of the electrons of the AB covalent bond should be assigned to one or the other nucleus.

(b) *Ab initio* quantum calculations are needed in order to explore questions about the linear structure, the role of the lone pair, the shape of the potential energy surface, and the extent to which the hydrogen bond has covalent sigma bond character. Yes, the hydrogen bond appears to have some sigma bond character. This was initially suggested by Linus Pauling in the 1930's and more recent experiments with Compton scattering of x-rays and NMR techniques indicate that the covalent character may provide as much as 20% of the hydrogen bond strength. A three-center molecular orbital model provides a degree of insight. A linear combination of an appropriate sigma orbital on A, the 1*s* hydrogen orbital, and an appropriate orbital for the lone pair on B yields a total of three molecular orbitals. One of the MOs is bonding, one is almost nonbonding, and the third is antibonding. Both bonding MO and the almost nonbonding orbital are occupied by two electrons (the sigma bonding electrons of A–H and the lone pair of B). The antibonding MO is empty. Thus, depending on the precise location of the almost nonbonding orbital, the nonbonding orbital may lower the total energy and account for the hydrogen bond.

Exercises

35.1(a) The O–H bond length of a water molecule is 95.85 pm and the Li^+ cation is 100 pm from the dipole center. Because these lengths are comparable, a calculation based on the assumption that the water dipole acts like a point dipole with a dipole length much shorter than the dipole-ion distance is unlikely to provide an accurate value of the dipole-ion interaction energy. However, such a calculation does provide an "order-of-magnitude" estimate. The minimum value of the dipole-ion interaction occurs with the dipole pointing toward the cation.

$$\begin{aligned} V_{\min} &\sim -\frac{\mu_{\mathrm{H_2O}}Q_{\mathrm{Li^+}}}{4\pi\varepsilon_0 r^2}\,[35.2] = -\frac{\mu_{\mathrm{H_2O}}e}{4\pi\varepsilon_0 r^2} \\ &\sim -\frac{(1.85\ \mathrm{D})\times(3.336\times10^{-30}\ \mathrm{C\,m\,D^{-1}})\times(1.602\times10^{-19}\ \mathrm{C})}{(1.113\times10^{-10}\ \mathrm{J^{-1}\,C^2\,m^{-1}})\times(100\times10^{-12}\ \mathrm{m})^2} \\ &\sim -8.88\times10^{-19}\ \mathrm{J} \end{aligned}$$

The interaction potential becomes a maximum upon flipping the dipole. This effectively changes the sign of the dipole in the previous calculation giving

$$V_{\text{max}} \sim 8.88\times10^{-19}\ \text{J}$$

The work w required to flip the dipole is the difference $V_{\text{max}} - V_{\text{min}}$.

$$w \sim V_{\text{max}} - V_{\text{min}} = 1.78\times10^{-18}\ \text{J}$$
$$w_{\text{m}} = wN_{\text{A}} \sim \boxed{1.07\times10^{3}\ \text{kJ mol}^{-1}}$$

35.2(a) The two linear quadrupoles are shown in Fig. 35.1 with a collinear configuration.

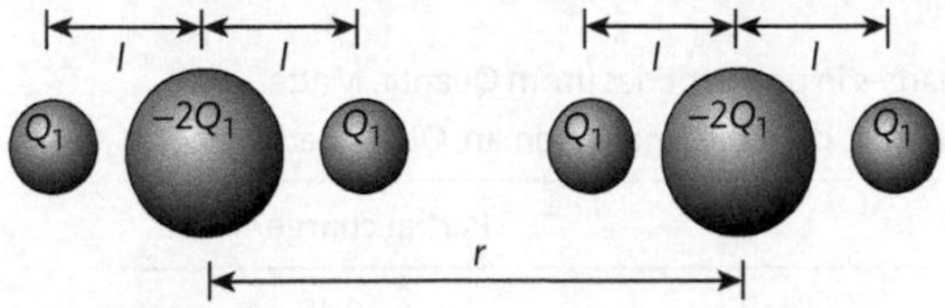

Figure 35.1

The total potential energy of the interaction between the quadrupoles is:

$$4\pi\varepsilon_0 V = \frac{Q_1^2}{r} - \frac{2Q_1^2}{r+l} + \frac{Q_1^2}{r+2l} - \frac{2Q_1^2}{r-l} + \frac{4Q_1^2}{r} - \frac{2Q_1^2}{r+l} + \frac{Q_1^2}{r-2l} - \frac{2Q_1^2}{r-l} + \frac{Q_1^2}{r}$$
$$\frac{4\pi\varepsilon_0 rV}{Q_1^2} = 6 - \frac{4}{1+x} - \frac{4}{1-x} + \frac{1}{1+2x} + \frac{1}{1-2x} \quad \text{where} \quad x = \frac{l}{r}$$

With the point quadrupole condition that $x << 1$ the last four terms in the above expression can be expanded with the Taylor series:

$$(1+z)^{-1} = 1 - z + z^2 - z^3 + z^4 - \cdots \quad \text{and} \quad (1-z)^{-1} = 1 + z + z^2 + z^3 + z^4 + \cdots$$

where z is either $2x$ or x.

$$\frac{4\pi\varepsilon_0 rV}{Q_1^2} = 6 - 4\{1 - x + x^2 - x^3 + x^4 - \cdots + 1 + x + x^2 + x^3 + x^4 + \cdots\}$$
$$+\{1-(2x)+(2x)^2-(2x)^3+(2x)^4-\cdots\}+\{1+(2x)+(2x)^2+(2x)^3+(2x)^4+\cdots\}$$
$$= -8x^2 - 8x^4 + 8x^2 + 32x^4 + \cdots$$
$$= 24x^4 + \text{higher order terms}$$

In the limit of small x values the higher order terms are negligibly small, thereby, leaving

$$V = \frac{6x^4 Q_1^2}{\pi\varepsilon_0 r} = \boxed{\frac{6l^4 Q_1^2}{\pi\varepsilon_0 r^5}}$$

Thus, $V \propto \frac{1}{r^5}$ for the quadrupole-quadrupole interaction. See Discussion question 34.2 and note that a quadrupole is n-pole array of charges with $n = 3$. So the above derivation demonstrates the general potential energy relation between an n-pole array and an m-pole array: $V \propto \frac{1}{r^{n+m-1}} = \frac{1}{r^{3+3-1}} = \frac{1}{r^5}$.

35.3(a) Using the partial charge presented in the table below, we estimate the partial charge on each hydrogen atom of a water molecule to be $Q_H = \delta e$ where $\delta = 0.42$. The electroneutrality of an H_2O molecule implies that the estimated partial charge on the oxygen atom is $Q_O = -2\delta e$. With a hydrogen bond length of 170 pm, the point charge model of the hydrogen bond in a vacuum estimates the potential of interaction to be

Partial charges in polypeptides (from Quanta, Matter, and Change; Atkins, de Paula, and Friedman, OUP, 1st ed. 2009)

Atom	Partial charge/e
C(=O)	+0.45
C(–CO)	+0.06
H(–C)	+0.02
H(–N)	+0.18
H(–O)	+0.42
N	–0.36
O	–0.38

$$V = \frac{Q_H Q_O}{4\pi\varepsilon_0 r} = -\frac{2(\delta e)^2}{4\pi\varepsilon_0 r} \quad [35.1a]$$

$$= -\frac{2(0.42\times1.60\times10^{-19}\ \mathrm{C})^2}{4\pi(8.85\times10^{-12}\ \mathrm{J^{-1}\ C^2\ m^{-1}})\times(170\times10^{-12}\ \mathrm{m})} = -4.8\times10^{-19}\ \mathrm{J}$$

The molar energy required to break these bonds is

$$E_m = -N_A V = -(6.022\times10^{23}\ \mathrm{mol^{-1}})\times(-4.8\times10^{-19}\ \mathrm{J}) = \boxed{28\bar{9}\ \mathrm{kJ\ mol^{-1}}}$$

Problems

35.1 The interaction is a dipole–induced-dipole interaction. The energy is given by eqn 35.7:

$$V = -\frac{\mu_1^2\alpha_2'}{4\pi\varepsilon_0 r^6}$$

$$= -\frac{[(1.3\,\mathrm{D})(3.336\times10^{-30}\,\mathrm{C\,m\,D^{-1}})]^2(1.04\times10^{-29}\,\mathrm{m^3})}{4\pi(8.854\times10^{-12}\,\mathrm{J^{-1}\,C^2\,m^{-1}})(4.0\times10^{-9}\,\mathrm{m})^6}$$

$$= \boxed{-4.3\times10^{-28}\,\mathrm{J} = -2.6\times10^{-4}\,\mathrm{J\,mol^{-1}}}$$

Comment: This value seems exceedingly small. The distance suggested in the problem may be too large compared to typical values. A distance of 0.40 nm, yields $V = -0.26\ \mathrm{kJ\,mol^{-1}}$.

35.3 The potential of the London dispersion interaction is

$$V = -\frac{C}{r^6}\ [35.8] \quad \text{where} \quad C = \tfrac{3}{2}\alpha_1'\alpha_2'\frac{I_1 I_2}{I_1 + I_2}$$

$$F = -\frac{\mathrm{d}V}{\mathrm{d}r} = -\frac{\mathrm{d}}{\mathrm{d}r}\left(-\frac{C}{r^6}\right) = C\frac{\mathrm{d}}{\mathrm{d}r}\left(r^{-6}\right) = \boxed{\frac{-6C}{r^7}}$$

The negative sign indicates that the force is attractive.

35.5 The number of molecules in a volume element $\mathrm{d}\tau$ is $\dfrac{N\mathrm{d}\tau}{V} = \mathcal{N}\mathrm{d}\tau$. The energy of interaction of these molecules with one at a distance R is $V(R)\mathcal{N}\mathrm{d}\tau$. The total interaction energy, taking into account the entire sample volume, is therefore

$$u = \int V(R)\mathcal{N}\mathrm{d}\tau = \mathcal{N}\int V(R)\mathrm{d}\tau \quad [V(R) \text{ is the interaction energy, not the volume}]$$

The total interaction energy of a sample of N molecules is $\tfrac{1}{2}Nu$ (the ½ is included to avoid double counting), and so the cohesive energy density is

$$\mathcal{U} = -\frac{U}{V} = \frac{-\tfrac{1}{2}Nu}{V} = -\frac{1}{2}\mathcal{N}u = -\frac{1}{2}\mathcal{N}^2\int V(R)\mathrm{d}\tau$$

For $V(R) = -\dfrac{C_6}{R^6}$ and $\mathrm{d}\tau = 4\pi R^2\,\mathrm{d}R$

$$-\frac{U}{V} = 2\pi\mathcal{N}^2 C_6\int_d^\infty \frac{\mathrm{d}R}{R^4} = \frac{2\pi}{3}\times\frac{\mathcal{N}^2 C_6}{d^3}$$

However, $\mathcal{N} = \dfrac{N_\mathrm{A}\rho}{M}$, where M is the molar mass; therefore

$$\mathcal{U} = \boxed{\left(\frac{2\pi}{3}\right)\times\left(\frac{N_\mathrm{A}\rho}{M}\right)^2\times\left(\frac{C_6}{d^3}\right)}$$

Topic 36 **Real Gases**

Discussion questions

36.1 Consider three temperature regions:

(1). $T < T_B$. At very low pressures, all gases show a compression factor, $Z \approx 1$. At high pressures, all gases have $Z > 1$, signifying that they have a molar volume greater than a perfect gas, which implies that repulsive forces are dominant. At intermediate pressures, most gases show $Z < 1$, indicating that attractive forces reducing the molar volume below the perfect value are dominant.

(2). $T \approx T_B$. $Z \approx 1$ at low pressures, slightly greater than 1 at intermediate pressures, and significantly greater than 1 only at high pressures. There is a balance between the attractive and repulsive forces at low to intermediate pressures, but the repulsive forces predominate at high pressures where the molecules are very close to each other.

(3). $T > T_B$. $Z > 1$ at all pressures because the frequency of collisions between molecules has increased with the higher temperature to a point for which repulsive forces dominate.

Exercises

36.1(a)

$$n_{N_2} = 3.15\ \text{g} \times \left(\frac{1\ \text{mol N}_2}{28.01\ \text{g}}\right) = 0.112\ \text{mol N}_2$$

$$V_m = \frac{V}{n_{N_2}} = \frac{2.05\ \text{dm}^3}{0.112\ \text{mol N}_2} = 0.0183\ \text{m}^3\ \text{mol}^{-1} = 1.83\times10^4\ \text{cm}^3\ \text{mol}^{-1}$$

Assuming that the dinitrogen virial equation of state has $C/V_m^{\,2} << B/V_m$,

$$p_{\text{virial}} = \frac{RT}{V_m}\left(1+\frac{B}{V_m}\right)\ [36.4b]$$

$$= \frac{(8.3145\ \text{J mol}^{-1}\ \text{K}^{-1})(273\ \text{K})}{0.0183\ \text{m}^3\ \text{mol}^{-1}}\left(1-\frac{10.5\ \text{cm}^3\ \text{mol}^{-1}}{1.83\times10^4\ \text{cm}^3\ \text{mol}^{-1}}\right)\ [\text{Table 36.1}]$$

$$= 1.24\times10^5\ \text{Pa} = \boxed{1.24\ \text{bar}}$$

According to the perfect (ideal) gas equation of state,

$$p_{\text{perfect}} = \frac{RT}{V_{\text{m}}} = \frac{(8.3145 \text{ J mol}^{-1} \text{ K}^{-1})(273 \text{ K})}{0.0183 \text{ m}^3 \text{ mol}^{-1}} = 1.24\times10^5 \text{ Pa} = \boxed{1.24 \text{ bar}}$$

The fact that $p_{\text{virial}} = p_{\text{perfect}}$ for nitrogen under these conditions results from the low pressure, high temperature conditions for which the molar volume is large enough to cause $B/V_{\text{m}} << 1$. The second coefficient of the virial equation of state is negligibly small under these conditions and the perfect gas equation of state provides accuracy.

36.2(a) The second virial coefficient equals zero at the Boyle temperature so the perfect gas law will accurately describe the pressure.

$$n_{\text{CO}_2} = 4.56 \text{ g} \times \left(\frac{1 \text{ mol CO}_2}{44.01 \text{ g}}\right) = 0.104 \text{ mol CO}_2$$

$$V_{\text{m}} = \frac{V}{n_{\text{CO}_2}} = \frac{2.25 \text{ dm}^3}{0.104 \text{ mol CO}_2} = 0.0216 \text{ m}^3 \text{ mol}^{-1}$$

$$p_{\text{perfect}} = \frac{RT_{\text{B}}}{V_{\text{m}}} = \frac{(8.3145 \text{ J mol}^{-1} \text{ K}^{-1})(714.8 \text{ K})}{0.0216 \text{ m}^3 \text{ mol}^{-1}} \text{ [Table 36.2]} = 2.75\times10^5 \text{ Pa} = \boxed{2.75 \text{ bar}}$$

36.3(a)

(a) The amount of xenon is $n = \dfrac{131 \text{ g}}{131 \text{ g mol}^{-1}} = 1.00 \text{ mol}$

$$V_{\text{m}} = \frac{V}{n} = \frac{1.0 \text{ dm}^3}{1.00 \text{ mol}} = 1.0 \text{ dm}^3 \text{ mol}^{-1}$$

$$p_{\text{perfect}} = \frac{RT}{V_{\text{m}}} = \frac{(8.3145 \text{ J K}^{-1}\text{mol}^{-1})\times(298.15 \text{ K})}{1.0\times10^{-3} \text{ m}^3 \text{ mol}^{-1}} = 25\times10^5 \text{ Pa} = \boxed{25 \text{ bar}}$$

(b) For xenon, Table 8.7 gives $a = 4.137 \text{ dm}^6 \text{ atm mol}^{-2}$ and $b = 5.16\times10^{-2} \text{ dm}^3 \text{ mol}^{-1}$

$$p_{\text{van der Waals}} = \frac{RT}{V_{\text{m}} - b} - \frac{a}{V_{\text{m}}^2}$$

$$= \frac{(8.3145 \text{ J K}^{-1}\text{mol}^{-1})\times(298.15 \text{ K})}{(1.0-0.0516)\times10^{-3} \text{ m}^3 \text{ mol}^{-1}} - \frac{4.137\times10^{-6} \text{ atm m}^6 \text{ mol}^{-2}}{(1.0\times10^{-3} \text{ m}^3 \text{ mol}^{-1})^2}\left(\frac{1.013\times10^5 \text{ Pa}}{1 \text{ atm}}\right)$$

$$= 22\times10^5 \text{ Pa} = \boxed{22 \text{ bar}}$$

36.4(a) The conversions needed are as follows:

$$1 \text{ atm} = 1.013\times10^5 \text{ Pa} \quad 1 \text{ Pa} = 1 \text{ kg m}^{-1}\text{s}^{-2} \quad 1 \text{ dm}^6 = 10^{-6} \text{ m}^6 \quad 1 \text{ dm}^3 = 10^{-3} \text{ m}^3$$

Therefore, $a = 0.751$ atm dm^6 mol^{-2} and $b = 0.0226$ dm^3 mol^{-1} become, after substitution of the conversions,

$$a = \boxed{7.61\times10^{-2}\ \text{kg m}^5\ \text{s}^{-2}\ \text{mol}^{-2}} \text{ and } b = \boxed{2.26\times10^{-5}\ \text{m}^3\ \text{mol}^{-1}}$$

36.5(a) The definition of Z is $Z = \frac{pV_m}{RT} = \frac{V_m}{V_m^\circ}$ [36.3a and b]. V_m is the actual molar volume, V_m° is the perfect gas molar volume. $V_m^\circ = \frac{RT}{p}$. Since V_m is 8 per cent smaller than that of a perfect gas, $V_m = 0.92\ V_m^\circ$.

(a) $Z = \frac{0.92\ V_m^\circ}{V_m^\circ} = \boxed{0.92}$

(b) $$V_m = \frac{ZRT}{p} = \frac{(0.92)\times(8.206\times10^{-2}\ \text{dm}^3\ \text{atm K}^{-1}\text{mol}^{-1})\times(250\ \text{K})}{12\ \text{atm}}$$
$$= \boxed{1.6\ \text{dm}^3\ \text{mol}^{-1}}$$

Since $V_m < V_m^\circ$, attractive forces dominate.

36.6(a) The amount of gas is first determined from its mass. Then, the van der Waals equation is used to determine its pressure at the working temperature. The initial conditions of 300 K and 100 atm are in a sense superfluous information.

$$n = \frac{92.4\ \text{kg}}{28.02\times10^{-3}\ \text{kg mol}^{-1}} = 3.30\times10^3\ \text{mol}$$

$$V = 1.000\ \text{m}^3 = 1.000\times10^3\ \text{dm}^3$$

$$p = \frac{nRT}{V-nb} - \frac{an^2}{V^2}\ [8.24\text{a}] = \frac{(3.30\times10^3\ \text{mol})\times(0.08206\ \text{dm}^3\ \text{atm K}^{-1}\text{mol}^{-1})\times(500\ \text{K})}{(1.000\times10^3\ \text{dm}^3)-(3.30\times10^3\ \text{mol})\times(0.0391\ \text{dm}^3\ \text{mol}^{-1})}$$
$$- \frac{(1.39\ \text{dm}^6\ \text{atm mol}^{-2})\times(3.30\times10^3\ \text{mol})^2}{(1.000\times10^3\ \text{dm}^3)^2}$$
$$= (155-15)\ \text{atm} = \boxed{140\ \text{atm}}$$

36.7(a) The Boyle temperature, T_B, is the temperature at which $B = 0$. In order to express T_B in terms of a and b, the van der Waals equation must be recast into the form of the virial equation.

$$p = \frac{RT}{V_m - b} - \frac{a}{V_m^2}\ [36.7\text{b}]$$

Factoring out $\dfrac{RT}{V_m}$ yields

$$p = \frac{RT}{V_m}\left\{\frac{1}{1-b/V_m} - \frac{a}{RTV_m}\right\}$$

So long as $b/V_m < 1$, the first term inside the brackets can be expanded using $(1-x)^{-1} = 1 + x + x^2 + \cdots$, which gives

$$p = \frac{RT}{V_m}\left\{1 + \left(b - \frac{a}{RT}\right)\times\left(\frac{1}{V_m}\right) + \cdots\right\}$$

We can now identify the second virial coefficient as $B = b - \dfrac{a}{RT}$.

Since at the Boyle temperature, $B = 0$, $T_B = \dfrac{a}{bR}$

From Table 36.3 $a = 6.260\ \text{dm}^6\ \text{atm mol}^{-2}$, $b = 5.42\times10^{-2}\ \text{dm}^3\ \text{mol}^{-1}$. Therefore,

$$T_B = \frac{6.260\ \text{dm}^6\ \text{atm mol}^{-2}}{(5.42\times10^{-2}\ \text{dm}^3\ \text{mol}^{-1})\times(8.206\times10^{-2}\ \text{dm}^3\ \text{atm K}^{-1}\ \text{mol}^{-1})} = \boxed{1.41\times10^3\ \text{K}}$$

36.8(a) Fig. 36.1 shows how the collision of two hard-sphere molecules establishes an excluded volume. The closest distance between two molecules of radius r, and volume $V_{\text{molecule}} = {}^4/_3\pi r^3$, is $2r$ so the volume excluded is ${}^4/_3\pi(2r)^3$ or $8V_{\text{molecule}}$. The volume excluded per molecule, a quantity that provides an estimate of the van der Waals coefficient b, is one-half this volume, or $4V_{\text{molecule}}$.

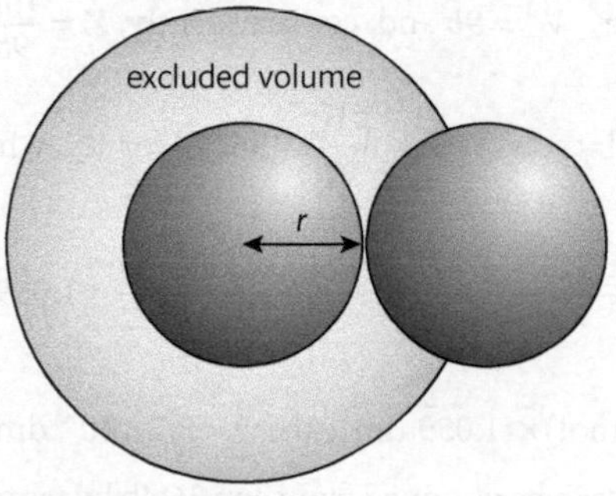

Figure 36.1

$b = 4V_{\text{molecule}}N_A = 4\left(\tfrac{4}{3}\pi r^3\right)N_A$ where $b_{Cl_2} = 5.42\times10^{-5}\ \text{m}^3\ \text{mol}^{-1}$ [Table 36.3]

$$r = \left(\frac{3b}{16\pi N_A}\right)^{1/3} = \left(\frac{3(5.42\times10^{-5}\ \text{m}^3\ \text{mol}^{-1})}{16\pi(6.022\times10^{23}\ \text{mol}^{-1})}\right)^{1/3} = \boxed{175\ \text{pm}}$$

36.9(a) The van der Waals equation [36.7b] is solved for b, which yields

$$b = V_m - \frac{RT}{\left(p + \frac{a}{V_m^2}\right)}$$

Substituting the data

$$b = 5.00\times10^{-4}\,\text{m}^3\,\text{mol}^{-1} - \frac{(8.3145\,\text{J K}^{-1}\,\text{mol}^{-1})\times(273\,\text{K})}{\left\{(3.0\times10^6\,\text{Pa}) + \left(\frac{0.50\,\text{m}^6\,\text{Pa mol}^{-2}}{(5.00\times10^{-4}\,\text{m}^3\,\text{mol}^{-1})^2}\right)\right\}}$$

$$= 0.46\times10^{-4}\,\text{m}^3\,\text{mol}^{-1} = \boxed{0.046\,\text{dm}^3\,\text{mol}^{-1}}$$

36.10(a) $Z = \frac{pV_m}{RT}\,[36.3\text{b}] = \frac{(3.0\times10^6\,\text{Pa})\times(5.00\times10^{-4}\,\text{m}^3\,\text{mol}^{-1})}{(8.3145\,\text{J K}^{-1}\,\text{mol}^{-1})\times(273\,\text{K})} = \boxed{0.66}$

Comment. The definition of Z involves the actual pressure, volume, and temperature and does not depend upon the equation of state used to relate these variables.

Problems

36.1 From the given equation of state $V_m = b + \frac{RT}{p} = b + V_m^\circ$. Thus,

$$Z = \frac{V_m}{V_m^\circ}\,[8.20] = \frac{b + V_m^\circ}{V_m^\circ} = \boxed{1 + \frac{b}{V_m^\circ}}.$$

For $V_m = 10b$, $10b = b + V_m^\circ$ or $V_m^\circ = 9b$ and, consenquently, $Z = \frac{10b}{9b} = \boxed{1.11}$

36.3 The molar volume is obtained by solving $Z = \frac{pV_m}{RT}$ [36.3b] for V_m, which yields

$$V_m = \frac{ZRT}{p} = \frac{(0.86)\times(0.08206\,\text{dm}^3\,\text{atm K}^{-1}\,\text{mol}^{-1})\times(300\,\text{K})}{20\,\text{atm}} = 1.0\overline{59}\,\text{dm}^3\,\text{mol}^{-1}$$

(a) Then, $V = nV_m = (8.2\times10^{-3}\,\text{mol})\times(1.0\overline{59}\,\text{dm}^3\,\text{mol}^{-1}) = 8.7\times10^{-3}\,\text{dm}^3 = \boxed{8.7\,\text{cm}^3}$

(b) An approximate value of B can be obtained from eqn 36.4b by truncation of the series after the second term, B/V_m. Solving for B gives

$$B = V_m\left(\frac{pV_m}{RT} - 1\right) = V_m\times(Z-1)$$

$$= (1.0\overline{59}\,\text{dm}^3\,\text{mol}^{-1})\times(0.86-1) = \boxed{-0.15\,\text{dm}^3\text{mol}^{-1}}$$

36.5 Since $B' = 0$ at the Boyle temperature (Topic 36.2): $B'(T_B) = a + be^{-c/T_B^2} = 0$

$$\text{Solving for } T_B : T_B = \sqrt{\frac{-c}{\ln\left(\frac{-a}{b}\right)}} = \sqrt{\frac{-(1131\,\text{K}^2)}{\ln\left[\frac{-(-0.1993\,\text{bar}^{-1})}{0.2002\,\text{bar}^{-1}}\right]}} = 501.0\,\text{K}$$

36.7 The van der Waals equation is a cubic equation in V_m so we solve for V_m using a numeric application on a scientific calculator or by using a numeric procedure in a math software package. Here's a Mathcad Prime 2 worksheet for the solution:

$a := 6.260 \cdot 10^{-6} \cdot \text{m}^6 \cdot \text{atm} \cdot \text{mol}^{-2}$ $b := 5.42 \cdot 10^{-5} \cdot \text{m}^3 \cdot \text{mol}^{-1}$ $R := 8.31447 \cdot \text{J} \cdot \text{K}^{-1} \cdot \text{mol}^{-1}$

$p := 150 \cdot 10^3 \cdot \text{Pa}$ $T := 250 \cdot \text{K}$

Guess Values

$$x := \frac{R \cdot T}{p}$$

Constraints

$$p = \frac{R \cdot T}{x - b} - \frac{a}{x^2}$$

Solver

$$\text{Find}(x) = 0.0136 \frac{\text{m}^3}{\text{mol}}$$

Thus, $V_m = \boxed{13.6\ \text{dm}^3\ \text{mol}^{-1}}$. Since $V_m < V_m^o = RT/p = 13.9\ \text{dm}^3\ \text{mol}^{-1}$, $\boxed{\text{attractive}}$ forces dominate at these conditions.

$$\text{Percentage Difference} = \frac{V_m - V_m^o}{V_m} \times 100\% = \frac{13.6 - 13.9}{13.6} \times 100\% = \boxed{-2\%}$$

36.9

$$p = \frac{RT}{V_m - b} - \frac{a}{V_m^2}\ [36.7\text{b}]$$

Factoring out $\frac{RT}{V_m}$ yields $p = \frac{RT}{V_m}\left\{\frac{1}{1 - b/V_m} - \frac{a}{RTV_m}\right\}$ and consequently

$$Z = \frac{pV_m}{RT}\ [36.3\text{b}] = \frac{1}{(1 - b/V_m)} - \frac{a}{RTV_m},$$

which upon the Taylor series expansion $\left(1 - \frac{b}{V_m}\right)^{-1} = 1 + \frac{b}{V_m} + \left(\frac{b}{V_m}\right)^2 + \cdots$ yields

$$Z = 1 + \left(b - \frac{a}{RT}\right) \times \left(\frac{1}{V_m}\right) + b^2\left(\frac{1}{V_m}\right)^2 + \cdots$$

We note that all terms beyond the second are necessarily positive, so only if

$$\frac{a}{RTV_m} > \frac{b}{V_m} + \left(\frac{b}{V_m}\right)^2 + \cdots$$

can Z be less than one. If we ignore terms beyond $\frac{b}{V_m}$, the conditions are simply stated as

$$Z < 1 \text{ when } \frac{a}{RT} > b \text{ and } Z > 1 \text{ when } \frac{a}{RT} < b$$

Thus, $Z < 1$ when attractive forces predominate, and $Z > 1$ when size effects (short-range repulsions) predominate.

Topic 37 Crystal structure

Discussion questions

37.1 A **space lattice** is the three-dimensional structural pattern formed by lattice points representing the locations of motifs which may be atoms, molecules, or groups of atoms, molecules, or ions within a crystal. All points of the space lattice have identical environments and they define the crystal structure. The **unit cell** is an imaginary parallelepiped from which the entire crystal structure can be generated, without leaving gaps, using translations of the unit cell alone. Each unit cell is defined in terms of lattice points and the unit cell is commonly formed by joining neighbouring lattice points by straight lines. The smallest possible unit cell is called the **primitive unit cell. Non-primitive unit cells** may exhibit lattice points within the cell, at the cell centre, on cell faces, or on cell edges.

37.3 We can use the Debye–Scherrer powder diffraction method with monochromic X-rays of wavelength λ, follow the procedure in Topics 37.2 and 37.3, and in particular look for systematic absences in the diffraction patterns. We can proceed through the following sequence:

1. Measure distances of the lines in the diffraction pattern from the centre.
2. From the known radius of the camera, convert the distances to incident angles θ for which the radiation exhibits constructive interference.
3. Calculate $\sin^2\theta$.
4. Based upon your experience, choose the cubic unit cell that you guess to be applicable and assign Miller indices (*hkl*) by $h^2+k^2+l^2$ values from lowest to highest to each incident angle. $h^2+k^2+l^2$ values equal to 7 and 15 are missing (see Brief illustration 37.4). When making the assignments, recognize that some indices will be absent depending upon the unit cell. Missing diffraction spots are summarized in Brief illustration 37.4 and Example 37.2. For identical atoms in a cubic P (primitive cubic) arrangement, the *hkl* all-odd have zero intensity. For identical atoms in a cubic I (body-centred cubic) arrangement, the $h+k+l$ odd have zero intensity. For cubic F (face-centred cubic), only lines for which h, k, and l are either all even or all odd will be present, others will be absent. The Miller indices for observed spots are summarized in the following figure (Quanta, Matter, and Change; Atkins, de Paula, and Friedman, OUP, 1st ed., Fig. 9.22, 2009).

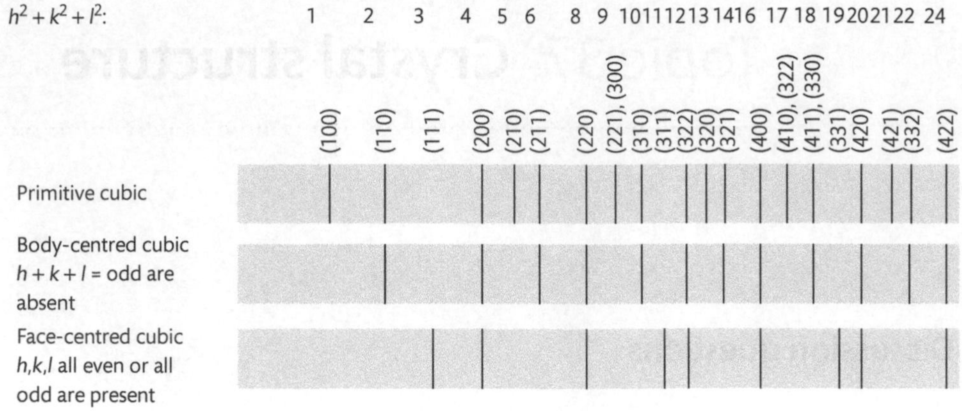

4. Find the factor $A = \lambda^2/4a^2 = \dfrac{\sin^2\theta}{h^2+k^2+l^2}$ for each spot [cubic unit cell, 37.1b and 37.2a with $n = 1$].
5. If one or more A values disagree with the majority, select another cubic unit cell and repeat Steps 3 and 4.
6. If A values agree to within experimental uncertainty, solve $A = \lambda^2/4a^2$ for a.

37.5 The **structure factor** F_{hkl} is the sum over all j atoms of terms each of which has a scattering factor f_j:

$$F_{hkl} = \sum_j f_j \mathrm{e}^{\mathrm{i}\phi_{hkl}(j)} \quad \text{where} \quad \phi_{hkl}(j) = 2\pi\left(hx_j + ky_j + lz_j\right) \quad [37.4]$$

The importance of the structure factor to the X-ray crystallographic method of structure determination is its relationship to the electron density distribution, $\rho(r)$, within the crystal:

$$\rho(\boldsymbol{r}) = \frac{1}{V}\sum_{hkl} F_{hkl}\mathrm{e}^{-2\pi\mathrm{i}(hx+ky+lz)} \quad [37.5]$$

Eqn 37.5 reveals that if the structure factors for the lattice planes can be measured then the electron density distribution can be calculated by performing the indicated sum. Therein lays the **phase problem**. Measurement detectors yield only the intensity of scattered radiation, which is proportional to $|F_{hkl}|^2$, and give no direct information about F_{hkl}. To see this, consider the structure factor form $F_{hkl} = |F_{hkl}|\mathrm{e}^{\mathrm{i}\alpha_{hkl}}$ where α_{hkl} is the phase of the hkl reflection plane. Then,

$$|F_{hkl}|^2 = \left\{|F_{hkl}|\mathrm{e}^{\mathrm{i}\alpha_{hkl}}\right\}^* \left\{|F_{hkl}|\mathrm{e}^{\mathrm{i}\alpha_{hkl}}\right\} = |F_{hkl}| \times |F_{hkl}|\mathrm{e}^{-\mathrm{i}\alpha_{hkl}}\mathrm{e}^{\mathrm{i}\alpha_{hkl}} = |F_{hkl}| \times |F_{hkl}|$$

and we see that all information about the phase is lost in an intensity measurement. It seems impossible to perform the sum of eqn 37.5 since we do not have the important factor $e^{i\alpha_{hkl}}$. Crystallographers have developed numerous methods to resolve the phase problem. In the **Patterson synthesis**, X-ray diffraction spot intensities are used to acquire separation and relative orientations of atom pairs. Another method uses the dominance of heavy atom scattering to deduce phase. **Heavy atom replacement** may be necessary for this type of application. **Direct methods** dominate modern X-ray diffraction analysis. These methods use statistical techniques, and the considerable computational capacity of the modern computer, to compute the probabilities that the phases have a particular value.

Exercises

37.1(a) $$\rho = \frac{\text{mass of unit cell}}{\text{volume of unit cell}} = \frac{m}{V}$$

$$m = nM = \frac{N}{N_A}M \quad [N \text{ is the number of formula units per unit cell}]$$

Then, $\rho = \dfrac{NM}{VN_A}$

$$N = \frac{\rho V N_A}{M}$$

$$= \frac{(3.9\times10^6\ \text{g m}^{-3})\times(634)\times(784)\times(516\times10^{-36}\ \text{m}^3)\times(6.022\times10^{23}\ \text{mol}^{-1})}{154.77\ \text{g mol}^{-1}} = 3.9$$

Therefore, $\boxed{N=4}$ and the true calculated density (in the absence of defects) is

$$\rho = \frac{(4)\times(154.77\ \text{g mol}^{-1})}{(634)\times(784)\times(516\times10^{-30}\ \text{cm}^3)\times(6.022\times10^{23})\text{mol}^{-1}} = \boxed{4.01\ \text{g cm}^{-3}}$$

37.2(a) The planes are sketched in Fig. 37.1. Expressed in multiples of the unit cell distances the planes are labeled (2, 3, 2) and (2, 2, ∞). Their Miller indices are the reciprocals of these multiples with all fractions cleared, thus

$$(2,3,2) \to \left(\tfrac{1}{2},\tfrac{1}{3},\tfrac{1}{2}\right) \to (3,2,3) \quad [\text{multiply by the lowest common denominator 6}]$$

$$(2,2,\infty) \to \left(\tfrac{1}{2},\tfrac{1}{2},0\right) \to (1,1,0) \quad [\text{multiply by the lowest common denominator 2}]$$

Dropping the commas, the planes are written $\boxed{(323) \text{ and } (110)}$.

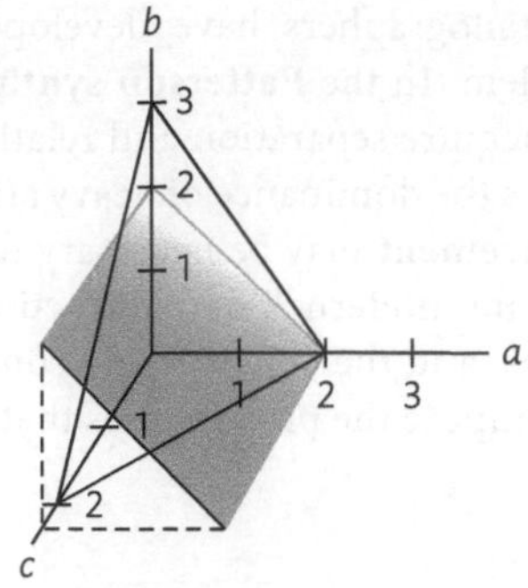

Figure 37.1

37.3(a) For the cubic unit cell: $d_{hkl} = \dfrac{a}{(h^2+k^2+l^2)^{1/2}}$ [37.1b]

$$d_{112} = \frac{a}{(1^2+1^2+2^2)^{1/2}} = \frac{562\ \text{pm}}{(6)^{1/2}} = \boxed{229\ \text{pm}}$$

$$d_{110} = \frac{a}{(1^2+1^2+0^2)^{1/2}} = \frac{562\ \text{pm}}{(2)^{1/2}} = \boxed{397\ \text{pm}}$$

$$d_{224} = \frac{a}{(2^2+2^2+4^2)^{1/2}} = \frac{562\ \text{pm}}{(24)^{1/2}} = \boxed{115\ \text{pm}}$$

37.4(a)

$$d_{hkl} = \left[\left(\frac{h}{a}\right)^2 + \left(\frac{k}{b}\right)^2 + \left(\frac{l}{c}\right)^2\right]^{-1/2} \quad [37.1c]$$

$$d_{321} = \left[\left(\frac{3}{812}\right)^2 + \left(\frac{2}{947}\right)^2 + \left(\frac{1}{637}\right)^2\right]^{-1/2} \text{pm} = \boxed{220\ \text{pm}}$$

37.5(a) Refer to Self-test 37.6 of the text and Discussion question 37.3 above. Systematic absences of diffraction lines from a cubic I lattice (body-centred cubic, bcc) correspond to $h+k+l=\text{odd}$. Hence the first three lines are from planes (110), (200), and (211) with $h^2+k^2+l^2$ values equal to 2, 4, and 6, respectively.

$$\sin\theta_{hkl} = \frac{\lambda}{2d_{hkl}} \text{ [37.2b] and } d_{hkl} = \frac{a}{(h^2+k^2+l^2)^{1/2}} \text{ [37.1b]},$$

$$\text{so} \quad \sin\theta_{hkl} = (h^2+k^2+l^2)^{1/2} \times \left(\frac{\lambda}{2a}\right)$$

In a bcc unit cell, the body diagonal of the cube is $4R$ where R is the atomic radius. The relationship of the side of the unit cell a to R is therefore (using the Pythagorean theorem twice, see Fig. 37.2):

$$a = 4R/3^{1/2} = \frac{4\times 126\text{ pm}}{3^{1/2}} = \boxed{291\text{ pm}} \quad \text{and} \quad \frac{\lambda}{2a} = \frac{72\text{ pm}}{(2)\times(291\text{ pm})} = 0.12\bar{4}$$

Thus,
$$\sin\theta_{110} = \sqrt{2}\times(0.12\bar{4}) = 0.17\bar{5} \qquad \theta_{110} = \sin^{-1}(0.17\bar{5}) = \boxed{10.1°}$$
$$\sin\theta_{200} = \sqrt{4}\times(0.12\bar{4}) = 0.24\bar{8} \qquad \theta_{200} = \sin^{-1}(0.24\bar{8}) = \boxed{14.4°}$$
$$\sin\theta_{211} = \sqrt{6}\times(0.12\bar{4}) = 0.30\bar{4} \qquad \theta_{211} = \sin^{-1}(0.30\bar{4}) = \boxed{17.7°}$$

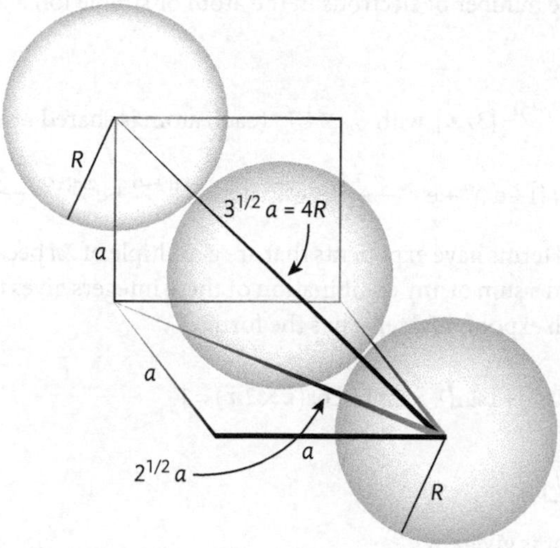

Figure 37.2

37.6(a) $$\theta_{hkl} = \arcsin\frac{\lambda}{2d_{hkl}}[37.2b] = \arcsin\left\{\frac{\lambda}{2}\left[\left(\frac{h}{a}\right)^2 + \left(\frac{k}{b}\right)^2 + \left(\frac{l}{c}\right)^2\right]^{1/2}\right\}[37.1c, \arcsin \equiv \sin^{-1}]$$

$$\theta_{100} = \arcsin\left\{\frac{154}{2}\times\left[\left(\frac{1}{542}\right)^2 + \left(\frac{0}{917}\right)^2 + \left(\frac{0}{645}\right)^2\right]^{1/2}\right\} = \boxed{8.16°}$$

$$\theta_{010} = \arcsin\left\{\frac{154}{2}\times\left[\left(\frac{0}{542}\right)^2 + \left(\frac{1}{917}\right)^2 + \left(\frac{0}{645}\right)^2\right]^{1/2}\right\} = \boxed{4.82°}$$

$$\theta_{111} = \arcsin\left\{\frac{154}{2}\times\left[\left(\frac{1}{542}\right)^2 + \left(\frac{1}{917}\right)^2 + \left(\frac{1}{645}\right)^2\right]^{1/2}\right\} = \boxed{11.7\bar{5}°}$$

37.7(a) $\theta = \arcsin\dfrac{\lambda}{2d}$ [37.2b, $\arcsin \equiv \sin^{-1}$]

$$\Delta\theta = \arcsin\frac{\lambda_2}{2d} - \arcsin\frac{\lambda_1}{2d} = \arcsin\left(\frac{154.433\text{ pm}}{(2)\times(77.8\text{ pm})}\right) - \arcsin\left(\frac{154.051\text{ pm}}{(2)\times(77.8\text{ pm})}\right)$$

$$= 1.07° = 0.0187\text{ rad}$$

Consequently, the difference in the glancing angles (2θ) is $\boxed{2.14°}$.

37.8(a) Justification 37.2 demonstrates that the scattering factor in the forward direction equals the number of electrons in the atom or simple ion. Consequently, $\boxed{f_{\text{Br}^-} = 36}$.

37.9(a) $F_{hkl} = \sum_j f_j e^{2\pi i(hx_j + ky_j + lz_j)}$ [37.4] with $f_j = \frac{1}{8}f$ (each atom is shared by eight cells).

Therefore, $F_{hkl} = \frac{1}{8}f\{1 + e^{2\pi ih} + e^{2\pi ik} + e^{2\pi il} + e^{2\pi i(h+k)} + e^{2\pi i(h+l)} + e^{2\pi i(k+l)} + e^{2\pi i(h+k+l)}\}$

All the exponential terms have exponents that are a multiple of 2π because h, k, and l are all integers and a sum of any combination of these integers gives the integer n. Consequently, each exponential term has the form:

$$e^{in\times 2\pi} = \cos(n\times 2\pi) + i\sin(n\times 2\pi) = \cos(n\times 2\pi) = 1$$

Therefore, $F_{hkl} = \boxed{f}$

37.10(a) The structure factor is given by

$$F_{hkl} = \sum_i f_i e^{i\phi_i} \quad \text{where} \quad \phi_i = 2\pi(hx_i + ky_i + lz_i) \text{ [37.4]}$$

All eight of the vertices of the cube are shared by eight cubes, so each vertex has a scattering factor of $f/8$ and their total contribution to the structure factor equals f (see exercise 37.9(a)). The two side-centred (C) ions of the unit cell are shared by two cells and in this problem each has a scattering factor of $2f$. Their coordinates are (½a, ½a, 0) and (½a, ½a, a). Thus, the contribution of the face-centred ions to the structure factor is

$$\sum_{i(\text{face-centred})} f_i e^{i\phi_i} = \tfrac{1}{2}(2f)e^{2\pi i(h/2+k/2)} + \tfrac{1}{2}(2f)e^{2\pi i(h/2+k/2+l)} = fe^{\pi i(h+k)}\left\{1 + e^{2\pi il}\right\}$$

But, according to Euler's identity ($e^{i\theta} = \cos\theta + i\sin\theta$),

$$e^{\pi i(h+k)} = \cos(\pi\{h+k\}) + i\sin(\pi\{h+k\}) = \cos(\pi\{h+k\}) = (-1)^{h+k}$$

and

$$e^{2\pi il} = \cos(2\pi l) + i\sin(2\pi l) = \cos(2\pi l) = 1$$

Thus, $\sum_{i(\text{face-centred})} f_i e^{i\phi_i} = f(-1)^{h+k}\{1+1\} = 2(-1)^{h+k} f$

and the structure factor becomes

$$F_{hkl} = \sum_{i(\text{vertices})} f_i e^{i\phi_i} + \sum_{i(\text{face-centred})} f_i e^{i\phi_i} = f + 2(-1)^{h+k} f = f\left\{1+2(-1)^{h+k}\right\}$$

$$F_{hkl} = \boxed{3f \text{ for } h+k \text{ even and } -f \text{ for } h+k \text{ odd}}$$

37.11(a) The electron density is given by

$$\rho(r) = \frac{1}{V}\sum_{hkl} F_{hkl} e^{-2\pi i(hx+ky+lz)} \quad [37.5]$$

The component along the x direction is

$$\rho(x) = \frac{1}{V}\sum_{h} F_h e^{-2\pi ihx}$$

Using the data of this problem, we sum from $h = -9$ to $+9$ and use the relationship $F_{-h} = F_h$. The following Mathcad computation of $\rho(x)$ and plot shown in Fig. 37.3 shows high electron density at the center of a unit cell edge.

$$F := \begin{pmatrix} 10 \\ -10 \\ 8 \\ -8 \\ 6 \\ -6 \\ 4 \\ -4 \\ 2 \\ -2 \end{pmatrix} \qquad \rho(x) := \sum_{h=-9}^{9} \left(F_{|h|} \cdot \cos(2\cdot\pi\cdot h\cdot x)\right)$$

$$\rho(.5) = 110$$

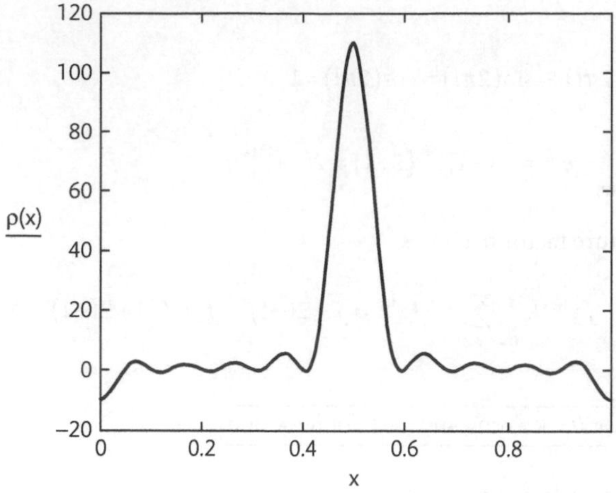

Figure 37.3

37.12(a) Using the information of exercise 37.11(a), the Mathcad computation of $P(x)$ is performed with eqn 37.6.

$$F := \begin{pmatrix} 10 \\ -10 \\ 8 \\ -8 \\ 6 \\ -6 \\ 4 \\ -4 \\ 2 \\ -2 \end{pmatrix} \qquad P(x) := \sum_{h=-9}^{9} \left[\left(F_{|h|}\right)^2 \cdot e^{-2\pi \cdot i \cdot h \cdot x} \right]$$

$$P(1.0) = 780$$

The Patterson synthesis $P(x)$ of Fig. 37.4 shows that atoms represented by this data are separated by 1 a unit along the x axis.

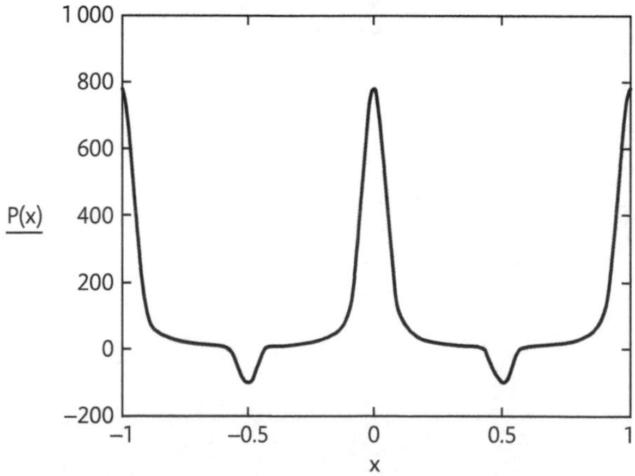

Figure 37.4

37.13(a) Draw points corresponding to the vectors joining each pair of atoms. Heavier atoms give more intense contribution than light atoms. Remember that there are two vectors joining any pair of atoms ($\overrightarrow{AB}$ and $\overleftarrow{AB}$); don't forget the AA zero vectors for the centre point of the diagram. See Fig. 37.5.

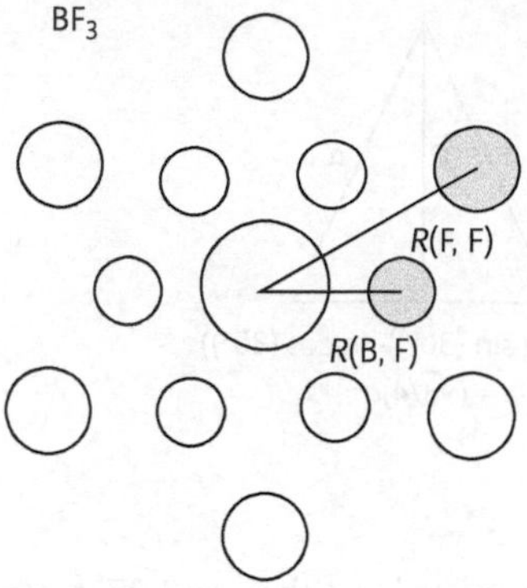

Fig 37.5

37.14(a) $\lambda = \dfrac{h}{p} = \dfrac{h}{mv}$

Hence, $v = \dfrac{h}{m\lambda} = \dfrac{6.626\times10^{-34}\ \text{J s}}{(1.675\times10^{-27}\ \text{kg})\times(65\times10^{-12}\ \text{m})} = \boxed{6.1\ \text{km s}^{-1}}$

37.15(a) $\lambda = \dfrac{h}{p} = \dfrac{h}{mv}$ Hence, $v = \dfrac{h}{m\lambda}$.

Combine $E = \frac{1}{2}kT$ and $E = \frac{1}{2}mv^2 = \dfrac{h^2}{2m\lambda^2}$, to obtain

$$\lambda = \frac{h}{(mkT)^{1/2}} = \frac{6.626\times10^{-34}\ \text{J s}}{[(1.675\times10^{-27}\ \text{kg})\times(1.381\times10^{-23}\ \text{J K}^{-1})\times(350\text{K})]^{1/2}} = \boxed{233\ \text{pm}}$$

Problems

37.1 The density of a face-centred cubic crystal is $4m/V$ where m is the mass of the unit hung on each lattice point and V is the volume of the unit cell. (The 4 comes from the fact that each of the cell's 8 vertices is shared by 8 cells, and each of the cell's 6 faces is shared by 2 cells.) So,

$$\rho = \frac{4m}{a^3} = \frac{4M}{N_A a^3} \text{ and } M = \frac{1}{4}\rho N_A a^3$$

$$M = \tfrac{1}{4}(1.287\ \text{g cm}^{-3})\times(6.022\times10^{23}\ \text{mol}^{-1})\times(12.3\times10^{-7}\ \text{cm})^3$$
$$= \boxed{3.61\times10^5\ \text{g mol}^{-1}}$$

37.3 The hexagonal unit cell is shown in Figure 37.6 from which we see that its volume V equals the value of c multiplied by the area of the hexagonal face. Since the area of the face equals 6 × (triangular area of Figure 37.6), we find that $\boxed{V=\left(3\sqrt{3}/2\right)a^2c}$.

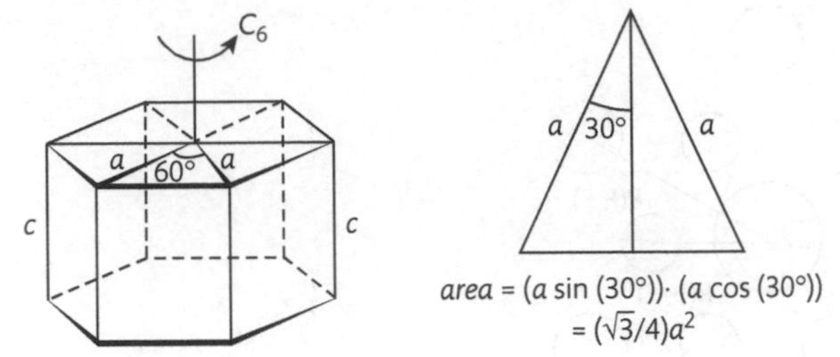

Figure 37.6

37.5 $V=abc\sin\beta$ and the information given tells us that $a=1.377b$, $c=1.436b$, and $\beta=122.82°$; hence

$$V=(1.377)\times(1.436\,b^3)\ \sin 122.82°=1.662\,b^3$$

Since $\rho=\dfrac{NM}{VN_{\rm A}}=\dfrac{2M}{1.662\,b^3N_{\rm A}}$, we find that

$$b=\left(\frac{2M}{1.662\rho N_{\rm A}}\right)^{1/3}$$

$$=\left(\frac{(2)\times(128.18\ \text{g mol}^{-1})}{(1.662)\times(1.152\times10^{6}\ \text{g m}^{-3})\times(6.022\times10^{23}\ \text{mol}^{-1})}\right)^{1/3}=605.8\,\text{pm}$$

Therefore, $a=\boxed{834\,\text{pm}}$, $b=\boxed{606\,\text{pm}}$, and $c=\boxed{870\,\text{pm}}$.

37.7 As shown in Problem 37.2, the volume of a primitive monoclinic unit cell is

$$V=abc\ \sin\beta=(1.0427\ \text{nm})\times(0.8876\ \text{nm})\times(1.3777\ \text{nm})\times\sin(93.254°)$$
$$=1.2730\ \text{nm}^3$$

The mass per unit cell is

$$m=\rho V=(2.024\,\text{g cm}^{-3})\times(1.2730\ \text{nm}^3)\times(10^{-7}\ \text{cm nm}^{-1})^3=2.577\times10^{-21}\ \text{g}$$

The monomer is $CuC_7H_{13}N_5O_8S$, so its molar mass is

$$M=[63.546+7(12.011)+13(1.008)+5(14.007)+8(15.999)+32.066]\ \text{g mol}^{-1}$$
$$=390.82\ \text{g mol}^{-1}$$

The number of monomer units, then, is the mass of the unit cell divided by the mass of the monomer

$$N = \frac{mN_A}{M} = \frac{(2.577\times10^{-21}\ \text{g})\times(6.022\times10^{23}\ \text{mol}^{-1})}{390.82\ \text{g mol}^{-1}} = 3.97 \text{ or } \boxed{4}$$

37.9 Consider for simplicity the two-dimensional lattice and planes shown in Fig. 37.7.

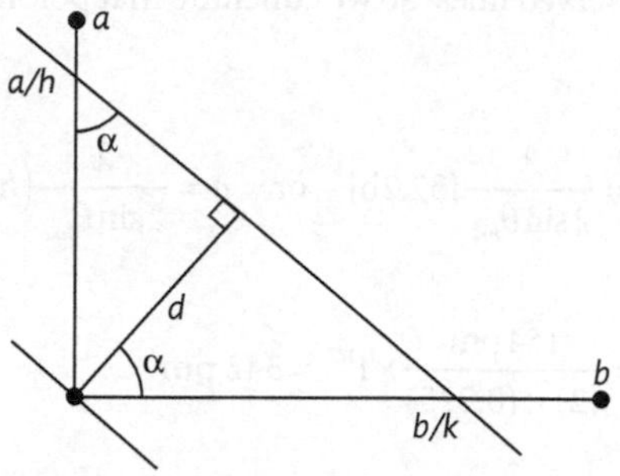

Fig 37.7

The (hk) planes cut the a and b axes at a/h and b/k, and we have

$$\sin\alpha = \frac{d}{(a/h)} = \frac{hd}{a} \quad \text{and} \quad \cos\alpha = \frac{d}{(b/k)} = \frac{kd}{b}$$

(i.e., $\sin(90° - \alpha) = \sin 90° \cos\alpha - \cos 90° \sin\alpha = \cos\alpha$)

Then, since $\sin^2\alpha + \cos^2\alpha = 1$, we can write

$$\left(\frac{hd}{a}\right)^2 + \left(\frac{kd}{b}\right)^2 = 1$$

and therefore

$$\frac{1}{d^2} = \left(\frac{h}{a}\right)^2 + \left(\frac{k}{b}\right)^2$$

The same argument extends by analogy (or further trigonometry) to three dimensions, to give

$$\boxed{\frac{1}{d^2} = \left(\frac{h}{a}\right)^2 + \left(\frac{k}{b}\right)^2 + \left(\frac{l}{c}\right)^2} \quad [37.1c]$$

37.11 Refer to Self-test 37.6 of the text and Discussion question 37.3 above. These references indicate systematic absences of diffraction lines from a cubic I lattice (body-centred cubic, bcc) corresponding to $h+k+l = \text{odd}$ thereby ruling out the presence of {100} planes in the diffraction pattern. These same planes are also ruled out of diffraction lines from a cubic F lattice (face-centred cubic, fcc) because the lines of this lattice have h, k, and l values which are either all even or all odd. These rules also indicate that {110} planes are not observed in the diffraction pattern cubic F lattices. Only the cubic P lattice exhibits diffraction lines from both {100} planes and {110} planes. The diffraction line from {111} planes of a cubic P lattice is expected to have low or zero intensity but, if observed, the separation between the sixth and seventh lines observed is larger than between the fifth and sixth lines because $\sin\theta_{hkl} \propto 1/d_{hkl} \propto \left(h^2+k^2+l^2\right)^{1/2}$, which corresponds to $5^{1/2}$, $6^{1/2}$, and $8^{1/2}$ for the 5th, 6th, and 7th observed lines. So we conclude that polonium has the $\boxed{\text{simple (primitive) cubic lattice}}$.

$$d_{hkl} = \frac{a}{\left(h^2+k^2+l^2\right)^{1/2}}\,[37.1\text{b}] = \frac{\lambda}{2\sin\theta_{hkl}}\,[37.2\text{b}] \quad \text{or} \quad a = \frac{\lambda}{2\sin\theta_{hkl}}\left(h^2+k^2+l^2\right)^{1/2}$$

Bragg reflection from {100}: $a = \dfrac{154\,\text{pm}}{(2)\times(0.225)} \times 1^{1/2} = 342\ \text{pm}$

$$\text{from } \{110\}:\quad a = \frac{154\,\text{pm}}{(2)\times(0.316)} \times 2^{1/2} = 345\ \text{pm}$$

$$\text{from } \{111\}:\quad a = \frac{154\,\text{pm}}{(2)\times(0.388)} \times 3^{1/2} = 344\ \text{pm}$$

Since these agree to within experimental uncertainty, we average to find that $\boxed{a = 344\ \text{pm}}$.

37.13 $\lambda = 2d_{100}\sin\theta_{100}$ [37.2b] and $d_{100} = a$ [37.1b, cubic lattice]

Therefore, $a = \dfrac{\lambda}{2\sin\theta_{100}}$ and

$$\frac{a(\text{KCl})}{a(\text{NaCl})} = \frac{\sin\theta_{100}(\text{NaCl})}{\sin\theta_{100}(\text{KCl})} = \frac{\sin 6°0'}{\sin 5°23'} = 1.114$$

Therefore, $a(\text{KCl}) = (1.114)\times(564\,\text{pm}) = \boxed{628\,\text{pm}}$.

The relative density ratio calculated from these unit cell dimensions is

$$\frac{\rho(\text{KCl})}{\rho(\text{NaCl})} = \left(\frac{M(\text{KCl})}{M(\text{NaCl})}\right)\times\left(\frac{a(\text{NaCl})}{a(\text{KCl})}\right)^3 = \left(\frac{74.55}{58.44}\right)\times\left(\frac{564\,\text{pm}}{628\,\text{pm}}\right)^3 = 0.924$$

whereas the ratio based upon macroscopic measurement of mass and volume is

$$\frac{\rho(\text{KCl})}{\rho(\text{NaCl})} = \frac{1.99\ \text{g cm}^{-3}}{2.17\ \text{g cm}^{-3}} = 0.917$$

The agreement of the two methods to within experiment uncertainties, $\boxed{\text{gave support}}$ to the X-ray method in the early period of its development.

37.15 The four values of $hx + ky + lz$ that occur in the exponential functions of F have the values 0, 5/2, 3, and 7/2, and so

$$F_{114} = f_{\rm I}\sum_j e^{i\phi_{114}(I_j)}\ [37.4] = f_{\rm I}\sum_j e^{2i\pi(x_j+y_j+4z_j)}$$
$$= f_{\rm I}\{1+e^{5i\pi}+e^{6i\pi}+e^{7i\pi}\} = f_{\rm I}\{1-1+1-1\} = \boxed{0}$$

37.17 (a) When there is only one pair of identical atoms, the Wierl equation reduces to

$$I(\theta) = f^2\frac{\sin sR}{sR} \quad \text{where} \quad s = \frac{4\pi}{\lambda}\sin\tfrac{1}{2}\theta$$

Using θ_{ex} and sR_{ex}, meaning $(sR)_{ex}$, to represent θ and sR values when $I(sR)$ is an extremum, we find an analytic formula for the extrema by taking the derivative of $I(sR)$ w/r/t sR, set it equal to zero, and solve for sR_{ex}. We find

$$sR_{ex} = \frac{\sin sR_{ex}}{\cos sR_{ex}} = \tan sR_{ex}$$

This equation may be solved numerically, as shown in the following Mathcad Prime 2 worksheet, to give the extrema. Alternatively, we can simply plot $I_{ff} \equiv I(sR)/f^2$ against sR to find sR values at which I_{ff} is a maximum or minimum. The graphical method is shown in Figure 37.8.

$sR_{ex_0} := 0 \quad n_{max} := 6 \quad n := 1..n_{max} \quad sR_{ex_n} := sR_{ex_{n-1}} + \pi + .001$

Solve block and results (Guess values are those above):

Solver Constraints Guess Values

$sR_{ex} = \tan(sR_{ex})$

$sR_{ex} := \text{Find}(sR_{ex})$

$$sR_{ex} = \begin{bmatrix} 0 \\ 4.493 \\ 7.725 \\ 10.904 \\ 14.006 \\ 17.221 \\ 20.371 \end{bmatrix} \qquad \tan(sR_{ex}) = \begin{bmatrix} 0 \\ 4.493 \\ 7.725 \\ 10.904 \\ 14.006 \\ 17.221 \\ 20.371 \end{bmatrix}$$

$$I_{ff}(sR) := \frac{\sin(sR)}{sR} \qquad I_{ff}(7.725) = 0.128 \qquad I_{ff}(14.066) = 0.071 \qquad I_{ff}(20.371) = 0.049$$

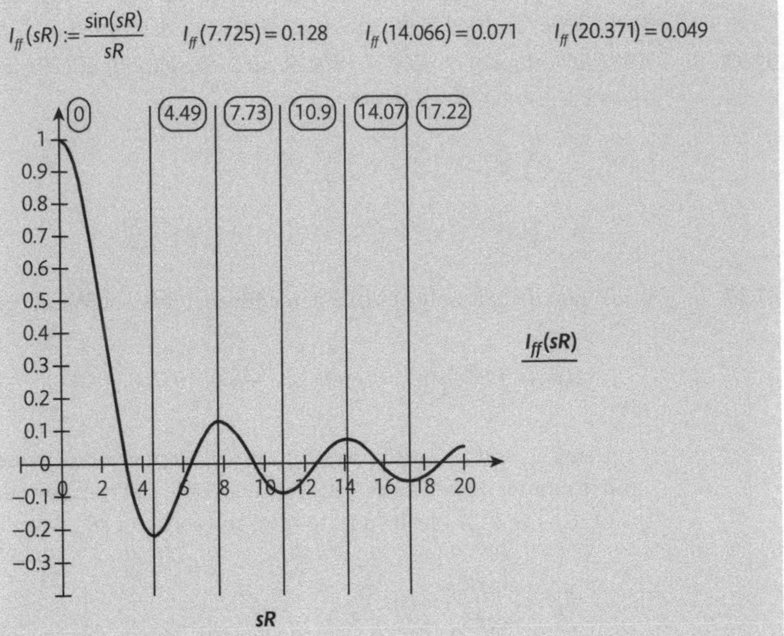

Figure 37.8

The angles of extrema are calculated using the Br_2 bond length of 228.3 pm, the equation $\theta_{ex} = 2\sin^{-1}\left(\dfrac{(sR)_{ex} \lambda}{4\pi R}\right)$ and sR extrema.

Neutron diffraction: $\theta_{1st\,max} = 0$

$$\theta_{1st\,min} = 2\sin^{-1}\left(\frac{(4.493)\times(78\text{ pm})}{4\pi(228.3\text{ pm})}\right) = \boxed{14.0°}$$

$$\theta_{2nd\,max} = 2\sin^{-1}\left(\frac{(7.725)\times(78\text{ pm})}{4\pi(228.3\text{ pm})}\right) = \boxed{24.2°}$$

Electron diffraction: $\theta_{1st\,max} = 0$

$$\theta_{1st\,min} = 2\sin^{-1}\left(\frac{(4.493)\times(4.0\text{ pm})}{4\pi(228.3\text{ pm})}\right) = \boxed{0.72°}$$

$$\theta_{2nd\,max} = 2\sin^{-1}\left(\frac{(7.725)\times(4.0\text{ pm})}{4\pi(228.3\text{ pm})}\right) = \boxed{1.23°}$$

(b) $$I=\sum_{i,j} f_i f_j \frac{\sin sR_{i,j}}{sR_{i,j}}, \qquad s=\frac{4\pi}{\lambda}\sin\tfrac{1}{2}\theta$$

$$=4f_\mathrm{C}f_\mathrm{Cl}\frac{\sin sR_\mathrm{CCl}}{sR_\mathrm{CCl}}+6f_\mathrm{Cl}^2\frac{\sin sR_\mathrm{ClCl}}{sR_\mathrm{ClCl}}$$

$$[4\ \mathrm{C{-}Cl\ pairs},\ 6\ \mathrm{Cl{-}Cl\ pairs},\ R_\mathrm{ClCl}=\left(\tfrac{8}{3}\right)^{1/2}R_\mathrm{CCl}]$$

$$=(4)\times(6)\times(17)\times(f^2)\times\left(\frac{\sin sR_\mathrm{CCl}}{sR_\mathrm{CCl}}\right)+(6)\times(17)^2\times(f^2)\times\frac{\sin\left(\tfrac{8}{3}\right)^{1/2}sR_\mathrm{CCl}}{\left(\tfrac{8}{3}\right)^{1/2}sR_\mathrm{CCl}}$$

$$\frac{I}{f^2}=(408)\times\frac{\sin sR_\mathrm{CCl}}{sR_\mathrm{CCl}}+(1\,061.85)\frac{\sin\left(\tfrac{8}{3}\right)^{1/2}sR_\mathrm{CCl}}{sR_\mathrm{CCl}}$$

A plot of this function is shown in Fig. 37.9.

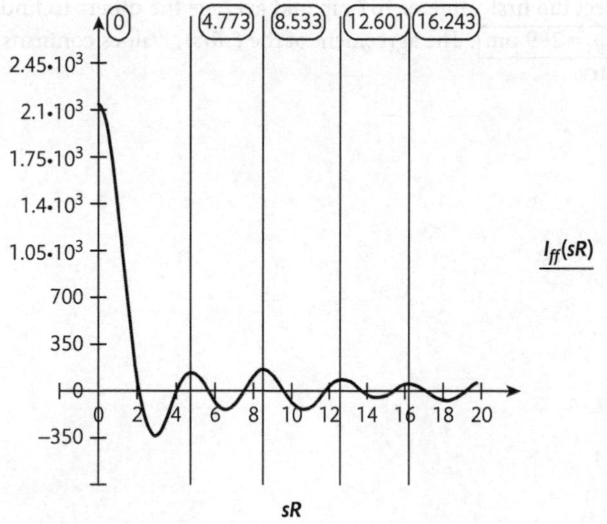

Figure 37.9

We find sR_ex from the graph, and s_ex from the data. Then, we take the ratio $(sR/s)_\mathrm{ex}$ to find the bond length R_CCl. The calculation of s requires the wavelength of the 10.0 keV electron beam.

$$\frac{p^2}{2m_\mathrm{e}}=eV \quad \text{or} \quad p=\left(2m_\mathrm{e}eV\right)^{1/2}$$

From the de Broglie relation,

$$\lambda = \frac{h}{p} = \frac{h}{(2m_e eV)^{1/2}}$$

$$= \frac{6.626\times10^{-34}\ \text{J s}}{2\times(9.109\times10^{-31}\ \text{kg})\times(1.609\times10^{-19}\ \text{C})\times(1.00\times10^{4}\ \text{V})}$$

$$= 12.2\ \text{pm}$$

We draw up the following table.

	Maxima			Minima			
θ_{ex}(exp)	3.17°	5.37°	7.90°	1.77°	4.10°	6.67°	9.17°
s_{ex}/pm^{-1}	0.0285	0.0483	0.0710	0.0156	0.0368	0.0599	0.0823
$(sR)_{ex}$	4.773	8.533	12.60	2.931	6.549	10.62	14.45
$(sR/s)_{ex}$/pm	167	177	177	173	178	177	176

We reject the first value of 167 pm and average the others to find $\boxed{R_{CCl} = 176\ \text{pm}}$ and $\boxed{R_{ClCl} = 289\ \text{pm}}$. The agreement of the $(sR/s)_{ex}$ values confirms the tetrahedral geometry.

Topic 38 **Bonding in solids**

Discussion questions

38.1 The majority of metals crystallize in structures which can be interpreted as the closest packing arrangements of hard spheres. These are the cubic close-packed (ccp) and hexagonal close-packed (hcp) structures. In these models, 74% of the volume of the unit cell is occupied by the atoms (packing fraction = 0.74). Most of the remaining metallic elements crystallize in the body-centered cubic (bcc) arrangement which is not too much different from the close-packed structures in terms of the efficiency of the use of space (packing fraction 0.68 in the hard sphere model). Polonium is an exception; it crystallizes in the simple cubic structure which has a packing fraction of 0.52. (See the solution to Exercise 38.2(a) for a derivation of all the packing fractions in cubic systems.) If atoms were truly hard spheres, we would expect that all metals would crystallize in either the ccp or hcp close-packed structures. The fact that a significant number crystallize in other structures is proof that a simple hard sphere model is an inaccurate representation of the interactions between the atoms. Covalent bonding between the atoms may influence the structure.

Exercises

38.1(a) The hatched area in Fig. 38.1 is $h\times 2R = 3^{1/2}R\times 2R = 2\sqrt{3}R^2$ where $h = 2R\cos 30°$. The net number of cylinders in a hatched area is 1, and the area of the cylinder's base is πR^2. The volume of the prism (of which the hatched area is the base) is $2\sqrt{3}R^2L$, and the volume occupied by the cylinders is $\pi R^2 L$. Hence, the packing fraction is

$$f = \frac{\pi R^2 L}{2\sqrt{3}R^2 L} = \frac{\pi}{2\sqrt{3}} = \boxed{0.9069}$$

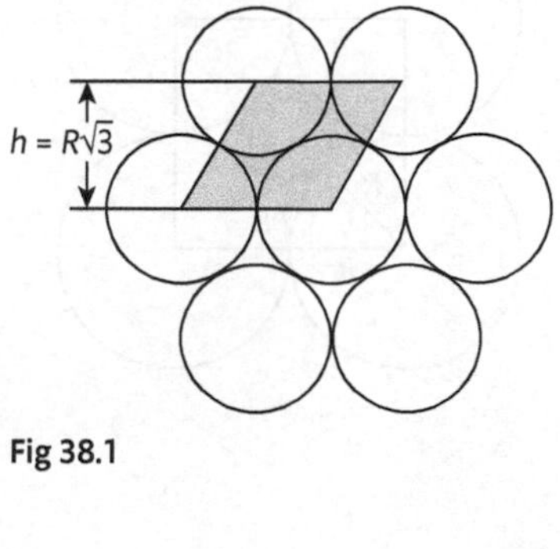

Fig 38.1

38.2(a) $f = \frac{NV_a}{V_c}$ where N is the number of atoms in each unit cell, V_a their individual volumes, and V_c the volume of the unit cell itself. Refer to Fig. 38.2 for a view of the primitive unit cell, the bcc unit cell, and the fcc unit cell.

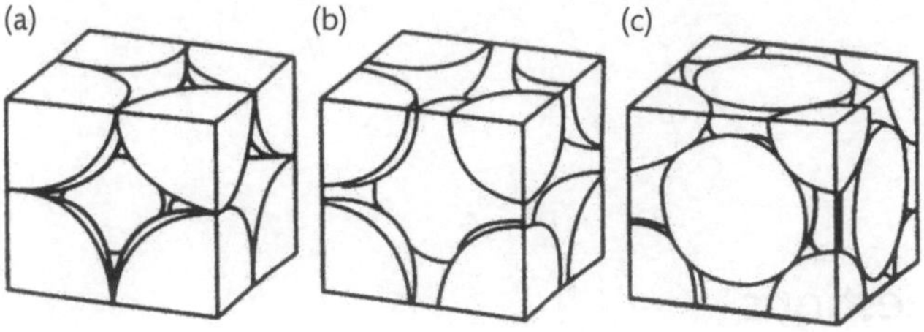

Fig 38.2

(a) For a primitive unit cell: $N = 1$, $V_a = \frac{4}{3}\pi R^3$, and $V_c = (2R)^3$

$$f = \frac{\left(\frac{4}{3}\pi R^3\right)}{(2R)^3} = \frac{\pi}{6} = \boxed{0.5236}$$

(b) For the bcc unit cell:

$$N = 2,\ V_a = \frac{4}{3}\pi R^3,\ \text{and}\ V_c = \left(\frac{4R}{\sqrt{3}}\right)^3 \quad \text{[body diagonal of a unit cube is } 4R\text{]}$$

$$f = \frac{2 \times \frac{4}{3}\pi R^3}{\left(\frac{4R}{\sqrt{3}}\right)^3} = \frac{\pi\sqrt{3}}{8} = \boxed{0.6802}$$

(c) For the fcc unit cell: $N = 4$, $V_a = \frac{4}{3}\pi R^3$, and $V_c = (2\sqrt{2}R)^3$

$$f = \frac{4 \times \frac{4}{3}\pi R^3}{(2\sqrt{2}R)^3} = \frac{\pi}{3\sqrt{2}} = \boxed{0.7405}$$

38.3(a) For sixfold coordination see Fig. 38.3.

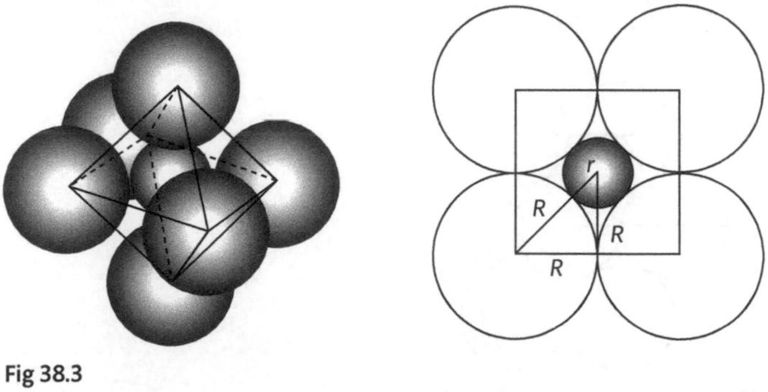

Fig 38.3

We assume that the larger spheres of radius R touch each other and that they also touch the smaller interior sphere. Hence, by the Pythagorean theorem

$$(R+r)^2 = 2(R)^2 \quad \text{or} \quad \left(1+\frac{r}{R}\right)^2 = 2$$

Thus, $\frac{r}{R}=0.414$ and $r=0.414R=(0.414)\times(181\text{ pm})=\boxed{74.9\text{ pm}}$

(b) For eightfold coordination see Fig. 38.4.

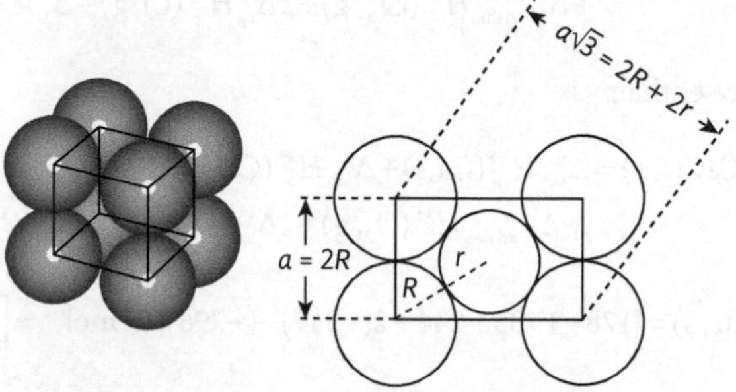

Fig 38.4

The body diagonal of a cube is $a\sqrt{3}$. Hence

$$a\sqrt{3}=2R+2r \quad \text{or} \quad \sqrt{3}R=R+r \quad [a=2R]$$

$$\frac{r}{R}=0.732 \quad \text{and} \quad r=0.732R=(0.732)\times(181\text{ pm})=\boxed{132\text{ pm}}$$

38.4(a) The volume change is a result of two partially counteracting factors: (1) different packing fraction (f), and (2) different radii.

$$\frac{V(\text{bcc})}{V(\text{hcp})}=\frac{f(\text{hcp})}{f(\text{bcc})}\times\frac{v(\text{bcc})}{v(\text{hcp})} \quad \text{where } v \text{ is atomic volume}$$

$$f(\text{hcp})=0.7405,\quad f(\text{bcc})=0.6802 \quad [\text{Exercise 38.2(a)}]$$

$$\frac{V(\text{bcc})}{V(\text{hcp})}=\frac{0.7405}{0.6802}\times\frac{(142.5)^3}{(145.8)^3}=1.016$$

Hence there is an $\boxed{\text{expansion}}$ of 1.6% in a transformation from hcp to bcc.

38.5(a) The lattice enthalpy is the difference in enthalpy between an ionic solid and the corresponding isolated ions. In this exercise, it is the enthalpy corresponding to the process

$$CaCl_2(s)\rightarrow Ca^{2+}(g)+2\,Cl^-(g) \quad \Delta_L H^{\ominus}(CaCl_2,s)$$

The standard lattice enthalpy can be computed from the standard enthalpies given in the exercise by considering the formation of $CaCl_2$(s) from its elements as occurring through the following steps: sublimation of Ca(s), removing two electrons from Ca(g), atomization of Cl_2(g), electron attachment to Cl(g), and formation of $CaCl_2$(s) lattice from gaseous ions. The formation reaction of $CaCl_2$(s) is

$$\mathrm{Ca(s)} + \mathrm{Cl_2(g)} \rightarrow \mathrm{CaCl_2(s)}$$

$$\Delta_f H^{\ominus}(\mathrm{CaCl_2, s}) = \Delta_{\mathrm{sub}} H^{\ominus}(\mathrm{Ca, s}) + \Delta_{\mathrm{ion}} H^{\ominus}(\mathrm{Ca, g}) + \Delta_{\mathrm{bond\ diss}} H^{\ominus}(\mathrm{Cl_2, g}) + 2\Delta_{\mathrm{eg}} H^{\ominus}(\mathrm{Cl, g}) - \Delta_{\mathrm{L}} H^{\ominus}(\mathrm{CaCl_2, s})$$

So the lattice enthalpy is

$$\Delta_{\mathrm{L}} H^{\ominus}(\mathrm{CaCl_2, s}) = \Delta_{\mathrm{sub}} H^{\ominus}(\mathrm{Ca, s}) + \Delta_{\mathrm{ion}} H^{\ominus}(\mathrm{Ca, g}) + \Delta_{\mathrm{bond\ diss}} H^{\ominus}(\mathrm{O_2, g}) + 2\Delta_{\mathrm{eg}} H^{\ominus}(\mathrm{Cl, g}) - \Delta_f H^{\ominus}(\mathrm{CaCl_2, s})$$

$$\Delta_L H^{\ominus}(\mathrm{CaCl_2, s}) = [178 + 1\,735 + 244 + 2(-349) - (-796)]\ \mathrm{kJ\ mol^{-1}} = \boxed{+2\,255.\ \mathrm{kJ\ mol^{-1}}}$$

Problems

38.1 Text Fig. 38.15 shows the diamond structure. Fig. 38.5 below is an easier to visualize form of the structure which shows the unit cell of diamond. The number of carbon atoms in the unit cell is $(8\times\frac{1}{8})+(6\times\frac{1}{2})+(4\times1)=8$ ($\frac{1}{8}$ for a corner atom, $\frac{1}{2}$ for a face-centered atom, and 1 for an atom entirely in the cell). The positions of the atoms are $(0,0,0)$, $(\frac{1}{2},\frac{1}{2},0)$, $(\frac{1}{2},0,\frac{1}{2})$, $(0,\frac{1}{2},\frac{1}{2})$, $(\frac{1}{4},\frac{1}{4},\frac{1}{4})$, $(\frac{1}{4},\frac{3}{4},\frac{3}{4})$, $(\frac{3}{4},\frac{1}{4},\frac{3}{4})$, $(\frac{3}{4},\frac{3}{4},\frac{1}{4})$, and as indicated in Fig. 38.6.

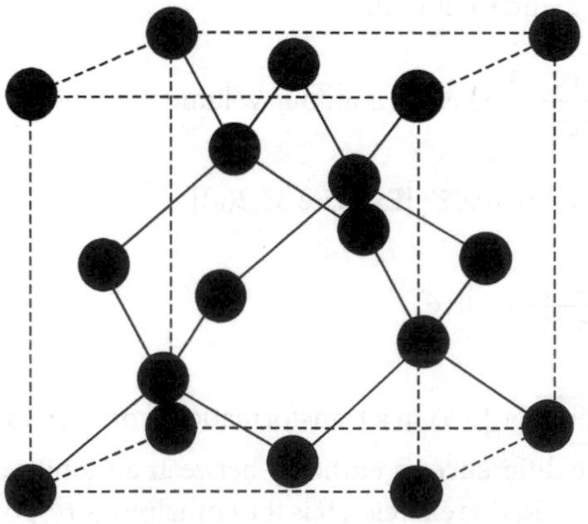

Fig 38.5

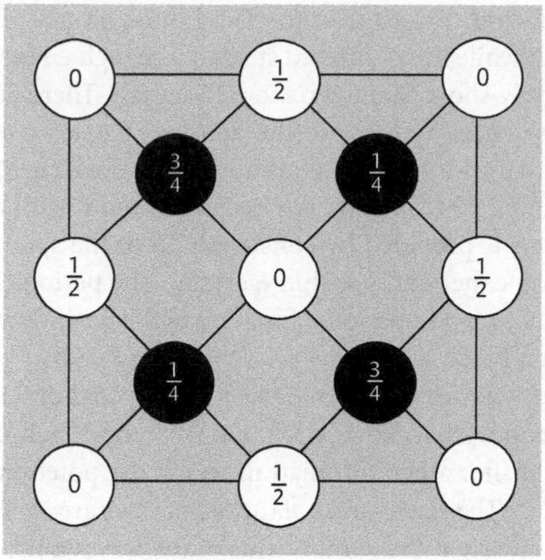

Fig 38.6

The fractions in Fig. 38.6 denote height above the base in units of the cube edge, a. Two atoms that touch lie along the body diagonal at $(0,0,0)$ and $(\frac{1}{4},\frac{1}{4},\frac{1}{4})$. Hence the distance $2r$ is one-fourth of the body diagonal which is $\sqrt{3}a$ in a cube. That is $2r=\sqrt{3}a/4$

The packing fraction is $\dfrac{\text{volume of atoms}}{\text{volume of unit cell}}=\dfrac{8V_a}{a^3}=\dfrac{(8)\times\frac{4}{3}\pi r^3}{(8r/\sqrt{3})^3}=\boxed{0.340}$

38.3 As demonstrated in Example 38.1 of the text and Exercise 38.2(a), close-packed spheres (cubic F or face-centred cubic) fill 0.7404 of the total volume of the crystal. Therefore 1 cm^3 of close-packed carbon atoms would contain

$$\frac{0.74040\ \text{cm}^3}{\left(\frac{4}{3}\pi r^3\right)}=3.838\times10^{23}\ \text{atoms}$$

$$\left(r=\left(\frac{154.45}{2}\right)\text{pm}=77.225\ \text{pm}=77.225\times10^{-10}\ \text{cm}\right)$$

Hence the close-packed density would be

$$\rho=\frac{\text{mass in }1\,\text{cm}^3}{1\ \text{cm}^3}=\frac{(3.838\times10^{23}\ \text{atom})\times(12.01\ \text{u/atom})\times(1.6605\times10^{-24}\ \text{g u}^{-1})}{1\ \text{cm}^3}$$

$$=\boxed{7.654\ \text{g cm}^{-3}}$$

The diamond structure is a very open structure which is dictated by the tetrahedral bonding of the carbon atoms. As a result many atoms that would be touching each other in a normal fcc structure do not in diamond; for example, the C atom in the center of a face does not touch the C atoms at the corners of the face. This reduces the density to the much lower 3.516 g cm^{-1}.

38.5 Permitted states at the low energy edge of the band must have a relatively long characteristic wavelength while the permitted states at the high energy edge of the band must have a relatively short characteristic wavelength. There are few wavefunctions that have these characteristics so the density of states is lowest at the edges. This is analogous to the MO picture that shows a few bonding MOs that lack nodes and few antibonding MOs that have the maximum number of nodes.

Another insightful view is provided by consideration of the spatially periodic potential that the electron experiences within a crystal. The periodicity demands that the electron wavefunction be a periodic function of the position vector $\vec{r}$. We can approximate it with a Bloch wave: $\psi \propto e^{i\vec{k}\cdot\vec{r}}$ where $\vec{k} = k_x\hat{i} + k_y\hat{j} + k_z\hat{k}$ is called the wave number vector. This is a bold, "free" electron approximation and in the spirit of searching for a conceptual explanation, not an accurate solution, suppose that the wavefunction satisfies a Hamiltonian in which the potential can be neglected: $\hat{H} = -(\hbar^2/2m)\nabla^2$. The eigenvalues of the Bloch wave are: $E = \hbar^2|\vec{k}|^2/2m$. The Bloch wave is periodic when the components of the wave number vector are multiples of a basic repeating unit. Writing the repeating unit as $2\pi/L$ where L is a length that depends upon the structure of the unit cell, we find: $k_x = 2n_x\pi/L$ where $n_x = 0, \pm1, \pm2, \ldots$. Similar equations can be written for k_y and k_z and with substitution the eigenvalues become: $E = (1/2m)(2\pi\hbar/L)^2(n_x^2 + n_y^2 + n_z^2)$. This equation suggests that the density of states for energy level E can be visually evaluated by looking at a plot of permitted n_x, n_y, n_z values as shown in the following graph of Fig. 38.7. The number of n_x, n_y, n_z values within a thin, spherical shell around the origin equals the density of states which have energy E. Three shells, labeled 1, 2, and 3, are shown in the graph. All have the same width but their energies increase with their distance from the origin. It is obvious that the low energy shell 1 has a much lower density of states than the intermediate energy shell 2. The sphere of

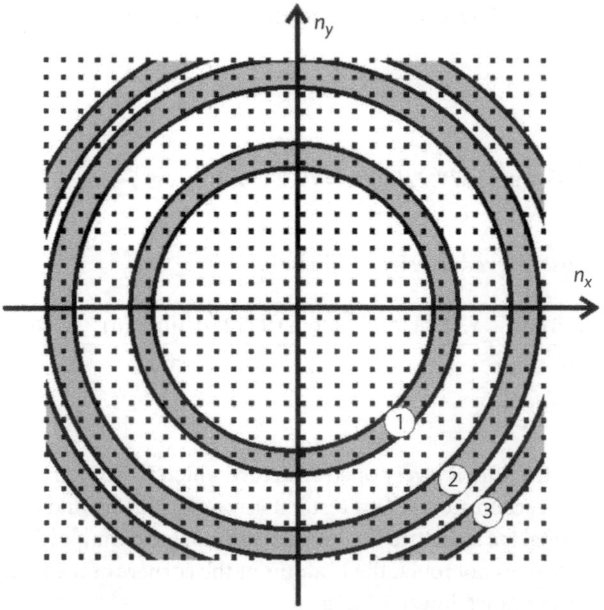

Figure 38.7

shell 3 has been cut into the shape determined by the periodic potential pattern of the crystal and, because of this phenomena, it also has a lower density of states than the intermediate energy shell 2. The general concept is that the low energy and high energy edges of a band have lower density of states than that of the band center.

38.7 The problem asks for an estimate of $\Delta_f H^\ominus(\text{CaCl, s})$. A Born–Haber cycle would envision formation of CaCl(s) from its elements as sublimation of Ca(s), ionization of Ca(g), atomization of Cl_2(g), electron gain of Cl(g), and formation of CaCl(s) from gaseous ions. Therefore

$$\Delta_f H^\ominus(\text{CaCl, s}) = \Delta_{\text{sub}} H^\ominus(\text{Ca, s}) + \Delta_{\text{ion}} H^\ominus(\text{Ca, g}) + \Delta_f H^\ominus(\text{Cl, g}) + \Delta_{\text{eg}} H^\ominus(\text{Cl, g}) - \Delta_L H^\ominus(\text{CaCl, s})$$

Before we can estimate the lattice enthalpy of CaCl, we select a lattice with the aid of the radius-ratio rule. The ionic radius for Cl^- is 181 pm (Table 38.2); use the ionic radius of K^+ (138 pm) for Cs^+.

$$\gamma = \frac{138\,\text{pm}}{181\,\text{pm}} = 0.762$$

suggesting the CsCl structure (Madelung constant $A = 1.763$, Table 38.3). We can interpret the Born–Mayer equation (eqn 38.7) as giving the negative of the lattice enthalpy.

$$\Delta_L H^\ominus \approx E_{\text{p,min}} = \frac{A|z_1 z_2|N_A e^2}{4\pi\varepsilon_0 d}\left(1 - \frac{d^*}{d}\right) \quad \text{where } d^* \text{ is taken to be 34.5 pm (common choice).}$$

The distance d is $d = (138 + 181)\,\text{pm} = 319\,\text{pm}$.

$$\Delta_L H^\ominus \approx \frac{(1.763)\times|(1)(-1)|\times(6.022\times10^{23}\,\text{mol}^{-1})\times(1.602\times10^{-19}\,\text{C})^2}{4\pi(8.854\times10^{-12}\,\text{J}^{-1}\,\text{C}^2\,\text{m}^{-1})\times(319\times10^{-12}\,\text{m})}\left(1 - \frac{34.5\,\text{pm}}{319\,\text{pm}}\right)$$
$$\approx 6.85\times10^5\,\text{J mol}^{-1} = 685\,\text{kJ mol}^{-1}$$

The enthalpy of formation, then, is

$$\Delta_f H^\ominus(\text{CaCl,s}) \approx [176 + 589.7 + 121.7 - 348.7 - 685]\,\text{kJ mol}^{-1} = \boxed{-146\,\text{kJ mol}^{-1}}.$$

Although formation of CaCl(s) from its elements is exothermic, formation of $CaCl_2$(s) is still more favoured energetically. Consider the disproportionation reaction 2 CaCl(s) → Ca(s) + $CaCl_2$(s) for which

$$\Delta H^\ominus = \Delta_f H^\ominus(\text{Ca, s}) + \Delta_f H^\ominus(\text{CaCl}_2\text{, s}) - 2\Delta_f H^\ominus(\text{CaCl, s})$$
$$\approx [0 - 795.8 - 2(-146)]\,\text{kJ mol}^{-1}$$
$$\approx -504\,\text{kJ mol}^{-1}$$

and the thermodynamic instability of CaCl(s) toward disproportionation to Ca(s) and $CaCl_2$(s) becomes apparent.

Note: Using the tabulated ionic radius of Ca (i.e., that of Ca^{2+}) would be less valid than using the atomic radius of a neighbouring monovalent ion, for the problem asks about a hypothetical compound of monovalent calcium. Predictions with the smaller Ca^{2+} radius (100 pm) differ from those listed above but the conclusion remains the same: the expected structure changes to rock-salt, the lattice enthalpy to 758 kJ mol^{-1}, $\Delta_f H^{\ominus}$(CaCl,s) to −219 kJ mol^{-1} and the disproportionation enthalpy to −358 kJ mol^{-1}.

38.9 The distance between adjacent ions on the two diagonal lines is d and, consequently, the distance between adjacent ions on the horizontal and vertical lines is $c = d\cos(45°) = d/2^{1/2}$. These four lines with their constant, close distance between adjacent ions determine the Madelung constant when there are very large numbers of ions in a line. The textbook figure shows that distances between ions on different lines quickly grow to a large value and, consequently, the total interaction between ions on different lines contributes very negligibly to the lattice energy per mole of ions. Thus, the Madelung constant of this ion arrangement is twice the Madelung constant for a line of alternating $+e$ cations and $-e$ anions separated by d plus twice the Madelung constant for a line of alternating $+e$ cations and $-e$ anions separated by c (see the one-dimensional model discussion in Topic 38.2(b) of the textbook).

$$E_P = 2\times(-2\ln 2)\times\left(\frac{N_A e^2}{4\pi\varepsilon_0 d}\right) + 2\times(-2\ln 2)\times\left(\frac{N_A e^2}{4\text{p}\varepsilon_0 c}\right)$$

$$= 2\times(-2\ln 2)\times\left(\frac{N_A e^2}{4\pi\varepsilon_0 d}\right) + 2\times(-2\ln 2)\times\left(\frac{2^{1/2} N_A e^2}{4\pi\varepsilon_0 d}\right)$$

$$= -2\times(2\ln 2)\times(1+2^{1/2})\times\left(\frac{N_A e^2}{4\pi\varepsilon_0 d}\right)$$

Comparing the above equation with eqn 38.5 yields the Madelung constant.

$$A = 2\times(2\ln 2)\times(1+2^{1/2}) = \boxed{6.694}$$

Topic 39 Electrical, optical, and magnetic properties of solids

Discussion questions

39.1 The **Fermi–Dirac distribution** is a version of the Boltzmann distribution that takes into account the effect of the Pauli exclusion principle. It can therefore be used to calculate the probability of occupancy of a state of energy E, $f(E)$, in a many-electron system at a temperature T:

$$f(E)=\frac{1}{e^{(E-\mu)/kT}+1} \quad [39.2a]$$

In this expression, μ is the '**chemical potential**', the energy of the level for which $f = ½$ provided that $T > 0$. For energies well above μ, the population resembles a Boltzmann distribution, decaying exponentially with increasing energy, i.e., the higher the temperature, the longer the exponential tail.

Exercises

39.1(a)

$$E_g = h\nu_{\min} = \frac{hc}{\lambda_{\max}} = \frac{(6.626\times10^{-34}\ \mathrm{J\ s})(2.998\times10^{8}\ \mathrm{m\ s^{-1}})}{350.\times10^{-9}\ \mathrm{m}}\left(\frac{1\ \mathrm{eV}}{1.602\times10^{-19}\ \mathrm{J}}\right)=\boxed{3.54\ \mathrm{eV}}$$

39.2(a) $m = g_e\{S(S+1)\}^{1/2}\mu_B$ [39.5, with S in place of s]

Therefore, because $m = 3.81\ \mu_B$ and $g_e \approx 2$,

$$S(S+1)=(\tfrac{1}{4})\times(3.81)^2=3.63$$

implying that $S = 1.47$. Because $S\approx\tfrac{3}{2}$, there must be $\boxed{\text{three unpaired spins}}$.

39.3(a) $\chi_m = \chi V_m [39.4] = \chi M/\rho = (-7.2\times10^{-7})\times(78.11\ \mathrm{g\ mol^{-1}})/(0.879\ \mathrm{g\ cm^{-3}})$

$$=\boxed{-6.4\times10^{-5}\ \mathrm{cm^3\ mol^{-1}}}=\boxed{-6.4\times10^{-11}\ \mathrm{m^3\ mol^{-1}}}$$

39.4(a) The molar susceptibility is given by

$$\chi_m = \frac{N_A g_e^2 \mu_0 \mu_B^2 S(S+1)}{3kT} \text{ [39.6]} \quad \text{so} \quad S(S+1) = \frac{3kT\chi_m}{N_A g_e^2 \mu_0 \mu_B^2}$$

$$S(S+1) = \frac{3(1.381\times10^{-23}\,\text{J K}^{-1})\times(294.53\,\text{K})\times(0.1463\times10^{-6}\,\text{m}^3\,\text{mol}^{-1})}{(6.022\times10^{23}\,\text{mol}^{-1})\times(2.0023)^2\times(4\pi\times10^{-7}\,\text{T}^2\,\text{J}^{-1}\,\text{m}^3)\times(9.27\times10^{-24}\,\text{J T}^{-1})^2}$$

$$= 6.84$$

$$S^2 + S - 6.841 = 0 \quad \text{and} \quad S = \frac{-1+\sqrt{1+4(6.841)}}{2} = 2.163$$

corresponding to $\boxed{4.326}$ effective unpaired spins. The theoretical number is $\boxed{5}$ corresponding to the $3d^5$ electronic configuration of Mn^{2+}.

Comment The discrepancy between the two values is accounted for by an antiferromagnetic interaction between the spins which alters χ_m from the form of eqn 39.6.

39.5(a) $$\chi_m = (6.3001\times10^{-6})\times\left(\frac{S(S+1)}{T/\text{K}}\,\text{m}^3\,\text{mol}^{-1}\right) \text{ [See Example 39.1]}$$

Since Cu(II) is a d^9 species, it has one unpaired spin, and so $S = s = \frac{1}{2}$. Therefore,

$$\chi_m = \frac{(6.3001\times10^{-6})\times\left(\frac{1}{2}\right)\times\left(\frac{3}{2}\right)}{298}\,\text{m}^3\,\text{mol}^{-1} = \boxed{+1.6\times10^{-8}\,\text{m}^3\,\text{mol}^{-1}}$$

Problems

39.1 The Fermi-Dirac distribution, *f*, can be transformed to the variables of text Figure 39.3 as follows.

$$\frac{E-\mu}{kT} = \left(\frac{\mu}{kT}\right)\times\left(\frac{E-\mu}{\mu}\right) \equiv z\times x$$

$$f(x,z) = \frac{1}{e^{z\times x}+1} \text{ [39.2a]} \quad \text{where} \quad x \equiv \frac{E-\mu}{\mu} \quad \text{and} \quad z \equiv \frac{\mu}{kT}$$

Text Figure 39.3 is replicated in the following Mathcad Prime 2 worksheet. The effect of changing values of z, the ratio of μ to kT, is readily explored by simply altering z values in the plot. Values $z = 0.10$, 1, and 10 are shown in the worksheet. It is seen that as $T\to\infty$, $z\to0$, and $f\to½$; that is all available energy states have an occupational probability of ½ (a state is filled with at most two electrons by the Pauli exclusion Principle). As $T\to0$, $z\to\infty$, and f becomes a step distribution for which $f = 1$ for $x < 0$ and $f = 0$ for $x > 0$ (see Brief illustration 39.1 of the text for the significance of this distribution).

Definitions for plot of Fermi-Dirac distribution, f: $= \dfrac{E-\mu}{\mu}$ $\quad z = \dfrac{\mu}{k \cdot T}$

$$f(x,z) := \frac{1}{e^{z \times x}+1} \qquad i := 0..300 \qquad x_i := -5.9999 + \frac{11.9999}{300} \cdot i$$

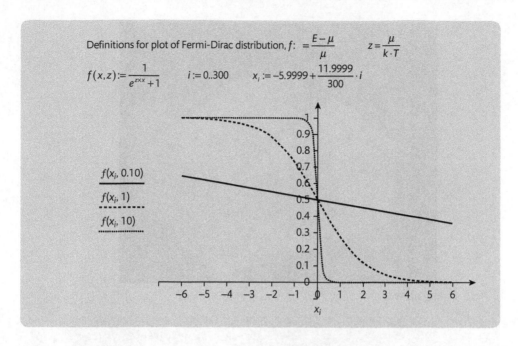

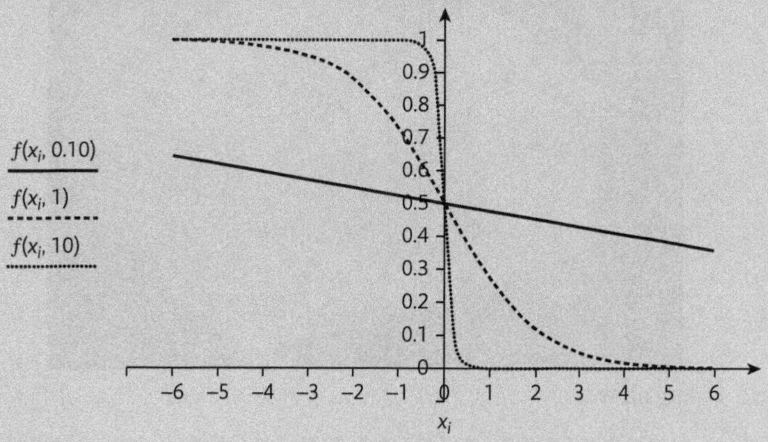

39.3 $$N_e = \int_{E=0}^{E=\infty} \mathrm{d}N(E) = \int_0^\infty \rho(E) \times f(E)\mathrm{d}E = \int_0^\infty \frac{\rho(E)}{e^{(E-\mu)/kT}+1} \mathrm{d}E \text{ [39.1 and 39.2a]}$$

In order for N to remain a constant of this equation as the temperature is raised, the exponential term $\exp\{(E-\mu)/kT\}$ must remain constant. It is apparent that, when $\exp\{E/kT\}$ gets smaller as T grows larger for each value of E, it must be multiplied by a larger value of $\exp\{-\mu/kT\}$, so $\mu(T)$ must decrease.

39.5 Tans and coworkers (S.J. Tans et al., *Nature*, **393**, 49 (1998)) have draped a semiconducting carbon nanotube (CNT) over metal (gold in Fig. 39.1) electrodes that are 400 nm apart atop a silicon surface coated with silicon dioxide. A bias voltage between the electrodes provides the source and drain of the molecular field-effect transistor(FET). The silicon serves as a gate electrode and the thin silicon oxide layer (at least 100 nm thick) insulates the gate from the CNT circuit. By adjusting the magnitude of an electric field applied to the gate, current flow across the CNT may be turned on and off.

Wind and coworkers (S.J. Wind et al., *Applied Physics Letters*, **80**(20, May 20), 3 817 (2002)) have designed (Fig. 39.2) a CNTFET of improved current carrying capability. The gate electrode is above the conduction channel and separated from the channel by a thin oxide dielectric. In this manner the CNT-to-air contact is eliminated, an arrangement that prevents the circuit from acting like a p-type transistor. This arrangement also reduces the gate oxide thickness to about 15 nm, allowing for much smaller gate voltages and a steeper subthreshold slope, which is a measure of how well a transistor turns on or off.

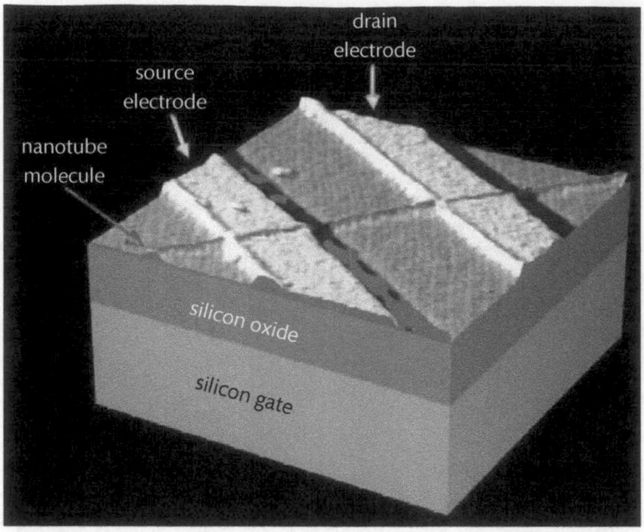

Figure 39.1

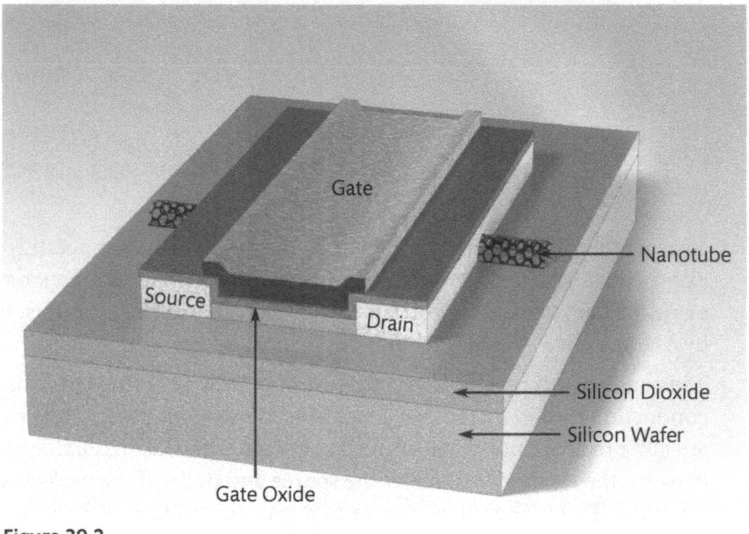

Figure 39.2

A single-electron transistor (SET) has been prepared by Cees Dekker and coworkers (*Science*, **293**, 76, (2001)) with a CNT. The SET is prepared by putting two bends in a CNT with the tip of an AFM (Fig. 39.3). Bending causes two buckles that, at a distance of 20 nm, serves as a conductance barrier. When an appropriate voltage is applied to the gate below the barrier, electrons tunnel one at a time across the barrier.

Weitz et al. (Phys. Stat. Sol. (b) **243**, 13, 3 394 (2006)) report on the construction of a single-wall CNT using a silane-based organic self-assembled monolayer (SAM) as a gate dielectric on top of a highly doped silicon wafer. The organic SAM is made of 18-phenoxyoctadecyltrichlorosilane. This ultrathin layer (Fig. 39.4)

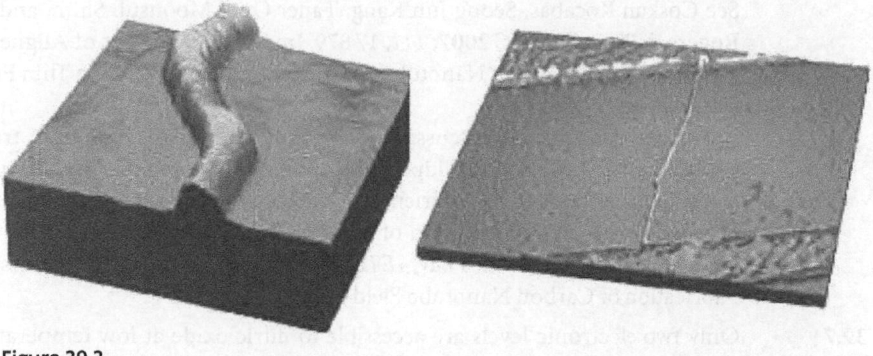

Figure 39.3

ensures strong gate coupling and therefore low operation voltages. Single-electron transistors (SETs) were obtained from individual metallic SWCNTs. Field-effect transistors made from individual semiconducting SWCNTs operate with gate-source voltages of −2 V, show good saturation, small hysteresis (200 mV) as well as a low subthreshold swing (290 mV/dec).

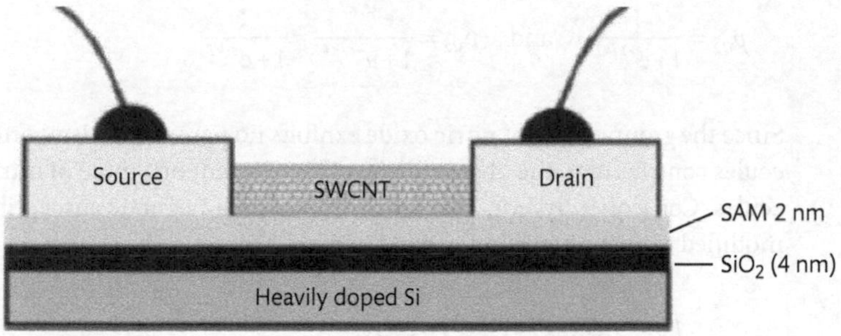

Figure 39.4

John Rodgers and researchers at the University of Illinois have reported a technique for producing near perfect alignment of CNT transistors (Fig. 39.5). The array is prepared by patterning thin strips of an iron catalyst on quartz crystals and then growing nanometer-wide CNTs along those strips using conventional carbon vapor deposition. The quartz crystal aligns the nanotubes. Transistor development then includes depositing source, drain, and gate electrodes using conventional photolithography. Transistors made with about 2,000 nanotubes can carry currents of one ampere. The research group also developed a technique for transferring the nanotube arrays onto any substrate, including silicon, plastic, and glass.

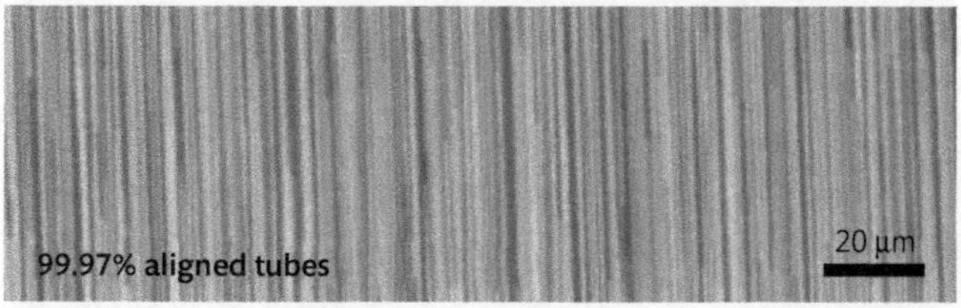

Figure 39.5

See Coskun Kocabas, Seong Jun Kang, Taner Ozel, Moonsub Shim, and John A. Rogers, *J. Phys. Chem. C* **2007,** *111,* 17 879, Improved Synthesis of Aligned Arrays of Single-Walled Carbon Nanotubes and Their Implementation in Thin Film Type Transistors.

Further background discussion of carbon nanotube field-effect transistors (CNTFET) can be found at wikipedia.org. For a review of the CNT catalytic growth technique, methods to grow oriented long CNTs with controlled diameters, and process steps for the fabrication of both back and top-grated CNTFET see K.C. Narasimhamurthy and R. Paily, *IETE Technical Review*, 2011, V 28, Issue 1, 57, Fabrication of Carbon Nanotube Field Effect Transistor.

39.7 Only two electronic levels are accessible to nitric oxide at low temperature. The ground state is a doubly degenerate $^2\Pi_{1/2}$ state while the excited state is a doubly degenerate $^2\Pi_{3/2}$ state that is 121.1 cm^{-1} above the ground state. These states originate from spin-orbital coupling of angular momentum. Let $\varepsilon = hc\tilde{\nu}$ be the energy separation between these levels, then the probabilities that a molecule is in one ($p_{1/2}$) or the other level ($p_{3/2}$) are given by the following equations, which are derived from the Boltzmann distribution in the note below.

$$p_{1/2} = \frac{1}{1+e^{-\varepsilon/kT}} \quad \text{and} \quad p_{3/2} = \frac{e^{-\varepsilon/kT}}{1+e^{-\varepsilon/kT}} = \frac{1}{1+e^{\varepsilon/kT}}$$

Since the ground state of nitric oxide exhibits no paramagnetism, only $p_{3/2}N_A$ molecules contribute to the observed magnetic moment of a mole of nitric oxide molecules. Consequently, eqn 39.6 for the molar paramagnetic susceptibility must be modified with the inclusion of a factor $p_{3/2}$.

$$\chi_m = \frac{p_{3/2}N_A g_e^2 \mu_0 \mu_B^2 S(S+1)}{3kT} \quad [39.6]$$

Substitution of $S(S+1) = (m/g_e\mu_B)^2$ [39.5] where m is the magnetic moment into the above expression gives

$$\chi_m = \frac{p_{3/2}N_A \mu_0 \mu_B^2 (m/\mu_B)^2}{3kT}$$

$$= \frac{N_A \mu_0 \mu_B^2 (m/\mu_B)^2}{3kT \times (1+e^{\varepsilon/kT})} \quad \text{where} \quad \varepsilon/k = hc\tilde{\nu}/k = hc \times (121.1 \text{ cm}^{-1})/k = 174.2 \text{ K}$$

Thus, with $m/\mu_B = 2$

$$\chi_m = \frac{6.286 \times 10^{-6} \text{ m}^3 \text{ mol}^{-1}}{(T/\text{K}) \times (1+e^{174.2/(T/\text{K})})}$$

This relation gives the molar paramagnetic susceptibility of NO as a function of temperature. For example, χ_m at 90 K is

$$\chi_m = \frac{6.286\times10^{-6}\ \text{m}^3\ \text{mol}^{-1}}{(90)\times\left(1+e^{174.2/(90)}\right)} = \boxed{8.81\times10^{-9}\ \text{m}^3\ \text{mol}^{-1}}$$

The mass paramagnetic susceptibility is

$$\chi_{\text{mass}} = \chi_m/M = \left(8.81\times10^{-9}\ \text{m}^3\ \text{mol}^{-1}\right)/\left(0.03001\ \text{kg mol}^{-1}\right)$$
$$= 2.94\times10^{-7}\ \text{m}^3\ \text{kg}^{-1}$$

Wishing to compare this with the value found in the older literature, we must convert the SI unit of susceptibility to the cgs (or emu) unit by dividing the SI unit by 4π, converting the m^3 to cm^3, and converting kg to g.

$$\chi_{\text{mass}}\ \text{in cgs} = \left(2.94\times10^{-4}\ \text{cm}^3\ \text{g}^{-1}\right)/4\pi = 23.4\times10^{-6}\ \text{cm}^3\ \text{g}^{-1}$$

This is in reasonable agreement with the accepted value of 19.8×10^{-6} cgs for the mass susceptibility of NO(s) at 90 K. Fig. 39.6 is a plot of the molar paramagnetic susceptibility, as modeled in this problem, against temperature below the normal fusion point (110 K) of nitric oxide. The curve is remarkably different than the $\chi_m(T)$ behavior of most paramagnetic substances. Paramagnetism is normally a property of the ground electronic state and, consequently, there is an inverse relation between χ_m and T [39.6] so that χ_m decreases with increasing T. Effective angular momentums of individual molecules align in a magnetic field at low temperature and become disoriented by thermal agitation as the temperature is increased. In the case of NO(s) it is the excited state that is paramagnetic so, when all molecules are in the ground state at absolute zero, $\chi_m = 0$. As T is increased from absolute zero, molecules are thermally promoted to the excited state and the observed paramagnetism increases as shown in Fig. 39.6.

Comment. The explanation of the magnetic properties of NO is more complicated and subtle than indicated by the solution here. In fact the full solution for this case was one of the important triumphs of the quantum theory of magnetism which was developed about 1930. See J. H. van Vleck, *The theory of electric and magnetic susceptibilities.* Oxford University Press (1932).

Note. The Boltzmann distribution indicates that the probability that a molecule is in the ground state energy level is given by $p_0 \propto g_0$ where g_0 is the degeneracy of the ground state while the probability that the molecule is in energy level "1" that is ε above the ground state is given by $p_1 \propto g_1 e^{-\varepsilon/kT}$. For a two-level system the constant of proportionality is provided by the normalization condition that $p_0 + p_1 = 1$. Thus, the constant of proportionality is $1/(g_0 + g_1 e^{-\varepsilon/kT})$ and the probabilities are

$$p_0 = g_0/(g_0 + g_1\,e^{-\varepsilon/kT}) \quad \text{and} \quad p_1 = g_1 e^{-\varepsilon/kT}/(g_0 + g_1 e^{-\varepsilon/kT})$$

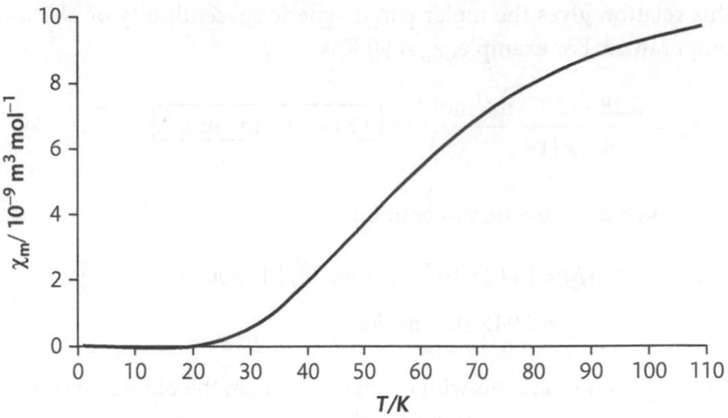

Figure 39.6

In the special case for which $g_0 = g_1$ the probabilities simplify to those given at the top.

Focus 8: Integrated activities

F8.1 Figure F8.1(a) shows a dark univalent probe cation in a vacancy within a two-dimensional square ionic lattice of grey univalent cations and white univalent anions. Let $d_0 = 200$ pm be the distance between nearest neighbors and let V_0 be the absolute value of the Coulombic interaction between nearest neighbors.

$$V_0 = \frac{e^2}{4\pi\varepsilon_0 d_0} = \frac{\left(1.602\times10^{-19}\ \text{C}\right)^2}{\left(1.113\times10^{-10}\ \text{J}^{-1}\ \text{C}^2\ \text{m}^{-1}\right)\times\left(200\times10^{-12}\ \text{m}\right)} = 1.153\times10^{-18}\ \text{J}$$

The symmetry of the lattice around the probe cation consists of four regions like that of Figure F8.1(b) so we calculate the total Coulombic interaction of the probe within the lattice quadrant of Figure F8.1(b) and multiply by 4. The calculation is pursued one column at a time and the column interactions are summed.

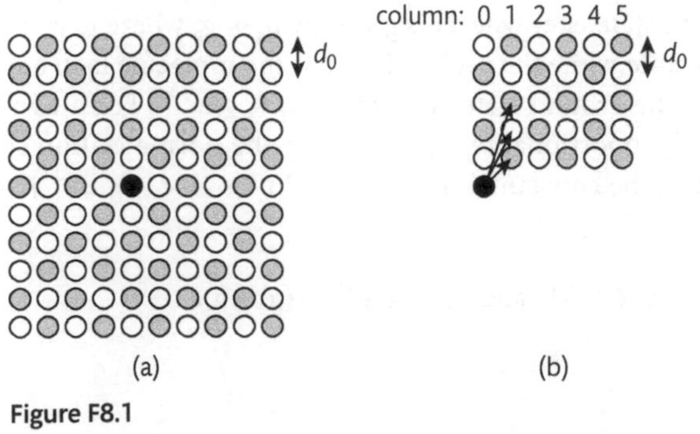

Figure F8.1

Probe-to-column 0 interaction:

$$V_{\text{column }0} = -V_0 \times \left(1 - \frac{1}{2} + \frac{1}{3} - \frac{1}{4} + \frac{1}{5} \cdots\right) = -V_0 \ln 2 = V_0 \sum_{n=1}^{\infty} \frac{(-1)^{n+0}}{\left(n^2 + 0^2\right)^{1/2}} \text{ (useful form)}$$

Probe-to-column 1 interaction using the Pythagorean theorem for the probe-ion distance:

$$V_{\text{column }1} = V_0 \times \left(\frac{1}{2^{1/2}} - \frac{1}{5^{1/2}} + \frac{1}{10^{1/2}} - \frac{1}{17^{1/2}} \cdots\right) = V_0 \sum_{n=1}^{\infty} \frac{(-1)^{n+1}}{\left(n^2 + 1^2\right)^{1/2}}$$

Similarly, the probe-to-column m interaction, using the Pythagorean theorem for the probe-ion distance, is

$$V_{\text{column }m} = V_0 \sum_{n=1}^{\infty} \frac{(-1)^{n+m}}{\left(n^2 + m^2\right)^{1/2}}$$

for $0 \le m < \infty$ (the sum is performed with a calculator or software)

The total interaction for the region shown in Fig. 19.15(b) is the sum of the above expression over all columns

$$\begin{aligned} V_{\text{Fig 19.15(b)}} &= V_0 \sum_{m=0}^{\infty} \sum_{n=1}^{\infty} \frac{(-1)^{n+m}}{\left(n^2 + m^2\right)^{1/2}} \\ &= -0.4038\, V_0 = (-0.4038) \times \left(1.153 \times 10^{-18}\text{ J}\right) \\ &= -4.656 \times 10^{-19}\text{ J} \end{aligned}$$

The total Coulombic interaction of the probe cation with the lattice is

$$V_{\text{total}} = 4V_{\text{Fig F8.1(b)}} = 4\left(-4.656 \times 10^{-19}\text{ J}\right) = \boxed{-1.862 \times 10^{-18}\text{ J}}$$

where the negative value indicates a net attraction.

Suppose that you interpret the problem to involve placement of the probe cation at the foot of the two-dimensional step shown in Fig. F8.2, you calculate the probe potential with

$$\begin{aligned} V_{\text{total}} &= 3V_{\text{Fig F8.1(b)}} - V_{\text{column }0} \\ &= 3 \times (-0.4038\, V_0) - (-V_0 \ln 2) \\ &= -0.5183\, V_0 \\ &= (-0.5183) \times \left(1.153 \times 10^{-18}\text{ J}\right) = \boxed{-5.975 \times 10^{-19}\text{ J}} \end{aligned}$$

Once again the negative value indicates a net attraction.

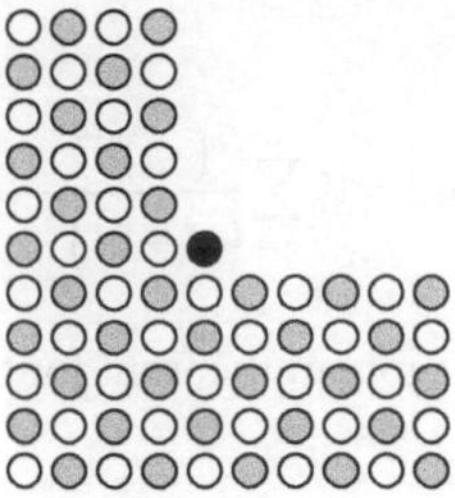

Figure F8.2

F8.3

(a) The classical perturbation energy for interaction between the vector dipole $\vec{\mu}$ and an applied vector electric field $\vec{\mathcal{E}}$ is given by the dot (scalar) product expression:

$$H_{\text{perturbation}} = -\vec{\mu}\cdot\vec{\mathcal{E}} = -\left\{\mu_x\mathcal{E}_x + \mu_y\mathcal{E}_y + \mu_z\mathcal{E}_z\right\} \quad \text{[extension of Justification 34.1]}$$

The dipole moment components of a one-dimensional (x) oscillator are (μ_x,0,0). For a perpendicular field the components are something like (0, $\mathcal{E}_y$,0). The dot product of these vectors is clearly zero meaning that a perpendicular field cannot perturb the one-dimensional oscillator. Eqn 34.5a, $\vec{\mu}^* = \alpha\vec{\mathcal{E}}$, addresses the question as to whether or not the applied field can induce a dipole. The induced response is in the direction of the applied field and the response is the same irrespective of the direction of the field. The constant of proportionality can be called the isotropic polarizability. In this problem, however, we ask whether $\mathcal{E}y$ can illicit the response μ_x^*. An affirmative answer requires an anisotropic material and polarizability, label it α_{xy}, for which $\mu_x^* = \alpha_{xy}\mathcal{E}_y$. For a one-dimensional harmonic oscillator, $\boxed{\alpha_{xy}\text{ must equal zero}}$ because there is no possible coupling of electron motion in the perpendicular directions.

(b) Eqn 34.7 is used to calculate the polarizability when the applied field is parallel to the one-dimensional oscillator, which is chosen to be the x direction in part (a).

$$\alpha = 2\sum_{n\neq 0}\frac{|\mu_{x,0n}|^2}{E_n^{(0)} - E_0^{(0)}} \sim 2\frac{|\mu_{x,01}|^2}{E_1^{(0)} - E_0^{(0)}}$$

where $\mu_{x,01} = e\left(\dfrac{\hbar}{2m\omega}\right)^{1/2}$ [given] and $E_1^{(0)} - E_0^{(0)} = \hbar\omega$ [12.5, harmonic oscillator].

Thus,

$$\alpha \sim \frac{2}{\hbar\omega}\left\{e\left(\frac{\hbar}{2m\omega}\right)^{1/2}\right\}^2 = \frac{e^2}{m\omega^2} = \frac{e^2}{m(k_{\mathrm{f}}/m)}\ [12.4] = \boxed{\frac{e^2}{k}}$$

The mass independence of the polarizability arises from the fact that the static (zero-frequency) polarizability is a response to the stationary electric field and does not depend on the inertial properties of the oscillator (the rate at which it responds to a changing force).

F8.5 Consider a single molecule surrounded by $N-1(\approx N)$ others in a container of volume V. The number of molecules in a spherical shell of thickness dR is $4\pi R^2\times\frac{N}{V}\mathrm{d}R$. Molecules cannot approach more closely than the molecular diameter d so $R\geq d$ and, therefore, the pair interaction energy is

$$u=\int_d^\infty 4\pi R^2\times\left(\frac{N}{V}\right)\times\left(\frac{-C_6}{R^6}\right)\mathrm{d}R=\frac{-4\pi NC_6}{V}\int_d^\infty\frac{\mathrm{d}R}{R^4}=\left(\frac{4\pi NC_6}{3V}\right)\times\left(\frac{1}{\infty^3}-\frac{1}{d^3}\right)$$

$$=\frac{-4\pi NC_6}{3Vd^3}$$

The mutual pairwise interaction energy of all N molecules is $U=\frac{1}{2}Nu$ (the $\frac{1}{2}$ appears because each pair must be counted only once, i.e. A with B but not A with B and B with A). Therefore,

$$U=\boxed{\frac{-2\pi N^2C_6}{3Vd^3}}$$

For a van der Waals gas, $\frac{n^2a}{V^2}=\left(\frac{\partial U}{\partial V}\right)_T=\frac{2\pi N^2C_6}{3V^2d^3}$ and therefore $a=\boxed{\frac{2\pi N_A^2C_6}{3d^3}}$ $[N=nN_A]$

F8.7 $\theta_{111}(100\,\mathrm{K})=22.0403°,\ \ \theta_{111}(300\,\mathrm{K})=21.9664°$

$\sin\theta_{111}(100\,\mathrm{K})=0.37526,\ \ \sin\theta_{111}(300\,\mathrm{K})=0.37406$

$$\frac{\sin\theta(300\,\mathrm{K})}{\sin\theta(100\,\mathrm{K})}=0.99681=\frac{a(100\,\mathrm{K})}{a(300\,\mathrm{K})}\quad\text{[37.2b, Bragg's law; see Problem 37.13]}$$

$$a(300\,\mathrm{K})=\frac{\lambda\sqrt{3}}{2\sin\theta_{111}}=\frac{(154.0562\,\mathrm{pm})\times\sqrt{3}}{(2)\times(0.37406)}=356.67\,\mathrm{pm}$$

$$a(100\,\mathrm{K})=(0.99681)\times(356.67\,\mathrm{pm})=355.53\,\mathrm{pm}$$

$$\frac{\delta a}{a}=\frac{356.67-355.53}{355.53}=3.206\times10^{-3}$$

$$\frac{\delta V}{V}=\frac{356.67^3-355.53^3}{355.53^3}=9.650\times10^{-3}$$

$$\alpha_{\text{volume}} = \frac{1}{V}\frac{\delta V}{\delta T} = \frac{9.560\times10^{-3}}{200\,\text{K}} = \boxed{4.8\times10^{-5}\,\text{K}^{-1}}$$

$$\alpha_{\text{linear}} = \frac{1}{a}\frac{\delta a}{\delta T} = \frac{3.206\times10^{-3}}{200\,\text{K}} = \boxed{1.6\times10^{-5}\,\text{K}^{-1}}$$

F8.9 1s hydrogenic radial wavefunction: $R_{1s}(\rho) = 2(Z/a_o)\mathrm{e}^{-\rho/2}$ where $\rho = (2Z/a_o)r$

We start by constructing a Gaussian function, $G_{1s}(\rho) = N\mathrm{e}^{-b\rho^2}$, that 'fits' or 'matches' $R_{1s}(\rho)$. Finding the normalization constant N requires evaluation of the normalization integral. This begins with the transformation $r \to \rho$.

$$\int_0^\infty G_{1s}(\rho)^2\mathrm{d}\tau = N^2\int_0^\infty \mathrm{e}^{-2b\rho^2}r^2\mathrm{d}r\ [18.5\text{b}] = N^2(a_o/2Z)^3\int_0^\infty \mathrm{e}^{-2b\rho^2}\rho^2\mathrm{d}\rho = 1$$

$$N = \left(\frac{(2Z/a_o)^3}{\int_0^\infty \mathrm{e}^{-2b\rho^2}\rho^2\mathrm{d}\rho}\right)^{1/2}$$

The integral in the denominator of the above expression is found in any math handbook. Here is the integration in the Mathcad Prime 2 worksheet used throughout this problem.

$$N(Z,b,a_o) := \left(\frac{2\cdot Z}{a_o}\right)^{\frac{3}{2}}\cdot\left(\frac{1}{\int_0^\infty \left(e^{-b\cdot\rho^2}\right)^2\cdot\rho^2\,d\rho}\right)^{\frac{1}{2}} \rightarrow \left(\frac{2\cdot Z}{a_o}\right)^{\frac{3}{2}}\cdot\sqrt{\frac{8\cdot\sqrt{2}\cdot b^{\frac{3}{2}}}{\sqrt{\pi}}}$$

To find the constant b, we calculate R_{1s} at 200 points in the range $0 \le r \le 3a_o$, setup the worksheet to calculate G_{1s} at each of these points, subtract the two, square the difference, and sum the result over all points. This gives the sum of squares (SS) for which we wish to vary b so as to minimize the sum, the total difference error (E). We call the sum *SSE* in the worksheet and Mathcad performs the minimization process with the 'minerr()' function within a solve block.

$$pm := 10^{-12}\cdot m \qquad a_o := 52.9177\cdot pm \qquad \rho(r,Z) := \frac{2\cdot Z\cdot r}{a_o}$$

$$R_{1s}(r,Z) := 2\cdot\left(\frac{Z}{a_o}\right)^{\frac{3}{2}}\cdot e^{-2^{-1}\cdot\rho(r,Z)} \qquad G_{1s}(r,Z,b) := N(Z,b,a_o)\cdot e^{-b\cdot\rho(r,Z)^2}$$

$$M := 200 \qquad i := 0..M \qquad r_{max} := 3\cdot a_o \qquad r_i := \frac{i}{M}\cdot r_{max}$$

$$SSE(Z,b) := \sum_i\left(R_{1s}(r_i,Z) - G_{1s}(r_i,Z,b)\right)^2$$

$Z := 1$

Guess Values

$b := 0.1$

Constraints

$SSE(Z, b) = 0$

Solver

$b := \text{minerr}(b)$

$b = 0.12795239$

Computed b values:

$Z = 1, b = 0.127952$
$Z = 2, b = 0.128470$
$Z = 3, b = 0.129258$

The following worksheet plot compares G_{1s} with R_{1s} over the range of r for $Z = 1$. The Gaussian wavefunction reduces the electron density at the nucleus and removes the sharp peak. The Gaussian function effectively replaces the cusp of the exponential decay near $r = 0$ with a broader, flatter function while simultaneously decreasing the long tail of the exponential. Physically, this is analogous to contraction of the orbital by increasing the nuclear charge Z. At intermediate r the Gaussian increases electron density but at large r the Gaussian approaches zero more rapidly. This comparison also applies for larger values of Z.

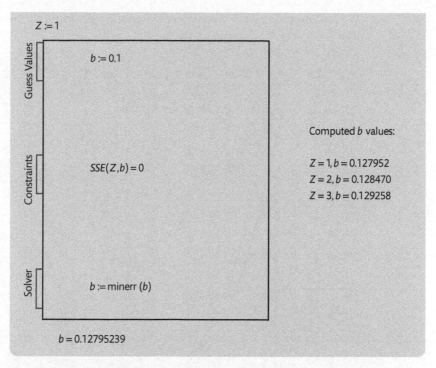

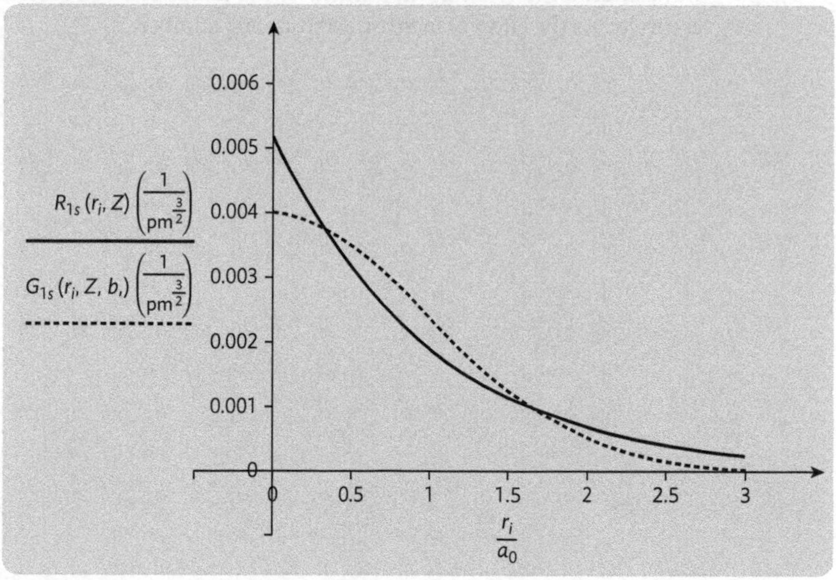

We can now proceed to a comparison of scattering factors for the two wavefunction over a range of $\xi \equiv \sin(\theta)/\lambda$ values.

$$f_{1s}(\xi, Z) := \frac{1}{4\cdot\pi\cdot\xi}\cdot\int_{0\text{pm}}^{100\cdot a_0} R_{1s}(r,Z)^2\cdot\sin(4\cdot\pi\cdot\xi\cdot r)\cdot r\,dr \qquad \text{where} \qquad \xi = \frac{\sin(\theta)}{\lambda}$$

$$f_{G1s}(\xi, Z) := \frac{1}{4\cdot\pi\cdot\xi}\cdot\int_{0\text{pm}}^{100\cdot a_0} G_{1s}(r,Z,b)^2\cdot\sin(4\cdot\pi\cdot\xi\cdot r)\cdot r\,dr$$

$$N := 200 \qquad \xi_{max} := 0.01\cdot\text{pm}^{-1} \qquad i := 0..N \qquad \xi_i := \frac{i}{N}\cdot\xi_{max}$$

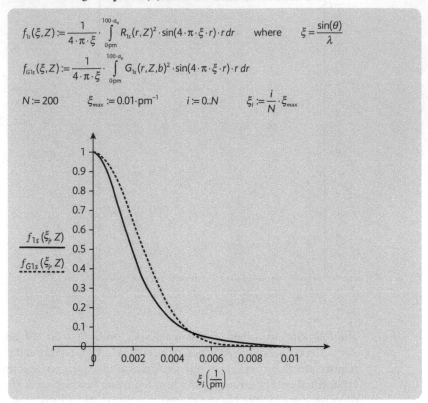

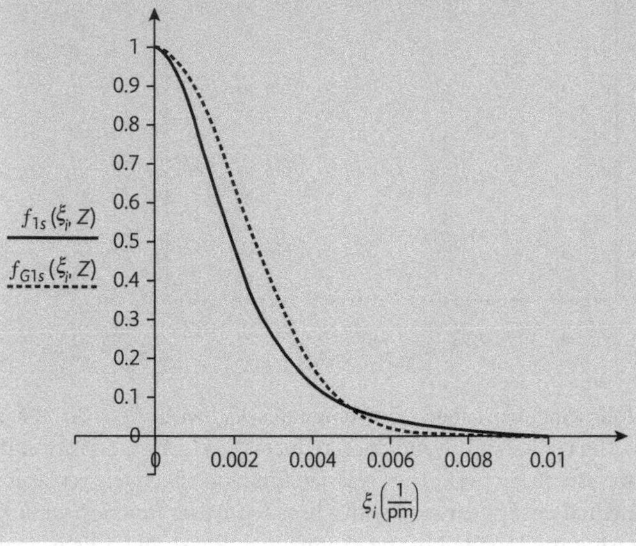

The above plot clearly shows that the replacement of the hydrogenic 1s wavefunction with a Gaussian wavefunction shifts the reflection angle away from the forward direction ($\theta = 0$). In the context of the discussion of Problem F8.8 this is equivalent to the effect of increasing the atomic number, Z.

Topic 40 **General features**

Discussion questions

40.1 Textbook schematics of absorption, emission, and Raman spectrometers are found in Figures 40.1, 40.12, and 40.13, respectively. A fundamental difference in the experimental arrangements is in the relative direction of the radiation exiting from the sample with respect to the incident radiation upon the sample. In absorption spectroscopy the angle between the two is 180°; that is, they are in line with each other and the analysed radiation is said to be **transmitted**. Identical beams incident upon the sample and reference cells are compared to eliminate small absorptions within the sample spectrum from components present in both cells, a solvent for example. In emission spectrometry the angle is 90°, which reduces the problem of distinguishing the relatively low intensity emission from the extremely intense incident radiation. Extremely low intensity Raman scattered radiation is collected with a curved mirror from angles in back-scattered directions and directed to the detector. This arrangement reduces problems caused by the high intensity incident radiation and increases the intensity of interest at the detector.

Exercises

40.1(a) The reduction in intensity obeys the Beer–Lambert law introduced in Section 11.2.

$$\log\frac{I}{I_0} = -\log\frac{I_0}{I} = -\varepsilon[\mathrm{J}]L \quad [40.8 \text{ and } 40.9]$$

$$= (-723\ \mathrm{dm^3\ mol^{-1}\ cm^{-1}})\times(4.25\times10^{-3}\ \mathrm{mol\ dm^{-3}})\times(0.250\ \mathrm{cm})$$

$$= -0.768$$

Hence, $\frac{I}{I_0} = 10^{-0.768} = 0.171$, and the reduction in intensity is $\boxed{82.9 \text{ per cent}}$.

40.2(a) $$\log\frac{I}{I_0} = -\log\frac{I_0}{I} = -\varepsilon[\mathrm{J}]L \quad [40.8 \text{ and } 40.9]$$

Hence, $\varepsilon = -\frac{1}{[\mathrm{J}]L}\log\frac{I}{I_0} = -\frac{\log(0.181)}{(1.39\times10^{-4}\ \mathrm{mol\ dm^{-3}})\times(1.00\ \mathrm{cm})}$

$$= \boxed{5.34\times10^3\ \mathrm{dm^3\ mol^{-1}\ cm^{-1}}}$$

40.3(a) $\log T = -A = -\varepsilon[\mathrm{J}]L$ [40.7–40.9]

$$[\mathrm{J}] = -\frac{1}{\varepsilon L}\log T = \frac{-\log(1-0.385)}{(386\ \mathrm{dm^3\ mol^{-1}\ cm^{-1}})\times(0.500\ \mathrm{cm})}$$

$$= \boxed{1.09\ \mathrm{mmol\ dm^{-3}}}$$

40.4(a) $$\mathcal{A} = \int_{\text{band}} \varepsilon(\tilde{\nu})\,\mathrm{d}\tilde{\nu}\ \ [40.10] = \int_{\tilde{\nu}_i}^{\tilde{\nu}_f} \varepsilon(\tilde{\nu})\,\mathrm{d}\tilde{\nu}$$

Since $\tilde{\nu} = \lambda^{-1}$ and $\tilde{\nu}/\mathrm{cm^{-1}} = 10^7/(\lambda/\mathrm{nm})$, the initial, peak, and final wavenumbers of the absorption band are:

$$\tilde{\nu}_i/\mathrm{cm^{-1}} = 10^7/(300) = 3.3\times10^4,\quad \tilde{\nu}_{\text{peak}}/\mathrm{cm^{-1}} = 10^7/(270) = 3.7\times10^4,$$
$$\text{and } \tilde{\nu}_f/\mathrm{cm^{-1}} = 10^7/(220) = 4.5\times10^4.$$

The positions of the wavenumber end points and peak (max) of the band are schematically presented in Fig. 40.1. It is apparent that, because of the relative position of the peak, the molar absorption coefficient is not symmetrically distributed around the peak wavenumber. The distribution is skewed toward higher wavenumbers. However, a reasonable estimate of the area under the curve may be approximated by adding the areas of triangle 1 and triangle 2 shown as dashed lines in the figure.

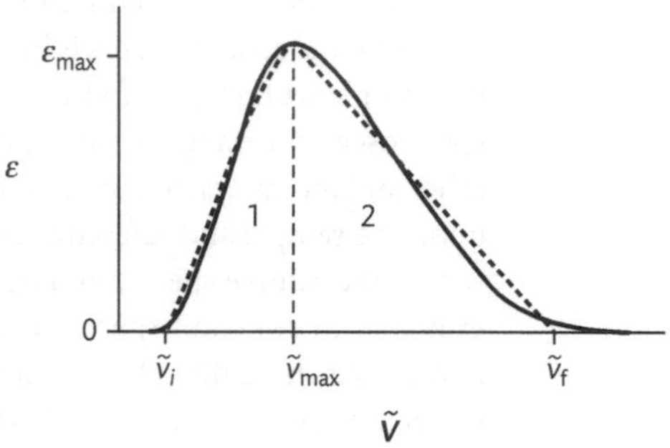

Figure 40.1

$$\mathcal{A} = \int_{\tilde{\nu}_i}^{\tilde{\nu}_f} \varepsilon(\tilde{\nu})\,\mathrm{d}\tilde{\nu} = \text{area 1} + \text{area 2}$$

$$= \frac{1}{2}(\tilde{\nu}_{\text{peak}} - \tilde{\nu}_i)\varepsilon_{\text{peak}} + \frac{1}{2}(\tilde{\nu}_f - \tilde{\nu}_{\text{peak}})\varepsilon_{\text{peak}} = \frac{1}{2}(\tilde{\nu}_f - \tilde{\nu}_i)\varepsilon_{\text{peak}}$$

$$= \frac{1}{2}\times(1.2\times10^4\ \mathrm{cm^{-1}})\times2.21\times10^4\ \mathrm{dm^3\ mol^{-1}\ cm^{-1}}$$

$$= \boxed{1.3\times10^8\ \mathrm{dm^3\ mol^{-1}\ cm^{-2}}}$$

40.5(a) $$\varepsilon = -\frac{1}{[\mathrm{J}]L}\log\frac{I}{I_0}\ \ [40.8, 40.9]\quad \text{with } L = 0.20\ \mathrm{cm}$$

We use this formula to draw up the following table.

$[\mathrm{Br_2}]/\mathrm{mol\ dm^{-3}}$	0.0010	0.0050	0.0100	0.0500	
I/I_0	0.814	0.356	0.127	3.0×10^{-5}	
$\varepsilon/(\mathrm{dm^3\ mol^{-1}\ cm^{-1}})$	447	449	448	452	mean: $44\bar{9}$

Hence, the molar absorption coefficient is $\varepsilon = \boxed{450\ \mathrm{dm^3\ mol^{-1}\ cm^{-2}}}$.

40.6(a)

$$\varepsilon = -\frac{1}{[\mathrm{J}]L}\log\frac{I}{I_0}\ [40.8, 40.9] = \frac{-1}{(0.010\ \mathrm{mol\ dm^{-3}})\times(0.20\ \mathrm{cm})}\log(0.48)$$

$$= \boxed{15\bar{9}\ \mathrm{dm^3\ mol^{-1}\ cm^{-1}}}$$

$$T = \frac{I}{I_0} = 10^{-[\mathrm{J}]\varepsilon L}\ [40.7]$$

$$= 10^{(-0.010\ \mathrm{mol\ dm^{-3}})\times(15\bar{9}\ \mathrm{dm^3\ mol^{-1}cm^{-1}})\times(0.40\ \mathrm{cm})} = 10^{-0.63\bar{6}} = 0.23,\ \text{or}\ \boxed{23\ \text{per cent}}$$

40.7(a) The Beer–Lambert law [40.7–40.9] is

$$\log\frac{I}{I_0} = -\varepsilon[\mathrm{J}]L \quad \text{so} \quad L = -\frac{1}{\varepsilon[\mathrm{J}]}\log\frac{I}{I_0}$$

For water, $[\mathrm{H_2O}] \approx \dfrac{1.00\ \mathrm{kg/dm^3}}{18.02\ \mathrm{g\ mol^{-1}}} = 55.5\ \mathrm{mol\ dm^{-3}}$

and $\varepsilon[\mathrm{J}] = (55.5\ \mathrm{mol\ dm^{-3}})\times(6.2\times10^{-5}\ \mathrm{dm^3\ mol^{-1}\ cm^{-1}}) = 3.4\times10^{-3}\ \mathrm{cm^{-1}}$,

so $\dfrac{1}{\varepsilon[\mathrm{J}]} = 2.9\ \mathrm{m}$

Hence, $L/\mathrm{m} = -2.9\times\log\dfrac{I}{I_0}$

(a) $\dfrac{I}{I_0} = 0.50,\ L = -2.9\ \mathrm{m}\times\log(0.50) = \boxed{0.87\ \mathrm{m}}$

(b) $\dfrac{I}{I_0} = 0.1,\ L = -2.9\ \mathrm{m}\times\log(0.10) = \boxed{2.9\ \mathrm{m}}$

40.8(a)

$$\nu_{\text{approach}} = \left(\frac{1+s/c}{1-s/c}\right)^{1/2}\nu\ [40.11]$$

or $\lambda_{\text{approach}} = \left(\dfrac{1-s/c}{1+s/c}\right)^{1/2}\lambda$ where $c = 2.9979\times10^8\ \mathrm{m\ s^{-1}} = 1.0793\times10^9\ \mathrm{km\ h^{-1}}$

$$= \left(\frac{1-60/1.0793\times10^9}{1+60/1.0793\times10^9}\right)^{1/2}\lambda$$

$$= \boxed{0.9999\times\lambda}$$

The coefficient in the above expression tells us that at such a slow speed there is an insignificant difference between the Doppler-shifted wavelength and the wavelength of the traffic light.

40.9(a) $\delta E \approx \hbar/\tau$ (eqn 16.9) so, since $E = h\nu$, $\delta\nu = (2\pi\tau)^{-1}$. Solving for τ:

$$\tau = (2\pi\delta\nu)^{-1} = (2\pi c\,\delta\tilde{\nu})^{-1} = \frac{5.309\ \mathrm{ps}}{\delta\tilde{\nu}/\mathrm{cm^{-1}}}$$

(a) $\tau = \dfrac{5.309\ \mathrm{ps}}{0.20} = \boxed{27\ \mathrm{ps}}$ (b) $\tau = \dfrac{5.309\ \mathrm{ps}}{2.0} = \boxed{2.7\ \mathrm{ps}}$

40.10(a) $\tau = \dfrac{5.309\ \text{ps}}{\delta\tilde{\nu}/\text{cm}^{-1}}$ [40.14, see exercise 40.9(a)]

(a) $\tau \approx 1.0\times10^{-13}\ \text{s} = 0.1\ \text{ps}$, implying that $\boxed{\delta\tilde{\nu} = 53\ \text{cm}^{-1}}$.

(b) $\tau \approx 100\times(1\times10^{-13}\ \text{s}) = 10\ \text{ps}$, implying that $\boxed{\delta\tilde{\nu} = 0.53\ \text{cm}^{-1}}$.

Problems

40.1 For a monochromatic beam of wavenumber $\tilde{\nu}$, the ratio of the output of a Michelson interferometer to the entering intensity I_0 is given by eqn 40.1 where p is the path length difference imposed by the position of mirror M_1 in text Figure 40.4.

$$y(x) = 1 + \cos 2\pi x \quad [40.1 \text{ with } y \equiv I(x)/I_0 \text{ and } x \equiv \tilde{\nu}p]$$

We examine the effect of varying the discrete step increment Δp of the instrument, which is equivalent to varying the step $\Delta x = \tilde{\nu}\cdot\Delta p$, in the one-period range $0 \le x \le 1$. Do this by dividing the range into N equal segments and treating N as a variable. This produces (x,y) pairs at each step and we plot the pairs for select values of N. Here's what the computational study looks like in a Mathcad Prime 2 worksheet:

$$f(N) := \left|\begin{array}{l} \text{for } i \in 0..N \\ \quad \left| \begin{array}{l} x_i \leftarrow \dfrac{i}{N} \\ y_i \leftarrow 1+\cos(2\cdot\pi\cdot x_i) \end{array}\right. \\ \text{return} \begin{bmatrix} x \\ y \end{bmatrix} \end{array}\right.$$

$x(N) := f(N)_{0,0}$

$y(N) := f(N)_{1,0}$

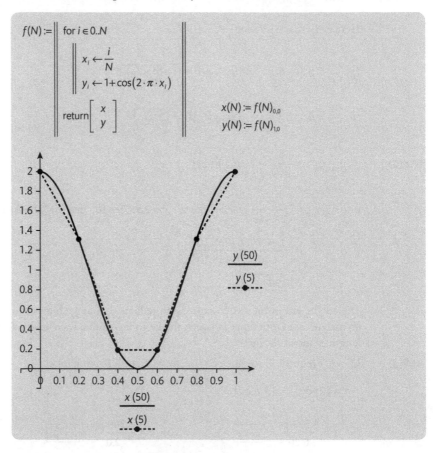

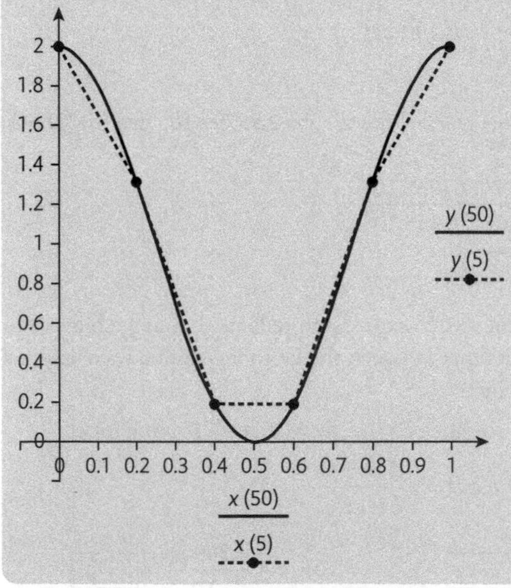

The plot clearly shows that when a large number of data pairs N are used, the Michelson interferometer output replicates the continuous variation of eqn 40.1 over the whole range. Even the selection of a small number of data pairs like $N = 5$ provides a good representation of the interferogram but one wonders whether important information is lost in regions between sampled points. To explore this question, we create a Prime 2 worksheet of an interferogram as a function of N when there are two spectrum lines of wavenumbers 100 cm^{-1} and 200 cm^{-1} of equal intensity (e.g., 1; the interferometer parameter p has the unit cm).

```
f(N) :=  for i ∈ 0..N
             p_i ← (i/N)·(2·π/200)
             y_i ← (1+cos(2·π·100·p_i)) + (1+cos(2·π·200·p_i))
         return [p
                 y]
```

$p(N) := f(N)_{0,0}$

$y(N) := f(N)_{1,0}$

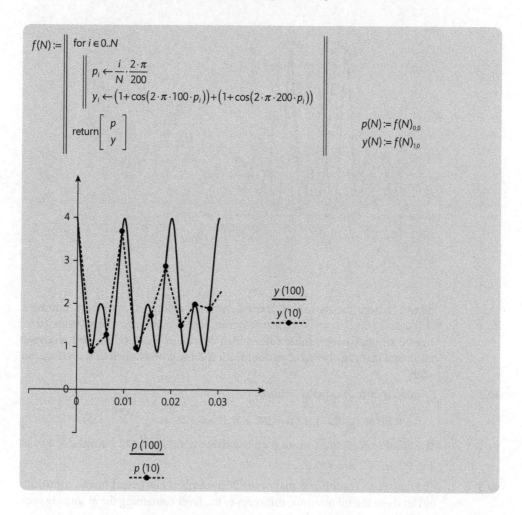

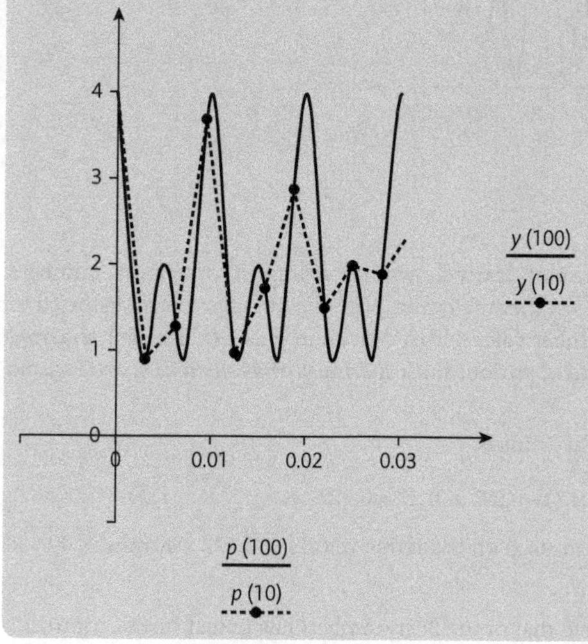

Once again, large values of N reproduce the interferogram of the continuous variation described by eqn 40.2. However, major features are omitted from the interferogram, or distorted, when either the range of p is smaller than a cycle in p or the discrete step value Δp is too large to represent the interferogram detail within a cycle. The plot of the data set $(p(10), y(10))$ is clearly a poor representation when the range is $0 \le p \le 0.03$ cm when the input consists of 100 cm^{-1} and 200 cm^{-1} lines.

We explore the effect of the distortions by performing a discrete Fourier transform with eqn 40.5 on the interferogram of the two-line spectrum. The worksheet uses the symbol ν to represent wavenumber and the plot divisions of 225 and 25 are inserted to normalize peaks to 1.

$$I(v,N) := 4\cdot\sum_{l=0}^{N}\left(y(N)_l - \frac{1}{2}\cdot y(N)_0\right)\cdot\cos(2\cdot\pi\cdot v\cdot p(N)_l)$$

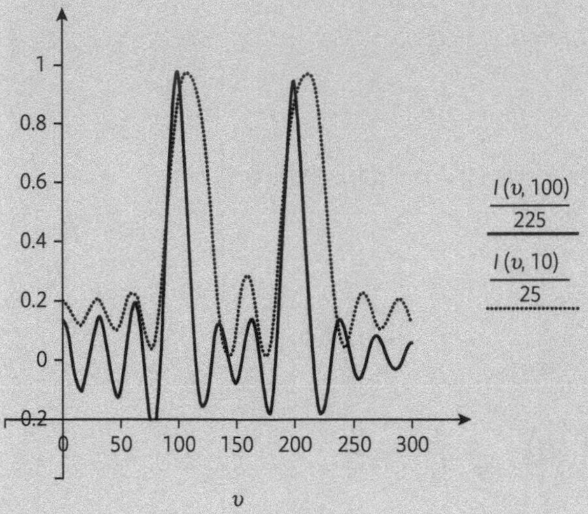

When N is large, the above plot clearly shows the two lines at correct wavenumbers of 100 cm^{-1} and 200 cm^{-1}. At the small value of N, the lines have been distorted to incorrectly high wavenumber values. This distortion problem is called **aliasing**, a problem that can also give spurious lines and must be avoided in spectroscopic work.

40.3 Fraction transmitted to the retina is

$$(1-0.30)\times(1-0.25)\times(1-0.09)\times 0.57 = 0.272$$

The number of photons focused on the retina in 0.1s is $0.272\times 40\ \text{mm}^2\times 0.1\ \text{s}\times 4\times 10^3\ \text{mm}^{-2}\ \text{s}^{-1} = \boxed{4.4\times 10^3}$.

40.5 The derivation is similar to that of the Beer-Lambert law in text Justification 40.1 but let dx be the infinitesimal thickness of the layer containing the absorbing species J. Then,

$$\mathrm{d}I = -\kappa[\mathrm{J}]I\mathrm{d}x$$

$$\frac{\mathrm{d}I}{I} = -\kappa[\mathrm{J}]\mathrm{d}x$$

$$\int_{I_0}^{I}\frac{\mathrm{d}I}{I}=-\kappa\int_0^L[\mathrm{J}]\mathrm{d}x=-\kappa[\mathrm{J}]_0\int_0^L \mathrm{e}^{-x/\lambda}\,\mathrm{d}x$$
$$\ln(I)\Big|_{I=I_0}^{I=I}=\kappa\lambda[\mathrm{J}]_0\,\mathrm{e}^{-x/\lambda}\Big|_{x=0}^{x=L}$$
$$-\ln\frac{I_0}{I}=-\kappa\lambda[\mathrm{J}]_0\left(1-\mathrm{e}^{-L/\lambda}\right)$$
$$(\ln 10)\log\frac{I_0}{I}=\kappa\lambda[\mathrm{J}]_0\left(1-\mathrm{e}^{-L/\lambda}\right)$$

Let $A=\log\frac{I_0}{I}$ [40.8] and $\varepsilon'=\kappa\lambda/(\ln 10)$. Then,

$$\boxed{A=\varepsilon'[\mathrm{J}]_0\left(1-\mathrm{e}^{-L/\lambda}\right)}$$

In the case for which $L/\lambda >> 1$, $\mathrm{e}^{-L/\lambda}=0$, $\boxed{A=\varepsilon'[\mathrm{J}]_0}$, and the absorbance is independent of the cell length.

In the case for which $L/\lambda << 1$, $\mathrm{e}^{-L/\lambda} \sim 1+(-L/\lambda)+(-L/\lambda)^2/2$ (Taylor series truncated after 2nd order).

Thus,

$$A=\varepsilon'[\mathrm{J}]\left(1-\mathrm{e}^{-L/\lambda}\right)$$
$$A=\varepsilon'[\mathrm{J}]\left(1-1+L/\lambda-(L/\lambda)^2/2\right)$$
$$A=\frac{\varepsilon' L[\mathrm{J}]}{\lambda}(1-L/2\lambda)$$
$$A=\varepsilon L[J]\times(1-L/2\lambda)\text{ when }L/\lambda<<1$$

40.7 Suppose that non-absorbing species B (pyridine in this problem) is progressively added to a solution of light absorbing species A (I_2 in this problem). Furthermore, suppose that A and B form a complex of light absorbing species AB. Suppose that all absorbance measurements are made at the equilibrium $\mathrm{A}+\mathrm{B}\rightleftharpoons\mathrm{AB}$. By the conservation of mass the sum of the equilibrium concentrations of the absorbing species, $[\mathrm{A}]_e + [\mathrm{AB}]_e$, is a constant for all additions of B. That is, the sum of concentrations of all species that contain A must equal the concentration of A present before any B is added, $[\mathrm{A}]_0$. The absorbance at any wavelength is

$$A_\lambda=\varepsilon_{\mathrm{A},\lambda}L[\mathrm{A}]_e+\varepsilon_{\mathrm{AB},\lambda}L[\mathrm{AB}]_e$$
$$=\varepsilon_{\mathrm{A},\lambda}L([\mathrm{A}]_0-[\mathrm{AB}]_e)+\varepsilon_{\mathrm{AB},\lambda}L[\mathrm{AB}]_e$$
$$=\varepsilon_{\mathrm{A},\lambda}L[\mathrm{A}]_0+\left(\varepsilon_{\mathrm{AB},\lambda}-\varepsilon_{\mathrm{A},\lambda}\right)L[\mathrm{AB}]_e$$

The first term to the right in the above equation is a constant for all additions of B; the second term is generally not constant and, consequently, A_λ generally varies with the addition of B. However, at any wavelength for which $\varepsilon_{\mathrm{AB},\lambda}=\varepsilon_{\mathrm{A},\lambda}$ the second term vanishes making A_λ a constant at all additions of B. The wavelength at which this happens is called the **isosbestic point**. It is characterized by the equilibrium between absorbing species and $\varepsilon_{\mathrm{AB},\lambda}=\varepsilon_{\mathrm{A},\lambda}$.

40.9 Gaussian distribution: $\varepsilon(\tilde{\nu})=\varepsilon_{\max}e^{-\frac{1}{2}\left(\frac{\tilde{\nu}-\mu}{\sigma}\right)^2}$ where μ is the mean of $\tilde{\nu}$ and σ is the standard deviation of the distribution. Dividing by $\varepsilon_{\max}$, taking the natural logarithm, and solving for $\tilde{\nu}-\mu$ gives

$$\tilde{\nu}-\mu=\pm\left(2\ln\frac{\varepsilon_{\max}}{\varepsilon}\right)^{1/2}\sigma$$

The width of the distribution at half-height, $\Delta\tilde{\nu}_{1/2}$, equals $2|\tilde{\nu}-\mu|$ evaluated at $\varepsilon=\varepsilon_{\max}/2$. Thus,

$$\Delta\tilde{\nu}_{1/2}=2(2\ln 2)^{1/2}\sigma \quad \text{or} \quad \sigma=\frac{\Delta\tilde{\nu}_{1/2}}{2(2\ln 2)^{1/2}}$$

We can now evaluate the integrated absorption coefficient, $\mathcal{A}$, in terms of $\varepsilon_{\max}$ and $\Delta\tilde{\nu}_{1/2}$.

$$\text{Let } x=\frac{\tilde{\nu}-\mu}{\sigma} \quad \text{and} \quad dx=\frac{1}{\sigma}d\tilde{\nu} \quad \text{and} \quad \varepsilon=\varepsilon_{\max}e^{-\frac{1}{2}x^2}$$

Then

$$\mathcal{A}=\int_{-\infty}^{\infty}\varepsilon d\tilde{\nu}\ [40.10]=\varepsilon_{\max}\sigma\int_{-\infty}^{\infty}e^{-\frac{1}{2}x^2}dx=(2\pi)^{1/2}\varepsilon_{\max}\sigma \quad \text{[standard integral]}$$

$$=\boxed{\frac{1}{2}\left(\frac{\pi}{\ln 2}\right)^{1/2}\varepsilon_{\max}\Delta\tilde{\nu}_{1/2}}$$

$$=1.064467\,\varepsilon_{\max}\Delta\tilde{\nu}_{1/2}$$

The Gaussian distribution is symmetric about the mean value of $\tilde{\nu}$, μ, which is the value of $\tilde{\nu}$ at the peak of the distribution. The absorption band of text Fig. F9.2 does not quite have this symmetry. It appears to be a skewed slightly toward the higher wavenumbers. Never the less, we estimate $\mathcal{A}$ by assuming that it can be approximated as a single Gaussian characterized by $\varepsilon_{\max}$ and $\Delta\tilde{\nu}_{1/2}$ values that are coarsely read off text Fig. F9.2.

$$\text{Coarse estimate: } \mathcal{A}=1.064467\,\varepsilon_{\max}\Delta\tilde{\nu}_{1/2}=1.064467\times(10\text{ dm}^3\text{ mol}^{-1}\text{ cm}^{-1})$$
$$\times(5.4\times10^3\text{ cm}^{-1})$$
$$=\boxed{5.7\times10^4\text{ dm}^3\text{ mol}^{-1}\text{ cm}^{-2}}$$

Let us now suppose that the slightly non-Gaussian shape exhibited by text Fig. F9.2 results from two separate absorption lines each of which has a molar absorption coefficient that is a Gaussian function of wavenumber. Fig. F9.2 is then an 'apparent' molar absorption coefficient that is the sum of two independent Gaussians each characterized by an amplitude A, a mean value μ, and a standard deviation σ. That's a total of 6 parameters to be adjusted to fit the data of the figure. We label the 3 parameters of the predominate Gaussian, the one with the lower mean wavenumber, with a '1'; the low amplitude, higher mean distribution is labeled with a '2'. A lot of $(\tilde{\nu},\varepsilon)$ data pairs are needed to determine precise values of the parameters so we expanded Fig. F9.2 and used Photoshop to read a total of 20 data pairs, several of which are displayed in the following Mathcad Prime 2 worksheet. Calling the sum of the two Gaussians G_{sum}, the worksheet uses guess values for the 6 parameters to calculate the difference $\varepsilon_{obs} - G_{sum}$ at each of the 20 $\tilde{\nu}_{obs}$, the difference is squared and summed over all data pairs, which the worksheet calls the 'sum of the squared errors' *SSE*. The idea is to systematically adjust the 6 parameters so as to minimize *SSE*. Mathcad performs the minimization process with the 'minerr()' function within a solve block. The symbol '*v*' is used to represent wavenumber within the worksheet.

$$\upsilon = \begin{bmatrix} 2.2\cdot 10^4 \\ 2.249\cdot 10^4 \\ 2.293\cdot 10^4 \\ \vdots \end{bmatrix} \text{cm}^{-1} \qquad \varepsilon = \begin{bmatrix} 0.412 \\ 0.7 \\ 1.317 \\ \vdots \end{bmatrix} \text{dm}^3\cdot\text{mol}^{-1}\cdot\text{cm}^{-1}$$

$$G_{sum}(\upsilon,\mu_1,\Delta\upsilon_1,A_1,\mu_2,\Delta\upsilon_2,A_2) := A_1\cdot e^{\frac{-1}{2}\left(\frac{\upsilon-\mu_1}{\Delta\upsilon_1}\right)^2} + A_2\cdot e^{\frac{-1}{2}\left(\frac{\upsilon-\mu_2}{\Delta\upsilon_2}\right)^2}$$

$$SSE(\mu_1,\Delta\upsilon_1,A_1,\mu_2,\Delta\upsilon_2,A_2) := \sum_{i=0}^{19}\left(\varepsilon_i - G_{sum}(\upsilon_i,\mu_1,\Delta\upsilon_1,A_1,\mu_2,\Delta\upsilon_2,A_2)\right)^2$$

Guess Values

$\mu_1 := 26\cdot 10^3\cdot\text{cm}^{-1}$ $\quad\mu_2 := 29\cdot 10^3\cdot\text{cm}^{-1}$

$\Delta\upsilon_1 := 2\cdot 10^3\cdot\text{cm}^{-1}$ $\quad\Delta\upsilon_2 := 2\cdot 10^3\cdot\text{cm}^{-1}$

$A_1 := 8\cdot\text{dm}^3\cdot\text{mol}^{-1}\cdot\text{cm}^{-1}$ $\quad A_2 := 3\cdot\text{dm}^3\cdot\text{mol}^{-1}\cdot\text{cm}^{-1}$

Constraints

$SSE(\mu_1,\Delta\upsilon_1,A_1,\mu_2,\Delta\upsilon_2,A_2) = 0$

Solver

Parameters := **minerr** $(\mu_1,\Delta\upsilon_1,A_1,\mu_2,\Delta\upsilon_2,A_2)$

$\mu_1 := Parameters_0 \qquad \mu_1 = (2.59\cdot 10^4)\frac{1}{\text{cm}}$

$\Delta\upsilon_1 := Parameters_1 \qquad \Delta\upsilon_1 = (1.597\cdot 10^3)\frac{1}{\text{cm}}$

$A_1 := Parameters_2 \qquad A_1 = 6.931\frac{\text{dm}^3}{\text{mol}\cdot\text{cm}}$

$\mu_2 := Parameters_3 \qquad \mu_2 = (2.856\cdot 10^4)\frac{1}{\text{cm}}$

$\Delta\upsilon_2 := Parameters_4 \qquad \Delta\upsilon_2 = (2.234\cdot 10^3)\frac{1}{\text{cm}}$

$A_2 := Parameters_5 \qquad A_2 = 5.131\frac{\text{dm}^3}{\text{mol}\cdot\text{cm}}$

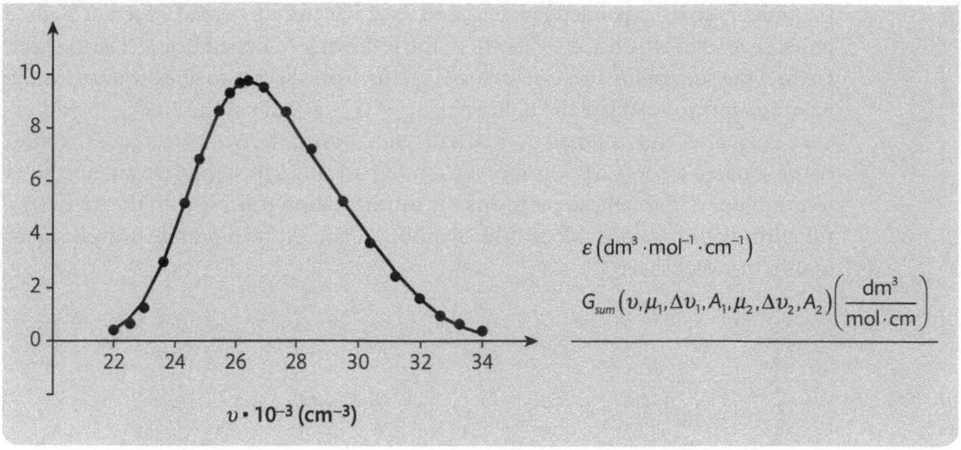

The plot shows that the sum of two Gaussians with adjusted parameters fits the data very nicely. The values of the six parameters are listed just above the plot. The following worksheet section uses the fitted function to calculate the integrated absorption coefficient with eqn 40.10. The earlier, coarse estimate is seen to be rather close to the more precise calculation.

$$\textit{integrated\_absorption\_coeff} := \int_{0\cdot\mathrm{cm}^{-1}}^{10^6\cdot\mathrm{cm}^{-1}} G_{sum}(\upsilon, \mu_1, \Delta\upsilon_1, A_1, \mu_2, \Delta\upsilon_2, A_2)\,d\upsilon$$

$$\textit{integrated\_absorption\_coeff} := (5.649 \cdot 10^4)\,\mathrm{dm}^3 \cdot \mathrm{mol}^{-1} \cdot \mathrm{cm}^{-2}$$

40.11 The light scattering equation

$$\frac{I_0}{I_\theta} = a + b \times \left(\frac{I_0}{I_\theta}\sin^2(\theta/2)\right) \quad \text{where} \quad a = (Kc_\mathrm{M}M)^{-1} \quad \text{and} \quad b = \frac{16\pi^2 R^2}{5\lambda^2}$$

has a linear form when I_0/I_θ is plotted against $\frac{I_0}{I_\theta}\sin^2(\theta/2)$. The intercept a and slope b are determined with a linear regression fit of this plot. Knowing a and b, the above equations can be used to calculate M and R. We begin by drawing up the requisite data table, preparing a plot (shown in Fig. 40.2), and calculating the linear regression fit.

$\theta/°$	15.0	45.0	70.0	85.0	90.0
$100 \times I_0/I_\theta$	4.20	4.37	4.63	4.83	4.90
$100 \times I_0/I_\theta \times \sin^2(\theta/2)$	0.0716	0.640	1.52	2.20	2.45

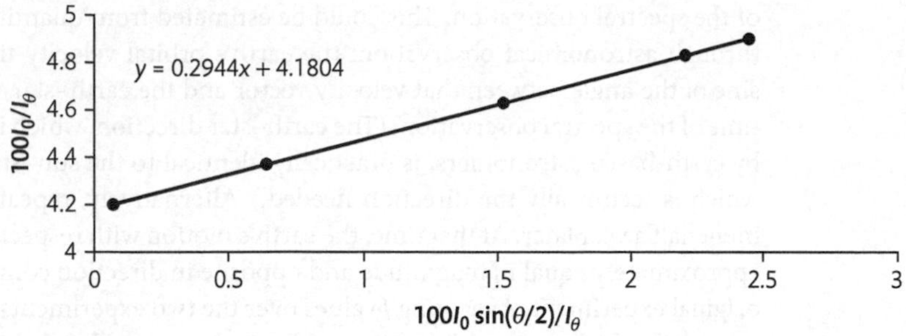

Figure 40.2

Thus, $a = 0.0418$ and $b = 0.294$.

$$M = (Kc_{\mathrm{M}}a)^{-1} = \{(2.40\times10^{-2}\ \mathrm{mol\ m^3\ kg^{-2}})\times(2.0\ \mathrm{kg\ m^{-3}})\times(0.0418)\}^{-1}$$
$$= \boxed{498\ \mathrm{kg\ mol^{-1}}}$$

$$R = (5b)^{1/2}\lambda/(4\pi) = (5\times0.294)^{1/2}\times(532\ \mathrm{nm})/(4\pi)$$
$$= \boxed{51.3\ \mathrm{nm}}$$

40.13 (a) Compute the ratios ν_{star}/ν for all three lines. We are given wavelength data, so we can use:

$$\frac{\nu_{\mathrm{star}}}{\nu} = \frac{\lambda}{\lambda_{\mathrm{star}}}.$$

The ratios are:

$$\frac{438.392\ \mathrm{nm}}{438.882\ \mathrm{nm}} = 0.998884,\quad \frac{440.510\ \mathrm{nm}}{441.000\ \mathrm{nm}} = 0.998889,\ \text{and}\ \frac{441.510\ \mathrm{nm}}{442.020\ \mathrm{nm}} = 0.998846.$$

The frequencies of the stellar lines are all less than those of the stationary lines, so we infer that the star is $\boxed{\text{receding}}$ from earth. The Doppler effect follows:

$$\nu_{\mathrm{receding}} = \nu f \quad \text{where } f = \left(\frac{1-s/c}{1+s/c}\right)^{1/2},\ \text{so}$$

$$f^2(1+s/c) = (1-s/c),\quad (f^2+1)s/c = 1-f^2,\quad s = \frac{1-f^2}{1+f^2}c$$

Our average value of f is 0.998873. (Note: the uncertainty is actually greater than the significant figures here imply, and a more careful analysis would treat uncertainty explicitly.) So the speed of recession with respect to the earth is:

$$s = \left(\frac{1-0.997747}{1+0.997747}\right)c = \boxed{1.128\times10^{-3}\ c} = \boxed{3.381\times10^{5}\ \mathrm{m\ s^{-1}}}$$

(b) One could compute the star's radial velocity with respect to the sun if one knew the earth's speed with respect to the sun along the sun–star vector at the time of the spectral observation. This could be estimated from quantities available through astronomical observation: the earth's orbital velocity times the cosine of the angle between that velocity vector and the earth–star vector at the time of the spectral observation. (The earth–star direction, which is observable by earth-based astronomers, is practically identical to the sun–star direction, which is technically the direction needed.) Alternatively, repeat the experiment half a year later. At that time, the earth's motion with respect to the sun is approximately equal in magnitude and opposite in direction compared to the original experiment. Averaging f values over the two experiments would yield f values in which the earth's motion is effectively averaged out.

40.15 Transform eqn 40.13 to the standard Gaussian form with the transformations $\sigma(T)=\left(kT/mc^2\right)^{1/2}$, which is the standard deviation, and $x(\nu_{\mathrm{obs}})=(\nu_{\mathrm{obs}}-\nu)/\nu$. σ^2 is the ratio of thermal energy to total, relativistic energy while x is a linear function of ν_{obs}. The study of $\mathrm{d}I/\mathrm{d}\nu_{\mathrm{obs}}$ and the effects of varying T are equivalent to study of $\mathrm{d}I/\mathrm{d}x$ and the effects of varying σ. Here's a Mathcad Prime 2 study.

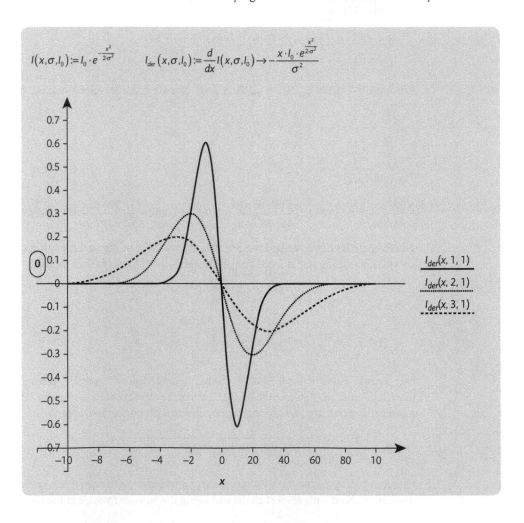

$I(x,\sigma,I_0) := I_0 \cdot e^{-\frac{x^2}{2\sigma^2}}$ $\qquad$ $I_{der}(x,\sigma,I_0) := \frac{d}{dx} I(x,\sigma,I_0) \rightarrow -\frac{x \cdot I_0 \cdot e^{\frac{x^2}{2\sigma^2}}}{\sigma^2}$

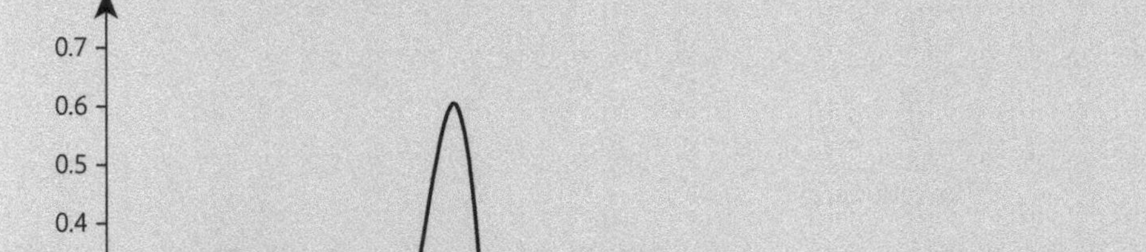

The plot indicates that as the standard deviation increases from 1 to 3, which is equivalent to a systematic increase in T, the extrema separate and the distribution of signal from the phase-sensitive detector broadens. An analytic determination of the dependence of the extrema separation upon σ involves finding $\mathrm{d}I_{\mathrm{der}}/\mathrm{d}x$, setting it equal to zero, and solving for x.

$$\frac{d}{dx}I_{der}(x,\sigma,I_0) \xrightarrow{\text{solve},x} \begin{bmatrix} \sigma \\ -\sigma \end{bmatrix}$$

Thus, the separation is $2\sigma(T) = \boxed{2(kT/mc^2)^{1/2}}$.

To finish the study, we calculate the standard deviation exhibited by an atomic hydrogen gas at the temperature of the surface of the sun (6 000 K).

$$\sigma_{T=6000\ \mathrm{K}} = \left(\frac{kT}{mc^2}\right)^{1/2} = \left(\frac{RT}{Mc^2}\right)^{1/2}$$

$$= \left(\frac{(8.314\ \mathrm{J\ K^{-1}\ mol^{-1}})\times(6\,000\ \mathrm{K})}{(1.0\times10^{-3}\ \mathrm{kg\ mol^{-1}})\times(3.00\times10^{8}\ \mathrm{m\ s^{-1}})^2}\right)^{1/2}$$

$$= 2.4\times10^{-5}$$

Topic 41 Molecular rotation

Note: The masses of nuclides are listed in Table 0.2 of the *Resource Section.*

Discussion questions

D41.1 Symmetric rotor: the energy depends on J and K^2 [eqn 41.13], hence each level except the $K = 0$ level is doubly degenerate. In addition, states of a given J have a component of their angular momentum along an external, laboratory-fixed axis, characterized by the quantum number M_J, that can take on $2J + 1$ values. The quantum number M_J does not affect the energy, consequently all $2J + 1$ orientations of the molecule have the same energy. It follows that a symmetric rotor level is $2(2J + 1)$-fold degenerate for $K \neq 0$, and $2J + 1$ degenerate for $K = 0$.

Linear rotor: A linear rotor has K fixed at 0, but there are still $2J + 1$ values of M_J, so the degeneracy is $2J + 1$.

Spherical rotor: A spherical rotor can be regarded as a version of a symmetric rotor in which $A = B$, and consequently, the energy is independent of the $2J + 1$ values that K can assume. Hence, there is a simultaneous degeneracy of $2J + 1$ in both K and M_J, resulting in a total degeneracy of $(2J + 1)^2$.

Asymmetric rotor: In an asymmetric rotor there is no longer a preferred direction in the molecule which carries out a simple rotation about J, thus for each value of J, there is only a degeneracy of $2J + 1$, as in a linear rotor.

If a decrease in rigidity affects the symmetry of the molecule, the rotational degeneracy could be affected also. For example, if a wobble resulted in a spherical rotor being transformed into an asymmetric rotor the rotational degeneracy of the molecule would change.

Exercises

E41.1(a) Ozone is an asymmetric rotor similar to H_2O; hence we follow the method of Example 41.1 and use $I = \sum_i m_i x_i^2$ [41.1] to calculate the moment of inertia, I_C about the C_2 axis.

$$I = m_O(x_O^2 + 0 + x_O^2) = 2m_O x_O^2 = 2m_O(R\sin\phi)^2,$$

where the bond angle is denoted 2ϕ and the bond length R. Substitution of the data gives

$$I = 2\times 15.9949\, m_u \times 1.66054\times 10^{-27}\,\text{kg}/m_u \times (1.28\times 10^{-10}\,\text{m}\times \sin 58.5^\circ)^2$$

$$= \boxed{6.33\times 10^{-46}\,\text{kg m}^2}$$

The corresponding rotational constant is

$$\tilde{A} = \frac{\hbar}{4\pi c I_c} = \frac{1.05447\times10^{-34}\ \text{J s}}{4\pi\times2.998\times10^{8}\ \text{m s}^{-1}\times6.33\times10^{-46}\ \text{kg m}^2}$$
$$= 44.21\ \text{m}^{-1} = \boxed{0.4421\ \text{cm}^{-1}}$$

E41.2(a) In order to conform to the symbols used in the first symmetric rotor figure of Table 41.1, we will use the molecular formula BA_4. $I_{\parallel}$ is along a bond and $I_{\perp}$ is perpendicular to both $I_{\parallel}$ and a molecular face that does not contain $I_{\parallel}$ (see the spherical rotor of text Figure 41.3). For our molecule, $R' = R$, $m_C = m_A$, and the equations of Table 41.1 simplify to

$$I_{\parallel}/m_A R^2 = 2(1-\cos\theta)$$
$$I_{\perp}/m_A R^2 = 1-\cos\theta+(m_A+m_B)\times(1+2\cos\theta)/m$$
$$+\left\{(3m_A+m_B)+6m_A\left[\tfrac{1}{3}(1+2\cos\theta)\right]^{1/2}\right\}/m \text{ where } m = 4m_A+m_B.$$

The $I_{\parallel}/m_A R^2$ moment of inertia ratio does not depend upon specific atomic masses so the plot of this ratio against θ, shown in Fig. 41.1 describes the angular dependence of all molecules having the formula BA_4. However, the $I_{\perp}/m_A R^2$ moment of inertia ratio does have a specifi c atomic mass dependency so we will plot its angular dependence for CH_4, an important fuel and powerful greenhouse gas. The computational equation is

$$I_{\perp}/m_A R^2 = 1-\cos\theta+13(1+2\cos\theta)/16+\left\{15+6\left[\tfrac{1}{3}(1+2\cos\theta)\right]^{1/2}\right\}/16$$

and its plot is also found in Fig. 41.1. As θ increases, atoms move away from the axis of $I_{\parallel}$, which causes the moment of inertia around this axis to increase. Atoms move toward the axis of $I_{\perp}$ as θ increases, thereby decreasing the moment of inertia around this axis.

Comment: Fig. 41.1 suggests that $I_{\parallel}(\theta_{\text{tetra}}) = I_{\perp}(\theta_{\text{tetra}})$ for all tetrahedral molecules where θ_{tetra} is the tetrahedral angle (approx. 109.471°). Can you provide an analytic proof that this is true? Hint: The exact value of θ_{tetra} is $\pi/2+\sin^{-1}(1/3)$ in radians where $\sin^{-1}()$ is the arcsin() function.

E41.3(a) (a) asymmetric (b) oblate symmetric (c) spherical (d) prolate symmetric

E41.4(a) The determination of two unknowns requires data from two independent experiments and the equation which relates the unknowns to the experimental data. In this exercise two independently determined values of B for two isotopically different HCN molecules are used to obtain the moments of inertia of the molecules and from these, by use of the equation for the moment of inertia of linear triatomic rotors (Table 41.1), the interatomic distances R_{HC} and R_{CN} are calculated.

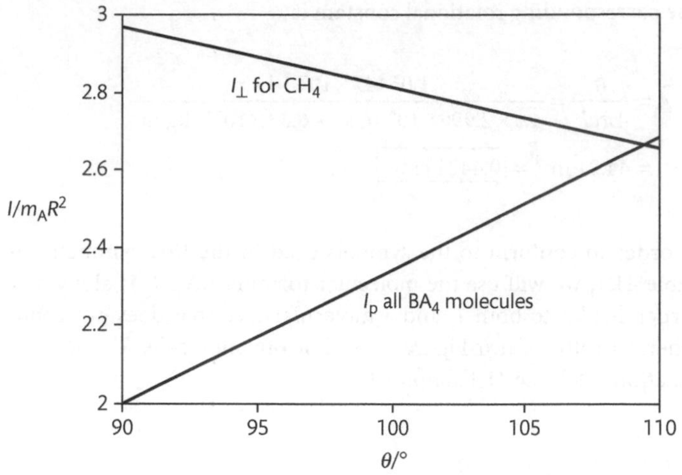

Figure 41.1

Rotational constants which are usually expressed in wavenumber (cm^{-1}) are sometimes expressed in frequency units (Hz). The conversion between the two is

$$B/\text{Hz} = c \times \tilde{B}/\text{cm}^{-1} \quad [c \text{ in cm s}^{-1}]$$

$$\text{Thus, } B(\text{in Hz}) = \frac{\hbar}{4\pi I} \text{ and } I = \frac{\hbar}{4\pi B}$$

Let, $^1\text{H} = \text{H}$, $^2\text{H} = \text{D}$, $R_{\text{HC}} = R_{\text{DC}} = R$, $R_{\text{CN}} = R'$. Then

$$I(\text{HCN}) = \frac{1.05457 \times 10^{-34}\ \text{J s}}{(4\pi) \times (4.4316 \times 10^{10}\ \text{s}^{-1})} = 1.8937 \times 10^{-46}\ \text{kg m}^2$$

$$I(\text{DCN}) = \frac{1.05457 \times 10^{-34}\ \text{Js}}{(4\pi) \times (3.6208 \times 10^{10}\ \text{s}^{-1})} = 2.3178 \times 10^{-46}\ \text{kg m}^2$$

and from Table 41.1 with isotope masses from the *Data Section*.

$$I(\text{HCN}) = m_{\text{H}}\text{R}^2 + m_N R'^2 - \frac{(m_{\text{H}}\text{R} - m_{\text{N}}\text{R}')^2}{m_{\text{H}} + m_{\text{C}} + m_N}$$

$$I(\text{HCN}) = \left[(1.0078R^2) + (14.0031R'^2) - \left(\frac{(1.0078R - 14.0031R')^2}{1.0078 + 12.0000 + 14.0031} \right) \right] \text{u}$$

Multiplying through by $m/m_{\text{u}} = (m_{\text{H}} + m_{\text{C}} + m_{\text{N}})/m_{\text{u}} = 27.0109$

$$27.0109 \times I(\text{HCN}) = \{27.0109 \times (1.0078R^2 + 14.0031R'^2) - (1.0078R - 14.0031R')^2\} m_{\text{u}}$$

$$\text{or}\quad \left(\frac{27.0109}{1.66054\times10^{-27}\,\text{kg}}\right)\times\left(1.8937\times10^{-46}\,\text{kg m}^2\right)=3.0804\times10^{-18}\,\text{m}^2$$

$$=\{27.0109\times\left(1.0078R^2+14.0031R'^2\right)-(1.0078R-14.0031R')^2\}\qquad\text{(a)}$$

In a similar manner we find for DCN

$$\left(\frac{28.0172}{1.66054\times10^{-27}\,\text{kg}}\right)\times\left(2.3178\times10^{-46}\,\text{kg m}^2\right)=3.9107\times10^{-18}\,\text{m}^2$$

$$=\left\{28.0172\times\left(2.0141R^2+14.0031R'^2\right)-(2.0141R-14.0031R')^2\right\}\qquad\text{(b)}$$

Thus there are two simultaneous quadratic equations (a) and (b) to solve for R and R'. These equations are most easily solved by readily available computer programs or by successive approximations. The results are

$$R=1.065\times10^{-10}\,\text{m}=\boxed{106.5\,\text{pm}}\quad\text{and}\quad R'=1.156\times10^{-10}\,\text{m}=\boxed{115.6\,\text{pm}}$$

These values are easily verified by direct substitution into the equations and agree well with the accepted values R_{HC} = 1.064 × 10^{-10} m and R_{CN} = 1.156 × 10^{-10} m.

Problems

P41.1 The center of mass of a diatomic molecule lies at a distance x from atom A and is such that the masses on either side of it balance

$$m_A x=m_B(R-x)$$

and it is at

$$x=\frac{m_B}{m}R\quad m=m_A+m_B$$

The moment of inertia of the molecule is

$$I=m_A x^2+m_B(R-x)^2[41.1]=\frac{m_A m_B^2 R^2}{m^2}+\frac{m_B m_A^2 R^2}{m^2}=\frac{m_A m_B}{m}R^2$$

$$=\boxed{m_{\text{eff}}R^2}\ \text{since}\ m_{\text{eff}}=\frac{m_A m_B}{m_A+m}$$

Topic 42 **Rotational spectroscopy**

Discussion questions

D42.1 Hydrogen molecules can exist in two forms: the *para*- form has antiparallel nuclear spins and the *ortho*- form has parallel nuclear spins. Because of these arrangements of the nuclear spins the *ortho*- form must have rotational wavefunctions restricted to odd J values only as discussed in detail in Section 42.3. *Ortho*-hydrogen cannot exist in the $J = 0$ state. Hence, the lowest energy level of *ortho*-hydrogen has $J = 1$ and therefore a zero-point energy. The conversion between the two forms is very slow.

D42.3 (1) *Rotational Raman spectroscopy*. The gross selection rule is that the molecule must be anisotropically polarizable, which is to say that its polarizability, α, depends upon the direction of the electric field relative to the molecule. Non-spherical rotors satisfy this condition. Therefore, linear and symmetric rotors are rotationally Raman active.

(2) *Vibrational Raman spectroscopy*. The gross selection rule is that the polarizability of the molecule must change as the molecule vibrates. All diatomic molecules satisfy this condition as the molecules swell and contract during a vibration, the control of the nuclei over the electrons varies, and the molecular polarizability changes. Hence both homonuclear and heteronuclear diatomics are vibrationally Raman active. In polyatomic molecules it is usually quite difficult to judge by inspection whether or not the molecule is anisotropically polarizable; hence group theoretical methods are relied on for judging the Raman activity of the various normal modes of vibration. The procedure is discussed in Section 44.4(b) and demonstrated in *Brief Illustration* 44.4.

Exercises

E42.1(a) Polar molecules show a pure rotational absorption spectrum. Therefore, select the polar molecules based on their well-known structures. Alternatively, determine the point groups of the molecules and use the rule that only molecules belonging to C_n, C_{nv}, and C_s may be polar, and in the case of C_n and C_{nv}, that dipole must lie along the rotation axis. Hence the polar molecules are

(b) HCl **(d)** CH_3Cl **(e)** CH_2Cl_2

Their point group symmetries are

(b) $C_{\infty v}$ **(d)** C_{3v} **(e)** $C_{2h}\,(trans), C_{2v}\,(cis)$

Therefore, these molecules show a pure rotational spectrum.

Comment. Note that the *cis* form of CH_2Cl_2 is polar, but the *trans* form is not.

E42.2(a) NO is a linear rotor, and we assume there is little centrifugal distortion; hence

$$\tilde{F}(J) = \tilde{B}J(J+1) \quad [41.9]$$

with $\tilde{B} = \dfrac{\hbar}{4\pi cI}$, $I = m_{\text{eff}}R^2$ [Table 41.1], and

$$m_{\text{eff}} = \frac{m_N m_O}{m_N + m_O} \text{[nuclide masses from the } \textit{Data Section}]$$

$$= \left(\frac{(14.003\,m_u)\times(15.995\,m_u)}{(14.003\,m_u)+(15.995\,m_u)}\right)\times(1.6605\times10^{-27}\text{ kg }m_u^{-1}) = 1.240\times10^{-26}\text{ kg}$$

Then, $I = (1.240\times10^{-26}\text{ kg})\times(1.15\times10^{-10}\text{ m})^2 = 1.64\bar{0}\times10^{-46}\text{ kg m}^2$

$$\text{and } \tilde{B} = \frac{1.0546\times10^{-34}\text{ J s}}{(4\pi)\times(2.998\times10^8\text{ m s}^{-1})\times(1.64\bar{0}\times10^{-46}\text{ kg m}^2)} = 170.\bar{7}\text{ m}^{-1} = 1.70\bar{7}\text{ cm}^{-1}$$

The wavenumber of the $J = 3 \leftarrow 2$ transition is

$$\tilde{\nu} = 2\tilde{B}(J+1)[42.8a] = 6\tilde{B}[J=2] = (6)\times(1.70\bar{7}\text{ cm}^{-1}) = 10.2\bar{4}\text{ cm}^{-1}$$

The frequency is

$$\nu = \tilde{\nu}c = (10.2\bar{4}\text{ cm}^{-1})\times\left(\frac{10^2\text{ m}^{-1}}{1\text{ cm}^{-1}}\right)\times(2.998\times10^8\text{ m s}^{-1}) = \boxed{3.07\times10^{11}\text{ Hz}}$$

When centrifugal distortion is taken into account the frequency decreases as can be seen by considering eqn 42.8b.

Question. What is the percentage change in these calculated values if centrifugal distortion is included?

E42.3(a) The wavenumber of the transition is related to the rotational constant by

$$hc\tilde{\nu} = \Delta E = hc\tilde{B}[J(J+1)-(J-1)J] = 2hc\tilde{B}J \quad [41.6, 41.8]$$

where J refer to the upper state $(J = 3)$. The rotational constant is related to molecular structure by

$$\tilde{B} = \frac{\hbar}{4\pi cI} \quad [41.7]$$

where I is moment of inertia. Putting these expressions together yields

$$\tilde{\nu} = 2\tilde{B}J = \frac{\hbar J}{2\pi cI} \quad \text{so} \quad I = \frac{\hbar J}{2\pi c\tilde{\nu}} = \frac{(1.0546\times10^{-34}\text{ J s})\times(3)}{2\pi(2.998\times10^{10}\text{ cm s}^{-1})\times(63.56\text{ cm}^{-1})}$$
$$= 2.642\times10^{-47}\text{ kg m}^2$$

The moment of inertia is related to the bond length by

$$I = m_{\text{eff}}R^2 \quad \text{so} \quad R = \sqrt{\frac{I}{m_{\text{eff}}}}$$

$$m_{\text{eff}}^{-1} = m_{\text{H}}^{-1} + m_{\text{Cl}}^{-1} = \frac{(1.0078\ m_{\text{u}})^{-1} + (34.9688\ m_{\text{u}})^{-1}}{1.66054\times10^{-27}\ \text{kg}\ m_{\text{u}}^{-1}} = 6.1477\times10^{26}\ \text{kg}^{-1}$$

and $R = \sqrt{(6.1477\times10^{26}\ \text{kg}^{-1})\times(2.642\times10^{-47}\ \text{kg m}^2)} = 1.274\times10^{-10}\ \text{m} = \boxed{127.4\ \text{pm}}$

E42.4(a) If the spacing of lines is constant, the effects of centrifugal distortion are negligible. Hence we may use for the wavenumbers of the transitions

$$\tilde{F}(J) - \tilde{F}(J-1) = 2\tilde{B}J \quad [41.10]$$

Since $J = 1, 2, 3, \ldots$, the spacing of the lines is $2\tilde{B}$

$$12.604\ \text{cm}^{-1} = 2\tilde{B}$$

$$\tilde{B} = 6.302\ \text{cm}^{-1} = 6.302\times10^2\ \text{m}^{-1}$$

$$I = \frac{\hbar}{4\pi c\tilde{B}} = m_{\text{eff}}R^2$$

$$\frac{\hbar}{4\pi c} = \frac{1.0546\times10^{-34}\ \text{J s}}{(4\pi)\times(2.9979\times10^8\ \text{m s}^{-1})} = 2.7993\times10^{-44}\ \text{kg m}$$

$$I = \frac{2.7993\times10^{-44}\ \text{kg m}}{6.302\times10^2\ \text{m}^{-1}} = \boxed{4.442\times10^{-47}\ \text{kg m}^2}$$

$$m_{\text{eff}} = \frac{m_{\text{Al}}m_{\text{H}}}{m_{\text{Al}} + m_{\text{H}}}$$

$$= \left(\frac{(26.98)\times(1.008)}{(26.98)+(1.008)}\right)\text{u}\times(1.6605\times10^{-27}\ \text{kg u}^{-1}) = 1.613\bar{6}\times10^{-27}\ \text{kg}$$

$$R = \left(\frac{I}{m_{\text{eff}}}\right)^{1/2} = \left(\frac{4.442\times10^{-47}\ \text{kg m}^2}{1.6136\times10^{-27}\ \text{kg}}\right)^{1/2} = 1.659\times10^{-10}\ \text{m} = \boxed{165.9\ \text{pm}}$$

E42.5(a) We select those molecules with an anisotropic polarizability. A practical rule to apply is that spherical rotors do not have anisotropic polarizabilities. Therefore $\boxed{\text{(c) CH}_4 \text{ is inactive}}$. All others are active.

E42.6(a) The Stokes lines appear at

$$\tilde{\nu}(J+2 \leftarrow J) = \tilde{\nu}_i - 2\tilde{B}(2J+3)[42.15] \quad \text{with} \quad J = 0,\ \tilde{\nu} = \tilde{\nu}_i - 6\tilde{B}$$

Since $\tilde{B} = 1.9987\ \text{cm}^{-1}$ (Table 43.1), the Stokes line appears at

$$\tilde{\nu} = (20487) - (6)\times(1.9987\ \text{cm}^{-1}) = \boxed{20\ 475\ \text{cm}^{-1}}$$

E42.7(a) The separation of lines is $4\tilde{B}$ [42.15], so $\tilde{B}=0.2438\ \text{cm}^{-1}$. Then we use

$$R=\left(\frac{\hbar}{4\pi m_{\text{eff}}c\tilde{B}}\right)^{1/2}$$

with $m_{\text{eff}}=\frac{1}{2}m(^{35}\text{Cl})=(\frac{1}{2})\times(34.9688\ m_u)=17.4844\ m_u$

Therefore

$$R=\left(\frac{1.05457\times10^{-34}\ \text{J s}}{(4\pi)\times(17.4844)\times(1.6605\times10^{-27}\ \text{kg})\times(2.9979\times10^{10}\ \text{cm s}^{-1})\times(0.2438\ \text{cm}^{-1})}\right)^{1/2}$$

$$=1.989\times10^{-10}\ \text{m}=\boxed{198.9\,\text{pm}}$$

E42.8(a) For diatomic molecules we can use eqn 42.16 to determine the statistical ratio of weights of populations. For chlorine-35 $I=3/2$, hence

$$\text{Ratio of (odd J/even J) weights of populations is } \frac{I+1}{I}=\boxed{\frac{5}{3}}$$

Problems

P42.1 Rotational line separations are $2\tilde{B}$ (in wavenumber units), $2\tilde{B}c$ (in frequency units), and $(2\tilde{B})^{-1}$ in wavelength units. Hence the transitions are separated by $\boxed{596\text{GHZ}}$, $\boxed{19.9\ \text{cm}^{-1}}$, and $\boxed{0.503\ \text{mm}}$.

Ammonia is a symmetric rotor (Section 41.2(b)) and we know that

$$\tilde{B}=\frac{\hbar}{4\pi cI_{\perp}} \quad [41.14]$$

and from Table 41.1,

$$I_{\perp}=m_AR^2(1-\cos\theta)+\left(\frac{m_Am_B}{m}\right)R^2(1+2\cos\theta)$$

$$m_A=1.6735\times10^{-27}\ \text{kg},\ m_B=2.3252\times10^{-26}\ \text{kg, and } m=2.8273\times10^{-26}\ \text{kg}$$

with $R=101.4$ pm and $\theta=106°47'$, which gives

$$I_{\perp}=(1.6735\times10^{-27}\ \text{kg})\times(101.4\times10^{-12}\ \text{m})^2\times(1-\cos106°47')$$
$$+\left(\frac{(1.6735\times10^{-27})\times(2.3252\times10^{-26}\ \text{kg}^2)}{2.8273\times10^{-26}\ \text{kg}}\right)$$
$$\times(101.4\times10^{-12}\ \text{m})^2\times(1+2\cos\ 106°47')$$
$$=2.815\bar{8}\times10^{-47}\ \text{kg m}^2$$

Therefore,

$$\tilde{B}=\frac{1.05457\times10^{-34}\ \text{J s}}{(4\pi)\times(2.9979\times10^{8}\ \text{m s}^{-1})\times(2.815\bar{8}\times10^{-47}\ \text{kg m}^2)}=994.1\ \text{m}^{-1}=\boxed{9.941\ \text{cm}^{-1}}$$

which is in accord with the data.

P42.3 $\tilde{\nu}=2\tilde{B}(J+1)$ [42.8a] $=2\tilde{B}$

Hence, $\tilde{B}(^{1}\text{HCl})=10.4392\ \text{cm}^{-1}$, $\tilde{B}(^{2}\text{HCl})=5.3920\ \text{cm}^{-1}$

$$\tilde{B}=\frac{\hbar}{4\pi cI}[41.7] \qquad I=m_{\text{eff}}R^2 \text{ [Table 41.1]}$$

$$R^2=\frac{\hbar}{4\pi c m_{\text{eff}}\tilde{B}} \qquad \frac{\hbar}{4\pi c}=2.79927\times10^{-44}\ \text{kg m}$$

$$m_{\text{eff}}(\text{HCl})=\left(\frac{(1.007825\,m_{\text{u}})\times(34.96885\,m_{\text{u}})}{(1.007825\,m_{\text{u}})+(34.96885\,m_{\text{u}})}\right)\times(1.66054\times10^{-27}\ \text{kg}\ m_{\text{u}}^{-1})$$
$$=1.62665\times10^{-27}\ \text{kg}$$

$$m_{\text{eff}}(\text{DCl})=\left(\frac{(2.0140\,m_{\text{u}})\times(34.96885\,m_{\text{u}})}{(2.0140\,m_{\text{u}})+(34.96885\,m_{\text{u}})}\right)\times(1.66054\times10^{-27}\ \text{kg}\ m_{\text{u}}^{-1})$$
$$=3.1622\times10^{-27}\ \text{kg}$$

$$R^2(\text{HCl})=\frac{2.79927\times10^{-44}\ \text{kg m}}{(1.62665\times10^{-27}\ \text{kg})\times(1.04392\times10^{3}\ \text{m}^{-1})}=1.64848\times10^{-20}\ \text{m}^2$$

$$R(\text{HCl})=1.28393\times10^{-10}\ \text{m}=\boxed{128.393\ \text{pm}}$$

$$R^2(^{2}\text{HCl})=\frac{2.79927\times10^{-44}\ \text{kg m}}{(3.1622\times10^{-27}\ \text{kg})\times(5.3920\times10^{2}\ \text{m}^{-1})}=1.6417\times10^{-20}\ \text{m}^2$$

$$R(^{2}\text{HCl})=1.2813\times10^{-10}\ \text{m}=\boxed{128.13\ \text{pm}}$$

Comment. Since the effects of centrifugal distortion have not been taken into account, the number of significant figures in the calculated values of R above should be no greater than 4, despite the fact that the data are precise to 6 figures.

P42.5 From the equation for a linear rotor in Table 41.1 it is possible to show that

$$I\times m=m_{\text{a}}m_{\text{c}}(R+R')^2+m_{\text{a}}m_{\text{b}}R^2+m_{\text{b}}m_{\text{c}}R'^2.$$

$$\text{Thus, } I(^{16}\text{O}^{12}\text{C}^{32}\text{S})=\left(\frac{m(^{16}\text{O})m(^{32}\text{S})}{m(^{16}\text{O}^{12}\text{C}^{32}\text{S}}\right)\times(R+R')^2+\left(\frac{m(^{12}\text{C})\{m(^{16}\text{O})R^2+m(^{32}\text{S})R'^2\}}{m(^{16}\text{O}^{12}\text{C}^{32}\text{S})}\right)$$

$$I(^{16}\text{O}^{12}\text{C}^{34}\text{S})=\left(\frac{m(^{16}\text{O})m(^{34}\text{S})}{m(^{16}\text{O}^{12}\text{C}^{34}\text{S}}\right)\times(R+R')^2+\left(\frac{m(^{12}\text{C})\{m(^{16}\text{O})R^2+m(^{34}\text{S})R'^2\}}{m(^{16}\text{O}^{12}\text{C}^{34}\text{S})}\right)$$

$m(^{16}\text{O})=15.9949\,m_{\text{u}}$, $m(^{12}\text{C})=12.0000\,m_{\text{u}}$, $m(^{32}\text{S})=31.9721\,m_{\text{u}}$, and $m(^{34}\text{S})=33.9679\,m_{\text{u}}$.

Hence,

$$I(^{16}\text{O}^{12}\text{C}^{32}\text{S})/m_{\text{u}}=(8.5279)\times(R+R')^2+(0.20011)\times(15.9949R^2+31.9721R'^2)$$

$$I(^{16}\text{O}^{12}\text{C}^{34}\text{S})/=(8.7684)\times(R+R')^2+(0.19366)\times(15.9949R^2+33.9679R'^2)$$

The spectral data provides the experimental values of the moments of inertia based on the relation $\nu=2c\tilde{B}(J+1)$ [42.8a] with $\tilde{B}=\dfrac{\hbar}{4\pi cI}$ [41.7]. These values are set equal to the above equations which are then solved for R and R'. The mean values of I obtained from the data are

$$I(^{16}\text{O}^{12}\text{C}^{32}\text{S}) = 1.37998\times10^{-45}\ \text{kg m}^2$$

$$I(^{16}\text{O}^{12}\text{C}^{34}\text{S}) = 1.41460\times10^{-45}\ \text{kg m}^2$$

Therefore, after conversion of the atomic mass units to kg, the equations we must solve are

$$1.37998\times10^{-45}\ \text{m}^2 = (1.4161\times10^{-26})\times(R+R')^2 + (5.3150\times10^{-27}R^2) + (1.0624\times10^{-26}R'^2)$$

$$1.41460\times10^{-45}\ \text{m}^2 = (1.4560\times10^{-26})\times(R+R')^2 + (5.1437\times10^{-27}R^2) + (1.0923\times10^{-26}R'^2)$$

These two equations may be solved for R and R'. They are tedious to solve by hand, but straightforward. Exercise 41.4(b) illustrates the details of the solution. Readily available mathematical software can be used to quickly give the result. The outcome is $R = \boxed{116.28\ \text{pm}}$ and $R' = \boxed{155.97\ \text{pm}}$. These values may be checked by direct substitution into the equations.

Comment. The starting point of this problem is the actual experimental data on spectral line positions. Exercise 41.4(b) is similar to this problem; it starts, however, from given values of the rotational constants B, which were themselves obtained from the spectral line positions. So the results for R and R' are expected to be essentially identical and they are.

Question. What are the rotational constants calculated from the data on the positions of the absorption lines?

P42.7 Plot frequency against J as in Fig. 42.1

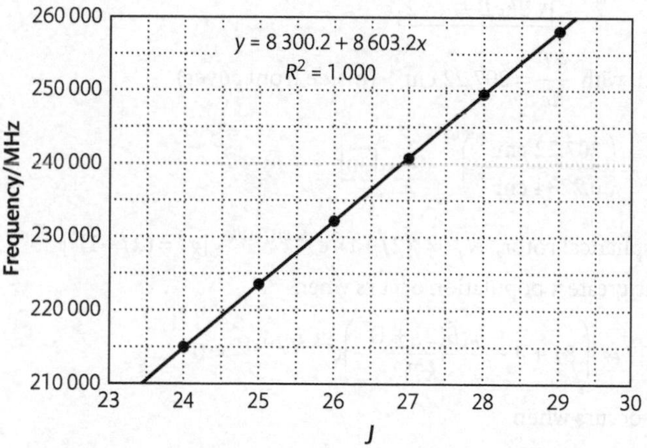

Fig 42.1

The rotational constant is related to the wavenumbers of observed transitions by

$$\tilde{\nu} = 2\tilde{B}(J+1) = \frac{\nu}{c} \quad \text{so} \quad \nu = 2\tilde{B}c(J+1)$$

A plot of ν versus J, then, has a slope of $2\tilde{B}c$. From Fig. 42.1, the slope is 8603 MHZ, so

$$\tilde{B}=\frac{8\,603\times10^{6}\,\text{s}^{-1}}{2(2.988\times10^{8}\,\text{m s}^{-1})}=\boxed{14.35\ \text{m}^{-1}}$$

The most highly populated energy level is roughly

$$J_{\text{max}}=\left(\frac{kT}{2hc\tilde{B}}\right)^{1/2}-\frac{1}{2}$$

$$\text{so } J_{\text{max}}=\left(\frac{(1.381\times10^{-23}\,\text{J K}^{-1})\times(298\,\text{K})}{(6.626\times10^{-34}\,\text{J s})\times(8\,603\times10^{6}\,\text{s}^{-1})}\right)^{1/2}-\frac{1}{2}=\boxed{26}\text{ at 298 K}$$

$$\text{and } J_{\text{max}}=\left(\frac{(1.381\times10^{-23}\,\text{J K}^{-1})\times(100\,\text{K})}{(6.626\times10^{-34}\,\text{J s})\times(8\,603\times10^{6}\,\text{s}^{-1})}\right)^{1/2}-\frac{1}{2}=\boxed{15}\text{ at 100 K}$$

P42.9 $N \propto ge^{-E/kT}$ [Boltzmann distribution, Topic 51]

$N_J \propto g_J e^{-E_J/kT} \propto (2J+1)e^{-hc\tilde{B}J(J+1)/kT}$ [$g_J=2J+1$ for a diatomic rotor]

The maximum population occurs when

$$\frac{\text{d}}{\text{d}J}N_J \propto \left\{2-(2J+1)^2\times\left(\frac{hc\tilde{B}}{kT}\right)\right\}e^{-hc\tilde{B}J(J+1)/kT}=0$$

and, since the exponential can never be zero at a finite temperature, then

$$(2J+1)^2\times\left(\frac{hc\tilde{B}}{kT}\right)=2$$

$$\text{or when } J_{\text{max}}=\boxed{\left(\frac{kT}{2hc\tilde{B}}\right)^{1/2}-\frac{1}{2}}$$

For ICl, with $\dfrac{kT}{hc}=207.22\ \text{cm}^{-1}$ (inside front cover)

$$J_{\text{max}}=\left(\frac{207.22\ \text{cm}^{-1}}{0.2284\ \text{cm}^{-1}}\right)^{1/2}-\frac{1}{2}=\boxed{30}$$

For a spherical rotor, $N_J \propto (2J+1)^2e^{-hc\tilde{B}J(J+1)/kT}$ [$g_J=(2J+1)^2$]

and the greatest population occurs when

$$\frac{\text{d}N_J}{\text{d}J}\propto\left(8J+4-\frac{hc\tilde{B}(2J+1)^3}{kT}\right)e^{-hc\tilde{B}J(J+1)/kT}=0$$

which occurs when

$$4(2J+1)=\frac{hc\tilde{B}(2J+1)^3}{kT}$$

$$\text{or at } J_{\text{max}}=\boxed{\left(\frac{kT}{hc\tilde{B}}\right)^{1/2}-\frac{1}{2}}$$

$$\text{For } CH_4,\ J_{max} = \left(\frac{207.22\ \text{cm}^{-1}}{5.24\ \text{cm}^{-1}}\right)^{1/2} - \frac{1}{2} = \boxed{6}$$

P42.11 *Temperature effects.* At extremely low temperatures (10 K) only the lowest rotational states are populated. No emission spectrum is expected for the cloud, and star light microwave absorptions by the cloud are by the lowest rotational states. At higher temperatures additional high-energy lines appear because higher energy rotational states are populated. Circumstellar clouds may exhibit infrared absorptions due to vibrational excitation as well as electronic transitions in the ultraviolet. Ultraviolet absorptions may indicate the photodissociation of carbon monoxide. High temperature clouds exhibit emissions.

Density effects. The density of an interstellar cloud may range from one particle to a billion particles per cm^3. This is still very much a vacuum compared to the laboratory high vacuum of a trillion particles per cm^3. Under such extreme vacuum conditions the half-life of any quantum state is expected to be extremely long and absorption lines should be very narrow. At the higher densities the vast size of nebulae obscures distant stars. High densities and high temperatures may create conditions in which emissions stimulate emissions of the same wavelength by molecules. A cascade of stimulated emissions greatly amplifies normally weak lines—the maser phenomena of Microwave Amplification by Stimulated Emission of Radiation.

Particle velocity effects. Particle velocity can cause Doppler broadening of spectral lines. The effect is extremely small for interstellar clouds at 10 K but is appreciable for clouds near high temperature stars. Outflows of gas from pulsing stars exhibit a red Doppler shift when moving away at high speed and a blue shift when moving toward us.

There will be many more transitions observable in circumstellar gas than in interstellar gas, because many more rotational states will be accessible at the higher temperatures. Higher velocity and density of particles in circumstellar material can be expected to broaden spectral lines compared to those of interstellar material by shortening collisional lifetimes. (Doppler broadening is not likely to be significantly different between circumstellar and interstellar material in the same astronomical neighborhood. The relativistic speeds involved are due to large-scale motions of the expanding universe, compared to which local thermal variations are insignificant.) A temperature of 1000 K is not high enough to significantly populate electronically excited states of CO; such states would have different bond lengths, thereby producing transitions with different rotational constants. Excited vibrational states would be accessible, though, and rovibrational transitions with P and R branches as detailed later in this chapter would be observable in circumstellar but not interstellar material. The rotational constant $\tilde{B}$ for $^{12}C^{16}O$ is 1.691 cm^{-1}. The first excited rotational energy level, $J=1$, with energy $J(J+1)hc\tilde{B} = 2hc\tilde{B}$, is thermally accessible at about 6 K (based on the rough equation of the rotational energy to thermal energy kT). In interstellar space, only two or three rotational lines would be observable; in circumstellar space (at about 1000 K) the number of transitions would be more like 20.

Topic 43 **Vibrational spectroscopy: Diatomic molecules**

Discussion questions

D43.1 True harmonic oscillation implies the existence of a parabolic potential energy. For low vibrational energies, near the bottom of the potential well, the assumption of a parabolic potential energy is a very good approximation. However, molecular vibrations are always anharmonic to a greater or lesser extent. At high excitation energies, the parabolic approximation is poor, and it is totally wrong near the dissociation limit. An advantage of the parabolic potential energy is that it allows for a relatively straight–forward solution of the Schrödinger equation for the vibrational motion. The Morse potential is a closer approximation to the true potential energy curve for molecular vibrations. It allows for the convergence of the energy levels at high values of the quantum numbers and for dissociation at large displacements. It fails at very short distances where it approaches a finite value. An advantage is that the Schrödinger equation can be solved exactly for the Morse potential.

D43.3 Rotational constants $\tilde{B}_v$ differ from one vibrational level, v, to another. The method of combination differences as described in Section 43.4(b) can be used to determine two rotational constants individually. The procedure is illustrated in detail in the solution to problem 43.9 where the values of the rotational constants $\tilde{B}_0$ and $\tilde{B}_1$ are determined for $^1H^{35}Cl$ from the combination difference equations, eqns 43.23 (a) and (b):

$$\Delta_0 = \tilde{\nu}_R(J-1) - \tilde{\nu}_P(J+1) = 4\tilde{B}_0(J+\tfrac{1}{2}), \text{ and}$$
$$\Delta_1 = \tilde{\nu}_R(J) - \tilde{\nu}_P(J) = 4\tilde{B}_1(J+\tfrac{1}{2})$$

Substituting values for $\tilde{\nu}_R$ and $\tilde{\nu}_P$ from data that gives the wavenumbers of the transitions for $J = 1, 2, 3, 4$ etc. we calculate the values for Δ_0 and Δ_1. Then $\tilde{B}_0$ and $\tilde{B}_1$ are calculated from the simultaneous equations for Δ_0 and Δ_1.

Exercises

E43.1(a) $$\omega = 2\pi\nu = \left(\frac{k}{m}\right)^{1/2}$$

$$k = 4\pi^2\nu^2 m = 4\pi^2 \times (2.0\ \text{s}^{-1})^2 \times (0.100\ \text{kg}) = 16\ \text{kg s}^{-2} = \boxed{16\ \text{N m}^{-1}}$$

E43.2(a) $\omega=\left(\dfrac{k}{m_{\mathrm{eff}}}\right)^{1/2}$ [43.7]

The fractional difference is

$$\frac{\omega'-\omega}{\omega}=\frac{\left(\dfrac{k}{m'_{\mathrm{eff}}}\right)^{1/2}-\left(\dfrac{k}{m_{\mathrm{eff}}}\right)^{1/2}}{\left(\dfrac{k}{m_{\mathrm{eff}}}\right)^{1/2}}=\frac{\left(\dfrac{1}{m'_{\mathrm{eff}}}\right)^{1/2}-\left(\dfrac{1}{m_{\mathrm{eff}}}\right)^{1/2}}{\left(\dfrac{1}{m_{\mathrm{eff}}}\right)^{1/2}}=\left(\frac{m_{\mathrm{eff}}}{m'_{\mathrm{eff}}}\right)^{1/2}-1$$

$$=\left(\frac{m(^{23}\mathrm{Na})m(^{35}\mathrm{Cl})\{m(^{23}\mathrm{Na})+m(^{37}\mathrm{Cl})\}}{\{m(^{23}\mathrm{Na})+m(^{35}\mathrm{Cl})\}m(^{23}\mathrm{Na})m(^{37}\mathrm{Cl})}\right)^{1/2}-1$$

$$=\left(\frac{m(^{35}\mathrm{Cl})}{m(^{37}\mathrm{Cl})}\times\frac{m(^{23}\mathrm{Na})+m(^{37}\mathrm{Cl})}{m(^{23}\mathrm{Na})+m(^{35}\mathrm{Cl})}\right)^{1/2}-1$$

$$=\left(\frac{34.9688}{36.9651}\times\frac{22.9898+36.9651}{22.9898+34.9688}\right)^{1/2}-1=-0.01077$$

Hence, the difference is $\boxed{1.077\ \text{percent}}$

E43.3(a) $\omega=\left(\dfrac{k}{m_{\mathrm{eff}}}\right)^{1/2}$ [43.7]; $\omega=2\pi\nu=2\pi\left(\dfrac{c}{\lambda}\right)=2\pi c\tilde{\nu}$

Therefore, $k=m_{\mathrm{eff}}\omega^2=4\pi^2m_{\mathrm{eff}}c^2\tilde{\nu}^2, m_{\mathrm{eff}}=\tfrac{1}{2}m(^{35}\mathrm{Cl})$

$$=(4\pi^2)\times\left(\frac{34.9688}{2}\right)\times(1.66054\times10^{-27}\ \mathrm{kg})$$
$$\times[(2.997924\times10^{10}\ \mathrm{cm\ s^{-1}})\times(564.9\ \mathrm{cm^{-1}})]^2$$
$$=\boxed{328.7\ \mathrm{N\ m^{-1}}}$$

E43.4(a) $\omega=\left(\dfrac{k}{m_{\mathrm{eff}}}\right)^{1/2}$ [43.7], so $k=m_{\mathrm{eff}}\omega^2=4\pi^2m_{\mathrm{eff}}c^2\tilde{\nu}^2$

$$m_{\mathrm{eff}}=\frac{m_1m_2}{m_1+m_2}[43.3]$$

$$m_{\mathrm{eff}}(\mathrm{H}^{19}\mathrm{F})=\frac{(1.0078)\times(18.9984)}{(1.0078)+(18.9984)}m_{\mathrm{u}}=0.9570\,m_{\mathrm{u}}$$

$$m_{\mathrm{eff}}(\mathrm{H}^{35}\mathrm{Cl})=\frac{(1.0078)\times(34.9688)}{(1.0078)+(34.9688)}m_{\mathrm{u}}=0.9796\,m_{\mathrm{u}}$$

$$m_{\mathrm{eff}}(\mathrm{H}^{81}\mathrm{Br})=\frac{(1.0078)\times(80.9163)}{(1.0078)+(80.9163)}m_{\mathrm{u}}=0.9954\,m_{\mathrm{u}}$$

$$m_{\text{eff}}\left(\text{H}^{127}\text{I}\right)=\frac{(1.0078)\times(126.9045)}{(1.0078)+(126.9045)}m_{\text{u}}=0.9999\,m_{\text{u}}$$

We draw up the following table

Question. Which ratio, $\frac{k}{B(\text{A}-\text{B})}$ or $\frac{\tilde{\nu}}{B(\text{A}-\text{B})}$, where $B(\text{A}-\text{B})$ are the bond energies, is the more nearly constant across the series of hydrogen halides? Why?

	HF	HCl	HBr	HI
$\tilde{\nu}$/cm^{-1}	4 141.3	2 988.9	2 649.7	2 309.5
$m_{\text{eff}}/m_{\text{u}}$	0.9570	0.9697	0.9954	0.9999
k/(N m^{-1})	$\boxed{967.0}$	$\boxed{515.6}$	$\boxed{411.8}$	$\boxed{314.2}$

Note the order of stiffness HF > HCl > HBr > HI.

E43.5(a) Data on three transitions are provided. Only two are necessary to obtain the value of $\tilde{\nu}$ and x_{e}. The third datum can then be used to check the accuracy of the calculated values.

$$\Delta\tilde{G}(v=1\leftarrow 0)=\tilde{\nu}-2\tilde{\nu}x_{\text{e}}=1\,556.22\ \text{cm}^{-1}[43.17]$$
$$\Delta\tilde{G}(v=2\leftarrow 0)=2\tilde{\nu}-6\tilde{\nu}x_{\text{e}}=3\,088.28\ \text{cm}^{-1}[43.18]$$

Multiply the first equation by 3, then subtract the second

$$\tilde{\nu}=(3)\times(1\,556.22\ \text{cm}^{-1})-(3\,088.28\ \text{cm}^{-1})=\boxed{1\,580.38\ \text{cm}^{-1}}$$

Then from the first equation

$$x_{\text{e}}=\frac{\tilde{\nu}-1\,556.22\,\text{cm}^{-1}}{2\tilde{\nu}}=\frac{(1\,580.38-1\,556.22)\text{cm}^{-1}}{(2)\times(1\,580.38\,\text{cm}^{-1})}=\boxed{7.644\times10^{-3}}$$

x_{e} data are usually reported as $x_{\text{e}}\tilde{\nu}$ which is

$$x_{\text{e}}\tilde{\nu}=12.08\ \text{cm}^{-1}$$

$$\Delta\tilde{G}(v=3\leftarrow 0)=3\tilde{\nu}-12\tilde{\nu}x_{\text{e}}$$
$$=(3)\times(1\,580.38\ \text{cm}^{-1})-(12)\times(12.08\ \text{cm}^{-1})=4\,596.18\ \text{cm}^{-1}$$

which is very close to the experimental value.

E43.6(a) See *Brief Illustration* 43.2 Select those molecules in which a vibration gives rise to a change in dipole moment. It is helpful to write down the structural formulas of the compounds. The infrared active compounds are

(b) HCl **(c)** CO_2 **(d)** H_2O

Comment: A more powerful method for determining infrared activity based on symmetry considerations is described in Section 44.4. Also see Exercises 44.5 and 44.6.

Problems

P43.1 The wavenumbers of the transitions with $\Delta v = +1$ are

$$\Delta G_{v+1/2} = \tilde{\nu} - 2(v+1)x_e\tilde{\nu} \ [43.17] \quad \text{and} \quad D_e = \frac{\tilde{\nu}^2}{4x_e\tilde{\nu}} [43.15]$$

A plot of $\Delta\tilde{G}(v)$ against $v+1$ should give a straight line with intercept $\tilde{\nu}$ at $v+1=0$ and slope $-2x_e\tilde{\nu}$.

Draw up the following table:

$v+1$	1	2	3
$\Delta\tilde{G}(v)$ / cm^{-1}	284.50	283.00	281.50

The points are plotted in Fig. 43.1.

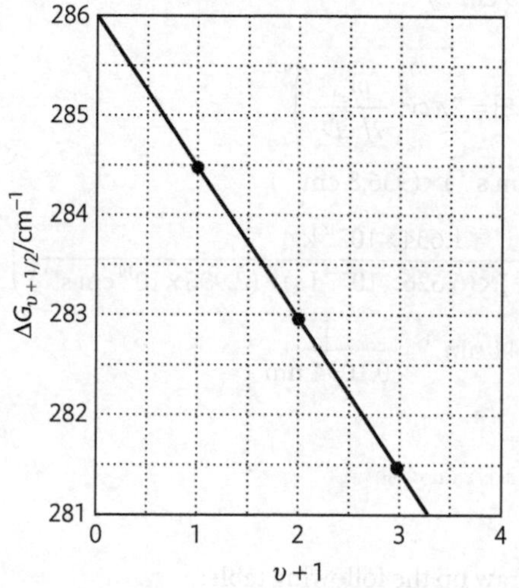

Fig 43.1

The intercept is at 286.0, so $\tilde{\nu} = 286\ \text{cm}^{-1}$. The slope is -1.50, so $x_e\tilde{\nu} = 0.750\ \text{cm}^{-1}$. It follows that

$$D_e = \frac{(286\ \text{cm}^{-1})^2}{(4)\times(0.750\ \text{cm}^{-1})} = 27\ 300\ \text{cm}^{-1}, \quad \text{or} \quad 3.38\ \text{eV}$$

The zero-point level lies at $\boxed{142.81\ \text{cm}^{-1}}$ and so $D_0 = \boxed{3.36\ \text{eV}}$. Since

$$m_{\text{eff}} = \frac{(22.99)\times(126.90)}{(22.99)+(126.90)}\,m_{\text{u}} = 19.46\bar{4}\,m_{\text{u}}$$

the force constant of the molecule is

$$k = 4\pi^2 m_{\text{eff}} c^2 \tilde{\nu}^2 \text{ [Exercise 43.3(a)]}$$
$$= (4\pi^2)\times(19.46\bar{4})\times(1.6605\times10^{-27}\ \text{kg})\times[(2.998\times10^{10}\ \text{cm s}^{-1})\times(286\ \text{cm}^{-1})]^2$$
$$= \boxed{93.8\ \text{N m}^{-1}}$$

P43.3 $V(R) = hc\tilde{D}_{\text{e}}\{1-\text{e}^{-a(R-R_{\text{e}})}\}^2$ [43.14]

$$\tilde{\nu} = \frac{\omega}{2\pi c} = 936.8\ \text{cm}^{-1} \quad x_{\text{e}}\tilde{\nu} = 14.15\ \text{cm}^{-1}$$

$$a = \left(\frac{m_{\text{eff}}}{2hc\tilde{D}_{\text{e}}}\right)^{1/2}\omega \quad x_{\text{e}} = \frac{\hbar a^2}{2m_{\text{eff}}\omega} \quad \tilde{D}_{\text{e}} = \frac{\tilde{\nu}}{4x_{\text{e}}}$$

$$m_{\text{eff}}(\text{RbH}) \approx \frac{(1.008)\times(85.47)}{(1.008)+(85.47)}\,m_{\text{u}} = 1.654\times10^{-27}\ \text{kg}$$

$$\tilde{D}_{\text{e}} = \frac{\tilde{\nu}^2}{4x_{\text{e}}\tilde{\nu}} = \frac{(936.8\ \text{cm}^{-1})^2}{(4)\times(14.15\ \text{cm}^{-1})} = 15\ 50\bar{5}\ \text{cm}^{-1}\ (1.92\ \text{eV})$$

$$a = 2\pi\nu\left(\frac{m_{\text{eff}}}{2hc\tilde{D}_{\text{e}}}\right)^{1/2} [43.8] = 2\pi c\tilde{\nu}\left(\frac{m_{\text{eff}}}{2hc\tilde{D}_{\text{e}}}\right)^{1/2}$$
$$= (2\pi)\times(2.998\times10^{10}\ \text{cm s}^{-1})\times(936.8\ \text{cm}^{-1})$$
$$\times\left(\frac{1.654\times10^{-27}\ \text{kg}}{(2)\times(15\ 505\ \text{cm}^{-1})\times(6.626\times10^{-34}\ \text{J s})\times(2.998\times10^{10}\ \text{cm s}^{-1})}\right)^{1/2}$$
$$= 9.144\times10^{9}\ \text{m}^{-1} = 9.44\ \text{nm}^{-1} = \frac{1}{0.1094\ \text{nm}}$$

Therefore, $\dfrac{V(R)}{hc\tilde{D}_{\text{e}}} = \{1-\text{e}^{-(R-R_{\text{e}})/(0.1094\ \text{nm})}\}^2$

with $R_{\text{e}} = 236.7$ pm. We draw up the following table:

R/pm	50	100	200	300	400	500	600	700	800
$V/(hc\tilde{D}_{\text{e}})$	20.4	6.20	0.159	0.193	0.601	0.828	0.929	0.971	0.988

These points are plotted in Fig 43.2 as the line labeled $J = 0$

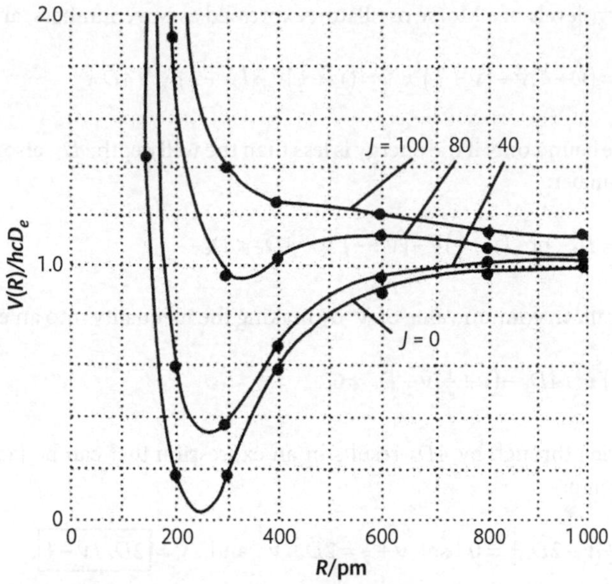

Fig 43.2

For the second part, we note that $\tilde{B} \propto \dfrac{1}{R^2}$ and write

$$V_J^* = V + hc\tilde{B}_e J(J+1)\times\left(\frac{R_e^2}{R^2}\right)$$

with $\tilde{B}_e$ the equilibrium rotational constant, $\tilde{B}_e = 3.020\ \text{cm}^{-1}$.

We then draw up the following table using the values of V calculated above:

R/pm	50	100	200	300	400	600	800	1 000
$\dfrac{R_e}{R}$	4.73	2.37	1.18	0.79	0.59	0.39	0.30	0.24
$\dfrac{V}{hc\tilde{D}_e}$	20.4	6.20	0.159	0.193	0.601	0.929	0.988	1.000
$\dfrac{V_{40}^*}{hc\tilde{D}_e}$	27.5	7.99	0.606	0.392	0.713	0.979	1.016	1.016
$\dfrac{V_{80}^*}{hc\tilde{D}_e}$	48.7	13.3	1.93	0.979	1.043	1.13	1.099	1.069
$\dfrac{V_{100}^*}{hc\tilde{D}_e}$	64.5	17.2	2.91	1.42	1.29	1.24	1.16	1.11

These points are also plotted in Fig 43.2

P43.5 The energy levels of a Morse oscillator, expressed as wavenumbers, are given by:

$$\tilde{G}(v)=(v+\tfrac{1}{2})\tilde{\nu}-(v+\tfrac{1}{2})^2x_e\tilde{\nu}=(v+\tfrac{1}{2})\tilde{\nu}-(v+\tfrac{1}{2})^2\tilde{\nu}^2/4\tilde{D}_e.$$

States are bound only if the energy is less than the well depth, $\tilde{D}_e$, also expressed as a wavenumber:

$$\tilde{G}(v)<\tilde{D}_e \quad \text{or} \quad (v+\tfrac{1}{2})\tilde{\nu}-(v+\tfrac{1}{2})^2\tilde{\nu}^2/4\tilde{D}_e<\tilde{D}_e$$

Solve for the maximum value of v by making the inequality into an equality:

$$(v+\tfrac{1}{2})^2\tilde{\nu}^2/4\tilde{D}_e-(v+\tfrac{1}{2})\tilde{\nu}+\tilde{D}_e=0$$

Multiplying through by $4\tilde{D}_e$ results in an expression that can be factored by inspection into:

$$\left[(v+\tfrac{1}{2})\tilde{\nu}-2\tilde{D}_e\right]^2=0 \quad \text{so} \quad v+\tfrac{1}{2}=2\tilde{D}_e/\tilde{\nu} \quad \text{and} \quad v=\boxed{2\tilde{D}_e/\tilde{\nu}-\tfrac{1}{2}}.$$

Of course, v is an integer, so its maximum value is really the greatest integer less than this quantity.

P43.7 $$\tilde{B}=\frac{\hbar}{4\pi cI}\ [41.7];\quad I=m_{\text{eff}}R^2;\quad R^2=\frac{\hbar}{4\pi cm_{\text{eff}}\tilde{B}}$$

$$m_{\text{eff}}=\frac{m_\text{C}m_\text{O}}{m_\text{C}+m_\text{O}}=\left(\frac{(12.0000\,m_\text{u})\times(15.9949\,m_\text{u})}{(12.0000\,m_\text{u})+(15.9949\,m_\text{u})}\right)\times(1.66054\times10^{-27}\text{ kg }m_\text{u}^{-1})$$
$$=1.13852\times10^{-26}\text{ kg}$$

$$\frac{\hbar}{4\pi c}=2.79932\times10^{-44}\text{ kg m}$$

$$R_0^2=\frac{2.79932\times10^{-44}\text{ kg m}}{(1.13852\times10^{-26}\text{ kg})\times(1.9314\times10^2\text{ m}^{-1})}=1.2730\bar{3}\times10^{-20}\text{ m}^2$$

$$R_0=1.1283\times10^{-10}\text{ m}=\boxed{112.83\text{ pm}}$$

$$R_1^2=\frac{2.79932\times10^{-44}\text{ kg m}}{(1.13852\times10^{-26}\text{ kg})\times(1.6116\times10^2\text{ m}^{-1})}=1.52565\times10^{-20}\text{ m}^2$$

$$R_1=1.2352\times10^{-10}\text{ m}=\boxed{123.52\text{ pm}}$$

Comment. The change in internuclear distance is roughly 10%, indicating that the rotations and vibrations of molecules are strongly coupled and that it is an oversimplification to consider them independently of each other.

P43.9 Examination of the wavenumbers of the transitions allows the following identifications to be made as shown in the table below. The J value in the table is the J value of the rotational states in the $v = 0$ vibrational level. The notation (R) and (P) after the J values refers to the R and P branches of the vibration–rotation spectrum.

$\tilde{\nu}$	2 998.05	2 981.05	2 963.35	2 944.99	2 925.92	2 906.25	2 865.14	2 843.63	2 821.59	2 799.00
J	5(R)	4(R)	3(R)	2(R)	1(R)	0(R)	1(P)	2(P)	3(P)	4(P)

The values of $\tilde{B}_0$ and $\tilde{B}_1$ are determined from the equations for the combination differences, eqns 43.23(a) and (b).

$$\Delta_0 = \tilde{\nu}_R(J-1) - \tilde{\nu}_P(J+1) = 4\tilde{B}_0(J+½), \text{ and}$$
$$\Delta_1 = \tilde{\nu}_R(J) - \tilde{\nu}_P(J) = 4\tilde{B}_1(J+½)$$

Substituting values for $\tilde{\nu}_R$ and $\tilde{\nu}_P$ from the above table for J = 1, 2, 3, and 4 we obtain the following values for Δ_0 and Δ_1.

J	1	2	3	4
$J+½$	3/2	5/2	7/2	9/2
Δ_0/cm^{-1}	62.62	104.33	145.99	
Δ_1/cm^{-1}	60.78	101.36	141.76	182.05
$\tilde{B}_0/\text{cm}^{-1}$	10.437	10.433	10.428	
$\tilde{B}_1/\text{cm}^{-1}$	10.130	10.136	10.126	10.113

An average of these values gives $\boxed{\tilde{B}_0 = 10.433\ \text{cm}^{-1}}$ and $\boxed{\tilde{B}_1 = 10.126\ \text{cm}^{-1}}$. Values for $\tilde{B}_0$ and $\tilde{B}_1$ may also be obtained by plotting the combination differences, Δ_0 and Δ_1, against $J+½$; the slopes give $4\tilde{B}_0$ and $4\tilde{B}_1$, respectively.

P43.11 The virial theorem states that if the potential energy of a particle has the form $V = ax^b$ then its mean potential and kinetic energies are related by $2\langle E_k\rangle = b\langle V\rangle$ [12.14]. For the harmonic oscillator potential energy b = 2 and $\langle E_k\rangle = ½E_v$ [12.13c] $= ½(v+½)\hbar\omega$ [12.13a]. Hence, $\langle V\rangle = \langle E_K\rangle = ½E_v = ½(v+½)\hbar\omega = ½k_f\langle x^2\rangle$ and

$\boxed{\langle x^2\rangle = \dfrac{1}{k_f}(v+½)\hbar\omega.}$ We see that as v increases, $\langle x^2\rangle$ increases and $\langle R^2\rangle = R_e^2 + \langle x^2\rangle$

[Solution to Problem 43.10] increases. As $\langle R^2\rangle$ increases, the moment of inertia I increases [Table 41.1]. As I increases the $\boxed{\text{rotational constant } B \text{ decreases}}$ [eqn 41.15] as the oscillator is excited to higher quantum states. Anharmonicity results in greater average values of R^2 as v increases [Fig. 43.6]; hence, $\boxed{B \text{ decreases with increased anharmonicity}}$.

P43.13 The set of peaks to the left of center are the P branch, those to the right are the R branch (with $\tilde{\nu}$ increasing to the right in the CO spectrum). Within the rigid rotor approximation the two sets are separated by $4\tilde{B}$. The effects of the interactions between vibration and rotation and of centrifugal distortion are least important for transitions with small J values hence the separation between the peaks immediately to the left and right of center will give good approximate values of B and bond length.

(a) $\tilde{\nu}_{\rm Q}(J) = \tilde{\nu}[43.21\text{b}] = 2143.26\ \text{cm}^{-1}$

(b) The zero–point energy is $\frac{1}{2}\tilde{\nu} = 1071.63\ \text{cm}^{-1}$. The molar zero–point energy in J mol^{-1} is

$$N_{\rm A}hc\times(1071.63\ \text{cm}^{-1}) = N_{\rm A}hc\times(1.07163\times10^{5}\ \text{m}^{-1})$$
$$= 1.28195\times10^{4}\ \text{J mol}^{-1} = \boxed{12.8195\ \text{kJ mol}^{-1}}$$

(c) $k = 4\pi^2\mu c^2\tilde{\nu}^2$

$$\mu(^{12}\text{C}^{16}\text{O}) = \frac{m_{\rm C}m_{\rm O}}{m_{\rm C}+m_{\rm O}} = \left(\frac{(12.0000\ m_{\rm u})\times(15.9949\ m_{\rm u})}{(12.0000\ m_{\rm u})+(15.9949\ m_{\rm u})}\right)\times(1.66054\times10^{-27}\ \text{kg}\ m_{\rm u}^{-1})$$
$$= 1.13852\times10^{-26}\ \text{kg}$$

$$k = 4\pi^2c^2\times(1.13852\times10^{-26}\ \text{kg})\times(2.14326\times10^{5}\ \text{m}^{-1})^2 = \boxed{1.85563\times10^{3}\text{N m}^{-1}}$$

(d) $4\tilde{B} \approx 7.655\ \text{cm}^{-1}$

$\tilde{B} \approx \boxed{1.91\ \text{cm}^{-1}}$ [4 significant figures not justified]

(e) $\tilde{B} = \dfrac{\hbar}{4\pi cI}[41.7] = \dfrac{\hbar}{4\pi c\mu R^2}$ [Table 41.1]

$$R^2 = \frac{\hbar}{4\pi c\mu\tilde{B}} = \frac{\hbar}{(4\pi c)\times(1.13852\times10^{-26}\ \text{kg})\times(191\ \text{m}^{-1})} = 1.28\overline{7}\times10^{-20}\ \text{m}^2$$

$R = 1.13\times10^{-10}\ \text{m} = \boxed{113\ \text{pm}}$

Topic 44 Vibrational spectroscopy: Polyatomic molecules

Discussion questions

D44.1 The gross selection rules tell us which are the allowed spectroscopic transitions. For both microwave and infrared spectroscopy, the allowed transitions depend on the existence of an oscillating dipole moment which can stir the electromagnetic field into oscillation (and vice versa for absorption). For microwave rotational spectroscopy, this implies that the molecule must have a permanent dipole moment, which is equivalent to an oscillating dipole when the molecule is rotating. See Fig. 42.1 of the text. In the case of infrared vibrational spectroscopy, the physical basis of the gross selection rule is that the molecule have a structure that allows for the existence of an oscillating dipole moment when the molecule vibrates. Polar molecules necessarily satisfy this requirement, but non-polar molecules may also have a fluctuating dipole moment upon vibration. See Fig. 43.3 of the text.

D44.3 The exclusion rule applies to the benzene molecule because it has a center of symmetry. Consequently, none of the normal modes of vibration of benzene can be both infrared and Raman active. If we wish to characterize all the normal modes we must obtain both kinds of spectra. See the solutions to Exercises 44.6(a) and 44.6(b) for specific illustrations of which modes are IR active and which are Raman active.

Exercises

E44.1(a) See *Brief Illustration* 44.3. Select those molecules in which a vibration gives rise to a change in dipole moment. It is helpful to write down the structural formulas of the compounds. The infrared active compounds are

(b) HCl **(c)** CO_2 **(d)** $H_2 0$

Comment: A more powerful method for determining infrared activity based on symmetry considerations is described in Section 44.4. Also see Exercises 44.6 and 44.7.

E44.2(a) The number of normal modes of vibration is given by (Section 44.4)

$$N_{\text{vib}} = \begin{Bmatrix} 3N-5 \text{ for linear molecules} \\ 3N-6 \text{ for nonlinear molecules} \end{Bmatrix}$$

where N is the number of atoms in the molecule. Hence, since none of these molecules are linear,

(a) 3 **(b)** 6 **(c)** 12

Comment: Even for moderately sized molecules the number of normal modes of vibration is large and they are usually difficult to visualize.

E44.3(a) This molecule is linear; hence the number of vibrational modes is $3N-5$. $N=44$ in this case; therefore, the number of vibrational modes is $\boxed{127}$

E44.4(a) $G_q(v)=(v+\frac{1}{2})\tilde{\nu}_q \qquad \tilde{\nu}_q=\frac{1}{2\pi c}\left(\frac{k_q}{m_q}\right)^{1/2}$ [44.1]

The lowest energy term is $\tilde{\nu}_2$ corresponding to the normal mode for bending. For this mode the oxygen atom may be considered to remain stationary and the effective mass is approximately $m_q=\frac{2m_{\mathrm{H}}m_{\mathrm{O}}}{2m_{\mathrm{H}}+m_{\mathrm{O}}}$. For the other modes the effective mass expression is more complicated and is beyond the scope of this text. However we know that in the ground vibrational state all normal modes have $v = 0$. Thus, since H_2O has the three normal modes shown in text Fig. 44.3, the ground vibrational term is the sum of eqn 44.1 normal mode terms:

$G_{\text{ground}}=G_1(0)+G_2(0)+G_3(0)=\boxed{\tfrac{1}{2}(\tilde{\nu}_1+\tilde{\nu}_2+\tilde{\nu}_3)}$

E44.5(a) See Figs. 44.3(H_2O, bent) and 44.2(CO_2, linear) of the text as well as the accompanying *Brief Illustration*. Decide which modes correspond to (i) a changing electric dipole moment, (ii) a changing polarizability, and take note of the exclusion rule.

(a) Nonlinear: all modes are both infrared and Raman active.

(b) Linear: the symmetric stretch is infrared inactive but Raman active.

The antisymmetric stretch is infrared active and (by the exclusion rule) Raman inactive. The two bending modes are infrared active and therefore Raman inactive.

E44.6(a) The uniform expansion is depicted in Fig. 44.1.

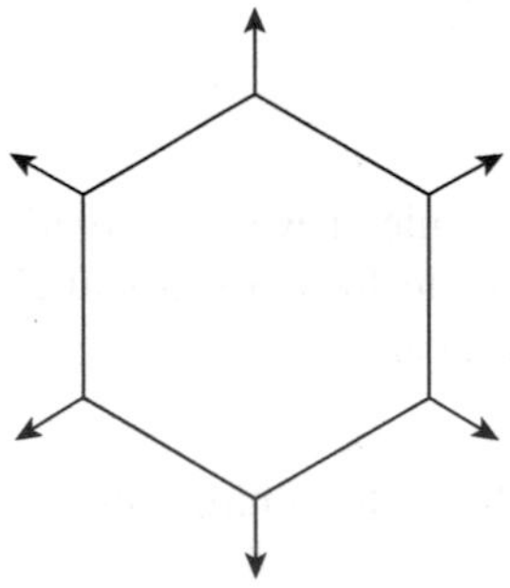

Benzene is centrosymmetric, and so the exclusion rule applies (Section 44.4). The mode is infrared inactive (symmetric breathing leaves the molecular dipole moment unchanged at zero), and therefore the mode may be $\boxed{\text{Raman active}}$ (and is). In group theoretical terms, the breathing mode has symmetry A_{1g} in D_{6h}, which is the point group for benzene, and quadratic forms x^2+y^2 and z^2 have this symmetry (see the character table for C_{6h}, a subgroup of D_{6h}). Hence, the mode is Raman active.

E44.7(a) Use the character table for the group C_{2v} (and see Example 44.3). The rotations span $A_2+B_1+B_2$. The translations span $A_1+B_1+B_2$. Hence the normal modes of vibration span the difference, $\boxed{4A_1+A_2+2B_1+2B_2}$

Comment: A_1, B_1 and B_2 are infrared active; all modes are Raman active.

E44.8(a) See the comment in the solution to Exercise 44.7(a). A_1, B_1 and B_2 are infrared active; all modes are Raman active.

Problems

P44.1 A sketch of $V(h)=V_0\left(1-e^{-bh^4}\right)$ is presented in Fig. 44.2.

(a) By analogy to the Morse potential and to the harmonic potential, we expect that

$$b^{1/4} \propto \nu \propto k^{1/2}$$

(b) Constructive interference of valence orbitals, shown in the adjacent sketch, should be the basic character of the ground-state wavefunction.

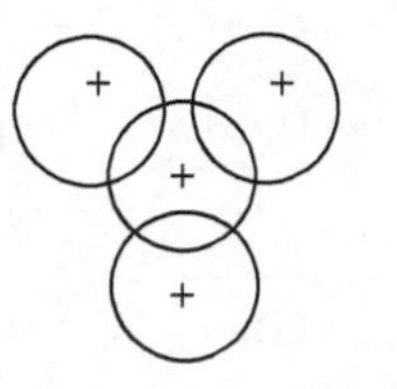

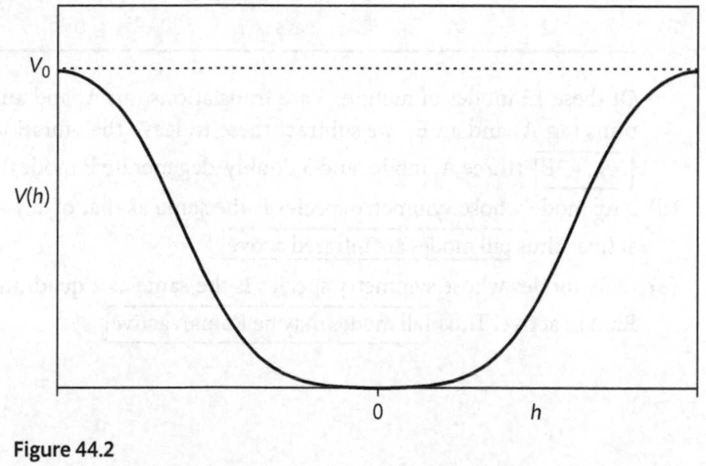

Figure 44.2

P44.3 (a) Follow the flow chart in Fig. 31.7 of the text. CH_3Cl is not linear, it has a C_3 axis (only one), it does not have C_2 axes perpendicular to C_3, it has no σ_h, but does have 3 σ_v planes; so it belongs to $\boxed{C_{3v}}$.

(b) The number of normal modes of a non-linear molecule is $3N-6$, where N is the number of atoms. So CH_3Cl has $\boxed{\text{nine}}$ normal modes

(c) To determine the symmetry of the normal modes, consider how the Cartesian axes of each atom are transformed under the symmetry operations of the C_{3v} group; the 15 Cartesian displacements constitute the basis here. All 15 cartesian axes are left unchanged under the identity, so the character of this operation is 15. Under a C_3 operation, the H atoms are taken into each other, so they do not contribute to the character of C_3. The z axes of the C and Cl atoms, are unchanged, so they contribute 2 to the character of C_3; for these two atoms

$$x \to -\frac{x}{2}+\frac{3^{1/2}y}{2} \text{ and } y \to -\frac{y}{2}+\frac{3^{1/2}x}{2}$$

so there is a contribution of −1/2 to the character from each of these coordinates in each of these atoms. In total, then $\chi = 0$ for C_3. To find the character of σ_v, call one of the σ_v planes the yz plane; it contains C, Cl, and one H atom. The y and z coordinates of these three atoms are unchanged, but the x coordinates are taken into their negatives, contributing 6 − 3 = 3 to the character for this operation; the other two atoms are interchanged, so they contribute nothing to the character. To find the irreducible representations that this basis spans, we multiply its characters by the characters of the irreducible representations, sum those products, and divide the sum by the order h of the group. The table below illustrates the procedure:

	E	$2C_2$	$3\sigma_v$		E	$2C_2$	$3\sigma_v$	sum / h
basis	15	0	3					
A_1	1	1	1	basis × A_1	15	0	3	4
A_2	1	1	−1	basis × A_2	15	0	−3	1
E	2	−1	0	basis × E	30	0	0	5

Of these 15 modes of motion, 3 are translations (an A_1 and an E) and 3 rotations (an A_2 and an E); we subtract these to leave the vibrations, which span $\boxed{3A_1 + 3E}$ (three A_1 modes and 3 doubly-degenerate E modes).

(d) Any mode whose symmetry species is the same as that of x, y, or z is infrared active. Thus $\boxed{\text{all modes are infrared active}}$.

(e) Only modes whose symmetry species is the same as a quadratic form may be Raman active. Thus $\boxed{\text{all modes may be Raman active}}$.

Topic 45 **Electronic spectroscopy**

Discussion questions

45.1 The ground electronic configuration of dioxygen, $1\sigma_g^{\,2}1\sigma_u^{\,2}2\sigma_g^{\,2}1\pi_u^{\,4}1\pi_g^{\,2}$, is discussed in Topic 24.1(c) and the determination of the term symbol, ${}^3\Sigma_g^-$, is described in Brief illustrations 45.1 and 45.2. The term symbol is Σ to represent a total orbital angular momentum of zero about the internuclear axis. This happens because for every π orbital electron with $\lambda = +1$ there is a π orbital electron with $\lambda = -1$. For example, the two $1\pi_g$ electrons are, according to Hund's rules, in separate degenerate orbitals for which one orbital has $\lambda = +1$ and the other has $\lambda = -1$. Except for the two $1\pi_g$ electrons, electrons have paired α and β spins, which results in zero contribution to the total spin angular momentum. According to Hund's rules, the two $1\pi_g$ electrons (i.e., $1\pi_g^{\,1}1\pi_g^{\,1}$) have parallel spins in the ground state. They provide a total spin angular momentum of $S = \frac{1}{2} + \frac{1}{2} = 1$ and a spin multiplicity of $2S + 1 = 3$, which appears as the left superscript 3. The term symbol indicates a gerade total symmetry because electrons are paired in the ungerade molecular orbitals and $u \times u = g$ and the resultant symmetry of electrons in different molecular orbitals must therefore be given by $g \times g = g$. The π orbitals change sign upon reflection in the plane that contains the internuclear axis. Consequently, the term symbol has the superscript − to indicate that the molecular wavefunction for O_2 changes sign upon reflection in the plane containing the nuclei.

45.3 A band head is the convergence of the frequencies of electronic transitions with increasing rotational quantum number, J. They result from the rotational structure superimposed on the vibrational structure of the electronic energy levels of the diatomic molecule. See Fig. 45.10. To understand how a band head arises, one must examine the equations describing the transition energies (eqn 45.8) where we see that convergence to an extreme in the energy absorption can only arise when terms in both $(B' - B)$ and $(B' + B)$ occur in the equation. The extreme occurs in the P branch when $B' > B$ but in is in the R branch when $B' < B$. Because only a term in $(B' - B)$ occurs for the Q branch, no band head can arise for that branch. The following Mathcad Prime 2 worksheet (in which v represents wavenumber) shows these branch heads by plotting $\Delta\tilde{\nu} \equiv \tilde{\nu}(J) - \tilde{\nu}$ [45.8a and c] against J.

$\Delta v_P(B, B', J) := -(B' + B)\cdot J + (B' - B)\cdot J^2$ $\qquad$ $i := 1..8$ $\qquad$ $J_i := i$

$\Delta v_R(B, B', J) := (B' + B)\cdot(J + 1) + (B' - B)\cdot(J + 1)^2$

$B := 1\cdot\text{cm}^{-1}$ $\qquad$ $B' := 1.3\cdot B$ $\qquad\qquad\qquad\qquad$ $B' := 1\cdot\text{cm}^{-1}$ $\qquad$ $B := 1.3\cdot B'$

$0.1\cdot\Delta v_R(B, B', J)\,(\text{cm}^{-1})$
$\Delta v_P(B, B', J)\,(\text{cm}^{-1})$

$\Delta v_R(B, B', J)\,(\text{cm}^{-1})$
$0.1\cdot\Delta v_P(B, B', J)\,(\text{cm}^{-1})$

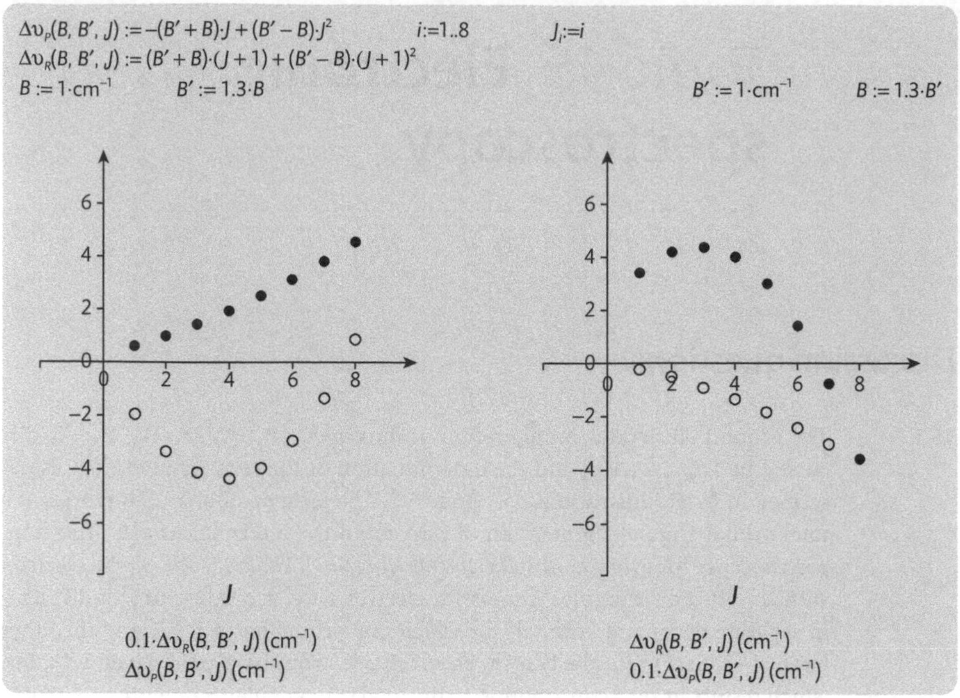

For the parameters chosen in the worksheet only the P branch shows an extreme $\Delta\tilde{v}$ value at $J_{\text{extremum}} = 4$ when $B' > B$ and only the R branch shows a branch head at $J_{\text{extremum}} = 3$ when $B > B'$. We see that $\Delta\tilde{v}$ values converge to the value at J_{extremum} from both low and high values of J as J approaches J_{extremum}. This is the branch head.

45.5 (a) The transition intensity is proportional to the square of the transition dipole moment. We initially suspect that the transition dipole moment should increase as the length L of the alternating carbon-to-carbon double/single/ double/single···· bond sequence of the polyene is increased and, consequently, the transition intensity should also increase as L increases. To test this hypothesis, consider that the polyene has N pi electrons which fill the first $n = N/2$ quantum states of the particle in a one-dimensional box of length $L = Nd = 2nd$ where d is the average carbon-to-carbon bond length. (The length choice $L = Nd$ adds half a bond length at each end to the distance between the two end-carbon nuclei.) The transition dipole moment of this model is

$$\mu_x = \int_0^L \psi_{n_f} x \psi_{n_i}\,dx$$

We quickly find a selection rule for transitions with the substitution $x = f(x) + L/2$ where $f(x) = x - L/2$ is a function of ungerade symmetry w/r/t inversion through the center of symmetry at $x = L/2$, and we note that the wavefunctions have alternating gerade and ungerade symmetry as n increases. Then,

$$\mu_x = \int_0^L \psi_{n_f}\{f(x) + L/2\}\psi_{n_i}\,dx = \int_0^L \psi_{n_f} f(x)\psi_{n_i}\,dx + (L/2)\int_0^L \psi_{n_f}\psi_{n_i}\,dx$$

The last integral vanishes because the wavefunctions of different energy levels are orthogonal which leaves

$$\mu_x = \int_0^L \psi_{n_f} f(x) \psi_{n_i} \, dx$$

Because the integrand factor $f(x)$ has ungerade symmetry, the product $\psi_{n_f}\psi_{n_i}$ of an allowed transition must also have ungerade symmetry so that the total symmetry of the integrand has u × u = g symmetry and the integral can be non-zero. Thus, the lowest energy transition is $n + 1 \leftarrow n$ where $n = n_i$. It is now convenient to return to the original transition dipole moment integral and substitute the wavefunctions for the final and initial states:

$$\mu_x = \frac{2}{L}\int_0^L \sin\left(\frac{(n+1)\pi x}{L}\right) x \sin\left(\frac{n\pi x}{L}\right) dx \quad [9.8a]$$

We need not evaluate the integral exactly because the development of a method to improve the intensity of a dye requires only a knowledge of the approximate relationship between u_x and either n or L (proportional properties for a polyene) so we recognize that $n+1 \sim n$ for a large polyene and we make the estimate

$$\mu_x \propto \frac{1}{L}\int_0^L x \sin^2\left(\frac{n\pi x}{L}\right) dx \text{ (The integral is found in standard mathematical tables.)}$$

$$\propto L$$

The model confirms the hypothesis that the transition dipole moment and, consequently, the transition intensity is increased by increasing the length of the polyene.

(b) Since $E_n = \dfrac{n^2 h^2}{8m_e L^2}$ [9.7b], $\Delta E = \dfrac{(2n+1)h^2}{8m_e L^2}[\Delta n = +1] \sim \dfrac{nh^2}{4m_e L^2}$ for large n.

But $L = 2nd$ is the length of the chain where d is the average carbon–carbon interatomic distance. Hence

$$\Delta E \propto \frac{1}{L}$$

Therefore, the transition moves toward the red as L is increased. When white light is used to illuminate the dyed object, the color absorbed is the **complementary color** to the reflected, observed color. Newton's color wheel, shown in Fig. 45.1, usefully displays complementary and observed colors. For example, draw a line from complementary violet through the circle centre to find the observed color of green-yellow. As the polyene length is increased, the complementary color progresses from violet to indigo, to blue, to green, etc. while the observed color progresses from green-yellow, to yellow-orange, to orange-red, to red-violet, to violet-indigo. We say that the apparent color of the dye shifts towards blue.

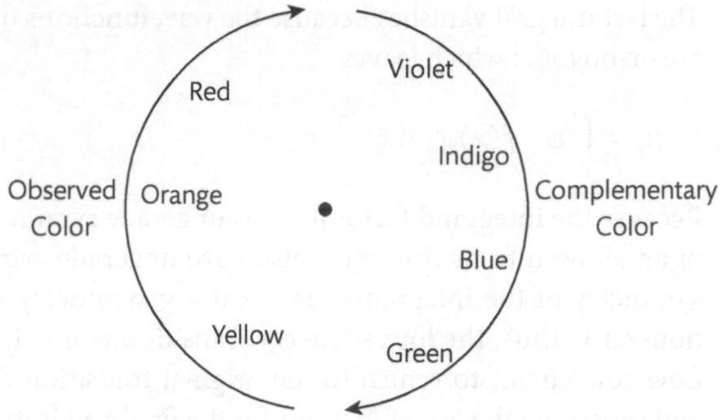

Figure 45.1

Exercises

45.1(a) The $1\sigma_g^2 1\sigma_u^2 1\pi_u^3 1\pi_g^1$ valence configuration has two unpaired electrons. Although Hund's rule does not apply to excited states, we examine the state of maximum spin multiplicity so $S = s_1 + s_2 = \frac{1}{2} + \frac{1}{2} = 1$ and the spin multiplicity is given by $2S+1=2(1)+1=\boxed{3}$. Because $u \times u = g$ and $g \times g = g$, the net parity of two electrons paired in an orbital is always gerade. Consequently, the overall parity is found by multiplying the parity of unpaired electrons. For this configuration, $u \times g = \boxed{u}$.

45.2(a) The electronic spectrum selection rules concerned with changes in angular momentum are (eqn 45.4): $\Delta\Lambda = 0, \pm1 \quad \Delta S = 0 \quad \Delta\Sigma = 0 \quad \Delta\Omega = 0, \pm1$ where $\Omega = \Lambda + \Sigma$. Λ gives the total orbital angular momentum about the internuclear axis and Σ gives the total spin angular momentum about the internuclear axis. The ± superscript selection rule for reflection in the plane along the internuclear axis is $+ \leftrightarrow +$ or $- \leftrightarrow -$ (i.e., $+ \leftrightarrow -$ is forbidden). The **Laporte selection rule** states that for a centrosymmetric molecule (those with a centre of inversion) the only allowed transitions are transitions that are accompanied by a change of parity: $u \leftrightarrow g$.

(a) The changes in the transition $^2\Pi \leftrightarrow {}^2\Pi$ are $\Delta\Lambda = 0$, $\Delta S = 0$, $\Delta\Sigma = 0$, and $\Delta\Omega = 0$ so the transition is $\boxed{\text{allowed}}$.

(b) The changes in the transition $^1\Sigma \leftrightarrow {}^1\Sigma$ are $\Delta\Lambda = 0$, $\Delta S = 0$, $\Delta\Sigma = 0$, and $\Delta\Omega = 0$ so the transition is $\boxed{\text{allowed}}$.

(c) The changes in the transition $\Sigma \leftrightarrow \Delta$ are $\Delta\Lambda = 2$ so the transition is $\boxed{\text{forbidden}}$.

(d) The transition $\Sigma^+ \leftrightarrow \Sigma^-$ is $\boxed{\text{forbidden}}$ because $+ \leftrightarrow -$ is forbidden.

(e) The transition $\Sigma^+ \leftrightarrow \Sigma^+$ is $\boxed{\text{allowed}}$ because $\Delta\Lambda = 0$ and $+ \leftrightarrow +$ is allowed.

45.3(a) We begin by evaluating the normalization constants N_0 and N_v.

$$N_0^2 = \frac{1}{\int_{-\infty}^{\infty} e^{-2ax^2}\,dx} = \left(\frac{2a}{\pi}\right)^{1/2} \text{ (standard integral); } N_0 = \left(\frac{2a}{\pi}\right)^{1/4}$$

$$\text{Likewise, } N_v^2 = \frac{1}{\int_{-\infty}^{\infty} e^{-2b(x-x_0)^2}\,dx} = \left(\frac{2b}{\pi}\right)^{1/2};\ N_v = \left(\frac{2b}{\pi}\right)^{1/4}$$

Furthermore, we can easily check that

$$ax^2 + b(x-x_0)^2 = z^2 + \frac{ab}{a+b}x_0^2 \quad \text{where} \quad z = (a+b)^{1/2}x - \frac{b}{(a+b)^{1/2}}x_0 \quad \text{and} \quad dx = \frac{1}{(a+b)^{1/2}}dz$$

Then, the vibration overlap integral between the vibrational wavefunction in the upper and lower electronic states is:

$$S(v,0)=\langle v|0\rangle=N_0N_v\int_{-\infty}^{\infty}\mathrm{e}^{-ax^2}\mathrm{e}^{-b(x-x_0)^2}\mathrm{d}x=N_0N_v\int_{-\infty}^{\infty}\mathrm{e}^{-\{ax^2+b(x-x_0)^2\}}\mathrm{d}x$$

$$=\frac{N_0N_v}{(a+b)^{1/2}}\int_{-\infty}^{\infty}\mathrm{e}^{-\left\{z^2+\frac{ab}{a+b}x_0^2\right\}}\mathrm{d}z=\frac{N_0N_v}{(a+b)^{1/2}}\mathrm{e}^{-\frac{ab}{a+b}x_0^2}\int_{-\infty}^{\infty}\mathrm{e}^{-z^2}\mathrm{d}z=N_0N_v\left(\frac{\pi}{a+b}\right)^{1/2}\mathrm{e}^{-\frac{ab}{a+b}x_0^2}$$

$$=\left(\frac{2a}{\pi}\right)^{1/4}\left(\frac{2b}{\pi}\right)^{1/4}\left(\frac{\pi}{a+b}\right)^{1/2}\mathrm{e}^{-\frac{ab}{a+b}x_0^2}=(4ab)^{1/4}\left(\frac{1}{a+b}\right)^{1/2}\mathrm{e}^{-\frac{ab}{a+b}x_0^2}$$

For the case $b=a/2$, this simplifies to

$$S(v,0)=\frac{2}{\left(3\sqrt{2}\right)^{1/2}}\mathrm{e}^{-ax_0^2/3}$$

The Franck-Condon factor is

$$|S(v,0)|^2=\boxed{\frac{2\sqrt{2}}{3}\mathrm{e}^{-2ax_0^2/3}}$$

45.4(a) $\psi_0=\left(\frac{2}{L}\right)^{1/2}\sin\left(\frac{\pi x}{L}\right)$ for $0\le x\le L$ and 0 elsewhere.

$\psi_v=\left(\frac{2}{L}\right)^{1/2}\sin\left\{\frac{\pi}{L}\left(x-\frac{L}{4}\right)\right\}$ for $\frac{L}{4}\le x\le\frac{5L}{4}$ and 0 elsewhere.

$$S(v,0)=\langle v|0\rangle=\frac{2}{L}\int_{L/4}^{L}\sin\left(\frac{\pi x}{L}\right)\sin\left\{\frac{\pi}{L}\left(x-\frac{L}{4}\right)\right\}\mathrm{d}x$$

The above integral is recognized as the standard integral (see math handbook):

$\int\sin(ax)\sin(ax+b)\mathrm{d}x=\frac{x}{2}\cos(b)-\frac{\sin(2ax+b)}{4a}$ with the transformations $a=\pi/L$ and $b=-\pi/4$. Thus,

$$S(v,0)=\frac{2}{L}\left[\frac{x}{2}\cos\left(-\frac{\pi}{4}\right)-\frac{\sin(2\pi x/L-\pi/4)}{4\pi/L}\right]_{x=L/4}^{x=L}=\frac{2}{L}\left[\frac{x}{2}\cos\left(\frac{\pi}{4}\right)-\frac{\sin(2\pi x/L-\pi/4)}{4\pi/L}\right]_{x=L/4}^{x=L}$$

$$=\left[x\cos\left(\frac{\pi}{4}\right)-\frac{\sin(2\pi x-\pi/4)}{2\pi}\right]_{x=1/4}^{x=1}$$

$$=\cos\left(\frac{\pi}{4}\right)-\frac{\sin(2\pi-\pi/4)}{2\pi}-\left[\frac{1}{4}\cos\left(\frac{\pi}{4}\right)-\frac{\sin(\pi/2-\pi/4)}{2\pi}\right]$$

$$=\frac{3}{4}\cos\left(\frac{\pi}{4}\right)-\frac{\sin(7\pi/4)}{2\pi}+\frac{\sin(\pi/4)}{2\pi}=\frac{3}{4}\cos\left(\frac{\pi}{4}\right)+\frac{\sin(\pi/4)}{2\pi}+\frac{\sin(\pi/4)}{2\pi}$$

$$=\frac{1}{4}\cos\left(\frac{\pi}{4}\right)\left\{3+\frac{4}{\pi}\right\}=\frac{\sqrt{2}}{8}\left\{3+\frac{4}{\pi}\right\}$$

The Franck-Condon factor is

$$|S(v,0)|^2 = \boxed{\frac{1}{32}\left(3+\frac{4}{\pi}\right)^2}$$

45.5(a) P branch ($\Delta J = -1$): $\tilde{\nu}_P(J) = \tilde{\nu} - (\tilde{B}' + \tilde{B})J + (\tilde{B}' - \tilde{B})J^2$ [45.8a]

When the bond is shorter in the excited state than in the ground state, $\tilde{B}' > \tilde{B}$ and $\tilde{B}' - \tilde{B}$ is positive. In this case, the lines of the P branch appear at successively decreasing energies as J increases, begin to converge, go through a head at J_{head}, begin to increase with increasing J, and become greater than $\tilde{\nu}$ when $J > (\tilde{B}' + \tilde{B})/(\tilde{B}' - \tilde{B})$ (see Discussion question 45.3); the quadratic shape of the $\tilde{\nu}_P$ against J curve is called the Fortrat parabola). This means that $\tilde{\nu}_P(J)$ is a minimum when $J = J_{\text{head}}$. It is reasonable to deduce that J_{head} is the closest integer to $\boxed{\tfrac{1}{2}(\tilde{B}' + \tilde{B})/(\tilde{B}' - \tilde{B})}$ because it takes twice as many J values to reach the minimum line of the P branch and to return to $\tilde{\nu}$. We can also find J_{head} by finding the minimum of the Fortrat parabola: $\mathrm{d}\tilde{\nu}_P/\mathrm{d}J = 0$ when $J = J_{\text{head}}$.

$$\frac{\mathrm{d}\tilde{\nu}_P}{\mathrm{d}J} = \frac{\mathrm{d}}{\mathrm{d}J}\left\{\tilde{\nu} - (\tilde{B}' + \tilde{B})J + (\tilde{B}' - \tilde{B})J^2\right\} = -(\tilde{B}' + \tilde{B}) + 2(\tilde{B}' - \tilde{B})J$$

$$-(\tilde{B}' + \tilde{B}) + 2(\tilde{B}' - \tilde{B})J_{\text{head}} = 0$$

$$J_{\text{head}} = \frac{(\tilde{B}' + \tilde{B})}{2(\tilde{B}' - \tilde{B})}$$

45.6(a) Since $\tilde{B}' < \tilde{B}$ and $\tilde{B}' - \tilde{B}$ is negative, the R branch shows a head at the closest integer to the value of $\boxed{\tfrac{1}{2}(\tilde{B}' + \tilde{B})/|(\tilde{B}' - \tilde{B})| - 1}$ (see Exercise 45.5(b)).

$$\frac{(\tilde{B}' + \tilde{B})}{2|(\tilde{B}' - \tilde{B})|} - 1 = \frac{(0.3101 + 0.3540)}{2|0.3101 - 0.3540|} - 1 = 6.6$$

$$J_{\text{head}} = \boxed{7}$$

45.7(a) When the R branch has a head, J_{head} is the closest integer to $\tfrac{1}{2}(\tilde{B}' + \tilde{B})/|(\tilde{B}' - \tilde{B})| - 1$ (see Exercise 45.5(b)). Thus, if we are only given that $J_{\text{head}} = 1$ and $\tilde{B} = 60.80\ \text{cm}^{-1}$, we know only that

$$0.5 < \tfrac{1}{2}(\tilde{B}' + \tilde{B})/(\tilde{B} - \tilde{B}') - 1 < 1.5$$

because the fractional value of a $\tfrac{1}{2}(\tilde{B}' + \tilde{B})/(\tilde{B}' - \tilde{B}) - 1$ calculation must be rounded-off to give the integer value J_{head}. Algebraic manipulation of the inequality yields

$$\frac{\{1 + 2(0.5)\}\tilde{B}}{\{3 + 2(0.5)\}} < \tilde{B}' < \frac{\{1 + 2(1.5)\}\tilde{B}}{\{3 + 2(1.5)\}}$$

$$\frac{\tilde{B}}{2} < \tilde{B}' < \frac{2\tilde{B}}{3}$$

$$\boxed{30.4\ \text{cm}^{-1} < \tilde{B}' < 40.5\ \text{cm}^{-1}}$$

When $\tilde{B}' < \tilde{B}$, the bond length in the electronically excited state is $\boxed{\text{greater}}$ than that in the ground state.

Here's an alternative solution that gives the same answer with insight into the band head concept: At the head of an R band, $\tilde{\nu}_{J_{\text{head}}} > \tilde{\nu}_{J_{\text{head}}-1}$ where $\tilde{\nu}_{J_{\text{head}}-1}$ is the transition $J_{\text{head}} \leftarrow J = J_{\text{head}} - 1$. Substitution of eqn 45.8(c) into this inequality yields the relation $\tilde{B}' > J_{\text{head}}\tilde{B}/(J_{\text{head}}+1)$. Similarly, $\tilde{\nu}_{J_{\text{head}}} > \tilde{\nu}_{J_{\text{head}}+1}$ where $\tilde{\nu}_{J_{\text{head}}+1}$ is the transition $J_{\text{head}}+2 \leftarrow J = J_{\text{head}}+1$. Substitution of eqn 45.8(c) into this inequality yields the relation $\tilde{B}' < (J_{\text{head}}+1)\tilde{B}/(J_{\text{head}}+2)$. Consequently, $J_{\text{head}}\tilde{B}/(J_{\text{head}}+1) < \tilde{B}' < (J_{\text{head}}+1)\tilde{B}/(J_{\text{head}}+2)$.

45.8(a) The transition wavenumber is $\tilde{\nu} = \dfrac{1}{\lambda} = \dfrac{1}{700\ \text{nm}} = 14 \times 10^3\ \text{cm}^{-1}$.

Water molecules are weak ligand field splitters, so we expect the d^5 electrons of Fe^{3+} to have the $t_{2g}^3e_g^2$ high spin ground state configuration in the octahedral $[Fe(OH_2)_6]^{3+}$ complex. The d-orbital electron spins are expected to be parallel with $S = 5/2$ and $2S + 1 = 6$ by Hund's maximum multiplicity rule. We also expect that $P > \Delta_o$ where P is the energy of repulsion for pairing two electrons in an orbital. A d–d transition to the $t_{2g}^4e_g^1$ octahedral excited state is expected to be both spin and parity forbidden and, therefore, have a very small molar absorption coefficient. This transition releases the energy Δ_O and requires the energy P needed to pair two electrons within a t_{2g} orbital. Thus, $\tilde{\nu} = P - \Delta_o$ and $\boxed{\Delta_o = P - \tilde{\nu}}$. Using the typical value $P \sim 28 \times 10^3\ \text{cm}^{-1}$ yields the estimate $\Delta_o \sim \boxed{14 \times 10^3\ \text{cm}^{-1}}$. See F.A. Cotton and G. Wilkinson, *Advanced Inorganic Chemistry*, 4th ed. (New York: Wiley-Interscience Publishers, 1980), 646, for electron-pairing energies.

45.9(a) The normalized wavefunctions are:

$$\psi_i = \left(\frac{1}{a}\right)^{1/2} \quad \text{for } 0 \le x \le a \text{ and 0 elsewhere.}$$

$$\psi_f = \left(\frac{1}{b - \tfrac{1}{2}a}\right)^{1/2} \quad \text{for } \tfrac{1}{2}a \le x \le b \text{ and 0 elsewhere.}$$

$$\int \psi_f x \psi_i \,\mathrm{d}x = \left(\frac{1}{a}\right)^{1/2}\left(\frac{1}{b-\tfrac{1}{2}a}\right)^{1/2}\int_{\frac{1}{2}a}^{a} x\,\mathrm{d}x = \left(\frac{1}{a}\right)^{1/2}\left(\frac{1}{b-\tfrac{1}{2}a}\right)^{1/2}\left.\frac{x^2}{2}\right|_{x=\frac{1}{2}a}^{x=a}$$

$$= \left(\frac{1}{a}\right)^{1/2}\left(\frac{1}{b-\tfrac{1}{2}a}\right)^{1/2}\left(\frac{3a^2}{8}\right) = \boxed{\frac{3}{8}\left(\frac{a^3}{b-\tfrac{1}{2}a}\right)^{1/2}}$$

45.10(a) The normalized wavefunctions are:

$$\psi_i = \left(\frac{1}{a\sqrt{\pi}}\right)^{1/2} \mathrm{e}^{-x^2/2a^2} \quad \text{for } -\infty \le x \le \infty \text{ and width } a.$$

$$\psi_f = \left(\frac{1}{a\sqrt{\pi}}\right)^{1/2} e^{-(x-a/2)^2/2a^2} \quad \text{for } -\infty \leq x \leq \infty \text{ and width } a.$$

$$\int \psi_f x \psi_i \, dx = \left(\frac{1}{a\sqrt{\pi}}\right) \int_{-\infty}^{\infty} x e^{-x^2/2a^2} e^{-(x-a/2)^2/2a^2} dx = \left(\frac{1}{a\sqrt{\pi}}\right) \int_{-\infty}^{\infty} x e^{-\{x^2+(x-a/2)^2\}/2a^2} dx$$

Since $x^2 + (x-a/2)^2 = \left(2^{1/2}x - \frac{a}{2^{3/2}}\right)^2 + \frac{a^2}{8}$, let $z = 2^{1/2}x - \frac{a}{2^{3/2}}$ and $x = 2^{-1/2}z + \frac{a}{4}$.
Then, $dx = 2^{-1/2} dz$ and substitution gives:

$$\int \psi_f x \psi_i \, dx = \left(\frac{e^{-1/16}}{a\sqrt{2\pi}}\right) \int_{-\infty}^{\infty} \left(2^{-1/2} z + \frac{a}{4}\right) e^{-z^2/2a^2} dz$$

$$= \left(\frac{e^{-1/16}}{a\sqrt{2\pi}}\right) \left\{ 2^{-1/2} \int_{-\infty}^{\infty} z e^{-z^2/2a^2} dz + \frac{a}{4} \int_{-\infty}^{\infty} e^{-z^2/2a^2} dz \right\}$$

The factors within the first integral have ungerade and gerade symmetry. Because u × g = u, the integrand has ungerade symmetry and the first integral is necessarily zero (the integral of an ungerade function over a symmetric interval equals zero).

$$\int \psi_f x \psi_i \, dx = \left(\frac{e^{-1/16}}{a\sqrt{2\pi}}\right) \left\{ \frac{a}{4} \int_{-\infty}^{\infty} e^{-z^2/2a^2} dz \right\} = \left(\frac{e^{-1/16}}{4\sqrt{2\pi}}\right) (2a^2\pi)^{1/2}$$

$$= \boxed{\tfrac{1}{4} e^{-1/16} a}$$

45.11(a) π-electrons in polyenes may be considered as particles in a one-dimensional box. Applying the Pauli exclusion principle, the N conjugated electrons will fill the levels, two electrons at a time, up to the level $n = N/2$. Since N is also the number of alkene carbon atom, Nd is the length of the box, with d the carbon–carbon interatomic distance. Hence

$$E_n = \frac{n^2 h^2}{8mN^2 d^2} \quad [9.8b]$$

For the lowest energy transition ($\Delta n = +1$)

$$\Delta E = h\nu = \frac{hc}{\lambda} = E_{N/2+1} - E_{N/2} = \frac{(N+1)h^2}{8mN^2 d^2}$$

Therefore, the larger N, the larger λ. Hence the absorption at 243 nm is due to the diene and that at 192 nm to the butene.

Question. How accurate is the formula derived above in predicting the wavelengths of the absorption maxima in these two compounds?

An alternative analysis uses simple Hückel theory of π molecular orbitals. The substituted butene is modeled with the ethene π energies of text Fig. 26.1 that shows a $\pi_{\text{LUMO}} \leftarrow \pi_{\text{HOMO}}$ transition energy of -2β (β is negative). The substituted 2,4-hexadiene is modeled with the butadiene π energies of text Fig. 26.2 that shows a $\pi_{\text{LUMO}} \leftarrow \pi_{\text{HOMO}}$ transition energy of -1.24β. Thus, the transition energy of the conjugated diene is lower and has a longer wavelength. This is an example of the general principle that the difference between neighboring energy levels becomes smaller as N becomes larger.

Problems

45.1 The term symbol $^2\Pi_g$ indicates that, since the term is a doublet, $S = ½$ and the excited state has one unpaired electron. The symbol Π tells us that there is $\hbar$ of orbital angular momentum around the internuclear axis (i.e., $|\Lambda| = 1$). From text Fig. 24.11 we see that the ground state of N_2^+ is $1\sigma_g^2 1\sigma_u^2 1\pi_u^4 2\sigma_g^1 1\pi_g^0$ for a $^2\Sigma_g^+$ term. Promoting the $2\sigma_g$ to the $1\pi_g$ MO maintains the spin and the gerade symmetry while increasing $|\Lambda|$ from 0 to 1. Thus, the excited state is $\boxed{1\sigma_g^2 1\sigma_u^2 1\pi_u^4 2\sigma_g^0 1\pi_g^1}$. We also note that, since the $1\pi_g$ MO changes sign upon reflection in the plane containing the internuclear axis, that the negative superscript may be included in the excited-state term to give $^2\Pi_g^-$. The transition $^2\Pi_g^- \leftarrow {}^2\Sigma_g^+$ will not appear in the electronic spectrum of N_2^+ because the parity has not changed and the reflection change $- \leftarrow +$ is not allowed.

45.3 The potential energy curves for the $X\,^3\Sigma_g^-$ and $B\,^3\Sigma_u^-$ electronic states of O_2 are represented schematically in Fig. 45.2 along with the notation used to represent the energy separation of this problem. Curves for the other electronic states of O_2 are not shown. Ignoring rotational structure and anharmonicity we may write

$$\tilde{\nu}_{00} \approx T_e + ½(\tilde{\nu}' - \tilde{\nu}) = 6.175\ \text{eV} \times \left(\frac{8\,065.5\ \text{cm}^{-1}}{1\ \text{eV}}\right) + ½(700 - 1\,580)\ \text{cm}^{-1}$$

$$\approx \boxed{49\,36\bar{4}\ \text{cm}^{-1}}$$

Comment: Note that the selection rule $\Delta v = \pm 1$ does not apply to vibrational transitions between different electronic states.

Question: What is the percentage change in $\tilde{\nu}_{00}$ if the anharmonicity constants $x_e\tilde{\nu}$ (Topic 43.3) 12.0730 cm^{-1} and 8.002 cm^{-1} for the ground and excited states, respectively, are included in the analysis?

45.5 (a) The H_2O^+ vibrational wavenumber (0.41 eV) of the photoelectron spectrum band at 12-13 eV corresponds to about 3 300 cm^{-1}, which is close to the 3 652 cm^{-1} symmetric stretching mode of the neutral ground state (see text Fig. 44.3). This suggests loss of a non-bonding electron. The absence of a long vibrational series in this band is compatible with an ionized equilibrium bond length that approximately equals that of the neutral ground state, thereby producing a large Franck-Condon factor for the adiabatic transition ($v' = 0 \leftarrow v = 0$) and smaller

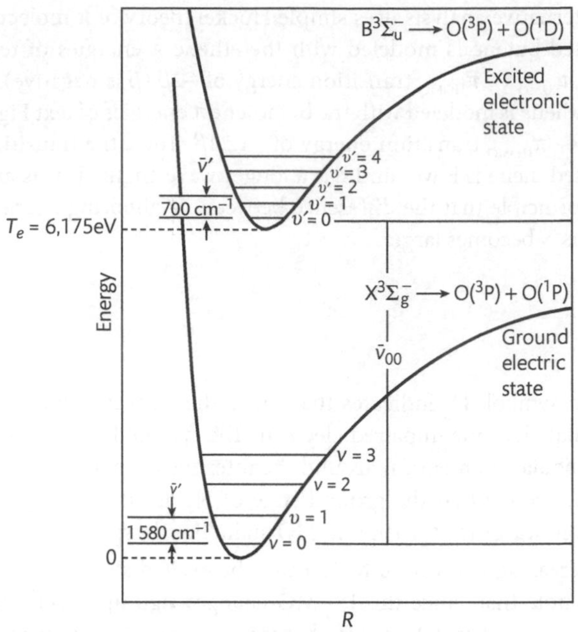

Figure 45.2

factors for other vibrational transitions ($\nu' = 1,2, \ldots \leftarrow \upsilon = 0$). These observations are compatible with the loss of a non-bonding electron because such a loss does not affect, or has little effect, on the bonding force constants, vibrational frequencies, or equilibrium bond lengths.

(b) The H_2O^+ vibrational wavenumber (0.125 eV) of the photoelectron spectrum band at 14-16 eV corresponds to about 1 000 cm^{-1}, which is very different than the 1 595 cm^{-1} bending mode of the neutral ground state (see text Fig. 44.3). This suggests loss of a σ bonding electron that severely reduces bond order, a bond force constant, and a vibrational frequency. Consequently, we expect the bond length of the ionized state to be longer than that of the neutral ground state, an observation that is compatible with the long vibrational series of the band because this yields many vertical transitions with significant Franck-Condon factors.

45.7 We need to establish whether the transition dipole moments

$$\mu_{\text{fi}} = \int \psi_{\text{f}}^* \mu \psi_{\text{i}} \, d\tau \text{ [16.8b, 42.2, Justification 43.1]}$$

connecting the states $n = 1$ and 2 and the states $n = 1$ and 3 are zero or nonzero. The particle in a box wavefunctions are

$$\psi_n = (2/L)^{1/2} \sin(n\pi x/L) \text{ [9.8a]}$$

Thus,

$$\mu_{2,1} \propto \int \sin\left(\frac{2\pi x}{L}\right) x \sin\left(\frac{\pi x}{L}\right) dx \propto \int x \left[\cos\left(\frac{\pi x}{L}\right) - \cos\left(\frac{3\pi x}{L}\right)\right] dx$$

and

$$\mu_{3,1} \propto \int \sin\left(\frac{3\pi x}{L}\right) x \sin\left(\frac{\pi x}{L}\right) dx \propto \int x\left[\cos\left(\frac{2\pi x}{L}\right) - \cos\left(\frac{4\pi x}{L}\right)\right] dx$$

having used $\sin\alpha\sin\beta = \frac{1}{2}\cos(\alpha-\beta) - \frac{1}{2}\cos(\alpha+\beta)$. Both of these integrals can be evaluated using the standard form

$$\int x(\cos ax)dx = \frac{1}{a^2}\cos ax + \frac{x}{a}\sin ax$$

$$\int_0^L x\cos\left(\frac{\pi x}{L}\right)dx = \frac{1}{(\pi/L)^2}\cos\left(\frac{\pi x}{L}\right)\Bigg|_0^L + \frac{x}{(\pi/L)}\sin\left(\frac{\pi x}{L}\right)\Bigg|_0^L = -2\left(\frac{L}{\pi}\right)^2 \neq 0$$

$$\int_0^L x\cos\left(\frac{3\pi x}{L}\right)dx = \frac{1}{(3\pi/L)^2}\cos\left(\frac{3\pi x}{L}\right)\Bigg|_0^L + \frac{x}{(3\pi/L)}\sin\left(\frac{3\pi x}{L}\right)\Bigg|_0^L = -2\left(\frac{L}{3\pi}\right)^2 \neq 0$$

Thus, $\mu_{2,1} \neq 0$. In a similar manner, $\mu_{3,1} = 0$.

Comment: A general formula for μ_{fi} applicable to all possible particle in a box transitions may be derived. The result is $(n = \mathrm{f}, m = \mathrm{i})$

$$\mu_{nm} = -\frac{eL}{\pi^2}\left[\frac{\cos(n-m)\pi - 1}{(n-m)^2} - \frac{\cos(n+m)\pi - 1}{(n+m)^2}\right]$$

For m and n both even or both odd numbers, $\mu_{\mathrm{nm}} = 0$; if one is even and the other odd, $\mu_{\mathrm{nm}} \neq 0$. Also, see Problem 45.9.

Question: Can you establish the general relation for μ_{nm} above?

45.9 $$S = \left[1 + \frac{R}{a_0} + \frac{1}{3}\left(\frac{R}{a_0}\right)^2\right]e^{-R/a_0} \quad [24.13]$$

$\mu = -eSR$ [given]

$\mu/(-ea_0) = (R/a_0) \times S$

$$= (R/a_0) \times \left[1 + \frac{R}{a_0} + \frac{1}{3}\left(\frac{R}{a_0}\right)^2\right]e^{-R/a_0}$$

We then draw up the following table:

R/a_0	0	1	2	3	4	5	6	7	8
$\mu/(-ea_0)$	0	0.858	1.173	1.046	0.757	0.483	0.283	0.155	0.081

These points are plotted in Fig. 45.3. The dipole-transition moment rises from zero at $R = 0$ to a peak at $R \sim 2.1\ a_0$ after which it declines to zero at infinity. The transition moment is zero at $R = 0$, because the initial state ($1s_A$) and the final state ($1s_A$)

are identical and, consequently, there is no dipole moment and no transition. As $R \to \infty$, the electron is confined to a single atom because its wavefunction does not extend to the other and, consequently, no transition is possible.

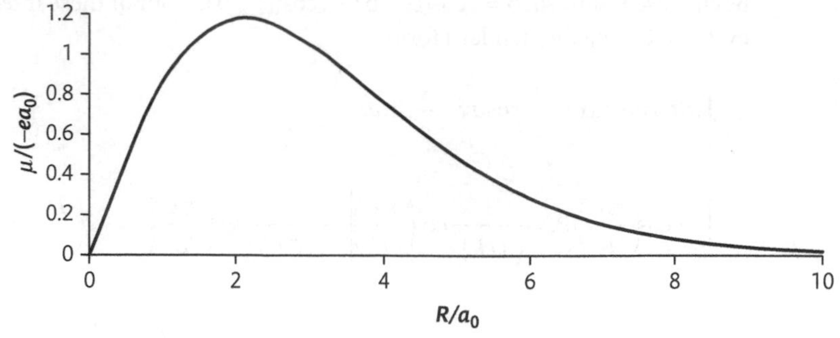

Figure 45.3

Topic 46 Decay of excited states

Discussion questions

46.1 The overall process associated with fluorescence involves the following steps. The molecule is first promoted from the vibrational ground state of a lower electronic level to a higher vibrational-electronic energy level by absorption of energy from a radiation field. Because of the requirements of the Franck-Condon principle, the transition is to excited vibrational levels of the upper electronic state. See text Fig. 46.2. Therefore, the absorption spectrum shows a vibrational structure characteristic of the upper state. The excited-state molecule can now lose energy to the surroundings through radiationless transitions and decay to the lowest vibrational level of the upper state. A spontaneous radiative transition now occurs to the lower electronic level and this fluorescence spectrum has a vibrational structure characteristic of the lower state. The fluorescence spectrum is not the mirror image of the absorption spectrum because the vibrational frequencies of the upper and lower states are different due to the difference in their potential energy curves.

46.3 (a) Continuous-wave (CW) laser emission is possible when heat is easily dissipated and population inversion can be continuously maintained by pumping. The red laser pointer is an example of a CW laser. Typically, the light amplification by stimulated emission of radiation is continuous in the optical cavity and one of the two mirrors, the output coupler, at the ends of the cavity is partially transparent so that only a fraction of the cavity radiation can continuously escape. The gain medium is pumped to the excited state by electricity, a flash lamp, or another laser.

(b) The pulsed laser periodically emits a pulse of high peak power radiation, which is much higher than can be achieved with a CW laser because the average laser power is released in a pulse of short duration. Pulses may be achieved by Q-switching or mode locking. In Q-switching, the laser cavity resonance characteristics are modified to make the cavity conditions unfavorable for lasing, during which time a healthy population inversion is achieved; the cavity is then suddenly brought to resonance, releasing the radiation pulse. The electro-optical Pockels cell or a saturable absorber may be used as Q-switching devices that give pulses of about 5 ns duration. Picosecond pulses can be achieved by the technique of mode locking in which a range of resonant modes of different frequency are phase locked and superimposed. Interference of the modes gives rise to short, regular bursts of radiation. Mode locking is achieved by varying the Q-factor of the laser cavity periodically at the frequency $c/2L$. The modulation can be achieved by linking a prism in the cavity to a transducer driven by a radiofrequency source at a frequency $c/2L$. The transducer sets up standing-wave vibrations in the prism and modulates the loss it introduces into the cavity.

Exercises

46.1(a)

(a) Vibrational energy spacings of the $\boxed{\text{lower}}$ state are determined by the spacing of the peaks of fluorescence spectrum of benzophenone, which emission intensity A of text Fig. F9.6 shows to be $\boxed{\tilde{\nu} \approx 1\,800\ \text{cm}^{-1}}$.

(b) The fluorescence spectrum gives $\boxed{\text{no information}}$ about the spacing of the upper vibrational levels (without a detailed analysis of the intensities of the lines).

46.2(a) When the steeply repulsive section of the O_2 potential energy curve for the excited state lies slightly toward the short side of the equilibrium bond length of the ground state and the minimum of the excited state lies to the longer side (as shown in text Fig. 46.8), a great many excited vibrational states overlap with the lowest energy vibration of the ground state. The **Franck-Condon factor** is appreciable for many vertical transitions (see text Fig. 45.7) and the absorption band is broad. Furthermore, predissociation to the unbound $^5\Pi_u$ state shortens the lifetime of excited vibrational states. This causes the high resolution lines of the corresponding vibrational-rotational transitions to be broad through the Heisenberg uncertainty principle $\Delta E \Delta t \geq \hbar/2$.

46.3(a) Referring to Example 46.1 of the text, we have

$$P_{\text{peak}} = E_{\text{pulse}}/t_{\text{pulse}} \quad \text{and} \quad P_{\text{average}} = E_{\text{total}}/t = E_{\text{pulse}} \times \nu_{\text{repetition}}$$

where $\nu_{\text{repetition}}$ is the pulse repetition rate.

$$t_{\text{pulse}} = E_{\text{pulse}}/P_{\text{peak}} = \frac{0.10\ \text{mJ}}{5.0\ \text{MW}} = \boxed{20\ \text{ps}}$$

$$\nu_{\text{repetition}} = P_{\text{average}}/E_{\text{pulse}} = \frac{7.0\ \text{kW}}{0.10\ \text{mJ}} = \boxed{70\ \text{MHz}}$$

Problems

46.1 The anthracene vapour fluorescence spectrum gives the vibrational splitting of the lower state. The wavelengths stated correspond to the wavenumbers 22 730, 24 390, 25 640, 27 030 cm^{-1}, indicating spacings of 1660, 1250, and 1390 cm^{-1}. The absorption spectrum spacing gives the separation of the vibrational levels of the upper state. The wavenumbers of the absorption peaks are 27 800, 29 000, 30 300, and 32 800 cm^{-1}. The vibrational spacings are therefore 1 200, 1 300, and 2 500 cm^{-1}. The data is compatible with the deactivation of the excited state vibrational modes before spontaneous emission returns the molecule to the ground electronic state. This produces a fluorescence band of lower energy than the absorption band. Furthermore, while the absorption band has a vibrational progression that depends upon vibrational modes of the excited state, the fluorescence band has a vibrational progression that depends upon vibrational modes of the ground state. The absorption and fluorescence spectra are not mirror images.

46.3 Assuming that the largest m/z value (9 912 g mol^{-1}) is an unfragmented polymer of 1+ charge, we expect that fragmentation should yield at least three peaks and the un-fragmented biopolymer of 2+ charge should give a peak at 4 956 g mol^{-1}. Because neither of these is observed, we conclude that the peak at 4 554 g mol^{-1} is a second distinct biopolymer.

Focus 9: Integrated activities

F9.1 Refer to the flow chart in Fig. 31.7. Yes at the first question (linear?) leads to linear point groups and therefore linear rotors. If the molecule is not linear, then yes at the next question (two or more C_n with $n > 2$?) leads to cubic and icosahedral groups and therefore spherical rotors. If the molecule is not a spherical rotor, yes at the next question leads to symmetric rotors if the highest C_n has $n > 2$; if not, the molecule is an asymmetric rotor.

(a) CH_4: not linear, but more than two $C_n\,(n>2)$, so $\boxed{\text{spherical rotor}}$.

(b) CH_3CN: not linear, C_3 (only one of them), so $\boxed{\text{symmetric rotor}}$.

(c) CO_2: linear, so $\boxed{\text{linear rotor}}$.

(d) CH_3OH: not linear, no C_n, so $\boxed{\text{asymmetric rotor}}$.

(e) Benzene: not linear, C_6, but only one high-order axis, so $\boxed{\text{symmetric rotor}}$.

(f) Pyridine: not linear, C_2, is highest rotational axis, so $\boxed{\text{asymmetric rotor}}$.

F9.3 (a) The problem states that the H_3^+ ion is an oblate symmetric rotor. Consequently, it must have the structure of an equilateral triangle with a C_3 principal axis. The rotational axis for $I_{\parallel}$, which is labelled I_C in this problem, is through the centre of the molecule, the location of the centre of mass, from which each atom is separated by the distance $R/\sqrt{3}$ where R is the equilibrium bond length. The rotational axis for $I_{\perp}$, which is labelled I_B in this problem, is on a line through one of the atoms and the centre of mass so the I_B axis bisects a H-H-H angle and two of the atoms are separated from the axis by the perpendicular distance $R/2$. The moment of inertia are given by

$$I_C = \sum_i m_H \times (\text{perpendicular distance from rotation axis})^2 \quad [41.1]$$

$$= 3m_H \times (\text{perpendicular distance from rotation axis})^2 = 3m_H \times \left(R/\sqrt{3}\right)^2$$
$$= m_H R^2$$

$$I_B = \sum_i m_H \times (\text{perpendicular distance from rotation axis})^2 \quad [41.1]$$

$$= 2m_H \times (\text{perpendicular distance from rotation axis})^2 = 2m_H \times (R/2)^2$$
$$= m_H R^2/2$$

Thus, $\boxed{I_C = 2I_B}$.

(b) $$\tilde{B} = \frac{\hbar}{4\pi c I_B}\ [41.14] = \frac{2\hbar}{4\pi c m_{\mathrm{H}} R^2} = \frac{\hbar}{2\pi c m_{\mathrm{H}} R^2}$$

$$R = \left(\frac{\hbar}{2\pi c m_{\mathrm{H}} \tilde{B}}\right)^{1/2} = \left(\frac{\hbar N_{\mathrm{A}}}{2\pi c M_{\mathrm{H}} \tilde{B}}\right)^{1/2}$$

$$= \left(\frac{(1.0546\times10^{-34}\ \mathrm{J\,s})\times(6.0221\times10^{23}\ \mathrm{mol^{-1}})}{2\pi\times(2.9979\times10^{10}\ \mathrm{cm\,s^{-1}})\times(1.0079\times10^{-3}\ \mathrm{kg\,mol^{-1}})\times(43.55\ \mathrm{cm^{-1}})}\right)^{1/2}$$

$$= 87.64\ \mathrm{pm}$$

Alternatively, the rotational constant $\tilde{C}$ can be used to calculate R.

$$\tilde{C} = \frac{\hbar}{4\pi c I_C}\ [41.14] = \frac{\hbar}{4\pi c m_{\mathrm{H}} R^2}$$

$$R = \left(\frac{\hbar}{4\pi c m_{\mathrm{H}} \tilde{C}}\right)^{1/2} = \left(\frac{\hbar N_{\mathrm{A}}}{4\pi c M_{\mathrm{H}} \tilde{C}}\right)^{1/2}$$

$$= \left(\frac{(1.0546\times10^{-34}\ \mathrm{J\,s})\times(6.0221\times10^{23}\ \mathrm{mol^{-1}})}{4\pi\times(2.9979\times10^{10}\ \mathrm{cm\,s^{-1}})\times(1.0079\times10^{-3}\ \mathrm{kg\,mol^{-1}})\times(20.71\ \mathrm{cm^{-1}})}\right)^{1/2}$$

$$= 89.87\ \mathrm{pm}$$

The two calculations differ slightly so we approximate the bond length as the average:

$$R_{\mathrm{av}} = \frac{(87.64+89.87)\ \mathrm{pm}}{2} = \boxed{88.8\ \mathrm{pm}}$$

(c) $$\tilde{B} = \frac{\hbar N_{\mathrm{A}}}{2\pi c M_{\mathrm{H}} R^2}$$

$$= \frac{(1.0546\times10^{-34}\ \mathrm{J\,s})\times(6.0221\times10^{23}\ \mathrm{mol^{-1}})}{2\pi\times(2.9979\times10^{10}\ \mathrm{cm\,s^{-1}})\times(1.0079\times10^{-3}\ \mathrm{kg\,mol^{-1}})\times(87.32\times10^{-12}\ \mathrm{m})^2}$$

$$= \boxed{43.87\ \mathrm{cm^{-1}}}$$

(d) $\dfrac{1}{m_{\mathrm{eff,H_3^+}}} = \sum_i \dfrac{1}{m_i} = \dfrac{3}{m_{\mathrm{H}}}$ or $m_{\mathrm{eff,H_3^+}} = \dfrac{m_{\mathrm{H}}}{3}$. Also, $m_{\mathrm{D}} = 2\,m_{\mathrm{H}}$ so $m_{\mathrm{eff,D_3^+}} = \dfrac{2m_{\mathrm{H}}}{3}$.

Assuming that the bonding force constants are the same in D_3^+ and H_3^+, eqn 43.8 relates the vibrational wavenumber to the effective mass.

$$\tilde{\nu}_{\mathrm{H_3^+}} = \frac{1}{2\pi c}\left(\frac{k_{\mathrm{f}}}{m_{\mathrm{eff,H_3^+}}}\right)^{1/2} \quad \text{and} \quad \tilde{\nu}_{\mathrm{D_3^+}} = \frac{1}{2\pi c}\left(\frac{k_{\mathrm{f}}}{m_{\mathrm{eff,D_3^+}}}\right)^{1/2}$$

Taking the ratio and solving for $\tilde{\nu}_{D_3^+}$ yields:

$$\tilde{\nu}_{D_3^+} = \left(\frac{m_{\text{eff},H_3^+}}{m_{\text{eff},D_3^+}}\right)^{1/2} \tilde{\nu}_{H_3^+} = \left(\frac{m_H/3}{2m_H/3}\right)^{1/2} \tilde{\nu}_{H_3^+} = \tilde{\nu}_{H_3^+}/\sqrt{2}$$

$$= \left(2\,521.6\text{ cm}^{-1}\right)/\sqrt{2}$$

$$= \boxed{1\,783.0\text{ cm}^{-1}}$$

Similarly,

$$\tilde{B}_{D_3^+} = \left(\frac{m_H}{m_D}\right)\tilde{B}_{H_3^+} = \tilde{B}_{H_3^+}/2 = \left(43.55\text{ cm}^{-1}\right)/2$$

$$= \boxed{21.78\text{ cm}^{-1}}$$

$$\tilde{C}_{D_3^+} = \left(\frac{m_H}{m_D}\right)\tilde{C}_{H_3^+} = \tilde{C}_{H_3^+}/2 = \left(20.71\text{ cm}^{-1}\right)/2$$

$$= \boxed{10.36\text{ cm}^{-1}}$$

F9.5 (a) SO_2 calculations are performed with Spartan '10 using the MP2 method with the 6-311++G** basis set. In a 6-311G basis set each atomic core basis function is expanded as a linear combination of six Gaussian functions. Valence orbitals are split into three basis set functions consisting of three, one, and one Gaussians. The 6-311++G basis set adds both an s-type and three p-type diffuse functions for each atom other than hydrogen and one s-type diffuse function for each hydrogen atom. The 6-311++G** basis set adds a set of five d-type polarization functions for each atom other than hydrogen and a set of three p-type polarization functions for each hydrogen atom.

The calculated equilibrium structure is shown in Figure F9.1 where we see that the two S–O bonds have equal length, thereby, justifying Lewis resonances structures; the Mulliken bond orders are found to be 1.47. The calculated bond length and angle agree with the experimental values of 143.21 pm and 119.54°. The calculated dipole is somewhat higher than the experimental value of 1.63 D. Greatly exaggerated stretches and bends of the three vibrational modes are also shown in Figure F9.1.

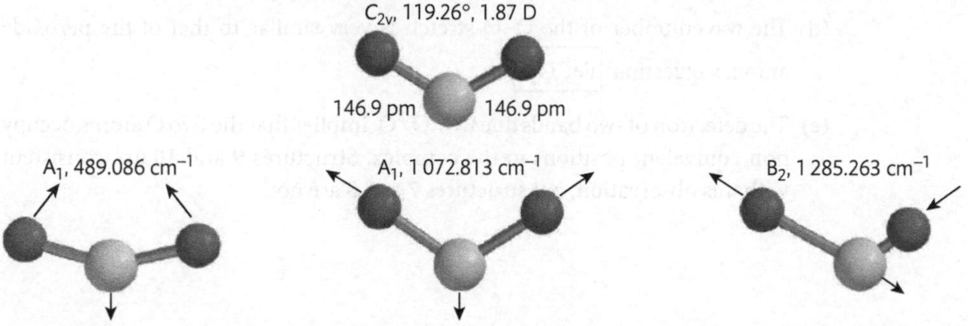

Figure F9.1

(b) The calculated values of the fundamental vibrational wavenumbers, $\boxed{489, 1073, \text{ and } 1285\text{ cm}^{-1}}$, are also reported in Figure F9.1. They correlate well with experimental values but are about 35-80 cm^{-1} lower. SCF calculations often yield systematically lower or higher values than experiment while approximately paralleling the experimental to within an additive constant.

F9.7 (a) Resonance Raman spectroscopy is preferable to vibrational spectroscopy for studying the O–O stretching mode because such a mode would be infrared inactive, or at best only weakly active. (The mode is sure to be inactive in free O_2, because it would not change the molecule's dipole moment. In a complex in which O_2 is bound, the O–O stretch may change the dipole moment, but it is not certain to do so at all, let alone strongly enough to provide a good signal.)

(b) Assuming that the two isotopic oxygen molecules have identical O–O bond strengths when bound to haemerythrin, the vibrational wavenumber depends on the effective mass alone.

$$\tilde{\nu} \propto \left(\frac{1}{m_{\text{eff}}}\right)^{1/2} \quad [43.8]$$

$$\frac{\tilde{\nu}(\text{bound }^{18}\text{O}_2)}{\tilde{\nu}(\text{bound }^{16}\text{O}_2)} = \left(\frac{m_{\text{eff}}(\text{bound }^{16}\text{O}_2)}{m_{\text{eff}}(\text{bound }^{18}\text{O}_2)}\right)^{1/2} = \left(\frac{16.0\text{ u}}{18.0\text{ u}}\right)^{1/2} = 0.943$$

and $\tilde{\nu}(\text{bound }^{18}\text{O}_2) = (0.943)\times(844\text{ cm}^{-1}) = \boxed{796\text{ cm}^{-1}}$

Note the assumption that the effective masses are proportional to the isotopic masses. This assumption is valid in the free molecule, where the effective mass of O_2 is equal to half the mass of the O atom; it is also valid if O_2 is strongly bound at one end, such that one atom is free and the other is essentially fixed to a very massive unit.

(c) The vibrational wavenumber is proportional to the square root of the force constant by eqn 43.8. The force constant is itself a measure of the strength of the bond (technically of its stiffness, which correlates with strength), which in turn is characterized by bond order. Simple molecular orbital analysis of O_2, O_2^-, and O_2^{2-} results in bond orders of $\boxed{2, 1.5, \text{ and } 1}$, respectively. Given decreasing bond order, one would expect decreasing vibrational wavenumbers (and vice versa).

(d) The wavenumber of the O–O stretch is very similar to that of the peroxide anion, suggesting $\boxed{Fe_2^{3+}O_2^{2-}}$.

(e) The detection of two bands due to $^{16}\text{O}^{18}\text{O}$ implies that the two O atoms occupy non-equivalent positions in the complex. Structures **9** and **10** are consistent with this observation, but structures **7** and **8** are not.

F9.9 In Fig. F9.2

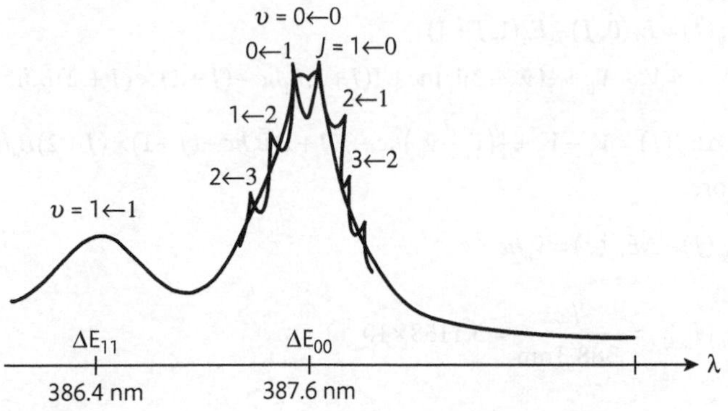

Fig F9.2

$$\Delta E_{11} = \frac{hc}{\lambda_{11}} = \frac{hc}{386.4\,\text{nm}} = 5.1409\times10^{-19}\,\text{J} = 3.2087\,\text{eV}$$

and

$$\Delta E_{00} = \frac{hc}{\lambda_{00}} = \frac{hc}{387.6\,\text{nm}} = 5.1250\times10^{-19}\,\text{J} = 3.1987\,\text{eV}$$

Energy of excited singlet, S_1: $E_1(\upsilon, J) = V_1 + (\upsilon + \tfrac{1}{2})\tilde{\nu}_1 hc + J(J+1)\tilde{B}_1 hc$
Energy of ground singlet, S_0: $E_0(\upsilon, J) = V_0 + (\upsilon + \tfrac{1}{2})\tilde{\nu}_0 hc + J(J+1)\tilde{B}_0 hc$
The midpoint of the 0–0 band corresponds to the forbidden Q branch ($\Delta J = 0$) with $J = 0$ and $\nu = 0 \leftarrow 0$.

$$\Delta E_{00} = E_1(0,0) - E_0(0,0) = (V_1 - V_0) + \tfrac{1}{2}(\tilde{\nu}_1 - \tilde{\nu}_0)hc \quad (1)$$

The midpoint of the 1–1 band corresponds to the forbidden Q branch ($\Delta J = 0$) with $J = 0$ and $\nu = 1 \leftarrow 1$.

$$\Delta E_{11} = E_1(1,0) - E_0(1,0) = (V_1 - V_0) + \tfrac{3}{2}(\tilde{\nu}_1 - \tilde{\nu}_0)hc \quad (2)$$

Multiplying eqn. 1 by three and subtracting eqn. 2 gives

$$3\Delta E_{00} - \Delta E_{11} = 2(V_1 - V_0)$$

$$\begin{aligned} V_1 - V_0 &= \tfrac{1}{2}(3\Delta E_{00} - \Delta E_{11}) \\ &= \tfrac{1}{2}\{3(5.1250) - (5.1409)\}\times10^{-19}\,\text{J} \\ &= 5.1171\times10^{-19}\,\text{J} = \boxed{3.1938\,\text{eV}} \quad (3) \end{aligned}$$

This is the potential energy difference between S_0 and S_1. Equations (1) and (3) may be solved for $\tilde{\nu}_1 - \tilde{\nu}_0$.

$$\begin{aligned} \tilde{\nu}_1 - \tilde{\nu}_0 &= 2\{\Delta E_{00} - (V_1 - V_0)\} \\ &= 2\{5.1250 - 5.1171\}\times10^{-19}\,\text{J}/hc \\ &= 1.5800\times10^{-21}\,\text{J} = 0.0098615\,\text{eV} = \boxed{79.538\,\text{cm}^{-1}} \end{aligned}$$

The $\tilde{\nu}_1$ value can be determined by analyzing the band head data for which $J+1 \leftarrow J$.

$$\Delta E_{10}(J) = E_1(0, J) - E_0(1, J+1)$$
$$= V_1 - V_0 + \tfrac{1}{2}(\tilde{\nu}_1 - 3\tilde{\nu}_0)hc + J(J+1)\tilde{B}_1 hc - (J+1)\times(J+2)\tilde{B}_0 hc$$

and $\Delta E_{00}(J) = V_1 - V_0 + \tfrac{1}{2}(\tilde{\nu}_1 - \tilde{\nu}_0)hc + J(J+1)\tilde{B}_1 hc - (J+1)\times(J+2)\tilde{B}_0 hc$

Therefore,

$$\Delta E_{00}(J) - \Delta E_{10}(J) = \tilde{\nu}_0 hc$$

$$\Delta E_{00}(J_{\text{head}}) = \frac{hc}{388.3\,\text{nm}} = 5.1158\times10^{-19}\,\text{J}$$

$$\Delta E_{10}(J_{\text{head}}) = \frac{hc}{421.6\,\text{nm}} = 4.7117\times10^{-19}\,\text{J}$$

$$\tilde{\nu}_0 = \frac{\Delta E_{00}(J) - \Delta E_{10}(J)}{hc}$$
$$= \frac{(5.1158 - 4.7117)\times10^{-19}\,\text{J}}{hc}$$
$$= \frac{4.0410\times10^{-20}\,\text{J}}{hc} = 0.25222\text{ eV} = \boxed{2\,034.3\text{ cm}^{-1}}$$

$$\tilde{\nu}_1 = \tilde{\nu}_0 + 79.538\,\text{cm}^{-1}$$
$$= (2\,034.3 + 79.538)\,\text{cm}^{-1} = \boxed{2\,113.8\text{ cm}^{-1} = \frac{4.1990\times10^{-20}\,\text{J}}{hc}}$$

$$\frac{I_{1-1}}{I_{0-0}} \approx \frac{e^{-E_1(1,0)/kT_{\text{eff}}}}{e^{-E_1(0,0)/kT_{\text{eff}}}} = e^{-(E_1(1,0)-E_1(0,0))/kT_{\text{eff}}}$$
$$\approx e^{-hc\tilde{\nu}_1/kT_{\text{eff}}}$$

$$\ln\left(\frac{I_{1-1}}{I_{0-0}}\right) = -\frac{hc\tilde{\nu}_1}{kT_{\text{eff}}}$$

$$T_{\text{eff}} = \frac{hc\tilde{\nu}_1}{k\ln\left(\dfrac{I_{0-0}}{I_{1-1}}\right)} = \frac{4.1990\times10^{-20}\,\text{J}}{(1.38066\times10^{-23}\,\text{J K}^{-1})\ln(10)} = \boxed{1\,321\text{ K}}$$

The relative population of the $v=0$ and $v=1$ vibrational states is the inverse of the relative intensities of the transitions from those states; hence $1/0.1 = \boxed{10}$.

It would seem that with such a high effective temperature more than eight of the rotational levels of the S_1 state should have a significant population. But the spectra of molecules in comets are never as clearly resolved as those obtained in the laboratory and that is most probably the reason why additional rotational structure does not appear in these spectra.

F9.11 (a) The molar absorption coefficient $\varepsilon(\tilde{\nu})$ is given by

$$\varepsilon(\tilde{\nu}) = \frac{A(\tilde{\nu})}{L[CO_2]} = \frac{RTA(\tilde{\nu})}{Lx_{CO_2}p} \quad \text{[Beer-Lambert law, perfect gas law, and Dalton's law]}$$

where $T = 298$ K, $L = 10$ cm, $p = 1$ bar, and $x_{CO_2} = 0.021$.

The absorption band originates with the $001 \leftarrow 000$ transition of the antisymmetric stretch vibrational mode at 2 349 cm^{-1} (text Fig. 42.2). The band is very broad because of accompanying rotational transitions and lifetime broadening of each individual absorption (also called collisional broadening or pressure broadening, Topic 40). The spectra reveals that the Q branch is missing so we conclude that the transition $\boxed{\Delta J = 0 \text{ is forbidden}}$ [42.1] for the $D_{\infty h}$ point group of CO_2. The P-branch ($\Delta J = -1$) is evident at lower energies and the R-branch ($\Delta J = +1$) is evident at higher energies. See Figs. F9.3(a) and (b).

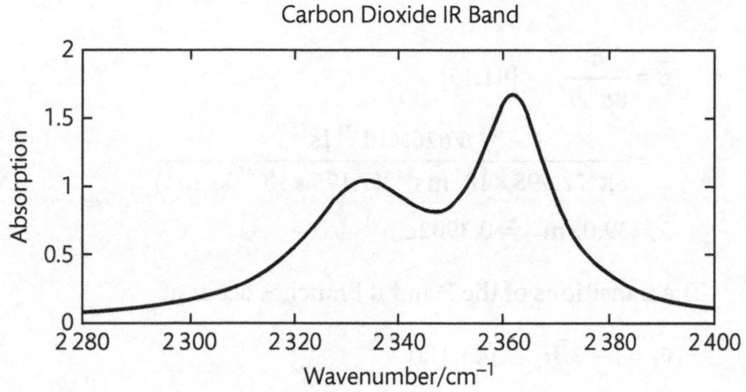

Fig F9.3(a)

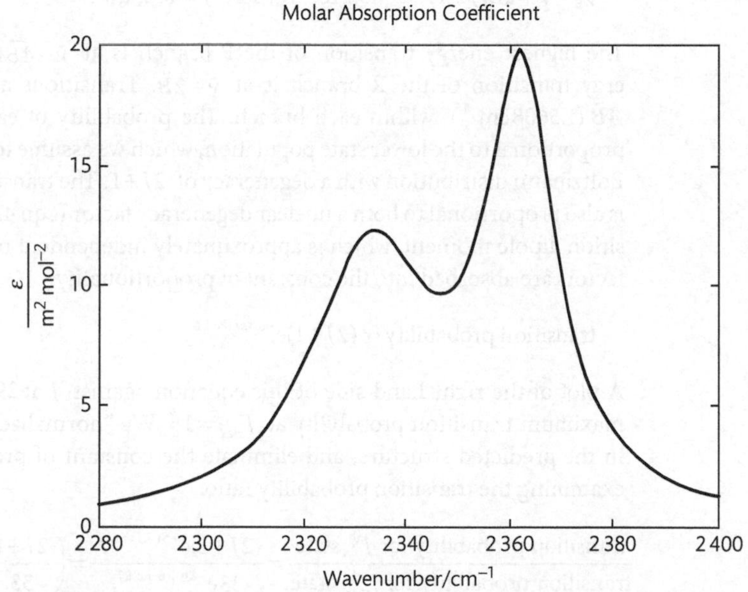

Fig F9.3(b)

(b) $^{16}O-^{12}C-^{16}O$ has two identical nuclei of zero spin so the CO_2 wavefunction must be symmetric w/r/t nuclear interchange and it must obey Bose–Einstein nuclear statistics. Consequently, J takes on even values only for the $\upsilon=0$ vibrational state and odd values only for the $\upsilon=1$ state. The (υ, J) states for this absorption band are $(1,J+1)\leftarrow(0,J)$ for $J=0,2,4,\ldots$. According to eqn 43.20, the energy of the $(0,J))$ state is

$$\tilde{S}(0,J)=\tfrac{1}{2}\tilde{\nu}+\tilde{B}J(J+1) \quad \text{where } \tilde{\nu}=2349\ \text{cm}^{-1}$$

$$\begin{aligned}I&=\frac{2M_{\text{O}}R^2}{N_{\text{A}}}\\&=\frac{2(0.01600\ \text{kg mol}^{-1})(116.2\times10^{-12}\ \text{m})^2}{6.022\times10^{23}\ \text{mol}^{-1}}\\&=7.175\times10^{-46}\ \text{kg m}^2 \qquad \text{(Table 41.1)}\end{aligned}$$

$$\begin{aligned}\tilde{B}&=\frac{h}{8\pi^2 cI} \qquad [41.15]\\&=\frac{6.626\times10^{-34}\ \text{J s}}{8\pi^2(2.998\times10^8\ \text{m s}^{-1})(7.175\times10^{-46}\ \text{kg m}^2)}\\&=39.02\ \text{m}^{-1}=0.3902\ \text{cm}^{-1}\end{aligned}$$

The transitions of the P and R branches occur at

$$\tilde{\nu}_{\text{P}}=\tilde{\nu}-2\tilde{B}J \qquad [43.21\text{a}]$$

and

$$\tilde{\nu}_{\text{R}}=\tilde{\nu}+2\tilde{B}(J+1) \qquad [43.21\text{c}] \quad \text{where } J=0,2,4,6\ldots$$

The highest energy transition of the P branch is at $\tilde{\nu}-4\tilde{B}$; the lowest energy transition of the R branch is at $\tilde{\nu}+2\tilde{B}$. Transitions are separated by $4\tilde{B}$ $(1.5608\ \text{cm}^{-1})$ within each branch. The probability of each transition is proportional to the lower state population, which we assume to be given by the Boltzmann distribution with a degeneracy of $2J+1$. The transition probability is also proportional to both a nuclear degeneracy factor (eqn 42.16) and a transition dipole moment, which is approximately independent of J. The former factors are absorbed into the constant of proportionality.

$$\text{transition probability} \propto (2J+1)\text{e}^{-S(0,J)hc/kT}$$

A plot of the right-hand-side of this equation against J at 298 K indicates a maximum transition probability at $J_{\text{max}}=16$. We "normalize" the maximum in the predicted structure, and eliminate the constant of proportionality by examining the transition probability ratio:

$$\frac{\text{transition probability for } J^{\text{th}} \text{ state}}{\text{transition probability for } J_{\text{max}} \text{ state}}=\frac{(2J+1)\text{e}^{-\tilde{S}(0,J)hc/kT}}{33\text{e}^{-\tilde{S}(0,16)hc/RT}}=\left(\frac{2J+1}{33}\right)\text{e}^{-(J^2+J-272)\tilde{B}hc/kT}$$

A plot, Fig. F9.4, of the above ratio against predicted wavenumbers can be compared to the ratio $A(\tilde{\nu})/A_{max}$ where A_{max} is the observed spectrum maximum (1.677). It shows a fair degree of agreement between the experimental and simple theoretical band shapes.

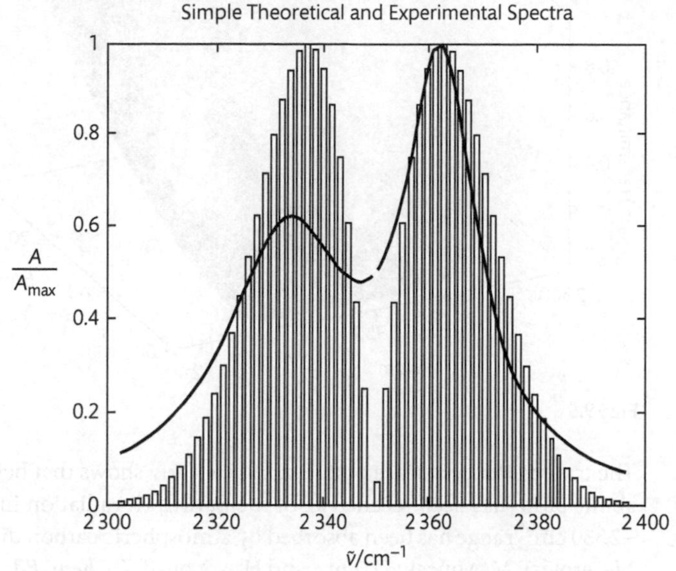

Fig F9.4

(c) Applying the Beer-Lambert law, we may write the relationship

$$A = \varepsilon(\tilde{\nu})\int_0^h [CO_2]\,dh$$

The strong absorption of the band suggests that h should not be a very great length and that $[CO_2]$ should be constant between the Earth's surface and h. Consequently, the integration gives

$$A = \varepsilon(\tilde{\nu})[CO_2]h$$

$$= \varepsilon(\tilde{\nu})h\left\{\frac{x_{CO_2}p}{RT}\right\} \quad \text{Dalton's law of partial pressures}$$

p and T are not expected to change much for modest values of h so we estimate that $p = 1$ bar and $T = 288\,K$.

$$A = \varepsilon(\tilde{\nu})h\left\{\frac{(3.3\times10^{-4})(1\times10^{5}\,Pa)}{(8.314\,J\,K^{-1}\,mol^{-1})(288\,K)}\right\}$$

$$= (0.0138\,m^{-3}\,mol)\times\varepsilon(\tilde{\nu})h$$

$$\text{Transmittance} = 10^{-A} = 10^{-(0.0138\,m^{-3}\,mol)\times\varepsilon(\tilde{\nu})h}$$

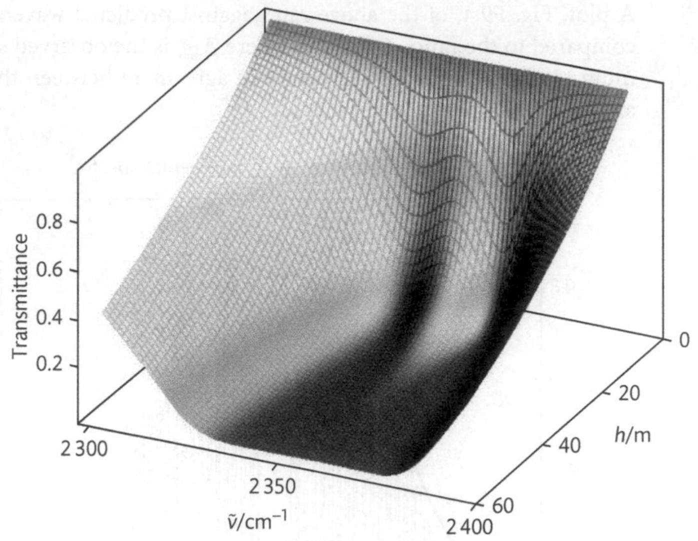

Fig F9.5

The transmittance surface plot, Fig. F9.5, clearly shows that before a height of about 30 m has been reached all of the Earth's IR radiation in the 2 320 cm^{-1} –2 380 cm^{-1} range has been absorbed by atmospheric carbon dioxide. See C.A. Meserole, F.M. Mulcahy, J. Lutz, and H.A. Yousif, *J. Chem. Ed.*, *74*, 316 (1997).

F9.13 (a) The calculations are perform with Spartan '10 using density functional theory (B3LYP) with the 6-31G* basis set. Equilibrium geometry characteristics for both the *trans* and *cis* form of (11) are summarized in the following table. The *trans* LUMO and HOMO are shown in Figures F9.6 and F9.7 while those of the *cis* form are shown in Figures F9.8 and F9.9.

Conformation	11-*trans* (11)	11-*cis* (11)
Total energy/eV	–23 783.28	–23 783.06
E_{LUMO}/eV	-5.78	-5.79
E_{HOMO}/eV	-7.91	-7.97
ΔE/eV	(a) 2.13	(b) 2.18
λ/nm	(a) 582	(b) 569
$C_5C_6C_7C_8$ torsion angle/°	170.97	170.26
$C_{11}C_{12}C_{13}C_{14}$ torsion angle/°	170.00	-179.73
C_1-C_2/pm	155.2	155.1
C_{11}-C_{12}/pm	139.0	139.6
C_{12}-C_{13}/pm	141.0	141.3
Dipole/D	13.58	12.27

The torsion angles are very close to the expected value of 180° with the exception of the torsion angle at C_{11}-***cis***, which is very close to the expected value of 0°. The aromatic character of the alternating π-bond system is evidenced by contrasting the computed bond lengths at a single bond away from the π -system (C_1–C_2), a double bond (C_{11}–C_{12}), and a single bond between doubles (C_{12}–C_{13}) within the Lewis structure. The former is a typical single C-C bond length. The latter two are approximately equal in both conformations and the value (~140 pm) is intermediate between a single and a double carbon bond, a characteristic of aromaticity. The dipole moment of the *cis* form is significantly smaller than that of the *trans* form and the difference may play an important role in the interaction of the chromophore with the surrounding protein matrix.

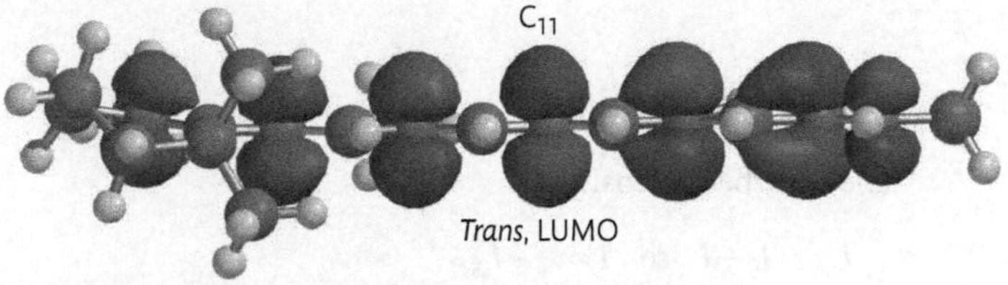

Trans, LUMO

Figure F9.6

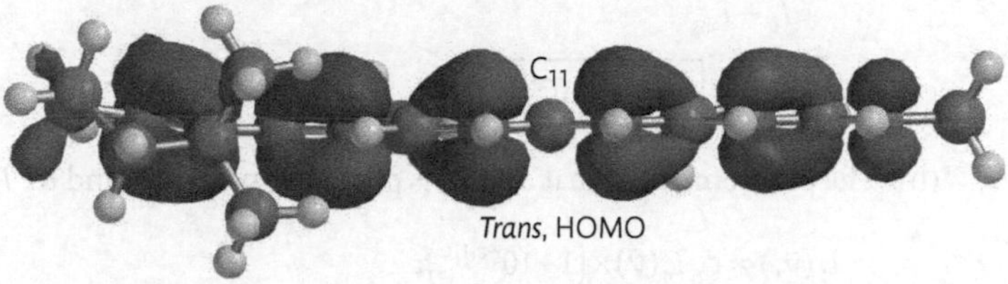

Trans, HOMO

Figure F9.7

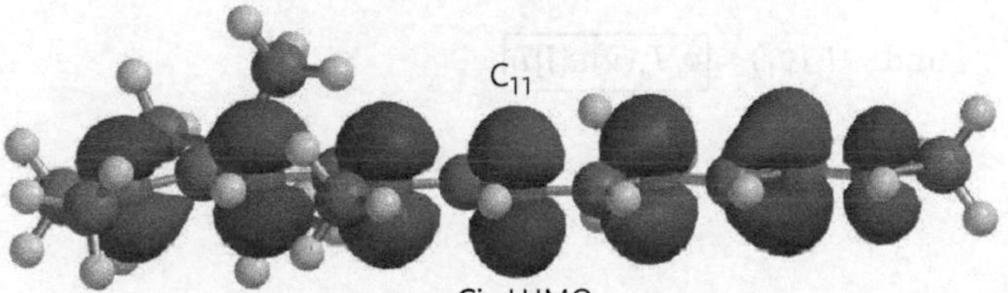

Cis, LUMO

Figure F9.8

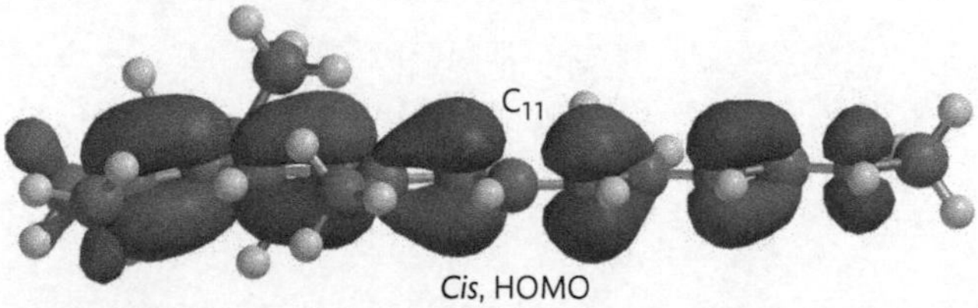

Cis, HOMO

Figure F9.9

(c) The lowest energy $\pi^* \leftarrow \pi$ transition is from the HOMO to the LUMO occurs in the visible near yellow for both confirmations. The transition energy appears to be very slightly lower for the *trans* conformation and at a slightly longer wavelength. However, this difference is so small that it may be an artifact of the computation and not of physical significance.

Question: The Figs. F9.7 and F9.9 show that both the *trans* and *cis* conformations have HOMO nodes at C_{11}. Does this affect the transition between the two conformations?

F9.15 (a) The Beer–Lambert Law is:

$$A = \log\frac{I_0}{I} = \varepsilon[\mathrm{J}]L \quad [40.6]$$

The absorbed intensity is:

$$I_{\mathrm{abs}} = I_0 - I \quad \text{so} \quad I = I_0 - I_{\mathrm{abs}}$$

Substitute this expression into the Beer–Lambert law and solve for I_{abs}:

$$\log\frac{I_0}{I_0 - I_{\mathrm{abs}}} = \varepsilon[\mathrm{J}]L \quad \text{so} \quad I_0 - I_{\mathrm{abs}} = I_0 \times 10^{-\varepsilon[\mathrm{J}]L}$$

and $I_{\mathrm{abs}} = \boxed{I_0 \times \left(1 - 10^{-\varepsilon[\mathrm{J}]L}\right)}$

(b) The problem states that $I_{\mathrm{f}}(\tilde{\nu}_{\mathrm{f}})$ is proportional to ϕ_{f} and to $I_{\mathrm{abs}}(\tilde{\nu})$, so:

$$I_{\mathrm{f}}(\tilde{\nu}_{\mathrm{f}}) \propto \phi_{\mathrm{f}}\, I_0(\tilde{\nu}) \times \left(1 - 10^{-\varepsilon[\mathrm{J}]L}\right)$$

If the exponent is small, we can expand $1 - 10^{-\varepsilon[\mathrm{J}]L}$ in a power series:

$$10^{-\varepsilon[\mathrm{J}]L} = (\mathrm{e}^{\ln 10})^{-\varepsilon[\mathrm{J}]L} \approx 1 - \varepsilon[\mathrm{J}]L\ln 10 + \cdots$$

and $I_{\mathrm{f}}(\tilde{\nu}_{\mathrm{f}}) \propto \boxed{\phi_{\mathrm{f}}\, I_0(\tilde{\nu})\varepsilon[\mathrm{J}]L}$

Topic 47 General principles

Discussion questions

D47.1 Refer to eqn 47.6, which describes the condition for resonance. The resonance field and resonance frequency are directly proportional to each other. High magnetic fields imply high frequencies and *vice versa*. The intensity of an NMR transition depends on a number of factors, two of which are seen in eqn 47.8a; one of these is the strength of the magnetic field itself and the other is the population difference between the energy levels of the α and β spins, which is also proportional to the magnetic field as shown by eqn 47.8b. Hence the intensity of the signal is proportional to $\mathcal{B}_0^2$. Consequently an increase in external field results in an even larger increase in signal intensity. See Section 47.1(b) and *Justification* 47.1 for a more detailed discussion. Chemical shifts as expressed in eqn 48.1 are also seen to be proportional to $\mathcal{B}_0$. Finally, as discussed in Section 48.3, the use of high magnetic fields simplifies the appearance of the complex spectra of macromolecules and allows them to be interpreted more readily. Although spin-spin splittings in NMR are not field dependent, strongly coupled (second order) spectra are simplified at high fields because the chemical shifts increase and individual groups of nuclei become identifiable again.

D47.3 The Larmor frequency is the rate of procession of a magnetic moment (electron or nuclear) in a magnetic field. Resonance occurs when the frequency of the applied radiation matches the Larmor frequency.

Exercises

E47.1(a) Work with eqn 47.4c. $\gamma_N\hbar = g_I\mu_N$, $\hbar$ has units J s, μ_N has units J T^{-1}; therefore γ_N has units

$$\frac{\text{J T}^{-1}}{\text{J s}} = \boxed{\text{s}^{-1}\ \text{T}^{-1}}$$

E47.2(a) The magnitude of the angular momentum is given by $\{I(I+1)\}^{1/2}\hbar$. For a proton $I = 1/2$, hence magnitude $= \dfrac{\sqrt{3}}{2}\hbar = \boxed{9.133\times10^{-35}\ \text{J s}}$. The components along the z–axis are $\pm\frac{1}{2}\hbar = \boxed{\pm 5.273\times 10^{-35}\ \text{J s.}}$ The angles that the projections of the angular momentum make with the z-axis are

$$\theta = \pm\cos^{-1}\frac{\hbar/2}{\sqrt{3}\hbar/2} = \boxed{\pm 0.9553\ \text{rad} = \pm 54.74^{\circ}}$$

E47.3(a) The resonance frequency is equal to the Larmor frequency of the proton and is given by

$$\nu = \nu_{\text{L}} = \frac{\gamma_{\text{N}}\mathcal{B}_0}{2\pi}[47.7]\ \text{with}\ \gamma_{\text{N}} = \frac{g_I\mu_{\text{N}}}{\hbar}[47.4\text{c}]$$

hence $\nu = \dfrac{g_I\mu_{\text{N}}\mathcal{B}_0}{h} = \dfrac{(5.5857)\times(5.0508\times10^{-27}\,J\,T^{-1})\times(13.5\ T)}{6.626\times10^{-34}\,J\,s} = \boxed{574\ \text{MHz}}$

E47.4(a) $E_{m_I} = -\gamma_{\text{N}}\hbar\mathcal{B}_0 m_I\ [47.4\text{c}] = -g_I\mu_{\text{N}}\mathcal{B}_0 m_I [47.4\text{c}, \gamma_{\text{N}}\hbar = g_I\mu_{\text{N}}]$

$m_I = \frac{3}{2}, \frac{1}{2}, -\frac{1}{2}, -\frac{3}{2}$

$E_{m_I} = (-0.4289)\times(5.051\times10^{-27}\ \text{J T}^{-1})\times(6.800\ \text{T}\times m_I) = \boxed{-1.473\times10^{-26}\ \text{J}\times m_I}$

E47.5(a) The energy level separation is

$$\Delta E = h\nu \quad \text{where} \quad \nu = \frac{\gamma_{\text{N}}\mathcal{B}_0}{2\pi}[47.6]$$

So

$$\nu = \frac{(6.73\times10^{7}\ \text{T}^{-1}\text{s}^{-1})\times 15.4T}{2\pi} = 1.65\times10^{8}\ \text{s}^{-1}$$
$$= 1.65\times10^{8}\ \text{Hz} = \boxed{165\ \text{MHz}}$$

E47.6(a) (a) By a calculation similar to that in Exercise 47.3(a) we can show that a 600-MHz NMR spectrometer operates in a magnetic field of 14.1 T. Thus

$$\Delta E = \gamma_{\text{N}}\hbar\mathcal{B}_0 = h\nu_{\text{L}} = h\nu \ \text{at resonance}$$
$$= (6.626\times10^{-34}\ \text{J s})\times(6.00\times10^{8}\ \text{s}^{-1}) = \boxed{3.98\times10^{-25}\ \text{J}}$$

(b) A 600-MHz NMR spectrometer means 600-MHz is the resonance frequency for protons for which the magnetic field is 14.1 T. In high-field NMRs, it is the field not the frequency which is fixed, so for the deuteron

$$\nu = \frac{g_I\mu_{\text{N}}\mathcal{B}_0}{h}\ [\text{Exercise 47.3(a)}]$$
$$= \frac{(0.8575)\times(5.051\times10^{-27}\ \text{J T}^{-1})\times(14.1\text{T})}{6.626\times10^{-34}\ \text{J s}} = 9.21\bar{6}\times10^{7}\ \text{Hz} = 92.1\bar{6}\ \text{MHz}$$

$$\Delta E = h\nu = (6.626\times10^{-34}\ \text{J s})\times(9.21\bar{6}\times10^{7}\ \text{s}^{-1}) = \boxed{6.11\times10^{-26}\ \text{J}}$$

Thus the separation in energy is larger for the proton $\boxed{\text{(a).}}$

E47.7(a) $\Delta E = h\nu = \gamma_N \hbar \mathcal{B} = g_I \mu_N \mathcal{B}$ [Solution to exercise 47.3(a)]

Hence, $$\mathcal{B} = \frac{h\nu}{g_I \mu_N} = \frac{(6.626\times10^{-34}\ \text{J Hz}^{-1})\times(50.0\times10^{6}\ \text{Hz})}{(0.40356)\times(5.051\times10^{-27}\ \text{J T}^{-1})} = \boxed{16.25\ \text{T}}$$

E47.8(a) In all cases the selection rule $\Delta m_I = \pm1$ is applied; hence

$$\mathcal{B}_0 = \frac{h\nu}{g_I \mu_N} = \frac{6.626\times10^{-34}\ \text{J Hz}^{-1}}{5.0508\times10^{-27}\ \text{J T}^{-1}} \times \frac{\nu}{g_I}$$

$$= (1.3119\times10^{-7})\times\frac{(\nu/\text{Hz})}{g_I}\text{T} = (0.13119)\times\frac{(\nu/\text{MHz})}{g_I}\text{T}$$

Refer to the following table:

$\mathcal{B}_0$ / T	(a)^{1}H	(b)^{2}H	(c)^{13}C
g_I	5.5857	0.85745	1.4046
(ii) 800 MHz	18.8	123	74.9
(i) 500 MHz	11.7	76.6	46.8

Comment. Magnetic fields above 30 T have not yet been obtained for use in NMR spectrometers. So, some of these calculated values have no current significance. As discussed in the solution to Exercises 47.6(a) and (b), it is the field, not the frequency, that is fixed in high-field NMR spectrometers. Thus an NMR spectrometer that is labeled a 500-MHz spectrometer refers to the resonance frequency for protons and has a magnetic field fixed at 11.7 T. In order to obtain resonance from other nuclei at the same fixed field as that used for protons you must construct a different probe that produces the frequency required for the particular nucleus under investigation.

E47.9(a) The ground state has

$$m_I + \tfrac{1}{2} = \alpha \text{ spin},\ m_I = -\tfrac{1}{2} = \beta \text{ spin}$$

Hence, with

$$\delta N = N_\alpha - N_\beta$$

$$\frac{\delta N}{N} = \frac{N_\alpha - N_\beta}{N_\alpha + N_\beta} = \frac{N_\alpha - N_\alpha e^{-\Delta E/kT}}{N_\alpha + N_\alpha e^{-\Delta E/kT}} \quad [\textit{Justification } 47.1]$$

$$= \frac{1-e^{-\Delta E/kT}}{1+e^{-\Delta E/kT}} \approx \frac{1-(1-\Delta E/kT)}{1+1} \approx \frac{\Delta E}{2kT} = \frac{g_I \mu_N \mathcal{B}_0}{2kT} \quad [\textit{for } \Delta E \ll kT]$$

That is, $$\frac{\delta N}{N} \approx \frac{g_I \mu_N \mathcal{B}_0}{2kT} = \frac{(5.5857)\times(5.0508\times10^{-27}\ \text{J T}^{-1})\times(\mathcal{B}_0)}{(2)\times(1.38066\times10^{-23}\ \text{J T}^{-1})\times(298\ \text{K})} \approx 3.43\times10^{-6}\,\mathcal{B}_0/\text{T}$$

(a) $\mathcal{B}_0 = 0.3$ T, $\quad \delta N/N = \boxed{1\times10^{-6}}$

(b) $\mathcal{B}_0 = 1.5$ T, $\quad \delta N/N = \boxed{5.1\times10^{-6}}$

(c) $\mathcal{B}_0 = 10$ T, $\quad \delta N/N = \boxed{3.4\times10^{-5}}$

E47.10(a) $$\delta N \approx \frac{Ng_I\mu_N\mathcal{B}_0}{2kT}[\text{Exercise47.9(a)}] = \frac{Nh\nu}{2kT}$$

Thus, $\delta N \propto \nu$

$$\frac{\delta N(800\text{ MHz})}{\delta N(60\text{ MHz})} = \frac{800\text{ MHz}}{60\text{ MHz}} = \boxed{13}$$

This ratio is not dependent on the nuclide as long as the approximation $\Delta E \ll kT$ holds (Exercise 47.9(a)).

E47.11(a) (a) $$\mathcal{B}_0 = \frac{h\nu}{g_I\mu_N} = \frac{(6.626\times10^{-34}\text{ J Hz}^{-1})\times(9\times10^9\text{ Hz})}{(5.5857)\times(5.051\times10^{-27}\text{ J T}^{-1})} = \boxed{2\times10^2\text{ T}}$$

(b) $$\mathcal{B}_0 = \frac{h\nu}{g_e\mu_B} = \frac{(6.626\times10^{-34}\text{ J Hz}^{-1})\times(300\times10^6\text{ Hz})}{(2.0023)\times(9.274\times10^{-24}\text{ J T}^{-1})} = \boxed{10\text{ mT}}$$

Comment. Because of the sizes (very large and very small) of these magnetic fields neither experiment seems feasible.

Question. What frequencies are required to observe electron resonance in the magnetic field of a 300 MHz NMR magnet and nuclear resonance in the field of a 9 GHz ($g = 2.00$) ESR magnet? Are these experiments feasible?

Problems

P47.1 (a) $\mu = g_I\mu_N|I|$ $[\mu_N = 5.05079\times10^{-27}\text{ J T}^{-1}]$

Using the formulas

$$\text{Sensitivity ratio}(\nu) = \frac{R_\nu(\text{nuclide})}{R_\nu(^1\text{H})} = \frac{2}{3}(I+1)\left[\frac{\mu(\text{nuclide})}{\mu(^1\text{H})}\right]$$

$$\text{Sensitivity ratio}(\mathcal{B}) = \frac{R_\mathcal{B}(\text{nuclide})}{R_\mathcal{B}(^1\text{H})} = \frac{1}{6}\left(\frac{I+1}{I^2}\right)\left[\frac{\mu(\text{nuclide})}{\mu(^1\text{H})}\right]^3$$

We construct the following table:

Nuclide	Spin I	μ/μ_N	Sensitivity ratio (ν)	Sensitivity ratio ($\mathcal{B}$)
^{2}H	1	0.85745	0.409	0.00965
^{13}C	$\frac{1}{2}$	0.7023	0.251	0.01590
^{14}N	1	0.40356	0.193	0.00101
^{19}F	$\frac{1}{2}$	2.62835	0.941	0.83350
^{31}P	$\frac{1}{2}$	1.1317	0.405	0.06654
^{1}H	$\frac{1}{2}$	2.79285		

(b) $\mu = \gamma_N\hbar I = g_I\mu_N|I|$

Hence $\gamma_N = \dfrac{\mu}{\hbar I}$

At constant frequency

$$R_\nu \propto (I+1)\mu\omega_0^2 \text{ or } R_\nu \propto (I+1)\mu \quad [\omega_0 \text{ is constant between the nuclei}]$$

Thus

$$\text{Sensitivity ratio}(\nu) = \frac{R_\nu(\text{nuclide})}{R_\nu(^1\text{H})}$$

$$= \tfrac{2}{3}(I+1)\left[\frac{\mu(\text{nuclide})}{\mu(^1\text{H})}\right] = \tfrac{2}{3}(I+1)\left[\frac{\mu(\text{nuclide})/\mu_\text{N}}{\mu(^1\text{H})/\mu_\text{N}}\right]$$

as above. Substituting $\omega_0 = \gamma_\text{N} B_0$ and $\gamma_\text{N} = \dfrac{\mu}{\hbar I}$, $\omega_0 = \dfrac{\mu B_0}{\hbar I}$ so

$$R_B \propto \frac{(I+1)\mu^3 B_0^2}{I^2}$$

$$\text{Sensitivity ratio}(\mathcal{B}) = \frac{R_B(\text{nuclide})}{R_B(^1\text{H})} = \tfrac{1}{6}\left(\frac{I+1}{I^2}\right)\left[\frac{\mu(\text{nuclide})}{\mu(^1\text{H})}\right]^3$$

$$= \tfrac{1}{6}\left(\frac{I+1}{I^2}\right)\left[\frac{\mu(\text{nuclide})/\mu_\text{N}}{\mu(^1\text{H})/\mu_\text{N}}\right]^3$$

as in part (a).

Topic 48 **Features of NMR spectra**

Discussion questions

D48.1 Detailed discussions of the origins of the local, neighboring group, and solvent contributions to the shielding constant can be found in Section 48.2. Here we will merely summarize the major features.

The local contribution is essentially the contribution of the electrons in the atom that contains the nucleus being observed. It can be expressed as a sum of a diamagnetic and paramagnetic parts, that is $\sigma(\text{local}) = \sigma_d + \sigma_p$. The diamagnetic part arises because the applied field generates a circulation of charge in the ground state of the atom. In turn, the circulating charge generates a magnetic field. The direction of this field can be found through Lenz's law which states that the induced magnetic field must be opposite in direction to the field producing it. Thus it shields the nucleus. The diamagnetic contribution is roughly proportional to the electron density on the atom and it is the only contribution for closed shell free atoms and for distributions of charge that have spherical or cylindrical symmetry. The local paramagnetic contribution is somewhat harder to visualize since there is no simple and basic principle analogous to Lenz's law that can be used to explain the effect. The applied field adds a term to the hamiltonian of the atom which mixes in excited electronic states into the ground state and any theoretical calculation of the effect requires detailed knowledge of the excited- state wavefunctions It is to be noted that the paramagnetic contribution does not require that the atom or molecule be paramagnetic. It is paramagnetic only in the sense that it results in an induced field in the same direction as the applied field.

The neighboring group contributions arise in a manner similar to the local contributions. Both diamagnetic and paramagnetic currents are induced in the neighboring atoms and these currents result in shielding contributions to the nucleus of the atom being observed. However, there are some differences: The magnitude of the effect is much smaller because the induced currents in neighboring atoms are much farther away. It also depends on the anisotropy of the magnetic susceptibility (see Topic 39) of the neighboring group as shown in eqn 48.12a. Only anisotropic susceptibilities result in a contribution.

Solvents can influence the local field in many different ways. Detailed theoretical calculations of the effect are difficult due to the complex nature of the solute-solvent interaction. Polar solvent–polar solute interactions are an electric field effect that usually causes deshielding of the solute protons. Solvent magnetic antisotropy can cause shielding or deshielding, for example, for solutes in benzene solution. In addition, there are a variety of specific chemical interactions between solvent and solute that can affect the chemical shift.

Exercises

E48.1(a)

(a) $\delta = \dfrac{\nu - \nu^\circ}{\nu^\circ} \times 10^6$ [48.4]

Since both ν and ν° depend upon the magnetic field in the same manner, namely

$$\nu = \frac{g_I \mu_N \mathcal{B}}{h} \quad \text{and} \quad \nu^\circ = \frac{g_I \mu_N \mathcal{B}_0}{h} \text{ [Exercise 47.3(a)]}$$

δ is $\boxed{\text{independent}}$ of both $\mathcal{B}$ and ν.

(b) Rearranging [48.4] we see $\nu - \nu^\circ = \nu^\circ \delta \times 10^{-6}$ and we see that the relative chemical shift is

$$\frac{\nu - \nu^\circ(800\text{ MHz})}{\nu - \nu^\circ(60\text{ MHz})} = \frac{(800\text{ MHz})}{(60\text{ MHz})} = \boxed{13}$$

Comment. This direct proportionality between $\nu - \nu^\circ$ and ν° is one of the major reasons for operating an NMR spectrometer at the highest frequencies possible.

E48.2(a) $\mathcal{B}_{\text{loc}} = (1-\sigma)\mathcal{B}_0$ [48.2]

$$|\Delta\mathcal{B}_{\text{loc}}| = |(\Delta\sigma)|\mathcal{B}_0 \approx |[\delta(\text{CH}_3) - \delta(\text{CHO})]|\mathcal{B}_0 \quad \left[|\Delta\sigma| \approx \left|\frac{\nu - \nu^\circ}{\nu^\circ}\right|\right]$$

$$= |(2.20 - 9.80)| \times 10^{-6}\mathcal{B}_0 = 7.60 \times 10^{-6}\mathcal{B}_0$$

(a) $\mathcal{B}_0 = 1.5$ T, $\quad |\Delta\mathcal{B}_{\text{loc}}| = 7.60 \times 10^{-6} \times 1.5\text{ T} = \boxed{11\ \mu\text{T}}$

(b) $\mathcal{B}_0 = 15$ T, $\quad |\Delta\mathcal{B}_{\text{loc}}| = \boxed{110\ \mu\text{T}}$

E48.3(a) $\nu - \nu^\circ = \nu^\circ \delta \times 10^{-6}$ [48.4]

$$|\Delta\nu| \equiv (\nu - \nu^\circ)(\text{CHO}) - (\nu - \nu^\circ)(\text{CH}_3)$$
$$= \nu^\circ[\delta(\text{CHO}) - \delta(\text{CH}_3)] \times 10^{-6}$$
$$= (9.80 - 2.20) \times 10^{-6}\nu^\circ = 7.60 \times 10^{-6}\nu^\circ$$

(a) $\nu^\circ = 250$ MHz, $|\Delta\nu| = 7.60 \times 10^{-6} \times 250\text{ MHz} = 1.90\text{ kHz}$

The spectrum is shown in Fig. 48.1 with the value of $|\Delta\nu|$ as calculated above.

(b) $\nu^\circ = 800$ MHz, $|\Delta\nu| = 6.08$ MHz

When the frequency is changed to 800 MHz, the $|\Delta\nu|$ changes to 6.08 kHz. The fine structure (the splitting within groups) remains the same as spin–spin splitting is unaffected by the strength of the applied field. However, the intensity of the lines increases by a factor of 3.2 because $\delta N/N \propto \nu$ (Exercise 47.10(a)).

The observed splitting pattern is that of an AX_3 (or A_3X) species, the spectrum of which is described in Section 48.3.

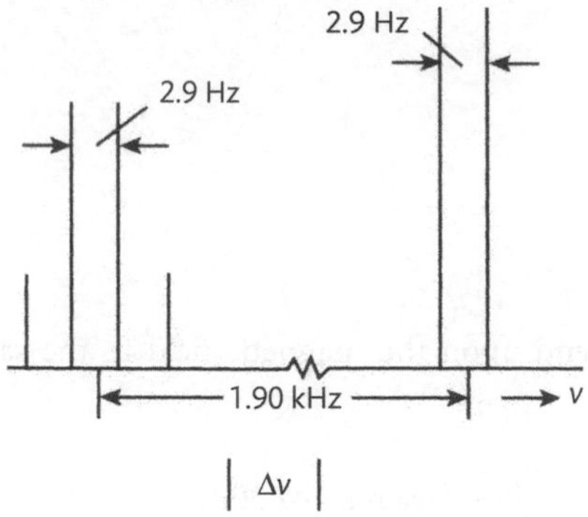

Fig 48.1

E48.4(a) The four equivalent ^{19}F nuclei ($I=\frac{1}{2}$) give a single line. However, the ^{10}B nucleus ($I=3$, 19.6 percent abundant) splits this line into $2\times3+1=7$ lines and the ^{11}B nucleus ($I=\frac{3}{2}$, 80.4 percent abundant) into $2\times\frac{3}{2}+1=4$ lines. The splitting arising from the ^{11}B nucleus will be larger than that arising from the ^{10}B nucleus (since its magnetic moment is larger, by a factor of 1.5, Table 47.2). Moreover, the total intensity of the four lines due to the ^{11}B nuclei will be greater (by a factor of $80.4/19.6\approx4$) than the total intensity of the seven lines due to the ^{10}B nuclei. The individual line intensities will be in the ratio $\frac{7}{4}\times4=7$ ($\frac{4}{7}$ the number of lines and about four times as abundant). The spectrum is sketched in Fig. 48.2.

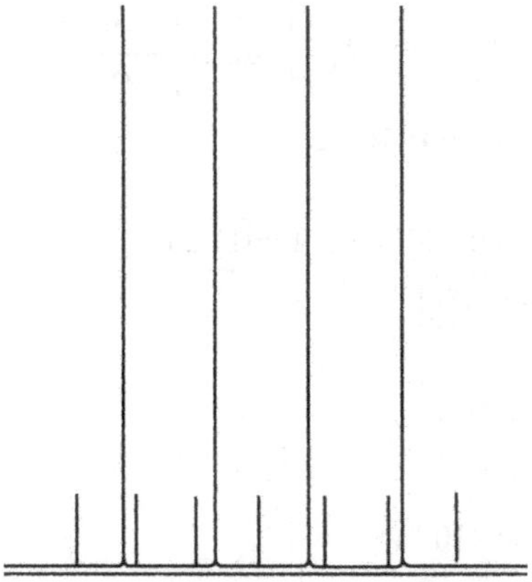

Fig 48.2

E48.5(a) $$\nu=\frac{g_I\mu_N\mathcal{B}}{h}\quad[\text{Solution to Exercise 47.3(a)}]$$

Hence, $$\frac{\nu(^{19}\mathbf{F})}{\nu(^{1}\mathbf{H})}=\frac{g(^{19}\mathbf{F})}{g(^{1}\mathbf{H})}$$

or $$\nu(^{19}\text{F})=\frac{5.2567}{5.5857}\times800\text{ MHz}=\boxed{753\text{ MHz}}$$

The proton resonance consists of 2 lines $(2\times\frac{1}{2}+1)$ and the ^{19}F resonance of 3 lines $[2\times(2\times\frac{1}{2})+1]$. The intensities are in the ratio 1:2:1 (Pascal's triangle for two equivalent spin $\frac{1}{2}$ nuclei, Section 48.3). The lines are spaced $\dfrac{5.5857}{5.2567}=1.06$ times greater in the fluorine region than the proton region. The spectrum is sketched in Fig. 48.3.

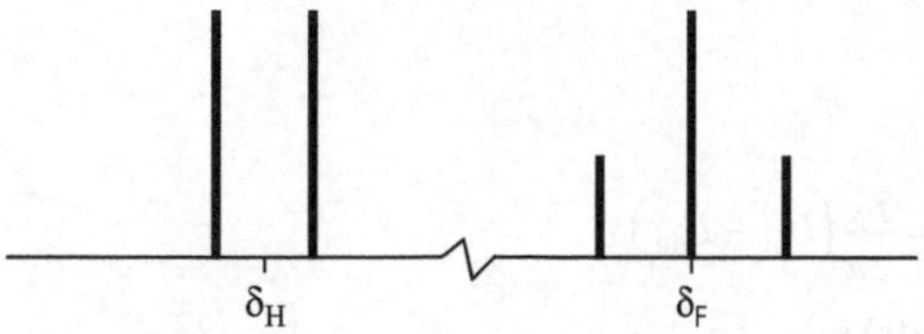

Figure 48.3

E48.6(a) The A, M, and X resonances lie in distinctively different groups. The A resonance is split into a 1:2:1 triplet by the M nuclei, and each line of that triplet is split into a 1:4:6:4:1 quintet by the X nuclei, (with $J_{AM}>J_{AX}$). The M resonance is split into a 1:3:3:1 quartet by the A nuclei and each line is split into a quintet by the X nuclei (with $J_{AM}>J_{MX}$). The X resonance is split into a quartet by the A nuclei and then each line is split into a triplet by the M nuclei (with $J_{AX}>J_{MX}$). The spectrum is sketched in Fig. 48.4.

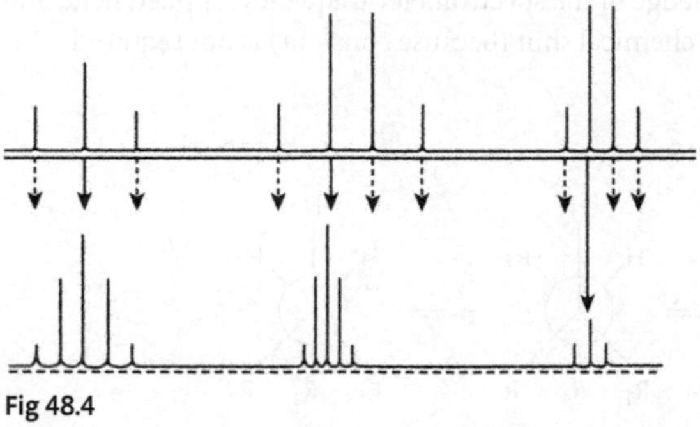

Fig 48.4

E48.7(a) $\tau\approx\dfrac{\sqrt{2}}{\pi\Delta\nu}$ [48.16, with $\delta\nu$ written as $\Delta\nu$]

$\Delta\nu=\nu^{\circ}(\delta'-\delta)\times10^{-6}$ [Exercise 48.3(a)]

Then $\tau\approx\dfrac{\sqrt{2}}{\pi\nu_0(\delta'-\delta)\times10^{-6}}$

$$\approx\frac{\sqrt{2}}{(\pi)\times(550\times10^{6}\text{ Hz})\times(4.8-2.7)\times10^{-6}}\approx3.9\times10^{-4}\text{ s}$$

Therefore, the signals merge when the lifetime of each isomer is less than about $\boxed{0.39\text{ ms}}$, corresponding to a conversion rate of about $\boxed{2.6\times10^{3}\text{ s}^{-1}}$

Problems

P48.1 We use $\nu = \dfrac{\gamma_N \mathcal{B}_{loc}}{2\pi} = \dfrac{\gamma_N}{2\pi}(1-\sigma)\mathcal{B}_0$ [48.3]
where $\mathcal{B}_0$ is the applied field.

Because shielding constants are quite small (a few parts per million) compared to 1, we may write for the purposes of this calculation

$$\nu = \frac{\gamma_N \mathcal{B}_0}{2\pi}$$

$$\nu_L - \nu_R = 100\ \text{Hz} = \frac{\gamma_N}{2\pi}\left(\mathcal{B}_{0L} - \mathcal{B}_{0R}\right)$$

$$\mathcal{B}_{0L} - \mathcal{B}_{0R} = \frac{2\pi \times 100\ \text{s}^{-1}}{\gamma_N}$$

$$= \frac{2\pi \times 100\ \text{s}^{-1}}{26.752\times 10^{7}\ \text{T}^{-1}\ \text{s}^{-1}} = 2.35\times 10^{-6}\ \text{T}$$

$$= 2.35\ \mu\text{T}$$

The field gradient required is then

$$\frac{2.35\ \mu\text{T}}{0.08\ \text{m}} = \boxed{29\ \mu\text{T m}^{-1}}$$

Note that knowledge of the spectrometer frequency, applied field, and the numerical value of the chemical shift (because constant) is not required.

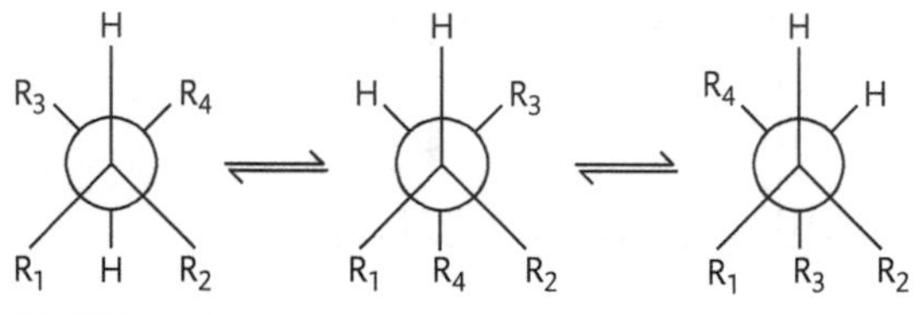

Fig 48.5

P48.3 It seems reasonable to assume that only staggered conformations can occur. Therefore the equilibria are as shown in Fig. 48.5.

When $R_3 = R_4 = H$, all three of the above conformations occur with equal probability:

$$^3J_{HH}(\text{methyl}) = \tfrac{1}{3}\left(^3J_t + 2\,^3J_g\right) \quad [\text{t} = \text{trans, g} = \text{gauche; CHR}_3\text{R}_4 = \text{methyl}]$$

The first conformation in the figure is trans, the second two are gauche.

Additional methyl groups will avoid being staggered between both R_1 and R_2. Therefore

$$^3J_{HH}(\text{ethyl}) = \tfrac{1}{2}(J_t + J_g) \quad [R_4 = H, R_3 = CH_3]$$
$$^3J_{HH}(\text{isopropyl}) = J_t \quad [R_3 = R_4 = CH_3]$$

We then have three simultaneous equations in two unknowns J_t and J_g.

$$\tfrac{1}{3}(^3J_t + 2\,^3J_g) = 7.3\ \text{Hz} \qquad (1)$$
$$\tfrac{1}{2}(^3J_t + ^3J_g) = 8.0\ \text{Hz} \qquad (2)$$
$$^3J_t = 11.2\ \text{Hz}$$

The two unknowns are overdetermined. The first two equations yield $^3J_t = 10.1$, $^3J_g = 5.9$. However, if we assume that $^3J_t = 11.2$ as measured directly in the isopropyl case then $^3J_g = 5.4$ (eqn 1) or 4.8 (eqn 2), with an average value of 5.1.

Using the original form of the Karplus equation

$$^3J_t = A\cos^2(180°) + B = 11.2$$
$$^3J_g = A\cos^2(60°) + B = 5.1$$

or

$$11.2 = A + B$$
$$5.1 = 0.25\,A + B$$

These simultaneous equations yield $A = 6.8$ Hz and $B = 4.8$ Hz. With these values of A and B, the original form of the Karplus equation fits the data exactly (at least to within the error in the values of ^3J_t and ^3J_g and in the measured values reported).

From the form of the Karplus equation in the text [48.14] we see that those values of A, B and C cannot be determined from the data given, as there are three constants to be determined from only two values of J. However, if we use the values of A, B and C given in the text, then

$$J_t = 7\ \text{Hz} + 1\ \text{Hz}(\cos 180°) + 5\ \text{Hz}(\cos 360°) = 11\ \text{Hz}$$
$$J_g = 7\ \text{Hz} + 1\ \text{Hz}(\cos 60°) + 5\ \text{Hz}(\cos 120°) = 5\ \text{Hz}$$

The agreement with the modern form of the Karplus equation is excellent, but not better than the original version. Both fit the data equally well. But the modern version is preferred as it is more generally applicable.

P48.5 $\quad ^3J_{HH} = A + B\cos\phi + C\cos 2\phi \quad [48.14]$

$$\frac{d}{d\phi}(^3J_{HH}) = -B\sin\phi - 2C\sin 2\phi = 0$$

This equation has a number of solutions:

$$\phi = 0,\ \phi = n\pi,\ \phi = \pi - \arccos\left(\frac{B}{4C}\right) = \arccos\left(\frac{B}{4C}\right)$$

The first two are trivial solutions.

$$\text{If } \phi = \arccos\left(\frac{B}{4C}\right) \text{ then, } \sin\phi = \sqrt{1 - \frac{B}{16C^2}}$$

$$\sin 2\phi = 2\sin\phi\cos\phi = 2\sqrt{1 - \frac{B^2}{16C^2}}\left(\frac{B}{4C}\right)$$

$$B\sin\phi + 2C\sin 2\phi = B\sqrt{1 - \frac{B^2}{16C^2}} + 4C\sqrt{1 - \frac{B^2}{16C^2}}\left(\frac{B}{4C}\right) = 0$$

So $\dfrac{B}{4C} = \cos\phi$ clearly satisfies the condition for an extremum.

The second derivative is

$$\frac{d^2}{d\phi^2}({}^3J_{HH}) = -B\cos\phi - 4C\cos 2\phi = -B\cos\phi - 4C(2\cos^2\phi - 1)$$

$$= -B\left(\frac{B}{4C}\right) - 4C\left(2\frac{B^2}{16C^2} - 1\right) = -\frac{B^2}{4C} - \frac{2B^2}{4C} + 4C$$

This quantity is positive if $16C^2 > 3B^2$

This is certainly true for typical values of B and C, namely $B = -1$ Hz and $C = 5$ Hz. Therefore the condition for a minimum is as stated, namely, $\boxed{\cos\phi = B/4C}$.

Topic 49 Pulse techniques in NMR

Discussion questions

D49.1 Before the application of a pulse the magnetization vector, ***M***, points along the direction of the static external magnetic field $\mathcal{B}_0$. There are more α spins than β spins. When we apply a rotating magnetic field $\mathcal{B}_1$ at right angles to the static field, the magnetization vector as seen in the rotating frame begins to precess about the $\mathcal{B}_1$ field with angular frequency $\omega_1 = \gamma_N \mathcal{B}_1$. The angle through which M rotates is $\theta = \gamma_N \mathcal{B}_1 t$, where t is the time for which the $\mathcal{B}_1$ pulse is applied. When $t = \pi / 2\gamma_N \mathcal{B}_1$, $\theta = \pi/2 = 90°$, and ***M*** has rotated into the *xy* plane. Now there are equal numbers of α and β spins. A 180° pulse applied for a time $\pi / \gamma_N \mathcal{B}_1$, rotates ***M*** antiparallel to the static field. Now there are more β spins than α spins. A population inversion has occurred.

D49.3 For example, at room temperature, the tumbling rate of benzene, the small molecule, in a mobile solvent, may be close to the Larmor frequency, and hence its spin-lattice relaxation time will be short. As the temperature increases, the tumbling rate may increase well beyond the Larmor frequency, resulting in an increased spin–lattice relaxation time.

For the large molecule (like a polymer) at room temperature, the tumbling rate may be well below the Larmor frequency, but with increasing temperature it will approach the Larmor frequency due to the increased thermal motion of the molecule combined with the decreased viscosity of the solvent. Therefore, the spin-lattice relaxation time may decrease

D49.5 The basic COSY experiment uses the simplest of all two-dimensional pulse sequences: a single 90° pulse to excite the spins at the end of the preparation period and a mixing period containing just a second 90° pulse.

The key to the COSY technique is the effect of the second 90° pulse, which can be illustrated by consideration of the four energy levels of an AX system (as shown in Figs. 49.12, 49.13, & 49.14 of the text). At thermal equilibrium, the population of the αAαX level is the greatest, and that of βAβX level is the smallest; the other two levels have the same energy and an intermediate population. After the first 90° pulse, the spins are no longer at thermal equilibrium. If a second 90° pulse is applied at a time t_1 that is short compared to the spin-lattice relaxation time T_1 the extra input of energy causes further changes in the populations of the four states. The changes in populations will depend on how far the individual magnetizations have precessed during the evolution period.

For simplicity, let us consider a COSY experiment in which the second 90° pulse is split into two selective pulses, one applied to X and one to A. Depending on the evolution time t_1, the 90° pulse that excites X may leave the population differences across each of the two X transitions unchanged, inverted, or somewhere in between. Consider the extreme case in which one population difference is inverted and the other unchanged. The 90° pulse that excites A will now generate an FID in which one of the two A transitions has increased in intensity, and the other has decreased. The overall effect is that precession of the X spins during the evolution period determines the amplitudes of the signals from the A spins obtained during the detection period. As the evolution time t_1 is increased, the intensities of the signals from A spins oscillate at rates determined by the frequencies of the two X transitions.

This transfer of information between spins is at the heart of two-dimensional NMR spectroscopy and leads to the correlation of different signals in a spectrum. In this case, information transfer tells us that there is a scalar coupling between A and X. If we conduct a series of experiments in which t_1 is incremented, Fourier transformation of the FIDs on t_2 yields a set of spectra $I(\nu_{1,}\ \nu_2)$ in which the A signal amplitudes oscillate as a function of t_1. A second Fourier transformation, this time on t_1, converts these oscillations into a two-dimensional spectrum $I(\nu_{1,}\ \nu_2)$. The signals are spread out in ν_1 according to their precession frequencies during the detection period. Thus, if we apply the COSY pulse sequence to our AX spin system, the result is a two-dimensional spectrum that contains four groups of signals centred on the two chemical shifts in ν_1 and ν_2. Each group will show fine structure, consisting of a block of four signals separated by J_{AX}. The diagonal peaks are signals centred on $(\delta_A\ \delta_A)$ and $(\delta_X\ \delta_X)$ and lie along the diagonal $\nu_1 = \nu_2$. They arise from signals that did not change chemical shift between t_1 and t_2. The cross peaks (or *off-diagonal peaks*) are signals centred on $(\delta_A\ \delta_X)$ and $(\delta_X\ \delta_A)$ and owe their existence to the coupling between A and X. Consequently, cross peaks in COSY spectra allow us to map the couplings between spins and to trace out the bonding network in complex molecules. Fig. 49.16 of the text shows a simple example of a proton COSY spectrum of 1-nitropropane.

Exercises

E49.1(a) Analogous to precession of the magnetization vector in the laboratory frame due to the presence of $\mathcal{B}_0$ that is

$$\nu_L = \frac{\gamma_N \mathcal{B}_0}{2\pi} \quad [47.6],$$

there is a precession in the rotating frame, due to the presence of $\mathcal{B}_1$, namely

$$\nu_L = \frac{\gamma_N \mathcal{B}_1}{2\pi} \quad \text{or} \quad \omega_1 = \gamma_N \mathcal{B}_1 \quad [\omega = 2\pi\nu]$$

Since ω is an angular frequency, the angle through which the magnetization vector rotates is

$$\theta = \gamma_N \mathcal{B}_1 t = \frac{g_I \mu_N}{\hbar} \mathcal{B}_1 t$$

$$\text{and } \mathcal{B}_1 = \frac{\theta \hbar}{g_I \mu_N t} = \frac{(\frac{\pi}{2})\times(1.055\times10^{-34}\ \text{J s})}{(5.586)\times(5.051\times10^{-27}\ \text{J T}^{-1})\times(1.0\times10^{-5}\ \text{s})} = \boxed{5.9\times10^{-4}\ \text{T}}$$

A 180° pulse requires $2\times10\ \mu\text{s} = \boxed{20\ \mu\text{s}}$

E49.2(a) The proton COSY spectrum of 1-nitropropane shows that (a) the C_a-H resonance with $\delta = 4.3$ shares a cross-peak with the C_b-H resonance at $\delta = 2.1$ and (b) the C_b-H resonance with $\delta = 2.1$ shares a cross-peak with the C_c-H resonance at $\delta = 1.1$. Off diagonal peaks indicate coupling between H's on various carbons. Thus peaks at (4,2) and (2,4) indicate that the H's on the adjacent CH_2 units are coupled. The peaks at (1,2) and (2,1) indicate that the H's on CH_3 and central CH_2 units are coupled.

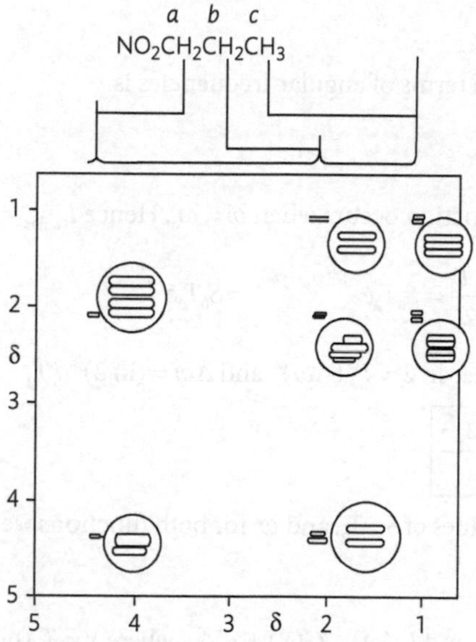

$= \boxed{0.58\ \text{mT}}$

Fig 49.1

Problems

P49.1 The envelopes of maxima and minima of the curve are determined by T_2 through eqn 49.1, but the time interval between the maxima of this decaying curve corresponds to the reciprocal of the frequency difference $\Delta\nu$ between the pulse frequency ν_0 and the Larmor frequency ν_L, that is $\Delta\nu = |\nu_0 - \nu_L|$

$$\Delta\nu = \frac{1}{0.12\ \text{s}} = 8.3\ \text{s}^{-1} = 8.3\ \text{Hz}$$

Therefore the Larmor frequency is $\boxed{400\times10^6\ \text{Hz} \pm 8\ \text{Hz}}$

According to eqns 49.1 and 49.5, the intensity of the maxima in the FID curve decays exponentially as e^{-t/T_2}. Therefore T_2 corresponds to the time at which the intensity has been reduced to $1/e$ of the original value. In the text figure, this corresponds to a time slightly before the fourth maximum has occurred, or about $\boxed{0.29\text{ s}}$

P49.3 (a) The *Lorentzian function* in terms of angular frequencies is

$$I_L(\omega)=\frac{S_0T_2}{1+T_2^2(\omega-\omega_0)^2}$$

The maximum in this function occurs when $\omega=\omega_0$. Hence $I_{L,max}=S_0T_2$ and

$$I_L(\Delta\omega_{1/2})=\frac{I_{L,max}}{2}=\frac{S_0T_2}{2}=\frac{S_0T_2}{1+T_2^2(\omega_{1/2}-\omega_0)^2}=\frac{S_0T_2}{1+T_2^2(\Delta\omega)^2}$$

where $\Delta\omega=\frac{1}{2}\Delta\omega_{½}$; hence $2=1+T_2^2(\Delta\omega)^2$ and $\Delta\omega=1/T_2$. Therefore

$$\boxed{\Delta\omega_{½}=\frac{2}{T_2}}$$

(b) The *Gaussian function* in terms of angular frequencies is

$$I_G(\omega)=S_0T_2e^{-T_2^2(\omega-\omega_0)^2}$$

The maximum in this function occurs when $\omega=\omega_0$. Hence $I_{G,max}=S_0T_2$ and

$$I_G(\Delta\omega_{1/2})=\frac{I_{G,max}}{2}=\frac{S_0T_2}{2}=S_0T_2e^{-T_2^2(\omega_{½}-\omega_0)^2}=S_0T_2e^{-T_2^2(\Delta\omega)^2}$$

where $\Delta\omega=\frac{1}{2}\Delta\omega_{½}$; hence $\ln 2=T_2^2(\Delta\omega)^2$ and $\Delta\omega=(\ln 2)^{½}/T_2$

Therefore $\boxed{\Delta\omega_{½}=\frac{2(\ln 2)^{½}}{T_2}}$

(c) If we choose the same values of S_0, T_2, and ω_0 for both functions we may rewrite them as

$$I_L(\omega)=L(x)\propto\frac{1}{1+x^2}\text{ and }I_G(\omega)=G(x)\propto e^{-x^2}\text{ where }x=T_2(\omega-\omega_0)$$

These functions are plotted against x in the following Mathcad worksheet. Note that the Lorentzian function is slightly sharper in the center, although this is difficult to discern with the scale of x used in the figure, and decreases much more slowly in the wings beyond the half amplitude points. Also note that the functions plotted in the figure are not normalized but are matched at their peak amplitude in order to more clearly display the differences in their shapes. If the curves had been normalized the areas under the two curves would be equal, but the peak height in the Lorentzian would be lower than the Gaussian peak height.

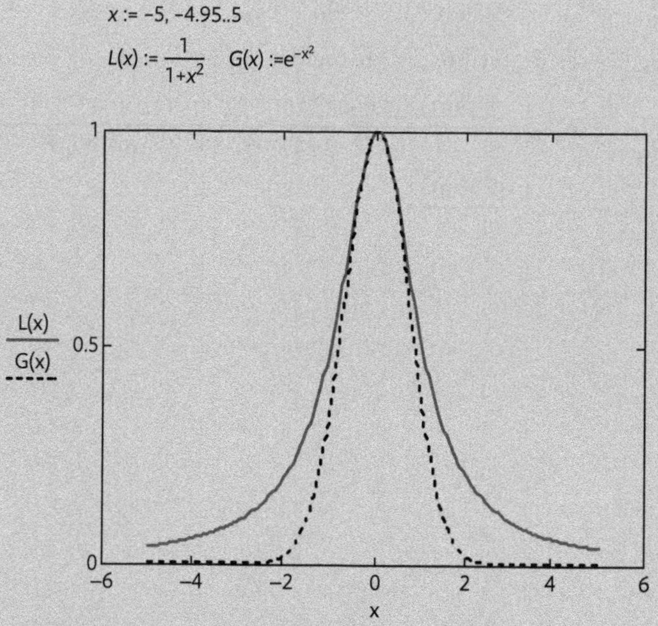

Figure 49.2

P49.5 We have seen (Problem 49.4) that if $G \propto \cos\omega_0 t$, then $I(\omega) \propto \dfrac{1}{\left[1+(\omega_0-\omega)^2\tau^2\right]}$ which peaks at $\omega \approx \omega_0$. Therefore, if

$$G(t) \propto a\cos\omega_1 t + b\cos\omega_2 t$$

we can anticipate that

$$I(\omega) \propto \frac{a}{1+(\omega_1-\omega)^2\tau^2} + \frac{b}{1+(\omega_2-\omega)^2\tau^2}$$

and explicit calculation shows this to be so. Therefore, $I(\omega)$ consists of two absorption lines, one peaking at $\omega \approx \omega_1$ and the other at $\omega \approx \omega_2$.

P49.7

$$\langle \mathcal{B}_{\text{nucl}} \rangle = \frac{-g_I\mu_N\mu_0 m_I}{4\pi R^3} \times \frac{\int_0^{\theta_{\max}} (1-3\cos^2\theta)\sin\theta\,d\theta}{\int_0^{\theta_{\max}} \sin\theta\,d\theta}$$

The denominator is the normalization constant and ensures that the total probability of being between 0 and $\theta_{\max}$ is 1.

$$\langle \mathcal{B}_{\text{nucl}} \rangle = \frac{-g_I\mu_N\mu_0 m_I}{4\pi R^3} \times \frac{\int_1^{x_{\max}} (1-3x^2)\,dx}{\int_1^{x_{\max}} dx} \quad [x_{\max} = \cos\theta_{\max}]$$

$$\langle \mathcal{B}_{\text{nucl}} \rangle = \frac{-g_I\mu_N\mu_0 m_I}{4\pi R^3} \times \frac{x_{\max}(1-x_{\max}^2)}{x_{\max}-1} = \frac{+g_I\mu_N\mu_0 m_I}{4\pi R^3}(\cos^2\theta_{\max} + \cos\theta_{\max})$$

If $\theta_{max} = \pi$ (complete rotation), $\cos\theta_{max} = -1$ and $\langle \mathcal{B}_{nucl} \rangle = 0$. If $\theta_{max} = 30°$,

$\cos^2\theta_{max} + \cos\theta_{max} = 1.616$ and

$$\langle \mathcal{B}_{nucl} \rangle = \frac{(5.5857)\times(5.0508\times10^{-27}\ \mathrm{J\,T^{-1}})\times(4\pi\times10^{-7}\ \mathrm{T^2\,J^{-1}\,m^3})\times(1.616)}{(4\pi)\times(1.58\times10^{-10}\ \mathrm{m})^3\times(2)}$$

$$= \boxed{0.58\ \mathrm{mT}}$$

Topic 50 **Electron paramagnetic resonance**

Discussion questions

D50.1 The hyperfine structure in the ESR spectrum of an atomic or molecular system is a result of two interactions: an anisotropic dipolar coupling between the net spin of the unpaired electrons and the nuclear spins and also an isotropic coupling due to the Fermi contact interaction. In solution, only the Fermi contact interaction contributes to the splitting as the dipolar contribution averages to zero in a rapidly tumbling system. In the case of π-electron radicals, such as $C_6H_6^-$, no hyperfine interaction between the unpaired electron and the ring protons might have been expected. The protons lie in the nodal plane of the molecular orbital occupied by the unpaired electron, so any hyperfine structure cannot be explained by a simple Fermi contact interaction which requires an unpaired electron density at the proton. However, an indirect spin polarization mechanism, similar to that used to explain spin–spin couplings in NMR, can account for the existence of proton hyperfine interactions in the ESR spectra of these systems. Refer to Fig. 50.6 of the text. Because of Hund's rule, the unpaired electron and the first electron in the C—H bond (the one from the C atom), will tend to align parallel to each other. The second electron in the C—H bond (the one from H) will then align antiparallel to the first by the Pauli principle, and finally the Fermi contact interaction will align the proton and electron on H antiparallel. The net result (parallel × antiparallel × antiparallel) is that the spins of the unpaired electron and the proton are aligned parallel. They have effectively detected each other.

Exercises

E50.1(a) $$g=\frac{h\nu}{\mu_B \mathcal{B}_0}\quad [50.2]$$

We shall often need the value

$$\frac{h}{\mu_B}=\frac{6.62608\times10^{-34}\ \mathrm{J\,Hz^{-1}}}{9.27402\times10^{-24}\ \mathrm{J\,T^{-1}}}=7.14478\times10^{-11}\ \mathrm{T\,Hz^{-1}}$$

Then, in this case

$$g = \frac{(7.14478\times10^{-11}\,\mathrm{T\,Hz^{-1}})\times(9.2231\times10^{9}\,\mathrm{Hz})}{329.12\times10^{-3}\,\mathrm{T}} = \boxed{2.0022}$$

E50.2(a) $a = \mathcal{B}(\text{line}\,3) - \mathcal{B}(\text{line}\,2) = \mathcal{B}(\text{line}\,2) - \mathcal{B}(\text{line}\,1)$

$$\left.\begin{aligned} \mathcal{B}_3 - \mathcal{B}_2 &= (334.8 - 332.5)\,\mathrm{mT} = 2.3\,\mathrm{mT} \\ \mathcal{B}_2 - \mathcal{B}_1 &= (332.5 - 330.2)\,\mathrm{mT} = 2.3\,\mathrm{mT} \end{aligned}\right\} a = \boxed{2.3\,\mathrm{mT}}$$

Use the centre line to calculate g

$$g = \frac{h\nu}{\mu_B \mathcal{B}_0} = (7.14478\times10^{-11}\,\mathrm{T\,Hz^{-1}})\times\frac{9.319\times10^{9}\,\mathrm{Hz}}{332.5\times10^{-3}\,\mathrm{T}} = \boxed{2.002\overline{5}}$$

E50.3(a) The centre of the spectrum will occur at $332.5\,\mathrm{mT}$. Proton 1 splits the line into two components with separation $2.0\,\mathrm{mT}$ and hence at $332.5\pm1.0\,\mathrm{mT}$. Proton 2 splits these two hyperfine lines into two, each with separation $2.6\,\mathrm{mT}$, and hence the lines occur at $332.5\pm1.0\pm1.3\,\mathrm{mT}$. The spectrum therefore consists of four lines of $\boxed{\text{equal intensity}}$ at the fields $\boxed{330.2\,\mathrm{mT}, 332.2\,\mathrm{mT}, 332.8\,\mathrm{mT}, 334.8\,\mathrm{mT}}$

E50.4(a) We construct Fig. 50.1(a) for CH_3 and Fig. 50.1(b) for CD_3. The predicted intensity distribution is determined by counting the number of overlapping lines of equal intensity from which the hyperfine line is constructed.

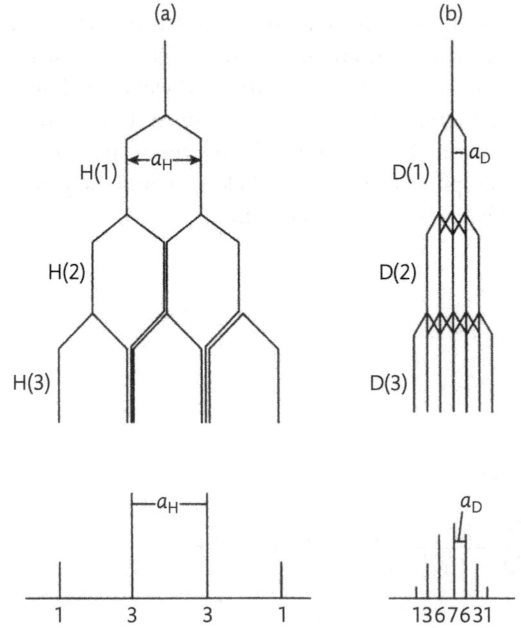

Fig 50.1

E50.5(a) $\mathcal{B}_0 = \dfrac{h\nu}{g\mu_B} = \dfrac{7.14478\times10^{-11}}{2.0025}\,\text{T Hz}^{-1}\times\nu[\text{Exercise 50.1(a)}] = 35.68\ \text{mT}\times(\nu/\text{GHz})$

(a) $\nu = 9.313\ \text{GHz},\quad \mathcal{B}_0 = \boxed{332.3\ \text{mT}}$

(b) $\nu = 33.80\ \text{GHz},\quad \mathcal{B}_0 = 1\,206\ \text{mT} = \boxed{1.206\ \text{T}}$

E50.6(a) Since the number of hyperfine lines arising from a nucleus of spin I is $2I+1$, we solve $2I+1=4$ and find that $\boxed{I=\frac{3}{2}}$.

Comment. Four lines of equal intensity could also arise from two inequivalent nuclei with $I=\frac{1}{2}$.

E50.7(a) The X nucleus produces six lines of equal intensity. The pair of H nuclei in XH_2 split each of these lines into a 1 : 2 : 1 triplet (Fig. 50.2(a)). The pair of D nuclei ($I=1$) in XD_2 split each line into a 1 : 2 : 3 : 2 : 1 quintet (Fig. 50.2(b)). The total number of hyperfine lines observed is then $6\times3=18$ in XH_2 and $6\times5=30$ in XD_2.

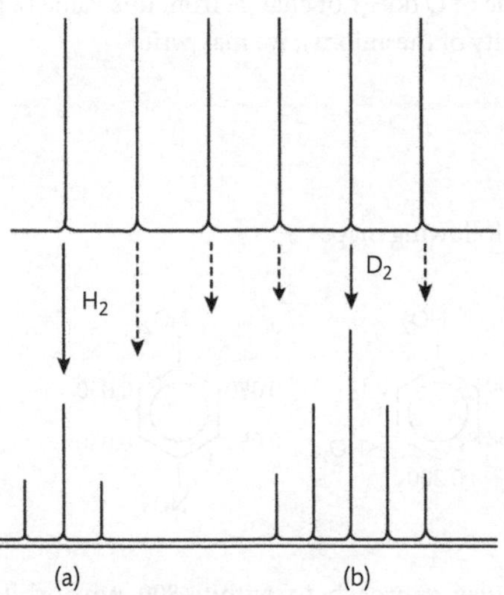

Fig 50.2(a) and 50.2(b)

Problems

P50.1 $\nu = \dfrac{g_e\mu_B\mathcal{B}}{h} = \dfrac{(2.00)\times(9.274\times10^{-24}\ \text{J T}^{-1})\times(1.0\times10^{3}\ \text{T})}{6.626\times10^{-34}\ \text{J s}} = \boxed{2.8\times10^{13}\ \text{Hz}}$

This frequency is in the infrared region of the electromagnetic spectrum and hence is comparable to the frequencies and energies of molecular vibrations; it is much greater than those of molecular rotations, and far less than those of molecular electronic motion.

P50.3 Refer to the figure in the solution to Exercise 50.4(a). The width of the CH_3 spectrum is $3a_H = \boxed{6.9\ \text{mT}}$. The width of the CD_3 spectrum is $6a_D$. It seems reasonable to assume, since the hyperfine interaction is an interaction of the magnetic moments of the nuclei with the magnetic moment of the electron, that the strength of the interactions is proportional to the nuclear moments.

$$\mu = g_I\mu_N I \quad \text{or} \quad \mu_z = g_I\mu_N m_I \quad [\text{eqns 47.3 \& 47.4}]$$

and thus nuclear magnetic moments are proportional to the nuclear g-values; hence

$$a_D \approx \frac{0.85745}{5.5857}\times a_H = 0.1535a_H = 0.35\ \text{mT}$$

Therefore, the overall width is $6a_D = \boxed{2.1\ \text{mT}}$

P50.5 For $C_6H_6^-$, $a = Q\rho$ with $Q = 2.25$ mT [50.5]

If we assume that the value of Q does not change from this value (a good assumption in view of the similarity of the anions), we may write

$$\rho = \frac{a}{Q} = \frac{a}{2.25\ \text{mT}}$$

and we can construct the following maps:

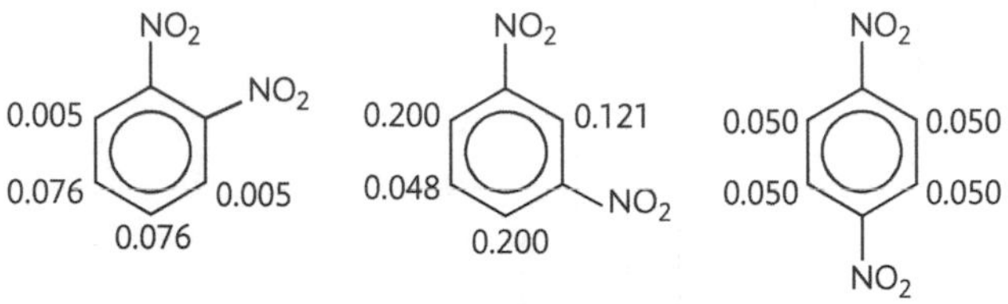

P50.7 When spin label molecules approach to within 800 pm, orbital overlap of the unpaired electrons and dipolar interactions between magnetic moments cause an exchange coupling interaction between the spins. The electron exchange process occurs at a rate that increases as concentration increases. Thus the process has a lifetime that is too long at low concentrations to affect the "pure" ESR signal. As the concentration increases, the linewidths increase until the triplet coalesces into a broad singlet. Further increase of the concentration decreases the exchange lifetime and therefore the linewidth of the singlet.

When spin labels within biological membranes are highly mobile, they may approach closely and the exchange interaction may provide the ESR spectra with information that mimics the moderate and high concentration signals above.

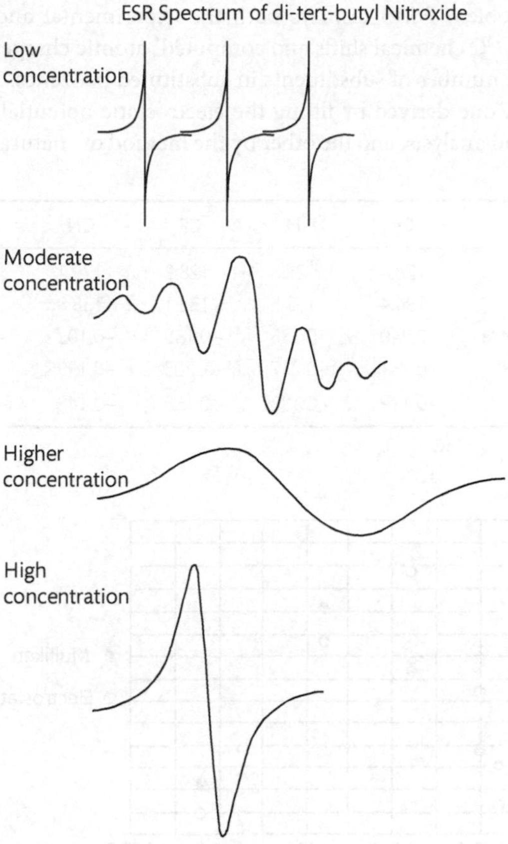

Fig 50.3

Focus 10: Integrated activities

F10.1 The first table below displays experimental ^{13}C chemical shifts and computed* atomic charges on the carbon atom *para* to a number of substituents in substituted benzenes. Two sets of charges are shown, one derived by fitting the electrostatic potential and the other by Mulliken population analysis.

Substituent	CH_3	H	CF_3	CN	NO_2
δ	128.4	128.5	128.9	129.1	129.4
Electrostatic charge/*e*	−0.1273	−0.0757	−0.0227	−0.0152	−0.0541
Mulliken charge/*e*	−0.1089	−0.1021	−0.0665	−0.0805	−0.0392

*Semi-empirical, PM3 level, PC Spartan Pro™

In Problem F6.15 we have recalculated net charges at a higher level than the semi-empirical PM3 level displayed above. The following table obtained from the solution to Problem F6.15 displays both the experimental and calculated* (HF-SCF/6-311G*) ^{13}C chemical shifts and computed* atomic charges on the carbon atom *para* to a number of substituents in substituted benzenes. Three sets of charges are shown, one derived by fitting the electrostatic potential, another by Mulliken population analysis, and the other by the method of "natural" charges.

Substituent	CH_3	H	CF_3	CN	NO_2
δ_{exp}	128.4	128.5	128.9	129.1	129.4
δ_{calc}*	134.4	133.5	132.1	138.8	141.8
Electrostatic charge*/e	−0.240	−0.135	−0.138	−0.102	−0.116
Mulliken charge*/e	−0.231	−0.217	−0..205	−0.199	−0.182
Natural charge*/e	−0.199	−0.183	−0.158	−0.148	−0.135

Spartan '06™; HF-SCF/6-311G

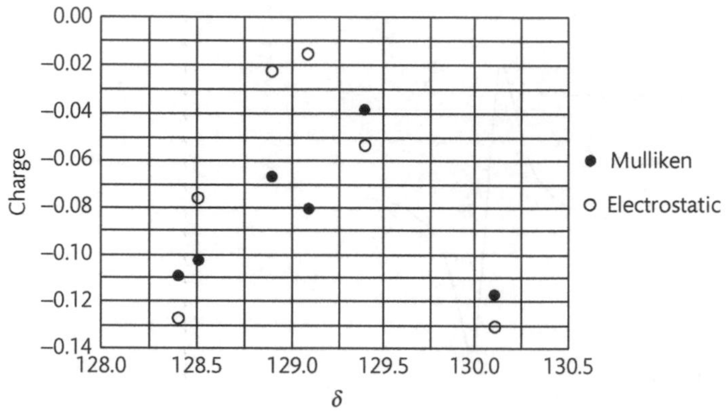

Fig F10.1

(a) Neither set of charges correlates well to the chemical shifts; however some correlation is apparent, particularly for the Mulliken Charges.

(b) The diamagnetic local contribution to shielding is roughly proportional to the electron density on the atom. The extent to which the *para*-carbon atom is affected by electron-donating or-withdrawing groups on the other side of the benzene ring is reflected in the net charge on the atom. If the diamagnetic local contribution dominated, then the more positive the atom, the greater the deshielding and the greater the chemical shift δ would be. That no good correlation is observed leads to several possible hypotheses: for example, the diamagnetic local contribution is not the dominant contribution in these molecules (or not in all of these molecules), or the computation is not sufficiently accurate to provide meaningful atomic charges. See the solution to Problem F6.15 for additional discussion.

F10.3

(a) The first figure displays spin densities computed by molecular modeling software (*ab initio*, density functional theory, Gaussian 98™).

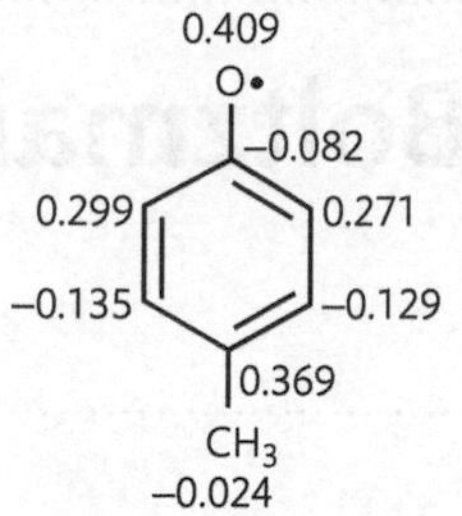

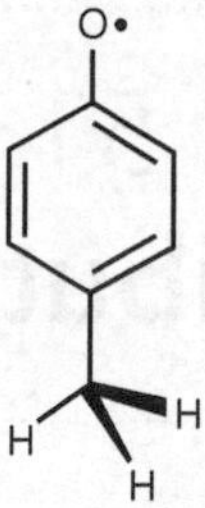

(b) First, note that the software assigned slightly different values to the two protons *ortho* to the oxygen and to the two protons *meta* to the oxygen. This is undoubtedly a computational artifact, a result of the minimum-energy structure having one methyl proton in the plane of the ring, which makes the right and left side of the ring slightly non-equivalent. (See second figure.) In fact, fast internal rotation makes the two halves of the ring equivalent. We will take the spin density at the *ortho* carbons to be 0.285 and those of the *meta* carbons to be −0.132. Predict the form of the spectrum by using the McConnell equation (50.5) for the splittings. The two *ortho* protons give rise to a 1:2:1 triplets with splitting $0.285 \times 2.25\ \mathrm{mT} = 0.64\ \mathrm{mT}$; these will in turn be split by the two *meta* protons into 1:2:1 triplets with splitting

$$0.132 \times 2.25\ \mathrm{mT} = 0.297\ \mathrm{mT} = 0.297\ \mathrm{mT}$$

And finally, these lines will be seen to be further split by the three methyl protons into 1:3:3:1 quartets with splittings 1.045 mT. Note that the McConnell relation cannot be applied to calculate these latter splittings, but the software generates them directly from calculated spin densities on the methyl hydrogens. The computed splittings agree well with experiment at the *ortho* positions (0.60 mT) and at the methyl hydrogens (1.19 mT) but less well at the *meta* positions (0.145 mT).

Topic 51 The Boltzmann distribution

Discussion questions

D51.1 The **population** of a state is the number of molecules of a sample that are *in* that state, on average. The population of a state is the number of molecules in the sample times the probability of the state. The **configuration** of a system is a list of populations in order of the energy of the corresponding states. For example, $\{N-3, 2, 1, 0, \ldots\}$ is a possible configuration of a system of N molecules in which all but three of the molecules are in the ground state, two molecules are in the next lowest state, one in the next state, etc. The **weight** of a configuration is the number of ways a given configuration can be comprised. In our example, the single molecule in the second excited state could be any of the system's N molecules, so there are N ways to arrange that state alone. The weight of the configuration $\{N_0, N_1, N_2, \ldots\}$ is

$$\mathcal{W} = \frac{N!}{N_0!N_1!N_2!\cdots} \quad (51.1)$$

When N is large (as it is for any macroscopic sample), the most probable configuration is so much more probable than all other possible configurations that it is the system's dominant configuration. See Topic 51.1.

D51.3 Because this chapter focuses on the application of statistics to the distribution of physical states in systems that contain a large number of atoms or molecules, we begin with a statistical answer: the thermodynamic temperature is the one quantity that determines the most probable populations of those states in systems at thermal equilibrium (Topic 51.1(b)). As a consequence, the temperature provides a necessary condition for thermal equilibrium; a system is at thermal equilibrium only if all of its sub-systems have the same temperature. Note that this is not a circular definition of temperature, for thermal equilibrium is not defined by uniformity of temperature: systems whose sub-systems can exchange energy tend toward thermal equilibrium. In this context, sub-systems can be different materials placed in contact (such as a block of copper in a beaker of water) or can be more abstract (such as rotational and vibrational modes of motion).

Finally, the equipartition theorem allows us to connect the temperature of statistical thermodynamics to the empirical concept of temperature developed long beforehand. Temperature is a measure of the intensity of thermal energy, directly proportional to the mean energy for each quadratic contribution to the energy (provided that the temperature is sufficiently high).

Exercises

E51.1(a) The weight is given by

$$\mathcal{W}=\frac{N!}{N_0!N_1!N_2!\cdots}=\frac{16!}{0!1!2!3!8!0!0!0!0!2!}$$

This can be simplified by removing the common factor of 8! from the numerator and denominator and noting that 0! = 1! = 1:

$$\mathcal{W}=\frac{16\times15\times14\times13\times12\times11\times10\times9}{2\times(3\times2)\times2}=\boxed{21\,621\,600}$$

E51.2(a) (a) $8!=8\times7\times6\times5\times4\times3\times2\times1=\boxed{40\,320}$ exactly.

(b) According to Stirling's simple approximation (eqn 51.2b),

$\ln x!\approx x\ln x-x$

so $\ln 8!\approx 8\ln 8-8=8.636$ and $8!\approx e^{8.636}=\boxed{5.63\times10^3}$

(c) According to Stirling's better approximation (eqn 51.2a),

$x!\approx(2\pi)^{1/2}x^{x+1/2}e^{-x}$ so $8!\approx(2\pi)^{1/2}8^{8.5}e^{-8}=\boxed{3.99\times10^4}$

E51.3(a) For two non-degenerate levels,

$$\frac{N_2}{N_1}=\frac{e^{-\beta\varepsilon_2}}{e^{-\beta\varepsilon_1}}=e^{-\beta(\varepsilon_2-\varepsilon_1)}=e^{-\beta\Delta\varepsilon}=e^{-\Delta\varepsilon/kT}\qquad\left[51.7\text{a with }\beta=\frac{1}{kT}\right]$$

Hence, as $T\to\infty$, $\frac{N_2}{N_1}=e^{-0}=\boxed{1}$. That is, the two levels would become equally populated.

E51.4(a) For two non-degenerate levels,

$$\frac{N_2}{N_1}=\frac{e^{-\beta\varepsilon_2}}{e^{-\beta\varepsilon_1}}=e^{-\beta(\varepsilon_2-\varepsilon_1)}=e^{-\beta\Delta\varepsilon}=e^{-\Delta\varepsilon/kT}\qquad\left[51.7\text{a with }\beta=\frac{1}{kT}\right]$$

$$\text{so}\quad \ln\frac{N_2}{N_1}=-\frac{\Delta\varepsilon}{kT}\quad\text{and}\quad T=-\frac{\Delta\varepsilon}{k\ln\frac{N_2}{N_1}}$$

$$\text{Thus}\quad T=-\frac{6.626\times10^{-34}\text{ J s}\times2.998\times10^{10}\text{ cm s}^{-1}\times400\text{ cm}^{-1}}{1.381\times10^{-23}\text{ J K}^{-1}\times\ln(1/3)}=\boxed{524\text{ K}}$$

E51.5(a) See Example 51.1. The ratio of populations of a particular *state* at the $J=5$ level to the population of the non-degenerate $J=0$ level is

$$\frac{N_2}{N_1}=\frac{e^{-\beta\varepsilon_2}}{e^{-\beta\varepsilon_1}}=e^{-\beta(\varepsilon_2-\varepsilon_1)}=e^{-\beta\Delta\varepsilon}=e^{-\Delta\varepsilon/kT}\qquad\left[51.7\text{a with }\beta=\frac{1}{kT}\right]$$

Because all of the states of a degenerate level are equally likely, the ratio of populations of a particular *level* is

$$\frac{N_5}{N_0}=\frac{g_5e^{-\beta\varepsilon_5}}{g_0e^{-\beta\varepsilon_0}}=\frac{g_5}{g_0}e^{-(\varepsilon_5-\varepsilon_0)/kT}$$

The degeneracy of linear rotor energy levels are

$$g_J = (2J+1)$$

and its energy levels are

$$\varepsilon_J = hc\tilde{B}J(J+1) \qquad \text{[Topic 41.2(c)]}$$

Thus, using $kT/hc = 207.224\ \text{cm}^{-1}$ at 298.15 K,

$$\frac{N_5}{N_0} = \frac{g_5}{g_0}\text{e}^{-5(5+1)hc\tilde{B}/kT} = \frac{(2\times5+1)}{(2\times0+1)}\text{e}^{-5(5+1)\times2.71\ \text{cm}^{-1}/207.224\ \text{cm}^{-1}} = \boxed{7.43}$$

E51.6(a) For two non-degenerate levels,

$$\frac{N_2}{N_1} = \frac{\text{e}^{-\beta\varepsilon_2}}{\text{e}^{-\beta\varepsilon_1}} = \text{e}^{-\beta(\varepsilon_2-\varepsilon_1)} = \text{e}^{-\beta\Delta\varepsilon} = \text{e}^{-\Delta\varepsilon/kT} \qquad \left[51.7\text{a with } \beta = \frac{1}{kT}\right]$$

so, assuming that other states (if any) are negligibly populated,

$$\ln\frac{N_2}{N_1} = -\frac{\Delta\varepsilon}{kT} \quad \text{and} \quad T = -\frac{\Delta\varepsilon}{k\ln\frac{N_2}{N_1}}$$

Thus $$T = -\frac{6.626\times10^{-34}\ \text{J s}\times2.998\times10^{10}\ \text{cm s}^{-1}\times540\ \text{cm}^{-1}}{1.381\times10^{-23}\ \text{J K}^{-1}\times\ln(10/90)} = \boxed{35\bar{4}\ \text{K}}$$

Problems

P51.1 We draw up the following table

0	ε	2ε	3ε	4ε	5ε	$\mathcal{W} = \dfrac{N!}{n_1!n_2!\ldots}$ [51.1]
4	0	0	0	0	1	5
3	1	0	0	1	0	20
3	0	1	1	0	0	20
2	2	0	1	0	0	30
2	1	2	0	0	0	30
1	3	1	0	0	0	20
0	5	0	0	0	0	1

(b) The most probable configurations are $\boxed{\{2, 2, 0, 1, 0, 0\}}$ and $\boxed{\{2, 1, 2, 0, 0, 0\}}$.

(a) There is no configuration in which the molecules are distributed evenly over the available states. The distribution closest to uniform would have two states with two molecules each and one state with one molecule. Those are precisely the two configurations identified in part (b).

P51.3 Listing all possible configurations for a 20-particle system would be very time-consuming and tedious indeed; however, listing representative configurations for a given total energy is manageable if done systematically. By "representative," list only one configuration that has a given weight.

For example, consider systems that have a total energy of 10ε, where ε is the separation between adjacent energy levels. There are many distinct configurations that have 17 particles in the ground state and one particle in each of three different states, including configurations in which the singly-occupied levels are $(\varepsilon, 2\varepsilon, 7\varepsilon)$, $(2\varepsilon, 3\varepsilon, 5\varepsilon)$, and $(\varepsilon, 4\varepsilon, 5\varepsilon)$. Any one of these configurations is "representative" of all of them, however, because they all have the same weight, namely

$$\mathcal{W} = \frac{N!}{N_0!N_1!N_2!\cdots}[51.1] = \frac{20!}{17!1!1!1!} = 20\times19\times18 = 6\,840$$

So it is sufficient to enumerate just one of the configurations whose occupancy numbers are 17, 1, 1, and 1 (to make sure that one such configuration exists consistent with the desired total energy). A systematic way of keeping track of representative configurations is to lower maximum occupancy numbers. The next set of occupancy numbers to look at would be 16, 4; then 16, 3, 1; 16, 2, 2; 16, 2, 1, 1; and 16, 1, 1, 1, 1. We would eliminate 16, 4, because no such occupancy numbers yield a total energy of 10ε.

Set up a spreadsheet to generate one of each kind of representative configuration, while keeping the total number of particles constant at 20 and total energy constant (at 10ε for the moment). Again, the most systematic way to do this is to give the highest occupancy numbers to the lowest-energy available states. (This rule also generates the most "exponential" configurations.) For example, out of several configurations corresponding to occupancy numbers of 16, 3, and 1, the one used is $N_0 = 16$, $N_1 = 3$, and $N_7 = 1$.

Next examine systems of total energy 10ε, 15ε, and 20ε. The most probable configurations and corresponding weights are:

$$10\varepsilon:\quad \{N_0 = 12, N_1 = 6, N_2 = 2, N_3 = 0, \ldots\}\quad \mathcal{W} = \frac{20!}{12!6!2!} = 3\,527\,160$$

$$15\varepsilon:\quad \{N_0 = 10, N_1 = 6, N_2 = 3, N_3 = 1, N_4 = 0, \ldots\}\quad \mathcal{W} = \frac{20!}{10!6!3!1!} = 155\,195\,040$$

$$20\varepsilon:\quad \{N_0 = 10, N_1 = 4, N_2 = 3, N_3 = 2, N_4 = 1, N_5 = 0, \ldots\}\quad \mathcal{W} = \frac{20!}{10!4!3!2!1!} = 2\,327\,925\,600$$

$$\text{and}\quad \{N_0 = 9, N_1 = 6, N_2 = 2, N_3 = 2, N_4 = 1, N_5 = 0, \ldots\}\quad \mathcal{W} = \frac{20!}{9!6!2!2!1!} = 2\,327\,925\,600$$

As the total energy increases, the most probable configuration has more occupied levels and occupancy of higher-energy levels.

The Boltzmann distribution would predict the following relative probabilities for equally spaced energy levels above the ground state:

$$\frac{p_j}{p_0} = \mathrm{e}^{-\beta j\varepsilon}\ [51.7\mathrm{a}]\quad \text{so}\quad \ln\frac{p_j}{p_0} = -\beta j\varepsilon$$

Thus, a plot of the natural log of relative probability vs. the ordinal number of the energy level (j) should be a straight line whose slope is $-\beta\varepsilon$. Using the non-zero occupancy numbers from the configuration, the plots are indeed roughly linear. Furthermore, the value of β decreases with increasing total energy, corresponding to an increase in temperature.

P51.5 Look immediately after *Justification* 51.1, to the expression for ln $\mathcal{W}$:

$$\ln \mathcal{W} = \ln N! - \sum_j \ln N_j!$$

Substitute the full version of Stirling's approximation:

$$\begin{aligned}\ln \mathcal{W} &= \ln\{(2\pi)^{1/2} N^{N+1/2} e^{-N}\} - \sum_j \ln\{(2\pi)^{1/2} N_j^{N_j+1/2} e^{-N_j}\} \\ &= \tfrac{1}{2}\ln(2\pi) + (N+\tfrac{1}{2})\ln N - N - \sum_j \left\{\tfrac{1}{2}\ln(2\pi) + (N_j+\tfrac{1}{2})\ln N_j - N_j\right\} \\ &= \tfrac{1}{2}\ln(2\pi) + (N+\tfrac{1}{2})\ln N - \sum_j \left\{\tfrac{1}{2}\ln(2\pi) + (N_j+\tfrac{1}{2})\ln N_j\right\}.\end{aligned}$$

With this expression in hand, we turn to Topic 51.2, to the derivation of the Boltzmann distribution. There the method of undetermined multipliers still yields

$$0 = \frac{\partial \ln \mathcal{W}}{\partial N_i} + \alpha - \beta\varepsilon_i \quad [51.9]$$

Evaluating the derivative is similar with the more accurate expression for $\mathcal{W}$ as with the more approximate form. First of all, derivatives of the 2π terms vanish, leaving

$$\frac{\partial \ln \mathcal{W}}{\partial N_i} = \frac{\partial\{(N+\frac{1}{2})\ln N\}}{\partial N_i} - \sum_j \frac{\partial\{(N_j+\frac{1}{2})\ln N_j\}}{\partial N_i}$$

The first term is

$$\begin{aligned}\frac{\partial\{(N+\frac{1}{2})\ln N\}}{\partial N_i} &= \left(\frac{\partial(N+\frac{1}{2})}{\partial N_i}\right)\ln N + (N+\tfrac{1}{2})\left(\frac{\partial \ln N}{\partial N_i}\right) \\ &= \ln N + \left(\frac{N+\frac{1}{2}}{N}\right)\left(\frac{\partial N}{\partial N_i}\right) = \ln N + \left(\frac{N+\frac{1}{2}}{N}\right),\end{aligned}$$

which differs from eqn 51.11 only insomuch as $\dfrac{N+\frac{1}{2}}{N}$ differs from 1. In macroscopic samples, N is so large for this difference to be utterly negligible. Similarly

$$\sum_j \frac{\partial\{(N_j+\frac{1}{2})\ln N_j\}}{\partial N_i} = \ln N_i + \left(\frac{N_i+\frac{1}{2}}{N_i}\right)$$

Again, in macroscopic samples, the last term differs from unity only negligibly *for any states that have a reasonable probability of occupation.* For those samples, the remainder of the derivation is exactly as in Topic 51.2. Any deviations are limited to states of exceedingly low occupancy—the extreme tail of the distribution.

P51.7 $$\frac{p(h)}{p(h_0)}=\frac{n(h)RT/V}{n(h_0)RT/V}=\mathrm{e}^{-\{(\varepsilon(h)-\varepsilon(h_0))/kT\}}[51.7\mathrm{a}]=\mathrm{e}^{-mg(h-h_0)/kT}$$

Defining $p(h=0)\equiv p_0$, we obtain the desired barometric formula:

$$\frac{p(h)}{p_0}=\mathrm{e}^{-mgh/kT}=\boxed{\mathrm{e}^{-Mgh/RT}}=\frac{\mathcal{N}(h)}{\mathcal{N}(0)}$$

Note that the result depends on the temperature, and it assumes that the temperature does not vary with height. To proceed, we must pick a temperature, so we use the standard temperature of 298 K—which is reasonable for the surface of the earth—but not for 8.0 km altitude.

For oxygen at 8.0 km,

$$\frac{M(\mathrm{O_2})gh}{RT}=\frac{(0.0320\ \mathrm{kg\ mol^{-1}})\times(9.81\ \mathrm{m\ s^{-2}})\times(8.0\times10^3\ \mathrm{m})}{(8.3145\ \mathrm{J\ K^{-1}\ mol^{-1}})\times(298\ \mathrm{K})}=1.01$$

so $$\frac{\mathcal{N}(8.0\ \mathrm{km})}{\mathcal{N}(0)}=\mathrm{e}^{-1.01}=\boxed{0.363}$$

For water,

$$\frac{M(\mathrm{H_2O})gh}{RT}=\frac{(0.0180\ \mathrm{kg\ mol^{-1}})\times(9.81\ \mathrm{m\ s^{-2}})\times(8.0\times10^3\ \mathrm{m})}{(8.3145\ \mathrm{J\ K^{-1}\ mol^{-1}})\times(298\ \mathrm{K})}=0.57$$

so $$\frac{\mathcal{N}(8.0\ \mathrm{km})}{\mathcal{N}(0)}=\mathrm{e}^{-0.57}=\boxed{0.57}$$

P51.9 Each protein binding site can be represented as a distinct box into which a ligand, L, may bind. All possible configurations are shown in the following table and the configuration count of *i* indistinguishable ligands being placed in *n* distinguishable sites is seen to be given by the combinatorial:

$$C(n,i)=\frac{n!}{(n-1)!\ i!}$$

				PL_0:	$C(4,0)=\frac{4!}{(4-0)!0!}=1$ configuration
L				PL_1:	$C(4,1)=\frac{4!}{(4-1)!1!}=4$ configurations
	L				
		L			
			L		
L	L			PL_2:	$C(4,2)=\frac{4!}{(4-2)!2!}=6$ configurations
L		L			
L			L		
	L	L			
	L		L		
		L	L		
L	L	L		PL_3:	$C(4,3)=\frac{4!}{(4-3)!3!}=4$ configurations
L	L		L		
L		L	L		
	L	L	L		
L	L	L	L	PL_4:	$C(4,4)=\frac{4!}{(4-4)!4!}=1$ configuration

Topic 52 **Partition functions**

Discussion questions

D52.1 The molecular partition is roughly equal to the number of physically distinct states thermally accessible to a molecule at a given temperature. At low temperatures, very little energy is available, so only the lowest-energy states of a molecule are accessible; therefore, as the temperature approaches absolute zero, the partition function approaches the degeneracy of the molecule's ground state. The higher the temperature, the greater the Boltzmann weighting factor $e^{-\beta\varepsilon}$ and the more accessible a state of energy ε becomes. Thus the number of accessible states increases with temperature. (The partition function is only "roughly equal" to the number of accessible states because at any non-zero temperature, each state is accessible with a finite probability proportional to $e^{-\beta\varepsilon}$. Thus, to state whether or not a given state is "accessible" at a given temperature is somewhat arbitrary, so an exactly counted number of accessible states is also arbitrary. Because the partition function is the sum of all of these fractional probabilities, it is a good estimate of this number.) See Topic 52.1.

D52.3 Distinct 'states' differ in one or more observable physical quantity such as energy or angular momentum. Thus, distinct states may have the same energy (may belong to the same 'energy level') if they differ in some other quantity; these are degenerate states. Each energy level has at least one state—more for degenerate levels. The partition function is a sort of count of **states**; it is a sum over all **states** of each state's Boltzmann factor. Because Boltzmann factors depend only on energy, the sum may practically be done as a sum over energy levels, as long as the degeneracy is included. That is, the sum over states within a given energy level is a sum of identical Boltzmann factors, equal to the degeneracy of the level times the Boltzmann factor.

Exercises

E52.1(a) (a) The thermal wavelength is

$$\Lambda = \frac{h}{(2\pi mkT)^{1/2}} \quad [52.7b]$$

We need the molecular mass, not the molar mass:

$$m = \frac{150\times10^{-3}\ \text{kg mol}^{-1}}{6.022\times10^{23}\ \text{mol}^{-1}} = 2.49\times10^{-25}\ \text{kg}$$

$$\text{So}\ \ \Lambda = \frac{6.626\times10^{-34}\ \text{J s}}{(2\pi\times2.49\times10^{-25}\ \text{kg}\times1.381\times10^{-23}\ \text{J K}^{-1}\times T)^{1/2}} = \frac{1.43\times10^{-10}\ \text{m}}{(T/\text{K})^{1/2}}$$

(i) $T = 300\ \text{K}$: $\Lambda = \dfrac{1.43\times10^{-10}\ \text{m}}{(300)^{1/2}} = \boxed{8.23\times10^{-12}\ \text{m}} = \boxed{8.23\ \text{pm}}$

(ii) $T = 3\,000\ \text{K}$: $\Lambda = \dfrac{1.43\times10^{-10}\ \text{m}}{(3\,000)^{1/2}} = \boxed{2.60\times10^{-12}\ \text{m}} = \boxed{2.60\ \text{pm}}$

(b) The translational partition function is

$$q^{\text{T}} = \frac{V}{\Lambda^3}\ [52.10\text{b}]$$

(i) $T = 300\ \text{K}$: $q^{\text{T}} = \dfrac{1.00\ \text{cm}^3}{(8.23\times10^{-12}\ \text{m})^3} = \dfrac{(1.00\times10^{-2}\ \text{m})^3}{(8.23\times10^{-12}\ \text{m})^3} = \boxed{1.79\times10^{27}}$

(ii) $T = 3\,000\ \text{K}$: $q^{\text{T}} = \dfrac{(1.00\times10^{-2}\ \text{m})^3}{(2.60\times10^{-12}\ \text{m})^3} = \boxed{5.67\times10^{28}}$

E52.2(a) $q^{\text{T}} = \dfrac{V}{\Lambda^3}$ [53.10b], implying that $\dfrac{q}{q'} = \left(\dfrac{\Lambda'}{\Lambda}\right)^3$

However, $\Lambda = \dfrac{h}{(2\pi mkT)^{1/2}}[52.7\text{b}] \propto \dfrac{1}{m^{1/2}}$ so $\dfrac{q}{q'} = \left(\dfrac{m}{m'}\right)^{3/2}$

Therefore, $\dfrac{q_{\text{H}_2}}{q_{\text{He}}} = \left(\dfrac{2\times1.008}{4.003}\right)^{3/2} = \boxed{0.3574}$

E52.3(a) The high-temperature expression for the rotational partition function of a linear molecule is

$$q^{\text{R}} = \frac{kT}{\sigma hc\tilde{B}}[52.13\text{b}],\quad \tilde{B} = \frac{\hbar}{4\pi cI}\ [41.7],\quad I = \mu R^2\ [\text{Table 41.1}]$$

$$\text{Hence}\quad q = \frac{8\pi^2 kTI}{\sigma h^2} = \frac{8\pi^2 kT\mu R^2}{\sigma h^2}$$

For O_2, $\mu = \frac{1}{2}m(\text{O}) = \frac{1}{2}\times16.00m_u = 8.00m_u$, and $\sigma = 2$; therefore

$$q = \frac{(8\pi^2)\times(1.381\times10^{-23}\ \text{J K}^{-1})\times(300\ \text{K})\times(8.00\times1.6605\times10^{-27}\ \text{kg})\times(1.2075\times10^{-10}\ \text{m})^2}{(2)\times(6.626\times10^{-34}\ \text{J s})^2}$$

$$= \boxed{72.2}$$

E52.4(a) The high-temperature expression for the rotational partition function of a nonlinear molecule is [52.14]

$$q^{\mathrm{R}} = \frac{1}{\sigma}\left(\frac{kT}{hc}\right)^{3/2}\left(\frac{\pi}{\tilde{A}\tilde{B}\tilde{C}}\right)^{1/2}$$

$$= \left(\frac{1.381\times10^{-23}\ \mathrm{J\,K^{-1}}\times T}{6.626\times10^{-34}\ \mathrm{J\,s}\times2.998\times10^{10}\ \mathrm{cm\,s^{-1}}}\right)^{3/2}\left(\frac{\pi}{3.1752\times0.3951\times0.3505\ \mathrm{cm^{-3}}}\right)^{1/2}$$

$$= 1.549\times(T/\mathrm{K})^{3/2}$$

(a) At 25°C, $q^{\mathrm{R}} = 1.549\times(298)^{3/2} = \boxed{7.97\times10^{3}}$

(b) At 100°C, $q^{\mathrm{R}} = 1.549\times(373)^{3/2} = \boxed{1.12\times10^{4}}$

E52.5(a) The rotational partition function of a nonsymmetrical linear molecule is

$$q^{\mathrm{R}} = \sum_J (2J+1)\mathrm{e}^{-hc\tilde{B}J(J+1)/kT} \qquad \left[52.11 \text{ with } \beta = \frac{1}{kT}\right]$$

Use $$\frac{hc\tilde{B}}{k} = \frac{6.626\times10^{-34}\ \mathrm{J\,s}\times2.998\times10^{10}\ \mathrm{cm\,s^{-1}}\times1.931\ \mathrm{cm^{-1}}}{1.381\times10^{-23}\ \mathrm{J\,K^{-1}}} = 2.778\ \mathrm{K},$$

so $$q^{\mathrm{R}} = \sum_J (2J+1)\mathrm{e}^{-2.778\ \mathrm{K}\times J(J+1)/T}$$

Use a spreadsheet or other mathematical software to evaluate the terms of the sum and to sum the terms until they converge. The high-temperature expression is

$$q^{\mathrm{R}} = \frac{kT}{hc\tilde{B}} = \frac{T}{2.778\ \mathrm{K}}$$

The explicit and high-temperature expressions are compared in Fig. 52.1. The high-temperature expression reaches 95% of the explicit sum at $\boxed{18\ \mathrm{K}}$.

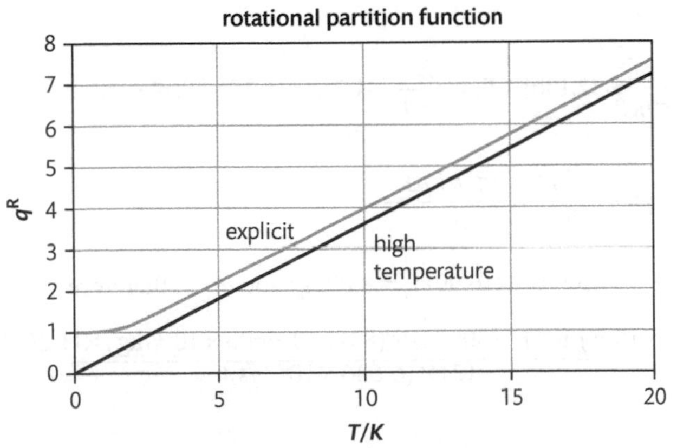

Fig 52.1

E52.6(a) The rotational partition function of a spherical rotor molecule, ignoring nuclear statistics, is

$$q^{\mathrm{R}} = \sum_J g_J \mathrm{e}^{-\varepsilon_J^{\mathrm{R}}/kT} \text{[52.1b]} = \sum_J (2J+1)^2 \mathrm{e}^{-hc\tilde{B}J(J+1)/kT} \text{[41.6]}.$$

Use $\dfrac{hc\tilde{B}}{k} = \dfrac{6.626\times10^{-34}\ \mathrm{J\,s}\times 2.998\times10^{10}\ \mathrm{cm\,s^{-1}}\times 5.241\ \mathrm{cm^{-1}}}{1.381\times10^{-23}\ \mathrm{J\,K^{-1}}} = 7.539\ \mathrm{K}$,

so $$q^{\mathrm{R}} = \sum_J (2J+1)^2 \mathrm{e}^{-7.539\ \mathrm{K}\times J(J+1)/T}$$

Use a spreadsheet or other mathematical software to evaluate the terms of the sum and to sum the terms until they converge. The high-temperature expression is eqn 52.14, neglecting σ and with $\tilde{A} = \tilde{B} = \tilde{C}$:

$$q^{\mathrm{R}} = \pi^{1/2}\left(\frac{kT}{hc\tilde{B}}\right)^{3/2} = \pi^{1/2}\left(\frac{T}{7.539\ \mathrm{K}}\right)^{3/2}$$

The explicit and high-temperature expressions are compared in Fig. 52.2. The high-temperature expression reaches 95% of the explicit sum at $\boxed{37\ \mathrm{K}}$.

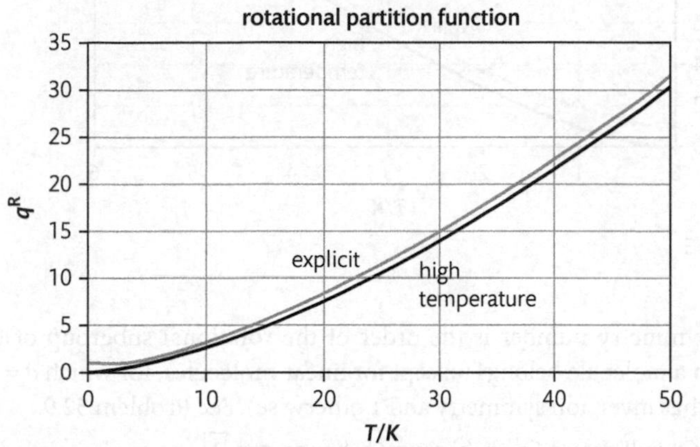

Fig 52.2

E52.7(a) The rotational partition function of a symmetric rotor molecule, ignoring nuclear statistics, is

$$q^{\mathrm{R}} = \sum_{J,K} g_{J,K} \mathrm{e}^{-\varepsilon_{J,K}^{\mathrm{R}}/kT} \text{[52.1b]} = \sum_{J=0} (2J+1)\mathrm{e}^{-hc\tilde{B}J(J+1)/kT}\left(1 + 2\sum_{K=1}^{J} \mathrm{e}^{-hc(\tilde{A}-\tilde{B})K^2/kT}\right) \text{[41.13]}$$

Use $\dfrac{hc\tilde{B}}{k} = \dfrac{6.626\times10^{-34}\ \mathrm{J\,s}\times 2.998\times10^{10}\ \mathrm{cm\,s^{-1}}\times 0.443\ \mathrm{cm^{-1}}}{1.381\times10^{-23}\ \mathrm{J\,K^{-1}}} = 0.637\ \mathrm{K}$, and

$$\frac{hc(\tilde{A}-\tilde{B})}{k} = \frac{6.626\times10^{-34}\ \mathrm{J\,s}\times 2.998\times10^{10}\ \mathrm{cm\,s^{-1}}\times(5.097-0.443)\ \mathrm{cm^{-1}}}{1.381\times10^{-23}\ \mathrm{J\,K^{-1}}} = 6.694\ \mathrm{K}$$

so $$q^{\mathrm{R}} = \sum_{J=0} (2J+1)\mathrm{e}^{-0.637\ \mathrm{K}\times J(J+1)/T}\left(1 + 2\sum_{K=1}^{J} \mathrm{e}^{-6.694\ \mathrm{K}\times K^2/T}\right)$$

Write a brief computer program or use other mathematical software to evaluate the terms of the sum and to sum the terms until they converge. Nested sums are straightforward to program in languages such as BASIC or FORTRAN, whereas spreadsheets are more unwieldy. Compare the results of the direct sum with the high-temperature expression, eqn 52.14, with $\tilde{B}=\tilde{C}$:

$$q^{\mathrm{R}}=\left(\frac{\pi}{\tilde{A}}\right)^{1/2}\left(\frac{kT}{hc}\right)^{3/2}\frac{1}{\tilde{B}}.$$

The explicit and high-temperature expressions are compared in Fig. 52.3. The high-temperature expression reaches 95% of the explicit sum at $\boxed{4.5\ \mathrm{K}}$.

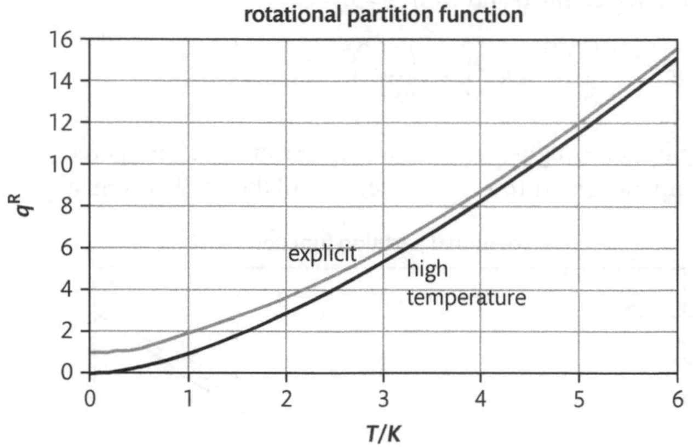

Fig 52.3

E52.8(a) The symmetry number is the order of the rotational subgroup of the group to which a molecule belongs (except for linear molecules, for which $\sigma=2$ if the molecule has inversion symmetry and 1 otherwise). See Problem 52.9.

(a) CO: Full group $C_{\infty v}$; subgroup C_1; hence $\sigma=\boxed{1}$

(b) O_2: Full group $D_{\infty h}$; subgroup C_2; $\sigma=\boxed{2}$

(c) H_2S: Full group C_{2v}; subgroup C_2; $\sigma=\boxed{2}$

(d) SiH_4: Full group T_d; subgroup T; $\sigma=\boxed{12}$

(e) $CHCl_3$: Full group C_{3v}; subgroup C_3; $\sigma=\boxed{3}$

E52.9(a) Ethene has four indistinguishable atoms that can be interchanged by rotations, so $\sigma=4$. The high-temperature expression for the rotational partition function of a non-linear molecule is [52.14]

$$q^{\mathrm{R}}=\frac{1}{\sigma}\left(\frac{kT}{hc}\right)^{3/2}\left(\frac{\pi}{\tilde{A}\tilde{B}\tilde{C}}\right)^{1/2}$$

$$=\frac{1}{4}\left(\frac{1.381\times10^{-23}\ \mathrm{J\,K^{-1}}\times298.15\ \mathrm{K}}{6.626\times10^{-34}\ \mathrm{J\,s}\times2.998\times10^{10}\ \mathrm{cm\,s^{-1}}}\right)^{3/2}\left(\frac{\pi}{4.828\times1.0012\times0.8282\ \mathrm{cm^{-3}}}\right)^{1/2}$$

$$=\boxed{660.6}$$

E52.10(a) The partition function for a mode of molecular vibration is

$$q^{\rm V} = \sum_v {\rm e}^{-vhc\tilde{\nu}/kT} = \frac{1}{1-{\rm e}^{-hc\tilde{\nu}/kT}} \quad [52.15 \text{ with } \beta = 1/kT]$$

Use $\dfrac{hc\tilde{\nu}}{k} = \dfrac{6.626\times10^{-34}\ {\rm J\,s}\times 2.998\times10^{10}\ {\rm cm\,s^{-1}}\times 323.2\ {\rm cm^{-1}}}{1.381\times10^{-23}\ {\rm J\,K^{-1}}} = 464.9\ {\rm K}$,

so $q^{\rm V} = \sum_v {\rm e}^{-vhc\tilde{\nu}/kT} = \dfrac{1}{1-{\rm e}^{-464.9\ {\rm K}/T}}$

The high-temperature expression is

$$q^{\rm V} = \frac{kT}{hc\tilde{\nu}} = \frac{T}{464.9\ {\rm K}} \quad [52.16]$$

The explicit and high-temperature expressions are compared in Fig. 52.4. The high-temperature expression reaches 95% of the explicit sum at $\boxed{4\,500\ {\rm K}}$.

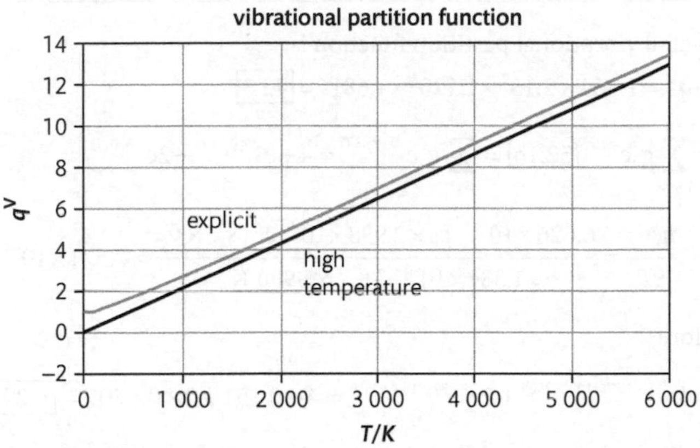

Fig 52.4

E52.11(a) The partition function for a mode of molecular vibration is

$$q^{\rm V} = \frac{1}{1-{\rm e}^{-hc\tilde{\nu}/kT}} \quad [52.15 \text{ with } \beta = 1/kT]$$

and the overall vibrational partition function is the product of the partition functions of the individual modes. (See Example 52.2.) We draw up the following table:

mode	1	2	3	4
$\tilde{\nu}/{\rm cm^{-1}}$	658	397	397	1 535
$hc\tilde{\nu}/kT$	1.893	1.142	1.142	4.416
$q^{\rm V}_{\rm mode}$	1.177	1.469	1.469	1.012

The overall vibrational partition function is

$$q^{\mathrm{V}} = 1.177 \times 1.469 \times 1.469 \times 1.012 = \boxed{2.571}$$

E52.12(a) The partition function for a mode of molecular vibration is

$$q^{\mathrm{V}} = \frac{1}{1-\mathrm{e}^{-hc\tilde{\nu}/kT}} \text{ [52.15 with } \beta = 1/kT]$$

and the overall vibrational partition function is the product of the partition functions of the individual modes. (See Example 52.2.) We draw up the following table, including the degeneracy of each level:

mode	1	2	3	4
$\tilde{\nu}/\mathrm{cm}^{-1}$	459	217	776	314
g_{mode}	1	2	3	3
$hc\tilde{\nu}/kT$	1.320	0.624	2.232	0.903
$q^{\mathrm{V}}_{\mathrm{mode}}$	1.364	2.15	1.120	1.681

The overall vibrational partition function is

$$q^{\mathrm{V}} = 1.364 \times 2.15^2 \times 1.120^3 \times 1.681^3 = \boxed{42.3}$$

E52.13(a) $$q = \sum_{\text{levels}} g_j \mathrm{e}^{-\beta\varepsilon_j} [52.1\mathrm{b}] = \sum_{\text{levels}} g_j \mathrm{e}^{-hc\tilde{\nu}_j/kT} = 4 + \mathrm{e}^{-hc\tilde{\nu}_1/kT} + 2\mathrm{e}^{-hc\tilde{\nu}_2/kT}$$

where $$\frac{hc\tilde{\nu}_j}{kT} = \frac{6.626 \times 10^{-34}\ \mathrm{J\,s} \times 2.998 \times 10^{10}\ \mathrm{cm\,s^{-1}} \times \tilde{\nu}_j}{1.381 \times 10^{-23}\ \mathrm{J\,K^{-1}} \times 1900\ \mathrm{K}} = 7.571 \times 10^{-4} \times (\tilde{\nu}_j/\mathrm{cm}^{-1})$$

Therefore

$$q = 4 + \mathrm{e}^{-7.571\times10^{-4}\times2500} + 2\mathrm{e}^{-7.571\times10^{-4}\times3500} = 4 + 0.151 + 2\times0.0707 = \boxed{4.292}$$

The individual terms in the last expression are the relative populations of the levels, namely 2×0.0707 to 0.151 to 4 (second excited level to first to ground) or $\boxed{0.0353 \text{ to } 0.0377 \text{ to } 1}$

Problems

P52.1 This problem can be carried out on a spreadsheet if care is taken with the layout. One may simply pick values of $\tilde{\nu}$; however, if one works in terms of the characteristic vibrational temperature, θ^{V}, one can employ more general dimensionless quantities as described below.

$$\theta^{\mathrm{V}} = \frac{hc\tilde{\nu}}{k}$$

We note that the energy levels of the Morse oscillator can be written as

$$E_v = (v+\tfrac{1}{2})k\theta^{\mathrm{V}} - (v+\tfrac{1}{2})^2 x_{\mathrm{e}} k\theta^{\mathrm{V}} = (v+\tfrac{1}{2})k\theta^{\mathrm{V}}\{1-(v+\tfrac{1}{2})x_{\mathrm{e}}\}$$

Thus, one can tabulate values of $E_v/k\theta^V$ without having to select a wavenumber. Similarly, one can employ the dimensionless temperature T/θ^V. As noted in the problem, Boltzmann factors require energies measured with respect to the ground state, so the energies in the exponents of the Boltzmann factors must be $E_v - E_0$. The energy expression for the Morse oscillator eventually reaches a maximum in v and then begins decreasing. Only the states up to and including the maximum energy are physically meaningful, so there are a finite number of Morse states. Thus, the partition function for a Morse oscillator is:

$$q = \sum_{v=0}^{v_{max}} \exp\left(-\frac{E_v - E_0}{kT}\right) = \sum_{v=0}^{v_{max}} \exp\left(-\frac{E_v - E_0}{k\theta^V} \times \frac{\theta^V}{T}\right)$$

To choose meaningful values of the anharmonicity, x_e, look up vibrational constants for some common diatomic molecules. x_e is about 0.03 for H_2 and about an order of magnitude smaller for I_2.

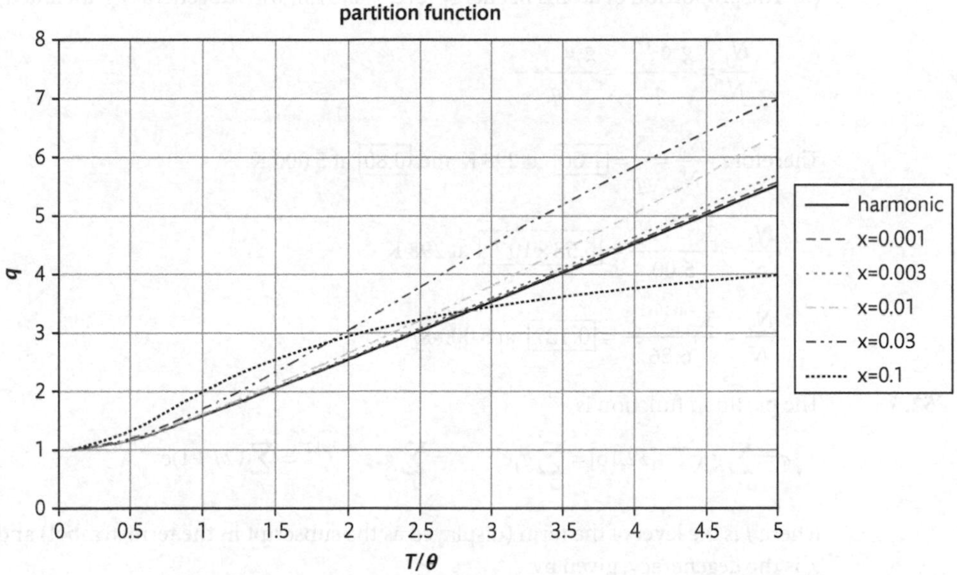

Fig 52.5

A plot of partition functions with various anharmonicities is shown in Fig. 52.5. For small values of x_e, the partition function closely resembles that of a harmonic oscillator (52.15). This provides a check on the calculation, for a Morse oscillator in the limit of small x_e **is** a harmonic oscillator. As x_e increases, the partition function gradually increases compared to the harmonic oscillator. This reflects the fact that the Morse oscillator energy levels become more closely spaced with increasing energy, so more levels are accessible at a given temperature. Eventually, however, the partition functions of highly anharmonic Morse oscillators fall below the harmonic curve at high temperatures. This reflects the fact that these Morse oscillators have a finite number of energy levels (indeed, a small number) so naturally a harmonic oscillator has more accessible levels at high temperature.

P52.3 (a) $q=\sum_j g_j \mathrm{e}^{-\beta\varepsilon_j}$ [52.1b] $=\sum_j g_j \mathrm{e}^{-hc\beta\tilde{\nu}_j}$

We use

$$hc\beta=\frac{6.626\times10^{-34}\ \mathrm{J\,s}\times2.998\times10^{10}\ \mathrm{cm\,s^{-1}}}{1.381\times10^{-23}\ \mathrm{J\,K^{-1}}\times298\ \mathrm{K}}=\frac{1}{207\ \mathrm{cm^{-1}}}\ \text{at 298 K}$$

and $\dfrac{1}{3476\ \mathrm{cm^{-1}}}$ at 5 000 K. Therefore,

$$q=5+\mathrm{e}^{-4707/207}+3\mathrm{e}^{-4751/207}+5\mathrm{e}^{-10559/207}$$

(i) $$=(5)+(1.3\times10^{-10})+(3.2\times10^{-10})+(3.5\times10^{-22})=\boxed{5.00}$$

$$q=5+\mathrm{e}^{-4707/3476}+3\mathrm{e}^{-4751/3476}+5\mathrm{e}^{-10559/3476}$$

(ii) $$=(5)+(0.26)+(0.76)+(0.24)=\boxed{6.26}$$

(b) The proportion of atoms in energy level j is [51.6, with degeneracy g_j included]

$$\frac{N_j}{N}=\frac{g_j\mathrm{e}^{-\beta\varepsilon_j}}{q}=\frac{g_j\mathrm{e}^{-hc\beta\tilde{\nu}_j}}{q}$$

Therefore, $\dfrac{N_0}{N}=\dfrac{5}{q}=\boxed{1.00}$ at 298 K and $\boxed{0.80}$ at 5 000 K.

$$\frac{N_2}{N}=\frac{3\mathrm{e}^{-4751/207}}{5.00}=\boxed{6.58\times10^{-11}}\ \text{at 298 K}$$

$$\frac{N_2}{N}=\frac{3\mathrm{e}^{-4751/3476}}{6.26}=\boxed{0.122}\ \text{at 5 000 K}$$

P52.5 The partition function is

$$q=\sum_J g_J\mathrm{e}^{-\beta\varepsilon_J}\,[52.1\mathrm{b}]=\sum_J g_J\mathrm{e}^{-hc\beta\tilde{\nu}_J}=\sum_J g_J\mathrm{e}^{-hc\tilde{\nu}_J/kT}=\sum_J(2J+1)\mathrm{e}^{-hc\tilde{\nu}_J/kT}$$

where J is the level of the term (displayed as the subscript in the term symbol) and g_J is the degeneracy, given by

$$g_J=2J+1.$$

At 298 K, $$\frac{hc\tilde{\nu}_j}{kT}=\frac{6.626\times10^{-34}\ \mathrm{J\,s}\times2.998\times10^{10}\ \mathrm{cm\,s^{-1}}\times\tilde{\nu}_j}{1.381\times10^{-23}\ \mathrm{J\,K^{-1}}\times298\ \mathrm{K}}=4.83\times10^{-3}\times\tilde{\nu}_j/\mathrm{cm^{-1}}$$

so $$q=1+3\mathrm{e}^{\,4.83\times10^{-3}\times557.1}+5\mathrm{e}^{-4.83\times10^{-3}\times1410.0}+5\mathrm{e}^{-4.83\times10^{-3}\times7125.3}+\mathrm{e}^{-4.83\times10^{-3}\times16367.3}=\boxed{1.209}.$$

At 1 000 K, $$\frac{hc\tilde{\nu}_j}{kT}=\frac{6.626\times10^{-34}\ \mathrm{J\,s}\times2.998\times10^{10}\ \mathrm{cm\,s^{-1}}\times\tilde{\nu}_j}{1.381\times10^{-23}\ \mathrm{J\,s}\times1\,000\ \mathrm{K}}=1.439\times10^{-3}\times\tilde{\nu}_j/\mathrm{cm^{-1}}$$

so $$q=1+3\mathrm{e}^{-1.439\times10^{-3}\times557.1}+5\mathrm{e}^{-1.439\times10^{-3}\times1\,410.0}+5\mathrm{e}^{-1.439\times10^{-3}\times7\,125.3}+\mathrm{e}^{-1.439\times10^{-3}\times16\,367.3}$$

$$=\boxed{3.004}.$$

P52.7 The partition function is

$$q=\sum_{v}e^{-\beta\varepsilon_v}\,[52.1b]=\sum_{v}e^{-hc\tilde{\nu}_v/kT}$$

(a) At 100 K, $\dfrac{hc\tilde{\nu}_j}{kT}=\dfrac{6.626\times10^{-34}\ \mathrm{J\,s}\times2.998\times10^{10}\ \mathrm{cm\,s^{-1}}\times\tilde{\nu}_j}{1.381\times10^{-23}\ \mathrm{J\,K^{-1}}\times100\ \mathrm{K}}=1.44\times10^{-2}\times\tilde{\nu}_j/\mathrm{cm^{-1}}$,

so $q=1+e^{-0.0144\times213.30}+e^{-0.0144\times425.39}+e^{-0.0144\times636.27}+e^{-0.0144\times845.93}=\boxed{1.049}$

(b) At 298 K, $\dfrac{hc\tilde{\nu}_j}{kT}=\dfrac{6.626\times10^{-34}\ \mathrm{J\,s}\times2.998\times10^{10}\ \mathrm{cm\,s^{-1}}\times\tilde{\nu}_j}{1.381\times10^{-23}\ \mathrm{J\,K^{-1}}\times298\ \mathrm{K}}=4.83\times10^{-3}\times\tilde{\nu}_j/\mathrm{cm^{-1}}$

so $q=1+e^{-4.83\times10^{-3}\times213.30}+e^{-4.83\times10^{-3}\times425.39}+e^{-4.83\times10^{-3}\times636.27}+e^{-4.83\times10^{-3}\times845.93}=\boxed{1.548}$

The fractions of molecules at the various levels are

$$\frac{N_v}{N}=\frac{e^{-\beta\varepsilon_v}}{q}[51.6]=\frac{e^{-hc\tilde{\nu}_v/kT}}{q}$$

So $\dfrac{N_0}{N}=\dfrac{1}{q}=$ (a) $\boxed{0.953}$, (b) $\boxed{0.645}$

$\dfrac{N_1}{N}=\dfrac{e^{-hc\tilde{\nu}_1/kT}}{q}=$ (a) $\boxed{0.044}$, (b) $\boxed{0.230}$

$\dfrac{N_2}{N}=\dfrac{e^{-hc\tilde{\nu}_2/kT}}{q}=$ (a) $\boxed{0.002}$, (b) $\boxed{0.083}$

Comment. Eqn 52.15 gives a closed-form expression for the vibrational partition function, based on an infinite ladder of harmonic oscillator states. The explicit sum will deviate slightly from this number because an actual molecule has a finite number of slightly anharmonic vibrational states below the bond dissociation energy.

P52.9 (a) Ethene has four indistinguishable atoms that can be interchanged by rotations, so $\sigma=4$. The high-temperature expression for the rotational partition function of a non-linear molecule is [52.14]

$$q^{\mathrm{R}}=\frac{1}{\sigma}\left(\frac{kT}{hc}\right)^{3/2}\left(\frac{\pi}{\tilde{A}\tilde{B}\tilde{C}}\right)^{1/2}$$
$$=\frac{1}{4}\left(\frac{1.381\times10^{-23}\ \mathrm{J\,K^{-1}}\times298.15\ \mathrm{K}}{6.626\times10^{-34}\ \mathrm{J\,s}\times2.998\times10^{10}\ \mathrm{cm\,s^{-1}}}\right)^{3/2}\left(\frac{\pi}{4.828\times1.0012\times0.8282\ \mathrm{cm^{-3}}}\right)^{1/2}$$
$$=\boxed{660.6}$$

(b) Pyridine belongs to the C_{2v} group, the same as water, so = 2.

$$q^{\mathrm{R}}=\frac{1}{\sigma}\left(\frac{kT}{hc}\right)^{3/2}\left(\frac{\pi}{\tilde{A}\tilde{B}\tilde{C}}\right)^{1/2}$$
$$=\frac{1}{2}\left(\frac{1.381\times10^{-23}\ \mathrm{J\,K^{-1}}\times298.15\ \mathrm{K}}{6.626\times10^{-34}\ \mathrm{J\,s}\times2.998\times10^{10}\ \mathrm{cm\,s^{-1}}}\right)^{3/2}\left(\frac{\pi}{0.2014\times0.1936\times0.0987\ \mathrm{cm^{-3}}}\right)^{1/2}$$
$$=\boxed{4.26\times10^{4}}$$

Topic 53 **Molecular energies**

Discussion question

D53.1 The equipartition theorem says that the average value of each quadratic contribution to the energy is the same and equal to $(1/2)kT = 1/(2\beta)$ This result agrees with the mean energy computed from a partition function, $\langle\varepsilon\rangle = -\dfrac{1}{q}\dfrac{\partial q}{\partial\beta}$ [53.3], at sufficiently high temperatures. The partition function yields the general result for mean energy. Mathematically, evaluating $\partial q/\partial\beta$ involves summing over states. In mathematical terms, the equipartition theorem is valid when the temperature is high enough for the discrete sum to be well-approximated by a continuous integral. In physical terms, this corresponds to temperatures sufficiently high for the difference between energy levels to be small compared to kT.

Exercises

E53.1(a) The mean energy is

$$\langle\varepsilon\rangle = \frac{1}{q}\sum_i \varepsilon_i e^{-\beta\varepsilon_i}\,[53.2] = \frac{\sum_i \varepsilon_i e^{-\beta\varepsilon_i}}{\sum_i e^{-\beta\varepsilon_i}} = \frac{\varepsilon e^{-\beta\varepsilon}}{1+e^{-\beta\varepsilon}} = \frac{\varepsilon}{1+e^{\beta\varepsilon}},$$

where the last expression specializes to two non-degenerate levels. Substitute

$$\varepsilon = hc\tilde{\nu} = 6.626\times10^{-34}\ \text{J s}\times2.998\times10^{10}\ \text{cm s}^{-1}\times500\ \text{cm}^{-1} = 9.93\times10^{-21}\ \text{J},$$

and $$\beta\varepsilon = \frac{\varepsilon}{kT} = \frac{9.93\times10^{-21}\ \text{J}}{1.381\times10^{-23}\ \text{J K}^{-1}\times298\ \text{K}} = 2.41,$$

so $$\langle\varepsilon\rangle = \frac{9.93\times10^{-21}\ \text{J}}{1+e^{2.41}} = \boxed{8.16\times10^{-22}\ \text{J}}$$

E53.2(a) The mean energy is

$$\langle\varepsilon\rangle = \frac{1}{q}\sum_{\text{states}} \varepsilon_i e^{-\beta\varepsilon_i}\,[53.2] = \frac{1}{q}\sum_{\text{levels}} g_i\varepsilon_i e^{-\beta\varepsilon_i} = \frac{1}{q}\sum_J (2J+1)\varepsilon_J e^{-\varepsilon_J/kT}$$

$$\varepsilon_J = hc\tilde{B}J(J+1) = 6.626\times10^{-34}\ \text{J s}\times2.998\times10^{10}\ \text{cm s}^{-1}\times1.931\ \text{cm}^{-1}\times J(J+1)$$
$$= J(J+1)\times3.836\times10^{-23}\ \text{J}$$

and $\dfrac{\varepsilon_J}{k} = \dfrac{J(J+1)\times 3.836\times 10^{-23}\ \text{J}}{1.381\times 10^{-23}\ \text{J K}^{-1}} = J(J+1)\times 2.778\ \text{K},$

so $\langle \varepsilon \rangle = \dfrac{1}{q}\sum_J J(J+1)(2J+1)\times 3.836\times 10^{-23}\ \text{J}\times \text{e}^{-J(J+1)\times 2.778\ \text{K}/T}$

Use a spreadsheet or other mathematical software to evaluate the terms of the sum and to sum the terms until they converge. For the partition function, see Exercise 52.5(a).

The equipartition value is simply kT (i.e., $kT/2$ for each rotational degree of freedom). The explicit and equipartition expressions are compared in Fig. 53.1. The explicit sum reaches 95% of the equipartition value at about $\boxed{18.5\ \text{K}}$.

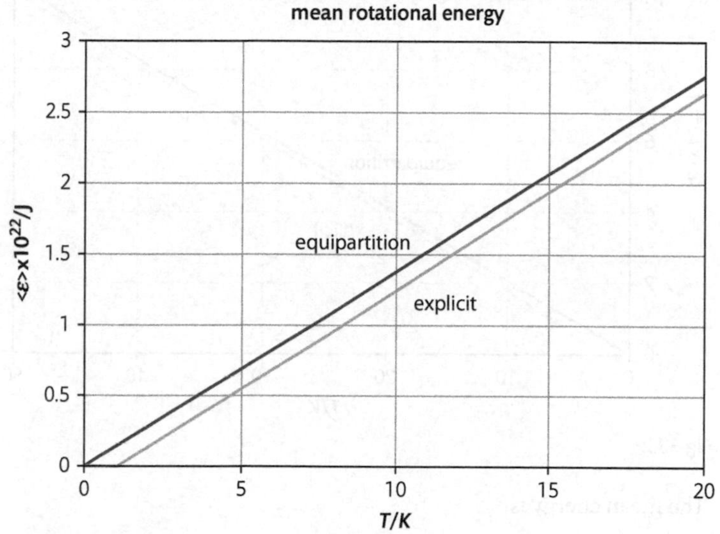

Fig 53.1

E53.3(a) The mean energy is

$$\langle \varepsilon \rangle = \frac{1}{q}\sum_{\text{states}} \varepsilon_i \text{e}^{-\beta\varepsilon_i}\ [53.2] = \frac{1}{q}\sum_{\text{levels}} g_i\varepsilon_i \text{e}^{-\beta\varepsilon_i} = \frac{1}{\sigma q}\sum_J (2J+1)^2 \varepsilon_J \text{e}^{-\varepsilon_J/kT}\quad [\text{Topic 41.2b}]$$

Note that the sum over levels is restricted by nuclear statistics; in order to avoid multiple counting, we sum over all J without restriction and divide the result by the symmetry number σ.

$$\varepsilon_J = hc\tilde{B}J(J+1) = 6.626\times 10^{-34}\ \text{J s}\times 2.998\times 10^{10}\ \text{cm s}^{-1}\times 5.241\ \text{cm}^{-1}\times J(J+1)$$
$$= J(J+1)\times 1.041\times 10^{-22}\ \text{J}$$

and $\dfrac{\varepsilon_J}{k} = \dfrac{J(J+1)\times 1.041\times 10^{-22}\ \text{J}}{1.381\times 10^{-23}\ \text{J K}^{-1}} = J(J+1)\times 7.539\ \text{K},$

so $\langle \varepsilon \rangle = \dfrac{1}{\sigma q}\sum_J J(J+1)(2J+1)^2\times 1.041\times 10^{-22}\ \text{J}\times \text{e}^{-J(J+1)\times 7.539\ \text{K}/T}$

Use a spreadsheet or other mathematical software to evaluate the terms of the sum and to sum the terms until they converge. For σq, see Exercise 52.6(a). The quantity evaluated explicitly in that exercise is σq, because we computed the partition function without taking the symmetry number into account; in effect, the sum evaluated here **and** the sum evaluated in the earlier exercise contain factors of σ, which cancel.

The equipartition value is simply $3kT/2$ (*i.e.*, $kT/2$ for each rotational degree of freedom). The explicit and equipartition expressions are compared in Fig. 53.2. The explicit sum reaches 95% of the equipartition value at about $\boxed{25\text{ K}}$.

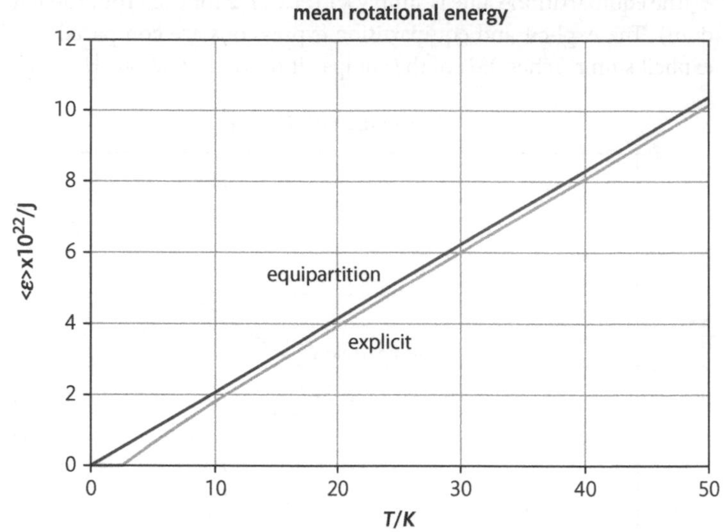

Fig 53.2

E53.4(a) The mean energy is

$$\langle\varepsilon\rangle = \frac{1}{q}\sum_{\text{states}} \varepsilon_i e^{-\beta\varepsilon_i} [53.2] = \frac{1}{q}\sum_{\text{levels}} g_i\varepsilon_i e^{-\beta\varepsilon_i}$$

$$= \frac{1}{\sigma q}\sum_{J=0}(2J+1)e^{-hc\tilde{B}J(J+1)/kT}\left(\sum_{K=-J}^{J}\varepsilon_{J,K}e^{-hc(\tilde{A}-\tilde{B})K^2/kT}\right) [41.13]$$

Note that the sum over levels is restricted by nuclear statistics; in order to avoid multiple counting, we sum over all J without restriction and divide the result by the symmetry number σ. (See Exercise 53.4(a).)

$$\varepsilon_{J,K} = hc\left\{\tilde{B}J(J+1)+\left(\tilde{A}-\tilde{B}\right)K^2\right\}$$

Use $hc\tilde{B} = 6.626\times10^{-34}\text{ J s}\times2.998\times10^{10}\text{ cm s}^{-1}\times0.443\text{ cm}^{-1} = 8.80\times10^{-24}\text{ J}$,

$$\frac{hc\tilde{B}}{k} = \frac{8.80\times10^{-24}\text{ J}}{1.381\times10^{-23}\text{ J K}^{-1}} = 0.637\text{ K},$$

$$hc\left(\tilde{A}-\tilde{B}\right) = 6.626\times10^{-34}\text{ J s}\times2.998\times10^{10}\text{ cm s}^{-1}\times(5.097-0.443)\text{ cm}^{-1}$$
$$= 9.245\times10^{-23}\text{ J},$$

and $\dfrac{hc(\tilde{A}-\tilde{B})}{k}=\dfrac{9.245\times10^{-23}\ \mathrm{J}}{1.381\times10^{-23}\ \mathrm{J\,K^{-1}}}=6.694\ \mathrm{K}$

so $$\langle\varepsilon\rangle=\frac{1}{\sigma q}\sum_{J=0}(2J+1)\mathrm{e}^{-0.637\ \mathrm{K}\times J(J+1)/T}$$

$$\times\sum_{K=-J}^{J}\{J(J+1)\times8.80\times10^{-24}\ \mathrm{J}+K^2\times9.245\times10^{-23}\ \mathrm{J}\}\mathrm{e}^{-6.694\ \mathrm{K}\times K^2/T}$$

Write a brief computer program or use other mathematical software to evaluate the terms of the sum and to sum the terms until they converge. Nested sums are straightforward to program in languages such as BASIC or FORTRAN, whereas spreadsheets are more unwieldy. For σq, see Exercise 52.7(a). The quantity evaluated explicitly in that exercise is σq, for there we computed the partition function without taking the symmetry number into account; in effect, the sum evaluated here **and** the sum evaluated in the earlier exercise contain factors of σ, which cancel. Compare the results of the direct sum with the equipartition value, namely $3kT/2$. The explicit and equipartition expressions are compared in Fig. 53.3. The explicit sum reaches 95% of the equipartition value at about $\boxed{4.5\ \mathrm{K}}$.

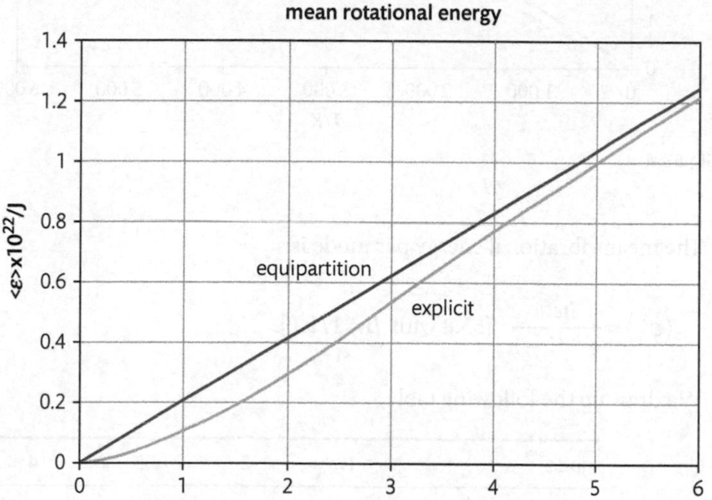

Fig 53.3

E53.5(a) The mean vibrational energy is

$$\langle\varepsilon^{\mathrm{V}}\rangle=\frac{hc\tilde{\nu}}{\mathrm{e}^{hc\tilde{\nu}/kT}-1}\quad[53.8\text{ with }\beta=1/kT].$$

Use $hc\tilde{\nu}=6.626\times10^{-34}\ \mathrm{J\,s}\times2.998\times10^{10}\ \mathrm{cm\,s^{-1}}\times323.2\ \mathrm{cm^{-1}}=6.420\times10^{-21}\ \mathrm{J}$

and $\dfrac{hc\tilde{\nu}}{k}=\dfrac{6.420\times10^{-21}\ \mathrm{J}}{1.381\times10^{-23}\ \mathrm{J\,K^{-1}}}=464.9\ \mathrm{K},$

so $\langle \varepsilon^{\rm V} \rangle = \dfrac{6.420\times10^{-21}\ {\rm J}}{{\rm e}^{464.9\ {\rm K}/T}-1}$

The equipartition value is simply kT for a single vibrational mode. The explicit and equipartition values are compared in Fig. 53.4. The explicit expression reaches 95% of the equipartition value at $\boxed{4\,600\ {\rm K}}$.

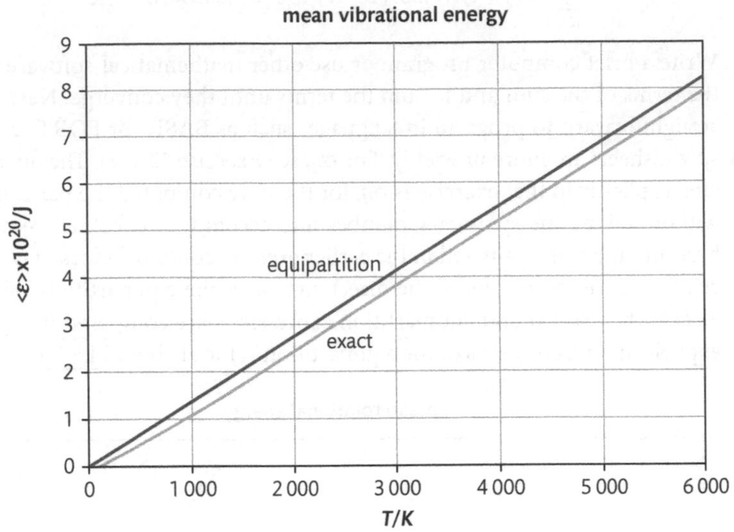

Fig 53.4

E53.6(a) The mean vibrational energy per mode is

$$\langle \varepsilon^{\rm V} \rangle = \frac{hc\tilde{\nu}}{{\rm e}^{hc\tilde{\nu}/kT}-1} \quad [53.8 \text{ with } \beta = 1/kT].$$

We draw up the following table

mode	1	2	3	4
$\tilde{\nu}$ / cm^{-1}	658	397	397	1535
$hc\tilde{\nu}$ / (10^{-21} J)	13.07	7.89	7.89	30.49
$(hc\tilde{\nu}/k)$ / K	946	571	571	2 208

So $\langle \varepsilon^{\rm V} \rangle = \dfrac{1.307\times10^{-20}\ {\rm J}}{{\rm e}^{946\ {\rm K}/T}-1} + 2\times\dfrac{7.89\times10^{-21}\ {\rm J}}{{\rm e}^{571\ {\rm K}/T}-1} + \dfrac{3.049\times10^{-20}\ {\rm J}}{{\rm e}^{2208\ {\rm K}/T}-1}$

The equipartition value is simply $4kT$, that is, kT per vibrational mode. The explicit and equipartition values are compared in Fig. 53.5. The explicit expression reaches 95% of the equipartition value at $\boxed{10\,500\ {\rm K}}$.

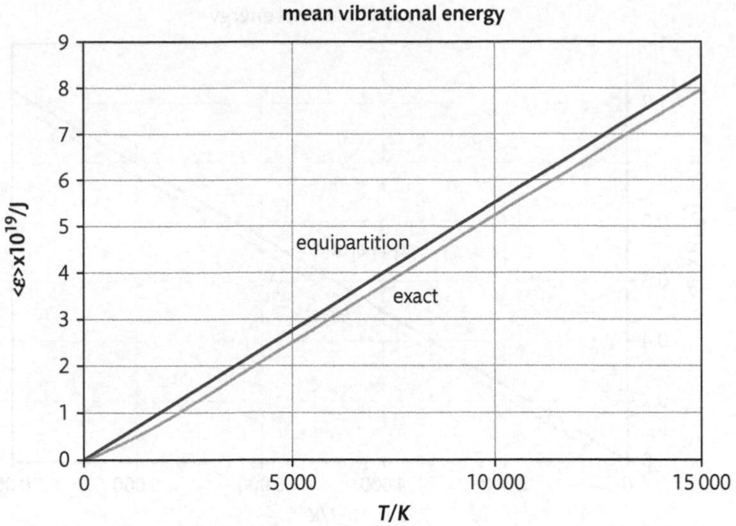

Fig 53.5

E53.7(a) The mean vibrational energy per mode is

$$\langle \varepsilon^{V} \rangle = \frac{hc\tilde{\nu}}{e^{hc\tilde{\nu}/kT}-1} \quad [53.8 \text{ with } \beta = 1/kT].$$

We draw up the following table

mode	1	2	3	4
$\tilde{\nu}$ / cm^{-1}	459	217	776	314
degeneracy	1	2	3	3
$hc\tilde{\nu}$ / (10^{-21} J)	9.12	4.31	15.42	6.24
$(hc\tilde{\nu}/k)$ / K	660	312	1 116	452

So $$\langle \varepsilon^{V} \rangle = \frac{9.12\times10^{-21}\text{ J}}{e^{660\text{ K}/T}-1} + 2\times\frac{4.31\times10^{-21}\text{ J}}{e^{312\text{ K}/T}-1} + 3\times\frac{1.542\times10^{-20}\text{ J}}{e^{1116\text{ K}/T}-1} + 3\times\frac{6.24\times10^{-21}\text{ J}}{e^{452\text{ K}/T}-1}$$

The equipartition value is simply $9kT$, that is, kT per vibrational mode. The explicit and equipartition values are compared in Figure 53.6. The explicit expression reaches 95% of the equipartition value at $\boxed{6\,500\text{ K}}$.

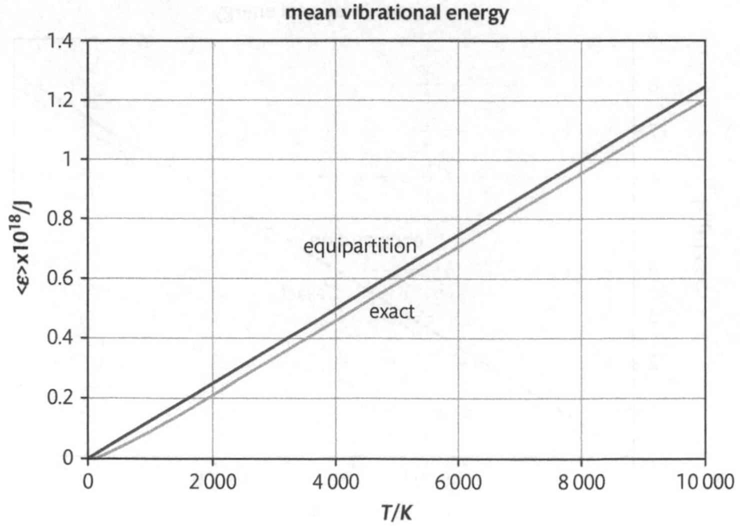

Fig 53.6

E53.8(a) $$\langle\varepsilon\rangle = -\frac{1}{q}\frac{\partial q}{\partial\beta}[53.4a] = -\frac{1}{q}\frac{\partial}{\partial\beta}\left(4+e^{-\beta\varepsilon_1}+2e^{-\beta\varepsilon_2}\right) = -\frac{1}{q}\left(-\varepsilon_1 e^{-\beta\varepsilon_1}-2\varepsilon_2 e^{-\beta\varepsilon_2}\right)$$

$$= \frac{hc}{q}\left(\tilde{\nu}_1 e^{-\beta hc\tilde{\nu}_1}+2\tilde{\nu}_2 e^{-\beta hc\tilde{\nu}_2}\right) = \frac{hc}{q}\left(\tilde{\nu}_1 e^{-hc\tilde{\nu}_1/kT}+2\tilde{\nu}_2 e^{-hc\tilde{\nu}_2/kT}\right)$$

Use $\dfrac{hc\tilde{\nu}_j}{kT} = 7.571\times10^{-4}\times(\tilde{\nu}_j/\text{cm}^{-1})$ and $q = 4.292$ [Exercise 52.13a]

$$\text{Thus } \langle\varepsilon\rangle = \frac{6.626\times10^{-34}\ \text{J s}\times2.998\times10^{10}\ \text{cm s}^{-1}}{4.292}$$
$$\times\left(2\,500\ \text{cm}^{-1}\times e^{-7.571\times10^{-4}\times2\,500}+2\times3\,500\ \text{cm}^{-1}\times e^{-7.571\times10^{-4}\times3\,500}\right)$$
$$= \boxed{4.033\times10^{-21}\ \text{J}}$$

Problems

P53.1 Follow the reasoning set out in Problem 52.1 to employ more general dimensionless quantities. The energy divided by the Boltzmann constant has dimensions of temperature, so define the characteristic temperature for this system by

$$\theta \equiv \frac{\hbar^2}{2m_e R^2 k}$$

With this definition, note that the energy levels for this spherical well can be written as

$$E_{nl} = X_{nl}^2 k\theta$$

Thus, one can tabulate values of $E_{nl}/k\theta$ without having to select a radius. Likewise, one can employ the dimensionless temperature T/θ . As noted in the problem, Boltzmann factors require energies measured with respect to the ground state, so the energies **in the exponents of the Boltzmann factors** must be $E_{nl} - E_{10}$. The expression for the partition function is [52.1]

$$q = \sum_{\text{states}} \exp\left(-\frac{E_{nl}-E_{10}}{kT}\right) = \sum_{\text{levels}} g_{nl}\exp\left(-\frac{E_{nl}-E_{10}}{kT}\right) = \sum_{\text{levels}} g_{nl}\exp\left(-\frac{E_{nl}-E_{10}}{k\theta}\times\frac{\theta}{T}\right)$$

Draw up the following table:

n	l	X_{nl}	$E_{nl}/k\theta$	$(E_{nl} - E_{10})/k\theta$	g_{nl}
1	0	3.142	9.872164	0	1
1	1	4.493	20.187049	10.314885	3
1	2	5.763	33.212169	23.340005	5
2	0	6.283	39.476089	29.603925	1
1	3	6.988	48.832144	38.95998	7
2	1	7.725	59.675625	49.803461	3

The partition function is plotted in Fig. 53.7(a). Note that at $T/\theta = 25$, the highest state already contributes more than 5% of the six-level sum. Thus, by this temperature, truncating the sum after six levels is becoming inaccurate.

$$\frac{\langle\varepsilon\rangle}{k\theta} = \frac{1}{q}\sum_{\text{levels}} \frac{E_{nl}}{k\theta} g_{nl}\exp\left(-\frac{E_{nl}-E_{10}}{k\theta}\times\frac{\theta}{T}\right) \quad [53.2]$$

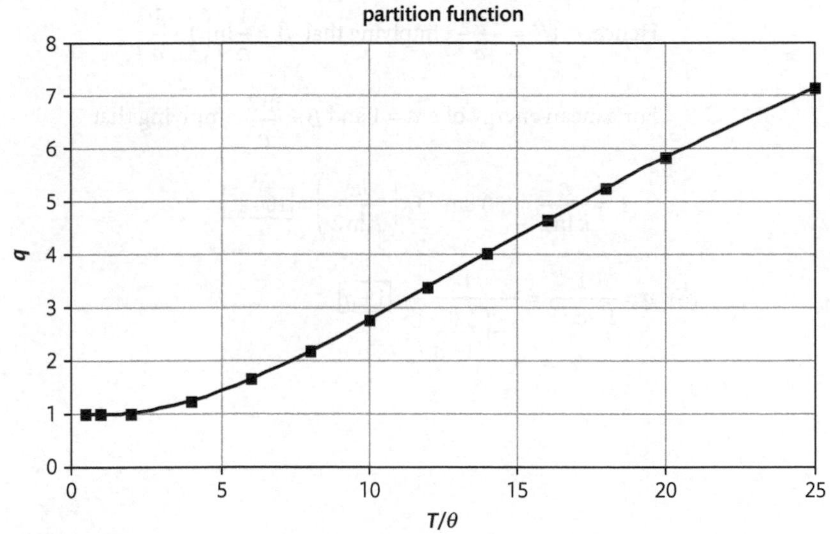

Fig 53.7(a)

The average energy (average of $E_{nl}/k\theta$, not of $(E_{nl} - E_{10})/k\theta$) is plotted in Fig. 53.7(b).

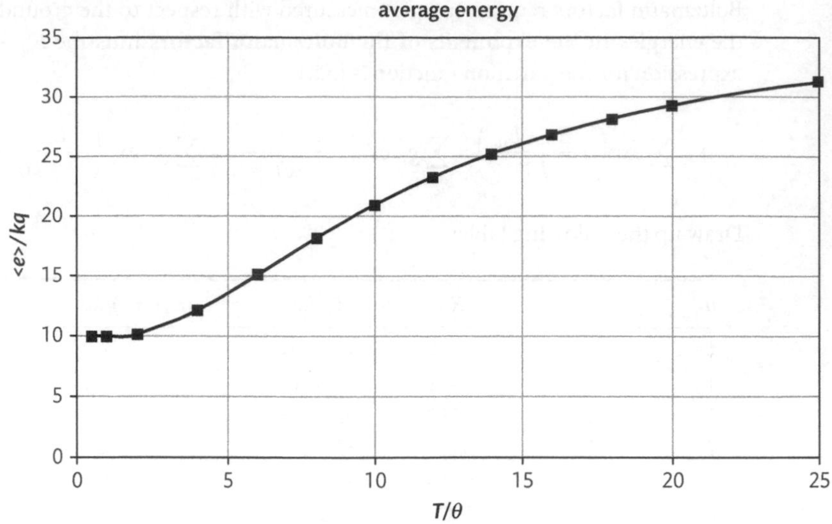

Fig 53.7(b)

P53.3 (a) $\langle\varepsilon\rangle = -\frac{1}{q}\frac{\partial q}{\partial\beta}$[53.4a], with $q = \frac{1}{1-e^{-\beta\varepsilon}}$[52.15]

so $\langle\varepsilon\rangle = -\frac{1}{q}\frac{\partial q}{\partial\beta} = \frac{\varepsilon e^{-\beta\varepsilon}}{1-e^{-\beta\varepsilon}} = \frac{\varepsilon}{e^{\beta\varepsilon}-1} = a\varepsilon$

Hence, $e^{\beta\varepsilon} = \frac{1+a}{a}$, implying that $\beta = \frac{1}{\varepsilon}\ln\left(1+\frac{1}{a}\right)$

For a mean energy of ε, $a = 1$ and $\beta = \frac{\ln 2}{\varepsilon}$, implying that

$$T = \frac{\varepsilon}{k\ln 2} = (50\ \text{cm}^{-1})\times\left(\frac{hc}{k\ln 2}\right) = \boxed{104\ \text{K}}$$

(b) $q = \frac{1}{1-e^{-\beta\varepsilon}} = \frac{1}{1-\left(\frac{a}{1+a}\right)} = \boxed{1+a}$

Topic 54 **The canonical ensemble**

Discussion questions

D54.1 Ensembles are needed to treat system of interacting particles (in contrast to the molecular partition functions that apply to independent particles). An ensemble is a set of a large number of imaginary replications of an actual physical (thermodynamic) system. Ensembles are useful in statistical thermodynamics because it is mathematically more tractable to perform an ensemble average to determine time averaged thermodynamic properties than it is to perform an average over time to determine these properties. (Macroscopic thermodynamic properties are averages over the time dependent properties of the particles that compose the macroscopic system.) The replications in an ensemble are identical in some respects but not in all respects. In the canonical ensemble, all replications have the same number of particles, the same volume, and the same temperature, but they need not have the same energy. Because they have the same temperature, they can exchange energy. Thus the canonical ensemble corresponds to a closed physical system at thermal equilibrium: averages using canonical ensembles apply to closed physical systems at thermal equilibrium

D54.3 Identical particles can be regarded as distinguishable when they are localized as in a crystal lattice where we can assign a set of coordinates to each particle. Strictly speaking, it is the lattice site that carries the set of coordinates, but as long as the particle is tethered to the site, it too can be considered distinguishable.

Exercise

E54.1(a) Inclusion of a factor of $1/N!$ is necessary when considering indistinguishable particles. Because of their translational freedom, gases are collections of indistinguishable particles. Solids are collections of particles that are distinguishable by their positions. The factor must be included in calculations on (a) $\boxed{\text{He gas}}$, (b) $\boxed{\text{Co gas}}$, and (d) $\boxed{H_2O \text{ vapor}}$, but not (c) solid CO.

Problem

P54.1 Basic thermodynamic relationships will be introduced in Foci 12 and 13. Among the key quantities is the internal energy, U, where

$$U = U(0) + \langle E \rangle \; [55.1b]$$

Here, $U(0)$ is the value of the energy at $T=0$. The relationship between the internal energy and other macroscopic thermodynamic functions is given by the fundamental equation of thermodynamics, eqn 66.1

$$dU = TdS - pdV,$$

where S is the entropy, a main subject of Focus 13. Already we can see a connection between U and the canonical partition function, Q, on the one hand (via eqn 54.5) and to macroscopic system variables like pressure and temperature on the other. However, the thermodynamic function most directly related to Q is the Helmholtz energy, A, defined in eqn 64.6

$$A = U - TS$$

The relationship of A to system variables, therefore, is

$$dA = dU - TdS - SdT = TdS - pdV - TdS - SdT = -pdV - SdT \quad [66.14],$$

so $$\left(\frac{\partial A}{\partial V}\right)_{T,N} = -p \quad [66.15]$$

The simple and direct relationship between A and Q is

$$A = A(0) - kT\ln Q \quad [64.11a]$$

Thus $$p = kT\left(\frac{\partial \ln Q}{\partial V}\right)_{T,N}$$

$$= kT\left(\frac{\partial \ln(q^N/N!)}{\partial V}\right)_{T,N}$$

$$= kT\left(\frac{\partial[N\ln q - \ln N!]}{\partial V}\right)_{T,N} = NkT\left(\frac{\partial \ln q}{\partial V}\right)_{T,N}.$$

The only contribution to the molecular partition function that depends on V is translation:

$$q = q^{\mathrm{T}}q^{\mathrm{R}}q^{\mathrm{V}}q^{\mathrm{E}} \ [52.6] = \frac{V}{\Lambda^3}q^{\mathrm{R}}q^{\mathrm{V}}q^{\mathrm{E}} \ [52.10\text{b}],$$

so $$p = NkT\left(\frac{\partial \ln(Vq^{\mathrm{R}}q^{\mathrm{V}}q^{\mathrm{E}}/\Lambda^3)}{\partial V}\right)_{T,N}$$

$$= NkT\left(\frac{\partial\left[\ln V + \ln\left(q^{\mathrm{R}}q^{\mathrm{V}}q^{\mathrm{E}}/\Lambda^3\right)\right]}{\partial V}\right)_{T,N} = NkT\left(\frac{\partial \ln V}{\partial V}\right)_{T,N}$$

$$= \frac{NkT}{V} \quad \text{or} \quad \boxed{pV = NkT = nRT}$$

See also Example 66.2.

Focus 11: Integrated activities

F11.1 The partition function is

$$q=\sum_i e^{-\beta\varepsilon_i}=\boxed{1+e^{-2\mu_B\beta\mathcal{B}}}\ \text{[energies measured from lower state]}$$

The mean energy is [53.4a]

$$\langle\varepsilon\rangle=-\frac{1}{q}\frac{dq}{d\beta}=\boxed{\frac{2\mu_B\mathcal{B}e^{-2\mu_B\beta\mathcal{B}}}{1+e^{-2\mu_B\beta\mathcal{B}}}}$$

Alternatively, if we measure energy with respect to zero magnetic field (rather than to ground state energy), then

$$\langle\varepsilon\rangle=\varepsilon_{gs}-\frac{1}{q}\frac{dq}{d\beta}=\boxed{-\mu_B\mathcal{B}+\frac{2\mu_B\mathcal{B}e^{-2\mu_B\beta\mathcal{B}}}{1+e^{-2\mu_B\beta\mathcal{B}}}}$$

We write $x=2\mu_B\beta\mathcal{B}$, so the partition function becomes

$$q=1+e^{-x}$$

and the mean energy, scaled by the energy separation, becomes

$$\frac{\langle\varepsilon\rangle}{2\mu_B\mathcal{B}}=-\frac{1}{2}+\frac{e^{-x}}{1+e^{-x}}=-\frac{1}{2}+\frac{1}{e^{x}+1}$$

The functions are plotted in Figs. F11.1(a) and (b).The effect of increasing the magnetic field is to concentrate population into the lower level.

The relative populations are given by eqn 51.8

$$\frac{N_+}{N_-}=e^{-\beta\Delta\varepsilon}=e^{-x}$$

$$x=2\mu_B\beta\mathcal{B}=\frac{2\mu_B\mathcal{B}}{kT}=\frac{2\times9.274\times10^{-24}\ \text{J T}^{-1}\times1.0\ \text{T}}{(1.381\times10^{-23}\ \text{J K}^{-1})T}=\frac{1.343}{T/\text{K}}$$

(a) $T=4K,\ \frac{N_+}{N_-}=e^{-1.343/4}=\boxed{0.71}$ (b) $T=298\ \text{K},\ \frac{N_+}{N_-}=e^{-1.343/298}=\boxed{0.996}$

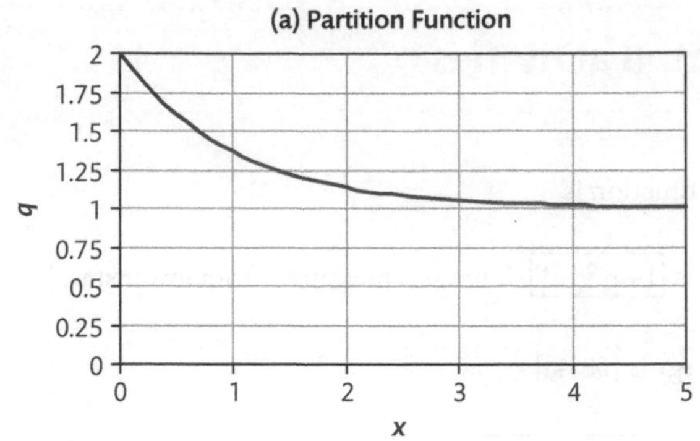

F11.1(a)

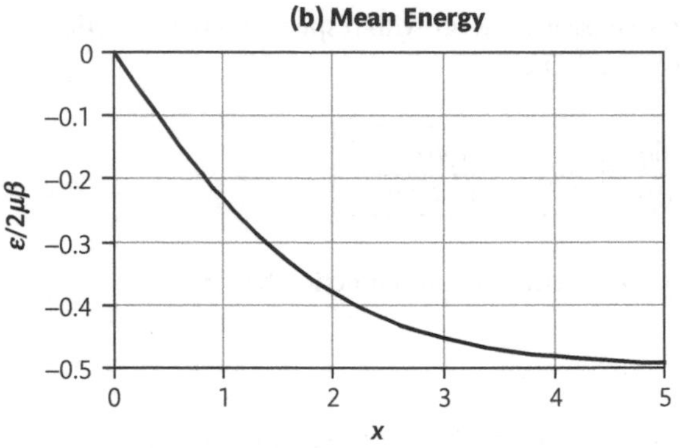

F11.1(b)

F11.3(a) One may simply pick values of the energy separation $\varepsilon = hc\tilde{\nu}$ and plot the partition function over a range of temperatures. However, if one works in terms of the characteristic vibrational temperature

$$\theta^{\mathrm{V}} = \frac{hc\tilde{\nu}}{k}$$

and a dimensionless temperature T/θ^{V}, one needs only a single plot, shown in Fig. F11.2. The partition function is

$$q^{\mathrm{V}} = \sum_{v} \mathrm{e}^{-vhc\tilde{\nu}/kT} = \frac{1}{1-\mathrm{e}^{-hc\tilde{\nu}/kT}} = \frac{1}{1-\mathrm{e}^{-\theta^{\mathrm{V}}/T}} \quad [52.15]$$

As noted in Topic 52.2(c), at high temperatures, q approaches T/θ^{V}. In the figure, which extends only to moderately high dimensionless temperatures, it is apparent that the partition function varies very nearly linearly with temperature with a slope very slightly less than one. In fact, q appears to approach a $\boxed{\text{limiting value of } T/\theta^{\mathrm{V}} + 0.5}$. At low temperatures, the partition function rises very slowly from unity.

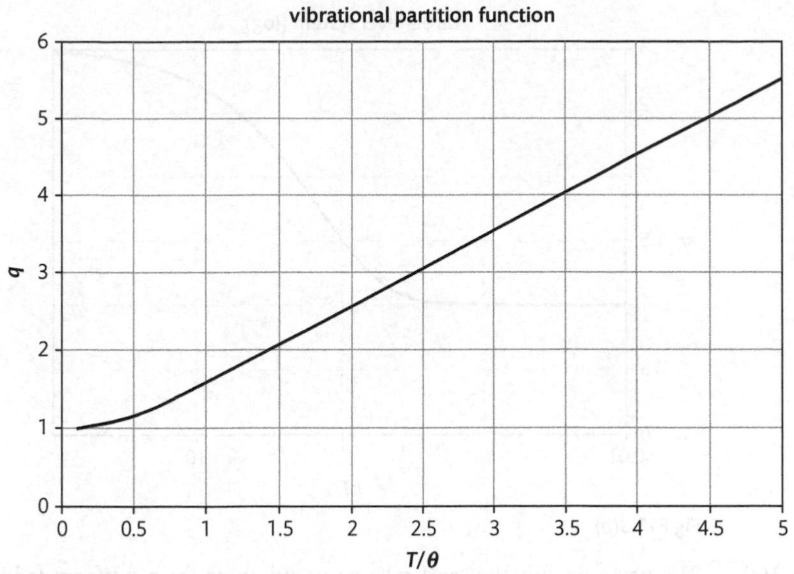

Fig F11.2

F11.3(b) The partition function is

$$q=\sum_i e^{-\beta\varepsilon_i}=\sum_i e^{-\varepsilon_i/kT}=1+e^{-\varepsilon/kT}+e^{-2\varepsilon/kT}$$

The partition function is plotted against kT/ε in Figs. F11.3(a) and F11.3(b). As can be seen in both of these figures, the partition function rises from a value of 1 at low temperatures (where only the ground state is occupied) to approach a value of 3 at high temperatures (where all three states are nearly equally populated). When the horizontal axis is plotted on a logarithmic scale (Fig. F11.3(b)), the sigmoidal character of the function is more apparent: the slope of the plot approaches zero in the limits of both low and high temperature.

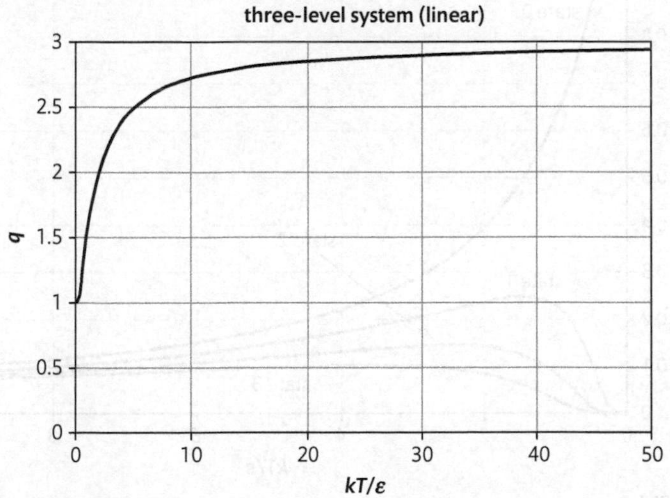

Fig F11.3(a)

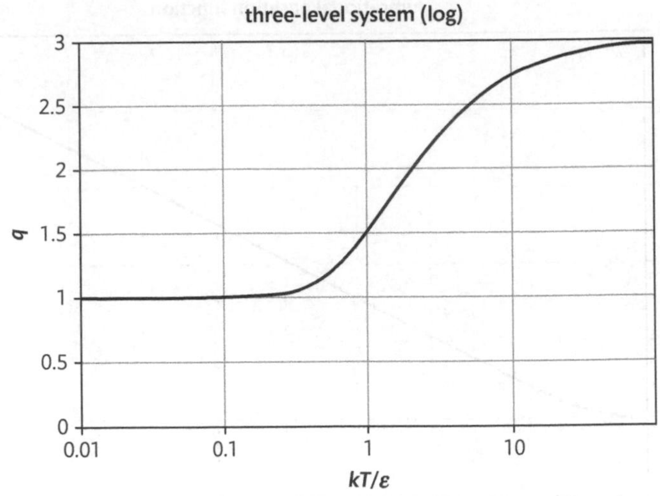

Fig F11.3(b)

F11.3(c) The partition function and relative populations for a uniform ladder of energy levels with separation ε are, respectively (eqn 52.2 with $\beta = 1/kT$):

$$q = \frac{1}{1 - e^{-\varepsilon/kT}} \quad \text{and} \quad p_i = \frac{N_i}{N} = \frac{e^{-\varepsilon_i/kT}}{q} = e^{-i\varepsilon/kT}(1 - e^{-\varepsilon/kT})$$

The relative populations of the lowest four states are plotted in Fig. F11.4. At low temperatures, the ground state is practically the only state occupied, so in that limit $p_0 = 1$ and $p_1 = p_2 = p_3 = 0$. At high temperatures, the populations of these four relatively low-lying states approach equality as they all decrease toward zero, because the number of accessible states increases without limit as the temperature increases without limit.

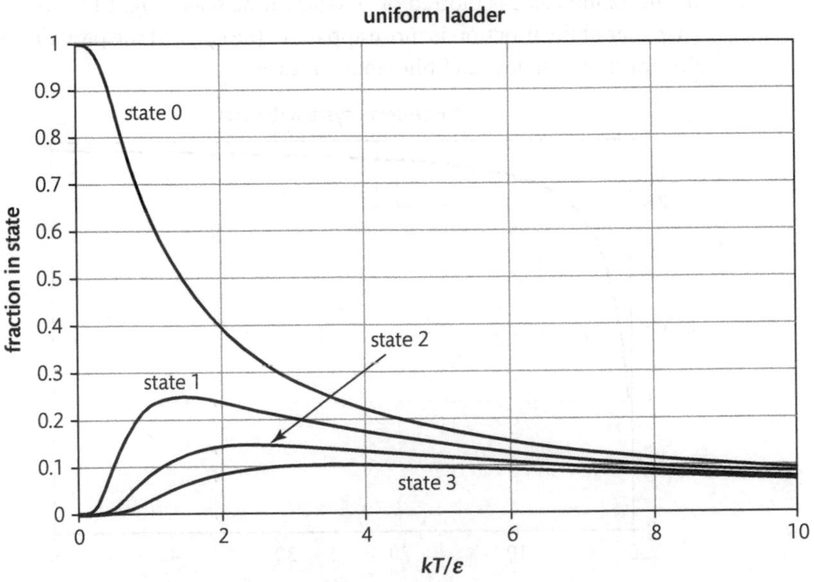

Fig F11.4

F11.3(d) The relative population of state i is

$$p_i = \frac{N_i}{N} = \frac{e^{-\varepsilon_i/kT}}{q} = \frac{e^{-i\varepsilon/kT}}{q}$$

The relative populations of the three states are plotted in Fig. F11.5. At low temperatures, the ground state is practically the only state occupied, so in that limit $p_0 = 1$ and $p_1 = p_2 = 0$. At high temperatures, the states are all practically equally occupied, so all three relative populations approach 1/3.

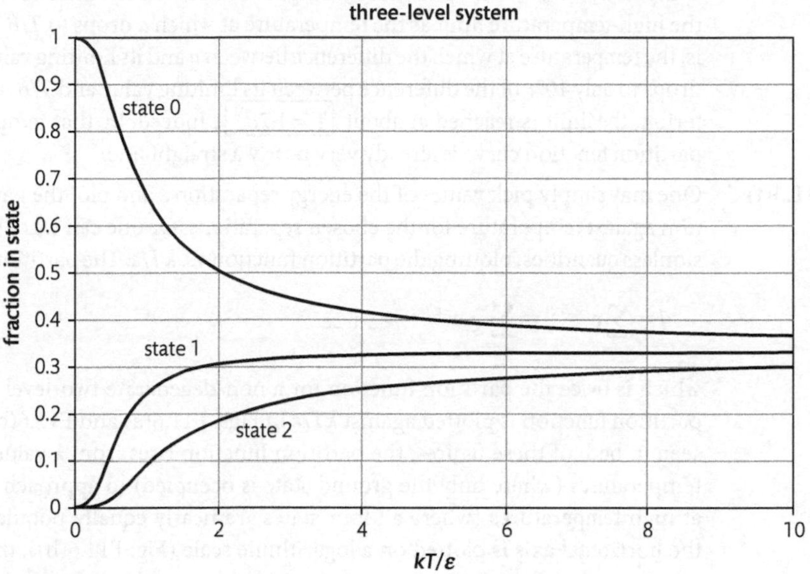

Fig F11.5

F11.3(e) One may simply pick values of the wavenumber $\tilde{\nu}$ and plot the partition function against temperature for the chosen wavenumbers. However, if one works in terms of the characteristic vibrational temperature

$$\theta^{\mathrm{V}} = \frac{hc\tilde{\nu}}{k}$$

and a dimensionless temperature T/θ^{V}, one needs only a single plot, shown in Fig. F11.2. The partition function is

$$q^{\mathrm{V}} = \sum_{v} e^{-vhc\tilde{\nu}/kT} = \frac{1}{1-e^{-hc\tilde{\nu}/kT}} = \frac{1}{1-e^{-\theta^{\mathrm{V}}/T}} \quad [52.15]$$

As noted in Topic 52.2(c), at high temperatures, q approaches T/θ^{V}. In the figure, which extends only to moderately high dimensionless temperatures, it is apparent that the partition function varies very nearly linearly with temperature with a slope very slightly less than one. In fact, q appears to approach a limiting value of $T/\theta^{\mathrm{V}} + 0.5$, and it approaches this value very quickly. Even at a dimensionless temperature of 1 (that is, at a temperature of $T = \theta^{\mathrm{V}}$), $q = 1.582$, not very far from its limiting value of $T/\theta^{\mathrm{V}} + 0.5 = 1.500$. As the temperature increases further, there are two distinct ways in which q approaches T/θ^{V}: one is that q continues slowly to get absolutely closer to $T/\theta^{\mathrm{V}} + 0.5$, and the other is that 0.5 becomes *relatively* small compared to T/θ^{V}.

One might arbitrarily define the temperature at which the high-temperature limit is reached is the temperature at which q drops to within 10% of T/θ^{V}. By that criterion, the limit is reached at about $T = 5.2\theta^{\mathrm{V}}$, because by that temperature, the 0.5 in the limiting value $T/\theta^{\mathrm{V}} + 0.5$ becomes relatively small (by our definition) compared to T/θ^{V}. Perhaps a more telling, but still arbitrary criterion would define the high-temperature limit as the temperature at which q drops to $T/\theta^{\mathrm{V}} + 0.55$ (that is, the temperature at which the difference between q and its limiting value $T/\theta^{\mathrm{V}} + 0.5$ drops to only 10% of the difference between its limiting value and T/θ^{V}). By that criterion, the limit is reached at about $\boxed{T = 1.7\theta^{\mathrm{V}}}$. Indeed, by that temperature, the partition function curve is already very nearly a straight line.

F11.3(f) One may simply pick values of the energy separation ε and plot the partition function against temperature for the chosen separations. Or one can work with dimensionless quantities, plotting the partition function vs. kT/ε. The partition function is

$$q = \sum_{\text{states}} e^{-\varepsilon_i/kT} = \sum_{\text{levels}} g_i e^{-\varepsilon_i/kT} = 2 + 2e^{-\varepsilon/kT}$$

which is twice the partition function for a non-degenerate two-level system. The partition function is plotted against kT/ε in Figs. F11.6(a) and F11.6(b). As can be seen in both of these figures, the partition function rises from a value of 2 at low temperatures (where only the ground state is occupied) to approach a value of 4 at high temperatures (where all four states are nearly equally populated). When the horizontal axis is plotted on a logarithmic scale (Fig. F11.6(b)), the sigmoidal character of the function is more apparent: the slope of the plot approaches zero in the limits of both low and high temperature. That slope begins to be significantly positive around the temperature $\boxed{T = 0.3\varepsilon/k}$.

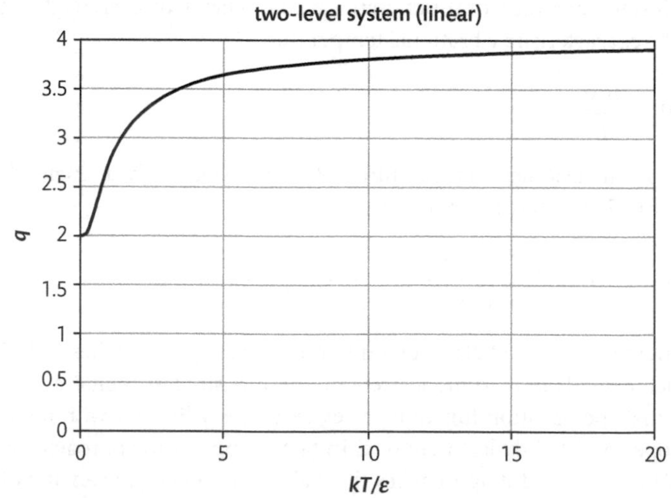

Fig F11.6(a)

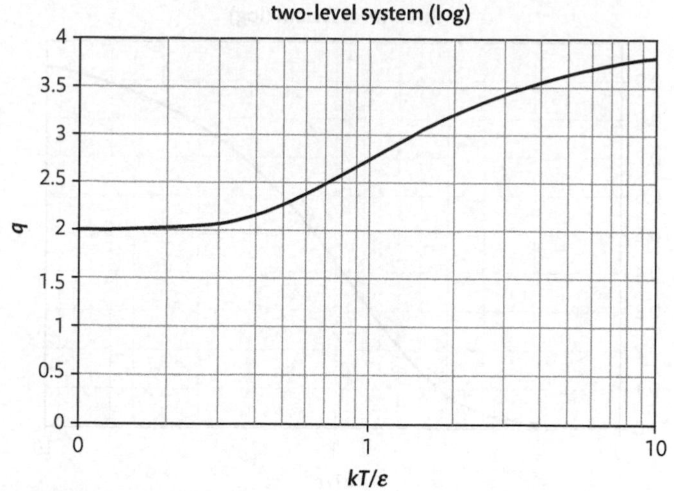

Fig F11.6(b)

F11.3(g) The mean energy is

$$\langle\varepsilon\rangle=\frac{1}{q}\sum_i \varepsilon_i e^{-\varepsilon_i/kT}=\frac{1}{q}\sum_i i\varepsilon e^{-i\varepsilon/kT}=\frac{\varepsilon e^{-\varepsilon/kT}+2\varepsilon e^{-2\varepsilon/kT}}{q}$$

The dimensionless mean energy, $\langle\varepsilon\rangle/\varepsilon$, is plotted against the dimensionless temperature kT/ε in Figs. F11.7(a) and F11.7(b). As can be seen in both of these figures, the mean energy rises from a value of 0 at low temperatures (where only the ground state is occupied) to approach a value of ε at high temperatures (where all three states are nearly equally populated). When the horizontal axis is plotted on a logarithmic scale (Fig. F11.7(b)), the sigmoidal character of the function is more apparent: the slope of the plot approaches zero in the limits of both low and high temperature.

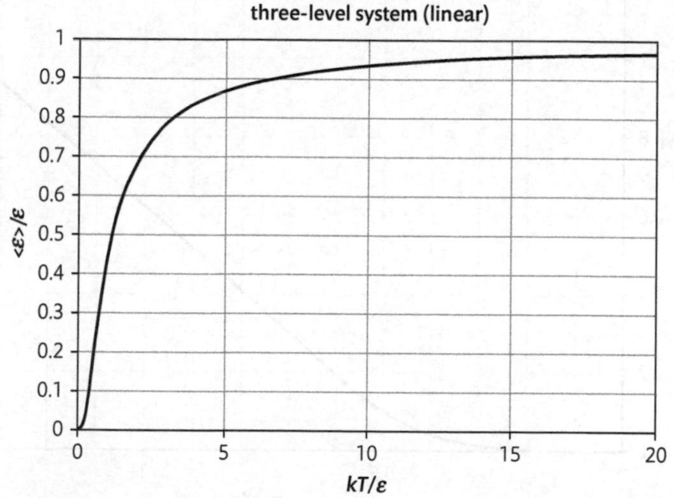

Fig F11.7(a)

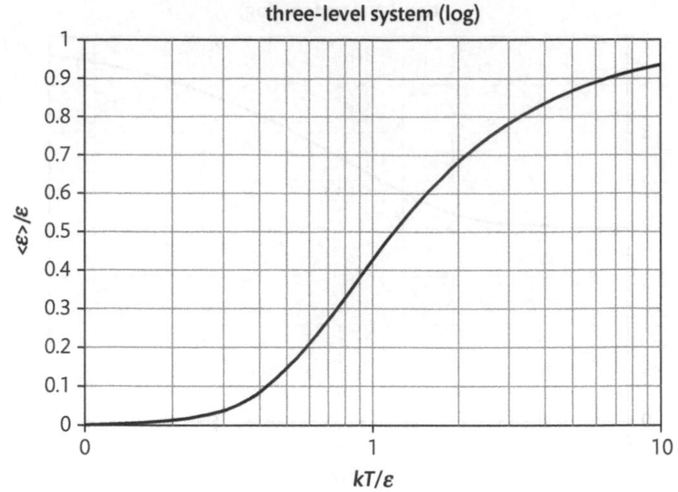

Fig F11.7(b)

F11.3(h) One may simply pick values of the rotational constant and plot the mean energy against temperature for the chosen values; however, if one works in terms of the characteristic rotational temperature, θ^{R}, one can employ more general dimensionless quantities as described below.

$$\theta^{R} = \frac{hc\tilde{B}}{k}.$$

A plot of the dimensionless mean energy, $\langle\varepsilon\rangle/hc\tilde{B} = \langle\varepsilon\rangle/k\theta^{R}$ vs. the dimensionless temperature T/θ^{R} for a non-symmetric linear rotor is shown in Fig. F11.8. The slope of the plot remains very small until about $T = 0.3\theta^{R}$. By about $\boxed{T = 0.6\theta^{R}}$, the slope of the plot has increased to nearly the constant value it has at high temperatures. In

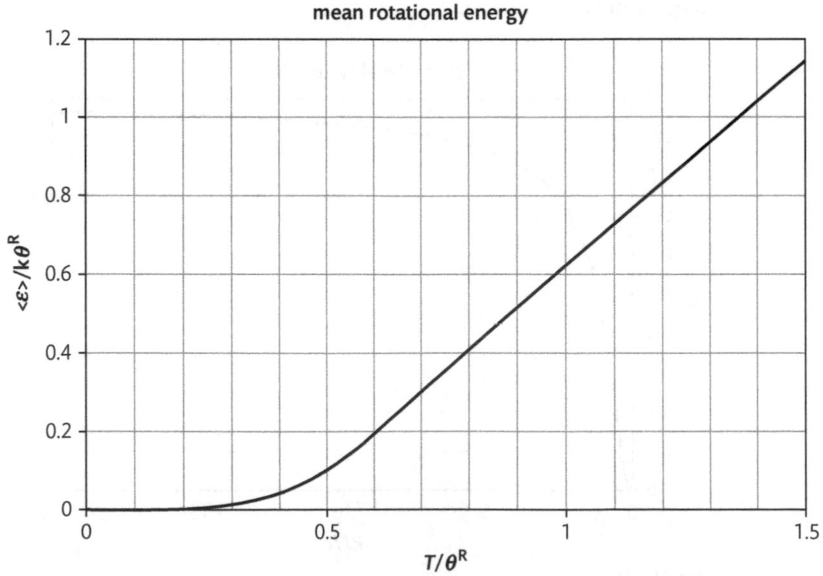

Fig F11.8

other words, at the former temperature, the mean energy begins to rise appreciably from zero, and by the latter temperature, it rises linearly with temperature as it will continue to do for still higher temperatures.

F11.3(i) One may simply pick values of the vibrational wavenumber and plot the mean energy against temperature for the chosen values; however, if one works in terms of the characteristic vibrational temperature, θ^{V}, one can employ more general dimensionless quantities as described below.

$$\theta^{V} = \frac{hc\tilde{\nu}}{k}.$$

A plot of the dimensionless mean energy, $\langle\varepsilon\rangle/hc\tilde{\nu} = \langle\varepsilon\rangle/k\theta^{V}$ vs. the dimensionless temperature T/θ^{V} for a harmonic oscillator is shown in Fig. F11.9. The slope of the plot remains very small until about $T = 0.2\theta^{V}$. By about $\boxed{T = 0.6\theta^{V}}$, the slope of the plot has increased to nearly the constant value it has at high temperatures. In other words, at the former temperature, the mean energy begins to rise appreciably from zero, and by the latter temperature, it rises linearly with temperature as it will continue to do for still higher temperatures.

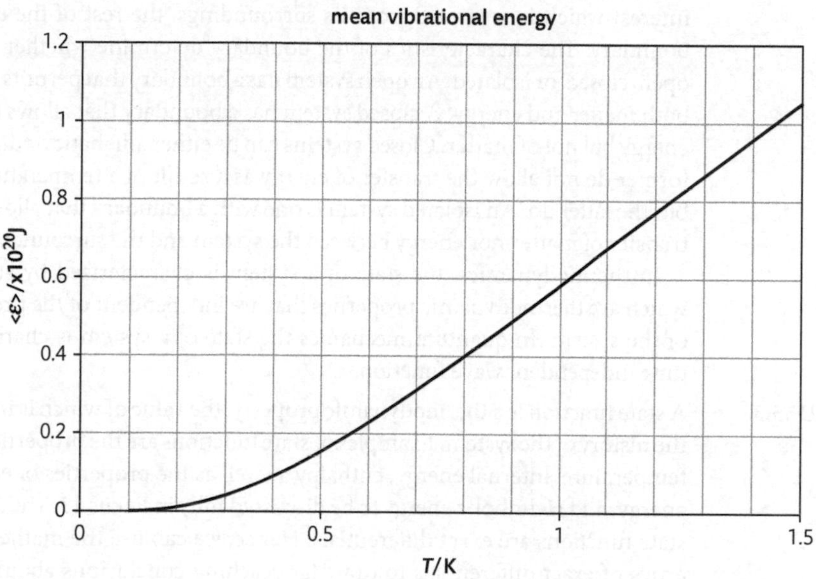

Fig F11.9

Topic 55 The First Law

Assume all gases are perfect unless stated otherwise. Unless otherwise stated, thermochemical data are for 298.15 K.

Discussion questions

D55.1 In physical chemistry, the universe is considered to be divided into two parts: the system and its surroundings. In thermodynamics, the system is the object of interest which is separated from its surroundings, the rest of the universe, by a boundary. The characteristics of the boundary determine whether the system is open, closed, or isolated. An open system has a boundary that permits the passage of both matter and energy. A closed system has a boundary that allows the passage of energy but not of matter. Closed systems can be either adiabatic or diathermic. The former do not allow the transfer of energy as a result of a temperature difference, but the latter do. An isolated system is one with a boundary that allows neither the transfer of matter nor energy between the system and the surroundings.

In thermodynamics, the state of a system is characterized by state functions which are thermodynamic properties that are independent of the previous history of the system. In quantum mechanics the state of a system is characterized by a time independent wave function.

D55.3 A state function is a thermodynamic property, the value of which is independent of the history of the system. Examples of state functions are the properties of pressure, temperature, internal energy, enthalpy as well as the properties of entropy, Gibbs energy, and Helmholtz energy to be discussed fully in Focus 13. The differentials of state functions are exact differentials. Hence, we can use the mathematical properties of exact differentials to draw far-reaching conclusions about the relations between physical properties and establish connections that were unexpected but turn out to be very significant. One practical importance of these results is that we can obtain the value of a property we require from the combination of measurements of other properties without actually having to measure the required property itself, the measurement of which might be very difficult.

Exercises

E55.1(a) $C_{V,\mathrm{m}} = \frac{1}{2}(3+\nu_{\mathrm{R}}^{*}+2\nu_{\mathrm{V}}^{*})R$ [58.15]

with a mode active if $T > \theta_{\mathrm{M}}$

(a) $\nu_R^* = 2,\ \nu_V^* \approx 1;$ hence $C_{V,m} = \frac{1}{2}(3+2+2\times 1)R = \boxed{\frac{7}{2}R}$ [experimental = 3.4R]

Hence the energy is

$$\frac{7}{2}RT = \frac{7}{2}\times 8.314\ \text{J K}^{-1}\ \text{mol}^{-1}\times 298.15\ \text{K} = 8671\ \text{J mol}^{-1} = \boxed{8.671\ \text{kJ mol}^{-1}}$$

Note that I_2 has quite a low vibrational wavenumber, so

$$\theta_V = \frac{hc\tilde{\nu}}{k} = \frac{(6.626\times 10^{-34}\ \text{J s})(2.998\times 10^{10}\ \text{cm s}^{-1})(214\ \text{cm}^{-1})}{1.381\times 10^{-23}\ \text{J K}^{-1}} = 308\ \text{K}$$

The temperature specified for this exercise is less than this but only slightly. Looking at Fig. 58.7 in the text suggests that the mode is closer to active than inactive when the temperature is even close to the vibrational temperature. Hence applying the equipartition theorem may not be adequate in this case.

(b) $\nu_R^* = 3,\ \nu_V^* \approx 0;$ hence $C_{V,m} = \frac{1}{2}(3+3+0)R = \boxed{3R}$ [experimental = 3.2R]

Hence the energy is

$$E = 3RT = 3\times 8.314\ \text{J K}^{-1}\ \text{mol}^{-1}\times 298.15\ \text{K} = 7436\ \text{J mol}^{-1} = \boxed{7.436\ \text{kJ mol}^{-1}}$$

(c) $\nu_R^* = 3,\ \nu_V^* \approx 4;$ hence $C_{V,m} = \frac{1}{2}(3+3+2\times 4)R = \boxed{7R}$ [experimental = 8.8R]

Hence the energy is

$$E = 7RT = 8\times 8.314\ \text{J K}^{-1}\ \text{mol}^{-1}\times 298.15\ \text{K} = \boxed{17.35\ \text{kJ mol}^{-1}}$$

Note that data from the book by G. H. Herzberg (*Molecular Spectra and Molecular Structure II. Infrared and Raman Spectra of Polyatomic Molecules*, D. van Nostrand Company, Inc., Princeton, NJ, 1945, pp 364–365) give low vibrational wavenumbers for four modes of benzene, and we have included these above. There are 26 more vibrational modes we have neglected, taking them to be inactive. Slight activity from these modes accounts for the difference of about 1.8R in the heat capacity between the experimental value and our estimate.

E55.2(a) (a) pressure, (b) temperature, and (d) enthalpy are state functions.

E55.3(a) The physical definition of work is $dw = -F\,dz$ [55.5]

In a gravitational field the force is the weight of the object, which is $F = mg$

If g is constant over the distance the mass moves, dw may be integrated to give the total work

$$w = -\int_{z_i}^{z_f} F\,dz = -\int_{z_i}^{z_f} mg\,dz = -mg(z_f - z_i) = -mgh \quad \text{where} \quad h = (z_f - z_i)$$

On earth: $w = -(60\,\text{kg})\times(9.81\,\text{m s}^{-2})\times(6.0\,\text{m}) = -3.5\times 10^3\ \text{J} = \boxed{3.5\times 10^3\ \text{J needed}}$

On the moon: $w = -(60\,\text{kg})\times(1.60\,\text{m s}^{-2})\times(6.0\,\text{m}) = -5.7\times 10^2\ \text{J}$

$= \boxed{5.7\times 10^2\ \text{J needed}}$

E55.4(a) This is an expansion against a constant external pressure; hence $w = -p_{ex}\Delta V$[55.8]

$$p_{ex} = (1.0\ \text{atm})\times(1.013\times 10^5\ \text{Pa atm}^{-1}) = 1.0\bar{1}\times 10^5\ \text{Pa}$$

The change in volume is the cross-sectional area times the linear displacement:

$$\Delta V = (50\,\text{cm}^2)\times(15\,\text{cm})\times\left(\frac{1\,\text{m}}{100\,\text{cm}}\right)^3 = 7.5\times10^{-4}\,\text{m}^3$$

so $\quad w = -(1.01\times10^5\,\text{Pa})\times(7.5\times10^{-4}\,\text{m}^3) = \boxed{-75\,\text{J}}$ as $1\,\text{Pa m}^3 = 1\,\text{J}$

E55.5(a) For all cases $\Delta U = 0$, since the internal energy of a perfect gas depends only on temperature. From the definition of enthalpy, $H = U + pV$, so $\Delta H = \Delta U + \Delta(pV) = \Delta U + \Delta(nRT)$ (perfect gas). Hence, $\Delta H = 0$ as well, at constant temperature for all processes in a perfect gas.

(a) $\boxed{\Delta U = \Delta H = 0}$

$$w = -nRT\ln\left(\frac{V_\text{f}}{V_\text{i}}\right)\ [55.12]$$

$$= -(1.00\,\text{mol})\times(8.314\,\text{J K}^{-1}\,\text{mol}^{-1})\times(293\,\text{K})\times\ln\left(\frac{30.0\,\text{dm}^3}{10.0\,\text{dm}^3}\right)$$

$$= -2.68\times10^3\,\text{J} = \boxed{-2.68\,\text{kJ}}$$

$$q = \Delta U - w\ [\text{First Law}] = 0 + 2.68\,\text{kJ} = \boxed{+2.68\,\text{kJ}}$$

(b) $\boxed{\Delta U = \Delta H = 0}$

$w = -p_\text{ex}\Delta V \qquad [55.8] \qquad \Delta V = (30.0 - 10.0)\,\text{dm}^3 = 20.0\,\text{dm}^3$

p_ex can be computed from the perfect gas law

$pV = nRT$

$$\text{so } p_\text{ex} = p_\text{f} = \frac{nRT}{V_\text{f}} = \frac{(1.00\,\text{mol})\times(0.08206\,\text{dm}^3\,\text{atm K}^{-1}\,\text{mol}^{-1})\times(293\,\text{K})}{30.0\,\text{dm}^3} = 0.801\,\text{atm}$$

$$w = -(0.801\,\text{atm})\times\left(\frac{1.013\times10^5\,\text{Pa}}{1\,\text{atm}}\right)\times(20.0\,\text{dm}^3)\times\left(\frac{1\,\text{m}^3}{10^3\,\text{dm}^3}\right)$$

$$= -1.62\times10^3\,\text{Pa m}^3 = -1.62\times10^3\,\text{J} = \boxed{-1.62\,\text{kJ}}$$

$$q = \Delta U - w = 0 + 1.62\,\text{kJ} = \boxed{+1.62\,\text{kJ}}$$

(c) $\boxed{\Delta U = \Delta H = 0}$

Free expansion is expansion against no force, so $\boxed{w = 0}$ and $q = \Delta U - w = 0 - 0 = \boxed{0}$

Comment: An isothermal free expansion of a perfect gas is also adiabatic.

E55.6(a) For a perfect gas at constant volume

$$\frac{p}{T} = \frac{nR}{V} = \text{constant}, \quad \text{hence,} \quad \frac{p_1}{T_1} = \frac{p_2}{T_2}$$

$$p_2 = \left(\frac{T_2}{T_1}\right)\times p_1 = \left(\frac{400\,\text{K}}{300\,\text{K}}\right)\times(1.00\,\text{atm}) = \boxed{1.33\,\text{atm}}$$

$$\Delta U = nC_{V,\mathrm{m}}\Delta T\ [2.16\mathrm{b}] = (n)\times(\tfrac{3}{2}R)\times(400\ \mathrm{K}-300\ \mathrm{K})$$
$$= (1.00\ \mathrm{mol})\times(\tfrac{3}{2})\times(8.314\ \mathrm{J\ K^{-1}\ mol^{-1}})\times(100\ \mathrm{K})$$
$$= 1.25\times10^{3}\ \mathrm{J} = \boxed{+1.25\ \mathrm{kJ}}$$

$\boxed{w=0}$ [constant volume] $\quad q = \Delta U - w$ [First Law] $= 1.25\ \mathrm{kJ} - 0 = \boxed{+1.25\ \mathrm{kJ}}$

Problems

P55.1 $\quad w = -\int_{V_1}^{V_2} p\,\mathrm{d}V \quad$ with $\quad p = \dfrac{nRT}{V-nb} - \dfrac{n^2a}{V^2}$ [van der Waals equation]

Therefore, $w = -nRT\int_{V_1}^{V_2}\dfrac{\mathrm{d}V}{V-nb} + n^2a\int_{V_1}^{V_2}\dfrac{\mathrm{d}V}{V^2} = \boxed{-nRT\ln\left(\dfrac{V_2-nb}{V_1-nb}\right) - n^2a\left(\dfrac{1}{V_2}-\dfrac{1}{V_1}\right)}$

This expression can be interpreted more readily if we assume $V \gg nb$, which is certainly valid at all but the highest pressures. Then using the first term of the Taylor series expansion,

$$\ln(1-x) = -x - \frac{x^2}{2}\ldots \qquad \text{for } |x| \ll 1$$

$$\ln(V-nb) = \ln V + \ln\left(1-\frac{nb}{V}\right) \approx \ln V - \frac{nb}{V}$$

and, after substitution

$$w \approx -nRT\ln\left(\frac{V_2}{V_1}\right) + n^2bRT\left(\frac{1}{V_2}-\frac{1}{V_1}\right) - n^2a\left(\frac{1}{V_2}-\frac{1}{V_1}\right)$$
$$\approx -nRT\ln\left(\frac{V_2}{V_1}\right) - n^2(a-bRT)\left(\frac{1}{V_2}-\frac{1}{V_1}\right)$$
$$\approx +w_0 - n^2(a-bRT)\left(\frac{1}{V_2}-\frac{1}{V_1}\right) = \text{Perfect gas value + van der Waals correction.}$$

w_0, the perfect gas value, is negative in expansion and positive in compression. Considering the correction term, in expansion $V_2 > V_1$, so $\left(\dfrac{1}{V_2}-\dfrac{1}{V_1}\right) < 0$. If attractive forces predominate, $a > bRT$ and the work done *by* the van der Waals gas is less in magnitude (less negative) than the perfect gas, the gas cannot easily expand. If repulsive forces predominate, $bRT > a$ and the work done *by* the van der Waals gas is greater in magnitude than the perfect gas, the gas easily expands. In the numerical calculations, consider a doubling of the initial volume.

(a) $w_0 = -nRT\ln\left(\dfrac{V_\mathrm{f}}{V_\mathrm{i}}\right) = (-1.0\ \mathrm{mol^{-1}})\times(8.314\ \mathrm{J\ K^{-1}\ mol^{-1}})\times(298\ \mathrm{K})\times\ln\left(\dfrac{2.0\ \mathrm{dm^3}}{1.0\ \mathrm{dm^3}}\right)$

$$w_0 = -1.7\bar{2}\times10^{3}\ \mathrm{J} = \boxed{-1.7\ \mathrm{kJ}}$$

(b) $w = w_0 - (1.0\,\text{mol})^2 \times [0 - (5.11\times10^{-2}\,\text{dm}^3\,\text{mol}^{-1}) \times (8.314\,\text{J K}^{-1}\text{mol}^{-1}) \times (298\,\text{K})]$

$$\times\left(\frac{1}{2.0\,\text{dm}^3} - \frac{1}{1.0\,\text{dm}^3}\right) = (-1.7\overline{2}\times10^3\,\text{J}) - (63\,\text{J}) = -1.7\overline{8}\times10^3\,\text{J} = \boxed{-1.8\,\text{kJ}}$$

(c) $w = w_0 - (1.0\,\text{mol})^2 \times (4.2\,\text{dm}^6\,\text{atm mol}^{-2}) \times \left(\frac{1}{2.0\,\text{dm}^3} - \frac{1}{1.0\,\text{dm}^3}\right)$

$$w = w_0 + 2.1\,\text{dm}^3\,\text{atm}$$

$$= (-1.7\overline{2}\times10^3\,\text{J}) + (2.1\,\text{dm}^3\,\text{atm}) \times \left(\frac{1\,\text{m}}{10\,\text{dm}}\right)^3 \times \left(\frac{1.01\times10^5\,\text{Pa}}{1\,\text{atm}}\right)$$

$$= (-1.7\overline{2}\times10^3\,\text{J}) + (0.21\times10^3\,\text{J}) = \boxed{-1.5\,\text{kJ}}$$

Schematically, the indicator diagrams for the cases (a), (b), and (c) would appear as in Fig. 55.1. For case (b) the pressure is always greater than the perfect gas pressure and for case (c) always less. Therefore,

$$\int_{V_1}^{V_2} p\,\mathrm{d}V(\text{c}) < \int_{V_1}^{V_2} p\,\mathrm{d}V(\text{a}) < \int_{V_1}^{V_2} p\,\mathrm{d}V(\text{b})$$

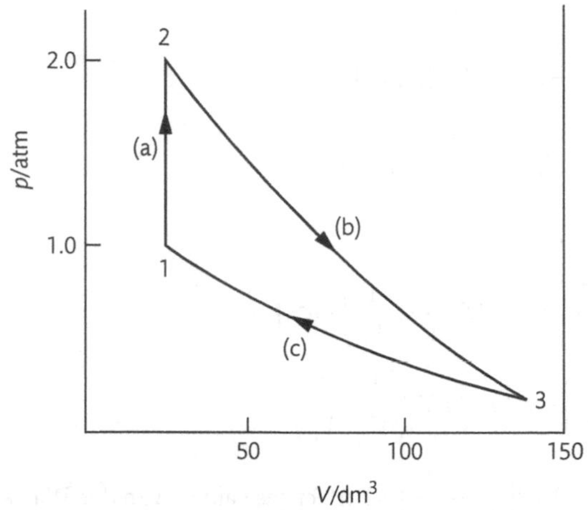

Fig 55.1

Topic 56 Enthalpy

Discussion questions

D56.1 An inversion temperature is the temperature at which the Joule-Thomson coefficient, μ, changes sign from negative to positive or vice-versa. For a perfect gas μ is always zero, thus it cannot have an inversion temperature. As explained in detail in Section 56.4, the existence of the Joule-Thomson effect depends upon intermolecular attractions and repulsions. A perfect gas has by definition no intermolecular attractions and repulsions, so it cannot exhibit the Joule-Thomson effect.

D56.3 The cooling of a real gas as it expands through a narrow throttle, without the gain or loss of heat to or from the environment, is called the Joule-Thomson effect. Attractive forces between molecules are reduced during the expansion with the requisite energy for separating the molecules being supplied by a reduction in molecular kinetic energy. The kinetic energy reduction manifests itself as a temperature reduction, since in the kinetic theory of gases, both real and ideal, temperature and kinetic energy are related, lesser kinetic energy corresponds to lower temperature.

After several successive expansions, the gas becomes so cold that it condenses to a liquid.

Exercises

E56.1(a) $C_p = \dfrac{q_p}{\Delta T}\,[56.4\text{c}] = \dfrac{229\,\text{J}}{2.55\,\text{K}} = 89.8\,\text{J K}^{-1}$

so $C_{\text{p,m}} = (89.8\,\text{J K}^{-1})/(3.0\,\text{mol}) = \boxed{30\,\text{J K}^{-1}\,\text{mol}^{-1}}$

For a perfect gas $C_{\text{p,m}} - C_{\text{v,m}} = R\,[56.6]$

$$C_{V,\text{m}} = C_{p,\text{m}} - R = (30-8.3)\,\text{J K}^{-1}\,\text{mol}^{-1} = \boxed{22\,\text{J K}^{-1}\,\text{mol}^{-1}}$$

E56.2(a) (a) $q = \Delta H$, since pressure is constant

$$\Delta H = \int_{T_i}^{T_f} \text{d}H, \quad \text{d}H = nC_{p,\text{m}}\text{d}T$$

$$\text{d}(H/\text{J}) = \{20.17 + 0.3665(T/\text{K})\}\text{d}(T/\text{K})$$

$$\Delta(H/\mathrm{J}) = \int_{T_i}^{T_f} (H/\mathrm{J}) = \int_{298}^{373} \{20.17 + 0.3665(T/\mathrm{K})\}\mathrm{d}(T/\mathrm{K})$$

$$= (20.17)\times(373-298) + \left(\frac{0.3665}{2}\right)\times\left(\frac{T}{\mathrm{K}}\right)^2\Bigg|_{298}^{373}$$

$$= (1.51\bar{3}\times10^3) + (9.22\bar{2}\times10^3) = 1.07\times10^4$$

$$q = \Delta H = \boxed{1.07\times10^4\,\mathrm{J}} = \boxed{+10.7\,\mathrm{kJ}} \text{ for 1.00 mol}$$

$$w = -p_{\mathrm{ex}}\Delta V \ [55.8] \quad \text{where} \quad p_{\mathrm{ex}} = p$$

$$w = -p\Delta V = -\Delta(pV)\ [\text{constant pressure}]$$

$$= -\Delta(nRT)\,[\text{perfect gas}] = -nR\Delta T$$
$$= (-1.00\,\mathrm{mol})\times(8.314\,\mathrm{J\,K^{-1}\,mol^{-1}})\times(373\,\mathrm{K}-298\,\mathrm{K})$$
$$= \boxed{-0.624\times10^3\,\mathrm{J}} = \boxed{-0.624\,\mathrm{kJ}}$$

$$\Delta U = q + w = (10.7\,\mathrm{kJ}) - (0.624\,\mathrm{kJ}) = \boxed{+10.1\,\mathrm{kJ}}$$

(b) The energy and enthalpy of a perfect gas depend on temperature alone; hence it does not matter whether the temperature change is brought about at constant volume or constant pressure; ΔH and ΔU are the same.

$$\Delta H = \boxed{+10.7\,\mathrm{kJ}}, \quad \Delta U = \boxed{+10.1\,\mathrm{kJ}}$$

Under constant volume, $w = \boxed{0}$

$$q = \Delta U - w = \boxed{+10.1\,\mathrm{kJ}}$$

E56.3(a) $q_p = C_p\Delta T\,[56.4c] = nC_{p,\mathrm{m}}\Delta T = (3.0\,\mathrm{mol})\times(29.4\,\mathrm{J\,K^{-1}\,mol^{-1}})\times(25\,\mathrm{K}) = \boxed{+2.2\,\mathrm{kJ}}$

$$\Delta H = q_p\,[56.2b] = \boxed{+2.2\,\mathrm{kJ}}$$

$$\Delta U = \Delta H - \Delta(pV)\,[\text{From}\,H \equiv U + pV] = \Delta H - \Delta(nRT)\,[\text{perfect gas}] = \Delta H - nR\Delta T$$
$$= (2.2\,\mathrm{kJ}) - (3.0\,\mathrm{mol})\times(8.314\,\mathrm{J\,K^{-1}\,mol^{-1}})\times(25\,\mathrm{K}) = (2.2\,\mathrm{kJ}) - (0.62\,\mathrm{kJ}) = \boxed{+1.6\,\mathrm{kJ}}$$

E56.4(a) The isothermal Joule-Thomson coefficient is

$$\mu_\mathrm{T} = \left(\frac{\partial H_\mathrm{m}}{\partial p}\right)_T = -\mu C_{p,\mathrm{m}}\ [56.10] = (-0.25\,\mathrm{K\,atm^{-1}})\times(29\,\mathrm{J\,K^{-1}\,mol^{-1}})$$
$$= \boxed{-7.2\,\mathrm{J\,atm^{-1}\,mol^{-1}}}$$

$$\mathrm{d}H = n\left(\frac{\partial H_\mathrm{m}}{\partial p}\right)_T \mathrm{d}p = -n\mu C_{p,\mathrm{m}}\,\mathrm{d}p$$

$$\Delta H = \int_{p_1}^{p_2}(-n\mu C_{p,\mathrm{m}})\mathrm{d}p = -n\mu C_{p,\mathrm{m}}(p_2 - p_1) \qquad [\mu \text{ and } C_p \text{ are constant}]$$

$$\Delta H = -(10.0\,\mathrm{mol})\times(+7.2\,\mathrm{J\,atm^{-1}\,mol^{-1}})\times(-85\,\mathrm{atm}) = +6.1\,\mathrm{kJ}$$

so $q(\text{supplied}) = +\Delta H = \boxed{+6.1\,\mathrm{kJ}}$

Problems

P56.1 The change in the molar enthalpy of formation of SO_2 as the temperature increases from 300K to 1500 K is calculated as follows:

$$\Delta(\Delta_\mathrm{f}H^\ominus) = \int_{300}^{1500} C^\ominus_{p,\mathrm{m}}(T)\mathrm{d}T$$

$C^\ominus_{p,\mathrm{m}}(T)$ can be obtained by fitting the heat capacity data to an equation of the form of eqn 56.5. When this is done using mathematical software we find

$$C^\ominus_{p,\mathrm{m}}(T)/(\mathrm{J\,K^{-1}\,mol^{-1}}) = 48.01 + 6.535\times10^{-3}T - 9.294\times10^{5}T^{-2}$$

Inserting this expression and performing the integration we obtain

$$\Delta(\Delta_\mathrm{f}H^\ominus) = \boxed{62.2\,\mathrm{kJ\,mol^{-1}}}$$

P56.3 Since the volume is fixed, $w = 0$

Since $\Delta U = q$ at constant volume, $\boxed{\Delta U = +2.35\,\mathrm{kJ}}$

$$\Delta H = \Delta U + \Delta(pV) = \Delta U + V\Delta p \ [\Delta V = 0]$$

From the van der Waals equation

$$p = \frac{RT}{V_\mathrm{m} - b} - \frac{a}{V_\mathrm{m}^2} \quad \text{so} \quad \Delta p = \frac{R\Delta T}{V_\mathrm{m} - b} \quad [\Delta V_m = 0 \text{ at constant volume}]$$

Therefore, $\Delta H = \Delta U + \dfrac{RV\Delta T}{V_\mathrm{m} - b}$

From the data,

$$V_m = \frac{15.0\,\mathrm{dm^3}}{2.0\,\mathrm{mol}} = 7.5\,\mathrm{dm^3\,mol^{-1}}, \Delta T = (341 - 300)\,\mathrm{K} = 41\,\mathrm{K}$$

$$V_\mathrm{m} - b = (7.5 - 4.3\times10^{-2})\,\mathrm{dm^3\,mol^{-1}} = 7.4\bar{6}\,\mathrm{dm^3\,mol^{-1}}$$

$$\frac{RV\Delta T}{V_\mathrm{m} - b} = \frac{(8.314\,\mathrm{J\,K^{-1}\,mol^{-1}})\times(15.0\,\mathrm{dm^3})\times(41\,\mathrm{K})}{7.4\bar{6}\,\mathrm{dm^3\,mol^{-1}}} = 0.68\,\mathrm{kJ}$$

Therefore, $\Delta H = (2.35\,\mathrm{kJ}) + (0.68\,\mathrm{kJ}) = \boxed{+3.03\,\mathrm{kJ}}$

P56.5 $\mu = \left(\frac{\partial T}{\partial p}\right)_H = -\frac{1}{C_p}\left(\frac{\partial H}{\partial p}\right)_T$ [56.8 and 56.10]

$$\mu = \frac{1}{C_p}\left\{T\left(\frac{\partial V}{\partial T}\right)_p - V\right\} \text{ [See eqns 56.9 \& 56.10]}$$

But $\quad V = \frac{nRT}{p} + nb \quad$ or $\quad \left(\frac{\partial V}{\partial T}\right)_p = \frac{nR}{p}$

Therefore,

$$\mu = \frac{1}{C_p}\left\{\frac{nRT}{p} - V\right\} = \frac{1}{C_p}\left\{\frac{nRT}{p} - \frac{nRT}{p} - nb\right\} = \frac{-nb}{C_p}$$

Since $b > 0$ and $C_p > 0$, we conclude that for this gas $\mu < 0$ or $\left(\frac{\partial T}{\partial p}\right)_H < 0$. This says that when the pressure drops during a Joule–Thomson expansion the temperature must increase.

P56.7 (a) The Joule–Thomson coefficient is related to the given data by

$$\mu = -(1/C_p)(\partial H/\partial p)_T = -(-3.29\times10^3 \text{ J mol}^{-1}\text{ MPa}^{-1})/(110.0 \text{ J K}^{-1}\text{ mol}^{-1})$$
$$= \boxed{29.9 \text{ K MPa}^{-1}}$$

(b) The Joule–Thomson coefficient is defined as

$$\mu = (\partial T/\partial p)_H \approx (\Delta T/\Delta p)_H$$

Assuming that the expansion is a Joule–Thomson constant-enthalpy process, we have

$$\Delta T = \mu\Delta p = (29.9 \text{ K MPa}^{-1})\times[(0.5-1.5)\times10^{-1}\text{ MPa}] = \boxed{-2.99 \text{ K}}$$

Topic 57 Thermochemistry

Discussion questions

D57.1 When a system is subjected to constant pressure conditions, and only expansion work can occur, the energy supplied as heat is the change in enthalpy of the system. Thus enthalpy changes in the system can be determined by measuring the amount of heat supplied. The calorimeters used for measuring these heats and hence the enthalpy changes of the systems being studied are constant pressure calorimeters. A simple example is a thermally insulated vessel (a coffee cup calorimeter) open to the atmosphere: the heat released in the reaction is determined by measuring the change in temperature of the contents. For a combustion reaction a constant-pressure flame calorimeter may be used as illustrated in Fig. 57.1 of the text where a certain amount of substance burns in a supply of oxygen and the rise in temperature is monitored. The most sophisticated way of measuring enthalpy changes is to use a differential scanning calorimeter which is described in detail in Section 57.1(b).

Exercises

E57.1(a) At constant pressure

$$q = \Delta H = n\Delta_{\text{vap}}H^{\ominus} = (0.75\ \text{mol})\times(30.0\ \text{kJ mol}^{-1}) = \boxed{22.5\ \text{kJ}}$$

and $w = -p\Delta V \approx -pV_{\text{vapor}} = -nRT = -(0.75\ \text{mol})\times(8.3145\ \text{J K}^{-1}\ \text{mol}^{-1})\times(250\ \text{K})$

$$w = -1.6\times10^{3}\ \text{J} = \boxed{-1.6\ \text{kJ}}$$

$$\Delta U = w + q = -1.6 + 22.5\ \text{kJ} = \boxed{20.9\ \text{kJ}}$$

Comment. Because the vapor is treated as a perfect gas, the specific value of the external pressure provided in the statement of the exercise does not affect the numerical value of the answer.

E57.2(a) The reaction is

$$C_6H_5C_2H_5(l) + \tfrac{21}{2}O_2(g) \rightarrow 8\ CO_2(g) + 5\ H_2O(l)$$

$$\begin{aligned}\Delta_c H^\ominus &= 8\Delta_f H^\ominus(CO_2,g)+5\Delta_f H^\ominus(H_2O,l)-\Delta_f H^\ominus(C_6H_5C_2H_5,l)\\ &=[(8)\times(-393.51)+(5)\times(-285.83)-(-12.5)]\ \text{kJ mol}^{-1}\\ &=\boxed{-4\,564.7\ \text{kJ mol}^{-1}}\end{aligned}$$

E57.3(a) First $\Delta_f H[(CH_2)_3,g]$ is calculated, and then that result is used to calculate $\Delta_r H$ for the isomerization

$$(CH_2)_3(g)+\tfrac{9}{2}O_2(g)\rightarrow 3\,CO_2(g)+3\,H_2O(l)\quad \Delta_c H=-2\,091\ \text{kJ mol}^{-1}$$

$$\begin{aligned}\Delta_f H[(CH_2)_3,g] &= -\Delta_c H+3\Delta_f H(CO_2,g)+3\Delta_f H(H_2O,g)\\ &=[+2\,091+(3)\times(-393.51)+(3)\times(-285.83)]\ \text{kJ mol}^{-1}\\ &=\boxed{+53\ \text{kJ mol}^{-1}}\end{aligned}$$

$$(CH_2)_3(g)\rightarrow C_3H_6(g)\quad \Delta_r H=?$$

$$\begin{aligned}\Delta_r H &= \Delta_f H(C_3H_6,g)-\Delta_f H[(CH_2)_3,g]\\ &=(20.42-53)\ \text{kJ mol}^{-1}=\boxed{-33\ \text{kJ mol}^{-1}}\end{aligned}$$

E57.4(a) Because $\Delta_f H^\ominus(H^+,aq)=0$, the whole of $\Delta_f H^\ominus(HCl,aq)$ is ascribed to $\Delta_f H^\ominus(Cl^-,aq)$. Therefore, $\Delta_f H^\ominus(Cl^-,aq)=\boxed{-167\ \text{kJ/mol}^{-1}}$

E57.5(a) For naphthalene the reaction is $C_{10}H_8(s)+12\,O_2(g)\rightarrow 10\,CO_2(g)+4\,H_2O(l)$
A bomb calorimeter gives $q_V=n\Delta_c U^\ominus$ rather than $q_p=n\Delta_c H^\ominus$; thus we need

$$\Delta_c U^\ominus=\Delta_c H^\ominus-\Delta n_g RT\ [57.3],\quad \Delta n_g=-2\ \text{mol}$$

$$\Delta_c H^\ominus=-5\,157\ \text{kJ mol}^{-1}\ [\text{Table 57.3}]$$

$$\begin{aligned}\Delta_c U^\ominus &= (-5\,157\ \text{kJ mol}^{-1})-(-2)\times(8.3\times10^{-3}\ \text{kJ K}^{-1}\ \text{mol}^{-1})\times(298\ \text{K})\\ &=\boxed{-5\,152\ \text{kJ mol}^{-1}}\end{aligned}$$

$$|q|=|q_V|=|n\Delta_c U^\ominus|=\left(\frac{120\times10^{-3}\ \text{g}}{128.18\ \text{g mol}^{-1}}\right)\times(5\,152\ \text{kJ mol}^{-1})=4.82\bar{3}\ \text{kJ}$$

$$C=\frac{|q|}{\Delta T}=\frac{4.82\bar{3}\ \text{kJ}}{3.05\ \text{K}}=\boxed{1.58\ \text{kJ K}^{-1}}$$

When phenol is used the reaction is

$$C_6H_5OH(s)+\tfrac{15}{2}O_2(g)\rightarrow 6\,CO_2(g)+3\,H_2O(l)$$

$$\Delta_c H^\ominus=-3\,054\ \text{kJ mol}^{-1}\ [\text{Table 57.3}]$$

$$\Delta_c U^\ominus = \Delta_c H^\ominus - \Delta n_g RT, \Delta n_g = -\frac{3}{2}$$
$$= (-3\,054 \text{ kJ mol}^{-1}) + (\tfrac{3}{2}) \times (8.314 \times 10^{-3} \text{ kJ K}^{-1} \text{ mol}^{-1}) \times (298 \text{ K})$$
$$= -3\,050 \text{ kJ mol}^{-1}$$

$$|q| = \frac{150 \times 10^{-3} \text{ g}}{94.12 \text{ g mol}^{-1}} \times (3\,050 \text{ kJ mol}^{-1}) = 4.86 \text{ kJ}$$

$$\Delta T = \frac{|q|}{C} = \frac{4.86 \text{ kJ}}{1.58 \text{ kJ K}^{-1}} = \boxed{+3.08 \text{ K}}$$

Comment. In this case $\Delta_c U^\ominus$ and $\Delta_c H^\ominus$ differed by about 0.1 per cent. Thus, to within 3 significant figures, it would not have mattered if we had used $\Delta_c H^\ominus$ instead of $\Delta_c U^\ominus$, but for very precise work it would.

E57.6(a) (a) reaction(3) = (−2) ′ reaction(1) + reaction(2) and $\Delta n_g = -1$

The enthalpies of reactions are combined in the same manner as the equations (Hess's law).

$$\Delta_r H^\ominus(3) = (-2) \times \Delta_r H^\ominus(1) + \Delta_r H^\ominus(2)$$
$$= [(-2) \times (-184.62) + (-483.64)] \text{ kJ mol}^{-1}$$
$$= \boxed{-114.40 \text{ kJ mol}^{-1}}$$

$$\Delta_r U^\ominus = \Delta_r H^\ominus - \Delta n_g RT[57.3] = (-114.40 \text{ kJ mol}^{-1}) - (-1) \times (2.48 \text{ kJ mol}^{-1})$$
$$= \boxed{-111.92 \text{ kJ mol}^{-1}}$$

(b) $\Delta_f H^\ominus$ refers to the formation of one mole of the compound, hence

$$\Delta_f H^\ominus(\text{J}) = \frac{\Delta_r H^\ominus(\text{J})}{\nu_\text{J}}$$

$$\Delta_f H^\ominus(\text{HCl,g}) = \frac{-184.62}{2} \text{ kJ mol}^{-1} = \boxed{-92.31 \text{ kJ mol}^{-1}}$$

$$\Delta_f H^\ominus(\text{H}_2\text{O,g}) = \frac{-483.64}{2} \text{ kJ mol}^{-1} = \boxed{-241.82 \text{ kJ mol}^{-1}}$$

E57.7(a) $\Delta_r H^\ominus = \Delta_r U^\ominus + \Delta n_g RT$ [57.3]; $\Delta n_g = +2$

$$= (-1\,373 \text{ kJ mol}^{-1}) + 2 \times (2.48 \text{ kJ mol}^{-1}) = \boxed{-1\,368 \text{ kJ mol}^{-1}}$$

Comment. As a number of these exercises have shown, the use of $\Delta_r H$ as an approximation for $\Delta_r U$ is often valid.

E57.8(a) (a) $\Delta_r H^\ominus = \sum_\text{Products} \nu \Delta_f H^\ominus - \sum_\text{Reactants} \nu \Delta_f H^\ominus$ [57.8]

$$\Delta_r H^\ominus(298\text{K}) = [(-110.53) - (-241.82)] \text{kJ mol}^{-1} = \boxed{+131.29 \text{kJ mol}^{-1}}$$

$$\Delta_r U^{\ominus}(298\text{K}) = \Delta_r H^{\ominus}(298\text{K}) - \Delta n_g RT \; [57.3]$$
$$= (131.29\,\text{kJ mol}^{-1}) - (1)\times(2.48\,\text{kJ mol}^{-1}) = \boxed{+128.81\,\text{kJ mol}^{-1}}$$

(b) $\Delta_r H^{\ominus}(478K) = \Delta_r H^{\ominus}(298K) + (T_2 - T_1)\Delta_r C_p^{\ominus}$ [*Brief Illustration* 57.5]

$$\Delta_r C_p^{\ominus} = C_{p,m}^{\ominus}(\text{CO,g}) + C_{p,m}^{\ominus}(\text{H}_2,\text{g}) - C_{p,m}^{\ominus}(\text{C,gr}) - C_{p,m}^{\ominus}(\text{H}_2\text{O,g})$$
$$= (29.14 + 28.82 - 8.53 - 33.58)\times 10^{-3}\ \text{kJ K}^{-1}\ \text{mol}^{-1}$$
$$= 15.85\times 10^{-3}\ \text{kJ K}^{-1}\ \text{mol}^{-1}$$

$$\Delta_r H^{\ominus}(478\text{K}) = (131.29\ \text{kJ mol}^{-1}) + (15.85\times 10^{-3}\ \text{kJ K}^{-1}\ \text{mol}^{-1})\times(180\ \text{K})$$
$$= (131.29 + 2.85)\ \text{kJ mol}^{-1} = \boxed{+134.14\ \text{kJ mol}^{-1}}$$

$$\Delta_r U^{\ominus}(478\ \text{K}) = \Delta_r H^{\ominus}(478\ \text{K}) - (1)\times(8.31\times 10^{-3}\ \text{kJ K}^{-1}\ \text{mol}^{-1})\times(478\ \text{K})$$
$$= (134.14 - 3.97)\ \text{kJ mol}^{-1} = \boxed{+130.17\ \text{kJ mol}^{-1}}$$

Comment. The differences in both $\Delta_r H^{\ominus}$ and $\Delta_r U^{\ominus}$ between the two temperatures are small and justify the use of the approximation that $\Delta_r C_p^{\ominus}$ is a constant.

E57.9(a) For the reaction $CH_4(g) + 2\,O_2(g) \rightarrow CO_2(g) + 2\,H_2O(g)$

$$\Delta_r H^{\ominus} = \Delta_f H^{\ominus}(\text{CO}_2,\text{g}) + 2\times\Delta_f H^{\ominus}(\text{H}_2\text{O,g}) - \Delta_f H^{\ominus}(\text{CH}_4,\text{g})$$

In order to calculate the enthalpy of reaction at 500 K we first calculate its value at 298 K using data in Tables 57.3 and 57.4.

$$\Delta_r H^{\ominus}(298\ \text{K}) = -393.51\ \text{kJ mol}^{-1} + 2\times(-241.82\ \text{kJ mol}^{-1}) - (-74.81\ \text{kJ mol}^{-1})$$
$$= -802.34\ \text{kJ mol}^{-1}$$

Then using data on the heat capacities of all the reacting substances we can calculate the change in enthalpy, ΔH, of each substance as the temperature increases from 298 K to 500 K. The enthalpy of reaction at 500 K could be obtained by adding all these enthalpy changes to the enthalpy of reaction at 298 K. This process is shown below:

$$\Delta_r H^{\ominus}(500\ \text{K}) = \Delta_r H^{\ominus}(298\ \text{K}) + \Delta H(\text{CO}_2,\text{g}) + 2\times\Delta H(\text{H}_2\text{O,g})$$
$$-\Delta H(\text{CH}_4,\text{g}) - 2\times\Delta H(\text{O}_2,\text{g})$$

However, we need not calculate each individual ΔH values. It is more efficient to proceed as follows using eqn 57.12:

$$\Delta_r H^{\ominus}(500\ \text{K}) = \Delta_r H^{\ominus}(298\ \text{K}) + \int_{298\text{K}}^{500\text{K}} \Delta_r C_p^{\ominus}\,\mathrm{d}T \; [57.12]$$

We will express the temperature dependence of the heat capacities in the form of the equation given in Problem 57.7 as data for the heat capacities of the substances involved in this reaction are available only in that form. They are not available for all the substances in the form of the equation of Table 56.1.

If $C_{p,\text{m}}^{\ominus} = \alpha + \beta T + \gamma T^2$ then

$$\Delta_r C_p^{\ominus} = \Delta\alpha + \Delta\beta T + \Delta\gamma T^2 \quad \text{with} \quad \Delta\alpha = \sum_J \nu_J \alpha_J \ \text{etc.}$$

Using the data given in Problem 57.7 we calculate

$$\Delta\alpha = (26.86+2\times30.36-14.16-2\times25.72)\ \mathrm{J\,K^{-1}\,mol^{-1}} = 21.98\ \mathrm{J\,K^{-1}\,mol^{-1}}$$

$$\Delta\beta = (6.97+2\times9.61-75.50-2\times12.98)\ \mathrm{mJ\,K^{-2}\,mol^{-1}} = -75.27\ \mathrm{mJ\,K^{-2}\,mol^{-1}}$$

$$\Delta\gamma = [-0.820+2\times1.184-(-17.99)-2\times(-3.862)]\ \mu\mathrm{J\,K^{-3}\,mol^{-1}} = 27.262\ \mu\mathrm{J\,K^{-3}\,mol^{-1}}$$

Integrating eqn 57.12 between T_1(298 K) and T_2(500 K) we obtain

$$\Delta_r H^\ominus(500\ \mathrm{K}) = \Delta_r H^\ominus(298\ \mathrm{K}) + \Delta\alpha(500\ \mathrm{K}-298\ \mathrm{K}) + \tfrac{1}{2}\Delta\beta[(500\ \mathrm{K})^2-(298\ \mathrm{K})^2]$$
$$+\tfrac{1}{3}\Delta\gamma[(500\ \mathrm{K})^3-(298\ \mathrm{K})^3] = \boxed{-803.07\ \mathrm{kJ\,mol^{-1}}}$$

E57.10(a) Since enthalpy is a state function, $\Delta_r H$ for the process (see Fig. 57.1)

$$\mathrm{Mg^{2+}(g)+2\,Cl(g)+2\,e^- \rightarrow MgCl_2(aq)}$$

is independent of path; therefore the change in enthalpy for the path on the left is equal to the change in enthalpy for the path on the right. All numerical values are in kJ mol⁻¹.

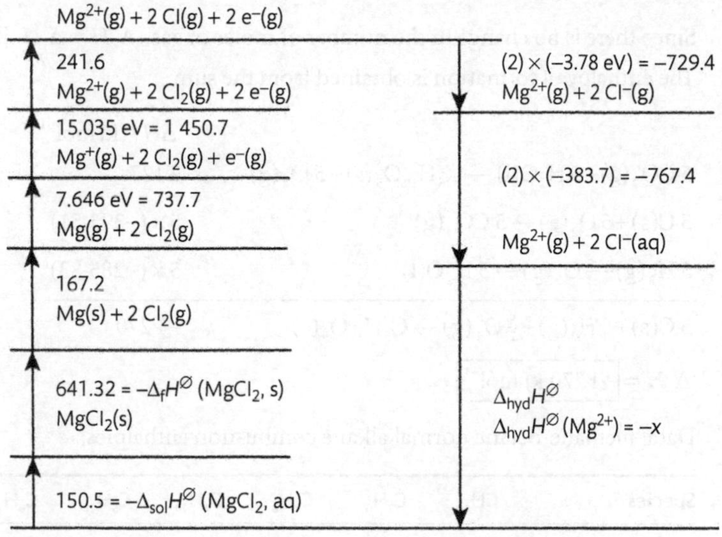

Fig 57.1

The cycle is the distance traversed upward along the left plus the distance traversed downward on the right. The sum of these distances is zero. Note that $E_{ea} = -\Delta_{eg}H$. Therefore, following the cycle up the left and down the right and using kJ units,

$$-(-150.5)-(-641.32)+(167.2)+(241.6)+(737.7+1\,450.7)$$
$$+2\times(-364.\bar{7})+2\times(-383.7)+\Delta_{\mathrm{hyd}}H(\mathrm{Mg^{2+}})=0$$

which yields $\Delta_{\mathrm{hyd}}H(\mathrm{Mg^{2+}}) = \boxed{-1\,892\ \mathrm{kJ\,mol^{-1}}}$

Problems

P57.1 The calorimeter is a constant-volume instrument as described in the text (Section 57.1); therefore $\Delta U = q_V$.

The calorimeter constant is determined from the data for the combustion of benzoic acid

$$\Delta U = \left(\frac{0.825\ \text{g}}{122.12\ \text{g mol}^{-1}}\right) \times (-3251\ \text{kJ mol}^{-1}) = -21.9\bar{6}\ \text{kJ}$$

Since $\Delta T = 1.940\ \text{K}$, $C = \dfrac{|q|}{\Delta T} = \dfrac{21.9\bar{6}\ \text{kJ}}{1.940\ \text{K}} = 11.3\bar{2}\ \text{kJ K}^{-1}$

For D-ribose, $\Delta U = -C\Delta T = -(11.3\bar{2}\ \text{kJ K}^{-1}) \times (0.910\ \text{K})$

Therefore,

$$\Delta_r U = \frac{\Delta U}{n} = -(11.3\bar{2}\ \text{kJ K}^{-1}) \times (0.910\ \text{K}) \times \left(\frac{150.13\ \text{g mol}^{-1}}{0.727\ \text{g}}\right) = -212\bar{7}\ \text{kJ mol}^{-1}$$

The combustion reaction for D-ribose is

$$C_5H_{10}O_5(s) + 5\,O_2(g) \rightarrow 5\,CO_2(g) + 5\,H_2O(l)$$

Since there is no change in the number of moles of gas, $\Delta_r H = \Delta_r U$ [57.3]

The enthalpy of formation is obtained from the sum

	$\Delta H/(\text{kJ mol}^{-1})$
$5\,CO_2(g) + 5\,H_2O(l) \rightarrow C_5H_{10}O_5(s) + 5\,O_2(g)$	2 127
$5\,C(s) + 5\,O_2(g) \rightarrow 5\,CO_2(g)$	$5 \times (-393.51)$
$5\,H_2(g) + \frac{5}{2}O_2(g) \rightarrow 5\,H_2O(l)$	$5 \times (-285.83)$
$5\,C(s) + 5\,H_2(g) + \frac{5}{2}O_2(g) \rightarrow C_5H_{10}O_5(s)$	−1 270

$\Delta_f H = \boxed{-1\,270\ \text{kJ mol}^{-1}}$

P57.3 Data: methane-octane normal alkane combustion enthalpies:

Species	CH_4	C_2H_6	C_3H_8	C_4H_{10}	C_5H_{12}	C_6H_{14}	C_8H_{18}
$\Delta_c H/(\text{kJ mol}^{-1})$	−890	−1 560	−2 220	−2 878	−3 537	−4 163	−5 471
$M/(\text{g mol}^{-1})$	16.04	30.07	44.10	58.13	72.15	86.18	114.23

Suppose $\Delta_c H = kM^n$. There are two methods by which a regression analysis can be used to determine the values of k and n. If you have a software package that can perform a "power fit" of the type $Y = aX^b$, the analysis is direct using $Y = \Delta_c H$ and $X = M$. Then, $k = a$ and $n = b$. Alternatively, taking the logarithm yields another equation—one of linear form

$$\ln|\Delta_c H| = \ln|k| + n \ln M \quad \text{where} \quad k < 0$$

This equation suggests a linear regression fit of $\ln(\Delta_c H)$ against $\ln M$ (Fig. 57.2). The intercept is $\ln k$ and the slope is n. Linear regression fit

$\ln|k| = 4.2112$, standard deviation = 0.0480 $k = -e^{4.2112} = \boxed{-67.44}$

$\boxed{n = 0.9253}$, standard deviation = 0.0121

$R = 1.000$

This is a good regression fit; essentially all of the variation is explained by the regression.

For decane the experimental value of $\Delta_c H$ equals $-6\,772.5$ kJ mol^{-1} (*CRC Handbook of Chemistry and Physics*, 81st Edition). The predicted value is

$$\Delta_c H = kM^n = -67.44(142.28)^{(0.9253)}\ \text{kJ mol}^{-1} = \boxed{-6\,625.5\ \text{kJ mol}^{-1}}$$

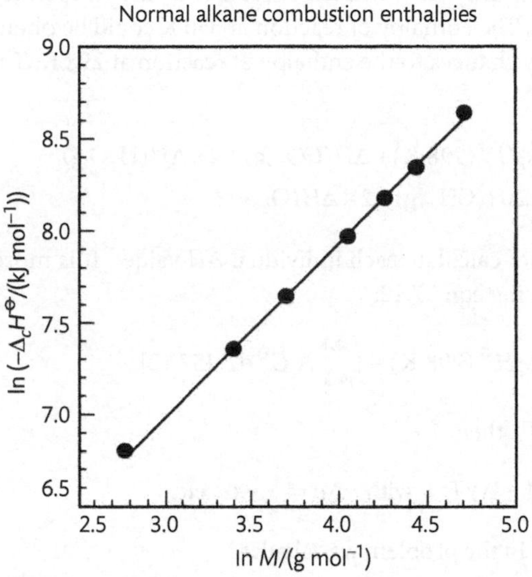

Fig 57.2

$$\text{Per cent error of prediction} = \left|\frac{-6\,772.5-(-6\,625.5)}{-6\,625.5}\right|\times 100$$

$$\text{Per cent error of prediction} = \boxed{2.17\ \text{per cent}}$$

P57.5 We must relate the formation of $DyCl_3$

$$Dy(s) + 1.5\,Cl_2(g) \rightarrow DyCl_3(s)$$

to the three reactions for which for which we have information. This reaction can be seen as a sequence of reaction (2), three times reaction (3), and the reverse of reaction (1), so

$$\Delta_f H^{\ominus}(DyCl_3,s) = \Delta_r H^{\ominus}(2) + 3\Delta_r H^{\ominus}(3) - \Delta_r H^{\ominus}(1),$$

$$\Delta_f H^{\ominus}(DyCl_3,s) = [-699.43 + 3(-158.31) - (-180.06)]\ \text{kJ mol}^{-1}$$

$$= \boxed{-994.30\ \text{kJ mol}^{-1}}$$

P57.7 This problem is essentially the same as Exercise 57.9(a). Here however we are asked to estimate the enthalpy of combustion at 350 K rather than 500 K.

For the reaction $CH_4(g)+2\,O_2(g)\rightarrow CO_2(g)+2\,H_2O(g)$

$$\Delta_r H^\ominus = \Delta_f H^\ominus(CO_2,g)+2\times\Delta_f H^\ominus(H_2O,g)-\Delta_f H^\ominus(CH_4,g)$$

In order to calculate the enthalpy of reaction at 500 K we first calculate its value at 298 K using data in Tables 57.3 and 57.4.

$$\Delta_r H^\ominus(298\text{ K}) = -393.51\text{ kJ mol}^{-1}+2\times(-241.82\text{ kJ mol}^{-1})-(-74.81\text{ kJ mol}^{-1})$$
$$= -802.34\text{ kJ mol}^{-1}$$

Then using data on the heat capacities of all the reacting substances we can calculate the change in enthalpy, ΔH, of each substance as the temperature increases from 298 K to 350 K. The enthalpy of reaction at 350 K could be obtained by adding all these enthalpy changes to the enthalpy of reaction at 298 K. This process is shown below:

$$\Delta_r H^\ominus(350\text{ K}) = \Delta_r H^\ominus(298\text{ K})+\Delta H(CO_2,g)+2\times\Delta H(H_2O,g)$$
$$-\Delta H(CH_4,g)-2\times\Delta H(O_2,g)$$

However, we need not calculate each individual ΔH values. It is more efficient to proceed as follows using eqn 57.12:

$$\Delta_r H^\ominus(350\text{ K}) = \Delta_r H^\ominus(298\text{ K})+\int_{298\text{ K}}^{500\text{ K}}\Delta_r C_p^\ominus \, dT \quad [57.12]$$

If $C_{p,m}^\ominus = \alpha+\beta T+\gamma T^2$ then

$$\Delta_r C_p^\ominus = \Delta\alpha+\Delta\beta T+\Delta\gamma T^2 \quad \text{with} \quad \Delta\alpha = \sum_J \nu_J\alpha_J \text{ etc.}$$

Using the data given in the problem we calculate

$$\Delta\alpha = (26.86+2\times30.36-14.16-2\times25.72)\text{ J K}^{-1}\text{ mol}^{-1} = 21.98\text{ J K}^{-1}\text{ mol}^{-1}$$

$$\Delta\beta = (6.97+2\times9.61-75.50-2\times12.98)\text{ mJ K}^{-2}\text{ mol}^{-1} = -75.27\text{ mJ K}^{-2}\text{ mol}^{-1}$$

$$\Delta\gamma = [-0.820+2\times1.184-(-17.99)-2\times(-3.862)]\ \mu\text{J K}^{-3}\text{ mol}^{-1}$$
$$= 27.262\ \mu\text{J K}^{-3}\text{ mol}^{-1}$$

Integrating eqn 57.12 between T_1(298 K) and T_2(350 K) we obtain

$$\Delta_r H^\ominus(350\text{ K}) = \Delta_r H^\ominus(298\text{ K})+\Delta\alpha(350\text{ K}-298\text{ K})+\frac{1}{2}\Delta\beta[(350\text{ K})^2-(298\text{ K})^2]$$
$$+\frac{1}{3}\Delta\gamma[(350\text{ K})^3-(298\text{ K})^3] = \boxed{-802.31\text{ kJ mol}^{-1}}$$

We note that in this example there was very little change in the enthalpy of combustion as the temperature increased from 298 K to 350 K. This is due to the coincidental cancellation of the enthalpy changes from the terms involving $\Delta\alpha$ and $\Delta\beta$.

P57.9 The needed data are the enthalpy of vaporization and heat capacity of water, available in the *Resource Section*.

$$C_{p,\mathrm{m}}(\mathrm{H_2O,l}) = 75.3\ \mathrm{J\ K^{-1}\ mol^{-1}} \qquad \Delta_{\mathrm{vap}}H^{\ominus}(\mathrm{H_2O}) = 44.0\ \mathrm{kJ\ mol^{-1}}$$

$$n(\mathrm{H_2O}) = \frac{65\ \mathrm{kg}}{0.018\ \mathrm{kg\ mol^{-1}}} = 3.6\times10^{3}\ \mathrm{mol}$$

From $\Delta H = nC_{p,\mathrm{m}}\Delta T$ we obtain

$$\Delta T = \frac{\Delta H}{nC_{p,\mathrm{m}}} = \frac{1.0\times10^{4}\ \mathrm{kJ}}{(3.6\times10^{3}\ \mathrm{mol})\times(0.0753\ \mathrm{kJ\ K^{-1}\ mol^{-1}})} = \boxed{+37\ \mathrm{K}}$$

From $\Delta H = n\Delta_{\mathrm{vap}}H^{\ominus} = \dfrac{m}{M}\Delta_{\mathrm{vap}}H^{\ominus}$

$$m = \frac{M\times\Delta H}{\Delta_{\mathrm{vap}}H^{\ominus}} = \frac{(0.018\ \mathrm{kg\ mol^{-1}})\times(1.0\times10^{4}\ \mathrm{kJ})}{44.0\ \mathrm{kJ\ mol^{-1}}} = \boxed{4.09\ \mathrm{kg}}$$

Comment. This estimate would correspond to about 30 glasses of water per day, which is much higher than the average consumption. The discrepancy may be a result of our assumption that evaporation of water is the main mechanism of heat loss.

Topic 58 The internal energy

Discussion questions

D58.1 One can use the general expression for π_T proved in Section 66.2(a), eqn 66.5) to derive its specific form for a van der Waals gas as given in Exercise 58.9(a), that is, $\pi_T = a/V_m^2$. (The derivation is carried out in Example 66.1.) For an isothermal expansion in a van der Waals gas $dU_m = (a/V_m)^2 dV_m$. Hence $\Delta U_m = -a(1/V_{m,2} - 1/V_{m,1})$. See this derivation in the solution to Exercise 58.9(a). This formula corresponds to what one would expect for a real gas. As the molecules get closer and closer the molar volume gets smaller and smaller, and the energy of attraction gets larger and larger.

D58.3 Equations of state can be thought of as expressions for the pressure of a gas in terms of the state functions, n, V, and T. They are obtained from the expression for the pressure in terms of the canonical partition function given in eqn 66.16 in Section 66.4.

$$p = kT\left(\frac{\partial \ln Q}{\partial V}\right)_T$$

Partition functions for perfect and imperfect gases are different. That for the perfect gas is given by $Q = q^N/N!$ with $q = V/\Lambda^3$. There is no one form for imperfect gases. One which can be shown to lead to the van der Waals equation of state is

$$Q = \frac{1}{N!}\left(\frac{2\pi mkT}{h^2}\right)^{3N/2}(V - Nb)e^{aN^2/kTV}$$

For the case of the perfect gas there are no molecular features in the partition function, but for imperfect gases there are repulsive and attractive features in the partition function which are related to the structure of the molecules.

Exercises

E58.1(a) The partition function is

$$q = \sum_j g_j e^{-\beta\varepsilon_j}$$

with degeneracies $\quad g_j = 2J + 1$

so $\quad q = 4 + 2e^{-\beta\varepsilon} \quad [g(^2P_{3/2}) = 4,\ g(^2P_{1/2}) = 2]$

$$U-U(0)=-\frac{N}{q}\frac{dq}{d\beta}=\frac{N\varepsilon e^{-\beta\varepsilon}}{2+e^{-\beta\varepsilon}}$$

$$C_V=\left(\frac{\partial U}{\partial T}\right)_V=-k\beta^2\left(\frac{\partial U}{\partial \beta}\right)_V=\frac{2R(\varepsilon\beta)^2 e^{-\beta\varepsilon}}{(2+e^{-\beta\varepsilon})^2}[N=N_A]$$

(a) Therefore, since at 500 K $\beta\varepsilon = 2.53\overline{5}$

$$C_{V,m}/R=\frac{(2)\times(2.53\overline{5})^2\times(e^{-2.53\overline{5}})}{(2+e^{-2.53\overline{5}})^2}=\boxed{0.236}$$

(b) At 900 K, when $\beta\varepsilon = 1.408$,

$$C_{V,m}/R=\frac{(2)\times(1.408)^2\times(e^{-1.408})}{(2+e^{-1.408})^2}=\boxed{0.193}$$

Comment. $C_{V,m}$ is smaller at 900 K than at 500 K, for then the temperature is higher than the peak in the "two-level" heat capacity curve.

E58.2(a) Assuming that all rotational modes are active we can draw up the following table for $C_{V,m}$, $C_{p,m}$ and γ with and without active vibrational modes.

	$C_{V,m}$	$C_{p,m}$	γ	Exptl	
$NH_3(v_V^*=0)$	$3R$	$4R$	1.33	1.31	$\boxed{\text{closer}}$
$NH_3(v_V^*=6)$	$9R$	$10R$	1.11		
$CH_4(v_V^*=0)$	$3R$	$4R$	1.33	1.31	$\boxed{\text{closer}}$
$CH_4(v_V^*=9)$	$12R$	$13R$	1.08		

The experimental values are obtained from Tables 57.3 and 57.4 assuming $C_{p,m}=C_{V,m}+R$. It is clear from the comparison in the above table that the vibrational modes are not active. This is confirmed by the experimental vibrational wavenumbers (see Herzberg, *Molecular Spectra and Molecular Structure II. Infrared and Raman Spectra of Polyatomic Molecules*, D. Van Nostrand Company, Inc., Princeton, NJ, 1945) all of which are much greater than kT at 298 K.

E58.3(a) We need to calculate $\Delta\varepsilon^2=\langle\varepsilon^2\rangle-\langle\varepsilon\rangle^2$ [58.13]. We assume that argon may be treated as a perfect gas and follow the results of Example 58.2. The root mean square deviation of the molecular energy is then

$$\sqrt{\Delta\varepsilon^2}=\sqrt{\tfrac{1}{2}k^2T^2}=\frac{kT}{\sqrt{2}}=\frac{1.381\times10^{-23}\ \mathrm{J\,K^{-1}}\times298\ \mathrm{K}}{\sqrt{2}}=\boxed{2.91\times10^{-21}\ \mathrm{J}}$$

E58.4(a) For reversible adiabatic expansion

$$T_f=T_i\left(\frac{V_i}{V_f}\right)^{1/c}\ [58.17]$$

where $c=\dfrac{C_{V,m}}{R}=\dfrac{C_{p,m}-R}{R}=\dfrac{(20.786-8.3145)\ \mathrm{J\,K^{-1}\,mol^{-1}}}{8.3145\ \mathrm{J\,K^{-1}\,mol^{-1}}}=1.500,$

so the final temperature is

$$T_f = (273.15\,\text{K}) \times \left(\frac{1.0\,\text{dm}^3}{3.0\,\text{dm}^3}\right)^{1/1.500} = \boxed{13\bar{1}\ \text{K}}$$

E58.5(a) In an adiabatic process, the initial and final pressures are related by (eqn 58.18) $p_f V_f^{\gamma} = p_i V_i^{\gamma}$

where $\gamma = \dfrac{C_p}{C_V} = \dfrac{C_V + nR}{C_V} = \dfrac{20.8\ \text{J K}^{-1} + (1.0\ \text{mol})(8.31\ \text{J K}^{-1}\ \text{mol}^{-1})}{20.8\ \text{J K}^{-1}} = 1.40$

Find V_i from the perfect gas law:

$$V_i = \frac{nRT_i}{p_i} = \frac{(1.0\ \text{mol}) \times (8.31\ \text{J K}^{-1}\ \text{mol}^{-1}) \times (300\ \text{K})}{4.25\ \text{atm}} \times \frac{1\ \text{atm}}{1.013 \times 10^5\ \text{Pa}}$$

$$V_i = 5.7\bar{9} \times 10^{-3}\ \text{m}^3$$

so $$V_f = V_i \left(\frac{p_i}{p_f}\right)^{1/\gamma} = (5.7\bar{9} \times 10^{-3}\ \text{m}^3) \times \left(\frac{4.25\ \text{atm}}{2.50\ \text{atm}}\right)^{1/1.40} = \boxed{0.0084\bar{6}\ \text{m}^3}$$

Find the final temperature from the perfect gas law:

$$T_f = \frac{p_f V_f}{nR} = \frac{(2.50\ \text{atm}) \times (0.0084\bar{6}\ \text{m}^3)}{(1.0\ \text{mol}) \times (8.31\ \text{J K}^{-1}\ \text{mol}^{-1})} \times \frac{1.013 \times 10^5\ \text{Pa}}{1\ \text{atm}}$$

$$T_f = \boxed{25\bar{7}\ \text{K}}$$

Adiabatic work is (eqn 58.16)

$$w_{ad} = C_V \Delta T = (20.8\ \text{J K}^{-1}) \times (25\bar{7} - 300)\ \text{K} = \boxed{-0.89 \times 10^3\ \text{J}}$$

E58.6(a) Reversible adiabatic work is

$$w_{ad} = C_V \Delta T\ [58.16] = n(C_{p,m} - R) \times (T_f - T_i)$$

where the temperatures are related by

$$T_f = T_i \left(\frac{V_i}{V_f}\right)^{1/c}\ [14.37a] \quad \text{where} \quad c = \frac{C_{V,m}}{R} = \frac{C_{p,m} - R}{R} = 3.463$$

So $$T_f = [(27.0 + 273.15)\text{K}] \times \left(\frac{500 \times 10^{-3}\ \text{dm}^3}{3.00\ \text{dm}^3}\right)^{1/3.463} = 179\text{K}$$

and $$w = \left(\frac{2.45\ \text{g}}{44.0\ \text{g mol}^{-1}}\right) \times [(37.11 - 8.3145)\ \text{J K}^{-1}\ \text{mol}^{-1}] \times (179 - 300)\ \text{K} = \boxed{-194\text{J}}$$

E58.7(a) For reversible adiabatic expansion

$$p_f V_f^{\gamma} = p_i V_i^{\gamma}\ [58.18] \quad \text{so} \quad p_f = p_i \left(\frac{V_i}{V_f}\right)^{\gamma} = (67.4\ \text{kPa}) \times \left(\frac{0.50\ \text{dm}^3}{2.0\ \text{dm}^3}\right)^{1.4} = \boxed{9.7\ \text{kPa}}$$

E58.8(a) See Exercise E58.9(a) and Problem P56.6. The internal pressure of a van der Waals gas is $\pi_T = a/V_m^2$.

The molar volume can be estimated from the perfect gas equation.

$$V_m = \frac{RT}{p} = \frac{0.08206\ \text{dm}^3\ \text{atm K}^{-1}\ \text{mol}^{-1} \times 400\ \text{K}}{1.00\ \text{bar} \times \left(\dfrac{1.000\ \text{atm}}{1.013\ \text{bar}}\right)} = 33.26\ \text{dm}^3\ \text{mol}^{-1}$$

$$\pi_T = \frac{a}{V_m^2} = \frac{5.464\ \text{atm dm}^6\ \text{mol}^{-2}}{(33.26\ \text{dm}^3\ \text{mol}^{-1})^2} = 4.96 \times 10^{-3}\ \text{atm} = \boxed{5.03\ \text{mbar}}$$

E58.9(a) The internal energy is a function of temperature and volume, $U_m = U_m(T, V_m)$, so

$$dU_m = \left(\frac{\partial U_m}{\partial T}\right)_{V_m} dT + \left(\frac{\partial U_m}{\partial V_m}\right)_T dV_m \qquad [\pi_T = (\partial U_m/\partial V_m)_T]$$

For an isothermal expansion $dT = 0$; hence

$$dU_m = \left(\frac{\partial U_m}{\partial V_m}\right)_T dV = \pi_T\, dV_m = \frac{a}{V_m^2} dV_m$$

$$\Delta U_m = \int_{V_{m,1}}^{V_{m,2}} dU_m = \int_{V_{m,2}}^{V_{m,1}} \frac{a}{V_m^2} dV_m = a\int_{1.00\ \text{dm}^3\ \text{mol}^{-1}}^{20.0\ \text{dm}^3\ \text{mol}^{-1}} \frac{dV_m}{V_m^2} = -\frac{a}{V_m}\bigg|_{1.00\ \text{dm}^3\ \text{mol}^{-1}}^{20.0\ \text{dm}^3\ \text{mol}^{-1}}$$

$$= -\frac{a}{20.0\ \text{dm}^3\ \text{mol}^{-1}} + \frac{a}{1.00\ \text{dm}^3\ \text{mol}^{-1}} = \frac{19.0a}{20.0\ \text{dm}^3\ \text{mol}^{-1}}$$

$$= 0.950\ a\ \text{mol dm}^{-3};$$

From Table 36.3, $a = 1.352\ \text{dm}^6\ \text{atm mol}^{-1}$

$$\Delta U_m = (0.950\ \text{mol dm}^{-3}) \times (1.352\ \text{dm}^6\ \text{atm mol}^{-2})$$

$$= (1.28\bar{4}\ \text{dm}^3\ \text{atm mol}^{-1}) \times \left(\frac{1\ \text{m}}{10\ \text{dm}}\right)^3 \times \left(\frac{1.013 \times 10^5\ \text{Pa}}{\text{atm}}\right) = \boxed{+130.\bar{1}\ \text{J mol}^{-1}}$$

$w = -\int p\, dV_m$ where $p = \dfrac{RT}{V_m - b} - \dfrac{a}{V_m^2}$ for a van der Waals gas. Hence,

$$w = -\int\left(\frac{RT}{V_m - b}\right) dV_m + \int \frac{a}{V_m^2} dV_m = -q + \Delta U_m$$

Therefore,

$$q = \int_{1.00\text{dm}^3\text{mol}^{-1}}^{20.0\text{dm}^3\text{mol}^{-1}} \left(\frac{RT}{V_m - b}\right) dV_m = RT\ln(V_m - b)\Big|_{1.00\text{dm}^3\text{mol}^{-1}}^{20.0\text{dm}^3\text{mol}^{-1}}$$

$$= (8.314\ \text{J K}^{-1}\ \text{mol}^{-1}) \times (298\ \text{K}) \times \ln\left(\frac{20.0 - 3.9 \times 10^{-2}}{1.00 - 3.9 \times 10^{-2}}\right) = \boxed{+7.52 \times 10^3\ \text{J mol}^{-1}}$$

and $w = -q + \Delta U_m = -(7.52 \times 10^3\ \text{J mol}^{-1}) + (130.\bar{1}\ \text{J mol}^{-1}) = \boxed{-7.39 \times 10^3\ \text{J mol}^{-1}}$

E58.10(a) $\alpha = \left(\frac{1}{V}\right)\left(\frac{\partial V}{\partial T}\right)_p$ [58.7]; $\alpha_{320} = \left(\frac{1}{V_{320}}\right)\left(\frac{\partial V}{\partial T}\right)_{p,320}$

$$\left(\frac{\partial V}{\partial T}\right)_p = V_{300}(3.9\times10^{-4}\,/\text{K} + 2.96\times10^{-6}\,T/\text{K}^2)$$

$$\left(\frac{\partial V}{\partial T}\right)_{p,320} = V_{300}(3.9\times10^{-4}\,/\text{K} + 2.96\times10^{-6}\times320/\text{K}) = 1.34\times10^{-3}\,\text{K}^{-1}V_{300}$$

$$V_{320} = V_{300}\{(0.75)+(3.9\times10^{-4})\times(320)+(1.48\times10^{-6})\times(320)^2\} = (V_{300})\times(1.02\bar{6})$$

$$\text{so}\quad \alpha_{320} = \left(\frac{1}{V_{320}}\right)\left(\frac{\partial V}{\partial T}\right)_{p,320} = \left(\frac{1}{1.02\bar{6}V_{300}}\right)\times(1.34\times10^{-3}\,\text{K}^{-1}\,V_{300})$$

$$\alpha_{320} = \frac{1.34\times10^{-3}\,\text{K}^{-1}}{1.02\bar{6}} = \boxed{1.31\times10^{-3}\,\text{K}^{-1}}$$

Comment. Knowledge of the density at 300 K is not required to solve this exercise, but it would be required to obtain numerical values of the volumes at the two temperatures.

E58.11(a) The isothermal compressibility is

$$\kappa_T = -\left(\frac{1}{V}\right)\left(\frac{\partial V}{\partial p}\right)_T \quad [58.8] \quad \text{so} \quad \left(\frac{\partial V}{\partial p}\right)_T = -\kappa_T V$$

At constant temperature

$$\text{d}V = \left(\frac{\partial V}{\partial p}\right)_T \text{d}p \quad \text{so} \quad \text{d}V = -\kappa_T V\,\text{d}p \quad \text{or} \quad \frac{\text{d}V}{V} = -\kappa_T\,\text{d}p$$

Substituting $V = \frac{m}{\rho}$ yields $\text{d}V = -\frac{m}{\rho^2}\text{d}\rho\,;\ \frac{\text{d}V}{V} = -\frac{\text{d}\rho}{\rho} = -\kappa_T \text{d}p$

Therefore, $\frac{\delta\rho}{\rho} \approx \kappa_T \delta p$

For $\frac{\delta\rho}{\rho} = 0.10\times10^{-2} = 1.0\times10^{-3}$, $\delta p \approx \frac{1.0\times10^{-3}}{\kappa_T} = \frac{1.0\times10^{-3}}{4.96\times10^{-5}\,\text{atm}^{-1}} = \boxed{2.0\times10^3\,\text{atm}}$

Problems

P58.1 The coefficient of thermal expansion is

$$\alpha = \frac{1}{V}\left(\frac{\partial V}{\partial T}\right)_p \approx \frac{\Delta V}{V\Delta T} \quad \text{so} \quad \Delta V \approx \alpha V \Delta T$$

This change in volume is equal to the change in height (sea level rise, Δh) times the area of the ocean (assuming that area remains constant). We will use α of pure water, although the oceans are complex solutions. For a 2°C rise in temperature

$$\Delta V = (2.1\times10^{-4}\,\text{K}^{-1})\times(1.37\times10^9\,\text{km}^3)\times(2.0\,\text{K}) = 5.8\times10^5\,\text{km}^3$$

$$\text{so} \quad \Delta h = \frac{\Delta V}{A} = 1.6\times10^{-3}\,\text{km} = \boxed{1.6\,\text{m}}$$

Since the rise in sea level is directly proportional to the rise in temperature, $\Delta T = 1°\text{C}$ would lead to $\Delta h = \boxed{0.80\text{ m}}$ and $\Delta T = 3.5°\text{C}$ would lead to $\Delta h = \boxed{2.8\text{ m}}$

Comment. More detailed models of climate change predict somewhat smaller rises but the same order of magnitude.

P58.3 $q^{\text{E}} = \sum_j g_j \text{e}^{-\beta\varepsilon_j} = 2 + 2\text{e}^{-\beta\varepsilon}, \quad \varepsilon = \Delta\varepsilon = 121.1\,\text{cm}^{-1}$

$$U_{\text{m}} - U_{\text{m}}(0) = -\frac{N_{\text{A}}}{q^{\text{E}}}\left(\frac{\partial q^{\text{E}}}{\partial \beta}\right)_V = \frac{2N_{\text{A}}\varepsilon \text{e}^{-\beta\varepsilon}}{q^{\text{E}}} = 1$$

$$C_{V,\text{m}} = -k\beta^2\left(\frac{\partial U_{\text{m}}}{\partial \beta}\right)_V \quad [14.26]$$

Let $x = \beta\varepsilon$, then $\text{d}\beta = \frac{1}{\varepsilon}\text{d}x$

$$C_{V,\text{m}} = -k\left(\frac{x}{\varepsilon}\right)^2 \varepsilon \frac{\partial}{\partial x}\left(\frac{N_{\text{A}}\varepsilon \text{e}^{-x}}{1+\text{e}^{-x}}\right) = -N_{\text{A}}kx^2 \times \frac{\partial}{\partial x}\left(\frac{\text{e}^{-x}}{1+\text{e}^{-x}}\right) = R\left(\frac{x^2\text{e}^{-x}}{(1+\text{e}^{-x})^2}\right)$$

Therefore

$$C_{V,\text{m}}/R = \frac{x^2\text{e}^{-x}}{(1+\text{e}^{-x})^2}, \quad x = \beta\varepsilon$$

We then draw up the following table

T / K	100	298	600
$(kT/hc)/\text{cm}^{-1}$	69.5	207	417
x	1.741	0.585	0.290
$C_{V,\text{m}}/R$	$\boxed{0.385}$	$\boxed{0.0786}$	$\boxed{0.0206}$
$C_{V,\text{m}}/(\text{J K}^{-1}\text{ mol}^{-1})$	3.20	0.654	0.171

Comment. Note that the double degeneracies do not affect the results because the two factors of 2 in q cancel when U is formed. In the range of temperature specified, the electronic contribution to the heat capacity decreases with increasing temperature.

P58.5 $q = 1 + 5\text{e}^{-\beta\varepsilon} \quad [g_J = 2J+1]$

$$\varepsilon = E(J=2) - E(J=0) = 6hcB \quad [E = hcBJ(J+1)]$$

$$\frac{U - U(0)}{N} = -\frac{1}{q}\frac{\partial q}{\partial \beta} = \frac{5\varepsilon \text{e}^{-\beta\varepsilon}}{1+5\text{e}^{-\beta\varepsilon}} \quad [N = N_{\text{A}}]$$

$$C_{V,\text{m}} = -k\beta^2\left(\frac{\partial U_{\text{m}}}{\partial \beta}\right)_V \quad [14.26]$$

$$C_{V,\text{m}}/R = \frac{5\varepsilon^2\beta^2\text{e}^{-\beta\varepsilon}}{(1+5\text{e}^{-\beta\varepsilon})^2} = \frac{180(hcB\beta)^2\text{e}^{-6hcB\beta}}{(1+5\text{e}^{-6hcB\beta})^2}$$

$$\frac{hcB}{k} = 1.4388\text{ cm K} \times 60.864\text{ cm}^{-1} = 87.571\text{ K}$$

Hence,

$$C_{V,\text{m}}/R = \frac{1.380\times10^6\,\text{e}^{-525.4\text{K}/T}}{(1+5\text{e}^{-525.4\text{K}/T})\times(T/\text{K})^2}$$

We draw up the following table:

T/K	50	100	150	200	250	300	350	400	450	500
$C_{V,\text{m}}/R$	0.02	0.68	1.40	1.35	1.04	0.76	0.56	0.42	0.32	0.26

These points are plotted in Fig. 58.1.

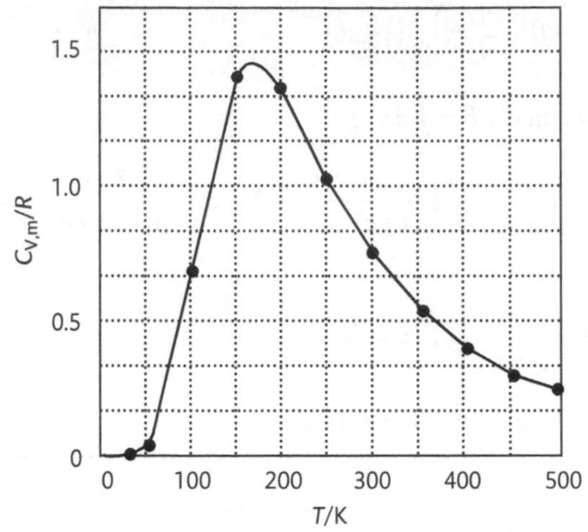

Fig 58.1

P58.7 (a) The probability distribution of rotational energy levels is the Boltzmann factor of each level, weighted by the degeneracy, over the partition function

$$p_J^{\text{R}}(T) = \frac{g(J)\text{e}^{-\varepsilon_J/kT}}{q^{\text{R}}} = \frac{(2J+1)\text{e}^{-hc\tilde{B}J(J+1)/kT}}{\sum_{J=0}(2J+1)\text{e}^{-hc\tilde{B}J(J+1)/kT}} \quad [52.11]$$

It is conveniently plotted against J at several temperatures using mathematical software. This distribution at 100 K is shown below as both a bar plot and a line plot.

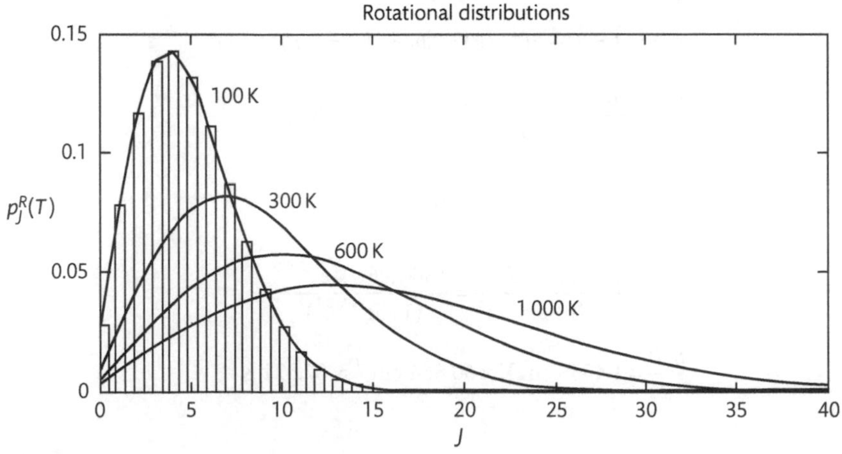

Fig 58.2

The plots show that higher rotational states become more heavily populated at higher temperature. Even at 100 K the most populated state has 4 quanta of rotational energy; it is elevated to 13 quanta at 1 000 K.

Values of the vibrational state probability distribution,

$$p_v^{\mathrm{V}}(T) = \frac{e^{-\varepsilon_j/kT}}{q^{\mathrm{V}}} = e^{-vhc\tilde{\nu}/kT}(1-e^{-hc\tilde{\nu}/kT}) \ [52.15]$$

are conveniently tabulated against v at several temperatures. Computations may be discontinued when values drop below some small number like 10^{-7}.

	$p_v^{\mathrm{V}}(T)$			
v	100 K	300 K	600 K	1 000 K
0	1	1	0.095	0.956
1	2.77×10^{-14}	3.02×10^{-5}	5.47×10^{-3}	0.042
2		9.15×10^{-10}	3.01×10^{-5}	1.86×10^{-3}
3			1.65×10^{-7}	8.19×10^{-5}
4				3.61×10^{-6}
5				1.59×10^{-7}

Only the state $v=0$ is appreciably populated below 1 000 K and even at 1 000 K only 4% of the molecules have 1 quanta of vibrational energy.

(b) The classical (equipartition) rotational partition function is

$$q_{\text{classical}}^{\mathrm{R}}(T) = \frac{kT}{hc\tilde{B}} = \frac{T}{\theta_{\mathrm{R}}} \ [52.13]$$

where q_{R} is the rotational temperature. We would expect the partition function to be well approximated by this expression for temperatures much greater than the rotational temperature.

$$\theta_{\mathrm{R}} = \frac{hc\tilde{B}}{k} = \frac{(6.626\times10^{-34}\ \mathrm{J\,s})\times(2.998\times10^{10}\ \mathrm{cm\,s^{-1}})\times(1.931\ \mathrm{cm^{-1}})}{1.381\times10^{-23}\ \mathrm{J\,K^{-1}}}$$

$$\theta_{\mathrm{R}} = 2.779\ \mathrm{K}$$

In fact $q_{\mathrm{R}} << T$ for all temperatures of interest in this problem (100 K or more). Agreement between the classical expression and the explicit sum is indeed good, as Fig. 58.3 confirms. The figure displays the percentage deviation $(q_{\text{classical}}^{\mathrm{R}} - q^{\mathrm{R}})100/q^{\mathrm{R}}$. The maximum deviation is about –0.9% at 100 K and the magnitude decreases with increasing temperature.

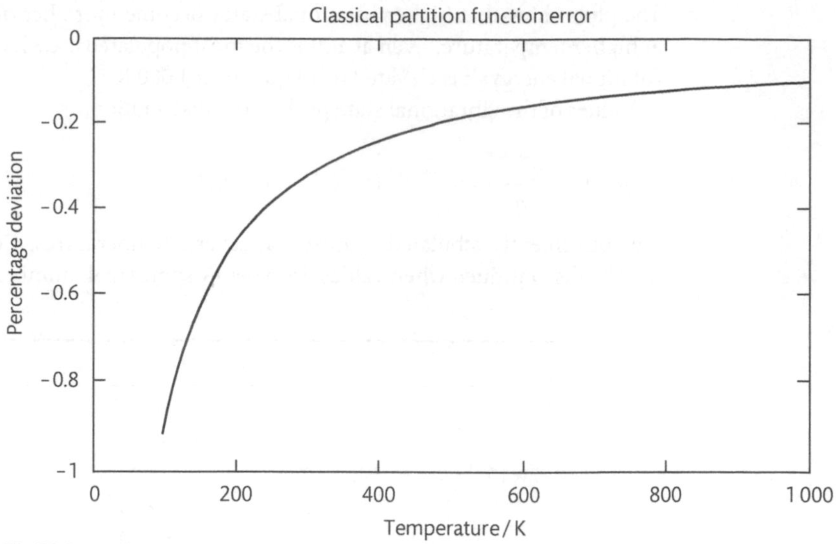

Fig 58.3

(c) The translational, rotational, and vibrational contributions to the total energy are specified by eqns 53.5b, 53.6b, and 53.8, respectively. As molar quantities, they are:

$$U^{\mathrm{T}} = \tfrac{3}{2}RT, \quad U^{\mathrm{R}} = RT, \quad U^{\mathrm{V}} = \frac{N_{\mathrm{A}}hc\tilde{\nu}}{\mathrm{e}^{hc\tilde{\nu}/kT} - 1}$$

The contributions to the difference in energy from its 100 K value are $\Delta U^{\mathrm{T}}(T) = U^{\mathrm{T}}(T) - U^{\mathrm{T}}(100\,\mathrm{K})$, etc. Fig. 58.4 shows the individual contributions to $\Delta U(T)$. Translational motion contributes 50% more than the rotational motion because it has 3 quadratic degrees of freedom compared to 2 quadratic degrees of freedom for rotation. Very little change occurs in the vibrational energy because very high temperatures are required to populate $v = 1, 2, \ldots$. states (see Part a).

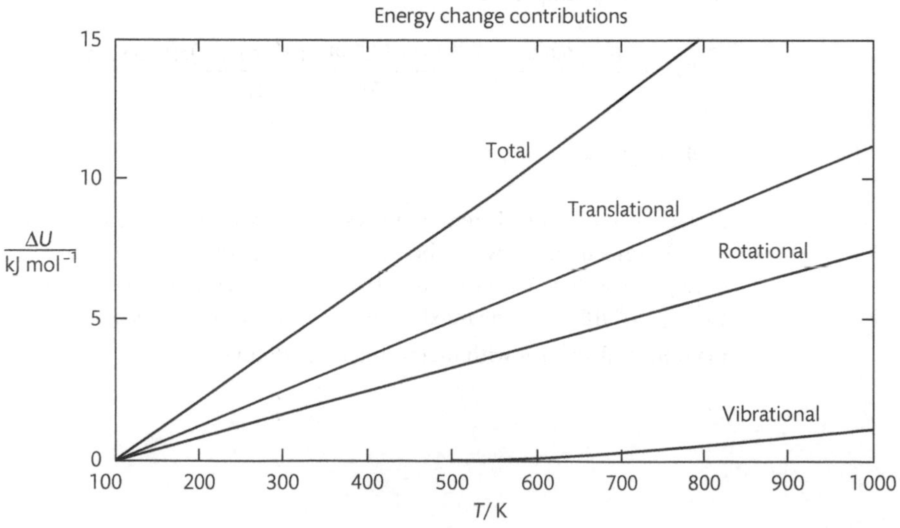

Fig 58.4

$$C_{V,\text{m}}(T)=\left(\frac{\partial U(T)}{\partial T}\right)_V=\left(\frac{\partial}{\partial T}\right)_V(U^\text{T}+U^\text{R}+U^\text{V})$$

$$=\frac{3}{2}R+R+\frac{\text{d}U^\text{V}}{\text{d}T}=\frac{5}{2}R+\frac{\text{d}U^\text{V}}{\text{d}T}$$

The derivative $\text{d}U^\text{V}/\text{d}T$ may be evaluated numerically with numerical software (we advise exploration of the technique) or it may be evaluated analytically using the equation for C_V in Example 58.3:

$$C_{V,\text{m}}^\text{V}=\frac{\text{d}U^\text{V}}{\text{d}T}=R\left\{\frac{\theta_\text{V}}{T}\left(\frac{\text{e}^{-\theta_\text{V}/2T}}{1-\text{e}^{-\theta_\text{V}/T}}\right)\right\}^2$$

where $\theta_\text{V}=hc\tilde{\nu}/k=3122$ K. Fig. 58.5 shows the ratio of the vibrational contribution to the sum of translational and rotational contributions. Below 300 K, vibrational motion makes a small, perhaps negligible, contribution to the heat capacity. The contribution is about 10% at 600 K and grows with increasing temperature.

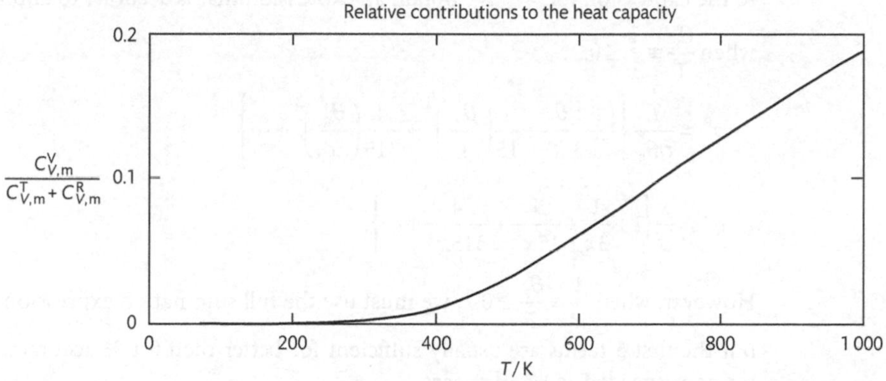

Fig 58.5

P58.9 We work with eqn 58.14 in the form $C_V=-\frac{N}{kT^2}\left(\frac{\partial\langle\varepsilon\rangle}{\partial\beta}\right)_V$ with $\langle\varepsilon\rangle$ given by

$\langle\varepsilon\rangle=\frac{\sum_i\varepsilon_i\text{e}^{-\beta\varepsilon_i}}{\sum_i\text{e}^{-\beta\varepsilon_i}}$. Differentiating this expression with respect to β we obtain

$$\left(\frac{\partial\langle\varepsilon\rangle}{\partial\beta}\right)_V=-\frac{\sum_i\varepsilon_i^2\text{e}^{-\beta\varepsilon_i}}{\sum_i\text{e}^{-\beta\varepsilon_i}}+\frac{\left(\sum_i\varepsilon_i\text{e}^{-\beta\varepsilon_i}\right)^2}{\sum_i\text{e}^{-2\beta\varepsilon_i}}\ \text{which can be rewritten as}$$

$$-\frac{\sum_i\varepsilon_i^2\text{e}^{-\beta\varepsilon_i}}{q}+\left(\frac{\sum_i\varepsilon_i\text{e}^{-\beta\varepsilon_i}}{q}\right)^2\ \text{or}$$

$$-\langle\varepsilon^2\rangle+\langle\varepsilon\rangle^2=-\Delta\varepsilon^2$$

After substituting into the expression above for C_V we obtain

$$C_V = -\frac{N}{kT^2}\left(\frac{\partial\langle\varepsilon\rangle}{\partial\beta}\right)_V = -\frac{N}{kT^2}(-\Delta\varepsilon^2) = \boxed{\frac{N\Delta\varepsilon^2}{kT^2}}$$, which was to be proved.

P58.11 We begin the derivation with the full expression for the molecular rotational partition function.

$$q^{\mathrm{R}} = \frac{1}{\sigma}\sum_J (2J+1)\mathrm{e}^{-\frac{hc\tilde{B}}{kT}J(J+1)}[52.11] = \frac{1}{\sigma}\sum_J (2J+1)\mathrm{e}^{-\frac{\theta_{\mathrm{R}}}{T}J(J+1)} \quad [\theta_{\mathrm{R}} = hc\tilde{B}/k]$$

We first note that no single analytical expression for this summation can be obtained that is valid at all temperatures. The summation process must be continued until additional terms change the sum to less than some pre-defined limit of accuracy, say 0.2 %. However, various approximate analytical expressions that are valid within certain temperature ranges can be derived. Taken together these expressions can give values that are valid within small error limits over the entire temperature range.

One of these approximations applies the Euler-Maclaurin summation formula to the expression for q^{R}. We obtain an expression that is accurate to about 0.1 % when $\frac{\theta_{\mathrm{R}}}{T} \equiv \frac{1}{x} \leq 0.7$.

$$q^{\mathrm{R}} = \frac{T}{\sigma\theta_{\mathrm{R}}}\left\{1 + \frac{1}{3}\frac{\theta_{\mathrm{R}}}{T} + \frac{1}{15}\left(\frac{\theta_{\mathrm{R}}}{T}\right)^2 + \frac{4}{315}\left(\frac{\theta_{\mathrm{R}}}{T}\right)^3 + \cdots\right\}$$

$$q^{\mathrm{R}} = \frac{x}{\sigma}\left\{1 + \frac{1}{3x} + \frac{1}{15x^2} + \frac{4}{315x^3} + \cdots\right\}.$$

However, when $\frac{1}{x} \equiv \frac{\theta_{\mathrm{R}}}{T} \geq 0.7$, we must use the full summation expression for q^{R}, but the first 5 terms are usually sufficient for better than 0.1 % accuracy. Fewer terms are needed as $1/x$ increases.

$$q^{\mathrm{R}} = \frac{1}{\sigma}(1 + 3\mathrm{e}^{-2/x} + 5\mathrm{e}^{-6/x} + 7\mathrm{e}^{-12/x} + 9\mathrm{e}^{-20/x} + \cdots)$$

The energy is calculated from $U_{\mathrm{m}} - U_{\mathrm{m}}(0) = kT^2\left[\frac{\partial \ln Q}{\partial T}\right]_V$

And the heat capacity from $C_{V,\mathrm{m}} = 2kT\left[\frac{\partial \ln Q}{\partial T}\right]_V + kT^2\left[\frac{\partial^2 \ln Q}{\partial T^2}\right]_V$

The canonical partition function for a linear rotor is $Q^{\mathrm{R}} = (q^{\mathrm{R}})^N$. Substituting this expression for Q^{R} into the expression for $C_{V,\mathrm{m}}$ and performing the differentiations term by term we obtain the final expressions for $C_{V,\mathrm{m}}$. These are: (1) using q^{R} from the Euler-Maclaurin expansion,

$$\boxed{C_{V,\mathrm{m}} = R\left\{1 + \frac{1}{45}\left(\frac{\theta_{\mathrm{R}}}{T}\right)^2 + \frac{16}{945}\left(\frac{\theta_{\mathrm{R}}}{T}\right)^3 + \ldots\right\}}$$, or

$$C_{V,\mathrm{m}} = R\left\{1+\frac{1}{45x^2}+\frac{16}{945x^3}+...\right\}\ (1),$$

(0.3% accuracy when $\frac{1}{x}\leq 0.65$, or $\frac{T}{\theta_\mathrm{R}}\geq 1.54$), and (2) using q^R from the full summation

$$C_{V,\mathrm{m}} = R\left\{\left(\frac{1}{xq^\mathrm{R}}\right)^2 12\mathrm{e}^{-2/x}(1+15\mathrm{e}^{-4/x}+20\mathrm{e}^{-6/x}+84\mathrm{e}^{-10/x}+175\mathrm{e}^{-12/x}+105\mathrm{e}^{-16/x}+\cdots)\right\}$$

$$C_{V,\mathrm{m}} = R\left\{\left(\frac{1}{xq^\mathrm{R}}\right)^2 12\mathrm{e}^{-2/x}(1+15\mathrm{e}^{-4/x}+20\mathrm{e}^{-6/x}+84\mathrm{e}^{-10/x}+175\mathrm{e}^{-12/x}+105\mathrm{e}^{-16/x}+\cdots)\right\}\ x\equiv\frac{T}{\theta_\mathrm{R}}\ (2)$$

(0.2% accuracy when $\frac{1}{x}\geq 0.65$ or $\frac{T}{\theta_\mathrm{R}}\leq 1.54$)

Examination of the data in Table 43.1 reveals that it is only in the case of H_2 and possibly the hydrogen halides that this latter expression, eqn (2) would be needed. The case of H_2 is treated separately in Problem 58.19.

An alternative, but equivalent, approach would be to use the expression derived in Problem 58.18. Again a decision would have to be made as to when to terminate the summation in order to obtain the desired level of accuracy.

In order to complete these calculations and to plot the heat capacity, a spread sheet program such as Excel®, or a computer algebra system such as MathCad® is very useful. See Fig. 58.6 for a plot of $C_\mathrm{m}(x)\equiv\frac{C_{V,\mathrm{m}}}{R}$ [eqn (2) above] against $x\equiv\frac{T}{\theta_\mathrm{R}}$.

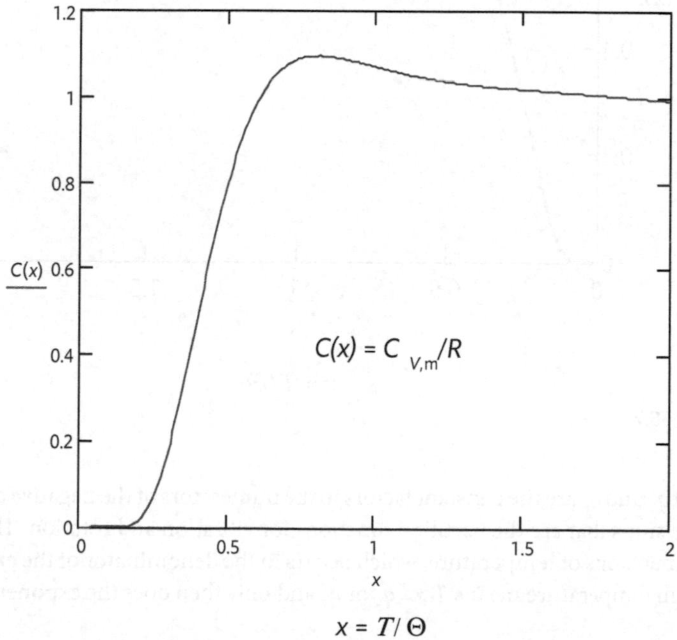

Figure 58.6

P58.13 The partition function for one dimension of vibration of an harmonic oscillator is $q^{\mathrm{V}} = \dfrac{1}{1-\mathrm{e}^{-\beta hc\tilde{\nu}}}$ [52.15] $= \dfrac{1}{1-\mathrm{e}^{-\theta_{\mathrm{E}}/T}}$. The energy is calculated from $U_{\mathrm{m}} - U_{\mathrm{m}}(0) = kT^2\left[\dfrac{\partial \ln Q}{\partial T}\right]_V$ with $Q^{\mathrm{V}} = (q^{\mathrm{V}})^N$. Performing the differentiation we obtain $U_{\mathrm{m}} - U_{\mathrm{m}}(0) = Nk\theta_{\mathrm{E}}\left(\dfrac{1}{\mathrm{e}^{\theta_{\mathrm{E}}/T}-1}\right)$. The heat capacity is calculated from $C_{V,\mathrm{m}} = \left(\dfrac{\partial U_{\mathrm{m}}}{\partial T}\right)_V$.

Performing this differentiation we obtain $C_{V,\mathrm{m}} = Nk\left\{\left(\dfrac{\theta_{\mathrm{E}}}{T}\right)^2 \dfrac{\mathrm{e}^{\theta_{\mathrm{E}}/T}}{(\mathrm{e}^{\theta_{\mathrm{E}}/T}-1)^2}\right\}$. For three independent dimensions of oscillation as in one mole of a crystalline solid, we multiply by three and set $Nk = R$, finally obtaining the Einstein formula which is $\boxed{C_{V,\mathrm{m}} = 3R\left\{\left(\dfrac{\theta_{\mathrm{E}}}{T}\right)^2 \dfrac{\mathrm{e}^{\theta_{\mathrm{E}}/T}}{(\mathrm{e}^{\theta_{\mathrm{E}}/T}-1)^2}\right\}}$. The heat capacity $C_{V,\mathrm{m}}/3R$ is plotted against $x = T/\theta_{\mathrm{E}}$ in Fig 58.7.

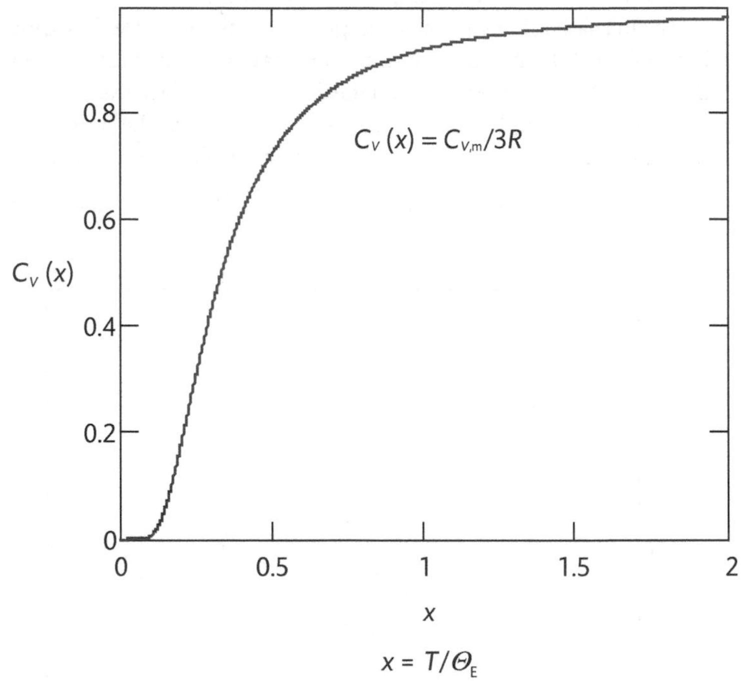

Fig 58.7

P58.15 (a) q_{V} and q_{R} are the constant factors in the numerators of the negative exponents in the sums that are the partition functions for vibration and rotation. They have the dimensions of temperature, which occurs in the denominator of the exponents. So high temperature means $T >> q_{\mathrm{V}}$ or q_{R} and only then does the exponential become substantial. Thus q_V and q_R are measures of the temperature at which higher vibrational and rotational states, respectively, become significantly populated.

$$\theta_R = \frac{hc\tilde{B}}{k} = \frac{(2.998\times10^{10}\,\text{cm s}^{-1})\times(6.626\times10^{-34}\,\text{J s})\times(60.864\,\text{cm}^{-1})}{(1.381\times10^{-23}\,\text{J K}^{-1})} = \boxed{87.55\,\text{K}}$$

$$\text{and}\quad \theta_V = \frac{hc\tilde{\nu}}{k} = \frac{(6.626\times10^{-34}\,\text{J s})\times(4\,400.39\,\text{cm}^{-1})\times(2.998\times10^{10}\,\text{cm s}^{-1})}{(1.381\times10^{-23}\,\text{J K}^{-1})} = \boxed{6\,330\,\text{K}}$$

(b) and (c) These parts of the solution were performed with Mathcad® and are reproduced on the following pages.

Objective: To calculate the equilibrium constant $K(T)$ and $C_p(T)$ for dihydrogen at high temperature for a system made with n mol H_2 at 1 bar.

$$H_2(g) \rightleftharpoons 2\,H(g)$$

At equilibrium the degree of dissociation, α, and the equilibrium amounts of H_2 and atomic hydrogen are related by the expressions

$$n_{H_2} = (1-\alpha)n \quad \text{and} \quad n_H = 2\alpha n$$

The equilibrium mole fractions are

$$x_{H_2} = (1-\alpha)n/\{(1-\alpha)n+2\alpha n\} = (1-\alpha)/(1+\alpha)$$
$$x_H = 2\alpha n/\{(1-\alpha)n+2\alpha n\} = 2\alpha/(1+\alpha)$$

The partial pressures are

$$p_{H_2} = (1-\alpha)p/(1+\alpha) \quad \text{and} \quad p_H = 2\alpha p/(1+\alpha)$$

The equilibrium constant is

$$K(T) = \frac{(p_H/p^\ominus)^2}{(p_{H_2}/p^\ominus)} = 4\alpha^2\frac{(p/p^\ominus)}{(1-\alpha^2)} = \frac{4\alpha^2}{(1-\alpha^2)} \quad \text{where } p = p^\ominus = 1\text{ bar}$$

The above equation is easily solved for a

$$\boxed{\alpha = (K/(K+4))^{1/2}}$$

The heat capacity at constant volume for the equilibrium mixture is

$$C_V(\text{mixture}) = n_H C_{V,m}(H) + n_{H_2} C_{V,m}(H_2)$$

The heat capacity at constant volume per mole of dihydrogen used to prepare the equilibrium mixture is

$$C_V = C_V(\text{mixture})/n = \{n_H C_{V,m}(H) + n_{H_2} C_{V,m}(H_2)\}/n$$
$$= \boxed{2\alpha C_{V,m}(H) + (1-\alpha)C_{V,m}(H_2)}$$

The formula for the heat capacity at constant pressure per mole of dihydrogen used to prepare the equilibrium mixture (C_p) can be deduced from the molar relationship

$$C_{p,\mathrm{m}} = C_{V,\mathrm{m}} + R$$

$$\begin{aligned} C_p &= \{n_\mathrm{H} C_{p,\mathrm{m}}(\mathrm{H}) + n_{\mathrm{H}_2} C_{p,\mathrm{m}}(\mathrm{H}_2)\}/n \\ &= \frac{n_\mathrm{H}}{n}\{C_{V,\mathrm{m}}(\mathrm{H}) + R\} + \frac{n_{\mathrm{H}_2}}{n}\{C_{V,\mathrm{m}}(\mathrm{H}_2) + R\} \\ &= \frac{n_\mathrm{H} C_{V,\mathrm{m}}(\mathrm{H}) + n_{\mathrm{H}_2} C_{V,\mathrm{m}}(\mathrm{H}_2)}{n} + R\left(\frac{n_\mathrm{H} + n_{\mathrm{H}_2}}{n}\right) \\ &= C_V + R(1+\alpha) \end{aligned}$$

Calculations

J = joule	s = second	kJ = 1 000 J
mol = mole	g = gram	bar = 1×10^5 Pa
$h = 6.62608\times10^{-34}$ J s	$c = 2.9979\times10^8$ m s^{-1}	$k = 1.38066\times10^{-23}$ J K^{-1}
$R = 8.31451$ J K^{-1} mol^{-1}	$N_\mathrm{A} = 60.2214\times10^{23}$ mol^{-1}	$p^\ominus = 1$ bar

Molecular properties of H_2

$\nu = 4\,400.39$ cm^{-1} $\quad \tilde{B} = 60.864$ cm^{-1} $\quad D = 432.1$ kJ mol^{-1}

$$m_\mathrm{H} = \frac{1\ \mathrm{g\ mol^{-1}}}{N_\mathrm{A}} \qquad m_{\mathrm{H}_2} = 2m_\mathrm{H}$$

$$\theta_\mathrm{V} = \frac{hc\tilde{\nu}}{k} \qquad \theta_\mathrm{R} = \frac{hc\tilde{B}}{k}$$

Computation of $K(T)$ and $\alpha(T)$

$$N = 200 \quad i = 0,\ldots,N \quad T_i = 500K + \frac{i\times5\,500\ \mathrm{K}}{N}$$

$$\Lambda_{\mathrm{H}i} = \frac{h}{(2\pi m_\mathrm{H} kT_i)^{1/2}} \qquad \Lambda_{\mathrm{H}_2 i} = \frac{h}{(2\pi m_{\mathrm{H}_2} kT_i)^{1/2}}$$

$$q_{\mathrm{V}i} = \frac{1}{1-\mathrm{e}^{-(\theta_\mathrm{V}/T_i)}} \qquad q_{\mathrm{R}i} = \frac{T_i}{2\theta_\mathrm{R}}$$

$$\boxed{K_{\mathrm{eq}i} = \frac{kT_i\left(\Lambda_{\mathrm{H}_2 i}\right)^3 \mathrm{e}^{-(D/RT_i)}}{p^\ominus q_{\mathrm{V}i} q_{\mathrm{R}i}\left(\Lambda_{\mathrm{H}i}\right)^6}} \qquad \alpha_i = \left(\frac{K_{\mathrm{eq}i}}{K_{\mathrm{eq}i}+4}\right)^{1/2}$$

See Fig. 58.8

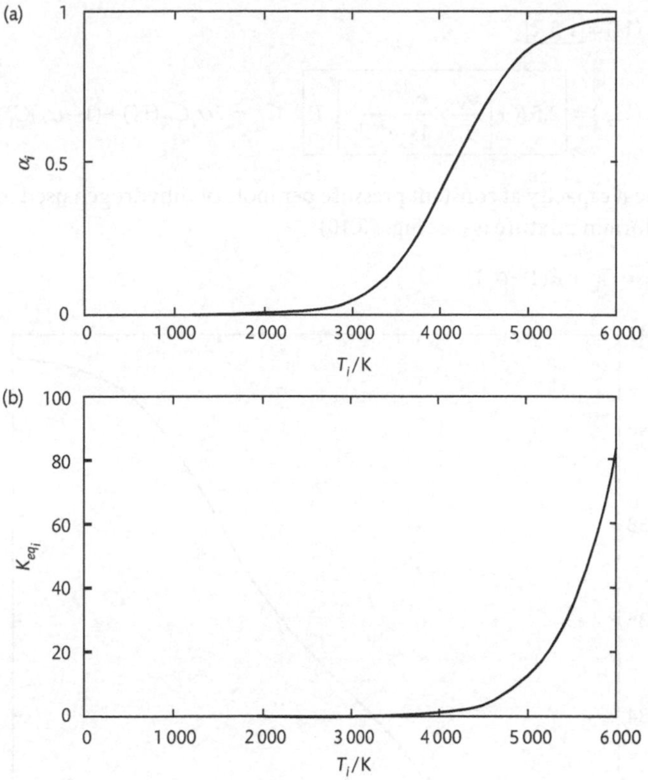

Fig 58.8

Heat capacity at constant volume per mole of dihydrogen used to prepare the equilibrium mixture is (see Fig. 58.9)

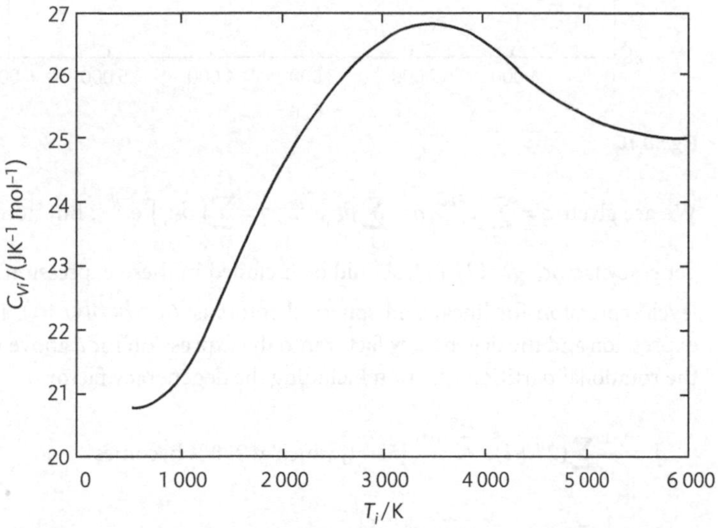

Fig 58.9

$$C_V(\mathrm{H}) = \boxed{1.5R}$$

$$C_V(\mathrm{H}_{2_i}) = \boxed{2.5R + \left[\frac{\theta_\mathrm{V}}{T_i} \times \frac{\mathrm{e}^{-(\theta_\mathrm{V}/2T_i)}}{1-\mathrm{e}^{\theta_\mathrm{V}/T_i}}\right]^2 R} \quad C_{V_i} = 2\alpha_i C_V(\mathrm{H}) + (1-\alpha_i)C_V(\mathrm{H}_{2_i})$$

The heat capacity at constant pressure per mole of dihydrogen used to prepare the equilibrium mixture is (see Fig. 58.10)

$$C_{p_i} = C_{V_i} + R(1+\alpha_i)$$

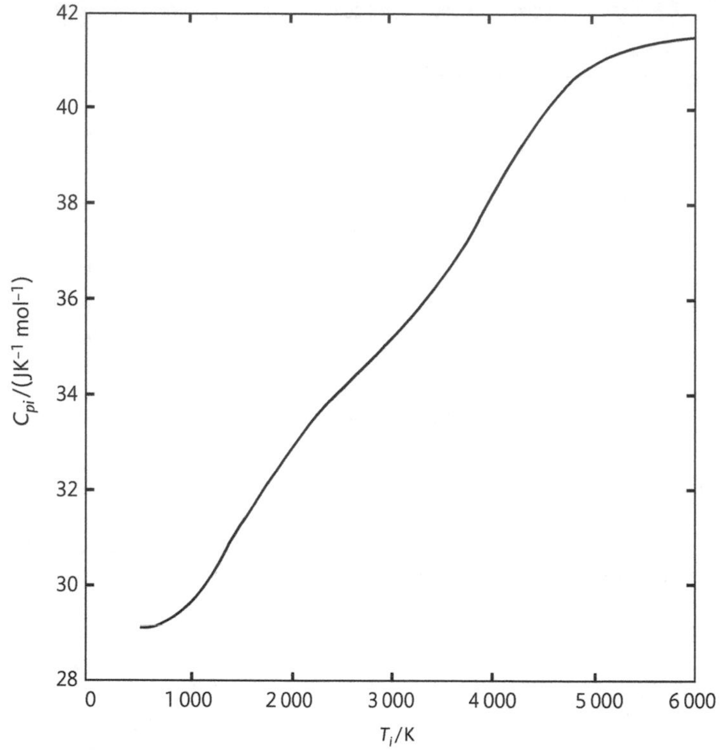

Fig 58.10

58.17 We are given $q = \sum_j \mathrm{e}^{-\beta\varepsilon_j}, \dot{q} = \sum_j \beta\varepsilon_j \mathrm{e}^{-\beta\varepsilon_j}, \ddot{q} = \sum_j (\beta\varepsilon_j)^2 \mathrm{e}^{-\beta\varepsilon_j}$. But note that the degeneracy factor, $g = (2J+1)$, should be included in these expressions. The energy level expression for linear and spherical rotors is $\varepsilon_J = hc\tilde{B}J(J+1)$. Inserting this expression and the degeneracy factor into the expression for q above we obtain for the rotational partition function including the degeneracy factor

$$q_\mathrm{R} = \frac{1}{\sigma}\sum_J (2J+1)\mathrm{e}^{-\frac{hc\tilde{B}}{kT}J(J+1)}$$ [52.11] which at 298 K becomes

$$q_\mathrm{R} = \frac{1}{\sigma}\sum_J (2J+1)\mathrm{e}^{-\frac{\tilde{B}}{207.22\ \mathrm{cm}^{-1}}J(J+1)}$$

The symmetry number for the molecule must be obtained from the point group of the molecule; it is equal to the number of distinct proper rotations of the molecule plus the identity operation. For HCl that is 1, but for CH_4 it is 12. For $\dot{q}^{R}$ and $\ddot{q}^{R}$ we obtain at 298 K

$$\dot{q}^{R}=\frac{1}{\sigma}\sum_{J}\left(\frac{\tilde{B}}{207.22\ \mathrm{cm}^{-1}}J(J+1)\right)(2J+1)\mathrm{e}^{-\frac{\tilde{B}}{207.22\ \mathrm{cm}^{-1}}J(J+1)}\quad\text{and}$$

$$\ddot{q}^{R}=\frac{1}{\sigma}\sum_{J}\left(\frac{\tilde{B}}{207.22\ \mathrm{cm}^{-1}}J(J+1)\right)^{2}(2J+1)\mathrm{e}^{-\frac{\tilde{B}}{207.22\ \mathrm{cm}^{-1}}J(J+1)}$$

(a) For HCl $\tilde{B}=10.593\ \mathrm{cm}^{-1}$. When this value is inserted into the expressions above the results of these summations are:

$$\boxed{q^{R}=19.899,\ \dot{q}^{R}=19.558,\ \text{and}\ \ddot{q}^{R}=576.536}$$

(b) For CCl_4 $\tilde{B}=5.797\ \mathrm{cm}^{-1}$. When this value is inserted into the expressions above along with $\sigma=12$, the results of these summations are:

$$\boxed{q^{R}=3.007,\ \dot{q}^{R}=2.979,\ \text{and}\ \ddot{q}^{R}=118.5}$$

P58.19 The partition functions for the *ortho-* and *para-* forms of H_2 are:

$$q^{R}_{\text{ortho}}=3\sum_{J=1,3,5\ldots}(2J+1)\mathrm{e}^{-\frac{\theta_R}{T}J(J+1)}\qquad[\theta_R=hc\tilde{B}/k=87.6\ \mathrm{K}]$$

$$q^{R}_{\text{para}}=\sum_{J=0,2,4\ldots}(2J+1)\mathrm{e}^{-\frac{\theta_R}{T}J(J+1)}$$

In order to conform to the notation of Problem 58.18 we may rewrite these expressions as

$$q^{R}_{\text{ortho}}=3\sum_{J=1,3,5\ldots}g(J)\mathrm{e}^{-\beta\varepsilon(J)}\ \text{with}\ g(J)=2J+1\ \text{and}\ \varepsilon(J)=hc\tilde{B}J(J+1)$$

$$q^{R}_{\text{para}}=\sum_{J=0,2,4\ldots}g(J)\mathrm{e}^{-\beta\varepsilon(J)}$$

To simplify the summation process, and in order to use the formulas derived in the solution to P14.32, we define a new quantum number K by $J=2K+1$ for the *ortho-* case and by $J=2K$ for the *para-* case. Then

$g_o(K)=4K+3$ and $\varepsilon_o(K)=hc\tilde{B}[(2K+1)(2K+2)]$ for *ortho*-hydrogen and

$g_p(K)=4K+1$ and $\varepsilon_p(K)=hc\tilde{B}[2K(2K+1)]$ for *para*-hydrogen.

With these modifications the partition functions can now be rewritten as

$$q^{R}_{\text{ortho}}=3\sum_{K=0,1,2\ldots}g_o(K)\mathrm{e}^{-\beta\varepsilon_o(K)}$$

$$q^{R}_{\text{para}}=\sum_{K=0,1,2\ldots}g_p(K)\mathrm{e}^{-\beta\varepsilon_p(K)}$$

Now the final expression for $C_{V,\mathrm{m}}/R$ derived in P58.18, with K in place of J, $g(K)$ in place of $g(J)$, and $\varepsilon(K)$ in place of $\varepsilon(J)$, can be used to calculate the heat capacities. With these substitutions that expression becomes

$$\frac{C_{V,\mathrm{m}}}{R}=\frac{1}{q}\sum_K g(K)\beta^2\varepsilon^2(K)\mathrm{e}^{-\beta\varepsilon(K)}-\frac{1}{q^2}\left(\sum_K g(K)\beta\varepsilon(K)\mathrm{e}^{-\beta\varepsilon(K)}\right)^2$$

This expression needs to be evaluated separately for both the *para*- and *ortho*-forms of hydrogen. It is most easily evaluated with a spreadsheet program such as Excel® or a CAS system such as MathCad®.

For the *ortho*- case $C_\mathrm{o}\equiv\dfrac{C_{V,\mathrm{m}}}{R}=\dfrac{1}{q_\mathrm{o}}\sum_K g_\mathrm{o}b_\mathrm{o}^2\mathrm{e}^{-b_\mathrm{o}}-\dfrac{1}{q_\mathrm{o}^2}\left(\sum_K g_\mathrm{o}b_\mathrm{o}\mathrm{e}^{-b_\mathrm{o}}\right)^2$ [eqn(1)], in which

$$b_\mathrm{o}=\beta\varepsilon_\mathrm{o}=\frac{\theta_\mathrm{R}}{T}(2K+1)(2K+2)=\frac{1}{x}(2K+1)(2K+2)\quad x\equiv\frac{T}{\theta_\mathrm{R}}$$

For the *para*- case $C_\mathrm{p}\equiv\dfrac{C_{V,\mathrm{m}}}{R}=\dfrac{1}{q_\mathrm{p}}\sum_K g_\mathrm{p}b_\mathrm{p}^2\mathrm{e}^{-b_\mathrm{p}}-\dfrac{1}{q_\mathrm{p}^2}\left(\sum_K g_\mathrm{p}b_\mathrm{p}\mathrm{e}^{-b_\mathrm{p}}\right)^2$ [eqn(2)], in which

$$b_\mathrm{p}=\beta\varepsilon_\mathrm{p}=\frac{\theta_\mathrm{R}}{T}[2K(2K+1)]=\frac{1}{x}[2K(2K+1)]\quad x\equiv T/\theta_\mathrm{R}$$

In the Mathcad® worksheet below the notation is as follows: $x\equiv T/\theta_\mathrm{R}$; for *ortho*-hydrogen, q_o is its partition function, $C_{\mathrm{o}1}$ and $C_{\mathrm{o}2}$ are the first and second terms on the right in eqn (1), and $C_\mathrm{o}=C_{V,\mathrm{m}}/R$ is its heat capacity in units of R. Similarly, for *para*-hydrogen, q_p is its partition function, $C_{\mathrm{p}1}$ and $C_{\mathrm{p}2}$ are the first and second terms on the right in eqn (2), and $C_\mathrm{p}=C_{V,\mathrm{m}}/R$ is its heat capacity in units of R.

$i:=0..24 \qquad x_{start}:=0.2 \qquad x_{end}:=5.0$

$x_i:=x_{start}+(x_{end}-x_{start})\cdot\dfrac{i}{24} \qquad x_i=T/\Theta$

$$q_{o_i}:=\sum_{K=0}^{6}\left[(4\cdot K+3)\cdot e^{-\frac{1}{x_i}\cdot(2\cdot K+1)\cdot(2\cdot K+2)}\right]$$

$$C_{o1_i}:=\left(\frac{1}{q_{o_i}}\right)\cdot\sum_{K=0}^{6}\left[(4\cdot K+3)\left[\left[\frac{1}{x_i}(2\cdot K+1)(2\cdot K+2)\right]^2\cdot e^{-\frac{1}{x_i}(2\cdot K+1)(2\cdot K+2)}\right]\right]$$

$$C_{o2_i}:=-\left(\frac{1}{q_{o_i}}\right)^2\left[\sum_{K=0}^{6}\left[(4\cdot K+3)\left[\frac{1}{x_i}(2\cdot K+1)(2\cdot K+2)\cdot e^{-\frac{1}{x_i}(2\cdot K+1)(2\cdot K+2)}\right]\right]\right]^2$$

$C_{o_i}:=C_{o1_i}+C_{o2_i}$

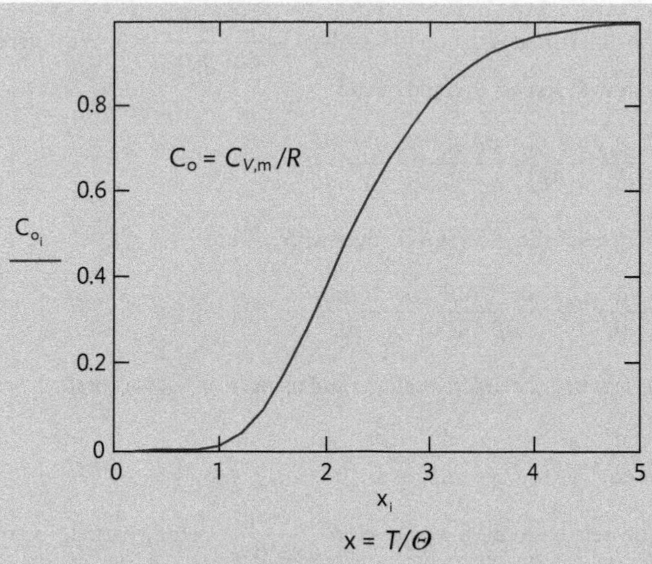

Fig 58.11

$$q_{p_i} := \sum_{K=0}^{6}\left[(4\cdot K+1)\cdot e^{-\frac{1}{x_i}\cdot(2\cdot K+1)\cdot(2\cdot K)}\right]$$

$$C_{p1_i} := \left(\frac{1}{q_{p_i}}\right)\cdot\sum_{K=0}^{6}\left[(4\cdot K+1)\left[\left[\frac{1}{x_i}(2\cdot K+1)(2\cdot K)\right]^2\cdot e^{-\frac{1}{x_i}(2\cdot K+1)(2\cdot K)}\right]\right]$$

$$C_{p2_i} := -\left(\frac{1}{q_{p_i}}\right)^2\left[\sum_{K=0}^{6}\left[(4\cdot K+1)\left[\frac{1}{x_i}(2\cdot K+1)(2\cdot K)\cdot e^{-\frac{1}{x_i}(2\cdot K+1)(2\cdot K)}\right]\right]\right]^2$$

$$C_{p_i} := C_{p1_i} + C_{p2_i}$$

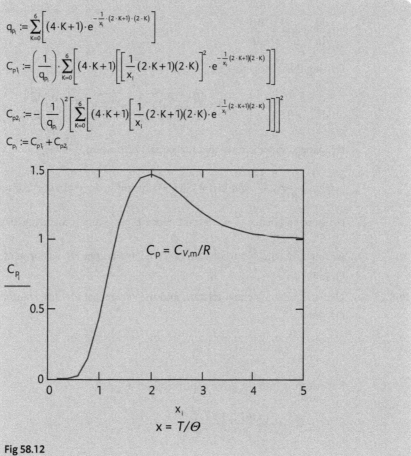

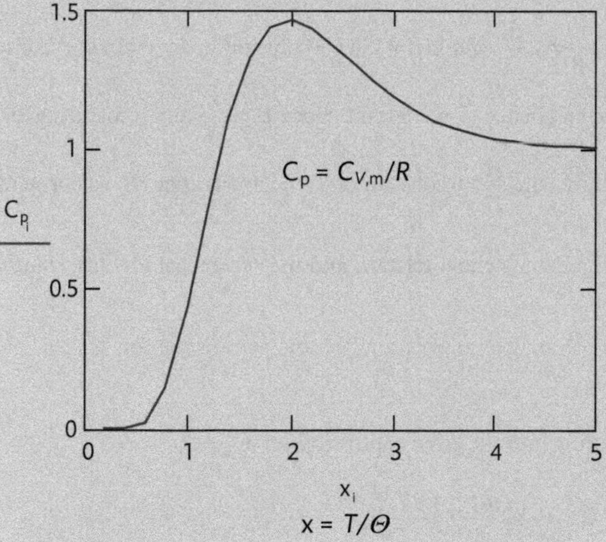

Fig 58.12

P58.19 The differential equation to be solved is $\frac{\mathrm{d}}{\mathrm{d}L}\left(\frac{\mathrm{e}^{-\beta\varepsilon_i}}{q(\beta)}\right)=0$. We differentiate this quotient with respect to L and obtain

$$\frac{\mathrm{d}}{\mathrm{d}L}\left(\frac{\mathrm{e}^{-\beta\varepsilon_i}}{q(\beta)}\right)=\frac{1}{q}\frac{\mathrm{d}(\mathrm{e}^{-\beta\varepsilon_i})}{\mathrm{d}L}-\frac{\mathrm{e}^{-\beta\varepsilon_i}}{q^2}\frac{\mathrm{d}q}{\mathrm{d}L}$$

In an adiabatic process β changes with L. Thus

$$\frac{\mathrm{d}}{\mathrm{d}L}\mathrm{e}^{-\beta\varepsilon_i}=\frac{\partial \mathrm{e}^{-\beta\varepsilon_i}}{\partial\beta}\frac{\mathrm{d}\beta}{\mathrm{d}L}+\frac{\partial \mathrm{e}^{-\beta\varepsilon_i}}{\partial\varepsilon_i}\frac{\mathrm{d}\varepsilon_i}{\mathrm{d}L}$$

Performing the differentiations and using $\varepsilon_i=\frac{\gamma_i}{L^2}$ we obtain

$$\frac{\mathrm{d}}{\mathrm{d}L}\mathrm{e}^{-\beta\varepsilon_i}=-\varepsilon_i\mathrm{e}^{-\beta\varepsilon_i}\frac{\mathrm{d}\beta}{\mathrm{d}L}+\frac{2\beta\varepsilon_i}{L}\mathrm{e}^{-\beta\varepsilon_i}$$

Now relative to the lowest level $q=\left(\frac{2\pi m}{h^2\beta}\right)^{\frac{1}{2}}L$, for a particle in a one-dimensional box. Differentiating q with respect to L, while recognizing again that β changes with L, we obtain (omitting a couple of algebraic steps)

$$\frac{\mathrm{d}q}{\mathrm{d}L}=\frac{q}{L}-\frac{q}{2\beta}\frac{\mathrm{d}\beta}{\mathrm{d}L}$$

Putting this all together we have

$$\frac{\mathrm{d}}{\mathrm{d}L}\left(\frac{\mathrm{e}^{-\beta\varepsilon_i}}{q(\beta)}\right)=\frac{1}{q}\left\{-\varepsilon_i\mathrm{e}^{-\beta\varepsilon_i}\frac{\mathrm{d}\beta}{\mathrm{d}L}+\frac{2\beta\varepsilon_i}{L}\mathrm{e}^{-\beta\varepsilon_i}\right\}-\frac{\mathrm{e}^{-\beta\varepsilon_i}}{q^2}\left\{\frac{q}{L}-\frac{q}{2\beta}\frac{\mathrm{d}\beta}{\mathrm{d}L}\right\}$$

By inspection we can see that this equals zero when $\frac{\mathrm{d}\beta}{\mathrm{d}L}=\frac{2\beta}{L}$

Thus $\frac{\mathrm{d}\beta}{\beta}=2\frac{\mathrm{d}L}{L}$ and $\ln\beta=2\ln L+\text{constant}$ or $\ln\beta=\ln L^2+$ constant, or $\ln\frac{\beta}{L^2}=$ constant and hence $\frac{\beta}{L^2}=\text{constant}$. Since $\beta=\frac{1}{kT}$ and recognizing that k is a constant we finally obtain $\frac{1}{TL^2}=\text{constant}$ or $TL^2=\text{constant}$ or $LT^{\frac{1}{2}}=\text{constant}$.

Q. E. D.

P58.23 Using the Euler's chain relation and the reciprocal identity [*Mathematical Background* 8].

$$\left(\frac{\partial p}{\partial T}\right)_V=-\left(\frac{\partial p}{\partial V}\right)_T\left(\frac{\partial V}{\partial T}\right)_p$$

Substituting into the given expression for C_p-C_V

$$C_p-C_V=-T\left(\frac{\partial p}{\partial V}\right)_T\left(\frac{\partial V}{\partial T}\right)_p^2$$

Using the reciprocal identity again

$$C_p - C_V = -\frac{T\left(\frac{\partial V}{\partial T}\right)_p^2}{\left(\frac{\partial V}{\partial p}\right)_T}$$

For a perfect gas, $pV = nRT$, so

$$\left(\frac{\partial V}{\partial T}\right)_p^2 = \left(\frac{nR}{p}\right)^2 \quad \text{and} \quad \left(\frac{\partial V}{\partial p}\right)_T = -\frac{nRT}{p^2}$$

so $$C_p - C_V = \frac{-T\left(\frac{nR}{p}\right)^2}{-\frac{nRT}{p^2}} = \boxed{nR}$$

P58.25 $$p = \frac{nRT}{V-nb} - \frac{n^2a}{V^2} \quad \text{[van der Waals equation]}$$

Hence $$\boxed{T = \left(\frac{p}{nR}\right)\times(V-nb) + \left(\frac{na}{RV^2}\right)\times(V-nb)}$$

$$\boxed{\left(\frac{\partial T}{\partial p}\right)_V = \frac{V-nb}{nR}} = \frac{V_m - b}{R} = \frac{1}{\left(\frac{\partial p}{\partial T}\right)_V}$$

For Euler's chain relation, we need to show that $\left(\frac{\partial T}{\partial p}\right)_V\left(\frac{\partial p}{\partial V}\right)_T\left(\frac{\partial V}{\partial T}\right)_p = -1$

Hence, in addition to $\left(\frac{\partial T}{\partial p}\right)_V$ we need $\left(\frac{\partial p}{\partial V}\right)_T$ and $\left(\frac{\partial V}{\partial T}\right)_p = \frac{1}{\left(\frac{\partial T}{\partial V}\right)_p}$

which can be found from

$$\left(\frac{\partial p}{\partial V}\right)_T = \frac{-nRT}{(V-nb)^2} + \frac{2n^2a}{V^3}$$

$$\left(\frac{\partial T}{\partial V}\right)_p = \left(\frac{p}{nR}\right) + \left(\frac{na}{RV^2}\right) - \left(\frac{2na}{RV^3}\right)\times(V-nb)$$

$$\left(\frac{\partial T}{\partial V}\right)_p = \left(\frac{T}{V-nb}\right) - \left(\frac{2na}{RV^3}\right)\times(V-nb)$$

Therefore,

$$\left(\frac{\partial T}{\partial p}\right)_V\left(\frac{\partial p}{\partial V}\right)_T\left(\frac{\partial V}{\partial T}\right)_p = \frac{\left(\frac{\partial T}{\partial p}\right)_V\left(\frac{\partial p}{\partial V}\right)_T}{\left(\frac{\partial T}{\partial V}\right)_p}$$

$$=\frac{\left(\frac{V-nb}{nR}\right)\times\left(\frac{-nRT}{(V-nb)^2}+\frac{2n^2a}{V^3}\right)}{\left(\frac{T}{V-nb}\right)-\left(\frac{2na}{RV^3}\right)\times(V-nb)}=\frac{\left(\frac{-T}{V-nb}\right)+\left(\frac{2na}{RV^3}\right)\times(V-nb)}{\left(\frac{T}{V-nb}\right)-\left(\frac{2na}{RV^3}\right)\times(V-nb)}$$

$$=-1$$

P58.27 $c=\left(\frac{RT\gamma}{M}\right)^{1/2}$, $p=\rho\frac{RT}{M}$, so $\frac{RT}{M}=\frac{p}{\rho}$; hence $\boxed{c=\left(\frac{\gamma p}{\rho}\right)^{1/2}}$

For argon, $\gamma=\frac{5}{3}$, so $c=\left(\frac{(8.314\,\mathrm{J\,K^{-1}\,mol^{-1}})\times(298\,\mathrm{K})\times\frac{5}{3}}{39.95\times10^{-3}\,\mathrm{kg\,mol^{-1}}}\right)^{1/2}=\boxed{322\,\mathrm{m\,s^{-1}}}$

Focus 12: Integrated activities

F12.1 $\Delta\varepsilon=\varepsilon=g\mu_B\mathcal{B}_0$ [50.2]

$q=1+\mathrm{e}^{-b\varepsilon}$

$C_{V,\mathrm{m}}/R=\frac{x^2\mathrm{e}^{-x}}{(1+\mathrm{e}^{-x})^2}$[Problem 58.3], $x=2\mu_B\mathcal{B}_0\beta$ [$g=2$ for electrons]

Therefore, if $\mathcal{B}_0=10.0\,\mathrm{T}$,

$$x=\frac{(2)\times(9.274\times10^{-24}\,\mathrm{J\,T^{-1}})\times(10.0\,\mathrm{T})}{(1.381\times10^{-23}\,\mathrm{J\,K^{-1}})\times T}=\frac{13.44}{T/\mathrm{K}}$$

(a) $T=100\,\mathrm{K}$, $x=0.134$, $C_V=4.47\times10^{-3}R$, implying that $C_V=3.7\times10^{-2}\,\mathrm{J\,K^{-1}\,mol^{-1}}$

Since the equipartition value is about $3R$ [$\nu_R^*=3, \nu_V^*\approx0$], the field brings about a change of appoximately $\boxed{0.1\text{ per cent.}}$

(b) $T=298\,\mathrm{K}$, $x=4.51\times10^{-2}$, $C_V=5.08\times10^{-4}R$, implying that $C_V=4.22\,\mathrm{mJ\,K^{-1}\,mol^{-1}}$, a change of about $\boxed{2\times10^{-2}\text{ percent}}$

Question. What percentage change would a magnetic field of 1 kT cause?

F12.3 The difference results from the definition $H=U+PV$; hence $\Delta H=\Delta U+\Delta(PV)$. As $\Delta(PV)$ is not usually zero, except for isothermal processes in a perfect gas, the difference between ΔH and ΔU is a non-zero quantity. Section 56.1 of the text shows that ΔH can be interpreted as the heat associated with a process at constant pressure, and ΔU as the heat at constant volume.

F12.5 (a) $U=U(T,V)$ so $\mathrm{d}U=\left(\frac{\partial U}{\partial T}\right)_V\mathrm{d}T+\left(\frac{\partial U}{\partial V}\right)_T\mathrm{d}V=C_V\,\mathrm{d}T+\left(\frac{\partial U}{\partial V}\right)_T\mathrm{d}V$

For $U=$ constant, $\mathrm{d}U=0$, and

$$C_V\,\mathrm{d}T=-\left(\frac{\partial U}{\partial V}\right)_T\mathrm{d}V \text{ or } C_V=-\left(\frac{\partial U}{\partial V}\right)_T\left(\frac{\mathrm{d}V}{\mathrm{d}T}\right)_U=-\left(\frac{\partial U}{\partial V}\right)_T\left(\frac{\partial V}{\partial T}\right)_U$$

This relationship is essentially Euler's chain relation [*MathematicalBackground* 8].

(b) $H = H(T,p)$ so $\mathrm{d}H = \left(\frac{\partial H}{\partial T}\right)_p \mathrm{d}T + \left(\frac{\partial H}{\partial p}\right)_T \mathrm{d}p = C_p\,\mathrm{d}T + \left(\frac{\partial H}{\partial p}\right)_T \mathrm{d}p$

According to Euler's chain relation

$$\left(\frac{\partial H}{\partial p}\right)_T \left(\frac{\partial p}{\partial T}\right)_H \left(\frac{\mathrm{d}T}{\mathrm{d}H}\right)_p = -1$$

so, using the reciprocal identity [*MathematicalBackground* 8].

$$\left(\frac{\partial H}{\partial p}\right)_T = -\left(\frac{\partial T}{\partial p}\right)_H \left(\frac{\mathrm{d}H}{\mathrm{d}T}\right)_p = -\mu C_p$$

F12.7 (a) $\mu = -\frac{1}{C_p}\left(\frac{\partial H}{\partial p}\right)_T = \frac{1}{C_p}\left\{T\left(\frac{\partial V_\mathrm{m}}{\partial T}\right)_p - V_\mathrm{m}\right\}$ [56.9, 56.10 and Problem 56.6]

$$V_\mathrm{m} = \frac{RT}{p} + aT^2 \quad \text{so} \quad \left(\frac{\partial V_\mathrm{m}}{\partial T}\right)_p = \frac{R}{p} + 2aT$$

$$\mu = \frac{1}{C_p}\left\{\frac{RT}{p} + 2aT^2 - \frac{RT}{p} - aT^2\right\} = \boxed{\frac{aT^2}{C_p}}$$

(b) $C_V = C_p - \alpha T V_\mathrm{m}\left(\frac{\partial p}{\partial T}\right)_V = C_p - T\left(\frac{\partial V_\mathrm{m}}{\partial T}\right)_p \left(\frac{\partial p}{\partial T}\right)_V$

But, $p = \frac{RT}{V_\mathrm{m} - aT^2}$

$$\left(\frac{\partial p}{\partial T}\right)_V = \frac{R}{V_\mathrm{m} - aT^2} - \frac{RT(-2aT)}{(V_\mathrm{m} - aT^2)^2}$$

$$= \frac{R}{(RT/p)} + \frac{2aRT^2}{(RT/p)^2} = \frac{p}{T} + \frac{2ap^2}{R}$$

Therefore

$$C_V = C_p - T\left(\frac{R}{p} + 2aT\right) \times \left(\frac{p}{T} + \frac{2ap^2}{R}\right)$$

$$= C_p - \frac{RT}{p}\left(1 + \frac{2apT}{R}\right) \times \left(1 + \frac{2apT}{R}\right) \times \left(\frac{p}{T}\right)$$

$$\boxed{C_V = C_p - R\left(1 + \frac{2apT}{R}\right)^2}$$

Topic 59 The Second Law

Assume that all gases are perfect and that data refer to 298.15 K unless otherwise stated.

Discussion questions

D59.1 We must remember that the second law of thermodynamics states only that the total entropy of both the system (here, the molecules organizing themselves into cells) and the surroundings (here, the medium) must increase in a naturally occurring process. It does not state that entropy must increase in a portion of the universe that interacts with its surroundings. In this case, the cells grow by using chemical energy from their surroundings (the medium) and in the process the increase in the entropy of the medium outweighs the decrease in entropy of the system. Hence, the second law is not violated.

Exercises

E59.1(a) All spontaneous processes are irreversible processes, which implies through eqn 61.9, the Clausius inequality, that $\Delta S_{tot} = \Delta S_{sys} + \Delta S_{surr} > 0$, for all spontaneous processes. In this case, $\Delta S_{tot} = 0$. Therefore, the process is $\boxed{\text{not spontaneous}}$.

Topic 60 The statistical entropy

Discussion questions

D60.1 For a thorough discussion of the relationship between the thermodynamic and statistical definitions of entropy, see Topics 60 and 61. We will not repeat all of that discussion here and will merely summarize the main points.

The thermodynamic entropy is defined in terms of the quantity $dS = \frac{dq_{rev}}{T}$ where dq_{rev} is the infinitesimal quantity of energy supplied as heat to the system reversibly at a temperature T.

The statistical entropy is defined in terms of the Boltzmann formula for the entropy: $S = k \ln W$ where k is the Boltzmann constant and W is the number of microstates, the total number of ways in which the molecules of the system can be arranged to achieve the same total energy of the system. These two definitions turn out to be equivalent provided the thermodynamic entropy is taken to be zero at $T = 0$.

The concept of the number of microstates makes quantitative the ill-defined qualitative concepts of 'disorder' and 'dispersal of matter and energy' that are used widely to introduce the concept of entropy: a more 'disorderly' distribution of energy and matter corresponds to a greater number of microstates associated with the same total energy. The more molecules that can participate in the distribution of energy, the more microstates there are for a given total energy and the greater the entropy than when the energy is confined to a smaller number of molecules.

The molecular interpretation of entropy given by the Boltzmann formula also suggests the thermodynamic definition. At high temperatures where the molecules of a system can occupy a large number of available energy levels, a small additional transfer of energy as heat will cause only a small change in the number of accessible energy levels, whereas at low temperatures the transfer of the same quantity of heat will increase the number of accessible energy levels and microstates significantly. Hence, the change in entropy upon heating will be greater when the energy is transferred to a cold body than when it is transferred to a hot body. This argument suggests that the change in entropy should be inversely proportional to the temperature at which the transfer takes place as in indicated in the thermodynamic definition.

60.3 See Section 52.2(a) for the derivation from the Boltzmann distribution of the formula for the partition function of a perfect gas, which is given by $q = \frac{V}{\Lambda^3}$, with $\Lambda = \frac{h}{(2\pi mkT)^{1/2}}$. The general expression for the entropy is

$S = \dfrac{U - U(0)}{T} + k\ln Q$ [60.5]. For indistinguishable, non-interacting particles, $Q = \dfrac{q^N}{N!}$. After insertion of this expression for Q, we obtain for the entropy $S = \dfrac{U - U(0)}{T} + Nk\ln\dfrac{q\mathrm{e}}{N}$ [60.3b]. From this expression, we derive the Sackur-Tetrode equation, $S_\mathrm{m} = R\ln\left(\dfrac{V_\mathrm{m}\mathrm{e}^{5/2}}{N_\mathrm{A}\Lambda^3}\right)$ [60.6b], for the entropy of a perfect gas. See Section 60.3(b) for the derivation. Because the molar volume appears in the numerator, the molar entropy increases in proportion to the natural log of the molar volume. In terms of the Boltzmann distribution this is natural: large containers have more closely spaced energy levels than small containers, so more states are thermally accessible. Because temperature appears in the numerator (the denominator of Λ), the molar entropy increases with temperature. The reason for this behavior from the point of view of the Boltzmann distribution is that more energy levels become accessible as temperature increases.

Exercises

E60.1(a) $S_\mathrm{m}^\ominus = R\ln\left(\dfrac{\mathrm{e}^{5/2}kT}{p^\ominus\Lambda^3}\right)$ [60.6b with $p = p^\ominus$]

(a) $$\Lambda = \frac{h}{(2\pi mkT)^{1/2}} = \frac{6.626\times10^{-34}\ \mathrm{J\,s}}{[(2\pi)\times(4.003)\times(1.6605\times10^{-27}\ \mathrm{kg})\times(1.381\times10^{-23}\ \mathrm{J\,K^{-1}}T)]^{1/2}}$$

$$= \frac{8.726\times10^{-10}\ \mathrm{m}}{(T/\mathrm{K})^{1/2}}$$

$$S_\mathrm{m}^\ominus = R\ln\left(\frac{(\mathrm{e}^{5/2})\times(1.381\times10^{-23}\ \mathrm{J\,K^{-1}}T)}{(1.013\times10^5\ \mathrm{Pa})\times(8.726\times10^{-10}\ \mathrm{m})^3}\right)\times\left(\frac{T}{\mathrm{K}}\right)^{3/2}$$

$$= R\ln(2.499\times(T/\mathrm{K})^{5/2})$$

$T = 298.15\ \mathrm{K}$, $S_\mathrm{m}^\ominus = (8.314\ \mathrm{J\,K^{-1}\,mol^{-1}})\times\ln(2.499\times(298)^{5/2})$

$$\boxed{= 126\ \mathrm{J\,K^{-1}\,mol^{-1}}}$$

(b) $$\Lambda = \frac{h}{(2\pi mkT)^{1/2}}$$

$$= \frac{6.626\times10^{-34}\ \mathrm{J\,s}}{[(2\pi)\times(131.29)\times(1.6605\times10^{-27}\ \mathrm{kg})\times(1.381\times10^{-23}\ \mathrm{J\,K^{-1}}T)]^{1/2}}$$

$$= \frac{1.524\times10^{-10}\ \mathrm{m}}{(T/\mathrm{K})^{1/2}}$$

$$S_\mathrm{m}^\ominus = R\ln\left(\frac{(\mathrm{e}^{5/2})\times(1.381\times10^{-23}\ \mathrm{J\,K^{-1}}\ T)}{(1.013\times10^5\ \mathrm{Pa})\times(1.524\times10^{-10}\ \mathrm{m})^3}\right)\times\left(\frac{T}{\mathrm{K}}\right)^{3/2}$$

$$= R\ln(469.1\times(T/\mathrm{K})^{5/2})$$

$T = 298.15\ \mathrm{K}$

$$S_\mathrm{m}^\ominus = (8.314\ \mathrm{J\,K^{-1}\,mol^{-1}})\times\ln(469.1\times(298)^{5/2}) \boxed{= 169\ \mathrm{J\,K^{-1}\,mol^{-1}}}$$

E60.2(a) From the solution to exercise 60.1(a) we have for helium

$S_m^{\ominus} = (8.314\ \mathrm{J\,K^{-1}\,mol^{-1}}) \times \ln(2.499 \times (T)^{5/2}) = 169\ \mathrm{J\,K^{-1}\,mol^{-1}}$. We solve for T.

$\boxed{T = 2.35 \times 10^3\ \mathrm{K}}$

E60.3(a) The rotational partition function of a non-linear molecule is (after substituting the numerical values of the constants in eqn 52.14)

$$q^R = \frac{1.0270}{\sigma}\frac{(T/\mathrm{K})^{3/2}}{(\tilde{A}\tilde{B}\tilde{C}/\mathrm{cm^{-3}})^{1/2}}\ [52.14]$$

$$= \frac{1.0270 \times 298^{3/2}}{(2) \times (27.878 \times 14.509 \times 9.287)^{1/2}}\ [\sigma = 2 \text{ from Table 43.1}] = \boxed{43.1}$$

$$\theta_R = \frac{hc(\tilde{A}\tilde{B}\tilde{C})^{1/3}}{k} = \frac{(6.626 \times 10^{-34}\ \mathrm{J\,s}) \times (2.998 \times 10^{10}\ \mathrm{cm\,s^{-1}}) \times [(27.878) \times (14.509) \times (9.287)\ \mathrm{cm^{-3}}]^{1/3}}{1.38 \times 10^{-23}\ \mathrm{J\,K^{-1}}}$$

$$= \boxed{22.36\ \mathrm{K}}$$

The high-temperature approximation is valid if $T > \theta_R$, where $\theta_R = \dfrac{hc(\tilde{A}\tilde{B}\tilde{C})^{1/3}}{k}$

Thus the high temperature approximation is valid.

$q^R = 43.1$ [Exercise 60.3(a) above]

All the rotational modes of water are fully active at 25°C; therefore

$$U_m^R - U_m^R(0) = E^R = \frac{3}{2}RT$$

$$S_m^R = \frac{E^R}{T} + R\ln q^R$$

$$= \frac{3}{2}R + R\ln 43.1 = \boxed{43.76\ \mathrm{J\,K^{-1}\,mol^{-1}}}$$

Comment. Division of q^R by $N_A!$ is not required for the internal contributions; internal motions may be thought of as localized (distinguishable). It is the overall canonical partition function, which is a product of internal and external contributions, that is divided by $N_A!$

E60.4(a) For CO_2, the high temperature approximation is valid and we may write

$$q_R = \frac{kT}{2hc\tilde{B}} = \frac{207.22\ \mathrm{cm^{-1}}}{2\tilde{B}} = \frac{207.22\ \mathrm{cm^{-1}}}{2 \times 0.3902\ \mathrm{cm^{-1}}} = 265.5$$

We can now use $S_m = \dfrac{U_m - U_m(0)}{T} + k\ln Q\ [60.5] = \dfrac{U_m - U_m(0)}{T} + R\ln q_R$ with

$U_m - U_m(0) = RT$. Therefore,

$$S_m = \frac{RT}{T} + R\ln q_R = 8.314\ \mathrm{J\,K^{-1}\,mol^{-1}} + 8.314\ \mathrm{J\,K^{-1}\,mol^{-1}} \times \ln 265.5$$

$$= \boxed{54.72\ \mathrm{J\,K^{-1}\,mol^{-1}}}$$

E60.5(a) We assume that the upper nine of the $(2\times\frac{9}{2}+1)=10$ spin-orbit states of the ion lie at an energy much greater than kT at 1 K; hence, since the spin degeneracy of Co^{2+} is 4 (the ion is a spin quartet), $q = 4$. The contribution to the entropy is

$$R\ln q = (8.314\ \mathrm{J\,K^{-1}mol^{-1}})\times(\ln 4) = \boxed{11.5\ \mathrm{J\,K^{-1}\,mol^{-1}}}$$

E60.6(a) The molar entropy of a collection of oscillators is given by

$$S_{\mathrm{m}} = \frac{U_{\mathrm{m}}-U_{\mathrm{m}}(0)}{T}+k\ln Q\ [60.5] = \frac{N_{\mathrm{A}}\langle\varepsilon\rangle}{T}+R\ln q$$

where $$\langle\varepsilon\rangle = \frac{hc\tilde{\nu}}{\mathrm{e}^{\beta hc\tilde{\nu}}-1} = k\frac{\theta}{\mathrm{e}^{\theta/T}-1}\ [53.8],\quad q=\frac{1}{1-\mathrm{e}^{-\beta hc\tilde{\nu}}}=\frac{1}{1-\mathrm{e}^{-\theta/T}}\ [52.15]$$

and θ_{V} is the vibrational temperature $hc\tilde{\nu}/k$. Thus

$$S_{\mathrm{m}} = \frac{R(\theta/T)}{\mathrm{e}^{\theta/T}-1} - R\ln(1-\mathrm{e}^{-\theta/T})$$

The vibrational entropy of formic acid is the sum of contributions of this form from each of its nine normal modes. The table below shows results from a spreadsheet programmed to compute S_{m}/R at a given temperature for the normal-mode wavenumbers of formic acid.

		T = 298 K		T = 500 K	
$\tilde{\nu}$ / cm^{-1}	θ_{V}/K	T/θ_{V}	S_{m}/R	T/θ_{V}	S_{m}/R
638	918.0	0.324635	1.624247	0.544689	4.340205
1033	1 486.3	0.200501	0.341767	0.33641	1.769583
625	899.3	0.331387	1.707377	0.556019	4.471129
1105	1 589.9	0.187436	0.254961	0.31449	1.500608
1229	1 768.3	0.168525	0.153044	0.28276	1.127318
1387	1 995.6	0.149327	0.079118	0.250549	0.779719
1770	2 546.7	0.117015	0.015427	0.196334	0.312684
2943	4 234.4	0.070376	8.53E-05	0.118081	0.016529
3570	5 136.5	0.058016	4.95E-06	0.097342	0.003239
			4.176032		14.32101

(a) At 298 K, $S_{\mathrm{m}} = 4.176\,R = \boxed{34.72\ \mathrm{J\,mol^{-1}\,K^{-1}}}$

(b) At 500 K, $S_{\mathrm{m}} = 14.32R = \boxed{119.06\ \mathrm{J\,mol^{-1}\,K^{-1}}}$

Comment. These calculated values are the vibrational contributions to the standard molar entropy of formic acid. The total molar entropy would also include translational and rotational contributions, but without knowledge of the rotational constants the total molar entropy cannot be calculated.

Problems

P60.1 The absorption lines are the values of differences in adjacent rotational terms. Using eqns 41.6, and 41.8, we have

$$F(J+1)-F(J)=\frac{E(J+1)-E(J)}{hc}=2\tilde{B}(J+1)$$

for $J = 0, 1, \ldots$. Therefore, we can find the rotational constant and reconstruct the energy levels from the data of Problem 52.6. To make use of all of the data, one would plot the wavenumbers, which represent $F(J+1)-F(J)$, vs. J; the slope of that linear plot is $2\tilde{B}$. However, in this case, plotting the data is not necessary because inspection of the data shows that the lines in the spectrum are equally spaced with a separation of 21.19 cm^{-1}, so that is the slope:

$$\text{slope} = 21.19\ \text{cm}^{-1} = 2\tilde{B} \quad \text{and hence} \quad \tilde{B} = 10.59\bar{5}\ \text{cm}^{-1}$$

The partition function is

$$q=\sum_{J=0}^{\infty}(2J+1)\mathrm{e}^{-\beta E(J)} \qquad \text{where} \qquad E(J) = hc\tilde{B}J(J+1)\ [41.6]$$

and the factor $(2J+1)$ is the degeneracy of the energy levels. Making these substitutions into the partition function q we obtain

$$q^{\mathrm{R}}=\sum_J(2J+1)\mathrm{e}^{-\frac{hc\tilde{B}}{kT}J(J+1)}\ [52.11]=\sum_J(2J+1)\mathrm{e}^{-\frac{\theta_{\mathrm{R}}}{T}J(J+1)} \quad [\theta_{\mathrm{R}}=hc\tilde{B}/k]$$

For HCl, $\theta_{\mathrm{R}}=hc\tilde{B}/k=15.244$ K. Defining $x\equiv\frac{T}{\theta_{\mathrm{R}}}$, q^{R} may be rewritten

$$q^{\mathrm{R}}=\sum_J(2J+1)\mathrm{e}^{-J(J+1)/x}$$

At temperatures above about 30 K the high temperature approximation for q^{R} would be adequate to calculate the molar entropy, but at lower temperatures the summation needs to be performed.

The molar entropy is calculated from $S_{\mathrm{m}}=\frac{U_{\mathrm{m}}-U_{\mathrm{m}}(0)}{T}+k\ln Q\ [60.5]=\frac{U_{\mathrm{m}}-U_{\mathrm{m}}(0)}{T}+R\ln q^{\mathrm{R}}$

and the molar energy from

$$U_{\mathrm{m}}-U_{\mathrm{m}}(0)=N_{\mathrm{A}}\langle\varepsilon^{\mathrm{R}}\rangle=-\frac{1}{q^{\mathrm{R}}}\left(\frac{\partial q^{\mathrm{R}}}{\partial\beta}\right)_V$$

$$U_{\mathrm{m}}-U_{\mathrm{m}}(0)=\frac{1}{q^{\mathrm{R}}}N_{\mathrm{A}}hc\tilde{B}\sum_{J=1}(2J+1)[J(J+1)]\mathrm{e}^{-J(J+1)/x} \quad \text{or since } hc\tilde{B}=k\theta_{\mathrm{R}}$$

$$U_{\mathrm{m}}-U_{\mathrm{m}}(0)=\frac{1}{q^{\mathrm{R}}}R\theta_{\mathrm{R}}\sum_{J=1}(2J+1)[J(J+1)]\mathrm{e}^{-J(J+1)/x}$$

Substituting into the expression for the entropy we obtain

$$S_m = \frac{1}{q^R} R \frac{1}{x} \sum_{J=1} (2J+1)[J(J+1)]e^{-J(J+1)/x} + R\ln q^R$$

S_m and q^R are best evaluated with a spreadsheet program such as Excel® or a CAS system such as Mathcad®. Here we have used Mathcad®. See the Mathcad® worksheet below. S_m/R is plotted as a function of $x \equiv \frac{T}{\theta_R}$ in Fig. 60.1.

$i = 0..47$

$x_{start} := 0.2 \qquad x_{end} := 9.6$

$x_i := x_{start} + (x_{end} - x_{start}) \cdot \frac{i}{47} \qquad x_i = T/\Theta$

$q_i := \sum_{J=0}^{6} \left[(2.J+1) \cdot e^{-J\frac{J+1}{x_i}} \right] \qquad q_i = qR$

$U_i := \left(\frac{1}{q_i}\right) \left[\frac{1}{x_i} \sum_{J=0}^{6} \left[(2 \cdot J+1) \cdot J \cdot (J+1) \cdot e^{-\left(J\frac{J+1}{x_i}\right)} \right] \right] \qquad U_i = U_m/RT$

$S_i := 8.314(U_i + \ln(q_i)) \qquad S_i = S_m/R$

$S_i = S_m/R$

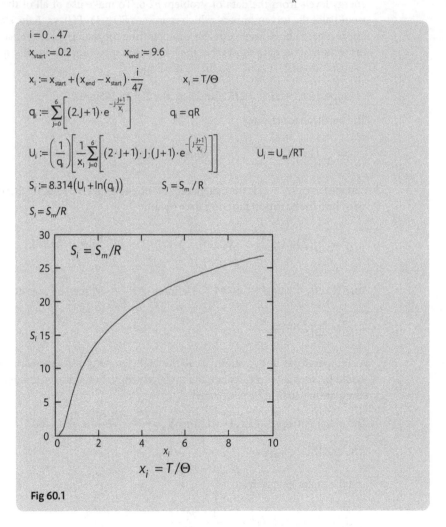

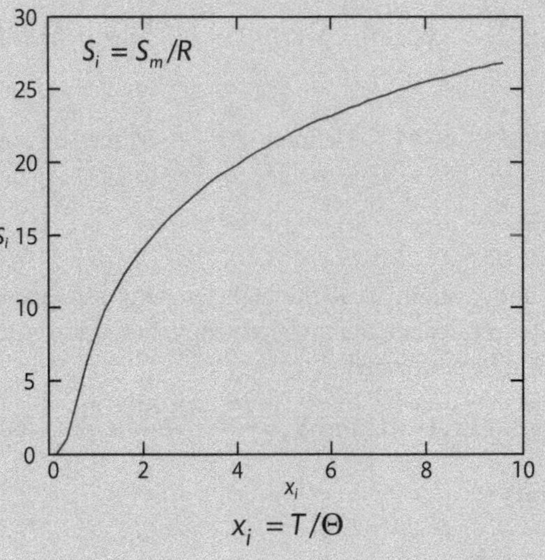

Fig 60.1

P60.3 The molar entropy is given by

$$S_m = \frac{U_m - U_m(0)}{T} + R\left(\ln\frac{q_m}{N_A} - 1\right) \quad \text{where} \frac{U_m - U_m(0)}{T} = -N_A\left(\frac{\partial \ln q}{\partial \beta}\right)_V$$

where $\dfrac{U_m - U_m(0)}{T} = -N_A\left(\dfrac{\partial \ln q}{\partial \beta}\right)_V$ and $\dfrac{q_m}{N_A} = \dfrac{q_m^T}{N_A} q^R q^V q^E$

The energy term $U_m - U_m(0)$ works out to be

$$U_m - U_m(0) = N_A[\langle \varepsilon^T \rangle + \langle \varepsilon^R \rangle + \langle \varepsilon^V \rangle + \langle \varepsilon^E \rangle]$$

Translation: [Section 52.2(a) and *Brief Illustration* 52.2] After some calculation we find

$$\frac{q_m^{T\ominus}}{N_A} = 2.561\times10^{-2}(T/K)^{5/2}\times(M/\text{g mol}^{-1})^{3/2}$$

$$= 2.561\times10^{-2}\times(298)^{5/2}\times(38.00)^{3/2} = 9.20\times10^6$$

and $\langle \varepsilon^T \rangle = \tfrac{3}{2}kT$

Rotation of a linear molecule: [Section 52.2(b)]. After some calculation we find

$$q^R = \frac{0.6950}{\sigma}\times\frac{T/K}{\tilde{B}/\text{cm}^{-1}}$$

The rotational constant is

$$\tilde{B} = \frac{\hbar}{4\pi c I} = \frac{\hbar}{4\pi c \mu R^2}$$

$$= \frac{(1.0546\times10^{-34}\text{ J s})\times(6.022\times10^{23}\text{ mol}^{-1})}{4\pi(2.998\times10^{10}\text{ cm s}^{-1})\times(\tfrac{1}{2}\times19.00\times10^{-3}\text{ kg mol}^{-1})\times(190.0\times10^{-12}\text{ m})^2}$$

$$= 0.4915\text{ cm}^{-1}$$

so $q^R = \dfrac{0.6950}{2}\times\dfrac{298}{0.4915} = 210.\bar{7}$

Also $\langle \varepsilon^R \rangle = kT$

Vibration:

$$q^V = \frac{1}{1-e^{-hc\tilde{\nu}/kT}} = \frac{1}{1-\exp\left(\dfrac{-1.4388(\tilde{\nu}/\text{cm}^{-1})}{T/K}\right)} = \frac{1}{1-\exp\left(\dfrac{-1.4388(450.0)}{298}\right)}$$

$$= 1.129$$

$$\langle \varepsilon^V \rangle = \frac{hc\tilde{\nu}}{e^{hc\tilde{\nu}/kT}-1} = \frac{(6.626\times10^{-34}\text{ J s})\times(2.998\times10^{10}\text{ cm s}^{-1})\times(450.0\text{ cm}^{-1})}{\exp\left(\dfrac{1.4388(450.0)}{298}\right)-1}$$

$$= 1.149\times10^{-21}\text{ J}$$

The Boltzmann factor for the lowest-lying excited electronic state is

$$\exp\left(\frac{-(1.609\text{ eV})\times(1.602\times10^{-19}\text{ J eV}^{-1})}{(1.381\times10^{-23}\text{ J K}^{-1})\times(298\text{ K})}\right) = 6\times10^{-28}$$

so we may take q^{E} to equal the degeneracy of the ground state, namely 2 and $\langle \varepsilon^{\mathrm{E}} \rangle$ to be zero. Putting it all together yields

$$\frac{U_{\mathrm{m}}-U_{\mathrm{m}}(0)}{T}=\frac{N_{\mathrm{A}}}{T}\left(\tfrac{3}{2}kT+kT+1.149\times10^{-21}\,\mathrm{J}\right)=\tfrac{5}{2}R+\frac{N_{\mathrm{A}}(1.149\times10^{-21}\,\mathrm{J})}{T}$$

$$=(2.5)\times(8.3145\,\mathrm{J\,mol^{-1}\,K^{-1}})+\frac{(6.022\times10^{23}\,\mathrm{mol^{-1}})\times(1.149\times10^{-21}\,\mathrm{J})}{298\,\mathrm{K}}$$

$$=23.11\,\mathrm{J\,mol^{-1}\,K^{-1}}$$

$$R\left(\ln\frac{q_{\mathrm{m}}}{N_{\mathrm{A}}}-1\right)=(8.3145\,\mathrm{J\,mol^{-1}\,K^{-1}})\times\{\ln[(9.20\times10^{6})\times(210.7)\times(1.129)\times(2)]-1\}$$

$$=176.3\,\mathrm{J\,mol^{-1}\,K^{-1}}\quad\text{and}\quad S_{\mathrm{m}}^{\circ}=\boxed{199.4\,\mathrm{J\,mol^{-1}\,K^{-1}}}$$

P60.5 The solution is provided in the MathCad® worksheet which is inserted below.

In this problem we will compare the entropy of a Morse oscillator with the entropy of a harmonic oscillator in the case that is characterized by the parameters:

Depth of potential minimum: $D_e := 50\cdot10^3\cdot\mathrm{cm}^{-1}$
Fundamental frequency: $v_e := 2000\ \mathrm{cm}^{-1}$
Anharmonicity constant: $x_e := 0.01$
Maximum quantum number of harmonic oscillator: $v_{harm} := 24$
Maximum quantum number of Morse oscillator: $v_{Morse} := 49$

Constants:

$h := 6.6260693\,10^{-34}\cdot \mathrm{joule}\cdot\mathrm{sec}$ $\quad c := 299\,792\,458\dfrac{\mathrm{m}}{\mathrm{sec}}$

$N_A := 6.0221415\,10^{23}\cdot\mathrm{mole}^{-1}$ $\quad k := 1.3806505\,10^{-23}\cdot\dfrac{\mathrm{joule}}{\mathrm{K}}$

Energy levels:

$G(v,x) := \left[(v+.5)\cdot v_e-(v+.5)^2\cdot x\cdot v_e\right]\cdot h\cdot c$ $\quad \varepsilon_{gs}(x) := G(0,x)$

Energy levels relative to zero for the lowest energy:

$G_{rel}(v,x) := G(v,x)-\varepsilon_{gs}(x)$

Molecular partition function:

$T := 5\cdot K, 25\cdot K..2000$ $\quad q(T,x,v_{max}) := \displaystyle\sum_{v=0}^{v_{max}} e^{\frac{-G_{rel}(v,x)}{k\cdot T}}$

Mean energy [53.4a]:

$$\varepsilon_{mean}(T,x,v_{max}) := \frac{k\cdot T^2}{q(T,x,v_{max})}\cdot\left(\frac{d}{dT}q(T,x,v_{max})\right)$$

Entropy for distinguishable oscillators [60.3a]:

$$S_m(T, x, v_{max}) := \frac{N_A \cdot \varepsilon_{mean}(T, x, v_{max})}{T} + N_A \cdot k \cdot \ln(q(T, x, v_{max}))$$

Molar entropy harmonic oscillator: $S_{harm}(T) := S_m(T, 0, v_{harm})$
Molar entropy Morse oscillator: $S_{Morse}(T) := S_m(T, x_e, v_{Morse})$

$T := 5K, 25K..2000K$

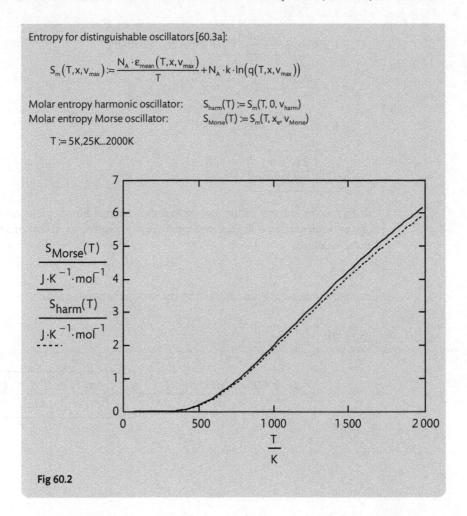

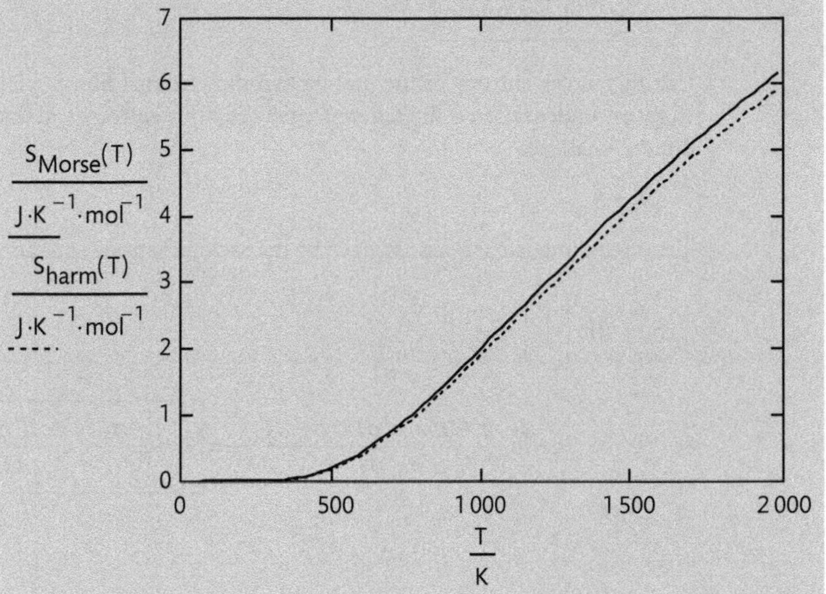

Fig 60.2

The above plot of the Morse oscillator entropy against temperature, when compared to the similar plot for the harmonic oscillator, shows that the Morse oscillator has the greater entropy. This happens because the Morse oscillator has the greater number of available energy states at any temperature. However, the difference is remarkably small.

P60.7 A Sackur-Tetrode type of equation describes the translational entropy of the gas. Here

$$q^T = q_x^T q_y^T \quad \text{with } q_x^T = \left(\frac{2\pi m}{\beta h^2}\right)^{1/2} X \ [52.7a]$$

where X is the length of the surface. Therefore,

$$q^T = \left(\frac{2\pi m}{\beta h^2}\right) XY = \frac{2\pi m\sigma}{\beta h^2}, \quad \sigma = XY$$

$$U_m - U_m(0) = -\frac{N_A}{q}\left(\frac{\partial q}{\partial \beta}\right) = RT \text{ [or by equipartition]}$$

$$S_m = \frac{U_m - U_m(0)}{T} + R(\ln q_m - \ln N_A + 1)\left[q_m = \frac{q}{n}\right]$$

$$= R + R\ln\left(\frac{eq_m}{N_A}\right) = R\ln\left(\frac{e^2 q_m}{N_A}\right)$$

$$= \boxed{R\ln\left(\frac{2\pi e^2 m\sigma_m}{h^2 N_A \beta}\right)}\left[\sigma_m = \frac{\sigma}{n}\right]$$

Call this molar entropy of the mobile two-dimensional film S_{m2}. The molar entropy of condensation is the difference between this entropy and that of a (three-dimensional) gas:

$$\Delta S_m = S_{m2} - S_{m3}$$

The three-dimensional value is given by the Sackur-Tetrode equation

$$S_m = R\ln\left\{e^{5/2}\left(\frac{2\pi m}{h^2\beta}\right)^{3/2}\frac{V_m}{N_A}\right\}$$

So $$\Delta S_m = R\ln\frac{e^2(2\pi m/h^2\beta)\times(\sigma_m/N_A)}{e^{5/2}(2\pi m/h^2\beta)^{3/2}\times(V_m/N_A)} = \boxed{R\ln\left\{\left(\frac{\sigma_m}{V_m}\right)\times\left(\frac{h^2\beta}{2\pi me}\right)^{1/2}\right\}}$$

Topic 61 The thermodynamic entropy

Discussion question

D61.1 For a thorough discussion of the relationship between the various formulations of the second law, see Sections 60.1 and 61.1. We will not repeat all of that discussion here and will merely summarize the main points.

There are two equivalent statements of the second law that are based on directly observable processes:

1. The **Kelvin statement:** No process is possible in which the sole result is the absorption of heat from a reservoir and its complete conversion into work.
2. The **Clausius statement:** No process is possible in which the sole result is the transfer of energy from a cooler to a hotter body.

It can be shown that these statements are equivalent and that they both lead to the existence of a state function of the system, S, called the entropy, defined through the relation $\mathrm{d}S = \frac{\mathrm{d}q_{\mathrm{rev}}}{T}$. d$S$ can be shown to be an exact differential, that is, $\oint \mathrm{d}S = 0$. Hence, S is a property of all systems. In Section 61.3, it is shown that the differential of the entropy as defined above leads to the Clausius inequality, that is $\mathrm{d}S \geq \frac{\mathrm{d}q}{T}$, where q represents the actual heat associated with a real, necessarily irreversible, process. We now suppose that the system is isolated from its surroundings, so that $\mathrm{d}q = 0$. The Clausius inequality then implies that $\mathrm{d}S \geq 0$, and we conclude that in an isolated system the entropy cannot decrease when a spontaneous change occurs. Since the universe as a whole can be considered an isolated system, this implies that $\Delta S_{\mathrm{universe}} = \Delta S_{\mathrm{sys}} + \Delta S_{\mathrm{surr}} \geq 0$, which is another version of the second law.

Exercises

E61.1(a) Efficiency, η, is $\frac{\text{work performed}}{\text{heat absorbed}} = \frac{|w|}{q_{\mathrm{h}}}$ [61.5] $= \frac{3.00\ \mathrm{kJ}}{10.00\ \mathrm{kJ}} = 0.300$. For an ideal heat engine we have $\eta_{\mathrm{rev}} = 1 - \frac{T_{\mathrm{c}}}{T_{\mathrm{h}}}$ [61.7] $= 0.300 = 1 - \frac{T_{\mathrm{c}}}{273.16\ \mathrm{K}}$. Solving for T_{c}, we obtain $\boxed{T_{\mathrm{c}} = 191.2\ \mathrm{K}}$ as the temperature of the organic liquid.

E61.2(a) Assume that the block is so large that its temperature does not change significantly as a result of the heat transfer. Then

$$\Delta S = \int_i^f \frac{dq_{rev}}{T} \; [61.2] = \frac{1}{T}\int_i^f dq_{rev} \; [\text{constant } T] = \frac{q_{rev}}{T}$$

(a) $\Delta S = \dfrac{100\times10^3\,\text{J}}{273.15\,\text{K}} = \boxed{366\,\text{J K}^{-1}}$ (b) $\Delta S = \dfrac{100\times10^3\,\text{J}}{323.15\,\text{K}} = \boxed{309\,\text{J K}^{-1}}$

Problems

P61.1 The isotherms correspond to T = constant, and the reversibly traversed adiabats correspond to S = constant. Thus we can represent the cycle as in Fig. 61.1.

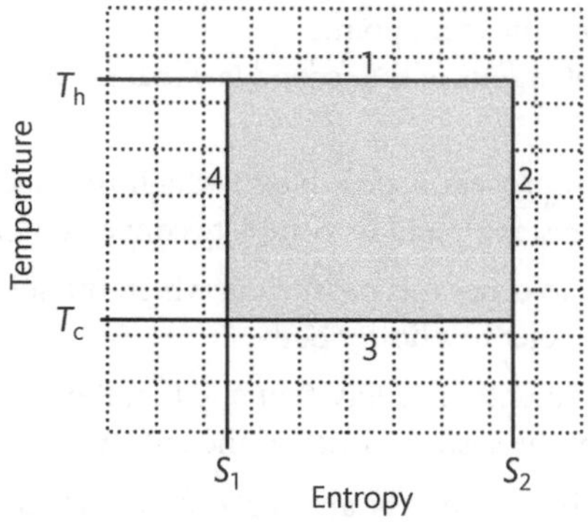

Fig 61.1

In this figure, paths 1, 2, 3, and 4 correspond to the four stages of the Carnot cycle listed in the text following Fig. 61.2. The area within the rectangle is

$$\text{Area} = \int_{\text{cyclic}} T\,dS = (T_h - T_c)\times(S_2 - S_1) = (T_h - T_c)\Delta S = (T_h - T_c)nR\ln\frac{V_B}{V_A}$$

(isothermal expansion from V_A to V_B, stage 1). But,

$$w(\text{cycle}) = \eta q_h = \left(\frac{T_h - T_c}{T_h}\right) nRT_h \ln\frac{V_B}{V_A} [\text{text Figure 61.2}] = nR(T_h - T_c)\ln\frac{V_B}{V_A}$$

Therefore, the area is equal to the net work done in the cycle.

P61.3 Paths A and B in Fig. 61.2 are the reversible adiabatic paths which are assumed to cross at state 1. Path C (dashed) is an isothermal path which connects the adiabatic paths at states 2 and 3. Now go round the cycle ($1 \rightarrow 2$, step 1; $2 \rightarrow 3$, step 2; $3 \rightarrow 1$, step 3).

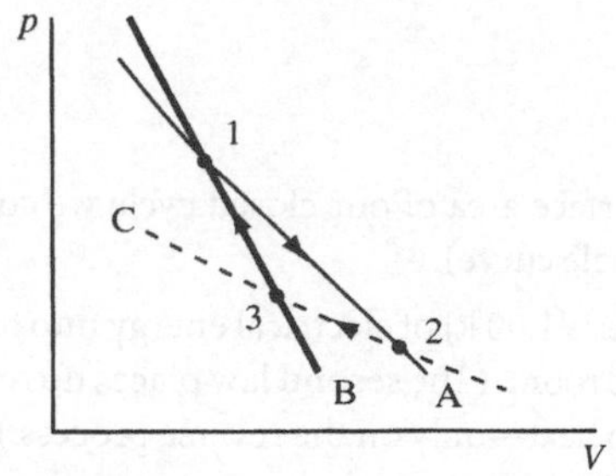

Fig 61.2

Step 1 $\Delta U_1 = q_1 + w_1 = w_1$ [$q_1 = 0$, adiabatic]
Step 2 $\Delta U_2 = q_2 + w_2 = 0$ [isothermal step, energy depends on temperature only]
Step 3 $\Delta U_3 = q_3 + w_3 = w_3$ [$q_3 = 0$, adiabatic]
For the cycle $\Delta U = 0 = w_1 + q_2 + w_2 + w_3$ or $w(\text{net}) = w_1 + w_2 + w_3 = -q_2$

But, $\Delta U_1 = -\Delta U_3$ [$\Delta T_1 = -\Delta T_2$]; hence $w_1 = -w_3$, and $w(\text{net}) = w_2 = -q_2$, or $-w(\text{net}) = q_2$

A net amount of work has been done by the system from heat obtained from a heat reservoir at the temperature of step 2, without at the same time transferring heat from a hot to a cold reservoir. This violates the Kelvin statement of the second law of thermodynamics. Therefore, the assumption that the two adiabatic reversible paths may intersect is disproved.

Question. May any adiabatic paths intersect, reversible or not?

Alternative solution not requiring the system to be a perfect gas

Note that step 2 above effectively requires the system to be a perfect gas. The following solution is more general:

Suppose that two adiabats cross at point 1. Consider the isotherm at T crossing both adiabats at points 2 and 3. We now can define a quasi-steady closed cycle along the two adiabats and the isotherm.

By definition of state variables:

$$\Delta U = \oint dU = 0$$

$$\Delta S = \oint dS = 0$$

Using the first law:

$$\Delta U = \oint dq + \oint dw = q + w$$
$$\Rightarrow q = -w$$

and the second law for reversible processes we have:

$$\Delta S = \oint \frac{dq}{T} = \frac{q}{T}$$
$$\Rightarrow q = 0$$

As T is finite we find:

$$w = -q = 0$$

As the work corresponds to the surface area of our closed cycle we conclude that the two adiabats coincide (are a single curve).

P61.5 In case (a), the electric heater converts 1.00 kJ of electrical energy into heat, providing $\boxed{1.00\text{ kJ}}$ of energy as heat to the room. (The second law places no restriction on the complete conversion of work to heat—only on the reverse process.) In case (b), we want to find the heat deposited in the room $|q_h|$:

$$|q_h| = |q_c| + |w| \qquad \text{where} \qquad \frac{|q_c|}{|w|} = c = \frac{T_c}{T_h - T_c}$$

so $$|q_c| = \frac{|w|T_c}{T_h - T_c} = \frac{1.00\text{ kJ} \times 260\text{ K}}{(295 - 260)\text{ K}} = 7.4\text{ kJ}$$

The heat transferred to the room is $|q_h| = |q_c| + |w| = 7.4\text{ kJ} + 1.00\text{ kJ} = \boxed{8.4\text{ kJ}}$. Most of the thermal energy the heat pump deposits into the room comes from outdoors. Difficult as it is to believe on a cold winter day, the intensity of thermal energy (that is, the absolute temperature) outdoors is a substantial fraction of that indoors. The work put into the heat pump is not simply converted to heat, but is "leveraged" to transfer additional heat from outdoors.

Topic 62 Entropy changes for specific processes

Discussion question

D62.1 The explanation of Trouton's rule is that $\dfrac{\Delta_{\text{vap}}H^{\ominus}}{T_{\text{b}}}$ is the standard entropy of vaporization, and we expect a comparable change in volume (with an accompanying comparable change in the number of accessible microstates) whenever an unstructured liquid forms a vapor. Hence, all unstructured liquids can be expected to have similar entropies of vaporization. Liquids that show significant deviations from Trouton's rule do so on account of strong molecular interactions that restrict molecular motion. As a result there is a greater dispersal of matter and energy when such liquids vaporize. Water is an example of a liquid with strong intermolecular interactions (hydrogen bonding) which tend to organize the molecules in the liquid, hence we expect its entropy of vaporization to be greater than 85 J K^{-1} mol^{-1}.

Exercises

E62.1(a) $\boxed{I_2(g)}$ will have the higher standard molar entropy at 298 K, primarily because ΔS_{fus} and ΔS_{vap} are greater for I_2. At 298 K, I_2 is a solid in its standard state.

E62.2(a)

$$\Delta S = nR\ln\left(\frac{V_{\text{f}}}{V_{\text{i}}}\right) \quad [62.2]$$

$$= \left(\frac{15\text{ g}}{44\text{ g/mol}}\right)\times 8.314\text{ J K}^{-1}\text{ mol}^{-1}\times\ln\left(\frac{3.0}{1.0}\right) = \boxed{3.1\text{ J K}^{-1}}$$

E62.3(a) Trouton's rule in the form $\Delta_{\text{vap}}H^{\ominus} = T_{\text{b}}\times 85\text{ J K}^{-1}\text{ mol}^{-1}$ can be used to obtain approximate enthalpies of vaporization. For benzene

$$\Delta_{\text{vap}}H^{\ominus} = (273.2+80.1)\text{K}\times 85\text{ J K}^{-1}\text{ mol}^{-1} = \boxed{30.0\text{ kJ/mol}^{-1}}$$

E62.4(a) $S_{\text{m}}(T_{\text{f}}) = S_{\text{m}}(T_{\text{i}}) + \int_{T_{\text{i}}}^{T_{\text{f}}} \frac{C_{V,\text{m}}}{T}\text{d}T$ [62.7, with $C_{V,\text{m}}$ in place of C_p]

If we assume that neon is a perfect gas then $C_{V,\mathrm{m}}$ may be taken to be constant and given by

$$C_{V,\mathrm{m}} = C_{p,\mathrm{m}} - R; \quad C_{p,\mathrm{m}} = 20.786\,\mathrm{J\,K^{-1}\,mol^{-1}}\,[\text{Table 57.4}]$$
$$= (20.786 - 8.314)\,\mathrm{J\,K^{-1}\,mol^{-1}}$$
$$= 12.472\,\mathrm{J\,K^{-1}\,mol^{-1}}$$

Integrating, we obtain

$$S_\mathrm{m}(500\,\mathrm{K}) = S_\mathrm{m}(298\,\mathrm{K}) + C_{V,\mathrm{m}} \ln \frac{T_\mathrm{f}}{T_\mathrm{i}}$$
$$= (146.22\,\mathrm{J\,K^{-1}\,mol^{-1}}) + (12.472\,\mathrm{J\,K^{-1}\,mol^{-1}}) \ln\left(\frac{500\,\mathrm{K}}{298\,\mathrm{K}}\right)$$
$$= (146.22 + 6.45)\,\mathrm{J\,K^{-1}\,mol^{-1}} = \boxed{152.67\,\mathrm{J\,K^{-1}\,mol^{-1}}}$$

E62.5(a) Since entropy is a state function, ΔS may be calculated from the most convenient path, which in this case corresponds to constant-pressure heating followed by constant-temperature compression.

$$\Delta S = nC_{p,\mathrm{m}} \ln\left(\frac{T_\mathrm{f}}{T_\mathrm{i}}\right) [62.8, \text{ at } p_\mathrm{i}] + nR\ln\left(\frac{V_\mathrm{f}}{V_\mathrm{i}}\right) [62.2, \text{ at } T_\mathrm{f}]$$

Since pressure and volume are inversely related (Boyle's law), $\frac{V_\mathrm{f}}{V_\mathrm{i}} = \frac{p_\mathrm{i}}{p_\mathrm{f}}$. Hence,

$$\Delta S = nC_{p,\mathrm{m}} \ln\left(\frac{T_\mathrm{f}}{T_\mathrm{i}}\right) - nR\ln\left(\frac{p_\mathrm{f}}{p_\mathrm{i}}\right) = (3.00\,\mathrm{mol}) \times \tfrac{5}{2} \times (8.314\,\mathrm{J\,K^{-1}\,mol^{-1}}) \times \ln\left(\frac{398\,\mathrm{K}}{298\,\mathrm{K}}\right)$$
$$- (3.00\,\mathrm{mol}) \times (8.314\,\mathrm{J\,K^{-1}\,mol^{-1}}) \times \ln\left(\frac{5.00\,\mathrm{atm}}{1.00\,\mathrm{atm}}\right)$$
$$= (18.0\bar{4} - 40.1\bar{4})\,\mathrm{J\,K^{-1}} = -22.1\,\mathrm{J\,K^{-1}}$$

Though ΔS (system) is negative, the process can still occur spontaneously if ΔS (total) is positive.

E62.6(a) For an adiabatic reversible process, $q = q_\mathrm{rev} = \boxed{0}$.

$$\Delta S = \int_\mathrm{i}^\mathrm{f} \frac{\mathrm{d}q_\mathrm{rev}}{T} = \boxed{0}$$

E62.7(a) Since the container is isolated, the heat flow is zero and therefore $\boxed{\Delta H = 0}$; since the masses of the blocks are equal, the final temperature must be their mean temperature, 25°C. Specific heat capacities are heat capacities per gram and are related to the molar heat capacities by

$$C_\mathrm{s} = \frac{C_\mathrm{m}}{M} \quad [C_{p,\mathrm{m}} \approx C_{V,\mathrm{m}} = C_\mathrm{m}]$$

So $nC_\mathrm{m} = mC_\mathrm{s}$ $[nM = m]$

$$\Delta S = mC_\mathrm{s} \ln\left(\frac{T_\mathrm{f}}{T_\mathrm{i}}\right) \quad [62.8]$$

$$\Delta S_1 = (1.00 \times 10^3\,\mathrm{g}) \times (0.385\,\mathrm{J\,K^{-1}\,g^{-1}}) \times \ln\left(\frac{298\,\mathrm{K}}{323\,\mathrm{K}}\right) = -31.0\,\mathrm{J\,K^{-1}}$$

$$\Delta S_2 = (1.00\times10^3\,\mathrm{g})\times(0.385\,\mathrm{J\,K^{-1}\,g^{-1}})\times\ln\left(\frac{298\,\mathrm{K}}{273\,\mathrm{K}}\right) = 33.7\,\mathrm{J\,K^{-1}}$$

$$\Delta S_{\mathrm{tot}} = \Delta S_1 + \Delta S_2 = \boxed{+2.7\,\mathrm{J\,K^{-1}}}$$

Comment. The positive value of ΔS_{tot} corresponds to a spontaneous process.

E62.8(a)

(a) $$\Delta S(\mathrm{gas}) = nR\ln\frac{V_{\mathrm{f}}}{V_{\mathrm{i}}}[62.2] = \left(\frac{14\,\mathrm{g}}{28.02\,\mathrm{g\,mol^{-1}}}\right)\times(8.314\,\mathrm{J\,K^{-1}\,mol^{-1}})\times(\ln 2) = \boxed{+2.9\,\mathrm{J\,K^{-1}}}$$

$\Delta S(\mathrm{surroundings}) = \boxed{-2.9\,\mathrm{J\,K^{-1}}}$ [overall zero entropy production]

$\Delta S(\mathrm{total}) = \boxed{0}$ [reversible process]

(b) $\Delta S(\mathrm{gas}) = \boxed{+2.9\,\mathrm{J\,K^{-1}}}$ [S a state function]

$\Delta S(\mathrm{surroundings}) = \boxed{0}$ [surroundings do not change]

$\Delta S(\mathrm{total}) = \boxed{+2.9\,\mathrm{J\,K^{-1}}}$

(c) $\Delta S(\mathrm{gas}) = \boxed{0}$ [$q_{\mathrm{rev}} = 0$]

$\Delta S(\mathrm{surroundings}) = \boxed{0}$ [no heat transferred to surroundings]

$\Delta S(\mathrm{total}) = \boxed{0}$

E62.9(a)

(a) $$\Delta_{\mathrm{vap}}S = \frac{\Delta_{\mathrm{vap}}H}{T_{\mathrm{b}}} = \frac{29.4\times10^3\,\mathrm{J\,mol^{-1}}}{334.88\,\mathrm{K}} = \boxed{+87.8\,\mathrm{J\,K^{-1}\,mol^{-1}}}$$

(b) If the vaporization occurs reversibly, $\Delta S_{\mathrm{tot}} = 0$, so $\Delta S_{\mathrm{surr}} = \boxed{-87.8\,\mathrm{J\,K^{-1}\,mol^{-1}}}$

E62.10(a) $$\Delta S = nC_{\mathrm{p}}(\mathrm{H_2O,s})\ln\frac{T_{\mathrm{f}}}{T_{\mathrm{i}}} + n\frac{\Delta_{\mathrm{fus}}H}{T_{\mathrm{fus}}} + nC_{\mathrm{p}}(\mathrm{H_2O,l})\ln\frac{T_{\mathrm{f}}}{T_{\mathrm{i}}} + n\frac{\Delta_{\mathrm{vap}}H}{T_{\mathrm{vap}}} + nC_{\mathrm{p}}(\mathrm{H_2O,g})\ln\frac{T_{\mathrm{f}}}{T_{\mathrm{i}}}$$

$$n = \frac{10.0\,\mathrm{g}}{18.02\,\mathrm{g\,mol^{-1}}} = 0.555\,\mathrm{mol}$$

$$\begin{aligned}\Delta S &= 0.555\,\mathrm{mol}\times38.02\,\mathrm{J\,K^{-1}\,mol^{-1}}\times\ln\frac{273.15}{263.15} + 0.555\,\mathrm{mol}\times\frac{6.008\,\mathrm{kJ/mol^{-1}}}{273.15\,\mathrm{K}}\\ &+0.555\,\mathrm{mol}\times75.291\,\mathrm{J\,K^{-1}\,mol^{-1}}\times\ln\frac{373.15}{273.15}\\ &+0.555\,\mathrm{mol}\times\frac{44.016\,\mathrm{kJ/mol^{-1}}}{373.15\,\mathrm{K}} + 0.555\,\mathrm{mol}\times33.58\,\mathrm{J\,K^{-1}\,mol^{-1}}\times\ln\frac{388.15}{373.15}\end{aligned}$$

$$\boxed{\Delta S = 92.2\,\mathrm{J\,K^{-1}}}$$

Comment. This calculation has been based on the assumption that the heat capacities remain constant over the range of temperatures involved and that the enthalpy of vaporization at 298.15 K given in Table 57.2 can be applied to the vaporization at 373.15 K. Neither one of these assumptions are strictly valid. Therefore, the calculated value is only approximate.

Problems

P62.1 (a) Because entropy is a state function $\Delta_{trs}S(l \rightarrow s, -5°C)$ may be determined indirectly from the following cycle

$$\begin{array}{ccc} H_2O(l,0°C) & \xrightarrow{\Delta_{trs}S(l \rightarrow s,0°C)} & H_2O(s,0°C) \\ \Delta S_l \uparrow & & \downarrow \Delta S_s \\ H_2O(l,-5°C) & \xrightarrow{\Delta_{trs}S(l \rightarrow s,-5°C)} & H_2O(s,-5°C) \end{array}$$

Thus $\quad \Delta_{trs}S(l \rightarrow s, -5°C) = \Delta S_l + \Delta_{trs}S(l \rightarrow s, 0°C) + \Delta S_s,$

where $\quad \Delta S_l = C_{p,m}(l)\ln\dfrac{T_f}{T} \quad [62.8;\ \theta_f = 0°C,\ \theta = -5°C]$

and $\quad \Delta S_s = C_{p,m}(s)\ln\dfrac{T}{T_f}$

$$\Delta S_l + \Delta S_s = -\Delta C_p \ln\frac{T}{T_f} \quad \text{with } \Delta C_p = C_{p,m}(l) - C_{p,m}(s) = +37.3\ \mathrm{J\,K^{-1}\,mol^{-1}}$$

$$\Delta_{trs}S(l \rightarrow s, T_f) = \frac{-\Delta_{fus}H}{T_f} \quad [62.5]$$

Thus, $\quad \Delta_{trs}S(l \rightarrow s, T) = \dfrac{-\Delta_{fus}H}{T_f} - \Delta C_p \ln\dfrac{T}{T_f}$

$$\Delta_{trs}S(l \rightarrow s, -5°C) = \frac{-6.01\times10^3\,\mathrm{J\,mol^{-1}}}{273\,\mathrm{K}} - (37.3\,\mathrm{J\,K^{-1}\,mol^{-1}})\times\ln\frac{268}{273}$$
$$= \boxed{-21.3\ \mathrm{J\,K^{-1}\,mol^{-1}}}$$

$$\Delta S_{sur} = \frac{\Delta_{fus}H(T)}{T}$$

$$\Delta_{fus}H(T) = -\Delta H_l + \Delta_{fus}H(T_f) - \Delta H_s$$

$$\Delta H_l + \Delta H_s = C_{p,m}(l)(T_f - T) + C_{p,m}(s)(T - T_f) = \Delta C_p(T_f - T)$$

$$\Delta_{fus}H(T) = \Delta_{fus}H(T_f) - \Delta C_p(T_f - T)$$

Thus, $\Delta S_{\text{sur}} = \dfrac{\Delta_{\text{fus}}H(T)}{T} = \dfrac{\Delta_{\text{fus}}H(T_{\text{f}})}{T} + \Delta C_p \dfrac{(T-T_{\text{f}})}{T}$

$$\Delta S_{\text{sur}} = \frac{6.01\,\text{kJ mol}^{-1}}{268\text{ K}} + (37.3\text{ J K}^{-1}\text{ mol}^{-1}) \times \left(\frac{268-273}{268}\right)$$

$$= \boxed{+21.7\text{ J K}^{-1}\text{ mol}^{-1}}$$

$$\Delta S_{\text{total}} = \Delta S_{\text{sur}} + \Delta S = (21.7-21.3)\text{ J K}^{-1}\text{ mol}^{-1} = \boxed{+0.4\text{ J K}^{-1}\text{mol}^{-1}}$$

Because $\Delta S_{\text{total}} > 0$, the transition $1 \rightarrow \text{s}$ is spontaneous at −5°C.

(b) A similar cycle and analysis can be set up for the transition liquid → vapour at 95°C. However, since the transformation here is to the high temperature state (vapour) from the low temperature state (liquid), which is the opposite of part (a), we can expect that the analogous equations will occur with a change of sign.

$$\Delta_{\text{trs}}S(1 \rightarrow \text{g}, T) = \Delta_{\text{trs}}S(1 \rightarrow \text{g}, T_{\text{b}}) + \Delta C_p \ln\frac{T}{T_{\text{b}}}$$

$$= \frac{\Delta_{\text{vap}}H}{T_{\text{b}}} + \Delta C_p \ln\frac{T}{T_{\text{b}}}, \qquad \Delta C_p = -41.9\text{ J K}^{-1}\text{ mol}^{-1}$$

$$\Delta_{\text{trs}}S(1 \rightarrow \text{g}, T) = \frac{40.7\text{ kJ mol}^{-1}}{373\text{ K}} - (41.9\text{ J K}^{-1}\text{ mol}^{-1}) \times \ln\left(\frac{368}{373}\right)$$

$$= \boxed{+109.7\text{ J K}^{-1}\text{ mol}^{-1}}$$

$$\Delta S_{\text{sur}} = \frac{-\Delta_{\text{vap}}H(T)}{T} = -\frac{\Delta_{\text{vap}}H(T_{\text{b}})}{T} - \frac{\Delta C_p(T-T_{\text{b}})}{T}$$

$$= \left(\frac{-40.7\text{ kJ mol}^{-1}}{368\text{ K}}\right) - (-41.9\text{ J K}^{-1}\text{ mol}^{-1}) \times \left(\frac{368-373}{368}\right)$$

$$= \boxed{-111.2\text{ J K}^{-1}\text{ mol}^{-1}}$$

$$\Delta S_{\text{total}} = (109.7-111.2)\text{J K}^{-1}\text{ mol}^{-1} = \boxed{-1.5\text{J K}^{-1}\text{ mol}^{-1}}$$

Since $\Delta S_{\text{total}} < 0$, the reverse transition, $\text{g} \rightarrow 1$, is spontaneous at 95°C.

P62.3 (a) $q(\text{total}) = q(\text{H}_2\text{O}) + q(\text{Cu}) = 0$, hence $-q(\text{H}_2\text{O}) = q(\text{Cu})$

$$q(\text{H}_2\text{O}) = n(-\Delta_{\text{vap}}H) + nC_{p,\text{m}}(\text{H}_2\text{O,l}) \times (\theta - 100°\text{C})$$

where θ is the final temperature of the water and copper.

$$q(\text{Cu}) = mC_{\text{s}}(\theta - 0) = mC_{\text{s}}\theta, \quad C_{\text{s}} = 0.385\text{ J K}^{-1}\text{g}^{-1}$$

Setting $-q(\text{H}_2\text{O}) = q(\text{Cu})$ allows us to solve for θ.

$$n(\Delta_{\text{vap}}H) - nC_{p,\text{m}}(\text{H}_2\text{O,l}) \times (\theta - 100°\text{C}) = mC_{\text{s}}\theta$$

Solving for θ yields:

$$\theta=\frac{n\{\Delta_{\text{vap}}H+C_{p,\text{m}}(\text{H}_2\text{O,l})\times 100^\circ\text{C}\}}{mC_\text{s}+nC_{p,\text{m}}(\text{H}_2\text{O,l})}$$

$$=\frac{(1.00\ \text{mol})\times(40.656\times10^3\ \text{J mol}^{-1}+75.3\ \text{J}\,^\circ\text{C}^{-1}\ \text{mol}^{-1}\times100^\circ\text{C})}{2.00\times10^3\ \text{g}\times0.385\ \text{J}\,^\circ\text{C}^{-1}\ \text{g}^{-1}+1.00\ \text{mol}\times75.3\ \text{J}\,^\circ\text{C}^{-1}\ \text{mol}^{-1}}$$

$$=57.0^\circ\text{C}=330.2\ \text{K}$$

$$q(\text{Cu})=(2.00\times10^3\,\text{g})\times(0.385\ \text{J K}^{-1}\,\text{g}^{-1})\times(57.0\text{K})=4.39\times10^4\,\text{J}=\boxed{43.9\ \text{kJ}}$$

$$q(\text{H}_2\text{O})=\boxed{-43.9\ \text{kJ}}$$

$$\Delta S(\text{total})=\Delta S(\text{H}_2\text{O})+\Delta S(\text{Cu})$$

$$\Delta S(\text{H}_2\text{O})=\frac{-n\Delta_{\text{vap}}H}{T_\text{b}}\ [62.5]+nC_{p,\text{m}}\ln\left(\frac{T_\text{f}}{T_\text{i}}\right)\ [62.8]$$

$$=-\frac{(1.00\ \text{mol})\times(40.656\times10^3\,\text{J mol}^{-1})}{373.2\ \text{K}}$$

$$+(1.00\ \text{mol})\times(75.3\ \text{J K}^{-1}\,\text{mol}^{-1})\times\ln\left(\frac{330.2\,\text{K}}{373.2\,\text{K}}\right)$$

$$=-108.\bar{9}\,\text{J K}^{-1}-9.22\ \text{J K}^{-1}=\boxed{-118.\bar{1}\ \text{J K}^{-1}}$$

$$\Delta S(\text{Cu})=mC_\text{s}\ln\frac{T_\text{f}}{T_\text{i}}=(2.00\times10^3\,\text{g})\times(0.385\ \text{J K}^{-1}\text{g}^{-1})\times\ln\left(\frac{330.2\,\text{K}}{273.2\,\text{K}}\right)$$

$$=\boxed{145.\bar{9}\,\text{J K}^{-1}}$$

$$\Delta S(\text{total})=-118.\bar{1}\,\text{J K}^{-1}+145.\bar{9}\,\text{J K}^{-1}=\boxed{28\,\text{J K}^{-1}}$$

This process is spontaneous since ΔS(surroundings) (surroundings) is zero and, hence,

$$\Delta S(\text{universe})=\Delta S(\text{total})>0$$

(b) The volume of the container may be calculated from the perfect gas law.

$$V=\frac{nRT}{p}=\frac{(1.00\ \text{mol})\times(0.08206\ \text{dm}^3\,\text{atm K}^{-1}\,\text{mol}^{-1})\times(373.2\ \text{K})}{1.00\ \text{atm}}=30.6\ \text{dm}^3$$

At 57°C the vapor pressure of water is 130 Torr (*Handbook of Chemistry and Physics*, 81st edition). The amount of water vapor present at equilibrium is then

$$n=\frac{pV}{RT}=\frac{(130\ \text{Torr})\times\left(\frac{1\text{atm}}{760\ \text{Torr}}\right)\times(30.6\ \text{dm}^3)}{(0.08206\ \text{dm}^3\text{atm K}^{-1}\text{mol}^{-1})\times(330.2\ \text{K})}=0.193\ \text{mol}$$

This is a substantial fraction of the original amount of water and cannot be ignored. Consequently the calculation needs to be redone taking into account the fact that only a part, n_l, of the vapor condenses into a liquid while the remainder (1.00 mol − n_l) remains gaseous. The heat flow involving water, then, becomes

$$q(\mathrm{H_2O}) = -n_l\Delta_{\mathrm{vap}}H + n_lC_{p,\mathrm{m}}(\mathrm{H_2O,l})\Delta T(\mathrm{H_2O}) + (1.00\ \mathrm{mol} - n_l)C_{p,\mathrm{m}}(\mathrm{H_2O,g})\Delta T(\mathrm{H_2O})$$

Because n_l depends on the equilibrium temperature through $n_l = 1.00\ \mathrm{mol} - \dfrac{pV}{RT}$, where p is the vapor pressure of water, we will have two unknowns (p and T) in the equation $-q(\mathrm{H_2O}) = q(\mathrm{Cu})$. There are two ways out of this dilemma: (1) p may be expressed as a function of T by use of the Clapeyron equation, or (2) by use of successive approximations. Redoing the calculation yields:

$$\theta = \frac{n_l\Delta_{\mathrm{vap}}H + n_lC_{p,\mathrm{m}}(\mathrm{H_2O,l})\times 100°\mathrm{C} + (1.00 - n_l)C_{p,\mathrm{m}}(\mathrm{H_2O,g})\times 100°\mathrm{C}}{mC_s + nC_{p,\mathrm{m}}(\mathrm{H_2O,l}) + (1.00 - n_l)C_{p,\mathrm{m}}(\mathrm{H_2O,g})}$$

With

$$n_l = (1.00\ \mathrm{mol}) - (0.193\ \mathrm{mol}) = 0.80\bar{7}\ \mathrm{mol}$$

(noting that $C_{p,\mathrm{m}}(\mathrm{H_2O,g})$ = 33.6 J mol^{-1} K^{-1} [Table 57.4]) θ = 47.2°C. At this temperature, the vapor pressure of water is 80.41 Torr, corresponding to

$$n_l = (1.00\ \mathrm{mol}) - (0.123\ \mathrm{mol}) = 0.87\bar{7}\ \mathrm{mol}$$

This leads to θ = 50.8°C. The successive approximations eventually converge to yield a value of $\boxed{49.9°\mathrm{C} = 323.1\,\mathrm{K}}$ for the final temperature. (At this temperature, the vapor pressure is 0.123 bar.) Using this value of the final temperature, the heat transferred and the various entropies are calculated as in part (a).

$$q(\mathrm{Cu}) = (2.00\times 10^3\ \mathrm{g})\times(0.385\ \mathrm{J\,K^{-1}\,g^{-1}})\times(49.9\ \mathrm{K}) = \boxed{38.4\ \mathrm{kJ}} = -q(\mathrm{H_2O})$$

$$\Delta S(\mathrm{H_2O}) = \frac{-n\Delta_{\mathrm{vap}}H}{T_\mathrm{b}} + nC_{p,\mathrm{m}}\ln\left(\frac{T_\mathrm{f}}{T_\mathrm{i}}\right) = \boxed{-119.\bar{8}\ \mathrm{J\,K^{-1}}}$$

$$\Delta S(\mathrm{Cu}) = mC_s\ln\frac{T_\mathrm{f}}{T_\mathrm{i}} = \boxed{129.\bar{2}\ \mathrm{J\,K^{-1}}}$$

$$\Delta S(\mathrm{total}) = -119.\bar{8}\ \mathrm{J\,K^{-1}} + 129.\bar{2}\ \mathrm{J\,K^{-1}} = \boxed{9\ \mathrm{J\,K^{-1}}}$$

P62.5 ΔS depends on only the initial and final states, so we can use $\Delta S = nC_{p,\mathrm{m}}\ln\dfrac{T_\mathrm{f}}{T_\mathrm{i}}$ [62.8]

Since $\quad q = nC_{p,\mathrm{m}}(T_\mathrm{f} - T_\mathrm{i}),\ T_\mathrm{f} = T_\mathrm{i} + \dfrac{q}{nC_{p,\mathrm{m}}} = T_\mathrm{i} + \dfrac{I^2Rt}{nC_{p,\mathrm{m}}}\ [q = ItV = I^2Rt]$

That is, $\Delta S = nC_{p,\mathrm{m}} \ln\left(1+\dfrac{I^2Rt}{nC_{p,\mathrm{m}}T_\mathrm{i}}\right)$

Since $n=\dfrac{500\ \mathrm{g}}{63.5\ \mathrm{g\ mol^{-1}}}=7.87\ \mathrm{mol}$

$$\Delta S=(7.87\ \mathrm{mol})\times(24.4\ \mathrm{J\ K^{-1}\ mol^{-1}})\times\ln\left(1+\frac{(1.00\ \mathrm{A})^2\times(1\ 000\ \Omega)\times(15.0\ \mathrm{s})}{(7.87)\times(24.4\ \mathrm{J\ K^{-1}})\times(293\ \mathrm{K})}\right)$$

$$=(192\ \mathrm{J\ K^{-1}})\times(\ln 1.27)=\boxed{+45.4\ \mathrm{J\ K^{-1}}}$$

$[1\ \mathrm{J}=1\ \mathrm{AVs}=1\ \mathrm{A^2\Omega\ s}]$

For the second experiment, no change in state occurs for the copper, hence, $\Delta S(\text{copper})=0$. However, for the water, considered as a large heat sink

$$\Delta S(\text{water})=\frac{q}{T}=\frac{I^2Rt}{T}=\frac{(1.00\ \mathrm{A})^2\times(1\ 000\ \Omega)\times(15.0\ \mathrm{s})}{293\ \mathrm{K}}=\boxed{+51.2\ \mathrm{J\ K^{-1}}}$$

P62.7 Let us write Newton's law of cooling as follows:

$$\frac{\mathrm{d}T}{\mathrm{d}t}=-A(T-T_s)$$

Where A is a constant characteristic of the system and T_S is the temperature of the surroundings. The negative sign appears because we assume $T > T_\mathrm{S}$. Separating variables

$\dfrac{\mathrm{d}T}{T-T_\mathrm{S}}=-A\mathrm{d}t$, and integrating, we obtain

$\ln(T-T_\mathrm{S})=-At+K$, where K is a constant of integration.

Let T_i be the initial temperature of the system when $t=0$, then

$K=\ln(T_\mathrm{i}-T_\mathrm{S})$ introducing this expression for K gives

$$\ln\left(\frac{T-T_\mathrm{S}}{T_\mathrm{i}-T_\mathrm{S}}\right)=-At \text{ or } T=T_\mathrm{S}+(T_\mathrm{i}-T_\mathrm{S})\mathrm{e}^{-At}$$

$$\frac{\mathrm{d}S}{\mathrm{d}t}=\frac{\mathrm{d}}{\mathrm{d}t}\left(C\ln\frac{T}{T_\mathrm{i}}\right)=\frac{\mathrm{d}}{\mathrm{d}t}(C\ln T)$$

From the above expression for T, we obtain $\ln T=\ln T_\mathrm{S}-At\ln(T_\mathrm{i}-T_\mathrm{S})$. Substituting $\ln t$ we obtain $\boxed{\dfrac{\mathrm{d}S}{\mathrm{d}t}=-CA\ln(T_\mathrm{i}-T_\mathrm{S})}$, where now T_i can be interpreted as any temperature T during the course of the cooling process.

Topic 63 **The Third Law**

Discussion question

D63.1 Because solutions of cations cannot be prepared in the absence of anions and *vice versa*, in order to assign numerical values to the entropies of ions in solution, we arbitrarily assign the value of zero to the standard entropy of H^+ ions in water at all temperatures, *i.e.*, $S^{\ominus}(H^+, aq) = 0$. With this choice, the entropies of ions in water are values relative to the hydrogen ion in water; hence they may be either positive or negative. Ion entropies vary as expected on the basis that they are related to the degree to which the ions order the water molecules around them in solution. Small, highly charged ions induce local structure in the surrounding water, and the disorder of the solution is decreased more than for the case of large, singly charged ions.

Exercises

E63.1(a) In each case $S_m = R \ln s$, where s is the number of orientations of about equal energy that the molecule can adopt. Therefore,

(a) $S_m = R\ln 3 = 8.3145\ \text{J K}^{-1}\ \text{mol}^{-1} \times \ln 3 = \boxed{9.13\ \text{J K}^{-1}\ \text{mol}^{-1}}$

(b) $S_m = R\ln 5 = 8.3145\ \text{J K}^{-1}\ \text{mol}^{-1} \times \ln 5 = \boxed{13.4\ \text{J K}^{-1}\ \text{mol}^{-1}}$

(c) $S_m = R\ln 6 = 8.3145\ \text{J K}^{-1}\ \text{mol}^{-1} \times \ln 6 = \boxed{14.9\ \text{J K}^{-1}\ \text{mol}^{-1}}$

E63.2(a) In each case $\quad \Delta_r S^{\ominus} = \sum_{\text{Products}} \nu S_m^{\ominus} - \sum_{\text{Products}} \nu S_m^{\ominus}$ [63.4]

with S_m values obtained from Tables 57.3 and 57.4.

(a) $\Delta_r S = 2\,S_m^{\ominus}(CH_3COOH, l) - 2\,S_m^{\ominus}(CH_3CHO, g) - S_m^{\ominus}(O_2, g)$

$= [(2\times 159.8) - (2\times 250.3) - 205.14]\ \text{J K}^{-1}\ \text{mol}^{-1} = \boxed{-386.1\ \text{J K}^{-1}\ \text{mol}^{-1}}$

(b) $\Delta_r S^{\ominus} = 2\,S_m^{\ominus}(AgBr, s) + S_m^{\ominus}(Cl_2, g) - 2\,S_m^{\ominus}(AgCl, s) - S_m^{\ominus}(Br_2, l)$

$= [(2\times 107.1) + (223.07) - (2\times 96.2) - (152.23)]\ \text{J K}^{-1}\ \text{mol}^{-1}$

$= \boxed{+92.6\ \text{J K}^{-1}\ \text{mol}^{-1}}$

(c) $\Delta_r S^{\ominus} = S_m^{\ominus}(HgCl_2, s) - S_m^{\ominus}(Hg, l) - S_m^{\ominus}(Cl_2, g)$

$= [146.0 - 76.02 - 223.07]\ \text{J K}^{-1}\ \text{mol}^{-1} = \boxed{-153.1\ \text{J K}^{-1}\ \text{mol}^{-1}}$

Problems

P63.1 $S_m^{\ominus}(T) = S_m^{\ominus}(298\text{ K}) + \Delta S$

$$\Delta S = \int_{T_1}^{T_2} C_{p,m}\frac{dT}{T} = \int_{T_1}^{T_2}\left(\frac{a}{T}+b+\frac{c}{T^3}\right)dT = a\ln\frac{T_2}{T_1} + b(T_2 - T_1) - \frac{1}{2}c\left(\frac{1}{T_2^2}-\frac{1}{T_1^2}\right)$$

(a) $$S_m^{\ominus}(373\text{ K}) = (192.45\text{ J K}^{-1}\text{ mol}^{-1}) + (29.75\text{ J K}^{-1}\text{ mol}^{-1})\times\ln\left(\frac{373}{298}\right)$$
$$+(25.10\times10^{-3}\text{ J K}^{-2}\text{ mol}^{-1})\times(75.0\text{ K})$$
$$+(\tfrac{1}{2})\times(1.55\times10^{5}\text{ J K}^{-1}\text{ mol}^{-1})\times\left(\frac{1}{(373.15)^2}-\frac{1}{(298.15)^2}\right)$$
$$= \boxed{200.7\text{ J K}^{-1}\text{ mol}^{-1}}$$

(b) $$S_m^{\ominus}(773\text{ K}) = (192.45\text{ J K}^{-1}\text{ mol}^{-1}) + (29.75\text{ J K}^{-1}\text{ mol}^{-1})\times\ln\left(\frac{773}{298}\right)$$
$$+(25.10\times10^{-3}\text{ J K}^{-2}\text{ mol}^{-1})\times(475\text{ K})$$
$$+(\tfrac{1}{2})\times(1.55\times10^{5}\text{ J K}^{-1}\text{ mol}^{-1})\times\left(\frac{1}{773^2}-\frac{1}{298^2}\right)$$
$$= \boxed{232.0\text{ J K}^{-1}\text{ mol}^{-1}}$$

P63.3 $$\Delta_r H^{\ominus} = \sum_{\text{products}} \nu_J \Delta_f H^{\ominus}(\text{J}) - \sum_{\text{reactants}} \nu_J \Delta_f H^{\ominus}(\text{J}) \quad [57.8]$$

$$\Delta_r H^{\ominus}(298\text{ K}) = 1\times\Delta_f H^{\ominus}(\text{CO,g}) + 1\times\Delta_f H^{\ominus}(\text{H}_2\text{O,g}) - 1\times\Delta_f H^{\ominus}(\text{CO}_2\text{,g})$$
$$= \{-110.53 - 241.82 - (-393.51)\}\text{ kJ mol}^{-1} = \boxed{+41.16\text{ kJ mol}^{-1}}$$

$$\Delta_r S^{\ominus} = \sum_{\text{products}} \nu_J S_m^{\ominus}(\text{J}) - \sum_{\text{reactants}} \nu_J S_m^{\ominus}(\text{J}) \quad [63.4]$$

$$\Delta_r S^{\ominus}(298\text{ K}) = 1\times S_m^{\ominus}(\text{CO, g}) + 1\times S_m^{\ominus}(\text{H}_2\text{O, g}) - 1\times S_m^{\ominus}(\text{CO}_2\text{, g}) - 1\times S_m^{\ominus}(\text{H}_2\text{, g})$$
$$= (197.67 + 188.83 - 213.74 - 130.684)\text{ kJ mol}^{-1}$$
$$= \boxed{+42.08\text{ J K}^{-1}\text{ mol}^{-1}}$$

$$\Delta_r H^{\ominus}(398\text{ K}) = \Delta_r H^{\ominus}(298\text{ K}) + \int_{298\text{ K}}^{398\text{ K}} \Delta_r C_p dT \quad [57.12]$$
$$= \Delta_r H^{\ominus}(298\text{ K}) + \Delta_r C_p \Delta T \quad [\text{heat capacities constant}]$$

$$\Delta_r C_p = 1\times C_{p,m}(\text{CO, g}) + 1\times C_{p,m}(\text{H}_2\text{O, g}) - 1\times C_{p,m}(\text{CO}_2\text{, g}) - 1\times C_{p,m}(\text{H}_2\text{, g})$$
$$= (29.14 + 33.58 - 37.11 - 28.824)\text{ J K}^{-1}\text{ mol}^{-1} = -3.21\text{ J K}^{-1}\text{ mol}^{-1}$$

$$\Delta_r H^{\ominus}(398\text{ K}) = (41.16\text{ kJ mol}^{-1}) + (-3.21\text{ J K}^{-1}\text{ mol}^{-1})\times(100\text{ K})$$
$$= \boxed{+40.84\text{ kJ mol}^{-1}}$$

For each substance in the reaction

$$\Delta S_{\mathrm{m}} = C_{p,\mathrm{m}} \ln\left(\frac{T_{\mathrm{f}}}{T_{\mathrm{i}}}\right) = C_{p,\mathrm{m}} \ln\left(\frac{398\ \mathrm{K}}{298\ \mathrm{K}}\right) \quad [62.7]$$

Thus

$$\Delta_{\mathrm{r}}S^{\ominus}(398\ \mathrm{K}) = \Delta_{\mathrm{r}}S^{\ominus}(298\ \mathrm{K}) + \sum_{\text{products}} \nu_{\mathrm{J}}C_{p,\mathrm{m}}(\mathrm{J})\ln\left(\frac{T_{\mathrm{f}}}{T_{\mathrm{i}}}\right) - \sum_{\text{reactants}} \nu_{\mathrm{J}}C_{p,\mathrm{m}}(\mathrm{J})\ln\left(\frac{T_{\mathrm{f}}}{T_{\mathrm{i}}}\right)$$

$$= \Delta_{\mathrm{r}}S^{\ominus}(298\ \mathrm{K}) + \Delta_{\mathrm{r}}C_p \ln\left(\frac{398\ \mathrm{K}}{298\ \mathrm{K}}\right)$$

$$= (42.01\ \mathrm{J\,K^{-1}\,mol^{-1}}) + (-3.21\ \mathrm{J\,K^{-1}\,mol^{-1}})\ln\left(\frac{398\ \mathrm{K}}{298\ \mathrm{K}}\right)$$

$$= (42.01 - 0.93)\ \mathrm{J\,K^{-1}\,mol^{-1}} = \boxed{+41.08\ \mathrm{J\,K^{-1}\,mol^{-1}}}$$

Comment. Both $\Delta_{\mathrm{r}}H^{\ominus}$ and $\Delta_{\mathrm{r}}S^{\ominus}$ changed little over 100 K for this reaction. This is not an uncommon result.

P63.5 Draw up the following table and proceed as in Problem 63.2:

T/K	14.14	16.33	20.03	31.15	44.08	64.81
$(C_{p,\mathrm{m}}/T)(\mathrm{J\,K^{-2}\,mol^{-1}})$	0.671	0.778	0.908	1.045	1.063	1.024

T/K	100.90	140.86	183.59	225.10	262.99	298.06
$(C_{p,\mathrm{m}}/T)(\mathrm{J\,K^{-2}\,mol^{-1}})$	0.942	0.861	0.787	0.727	0.685	0.659

Plot $C_{\mathrm{p,m}}$ against T (Fig. 63.1(a)) and $C_{\mathrm{p,m}}/T$ against T (Fig. 63.1(b)), extrapolating to $T = 0$ with $C_{\mathrm{p,m}} = aT^3$ fitted at $T = 14.14$ K, which gives $a = 3.36\ \mathrm{mJ\,K^{-1}\,mol^{-1}}$. Integration by determining the area under the curve then gives

$$H_{\mathrm{m}}^{\ominus}(298\mathrm{K}) - H_{\mathrm{m}}^{\ominus}(0) = \int_0^{298\mathrm{K}} C_{p,\mathrm{m}}\,\mathrm{d}T = \boxed{34.4\ \mathrm{kJ\,mol^{-1}}}$$

(a)

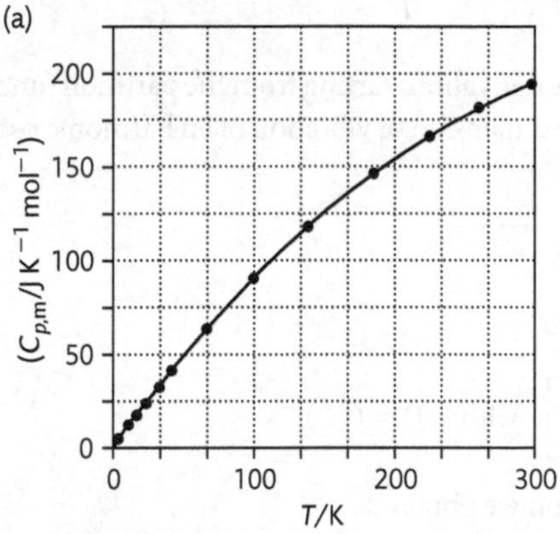

Fig 63.1(a)

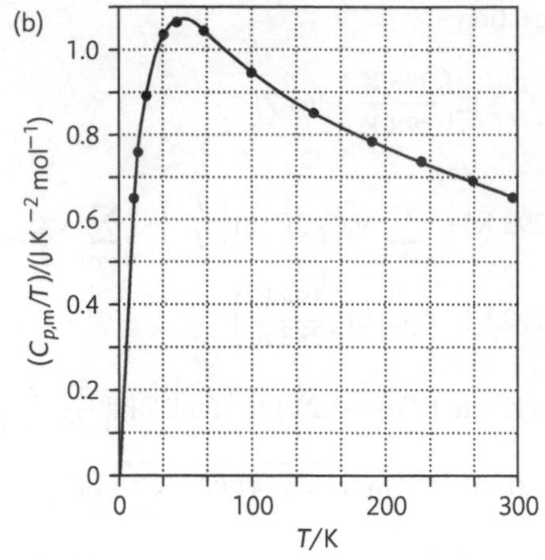

Fig 63.1(b)

$$S_m(298\ \text{K}) = S_m(0) + \int_0^{298\ \text{K}} \frac{C_{p,m}}{T}\text{d}T = S_m(0) + \boxed{243\ \text{J K}^{-1}\ \text{mol}^{-1}}$$

P63.7 Einstein solid: $C_{V,m}(T) = 3Rf_E(T)$ $f_E(T) = \left(\frac{\theta_E}{T}\right)^2\left(\frac{e^{\theta/2T}}{e^{\theta/T}-1}\right)^2$

$$S_m = \int_0^T (C_{V,m}/T)\text{d}T = 3R\int_0^T [f_E(T)/T]\text{d}T$$

$$= 3R\int_0^T \left(\frac{\theta_E}{T}\right)^2\left(\frac{1}{T}\right)\left\{\frac{e^{\theta/T}}{(1-e^{\theta/T})^2}\right\}\text{d}T$$

Set $x = \theta_E/T$, then $\text{d}T = -T^2\text{d}x/\theta_E$; hence

$\frac{S_m}{3R} = \int_{\theta_E/T}^{\infty}\left[\frac{xe^x}{(1-e^x)^2}\right]\text{d}x$. This integral can be evaluated analytically. The result is

$$\frac{S_m}{3R} = \left[\frac{x}{e^x-1} - \ln(1-e^{-x})\right] \quad \text{with } x \equiv \frac{\theta_E}{T}$$

A more instructive alternative derivation starting from the partition function follows. The partition function or one dimension of vibration of an harmonic oscillator is

$$q^V = \frac{1}{1-e^{-\beta hc\tilde{\nu}}}\ [52.15] = \frac{1}{1-e^{-\theta_E/T}}$$

The energy is calculated from

$$U_m - U_m(0) = kT^2\left[\frac{\partial \ln Q}{\partial T}\right]_V \quad \text{with } Q^V = (q^V)^N$$

Performing the differentiation we obtain

$U_m - U_m(0) = R\theta_E\left(\frac{1}{e^{\theta_E/T}-1}\right)$. We have used $N = N_A$ and $N_A k = R$

The molar entropy is calculated from

$$S_m = \frac{U_m - U_m(0)}{T} + k\ln Q\ [60.5] = \frac{U_m - U_m(0)}{T} + R\ln q^V$$

Substituting for $U_m - U_m(0)$ and q^V and multiplying by 3 for the 3 dimensions of vibration in a crystal we obtain $\frac{S_m}{3R} = \left[\frac{x}{e^x - 1} - \ln(1 - e^{-x})\right]$ with $x \equiv \frac{\theta_E}{T}$. Note that this is the same expression obtained above. We now plot this result against $y \equiv \frac{1}{x} = \frac{T}{\Theta_E}$

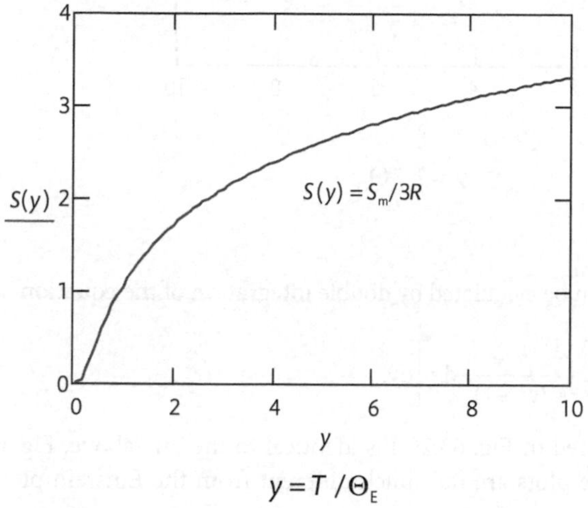

Fig 63.2(a)

Debye solid: $C_{V,m}(T) = 3Rf_D(T)\quad f_D(T) = 3\left(\frac{T}{\theta_D}\right)^3 \int_0^{\theta_D/T} \frac{x^4 e^x}{(e^x - 1)^2} dx$

$$S_m = \int_0^T (C_{V,m}/T)dT = 3R\int_0^T [f_D(T)/T]dT$$

The function $f_D(T)$ cannot be integrated analytically, only numerically, the integration yielding a table of numbers, one for each value of $\frac{\theta_D}{T}$. The entropy requires a second integration which would be the area under a plot of $C_{V,m}$ against T. This double integration can certainly be accomplished rather easily with available software such as MathCad® and we do this below. However we could again proceed as we did in the alternative derivation of the entropy equation for the Einstein solid, starting from the partition function for the Debye solid. This derivation is complex, but in essence is conceptually the same as what we did above for the case of the Einstein solid. The derivation finally yields an equation for the entropy that involves only a single numerical integration.

$$\frac{S}{3R} = \frac{3}{u^3}\int_0^u \left[\frac{x}{e^x - 1} - \ln(1 - e^{-x})\right] x^2 dx \text{ with } u \equiv \frac{\theta_D}{T}$$

In the graph below, we plot this function against $y = \frac{1}{u} = \frac{T}{\Theta_D}$

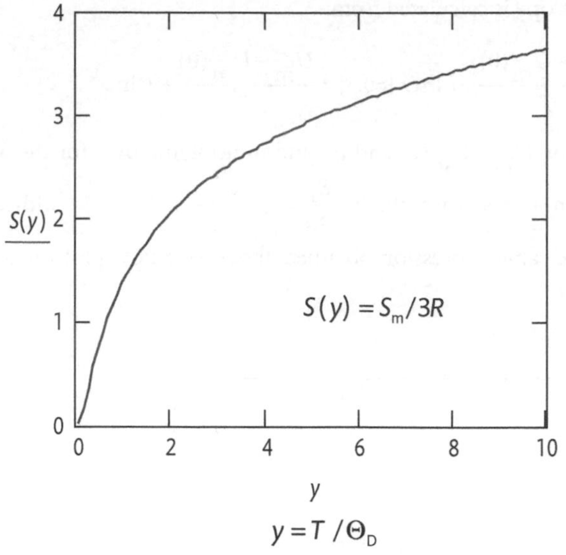

Fig 63.2(b)

The entropy can be calculated by double integration of the equation

$$\frac{S_m}{3R} = \int_0^y 3y^2 \left[\int_0^{\frac{1}{y}} \frac{x^4 e^x}{(e^x - 1)^2} dx \right] dy.$$

The plot obtained in Fig. 63.2(c) is identical to the one above, Fig. 63.2(b). Also note the Debye plots are not much different from the Einstein plot. It is a little steeper at low values of y.

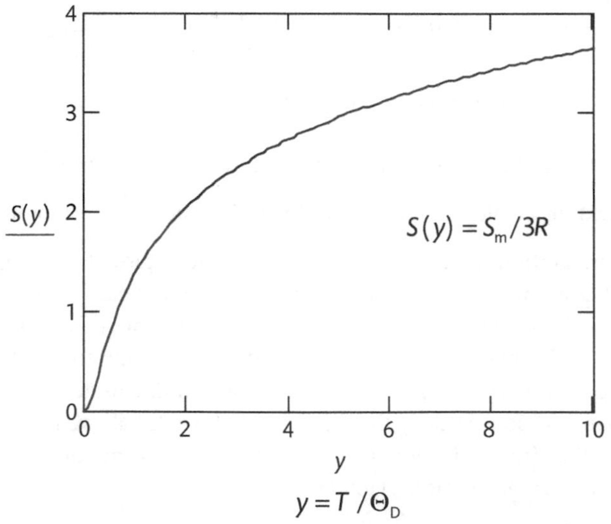

Fig 63.2(c)

Topic 64 Spontaneous processes

Discussion questions

D64.1 All of these expressions are obtained from a combination of the first law of thermodynamics with the second law in the form of the Clausius inequality, $TdS \geq dq$, as was done at the start of *Justification* 64.1.) It may be written as

$$-dU - p_{ex}dV + dw_{add} + TdS \geq 0$$

where we divided the work into pressure-volume work and additional work.

Under conditions of constant energy and volume and no additional work, that is, an isolated system, this relation reduces to $dS \geq 0$ which is equivalent to $\Delta S_{tot} = \Delta S_{universe} \geq 0$. (The universe is an isolated system.)

Under conditions of constant entropy and volume and no additional work, the fundamental relation reduces to $dU \leq 0$.

Under conditions of constant temperature and volume, with no additional work, the relation reduces to $dA \leq 0$, where A is defined as $U - TS$.

Under conditions of constant temperature and pressure, with no additional work, the relation reduces to $dG \leq 0$, where G is defined as $U + pV - TS = H - TS$.

In all of these relations, choosing the inequality provides the criteria for *spontaneous change*. Choosing the equal sign gives us the criteria for **equilibrium** under the conditions specified.

Exercises

E64.1(a) In each case we use

$$\Delta_r G^{\ominus} = \Delta_r H^{\ominus} - T\Delta_r S^{\ominus} \text{ [Example 64.2]}$$

along with

$$\Delta_r H^{\ominus} = \sum_{\text{Products}} \nu\Delta_f H^{\ominus} - \sum_{\text{Reactants}} \nu\Delta_f H^{\ominus} \text{[57.10]}$$

(a) $\Delta_r H^{\ominus} = 2\,\Delta_f H^{\ominus}(CH_3COOH, l) - 2\,\Delta_f H^{\ominus}(CH_3CHO, g)$

$= [2\times(-484.5) - 2\times(-166.19)]\text{ kJ mol}^{-1} = -636.6\overline{2}\text{ kJ mol}^{-1}$

$$\Delta_r G^\ominus = -636.6\bar{2}\ \text{kJ mol}^{-1} - (298.15\ \text{K}) \times (-386.1\ \text{J K}^{-1}\ \text{mol}^{-1})$$
$$= \boxed{-521.5\ \text{kJ mol}^{-1}}$$

(b) $\Delta_r H^\ominus = 2\,\Delta_f H^\ominus(\text{AgBr, s}) - 2\,\Delta_f H^\ominus(\text{AgCl, s})$
$$= [2\times(-100.37) - 2\times(-127.07)]\ \text{kJ mol}^{-1} = +53.40\ \text{kJ mol}^{-1}$$

$$\Delta_r G^\ominus = +53.40\ \text{kJ mol}^{-1} - (298.15\ \text{K}) \times (+92.6)\ \text{J K}^{-1}\ \text{mol}^{-1} = \boxed{+25.8\ \text{kJ mol}^{-1}}$$

(c) $\Delta_r H^\ominus = \Delta_f H^\ominus(\text{HgCl}_2,\text{s}) = -224.3\ \text{kJ mol}^{-1}$

$$\Delta_r G^\ominus = -224.3\ \text{kJ mol}^{-1} - (298.15\ \text{K}) \times (-153.1\ \text{J K}^{-1}\ \text{mol}^{-1}) = \boxed{-178.7\ \text{kJ mol}^{-1}}$$

E64.2(a) $\Delta_r G^\ominus = \Delta_r H^\ominus - T\Delta_r S^\ominus$ [65.2] $\Delta_r H^\ominus = \sum_{\text{Products}} \nu \Delta_f H^\ominus - \sum_{\text{Reactants}} \nu \Delta_f H^\ominus$ [57.10]

$$\Delta_r S^\ominus = \sum_{\text{Products}} \nu S_m^\ominus - \sum_{\text{Reactants}} \nu S_m^\ominus \ [63.4]$$

$$\Delta_r H^\ominus = 2\,\Delta_f H^\ominus(\text{H}_2\text{O, l}) - 4\,\Delta_f H^\ominus(\text{HI, g})$$
$$= \{2\times(-285.83) - 4\times(+26.48)\}\ \text{kJ mol}^{-1}$$
$$= -677.58\ \text{kJ mol}^{-1}$$

$$\Delta_r S = 2\,S_m^\ominus(\text{I}_2,\text{s}) + 2\,S_m^\ominus(\text{H}_2\text{O, l}) - 4\,S_m^\ominus(\text{HI, g}) - S_m^\ominus(\text{O}_2,\text{g})$$
$$= [(2\times116.135) + (2\times69.91) - (4\times206.59) - (205.14)]\ \text{J K}^{-1}\ \text{mol}^{-1}$$
$$= -659.41\ \text{J K}^{-1}\ \text{mol}^{-1} = -0.65941\ \text{kJ K}^{-1}\ \text{mol}^{-1}$$

$$\Delta_r G^\ominus = -677.58\ \text{kJ mol}^{-1} - (298.15\ \text{K}) \times (-0.65941\ \text{kJ K}^{-1}\ \text{mol}^{-1})$$
$$= \boxed{-480.98\ \text{kJ mol}^{-1}}$$

Question. Repeat the calculation based on $\Delta_f G$ data of Table 57.4. What difference, if any, is there from the value above?

E64.3(a) $\text{CH}_4(\text{g}) + 2\,\text{O}_2(\text{g}) \rightarrow \text{CO}_2(\text{g}) + 2\,\text{H}_2\text{O(l)}$

$$\Delta_r G^\ominus = \sum_{\text{Products}} \nu \Delta_f G^\ominus - \sum_{\text{Rectants}} \nu \Delta_f G^\ominus \ [65.3]$$

$$\Delta_r G^\ominus = \Delta_f G^\ominus(\text{CO}_2,\text{g}) + 2\Delta_f G^\ominus(\text{H}_2\text{O,l}) - \Delta_f G^\ominus(\text{CH}_4,\text{g})$$
$$= \{-394.36 + (2\times-237.13) - (-50.72)\}\ \text{kJ mol}^{-1} = -817.90\ \text{kJ mol}^{-1}$$

Therefore, the maximum non-expansion work is $\boxed{817.90\ \text{kJ mol}^{-1}}$ since $|w_{\text{add}}| = |\Delta G|$

E64.4(a) In each case the contribution to G is given by

$G - G(0) = -nRT\ln q$ [64.12b; but see Comment to the solution to Exercise 60.3(a)]

Therefore, we first evaluate q^{R} and q^{V}

$$q^{R} = \frac{0.6950}{\sigma}\frac{T/K}{\tilde{B}/\text{cm}^{-1}} [52.13, \sigma = 2]$$

$$= \frac{0.6950 \times (298)}{(2)\times(0.3902)} = 265$$

$$q^{V} = \left(\frac{1}{1-e^{-a}}\right)\times\left(\frac{1}{1-e^{-b}}\right)^{2}\times\left(\frac{1}{1-e^{-c}}\right) [52.15]$$

with

$$a = \frac{(1.4388)\times(1388.2)}{298} = 6.70\bar{2}$$

$$b = \frac{(1.4388)\times(667.4)}{298} = 3.22\bar{2}$$

$$c = \frac{(1.4388)\times(2349.2)}{298} = 11.3\bar{4}$$

Hence

$$q^{V} = \frac{1}{1-e^{-6.702}}\times\left(\frac{1}{1-e^{-3.222}}\right)^{2}\times\frac{1}{1-e^{-11.34}} = 1.08\bar{6}$$

Therefore, the rotational contribution to the molar Gibbs energy is

$$-RT\ln q^{R} = -8.314\ \text{J K}^{-1}\ \text{mol}^{-1}\times 298\ \text{K}\times\ln 265$$

$$= \boxed{-13.8\ \text{kJ mol}^{-1}}$$

and the vibrational contribution is

$$-RT\ln q^{V} = -8.314\ \text{J K}^{-1}\ \text{mol}^{-1}\times 298\ \text{K}\times\ln 1.08\bar{6} = \boxed{-0.20\ \text{kJ mol}^{-1}}$$

E64.5(a) The partition function is

$$q = \sum_{j} g_{j}e^{-\beta\varepsilon_{j}}$$

with degeneracies $g_{j} = 2J+1$

so $q = 4 + 2e^{-\beta\varepsilon} [g(^{2}P_{3/2}) = 4,\ g(^{2}P_{1/2}) = 2]$

(a) At 500 K

$$\beta\varepsilon = \frac{(1.4388\ \text{cm K})\times(881\ \text{cm}^{-1})}{500\ \text{K}} = 2.53\bar{5}$$

Therefore, the contribution to G_{m} is

$$G_{m} - G_{m}(0) = -RT\ \ln q \quad [64.12b, \text{but see the Comment to Exercise 60.3(a)}]$$

$$-RT\ \ln q = -(8.314\ \mathrm{J\ K^{-1}\ mol^{-1}})\times(500\mathrm{K})\times\ln(4+2\times e^{-2.535})$$
$$= -(8.314\ \mathrm{J\ K^{-1}\ mol^{-1}})\times(500\mathrm{K})\times(\ln 4+0.158) = \boxed{-6.42\ \mathrm{kJ\ mol^{-1}}}$$

(b) At 900 K

$$\beta\varepsilon = \frac{(1.4388\ \mathrm{cm\ K})\times(881\ \mathrm{cm^{-1}})}{900\ \mathrm{K}} = 1.40\bar{8}$$

Therefore, the contribution to G_m is

$$G_m - G_m(0) = -RT\ \ln q$$
$$-RT\ \ln q = -(8.314\ \mathrm{J\ K^{-1}\ mol^{-1}})\times(900\ \mathrm{K})\times\ln(4+2\times e^{-1.408})$$
$$= -(8.314\ \mathrm{J\ K^{-1}\ mol^{-1}})\times(900\ \mathrm{K})\times(\ln 4+0.489) = \boxed{-14.0\ \mathrm{kJ\ mol^{-1}}}$$

Problems

P64.1 First, determine the final state in each section. In section B, the volume was halved at constant temperature, so the pressure was doubled: $p_{B,f} = 2p_{B,i}$. The piston ensures that the pressures are equal in both chambers, so $p_{A,f} = 2p_{B,i} = 2p_{A,i}$. From the perfect gas law

$$\frac{T_{A,f}}{T_{A,i}} = \frac{p_{A,f}V_{A,f}}{p_{A,i}V_{A,i}} = \frac{(2p_{A,i})\times(3.00\ \mathrm{dm^3})}{(p_{A,i})\times(2.00\ \mathrm{dm^3})} = 3.00 \qquad \text{so} \qquad T_{A,f} = 900\ \mathrm{K}.$$

(a) $\Delta S_A = nC_{V,m}\ln\left(\frac{T_{A,f}}{T_{A,i}}\right)$[62.8 with C_V in place of C_p]$+nR\ln\left(\frac{V_{A,f}}{V_{A,i}}\right)$[62.2]

$$\Delta S_A = (2.0\ \mathrm{mol})\times(20\ \mathrm{J\ K^{-1}\ mol^{-1}})\times\ln 3.00$$
$$+(2.00\ \mathrm{mol})\times(8.314\ \mathrm{J\ K^{-1}\ mol^{-1}})\times\ln\left(\frac{3.00\ \mathrm{dm^3}}{2.00\ \mathrm{dm^3}}\right)$$
$$= \boxed{50.7\ \mathrm{J\ K^{-1}}}$$

$$\Delta S_B = nR\ln\left(\frac{V_{B,f}}{V_{B,i}}\right) = (2.00\ \mathrm{mol})\times(8.314\ \mathrm{J\ K^{-1}\ mol^{-1}})\times\ln\left(\frac{1.00\ \mathrm{dm^3}}{2.00\ \mathrm{dm^3}}\right)$$
$$= \boxed{-11.5\ \mathrm{J\ K^{-1}}}$$

(b) The Helmholtz free energy is defined as $A = U - TS$. Because section B is isothermal, $\Delta U = 0$ and $\Delta(TS) = T\Delta S$, so

$$\Delta A_B = -T_B\Delta S_B = -(300\ \mathrm{K})(-11.5\ \mathrm{J\ K^{-1}}) = 3.46\times10^3\ \mathrm{J} = \boxed{+3.46\ \mathrm{kJ}}$$

In section A, we cannot compute $\Delta(TS)$, so we cannot compute ΔU. ΔA $\boxed{\text{is indeterminate}}$ in both magnitude and sign. We know that in a perfect gas, U depends only on temperature; moreover, $U(T)$ is an increasing function of

T, for $\dfrac{\partial U}{\partial T} = C$ (heat capacity), which is positive; since $\Delta T > 0, \Delta U > 0$ as well. But $\Delta(TS) > 0$ too, since both the temperature and the entropy increase.

(c) Likewise, under constant-temperature conditions

$$\Delta G = \Delta H - T\Delta S$$

In section B, $\Delta H_B = 0$ (constant temperature, perfect gas), so

$$\Delta G_B = -T_B \Delta S_B = -(300\text{ K})\times(-11.5\text{ J K}^{-1}) = \boxed{3.46\times10^3\text{ J}}$$

ΔG_A is $\boxed{\text{indeterminate}}$ in both magnitude and sign.

(d) $\Delta S(\text{total system}) = \Delta S_A + \Delta S_B = (50.7 - 11.5)\text{ J K}^{-1} = \boxed{+39.2\text{ J K}^{-1}}$

If the process has been carried out reversibly as assumed in the statement of the problem we can say

$$\Delta S(\text{system}) + \Delta S(\text{surroundings}) = 0$$

Hence, $\Delta S(\text{surroundings}) = \boxed{-39.2\text{ J K}^{-1}}$

Question. Can you design this process such that heat is added to section A reversibly?

P64.3 (a) At constant temperature,

$$\Delta_r G = \Delta_r H - T\Delta_r S \quad \text{so} \quad \Delta_r S = \frac{\Delta_r H - \Delta_r G}{T}$$

and $\Delta_r S = \dfrac{[-20-(-31)]\text{ kJ mol}^{-1}}{310\text{ K}} = +0.035\text{ kJ K}^{-1}\text{ mol}^{-1} = \boxed{+35\text{ J K}^{-1}\text{ mol}^{-1}}$

(b) The power density P is

$$P = \frac{|\Delta_r G|\, n}{V}$$

where n is the number of moles of ATP hydrolyzed per second

$$n = \frac{N}{N_A} = \frac{10^6\text{ s}^{-1}}{6.02\times10^{23}\text{ mol}^{-1}} = 1.6\bar{6}\times10^{-18}\text{ mol s}^{-1}$$

and V is the volume of the cell

$$V = \tfrac{4}{3}\pi r^3 = \tfrac{4}{3}\pi(10\times10^{-6}\text{ m})^3 = 4.1\bar{9}\times10^{-15}\text{ m}^3$$

Thus $P = \dfrac{|\Delta_r G|\, n}{V} = \dfrac{(31\times10^3\text{ J mol}^{-1})\times(1.6\bar{6}\times10^{-18}\text{ mol s}^{-1})}{4.1\bar{9}\times10^{-15}\text{ m}^3} = \boxed{12\text{ W m}^{-3}}$

This is orders of magnitude less than the power density of a computer battery, which is about

$$P_{\text{battery}} = \frac{15\text{ W}}{100\text{ cm}^3}\times\left(\frac{100\text{ cm}}{1\text{ m}}\right)^3 = \boxed{1.5\times10^4\text{ W m}^{-3}}$$

(c) Simply make a ratio of the magnitudes of the free energies

$$\frac{14.2\text{ kJ (mol glutamine)}^{-1}}{31\text{ kJ (mol ATP)}^{-1}} = \boxed{0.46\frac{\text{mol ATP}}{\text{mol glutamine}}}$$

Topic 65 **Standard Gibbs energies**

Discussion questions

D65.1 The Gibbs energy of formation refers to the change in Gibbs energy associated with a formation reaction. A standard formation reaction is the reaction of elements in their standard states to form a product in its standard state. In that sense, the methods of determination of standard Gibbs energies of formation are the same as for determining the Gibbs energy change for chemical reactions in general. Here we list, but do not describe in detail, a few of the more important methods that are used.

(1) From the equilibrium constant of the formation reaction. $\Delta_f G^{\ominus} = -RT\ln K$. See focus 15.

(2) From the potential of an electrochemical cell in which the reaction generating the potential is a formation reaction. $\Delta_f G^{\ominus} = -\nu FE_{cell}$. This method gives accurate values of $\Delta_f G^{\ominus}$. See focus 15.

(3) From the partition functions of the elements and compound involved in the formation reaction. The partition functions can be determined from spectroscopic data as illustrated in Example 65.1

Exercises

E65.1(a) In each case $\Delta_r G^{\ominus} = \sum_{\text{Products}} \nu\Delta_f G^{\ominus} - \sum_{\text{Reactants}} \nu\Delta_f G^{\ominus}$ [65.3]

with $\Delta_f G^{\ominus}$(J) values from Tables 57.3 and 57.4.

(a) $\Delta_r G^{\ominus} = 2\,\Delta_f G^{\ominus}(CH_3COOH, l) - 2\,\Delta_f G^{\ominus}(CH_3CHO, g)$

$$= [2\times(-389.9) - 2\times(-128.86)]\ \text{kJ mol}^{-1} = \boxed{-522.1\ \text{kJ mol}^{-1}}$$

(b) $\Delta_r G^{\ominus} = 2\Delta_f G^{\ominus}(AgBr, s) - 2\Delta_f G^{\ominus}(AgCl, s) = [2\times(-96.90)$

$$-2\times(-109.79)]\ \text{kJ mol}^{-1}$$

$$= \boxed{+25.78\ \text{kJ mol}^{-1}}$$

(c) $\Delta_r G^{\ominus} = \Delta_f G^{\ominus}(HgCl_2, s) = \boxed{-178.6\ \text{kJ mol}^{-1}}$

Comment. In each case these values of $\Delta_r G^{\ominus}$ agree closely with the calculated values in Exercise 64.1(a).

E65.2(a) The formation reaction for ethyl acetate is

$$4\,C(s) + 4\,H_2(g) + O_2(g) \rightarrow CH_3COOC_2H_5\,(l)$$

$$\Delta_f G^{\ominus} = \Delta_f H^{\ominus} - T\Delta_f S^{\ominus} \text{ [Example 64.2]}$$

$\Delta_f H^{\ominus}$ is to be obtained from $\Delta_c H^{\ominus}$ for ethyl acetate and data from Tables 57.3 and 57.4.

Thus

$$CH_3COOC_2H_5(l) + 5\,O_2(g) \rightarrow 4\,CO_2(g) + 4\,H_2O(l)$$

$$\Delta_c H^{\ominus} = 4\Delta_f H^{\ominus}(CO_2, g) + 4\Delta_f H^{\ominus}(H_2O, l) - \Delta_f H^{\ominus}(CH_3COOC_2H_5(l))$$

$$\begin{aligned}\Delta_f H^{\ominus}(CH_3COOC_2H_5(l)) &= 4\Delta_f H^{\ominus}(CO_2, g) + 4\Delta_f H^{\ominus}(H_2O, l) - \Delta_c H^{\ominus}\\ &= [4\times(-393.51) + 4\times(-285.83) - (-2231]\,\text{kJ mol}^{-1}\\ &= -486.\bar{4}\,\text{kJ mol}^{-1}\end{aligned}$$

$$\Delta_r S^{\ominus} = \sum_{\text{Products}} \nu S_m^{\ominus} - \sum_{\text{Reactants}} \nu S_m^{\ominus} \text{ [63.4]}$$

$$\begin{aligned}\Delta_f S &= S_m^{\ominus}(CH_3COOC_2H_5(l)) - 4S_m^{\ominus}(C, s) - 4S_m^{\ominus}(H_2, g) - S_m^{\ominus}(O_2, g)\\ &= [259.4 - (4\times 5.740) - (4\times 130.68) - (205.14)]\,\text{J K}^{-1}\text{mol}^{-1}\\ &= -491.4\,\text{J K}^{-1}\text{mol}^{-1}\end{aligned}$$

Hence $\Delta_r G^{\ominus} = -486.\bar{4}\,\text{kJ mol}^{-1} - (298.15\,\text{K})\times(-491.\bar{4}\,\text{J K}^{-1}\text{mol}^{-1}) = \boxed{-340\,\text{kJ mol}^{-1}}$

Problem

P65.1 The standard molar Gibbs energy is given by

$$G_m^{\ominus} - G_m^{\ominus}(0) = RT\ln\frac{q_m^{\ominus}}{N_A} \quad \text{where} \frac{q_m^{\ominus}}{N_A} = \frac{q_m^{T\ominus}}{N_A} q^R q^V q^E \text{ [See Sections 64.2(b) and 52.2]}$$

Translation: (See Section 52.2 and the solution to Problems 60.2 and 64.4 for all partition functions)

$$\begin{aligned}\frac{q_m^{T\ominus}}{N_A} &= 2.561\times10^{-2}(T/\text{K})^{5/2}(M/\text{g mol}^{-1})^{3/2}\\ &= (2.561\times10^{-2})\times(2000)^{5/2}\times(38.90)^{3/2} = 1.111\times10^{9}\end{aligned}$$

Rotation of a linear molecule:

$$q^R = \frac{kT}{\sigma hc\tilde{B}} = \frac{0.6950}{\sigma}\times\frac{T/\text{K}}{\tilde{B}/\text{cm}^{-1}}$$

The rotational constant is

$$\tilde{B} = \frac{\hbar}{4\pi cI} = \frac{\hbar}{4\pi c m_{\text{eff}} R^2}$$

where $m_{\text{eff}} = \dfrac{m_{\text{B}}m_{\text{Si}}}{m_{\text{B}}+m_{\text{Si}}} = \dfrac{(10.81)\times(28.09)}{10.81+28.09}\times\dfrac{10^{-3}\,\text{kg mol}^{-1}}{6.022\times10^{23}\,\text{mol}^{-1}} = 1.296\times10^{-26}\,\text{kg}$

$$\tilde{B} = \frac{1.0546\times10^{-34}\,\text{J s}}{4\pi(2.998\times10^{10}\,\text{cm s}^{-1})\times(1.296\times10^{-26}\,\text{kg})\times(190.5\times10^{-12}\,\text{m})^2} = 0.5952\ \text{cm}^{-1}$$

so $q^{\text{R}} = \dfrac{0.6950}{1}\times\dfrac{2000}{0.5952} = 2335$

Vibration:

$$q^{\text{V}} = \frac{1}{1-\text{e}^{-hc\tilde{\nu}/kT}} = \frac{1}{1-\exp\left(\dfrac{-1.4388(\tilde{\nu}/\text{cm}^{-1})}{T/\text{K}}\right)} = \frac{1}{1-\exp\left(\dfrac{-1.4388(772)}{2000}\right)}$$

$$= 2.467$$

The Boltzmann factor for the lowest-lying electronic excited state is

$$\exp\left(\frac{-(1.4388)\times(8000)}{2000}\right) = 3.2\times10^{-3}$$

The degeneracy of the ground level is 4 (spin degeneracy = 4, orbital degeneracy = 1), and that of the excited level is also 4 (spin degeneracy = 2, orbital degeneracy = 2), so

$$q^{\text{E}} = 4(1+3.2\times10^{-3}) = 4.013$$

Putting it all together yields

$$\begin{aligned} G_{\text{m}}^{\ominus} - G_{\text{m}}^{\ominus}(0) &= (8.3145\,\text{J mol}^{-1}\,\text{K}^{-1})\times(2000\,\text{K}) \\ &\quad\times\ln[(1.111\times10^{9})\times(2335)\times(2.467)\times(4.013)] \\ &= 5.135\times10^{5}\,\text{J mol}^{-1} = \boxed{513.5\,\text{kJ mol}^{-1}} \end{aligned}$$

Topic 66 Combining the First and Second Laws

Discussion question

D66.1 The relation $(\partial G/\partial T)_p = -S$ shows that the Gibbs function of a system decreases with T at constant p in proportion to the magnitude of its entropy. This makes good sense when one considers the definition of G, which is $G = U + pV - TS$. Hence, G is expected to decrease with T in proportion to S when p is constant. Furthermore, an increase in temperature causes entropy to increase according to

$$\Delta S = \int_i^f \mathrm{d}q_{\mathrm{rev}}/T$$

The corresponding increase in molecular disorder causes a decline in the Gibbs energy. (Entropy is always positive.)

Exercises

E66.1(a) $$\Delta G = nRT\ln\left(\frac{p_f}{p_i}\right)[66.12] = nRT\ln\left(\frac{V_i}{V_f}\right) \text{ [Boyle's law]}$$

$$\Delta G = (2.5\times10^{-3}\,\mathrm{mol})\times(8.314\ \mathrm{J\ K^{-1}\,mol^{-1}})\times(300\,\mathrm{K})\times\ln\left(\frac{42}{600}\right) = \boxed{-17\,\mathrm{J}}$$

E66.2(a) $$\left(\frac{\partial G}{\partial T}\right)_p = -S\,[66.7];\ \text{hence}\ \left(\frac{\partial G_f}{\partial T}\right)_p = -S_f,\ \text{and}\ \left(\frac{\partial G_i}{\partial T}\right)_p = -S_i$$

$$\Delta S = S_f - S_i = -\left(\frac{\partial G_f}{\partial T}\right)_p + \left(\frac{\partial G_i}{\partial T}\right)_p = -\left(\frac{\partial(G_f - G_i)}{\partial T}\right)_p$$

$$= -\left(\frac{\partial \Delta G}{\partial T}\right)_p = -\frac{\partial}{\partial T}\left(-85.40\,\mathrm{J} + 36.5\,\mathrm{J}\times\frac{T}{\mathrm{K}}\right)$$

$$= \boxed{-36.5\,\mathrm{J\,K^{-1}}}$$

E66.3(a) We will assume that the volume and molar volume of octane changes little over the range of pressures given and that, therefore, equation 66.11 which applies to incompressible substances can be used to solve this exercise. The change in Gibbs energy for this sample is then given by

$$\Delta G = nV_m\Delta p\,[66.11] = V\Delta p$$

$$\Delta G = (1.0\,\mathrm{dm^3})\times\left(\frac{1\,\mathrm{m^3}}{10^3\,\mathrm{dm^3}}\right)\times(99\,\mathrm{atm})\times(1.013\times10^5\,\mathrm{Pa\,atm^{-1}})$$

$$= 10\ \mathrm{kPa\,m^3} = \boxed{+10\,\mathrm{kJ}}$$

In order to calculate the change in Gibbs energy per mole we calculate the molar volume

$$V_{\rm m} = \frac{M}{\rho(\text{density})} = \frac{114.23\ \text{g mol}^{-1}}{0.703\ \text{g cm}^{-3}} \times \frac{10^{-6}\ \text{m}^3}{\text{cm}^3} = 1.625\times10^{-4}\ \text{m}^3$$

$$\Delta G_{\rm m} = V_{\rm m}\Delta p\,[66.11] = 1.625\times10^{-4}\ \text{m}^3\text{mol}^{-1}\times(99\ \text{atm})\times(1.013\times10^5\ \text{Pa atm}^{-1})$$
$$= \boxed{1.6\ \text{kJ mol}^{-1}}$$

E66.4(a) $\Delta G_{\rm m} = RT\ln\frac{p_{\rm f}}{p_{\rm i}}\,[66.12] = (8.314\ \text{J K}^{-1}\,\text{mol}^{-1})\times(298\text{K})\times\ln\left(\frac{100.0}{1.0}\right) = \boxed{+11\ \text{kJ mol}^{-1}}$

E66.5(a) $dG = -S\,dT + V\,dp$ [66.6]; at constant T, $dG = V\,dp$; therefore

$$\Delta G = \int_{p_{\rm i}}^{p_{\rm f}} V\,dp$$

The change in volume of a condensed phase under isothermal compression is given by the isothermal compressibility (eqn 58.8).

$$\kappa_T = -\frac{1}{V}\left(\frac{\partial V}{\partial p}\right)_T = 76.8\times10^{-6}\ \text{atm}^{-1}\quad\text{[Table 58.1]}$$

This small isothermal compressibility (typical of condensed phases) tells us that we can expect a small change in volume from even a large increase in pressure. So we can make the following approximations to obtain a simple expression for the volume as a function of the pressure

$$\kappa_T \approx -\frac{1}{V}\left(\frac{V-V_{\rm i}}{p-p_{\rm i}}\right) \approx -\frac{1}{V_{\rm i}}\left(\frac{V-V_{\rm i}}{p}\right)\quad\text{so}\quad V = V_{\rm i}(1-\kappa_T p),$$

where $V_{\rm i}$ is the volume at 1 atm, namely the sample mass over the density, m/ρ.

$$\Delta G = \int_{1\,\text{atm}}^{3\,000\,\text{atm}} \frac{m}{\rho}(1-\kappa_T p)\,dp$$
$$= \frac{m}{\rho}\left(\int_{1\,\text{atm}}^{3\,000\,\text{atm}} dp - \kappa_T\int_{1\,\text{atm}}^{3\,000\,\text{atm}} p\,dp\right)$$
$$= \frac{m}{\rho}\left(p\Big|_{1\,\text{atm}}^{3\,000\,\text{atm}} - \kappa_T\frac{p^2}{2}\Big|_{1\,\text{atm}}^{3\,000\,\text{atm}}\right)$$
$$= \frac{35\ \text{g}}{0.789\ \text{g cm}^{-3}}\left(2\,999\ \text{atm} - (76.8\times10^{-6}\ \text{atm}^{-1})\times(9.00\times10^6\ \text{atm}^2)/2\right)$$
$$= 44.\bar{4}\ \text{cm}^3\times\left(\frac{1\text{m}}{100\ \text{cm}}\right)^3\times 2\,653\ \text{atm}\times(1.013\times10^5\ \text{Pa atm}^{-1})$$
$$= 1.2\times10^4\ \text{J} = \boxed{12\ \text{kJ}}$$

Problems

P66.1 The Gibbs–Helmholtz equation [66.9] may be recast into an analogous equation involving ΔG and ΔH, since

$$\left(\frac{\partial \Delta G}{\partial T}\right)_p=\left(\frac{\partial G_{\mathrm{f}}}{\partial T}\right)_p-\left(\frac{\partial G_{\mathrm{i}}}{\partial T}\right)_p$$

and $\Delta H=H_{\mathrm{f}}-H_{\mathrm{i}}$

Thus, $\left(\frac{\partial}{\partial T}\frac{\Delta_{\mathrm{r}}G^{\ominus}}{T}\right)_p=-\frac{\Delta_{\mathrm{r}}H^{\ominus}}{T^2}$

$$\mathrm{d}\left(\frac{\Delta_{\mathrm{r}}G^{\ominus}}{T}\right)=\left(\frac{\partial}{\partial T}\frac{\Delta_{\mathrm{r}}G^{\ominus}}{T}\right)_p\mathrm{d}T[\text{constant pressure}]=-\frac{\Delta_{\mathrm{r}}H^{\ominus}}{T^2}\mathrm{d}T$$

$$\Delta\left(\frac{\Delta_{\mathrm{r}}G^{\ominus}}{T}\right)=-\int_{T_{\mathrm{c}}}^{T}\frac{\Delta_{\mathrm{r}}H^{\ominus}\,\mathrm{d}T}{T^2}$$

$$\approx-\Delta_{\mathrm{r}}H^{\ominus}\int_{T_{\mathrm{c}}}^{T}\frac{\mathrm{d}T}{T^2}=\Delta_{\mathrm{r}}H^{\ominus}\left(\frac{1}{T}-\frac{1}{T_{\mathrm{c}}}\right)\quad[\Delta_{\mathrm{r}}H^{\ominus}\text{ assumed constant}]$$

Therefore, $\frac{\Delta_{\mathrm{r}}G^{\ominus}(T)}{T}-\frac{\Delta_{\mathrm{r}}G^{\ominus}(T_{\mathrm{c}})}{T_{\mathrm{c}}}\approx\Delta_{\mathrm{r}}H^{\ominus}\left(\frac{1}{T}-\frac{1}{T_{\mathrm{c}}}\right)$

and so $\Delta_{\mathrm{r}}G^{\ominus}(T)=\frac{T}{T_{\mathrm{c}}}\Delta_{\mathrm{r}}G^{\ominus}(T_{\mathrm{c}})+\left(1-\frac{T}{T_{\mathrm{c}}}\right)\Delta_{\mathrm{r}}H^{\ominus}(T_{\mathrm{c}})$

$$=\tau\Delta_{\mathrm{r}}G^{\ominus}(T_{\mathrm{c}})+(1-\tau)\Delta_{\mathrm{r}}H^{\ominus}(T_{\mathrm{c}})\quad\text{where}\quad\tau=\frac{T}{T_{\mathrm{c}}}$$

For the reaction

$$2\,\mathrm{CO(g)}+\mathrm{O_2(g)}\rightarrow 2\,\mathrm{CO_2(g)}$$

$$\Delta_{\mathrm{r}}G^{\ominus}(T_{\mathrm{c}})=2\,\Delta_{\mathrm{f}}G^{\ominus}(\mathrm{CO_2,g})-2\,\Delta_{\mathrm{f}}G^{\ominus}(\mathrm{CO,g})$$
$$=[2\times(-394.36)-2\times(-137.17)]\,\mathrm{kJ\,mol^{-1}}=-514.38\,\mathrm{kJ\,mol^{-1}}$$

$$\Delta_{\mathrm{r}}H^{\ominus}(T_{\mathrm{c}})=2\,\Delta_{\mathrm{f}}H^{\ominus}(\mathrm{CO_2,g})-2\,\Delta_{\mathrm{f}}H^{\ominus}(\mathrm{CO,g})$$
$$=[2\times(-393.51)-2\times(-110.53)]\,\mathrm{kJ\,mol^{-1}}=-565.96\,\mathrm{kJ\,mol^{-1}}$$

Therefore, since $\tau=\frac{375}{298.15}=1.25\overline{8}$

$$\Delta_{\mathrm{r}}G^{\ominus}(375\,\mathrm{K})=\{(1.25\overline{8})\times(-514.38)+(1-1.25\overline{8})\times(-565.96)\}\,\mathrm{kJ\,mol^{-1}}$$
$$=\boxed{-501\,\mathrm{kJ\,mol^{-1}}}$$

P66.3 $w_{\text{add, max}} = \Delta_r G$ [64.17]

$$\Delta_r G^{\ominus}(37°\text{C}) = \tau\,\Delta_r G^{\ominus}(T_c) + (1-\tau)\Delta_r H^{\ominus}(T_c) \quad \left[\text{Problem 66.2, } \tau = \frac{T}{T_c}\right]$$

$$= \left(\frac{310\text{ K}}{298.15\text{ K}}\right) \times (-6\,333\text{ kJ mol}^{-1}) + \left(1 - \frac{310\text{ K}}{298.15\text{ K}}\right) \times (-5\,797\text{ kJ mol}^{-1})$$

$$= -6\,354\text{ kJ mol}^{-1}$$

The difference is

$\Delta_r G^{\ominus}(37°\text{C}) - \Delta_r G^{\ominus}(T_c) = \{-6\,354 - (-6\,333)\}\text{ kJ mol}^{-1} = \boxed{-21\text{ kJ mol}^{-1}}$ Therefore an additional 21 kJ mol^{-1} of non-expansion work may be done at the higher temperature.

Comment. As shown by Problem 66.1, increasing the temperature does not necessarily increase the maximum non-expansion work. The relative magnitude of $\Delta_r G^{\ominus}$ and $\Delta_r H^{\ominus}$ is the determining factor.

P66.5 $H \equiv U + pV$

$$dH = dU + p\,dV + V\,dp = T\,dS - p\,dV\ [3.43] + p\,dV + V\,dp = T\,dS + V\,dp$$

Since H is a state function, dH is exact, and it follows that

$$\left(\frac{\partial H}{\partial S}\right)_p = T \quad \text{and} \quad \boxed{\left(\frac{\partial V}{\partial S}\right)_p = \left(\frac{\partial T}{\partial p}\right)_S}$$

Similarly, $A \equiv U - TS$

$$dA = dU - T\,dS - S\,dT = T\,dS - p\,dV\ [3.43] - T\,dS - S\,dT = -p\,dV - S\,dT$$

Since dA is exact,

$$\boxed{\left(\frac{\partial S}{\partial V}\right)_T = \left(\frac{\partial p}{\partial T}\right)_V}$$

P66.7 If $S = S(T,p)$

then $$dS = \left(\frac{\partial S}{\partial T}\right)_p dT + \left(\frac{\partial S}{\partial p}\right)_T dp$$

$$T\,dS = T\left(\frac{\partial S}{\partial T}\right)_p dT + T\left(\frac{\partial S}{\partial p}\right)_T dp$$

Use $$\left(\frac{\partial S}{\partial T}\right)_p = \left(\frac{\partial S}{\partial H}\right)_p \left(\frac{\partial H}{\partial T}\right)_p = \frac{1}{T} \times C_p \quad \left[\left(\frac{\partial H}{\partial S}\right)_p = T, \text{ Problem 66.5}\right]$$

$$\left(\frac{\partial S}{\partial p}\right)_T = -\left(\frac{\partial V}{\partial T}\right)_p \quad \text{[Maxwell relation]}$$

Hence, $T\,dS = C_p\,dT - T\left(\frac{\partial V}{\partial T}\right)_p dp = \boxed{C_p dT - \alpha TV\,dp}$

For reversible, isothermal compression, $TdS = dq_{rev}$ and $dT = 0$; hence

$$dq_{rev} = -\alpha TV\,dp$$

$$q_{rev} = \int_{p_i}^{p_f} -\alpha TV\,dp = \boxed{-\alpha TV\Delta p}\ [\alpha \text{ and } V \text{ assumed constant}]$$

For mercury

$$q_{rev} = (-1.82\times10^{-4}\ \text{K}^{-1})\times(273\ \text{K})\times(1.00\times10^{-4}\ \text{m}^{-3})\times(1.0\times10^{8}\ \text{Pa}) = \boxed{-0.50\ \text{kJ}}$$

P66.9 It is most efficient to first explore how the molar Gibbs energy depends on pressure using the full van der Waals expression, and then in turn to set a to zero and finally both a and b to zero in the expression and repeat the calculation. We use Mathcad® to perform the calculation and plot the graphs. We also use data for CO_2 as our sample gas.

The general solution of this problem requires that we develop a function that can calculate the molar volume for given values of pressure, temperature, and van der Waals constants.

Misc. definitions: $R := 8.31447\cdot J\cdot K^{-1}\cdot mol^{-1}$ $dm := 10^{-1}\cdot m$ $kJ := 10^{3}\cdot J$ $TOL := 10^{-3}$

Function for calculation of the molar volume:

V_m estimate for Given/Find solve block: $V := 0.1\cdot dm^3\cdot mol^{-1}$

Given $p = \frac{R\cdot T}{V-b} - \frac{a}{V^2}$ $V_m(p, T, a, b) := \text{Find}(V)$

$a := 3.610\cdot atm\cdot dm^6\cdot mol^{-2}$ $b := 4.29\cdot10^{-2}\cdot dm3\cdot mol^{-1}$

To see the effect of first reducing *a* to zero and then the effect of reducing *b* to zero we define a_1 and b_1 as follows.

$a_1 := 0\cdot atm\cdot dm^6\cdot mol^{-2}$

$b_1 := 0\cdot dm^3\cdot mol^{-1}$

$V_m(250\cdot bar, 325\cdot K, a, b) = 7.004\times10^{-2}\ dm^3\cdot mol^{-1}$

$V_m(250\cdot bar, 325\cdot K, a_1, b) = 1.510\times10^{-1}\ dm^3\cdot mol^{-1}$

Function for calculation of the molar volume:

$V_m(250\cdot bar, 325\cdot K, a_1, b_1) = 1.081\times10^{-1}\ dm^3\cdot mol^{-1}$

We note that when $b = 0$, for this choice of pressure and temperature, there is no real solution for the molar volume.

We can now use eqn 66.10b to calculate the molar Gibbs energy change when pressure changes at constant temperature. Let $\Delta G_m(p,T) = G_m(p,T) - G_m(1\ \text{bar},T)$, then we calculate the following integrals. The first integral has both *a* and *b* nonzero, the second has $a = 0$, the third $b = 0$.

$$\Delta G_m(p, T) := \int_{1\cdot bar}^{p} V_m(p, T, a, b)\,dp$$

$$\Delta G_{ma}(p, T) := \int_{1\cdot bar}^{p} V_m(p, T, a_1, b)\,dp$$

$$\Delta G_{mb}(p, T) := \int_{1\cdot bar}^{p} V_m(p, T, a, b_1)\,dp$$

This last integral cannot be evaluated over the complete range of pressures from 1 bar to 300 bar because the molar volume is not real over all of this range. Let us then calculate the following integral which has both *a* and *b* equal zero, which is the case of the perfect gas.

$$\Delta G_{id}(p, T) := \int_{1 \cdot bar}^{p} V_m(p, T, a_1, b_1)\,dp$$

For our carbon dioxide example the Gibbs energy change that occurs upon increasing the pressure from 1 bar to 300 bar at 325 K is:

$$\Delta G_M(300 \cdot bar, 325 \cdot K) = 1.283 \times 10^1\,kj \cdot mol^{-1}$$

$$\Delta G_{MA}(300 \cdot bar, 325 \cdot K) = 1.670 \times 10^1\,kj \cdot mol^{-1}$$

$$\Delta G_{id}(300 \cdot bar, 325 \cdot K) = 1.541 \times 10^1\,kj \cdot mol^{-1}$$

We plot the change of the Gibbs energy over a chosen pressure range at the particular temperature of 325 K.

$p := 1 \cdot bar, 2 \cdot bar.. 300 \cdot bar$

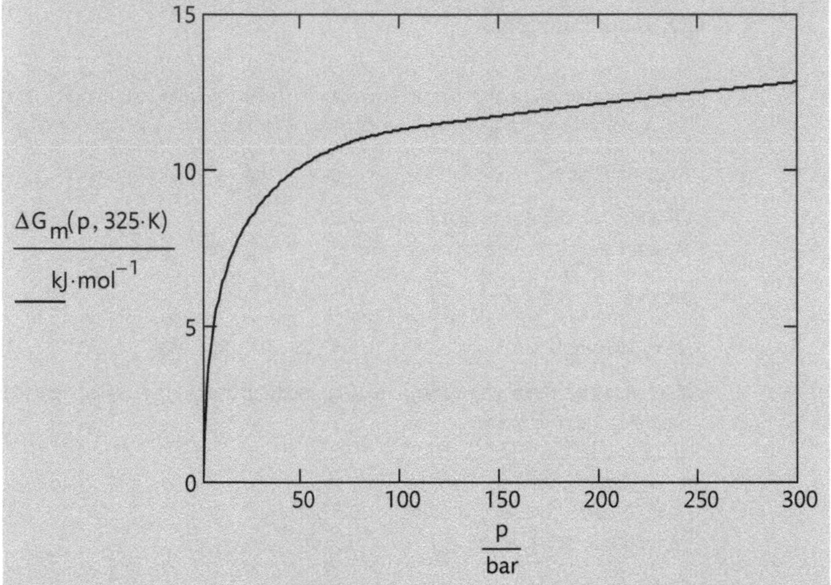

Fig 66.1(a)

To see the effect of reducing *a* to zero, and of both *a* and *b* to zero (the perfect gas), we plot on the same graph below, Figure 66.1(b), ΔG_m, ΔG_{ma}(with $a = 0$), and ΔG_{id}(both *a* and *b* equal zero).

Examination of Figure 66.1(b) shows that when intermolecular attractions are absent, but repulsions present (the top curve), the Gibbs energy increases most rapidly with pressure. When both attractions and repulsions are present (the bottom curve, same as Figure 66.1(a)), the Gibbs energy increases least rapidly with pressure, at least for the case of carbon dioxide, but since the rate of increase is dependent upon the magnitude of *a* and *b*, so it would not necessarily be the same for a different gas.

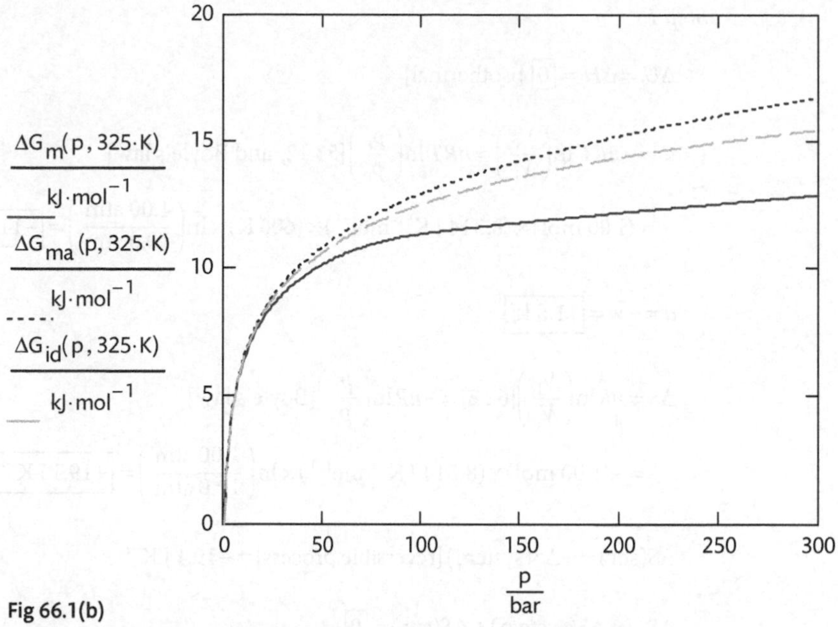

Fig 66.1(b)

Focus 13: Integrated activities

F13.1 We evaluate β by comparing calculated and experimental values for thermodynamic properties. The calculated values are obtained from the theoretical formulas for these properties, all of which are expressed in terms of the parameter β. So there can be many ways of identifying β, as many as there are thermodynamic properties. One way is through the energy as shown in Section 51.19(b). Another is through the pressure as demonstrated in Section 66.3. Another yet is through the entropy, Sections 59.3 and 66.3, and this approach to the identification may be the most fundamental.

When plotted against T, in model systems, such as a two-level system, q and U show sharp discontinuities upon passing through $T=0$. Only S is continuous. However, when plotted against β, all three are continuous upon passing through $\beta=0$. See the solution to problem 58.28. Hence β seems a more natural variable than T. Another way of looking at this is to consider τ, defined as $\tau=1/\beta=kT$, to be the fundamental temperature, and then the "fundamental" constant k appears merely as a scale factor.

F13.3

	Step1	Step2	Step3	Step4	Cycle
q	+11.5 kJ	0	−5.74 kJ	0	+5.8 kJ
w	−11.5 kJ	−3.74 kJ	+5.74 kJ	+3.74 kJ	−5.8 kJ
ΔU	0	−3.74 kJ	0	+3.74 kJ	0
ΔH	0	−6.23 kJ	0	+6.23 kJ	0
ΔS	+19.1 J K^{-1}	0	−19.1 J K^{-1}	0	0
ΔS_{tot}	0	0	0	0	0
ΔG	−11.5 kJ	?	+5.74 kJ	?	0

Step 1

$$\Delta U = \Delta H = \boxed{0} \text{ [isothermal]}$$

$$w = -nRT\ln\left(\frac{V_f}{V_i}\right) = nRT\ln\left(\frac{p_f}{p_i}\right) \text{[55.12, and Boyle's law]}$$

$$= (1.00 \text{ mol})\times(8.314 \text{ J K}^{-1} \text{ mol}^{-1})\times(600 \text{ K})\times\ln\left(\frac{1.00 \text{ atm}}{10.0 \text{ atm}}\right) = \boxed{-11.5 \text{ kJ}}$$

$$q = -w = \boxed{11.5 \text{ kJ}}$$

$$\Delta S = nR\ln\left(\frac{V_f}{V_i}\right)\text{[62.2]} = -nR\ln\left(\frac{p_f}{p_i}\right)\text{[Boyle's law]}$$

$$= -(1.00 \text{ mol})\times(8.314 \text{ J K}^{-1} \text{ mol}^{-1})\times\ln\left(\frac{1.00 \text{ atm}}{10.0 \text{ atm}}\right) = \boxed{+19.1 \text{ J K}^{-1}}$$

$$\Delta S(\text{sur}) = -\Delta S(\text{system}) \text{ [reversible process]} = -19.1 \text{ J K}^{-1}$$

$$\Delta S_{\text{tot}} = \Delta S(\text{system}) + \Delta S(\text{sur}) = \boxed{0}$$

$$\Delta G = \Delta H - T\Delta S = 0 - (600 \text{ K})\times(19.1 \text{ J K}^{-1}) = \boxed{-11.5 \text{ kJ mol}^{-1}}$$

Step 2

$$q = \boxed{0} \text{ [adiabatic]}$$

$$\Delta U = nC_{V,\text{m}}\Delta T \text{ [55.18b]}$$

$$= (1.00 \text{ mol})\times\left(\frac{3}{2}\right)\times(8.314 \text{ J K}^{-1} \text{ mol}^{-1})\times(300 \text{ K} - 600 \text{ K}) = \boxed{-3.74 \text{ kJ}}$$

$$w = \Delta U = \boxed{-3.74 \text{ kJ}}$$

$$\Delta H = \Delta U + \Delta(pV) = \Delta U + nR\Delta T$$

$$= (-3.74 \text{ kJ}) + (1.00 \text{ mol})\times(8.314 \text{ J K}^{-1} \text{ mol}^{-1})\times(-300 \text{ K})$$

$$= \boxed{-6.23 \text{ kJ}}$$

$$\Delta S = \Delta S(\text{sur}) = \boxed{0} \text{ [reversible adiabatic process]}$$

$$\Delta S_{\text{tot}} = \boxed{0}$$

$$\Delta G = \Delta(H - TS) = \Delta H - S\Delta T \text{ [no change in entropy]}$$

Although the change in entropy is known to be zero, the entropy itself is not known, so ΔG is $\boxed{\text{indeterminate}}$.

Step 3

These quantities may be calculated in the same manner as for *Step 1* or more easily as follows

$$\Delta U = \Delta H = \boxed{0} \text{ [isothermal]}$$

$$\varepsilon_{\text{rev}} = 1 - \frac{T_c}{T_h} [15.17] = 1 - \frac{300\text{ K}}{600\text{ K}} = 0.500 = 1 + \frac{q_c}{q_h} [15.16]$$

$$q_c = -0.500 q_h = -(0.500) \times (11.5\text{ kJ}) = -5.74\text{ kJ}$$

$$q_c = \boxed{-5.74\text{ kJ}} \qquad w = -q_c = \boxed{5.74\text{ kJ}}$$

$$\Delta S = \frac{q_{\text{rev}}}{T} \text{ [isothermal]} = \frac{-5.74 \times 10^3\text{ J}}{300\text{ K}} = \boxed{-19.1\text{ J K}^{-1}}$$

$$\Delta S(\text{sur}) = -\Delta S(\text{system}) = +19.1\text{ J K}^{-1}$$

$$\Delta S_{\text{tot}} = \boxed{0}$$

$$\Delta G = \Delta H - T\Delta S = 0 - (300\text{ K}) \times (-19.1\text{ J K}^{-1}) = \boxed{+5.74\text{ kJ}}$$

Step 4

ΔU and ΔH are the negative of their values in *Step 2*. (Initial and final temperatures reversed.)

$$\Delta U = \boxed{+3.74\text{ kJ}}, \quad \Delta H = \boxed{+6.23\text{ kJ}}, \quad q = \boxed{0} \text{ [adiabatic]}$$

$$w = \Delta U = \boxed{+3.74\text{ kJ}}$$

$$\Delta S = \Delta S(\text{sur}) = \boxed{0} \text{ [reversible adiabatic process]}$$

$$\Delta S_{\text{tot}} = \boxed{0}$$

Again $\Delta G = \Delta(H - TS) = \Delta H - S\Delta T$ [no change in entropy]
but S is not known, so ΔG is $\boxed{\text{indeterminate}}$.

Cycle

$$\Delta U = \Delta H = \Delta S = \Delta G = \boxed{0} \quad [\Delta(\text{state function}) = 0 \text{ for any cycle}]$$

$$\Delta S(\text{sur}) = 0 \text{[all reversible processes]}$$

$$\Delta S_{\text{tot}} = \boxed{0}$$

$$q(\text{cycle}) = (11.5 - 5.74)\text{ kJ} = \boxed{5.8\text{ kJ}} \quad w(\text{cycle}) = -q(\text{cycle}) = \boxed{-5.8\text{ kJ}}$$

F13.5 The fundamental equation of thermodynamics is $dU = TdS - pdV$ [66.1]. The required result follows from this relation at constant V, that is when $dV = 0$. Then $\left(\frac{\partial U}{\partial S}\right)_V = T$. Thus the temperature can be identified as the slope of a plot U

against S. The problem suggests we create this plot based on the calculation of Problem 60.6 which applies to a two level system at both positive and negative temperatures.

The partition function for a two-level system with energy separation ε is

$$q = 1 + e^{-\varepsilon/kT} = 1 + e^{-\beta\varepsilon}$$

The entropy is calculated from

$$S = \frac{U}{T} + Nk\ln q = k\beta U + Nk\ln q$$

$$U = -\frac{N}{q}\frac{dq}{d\beta} = \frac{N\varepsilon}{1+e^{\beta\varepsilon}}$$

$$\frac{U}{N\varepsilon} = \frac{1}{1+e^{\beta\varepsilon}}$$

$$S = Nk\left[\frac{\beta\varepsilon}{1+e^{\beta\varepsilon}} + \ln(1+e^{-\beta\varepsilon})\right]$$

$$\frac{S}{Nk} = \left[\frac{\beta\varepsilon}{1+e^{\beta\varepsilon}} + \ln(1+e^{-\beta\varepsilon})\right]$$

We now use MathCad® to plot $U(x) \equiv \frac{U}{N\varepsilon}$ against $S(x) \equiv \frac{S}{Nk}$ (and vice versa) with $x \equiv \beta\varepsilon = \frac{\varepsilon}{kT}$ as in Problem 60.6.

$$\mathbf{S}(x) := \left(\frac{x}{1+e^{x}}\right) + \ln\left(1+e^{-x}\right) \qquad \mathbf{U}(x) := \frac{1}{1+e^{x}}$$

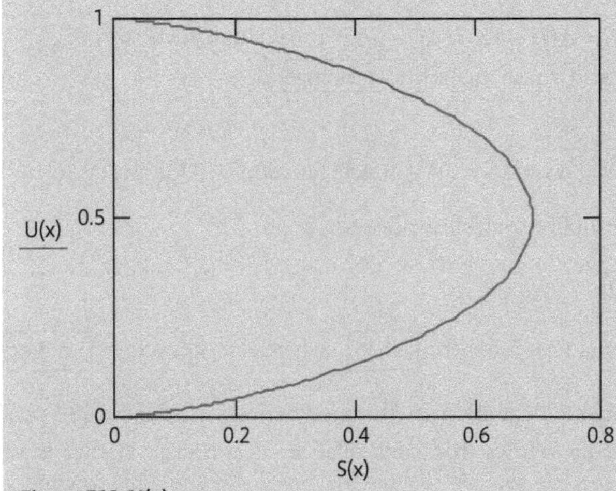

Figure F13.1I(a)

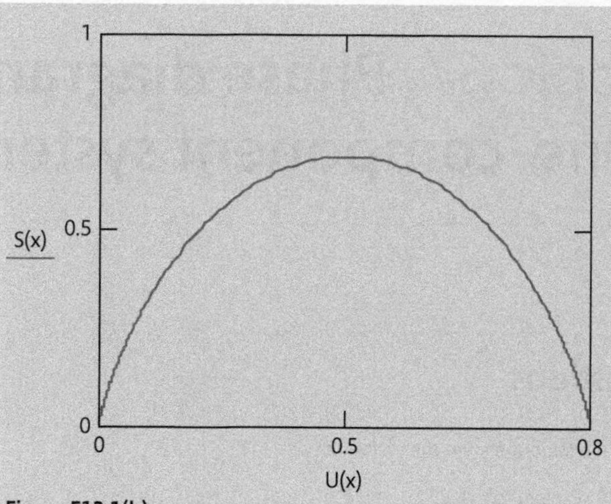

Figure F13.1(b)

When we examine Figure F13.1(a) we identify that portion of the curve that has a negative slope with negative temperatures and the portion with a positive slope with positive temperatures. Considering Figures F13.1(a) & (b) we see that entropy rises as energy is supplied to the system when $T > 0$, but decreases when energy is supplied for $T < 0$.

Topic 67 Phase diagrams: One-component systems

Discussion questions

D67.1

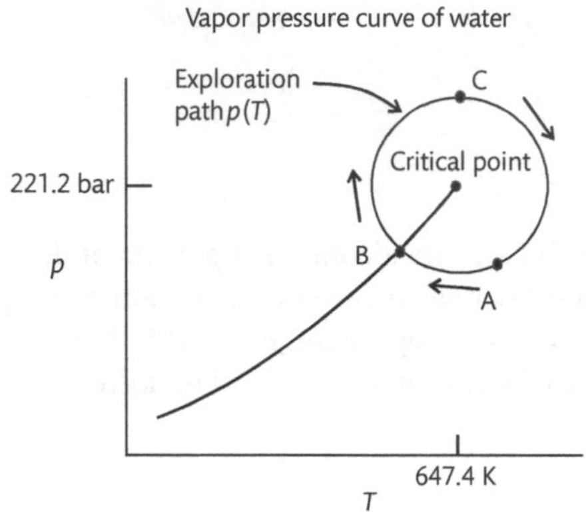

Fig 67.1

Refer to Fig. 67.1 above and Fig. 67.7 in the main text. Starting at point A and continuing clockwise on path $p(T)$ toward point B, we see only a gaseous phase within the container with water at pressures and temperatures $p(T)$. Upon reaching point B on the vapor pressure curve, liquid appears on the bottom of the container and a phase boundary or meniscus is evident between the liquid and less dense gas above it. The liquid and gaseous phases are at equilibrium at this point. Proceeding clockwise away from the vapor pressure curve the meniscus disappears, and the system becomes wholly liquid. Continuing along $p(T)$ to point C at the critical temperature no abrupt changes are observed in the isotropic fluid. Before point C is reached, it is possible to return to the vapor pressure curve and a liquid-gas equilibrium by reducing the pressure isothermally. Continuing clockwise from point C along path $p(T)$ back to point A, no phase boundary is observed even though we now consider the water to have returned to the gaseous state. Additionally, if the pressure is isothermally reduced at any point after point C, it is impossible to return to a liquid-gas equilibrium.

When the path $p(T)$ is chosen to be very close to the critical point, the water appears opaque. At near-critical conditions, densities and refractive indices of the liquid and gas phases are nearly identical. Furthermore, molecular fluctuations cause spatial variations of densities and refractive indices on a scale large enough to strongly scatter visible light. This is called critical opalescence.

D67.3 Phase: a state of matter that is uniform throughout, not only in chemical composition but also in physical state.
Constituent: any chemical species present in the system.
Component: a chemically independent constituent of the system. It is best understood in relation to the phrase "number of components" which is the minimum number of independent species necessary to define the composition of all the phases present in the system.

Degree of freedom (or variance): the number of intensive variables that can be changed without disturbing the number of phases in equilibrium.

Exercise

E67.1(a) Point a is far from any phase boundaries, so it is in a single-phase region: $\boxed{P=1}$.
Point b is a triple point, at the intersection of three phase boundaries: $\boxed{P=3}$.

Problem

P67.1 (i) Below a denaturant concentration of 0.1 only the native and unfolded forms are stable; the "molten globule" form is not stable.

(ii) At denaturant concentration of 0.15 only the native form is stable below a temperature of about 0.65. At temperature 0.65 the native and molten-globule forms are at equilibrium. Heating above 0.65 causes all native forms to become molten-globules. At temperature 0.85, equilibrium between molten-globule and unfolded protein is observed and above this temperature only the unfolded form is stable.

Topic 68 Phase diagrams: Two-component systems

Discussion question

D68.1 The principal factor is the shape of the two-phase liquid-vapor region in the phase diagram (usually a temperature-composition diagram). The closer the liquid and vapor lines are to each other, the more steps of the sort described in Section 68.1(b) are needed to move from a given mixed composition to an acceptable enrichment in one of the components. (See Fig. 68.3 of the main text.) But the presence of an azeotrope could prevent the desired degree of separation from being achieved. Incomplete miscibility of the components at specific concentrations could also affect the number of plates required.

Exercises

E68.1(a) See Figure 68.1.

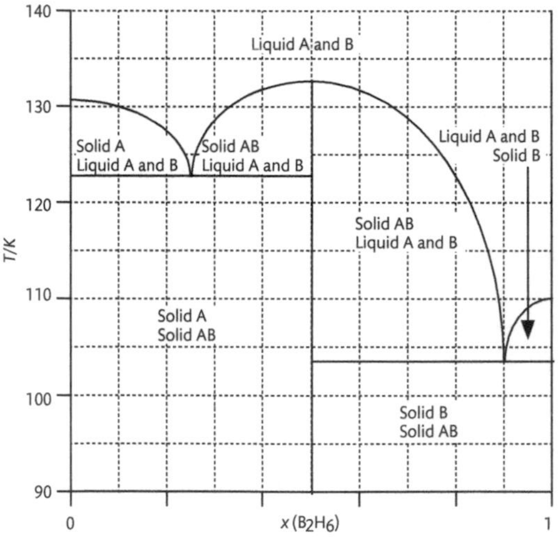

Figure 68.1

E68.2(a) Refer to Fig. 68.11 of the main text. At the lowest temperature shown on the phase diagram, there are two liquid phases, a water-rich phase ($x_B = 0.05$) and a methylpropanol-rich phase ($x_B = 0.88$). The latter phase is about 10 times as abundant as the former according to the lever rule, because it is about 10 times closer in composition (0.88 – 0.80 vs. 0.80 – 0.05). On heating, the compositions of the two phases change, the water-rich phase increasing significantly in methylpropanol and the methylpropanol-rich phase more gradually increasing in water. (Note how the composition of the left side of the diagram changes more with temperature than the right.) The relative proportions of the phases continue to be given by the lever rule. Just before the isopleth intersects the phase boundary, the methylpropanol-rich phase ($x_B = 0.8$) is in equilibrium with a vanishingly small water-rich phase ($x_B = 0.36$). Then, at T_1, the phases merge, and the single-phase region is encountered with $x_B = 0.8$.

Problems

P68.1 The data are plotted in Fig. 68.2. From the graph, the vapor in equilibrium with a liquid of composition . . .

(a) $x_M = 0.25$ is determined from the tie line labelled *a* in the figure extending from $x_M = 0.25$ to $\boxed{y_M = 0.36}$.

(b) $x_O = 0.25$ is determined from the tie line labelled *b* in the figure extending from $x_M = 0.75$ to $\boxed{y_M = 0.81}$.

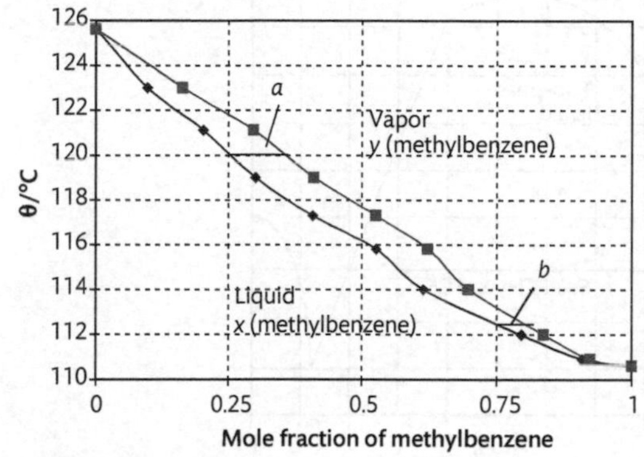

Fig 68.2

P68.3 The phase diagram is shown in Fig. 68.3. Point symbols are plotted at the given data points. The lines are schematic at best.

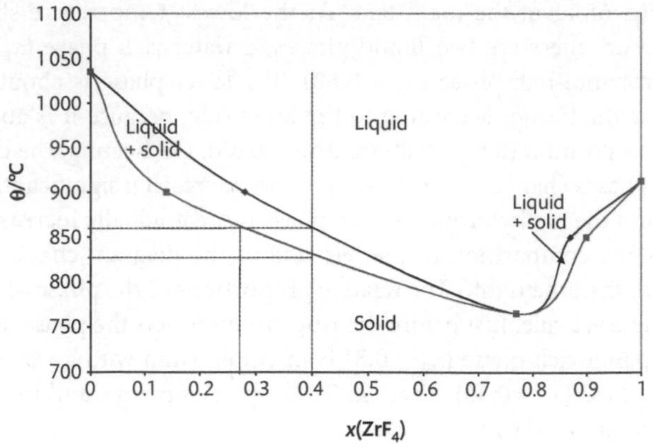

Fig 68.3

At 860°C, a solid solution with $x(ZrF_4) = 0.27$ appears. The solid solution continues to form, and its ZrF_4 content increases until it reaches $x(ZrF_4) = 0.40$ at 830°C. At that temperature and below, the entire sample is solid.

P68.5 The phase diagram is drawn in Fig. 68.4.

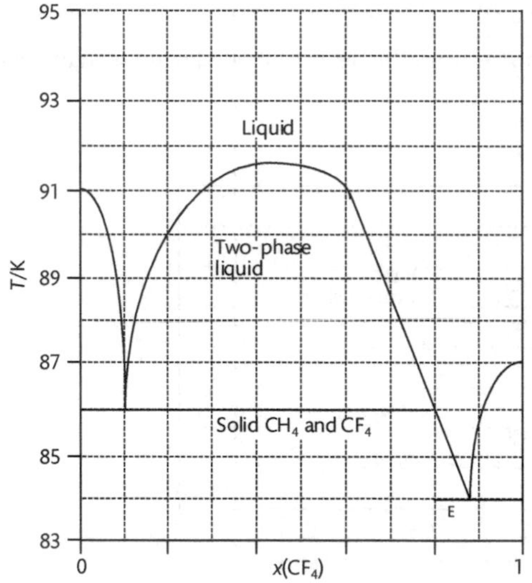

Fig 68.4

Topic 69 **Physical transformations**

Discussion questions

D69.1 Water is one of relatively few substances whose melting point *decreases* as the pressure increases. Carbon dioxide, on the other hand, exhibits the more common behavior of a melting point that *increases* with increasing pressure. The behavior of both substances is commonly explained in terms of higher pressure favoring the denser state. For most substances, the solid is denser than the liquid, so increasing the pressure from a point on the solid-liquid coexistence curve brings one into the solid-only region because the phase boundary veers toward higher temperature. For water, the liquid is denser than the solid, so increasing the pressure takes one into the denser state (this time the liquid) because the phase boundary veers toward lower temperature.

So far the explanation is in terms of strictly macroscopic properties. Microscopic (molecular) considerations can explain why liquid water is denser than ice. The crystal structure of ice is unusually open because of hydrogen bonding and the shape of the water molecule. The geometric constraints that allow both hydrogen atoms in a given water molecule to form hydrogen bonds with neighboring molecules keep the molecules from approaching as closely as they could in the absence of hydrogen bonds.

D69.3 Eqn 69.11 gives the temperature dependence of the chemical potential. It tells us that the instantaneous slope of a plot of chemical potential, μ, versus temperature, T, is equal to minus the molar entropy, S_m. Because S_m is positive for all pure substances at temperatures above absolute zero, we immediately see that μ decreases with increasing temperature. Furthermore, our association of entropy with disorder on the molecular scale allows us to say that S_m is greater for gases than for liquids than for solids. Thus, μ decreases more strongly with temperature in gases than in liquids than in solids. If we begin at a temperature so low that μ is lower for the solid than for the liquid than for the gas, then we can anticipate that there will exist temperatures at which the curves of μ vs. T of the various phases will cross: the decreasing curve of μ_{solid} does not decrease as strongly as that of μ_{liquid}, and eventually the curves cross. At that temperature, the solid and liquid are in equilibrium; at lower temperatures $\mu_{solid} < \mu_{liquid}$, so the solid is more stable; at higher temperatures $\mu_{solid} > \mu_{liquid}$, and the liquid is more stable. There will be similar crossing points between μ_{liquid} and μ_{gas} and between μ_{solid} and μ_{gas}.

Fig. 69.6 of the main text illustrates the matter schematically. The crossing points of solid and liquid chemical potentials and of liquid and vapor chemical potentials are the melting and boiling temperatures respectively at the pressure that applies to the figure. If one followed the corresponding phase diagram at the same pressure, one would move from a solid-only region to a liquid-only region to a vapor-only region as one increased the temperature; the transition points in the phase diagram are the crossing points in the chemical potential diagram.

Question: Fig. 69.6 implies that sublimation does not occur for the substance depicted, at least not at the pressure that applies to the figure. Why not? Can you draw a figure like Fig. 69.6 that is consistent with sublimation?

D69.5 Mathematically, the reason is clear. The dependence of chemical potential on pressure, expressed in eqn 69.12, follows from the chemical potential's status as a molar Gibbs energy and from the dependence of the Gibbs energy on conditions as expressed in the fundamental equation of chemical thermodynamics (eqn 69.7).

The question implies that the dependence of chemical potential on pressure is not intuitive for an incompressible phase. Perhaps the dependence is more comprehensible for a compressible phase, because increasing pressure does work on a compressible phase. In that case, one can imagine a compressible phase (a vapor, for example) in equilibrium with an incompressible phase (say a liquid) in a closed container. As the pressure increases, the chemical potential of the vapor increases, yet the chemical potentials of the two phases in equilibrium remain equal; therefore, the chemical potential of the liquid must also increase with pressure.

Exercises

E69.1(a) At equilibrium, the chemical potential of a substance is the same throughout a sample, regardless of how many phases are present. [Topic 69.2(b)] Thus

$$\boxed{\mu_W(s) = \mu_W(l)} \text{ and } \boxed{\mu_E(s) = \mu_E(l)}.$$

E69.2(a) The Gibbs-Duhem equation relates changes in the chemical potential of a mixture's components to the composition:

$$d\mu_E = -\frac{n_W}{n_E} d\mu_W \ [69.10]$$

For a small macroscopic change,

$$\delta\mu_E = -\frac{n_W}{n_E}\delta\mu_W = -\frac{0.60}{0.40} \times 0.25 \text{ J mol}^{-1} = \boxed{-0.38 \text{ J mol}^{-1}}$$

E69.3(a) The temperature dependence of the chemical potential is given by

$$\left(\frac{\partial\mu}{\partial T}\right)_p = -S_m \ [69.11]$$

For a small macroscopic change,

$$\delta\mu = -S_m\delta T = (-69.91 \text{ J mol}^{-1} \text{ K}^{-1}) \times (5 \text{ K}) = \boxed{-3\times10^2 \text{ J mol}^{-1}}$$

E69.4(a) The pressure dependence of the chemical potential is given by

$$\left(\frac{\partial \mu}{\partial p}\right)_T = V_m \quad [69.12]$$

So $\Delta\mu = \int_{1.0\text{ bar}}^{100\text{ kbar}} \left(\frac{\partial \mu}{\partial p}\right)_T dp = \int_{1.0\text{ bar}}^{100\text{ kbar}} V_m\, dp = V_m \Delta p$

$$= \frac{18.0\text{ g}}{\text{mol}} \times \frac{\text{cm}^3}{0.997\text{ g}} \times \left(\frac{1\text{ m}}{100\text{ cm}}\right)^3 \times 99\text{ kbar} \times \frac{10^3 \times 10^5\text{ Pa}}{1\text{ kbar}}$$

$$= \boxed{1.79\times10^5\text{ J mol}^{-1}}.$$

E69.5(a) The Clapeyron equation relates the pressure and temperature of phase boundaries to state functions:

$$\frac{dp}{dT} = \frac{\Delta_{trs} S}{\Delta_{trs} V} \quad [69.14]$$

so $\Delta_{fus} S = \Delta_{fus} V \times \left(\frac{dp}{dT}\right) \approx \Delta_{fus} V \times \frac{\Delta p}{\Delta T}$,

where the approximation holds if $\Delta_{fus} S$ and $\Delta_{fus} V$ are independent of temperature.

$$\Delta_{fus} S = [(163.3-161.0)\times(10^{-2}\text{ m})^3\text{ mol}^{-1}] \times \left(\frac{(100-1)\text{ atm}\times(1.013\times10^5\text{ Pa atm}^{-1})}{(351.26-350.75)\text{K}}\right)$$

$$= \boxed{+45\text{ J K}^{-1}\text{mol}^{-1}}.$$

$$\Delta_{fus} H = T_f \Delta_{fus} S = (350.75\text{ K})\times(45\text{ J K}^{-1}\text{mol}^{-1})$$

$$= \boxed{+1.59\times10^4\text{ J mol}^{-1}} = \boxed{+15.9\text{ kJ mol}^{-1}}.$$

E69.6(a) The Clausius-Clapeyron equation gives the variation of vapor pressure with temperature, if the vapor is a perfect gas:

$$\frac{d\ln p}{dT} = \frac{\Delta_{vap} H}{RT^2} \quad [69.16]$$

so $\int d\ln p = \int \frac{\Delta_{vap} H}{RT^2} dT$ and $\ln p = \text{constant} - \frac{\Delta_{vap} H}{RT}$

Therefore, $\Delta_{vap} H = (1\,501.8\text{ K})\times R = (1\,501.8\text{ K})\times(8.3145\text{ J mol}^{-1}\text{ K}^{-1})$,

$$\Delta_{vap} H = \boxed{+12\,487\text{ J mol}^{-1}} = \boxed{+12.487\text{ kJ mol}^{-1}}$$

E69.7(a) Use eqn 69.15(a) and assume that $\Delta_{fus} H$ and $\Delta_{fus} V$ are independent of temperature:

$$\frac{dp}{dT} = \frac{\Delta_{fus} H}{T_f \Delta_{fus} V} \approx \frac{\Delta p}{\Delta T}$$

Thus $\Delta T \approx \dfrac{T_{\mathrm{f}}\Delta_{\mathrm{fus}}V}{\Delta_{\mathrm{fus}}H}\times \Delta p$

Because the change in molar volume upon fusion (melting) is

$$\Delta_{\mathrm{fus}}V = (78.11\ \mathrm{g\,mol^{-1}})\times\left(\frac{1}{0.879\ \mathrm{g\,cm^{-3}}}-\frac{1}{0.891\ \mathrm{g\,cm^{-3}}}\right)\times\left(\frac{\mathrm{m}}{100\ \mathrm{cm}}\right)^{3}$$
$$=1.2\bar{0}\times10^{-6}\ \mathrm{m^3\,mol^{-1}},$$

the change in melting temperature is

$$\Delta T \approx \frac{(278.6\bar{5}\ \mathrm{K})\times(1.2\bar{0}\times10^{-6}\ \mathrm{m^3\,mol^{-1}})}{10.59\times10^{3}\ \mathrm{J\,mol^{-1}}}\times(10.0\times10^{3}-1)\ \mathrm{bar}\times\frac{10^{5}\ \mathrm{Pa}}{\mathrm{bar}}$$
$$=31.\bar{5}\ \mathrm{K}.$$

Therefore, at 10.0 kbar, $T_{\mathrm{f}} = (5.5+31.\bar{5})°\mathrm{C} = \boxed{37°\mathrm{C}}$

E69.8(a) (a) According to Trouton's rule (*Brief illustration* 62.2)

$$\Delta_{\mathrm{vap}}H = (85\ \mathrm{J\,K^{-1}\,mol^{-1}})\times T_{\mathrm{b}} = (85\ \mathrm{J\,K^{-1}\,mol^{-1}})\times(76.8+273.15)\ \mathrm{K}$$
$$=\boxed{2.9\bar{7}\times10^{4}\ \mathrm{J\,mol^{-1}}}=\boxed{29.\bar{7}\ \mathrm{kJ\,mol^{-1}}}.$$

(b) Integrating the Clausius–Clapeyron equation [69.16] yields

$$\int \mathrm{d}\ln p = \int \frac{\Delta_{\mathrm{vap}}H}{RT^2}\mathrm{d}T.$$

Assuming that $\Delta_{\mathrm{vap}}H$ is independent of temperature

$$\ln\frac{p_2}{p_1} = \frac{\Delta_{\mathrm{vap}}H}{R}\left(\frac{1}{T_1}-\frac{1}{T_2}\right)$$

Let T_1 be the normal boiling temperature, 350.0 K; then p_1 = 1 bar. Thus at 25°C

$$\ln\frac{p_2}{\mathrm{bar}} = \left(\frac{2.9\bar{7}\times10^{3}\ \mathrm{J\,mol^{-1}}}{8.3145\ \mathrm{J\,K^{-1}\,mol^{-1}}}\right)\times\left(\frac{1}{350.0\ \mathrm{K}}-\frac{1}{298\ \mathrm{K}}\right)=-1.77\bar{6}$$

so $p_2 = \boxed{0.169\ \mathrm{bar}}$

At 70°C, $\ln\dfrac{p_2}{\mathrm{bar}} = \left(\dfrac{2.9\bar{7}\times10^{3}\ \mathrm{J\,mol^{-1}}}{8.3145\ \mathrm{J\,K^{-1}\,mol^{-1}}}\right)\times\left(\dfrac{1}{350.0\ \mathrm{K}}-\dfrac{1}{343\ \mathrm{K}}\right)=-0.20\bar{3}$

so $p_2 = \boxed{0.82\ \mathrm{bar}}$

E69.9(a) Use eqn 69.15(a) and assume that $\Delta_{\mathrm{fus}}H$ and $\Delta_{\mathrm{fus}}V$ are independent of temperature:

$$\frac{\mathrm{d}p}{\mathrm{d}T} = \frac{\Delta_{\mathrm{fus}}H}{T_{\mathrm{f}}\Delta_{\mathrm{fus}}V} \approx \frac{\Delta p}{\Delta T}$$

Thus $\Delta T \approx \dfrac{T_{\mathrm{f}}\Delta_{\mathrm{fus}}V}{\Delta_{\mathrm{fus}}H}\times \Delta p$

$$\Delta_{\text{fus}}V = (18.02\ \text{g mol}^{-1})\times\left(\frac{1}{1.00\ \text{g cm}^{-3}}-\frac{1}{0.92\ \text{g cm}^{-3}}\right)\times\left(\frac{\text{m}}{100\ \text{cm}}\right)^3$$

$$= -1.5\bar{7}\times10^{-6}\ \text{m}^3\ \text{mol}^{-1}\ ,$$

$$\text{so}\quad \Delta T \approx \frac{(273.15\ \text{K})\times(-1.5\bar{7}\times10^{-6}\ \text{m}^3\ \text{mol}^{-1})}{6008\ \text{J mol}^{-1}}\times(50-1)\ \text{bar}\times\frac{10^5\ \text{Pa}}{\text{bar}} = -0.35\ \text{K}$$

Therefore, at 50 bar, $T_f = \boxed{-0.35°\text{C}} = (273.15-0.35)\ \text{K} = \boxed{272.80\ \text{K}}$

Problems

P69.1 At the triple point, T_3, the vapor pressures of liquid and solid are equal, hence

$$10.5916-\frac{1\,871.2\ \text{K}}{T_3} = 8.3186-\frac{1\,425.7\ \text{K}}{T_3};\quad T_3 = \boxed{196.0\ \text{K}}$$

$$\text{and}\quad \log(p_3/\text{Torr}) = \frac{-1871.2\ \text{K}}{196.0\ \text{K}}+10.5916 = 1.044\bar{7};\quad p_3 = \boxed{11.1\ \text{Torr}}$$

P69.3 Use the Clapeyron equation in the form of eqn 69.15(a),

$$\frac{dp}{dT} = \frac{\Delta_{\text{fus}}H}{T\Delta_{\text{fus}}V}$$

Assume that $\Delta_{\text{fus}}H$ and $\Delta_{\text{fus}}V$ are approximately constant in the neighborhood of the standard melting point and integrate:

$$dp = \frac{dT}{T}\frac{\Delta_{\text{fus}}H}{\Delta_{\text{fus}}V}\quad \text{so}\quad p_2-p_1 = \frac{\Delta_{\text{fus}}H}{\Delta_{\text{fus}}V}\ln\frac{T_2}{T_1}$$

Solve for T_2:

$$T_2 = T_1\exp\left(\frac{(p_2-p_1)\Delta_{\text{fus}}V}{\Delta_{\text{fus}}H}\right)$$

The difference in pressure is

$$p_2-p_1 = 10.0\ \text{m Hg}\times\frac{1\,000\ \text{mm Hg}}{1\ \text{m Hg}}\times\frac{133.3\ \text{Pa}}{1\ \text{mm Hg}}-1\ \text{bar}\times\frac{10^5\ \text{Pa}}{1\ \text{bar}} = 1.23\times10^6\ \text{Pa}$$

$$\text{so}\quad T_2 = 234.3\ \text{K}\times\exp\left(\frac{1.23\times10^6\ \text{Pa}\times0.517\times(10^{-2}\ \text{m})^3\ \text{mol}^{-1}}{2.292\times10^3\ \text{J mol}^{-1}}\right)$$

$$= \boxed{234.4\ \text{K}}.$$

P69.5 The equations describing the coexistence curve on the solid–liquid boundary is

$$\text{(a)}\quad p = p^* + \frac{\Delta_{\text{fus}}H}{\Delta_{\text{fus}}V}\ln\frac{T}{T^*}\quad \text{[Problem 69.3]}$$

The liquid–vapor and solid-vapor boundaries can be obtained by integrating the Clausius-Clapeyron equation (69.16) away from the triple point. For the liquid-vapor boundary,

$$\mathrm{d}\ln p = \frac{\mathrm{d}T}{T^2}\frac{\Delta_{\mathrm{vap}}H}{R} \quad \text{so} \quad \ln\frac{p}{p^*} = \frac{\Delta_{\mathrm{vap}}H}{R}\left(\frac{1}{T^*}-\frac{1}{T}\right) \text{ and}$$

$$\text{(b)}\quad p = p^*\exp\left\{\frac{\Delta_{\mathrm{vap}}H}{R}\left(\frac{1}{T^*}-\frac{1}{T}\right)\right\}$$

Similarly, the solid-vapor boundary is given by

$$\text{(c)}\quad p = p^*\exp\left\{\frac{\Delta_{\mathrm{sub}}H}{R}\left(\frac{1}{T^*}-\frac{1}{T}\right)\right\}$$

We need $\Delta_{\mathrm{sub}}H = \Delta_{\mathrm{fus}}H + \Delta_{\mathrm{vap}}H = 41.4\,\mathrm{kJ\,mol^{-1}}$

$$\Delta_{\mathrm{fus}}V = M\times\left(\frac{1}{\rho(\mathrm{l})}-\frac{1}{\rho(\mathrm{s})}\right) = \left(\frac{78.11\,\mathrm{g\,mol^{-1}}}{\mathrm{g\,cm^{-3}}}\right)\times\left(\frac{1}{0.879}-\frac{1}{0.891}\right)$$
$$= +1.19\bar{7}\,\mathrm{cm^3\,mol^{-1}}$$

After insertion of these numerical values into the above equations, we obtain

$$\text{(a)}\quad p = p^* + \left(\frac{10.6\times10^3\,\mathrm{J\,mol^{-1}}}{1.197\times10^{-6}\,\mathrm{m^3\,mol^{-1}}}\right)\ln\frac{T}{T^*}$$
$$= p^* + 8.85\bar{5}\times10^9\,\mathrm{Pa}\times\ln\frac{T}{T^*}$$
$$= p^* + (6.64\times10^7\,\mathrm{Torr})\ln\frac{T}{T^*}\ [1\,\mathrm{Torr} = 133.322\,\mathrm{Pa}]$$

This line is plotted as *a* in Fig. 69.1, starting from the triple point, (p^*, T^*) = (36 Torr, 5.50°C (278.65 K)).

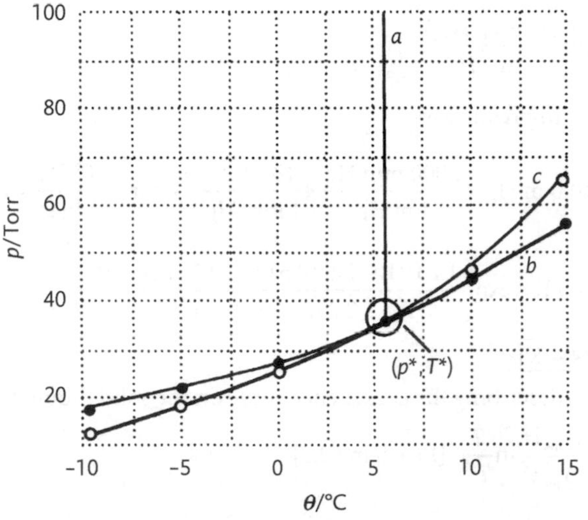

Fig 69.1

(b) $$p = p^* \exp\left\{\frac{30.8\times10^3\ \mathrm{J\,mol^{-1}}}{8.314\ \mathrm{J\,K^{-1}\,mol^{-1}}}\times\left(\frac{1}{T^*}-\frac{1}{T}\right)\right\} = p^*\exp\left\{370\bar{5}\ \mathrm{K}\times\left(\frac{1}{T^*}-\frac{1}{T}\right)\right\}$$

This equation is plotted as line b in Fig. 69.1, also starting from the triple point.

(c) $$p = p^*\exp\left\{\frac{41.4\times10^3\ \mathrm{J\,mol^{-1}}}{8.314\ \mathrm{J\,K^{-1}\,mol^{-1}}}\times\left(\frac{1}{T^*}-\frac{1}{T}\right)\right\} = p^*\exp\left\{49\overline{80}\ \mathrm{K}\times\left(\frac{1}{T^*}-\frac{1}{T}\right)\right\}$$

These points are plotted as line c in Fig. 69.1.

The actual phase boundary is curve b above the triple point and curve c below it; that is, the actual phase boundary is whichever curve (b or c) gives the lower pressure.

P69.7 The general condition of equilibrium in an isolated system is $\mathrm{d}S = 0$. If α and β are in thermal contact with each other and if they comprise an isolated system, then

$$\mathrm{d}S = \mathrm{d}S_\alpha + \mathrm{d}S_\beta = \frac{\mathrm{d}q_{\mathrm{rev},\alpha}}{T_\alpha} + \frac{\mathrm{d}q_{\mathrm{rev},\beta}}{T_\beta} \quad [61.1]$$

Since α and β are in thermal contact with each other and with nothing else, the heat flow into one of the parts is equal and opposite the heat flow into the other—whether reversible or not:

$$\mathrm{d}q_{\mathrm{rev},\alpha} = -\mathrm{d}q_{\mathrm{rev},\beta},$$

so $$\mathrm{d}S = \mathrm{d}q_{\mathrm{rev},\alpha}\left(\frac{1}{T_\alpha}-\frac{1}{T_\beta}\right)$$

If $\mathrm{d}S$ is to vanish, either the temperatures must be equal or there must be no heat flow; however, heat **does** flow between subsystems at different temperatures in thermal contact. (Heat can even flow between subsystems in thermal contact at the **same** temperature, but it cannot vanish **unless** the temperatures are the same.) The result, whether or not the heat flow is zero, $\boxed{T_\alpha = T_\beta}$

P69.9 $\mathrm{d}H = C_p\,\mathrm{d}T + V\,\mathrm{d}p$implying that $\mathrm{d}\Delta H = \Delta C_p\,\mathrm{d}T + \Delta V\,\mathrm{d}p$

However, along a phase boundary $\mathrm{d}p$ and $\mathrm{d}T$ are related by

$$\frac{\mathrm{d}p}{\mathrm{d}T} = \frac{\Delta H}{T\Delta V} \quad \text{[Clapeyron equation]}$$

Therefore,

$$\mathrm{d}\Delta H = \left(\Delta C_p + \Delta V \times \frac{\Delta H}{T\Delta V}\right)\mathrm{d}T = \left(\Delta C_p + \frac{\Delta H}{T}\right)\mathrm{d}T \quad \text{and} \quad \frac{\mathrm{d}\Delta H}{\mathrm{d}T} = \Delta C_p + \frac{\Delta H}{T}$$

Then, since

$$\frac{\mathrm{d}}{\mathrm{d}T}\left(\frac{\Delta H}{T}\right) = \frac{1}{T}\frac{\mathrm{d}\Delta H}{\mathrm{d}T} - \frac{\Delta H}{T^2} = \frac{1}{T}\left(\frac{\mathrm{d}\Delta H}{\mathrm{d}T} - \frac{\Delta H}{T}\right)$$

substituting the first result gives

$$\frac{\mathrm{d}}{\mathrm{d}T}\left(\frac{\Delta H}{T}\right) = \frac{\Delta C_p}{T}$$

Therefore,

$$\mathrm{d}\left(\frac{\Delta H}{T}\right)=\frac{\Delta C_p\,\mathrm{d}T}{T}=\boxed{\Delta C_p\,\mathrm{d}\ln T}$$

P69.11 In each phase the slopes are given by

$$\left(\frac{\partial\mu}{\partial T}\right)_p=-S_{\mathrm{m}}\quad[69.11]$$

The curvature of a plot of μ against T is given by

$$\left(\frac{\partial^2\mu}{\partial T^2}\right)_p=-\left(\frac{\partial S_{\mathrm{m}}}{\partial T}\right)_p$$

This derivative is readily obtained by looking at the expression for the enthalpy analogous to the fundamental equation of thermodynamics (eqn 66.1). That expression is for the internal energy:

$$\mathrm{d}U=T\mathrm{d}S-p\mathrm{d}V$$

Adding $\mathrm{d}(pV)$ to both sides yields an analogous expression for $\mathrm{d}H$:

$$\mathrm{d}U+\mathrm{d}(pV)=\mathrm{d}H=T\mathrm{d}S-p\mathrm{d}V+p\mathrm{d}V+V\mathrm{d}p=T\mathrm{d}S+V\mathrm{d}p$$

from which it follows that

$$\left(\frac{\partial H}{\partial T}\right)_p=T\left(\frac{\partial S}{\partial T}\right)_p=C_p\quad[56.3]$$

Dividing both sides by n (to get molar quantities) and by $-T$ yields

$$-\left(\frac{\partial S_{\mathrm{m}}}{\partial T}\right)_p=\boxed{-\frac{1}{T}\times C_{p,\mathrm{m}}}=\left(\frac{\partial^2\mu}{\partial T^2}\right)_p$$

Because $C_{\mathrm{p,m}}$ is necessarily positive, the curvatures in all states of matter are necessarily negative; plots of μ vs. T are concave down. $C_{\mathrm{p,m}}$ is often largest for the liquid state, though not always; but it is the ratio $C_{\mathrm{p,m}}/T$ that determines the magnitude of the curvature, so no precise answer can be given for the state with the greatest curvature. It depends upon the substance.

Topic 70 **Ideal mixtures**

Discussion question

D70.1 Raoult's Law (eqn 70.6) defines the behavior of ideal solutions. Like ideal gases, what makes the behavior ideal can be expressed in terms of intermolecular interactions. Unlike ideal gases, however, the interactions in an ideal solution cannot be neglected. Instead, ideal behavior amounts to having the same interactions among molecules of the mixture's different components as molecules of each component have with other molecules of that same component. In shorthand, ideal behavior consists of A-B interactions being the same as A-A and B-B interactions. If that is the case, then the cohesive forces that would keep a molecule in the liquid phase would be the same in a solution as in a pure liquid, and the vapor pressure of a component will differ from that of a pure liquid only in proportion to its abundance (mole fraction). Thus, we expect Raoult's law to be valid for mixtures of components that have very similar chemical structures. Similar structures imply both similar intermolecular interactions (governed largely by polarity) and similar sizes (implying that the mole fraction is a good approximation to the relative proportion of the surface area occupied by each component—a factor that is relevant to rates of evaporation and condensation).

Exercises

E70.1(a) Assume perfect gas behavior.

$$pV = nRT = \frac{mRT}{M} \quad \text{so} \quad m = \frac{pVM}{RT},$$

where $V = 4.0\ \text{m} \times 4.0\ \text{m} \times 3.0\ \text{m} = 48\ \text{m}^3$.

(a) $$m = \frac{(3.2\times10^3\ \text{Pa})\times(48\ \text{m}^3)\times(18.02\ \text{g mol}^{-1})}{(8.3145\ \text{J K}^{-1}\ \text{mol}^{-1})\times(298.15\ \text{K})} = \boxed{1.1\times10^3\ \text{g}} = \boxed{1.1\ \text{kg}}$$

(b) $$m = \frac{(13.1\times10^3\ \text{Pa})\times(48\ \text{m}^3)\times(78.11\ \text{g mol}^{-1})}{(8.3145\ \text{J K}^{-1}\ \text{mol}^{-1})\times(298.15\ \text{K})} = \boxed{2.0\times10^4\ \text{g}} = \boxed{20.\ \text{kg}}$$

E70.2(a) $\Delta_{mix}G = nRT(x_A \ln x_A + x_B \ln x_B)$ [70.1].

The mole fractions are equal, so $x_A = x_B = 0.50$. Because the gases are perfect, $pV = nRT$. Therefore,

$$\Delta_{mix}G = (pV)\times(0.50\ln 0.50 + 0.50\ln 0.50) = pV\ln 0.50$$

$$= (1.5\text{ bar})\times\left(\frac{10^5\text{ Pa}}{\text{bar}}\right)\times 10.0\times(10^{-1}\text{ m})^3\times(\ln 0.50)$$

$$= -1.0\bar{4}\times 10^3\text{ J} = \boxed{-1.0\bar{4}\text{ kJ}}$$

$$\Delta_{mix}S = -nR(x_A\ln x_A + x_B\ln x_B)\ [70.2] = \frac{-\Delta_{mix}G}{T} = -\frac{-1.0\bar{4}\times 10^3\text{ J}}{298\text{ K}}$$

$$= \boxed{+3.5\text{ J K}^{-1}}$$

E70.3(a) $\Delta_{mix}G = nRT(x_A \ln x_A + x_B \ln x_B + x_C \ln x_C)$ [70.1 with three components].

We need to determine mole fractions from the given mass percentages. Assume a 100-g sample:

$$n_N = \frac{76\text{ g}}{28.02\text{ g mol}^{-1}} = 2.71\text{ mol},,\ n_O = \frac{23\text{ g}}{32.00\text{ g mol}^{-1}} = 0.72\text{ mol},,$$

$$n_{Ar} = \frac{1\text{ g}}{39.95\text{ g mol}^{-1}} = 0.03\text{ mol, and } n_{total} = 3.46\text{ mol}$$

Thus $x_N = n_N/n_{total} = 2.71/3.46 = 0.78$; similarly $x_O = 0.21$ and $x_{Ar} = 0.007$, and for one mole of air at 298 K

$$\Delta_{mix}G = (8.3145\text{ J mol}^{-1}K^{-1})\times(298\text{ K})$$
$$\times(0.78\ln 0.78 + 0.21\ln 0.21 + 0.007\ln 0.007),$$

$$\Delta_{mix}G = \boxed{-1.37\times 10^3\text{ J mol}^{-1}} = \boxed{-1.37\text{ kJ mol}^{-1}}$$

$$\Delta_{mix}S = \frac{-\Delta_{mix}G}{T}\ [\text{Exercise 70.2(a)}] = -\frac{-1.37\times 10^3\text{ J mol}^{-1}}{298\text{ K}} = \boxed{+4.6\text{ J K}^{-1}\text{ mol}^{-1}}$$

For a perfect gas, $\boxed{\Delta_{mix}H = 0}$ [no intermolecular interactions before or after mixing].

E70.4(a) Hexane and heptane form nearly ideal solutions, therefore eqn 70.2 applies:

$$\Delta_{mix}S = -nR(x_A\ln x_A + x_B\ln x_B)$$

(a) To maximize $\Delta_{mix}S$, differentiate the equation with respect to x_A and look for the value of x_A which makes the derivative vanish. Since $x_B = 1 - x_A$, we need to differentiate

$$\Delta_{mix}S = -nR\{x_A\ln x_A + (1-x_A)\ln(1-x_A)\}$$

This yields

$$\frac{d\Delta_{mix}S}{dx_A} = -nR\left\{\ln x_A + \frac{x_A}{x_A} - \ln(1-x_A) - \frac{(1-x_A)}{(1-x_A)}\right\} = -nR\ln\frac{x_A}{1-x_A}$$

which is zero when $x_A = \boxed{\frac{1}{2}} = x_B$. Hence, the maximum entropy of mixing occurs for the preparation of a mixture that contains equal mole fractions of the two components.

(b) An equimolar mixture still maximizes the entropy of mixing; we only have to express the concentrations by mass fraction rather than mole fraction:

$$n_A = n_B \quad \text{so} \quad \frac{m_A}{M_A} = \frac{m_B}{M_B} = \frac{m_{\text{total}} - m_A}{M_B}$$

Solve for m_A in terms of m_{total}:

$$m_A = m_{\text{total}}\left(\frac{M_A}{M_A + M_B}\right)$$

But m_A/m_{total} is just the mass fraction of A (hexane, say). So the mass fractions are

$$\frac{m_{\text{hex}}}{m_{\text{total}}} = \frac{86.17 \text{ g mol}^{-1}}{(86.17+100.20) \text{ g mol}^{-1}} = \boxed{0.4624} \text{ and } \frac{m_{\text{hep}}}{m_{\text{total}}} = 1 - 0.4624 = \boxed{0.5376}.$$

E70.5(a) Check whether $p_{\text{HCl}}/x_{\text{HCl}}$ is constant; if so, that is the Henry's law constant. [70.8]

x_{HCl}	0.005	0.012	0.019
$(p_{\text{HCl}}/\text{kPa})/x_{\text{HCl}}$	6.4×10^3	6.4×10^3	6.4×10^3

Hence, $K_B = \boxed{6.4\times10^3 \text{ kPa}}$

E70.6(a) The vapor pressures of components A and B may be expressed in terms of their compositions in the liquid (mole fractions x_A and x_B) or in the vapor (mole fractions y_A and y_B). The pressures are the same whatever the expression; hence the expressions can be set equal to each other and solved for the composition.

$$p_A = y_A p = 0.350p = x_A p_A{}^* = x_A \times (76.7 \text{ kPa})$$

and $p_B = y_B p = (1-y_A)p = 0.650p = x_B p_B{}^* = (1-x_A) \times (52.0 \text{ kPa})$

Therefore, $\dfrac{y_A p}{y_B p} = \dfrac{x_A p_A{}^*}{x_B p_B{}^*}$

Hence $\dfrac{0.350}{0.650} = \dfrac{76.7 x_A}{52.0(1-x_A)}$

which solves to $x_A = \boxed{0.268}$ and $x_B = 1 - x_A = \boxed{0.732}$

Also, since $0.350p = x_A p_A{}^*$

$$p = \frac{x_A p_A{}^*}{0.350} = \frac{(0.268)\times(76.7 \text{ kPa})}{0.350} = \boxed{58.6 \text{ kPa}}$$

E70.7(a) In Exercise 70.5(a), the Henry's law constant, K_B, was determined for concentrations expressed in mole fractions. To use that value, $K_B = 6.4 \times 10^3$ kPa, we must express the molality given here as a mole fraction. For convenience, assume a sample with 1 kg solvent ($GeCl_4$). Thus

$$n(\mathrm{GeCl_4}) = \frac{1000\,\mathrm{g}}{214.44\ \mathrm{g\,mol^{-1}}} = 4.663\ \mathrm{mol} \quad \text{and} \quad n(\mathrm{HCl}) = 0.15\ \mathrm{mol}$$

Therefore, $x_{\mathrm{HCl}} = \dfrac{0.15\ \mathrm{mol}}{0.15\ \mathrm{mol} + 4.663\ \mathrm{mol}} = 0.031$

$$p_{\mathrm{HCl}} = (0.031) \times (6.4 \times 10^3\,\mathrm{kPa}) = \boxed{2.0 \times 10^2\,\mathrm{kPa}}$$

E70.8(a) We assume that the solvent, benzene (A), is ideal and obeys Raoult's law; then

$$p_A = x_A p_A^* \quad \text{and} \quad x_A = \frac{n_A}{n_A + n_B}$$

Hence $p_A = \dfrac{n_A p_A^*}{n_A + n_B}$; which solves to

$$n_B = \frac{n_A (p_A^* - p_A)}{p_A}$$

Then, since $n_B = \dfrac{m_B}{M_B}$, where m_B is the mass of B (solute) present,

$$M_B = \frac{m_B p_A}{n_A (p_A^* - p_A)} = \frac{m_B M_A p_A}{m_A (p_A^* - p_A)}$$

From the data

$$M_B = \frac{(19.0\mathrm{g}) \times (78.11\ \mathrm{g\,mol^{-1}}) \times (51.5\ \mathrm{kPa})}{(500\mathrm{g}) \times (53.3 - 51.5)\mathrm{kPa}} = \boxed{85\ \mathrm{g\,mol^{-1}}}$$

E70.9(a) With concentrations expressed in molality, Henry's law is $p_B = b_B K_B$ [70.8b].

Solving for b, the solubility, we have $b_B = \dfrac{p_B}{K_B}$

where K_B can be found in Table 70.1 and p_B, according to Dalton's law of partial pressures is

$$p_B = x_B p_{\mathrm{total}} = x_B \times 1\ \mathrm{atm} = x_B \times 101\ \mathrm{kPa}$$

(a) $b = \dfrac{0.10\ \mathrm{atm}}{3.01 \times 10^3\ \mathrm{kPa\,kg\,mol^{-1}}} \times \dfrac{101\ \mathrm{kPa}}{1\ \mathrm{atm}} = \boxed{3.4 \times 10^{-3}\ \mathrm{mol\,kg^{-1}}}$

(b) $b = \dfrac{1.00\ \mathrm{atm}}{3.01 \times 10^3\ \mathrm{kPa\,kg\,mol^{-1}}} \times \dfrac{101\ \mathrm{kPa}}{1\ \mathrm{atm}} = \boxed{3.4 \times 10^{-2}\ \mathrm{mol\,kg^{-1}}}$

Problem

P70.1 $p_A = y_A p$ and $p_B = y_B p$ [Dalton's law]. Hence, draw up the following table:

p_A/kPa	0	1.399	3.566	5.044	6.996	7.940	9.211	10.105	11.287	12.295
x_A	0	0.0898	0.2476	0.3577	0.5194	0.6036	0.7188	0.8019	0.9105	1
y_A	0	0.0410	0.1154	0.1762	0.2772	0.3393	0.4450	0.5435	0.7284	1

p_B/kPa	0	4.209	8.487	11.487	15.462	18.243	23.582	27.334	32.722	36.066
x_B	0	0.0895	0.1981	0.2812	0.3964	0.4806	0.6423	0.7524	0.9102	1
y_B	0	0.2716	0.4565	0.5550	0.6607	0.7228	0.8238	0.8846	0.9590	1

The data are plotted in Fig. 70.1.

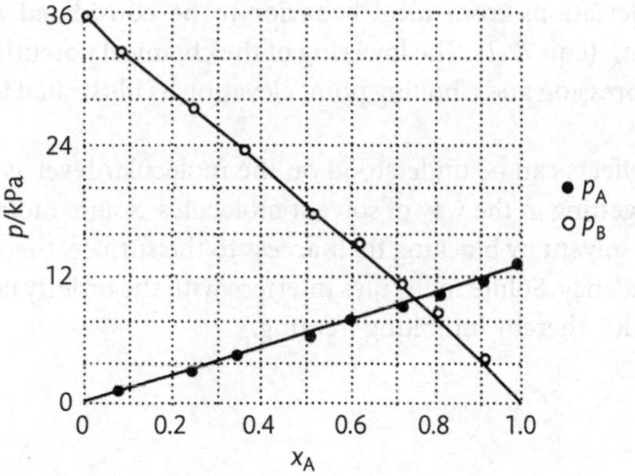

Fig 70.1

We can assume, at the lowest concentrations of both A and B, that Henry's law will hold. The Henry's law constants are then given by

$$K_A = \frac{p_A}{x_A} = \boxed{15.58\,\text{kPa}} \text{ from the point at } x_A = 0.0898$$

and $$K_B = \frac{p_B}{x_B} = \boxed{47.03\,\text{kPa}} \text{ from the point at } x_B = 0.0895$$

Topic 71 **Colligative properties**

Discussion question

D71.1 All the colligative properties are a result of the lowering of the chemical potential of the solvent due to the presence of the solute. This reduction takes the form $\mu_A = \mu_A{}^* + RT\ln x_A$ (eqn 70.7) if the solution can be considered ideal, or anticipating deviations from ideal behavior to be considered in Topic 72, $\mu_A = \mu_A{}^* + RT\ln a_A$ (eqn 72.1). The lowering of the chemical potential results in a freezing point depression and a boiling point elevation as illustrated in Fig. 71.1 of the main text.

Both of these effects can be understood on the molecular level as the result of solute molecules getting in the way of solvent molecules. Solute molecules inhibit the boiling of the solvent by blocking their access to the surface, thereby reducing their escaping tendency. Solute molecules interfere with the orderly crystallization of solvent molecules, thereby inhibiting freezing.

Exercises

E71.1(a) The best value of the molar mass is obtained from values of the data extrapolated to zero concentration, since it is under this condition that the van't Hoff equation applies.

$$\Pi = [\mathrm{B}]RT = \frac{n_B RT}{V} \;[71.3] \quad \text{so} \quad \Pi = \frac{m_B RT}{M_B V} = \frac{cRT}{M_B} \quad \text{where} \quad c = \frac{m}{V}$$

(Drop subscript B from here on; masses, etc., will be understood to refer to solutes.) At the same time, the osmotic pressure is the hydrostatic pressure:

$$\Pi = \rho g h, \quad \text{so} \quad h = \left(\frac{RT}{\rho g M_B}\right)c$$

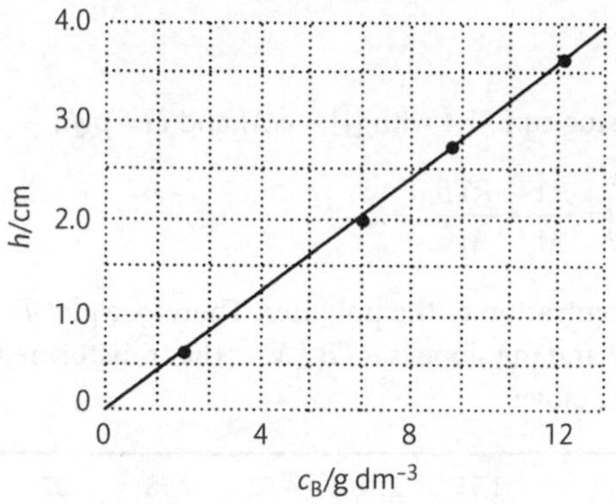

Fig 71.1

Thus, a plot of h against c should be a straight line with slope $\dfrac{RT}{\rho g M_B}$. Fig. 71.1 shows that plot. The slope of the best-fit line is

$$\frac{RT}{\rho g M_B} = \frac{0.2854\ \text{cm}}{\text{g dm}^{-3}} = \frac{0.2854 \times (10^{-2}\ \text{m})}{(10^{-3}\ \text{kg}) \times (10^{-1}\ \text{m})^{-3}} = 2.854 \times 10^{-3}\ \text{m}^4\ \text{kg}^{-1}$$

Therefore,

$$M_B = \frac{RT}{(\rho g) \times \text{slope}}$$

$$= \frac{(8.3145\ \text{J K}^{-1}\ \text{mol}^{-1}) \times (298.15\text{K})}{(1.004 \times 10^3\ \text{kg m}^{-3}) \times (9.807\ \text{m s}^{-2}) \times (2.854 \times 10^{-3}\ \text{m}^4\ \text{kg}^{-1})}$$

$$= \boxed{88.2\ \text{kg mol}^{-1}}.$$

E71.2(a) See Example 71.2.

(a) The number-average molar mass is

$$\overline{M}_n = \frac{N_1 M_1 + N_2 M_2}{N_1 + N_2} = \frac{\left(\dfrac{m_1}{M_1}\right) M_1 + \left(\dfrac{m_2}{M_2}\right) M_2}{\left(\dfrac{m_1}{M_1}\right) + \left(\dfrac{m_2}{M_2}\right)} = \frac{m_1 + m_2}{\left(\dfrac{m_1}{M_1}\right) + \left(\dfrac{m_2}{M_2}\right)}$$

Assume 100 g (a convenient sample size for mass percent information).

$$\overline{M}_n = \frac{100\ \text{g}}{\left(\dfrac{30\ \text{g}}{30\ \text{kg mol}^{-1}}\right) + \left(\dfrac{70\ \text{g}}{15\ \text{kg mol}^{-1}}\right)} = \boxed{18\ \text{kg mol}^{-1}}$$

(b) The weight-average molar mass is

$$\overline{M}_w = \frac{m_1 M_1 + m_2 M_2}{m_1 + m_2} = \frac{(30) \times (30) + (70) \times (15)}{100}\ \text{kg mol}^{-1} = \boxed{20\ \text{kg mol}^{-1}}$$

(c) The heterogeneity index is $\dfrac{\overline{M}_w}{\overline{M}_n} = \dfrac{20}{18} = \boxed{1.1}$

Problems

P71.1 Refer to Example 71.1, using eqn 71.4 with [J] = c/M and $\Pi = \rho g h$:

$$\frac{\Pi}{c}=\frac{RT}{M}\left(1+\frac{B}{M}c+\ldots\right)=\frac{RT}{M}+\frac{RTB}{M^2}c+\ldots$$

where c is the mass concentration of the polymer. Therefore plot Π / c against c. The y-intercept is RT/M and the slope is RTB/M^2. The transformed data to plot are given in the following table:

c / (mg cm^{-3})	1.33	2.10	4.52	7.18	9.87
Π / Pa	30	51	132	246	390
(Π/c) / (Pa mg^{-1} cm^3)	22.6	24.3	29.2	34.3	39.5

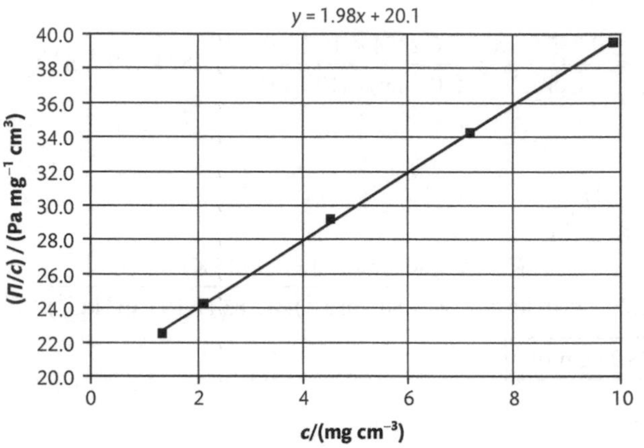

Fig 71.2

The plot is shown in Fig. 71.2. The y-intercept is 20.1 Pa mg^{-1} cm^3. The slope is 1.98 Pa mg^{-2} cm^6. Therefore

$$M=\frac{RT}{20.1\ \text{Pa mg}^{-1}\ \text{cm}^3}=\frac{8.3145\ \text{J K}^{-1}\ \text{mol}^{-1}\times 303\ \text{K}}{20.1\ \text{Pa mg}^{-1}\ \text{cm}^3}\times\frac{10^{-3}\ \text{g}}{1\ \text{mg}}\times\left(\frac{1\ \text{cm}}{10^{-2}\ \text{m}}\right)^3$$

$$=\boxed{1.25\times10^5\ \text{g mol}^{-1}}$$

$$B=\frac{M^2}{RT}\times1.98\ \text{Pa mg}^{-2}\ \text{cm}^6=\frac{M}{\left(\frac{RT}{M}\right)}\times1.98\ \text{Pa mg}^{-2}\ \text{cm}^6$$

$$=\frac{1.25\times10^5\ \text{g mol}^{-1}\times1.98\ \text{Pa mg}^{-2}\ \text{cm}^6}{20.1\ \text{Pa mg}^{-1}\ \text{cm}^3}$$

$$=1.23\times10^4\ \text{g mol}^{-1}\ \text{mg}^{-1}\ \text{cm}^3\times\frac{1\ \text{mg}}{10^{-3}\ \text{g}}\times\left(\frac{10^{-1}\ \text{dm}}{1\ \text{cm}}\right)^3=\boxed{1.23\times10^4\ \text{dm}^3\ \text{mol}^{-1}}$$

P71.3 At the boiling point of the solution (T_b) the chemical potential of the liquid solvent is equal to that of the solvent vapor:

$$\mu_A(l,T_b) = \mu_A(g,T_b)$$

Substituting the ideal-solution expressions for these chemical potentials yields:

$$\mu_A^{\ominus}(l,T_b) + RT_b \ln x_A = \mu_A^{\ominus}(g,T_b) + RT_b \ln(p_A/p^{\ominus})$$

The last term vanishes because $p_A = p°$ at the boiling point of the solution. We are interested in comparing T_b to T_b^*, the boiling point of the pure solvent, so we apply eqn 69.11 to both pure-substance chemical potentials:

$$\left(\frac{\partial \mu}{\partial T}\right)_p = -S_m \quad \text{so} \quad \mu(T_b) \approx \mu(T_b^*) - (T_b - T_b^*)S_m \quad \text{if } T_b - T_b^* \text{ is small}$$

Thus we can express chemical potential at the boiling point of the **solution** (T_b), in terms of differences from the chemical potential at the boiling point of the **pure substance** (T_b^*):

$$\mu_A^{\ominus}(l,T_b^*) - (T_b - T_b^*)S_m(l) + RT_b \ln x_A = \mu_A^{\ominus}(g,T_b^*) - (T_b - T_b^*)S_m(g)$$

Now $\mu_A^{\ominus}(l,T_b^*) = \mu_A^{\ominus}(g,T_b^*)$ [pure liquid and vapor are in equilibrium at T_b^*], so eliminating them from the equation and gathering the entropy terms on the left, we have

$$(T_b - T_b^*)\{S_m(g) - S_m(l)\} \approx -RT_b \ln x_A = -RT_b \ln(1-x_B) \approx RT_b x_B \ [x_B \text{ small}]$$

The differences on the left side of the equation are simply ΔT_b and $\Delta_{vap}S$, so we have

$$\Delta T_b \approx \frac{RT_b}{\Delta_{vap}S} x_B \approx \frac{RT_b^*}{\Delta_{vap}S} x_B [T_b^* \approx T_b] \approx \boxed{\frac{R(T_b^*)^2}{\Delta_{vap}H} x_B} \ [\Delta_{vap}H = T_b^* \Delta_{vap}S]$$

Topic 72 **Real Solutions**

Discussion question

D72.1 A regular solution has excess entropy of zero but an excess enthalpy that is non-zero and dependent on composition, perhaps in the manner of eqn 72.13. We can think of a regular solution as one in which the different molecules of the solution are distributed randomly, as in an ideal solution, but have different energies of interaction with each other. The parameter ξ is a measure of the difference between A-B interactions and A-A interactions. If $\xi < 0$, A-B interactions are energetically more favorable than A-A interactions; in that case, mixing is exothermic and even more spontaneous in all proportions than for an ideal solution. If, on the other hand, $\xi > 0$, A-B interactions are energetically less favorable than A-A interactions. In this case, mixing is endothermic. If ξ is not too large, mixing can still be spontaneous in all proportions. If ξ is larger, mixing is only spontaneous when one component dominates the composition. (See eqn 72.14.)

Exercises

E72.1(a) The excess enthalpy in a regular solution is (eqn 72.13)

$$H^{\mathrm{E}} = n\xi RTx_{\mathrm{A}}x_{\mathrm{B}} = n\xi RTx_{\mathrm{A}}(1-x_{\mathrm{A}})$$

The maximum value occurs at the composition where $x_{\mathrm{A}}(1-x_{\mathrm{A}})$ is at a maximum. That maximum occurs at $x_{\mathrm{A}} = x_{\mathrm{B}} = {}^{1}/_{2}$. Now we can find ξ:

$$\xi = \frac{H^{\mathrm{E}}_{\max}/n}{RTx_{\mathrm{A}}x_{\mathrm{B}}} = \frac{1.8\times10^{3}\ \mathrm{J\,mol^{-1}}}{(8.3145\ \mathrm{J\,K^{-1}\ mol^{-1}})(313\ \mathrm{K})(0.500)^{2}} = \boxed{2.8}$$

Phase separation occurs if $\xi > 2$, so $\boxed{\text{phase separation does occur}}$ at this temperature.

E72.2(a) For A (Raoult's law basis; concentration in mole fraction)

$$a_{\mathrm{A}} = \frac{p_{\mathrm{A}}}{p_{\mathrm{A}}^{*}}[72.2] = \frac{250\ \mathrm{Torr}}{300\ \mathrm{Torr}} = \boxed{0.833} \qquad \gamma_{\mathrm{A}} = \frac{a_{\mathrm{A}}}{x_{\mathrm{A}}} = \frac{0.833}{0.9} = \boxed{0.9\overline{3}}$$

For B (Henry's law basis; concentration in mole fraction)

$$a_{\mathrm{B}} = \frac{p_{\mathrm{B}}}{K_{\mathrm{B}}}[72.9] = \frac{25\ \mathrm{Torr}}{200\ \mathrm{Torr}} = \boxed{0.125} \qquad \gamma_{\mathrm{B}} = \frac{a_{\mathrm{B}}}{x_{\mathrm{B}}} = \frac{0.125}{0.1} = \boxed{1.\overline{25}}$$

For B (Henry's law basis; concentration in molality), we use an equation analogous to 72.9 but with a modified Henry's law constant K'_B:

$$a_B = \frac{p_B}{K'_B b^\ominus} \quad \text{with } p_B = b_B K'_B \text{ analogous to } p_B = x_B K_B.$$

K'_B and K_B are related by equating the two expressions for p_B:

$$p_B = b_B K'_B = x_B K_B \quad \text{so}$$

$$K'_B = \frac{x_B K_B}{b_B} = \frac{0.1 \times 200\ \text{Torr}}{2.22\ \text{mol kg}^{-1}} = 9.\bar{0}\ \text{Torr kg mol}^{-1}$$

Then, $a_B = \dfrac{25\ \text{Torr}}{9.\bar{0}\ \text{Torr}} = \boxed{2.\bar{8}} \qquad \gamma_B = \dfrac{a_B}{b_B/b^\ominus} = \dfrac{2.8}{2.22} = \boxed{1.\bar{2}\bar{5}}$

Comment. The two methods for the "solute" B give different values for the activities. This is reasonable since the reference states are different and therefore the chemical potentials in the reference states are also different.

Question. What are the activity and activity coefficient of B in the ideal solution (Raoult's law) basis?

E72.3(a) Activities and activity coefficients on the ideal-solution (Raoult's law) basis are defined by

$$a_A = \frac{p_A}{p_A^*} \quad \text{and} \quad \gamma_A = \frac{a_A}{x_A}$$

So we need partial pressures. Dalton's law relates the vapor-phase mole fractions to partial pressures

$$y_A = \frac{p_A}{p_A + p_M} = \frac{p_A}{101.3\ \text{kPa}} = 0.516,$$

So $p_A = 0.516 \times 101.3\ \text{kPa} = 52.3\ \text{kPa}$ and $p_M = (101.3 - 52.3)\ \text{kPa} = 49.0\ \text{kPa}$

$$a_A = \frac{p_A}{p_A^*} = \frac{52.3\ \text{kPa}}{105\ \text{kPa}} = \boxed{0.498} \qquad a_M = \frac{p_M}{p_M^*} = \frac{49.0\ \text{kPa}}{73.5\ \text{kPa}} = \boxed{0.667}$$

$$\gamma_A = \frac{a_A}{x_A} = \frac{0.498}{0.400} = \boxed{1.24} \qquad \gamma_M = \frac{a_M}{x_M} = \frac{0.667}{0.600} = \boxed{1.11}.$$

E72.4(a) The excess enthalpy in a regular solution is (eqn 72.13)

$$H^E = n\xi RTx_A x_B = n\xi RTx_A(1-x_A)$$

The maximum excess enthalpy occurs at $x_A = x_B = {}^1/_2$ [Exercise 72.1(a)]

$$\xi = \frac{H^E_{\text{max}}/n}{RTx_A x_B} = \frac{800\ \text{J mol}^{-1}}{(8.3145\ \text{J K}^{-1}\ \text{mol}^{-1})(293\ \text{K})(0.500)^2} = 1.31$$

The Margules equations [72.15] give the natural logarithms of activity coefficients:

$$\ln \gamma_A = \xi x_B^2 \quad \text{and} \quad \ln \gamma_B = \xi x_A^2$$

Because $x_A = x_B$, $\ln \gamma_A = \ln \gamma_B = 1.31 \times (0.500)^2 = 0.328$ and $\gamma_A = \gamma_B = 1.39$

Finally, $a_A = x_A \gamma_A = 0.500 \times 1.39 = \boxed{0.694} = a_B$

E72.5(a) $I = \frac{1}{2}\sum_i (b_i / b^\ominus) z_i^2$ [72.27]

For a salt of formula MpXq, $b_M = pb$ and $b_X = qb$,

so $I = \frac{1}{2}(pz_+^2 + qz_-^2)\left(\frac{b}{b^\ominus}\right)$

$$I(\mathrm{KCl}) = \frac{1}{2}(1\times 1 + 1\times 1)\left(\frac{b_{\mathrm{KCl}}}{b^\ominus}\right) = \frac{b_{\mathrm{KCl}}}{b^\ominus}$$

$$I(\mathrm{CuSO_4}) = \frac{1}{2}(1\times 2^2 + 1\times 2^2)\left(\frac{b_{\mathrm{CuSO_4}}}{b^\ominus}\right) = 4\left(\frac{b_{\mathrm{CuSO_4}}}{b^\ominus}\right)$$

$$I = I(\mathrm{KCl}) + I(\mathrm{CuSO_4}) = \frac{b_{\mathrm{KCl}}}{b^\ominus} + 4\left(\frac{b_{\mathrm{CuSO_4}}}{b^\ominus}\right) = 0.15 + 4\times 0.35 = \boxed{1.55}.$$

E72.6(a) The ionic strength of the original solution is $I(\mathrm{KNO_3}) = 0.250$. [$b/b^\ominus$ for a singly-charged salt.] So the added salt must contribute an additional ionic strength of 0.200.

(a) $I(\mathrm{Ca(NO_3)_2}) = \frac{1}{2}(2^2 + 2\times 1^2)\frac{b}{b^\ominus} = 3\frac{b}{b^\ominus} = 0.200$

so $b = \frac{0.200}{3} b^\ominus = 0.0667\,\mathrm{mol\,kg^{-1}}$

and $m(\mathrm{Ca(NO_3)_2}) = 0.0667\ \mathrm{mol\ kg^{-1}} \times 0.800\ \mathrm{kg} \times 261.32\ \mathrm{g\ mol^{-1}} = \boxed{8.75\ \mathrm{g}}$

(b) An additional 0.200 mol NaCl must be added per kg of solvent, so

$m(\mathrm{NaCl}) = 0.200\ \mathrm{mol\ kg^{-1}} \times 0.800\ \mathrm{kg} \times 58.44\ \mathrm{g\ mol^{-1}} = \boxed{9.35\ \mathrm{g}}$

E72.7(a) The solution is dilute, so use the Debye–Hückel limiting law.

$\log \gamma_\pm = -|z_+ z_-| A I^{1/2}$ [72.26]

$$I = \frac{1}{2}\sum_i z_i^2\left(\frac{b_i}{b^\ominus}\right) = \frac{1}{2}\{(2^2\times 0.0050) + (1\times 0.0050\times 2) + (1\times 0.0040) + (1\times 0.0040)\}$$

$= \boxed{0.0190}$.

$\mathrm{CaCl_2}$: $\log \gamma_\pm = -2\times 1\times 0.509\times (0.0190)^{1/2} = -0.140$ so $\gamma_\pm = \boxed{0.72}$

NaF: $\log \gamma_\pm = -1\times 1\times 0.509\times (0.0190)^{1/2} = -0.070$ so $\gamma_\pm = \boxed{0.85}$

E72.8(a) The extended Debye–Hückel law (eqn 72.59 with $C = 0$), is

$$\log \gamma_\pm = -\frac{A|z_+ z_-| I^{1/2}}{1 + BI^{1/2}}$$

Solve for B.

$$B = -\left(\frac{1}{I^{1/2}} + \frac{A|z_+ z_-|}{\log \gamma_\pm}\right) = -\left(\frac{1}{(b/b^\ominus)^{1/2}} + \frac{0.509}{\log \gamma_\pm}\right)$$

Draw up the following table

b/(mol kg^{-1})	5.0×10^{-3}	10.0×10^{-3}	20.0×10^{-3}
$\gamma_\pm$	0.930	0.907	0.879
B	2.01	2.01	2.02

The values of B are constant, illustrating that the extended law fits these activity coefficients with $\boxed{B = 2.01}$.

Problems

P72.1 (a) On a Raoult's law basis, the following relations apply to both components:

$$p_J = a_J p_J \ [72.2] = \gamma_J x_J p_J \ [72.4] \quad \text{so} \quad \gamma_J = \frac{p_J}{x_J p_J^*}$$

The vapor pressures of the pure components are given in the table of data. p_J^* is the value of p_J at $x_J = 1$: $p_I^* = 47.12$ kPa and $p_A^* = 37.38$ kPa.

(b) On a Henry's law basis, the following relations apply to the solute (*i.e.*, in this problem, iodoethane):

$$a_B = \frac{p_B}{K_B} \ [72.9] \quad \text{so} \quad \gamma_B = \frac{p_B}{x_B K_B}$$

(The solvent activity remains based on Raoult's law, so γ_A would remain as in part (a).) Henry's law constants are determined by plotting the data and extrapolating the low concentration data (*i.e.*, the best straight line in the neighborhood of $x_B = 0$) all the way across the plot to $x_B = 1$. The data are plotted in Fig. 72.1, which also displays p_I^*, p_A^*, K_I, and K_A. $K_I = 64.4$ kPa.

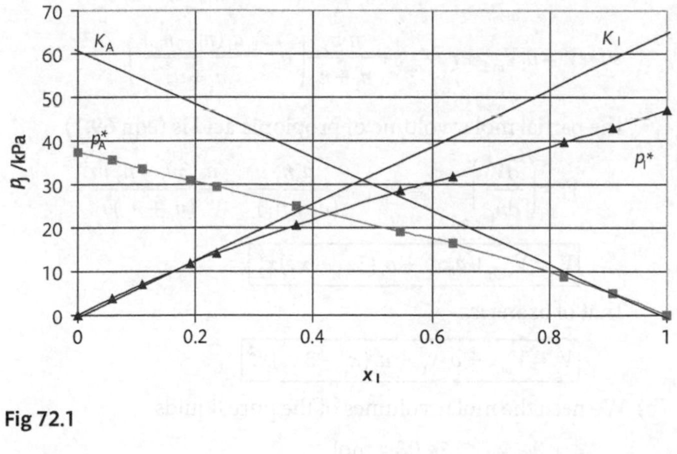

Fig 72.1

x_I	p_I/kPa	p_A/kPa	γ_I(Raoult)	γ_A(Raoult)	γ_I(Henry)
0	0	37.38†	–	1	1
0.0579	3.73	35.48	1.367	1.008	1.000
0.1095	7.03	33.64	1.362	1.011	0.997
0.1918	11.7	30.85	1.295	1.021	0.947
0.2353	14.05	29.44	1.267	1.030	0.927
0.3718	20.72	25.04	1.183	1.066	0.865
0.5478	28.44	19.23	1.102	1.138	0.806
0.6349	31.88	16.39	1.066	1.201	0.779
0.8253	39.58	8.88	1.018	1.360	0.744
0.9093	43.00	5.09	1.004	1.501	0.734
1.0000	47.12‡	0	1	–	0.731

†the value of p_A^*; ‡the value of p_I^*

Question. In this problem both I and A were treated as solvents, but only I as a solute. Extend the table by including a column for γ_A(Henry).

P72.3 (a) The volume of an ideal mixture is

$$V_{\text{ideal}} = n_1 V_{\text{m},1} + n_2 V_{\text{m},2},$$

so the volume of a real mixture is

$$V = V_{\text{ideal}} + V^{\text{E}}.$$

We have an expression for excess molar volume in terms of mole fractions. To compute partial molar volumes, we need an expression for the actual excess volume as a function of moles

$$V^{\text{E}} = (n_1 + n_2)V_{\text{m}}^{\text{E}} = \frac{n_1 n_2}{n_1 + n_2}\left(a_0 + \frac{a_1(n_1 - n_2)}{n_1 + n_2}\right)$$

$$\text{so} \quad V = n_1 V_{\text{m},1} + n_2 V_{\text{m},2} + \frac{n_1 n_2}{n_1 + n_2}\left(a_0 + \frac{a_1(n_1 - n_2)}{n_1 + n_2}\right)$$

The partial molar volume of propionic acid is (eqn 69.1)

$$V_1 = \left(\frac{\partial V}{\partial n_1}\right)_{p,T,n_2} = V_{\text{m},1} + \frac{a_0 n_2^2}{(n_1 + n_2)^2} + \frac{a_1(3n_1 - n_2)n_2^2}{(n_1 + n_2)^3}$$

$$\boxed{V_1 = V_{\text{m},1} + a_0 x_2^2 + a_1(3x_1 - x_2)x_2^2}$$

That of oxane is

$$\boxed{V_2 = V_{\text{m},2} + a_0 x_1^2 + a_1(x_1 - 3x_2)x_1^2}$$

(b) We need the molar volumes of the pure liquids

$$V_{\text{m},1} = \frac{M_1}{\rho_1} = \frac{74.08\ \text{g mol}^{-1}}{0.97174\ \text{g cm}^{-3}} = 76.23\ \text{cm}^3\ \text{mol}^{-1}$$

and $V_{m,2} = \dfrac{86.13\text{ g mol}^{-1}}{0.86398\text{ g cm}^{-3}} = 99.69\text{ cm}^3\text{ mol}^{-1}$

In an equimolar mixture, the partial molar volume of propionic acid is

$$V_1 = 76.23 + (-2.4697)\times(0.5)^2 + (0.0608)\times\{3(0.5)-0.5\}\times(0.5)^2\text{ cm}^3\text{ mol}^{-1}$$

$$= \boxed{75.63\text{ cm}^3\text{ mol}^{-1}}.$$

and that of oxane is

$$V_2 = 99.69 + (-2.4697)\times(0.5)^2 + (0.0608)\times\{0.5-3(0.5)\}\times(0.5)^2\text{ cm}^3\text{ mol}^{-1}$$

$$= \boxed{99.06\text{ cm}^3\text{ mol}^{-1}}.$$

P72.5 According to the Debye–Hückel limiting law

$$\log\gamma_\pm = -0.509\,|z_+z_-|\,I^{1/2}\ [72.26] = -0.509\left(\frac{b}{b^\ominus}\right)^{1/2}$$

We draw up the following table:

b/(mmol kg^{-1})	1.0	2.0	5.0	10.0	20.0
$I^{1/2}$	0.032	0.045	0.071	0.100	0.141
$\gamma_\pm$(calc)	0.964	0.949	0.920	0.889	0.847
$\gamma_\pm$(exp)	0.9649	0.9519	0.9275	0.9024	0.8712
log $\gamma_\pm$(calc)	−0.0161	−0.0228	−0.0360	−0.0509	−0.0720
log $\gamma_\pm$(calc)	−0.0155	−0.0214	−0.0327	−0.0446	−0.0599

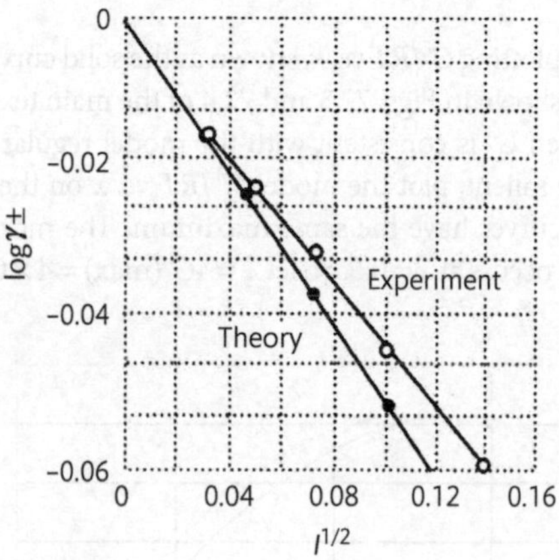

Fig 72.2

The points are plotted against $I^{1/2}$ in Fig. 72.2. Note that the limiting slopes of the calculated and experimental curves coincide. We use the extended Debye–Hückel law (eqn 72.29 with $C = 0$). A sufficiently good value of B in the extended law may be obtained by assuming that the constant A in the extended law is the same as A in the limiting law. Using the data at 20.0 mmol kg^{-1} we may solve for B.

$$B = -\frac{A}{\log\gamma_\pm} - \frac{1}{I^{1/2}} = -\frac{0.509}{(-0.0599)} - \frac{1}{0.141} = 1.40\overline{5}$$

Thus,

$$\log\gamma_\pm = -\frac{0.509 I^{1/2}}{1 + 1.40\overline{5} I^{1/2}}$$

In order to determine whether or not the fit is improved, we use the data at 10.0 mmol kg^{-1}:

$$\log\gamma_\pm = \frac{-(0.509)\times(0.100)}{(1)+(1.405)\times(0.100)} = -0.0446$$

which fits the data almost exactly. The fits to the other data points will also be almost exact.

P72.7 $G^E = RTx(1-x)\{0.4857 - 0.1077(2x-1) + 0.0191(2x-1)^2\}$

A regular solution is defined by $S^E = 0$. Thus, a solution is regular if $G^E = H^E$. Because an ideal solution has zero enthalpy of mixing, $H^E = \Delta_{mix}H$; therefore, H^E is readily measured experimentally by calorimetry. Thus, the test of whether the solution is regular amounts to measuring $\Delta_{mix}H$ and determining how closely it matches the expression for G^E.

To test whether the given expression is consistent with the regular solution **model** given by eqn 72.13, we must compare the above expression for G^E with the model expression for H^E:

$$H^E = \xi RTx(x-1)$$

Begin the comparison by plotting G^E/RT vs. x, shown as the solid curve in Fig. 72.3, which resembles the curves shown in Figs. 72.3 and 72.4 of the main text. The resemblance suggests that the given G^E is consistent with the model regular solution. To make the comparison more salient, plot the model H^E/RT vs. x on the same graph, selecting ξ so that the two curves have the same maximum. The maximum in the model function is $\xi/4$, and it occurs at $x = 0.5$. So let $\xi = 4G^E(\max) = 4\times 0.123 = 0.497$.

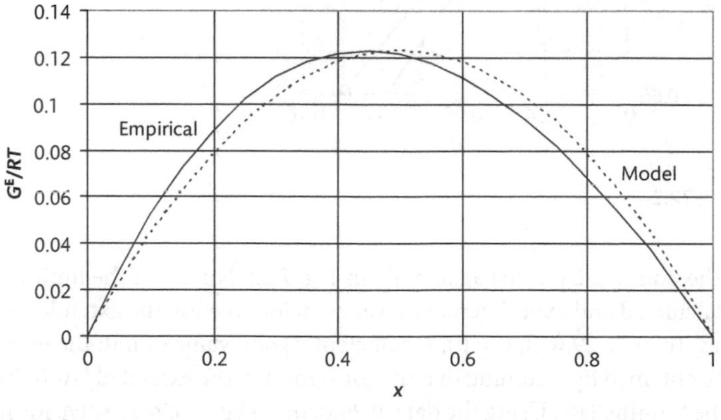

Fig 72.3

G^E is reasonably consistent with the model regular solution over the entire range of composition.

P72.9 The analogous integral is

$$V_P(x_T) - V_P(0) = -\int_{V_T(0)}^{V_T(x_T)} \frac{x_T}{1-x_T} \mathrm{d}V_T$$

We should now plot $x_T/(1-x_T)$ against V_T and estimate the integral. We must integrate up to $x_T = 0.5$, which corresponds to $x_T/(1-x_T) = 1$. So draw up the following table including all data points with $x_T < 0.5$ and one point beyond so as to define the plot of the integrand over the interval of integration.

x_T	0	0.194	0.385	0.559
V_T (cm^3 mol^{-1})	73.99	75.29	76.50	77.55
$x_A/(1-x_A)$	0	0.241	0.626	1.266

The points are plotted in Fig. 72.4. In order to estimate the upper limit, we draw a smooth curve through the data points. The value of V_T where $x_T = 0.5$ is 77.1 cm^3 mol^{-1}

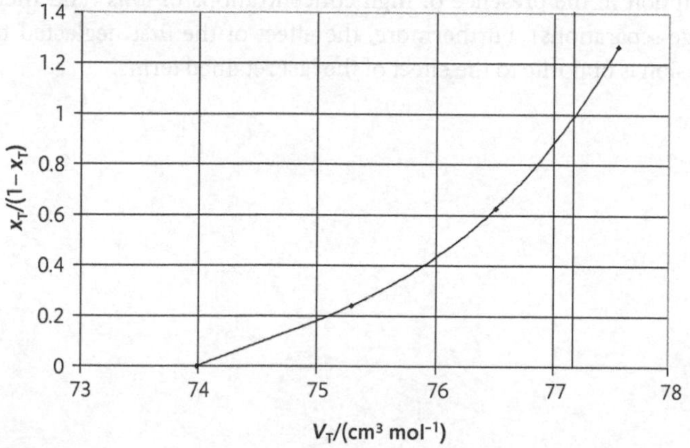

Fig 72.4

The integral is the area under the curve. Applying the trapezoid rule yields 1.39 cm^3 mol^{-1} for the integral, so

$$V_P(x_T = 0.5) = V_P(x_T = 0) - 1.39 \text{ cm}^3 \text{ mol}^{-1}$$

We need $V_P(x_T = 0)$, that is, the partial molar volume of propanone in a "mixture" containing no trichloromethane. $V_P(x_T = 0)$ is the partial molar volume of pure propanone, *i.e.*, the molar volume of propanone:

$$V_m(\text{propanone}) = \frac{M}{\rho} = \frac{58.08 \text{ g mol}^{-1}}{0.7857 \text{ g cm}^{-3}} = 73.92 \text{ cm}^3 \text{ mol}^{-1}$$

$$V_P(x_T = 0.5) = (73.92 - 1.39) \text{ cm}^3 \text{ mol}^{-1} = \boxed{72.53 \text{ cm}^3 \text{ mol}^{-1}}$$

P72.11 Le Chatelier's principle states that a system at equilibrium responds to a disturbance of that equilibrium by minimizing the effect of the disturbance. Here the disturbance is the addition of ions not directly involved in the equilibrium. Their effect is to increase the ionic strength of the system, and through the ionic strength to change the activity of the ions. According to the Debye-Hückel limiting law (eqn 72.26), the activity coefficients of the ions decrease with increasing ionic strength and therefore the activities of the ions likewise decrease. The equilibrium would then shift in such a way as to minimize this decrease of ionic activity by increasing the concentration of the charged protein and counter ions. This is the salting-in effect. The extended Debye-Hückel law (eqn 72.29 with $C=0$), however, suggests that increasing the ionic strength still further eventually causes the activity coefficients to increase. In this regime, the concentrations of the charged protein and counter ions would decrease as they precipitate out; this is the salting-out effect.

The extended Debye-Hückel law can tell us ***how*** activity coefficients vary with ionic strength, but it does not give us insight into why. Examination of a derivation of the limiting law can help us. Part of that derivation includes applying a power-series approximation to an exponential function, truncating it after the term linear in the energy of electrostatic interaction. That approximation depends on the energy of electrostatic interaction being small compared to kT. That is a questionable assumption in the presence of high concentrations of ions (and therefore small average separations). Furthermore, the effect of the first neglected terms in the expansion is opposite to the effect of the last retained terms.

Focus 14: Integrated activities

F14.1 The data are plotted in Fig. 14.1(a). The vapor pressure curves define reasonably straight lines at either end of the composition range. That the benzene vapor pressure curve is non-linear is quite apparent, though: Henry's law provides a good approximation to p_B only for x_B less than about 0.15, and Raoult's law approximates p_B well only for very nearly pure benzene. (The benzene vapor pressure curve ***is*** reasonably linear for large x_B, but the best-fit line to those data clearly does not pass through $x_B = 0$, as would be required for Raoult's law.) The acetic acid vapor pressure curve is closer than the benzene curve to linear, even when plotted on a comparable scale.

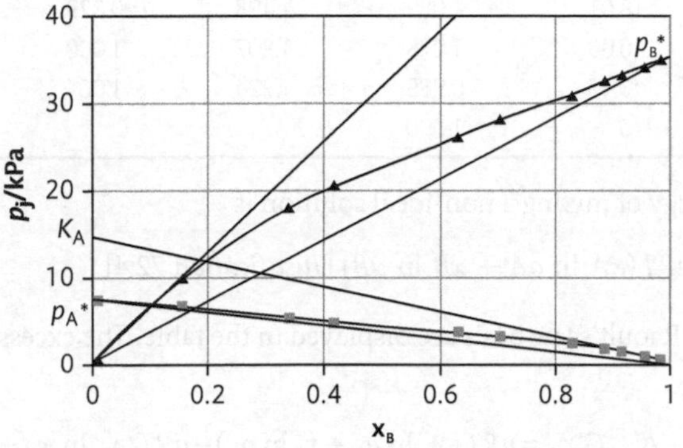

Fig F14.1(a)

(a) On a Raoult's law basis, the following relations apply to both components:

$$p_J = a_J p_J^* \ [72.2] = \gamma_J x_J p_J^* \ [72.4] \quad \text{so} \quad \gamma_J = \frac{p_J}{x_J p_J^*}$$

The vapor pressures of the pure components are obtained by linear extrapolation of the vapor pressure data in the neighborhood of each component as a pure substance (that is, near $x_B = 1$ for B and near $x_B = 0$ for A). Note, for the purposes of extrapolating the best estimate of the pure-substance vapor pressures, I did not use Raoult's law (which would force the straight line through zero at the opposite end of the composition range) but an empirical linear fit. The results are $p_A^* = 7.45$ kPa and $p_B^* = 35.41$ kPa. The activity coefficients are shown in the table below. In that table, the limiting pressures are shown in italics.

(b) On a Henry's law basis, the following relations apply to the solute (*i.e.*, in this problem, benzene):

$$a_B = \frac{p_B}{K_B} \ [72.9] \quad \text{so} \quad \gamma_B = \frac{p_B}{x_B K_B}$$

The Henry's law constant is the slope of a best-fit line in the neighborhood of $x_J = 0$ and constrained to pass through $p_J = 0$ at $x_J = 0$. The Henry's law constants obtained from these data are $K_B = 63.99$ kPa and $K_A = 19.42$ kPa.

x_B	p_A/kPa	p_B/kPa	γ_A(Raoult)	γ_B(Raoult)	γ_B(Henry)	$G^E{}_m$/kJ mol^{-1}
1	0	35.41		1.000	0.553	0.0
0.984	0.484	35.05	4.060	1.006	0.557	75.9
0.9561	0.967	34.29	2.957	1.013	0.560	160.6
0.9165	1.535	33.28	2.468	1.025	0.567	264.6
0.8862	1.89	32.64	2.229	1.040	0.576	338.8
0.8286	2.45	30.90	1.919	1.053	0.583	415.4
0.7027	3.31	28.16	1.494	1.132	0.626	554.5
0.6304	3.83	26.08	1.391	1.168	0.647	591.2
0.4166	4.84	20.42	1.114	1.384	0.766	532.6
0.3396	5.36	18.01	1.089	1.498	0.829	520.5
0.1563	6.76	10.00	1.075	1.807	1.000	413.4
0.0069	7.29	0.47	0.985	1.924	1.064	−27.3
0	7.45	0	1.000			0.0

The Gibbs energy of mixing a non-ideal solution is

$$\Delta mixG = nRT(xA \ln aA + xB \ln aB) \text{ [\textit{Justification} 72.4]}$$

Activities on a Raoult's law basis are displayed in the table. The excess Gibbs function is

$$G^E = \Delta_{mix} G - \Delta_{mix} G_{ideal} = nRT(x_A \ln a_A + x_B \ln a_B) - nRT(x_A \ln x_A + x_B \ln x_B)$$
[70.9a]

so $$G^E = nRT\left(x_A \ln\frac{a_A}{x_A} + x_B \ln\frac{a_B}{x_B}\right) = nRT(x_A \ln\gamma_A + x_B \ln\gamma_B)$$

The molar excess Gibbs function computed on the basis of Raoult's law activity coefficients is plotted against composition in Fig. 14.1(b).

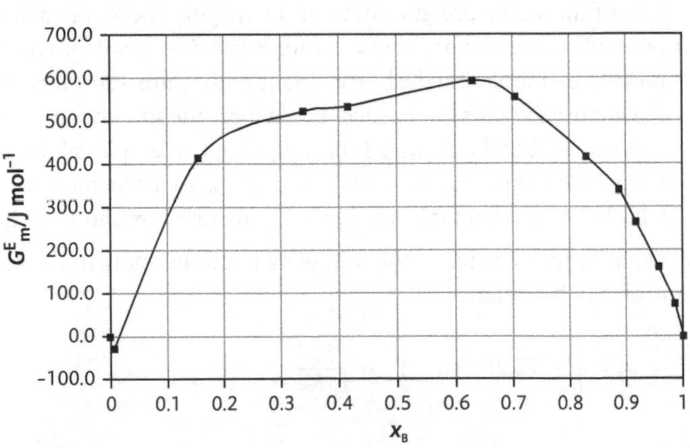

Fig 14.1(b)

F14.3 The partial pressure of a solvent is related to its activity by

$$p_A = a_A p_A \ [72.2] = \gamma_A x_A p_A \ [72.4] \quad \text{so} \quad \gamma_A = \frac{p_A}{x_A p_A^*} = \frac{y_A p}{x_A p_A^*}$$

Sample calculation at 80 K:

$$\gamma(O_2, 80\,K) = \frac{0.11 \times 100\ \text{kPa}}{0.34 \times 225\ \text{Torr}} \times \frac{760\ \text{Torr}}{101.325\ \text{kPa}} = 1.079$$

Summary

T/K	77.3	78	80	82	84	86	88	90.2
$\gamma(O_2)$	–	0.877	1.079	1.039	0.995	0.993	0.990	0.987

To within the experimental uncertainties the solution appears to be ideal ($\gamma \approx 1$). The low value at 78 K may be caused by nonideality; however, the larger relative uncertainty in $y(O_2)$ is probably the origin of the low value. A temperature–composition diagram is shown in Fig. 14.2(a). The near ideality of this solution is, however, best shown in the pressure–composition diagram of Fig. 14.2(b). There the liquid line is essentially a straight line as predicted for an ideal solution.

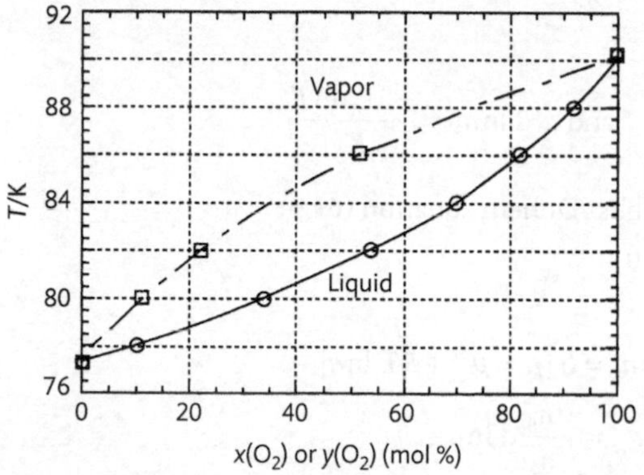

(a)

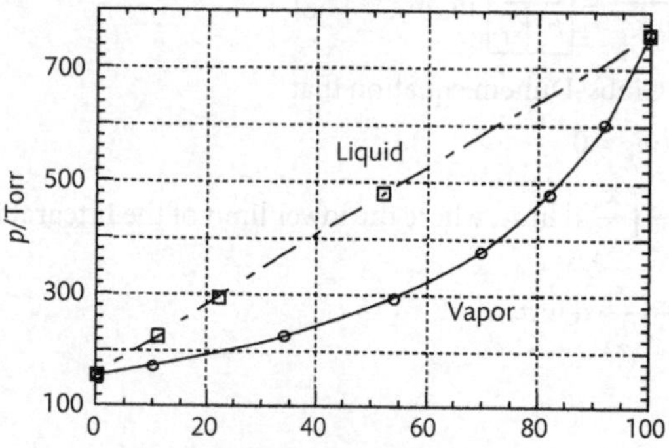

(b)

Fig F14.2

F14.5 At equilibrium the chemical potentials of the approximately pure solid and of the solvent in solution are equal:

$$\mu_A^*(s) = \mu_A^*(l) + RT\ln a_A$$

$$\Delta_{fus}G = \mu_A^*(l) - \mu_A^*(s) = -RT\ln a_A$$

Hence, $\ln a_A = \dfrac{-\Delta_{fus}G}{RT}$

$$\frac{d\ln a_A}{dT} = -\frac{1}{R}\frac{d}{dT}\left(\frac{\Delta_{fus}G}{T}\right) = \frac{\Delta_{fus}H}{RT^2} \text{ [Gibbs-Helmholtz eqn (66.8b)]}$$

Since $\Delta T = T_f^* - T, d\Delta T = -dT$

and $\dfrac{d\ln a_A}{d\Delta T} = \dfrac{-\Delta_{fus}H}{RT^2} \approx \dfrac{-\Delta_{fus}H}{RT_f^2} = -\dfrac{1}{K_f'}$

where K_f' is the freezing point depression constant from Problem 72.2. It corresponds to the mole-fraction-based expression

$$\Delta T = K_f' x_B$$

It is related to the molality-based expression

$$\Delta T = K_f b_B$$

by $K_f = \dfrac{K_f' x_B}{b_B} = \dfrac{K_f' x_B m_A}{n_B} \approx K_f' M_A \quad [n_A \approx n_{total}]$

Therefore,

$$\frac{d\ln a_A}{d\Delta T} = \boxed{-\frac{M_A}{K_f}} \quad \text{and} \quad d\ln a_A = \frac{-M_A d\Delta T}{K_f}$$

According to the Gibbs–Duhem equation (69.9)

$$n_A d\mu_A + n_B d\mu_B = 0$$

which implies that

$$n_A d\ln a_A + n_B d\ln a_B = 0 \ [\mu = \mu^\ominus + RT\ln a]$$

and hence that $d\ln a_A = -\dfrac{n_B}{n_A} d\ln a_B$

Hence, $\dfrac{d\ln a_B}{d\Delta T} = \dfrac{n_A M_A}{n_B K_f} = \boxed{\dfrac{1}{b_B K_f}} \quad [n_A M_A = 1\text{ kg}]$

We know from the Gibbs-Duhem equation that

$$x_A d\ln a_A + x_B d\ln a_B = 0$$

and that $\int d\ln a_A = -\int \dfrac{x_B}{x_A} d\ln a_B$, where the lower limit of the integral is pure A.

Therefore $\ln a_A = -\int \dfrac{x_B}{x_A} d\ln a_B$

The osmotic coefficient was defined in Problem 72.10 as

$$\phi = -\frac{1}{r}\ln a_A = -\frac{x_A}{x_B}\ln a_A$$

Therefore,

$$\phi = \frac{x_A}{x_B}\int \frac{x_B}{x_A}\mathrm{d}\ln a_B \approx \frac{1}{b}\int_0^b b\,\mathrm{d}\ln a_B = \frac{1}{b}\int_0^b b\,\mathrm{d}\ln \gamma b = \frac{1}{b}\int_0^b b\,\mathrm{d}\ln b + \frac{1}{b}\int_0^b b\,\mathrm{d}\ln \gamma$$

$$= 1 + \frac{1}{b}\int_0^b b\,\mathrm{d}\ln \gamma.$$

From the Debye-Hückel limiting law (eqn 72.26),

$$\ln\gamma = -A'I^{1/2} = -A'fb^{1/2} \qquad [A' = 2.303A]$$

where f is the ratio between b and I, which depends on the charges of the electrolyte.

Hence, $\mathrm{d}\ln\gamma = -\frac{A'f}{2}b^{-1/2}\,\mathrm{d}b$ and so

$$\phi = 1 + \frac{1}{b}\left(-\frac{A'f}{2}\right)\int_0^b b^{1/2}\,\mathrm{d}b = 1 - \frac{1}{2}\left(\frac{A'f}{b}\right)\times\frac{2}{3}b^{3/2} = \boxed{1 - \frac{1}{3}A'I^{1/2}}$$

F14.7 The question asks us to express derivatives of the chemical potential in statistical thermodynamic terms, so we start with the chemical potential itself, which is the partial molar Gibbs function.

$$G = G(0) - nRT\ln\frac{q}{N}\ [64.12b],$$

$$\text{so}\quad \mu = \left(\frac{\partial G}{\partial n}\right)_{p,T} = \frac{\partial G(0)}{\partial n} - RT\ln\frac{q}{N} - nRT\frac{\partial}{\partial n}\left(\ln\frac{q}{N}\right)$$

How does q/N depend on n (at constant p and T)? $N = nN_A$. q depends on molecular parameters and temperature and (in the translational partition function) volume. Recall that

$$q^T = \frac{V}{\Lambda^3}\ [52.10b]$$

$$\text{so}\quad \frac{q}{N} = \frac{q^T q^R q^V q^E}{N} = \frac{Vq^R q^V q^E}{N\Lambda^3} = \frac{kTq^R q^V q^E}{p\Lambda^3}$$

where we used the perfect gas law in the form $pV = NkT$ in the last step. Thus, for a perfect gas, the partial derivative of $\ln q/N$ with respect to n at constant p and T vanishes, and

$$\mu = \left(\frac{\partial G}{\partial n}\right)_{p,T} = \frac{\partial G(0)}{\partial n} - RT\ln\frac{q}{N}$$

(a) We are to show that eqn 69.11,

$$\left(\frac{\partial \mu}{\partial T}\right)_{p,n} = -S_{\mathrm{m}}$$

is consistent with eqns 64.12b (above) and eqn 60.3,

$$S = \frac{U-U(0)}{T} + Nk\ln\frac{qe}{N} = \frac{U-U(0)}{T} + nR + nR\ln\frac{q}{N}$$

We differentiate our statistical-thermodynamic expression of the chemical potential:

$$\left(\frac{\partial \mu}{\partial T}\right)_{p,n} = \frac{\partial}{\partial T}\left(\frac{\partial G(0)}{\partial n} - RT\ln\frac{q}{N}\right)_{p,n} = -R\left(\ln\frac{q}{N}\right) - RT\frac{\partial}{\partial T}\left(\ln\frac{q}{N}\right)_{p,n}$$

$$= -R\ln\frac{q}{N} - RT\left(\frac{\partial \ln q}{\partial T}\right)_{p,n}$$

So we must show that this equals S/n. To bridge the gap between our expressions, we need a statistical-thermodynamic expression for internal energy. For a collection of independent particles

$$U-U(0) = N\langle\varepsilon\rangle = -N\left(\frac{\partial \ln q}{\partial \beta}\right)_V \text{ [53.4b]} = -N\left(\frac{\partial \ln q}{\partial T}\right)_V\left(\frac{\partial T}{\partial \beta}\right)_V$$

$$= -N\left(\frac{\partial \ln q}{\partial T}\right)_V\left(\frac{\partial \beta}{\partial T}\right)^{-1} = NkT^2\left(\frac{\partial \ln q}{\partial T}\right)_V$$

Note that what is held constant here is volume, not pressure as in the chemical potential derivative, so we need the following property of partial derivatives based on eqn MB 8.3:

$$\left(\frac{\partial \ln q}{\partial T}\right)_V = \left(\frac{\partial \ln q}{\partial T}\right)_p + \left(\frac{\partial \ln q}{\partial p}\right)_T\left(\frac{\partial p}{\partial T}\right)_V$$

$$= \left(\frac{\partial \ln q}{\partial T}\right)_p + \left\{\frac{\partial}{\partial p}\left(\ln\frac{nRTq^{\mathrm{R}}q^{\mathrm{V}}q^{\mathrm{E}}}{p\Lambda^3}\right)\right\}_T\left\{\frac{\partial}{\partial T}\left(\frac{nRT}{V}\right)\right\}_V$$

$$= \left(\frac{\partial \ln q}{\partial T}\right)_p - \frac{1}{p}\times\frac{nR}{V} = \left(\frac{\partial \ln q}{\partial T}\right)_p - \frac{1}{T}$$

Thus $\dfrac{U-U(0)}{T} = nRT\left(\dfrac{\partial \ln q}{\partial T}\right)_p - nR$

which allows us to identify our statistical-thermodynamic expression for $\partial\mu/\partial T$:

$$\left(\frac{\partial \mu}{\partial T}\right)_{p,n} = -R\ln\frac{q}{N} - \frac{U-U(0)}{nT} + R = -\frac{S}{n} = \boxed{-S_{\mathrm{m}}}$$

(b) We are to show that eqn 69.12

$$\left(\frac{\partial \mu}{\partial p}\right)_{T,n} = V_{\mathrm{m}}$$

is consistent with statistical thermodynamic results based on eqn 64.12b. So we differentiate

$$\left(\frac{\partial \mu}{\partial p}\right)_{T,n} = \frac{\partial}{\partial p}\left(\frac{\partial G(0)}{\partial n} - RT\ln\frac{q}{N}\right)_{T,n} = -RT\left(\frac{\partial}{\partial p}(\ln q - \ln N)\right)_{T,n}$$

$$= -RT\left(\frac{\partial \ln q}{\partial p}\right)_{T,n} = -RT \times \left(-\frac{1}{p}\right) = \frac{V}{n} = \boxed{V_{\mathrm{m}}}$$

Topic 73 Chemical transformations

Discussion questions

73.1 The position of equilibrium is always determined by the condition that the reaction quotient, Q, must equal the equilibrium constant, K. If the mixing in of an additional amount of reactant or product destroys that equality, then the reacting system will shift in such a way as to restore the equality. That implies that some of the added reactant or product must be removed by the reacting system and the amounts of other components will also be affected. These adjustments restore the concentrations to their (new) equilibrium values.

73.3 The thermodynamic equilibrium constant involves activities rather than pressures (see eqn 73.12 and Example 73.1). At low pressures, the activities of gases may be replaced with their partial pressures with little error, but at high pressures that is not a good approximation. The difference between the equilibrium constant expressed in activities and the constant expressed in pressures is dependent upon two factors: the stoichiometry of the reaction and the magnitude of the partial pressures. Thus there is no one answer to this question. For the example of the ammonia synthesis reaction, in a range of pressures where the activity coefficients are greater than one, an increase in pressure results in a greater shift to the product side than would be predicted by the constant expressed in partial pressures.

The activity coefficients of real gases give an indication of dominant intermolecular forces. Coefficients greater than 1 are observed when repulsions dominate; coefficients less than one dominate when attractions dominate. In the limit of zero pressure all gases behave as perfect gases with activity coefficients equal to 1 and, therefore, $a_{\mathrm{J(gas)}} = \gamma_\mathrm{J} P_\mathrm{J} / p^{\ominus} = P_\mathrm{J} / p^{\ominus}$ in the perfect gas case only. Thus, we conclude that intermolecular forces cause the thermodynamic equilibrium constant to respond differently to changes in pressure from the equilibrium constant expressed in terms of partial pressures.

For an exothermic reaction, such as the ammonia synthesis, an increase in temperature will shift the reaction to the reactant side, but the relative shift is independent of the activity coefficients. The ratio $\ln (K_2/K_1)$ depends only on $\Delta_\mathrm{r} H^{\ominus}$ (see eqn 75.3 and 75.5).

Exercises

73.1(a) Eqn 73.12 provides equilibrium constants in terms of activities for part (i) with the activities of pure solids and liquids being equal to 1. For part (ii), substitute $a_{solute} = \gamma_{solute} b_{solute}/b^{\ominus}$ and assume perfect gas behavior with the substitution $a_{gas} = P_{gas}/p^{\ominus}$.

(a) (i) $K = \dfrac{a_{COCl(g)}a_{Cl(g)}}{a_{CO(g)}a_{Cl_2(g)}}$ (ii) $K = \dfrac{p_{COCl}p_{Cl}}{p_{CO}p_{Cl_2}}$

(b) (i) $K = \dfrac{a^2_{SO_3(g)}}{a^2_{SO_2(g)}a_{O_2(g)}}$ (ii) $K = \dfrac{p^2_{SO_3}p^{\ominus}}{p^2_{SO_2}p_{O_2}}$

(c) (i) $K = \dfrac{a_{FeSO_4(aq)}}{a_{PbSO_4(aq)}}$ (ii) $K = \dfrac{\gamma_{FeSO_4}}{\gamma_{PbSO_4}} \times \dfrac{b_{FeSO_4}}{b_{PbSO_4}}$

(d) (i) $K = \dfrac{a^2_{HCl(aq)}}{a_{H_2(g)}}$ (ii) $K = \dfrac{\gamma^2_{HCl(aq)}b^2_{HCl(aq)}}{p_{H_2(g)}} \times \dfrac{p^{\ominus}}{(b^{\ominus})^2}$

(e) (i) $K = \dfrac{a_{CuCl_2(aq)}}{a^2_{CuCl(aq)}}$ (ii) $K = \dfrac{\gamma_{CuCl_2}}{\gamma^2_{CuCl}} \times \dfrac{b_{CuCl_2}b^{\ominus}}{b^2_{CuCl}}$

73.2(a) $\nu_{Hg_2Cl_2} = -1,\ \nu_{H_2} = -1,\ \nu_{HCl} = 2,\ \nu_{Hg} = 2 \quad (\Delta\nu_g = -1\ [57.3])$

73.3(a) Let B = borneol and I = isoborneol; B ⇌ I

$$\Delta_r G = \Delta G^{\ominus} + RT\ln Q \quad \text{where} \quad Q = \frac{p_I}{p_B}\ [73.5]$$

$$p_B = x_B p = \frac{0.15\ \text{mol}}{0.15\ \text{mol} + 0.30\ \text{mol}} \times 600\ \text{Torr} = 200\ \text{Torr}; \qquad p_I = p - p_B = 400\ \text{Torr}$$

$$Q = \frac{400\ \text{Torr}}{200\ \text{Torr}} = 2.00$$

$$\Delta_r G = (+9.4\ \text{kJ mol}^{-1}) + (8.314\ \text{J K}^{-1}\ \text{mol}^{-1}) \times (503\ \text{K}) \times (\ln 2.00) = \boxed{+12.3\ \text{kJ mol}^{-1}}$$

This mixture reacts spontaneously to the left (toward the formation of borneol).

73.4(a) The formation reaction is: 2 Ag(s) + ½ O_2(g) ⇌ Ag_2O(s)

$$K = \frac{1}{a^{1/2}_{O_2(g)}} = \left(\frac{p^{\ominus}}{p_{O_2}}\right)^{1/2}$$

[73.12; $a_{Ag(s)} = a_{Ag_2O(s)} = 1$ and, assuming perfect gas behavior, $a_{O_2(g)} = p_{O_2}/p^{\ominus}$]

$$= \left(\frac{10^5\ \text{Pa}}{11.85\ \text{Pa}}\right)^{1/2} = 91.86$$

$$\Delta_r G^{\ominus} = -RT\ln K \text{ [73.13]}$$
$$= -(8.3145\text{ J K}^{-1}\text{ mol}^{-1})\times(298\text{ K})\times(\ln 91.86)$$
$$= \boxed{-11.20\text{ kJ mol}^{-1}}$$

73.5(a) $CaF_2(s) \rightleftharpoons Ca^{2+}(aq) + 2F^{-}(aq) \quad K = 3.9\times10^{-11}$

$$\Delta_r G^{\ominus} = -RT\ln K \text{ [73.13]}$$
$$= -(8.3145\text{ J K}^{-1}\text{ mol}^{-1})\times(298.15\text{ K})\times\ln(3.9\times10^{-11}) = +59.4\text{ kJ mol}^{-1}$$
$$= \Delta_f G^{\ominus}(CaF_2,\text{ aq}) - \Delta_f G^{\ominus}(CaF_2,\text{ s}) \text{ [73.10]}$$

$$\Delta_f G^{\ominus}(CaF_2,\text{aq}) = \Delta_r G^{\ominus} + \Delta_f G^{\ominus}(CaF_2,\text{s})$$
$$= [59.4 - 1167]\text{ kJ mol}^{-1} = \boxed{-1108\text{ kJ mol}^{-1}}$$

73.6(a) Draw up the following equilibrium table for the gas-phase reaction: $2\text{ A} + \text{B} \rightleftharpoons 3\text{ C} + 2\text{ D}$

	A	B	C	D	Total
Initial amounts/mol	1.00	2.00	0	1.00	4.00
Stated change/mol			+0.90		
Implied change/mol	−0.60	−0.30	+0.90	+0.60	
Equilibrium amounts/mol	0.40	1.70	0.90	1.60	4.60
Mole fractions	0.087	0.370	0.196	0.348	1.001

(a) Equilibrium mole fractions are given in the table.

(b) $K_x = \prod_J x_J^{\nu_J}$

$$K_x = \frac{(0.196)^3\times(0.348)^2}{(0.087)^2\times(0.370)} = 0.32\bar{6} = \boxed{0.33}$$

(c) $p_J = x_J p, \quad p = 1\text{ bar}, \quad p^{\ominus} = 1\text{ bar}$

Assuming that the gases are perfect, $a_J = p_J/p^{\ominus}$, hence

$$K = \frac{(p_C/p^{\ominus})^3\times(p_D/p^{\ominus})^2}{(p_A/p^{\ominus})^2\times(p_B/p^{\ominus})} \text{ [73.12]}$$
$$= \frac{x_C^3 x_D^2}{x_A^2 x_B}\times\left(\frac{p}{p^{\ominus}}\right)^2 = K_x \text{ (when } p = 1.00\text{ bar)} = \boxed{0.33}$$

(d) $\Delta_r G^{\ominus} = -RT\ln K \text{ [73.13]} = -(8.3145\text{ J K}^{-1}\text{ mol}^{-1})\times(298\text{ K})\times(\ln 0.32\bar{6})$

$$= \boxed{+2.8\text{ kJ mol}^{-1}}$$

73.7(a) $ATP^{4-}(aq)+H_2O(l)\rightarrow ADP^{3-}(aq)+HPO_4^{2-}(aq)+H_3O^+(aq)\ \Delta_r G^\ominus=+10\ \text{kJ mol}^{-1}$

The biological standard state applies to the state where $a_{H^+}=10^{-7}$ and all other activities are 1 (i.e., the biological standard state, $\oplus$, has pH = 7). Hence $Q^\oplus=1\times10^{-7}$

$$\begin{aligned}\Delta_r G^\oplus &= \Delta_r G^\ominus + RT\ln Q^\oplus \quad [73.9]\\ &= 10\ \text{kJ mol}^{-1}+\left(8.3145\ \text{J K}^{-1}\ \text{mol}^{-1}\right)\times(298\ \text{K})\times\ln\left(1\times10^{-7}\right)\\ &= \boxed{-30\ \text{kJ mol}^{-1}}\end{aligned}$$

Problems

73.1 (a) $I_2(s)+Br_2(g)\rightleftharpoons 2\,IBr(g)$

$$\begin{aligned}\Delta_r G^\ominus &= -RT\ln K\ [73.13]\\ &= -(8.3145\ \text{J K}^{-1}\text{mol}^{-1})\times(298\ \text{K})\times(\ln 0.164)=4.48\times10^3\ \text{J mol}^{-1}\\ &= \boxed{+4.48\ \text{kJ mol}^{-1}}\end{aligned}$$

(b) $K=\prod_J a_J^{\nu_J}\ [73.12]=\dfrac{(p_{IBr}/p^\ominus)^2}{p_{Br_2}/p^\ominus}$ [perfect gases]

Assuming that the sublimation vapour pressure of the iodine, p_{I_2}, is negligibly small:

$$p=p_{I_2}+p_{Br_2}+p_{IBr}=p_{Br_2}+p_{IBr}\quad\text{or}\quad p_{Br_2}=p-p_{IBr}$$

Thus, $K\times\left(\dfrac{p-p_{IBr}}{p^\ominus}\right)=\left(p_{IBr}/p^\ominus\right)^2$

$$\left(p_{IBr}/p^\ominus\right)^2+K\times\left(p_{IBr}/p^\ominus\right)-K\times p/p^\ominus=0$$

$$p_{IBr}/p^\ominus=\frac{-K+\sqrt{K^2+4K\times p/p^\ominus}}{2}\quad\text{[positive solution to quadratic eqn]}$$

$$p_{IBr}=\frac{-0.164+\sqrt{(0.164)^2+4\times(0.164)\times\left(\dfrac{0.164\ \text{atm}}{1\ \text{bar}}\right)\times\left(\dfrac{1.01325\ \text{bar}}{1\ \text{atm}}\right)}}{2}\times(1\ \text{bar})$$

$$=\boxed{0.102\ \text{bar}}$$

$$\begin{aligned}p_{Br_2}&=p-p_{IBr}\\ &=0.164\ \text{atm}\times\left(\frac{1.01325\ \text{bar}}{1\ \text{atm}}\right)-0.102\ \text{bar}=0.064\ \text{bar}\end{aligned}$$

(c) In order to include the effect of iodine pressure, we must search the literature of physical data to find the sublimation vapour pressure of iodine at 25°. It is $p_{I_2} = 0.305$ Torr. Now, the following two eqns in the two unknowns (p_{Br_2} and p_{IBr}) can be solved for the unknowns.

$$p = p_{I_2} + p_{Br_2} + p_{IBr} \quad \text{and} \quad K = \left(p_{IBr}/p^{\ominus}\right)^2 \Big/ \left(p_{Br_2}/p^{\ominus}\right)$$

We can use an algebraic method like that of part (b) or a numeric solver of a mathematical software program to find the unknowns. However, we first check that the sublimation vapour pressure is significant. Rearranging Dalton's law to solve for $p_{Br_2} + p_{IBr}$:

$$\begin{aligned} p_{Br_2} + p_{IBr} &= p - p_{I_2} = 0.164 \text{ atm} - 0.305 \text{ Torr} = 0.164 \text{ atm} - 0.000401 \text{ atm} \\ &= 0.164 \text{ atm} \end{aligned}$$

This shows that the sum $p_{Br_2} + p_{IBr}$ equals the value of part (b) to within the significant figures of the data. Thus, there is no need to recalculate. The answer remains that of part (b). If you're interested in how the solution proceeds with software, survey the following Mathcad Prime 2 worksheet, which uses the symbols *A*, *B*, and *C* to represent the partial pressures of p_{I_2}, p_{Br_2} and p_{IBr}, respectively. The solution shows that the effect of the iodine sublimation vapour pressure is smaller than the uncertainty of the provided experimental data. That is, the solution is equivalent to that of part (b) when rounded to significant figures. Increase the value of *A* to find a magnitude at which the sublimation vapour pressure becomes significant in this problem (~1 Torr).

$p_{std} := 1 \cdot \text{bar}$ $\quad A := 0.305 \cdot \text{torr}$ $\quad K := 0.164$ $\quad p := 0.164 \cdot \text{atm}$

Guess Values

$B := 0.06 \cdot \text{bar}$

$C := 0.1 \cdot \text{bar}$

Constraints

$p = A + B + C$

$K \cdot p_{std} = \dfrac{B^2}{C}$

Solver

$\begin{bmatrix} B \\ C \end{bmatrix} := \text{Find}(B, C)$ $\quad B = 0.10215 \text{ bar}$ $\quad C = 0.06362 \text{ bar}$

73.3 $\quad H_2O(g) \rightarrow H_2(g) + \tfrac{1}{2}\, O_2(g) \qquad \Delta_r G^{\ominus} = +118.08 \text{ kJ mol}^{-1}$ at 2 300 K

$$K = e^{-\Delta_r G^{\ominus}/RT} \quad [73.13]$$

$$= e^{-(118\,080 \text{ kJ mol}^{-1})/(8.3145 \text{ J K}^{-1} \text{ mol}^{-1}) \times (2\,300 \text{ K})} = 0.002082$$

Draw up the following equilibrium table:

	$H_2O(g)$	$H_2(g)$	$O_2(g)$	Totals
Amounts at Start	n	—	—	n
Amounts at Equilibrium	$(1-\alpha)n$	αn	$\frac{1}{2}\alpha n$	$(1+\frac{1}{2}\alpha)n$
Mole fractions at Equilibrium	$\dfrac{(1-\alpha)}{(1+\frac{1}{2}\alpha)}$	$\dfrac{\alpha}{(1+\frac{1}{2}\alpha)}$	$\dfrac{\frac{1}{2}\alpha}{(1+\frac{1}{2}\alpha)}$	1

$$K=\prod_J a_J^{\nu_J}\ [17.12]=\frac{(p_{H_2}/p^{\ominus})(p_{O_2}/p^{\ominus})^{1/2}}{(p_{H_2O}/p^{\ominus})}\ \text{[perfect gases]}$$

$$=\frac{(x_{H_2}p/p^{\ominus})(x_{O_2}p/p^{\ominus})^{1/2}}{(x_{H_2O}p/p^{\ominus})}\ [p_J=x_Jp]$$

$$=\frac{(x_{H_2})(x_{O_2})^{1/2}}{(x_{H_2O})}\ [p=p^{\ominus}=1\text{ bar}]$$

$$=\left(\frac{\alpha}{1+\frac{1}{2}\alpha}\right)\times\left(\frac{\frac{1}{2}\alpha}{1+\frac{1}{2}\alpha}\right)^{1/2}\times\left(\frac{1+\frac{1}{2}\alpha}{1-\alpha}\right)$$

$$=(\tfrac{1}{2})^{1/2}\alpha^{3/2}\times\left(\frac{1}{1+\frac{1}{2}\alpha}\right)^{1/2}\times\left(\frac{1}{1-\alpha}\right)\qquad\text{(i)}$$

The equilibrium constant is small and, consequently, it is reasonable to estimate that $\alpha << 1$. In this case, the factors $1+\frac{1}{2}\alpha$ and $1-\alpha$ in Equation (i) are approximately equal to 1.

$$K\simeq(\tfrac{1}{2})^{1/2}\alpha^{3/2}$$

$$\alpha\simeq\left\{(\tfrac{1}{2})^{-1/2}K\right\}^{2/3}$$

$$\simeq\left\{(\tfrac{1}{2})^{-1/2}(0.002082)\right\}^{2/3}$$

$$\simeq\boxed{0.02054}$$

Alternatively, the numeric solver of a scientific calculator or software package like Mathcad can be used to acquire the numeric solution to Equation (i). It is $\alpha=0.02033$.

73.5 Draw up the following table for the reaction: $H_2(g)+I_2 \rightleftharpoons 2HI(g)\quad K=870$.

	H_2	I_2	H_I	Total
Initial amounts/mol	0.300	0.400	0.200	0.900
Change / mol	$-x$	$-x$	$+2x$	
Equilibrium amounts/mol	$0.300-x$	$0.400-x$	$0.200+2x$	0.900
Mole fraction	$\dfrac{0.300-x}{0.900}$	$\dfrac{0.400-x}{0.900}$	$\dfrac{0.200+2x}{0.900}$	1

$$K=\frac{(p_{\mathrm{HI}}/p^{\ominus})^2}{(p_{\mathrm{H_2}}/p^{\ominus})(p_{\mathrm{I_2}}/p^{\ominus})}\ \text{[perfect gases]}$$

$$=\frac{(x_{\mathrm{HI}})^2}{(x_{\mathrm{H_2}})(x_{\mathrm{I_2}})}\ [p_{\mathrm{J}}=x_{\mathrm{J}}p]=\frac{(0.200+2x)^2}{(0.300-x)(0.400-x)}=870\ \text{[given]}$$

Therefore,

$$(0.0400)+(0.800x)+4x^2=(870)\times(0.120-0.700x+x^2)\ \text{or}\ 866x^2-609.80x+104.36=0$$

which solves to $x=0.293$ [$x=0.411$ is excluded because x cannot exceed 0.300]. The final composition is therefore $\boxed{0.007\ \text{mol H}_2,}$ $\boxed{0.107\ \text{mol I}_2}$, and $\boxed{0.786\ \text{mol HI}}$

73.7 The formation reaction is $\mathrm{Si(s)+H_2(g)\rightleftharpoons SiH_2(g)}$ with the equilibrium constant

$$K=\exp(-\Delta_f G^{\ominus}/RT)=\exp(-\Delta_f H^{\ominus}/RT)\exp(\Delta_f S^{\ominus}/R).$$

Let h be the uncertainty in $\Delta_f H^{\ominus}$ so that the high value is the low value plus h. The K based on the low value is

$$K_{\mathrm{low}H}=\exp\left(\frac{-\Delta_f H^{\ominus}_{\mathrm{low}}}{RT}\right)\exp\left(\frac{\Delta_f S^{\ominus}}{R}\right)=\exp\left(\frac{-\Delta_f H^{\ominus}_{\mathrm{high}}}{RT}\right)\exp\left(\frac{h}{RT}\right)\exp\left(\frac{\Delta_f S^{\ominus}}{R}\right)$$

$$=\exp\left(\frac{h}{RT}\right)K_{\mathrm{high}H}$$

So, $$\frac{K_{\mathrm{low}\,H}}{K_{\mathrm{high}\,H}}=\exp\left(\frac{h}{RT}\right)$$

(a) At 298 K, $$\frac{K_{\mathrm{low}H}}{K_{\mathrm{high}H}}=\exp\left(\frac{(289-243)\ \mathrm{kJ\ mol^{-1}}}{(8.3145\times10^{-3}\ \mathrm{kJ\ K^{-1}\ mol^{-1}})\times(298\ \mathrm{K})}\right)=\boxed{1.2\times10^{8}}$$

(b) At 700 K, $$\frac{K_{\mathrm{low}H}}{K_{\mathrm{high}H}}=\exp\left(\frac{(289-243)\ \mathrm{kJ\ mol^{-1}}}{(8.3145\times10^{-3}\ \mathrm{kJ\ K^{-1}\ mol^{-1}})\times(700\ \mathrm{K})}\right)=\boxed{2.7\times10^{3}}$$

73.9 $a\,\mathrm{A(g)}+b\,\mathrm{B(g)}\rightleftharpoons c\,\mathrm{C(g)}+d\,\mathrm{D(g)}$

$$K=\frac{a^c_{\mathrm{C(g)}}a^d_{\mathrm{D(g)}}}{a^a_{\mathrm{A(g)}}a^b_{\mathrm{B(g)}}}$$

$$=\frac{(p_{\mathrm{C}}/p^{\ominus})^c(p_{\mathrm{D}}/p^{\ominus})^d}{(p_{\mathrm{A}}/p^{\ominus})^a(p_{\mathrm{B}}/p^{\ominus})^b}\ [\text{perfect gas},\ a_{\mathrm{J}}=p_{\mathrm{J}}/p^{\ominus}]$$

$$=\frac{p^c_{\mathrm{C}}p^d_{\mathrm{D}}}{p^a_{\mathrm{A}}p^b_{\mathrm{B}}}(p^{\ominus})^{-\Delta\nu_g}\quad \text{where}\quad \Delta\nu_g=c+d-a-b$$

$$=\frac{[\mathrm{C}]^c[\mathrm{D}]^d}{[\mathrm{A}]^a[\mathrm{B}]^b}\left(\frac{RT}{p^{\ominus}}\right)^{\Delta\nu_g}\quad [p_{\mathrm{J}}=n_{\mathrm{J}}RT/V=[J]RT]$$

$$=\frac{([\mathrm{C}]/c^{\ominus})^c([\mathrm{D}]/c^{\ominus})^d}{([\mathrm{A}]/c^{\ominus})^a([\mathrm{B}]/c^{\ominus})^b}\left(\frac{c^{\ominus}RT}{p^{\ominus}}\right)^{\Delta\nu_g} \quad [c^{\ominus}=1\ \mathrm{mol\ dm^{-3}}]$$

$$=K_c\left(c^{\ominus}RT/p^{\ominus}\right)^{\Delta\nu_g} \quad \text{where} \quad K_c=\frac{([\mathrm{C}]/c^{\ominus})^c([\mathrm{D}]/c^{\ominus})^d}{([\mathrm{A}]/c^{\ominus})^a([\mathrm{B}]/c^{\ominus})^b}$$

73.11 The fractional saturation, s, the fraction of Mb molecules (or Hb molecules) that are oxygenated, is related to the partial pressure of oxygen, p $(=p/1\ \text{Torr})$, through the Hill equation:

$$\log\frac{s}{1-s}=\upsilon_{\mathrm{H}}\log p-\upsilon_{\mathrm{H}}\log K_{\mathrm{H}} \quad \text{or} \quad p=K_{\mathrm{H}}\times\left(\frac{s}{1-s}\right)^{1/\upsilon_{\mathrm{H}}} \quad \text{or} \quad s=\frac{(p/K_{\mathrm{H}})^{\upsilon_{\mathrm{H}}}}{1+(p/K_{\mathrm{H}})^{\upsilon_{\mathrm{H}}}}$$

where ν_{H} and K_{H} are the Hill coefficients.

(a) Let $p_{½}$ be the oxygen pressure when $s=½$. Then,

$$p_{½}=K_{\mathrm{H}}\times\left(\frac{½}{1-½}\right)^{1/\upsilon_{\mathrm{H}}}$$
$$=K_{\mathrm{H}}$$

Thus, measurement of $p_{½}$ directly yields K_{H}. The plots of Problem 73.10 for $K=0.5$ give $p_{½}=2.0$ Torr for Mb and $p_{½}=36.0$ Torr for Mb and we conclude that $\boxed{K_{\mathrm{H}}(\mathrm{Mb})=2.0}$ and $\boxed{K_{\mathrm{H}}(\mathrm{Hb})=36.0}$. The fractional saturation of Mb and Hb at a given p is calculated with the equation:

$$s=\frac{(p/K_{\mathrm{H}})^{\upsilon_{\mathrm{H}}}}{1+(p/K_{\mathrm{H}})^{\upsilon_{\mathrm{H}}}}$$

The values of K_{Hill} are reported above and the values of ν_{H} are given as $\nu_{\mathrm{H}}(\mathrm{Mb})=1$ and $\nu_{\mathrm{H}}(\mathrm{Hb})=2.8$. We draw a table:

Oxygen partial pressure		s	
p/kPa	p/Torr	Mb	Hb
1.0	7.5	0.79	0.01
1.5	11.3	0.85	0.04
2.5	18.8	0.90	0.14
4.0	30.0	0.94	0.38
8.0	60.0	0.97	0.81

(b) Using the $K_{\mathrm{H}}(\mathrm{Hb})$ value for Hb reported above and the value $\nu_{\mathrm{H}}(\mathrm{Hb})=4$ in the Hill equation yields the fractional saturations reported in the following table.

p/kPa	p/Torr	s(Hb)
1.0	7.5	0.002
1.5	11.3	0.01
2.5	18.8	0.07
4.0	30.0	0.33
8.0	60.0	0.89

73.13 For the ATP hydolysis at 37°C:

$ATP(aq) + H_2O(l) \rightarrow ADP(aq) + P_i^-(aq) + H_3O^+(aq)\ \Delta_r G^{\oplus} = -31\ kJ\ mol^{-1}$

[Brief illustration 73.7]

$$\Delta_r G = \Delta_r G^{\ominus} + RT\ln Q\ \ [73.9]$$

$$\Delta_r G^{\oplus} = \Delta_r G^{\ominus} + RT\ln Q^{\oplus}$$

In Equation 73.9 molar solution concentrations are used with 1 M standard states so the standard state ($\ominus$) pH equals zero in contrast to the biological standard state ($\oplus$) of pH 7, which corresponds to $[H_3O+] = 10^{-7}$ M.

So, $$Q^{\oplus} = \frac{10^{-7}\ \text{M}}{1\ \text{M}} = 10^{-7}$$

$$\begin{aligned}\Delta_r G^{\ominus} &= \Delta_r G^{\oplus} - RT\ln Q^{\oplus}\\ &= -31\ \text{kJ mol}^{-1} - (8.3145\ \text{J K}^{-1}\ \text{mol}^{-1})\times(310\ \text{K})\ln(10^{-7})\\ &= +11\ \text{kJ mol}^{-1}\end{aligned}$$

This calculation shows that under standard conditions the hydrolysis of ATP is not spontaneous! It is endergonic. The calculation of the ATP hydrolysis free energy with the cell conditions pH = 7, $[ATP] = [ADP] = [P_i^-] = 1.0\times10^{-6}$ M, is interesting.

$$\begin{aligned}\Delta_r G &= \Delta_r G^{\ominus} + RT\ln Q = \Delta_r G^{\ominus} + RT\ln\left(\frac{[\text{ADP}]\times[\text{P}_i^-]\times[\text{H}^+]}{[\text{ATP}]\times(1\ \text{M})^2}\right)\\ &= +11\ \text{kJ mol}^{-1} + (8.3145\ \text{J K}^{-1}\ \text{mol}^{-1})\times(310\ \text{K})\ln(10^{-6}\times10^{-7})\\ &= -66\ \text{kJ mol}^{-1}\end{aligned}$$

The concentration conditions in biological cells make the hydrolysis of ATP spontaneous and very exergonic. A maximum of 66 kJ of work is available to drive coupled chemical reactions when a mole of ATP is hydrolyzed.

73.15 Yes, a bacterium can evolve to utilize the ethanol/nitrate pair to exergonically release the free energy needed for ATP synthesis. The ethanol reductant may yield any of the following products:

$$\underset{\text{ethanol}}{CH_3CH_2OH} \rightarrow \underset{\text{ethanal}}{CH_3CHO} \rightarrow \underset{\text{ethanoic acid}}{CH_3COOH} \rightarrow CO_2 + H_2O$$

The nitrate oxidant may receive electrons to yield any of the following products:

$$\underset{\text{nitrate}}{NO_3^-} \rightarrow \underset{\text{nitrite}}{NO_2^-} \rightarrow \underset{\text{dinitrogen}}{N_2} \rightarrow \underset{\text{ammonia}}{NH_3}$$

Oxidation of two ethanol molecules to carbon dioxide and water can transfer 8 electrons to nitrate during the formation of ammonia. The half-reactions and net reaction are:

$$2\,[CH_3CH_2OH(l) \rightarrow 2\,CO_2(g) + H_2O(l) + 4\,H^+(aq) + 4\,e^-]$$

$$NO_3^-(aq) + 9\,H^+(aq) + 8\,e^- \rightarrow NH_3(aq) + 3\,H_2O(l)$$

$$2\,CH_3CH_2OH(l) + H^+(aq) + NO_3^-(aq) \rightarrow 4\,CO_2(g) + 5\,H_2O(l) + NH_3(aq)$$

$\Delta_r G^\ominus = -2331.29$ kJ for the reaction as written (data of the resource section). Of course, enzymes must evolve that couple this exergonic redox reaction to the production of ATP, which would then be available for carbohydrate, protein, lipid, and nucleic acid synthesis.

Topic 74 The statistical description of equilibrium

Discussion question

74.1 See Justification 74.1 for a derivation of the general expression (eqn 74.4b) for the equilibrium constant in terms of the partition functions and difference in molar energy, $\Delta_r E_0$, of the products and reactants in a chemical reaction. The partition functions are functions of temperature and the ratio of partition functions in eqn 74.4b will therefore vary with temperature. However, the most direct effect of temperature on the equilibrium constant is through the exponential term $e^{-\Delta_r E_0/RT}$. The manner in which both factors affect the magnitudes of the equilibrium constant and its variation with temperature is described in detail for a simple $R \rightleftharpoons P$ gas phase equilibrium in Topic 74.2 and Justification 74.2.

Exercises

74.1(a) $I_2(g) \rightarrow 2\,I(g)$

Data : $\tilde{\nu} = 214.36\ \text{cm}^{-1}, \tilde{B} = 0.0373\ \text{cm}^{-1}, D_e = 1.5422\ \text{eV}, T = 1\,000\ \text{K}$

$$kT/p^{\ominus} = k \times (1\,000\ \text{K})/p^{\ominus} = 1.3807 \times 10^{-25}\ \text{m}^3$$

$$D_0/RT = \left(D_e - \tfrac{1}{2}hc\tilde{\nu}\right)N_A/RT = 17.7422$$

$\sigma = 2$ for a homonuclear diatomic molecule

$g_e(\text{I}) = 4$ and $g_e(\text{I}_2) = 1$

Thermal wavelengths [52.7b]

$$\Lambda(\text{I}) = h/\left(2\pi m_1 kT\right)^{1/2} = 4.9008 \times 10^{-12}\ \text{m}$$

$$\Lambda(\text{I}_2) = h/\left(2\pi m_{\text{I}_2} kT\right)^{1/2} = 3.4654 \times 10^{-12}\ \text{m}$$

Partition functions:

$$q^{\text{R}}(\text{I}_2) = kT/(\sigma hc\tilde{B})\ [52.13\text{b}] = 9\,317$$

$$q^{\text{V}}(\text{I}_2) = (1 - e^{-hc\tilde{\nu}/kT})^{-1}\ [52.15] = 3.7681$$

Equilibrium constant:

$$K = \frac{kT}{p^{\ominus}}\left(\frac{g_e(X)^2}{g_e(X_2)}\right)\left(\frac{1}{q^V(X_2)q^R(X_2)}\right)\left(\frac{\Lambda(X_2)}{\Lambda(X)^2}\right)^3 e^{-D_0(X-X)/RT} \quad [74.6]$$

$$= (1.3807\times10^{-25}\ m^3)\times\left(\frac{4^2}{1}\right)\times\left(\frac{1}{(3.7681)\times(9\,317)}\right)\times\left(\frac{3.4654\times10^{-12}\ m}{(4.9008\times10^{-12}\ m)^2}\right)^3 e^{-17.7422}$$

$$= \boxed{0.003725}$$

Problems

74.1 $CD_4(g) + HCl(g) \rightarrow CHD_3(g) + DCl(g)$

$$K = \frac{q_m^{\ominus}(CHD_3)q_m^{\ominus}(DCl)}{q_m^{\ominus}(CD_4)q_m^{\ominus}(HCl)} e^{-\Delta_r E_0/kT} \quad [74.4b;\ N_A \text{ factors cancel}]$$

Each of these partition functions is a product $q_m^{\ominus} = q_m^T q^R q^V q^E$ with all $q^E = 1$. The T, R, and V partition functions are given by

$$q^T = \frac{V}{\Lambda^3} \quad [52.10b] \quad \text{where } \Lambda = \frac{h}{(2\pi mkT)^{1/2}} \quad [52.7b]$$

$q^R = kT/(\sigma hc\tilde{B})$ [52.13b] where $\sigma_{XY} = 1$ and $\sigma_{XX} = 2$ for a linear rotor

$$q^R = \frac{1}{\sigma}\left(\frac{kT}{hc}\right)^{3/2}\left(\frac{\pi}{\tilde{A}\tilde{B}^2}\right)^{1/2}$$ for a symmetrical rotor (CHD_3, $\sigma = 3$, Justification 52.3, [52.14])

$$q^R = \frac{1}{\sigma}\left(\frac{kT}{hc}\right)^{3/2}\left(\frac{\pi}{\tilde{B}^3}\right)^{1/2}$$ for a spherical rotor (CD_4, $\sigma = 12$, [52.12b], [52.14])

$$q^V = (1 - e^{-hc\tilde{\nu}/kT})^{-1} \quad [52.15]$$

$\Delta_r E_0$ equals the difference in zero-point energies.

$$\frac{\Delta_r E_0}{hc} = \frac{1}{2}\left\{\begin{matrix}(2\,993 + 2\,142 + 3\times1\,003 + 2\times1\,291 + 2\times1\,036 + 2\,145) \\ -(2\,109 + 2\times1\,092 + 3\times2\,259 + 3\times996 + 2\,991)\end{matrix}\right\} cm^{-1}$$

$$= -1\,053\ cm^{-1}$$

and $$-\frac{\Delta_r E_0}{kT} = -\frac{hc}{k}\times\frac{\Delta_r E_0}{hc}\times\frac{1}{T} = -\frac{1.4388\times(-1\,053)}{T/K} = +\frac{1515}{T/K}$$

The ratio of translational partition functions is

$$\frac{q_m^T(CHD_3)q_m^T(DCl)}{q_m^T(CD_4)q_m^T(HCl)} = \left(\frac{M(CHD_3)M(DCl)}{M(CD_4)M(HCl)}\right)^{3/2} = \left(\frac{19.06\times37.46}{20.07\times36.46}\right)^{3/2} = 0.964$$

The ratio of rotational partition functions is

$$\frac{q^{R}(CHD_3)q^{R}(DCl)}{q^{R}(CD_4)q^{R}(HCl)} = \frac{\sigma(CD_4)}{\sigma(CHD_3)}\frac{(\tilde{B}(CD_4)/cm^{-1})^{3/2}\,\tilde{B}(HCl)/cm^{-1}}{(\tilde{A}(CHD_3)\tilde{B}(CHD_3)^2/cm^{-3})^{1/2}\,\tilde{B}(DCl)/cm^{-1}}$$

$$= \frac{12}{3}\times\frac{2.63^{3/2}\times 10.59}{(2.63\times 3.28^2)^{1/2}\times 5.445} = 6.24$$

The ratio of vibrational partition functions (Call it Q^{V} for convenience below) is

$$Q^{V} = \frac{q^{V}(CHD_3)q^{V}(DCl)}{q^{V}(CD_4)q^{V}(HCl)} = \frac{q(2\,993)q(2\,142)q(1\,003)^3 q(1\,291)^2 q(1\,036)^2 q(2\,145)}{q(2\,109)q(1\,092)^2 q(2\,259)^3 q(996)^3 q(2\,991)}$$

where $q(x) = \dfrac{1}{1-e^{-1.4388x/(T/K)}}$

The equilibrium constant is

$$K = 0.964\times 6.24\times Q^{V}e^{+1515/(T/K)} = 6.02Q^{V}e^{+1515/(T/K)}$$

We can now evaluate K (on a computer), and obtain the following values:

T/K	300	400	500	600	700	800	900	1 000
K	945	273	132	83	61	49	42	37

The values of $K(T)$ are plotted in Fig. 74.1.

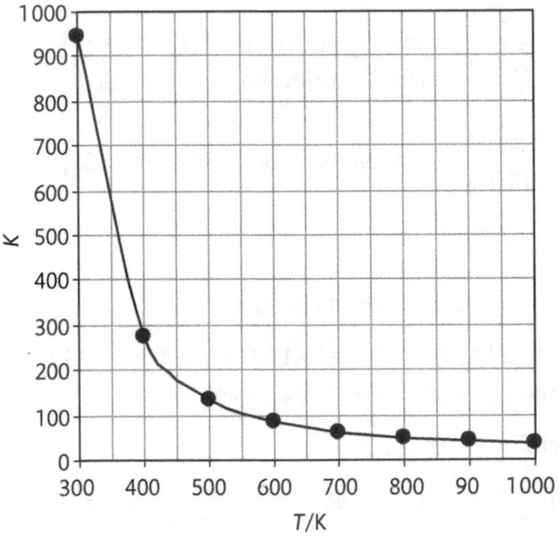

Fig 74.1

Topic 75 The response of equilibria to the conditions

Discussion questions

75.1 (1) Response to change in pressure. The equilibrium constant is independent of pressure, but the individual partial pressures can change as the total pressure changes. This will happen when there is a difference, $\Delta\nu_g$, between the sums of the number of moles of gases on the product and reactant sides of the balanced chemical reaction equation:

$$\Delta\nu_g = \sum_{J=\text{product gases}} \nu_J - \sum_{J=\text{reactant gases}} |\nu_J|$$

The requirement of an unchanged equilibrium constant implies that the side with the smaller number of moles of gas be favored as pressure increases. To see this, we examine the general reaction equation $0 = \sum_J \nu_J J$ [73.8] in the special case for which all reactants and products are perfect gases. In this case the activities of eqn 73.12 equal the partial pressure of the gaseous species and, therefore,

$$a_{J(\text{gas})} = p_J/p^\ominus = x_J p/p^\ominus$$

where x_J is the mole fraction of gaseous species J. Substitution into eqn 73.12 and simplification yields a useful equation:

$$K = \left(\prod_J a_J^{\nu_J}\right)_{\text{equilibrium}} = \left(\prod_J x_J^{\nu_J}\left(p/p^\ominus\right)^{\nu_J}\right)_{\text{equilibrium}}$$

$$= \left(\prod_J x_J^{\nu_J}\right)_{\text{equilibrium}}\left(\prod_J \left(p/p^\ominus\right)^{\nu_J}\right)_{\text{equilibrium}} = \left(\prod_J x_J^{\nu_J}\right)_{\text{equilibrium}}\left(p/p^\ominus\right)^{\Delta\nu_g}$$

$$= K_x\left(p/p^\ominus\right)^{\Delta\nu_g} \quad \text{where} \quad K_x = \left(\prod_J x_J^{\nu_J}\right)_{\text{equilibrium}}$$

K_x is not an equilibrium constant. It is a ratio of product and reactant concentration factors that has a form analogous to the equilibrium constant K.

However, whereas K depends upon temperature alone, the concentration ratio K_x depends upon both temperature and pressure. Solving for K_x provides an equation that directly indicates its pressure dependence.

$$K_x = K\left(p/p^\ominus\right)^{-\Delta\nu_g}$$

This equation indicates that, if $\Delta\nu_g = 0$ (an equal number of gas moles on both sides of the balanced reaction equation), $K_x = K$ and the concentration ratio has no pressure dependence. An increase in pressure causes no change in K_x and no shift in the concentration equilibrium is observed upon a change in pressure. However this equation indicates that, if $\Delta\nu_g < 0$ (fewer moles of gas on the product side of the balanced reaction equation), $K_x = K\left(p/p^\ominus\right)^{|\Delta\nu_g|}$. Because p is raised to a positive power in this case, an increase in pressure causes K_x to increase. This means that the numerator concentrations (products) must increase while the denominator concentrations (reactants) decrease. The concentrations shift to the product side to reestablish equilibrium when an increase in pressure has stressed the reaction equilibrium. Similarly, if $\Delta\nu_g >$ 0 (fewer moles of gas on the reactant side of the balanced reaction equation), $K_x = K\left(p/p^\ominus\right)^{-|\Delta\nu_g|}$. Because p is raised to a negative power in this case, the concentrations now shift to the reactant side to reestablish equilibrium when an increase in pressure has stressed the reaction equilibrium.

(2) Response to change in temperature. Equation 75.3 shows that K decreases with increasing temperature when the reaction is exothermic; thus the reaction shifts to the left, the opposite occurs in endothermic reactions.

Failure of Le Chatelier's principle is very unusual. The expected response to pressure may prove wrong should the gas components be non-perfect, real gases. Should the gases of the reaction side that is favored by Le Chatelier's principle strongly repel while the gas molecules of the other side strongly attract, a compression may shift the reaction in the direction opposite to that expected for perfect gases. This is reflected in the analysis of eqn 73.12 for the special case in which all products and gases are real gases having activity coefficients that do not equal 1. Coefficients greater than 1 are observed when intermolecular repulsions dominate, and they are less than 1 when attractive forces dominate. In this case, $a_{J(gas)} = \gamma_J p_J/p^\ominus = \gamma_J x_J p/p^\ominus$ where γ_J is the activity coefficient of gas J. Substitution and simplification yields

$$K_x = K\left(p/p^\ominus\right)^{-\Delta\nu_g}/K_\gamma \quad \text{where} \quad K_\gamma = \prod_J \gamma_J^{\nu_J}$$

If $\Delta\nu_g < 0$ (fewer moles of non-perfect gas on the product side of the balanced reaction equation), a pressure increase will not shift the reaction equilibrium toward the products, which is predicted by Le Chatelier's principle, provided that the pressure increase causes K_γ to increase to the extent that the ratio $\left(p/p^\ominus\right)^{-\Delta\nu_g}/K_\gamma$ decreases.

Exercises

75.1(a) For the reaction $H_2CO(g) \rightarrow CO(g) + H_2(g)$, $\Delta\nu_g = 1$

Assuming that all reactants and products are perfect gases, the activities of eqn 73.12 equal the partial pressure of the gaseous species and, therefore,

$$a_{J(gas)} = p_J/p^{\ominus} = x_J p/p^{\ominus}$$

where x_J is the mole fraction of gaseous species J. Substitution into eqn 73.12 and simplification yields a useful equation:

$$K = \left(\prod_J a_J^{\nu_J}\right)_{\text{equilibrium}} = \left(\prod_J x_J^{\nu_J}\left(p/p^{\ominus}\right)^{\nu_J}\right)_{\text{equilibrium}}$$

$$= \left(\prod_J x_J^{\nu_J}\right)_{\text{equilibrium}} \left(\prod_J \left(p/p^{\ominus}\right)^{\nu_J}\right)_{\text{equilibrium}} = \left(\prod_J x_J^{\nu_J}\right)_{\text{equilibrium}} \left(p/p^{\ominus}\right)^{\Delta\nu_g}$$

$$= K_x\left(p/p^{\ominus}\right)^{\Delta\nu_g} \quad \text{where } K_x = \left(\prod_J x_J^{\nu_J}\right)_{\text{equilibrium}} \quad \text{and } \Delta\nu_g = \sum_{\text{J=product gases}} \nu_J - \sum_{\text{J=reactant gases}} |\nu_J|$$

K_x is not an equilibrium constant. It is a ratio of product and reactant concentration factors that has a form analogous to the equilibrium constant K. Therefore,

$$K_x = K\left(\frac{p}{p^{\ominus}}\right)^{-\Delta\nu_g}$$

Since K depends upon temperature alone,

$$\frac{K_x(p_2)}{K_x(p_1)} = \frac{K\left(p_2/p^{\ominus}\right)^{-\Delta\nu_g}}{K\left(p_1/p^{\ominus}\right)^{-\Delta\nu_g}} = \left(\frac{p_2}{p_1}\right)^{-\Delta\nu_g}$$

$$\frac{K_x(3.0\text{ bar})}{K_x(1.0\text{ bar})} = \left(\frac{3.0\text{ bar}}{1.0\text{ bar}}\right)^{-1} = 0.33$$

Thus, $\boxed{K_x \text{ is reduced by } 67\,\%}$ when the pressure is increased from 1.0 bar to 3.0 bar. This is quantification of the Le Chatelier's principle that the equilibrium shifts to the left (side that has smaller number of gas moles) to reduce the stress of increased pressure.

75.2(a) At 1 280 K, $\Delta_r G^{\ominus} = +33\times10^3\text{ J mol}^{-1}$; thus

$$\ln K_1(1\,280\text{K}) = -\frac{\Delta_r G^{\ominus}}{RT} = -\frac{33\times10^3\text{ J mol}^{-1}}{(8.3145\text{ J K}^{-1}\text{mol}^{-1})\times(1\,280\text{ K})} = -3.1\bar{0}$$

$$K_1 = \boxed{0.045}$$

$$\ln K_2 = \ln K_1 - \frac{\Delta_r H^{\ominus}}{R}\left(\frac{1}{T_2} - \frac{1}{T_1}\right) \quad [75.5]$$

We look for the temperature T_2 that corresponds to $\ln K_2 = \ln(1) = 0$. This is the crossover temperature. Solving for T_2 from eqn 75.5 with $\ln K_2 = 0$, we obtain

$$\frac{1}{T_2} = \frac{R\ln K_1}{\Delta_r H^\ominus} + \frac{1}{T_1} = \left(\frac{(8.3145\ \mathrm{J\,K^{-1}\,mol^{-1}})\times(-3.1\bar{0})}{224\times10^3\ \mathrm{J\,mol^{-1}}}\right) + \left(\frac{1}{1280\ \mathrm{K}}\right)$$

$$= 6.6\bar{6}\times10^{-4}\ \mathrm{K^{-1}}$$

$$T_2 = \boxed{1\bar{5}\bar{0}0\ \mathrm{K}}$$

75.3(a) Given $\ln K = -1.04 - \frac{1088\ \mathrm{K}}{T} + \frac{1.51\times10^5\ \mathrm{K^2}}{T^2}$,

and since $\frac{\mathrm{d}\ln K}{\mathrm{d}(1/T)} = \frac{-\Delta_r H^\ominus}{R}$ [75.3]

$$\frac{-\Delta_r H^\ominus}{R} = -1088\ \mathrm{K} + \frac{(2)\times(1.51\times10^5\ \mathrm{K^2})}{T}$$

Then, at 450 K

$$\Delta_r H^\ominus = \left(1088\ \mathrm{K} - \frac{3.02\times10^5\ \mathrm{K^2}}{450\ \mathrm{K}}\right)\times(8.3145\ \mathrm{J\,K^{-1}\,mol^{-1}}) = \boxed{+3.47\ \mathrm{kJ\,mol^{-1}}}$$

$$\Delta_r G^\ominus = -RT\ln K\ [17.13] = RT\times\left(1.04 + \frac{1088\ \mathrm{K}}{T} - \frac{1.51\times10^5\ \mathrm{K^2}}{T^2}\right)$$

$$= (8.3145\ \mathrm{J\,K^{-1}\,mol^{-1}})\times(450\ \mathrm{K})\times\left(1.04 + \frac{1088\ \mathrm{K}}{450\ \mathrm{K}} - \frac{1.51\times10^5\ \mathrm{K^2}}{(450\ \mathrm{K})^2}\right) = +10.15\ \mathrm{kJ\,mol^{-1}}$$

$$= \Delta_r H^\ominus - T\Delta_r S^\ominus\ [15.50]$$

Therefore, $\Delta_r S^\ominus = \frac{\Delta_r H^\ominus - \Delta_r G^\ominus}{T} = \frac{3.47\ \mathrm{kJ\,mol^{-1}} - 10.15\ \mathrm{kJ\,mol^{-1}}}{450\ \mathrm{K}}$

$$= \boxed{-14.8\ \mathrm{J\,K^{-1}\,mol^{-1}}}$$

75.4(a) $\ln\frac{K_2}{K_1} = -\frac{\Delta_r H^\ominus}{R}\left(\frac{1}{T_2} - \frac{1}{T_1}\right)$ [75.5]

Therefore, $\Delta_r H^\ominus = -\frac{R\ln\left(\frac{K_2}{K_1}\right)}{\left(\frac{1}{T_2} - \frac{1}{T_1}\right)}$

$T_2 = 308$ K; hence, with the substitution $K_2/K_1 = \kappa$

$$\Delta_r H^\ominus = -\frac{(8.3145\ \mathrm{J\,K^{-1}\,mol^{-1}})\times(\ln\kappa)}{\left(\frac{1}{308\ \mathrm{K}} - \frac{1}{298\ \mathrm{K}}\right)} = 76.3\ \mathrm{kJ\,mol^{-1}}\times\ln\kappa$$

Therefore,

(a) $\kappa = 2,\ \Delta_r H^\ominus = (76.3\ \text{kJ mol}^{-1})\times\ln 2 = \boxed{+53\ \text{kJ mol}^{-1}}$

(b) $\kappa = \tfrac{1}{2},\ \Delta_r H^\ominus = (76.3\ \text{kJ mol}^{-1})\times\ln\tfrac{1}{2} = \boxed{-53\ \text{kJ mol}^{-1}}$

75.5(a) The decomposition reaction is $CaCO_3(s) \rightleftharpoons CaO(s) + CO_2(g)$

For the purposes of this exercise we may assume that the required temperature is that temperature at which $K = 1$, which corresponds to a pressure of 1 bar for the gaseous product, $\ln K = 0$, and $\Delta_r G^\ominus = 0$.

$$\Delta_r G^\ominus = \Delta_r H^\ominus - T\Delta_r S^\ominus = 0 \quad \text{when}\ \Delta_r H^\ominus = T\Delta_r S^\ominus$$

Therefore, the decomposition temperature (when $K = 1$) is

$$T = \frac{\Delta_r H^\ominus}{\Delta_r S^\ominus}$$

$$\Delta_r H^\ominus = \{(-635.09) - (393.51) - (-1\,206.9)\}\ \text{kJ mol}^{-1} = +178.3\ \text{kJ mol}^{-1}$$

$$\Delta_r S^\ominus = \{(39.75) + (213.74) - (92.9)\}\ \text{J K}^{-1}\ \text{mol}^{-1} = +160.6\ \text{J K}^{-1}\ \text{mol}^{-1}$$

$$T = \frac{178.3\times10^3\ \text{J mol}^{-1}}{160.6\ \text{J K}^{-1}\ \text{mol}^{-1}} = \boxed{1110\ \text{K}}\ (840\ ^\circ\text{C})$$

75.6(a) The reaction equation is: $PbO(s) + CO(g) \rightleftharpoons Pb(s) + CO_2(g)$.

$$\nu_{Pb} = 1, \quad \nu_{CO_2} = 1, \quad \nu_{PbO} = -1, \quad \nu_{CO} = -1$$

(a) $\Delta_r G^\ominus = \sum_J \nu_J\, \Delta_f G^\ominus(J)$ [73.10]

$$\begin{aligned}\Delta_r G^\ominus &= \Delta_f G^\ominus(\text{Pb, s}) + \Delta_f G^\ominus(\text{CO}_2,\ \text{g}) - \Delta_f G^\ominus(\text{PbO, s, red}) - \Delta_f G^\ominus(\text{CO, g})\\ &= (-394.36\ \text{kJ mol}^{-1}) - (-188.93\ \text{kJ mol}^{-1}) - (-137.17\ \text{kJ mol}^{-1})\\ &= \boxed{-68.26\ \text{kJ mol}^{-1}}\end{aligned}$$

$$\ln K = \frac{-\Delta_r G^\ominus}{RT}\ [73.13] = \frac{+68.26\times10^3\ \text{J mol}^{-1}}{(8.3145\ \text{J K}^{-1}\ \text{mol}^{-1})\times(298\ \text{K})} = 27.55; \quad K = \boxed{9.2\times10^{11}}$$

(b)
$$\begin{aligned}\Delta_r H^\ominus &= \Delta_f H^\ominus(\text{Pb, s}) + \Delta_f H^\ominus(\text{CO}_2,\ \text{g}) - \Delta_f H^\ominus(\text{PbO, s, red}) - \Delta_f H^\ominus(\text{CO, g})\\ &= (-393.51\ \text{kJ mol}^{-1}) - (-218.99\ \text{kJ mol}^{-1}) - (-110.53\ \text{kJ mol}^{-1})\\ &= \boxed{-63.99\ \text{kJ mol}^{-1}}\end{aligned}$$

$$\begin{aligned}\ln K(400\ \text{K}) &= \ln K(298\ \text{K}) - \frac{\Delta_r H^\ominus}{R}\left(\frac{1}{400\ \text{K}} - \frac{1}{298\ \text{K}}\right)\ [75.5]\\ &= 27.55 - \left(\frac{-63.99\times10^3\ \text{J mol}^{-1}}{8.314\ \text{J K}^{-1}\ \text{mol}^{-1}}\right)\times(-8.55\overline{7}\times10^{-4}\ \text{K}^{-1}) = 20.9\overline{6}\end{aligned}$$

$$K(400\ \text{K}) = \boxed{1.3\times10^9}$$

$$\Delta_r G^{\ominus}(400\ \text{K}) = -RT\ln K(400\ \text{K})\ [17.13] = -(8.3145\ \text{J K}^{-1}\ \text{mol}^{-1})\times(400\ \text{K})\times(20.9\bar{6})$$
$$= -6.97\times10^4\ \text{J mol}^{-1} = \boxed{-69.7\ \text{kJ mol}^{-1}}$$

Problems

75.1 $NH_4Cl(s) \rightleftharpoons NH_3(g) + HCl(g)$

$$p = p_{NH_3} + p_{HCl} = 2p_{NH_3} \quad [p_{NH_3} = p_{HCl}]$$

(a) $K = \prod_J a_J^{\nu_J}$ [73.12]; $a_{J(g)} = p_J/p^{\ominus}$ for perfect gases; $a_{NH_4Cl(s)} = 1$

$$= \left(\frac{p_{NH_3}}{p^{\ominus}}\right)\times\left(\frac{p_{HCl}}{p^{\ominus}}\right) = \frac{p_{NH_3}^2}{p^{\ominus 2}} = \frac{1}{4}\times\left(\frac{p}{p^{\ominus}}\right)^2$$

At 427 °C (700 K), $K = \frac{1}{4}\times\left(\frac{608\ \text{kPa}}{100\ \text{kPa}}\right)^2 = \boxed{9.24}$

At 459 °C (732 K), $K = \frac{1}{4}\times\left(\frac{1\,115\ \text{kPa}}{100\ \text{kPa}}\right)^2 = \boxed{31.08}$

(b) $\Delta_r G^{\ominus} = -RT\ln K$ [73.13]

$$= (-8.3145\ \text{J K}^{-1}\ \text{mol}^{-1})\times(700\ \text{K})\times\ln(9.24)$$
$$= \boxed{-12.9\ \text{kJ mol}^{-1}}\ \text{at } 427\,°\text{C}$$

(c) $\Delta_r H^{\ominus} = \dfrac{R\ln\frac{K_2}{K_1}}{\left(\frac{1}{T_1}-\frac{1}{T_2}\right)}$ [75.5]

$$= \frac{(8.3145\ \text{J K}^{-1}\ \text{mol}^{-1})\times\ln(31.08/9.24)}{\left(\frac{1}{700\ \text{K}}-\frac{1}{732\ \text{K}}\right)} = \boxed{+161\ \text{kJ mol}^{-1}}$$

(d) $\Delta_r S^{\ominus} = \dfrac{\Delta_r H^{\ominus} - \Delta_r G^{\ominus}}{T} = \dfrac{(161\ \text{kJ mol}^{-1}) - (-12.9\ \text{kJ mol}^{-1})}{700\ \text{K}} = \boxed{+248\ \text{J K}^{-1}\ \text{mol}^{-1}}$

75.3 $U(s) + \tfrac{3}{2}\,H_2(g) \rightleftharpoons UH_3(s)$

$$K = a_{H_2}^{-3/2} = (p_{H_2}/p^{\ominus})^{-3/2}\ [\text{perfect gas and } p_{H_2} = p]$$
$$= (p/p^{\ominus})^{-3/2}$$

$$\Delta_f H^{\ominus} = RT^2 \frac{\mathrm{d}\ln K}{\mathrm{d}T} \ [75.3] = RT^2 \frac{\mathrm{d}}{\mathrm{d}T}\ln\left(p/p^{\ominus}\right)^{-3/2} = -\tfrac{3}{2}RT^2 \frac{\mathrm{d}}{\mathrm{d}T}\left(\ln p - \ln p^{\ominus}\right)$$

$$= -\tfrac{3}{2}RT^2 \frac{\mathrm{d}}{\mathrm{d}T}(\ln p)$$

$$= -\tfrac{3}{2}RT^2 \frac{\mathrm{d}}{\mathrm{d}T}(A + B/T + C\ln(T/\mathrm{K})) = -\tfrac{3}{2}RT^2 \times \left(\frac{-B}{T^2} + \frac{C}{T}\right)$$

$$= \boxed{\tfrac{3}{2}R \times (B - CT)} \quad \text{where} \quad B = -1.464 \times 10^4 \text{ K} \quad \text{and} \quad C = -5.65$$

$$\Delta_r C_p^{\ominus} = \left(\frac{\partial \Delta_f H^{\ominus}}{\partial T}\right)_p \ [\text{from } 56.3] = -\frac{3}{2}CR = \boxed{70.5 \text{ J K}^{-1} \text{ mol}^{-1}}$$

75.5 $\mathrm{CaCl_2 \cdot NH_3(s) \rightleftharpoons CaCl_2(s) + NH_3(g)}$ $\quad K = \frac{p}{p^{\ominus}}$ and $\Delta_r H^{\ominus} = +78 \text{ kJ mol}^{-1}$

$$\Delta_r G^{\ominus} = -RT\ln K = -RT\ln\frac{p}{p^{\ominus}}$$

$$\Delta_r G^{\ominus}(400\text{ K}) = -(8.3145 \text{ J K}^{-1}\text{ mol}^{-1}) \times (400\text{ K}) \times \ln\left(\frac{1.71\text{ kPa}}{100.0\text{ kPa}}\right) \quad [p^{\ominus} = 1\text{ bar} = 100.0\text{ kPa}]$$

$$= +13.5 \text{ kJ mol}^{-1}$$

$$\frac{\Delta_r G_{\ominus}(T_2)}{T_2} - \frac{\Delta_r G^{\ominus}(T_1)}{T_1} = \Delta_r H^{\ominus}\left(\frac{1}{T_2} - \frac{1}{T_1}\right) \ [75.5 \text{ and } 73.13]$$

Therefore, taking $T_1 = 400$ K and letting $T = T_2$ be any temperature in the range 350 K to 470 K,

$$\Delta_r G^{\ominus}(T) = \left(\frac{T}{400\text{ K}}\right) \times (13.5 \text{ kJ mol}^{-1}) + (78 \text{ kJ mol}^{-1}) \times \left(1 - \frac{T}{400\text{ K}}\right)$$

$$= (78 \text{ kJ mol}^{-1}) + \left(\frac{(13.5 - 78)\text{ kJ mol}^{-1}}{400}\right) \times \left(\frac{T}{\text{K}}\right)$$

That is, $\Delta_r G^{\ominus}(T)/(\text{kJ mol}^{-1}) = \boxed{78 - 0.161 \times (T/\text{K})}$

75.7 The equilibrium we need to consider is $\mathrm{I_2(g) \rightleftharpoons 2\,I(g)}$ ($M_\mathrm{I} = 126.90 \text{ g mol}^{-1}$). It is convenient to express the equilibrium constant in terms of α, the degree of dissociation of $\mathrm{I_2}$, which is the predominant species at low temperatures. Recognizing that the data $n_{\mathrm{I_2}}$ is related to the total iodine mass, m_I, by $n_{\mathrm{I_2}} = m_\mathrm{I}/M_{\mathrm{I_2}}$ we draw the following table.

	I	$\mathrm{I_2}$	**Total**
Equilibrium amounts	$2\alpha n_{\mathrm{I_2}}$	$(1-\alpha)n_{\mathrm{I_2}}$	$(1+\alpha)n_{\mathrm{I_2}}$
Mole fraction	$\frac{2\alpha}{1+\alpha}$	$\frac{1-\alpha}{1+\alpha}$	1
Partial pressure	$\frac{2\alpha p}{1+\alpha}$	$\left(\frac{1-\alpha}{1+\alpha}\right)p$	p

The equilibrium constant for the dissociation is

$$K = \frac{(p_I/p^{\ominus})^2}{p_{I_2}/p^{\ominus}} = \frac{p_I^2}{p_{I_2}p^{\ominus}} = \frac{4\alpha^2(p/p^{\ominus})}{1-\alpha^2}$$

We also know that

$$pV = n_{\text{total}}RT = (1+\alpha)n_{I_2}RT$$

Implying that $\alpha = \dfrac{pV}{n_{I_2}RT} - 1$ where $V = 342.68\ \text{cm}^3$. The provided data along with calculated values of α and $K(T)$ are summarized in the following table.

T/K	973	1 073	1 173
p/atm	0.06244	0.07500	0.09181
$10^4\ n_{I_2}$/ mol	2.4709	2.4555	2.4366
α	0.08459	0.1887	0.3415
K	1.82×10^{-3}	1.12×10^{-2}	4.91×10^{-2}

Since $\Delta_r H^{\ominus}$ is expected to be approximately a constant over this temperature range and since $\Delta_r H^{\ominus} = -R\left(\dfrac{\mathrm{d}\ln K}{\mathrm{d}(1/T)}\right)$ [75.3], a plot of $\ln K$ against $1/T$ should be linear with slope $= -\Delta_r H^{\ominus}/R$. The linear regression fit to the plot is found to be $\ln K = 13.027-(18\,809\ \text{K})/T$ with $R^2 = 0.999969$. Thus,

$$\Delta_r H^{\ominus} = -(-18\ 809\ \text{K})R = \boxed{+156\ \text{kJ mol}^{-1}}$$

75.9 $$\Delta_r H(T') = \Delta_r H(T) + \int_T^{T'} \Delta_r C_p(T)\mathrm{d}T \quad [57.12;\ \Delta_r C_p(T) = \sum_J \nu_J C_{p,J}]$$

$$\Delta_r S(T') = \Delta_r S(T) + \int_T^{T'} \frac{\Delta_r C_p(T)}{T}\mathrm{d}T \quad \text{[63.1 applied to reaction equations]}$$

$$\begin{aligned}\Delta_r G(T') &= \Delta_r H(T') - T'\Delta_r S(T')\\ &= \left\{\Delta_r H(T) + \int_T^{T'} \Delta_r C_p(T)\mathrm{d}T\right\} - T'\left\{\Delta_r S(T) + \int_T^{T'} \frac{\Delta_r C_p(T)}{T}\mathrm{d}T\right\}\\ &= \Delta_r H(T) - T\Delta_r S(T) - (T'-T)\Delta_r S(T) + \int_T^{T'} \Delta_r C_p(T)\mathrm{d}T - T'\int_T^{T'} \frac{\Delta_r C_p(T)}{T}\mathrm{d}T\\ &= \Delta_r G(T) - (T'-T)\Delta_r S(T) + \int_T^{T'} \Delta_r C_p(T)\mathrm{d}T - T'\int_T^{T'} \frac{\Delta_r C_p(T)}{T}\mathrm{d}T\\ &= \Delta_r G(T) - (T'-T)\Delta_r S(T) + \int_T^{T'} \left(1-\frac{T'}{T}\right)\Delta_r C_p(T)\mathrm{d}T\end{aligned}$$

$$\Delta_r C_p(T)) = \Delta a + T\Delta b + \frac{\Delta c}{T^2}$$

$$\left(1-\frac{T'}{T}\right)\Delta_r C_p(T) = \Delta a + T\Delta b + \frac{\Delta c}{T^2} - \frac{T'\Delta a}{T} - T'\Delta b - \frac{T'\Delta c}{T^3}$$

$$= \Delta a - T'\Delta b + T\Delta b - \frac{T'\Delta a}{T} + \frac{\Delta c}{T^2} - \frac{T'\Delta c}{T^3}$$

$$\int_T^{T'}\left(1-\frac{T'}{T}\right)\Delta_r C_p(T)\mathrm{d}T = (\Delta a - T'\Delta b)(T'-T) + \tfrac{1}{2}(T'^2 - T^2)\Delta b - T'\Delta a \ln\frac{T'}{T} + \Delta c\left(\frac{1}{T}-\frac{1}{T'}\right) - \tfrac{1}{2}T'\Delta c\left(\frac{1}{T^2}-\frac{1}{T'^2}\right)$$

Therefore,

$$\boxed{\Delta_r G(T') = \Delta_r G(T) + (T-T')\Delta_r S(T) + \alpha(T',T)\times\Delta a + \beta(T',T)\times\Delta b + \gamma(T',T)\times\Delta c}$$

where $\alpha(T',T) = T' - T - T'\ln\frac{T'}{T}$,

$\beta(T',T) = \frac{1}{2}(T'^2 - T^2) - T'(T'-T)$, and $\gamma(T',T) = \frac{1}{T} - \frac{1}{T'} + \tfrac{1}{2}T'\left(\frac{1}{T'^2} - \frac{1}{T^2}\right)$

Applying this to the formation reaction of water:

$$H_2(g) + \tfrac{1}{2}O_2(g) \rightarrow H_2O(l)$$

$$\Delta_f G^{\ominus}(298\ \text{K}) = -237.13\ \text{kJ mol}^{-1} \ \text{and}\ \Delta_r S^{\ominus}(298\ \text{K}) = -163.34\ \text{J K}^{-1}\ \text{mol}^{-1}$$

$$\Delta a = a(H_2O) - a(H_2) - \tfrac{1}{2}a(O_2) = (75.29 - 27.88 - 14.98)\ \text{J K}^{-1}\ \text{mol}^{-1}$$

$$= +33.03\ \text{J K}^{-1}\ \text{mol}^{-1}$$

$$\Delta b = [(0) - (3.26\times10^{-3}) - (2.09\times10^{-3})]\ \text{J K}^{-2}\ \text{mol}^{-1} = -5.35\times10^{-3}\ \text{J K}^{-2}\ \text{mol}^{-1}$$

$$\Delta c = [(0) - (0.50\times10^5) + (0.83\times10^5)]\ \text{J K mol}^{-1} = +0.33\times10^5\ \text{J K mol}^{-1}$$

For $T = 298$ K and $T' = 372$ K,

$$\alpha = -8.5\ \text{K}, \quad \beta = -2738\ \text{K}^2, \quad \gamma = -8.288\times10^{-5}\ \text{K}^{-1}$$

and so

$$\begin{aligned}\Delta_f G^{\ominus}(372\ \text{K}) &= (-237.13\ \text{kJ mol}^{-1}) + (-74\ \text{K})\times(-163.34\ \text{J K}^{-1}\ \text{mol}^{-1})\\ &\quad + (-8.5\ \text{K})\times(33.03\times10^{-3}\ \text{kJ K}^{-1}\ \text{mol}^{-1})\\ &\quad + (-2738\ \text{K}^2)\times(-5.35\times10^{-6}\ \text{kJ K}^{-2}\ \text{mol}^{-1})\\ &\quad + (-8.288\times10^{-5}\ \text{K}^{-1})\times(0.33\times10^2\ \text{kJ K mol}^{-1})\\ &= [(-237.13) + (12.09) - (0.28) + (0.015) - (0.003)]\ \text{kJ mol}^{-1}\\ &= \boxed{-225.31\ \text{kJ mol}^{-1}}\end{aligned}$$

Note that the β and γ terms are not significant (for this reaction and temperature range).

Topic 76 Electrochemical cells

Discussion questions

76.1 A galvanic cell uses a spontaneous chemical reaction to generate a potential difference and deliver an electric current to an external device. An electrolytic cell uses an external potential difference to drive a chemical reaction in the cell that is by itself non-spontaneous. In their essential features, these two kinds of cells can be considered opposites of each other, in the sense that an electrolytic cell can be thought of as a galvanic cell operating in the reverse direction. For some electrochemical cells, this is easy to accomplish. We say they are rechargeable. The most common example is the lead-acid battery used in automobiles. For many other cells, however, this kind of reversibility cannot be achieved. A fuel cell, like the galvanic cell, uses a spontaneous chemical reaction to generate a potential difference and deliver an electric current to an external device. Unlike the galvanic cell, the fuel cell must receive reactants from an external storage tank.

Exercise

76.1(a) The cell notation specifies the right and left electrodes. Note that for proper cancellation we must equalize the number of electrons in half-reactions being combined. To calculate the standard cell potential (emf) of the cell we have used $E^\ominus = E^\ominus_R - E^\ominus_L$, with standard electrode potentials from Table 77.1.

	$E^\ominus$
(a) R: $2\,Ag^+(aq) + 2e^- \rightarrow 2\,Ag(s)$	+0.80 V
L: $Zn^+(aq) + 2e^- \rightarrow Zn(s)$	−0.76 V
Overall (R − L): $2\,Ag^+(aq) + Zn(s) \rightarrow 2\,Ag(s) + Zn^{2+}(aq)$	+1.56 V
(b) R: $2\,H^+(aq) + 2e^- \rightarrow H_2(g)$	0
L: $Cd^{2+}(aq) + 2e^- \rightarrow Cd(s)$	−0.40 V
Overall (R − L): $Cd(s) + 2\,H^+(aq) \rightarrow Cd^{2+}(aq) + H_2(g)$	+0.40 V

(c) R: $Cr^{3+}(aq)+3e^- \rightarrow Cr(s)$ -0.74 V

L: $3[Fe(CN)_6]^{3-}(aq)+3e^- \rightarrow 3[Fe(CN)_6]^{4-}(aq)$ $+0.36$ V

Overall (R−L): $Cr^{3+}(aq)+3[Fe(CN)_6]^{4-}(aq) \rightarrow Cr(s)+3[Fe(CN)_6]^{3-}(aq)$ -1.10 V

(d) R: $Sn^{4+}(aq)+2e^- \rightarrow Sn^{2+}(aq)$ $+0.15$ V

L: $2\,Fe^{3+}(aq)+2\,e^- \rightarrow 2\,Fe^{2+}(aq)$ $+0.77$ V

Overall (R−L): $Sn^{2+}(aq)+2\,Fe^{2+}(aq) \rightarrow Sn^{2+}(aq)+2\,Fe^{2+}(aq)$ -0.62 V

Comment: Those cells for which $E^{\ominus}>0$ may operate as spontaneous galvanic cells under standard conditions. Those for which $E^{\ominus}<0$ may operate as non-spontaneous electrolytic cells. Recall that $E^{\ominus}$ informs us of the spontaneity of a cell under standard conditions only. For other conditions we require E.

Problems

76.1 (a) The cell reaction is $H_2(g)+\frac{1}{2}O_2(g) \rightarrow H_2O(l)$ $\quad \nu=2$

$$\Delta_r G^{\ominus}=\Delta_f G^{\ominus}(H_2O,l)=-237.13\ \text{kJ mol}^{-1}\ \text{[Table 57.4]}$$

$$E^{\ominus}=-\frac{\Delta_r G^{\ominus}}{\nu F}[76.3]=\frac{+237.13\ \text{kJ mol}^{-1}}{(2)\times(96.485\ \text{kC mol}^{-1})}=\boxed{+1.23\ \text{V}}$$

(b) $C_6H_6(l)+\frac{15}{2}O_2(g) \rightarrow 6\,CO_2(g)+3\,H_2O(l)$

$$\begin{aligned}\Delta_f G^{\ominus}&=6\,\Delta_f G^{\ominus}(CO_2,g)+3\,\Delta_f G^{\ominus}(H_2O,l)-\Delta_f G^{\ominus}(C_6H_6,l)\\&=(6)\times(-394.36)+(3)\times(-237.13)-(+124.3.03)]\ \text{kJ mol}^{-1}\\&\quad\text{[Tables 57.3 and 57.4]}\\&=-3\,201.58\ \text{kJ mol}^{-1}\end{aligned}$$

To find the value of ν for this reaction, recognize that a carbon atom in benzene has a −1 oxidation state while a carbon atom in carbon dioxide has a + 4 oxidation state. This is a change of + 5 per carbon or + 30 per benzene molecule. Thus, $\nu=30$ for the reaction equation. Alternatively, to find ν we break the cell reaction down into half-reactions as follows.

R: $\frac{15}{2}O_2(g)+30\,e^-+30\,H^+(aq) \rightarrow 15\,H_2O(l)$

L: $6\,CO_2(g)+30\,e^-+30\,H^+(aq) \rightarrow C_6H_6(l)+12\,H_2O(1)$

R−L: $C_6H_6(l)+\frac{15}{2}O_2(g) \rightarrow 6\,CO_2(g)+3\,H_2O(l)$

Once again, we find that $\nu=30$.

Therefore, $E^{\ominus}=\dfrac{-\Delta G^{\ominus}}{\nu F}\ [17.30]=\dfrac{+3\,201.58\ \text{kJ mol}^{-1}}{(30)\times(96.485\ \text{kC mol}^{-1})}=\boxed{+1.11\ \text{V}}$

Topic 77 Standard electrode potentials

Discussion questions

77.1 The **electrochemical series** lists metallic elements and hydrogen in the order of their reducing power as measured by their standard potentials in aqueous solution (see Table 77.1). It is used to quickly determine whether one metal can spontaneously displace another from solution at 298 K. The application rule is: a low element in the series will displace the cation of a high element in the series. For example, zinc is lower than hydrogen in the series so zinc will spontaneously react with the hydronium cation, $H^+(aq)$, to form the zinc cation and hydrogen gas: $Zn(s) + 2\,H^+(aq) \rightarrow Zn^{2+}(aq) + H_2(g)$.

77.3 Eqn 77.8:

$$\Delta_r H^{\ominus} = -\nu F\left(E^{\ominus}_{cell} - T\frac{dE^{\ominus}_{cell}}{dT}\right) \quad [77.8],$$

provides a non-calorimetric method for measuring $\Delta_r H^{\ominus}$. $E^{\ominus}_{cell}$ must be measured over a range of temperatures so that the temperature derivative, $dE^{\ominus}_{cell}/dT$, may be evaluated by curve fitting the $E^{\ominus}_{cell}(T)$ data. Substitution of the $E^{\ominus}_{cell}$ and $dE^{\ominus}_{cell}/dT$ values into eqn 77.8 yields the standard reaction enthalpy. The procedure is illustrated in text Example 77.3. In the special case for which $dE^{\ominus}_{cell}/dT$ is negligibly small, we may conclude that $\Delta_r S^{\ominus} \cong 0$ (see eqn 77.7) and $\Delta_r H^{\ominus} \cong \Delta_r G^{\ominus} = -\nu F E^{\ominus}_{cell}$

Exercises

77.1(a) If half-reaction (a) is the direct sum of half-reactions (b) and (c), eqn 77.4 gives

$$E^{\ominus}(a) = \frac{\nu_b E^{\ominus}(b) + \nu_c E^{\ominus}(c)}{\nu_a}$$

We observe that Ce^{4+}/Ce is the sum of Ce^{3+}/Ce and Ce^{4+}/Ce^{3+} so we write

$$E^{\ominus}(Ce^{4+}/Ce) = \frac{E^{\ominus}(Ce^{4+}/Ce^{3+}) + 3E^{\ominus}(Ce^{3+}/Ce)}{4} = \frac{(+1.61\text{ V}) + 3(-2.48\text{ V})}{4} = \boxed{-1.46\text{ V}}$$

77.2(a) Zn is lower than mercury in the electrochemical series of Table 77.1 and Table 77.2. Consequently, metallic zinc displaces the mercury(II) cation from aqueous solution and $\boxed{\text{elemental mercury cannot spontaneously displace the zinc(II) cation from solution under standard conditions}}$. An alternative view observes that, since $E^{\ominus}(Zn^{2+}/Zn) - E^{\ominus}(Hg^{2+}/Hg) < 0$, the reaction Hg(l) + Zn^{2+} (aq) → Hg^{2+} (aq) + Zn(s) is not spontaneous.

77.3(a)

$$\left.\begin{array}{ll} R: Ag^{+}(aq) + e^{-} \rightarrow Ag(s) & +0.80\text{ V} \\ L: AgI(s) + e^{-} \rightarrow Ag(s) + I^{-}(aq) & -0.15\text{ V} \end{array}\right\} E^{\ominus} = E^{\ominus}_{R} - E^{\ominus}_{L} = +0.95\text{ V}$$

$$\text{Overall}: Ag^{+}(aq) + I^{-}(aq) \rightarrow AgI(s) \qquad \nu = 1$$

$$\ln K = \frac{\nu F E^{\ominus}}{RT}\,[76.5] = \frac{0.95\text{ V}}{25.693\times10^{-3}\text{ V}} = 37$$

$$K = 1.1\bar{7}\times10^{16}$$

$$\text{(a)}\ K = \frac{a_{AgI(s)}}{a_{Ag^{+}(aq)}a_{I^{-}(aq)}} = \frac{1}{[Ag^{+}][I^{-}]} = \frac{1}{[Ag^{+}]^{2}} = 1.1\bar{7}\times10^{16}$$

In the above equation the activity of the solid equals 1 and, since the solution is extremely dilute, the activity coefficients of dissolved ions also equal 1. Solving for the molar ion concentration gives [Ag^{+}] = [I^{-}] = 9.2 × 10^{-9} M. AgI has a solubility equal to $\boxed{9.2\times10^{-9}\text{ M}}$.

(b) The solubility equilibrium is written as the reverse of the cell reaction. Therefore,

$$K_{S} = K^{-1} = 1/1.2\times10^{16} = \boxed{8.3\times10^{-17}}.$$

77.4(a) $Hg_2Cl_2(s) + H_2(g) \rightarrow 2\,Hg(l) + 2\,HCl(aq)$ and $\nu = 2$

Within this small 10°C temperature range we may assume that the standard reduction potential is linear in temperature. The two-point method provides the slope.

$$\text{slope, } a = \frac{dE^{\ominus}_{cell}}{dT} = \frac{\Delta E^{\ominus}_{cell}}{\Delta T} = \frac{(0.2669 - 0.2699)\text{ V}}{(303 - 293)\text{K}} = -3.0\times10^{-4}\text{ V K}^{-1}$$

The linear expression is

$$E^{\ominus}_{cell}(T) = E^{\ominus}_{1} + a\times(T - T_{1}) = 0.2699\text{ V} + a\times(T - 293\text{ K})$$

$$= aT + b \quad \text{where} \quad a = -3.0\times10^{-4}\text{ V K}^{-1} \quad \text{and} \quad b = 0.3578\text{ V}$$

Thus,

$$\Delta_r G^{\ominus}(298\text{ K}) = -\nu F E^{\ominus}_{cell}(298\text{K})\ [76.2]$$

$$= -2\times(96485\text{ C mol}^{-1})\times\{(-3.0\times10^{-4}\text{ V K}^{-1})\times(298\text{ K}) + 0.3578\text{ V}\}$$

$$= \boxed{-52\text{ kJ mol}^{-1}}$$

$$\Delta_r S^{\ominus} = \nu F\frac{dE^{\ominus}_{cell}}{dT}\ [77.7]$$

$$= 2\times(96\,485\text{ C mol}^{-1})\times(-3.0\times10^{-4}\text{ V K}^{-1})$$

$$= \boxed{-58\text{ J K}^{-1}\text{ mol}^{-1}}$$

$$\begin{aligned}\Delta_r H^\ominus &= \Delta_r G^\ominus + T\Delta_r S^\ominus \ [65.2]\\ &= (-52\ \text{kJ mol}^{-1}) + (298\ \text{K}) \times (-58\ \text{J K}^{-1}\ \text{mol}^{-1})\\ &= \boxed{-69\ \text{kJ mol}^{-1}}\end{aligned}$$

77.5(a) We calculate the standard Gibbs energy with

$$\Delta_r G^\ominus_{\text{cell}} = -\nu F E^\ominus_{\text{cell}}\ [76.2] = -\nu F\left(E^\ominus_{\text{R}} - E^\ominus_{\text{L}}\right)\ [77.1, \text{Table 77.1}]$$

	ν	$E^\ominus_R$/V	$E^\ominus_L$/V	$E^\ominus_{\text{cell}}$/V	$\Delta_r G^\ominus_{\text{cell}}$/kJ mol^{-1}
(a)	2	+0.80	−0.76	+1.56	−301
(b)	2	0.00	−0.40	+0.40	−77
(c)	3	−0.74	+0.36	−1.10	+318
(d)	2	+0.15	+0.77	−0.62	+120

77.6(a) The specification of the right and left electrodes is determined by the direction of the reaction as written. As always, in combining half-reactions to form an overall cell reaction we must write the half-reactions with equal number of electrons to ensure proper cancellation. We first identify the half-reactions and then set up the corresponding cell. To calculate the standard cell potential (emf) of the cell we have used $E^\ominus_{\text{cell}} = E^\ominus_R - E^\ominus_L$, with standard electrode potentials from Table 77.1.

		$E^\ominus$
(a)	R: $Pb^{2+}(aq) + 2e^- \rightarrow Pb(s)$	−0.13 V
	L: $Fe^{2+}(aq) + 2e^- \rightarrow Fe(s)$	−0.44 V
	Hence the cell is	
	$Fe(s)\vert FeSO_4(aq)\Vert PbSO_4(aq)\vert Pb(s)$	+0.31 V
(b)	R: $Hg_2Cl_2(s) + 2e^- \rightarrow 2Hg(l) + 2Cl^-(aq)$	+0.27 V
	L: $H^+(aq) + e^- \rightarrow \frac{1}{2}H_2(g)$	0
	and the cell is	
	$Pt\vert H_2(g)\vert H^+(aq)\vert Hg_2Cl_2(s)\vert Hg(l)$	
	or $Pt\vert H_2(g)\vert HCl(aq)\vert Hg_2Cl_2(s)\vert Hg(l)$	+0.27 V
(c)	R: $O_2(g) + 4H^+(aq) + 4e^- \rightarrow 2H_2O(l)$	+1.23 V
	L: $2H^+(aq) + 2e^- \rightarrow H_2(g)$	0
	and the cell is	
	$Pt\vert H_2(g)\vert H^+(aq)\vert O_2(g)\vert Pt$	+1.23 V

Comment: All of these cells have $E^\ominus > 0$, corresponding to a spontaneous cell reaction under standard conditions. If $E^\ominus$ had turned out to be negative, the spontaneous reaction would have been the reverse of the one given, with the right and left electrodes of the cell also reversed.

77.7(a) $Ag|AgBr(s)|KBr(aq, 0.050\text{ mol kg}^{-1})||Cd(NO_3)_2(aq, 0.010\text{ mol kg}^{-1})|Cd(s)$

(a) $\text{R: } Cd^{2+}(aq)+2e^- \rightarrow Cd(s) \qquad E^\ominus = -0.40\text{ V (Table 17.2)}$

$\text{L: } AgBr(s)+e^- \rightarrow Ag(s)+Br^-(aq) \qquad E^\ominus = +0.0713\text{ V}$

$\text{R}-\text{L: } Cd^{2+}(aq)+2Br^-(aq)+2Ag(s) \rightarrow Cd(s)+2AgBr(s) \qquad E^\ominus_{cell} = -0.47\text{ V}$

The cell reaction is not spontaneous toward the right under standard conditions because $E^\ominus_{cell} < 0$

(b) The Nernst equation for the above cell reaction is:

$$E_{cell} = E^\ominus_{cell} - \frac{RT}{\nu F}\ln Q \text{ [76.4]}$$

where $\nu = 2$ and $Q = \dfrac{1}{a_{Cd^{2+}}a^2_{Br^-}} = \dfrac{1}{\gamma_{Cd^{2+}}\gamma^2_{Br^-}} \times \dfrac{(b^\ominus)^3}{b_{Cd^{2+}}b^2_{Br^-}}$

$$= \frac{1}{\gamma_{\pm,R}\gamma^2_{\pm,L}} \times \frac{(b^\ominus)^3}{b_{Cd^{2+}}b^2_{Br^-}} \text{ [73.11]}$$

$b_{Cd^{2+}} = 0.010\text{ mol kg}^{-1}$ for the right-hand electrode

$b_{Br^-} = 0.050\text{ mol kg}^{-1}$ for the left-hand electrode

(c) The ionic strength and mean activity coefficient at the right-hand electrode are:

$$I_R = \tfrac{1}{2}\sum_i z_i^2(b_i/b^\ominus) \text{ [72.27]} = \tfrac{1}{2}\{4(0.010)+1(.020)\} = 0.030$$

$$\log\gamma_{\pm,R} = -|z_+z_-|AI^{1/2} \text{ [72.26]} = -2\times(0.509)\times(0.030)^{1/2} = -0.176$$

$$\gamma_{\pm,R} = 0.666$$

The ionic strength and mean acitivity coefficient at the left-hand electrode are:

$$I_L = \tfrac{1}{2}\sum_i z_i^2(b_i/b^\ominus) \text{ [72.27]} = \tfrac{1}{2}\{1(0.050)+1(.050)\} = 0.050$$

$$\log\gamma_{\pm,L} = -|z_+z_-|AI^{1/2} \text{ [72.26]} = -1\times(0.509)\times(0.050)^{1/2} = -0.114$$

$$\gamma_{\pm,L} = 0.769$$

Therefore,

$$Q = \left(\frac{1}{(0.666)\times(0.769)^2}\right)\times\left(\frac{1}{(0.010)\times(0.050)^2}\right) = 1.02\times10^5$$

and $E_{cell} = -0.47\text{ V} - \left(\dfrac{25.693\times10^{-3}\text{ V}}{2}\right)\ln(1.02\times10^5)$

$$= \boxed{-0.62\text{ V}}$$

77.8(a) We first recognize the reduction and oxidation processes within the reaction equation and calculate $E^\ominus_{cell}$ using $E^\ominus = E^\ominus(\text{reduction couple}) - E^\ominus(\text{oxidation couple})$ [77.1] with the standard reduction potentials of Table 77.1. Inspection of the balanced reaction equation and comparison with the redox couples also gives the stoichiometric coefficient of the electrons transferred, ν. The equilibrium constant is calculated with

$$K = e^{\nu FE^{\ominus}/RT} \quad [76.5] \quad \text{where} \quad RT/F = 25.693\ \text{mV at } 25\ °\text{C}.$$

(a) $E^{\ominus} = E^{\ominus}(Sn^{4+}/Sn^{2+}) - E^{\ominus}(Sn^{2+}/Sn)$ and $\nu = 2$

$$E^{\ominus} = +0.15\ \text{V} - (-0.14\ \text{V}) = +0.29\ \text{V}$$

$$K = e^{2\times(0.29)/(0.025693)} = \boxed{6.4\times10^{9}}$$

(b) $E^{\ominus} = E^{\ominus}(Hg^{2+}/Hg) - E^{\ominus}(Fe^{2+}/Fe)$ and $\nu = 2$

$$E^{\ominus} = +0.86\ \text{V} - (-0.44\ \text{V}) = +1.30\ \text{V}$$

$$K = e^{2\times(1.30)/(0.025693)} = \boxed{8.9\times10^{43}}$$

Problems

77.1 (a) $I = \tfrac{1}{2}\{z_+^2 b_+ + z_-^2 b_-\}/b^{\ominus}$ [72.27] $= 4b/b^{\ominus}$

For $CuSO_4$, $I = (4)\times(1.0\times10^{-3}) = \boxed{4.0\times10^{-3}}$

For $ZnSO_4$, $I = (4)\times(3.0\times10^{-3}) = \boxed{1.2\times10^{-2}}$

(b) $\log\gamma_{\pm} = -|z_+z_-|AI^{1/2}$ [72.26]

$$\log\gamma_{\pm}(CuSO_4) = -(4)\times(0.509)\times(4.0\times10^{-3})^{1/2} = -0.12\overline{88}$$

$$\gamma_{\pm}(CuSO_4) = \boxed{0.74}$$

$$\log\gamma_{\pm}(ZnSO_4) = -(4)\times(0.509)\times(1.2\times10^{-2})^{1/2} = -0.22\overline{30}$$

$$\gamma_{\pm}(ZnSO_4) = \boxed{0.60}$$

(c) The reaction in the Daniell cell is

$$Cu^{2+}(aq) + SO_4^{2-}(aq) + Zn(s) \rightarrow Cu(s) + Zn^{2+}(aq) + SO_4^{2-}(aq) \quad \nu = 2$$

Hence, $$Q = \frac{a(Zn^{2+})a(SO_4^{2-},R)}{a(Cu^{2+})a(SO_4^{2-},L)}$$

$$= \frac{\gamma_+ b_+(Zn^{2+})\gamma_- b_-(SO_4^{2-},R)}{\gamma_+ b_+(Cu^{2+})\gamma_- b_-(SO_4^{2-},L)} \left[b \equiv \frac{b}{b^{\ominus}}\text{ here and below}\right]$$

where the designations R and L refer to the right and left sides of the equation for the cell reaction and all b are assumed to be unitless, that is, $b/b^{\ominus}$

$$b_+(Zn^{2+}) = b_-(SO_4^{2-},R) = b(ZnSO_4)$$

$$b_+(Cu^{2+}) = b_-(SO_4^{2-},L) = b(CuSO_4)$$

Therefore,

$$Q = \frac{\gamma_{\pm}^2(ZnSO_4)b^2(ZnSO_4)}{\gamma_{\pm}^2(CuSO_4)b^2(CuSO_4)} = \frac{(0.60)^2\times(3.0\times10^{-3})^2}{(0.74)^2\times(1.0\times10^{-3})^2} = 5.9\overline{2} = \boxed{5.9}$$

(d) $E^{\ominus} = -\dfrac{\Delta_r G^{\ominus}}{\nu F}$ [76.3] $= \dfrac{-(-212.7\times10^3\ \text{J mol}^{-1})}{(2)\times(9.6485\times10^4\ \text{C mol}^{-1})} = \boxed{+1.102\ \text{V}}$

(e) $E = E^{\ominus} - \dfrac{25.693\times10^{-3}\ \text{V}}{\nu}\ln Q$ [76.4] $= (1.102\ \text{V}) - \left(\dfrac{25.693\times10^{-3}\ \text{V}}{2}\right)\ln(5.9\bar{2})$

$= (1.102\ \text{V}) - (0.023\ \text{V}) = \boxed{+1.079\ \text{V}}$

77.3 $Hg_2Cl_2(s) + Zn(s) \rightarrow 2\ Hg(l) + ZnCl_2(aq)$ and $\nu = 2$

(a) $E = E^{\ominus} - \dfrac{25.693\ \text{mV}}{\nu}\ln Q$ [76.4, 25 °C]

$Q = a(Zn^{2+})a^2(Cl^-)$

$= \{\gamma_+ b(Zn^{2+})/b^{\ominus}\}\times\gamma_-^2\{b(Cl^-)/b^{\ominus}\}^2$

where $b(Zn^{2+}) = b$, $b(Cl^-) = 2b$, and $\gamma_+\gamma_-^2 = \gamma_\pm^3$

Therefore, $Q = \gamma_\pm^3 \times 4b^3$ $\left[b \equiv b/b^{\ominus}\ \text{here and below}\right]$

and $E = E^{\ominus} - \dfrac{25.693\ \text{mV}}{2}\ln(4b^3\gamma_\pm^3)$

$= E^{\ominus} - \tfrac{3}{2}\times(25.693\ \text{mV})\times\ln(4^{1/3}b\gamma_\pm)$

$= \boxed{E^{\ominus} - (38.54\ \text{mV})\times\ln(4^{1/3}b) - (38.54\ \text{mV})\ln(\gamma_\pm)}$

(b) $E^{\ominus}_{\text{cell}} = E^{\ominus}(Hg_2^{2+}/Hg) - E^{\ominus}(Zn^{2+}/Zn)$

$= +0.2676\ \text{V} - (-0.7628\ \text{V}) = \boxed{1.0304\ \text{V}}$

(c) $\Delta_r G = -\nu FE = -(2)\times(9.6485\times10^4\ \text{C mol}^{-1})\times(1.2272\ \text{V}) = \boxed{-236.81\ \text{kJ mol}^{-1}}$

$\Delta_r G^{\ominus} = -\nu FE^{\ominus} = -(2)\times(9.6485\times10^4\ \text{C mol}^{-1})\times(1.0304\ \text{V})$

$= \boxed{-198.84\ \text{kJ mol}^{-1}}$

$\ln K = -\dfrac{\Delta_r G^{\ominus}}{RT} = \dfrac{1.9884\times10^5\ \text{J mol}^{-1}}{(8.3145\ \text{J K}^{-1}\ \text{mol}^{-1})\times(298.15\ \text{K})}$

$= 80.211$ so $K = \boxed{6.84\times10^{34}}$

(d) From part (a)

$1.2272\ \text{V} = 1.0304\ \text{V} - (38.54\ \text{mV})\times\ln(4^{1/3}\times0.0050) - (38.54\ \text{mV})\times\ln\gamma_\pm$

$\ln\gamma_\pm = -\dfrac{(1.2272\ \text{V}) - (1.0304\ \text{V}) - (0.186\bar{4}\ \text{V})}{0.03854\ \text{V}} = -0.269\bar{8}$ so $\gamma_\pm = \boxed{0.763}$

(e) $\log\gamma_\pm = -|z_-z_+|AI^{1/2}$ [72.26]

$I = \tfrac{1}{2}\sum_i z_i^2(b_i/b^{\ominus})$ [72.27]

where $b(\mathrm{Zn}^{2+}) = b = 0.0050\ \mathrm{mol\,kg^{-1}}$ and $b(\mathrm{Cl^-}) = 2b = 0.010\ \mathrm{mol\,kg^{-1}}$

Thus, $I = \tfrac{1}{2}[(4)\times(0.0050)+(0.010)] = 0.015$

$$\log \gamma_\pm = -(2)\times(0.509)\times(0.015)^{1/2} = -0.12\bar{5} \quad \text{so} \quad \gamma_\pm = \boxed{0.75}$$

This compares remarkably well to the value obtained from experimental data in part (d).

(f) $\Delta_r S = -\left(\dfrac{\partial \Delta_r G}{\partial T}\right)_p$ [66.7]

$$= \nu F\left(\frac{\partial E}{\partial T}\right)_p [76.3] = (2)\times(9.6485\times10^4\ \mathrm{C\,mol^{-1}})\times(-4.52\times10^{-4}\ \mathrm{V\,K^{-1}})$$

$$= \boxed{-87.2\ \mathrm{J\,K^{-1}\,mol^{-1}}}$$

$$\Delta_r H = \Delta_r G + T\Delta_r S \quad [65.2]$$

$$= (-236.81\ \mathrm{kJ\,mol^{-1}}) + (298.15\ \mathrm{K})\times(-87.2\ \mathrm{J\,K^{-1}\,mol^{-1}})$$

$$= \boxed{-262.4\ \mathrm{kJ\,mol^{-1}}}$$

77.5 $\mathrm{Pt|H_2(g,}p^\ominus\mathrm{)|NaOH(aq, 0.01000\ mol\,kg^{-1}), NaCl(aq, 0.01125\ mol\,kg^{-1})}$
$\mathrm{|AgCl(s)|Ag(s)}$

$$\mathrm{H_2(s,}p^\ominus\mathrm{)} + 2\,\mathrm{AgCl(s)} \rightarrow 2\,\mathrm{Ag(s)} + 2\mathrm{Cl^-(aq)} + 2\,\mathrm{H^+(aq)} \quad \text{where} \quad \nu = 2$$

$$E = E^\ominus - \frac{RT}{2F}\ln\{a(\mathrm{H^+})a(\mathrm{Cl^-})\}^2 \quad [76.4]$$

$$= E^\ominus - \frac{RT}{F}\ln\{a(\mathrm{H^+})a(\mathrm{Cl^-})\} = E^\ominus - \frac{RT}{F}\ln\frac{K_w a(\mathrm{Cl^-})}{a(\mathrm{OH^-})}$$

$$= E^\ominus - \frac{RT}{F}\ln\frac{K_w\gamma_\pm b(\mathrm{Cl^-})}{\gamma_\pm b(\mathrm{OH^-})}$$

$$= E^\ominus - \frac{RT}{F}\ln\frac{K_w b(\mathrm{Cl^-})}{b(\mathrm{OH^-})} = E^\ominus - \frac{RT}{F}\ln K_w - \frac{RT}{F}\ln\frac{b(\mathrm{Cl^-})}{b(\mathrm{OH^-})}$$

$$= E^\ominus + (2.303)\frac{RT}{F}\times \mathrm{p}K_w - \frac{RT}{F}\ln\frac{b(\mathrm{Cl^-})}{b(\mathrm{OH^-})} \quad \left(\mathrm{p}K_w = -\log K_w = \frac{-\ln K_w}{2.303}\right)$$

Hence,
$$\mathrm{p}K_w = \frac{E - E^\ominus}{2.303RT/F} + \frac{\ln\left(\dfrac{b(\mathrm{Cl^-})}{b(\mathrm{OH^-})}\right)}{2.303} = \frac{E - E^\ominus}{2.303RT/F} + 0.05114$$

Using Table 77.1, we find that

$$E^\ominus = E_R^\ominus - E_L^\ominus = E^\ominus(\mathrm{AgCl, Ag}) - E^\ominus(\mathrm{H^+/H_2}) = +0.22\ \mathrm{V} - 0 = +0.22\ \mathrm{V}$$

This value does not have the precision needed for computations with the high precision data of this problem. Consequently, we will use the more precise value found in the *CRC Handbook of Chemistry and Physics*(71st ed): $E^{\ominus} = 0.22233$ V. We then draw up the following table:

θ / °C	20.0	25.0	30.0
E/V	1.04774	1.04864	1.04942
$\ln(10)RTF^{-1}$ / V	0.058168	0.059160	0.060152
pK_w	14.24	14.02	13.80

A plot of pK_w against T is seen to be linear with a regression fit of $pK_w = 0.9039 + 3\,910\ \text{K} \times (1/T)$ with $R^2 = 0.99991$. So, $d(pK_w)/d(1/T) = 3\,910$ K.

$$\Delta_r H^{\ominus} = -R\frac{d\ln K_w}{d(1/T)} \quad [75.3]$$

$$= (2.303)R\frac{d(pK_w)}{d(1/T)}$$

$$= (2.303)\times(8.3145\ \text{J K}^{-1}\ \text{mol}^{-1})\times(3910\ \text{K})$$

$$= \boxed{+74.9\ \text{kJ mol}^{-1}}$$

$$\Delta_r G^{\ominus}(298.15\ \text{K}) = -RT\ln K_w\ [73.13] = 2.303RT\times pK_w$$

$$= 2.303RT\times\left(0.9039 + 3910\ \text{K}/T\right)$$

$$= 2.303\times(8.3145\ \text{J K}^{-1}\ \text{mol}^{-1})\times(298.15\ \text{K})$$

$$\times\left(0.9039 + 3910\ \text{K}/(298.15\ \text{K})\right)$$

$$= \boxed{+80.0\ \text{kJ mol}^{-1}}$$

$$\Delta_r S^{\ominus} = \frac{\Delta_r H^{\ominus} - \Delta_r G^{\ominus}}{T} \quad [65.2]$$

$$= \frac{(74.9 - 80.0)\text{kJ mol}^{-1}}{298.15\ \text{K}}$$

$$= \boxed{-17.1\ \text{J K}^{-1}\ \text{mol}^{-1}}$$

77.7 The method of the solution is first to determine $\Delta_r G^{\ominus}, \Delta_r H^{\ominus}$, and $\Delta_r S^{\ominus}$ for the cell reaction

$$\tfrac{1}{2}H_2(g) + AgCl(s) \rightarrow Ag(s) + HCl(aq) \qquad \nu = 1$$

and then, from the values of these quantities and the known values of $\Delta_f G^{\ominus}$, $\Delta_f H^{\ominus}$, and $S^{\ominus}$, for all the species other than $Cl^-(aq)$, to calculate $\Delta_f G^{\ominus}$, $\Delta_f H^{\ominus}$, and $S^{\ominus}$ for $Cl^-(aq)$.

$$\Delta_r G^{\ominus} = -\nu F E^{\ominus} \quad [76.3] \text{ at } 298.15\ \text{K}\ (25.00\ °\text{C})$$

$$E^{\ominus}/\text{V} = (0.23659) - (4.8564\times10^{-4})\times(25.00) - (3.4205\times10^{-6})$$

$$\times(25.00)^2 + (5.869\times10^{-9})\times(25.00)^3$$

$$= +0.22240\ \text{V}.$$

Therefore, $\Delta G^{\ominus} = -(96.485\ \text{kC mol}^{-1}) \times (0.22240\ \text{V}) = -21.46\ \text{kJ mol}^{-1}$

$$\Delta_r S^{\ominus} = -\left(\frac{\partial \Delta_r G^{\ominus}}{\partial T}\right)_p \quad [66.7] = \nu F \left(\frac{\partial E^{\ominus}}{\partial T}\right)_p$$

$$= \nu F \left(\frac{\partial E^{\ominus}}{\partial \theta}\right)_p \frac{{}^\circ\text{C}}{\text{K}} \quad [d\theta/{}^\circ\text{C} = dT/\text{K}] \qquad \text{(a)}$$

$$\left(\partial E^{\ominus}/\partial \theta\right)_p/\text{V} = \left(-4.8564\times10^{-4}/{}^\circ\text{C}\right) - (2)\times\left(3.4205\times10^{-6}\,\theta/({}^\circ\text{C})^2\right) + (3) \times\left(5.869\times10^{-9}\,\theta^2/({}^\circ\text{C})^3\right)$$

$$\left(\partial E^{\ominus}/\partial \theta\right)_p/\left(\text{V}\,{}^\circ\text{C}^{-1}\right) = \left(-4.8564\times10^{-4}\right) - \left(6.8410\times10^{-6}\left(\theta/{}^\circ\text{C}\right)\right) + \left(1.7607\times10^{-8}\left(\theta^\circ/C\right)^2\right)$$

Therefore, at 25 °C,

$$\left(\partial E^{\ominus}/\partial \theta\right)_p = -6.4566\times10^{-4}\ \text{V}/{}^\circ\text{C}$$

and

$$\left(\partial E^{\ominus}/\partial \theta\right)_p = (-6.4566\times10^{-4}\ \text{V}/{}^\circ\text{C})\times({}^\circ\text{C/K}) = -6.4566\times10^{-4}\ \text{V K}^{-1}$$

Hence, from equation (a)

$$\Delta_r S^{\ominus} = \left(-96.485\ \text{kC mol}^{-1}\right)\times\left(6.4566\times10^{-4}\ \text{V K}^{-1}\right) = -62.30\ \text{J K}^{-1}\ \text{mol}^{-1}$$

and $\Delta_r H^{\ominus} = \Delta_r G^{\ominus} + T\Delta_r S^{\ominus}$

$$= -(21.46\ \text{kJ mol}^{-1}) + (298.15\ \text{K})\times(-62.30\ \text{J K}^{-1}\ \text{mol}^{-1})$$

$$= -40.03\ \text{kJ mol}^{-1}$$

For the cell reaction $\tfrac{1}{2}H_2(g) + AgCl(s) \rightarrow Ag(s) + HCl(aq)$

$$\Delta_r G^{\ominus} = \Delta_f G^{\ominus}(\text{H}^+) + \Delta_f G^{\ominus}(\text{Cl}^-) - \Delta_f G^{\ominus}(\text{AgCl})$$

$$= \Delta_f G^{\ominus}(\text{Cl}^-) - \Delta_f G^{\ominus}(\text{AgCl}) \quad \left[\Delta_f G^{\ominus}(\text{H}^+) = 0\right]$$

Hence, $\Delta_f G^{\ominus}(\text{Cl}^-) = \Delta_r G^{\ominus} + \Delta_f G^{\ominus}(\text{AgCl}) = (-21.46 - 109.79)\ \text{kJ mol}^{-1}$

$$= \boxed{-131.25\ \text{kJ mol}^{-1}}$$

Similarly, $\Delta_f H^{\ominus}(\text{Cl}^-) = \Delta_r H^{\ominus} + \Delta_f H^{\ominus}(\text{AgCl}) = (-40.03 - 127.07)\ \text{kJ mol}^{-1}$

$$= \boxed{-167.10\ \text{kJ mol}^{-1}}$$

For the entropy of Cl^- in solution we use

$$\Delta_r S^{\ominus} = S^{\ominus}(\text{Ag}) + S^{\ominus}(\text{H}^+) + S^{\ominus}(\text{Cl}^-) - \tfrac{1}{2}S^{\ominus}(\text{H}_2) - S^{\ominus}(\text{AgCl})$$

with $S^\ominus(\mathrm{H^+})=0$. Then,

$$\begin{aligned} S^\ominus(\mathrm{Cl^-}) &= \Delta_r S^\ominus - S^\ominus(\mathrm{Ag}) + \tfrac{1}{2}S^\ominus(\mathrm{H_2}) + S^\ominus(\mathrm{AgCl}) \\ &= \{(-62.30)-(42.55)+\tfrac{1}{2}\times(130.68)+(96.2)\}\ \mathrm{J\,K^{-1}\,mol^{-1}} \\ &= \boxed{+56.7\ \mathrm{J\,K^{-1}\,mol^{-1}}} \end{aligned}$$

77.9 $\mathrm{MX(s)} \rightleftharpoons \mathrm{M^+(aq)+X^-(aq)}$ and $K_s = a(\mathrm{M^+})a(\mathrm{X^-}) = b(\mathrm{M^+})b(\mathrm{X^-})\gamma_\pm^2\ [b \equiv b/b^\ominus]$

The concentration of the soluble salt, $b(\mathrm{M^+})=S'$, is very low compared to the concentration of the freely soluble salt, [NX] = C. Consequently, $b(\mathrm{X^-})=S'+C=C$ and the ionic strength depends upon C alone.

$$I = \tfrac{1}{2}\left(b(\mathrm{N^+})+b(\mathrm{X^-})\right)\ [72.27] = C$$

$$\ln\gamma_\pm = 2.303\log\gamma_\pm = -2.303AI^{1/2}\ [72.26] \quad \text{where} \quad A=0.509$$

$$\gamma_\pm = \mathrm{e}^{-2.303AI^{1/2}} \quad \text{and} \quad \gamma_\pm^2 = \mathrm{e}^{-4.606AI^{1/2}}$$

Hence, $K_s = S'C\,\mathrm{e}^{-4.606AI^{1/2}}$ or $S' = K_s\mathrm{e}^{4.606AI^{1/2}}/C$

77.11 The half-reactions involved are:

$$\mathrm{R:}\ \mathrm{cyt_{ox}} + \mathrm{e^-} \rightarrow \mathrm{cyt_{red}} \qquad E^\ominus_{\mathrm{cyt}}$$

$$\mathrm{L:}\ \mathrm{D_{ox}} + \mathrm{e^-} \rightarrow \mathrm{D_{red}} \qquad E^\ominus_{\mathrm{D}}$$

The overall cell reaction is:

$$\mathrm{R-L:}\ \mathrm{cyt_{ox}} + \mathrm{D_{red}} \rightleftharpoons \mathrm{cyt_{red}} + \mathrm{D_{ox}} \qquad E^\ominus = E^\ominus_{\mathrm{cyt}} - E^\ominus_{\mathrm{D}}$$

(a) The Nernst equation [76.4] for the cell reaction is

$$E = E^\ominus - \frac{RT}{F}\ln\frac{[\mathrm{cyt_{red}}][\mathrm{D_{ox}}]}{[\mathrm{cyt_{ox}}][\mathrm{D_{red}}]}.$$

At equilibrium, $E=0$. Therefore,

$$\ln\frac{[\mathrm{cyt_{red}}]_{\mathrm{eq}}[\mathrm{D_{ox}}]_{\mathrm{eq}}}{[\mathrm{cyt_{ox}}]_{\mathrm{eq}}[\mathrm{D_{red}}]_{\mathrm{eq}}} = \frac{F}{RT}\left(E^\ominus_{\mathrm{cyt}} - E^\ominus_{\mathrm{D}}\right)$$

$$\ln\left(\frac{[\mathrm{D_{ox}}]_{\mathrm{eq}}}{[\mathrm{D_{red}}]_{\mathrm{eq}}}\right) = \ln\left(\frac{[\mathrm{cyt}]_{\mathrm{ox}}}{[\mathrm{cyt}]_{\mathrm{red}}}\right) + \frac{F}{RT}\left(E^\ominus_{\mathrm{cyt}} - E^\ominus_{\mathrm{D}}\right)$$

Therefore, a plot of $\ln\left(\frac{[\mathrm{D_{ox}}]_{\mathrm{eq}}}{[\mathrm{D_{red}}]_{\mathrm{eq}}}\right)$ against $\ln\left(\frac{[\mathrm{cyt}]_{\mathrm{ox}}}{[\mathrm{cyt}]_{\mathrm{red}}}\right)$ is linear with a slope of one and an intercept of $\frac{F}{RT}\left(E^\ominus_{\mathrm{cyt}} - E^\ominus_{\mathrm{D}}\right)$

(b) Draw up the following table.

$\ln\left(\frac{[D_{ox}]_{eq}}{[D_{red}]_{eq}}\right)$	−5.882	−4.776	−3.661	−3.002	−2.593	−1.436	−0.6274
$\ln\left(\frac{[cyt_{ox}]_{eq}}{[cyt_{red}]_{eq}}\right)$	−4.547	−3.772	−2.415	−1.625	−1.094	−0.2120	−0.3293

The plot of $\ln\left(\frac{[D_{ox}]_{eq}}{[D_{red}]_{eq}}\right)$ against $\ln\left(\frac{[cyt_{ox}]_{eq}}{[cyt_{red}]_{eq}}\right)$ is shown in Fig. 77.1. The intercept is −1.2124. Hence

$$E^{\ominus}_{cyt} = \frac{RT}{F}\times(-1.2124)+0.237\,V$$
$$=0.0257\,V\times(-1.2124)+0.237\,V$$
$$=\boxed{+0.206\,V}.$$

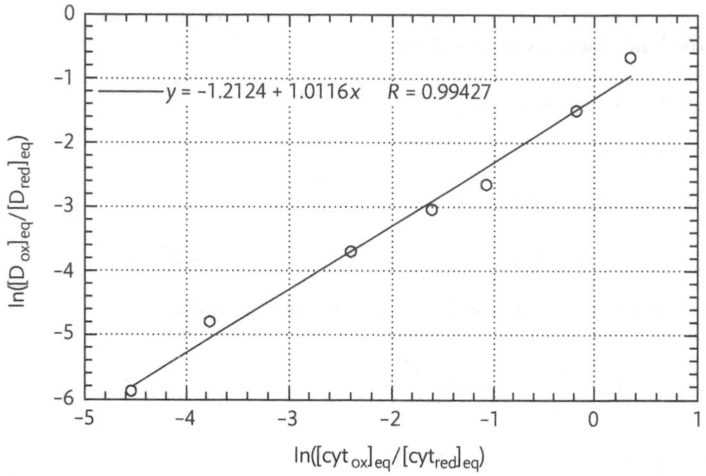

Fig 77.1

Focus 15: Integrated activities

F15.1 $N_2O_4(g) \overset{K}{\rightleftharpoons} 2\,NO_2(g)$

$(1-\alpha)n$	$2\alpha n$	amounts
$\frac{1-\alpha}{1+\alpha}$	$\frac{2\alpha}{1+\alpha}$	mole fractions
$\left(\frac{1-\alpha}{1+\alpha}\right)p$	$\left(\frac{2\alpha}{1+\alpha}\right)p$	partial pressures [$p \equiv p/p^{\ominus}$ here]

$$K=\frac{(2\alpha/1+\alpha)^2 p}{(1-\alpha/1+\alpha)}=\frac{4\alpha^2}{1-\alpha^2}p$$

Now solve for α.

$$\alpha^2=\frac{K}{4p+K},\qquad \alpha=\left(\frac{K}{4p+K}\right)^{1/2}$$

The degree of dimerization is $d=1-\alpha=1-\left(\frac{K}{4p+K}\right)^{1/2}=\boxed{1-\left(\frac{1}{4(p/K)+1}\right)^{1/2}}$

The susceptibility varies in proportion to $\alpha=1-d$. As pressure increases, α decreases, and the susceptibility $\boxed{\text{decreases}}$.

To determine the effect of temperature we need $\Delta_r H\approx\Delta_r H^{\ominus}$ for the reaction above.

$$\Delta_r H^{\ominus}=2\times(33.18\text{ kJ mol}^{-1})-9.16\text{ kJ mol}^{-1}=+57.2\text{ kJ mol}^{-1}$$

A positive $\Delta_r H^{\ominus}$ indicates that $NO_2(g)$ is favoured as the temperature increases; hence the susceptibility $\boxed{\text{increases}}$ with temperature.

Topic 78 The kinetic theory of gases

Discussion questions

78.1 The assumptions of the kinetic model of gases and discussion of conditions for which the fundamental assumptions may not be applicable:

The gas consists of molecules of mass *m* in ceaseless random motion obeying the laws of classical mechanics. Critique: The laws of classical mechanics include the conservation of particulate total energy, linear momentum, and angular momentum. These laws are predicated upon the assumption that gaseous matter exhibits the particulate behavior alone to the exclusion of electron wave behavior that is required to explain chemical reactions that may occur during particle collisions. Furthermore, the assumption of ceaseless random motion means that the simple kinetic model does not include a description of convective flow or diffusion (this restriction is relieved by dividing the gas into very small adjacent volumes in which each separate volume is characterized by its own temperature, pressure, and composition).

The size of the molecules is negligible, in the sense that their diameters are much smaller than the average distance travelled between collisions. Critique: A molecular gas at low molar volumes occupies a significant percentage of available space. The negligible molecular volume restriction limits the applicability of the simple kinetic model to the limit of zero pressure and prevents formation of a description of the van der Waals constants. However, the restriction can be relieved by treating each molecule as a ball bearing or billiard ball of fixed radius a and diameter d ($= 2a$). The resultant hard-sphere model, which is a kind of central-force model, allows for the estimation of the collision frequency and mean free path; it does not account for collision effects from dipole forces that are not centered on the center-of-mass.

The molecules interact only through brief elastic collisions (i.e., collisions in which the total translational kinetic energy of the molecules is conserved). Critique: This assumption is inherent in the analysis of collision frequency, mean free path, and the collision frequency with walls and surfaces. The model does not account for unevenness of surfaces, which creates surface pockets that may trap a gas molecule during collision, or cause a gas molecule to reflect at an angle other than the collision angle. The successful description of the simple kinetic model to wall collisions and effusion does suggest that it describes these phenomena as an average of many possibilities. This basic assumption of the model does not account for **inelastic collisions** in which molecule collisions at high temperature can exchange energy between translational, vibrational, and rotational degrees of freedom. This complication becomes important when the mean translational energy is comparable to the energy difference between the ground and the first excited state.

Exercises

78.1(a) $$\upsilon_{\text{mean}} = \left(\frac{8RT}{\pi M}\right)^{1/2} \ [78.7] \quad \text{and} \quad \upsilon_{\text{rms}} = \langle \upsilon^2 \rangle^{1/2} = \left(\frac{3RT}{M}\right)^{1/2} \ [78.2]$$

Mean translational kinetic energy:

$$\langle \varepsilon_{\text{k}} \rangle = \langle \tfrac{1}{2} m\upsilon^2 \rangle N_{\text{A}} = \tfrac{1}{2} m N_{\text{A}} \langle \upsilon^2 \rangle = \tfrac{1}{2} M \upsilon_{\text{rms}}^2 = \tfrac{1}{2} M \times \left(\frac{3RT}{M}\right) = \tfrac{3}{2} RT$$

The ratios of species 1 to species 2 at the same temperature are:

$$\frac{\upsilon_{\text{mean1}}}{\upsilon_{\text{mean2}}} = \left(\frac{M_2}{M_1}\right)^{1/2} \quad \text{and} \quad \frac{\langle \varepsilon_{\text{k}} \rangle_{\text{m1}}}{\langle \varepsilon_{\text{k}} \rangle_{\text{m2}}} = 1$$

(a) $$\frac{\upsilon_{\text{meanH}_2}}{\upsilon_{\text{meanHg}}} = \left(\frac{200.6}{2.016}\right)^{1/2} = \boxed{9.975}$$

(b) The mean translation kinetic energy is independent of molecular mass and depends upon temperature alone! Consequently, because the mean translational kinetic energy for a gas is proportional to T, the ratio of mean translational kinetic energies for gases at the same temperature always equals $\boxed{1}$.

78.2(a) $$\upsilon_{\text{rms}} = \langle \upsilon^2 \rangle^{1/2} = \left(\frac{3RT}{M}\right)^{1/2} \ [78.2]$$

$$\upsilon_{\text{rmsH}_2} = \left\{\frac{3\times(8.3145\ \text{J mol}^{-1}\ \text{K}^{-1})\times(293\ \text{K})}{2.016\times10^{-3}\ \text{kg mol}^{-1}}\right\}^{1/2} \quad (1\ \text{J} = 1\ \text{kg m}^2\ \text{s}^{-2})$$

$$= \boxed{1904\ \text{m s}^{-1}}$$

$$\upsilon_{\text{rmsO}_2} = \left\{\frac{3\times(8.3145\ \text{J mol}^{-1}\ \text{K}^{-1})\times(293\ \text{K})}{32.00\times10^{-3}\ \text{kg mol}^{-1}}\right\}^{1/2}$$

$$= \boxed{478\ \text{m s}^{-1}}$$

78.3(a) Scientific calculators easily compute both the Maxwell distribution and the requisite integral. Here's a short Mathcad Prime 2 calculation:

$M := 28.03 \cdot \text{gm} \cdot \text{mol}^{-1}$ $\quad T := 400 \cdot \text{K}$ $\quad R := 8.31447 \cdot \text{J} \cdot \text{K}^{-1} \cdot \text{mol}^{-1}$

$f(\upsilon) := 4 \cdot \pi \cdot \left(\dfrac{M}{2 \cdot \pi \cdot R \cdot T}\right)^{\frac{3}{2}} \cdot \upsilon^2 \cdot e^{\frac{M\upsilon^2}{2 \cdot R \cdot T}}$ [78.4] $\quad F(\upsilon_1 \cdot \upsilon_2) := \displaystyle\int_{\upsilon_1}^{\upsilon_2} f(\upsilon)\, d\upsilon$ [78.5]

$F(200 \cdot \text{m} \cdot \text{s}^{-1} \cdot 210 \cdot \text{m} \cdot \text{s}^{-1}) = 6.873 \cdot 10^{-3}$

Thus, $\boxed{0.69\%}$ of the molecules have a speed in this range.

78.4(a) $$\upsilon_{mp} = \left(\frac{2RT}{M}\right)^{1/2} \quad [78.8]$$

$$= \left\{\frac{2\times(8.3145\ \text{J mol}^{-1}\ \text{K}^{-1})\times(293\ \text{K})}{44.01\times10^{-3}\ \text{kg mol}^{-1}}\right\}^{1/2} \qquad (1\ \text{J} = 1\ \text{kg m}^2\ \text{s}^{-2})$$

$$= \boxed{333\ \text{m s}^{-1}}$$

$$\upsilon_{mean} = \left(\frac{8RT}{\pi M}\right)^{1/2} \quad [78.7]$$

$$= \left\{\frac{8\times(8.3145\ \text{J mol}^{-1}\ \text{K}^{-1})\times(293\ \text{K})}{\pi\times(44.01\times10^{-3}\ \text{kg mol}^{-1})}\right\}^{1/2}$$

$$= \boxed{375\ \text{m s}^{-1}}$$

$$\upsilon_{rel} = 2^{1/2}\upsilon_{mean} \quad [78.9a]$$

$$= 2^{1/2}\times(375\ \text{m s}^{-1}) = \boxed{530\ \text{m s}^{-1}}$$

78.5(a) (a) $$\upsilon_{mean} = \left(\frac{8RT}{\pi M}\right)^{1/2} \quad [78.7]$$

$$= \left\{\frac{8\times(8.3145\ \text{J mol}^{-1}\ \text{K}^{-1})\times(298\ \text{K})}{\pi\times(28.02\times10^{-3}\ \text{kg mol}^{-1})}\right\}^{1/2}$$

$$= \boxed{475\ \text{m s}^{-1}}$$

(b) $$\lambda = \frac{kT}{\sigma p}\ [78.12] = \frac{kT}{\pi d^2 p} \quad [\sigma = \pi d^2]$$

$$= \frac{(1.381\times10^{-23}\ \text{J K}^{-1})\times(298\ \text{K})}{\pi\times(395\times10^{-12}\ \text{m})^2\times(1.01325\times10^5\ \text{Pa})} \quad \left[\begin{array}{l}1\ \text{atm} = 1.01325\ \text{bar}\\ = 1.01325\times10^5\ \text{Pa}\end{array}\right]$$

$$= 8.29\times10^{-8}\ \text{m}\ [1\ \text{Pa} = 1\ \text{J m}^{-3}] = \boxed{82.9\ \text{nm}}$$

(c) $$z = \frac{\upsilon_{rel}}{\lambda}\ [78.11] = \frac{\sqrt{2}\,\upsilon_{mean}}{\lambda} \quad [78.9a]$$

$$= \frac{\sqrt{2}(475\ \text{m s}^{-1})}{82.9\times10^{-9}\ \text{m}} = \boxed{8.10\times10^9\ \text{s}^{-1}}$$

78.6(a) The volume and radius of a spherical container are related by

$V = \tfrac{4}{3}\pi r^3$ or $r = (3V/4\pi)^{1/3}$

Thus, $r = \{3\times(1.00^{-4}\ \text{m}^3)/4\pi\}^{1/3} = 0.0288\ \text{m}$

$$\lambda = \frac{kT}{\sigma p}\ [78.12] \quad \text{or} \quad p = \frac{kT}{\sigma\lambda} = \frac{kT}{2r\sigma} \quad [\text{when } \lambda = 2r]$$

$$p = \frac{(1.381\times10^{-23}\ \text{J K}^{-1})\times(293\ \text{K})}{2\times(0.0288\ \text{m})\times(0.36\times10^{-18}\ \text{m}^2)} = \boxed{0.195\ \text{Pa}} \quad [1\ \text{J} = 1\ \text{Pa m}^3]$$

78.7(a) $\lambda = \dfrac{kT}{\sigma p}$ [78.12]

$$= \frac{(1.381\times10^{-23}\ \mathrm{J\,K^{-1}})\times(217\ \mathrm{K})}{(0.43\times10^{-18}\ \mathrm{m^2})\times(0.050)\times(1.013\times10^5\ \mathrm{Pa})}$$

$$= \boxed{1.38\times10^{-6}\ \mathrm{m}}$$

78.8(a) $z = \dfrac{\sigma p}{kT}v_{\mathrm{rel}}$ [78.10] $= \dfrac{\sqrt{2}\,\sigma p}{kT}v_{\mathrm{mean}}$ [78.9a] $= \dfrac{\sqrt{2}\,\sigma N_{\mathrm{A}} p}{RT}\times\left(\dfrac{8RT}{\pi M}\right)^{1/2}$

$$= \sqrt{2}\,\sigma N_{\mathrm{A}} p\times\left(\frac{8}{\pi MRT}\right)^{1/2}$$

$$= \sqrt{2}\,(0.38\times10^{-18}\ \mathrm{m^2})\times(6.022\times10^{23}\ \mathrm{mol^{-1}})$$

$$\times\left(\frac{8}{\pi(39.95\times10^{-3}\ \mathrm{kg\,mol^{-1}})\times(8.3145\ \mathrm{J\,K^{-1}\,mol^{-1}})\times(298\ \mathrm{K})}\right)^{1/2} p$$

$$= (5.2\times10^4\ \mathrm{s^{-1}\,Pa^{-1}})\times(1.013\times10^5\ \mathrm{Pa})(p/\mathrm{atm})$$

$$= (5.3\times10^9\ \mathrm{s^{-1}})\times(p/\mathrm{atm})$$

(a) $p = 10$ atm, $z = \boxed{5.3\times10^{10}}\ \mathrm{s^{-1}}$

(b) $p = 1.0$ atm, $z = \boxed{5.3\times10^{9}}\ \mathrm{s^{-1}}$

(c) $p = 1.0$ μatm, $z = \boxed{5.3\times10^{3}}\ \mathrm{s^{-1}}$

78.9(a) $A = (5.0\ \mathrm{mm})\times(4.0\ \mathrm{mm}) = 2.0\times10^{-5}\ \mathrm{m^2}$

The collision frequency of the Ar gas molecules with surface area A equals $Z_{\mathrm{W}}A$

$$Z_{\mathrm{W}}A = \frac{p}{(2\pi MkT/N_{\mathrm{A}})^{1/2}}A \quad [78.13;\ m = M/N_{\mathrm{A}}]$$

$$= \frac{(25\ \mathrm{Pa})\times(2.0\times10^{-5}\ \mathrm{m^2})}{\left\{\begin{array}{l}2\pi(39.95\times10^{-3}\ \mathrm{kg\,mol^{-1}})\times(1.381\times10^{-23}\ \mathrm{J\,K^{-1}})\\ \times(300\ \mathrm{K})/(6.022\times10^{23}\ \mathrm{mol^{-1}})\end{array}\right\}^{1/2}}$$

$$= 1.2\times10^{19}\ \mathrm{s^{-1}}$$

The number of argon molecule collisions within A in time interval t equals $Z_{\mathrm{W}}At$ if p does not change significantly during the period t.

$$Z_{\mathrm{W}}At = (1.2\times10^{19}\ \mathrm{s^{-1}})\times(100\ \mathrm{s}) = \boxed{1.2\times10^{21}}$$

78.10(a) If p does not change significantly during the period t: $t_{\mathrm{O_2}} = \dfrac{(\text{rate of effusion})_{\mathrm{H_2}}}{(\text{rate of effusion})_{\mathrm{O_2}}}t_{\mathrm{H_2}}$

The rate of effusion at constant temperature, pressure, and molecule count is proportional to $M^{-1/2}$ [78.14]. Consequently, the above expression conveniently simplifies to

$$t_{\mathrm{O_2}} = \left(\frac{M_{\mathrm{O_2}}}{M_{\mathrm{H_2}}}\right)^{1/2} t_{\mathrm{H_2}}$$

$$= \left(\frac{32.00}{2.02}\right)^{1/2}\times(135\ \mathrm{s}) = \boxed{537\ \mathrm{s}}$$

78.11(a) If p does not change significantly during the period t, the mass loss equals the effusion mass loss multiplied by the time period t: m_{loss} = (rate of effusion) × t × m = (rate of effusion) × t × M/N_A

$$m_{\text{loss}} = \left(\frac{pA_0N_A}{(2\pi MRT)^{1/2}}\right)\times\left(\frac{Mt}{N_A}\right) \quad [78.14]$$

$$= pA_0t\times\left(\frac{M}{2\pi RT}\right)^{1/2}$$

$$= (0.735\ \text{Pa})\times\left\{\pi\left(0.75\times10^{-3}\ \text{m}\right)^2\right\}\times(3600\ \text{s})$$

$$\times\left(\frac{0.300\ \text{kg mol}^{-1}}{2\pi\left(8.3145\ \text{J mol}^{-1}\ \text{K}^{-1}\right)\times(500\ \text{K})}\right)^{1/2}$$

$$= \boxed{16\ \text{mg}}$$

78.12(a) The pressure of this exercise changes significantly during time period t so it is useful to spend a moment finding an expression for $p(t)$. Mathematically, the rate of effusion is the derivative–$\mathrm{d}N/\mathrm{d}t$. Substitution of the perfect gas law for N, $N = pVN_A/RT$ where V and T are constants, reveals that the rate of effusion can be written as–$(N_AV/RT)\mathrm{d}p/\mathrm{d}t$. This formulation of the rate of effusion, along with eqn 78.14, is used to find $p(t)$.

$$-\left(\frac{N_AV}{RT}\right)\frac{\mathrm{d}p}{\mathrm{d}t} = \frac{pA_0N_A}{(2\pi MRT)^{1/2}} \quad [78.14]$$

$$\frac{\mathrm{d}p}{\mathrm{d}t} = -\frac{pA_0}{V}\left(\frac{RT}{2\pi M}\right)^{1/2}$$

$$\frac{\mathrm{d}p}{p} = -\frac{\mathrm{d}t}{\tau} \quad \text{where} \quad \tau = \frac{V}{A_0}\left(\frac{2\pi M}{RT}\right)^{1/2}$$

$$\int_{p_0}^{p}\frac{\mathrm{d}p}{p} = -\frac{1}{\tau}\int_0^t \mathrm{d}t \quad \text{where } p_0 \text{ is the initial pressure}$$

$$\ln\frac{p}{p_0} = -\frac{t}{\tau} \quad \text{or} \quad p(t) = p_0\mathrm{e}^{-t/\tau}$$

The nitrogen gas and unknown gas data can be used to determine the relaxation time, τ, for each.

$$\tau_{N_2} = \frac{t_{N_2}}{\ln(p_0/p)_{N_2}} = \frac{45\ \text{s}}{\ln(74/20)} = 34\ \text{s}$$

$$\tau_{\text{unk}} = \frac{t_{\text{unk}}}{\ln(p_0/p)_{\text{unk}}} = \frac{152\ \text{s}}{\ln(74/20)} = 116\ \text{s}$$

The above definition of τ shows that it is proportional to $M^{1/2}$. Since the ratio of the relaxation times cancels the constant of proportionality,

$$\left(\frac{M_{\text{unk}}}{M_{\text{N}_2}}\right)^{1/2}=\frac{\tau_{\text{unk}}}{\tau_{\text{N}_2}}$$

$$M_{\text{unk}}=\left(\frac{\tau_{\text{unk}}}{\tau_{\text{N}_2}}\right)^2 M_{\text{N}_2}$$

$$=\left(\frac{116}{34.\bar{4}}\right)^2\times(28.02\ \text{g mol}^{-1})=\boxed{31\bar{9}\ \text{g mol}^{-1}}$$

78.13(a) In Exercise 78.12(a) it is shown that

$$\ln\frac{p}{p_0}=-\frac{t}{\tau}\quad\text{or}\quad p(t)=p_0\mathrm{e}^{-t/\tau}\quad\text{where}\quad \tau=\frac{V}{A_0}\left(\frac{2\pi M}{RT}\right)^{1/2}$$

The relaxation time, τ, of oxygen is calculated with the data.

$$\tau=\left(\frac{3.0\ \text{m}^3}{\pi(0.10\times10^{-3}\ \text{m})^2}\right)\times\left\{\frac{2\pi(32.00\times10^{-3}\ \text{kg mol}^{-1})}{(8.3145\ \text{J mol}^{-1}\ \text{K}^{-1})\times(298\ \text{K})}\right\}^{1/2}$$

$$=8.6\times10^5\ \text{s}=10.\bar{0}\ \text{days}$$

The time required for the specified pressure decline is calculated with the above eqn

$$t=\tau\ln(p_0/p)=(10.\bar{0}\ \text{days})\times\ln(80/70)=\boxed{1.3\ \text{days}}$$

Problems

78.1 The time in seconds for a disk to rotate 360° is the inverse of the frequency. The time for it to advance 2° is $\left(\frac{2^\circ}{360^\circ}\right)\times\frac{1}{\nu}$. This is the time required for slots in neighboring disks to coincide along the atomic beam. For an atom to pass through all neighboring slots it must have the speed

$$v_x=\text{disk spacing / alignment time}=(1.0\ \text{cm})/\left\{\left(\frac{2^\circ}{360^\circ}\right)\times\frac{1}{\nu}\right\}$$

$$=180\nu\,\text{cm}=180\times(\nu/\text{Hz})\ \text{cm s}^{-1}$$

Theoretically, the velocity distribution in the *x*-direction is

$$f(v_x)=\left(\frac{m}{2\pi kT}\right)^{1/2}\mathrm{e}^{-mv_x^2/2kT}\quad[\text{Justification 78.2 with } M/R=m/k]$$

where

$$mv_x^2/2kT=\frac{83.80\times(1.6605\times10^{-27}\ \text{kg})\times\{1.80\times(\nu/\text{Hz})\ \text{m s}^{-1}\}^2}{2\times(1.381\times10^{-23}\ \text{J K}^{-1})\times T}$$

$$=\frac{(1.632\times10^{-2})\times(\nu/\text{Hz})^2}{T/\text{K}}$$

Therefore, as $I \propto f$, $I \propto T^{-1/2}e^{-mv_x^2/2kT}$ we can write $I \propto (T/\mathrm{K})^{-1/2}\, e^{-1.632\times10^{-2}\times(\nu/\mathrm{Hz})^2/(T/\mathrm{K})}$ and eliminate the constant of proportionality by taking the ratio of intensities at the two temperatures.

$$I(T_1)/I(T_2) = (T_1/T_2)^{-1/2}\, e^{-1.632\times10^{-2}\times(\nu/\mathrm{Hz})^2\times\{(T_1/\mathrm{K})^{-1}-(T_2)^{-1}\}}$$

$$I(40\ \mathrm{K})/I(100\ \mathrm{K}) = (40/100)^{-1/2}\, e^{-1.632\times10^{-2}\times(\nu/\mathrm{Hz})^2\times\{(40)^{-1}-(100)^{-1}\}}$$

$$= 1.581\times e^{-2.448\times10^{-4}\times(\nu/\mathrm{Hz})^2}$$

Draw up the following table.

ν/Hz	20	40	80	100	120
I_{exp} (40 K)	0.846	0.513	0.069	0.015	0.002
I_{exp} (100 K)	0.592	0.485	0.217	0.119	0.057
$\{I\,(40\ \mathrm{K})/I\,(100\ \mathrm{K})\}_{\mathrm{exp}}$	1.43	1.06	0.32	0.13	0.035
$\{I\,(40\ \mathrm{K})/I\,(100\ \mathrm{K})\}_{\mathrm{calc}}$	1.43	1.07	0.33	0.14	0.047

The calculated intensity ratios are in excellent agreement with the experimental ratios.

78.3 The time constant for the exponential mass loss is

$$\tau = \left(\frac{2\pi M}{RT}\right)^{1/2}\frac{V}{A}$$

$$= \left\{\frac{2\pi\times(137.33\times10^{-3}\ \mathrm{kg\ mol^{-1}})}{(8.3145\ \mathrm{J\ K^{-1}\ mol^{-1}})\times(1573\ \mathrm{K})}\right\}^{1/2}\times\left(\frac{100\times10^{-6}\ \mathrm{m^3}}{0.10\times10^{-6}\ \mathrm{m^2}}\right) = 8.12\ \mathrm{s}$$

The expression for the exponential pressure decay is

$p(t) = p_0 e^{-t/\tau}$ or $t = \tau\ln(p_0/p)$ where p_0 is the initial pressure.

Thus, the time required for the pressure to drop to 10% of the initial value is

$$t = (8.12\ \mathrm{s})\times\ln(10) = \boxed{18.9\ \mathrm{s}}$$

78.5 The most probable speed of a gas molecule corresponds to the condition that the Maxwell distribution be a maximum (it has no minimum); hence we find it by setting the first derivative of the function equal to zero and solving for the value of υ for which this condition holds.

$$f(\upsilon) = 4\pi\left(\frac{m}{2\pi kT}\right)^{3/2}\upsilon^2 e^{-m\upsilon^2/2kT}\ [78.4] = \mathrm{const}\times\upsilon^2 e^{-m\upsilon^2/2kT}\quad [M/R = m/k]$$

$$\frac{\mathrm{d}f(v)}{\mathrm{d}v} = 0\ \text{ when }\ \left(2-\frac{mv^2}{kT}\right) = 0$$

So, $\boxed{\upsilon(\text{most probable}) = \upsilon_{\mathrm{mp}} = \left(\frac{2kT}{m}\right)^{1/2} = \left(\frac{2RT}{M}\right)^{1/2}}$ [78.8]

The average kinetic energy corresponds to the average of $\frac{1}{2}m\upsilon^2$. The average is obtained by determining

$$\langle\upsilon^2\rangle = \int_0^\infty \upsilon^2 f(\upsilon)\mathrm{d}\upsilon\ [78.6] = 4\pi\left(\frac{m}{2\pi kT}\right)^{3/2}\int_0^\infty \upsilon^4 \mathrm{e}^{-m\upsilon^2/2kT}\,\mathrm{d}\upsilon\ [78.4]$$

The integral evaluates to $\dfrac{3\pi^{1/2}}{8}\left(\dfrac{m}{2kT}\right)^{-5/2}$.

Then, $\langle\upsilon^2\rangle = 4\pi\left(\dfrac{m}{2\pi kT}\right)^{3/2}\times\dfrac{3\pi^{1/2}}{8}\left(\dfrac{m}{2kT}\right)^{-5/2} = \dfrac{3kT}{m}$

Thus, $\langle\varepsilon\rangle = \frac{1}{2}m\langle\upsilon^2\rangle = \frac{3}{2}kT$ and we conclude that kinetic theory validates the equipartition theorem for translational kinetic energy.

78.7 We proceed as in Justification 78.2 except that instead of taking a product of three one-dimensional distributions in order to get the three-dimensional distribution, we make a product of two one-dimensional distributions.

$$f(\upsilon_x,\upsilon_y)\mathrm{d}\upsilon_x\mathrm{d}\upsilon_y = f(\upsilon_x^2)f(\upsilon_y^2)\mathrm{d}\upsilon_x\mathrm{d}\upsilon_y = \left(\frac{m}{2\pi kT}\right)\mathrm{e}^{-m\upsilon^2/2kT}\mathrm{d}\upsilon_x\mathrm{d}\upsilon_y$$

where $\upsilon^2 = \upsilon_x^2 + \upsilon_y^2$. The probability $f(\upsilon)\mathrm{d}\upsilon$ that the molecules have a two-dimensional speed, υ, in the range υ to $\upsilon+\mathrm{d}\upsilon$ is the sum of the probabilities that it is in any of the area elements $\mathrm{d}\upsilon_x\mathrm{d}\upsilon_y$ in the circular shell of radius υ. The sum of the area elements is the area of the circular shell of radius υ and thickness $\mathrm{d}\upsilon$ which is $\pi(\upsilon+\mathrm{d}\pi)^2 - \pi\upsilon^2 = 2\pi\upsilon\mathrm{d}\upsilon$. Therefore,

$$\boxed{f(\upsilon) = \left(\frac{m}{kT}\right)\upsilon\mathrm{e}^{-m\upsilon^2/2kT}}\quad\left[\frac{M}{R} = \frac{m}{k}\right]$$

The mean speed is determined as

$$\upsilon_{\text{mean}} = \int_0^\infty \upsilon f(\upsilon)\mathrm{d}\upsilon = \left(\frac{m}{kT}\right)\int_0^\infty \upsilon^2\mathrm{e}^{-m\upsilon^2/2kT}\,\mathrm{d}\upsilon = \left(\frac{m}{kT}\right)\times\left(\frac{\pi^{1/2}}{4}\right)\times\left(\frac{2kT}{m}\right)^{3/2}$$

[standard integral]

$$= \boxed{\left(\frac{\pi kT}{2m}\right)^{1/2}\text{ or }\left(\frac{\pi RT}{2M}\right)^{1/2}}$$

78.9 Rewriting eqn 78.4 with $M/R = m/k$,

$$f(\upsilon) = 4\pi\left(\frac{m}{2\pi kT}\right)^{3/2}\upsilon^2\mathrm{e}^{-m\upsilon^2/2kT}$$

The proportion of molecules with speeds less than c (value to be selected latter) is

$$P = \int_0^c f(\upsilon)\mathrm{d}\upsilon = 4\pi\left(\frac{m}{2\pi kT}\right)^{3/2}\int_0^c \upsilon^2\mathrm{e}^{-m\upsilon^2/2kT}\,\mathrm{d}\upsilon$$

Defining $a \equiv m/2kT$,

$$P = 4\pi\left(\frac{a}{\pi}\right)^{3/2}\int_0^c \upsilon^2\mathrm{e}^{-a\upsilon^2}\,\mathrm{d}\upsilon = -4\pi\left(\frac{a}{\pi}\right)^{3/2}\frac{\mathrm{d}}{\mathrm{d}a}\int_0^c \mathrm{e}^{-a\upsilon^2}\,\mathrm{d}\upsilon$$

Defining $\chi^2 \equiv av^2$. Then, $\mathrm{d}\upsilon = a^{-1/2}\mathrm{d}\chi$ and

$$P = -4\pi\left(\frac{a}{\pi}\right)^{3/2}\frac{\mathrm{d}}{\mathrm{d}a}\left\{\frac{1}{a^{1/2}}\int_0^{ca^{1/2}} \mathrm{e}^{-\chi^2}\,\mathrm{d}\chi\right\}$$

$$= -4\pi\left(\frac{a}{\pi}\right)^{3/2}\left\{-\frac{1}{2}\left(\frac{1}{a}\right)^{3/2}\int_0^{ca^{1/2}} \mathrm{e}^{-\chi^2}\,\mathrm{d}\chi + \left(\frac{1}{a}\right)^{1/2}\frac{\mathrm{d}}{\mathrm{d}a}\int_0^{ca^{1/2}} \mathrm{e}^{-\chi^2}\,\mathrm{d}\chi\right\}.$$

Then we use $\int_0^{ca^{1/2}} \mathrm{e}^{-\chi^2}\,\mathrm{d}\chi = (\pi^{1/2}/2)\mathrm{erf}(ca^{1/2})$

$$\frac{\mathrm{d}}{\mathrm{d}a}\int_0^{ca^{1/2}} \mathrm{e}^{-\chi^2}\,\mathrm{d}\chi = \left(\frac{\mathrm{d}ca^{1/2}}{\mathrm{d}a}\right)\times(\mathrm{e}^{-c^2a}) = \frac{1}{2}\left(\frac{c}{a^{1/2}}\right)\mathrm{e}^{-c^2a}$$

where we have used $\frac{\mathrm{d}}{\mathrm{d}z}\int_0^z f(y)\mathrm{d}y = f(z)$.

Substituting and cancelling we obtain $P = \mathrm{erf}(ca^{1/2}) - (2ca^{1/2}/\pi^{1/2})\mathrm{e}^{-c^2a}$

Now, let c be the root mean square speed, $c = \upsilon_{\mathrm{rms}} = (3kT/m)^{1/2}$ [78.3], so

$$ca^{1/2} = (3kT/m)^{1/2} \times (m/2kT)^{1/2} = (3/2)^{1/2}$$

and $P = \mathrm{erf}\left(\sqrt{\frac{3}{2}}\right) - \left(\frac{6}{\pi}\right)^{1/2}\mathrm{e}^{-3/2} = 0.92 - 0.31 = 0.61$

(a) The fraction of molecules having a speed greater than υ_{rms} is $1-P = 0.39$; that is $\boxed{39\%}$.

(b) The fraction of molecules having a speed less than υ_{rms} is $P = 0.61$; that is $\boxed{61\%}$.

(c) For the proportions in terms of the mean speed, υ_{mean}, replace c by

$$\upsilon_{\mathrm{mean}} = (8kT/\pi m)^{1/2}\ [78.7] = (8/3\pi)^{1/2}\,\upsilon_{\mathrm{rms}}\ [78.3]$$

So, $ca^{1/2} = \upsilon_{\mathrm{mean}}a^{1/2} = (8/3\pi)^{1/2}\upsilon_{\mathrm{rms}}a^{1/2} = (8/3\pi)^{1/2} \times (3/2)^{1/2} = 2/\pi^{1/2}$

Then, $P = \mathrm{erf}(\upsilon_{\mathrm{mean}}a^{1/2}) - (2\upsilon_{\mathrm{mean}}a^{1/2}/\pi^{1/2}) \times (\mathrm{e}^{-\upsilon_{\mathrm{mean}}^2 a})$

$$= \mathrm{erf}(2/\pi^{1/2}) - (4/\pi)\mathrm{e}^{-4/\pi}$$

$$= 0.889 - 0.356 = \boxed{0.533}$$

That is, $\boxed{53\%}$ of the molecules have a speed less than the mean, and $\boxed{47\%}$ have a speed greater than the mean.

78.11 $$\langle \upsilon^n \rangle = \int_0^\infty \upsilon^n f(\upsilon)\mathrm{d}\upsilon\ [78.6] = 4\pi\left(\frac{m}{2\pi kT}\right)^{3/2}\int_0^\infty \upsilon^{n+2}\mathrm{e}^{-m\upsilon^2/2kT}\,\mathrm{d}\upsilon\ [78.4]$$

The integral of the above expression must be handled in a manner that depends upon whether n is odd or even. When n is an odd, positive integer ($n = 1, 3, 5, \ldots$), let $n + 2 = 2p + 1$ where $p = 0, 1, 2, 3, \ldots$ and the integral is given by

$$\int_0^\infty \upsilon^{n+2}\mathrm{e}^{-m\upsilon^2/2kT}\,\mathrm{d}\upsilon = \int_0^\infty \upsilon^{2p+1}\mathrm{e}^{-m\upsilon^2/2kT}\,\mathrm{d}\upsilon = \tfrac{1}{2}p!\left(\frac{2kT}{m}\right)^{p+1}\ \text{[standard integral]}$$

$$= \tfrac{1}{2}\left(\frac{n+1}{2}\right)!\left(\frac{2kT}{m}\right)^{\frac{n+3}{2}}$$

Thus,

$$\langle v^n \rangle = 4\pi\left(\frac{m}{2\pi kT}\right)^{3/2} \times \tfrac{1}{2}\left(\frac{n+1}{2}\right)!\left(\frac{2kT}{m}\right)^{\frac{n+3}{2}}$$

$$= \boxed{\frac{2}{\pi^{1/2}}\left(\frac{n+1}{2}\right)!\left(\frac{2kT}{m}\right)^{n/2}} \text{ when } n \text{ is an odd, positive integer.}$$

When n is an even, positive integer ($n = 0, 2, 4, \ldots$), let $n + 2 = 2p$ where $p = 0, 1, 2, 3, \ldots$ and the integral is given by

$$\int_0^\infty v^{n+2} e^{-mv^2/2kT}\,dv = \int_0^\infty v^{2p} e^{-mv^2/2kT}\,dv$$

$$= \frac{(2p-1)!!}{2^{p+1}}\left(\frac{2kT}{m}\right)^p\left(\frac{2\pi kT}{m}\right)^{1/2} \quad \left[\begin{array}{l}\text{standard integral, } (2p-1)!!\\ = 1\times3\times5\cdots\times(2p-1)\end{array}\right]$$

$$= \frac{\pi^{1/2}(n+1)!!}{2^{\frac{n+4}{2}}}\left(\frac{2kT}{m}\right)^{\frac{n+3}{2}} \quad [(n+1)!! = 1\times3\times5\cdots\times(n+1)]$$

Thus,

$$\langle v^n \rangle = 4\pi\left(\frac{m}{2\pi kT}\right)^{3/2} \times \frac{\pi^{1/2}(n+1)!!}{2^{\frac{n+4}{2}}}\left(\frac{2kT}{m}\right)^{\frac{n+3}{2}}$$

$$= \boxed{\frac{(n+1)!!}{2^{n/2}}\left(\frac{2kT}{m}\right)^{n/2}} \text{ when } n \text{ is an even, positive integer.}$$

78.13 The work required for a mass, m, to go from a distance R from the center of a planet of mass m_p to infinity is $w = \int_R^\infty F\,dr$ where F is the force of gravity. The force and is given by Newton's law of universal gravitation, which is

$$F = \frac{Gmm_p}{r^2} \text{ where } r \geq R \text{ and } G \text{ is the gravitational constant}$$

Then, $w = \int_R^\infty \frac{Gmm_p}{r^2}\,dr = \frac{Gmm_p}{R}$

According to Newton's second law of motion, $F_{surface} = mg$ where $g = \frac{Gm_p}{R^2}$

Thus, $w = gRm$. This work is the minimum kinetic energy that the particle must have in order to escape the planet's gravitational attraction from a distance R from the planet's center. Hence, the escape velocity, v_e, is given by

$$w = \tfrac{1}{2}mv_e^2 = mgR$$

$$v_e = (2gR)^{1/2}$$

This equation indicates that, although the escape velocity depends upon the gravitational properties of the planet, it is independent of the mass of the escaping particle. A molecule has the same escape velocity as a macroscopic rocket.

What proportion of the molecules have enough speed to escape when the temperature is T? In order to calculate the proportion of molecules that have speeds exceeding the escape velocity we must integrate the Maxwell distribution [78.4] from v_e to infinity. The integration gives the probability that a molecule has a speed greater than or equal to v_e and is easily converted to the proportion of molecules having speed in this range. This integral cannot be evaluated analytically but, if desired, it can be expressed in terms of the error function. However, scientific calculators and computer software easily perform a numerical calculation. Here is a Mathcad Prime 2 worksheet for calculating the probability P:

$$R := 8.3145 \cdot \mathrm{J \cdot K^{-1} \cdot mol^{-1}} \qquad f(v,T,M) := 4 \cdot \pi \cdot \left(\frac{M}{2 \cdot \pi \cdot R \cdot T}\right)^{\frac{3}{2}} \cdot v^2 \cdot e^{\frac{-M \cdot v^2}{2 \cdot R \cdot T}}$$

$$P(v_e, T, M) := \int_{v_e}^{\infty \cdot \mathrm{m \cdot s^{-1}}} f(v,T,M)\,dv$$

(a) Earth: $g_{\mathrm{Earth}} = \dfrac{Gm_{\mathrm{Earth}}}{R^2_{\mathrm{Earth}}} = 9.81\ \mathrm{m\ s^{-2}}$

$$v_e = \left[2 \times (9.81\ \mathrm{m\ s^{-2}}) \times (6.37 \times 10^6\ \mathrm{m})\right]^{1/2} = \boxed{11.2\ \mathrm{km\ s^{-1}}}$$

At what temperatures do H_2, He, and O_2 molecules have mean speeds equal to the escape speeds?

$$v_{\mathrm{mean}} = \left(\frac{8RT}{\pi M}\right)^{1/2} \quad [78.7]$$

$$T = \frac{\pi M v^2_{\mathrm{mean}}}{8R} = \frac{\pi M v_e^2}{8R}$$

$$= \frac{\pi \times (11.2 \times 10^3\ \mathrm{m\ s^{-1}})^2 \times (\mathrm{kg\ mol^{-1}})}{8 \times (8.3145\ \mathrm{J\ K^{-1}\ mol^{-1}})} \times \frac{M}{\mathrm{kg\ mol^{-1}}}$$

$$= (5.92 \times 10^6\ \mathrm{K}) \times \frac{M}{\mathrm{kg\ mol^{-1}}}$$

Thus, $T_{H_2} = (5.92 \times 10^6\ \mathrm{K}) \times (2.016 \times 10^{-3}) = \boxed{1.19 \times 10^4\ \mathrm{K}}$

$$T_{\mathrm{He}} = (5.92 \times 10^6\ \mathrm{K}) \times (4.00 \times 10^{-3}) = \boxed{2.37 \times 10^4\ \mathrm{K}}$$

$$T_{O_2} = (5.92 \times 10^6\ \mathrm{K}) \times (32.00 \times 10^{-3}) = \boxed{1.89 \times 10^5\ \mathrm{K}}$$

What is the probability that the molecules have enough speed to escape when the temperature is T?

For dihydrogen at 240 K and 1 500 K: $P(11.2 \cdot \mathrm{km \cdot s^{-1}}, 240 \cdot \mathrm{K}, 2.016 \cdot \mathrm{gm \cdot mol^{-1}}) = 2.498 \cdot 10^{-27}$

$P(11.2 \cdot \mathrm{km \cdot s^{-1}}, 1\,500 \cdot \mathrm{K}, 2.016 \cdot \mathrm{gm \cdot mol^{-1}}) = 1.487 \cdot 10^{-4}$

For helium at 240 K and 1 500 K: $P(11.2 \cdot \mathrm{km \cdot s^{-1}}, 240 \cdot \mathrm{K}, 4.00 \cdot \mathrm{gm \cdot mol^{-1}}) = 3.21 \cdot 10^{-54}$

$P(11.2 \cdot \mathrm{km \cdot s^{-1}}, 1\,500 \cdot \mathrm{K}, 4.00 \cdot \mathrm{gm \cdot mol^{-1}}) = 1.202 \cdot 10^{-8}$

For dioxygen at 240 K and 1 500 K: $P(11.2 \cdot \mathrm{km \cdot s^{-1}}, 240 \cdot \mathrm{K}, 32.00 \cdot \mathrm{gm \cdot mol^{-1}}) = 0$

$P(11.2 \cdot \mathrm{km \cdot s^{-1}}, 1\,500 \cdot \mathrm{K}, 32.00 \cdot \mathrm{gm \cdot mol^{-1}}) = 1.888 \cdot 10^{-69}$

(b) Mars: $g_{\text{Mars}} = \dfrac{m_{\text{Mars}}}{m_{\text{Earth}}} \times \dfrac{R^2_{\text{Earth}}}{R^2_{\text{Mars}}} \times g_{\text{Earth}}$

$$= (0.108)\times\left(\frac{6.37}{3.38}\right)^2 \times (9.81\,\text{m}\,\text{s}^{-2}) = 3.76\,\text{m}\,\text{s}^{-2}$$

$$v_e = \left[2\times(3.76\text{ m s}^{-2})\times(3.38\times10^6\text{ m})\right]^{1/2} = \boxed{5.04\text{ km s}^{-1}}$$

At what temperatures do H_2, He, and O_2 molecules have mean speeds equal to the escape speeds?

$$v_{\text{mean}} = \left(\frac{8RT}{\pi M}\right)^{1/2} \quad [78.7]$$

$$T = \frac{\pi M v^2_{\text{mean}}}{8R} = \frac{\pi M v_e^2}{8R}$$

$$= \frac{\pi\times(5.04\times10^3\text{ m s}^{-1})^2\times(\text{kg mol}^{-1})}{8\times(8.3145\text{ J K}^{-1}\text{ mol}^{-1})}\times\frac{M}{\text{kg mol}^{-1}}$$

$$= (1.20\times10^6\text{ K})\times\frac{M}{\text{kg mol}^{-1}}$$

Thus, $T_{H_2} = (1.20\times10^6\text{ K})\times(2.016\times10^{-3}) = \boxed{2.42\times10^3\text{ K}}$

$$T_{\text{He}} = (1.20\times10^6\text{ K})\times(4.00\times10^{-3}) = \boxed{4.80\times10^3\text{ K}}$$

$$T_{O_2} = (1.20\times10^6\text{ K})\times(32.00\times10^{-3}) = \boxed{3.84\times10^4\text{ K}}$$

What is the probability that the molecules have enough speed to escape when the temperature is *T*?

For dihydrogen at 240 K and 1 500 K:	$P(5.04\cdot\text{km}\cdot\text{s}^{-1}, 240\cdot\text{K}, 2.016\cdot\text{gm}\cdot\text{mol}^{-1}) = 1.122\cdot10^{-5}$
	$P(5.04\cdot\text{km}\cdot\text{s}^{-1}, 1\,500\cdot\text{K}, 2.016\cdot\text{gm}\cdot\text{mol}^{-1}) = 2.502\cdot10^{-1}$
For helium at 240 K and 1 500 K:	$P(5.04\cdot\text{km}\cdot\text{s}^{-1}, 240\cdot\text{K}, 4.00\cdot\text{gm}\cdot\text{mol}^{-1}) = 4.611\cdot10^{-11}$
	$P(5.04\cdot\text{km}\cdot\text{s}^{-1}, 1\,500\cdot\text{K}, 4.00\cdot\text{gm}\cdot\text{mol}^{-1}) = 4.307\cdot10^{-2}$
For dioxygen at 240 K and 1 500 K:	$P(5.04\cdot\text{km}\cdot\text{s}^{-1}, 240\cdot\text{K}, 32.00\cdot\text{gm}\cdot\text{mol}^{-1}) = 5.659\cdot10^{-88}$
	$P(5.04\cdot\text{km}\cdot\text{s}^{-1}, 1\,500\cdot\text{K}, 32.00\cdot\text{gm}\cdot\text{mol}^{-1}) = 4.246\cdot10^{-14}$

Based on these number alone, it would appear that H_2 and He would be depleted from the atmosphere of both Earth and Mars only after many (millions?) years; that the rate on Mars, though still slow, would be many orders of magnitude larger than on Earth; that O_2 would be retained on Earth indefinitely; and that the rate of O_2 depletion on Mars would be very slow (billions of years?), though not totally negligible. The temperatures of both planets may have been higher in the past than they are now. In the analysis of the data, we must remember that the probabilities, *P*, are not rates of depletion, though the rates should be roughly proportional to *P*.

78.15 Dry atmospheric air is 78.08% N_2, 20.95% O_2, 0.93% Ar, 0.03%CO_2, plus traces of other gases. Nitrogen, oxygen, and carbon dioxide contribute 99.06% of the molecules in a volume with each molecule contributing an average rotational energy equal to kT. The rotational energy density is given by

$$\rho_R = \frac{E_R}{V} = \frac{0.9906N \times \varepsilon^R}{V} = \frac{0.9906 \times \varepsilon^R \times pN_A}{RT}$$
$$= \frac{0.9906\,kTpN_A}{RT} = 0.9906p$$
$$= 0.9906 \times (1.013 \times 10^5 \text{ Pa}) = 0.1004 \text{ J cm}^{-1}$$

The total energy density (translational plus rotational) is

$$\rho_T = \rho_K + \rho_R = 0.15 \text{ J cm}^{-3} + 0.10 \text{ J cm}^{-3} = \boxed{0.25 \text{ J cm}^{-3}}$$

Topic 79 **Transport properties of gases**

Discussion questions

79.1 Fick's first law, $J_z = -D\dfrac{\mathrm{d}\mathcal{N}}{\mathrm{d}z}$ [79.3], arises in a non-equilibrium gas in which the number density $\mathcal{N}$ is a smooth, continuous function of the spatial coordinates. As illustrated in Justification 79.1, the net flux of matter from one infinitesimally small volume of space to an adjacent volume is the difference the matter flux from density to low density and the matter flux from the low density to the high density. This difference is non-zero and proportional to the density (concentration) gradient in the absence of equilibrium. The constant of proportionality is the diffusion constant D and the net flux is always positive down the concentration gradient. That is, there is a net flow of mass from high to low concentration.

79.3 As a hypothesis, we speculate that gaseous molecular attraction effectively reduces the magnitude of the collision cross-section σ. Since the transport properties of diffusion, thermal energy transport, and viscosity in the approximate regime $\sigma << \lambda <<$ *container size* are all inversely proportional to σ, the hypothesis suggests an increase in these transport properties when molecular attraction becomes relevant. Stronger molecular repulsions effectively increase the molecular size parameter σ, thereby, decreasing the magnitude of the transport coefficients, which decreases the gaseous transport properties.

Exercises

79.1(a) For a perfect argon gas: $\lambda = \dfrac{kT}{\sigma p}$ [78.12] $= \dfrac{RT}{\sigma N_A p} = \dfrac{1}{\sigma N_A[\mathrm{Ar}]}$. Thus,

$$\kappa = \tfrac{1}{3}\lambda v_{\mathrm{mean}} C_{V,\mathrm{m}}[\mathrm{Ar}] \ [79.8\mathrm{a}] = \tfrac{1}{3}\left(\frac{1}{\sigma N_A[\mathrm{Ar}]}\right) v_{\mathrm{mean}} C_{V,\mathrm{m}}[\mathrm{Ar}] = \tfrac{1}{3}\left(\frac{C_{V,\mathrm{m}}}{\sigma N_A}\right) v_{\mathrm{mean}}$$

$$= \tfrac{1}{3}\left(\frac{C_{V,\mathrm{m}}}{\sigma N_A}\right) \times \left(\frac{8RT}{\pi M}\right)^{1/2} \quad [78.7]$$

$$= \tfrac{1}{3}\left(\frac{12.5\ \mathrm{J\,K^{-1}\,mol^{-1}}}{(0.36\times10^{-18}\ \mathrm{m^2})\times(6.022\times10^{23}\ \mathrm{mol^{-1}})}\right)$$

$$\times\left(\frac{8\times(8.3145\ \mathrm{J\,K^{-1}\,mol^{-1}})\times(298\ \mathrm{K})}{\pi\times(39.95\times10^{-3}\ \mathrm{kg\,mol^{-1}})}\right)^{1/2}$$

$$= \boxed{7.6\times10^{-3}\ \mathrm{J\,K^{-1}\,m^{-1}\,s^{-1}}}$$

Comment. This calculated value does not agree well with the value of κ listed in Table 79.1.

Question. Can the differences between the calculated and experimental values of κ be accounted for by the difference in temperature (298 K here, 273 K in Table 79.1)? If not, what might be responsible for the difference?

79.2(a)

$$D = \tfrac{1}{3}\lambda v_{\text{mean}} \ [79.7]$$

$$= \tfrac{1}{3}\left(\frac{v_{\text{rel}}}{z}\right)v_{\text{mean}} \ [78.11] = \tfrac{1}{3}\left(\frac{v_{\text{rel}}}{\sigma v_{\text{rel}}p/kT}\right)v_{\text{mean}} \ [78.10]$$

$$= \tfrac{1}{3}\left(\frac{kT}{\sigma p}\right)\times\left(\frac{8RT}{\pi M}\right)^{1/2} \ [78.7]$$

$$= \tfrac{1}{3}\left(\frac{(1.381\times10^{-23}\ \text{J K}^{-1})\times(293.15\ \text{K})}{(0.36\times10^{-18}\ \text{m}^2)\times\text{Pa}}\right)$$

$$\times\left(\frac{8\times(8.3145\ \text{J K}^{-1}\ \text{mol}^{-1})\times(293.15\ \text{K})}{\pi\times(39.95\times10^{-3}\ \text{kg mol}^{-1})}\right)^{1/2}\times\frac{1}{p/\text{Pa}}$$

$$= (1.4\bar{8}\ \text{m}^2\ \text{s}^{-1})\times\frac{1}{p/\text{Pa}}$$

$$J_z/N_A = -\frac{D}{N_A}\frac{d\mathcal{N}}{dz} \ [79.3] = -\frac{D}{N_A}\frac{d}{dz}\left(\frac{N_A p}{RT}\right)$$

$$= -\left(\frac{D}{RT}\right)\frac{dp}{dz}$$

$$= -\left(\frac{1}{(8.3145\ \text{J K}^{-1}\ \text{mol}^{-1})\times(293.15\ \text{K})}\right)\times\left(\frac{1.4\bar{8}\ \text{m}^2\ \text{s}^{-1}}{p/\text{Pa}}\right)$$

$$\times(1.0\times10^5\ \text{Pa m}^{-1})$$

$$= -\frac{60.\bar{7}\ \text{mol m}^{-2}\ \text{s}^{-1}}{p/\text{Pa}}$$

(a) $p = 1.00$ Pa, $\boxed{D = 1.5\ \text{m}^2\ \text{s}^{-1},\ J_z/N_A = -61\ \text{mol m}^{-2}\ \text{s}^{-1}}$

(b) $p = 100$ kPa, $\boxed{D = 1.5\times10^{-5}\ \text{m}^2\ \text{s}^{-1},\ J_z/N_A = -6.1\times10^{-4}\ \text{mol m}^{-2}\ \text{s}^{-1}}$

(c) $p = 10.0$ MPa, $\boxed{D = 1.5\times10^{-7}\ \text{m}^2\ \text{s}^{-1},\ J_z/N_A = -6.1\times10^{-6}\ \text{mol m}^{-2}\ \text{s}^{-1}}$

79.3(a) For a perfect argon gas: $\lambda = \dfrac{kT}{\sigma p}\ [78.12] = \dfrac{RT}{\sigma N_A p} = \dfrac{1}{\sigma N_A[\text{Ar}]}$. Thus,

$$\kappa = \tfrac{1}{3}\lambda v_{\text{mean}}C_{V,\text{m}}[\text{Ar}] \ [79.8\text{a}] = \tfrac{1}{3}\left(\frac{1}{\sigma N_A[\text{Ar}]}\right)v_{\text{mean}}C_{V,\text{m}}[\text{Ar}] = \tfrac{1}{3}\left(\frac{C_{V,\text{m}}}{\sigma N_A}\right)v_{\text{mean}}$$

$$= \tfrac{1}{3}\left(\frac{C_{V,\text{m}}}{\sigma N_A}\right)\times\left(\frac{8RT}{\pi M}\right)^{1/2} \ [78.7]$$

$$= \tfrac{1}{3}\left(\frac{12.5\ \mathrm{J\,K^{-1}\,mol^{-1}}}{(0.36\times10^{-18}\ \mathrm{m^2})\times(6.022\times10^{23}\ \mathrm{mol^{-1}})}\right)$$

$$\times\left(\frac{8\times(8.3145\ \mathrm{J\,K^{-1}\,mol^{-1}})\times(280\ \mathrm{K})}{\pi\times(39.95\times10^{-3}\ \mathrm{kg\,mol^{-1}})}\right)^{1/2}$$

$$= 7.4\times10^{-3}\ \mathrm{J\,K^{-1}\,m^{-1}\,s^{-1}}$$

$$J = -\kappa\frac{\mathrm{d}T}{\mathrm{d}z}\ [79.4]$$

$$= -(7.4\times10^{-3}\ \mathrm{J\,K^{-1}\,m^{-1}\,s^{-1}})\times(10.5\ \mathrm{K\,m^{-1}})$$

$$= \boxed{-0.078\ \mathrm{J\,m^{-2}\,s^{-1}}}$$

79.4(a) For a perfect neon gas: $C_{V,\mathrm{m}} = C_{p,\mathrm{m}} - R = 20.786\ \mathrm{J\,K^{-1}\,mol^{-1}} - R = 12.472\ \mathrm{J\,K^{-1}\,mol^{-1}}$

$$\kappa = \tfrac{1}{3}\lambda v_{\mathrm{mean}} C_{V,\mathrm{m}}[\mathrm{Ne}]\ [79.8a] = \tfrac{1}{3}\left(\frac{1}{\sigma N_{\mathrm{A}}[\mathrm{Ne}]}\right)v_{\mathrm{mean}} C_{V,\mathrm{m}}[\mathrm{Ne}] = \tfrac{1}{3}\left(\frac{C_{V,\mathrm{m}}}{\sigma N_{\mathrm{A}}}\right)v_{\mathrm{mean}}$$

$$= \tfrac{1}{3}\left(\frac{C_{V,\mathrm{m}}}{\sigma N_{\mathrm{A}}}\right)\times\left(\frac{8RT}{\pi M}\right)^{1/2}\quad [78.7]$$

Solve for σ.

$$\sigma = \tfrac{1}{3}\left(\frac{C_{V,\mathrm{m}}}{\kappa N_{\mathrm{A}}}\right)\times\left(\frac{8RT}{\pi M}\right)^{1/2}$$

$$= \tfrac{1}{3}\left(\frac{12.472\ \mathrm{J\,K^{-1}\,mol^{-1}}}{(46.5\times10^{-3}\ \mathrm{J\,K^{-1}\,m^{-1}\,s^{-1}})\times(6.022\times10^{23}\ \mathrm{mol^{-1}})}\right)$$

$$\times\left(\frac{8\times(8.3145\ \mathrm{J\,K^{-1}\,mol^{-1}})\times(273\ \mathrm{K})}{\pi\times(20.18\times10^{-3}\ \mathrm{kg\,mol^{-1}})}\right)^{1/2}$$

$$= 7.95\times10^{-20}\ \mathrm{m^2} = \boxed{0.0795\ \mathrm{nm^2}}$$

The value reported in Table 78.1 is $0.24\ \mathrm{nm^2}$. Question: What approximations inherent in the equation used in the solution to this exercise are likely to cause a factor of 3 difference?

79.5(a) The thermal energy flux ('heat' flux) is described by: $J(\text{energy}) = -\kappa\dfrac{dT}{dz}$ [79.4]

where the negative sign indicates flow toward lower temperature. This is the rate of energy transfer per unit area. The total rate of energy transfer across area A is

$$\frac{\mathrm{d}E}{\mathrm{d}t} = AJ(\text{energy}) = -\kappa A\frac{\mathrm{d}T}{\mathrm{d}z}$$

To calculate the temperature gradient with the given data, we assume that the gradient is in a steady-state. Then, recognizing that temperature differences have identical magnitude in Celsius or Kelvin units,

$$\frac{dT}{dz} = \frac{\Delta T}{\Delta z} = \frac{\{(-15)-(28)\}\ \text{K}}{1.0\times10^{-2}\ \text{m}} = -4.3\times10^{3}\ \text{K m}^{-1}.$$

We now assume that the coefficient of thermal conductivity of the gas between the window panes is comparable to that of nitrogen given in Table 79.1: $\kappa \approx 0.0240\ \text{J K}^{-1}\ \text{m}^{-1}\ \text{s}^{-1}$. Then, the rate of outward energy transfer is

$$\frac{dE}{dt} \approx -(0.0240\ \text{J K}^{-1}\ \text{m}^{-1}\ \text{s}^{-1})\times(1.0\ \text{m}^2)\times(-4.3\times10^{3}\ \text{K m}^{-1})$$

$$\approx 10\bar{3}\ \text{J s}^{-1} \quad \text{or} \quad \boxed{10\bar{3}\ \text{W}}$$

A $10\bar{3}$ W heater is needed to balance this rate of heat loss.

79.6(a)

$$\eta = \frac{pMD}{RT}\ [79.9] = \left(\frac{pM}{RT}\right)\times\left(\frac{kT}{3\sigma p}\right)\times\left(\frac{8RT}{\pi M}\right)^{1/2}$$

$$= \left(\frac{k}{3\sigma}\right)\times\left(\frac{8MT}{\pi R}\right)^{1/2}$$

Solve for σ.

$$\sigma = \left(\frac{k}{3\eta}\right)\times\left(\frac{8MT}{\pi R}\right)^{1/2}$$

$$= \tfrac{1}{3}\left(\frac{1.381\times10^{-23}\ \text{J K}^{-1}}{298\times10^{-6}\times10^{-1}\ \text{kg m}^{-1}\ \text{s}^{-1}}\right)\times\left(\frac{8\times(20.18\times10^{-3}\ \text{kg mol}^{-1})\times(273\ \text{K})}{\pi\times(8.3145\ \text{J K}^{-1}\ \text{mol}^{-1})}\right)^{1/2}$$

$$= 2.01\times10^{-19}\ \text{m}^2 = \boxed{0.201\ \text{nm}^2}$$

79.7(a)

$$\eta = \frac{pMD}{RT}\ [79.9\text{c}] = \left(\frac{pM}{RT}\right)\times\left(\frac{kT}{3\sigma p}\right)\times\left(\frac{8RT}{\pi M}\right)^{1/2}\ [79.7] = \left(\frac{k}{3\sigma}\right)\times\left(\frac{8MT}{\pi R}\right)^{1/2}$$

$$= \tfrac{1}{3}\left(\frac{1.381\times10^{-23}\ \text{J K}^{-1}}{0.40\times10^{-18}\ \text{m}^2}\right)\times\left(\frac{8\times(29.0\times10^{-3}\ \text{kg mol}^{-1})\times T}{\pi\times(8.3145\ \text{J K}^{-1}\ \text{mol}^{-1})}\right)^{1/2}$$

$$= (1.08\times10^{-6}\ \text{kg m}^{-1}\ \text{s}^{-1})\times(T/\text{K})^{1/2} = (1.08\times10^{-5}\ \text{P})\times(T/\text{K})^{1/2}$$

(a) At 273 K, $\eta = 178\ \mu\text{P}$

(b) At 298 K, $\eta = 186\ \mu\text{P}$

(c) At 1000 K, $\eta = 342\ \mu\text{P}$

Problems

79.1 Viscosity is independent of pressure in simple molecular kinetic theory.

$$\eta = \frac{pMD}{RT}\ [79.9\text{c}] = \left(\frac{pM}{RT}\right)\times\left(\frac{kT}{3\sigma p}\right)\times\left(\frac{8RT}{\pi M}\right)^{1/2}$$

$$= \left(\frac{k}{3\sigma}\right)\times\left(\frac{8MT}{\pi R}\right)^{1/2}$$

Solve for σ.

$$\sigma=\left(\frac{k}{3\eta}\right)\times\left(\frac{8MT}{\pi R}\right)^{1/2}$$

$$=\tfrac{1}{3}\left(\frac{1.381\times10^{-23}\ \text{J K}^{-1}}{(10^{-7}\ \text{kg m}^{-1}\ \text{s}^{-1})\times(\eta/\mu\text{P})}\right)\times\left(\frac{8\times(17.03\times10^{-3}\ \text{kg mol}^{-1})\times\text{K}}{\pi\times(8.3145\ \text{J K}^{-1}\ \text{mol}^{-1})}\right)^{1/2}$$

$$\times(T/\text{K})^{1/2}$$

$$=(3.325\times10^{-18}\ \text{m}^2)\times\frac{(T/\text{K})^{1/2}}{\eta/\mu\text{P}}=(3.325\ \text{nm}^2)\times\frac{(T/\text{K})^{1/2}}{\eta/\mu\text{P}}$$

(a) At 270 K, $\eta=90.8\ \mu\text{P}$, $\boxed{\sigma=0.602\ \text{nm}^2}$, $\boxed{d=(\sigma/\pi)^{1/2}=438\ \text{pm}}$

(b) At 490 K, $\eta=174.9\ \mu\text{P}$, $\boxed{\sigma=0.421\ \text{nm}^2}$, $\boxed{d=(\sigma/\pi)^{1/2}=366\ \text{pm}}$

The smaller molecular diameter at higher temperature is consistent with the idea that, at higher temperatures, more forceful collisions contract a molecule's perimeter.

79.3 $\lambda=kT/(\sigma p)$ [78.12] or, since $T/p=1/k\mathcal{N}$ where $\mathcal{N}$ is the number density, $\lambda=1/\sigma\mathcal{N}$

$$D=\tfrac{1}{3}\lambda v_{\text{mean}}\ [79.7]$$

$$=\tfrac{1}{3}\left(\frac{1}{\sigma\mathcal{N}}\right)v_{\text{mean}}\ [78.11]=\tfrac{1}{3}\left(\frac{1}{\pi(2a_0)^2\mathcal{N}}\right)\times\left(\frac{8RT}{\pi M}\right)^{1/2}\ [78.7]$$

$$=\tfrac{1}{3}\left(\frac{1}{\pi\times(2\times5.29\times10^{-11}\ \text{m})^2\times(1.0\times10^6\ \text{m}^{-3})}\right)$$

$$\times\left(\frac{8\times(8.3145\ \text{J K}^{-1}\ \text{mol}^{-1})\times(1.00\times10^4\ \text{K})}{\pi\times(1.0079\times10^{-3}\ \text{kg mol}^{-1})}\right)^{1/2}$$

$$=\boxed{2.37\times10^{17}\ \text{m}^2\ \text{s}^{-1}}$$

$$\kappa=\tfrac{1}{3}\lambda v_{\text{mean}}C_{V,\text{m}}[\text{H}]\ [79.8\text{a}]=\tfrac{1}{3}\left(\frac{1}{\sigma N_\text{A}[\text{H}]}\right)v_{\text{mean}}C_{V,\text{m}}[\text{H}]=\tfrac{1}{3}\left(\frac{C_{V,\text{m}}}{\sigma N_\text{A}}\right)v_{\text{mean}}$$

$$=\tfrac{1}{3}\left(\frac{\tfrac{3}{2}R}{\sigma N_\text{A}}\right)\times\left(\frac{8RT}{\pi M}\right)^{1/2}\ \text{[78.7 and equipartition theorm]}$$

$$=\tfrac{1}{2}\left(\frac{k}{\pi(2a_0)^2}\right)\times\left(\frac{8RT}{\pi M}\right)^{1/2}$$

$$=\tfrac{1}{2}\left(\frac{1.381\times10^{-23}\ \text{J K}^{-1}}{\pi\times(2\times5.29\times10^{-11}\ \text{m})^2}\right)\times\left(\frac{8\times(8.3145\ \text{J K}^{-1}\ \text{mol}^{-1})\times(1.00\times10^4\ \text{K})}{\pi\times(1.0079\times10^{-3}\ \text{kg mol}^{-1})}\right)^{1/2}$$

$$=\boxed{2.85\ \text{J K}^{-1}\ \text{m}^{-1}\ \text{s}^{-1}}$$

The validity of these calculations is in doubt because the kinetic theory of gases assumes the Maxwell–Boltzmann distribution, essentially an equilibrium distribution. In such a dilute medium, the timescales on which particles exchange energy by collision make an assumption of equilibrium unwarranted. It is especially dubious considering that atoms are more likely to interact with photons from stellar radiation than with other atoms.

Topic 80 **Motion in liquids**

Discussion questions

80.1 The **ionic radius**, as assigned according to the distances between ions in a crystal, is a measure of ion size. The **hydrodynamic radius** (or **Stokes radius**) of an ion is its effective radius in solution taking into account all the water molecules it carries in its **hydration shell**. A hydrodynamic radius of a small ion is typically much larger than the ionic radius. This happens because small ions give rise to stronger electric fields than large ones so the small ions are more extensively solvated than big ones. Thus, an ion of small ionic radius may have a large hydrodynamic radius because it drags many solvent molecules through the solution as it migrates.

80.3 Eqn 80.12 indicates that, for the passage of a solvated ion sphere through a continuous media, ion mobility is inversely proportional to viscosity. To test the thought that this transport mechanism explains observed differences between proton mobility in water and in liquid ammonia, we look up the respective viscosities and find that the viscosity of ammonia is 0.276×10^{-3} kg m^{-1} s^{-1} at -40°C and the viscosity of water is 0.891×10^{-3} kg m^{-1} s^{-1} at 25°C. This mechanism predicts a larger proton mobility in ammonia, the reverse of fact. This, and the fact that proton mobility in water is extraordinarily large, forces us to look for another mechanism of proton transport. Thus, we turn to text Fig. 80.3.

Text Figure 80.3 is a schematic illustration of fast proton transfer along a chain of hydrogen bonded water molecules. Each step along the chain of events has the basic form illustrated in Fig. 80.1.

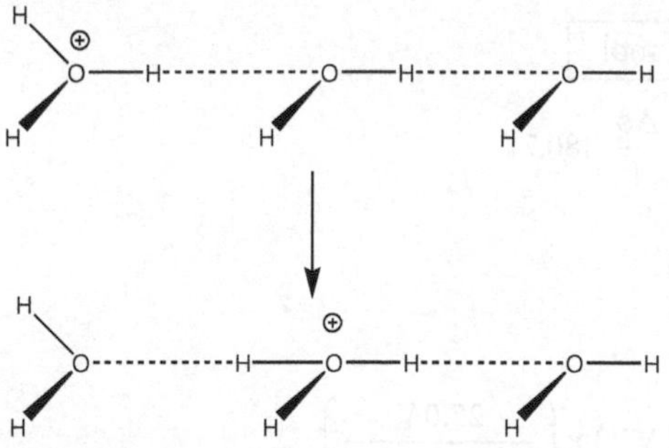

Figure 80.1

There is a strong hydrogen bond in liquid water and the speed of proton transfer suggests a low activation energy for the proton jump. This activation energy is readily available via molecule thermal motions. Proton mobility in liquid ammonia may occur by a similar mechanism but, whereas the protonated species in water is viewed as the simplified H_3O^+ species, the protonated species in ammonia is NH_4^+. The hydrogen bond between ammonia molecules is weaker than the hydrogen bond in water but the lower charge mobility suggests a larger ratio of activation energy to available thermal energy for the proton jump. The collision frequency is comparable between water and ammonia; however, fewer of these thermal collisions have the requisite energy to promote the proton jump in liquid ammonia.

Exercises

80.1(a) We take the natural logarithm of eqn 80.2 and solve for the activation energy, E_a.

$$\eta = \eta_0 e^{E_a/RT} \quad [80.2]$$
$$\ln \eta = \ln \eta_0 + E_a / RT$$
$$\ln \eta_{T_1} - \ln \eta_{T_2} = \frac{E_a}{R} \times \left(\frac{1}{T_1} - \frac{1}{T_2}\right)$$

Therefore,

$$E_a = \frac{R \ln\left(\eta_{T_1} / \eta_{T_2}\right)}{\left(\dfrac{1}{T_1} - \dfrac{1}{T_2}\right)}$$
$$= \frac{\left(8.3145 \text{ J K}^{-1} \text{ mol}^{-1}\right) \times \ln(1.002/0.7975)}{\left(\dfrac{1}{293 \text{ K}} - \dfrac{1}{303 \text{ K}}\right)} = \boxed{16.8 \text{ J mol}^{-1}}$$

80.2(a) Molar ionic conductivity is related to mobility by

$$\lambda = zuF \quad [80.13]$$
$$= 1 \times \left(7.91 \times 10^{-8} \text{ m}^2 \text{ s}^{-1} \text{ V}^{-1}\right) \times \left(96485 \text{ C mol}^{-1}\right)$$
$$= \boxed{7.63 \times 10^{-3} \text{ S m}^2 \text{ mol}^{-1}}$$

80.3(a) $s = u\mathcal{E}$ [80.11] and $\mathcal{E} = \dfrac{\Delta\phi}{l}$ [80.7]

Therefore,

$$s = u\left(\frac{\Delta\phi}{l}\right)$$
$$= (7.92 \times 10^{-8} \text{ m}^2 \text{ s}^{-1} \text{ V}^{-1}) \times \left(\frac{25.0 \text{ V}}{7.00 \times 10^{-3} \text{ m}}\right)$$
$$= 2.83 \times 10^{-4} \text{ m s}^{-1} \quad \text{or} \quad \boxed{283 \text{ μm s}^{-1}}$$

80.4(a) The basis for the solution is the law of independent migration of ions (eqn 80.6). Switching counterions does not affect the mobility of the remaining other ion at infinite dilution.

$$\Lambda_m^o = \nu_+\lambda_+ + \nu_-\lambda_- \quad [80.6]$$

$$\Lambda_m^o(\text{NaI}) = \lambda(\text{Na}^+) + \lambda(\text{I}^-) = 12.69 \text{ mS m}^2 \text{ mol}^{-1}$$

$$\Lambda_m^o(\text{NaNO}_3) = \lambda(\text{Na}^+) + \lambda(\text{NO}_3^-) = 12.16 \text{ mS m}^2 \text{ mol}^{-1}$$

$$\Lambda_m^o(\text{AgNO}_3) = \lambda(\text{Ag}^+) + \lambda(\text{NO}_3^-) = 13.34 \text{ mS m}^2 \text{ mol}^{-1}$$

Hence,

$$\Lambda_m^o(\text{AgI}) = \Lambda_m^o(\text{AgNO}_3) + \Lambda_m^o(\text{NaI}) - \Lambda_m^o(\text{NaNO}_3)$$
$$= (13.34 + 12.69 - 12.16) \text{ mS m}^2 \text{ mol}^{-1} = \boxed{13.87 \text{ mS m}^2 \text{ mol}^{-1}}$$

Question. How well does this result agree with the value calculated directly from the data of Table 80.2?

80.5(a) $u = \dfrac{\lambda}{zF}$ [80.17] ; $z = 1$; $1\text{ S} = 1\,\Omega^{-1} = 1\text{ C V}^{-1}\text{ s}^{-1}$

$$u(\text{Li}^+) = \frac{3.87 \text{ mS m}^2 \text{ mol}^{-1}}{9.6485 \times 10^4 \text{ C mol}^{-1}} = 4.01 \times 10^{-5} \text{ mS C}^{-1} \text{ m}^2$$
$$= \boxed{4.01 \times 10^{-8} \text{ m}^2 \text{ V}^{-1} \text{ s}^{-1}}$$

$$u(\text{Na}^+) = \frac{5.01 \text{ mS m}^2 \text{ mol}^{-1}}{9.6485 \times 10^4 \text{ C mol}^{-1}} = \boxed{5.19 \times 10^{-8} \text{ m}^2 \text{ V}^{-1} \text{ s}^{-1}}$$

$$u(\text{K}^+) = \frac{7.35 \text{ mS m}^2 \text{ mol}^{-1}}{9.6485 \times 10^4 \text{ C mol}^{-1}} = \boxed{7.62 \times 10^{-8} \text{ m}^2 \text{ V}^{-1} \text{ s}^{-1}}$$

80.6(a) $a = \dfrac{kT}{6\pi\eta D}$ [80.19*b*]; $1\text{ P} = 10^{-1} \text{ kg m}^{-1} \text{ s}^{-1}$

$$a = \frac{(1.381 \times 10^{-23} \text{ J K}^{-1}) \times (298 \text{ K})}{6\pi \times (1.00 \times 10^{-3} \text{ kg m}^{-1} \text{ s}^{-1}) \times (5.2 \times 10^{-10} \text{ m}^2 \text{ s}^{-1})}$$
$$= 4.2 \times 10^{-10} \text{ m} \quad \text{or} \quad \boxed{420 \text{ pm}}$$

Problems

80.1 We take the natural logarithm of eqn 80.2

$$\eta \propto e^{E_a/RT} \quad [80.2]$$
$$\ln \eta = \text{constant} + E_a/RT$$

and recognize that a plot of ln η against $1/T$ has a slope equal to E_a/R. Thus, a linear regression fit of ln η against $1/T$, shown in Fig. 80.2, yields the slope from which we calculate E_a with the expression $E_a = \text{slope} \times R$.

$$E_a = (1222\ \text{K}) \times \left(8.3145\ \text{J K}^{-1}\ \text{mol}^{-1}\right) = \boxed{10.2\ \text{kJ mol}^{-1}}$$

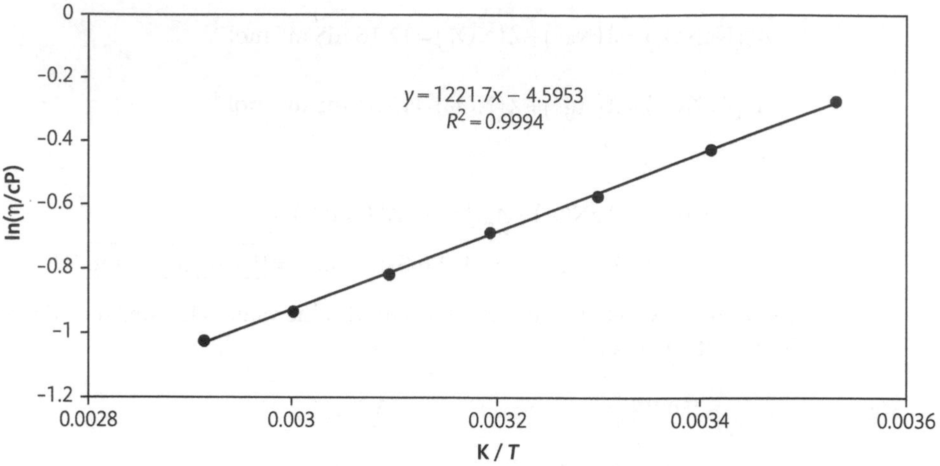

Figure 80.2

80.3 The molar conductivity, Λ_m, is related to the conductivity, κ, by $\Lambda_m = \kappa/c$ [80.4] and the **Kohlrausch law** [80.5] indicates that molar conductivity is linear in $c^{1/2}$.

$$\Lambda_m = \Lambda_m^o - \mathcal{K} c^{1/2} \quad [80.5]$$

We draw a data table and calculate values for a plot of Λ_m against $c^{1/2}$:

$c/(\text{mol dm}^{-3})$	1.334	1.432	1.529	1.672	1.725
$\kappa/(\text{mS cm}^{-1})$	131	139	147	156	164
$c^{1/2}/(\text{mol dm}^{-3})^{1/2}$	1.155	1.197	1.237	1.293	1.313
$\Lambda_m/(\text{mS m}^2\ \text{mol}^{-1})$	9.82	9.71	9.61	9.33	9.51

The plot, shown in Fig. 80.3, is linear and the linear regression fit yields the Kohlrausch parameters:

$$\Lambda_m{}^\circ = \boxed{12.78\ \text{mS m}^2\ \text{mol}^{-1}}$$

$$\mathcal{K} = \boxed{2.57\ \text{mS m}^2\ (\text{mol dm}^{-1})^{-3/2}}$$

80.5 The molar conductivity, Λ_m, is related to the conductivity, κ, by $\Lambda_m = \kappa/c$ [80.4] $= C/Rc$ where the cell constant is $C = 0.2063\ \text{cm}^{-1}$. The Kohlrausch law [80.5] indicates that molar conductivity is linear in $c^{1/2}$.

$$\Lambda_m = \Lambda_m{}^\circ - \mathcal{K} c^{1/2} \quad [80.5]$$

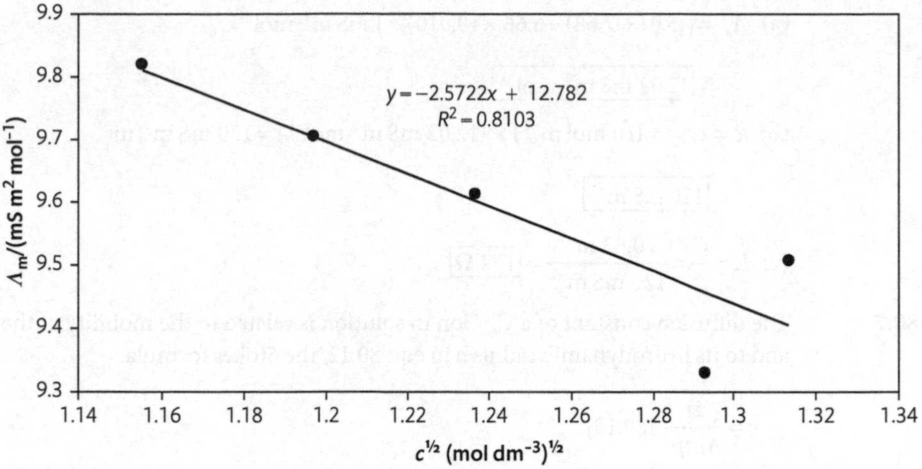

Figure 80.3

We draw a data table and calculate values for a plot of Λ_m against $c^{1/2}$:

$c/(\text{mol dm}^{-3})$	0.00050	0.0010	0.0050	0.010	0.020	0.050
R/Ω	3314	1669	342.1	174.1	89.08	37.14
$c^{1/2}/(\text{mol dm}^{-3})^{1/2}$	0.0224	0.0316	0.0707	0.100	0.141	0.224
$\Lambda_m/(\text{mS m}^2\text{ mol}^{-1})$	12.45	12.36	12.06	11.85	11.58	11.11

The plot, shown in Fig. 80.3, is linear and the linear regression fit yields the intercept and slope. The intercept is the limiting molar conductivity and the slope is the negative of the Kohlrausch parameter $\mathcal{K}$;

$$\Lambda_m^\circ = \boxed{12.6 \text{ mS m}^2 \text{ mol}^{-1}}$$

$$\mathcal{K} = \boxed{6.66 \text{ mS m}^2 \text{ (mol dm}^{-1})^{-3/2}}$$

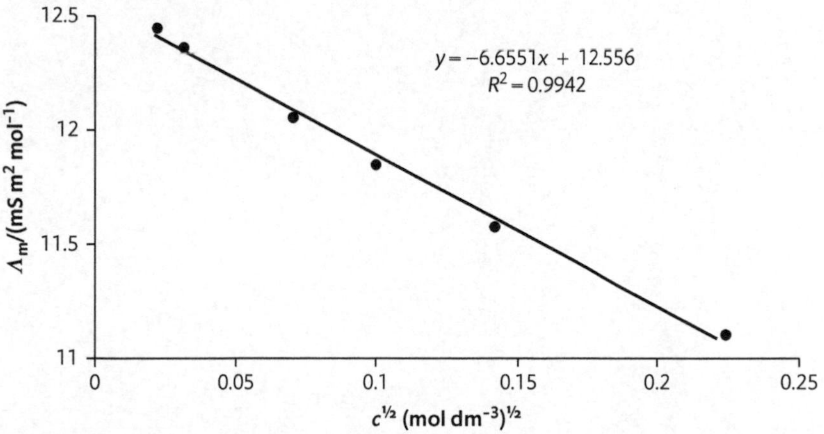

Figure 80.4

(a) $\Lambda_m = \left((5.01+7.68)-6.66\times(0.010)^{1/2}\right)\ \text{mS m}^2\ \text{mol}^{-1}$

$= \boxed{12.02\ \text{mS m}^2\ \text{mol}^{-1}}$

(b) $\kappa = c\Lambda_m = (10\ \text{mol m}^{-3})\times(12.02\ \text{mS m}^2\ \text{mol}^{-1}) = 120\ \text{mS m}^2\ \text{m}^{-3}$

$= \boxed{120\ \text{mS m}^{-1}}$

(c) $R = \dfrac{C}{\kappa} = \dfrac{20.63\ \text{m}^{-1}}{120\ \text{mS m}^{-1}} = \boxed{172\ \Omega}$

80.7 The diffusion constant of a C_{60}^- ion in solution is related to the mobility of the ion and to its hydrodynamic radius a in eqn 80.12, the Stokes formula:

$$a = \frac{ze}{6\pi\eta u}\ [80.12]$$

$$= \frac{(1)\times\left(1.602\times10^{-19}\ \text{C}\right)}{6\pi\left(0.93\times10^{-3}\ \text{kg m}^{-1}\text{s}^{-1}\right)\times\left(1.1\times10^{-8}\ \text{m}^2\ \text{V}^{-1}\ \text{s}^{-1}\right)} = 8.3\times10^{-10}\ \text{m} = \boxed{0.83\ \text{nm}}$$

This is substantially larger than the 0.5 nm van der Waals radius of a Buckminsterfullerene (C_{60}) molecule because the anion attracts a considerable hydration shell through the London dispersion attraction to the non-polar solvent molecules and through the ion-induced dipole interaction. The Stokes radius reflects the larger effective radius of the combined anion and its hydration shell.

80.9 Since $D \propto \eta^{-1}$ [80.16 and 80.12] and $\eta \propto e^{E_a/RT}$ [80.2], we expect that $D \propto e^{-E_a/RT}$

Therefore, we take the ratio of D at two different temperatures so as to eliminate the constant of proportionality, and solve for E_a to find:

$$E_a = -\frac{R\ln\left(D_{T_1}/D_{T_2}\right)}{\left(\dfrac{1}{T_1}-\dfrac{1}{T_2}\right)} = -\frac{\left(8.3145\ \text{J K}^{-1}\ \text{mol}^{-1}\right)\times\ln(2.89/2.05)}{\dfrac{1}{298\ \text{K}}-\dfrac{1}{273\ \text{K}}} = 9.3\ \text{kJ mol}^{-1}$$

That is, the activation energy for diffusion is $\boxed{9.3\ \text{kJ mol}^{-1}}$

Topic 81 Diffusion

Discussion questions

81.1 The **thermodynamic force** $\mathcal{F}$ is defined in eqn 81.1.

$$\mathcal{F}=-\left(\frac{\partial \mu}{\partial x}\right)_{p,T} \quad [81.1]$$

This expression is a summary of the second thermodynamic law that molecules move in the direction that minimizes the chemical potential of the molecules when p and T are local constants. $\mathcal{F}$ is not one of the "real" forces such as gravity or electromagnetism; it is the negative gradient of the chemical potential, which has a balance of terms involving enthalpy and entropy:

$$\mu_{\rm J}=\left(\frac{\partial G}{\partial n_{\rm J}}\right)_{p,T,n'}[69.4]=\left(\frac{\partial}{\partial n_{\rm J}}\right)_{p,T,n'}(H-TS)$$

Thus, the thermodynamic force moves molecules so as to minimize the enthalpy, to which molecular interactions provide a great contribution, while simultaneously attempting to maximize entropy. Often times one, or the other, of these tendencies predominate. For an ideal solution the gradient of the molar enthalpy is zero so the force represents the spontaneous tendency for molecules to disperse so that entropy is maximized.

Exercises

81.1(a) Eqn 81.14, $\langle x^2\rangle=2Dt$, gives the mean square distance travelled in any one dimension in time t. We need the distance travelled from a point in any direction. The distinction here is the distinction between the one-dimensional and three-dimensional diffusion. The mean square three dimensional distance can be obtained from the one dimensional mean square distance since motions in the three directions are independent.

$$r^2=x^2+y^2+z^2 \text{ [Pythagorean theorem]},$$

$$\begin{aligned}\langle r^2\rangle&=\langle x^2\rangle+\langle y^2\rangle+\langle z^2\rangle=3\langle x^2\rangle \text{ [independent motion]}\\&=3\times 2Dt \text{ [81.14 for } \langle x^2\rangle]\\&=6Dt\end{aligned}$$

$$\text{Therefore,}\quad t=\frac{\langle r^2\rangle}{6D}=\frac{(5.0\times10^{-3}\ \text{m})^2}{(6)\times(6.73\times10^{-10}\ \text{m}^2\,\text{s}^{-1})}=\boxed{6.2\times10^3\ \text{s}}$$

81.2(a) The diffusion equation solution for these boundary conditions is provided in eqn 81.11 with

$$n_0=(20.0\ \text{g})\times\left(\frac{1\ \text{mol sucrose}}{342.30\ \text{g}}\right)=5.84\times10^{-2}\ \text{mol sucrose},$$

$A=5.0\ \text{cm}^2$, $D=5.216\times10^{-9}\ \text{m}^2\text{s}^{-1}$, and $x\ =\ 10$ cm.

$$c(x,t)=\frac{n_0}{A(\pi Dt)^{1/2}}\text{e}^{-x^2/4Dt}\ [81.11]$$

$$c(10\ \text{cm},t)=\frac{5.84\times10^{-2}\ \text{mol}}{(5.0\times10^{-4}\ \text{m}^2)\times\{\pi(5.216\times10^{-9}\ \text{m}^2\ \text{s}^{-1})\}^{1/2}\,t^{1/2}}\text{e}^{-(0.10\ \text{m})^2/4\times(5.216\times10^{-9}\ \text{m}^2\ \text{s}^{-1})\times t}$$
$$=(9.12\times10^5\ \text{mol m}^{-3})\times(t/\text{s})^{-1/2}\,\text{e}^{-(4.79\times10^5)/(t/\text{s})}$$

(a) $t=10$ s:

$$c(10\ \text{cm},10\ \text{s})=(9.12\times10^5\ \text{mol m}^{-3})\times(10)^{-1/2}\,\text{e}^{-(4.79\times10^5)/(10)}=\boxed{0.00\ \text{mol dm}^{-3}}$$

(b) $t=24\times3\,600\ \text{s}=8.64\times10^4$ s:

$$c(10\ \text{cm},8.64\times10^4\ \text{s})=(9.12\times10^5\ \text{mol m}^{-3})\times(8.64\times10^4)^{-1/2}\,\text{e}^{-(4.79\times10^5)/(8.64\times10^4)}$$
$$=12.1\ \text{mol m}^{-3}=\boxed{0.0121\ \text{mol dm}^{-3}}$$

Problems

81.1

$$\mathcal{F}=-\frac{RT}{c}\times\frac{\text{d}c}{\text{d}x}\ [81.2\text{b}]$$

$$\frac{\text{d}c}{\text{d}x}=\frac{(0.05-0.10)\ \text{mol dm}^{-3}}{0.10\ \text{m}}=-0.50\ \text{mol dm}^{-3}\ \text{m}^{-1}\ [\text{linear gradation}]$$

$$RT=2.48\times10^3\ \text{J mol}^{-1}=2.48\times10^3\ \text{N m mol}^{-1}$$

(a) $$\mathcal{F}=-\left(\frac{2.48\times10^3\ \text{N m mol}^{-1}}{0.10\ \text{mol dm}^{-3}}\right)\times(-0.50\ \text{mol dm}^{-3}\ \text{m}^{-1})$$
$$=\boxed{12\ \text{kN mol}^{-1}}\ \text{or}\ \boxed{2.0\times10^{-20}\ \text{N molecule}^{-1}}$$

(b) $$\mathcal{F}=-\left(\frac{2.48\times10^3\ \text{N m mol}^{-1}}{0.075\ \text{mol dm}^{-3}}\right)\times(-0.50\ \text{mol dm}^{-3}\ \text{m}^{-1})$$
$$=\boxed{16.\bar{5}\ \text{kN mol}^{-1}}\ \text{or}\ \boxed{2.7\times10^{-20}\ \text{N molecule}^{-1}}$$

(c) $$\mathcal{F}=-\left(\frac{2.48\times10^3\ \text{N m mol}^{-1}}{0.050\ \text{mol dm}^{-3}}\right)\times(-0.50\ \text{mol dm}^{-3}\ \text{m}^{-1})$$
$$=\boxed{24.\bar{8}\ \text{kN mol}^{-1}}\ \text{or}\ \boxed{4.1\times10^{-20}\ \text{N molecule}^{-1}}$$

81.3 $\mathcal{F} = -\dfrac{RT}{c} \times \dfrac{dc}{dx}$ [81.2b] with the axis origin at the center of the tube

$RT = 2.48 \times 10^3\ \text{J mol}^{-1} = 2.48 \times 10^3\ \text{N m mol}^{-1}$

$c = c(x) = c_0\left(1-e^{-ax^2}\right)$ where $c_0 = 0.100\ \text{mol dm}^{-3}$ and $a = 0.10\ \text{cm}^{-2}$

$$\frac{dc}{dx} = 2axc_0 e^{-ax^2}$$

$$\mathcal{F} = -2aRTxe^{-ax^2}\left(1-e^{-ax^2}\right)^{-1} = 2aRTx\left(1-e^{ax^2}\right)^{-1} = \left(50.\ \text{kN cm}^{-1}\ \text{mol}^{-1}\right)x\left(1-e^{ax^2}\right)^{-1}$$

A plot of the force per mole against x is shown in Fig. 81.1. It demonstrates that mass is pushed by the thermodynamic force toward the center of the tube where the concentration is lowest; a negative force pushes toward the left, positive force pushes toward the right. The force per molecule is calculated with the equation:

$$\mathcal{F} = 2aRTx\left(1-e^{ax^2}\right)^{-1}/N_A = \left(8.2 \times 10^{-23}\ \text{kN cm}^{-1}\ \text{molecule}^{-1}\right)x\left(1-e^{ax^2}\right)^{-1}$$

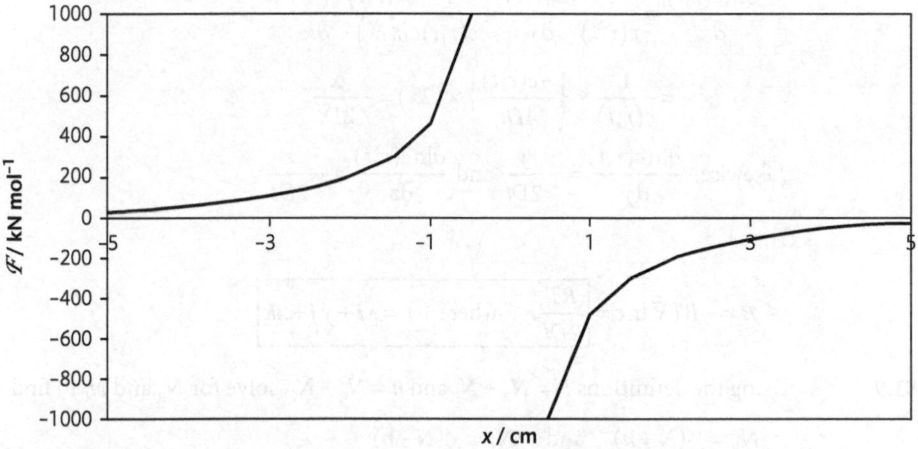

Figure 81.1

81.5 The diffusion equation is: $\dfrac{\partial c}{\partial t} = D\dfrac{\partial^2 c}{\partial x^2}$ [81.7].

We confirm that $c(x,t) = \dfrac{a}{t^{1/2}}e^{-bx^2/t}$ [81.11], where $a = \dfrac{n_0}{A(\pi D)^{1/2}}$ and $b = \dfrac{1}{4D}$, is a solution to the diffusion equation by taking both the partial derivative w/r/t time and the second partial w/r/t position to find whether they are proportional.

$$\frac{\partial c}{\partial t} = -\left(\tfrac{1}{2}\right)\times\left(\frac{a}{t^{3/2}}\right)e^{-bx^2/t} + \left(\frac{a}{t^{1/2}}\right)\times\left(\frac{bx^2}{t^2}\right)e^{-bx^2/t} = -\frac{c}{2t} + \frac{bx^2}{t^2}c$$

$$\frac{\partial c}{\partial x} = \left(\frac{a}{t^{1/2}}\right)\times\left(\frac{-2bx}{t}\right)e^{-bx^2/t}$$

$$\frac{\partial^2 c}{\partial x^2} = -\left(\frac{2b}{t}\right)\times\left(\frac{a}{t^{1/2}}\right)e^{-bx^2/t} + \left(\frac{a}{t^{1/2}}\right)\times\left(\frac{2bx}{t}\right)^2 e^{-bx^2/t} = -\left(\frac{2b}{t}\right)c + \left(\frac{2bx}{t}\right)^2 c$$

$$= -\left(\frac{1}{2Dt}\right)c + \left(\frac{bx^2}{Dt^2}\right)c$$

$$= \frac{1}{D}\frac{\partial c}{\partial t} \text{ as required.}$$

Initially the material is concentrated at $x=0$. Note that $c=0$ for $x>0$ when $t=0$ on account of the very strong exponential factor $\left(e^{-bx^2/t} \to 0 \text{ more strongly than } \frac{1}{t^{1/2}} \to \infty\right)$. When $x=0$, $e^{-x^2/4Dt}=1$. We confirm the correct behavior by noting that $\langle x\rangle = 0$ and $\langle x^2\rangle = 0$ at $t=0$ [81.3 and 81.14], and so all the material must be at $x=0$ at $t=0$.

81.7 $c(r,t) = \dfrac{n_0}{8(\pi Dt)^{3/2}} e^{-r^2/4Dt}$ [81.12] where $r^2 = x^2+y^2+z^2$

$$\frac{\partial \ln c(r,t)}{\partial x} = \frac{1}{c(r,t)}\frac{\partial c(r,t)}{\partial x} = \frac{1}{c(r,t)}\frac{\partial c(r,t)}{\partial (r^2)}\frac{\partial (r^2)}{\partial x}$$

$$= \frac{1}{c(r,t)}\times\left\{\frac{-c(r,t)}{4Dt}\right\}\times(2x) = -\frac{x}{2Dt}$$

Likewise, $\dfrac{\partial \ln c(r,t)}{\partial y} = -\dfrac{y}{2Dt}$ and $\dfrac{\partial \ln c(r,t)}{\partial z} = -\dfrac{z}{2Dt}$

Thus,

$$\mathcal{F} = -RT\nabla \ln c = \boxed{\frac{RT}{2Dt}\boldsymbol{r} \quad \text{where} \quad \boldsymbol{r} = x\boldsymbol{i} + y\boldsymbol{j} + z\boldsymbol{k}}$$

81.9 Using the definitions $N = N_R + N_L$ and $n = N_R - N_L$, solve for N_R and N_L to find

$$N_R = ½(N+n) \quad \text{and} \quad N_L = ½(N-n)$$

Following the discussion of Justification 81.2, we then have

$$P(n\lambda) = \frac{\text{number of paths with } N_R \text{ steps to the right}}{\text{total number of paths}} = \frac{N!}{(N_L)!N_R!\,2^N}$$

$$= \frac{N!}{\{½(N+n)\}!\{½(N-n)\}!\,2^N}$$

This is the "exact" random walk probability. After application of Stirling's approximation we have the "approximate" probability. P_{Approx} may be written in terms of the variables (x,t) or (n,N) because $x = n\lambda$ and $t = N\tau$.

$$P_{\text{Approx}} = \left(\frac{2\tau}{\pi t}\right)^{1/2} e^{-x^2\tau/2t\lambda^2} \text{ [81.15]} = \left(\frac{2}{\pi N}\right)^{1/2} e^{-n^2/2N}$$

We calculate the probability P of being at $x = 6\lambda$ for $N = 4, 6, \ldots 180$ using Mathcad Prime 2 and we plot P_{Exact} against N. We include a plot of the fractional deviation of P_{Approx} against N from which we see that the deviation drops below 0.1% when $\boxed{N > 60}$.

$$n := 6 \quad P_{Exact}(N) := \frac{N!}{\left(\frac{1}{2}\cdot(N+n)\right)!\left(\frac{1}{2}\cdot(N-n)\right)!2^N} \quad P_{Approx}(N) := \left(\frac{2}{\pi\cdot N}\right)^{\frac{1}{2}}\cdot e^{-\frac{n^2}{2\cdot N}}$$

$N := 6, 8..180$

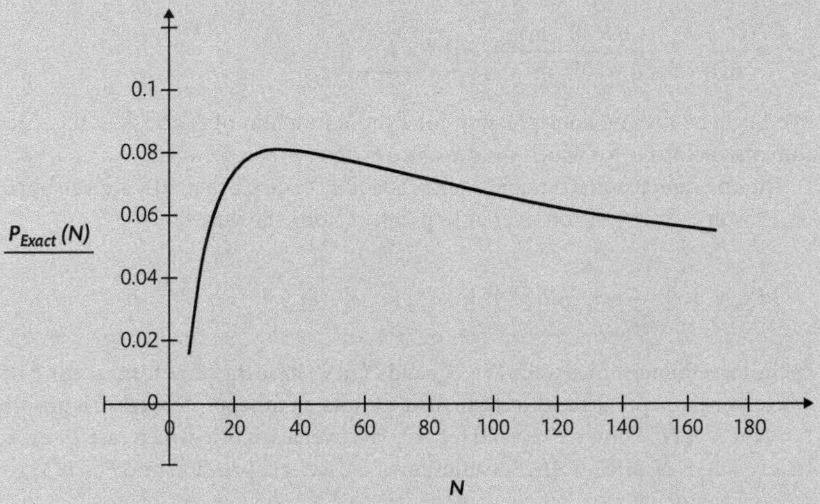

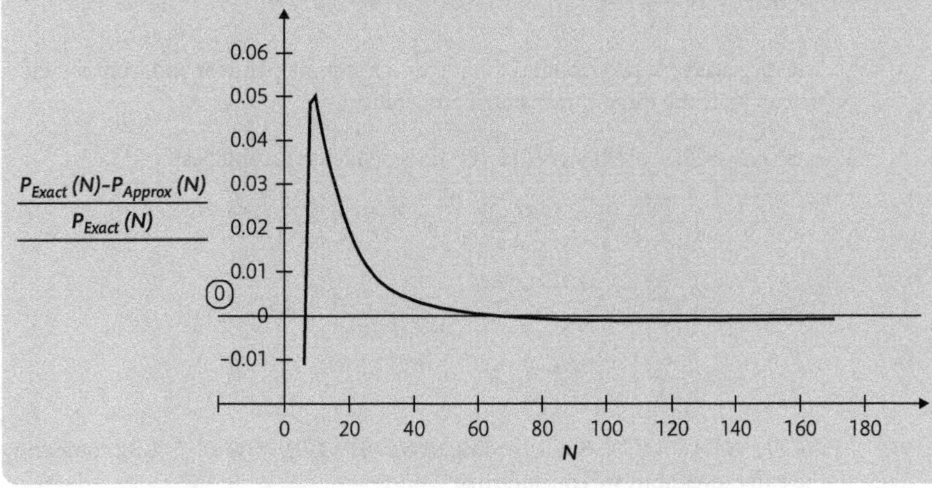

81.11 Eqn 81.14, $\langle x^2 \rangle = 2Dt$, gives the mean square distance traveled in any one dimension in time t. We need the distance traveled from a point in any direction. The distinction here is the distinction between the one-dimensional and three-dimensional diffusion. The mean square three-dimensional distance can be obtained from the one-dimensional mean square distance since motions in the three directions are independent.

$$r^2 = x^2 + y^2 + z^2 \text{ [Pythagorean theorem]}$$
$$\langle r^2\rangle = \langle x^2\rangle + \langle y^2\rangle + \langle z^2\rangle = 3\langle x^2\rangle \text{ [independent motion]}$$
$$= 3\times 2Dt \text{ [81.14 for } \langle x^2\rangle] = 6Dt$$

Therefore,

$$t = \frac{\langle r^2\rangle}{6D} = \frac{(1.0\times 10^{-6}\text{ m})^2}{6(1.0\times 10^{-11}\text{ m}^2\text{ s}^{-1})} = \boxed{1.7\times 10^{-2}\text{ s}}$$

81.13 We begin by finding an expression for Dt as a function of N and λ in the three-dimensional space for which $-\infty < x < \infty, -\infty < y < \infty, -\infty < z < \infty$.

The one-dimensional random walk probability of displacement x is given by eqn 81.15 with N_x being the number of steps taken along the x-axis.

$$P(x, N_x) = \left(\frac{2}{\pi N_x}\right)^{1/2} e^{-x^2/2N_x\lambda^2} \quad [81.15]$$

In the three-dimensional walk, $P(y, N_y)$ and $P(z, N_z)$ have the same form as eqn 81.15 when there is no preferred direction. Also, the total number of steps taken is given by $N = N_x + N_y + N_z$. Eqn 81.15 is valid for very large values of N in which case we expect that $N_x = N_y = N_z$ or $N_x = \frac{1}{3}N$. Substitution of the latter expression into $P(x, N_x)$ gives

$$P(x, N) = \left(\frac{6}{\pi N}\right)^{1/2} e^{-3x^2/2N\lambda^2}$$

The displacement probabilities for x, y, and z are independent and, consequently, we can write the three-dimensional probability as

$$P(r, N) = P(x, N)P(y, N)P(z, N) \text{ [independent probabilities]}$$
$$= \left(\frac{6}{\pi N}\right)^{1/2} e^{-3x^2/2N\lambda^2}\left(\frac{6}{\pi N}\right)^{1/2} e^{-3y^2/2N\lambda^2}\left(\frac{6}{\pi N}\right)^{1/2} e^{-3z^2/2N\lambda^2}$$
$$= \left(\frac{6}{\pi N}\right)^{3/2} e^{-3(x^2+y^2+z^2)/2N\lambda^2}$$
$$= \left(\frac{6}{\pi N}\right)^{3/2} e^{-3r^2/2N\lambda^2}$$

So, $P(r, N) \propto e^{-3r^2/2N\lambda^2}$. But according to eqn 81.12 $P(r, t) \propto e^{-r^2/4Dt}$. By comparing the two expressions we conclude that

$$\boxed{Dt = N\lambda^2/6}$$

Now, we turn to solving the current problem, which places the object at the coordinate origin at $t = 0$ and determines r_{mean} after N steps when the object has access to these regions of hemispherical space: $-\infty < x < \infty, -\infty < y < \infty, 0 \le z < \infty$. Eqn 81.12

indicates that the probability density for r is proportional to $\exp(-r^2/4Dt)$. The normalization constant $C_{\text{normalization}}$ for the hemispherical space is:

$$C_{\text{normalization}}^{-1} = \int_0^{\infty}\int_{-\infty}^{\infty}\int_{-\infty}^{\infty} e^{-r^2/4Dt}\,dx\,dy\,dz = \int_0^{\infty}\int_0^{2\pi}\int_0^{\pi/2} e^{-r^2/4Dt} r^2 \sin\theta\,d\theta\,d\phi\,dr$$

$$= 2\pi\int_0^{\infty} e^{-r^2/4Dt} r^2\,dr = 2\pi \times \left(2\pi^{1/2}(Dt)^{3/2}\right) = 4(\pi Dt)^{3/2} \text{ [standard integral]}$$

$$C_{\text{normalization}} = \frac{1}{4(\pi Dt)^{3/2}}$$

$$\langle r\rangle = C_{\text{normalization}}\int_0^{\infty}\int_{-\infty}^{\infty}\int_{-\infty}^{\infty} re^{-r^2/4Dt}\,dx\,dy\,dz = C_{\text{normalization}}\int_0^{\infty}\int_0^{2\pi}\int_0^{\pi/2} e^{-r^2/4Dt} r^3 \sin\theta\,d\theta\,d\phi\,dr$$

$$= \left(\frac{1}{4(\pi Dt)^{3/2}}\right) \times 2\pi \times \left(8D^2t^2\right)$$

$$= 4\left(\frac{Dt}{\pi}\right)^{1/2} = 4\left(\frac{N\lambda^2}{6\pi}\right)^{1/2}$$

$$= \boxed{\lambda\left(\frac{8N}{3\pi}\right)^{1/2}}$$

Focus 16: Integrated activities

F16.1 $\langle x^2\rangle = 2Dt$ [81.14], $D = \dfrac{kT}{6\pi\eta a}$ [80.19b]

$$\eta = \frac{kT}{6\pi Da} = \frac{kTt}{3\pi a\langle x^2\rangle} = \frac{(1.381\times10^{-23}\text{ J K}^{-1})\times(298.15\text{ K})\times t}{(3\pi)\times(2.12\times10^{-7}\text{ m})\times\langle x^2\rangle}$$

$$= (2.06\times10^{-15}\text{ J m}^{-1})\times\left(\frac{t}{\langle x^2\rangle}\right) = (2.06\times10^{-3}\text{ J m}^{-3}\text{ s})\times\left(\frac{t/\text{s}}{\langle x^2\rangle/10^{-12}\text{ m}^2}\right)$$

$$= (2.06\times10^{-3}\text{ kg m}^{-1}\text{ s}^{-1})\times\left(\frac{t/\text{s}}{\langle x^2\rangle/10^{-12}\text{ m}^2}\right)$$

We draw up the following table:

t/s	30	60	90	120
$\langle x^2\rangle/10^{-12}$ m^2	88.2	113.5	128	144
$\eta/10^{-3}$ kg m^{-1} s^{-1}	0.701	1.09	1.45	1.72

Hence, the mean value is 1.2×10^{-3} kg m^{-1} s^{-1} with a standard deviation of 0.4×10^{-3} kg m^{-1} s^{-1}.

Alternatively, we can view the media as having an effective viscosity that depends upon size of the latex spheres. Then, a plot of viscosity against $\langle x^2\rangle$ reveals a linear relation with a least squares regression fit of

$$\boxed{\text{Effective viscosity}/(10^{-3}\text{ kg m}^{-1}\text{ s}^{-1}) = 0.01860\times\langle x^2\rangle/(10^{-12}\text{ m}^2) - 0.9626}$$

$R^2 = 0.9915$

Topic 82 **Reaction rates**

Discussion question

D82.1 As far as this topic is concerned, the order of a reaction with respect to a species (a product or a reactant) is the power to which the concentration of that species is raised in a power-law type of rate law (eqn 82.7). The **overall order** of a reaction is the sum of the orders with respect to all of the species that appear in the rate law. Reaction orders are empirical quantities, and they need not be integers or half-integers. In Topic 86, we will see that integer reaction orders are common, though, and we will see how they come about.

The relevance of a pseudofirst-order reaction or of any pseudo- reaction order for that matter, is not so much what the order appears to be but rather the strategy involved that allows for the empirical determination of the rate law. In the technique called the **isolation method**, reaction conditions can often be manipulated such that the concentrations of all reactants but one are in such excess that their concentrations can be considered constant over the course of a kinetics experiment. In that case, a rate law of the form

$$v = k_r[A]^a[B]^b \ldots$$

would appear to simplify to

$$v = k_r'[A]^a$$

where the apparent rate constant, k_r' includes the actual rate constant and nearly constant concentration terms:

$$k_r' = k_r[B]_0^b \ldots$$

A pseudofirst-order reaction, then, is a reaction that is first-order with respect to one reactant run under conditions such that the overall order of the reaction appears to be first order. (That is, in the example above, $a = 1$.)

Topic 83 examines the mathematical properties of various rate laws in some detail. Refer to Table 83.3. We will consider only reactions whose rates depend on the concentration of a single reactant (so that the overall order is the order with respect to that one reactant).

In a first-order reaction, the rate of reaction is directly proportional to the concentration of the reactant:

$$v = k_r[A] \qquad \text{and} \qquad \ln\,[A] = \ln\,[A]_0 - k_r t \ [83.1b]$$

A plot of the logarithm of reactant concentration against time is a straight line.

In a zero-order reaction, the rate of reaction is constant, independent of the reactant concentration:

$$v = k_r \qquad \text{and} \qquad [A] = [A]_0 - k_r t \quad \text{[based on Table 83.3]}$$

A plot of the reactant concentration itself against time is a straight line.

In a second-order reaction, the rate of reaction is proportional to the square of the reactant concentration:

$$v = k_r[A]^2 \qquad \text{and} \qquad \frac{1}{[A]} = \frac{1}{[A]_0} + k_r t \quad [83.4b]$$

A plot of the reciprocal of reactant concentration against time is a straight line.

Exercises

E82.1(a) Let the initial amount of ICl be n_{ICl} and the initial amount of H_2 be n_H; the initial amounts of I_2 and HCl are assumed to be zero. Thus, the initial total quantity of gas is $n_{ICl} + n_H$. Let the amount of I_2 formed at any given time be n. In that case, the amount of HCl is $2n$, that of H_2 is $n_H - n$, and the amount of ICl is $n_{ICl} - 2n$. At any given time, then, the total quantity of gas is

$$n_{total} = n_{ICl} - 2n + n_H - n + n + 2n = n_{ICl} + n_H = n_{initial}$$

Thus there is no change in the amount of gas during the course of the reaction. Since there is no change in volume or temperature either, there is $\boxed{\text{no change in pressure}}$.

Comment. Measuring the pressure would **not** be a practical way of monitoring the progress of this reaction.

E82.2(a) $v = \frac{1}{\nu_J}\frac{d[J]}{dt}$ [82.3b] so $\frac{d[J]}{dt} = \nu_J v$

Rate of formation of $C = 3v = \boxed{8.1\ \text{mol dm}^{-3}\ \text{s}^{-1}}$

Rate of formation of $D = v = \boxed{2.7\ \text{mol dm}^{-3}\ \text{s}^{-1}}$

Rate of consumption of $A = v = \boxed{2.7\ \text{mol dm}^{-3}\ \text{s}^{-1}}$

Rate of consumption of $B = 2v = \boxed{5.4\ \text{mol dm}^{-3}\ \text{s}^{-1}}$

E82.3(a) $v = \frac{1}{\nu_J}\frac{d[J]}{dt}\ [82.3b] = \frac{1}{2}\frac{d[C]}{dt} = \frac{1}{2} \times (2.7\ \text{mol dm}^{-3}\ \text{s}^{-1}) = \boxed{1.3\overline{5}\ \text{mol dm}^{-3}\ \text{s}^{-1}}$

Rate of formation of $D = 3v = \boxed{4.0\overline{5}\ \text{mol dm}^{-3}\ \text{s}^{-1}}$

Rate of consumption of $A = 2v = \boxed{2.7\ \text{mol dm}^{-3}\ \text{s}^{-1}}$

Rate of consumption of $B = v = \boxed{1.3\overline{5}\ \text{mol dm}^{-3}\ \text{s}^{-1}}$

E82.4(a) The rate is expressed in mol dm^{-3} s^{-1}; therefore

$$\text{mol dm}^{-3}\ \text{s}^{-1} = [k_r] \times (\text{mol dm}^{-3}) \times (\text{mol dm}^{-3}),$$

where $[k_r]$ denotes units of k_r, requires the units to be $\boxed{\text{dm}^3\ \text{mol}^{-1}\ \text{s}^{-1}}$

(a) Rate of formation of $A = v = \boxed{k_r[A][B]}$

(b) Rate of consumption of $C = 3v = \boxed{3k_r[A][B]}$

E82.5(a) Given $\dfrac{d[C]}{dt} = k_r[A][B][C]$,

the rate of reaction is [82.3b]

$$v = \frac{1}{\nu_J}\frac{d[J]}{dt} = \frac{1}{2}\frac{d[C]}{dt} = \boxed{\frac{1}{2}k_r[A][B][C]}$$

The units of k_r, $[k_r]$, must satisfy

$$\text{mol dm}^{-3}\ \text{s}^{-1} = [k_r]\times(\text{mol dm}^{-3})\times(\text{mol dm}^{-3})\times(\text{mol dm}^{-3})$$

Therefore, $[k_r] = \boxed{\text{dm}^6\ \text{mol}^{-2}\ \text{s}^{-1}}$

E82.6(a) (a) For a second-order reaction, denoting the units of k_r by $[k_r]$

$$\text{mol dm}^{-3}\ \text{s}^{-1} = [k_r]\times(\text{mol dm}^{-3})^2;\ \text{therefore}\ \boxed{[k_r] = \text{dm}^3\ \text{mol}^{-1}\ \text{s}^{-1}}$$

For a third-order reaction

$$\text{mol dm}^{-3}\ \text{s}^{-1} = [k_r]\times(\text{mol dm}^{-3})^3;\ \text{therefore}\ \boxed{[k_r] = \text{dm}^6\ \text{mol}^{-2}\ \text{s}^{-1}}$$

(b) For a second-order reaction

$$\text{kPa s}^{-1} = [k_r]\times\text{kPa}^2;\ \text{therefore}\ \boxed{[k_r] = \text{kPa}^{-1}\ \text{s}^{-1}}$$

For a third-order reaction

$$\text{kPa s}^{-1} = [k_r]\times\text{kPa}^3;\ \text{therefore}\ \boxed{[k_r] = \text{kPa}^{-2}\ \text{s}^{-1}}$$

Problem

P82.1 The rate law is

$$v = k_r[A]^a \propto p_A^a = \{p_{A,0}(1-f)\}^a$$

where f is the fraction reacted. Thus

$$\frac{v_1}{v_2} = \frac{p_{A,1}^a}{p_{A,2}^a} = \left(\frac{1-f_1}{1-f_2}\right)^a$$

Taking logarithms

$$\ln\left(\frac{v_1}{v_2}\right) = a\ln\left(\frac{1-f_1}{1-f_2}\right)$$

$$\text{so}\quad a = \frac{\ln\left(\dfrac{v_1}{v_2}\right)}{\ln\left(\dfrac{1-f_1}{1-f_2}\right)} = \frac{\ln\left(\dfrac{9.71}{7.67}\right)}{\ln\left(\dfrac{0.90}{0.80}\right)} = 2.0$$

The reaction is $\boxed{\text{second order}}$.

Comment. Knowledge of the initial pressure is not required for the solution to this exercise. The ratio of pressures was computed using fractions of the initial pressure.

Topic 83 **Integrated rate laws**

Discussion question

D83.1 The determination of a rate law is simplified by the isolation method in which the concentrations of all the reactants except one are in large excess. If B is in large excess, for example, then to a good approximation its concentration is constant throughout the reaction. Although the true rate law might be $v = k_r[A]^a[B]^b$, we can approximate [B] by $[B]_0$ and write

$$v = k_r'[A]^a, \text{ where } k_r' = k_r[B]_0^{\,b} \quad \text{[analogous to 82.10]}$$

which depends on the concentration of only one reactant. The dependence of the rate on the concentration of each of the reactants may be found by isolating them in turn (by having all the other substances present in large excess), and so constructing the overall rate law.

In the method of initial rates, which is often used in conjunction with the isolation method, the rate is measured at the beginning of the reaction for several different initial concentrations of reactants. We shall suppose that the rate law for a reaction with A isolated is $v = k_r'[A]^a$; then its initial rate, v_0, is given by the initial values of the concentration of A, and we write $v_0 = k_r'[A]_0^{\,a}$. Taking logarithms gives

$$\log v_0 = \log k_r' + a \log [A]_0 \quad [82.11]$$

For a series of initial concentrations, a plot of the logarithms of the initial rates against the logarithms of the initial concentrations of A should be a straight lime with slope *a*.

The method of initial rates might not reveal the full rate law, for the products may participate in the reaction and affect the rate. For example, products participate in the synthesis of HBr, where the full rate law depends on the concentration of HBr. To avoid this difficulty, the rate law should be fitted to the data throughout the reaction. The fitting may be done, in simple cases at least, by using a proposed rate law to predict the concentration of any component at any time, and comparing it with the data.

Because rate laws are differential equations, we must integrate them if we want to find the concentrations as a function of time. Even the most complex rate laws may be integrated numerically. However, in a number of simple cases analytical solutions are easily obtained and prove to be very useful. These are summarized in Table 83.3. In order to determine the rate law, one plots the right hand side of the integrated rate laws shown in the table against t in order to see which of them results in a straight line through the origin. The one that does is the correct rate law. One must be careful to plot the data for longer than the initial rate, however. If concentration data are available only over a relatively short time span, then more than one integrated rate law could conceivably fit the data.

Exercises

E83.1(a) Table 83.3 gives a general expression for the half-life of a reaction of the type $A \rightarrow P$ for orders other than 1:

$$t_{1/2} = \frac{2^{n-1}-1}{(n-1)k_r[A]_0^{n-1}} \propto [A]_0^{1-n} \propto p_0^{1-n}$$

where the proportionality constants may be functions of the reaction order, the rate constant, or even the temperature, but not of the concentration. Form a ratio of the half-lives at different initial pressures:

$$\frac{t_{1/2}(p_{0,1})}{t_{1/2}(p_{0,2})} = \left(\frac{p_{0,1}}{p_{0,2}}\right)^{1-n} = \left(\frac{p_{0,2}}{p_{0,1}}\right)^{n-1}$$

Hence $\ln\left(\frac{t_{1/2}(p_{0,1})}{t_{1/2}(p_{0,2})}\right) = (n-1)\ln\left(\frac{p_{0,2}}{p_{0,1}}\right)$

or $(n-1) = \dfrac{\ln\left(\frac{410\text{ s}}{880\text{ s}}\right)}{\ln\left(\frac{169\text{ Torr}}{363\text{ Torr}}\right)} = 0.999 \approx 1$

Therefore, $\boxed{n=2}$

E83.2(a) $2\,N_2O_5 \rightarrow 4\,NO_2 + O_2 \quad v = k_r[N_2O_5]$

Therefore, rate of consumption of $N_2O_5 = 2v = 2k_r[N_2O_5]$

$$\frac{d[N_2O_5]}{dt} = -2k_r[N_2O_5] \quad \text{so} \quad [N_2O_5] = [N_2O_5]_0 e^{-2k_r t}$$

Solve this for t:

$$t = \frac{1}{2k_r}\ln\frac{[N_2O_5]_0}{[N_2O_5]}$$

Therefore, the half life is:

$$t_{1/2} = \frac{1}{2k_r}\ln 2 = \frac{\ln 2}{(2)\times(3.38\times10^{-5}\text{ s}^{-1})} = \boxed{1.03\times10^4\text{ s}}$$

Since the partial pressure of N_2O_5 is proportional to its concentration

$$p(N_2O_5) = p_0(N_2O_5)e^{-2k_r t}$$

(a) $p(N_2O_5) = (500\,\text{Torr}) \times \left(e^{-(2\times3.38\times10^{-5}/\text{s})\times(50\,\text{s})}\right) = \boxed{498\,\text{Torr}}$

(b) $p(N_2O_5) = (500\,\text{Torr}) \times \left(e^{-(2\times3.38\times10^{-5}/\text{s})\times(20\times60\,\text{s})}\right) = \boxed{461\,\text{Torr}}$

Comment. The half-life formula in Table 83.3 is based on a rate constant for the rate of change of the reactant, that is, based on the assumption that

$$-\frac{d[A]}{dt} = k_r[A]$$

Our expression for the rate of consumption has $2k_r$ instead of k_r, and our expression for $t_{1/2}$ does likewise.

E83.3(a) The integrated second-order rate law for a reaction of the type A + B → products is

$$k_r t = \frac{1}{[B]_0 - [A]_0}\ln\left(\frac{[B]/[B]_0}{[A]/[A]_0}\right) \quad [83.8]$$

Introducing $[B] = [B]_0 - x$ and $[A] = [A]_0 - x$ and rearranging we obtain

$$k_r t = \left(\frac{1}{[B]_0 - [A]_0}\right)\ln\left(\frac{[A]_0([B]_0 - x)}{([A]_0 - x)[B]_0}\right)$$

Solving for x yields, after some rearranging,

$$x = \frac{[A]_0[B]_0\left(e^{k_r([B]_0-[A]_0)t} - 1\right)}{[B]_0 e^{([B]_0-[A]_0)k_r t} - [A]_0} = \frac{(0.060)\times(0.110\,\text{mol dm}^{-3})\times\left(e^{(0.110-0.060)\times0.11\times t/\text{s}} - 1\right)}{(0.110)\times e^{(0.110-0.060)\times0.11\times t/\text{s}} - 0.060}$$

$$= \frac{(0.060\,\text{mol dm}^{-3})\times(e^{0.0055t/\text{s}} - 1)}{e^{0.0055t/\text{s}} - 0.55}$$

(a) After 20 s

$$x = \frac{(0.060\,\text{mol dm}^{-3})\times(e^{0.0055\times20} - 1)}{e^{0.0055\times20} - 0.55} = 0.0122\,\text{mol dm}^{-3}$$

which implies that

$$[CH_3COOC_2H_5] = (0.110 - 0.0122)\,\text{mol dm}^{-3} = \boxed{0.098\,\text{mol dm}^{-3}}$$

(b) After 15 min = 900 s,

$$x = \frac{(0.060\,\text{mol dm}^{-3})\times(e^{0.0055\times900} - 1)}{e^{0.0055\times900} - 0.55} = 0.060\,\text{mol dm}^{-3}$$

$$\text{so}\quad [\mathrm{CH_3COOC_2H_5}] = (0.110 - 0.060)\,\mathrm{mol\,dm^{-3}} = \boxed{0.050\,\mathrm{mol\,dm^{-3}}}$$

E83.4(a) The rate of consumption of A is

$$-\frac{\mathrm{d[A]}}{\mathrm{d}t} = 2v = 2k_\mathrm{r}[\mathrm{A}]^2 \quad [\nu_\mathrm{A} = -2]$$

which integrates to $\dfrac{1}{[\mathrm{A}]} - \dfrac{1}{[\mathrm{A}]_0} = 2k_\mathrm{r}t$ [83.4b with k_r replaced by $2k_\mathrm{r}$]

Therefore, $t = \dfrac{1}{2k_\mathrm{r}}\left(\dfrac{1}{[\mathrm{A}]} - \dfrac{1}{[\mathrm{A}]_0}\right)$

$$t = \left(\frac{1}{2\times 4.30\times 10^{-4}\,\mathrm{dm^3\,mol^{-1}\,s^{-1}}}\right)\times\left(\frac{1}{0.010\,\mathrm{mol\,dm^{-3}}} - \frac{1}{0.210\,\mathrm{mol\,dm^{-3}}}\right)$$
$$= \boxed{1.11\times 10^5\,\mathrm{s}} = \boxed{1.28\ \text{days}}$$

Problems

P83.1 The concentration of A varies with time as

$$[\mathrm{A}] = [\mathrm{A}]_0 \mathrm{e}^{-k_\mathrm{r}t} \quad [83.1\mathrm{b}]$$

Dimensionless concentrations $[\mathrm{A}]/[\mathrm{A}]_0$ and $[\mathrm{B}]/[\mathrm{A}]_0$ are plotted against the dimensionless time $k_\mathrm{r}t$ in Fig. 83.1(a). The same variables are plotted against a logarithmic horizontal axis in Fig. 83.1(b). The latter graph shows that most of the change happens in the interval $0.3 < k_\mathrm{r}t < 3$.

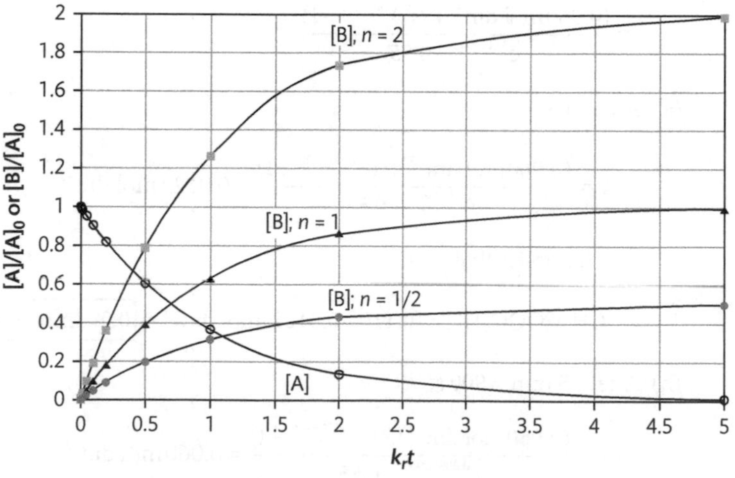

Fig 83.1(a)

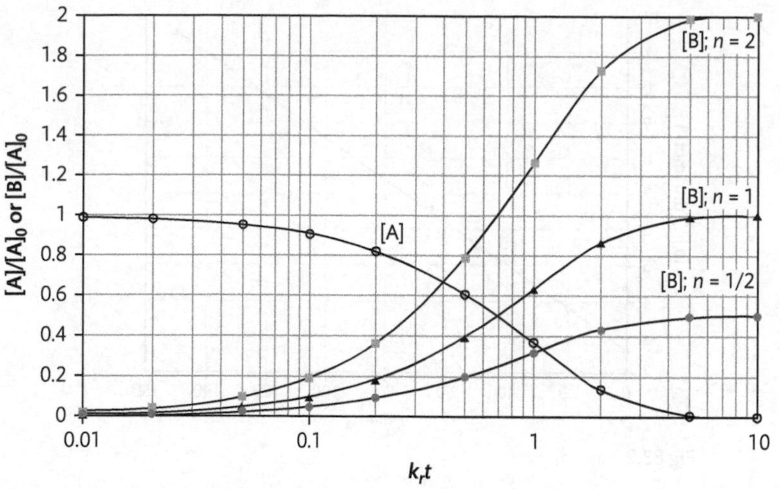

Fig 83.1(b)

P83.3 A simple but practical approach is to make an initial guess at the order by observing whether the half-life of the reaction appears to depend on concentration. If it does not, the reaction is first-order; if it does, refer to Table 83.3 for an expression for the half-life of a reaction of the type $A \rightarrow P$ for orders other than 1:

$$t_{1/2} = \frac{2^{n-1}-1}{(n-1)k_r[A]_0^{n-1}} \propto [A]_0^{1-n}$$

Examination of the data shows that the first half-life is roughly 45 minutes and the second is about double the first. (Compare the 0-50 minute data to the 50-150 minute data.) That is, the half-life starting from **half** of the initial concentration is about **twice** the initial half-life, suggesting that the half-life is inversely proportional to initial concentration:

$$t_{1/2} \propto [A]_0^{-1} = [A]_0^{1-n} \text{ with } n = 2$$

Confirm this suggestion by plotting $1/[A]$ against time. A second-order will obey

$$\frac{1}{[A]} = k_r t + \frac{1}{[A]_0} \quad [83.4b]$$

We draw up the following table ($A = NH_4CNO$)

t/min	0	20.0	50.0	65.0	150
m(urea)/g	0	7.0	12.1	13.8	17.7
m(A)/g	22.9	15.9	10.8	9.1	5.2
[A]/(mol dm^{-3})	0.381	0.265	0.180	0.152	0.0866
[A]$^{-1}$/(dm^3 mol^{-1})	2.62	3.78	5.56	6.60	11.5

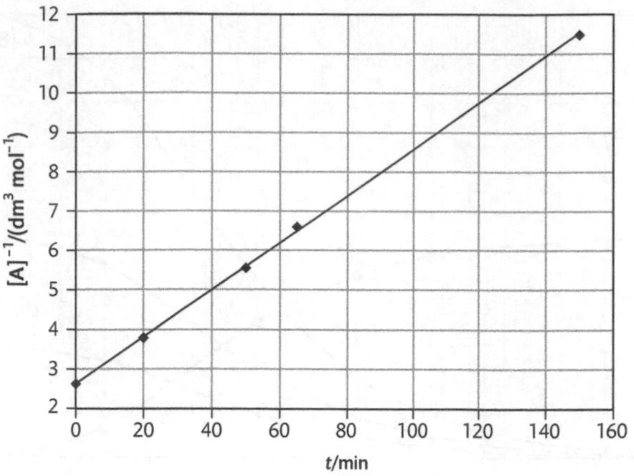

Fig 83.2

The data are plotted in Fig. 83.2 and fit closely to a straight line. Hence, the reaction is indeed $\boxed{\text{second-order}}$. The rate constant is the slope: $\boxed{k_r = 0.059\bar{4}\,\text{dm}^3\,\text{mol}^{-1}\,\text{min}^{-1}}$. To find [A] at 300 min, use eqn 83.4c:

$$[\text{A}] = \frac{[\text{A}]_0}{1 + k_r t[\text{A}]_0} = \frac{0.381\,\text{mol dm}^{-3}}{1 + (0.059\bar{4}) \times (300) \times (0.381)} = 0.048\bar{9}\,\text{mol dm}^{-3}$$

The mass of NH_4CNO left after 300 minutes is

$$m = (0.048\bar{9}\,\text{mol dm}^{-3}) \times (1.00\,\text{dm}^3) \times (60.06\,\text{g mol}^{-1}) = \boxed{2.94\,\text{g}}$$

P83.5 Use the procedure adopted in the solutions to problems 83.3 and 83.4: is the half-life (or any other similarly defined "fractional life") constant, or does it vary over the course of the reaction? The data are not quite so clear-cut. The half-life appears to be approximately constant at about 10 minutes: in the interval 0-10 the initial concentration drops by just over half, while in the interval 2-12 the concentration drops by slightly less than half. Another measure would compare the fractional consumption in two equal time intervals. The fractional consumption in the 0-2 interval is about 1/6, while that in the 10-12 interval is about 1/10—suggesting that the fractional consumption is **not** constant over the time the reaction was monitored. We draw up the following table (A = nitrile) in order to examine both first-order and second-order plots. The former is a plot of $\ln\left(\frac{[\text{A}]}{[\text{A}]_0}\right)$ against time (eqn 83.1b); the latter is a plot of 1/[A] against time:

$t/(10^3\,\text{s})$	0	2.00	4.00	6.00	8.00	10.00	12.00
$[\text{A}]/(\text{mol dm}^{-3})$	1.50	1.26	1.07	0.92	0.81	0.72	0.65
$\frac{[\text{A}]}{[\text{A}]_0}$	1.00	0.840	0.713	0.613	0.540	0.480	0.433
$\ln\left(\frac{[\text{A}]}{[\text{A}]_0}\right)$	0	−0.174	−0.338	−0.489	−0.616	−0.734	−0.836
$(1/[\text{A}])/(\text{dm}^3\,\text{mol}^{-1})$	0.667	0.794	0.935	1.09	1.23	1.39	1.54

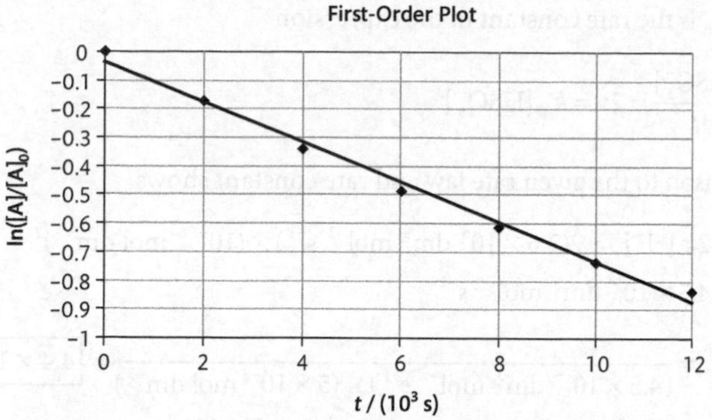

Fig 83.3(a)

The first-order plot is not bad: the correlation coefficient is 0.991. The corresponding first-order rate constant is $k_r = -\text{slope} = \boxed{7.0 \times 10^{-5}\ \text{s}^{-1}}$

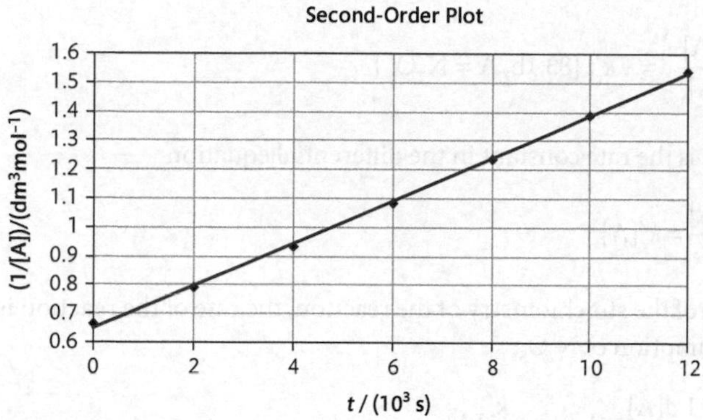

Fig 83.3(b)

The second-order plot looks even better: the correlation coefficient is 0.999. The corresponding second-order rate constant is $k_r = \text{slope} = \boxed{7.3 \times 10^{-5}\ \text{dm}^3\ \text{mol}^{-1}\ \text{s}^{-1}}$

Comment. Based on the given data, the reaction appears to be closer to second-order than to first-order. (The reaction order need not be an integer.) This conclusion is a tenuous one, though, based on the assumption that the data contain little experimental error. The best course of action for an investigator seeking to establish the reaction order would be further experimentation, following the reaction over a wider range of concentrations.

P83.7 The initial rate is

$$v_0 = (3.6 \times 10^6\ \text{dm}^9\ \text{mol}^{-3}\ \text{s}^{-1}) \times (5 \times 10^{-5}\,\text{mol dm}^{-3})^2 \times (10^{-5.6}\ \text{mol dm}^{-3})^2$$

$$= \boxed{6 \times 10^{-14}\,\text{mol dm}^{-3}\ \text{s}^{-1}}$$

The half-life for a second-order reaction is

$$t_{1/2} = \frac{1}{k_{\text{eff}}[\text{HSO}_3^-]_0} \quad [83.5]$$

where k_{eff} is the rate constant in the expression

$$-\frac{\mathrm{d}[\mathrm{HSO_3^-}]}{\mathrm{d}t} = 2v = k_{\text{eff}}[\mathrm{HSO_3^-}]^2$$

Comparison to the given rate law and rate constant shows

$$k_{\text{eff}} = 2k_r[\mathrm{H^+}]^2 = 2(3.6\times10^6\ \mathrm{dm^9\ mol^{-3}\ s^{-1}})\times(10^{-5.6}\ \mathrm{mol\ dm^{-3}})^2$$
$$= 4.5\times10^{-5}\ \mathrm{dm^3\ mol^{-1}\ s^{-1}}$$

and $$t_{1/2} = \frac{1}{(4.5\times10^{-5}\ \mathrm{dm^3\ mol^{-1}\ s^{-1}})\times(5\times10^{-5}\ \mathrm{mol\ dm^{-3}})} = \boxed{4.\bar{4}\times10^8\ \mathrm{s} = 14\ \mathrm{yr}}$$

P83.9 Examination of the data shows that the half-life remains constant at about 2 minutes. Therefore, the reaction is $\boxed{\text{first-order}}$. This can be confirmed by fitting any two pairs of data to the integrated first-order rate law, solving for the rate constant from each pair, and checking to see that they are the same to within experimental error.

$$\ln\left(\frac{[\mathrm{A}]}{[\mathrm{A}]_0}\right) = -k_r't\ [83.1\mathrm{b},\ \mathrm{A}=\mathrm{N_2O_5}]$$

Note: k_r' is the rate constant in the differential equation

$$-\frac{\mathrm{d}[\mathrm{A}]}{\mathrm{d}t} = k_r'[\mathrm{A}]$$

Because of the stoichiometry of the reaction, the rate of the reaction is half the rate of consumption of N_2O_5:

$$v = -\frac{1}{2}\frac{\mathrm{d}[\mathrm{A}]}{\mathrm{d}t} = k_r[\mathrm{A}] = \frac{k_r'}{2}[\mathrm{A}]$$

Solving for k_r',

$$k_r' = \frac{\ln\left(\frac{[\mathrm{A}]_0}{[\mathrm{A}]}\right)}{t}$$

At $t = 1.00$ min, $[\mathrm{A}] = 0.705\ \mathrm{mol\ dm^{-3}}$ and

$$k_r' = \frac{\ln\left(\frac{1.000}{0.705}\right)}{1.00\ \mathrm{min}} = 0.350\ \mathrm{min^{-1}} = 5.83\times10^{-3}\ \mathrm{s^{-1}}$$

At $t = 3.00$ min, $[\mathrm{A}] = 0.349\ \mathrm{mol\ dm^{-3}}$ and

$$k_r' = \frac{\ln\left(\frac{1.000}{0.349}\right)}{3.00\ \mathrm{min}} = 0.351\ \mathrm{min^{-1}} = 5.85\times10^{-3}\ \mathrm{s^{-1}}$$

Values of k_r' may be determined in a similar manner at all other times. The average value of k_r' obtained is $\boxed{5.84\times10^{-3}\ \mathrm{s^{-1}}}$ (which makes $\boxed{k_r = 2.92\times10^{-3}\ \mathrm{s^{-1}}}$). The constancy of k_r', which varies only between 5.83 and 5.85 × 10^{-3} s^{-1} confirms that the reaction is $\boxed{\text{first-order}}$. A linear regression of ln[A] against t yields the same result. The half-life is (eqn 83.2)

$$t_{1/2} = \frac{\ln 2}{k_r'} = \frac{0.693}{5.84 \times 10^{-3}\ \text{s}^{-1}} = 118.\bar{7}\ \text{s} = \boxed{1.98\ \text{min}}\ .$$

P83.11 The data for this experiment do not extend much beyond one half-life. Therefore the half-life method of predicting the order of the reaction as described in the solutions to Problems 83.3 and 83.4 cannot be used here. However, a similar method based on "three-quarters lives" will work. For a first-order reaction, we may write (analogous to the derivation of eqn 83.2)

$$k_r t_{3/4} = -\ln\frac{\frac{3}{4}[\text{A}]_0}{[\text{A}]_0} = -\ln\frac{3}{4} = \ln\frac{4}{3} = 0.288 \quad \text{or} \quad t_{3/4} = \frac{0.288}{k_r}$$

Thus the three-quarters life (or any given fractional life) is also independent of concentration for a first-order reaction. Examination of the data shows that the first three-quarters life (time to [A] = 0.237 mol dm^{-3}) is about 80 min and by interpolation the second (time to [A] = 0.178 mol dm^{-3}) is also about 80 min. Therefore the reaction is first-order and the rate constant is approximately

$$k_r = \frac{0.288}{t_{3/4}} \approx \frac{0.288}{80\,\text{min}} = 3.6 \times 10^{-3}\ \text{min}^{-1}$$

A least-squares fit of the data to the first-order integrated rate law [83.1b] gives the slightly more accurate result, $k_r = \boxed{3.65 \times 10^{-3}\ \text{min}^{-1}}$. The half life is

$$t_{1/2} = \frac{\ln 2}{k_r} = \frac{\ln 2}{3.65 \times 10^{-3}\ \text{min}^{-1}} = \boxed{190\ \text{min}}$$

P83.13 The data do not extend much beyond one half-life; therefore, we cannot see whether the half-life is constant over the course of the reaction as a preliminary step in guessing a reaction order. In a first-order reaction, however, not only the half-life but any other similarly defined fractional lifetime remains constant. (See Problem 83.11.) In this problem, we can see that the ⅔-life is **not** constant. (It takes less than 1.6 ms for [ClO] to drop from the first recorded value (8.49 μmol dm^{-3}) by more than ⅓ of that value (to 5.79 μmol dm^{-3}); it takes more than 4.0 more ms for the concentration to drop by not even ⅓ of **that** value (to 3.95 μmol dm^{-3}). So our working assumption is that the reaction is not first-order but second-order. Draw up the following table:

t/ms	[ClO]/(μmol dm^{-3})	(1/[ClO])/(dm^3 μmol^{-1})
0.12	8.49	0.118
0.62	8.09	0.124
0.96	7.10	0.141
1.60	5.79	0.173
3.20	5.20	0.192
4.00	4.77	0.210
5.75	3.95	0.253

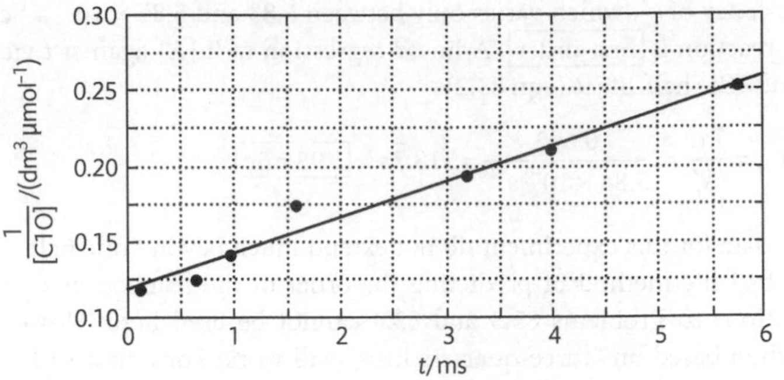

Fig 83.4

The plot of 1/[ClO] vs. t yields a reasonable straight line; the linear least-squares fit is:

$$(1/[\mathrm{ClO}])/(\mathrm{dm^3\ \mu mol^{-1}}) = 0.118 + 0.0237(t\ /\ \mathrm{ms}) \quad R^2 = 0.974$$

The rate constant is equal to the slope

$$k_r' = 0.0237\,\mathrm{dm^3\ \mu mol^{-1}\ ms^{-1}} = \boxed{2.37\times 10^7\ \mathrm{dm^3\ mol^{-1}\ s^{-1}}}$$

Note: k_r' is the rate constant in the differential equation

$$-\frac{\mathrm{d[ClO]}}{\mathrm{d}t} = k_r'[\mathrm{ClO}]$$

Because of the stoichiometry of the reaction, the rate of the reaction is half the rate of consumption of ClO:

$$v = -\frac{1}{2}\frac{\mathrm{d[ClO]}}{\mathrm{d}t} = k_r[\mathrm{ClO}] = \frac{k_r'}{2}[\mathrm{ClO}]$$

so $\boxed{k_r = 1.18\times 10^7\ \mathrm{dm^3\ mol^{-1}\ s^{-1}}}$

The half-life depends on the initial concentration (eqn 83.5):

$$t_{1/2} = \frac{1}{k_r[\mathrm{ClO}]_0} = \frac{1}{(2.37\times10^{-7}\ \mathrm{dm^3\ mol^{-1}\,s^{-1}})(8.47\times10^{-6}\ \mathrm{mol\ dm^{-3}})} = \boxed{4.98\times10^{-3}\ \mathrm{s}}$$

P83.15 $\mathrm{A+B\rightarrow P}$, $\dfrac{\mathrm{d[P]}}{\mathrm{d}t} = k_r[\mathrm{A}]^m[\mathrm{B}]^n$

and for a short interval δt,

$$\delta[\mathrm{P}] \approx k_r[\mathrm{A}]^m[\mathrm{B}]^n\,\delta t$$

Therefore, since $\delta[\mathrm{P}] = [\mathrm{P}]_t - [\mathrm{P}]_0 = [\mathrm{P}]_t$

$$\frac{[\mathrm{P}]}{[\mathrm{A}]} = k_r[\mathrm{A}]^{m-1}[\mathrm{B}]^n\,\delta t$$

$\dfrac{[\text{Chloropropane}]}{[\text{Propene}]}$ is independent of [Propene], implying that $m = 1$.

$$\frac{[\text{Chloropropane}]}{[\text{HCl}]} = \begin{cases} p(\text{HCl}) & 10 \quad 7.5 \quad 5.0 \\ & 0.05 \quad 0.03 \quad 0.01 \end{cases}$$

These results suggest that the ratio is roughly proportional to $p(\text{HCl})^2$, and therefore that $m = 3$ when A is identified with HCl. The rate law is therefore

$$\frac{\mathrm{d}[\text{Chloropropane}]}{\mathrm{d}t} = k_r[\text{Propene}][\text{HCl}]^3$$

and the reaction is $\boxed{\text{first-order}}$ in propene and $\boxed{\text{third-order}}$ in HCl.

P83.17 $v = \dfrac{d[\text{P}]}{\mathrm{d}t} = k_r[\text{A}][\text{B}]$

Let the initial concentrations be $[\text{A}]_0 = A_0$, $[\text{B}]_0 = B_0$, and $[\text{P}]_0 = 0$. Then, when P is formed in concentration x, the concentration of A changes to $A_0 - 2x$ and that of B changes to $B_0 - 3x$. Therefore

$$\frac{\mathrm{d}[\text{P}]}{\mathrm{d}t} = \frac{\mathrm{d}x}{\mathrm{d}t} = k_r(A_0 - 2x)(B_0 - 3x) \quad \text{with} \quad x = 0 \text{ at } t = 0$$

$$\int_0^t k_r \mathrm{d}t = \int_0^x \frac{\mathrm{d}x}{(A_0 - 2x)\times(B_0 - 3x)}$$

Apply partial fractions decomposition to the integrand on the right.

$$\int_0^t k_r \mathrm{d}t = \int_0^x \left(\frac{6}{2B_0 - 3A_0}\right) \times \left(\frac{1}{3(A_0 - 2x)} - \frac{1}{2(B_0 - 3x)}\right) \mathrm{d}x$$

$$= \left(\frac{-1}{(2B_0 - 3A_0)}\right) \times \left(\int_0^x \frac{\mathrm{d}x}{x - (1/2)A_0} - \int_0^x \frac{\mathrm{d}x}{x - (1/3)B_0}\right)$$

$$k_r t = \left(\frac{-1}{(2B_0 - 3A_0)}\right) \times \left[\ln\left(\frac{x - \frac{1}{2}A_0}{-\frac{1}{2}A_0}\right) - \ln\left(\frac{x - \frac{1}{3}B_0}{-\frac{1}{3}B_0}\right)\right]$$

$$= \left(\frac{-1}{2B_0 - 3A_0}\right) \ln\left(\frac{(2x - A_0)B_0}{A_0(3x - B_0)}\right)$$

$$= \boxed{\left(\frac{1}{3A_0 - 2B_0}\right) \ln\left(\frac{(2x - A_0)B_0}{A_0(3x - B_0)}\right)}$$

P83.19 The rate law $\dfrac{\mathrm{d}[\text{A}]}{\mathrm{d}t} = -k_r[\text{A}]^n$ for $n \neq 1$ integrates to

$$k_r t = \left(\frac{1}{n-1}\right) \times \left(\frac{1}{[\text{A}]^{n-1}} - \frac{1}{[\text{A}]_0^{n-1}}\right) \quad \text{[Table 83.3]}$$

At $t = t_{1/2}$, $$kt_{1/2} = \left(\frac{1}{n-1}\right)\left[\left(\frac{2}{[A]_0}\right)^{n-1} - \left(\frac{1}{[A]_0}\right)^{n-1}\right]$$

At $t = t_{3/4}$, $$kt_{3/4} = \left(\frac{1}{n-1}\right)\left[\left(\frac{4}{3[A]_0}\right)^{n-1} - \left(\frac{1}{[A]_0}\right)^{n-1}\right]$$

Hence, $$\frac{t_{1/2}}{t_{3/4}} = \boxed{\frac{2^{n-1}-1}{\left(\frac{4}{3}\right)^{n-1}-1}}$$

Topic 84 Reactions approaching equilibrium

Discussion question

D84.1 A temperature-jump experiment makes use of the fact that large quantities of energy can be delivered to a system in a very short time by electrical discharge or laser, with the result that a system's temperature can be changed in a very short time. Changing the temperature of an equilibrium mixture suddenly disturbs the equilibrium, since equilibrium constants and therefore equilibrium compositions depend on temperature. By changing the temperature suddenly, a system that was at equilibrium at its initial temperature is no longer at equilibrium. Monitoring the concentration of one or more components of the equilibrium as the concentrations adjust to establish equilibrium at the new temperature can yield information on the rate of reaction at the new temperature. "Relaxation" (adjustment toward equilibrium) occurs with a first-order rate constant equal to the sum of the forward and reverse rate constants of the equilibrium reaction. See Topic 84.2 and eqn 84.9.

Exercise

E84.1(a) The reactions whose rate constants are sought are the forward and reverse reactions in the following equilibrium.

$$NH_3(aq)+H_2O(l) \underset{k_r'}{\overset{k_r}{\rightleftharpoons}} NH_4^+(aq)+OH^-(aq)$$

The rate constants are related by

$$K_b = \frac{k_r}{k_r'} = \frac{[NH_4^+][OH^-]}{[NH_3]} = 1.78\times10^{-5}\ \text{mol dm}^{-3}$$

where the concentrations are equilibrium concentrations. (We assign units to K_b, which technically is a pure number, to help us keep track of units in the rate constants. Keeping track of the units makes us realize that k_r is a pseudo first-order protonation of NH_3 in excess water, for water does not appear in the above expression.) We need one more relationship between the constants, which we can obtain by proceeding as in Example 84.1:

$$\frac{1}{\tau} = k_r + k_r'([NH_4^+]+[OH^-])$$

Substitute into this expression

$$k_r = K_b k_r' \quad \text{and} \quad [NH_4^+] = [OH^-] = (K_b[NH_3])^{1/2}$$

hence $\dfrac{1}{\tau} = K_b k_r' + 2k_r'(K_b[NH_3])^{1/2} = k_r'\{K_b + 2(K_b[NH_3])^{1/2}\}$

So the reverse rate constant is

$$k_r' = \frac{1}{\tau\{K_b + 2(K_b[NH_3])^{1/2}\}}$$

$$= \frac{1}{7.61\times10^{-9}\ \text{s}\{1.78\times10^{-5}\ \text{mol dm}^{-3} + 2(1.78\times10^{-5}\times0.15)^{1/2}\ \text{mol dm}^{-3}\}}$$

$$= \boxed{4.0\times10^{10}\ \text{dm}^{-3}\ \text{mol s}^{-1}}$$

and the forward constant is

$$k_r = K_b k_r' = 1.78\times10^{-5}\ \text{mol dm}^{-3}\times 4.0\times10^{10}\ \text{dm}^3\ \text{mol}^{-1}\ \text{s}^{-1} = \boxed{7.1\times10^5\ \text{s}^{-1}}$$

Recall that k_r is the rate constant in the pseudo first-order rate law

$$-\frac{d[NH_3]}{dt} = k_r[NH_3]$$

Let us call k_2 the rate constant in the overall second-order rate law

$$-\frac{d[NH_3]}{dt} = k_2[NH_3][H_2O]$$

Setting these two expressions equal to each other yields

$$k_2 = \frac{k_r}{[H_2O]} = \frac{7.1\times10^5\ \text{s}^{-1}}{(1000\ \text{g dm}^{-3})/(18.02\ \text{g mol}^{-1})} = \boxed{1.28\times10^4\ \text{dm}^3\ \text{mol}^{-1}\ \text{s}^{-1}}$$

Problems

P84.1 Differentiate eqn 84.4:

$$\frac{d[A]}{dt} = \frac{d}{dt}\left(\frac{k_r' + k_r e^{-(k_r+k_r')t}}{k_r' + k_r}\right)[A]_0$$

The only time dependence in the expression is in the exponential, so

$$\frac{d[A]}{dt} = \frac{[A]_0 k_r}{k_r' + k_r}\frac{d}{dt}e^{-(k_r+k_r')t} = -[A]_0 k_r e^{-(k_r+k_r')t}$$

According to eqn 84.3, this expression should be equal to $-(k_r + k_r')[A] + k_r'[A]_0$, so we expand the latter by substituting eqn 84.4 for A:

$$-(k_r + k_r')[A] + k_r'[A]_0 = -(k_r + k_r')\left(\frac{k_r' + k_r e^{-(k_r+k_r')t}}{k_r' + k_r}\right)[A]_0 + k_r'[A]_0$$

$$= -(k_r' + k_r e^{-(k_r+k_r')t})[A]_0 + k_r'[A]_0 = -k_r e^{-(k_r+k_r')t}[A]_0$$

Indeed, this is equal to the result of the differentiation above.

P84.3 We proceed as in Topic 84.1. Stoichimetry says that any change in the amount of A is equal in magnitude and opposite in sign from the change in amount of B. Thus

$$[A]_0 - [A] = -\{[B]_0 - [B]\} \quad \text{and} \quad [B] = [A]_0 + [B]_0 - [A]$$

Eqn 84.3 becomes

$$\frac{d[A]}{dt} = -k_r[A] + k_r'\{[A]_0 + [B]_0 - [A]\} = -(k_r + k_r')[A] + k_r'\{[A]_0 + [B]_0\}$$

and its solution becomes

$$\boxed{[A] = \left(\frac{k_r' + k_r e^{-(k_r + k_r')t}}{k_r' + k_r}\right)\{[A]_0 + [B]_0\}}$$

The equilibrium compositions are

$$[A]_{eq} = \boxed{\frac{k_r'\{[A]_0 + [B]_0\}}{k_r + k_r'}} \quad \text{and} \quad [B]_{eq} = \boxed{\frac{k_r\{[A]_0 + [B]_0\}}{k_r + k_r'}}$$

(That is, just as $[A]_0 + [B]_0$ replaces $[A]_0$ in eqn 84.3, it also replaces it in eqns 84.4 and 84.5.)

P84.5 (a) First, find an expression for the relaxation time, using Example 84.1 as a model:

$$\frac{d[A]}{dt} = -2k_a[A]^2 + 2k_a'[A_2]$$

Rewrite the expression in terms of a difference from equilibrium values, $[A] = [A]_{eq} + x$

$$\frac{d[A]}{dt} = \frac{d([A]_{eq} + x)}{dt} = \frac{dx}{dt} = -2k_a([A]_{eq} + x)^2 + 2k_a'([A_2]_{eq} - \tfrac{1}{2}x)$$

$$\frac{dx}{dt} = -2k_a[A]_{eq}^2 - 4k_a[A]_{eq}\,x - 2k_a x^2 + 2k_a'[A_2]_{eq} - k_a'x$$

Neglect powers of x greater than x^1, and use the fact that at equilibrium the forward and reverse rates are equal,

$$k_a[A]_{eq}^2 = k_a'[A_2]_{eq}$$

to obtain

$$\frac{dx}{dt} \approx -(4k_a[A]_{eq} + k_a')x \quad \text{so} \quad \frac{1}{\tau} \approx 4k_a[A]_{eq} + k_a'$$

To get the desired expression, square the reciprocal relaxation time,

$$(*)\ \frac{1}{\tau^2} \approx 16k_a^2[A]_{eq}^2 + 8k_a k_a'[A]_{eq} + (k_a')^2$$

introduce $[A]_{tot} = [A]_{eq} + 2[A_2]_{eq}$ into the middle term,

$$\frac{1}{\tau^2} \approx 16k_a^2[\text{A}]_{\text{eq}}^2 + 8k_a k_a'([\text{A}]_{\text{tot}} - 2[\text{A}_2]_{\text{eq}}) + (k_a')^2$$

$$\approx 16k_a^2[\text{A}]_{\text{eq}}^2 + 8k_a k_a'[\text{A}]_{\text{tot}} - 16k_a k_a'[\text{A}_2]_{\text{eq}} + (k_a')^2 = \boxed{8k_a k_a'[\text{A}]_{\text{tot}} + (k_a')^2}$$

and use the equilibrium condition again to see that the remaining equilibrium concentrations cancel each other.

Comment. Introducing $[\text{A}]_{\text{tot}}$ into just one term of eqn * above is a permissible step but not a very systematic one. It is worth trying because of the resemblance between eqn * and the desired expression: We would be finished if we could get $[\text{A}]_{\text{tot}}$ into the middle term and somehow get the first term to disappear! A more systematic but messier approach would be to express $[\text{A}]_{\text{eq}}$ in terms of the desired $[\text{A}]_{\text{tot}}$ by using the equilibrium condition and $[\text{A}]_{\text{tot}} = [\text{A}]_{\text{eq}} + 2[\text{A}_2]_{\text{eq}}$: Solve both of those equations for $[\text{A}_2]_{\text{eq}}$, set the two resulting expressions equal to each other, solve for $[\text{A}]_{\text{eq}}$ in terms of the desired $[\text{A}]_{\text{tot}}$, and substitute **that** expression for $[\text{A}]_{\text{eq}}$ everywhere in eqn *.

(b) Plot $\frac{1}{\tau^2}$ vs. $[\text{A}]_{\text{tot}}$

The resulting curve should be a straight line whose y-intercept is $(k_a')^2$ and whose slope is $8k_a k_a'$.

(c) Draw up the following table:

$[\text{A}]_{\text{tot}}/(\text{mol dm}^{-3})$	0.500	0.352	0.251	0.151	0.101
τ/ns	2.3	2.7	3.3	4.0	5.3
$1/(\tau/\text{ns})^2$	0.189	0.137	0.092	0.062	0.036

The plot is shown in Fig. 84.1.

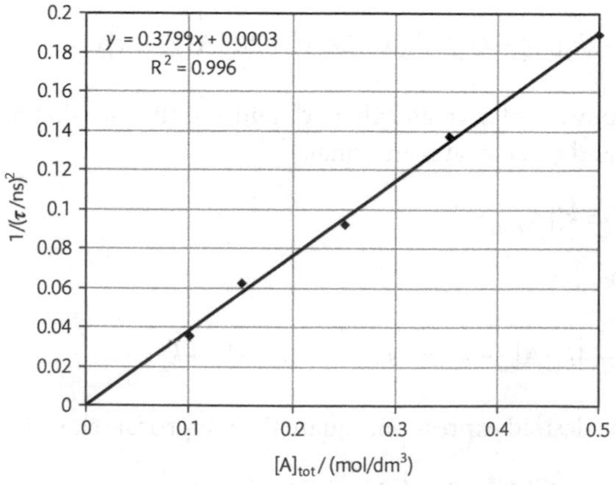

Fig 84.1

The y-intercept is 0.0003 ns^{-2} and the slope is 0.38 ns^{-2} dm^3 mol^{-1}, so

$$k_a' = \{3\times10^{-4}\times(10^{-9}\ \mathrm{s})^{-2}\}^{1/2} = (3\times10^{14}\ \mathrm{s}^{-2})^{1/2} = \boxed{1.\overline{7}\times10^{7}\ \mathrm{s}^{-1}}$$

$$k_a = \frac{0.38\times(10^{-9}\ \mathrm{s})^{-2}\ \mathrm{dm}^3\ \mathrm{mol}^{-1}}{8\times(1.\overline{7}\times10^{7}\ \mathrm{s}^{-1})} = \boxed{2.\overline{7}\times10^{9}\ \mathrm{dm}^3\ \mathrm{mol}^{-1}\ \mathrm{s}^{-1}}$$

and $$K = \frac{k_a/\mathrm{dm}^3\ \mathrm{mol}^{-1}\ \mathrm{s}^{-1}}{k_a'/\mathrm{s}^{-1}} = \frac{2.\overline{7}\times10^{9}}{1.\overline{7}\times10^{7}} = \boxed{1.\overline{6}\times10^{2}}$$

Comment. The data define a good straight line, as the correlation coefficient $R^2 = 0.996$ shows. That straight line appears to go through the origin, but the best-fit equation gives a small non-zero y-intercept. Inspection of the plot shows that several of the data points lie about as far from the fit line as the y-intercept does from zero. This suggests that y-intercept has a fairly high relative uncertainty and so do the rate constants.

Topic 85 The Arrhenius equation

Discussion question

D85.1 In eqn 85.1, k_r is the **rate constant**, A the **pre-exponential factor** (a name that makes more sense in light of the form of the Arrhenius equation given in eqn 85.4), and E_a the **activation energy**. The pre-exponential factor is sometimes called the frequency factor or even the Arrhenius A-factor. The Arrhenius equation is both an empirical expression that often provides a good summary of the temperature dependence of reaction rates and an expression whose parameters have sound physical significance. As an empirical expression, the equation is good, but neither unique nor universal. Over relatively small temperature ranges, other expressions can fit rate data; for example, plots of ln k_r vs. ln T or vs. T are often linear. Over large temperature ranges, plots of ln k_r vs. $1/T$ sometimes exhibit curvature, and kineticists may then turn to three- or four-parameter fits for greater accuracy. Eqn 85.4 bears some resemblance to the Boltzmann distribution, and the activation energy is the minimum energy required for an encounter between reactant molecules to lead to reaction. The pre-exponential factor can be interpreted in terms of the frequency of collisions between reactant molecules that have the right orientation for reaction to take place. More detailed interpretation of the Arrhenius parameters will be seen in Focus 18, Reaction Dynamics.

Exercises

E85.1(a) Eqn 85.2 can be rearranged to yield the activation energy:

$$E_a = \frac{R\ln\dfrac{k_{r,2}}{k_{r,1}}}{\left(\dfrac{1}{T_1}-\dfrac{1}{T_2}\right)} = \frac{(8.3145\ \text{J K}^{-1}\ \text{mol}^{-1})\ln\dfrac{2.67\times10^{-2}}{3.80\times10^{-3}}}{\left(\dfrac{1}{(273+35)\ \text{K}}-\dfrac{1}{(273+50)\ \text{K}}\right)}$$

$$= \boxed{1.08\times10^{5}\ \text{J mol}^{-1}} = \boxed{108\ \text{kJ mol}^{-1}}$$

With the activation energy in hand, the pre-exponential factor can be found from either rate constant by rearranging eqn 85.4.

$$A = k_r e^{E_a/RT} = (3.80\times10^{-3}\ \text{dm}^3\ \text{mol}^{-1}\ \text{s}^{-1})e^{1.08\times10^5\ \text{J mol}^{-1}/(8.3145\ \text{J K}^{-1}\ \text{mol}^{-1})(273+35)\ \text{K}}$$

$$A = \boxed{6.50\times10^{15}\ \text{dm}^3\ \text{mol}^{-1}\ \text{s}^{-1}}$$

Computing A from both provides a useful check on the calculation.

$$A = k_r e^{E_a/RT} = (2.67\times10^{-2}\ \text{dm}^3\ \text{mol}^{-1}\ \text{s}^{-1})e^{1.08\times10^5\ \text{J mol}^{-1}/(8.3145\ \text{J K}^{-1}\ \text{mol}^{-1})(273+50)\ \text{K}}$$

$$A = \boxed{6.50\times10^{15}\ \text{dm}^3\ \text{mol}^{-1}\ \text{s}^{-1}}$$

E85.2(a) Eqn 85.2 can be rearranged to yield the activation energy:

$$E_a = \frac{R\ln\dfrac{k_{r,2}}{k_{r,1}}}{\left(\dfrac{1}{T_1}-\dfrac{1}{T_2}\right)} = \frac{(8.3145\ \text{J K}^{-1}\ \text{mol}^{-1})\ln 3.00}{\left(\dfrac{1}{(273+24)\ \text{K}}-\dfrac{1}{(273+49)\ \text{K}}\right)}$$

$$= \boxed{3.49\times10^4\ \text{J mol}^{-1}} = \boxed{34.9\ \text{kJ mol}^{-1}}$$

Problems

P85.1 Eqn 85.3 defines the activation energy whether or not that quantity is a constant with temperature:

$$E_a = RT^2\left(\frac{d\ln k_r}{dT}\right)$$

Multiply both sides by dT and divide by RT^2:

$$\frac{E_a dT}{RT^2} = d\ln k_r$$

Now integrate both sides, taking E_a to be constant:

$$\int\frac{E_a dT}{RT^2} = \frac{E_a}{R}\int\frac{dT}{T^2} = -\frac{E_a}{RT} + c = \int d\ln k_r = \ln k_r$$

This is eqn 85.1 once we identify the constant of integration c as $\ln A$.

P85.3 If the rate constant obeys the Arrhenius equation, a plot of $\ln k_r$ against $1/T$ should yield a straight line with slope $-E_a/R$ (eqn 85.1). Construct a table as follows:

k_r / dm^3 mol^{-1} s^{-1}	T/K	1000 K/T	ln (k_r / s^{-1})
1.44×10^7	300.3	3.330	16.5
3.03×10^7	341.2	2.931	17.2
6.90×10^7	392.2	2.550	18.0

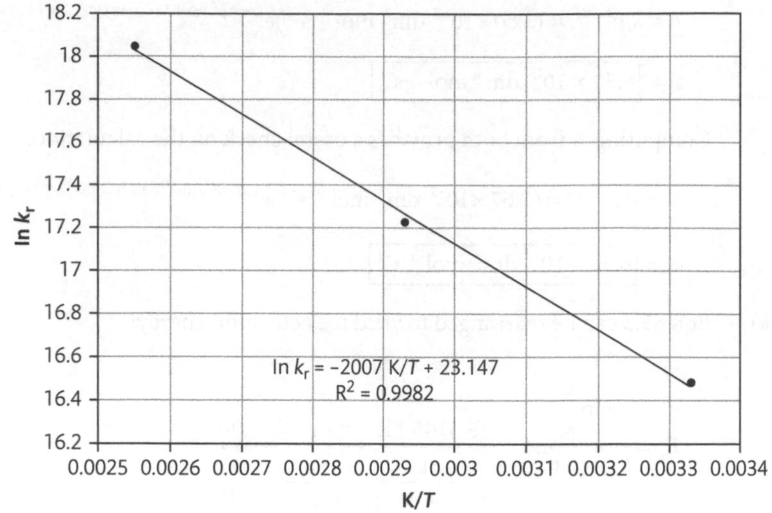

Fig 85.1

The plot is shown in Fig. 85.1. The best-fit straight line fits the three data points very well:

$$\ln (k_r/\mathrm{dm^3\ mol^{-1}\ s^{-1}}) = -2.01\times 10^3\ \mathrm{K}/T + 23.1$$

so $E_a = -(-2.01\times 10^3\ \mathrm{K})\times(8.3145\ \mathrm{J\ mol^{-1}\ K^{-1}}) = 1.67\times 10^4\ \mathrm{J\ mol^{-1}} = \boxed{16.7\ \mathrm{kJ\ mol^{-1}}}$

Compute A from each rate constant, rearranging eqn 85.4:

$$A = k_r e^{E_a/RT}$$

T/K	300.3	341.2	392.2
$10^{-7}\ k_r/(\mathrm{dm^3\ mol^{-1}\ s^{-1}})$	1.44	3.03	6.9
E_a / RT	6.69	5.89	5.12
$10^{-10}\ A/(\mathrm{dm^3\ mol^{-1}\ s^{-1}})$	1.16	1.10	1.16

The mean is $\boxed{1.14\times 10^{10}\ \mathrm{dm^3\ mol^{-1}\ s^{-1}}}$

P85.5 The Arrhenius expression for the rate constant is (eqn 85.1)

$$\ln k_r = \ln A - E_a/RT$$

A plot of $\ln k_r$ versus $1/T$ will have slope $-E_a/R$ and y-intercept $\ln A$. The transformed data and plot (Fig. 85.2) follow:

T/K	295	223	218	213	206	200	195
$10^{-6}k_r/(\mathrm{dm^3\ mol^{-1}\ s^{-1}})$	3.55	0.494	0.452	0.379	0.295	0.241	0.217
$\ln k_r/(\mathrm{dm^3\ mol^{-1}\ s^{-1}})$	15.08	13.11	13.02	12.85	12.59	12.39	12.29
10^{-3} K/T	3.39	4.48	4.59	4.69	4.85	5.00	5.13

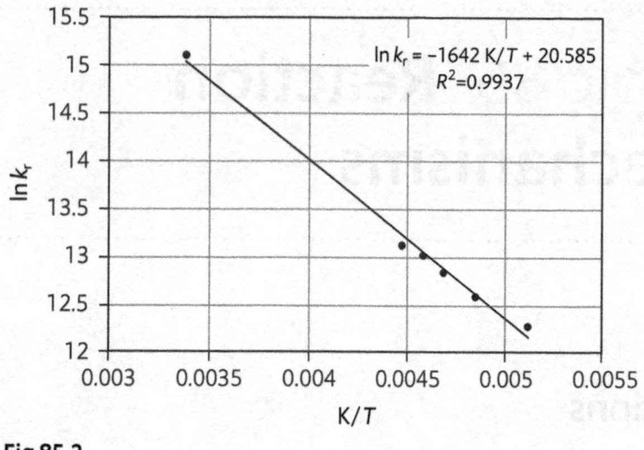

Fig 85.2

So $E_a = -(8.3145\ \text{J K}^{-1}\ \text{mol}^{-1}) \times (-1642\ \text{K}) = 1.37 \times 10^4\ \text{J mol}^{-1} = \boxed{13.7\ \text{kJ mol}^{-1}}$

and $A = e^{20.585}\ \text{dm}^3\ \text{mol}^{-1}\ \text{s}^{-1} = \boxed{8.7 \times 10^8\ \text{dm}^3\ \text{mol}^{-1}\ \text{s}^{-1}}$

Topic 86 **Reaction mechanisms**

Discussion questions

D86.1 The overall reaction order is the sum of the powers of the concentrations of all of the substances appearing in the *experimental* rate law for the reaction (eqn 82.7); hence, it is the sum of the individual orders (exponents) associated with a given reactant (or, occasionally, product). Reaction order is an experimentally determined, *not theoretical*, quantity, although theory may attempt to predict or explain it. Molecularity is the number of reactant molecules participating in an *elementary* reaction. Molecularity has meaning only for an elementary reaction, but reaction order applies to any reaction. In general, reaction order bears no necessary relation to the stoichiometry of the reaction, with the exception of elementary reactions, where the order of the reaction corresponds to the number of molecules participating in the reaction; that is, to its molecularity. Thus for an elementary reaction (but *only* for an elementary reaction), overall order and molecularity are the same and are determined by the stoichiometry.

D86.3 The steady-state approximation is the assumption that the rate of change of the concentrations of intermediates in consecutive chemical reactions is negligibly small. It is a good approximation when at least one of the reaction steps involving the intermediate is very fast, that is, has a large rate constant relative to other steps. See Topic 86.3. A pre-equilibrium approximation is similar in that it is a good approximation when the rate of formation of the intermediate from the reactants and the rate of its reversible decay back to the reactions are both very fast in comparison to the rate of formation of the product from the intermediate. This results in the intermediate being in approximate equilibrium with the reactants over relatively long time periods (though short compared to the overall time scale of the reaction). Thus, the concentration of the intermediate remains approximately constant over the time period that the equilibrium can be considered to be maintained. This allows one to relate the rate constants and concentrations to each other through a constant (the pre-equilibrium constant). See Topic 86.5. The approximations differ in ***how*** the intermediate concentration is kept nearly constant. In the steady-state case, it is because formation of the intermediate is slow, limiting the reactions that come afterward. In the pre-equilibrium case, it is because formation of the intermediate and dissociation back to reactants are both fast compared to the reactions that follow.

D86.5 Yes, a negative activation energy is quite possible for composite reactions. The rate constant of a composite reaction can be a product or ratio of rate constants and equilibrium constants of elementary reactions that contribute to the composite reaction. In general, elementary reactions that have a positive activation energy whose rate constants appear in the denominator of a composite rate constant tend to reduce the activation energy of the overall reaction, as illustrated in eqn 86.13. There is no reason why that reduction cannot be to a negative value.

The most common molecular interpretation of the activation energy is as the "height" of an energy barrier that must be overcome by reactants in order to form products, as mentioned in Topic 85. That interpretation is most straightforward for elementary reactions. In a complicated potential energy surface in which reactants and products are linked by several intermediates and possibly several reaction pathways, it is no surprise that this simple interpretation fails.

Question. Show that the following mechanism leads to an overall negative activation energy for the rate of formation of P if $E_a(2) > E_a(3)$.

$$\mathrm{A} \rightarrow \mathrm{I} \quad k_1 \text{ (slow)}$$

$$\mathrm{I} \rightarrow \mathrm{B} \quad k_2$$

$$\mathrm{I} \rightarrow \mathrm{P} \quad k_3$$

Exercises

E86.1(a) The rate of the overall reaction is

$$v = \frac{\mathrm{d[P]}}{\mathrm{d}t} = k_2[\mathrm{A}][\mathrm{B}]$$

however, we cannot have the concentration of an intermediate in the overall rate law.

(i) Assume a pre-equilibrium with

$$K = \frac{[\mathrm{A}]^2}{[\mathrm{A_2}]}, \quad \text{implying that} \quad [\mathrm{A}] = K^{1/2}[\mathrm{A_2}]^{1/2}$$

and $v = \boxed{k_2 K^{1/2}[\mathrm{A_2}]^{1/2}[\mathrm{B}]} = k_{\mathrm{eff}}[\mathrm{A_2}]^{1/2}[\mathrm{B}]$

where $k_{\mathrm{eff}} = k_2 K^{1/2}$

(ii) Apply the steady-state approximation:

$$\frac{\mathrm{d[A]}}{\mathrm{d}t} \approx 0 = 2k_1[\mathrm{A_2}] - 2k_1'[\mathrm{A}]^2 - k_2[\mathrm{A}][\mathrm{B}]$$

This is a quadratic equation in [A]

$$[\mathrm{A}] = \frac{-b \pm (b^2 - 4ac)^{1/2}}{2a} = \frac{k_2[\mathrm{B}] \pm \{k_2^2[\mathrm{B}]^2 + 16k_1'k_1[\mathrm{A_2}]\}^{1/2}}{-4k_1'}$$

$$= \frac{k_2[\mathrm{B}]}{4k_1'}\left\{\left(1 + \frac{16k_1'k_1[\mathrm{A_2}]}{k_2^2[\mathrm{B}]^2}\right)^{1/2} - 1\right\}$$

(In the last step, choose the sign that gives a positive quantity for [A].) Thus the rate law is

$$v = k_2[\mathrm{A}][\mathrm{B}] = \boxed{\frac{k_2^2[\mathrm{B}]^2}{4k_1'}\left\{\left(1+\frac{16k_1'k_1[\mathrm{A}_2]}{k_2^2[\mathrm{B}]^2}\right)^{1/2}-1\right\}}$$

This is a perfectly good rate law, albeit a complicated one. It is not in typical power-law form, but it is a function of reactant concentrations only, with no intermediates. This law simplifies under certain conditions. If $16k_1k_1'[\mathrm{A}_2] \gg k_2^2[\mathrm{B}]^2$, then

$$\left(1+\frac{16k_1'k_1[\mathrm{A}_2]}{k_2^2[\mathrm{B}]^2}\right)^{1/2}-1 \approx \frac{4\{k_1'k_1[\mathrm{A}_2]\}^{1/2}}{k_2[\mathrm{B}]}$$

and $$v \approx \frac{k_2^2[\mathrm{B}]^2}{4k_1'}\times\frac{4\{k_1'k_1[\mathrm{A}_2]\}^{1/2}}{k_2[\mathrm{B}]} = k_2[\mathrm{B}]\left(\frac{k_1[\mathrm{A}_2]}{k_1'}\right)^{1/2} = \boxed{k_2K^{1/2}[\mathrm{A}_2]^{1/2}[\mathrm{B}]}$$

recovering the pre-equilibrium rate law. If, on the other hand, $16k_1k_1'[\mathrm{A}_2] \ll k_2^2[\mathrm{B}]^2$, we expand the square root

$$\left(1+\frac{16k_1'k_1[\mathrm{A}_2]}{k_2^2[\mathrm{B}]^2}\right)^{1/2} \approx 1+\frac{8k_1'k_1[\mathrm{A}_2]}{k_2^2[\mathrm{B}]^2}$$

and $$v \approx \frac{k_2^2[\mathrm{B}]^2}{4k_1'}\left(1+\frac{8k_1'k_1[\mathrm{A}_2]}{k_2^2[\mathrm{B}]^2}-1\right) = \boxed{2k_1[\mathrm{A}_2]}$$

Comment. If the equilibrium is "fast", the latter condition will not be fulfilled. In fact, this special case amounts to having the first step rate-limiting. Note that the full (messy) steady-state approximation is less severe than either the pre-equilibrium or the rate-limiting step approximations, for it includes both as special cases.

E86.2(a) Let the steps be

$$\mathrm{A}+\mathrm{B} \rightleftharpoons \mathrm{I} \qquad (\text{fast}: k_\mathrm{a}, k_\mathrm{a}')$$

and $$\mathrm{I} \rightarrow \mathrm{P} \qquad (k_\mathrm{b})$$

Then the rate of reaction is

$$v = \frac{\mathrm{d}[\mathrm{P}]}{\mathrm{d}t} = k_\mathrm{b}[\mathrm{I}]$$

Applying the pre-equilibrium approximation yields

$$\frac{[\mathrm{I}]}{[\mathrm{A}][\mathrm{B}]} = K = \frac{k_\mathrm{a}}{k_\mathrm{a}'} \quad \text{so} \quad [\mathrm{I}] = \frac{k_\mathrm{a}[\mathrm{A}][\mathrm{B}]}{k_\mathrm{a}'}$$

and $$v = \frac{k_\mathrm{a}k_\mathrm{b}[\mathrm{A}][\mathrm{B}]}{k_\mathrm{a}'} = k_\mathrm{r}[\mathrm{A}][\mathrm{B}] \quad \text{with} \quad k_\mathrm{r} = \frac{k_\mathrm{a}k_\mathrm{b}}{k_\mathrm{a}'}$$

Thus $E_\mathrm{a} = E_\mathrm{a}(a)+E_\mathrm{a}(b)-E_\mathrm{a}'(a)\,[86.13] = (25+10-38)\ \mathrm{kJ\ mol^{-1}} = \boxed{-3\ \mathrm{kJ\ mol^{-1}}}$

Comment. Activation energies are rarely negative; however, some composite reactions are known to have small, negative activation energies.

Problems

P86.1 Using spreadsheet software to evaluate eqn 86.4b, one can draw up a plot like that in Fig. 86.1. The curves in this plot represent the concentration of the intermediate [I] as a function of time. They are labeled with the ratio k_a/k_b, where $k_b = 1\ s^{-1}$ for all curves and k_a varies. The darkest curve, labeled 10, corresponds to $k_a = 10\ s^{-1}$, as specified in part (a) of the problem. As the ratio k_a/k_b gets smaller (or, as the problem puts it, the ratio k_b/k_a gets larger), the concentration profile for I becomes lower, broader, and flatter; that is, [I] becomes more nearly constant over a longer period of time. This is the nature of the steady-state approximation, which becomes more and more valid as consumption of the intermediate becomes fast compared with its formation.

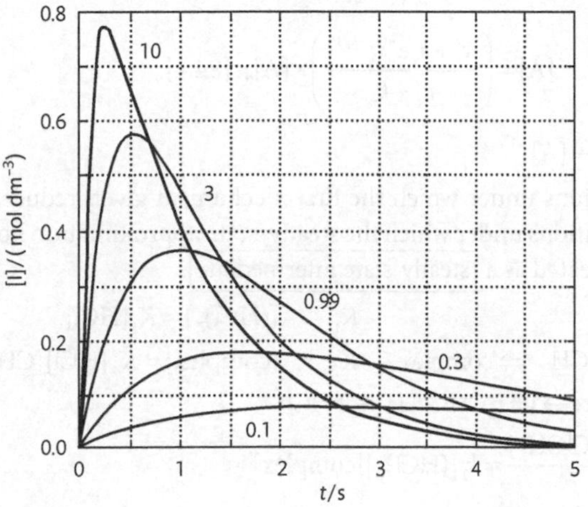

Fig 86.1

P86.3 The rate equations are

$$\frac{d[A]}{dt} = -k_a[A] + k_a'[B]$$

$$\frac{d[B]}{dt} = k_a[A] - k_a'[B] - k_b[B] + k_b'[C]$$

$$\frac{d[C]}{dt} = k_b[B] - k_b'[C]$$

These equations are a set of coupled differential equations. Although it is not immediately apparent, they have a closed-form general solution; however, we are looking for the particular circumstances under which the mechanism reduces to the second form given. Since the reaction involves an intermediate, let us explore the result of applying the steady-state approximation to it. Then

$$\frac{d[B]}{dt} = k_a[A] - k_a'[B] - k_b[B] + k_b'[C] \approx 0$$

and $[B] \approx \dfrac{k_a[A]+k_b'[C]}{k_a'+k_b}$

Therefore, $\dfrac{d[A]}{dt} = -\dfrac{k_a k_b}{k_a'+k_b}[A] + \dfrac{k_a' k_b'}{k_a'+k_b}[C]$

This rate expression may be compared to that given in the text (eqn 84.2) for a first-order reaction approaching equilibrium

$$A \underset{k_r'}{\overset{k_r}{\rightleftarrows}} C$$

Here $k_r = \dfrac{k_a k_b}{k_a'+k_b}$ and $k_r' = \dfrac{k_a' k_b'}{k_a'+k_b}$

The solutions are $[A] = \left(\dfrac{k_r' + k_r e^{-(k_r'+k_r)t}}{k_r'+k_r}\right) \times [A]_0$ [84.4]

and $[C] = [A]_0 - [A]$

Thus, the conditions under which the first mechanism given reduces to the second are the conditions under which the steady-state approximation holds, namely, when B can be treated as a **steady-state intermediate**.

P86.5 $2\,HCl \rightleftharpoons (HCl)_2$ $\quad K_1 \quad [(HCl)_2] = K_1[HCl]^2$

$HCl + CH_3CH{=}CH_2 \rightleftharpoons \text{complex}$ $\quad K_2 \quad [\text{complex}] = K_2[HCl][CH_3CH{=}CH_2]$

$(HCl)_2 + \text{complex} \rightarrow CH_3CHClCH_3 + 2\,HCl$ $\quad k_r$

$$v = \frac{d[CH_3CHClCH_3]}{dt} = k_r[(HCl)_2][\text{complex}]$$

Both $(HCl)_2$ and the complex are intermediates, so substitute for them using the equilibrium expressions:

$$v = k_r[(HCl)_2][\text{complex}] = k_r(K_1[HCl]^2)(K_2[HCl][CH_3CH{=}CH_2])$$
$$= \boxed{k_r K_1 K_2 [HCl]^3 [CH_3CH{=}CH_2]},$$

which is third-order in HCl and first-order in propene. One approach to experimental verification is to look for evidence of proposed intermediates, using infrared spectroscopy to search for $(HCl)_2$, for example.

Focus 17: Integrated activities

F17.1 (a) The rates of change are

$$(1)\quad \frac{d[F_2O]}{dt} = -k_a[F_2O]^2 - k_b[F][F_2O]$$

$$(2)\quad \frac{d[F]}{dt} = k_a[F_2O]^2 - k_b[F][F_2O] + 2k_c[OF]^2 - 2k_d[F]^2[F_2O] \approx 0 \text{ [steady-state]}$$

$$(3)\quad \frac{d[OF]}{dt} = k_a[F_2O]^2 + k_b[F][F_2O] - 2k_c[OF]^2 \approx 0 \text{ [steady-state]}$$

Applying the steady-state approximation to both [F] and [OF] and adding the equations (2) and (3) gives

$$\begin{aligned} k_a[F_2O]^2 - k_b[F][F_2O] + 2k_c[OF]^2 - 2k_d[F]^2[F_2O] &= 0 \\ k_a[F_2O]^2 + k_b[F][F_2O] - 2k_c[OF]^2 \qquad\qquad &= 0 \\ \hline 2k_a[F_2O]^2 \qquad\qquad\qquad - 2k_d[F]^2[F_2O] &= 0 \end{aligned}$$

Solving for [F] gives

$$[F] = \left(\frac{k_a}{k_d}[F_2O]\right)^{1/2}$$

Substituting this result into (1) gives

$$\frac{d[F_2O]}{dt} = -k_a[F_2O]^2 - k_b\left(\frac{k_a}{k_d}\right)^{1/2}[F_2O]^{3/2}$$

$$\text{or}\quad -\frac{d[F_2O]}{dt} = \boxed{k_a[F_2O]^2 + k_b\left(\frac{k_a}{k_d}\right)^{1/2}[F_2O]^{3/2}}$$

Comparison with the experimental rate law reveals that they are consistent when we make the following identifications.

$$k_r = k_a = 7.8\times10^{13}\,e^{-E_1/RT}\,\text{dm}^3\,\text{mol}^{-1}\,\text{s}^{-1}$$

$$E_a = (19\,350\text{ K})R = 160.9\text{ kJ mol}^{-1} = \text{activation energy for step (1)},$$

$$k_r' = k_b\left(\frac{k_a}{k_d}\right)^{1/2} = 2.3\times10^{10}\,e^{-E_a'/RT}\,\text{dm}^3\,\text{mol}^{-1}\,\text{s}^{-1}$$

$$E_a' = (16\,910\text{ K})R = 140.6\text{ kJ mol}^{-1}$$

(b) $\frac{1}{2}O_2 + F_2 \rightarrow F_2O \qquad \Delta_f H(F_2O) = 24.41\ \text{kJ mol}^{-1}$

$2\ F \rightarrow F_2 \qquad \Delta H = -D(F\text{—}F) = -160.6\ \text{kJ mol}^{-1}$

$O \rightarrow \frac{1}{2}O_2 \qquad \Delta H = -\frac{1}{2}D(O\text{—}O) = -249.1\ \text{kJ mol}^{-1}$

$2\ F + O \rightarrow F_2O$

$$\Delta H(FO-F) + \Delta H(O-F) = -\left[\Delta_f H(F_2O) - D(F-F) - \tfrac{1}{2}D(O-O)\right]$$
$$= -(24.41 - 160.6 - 249.1)\ \text{kJ mol}^{-1}$$
$$= 385.3\ \text{kJ mol}^{-1}$$

We estimate that $\Delta H(FO-F) \approx E_1 = \boxed{160.9\ \text{kJ mol}^{-1}}$

Then $\Delta H(O-F) \approx (385.3 - 160.9)\ \text{kJ mol}^{-1} \approx \boxed{224.4\ \text{kJ mol}^{-1}}$

In order to determine the activation energy of reaction (2) we assume that each rate be expressed in Arrhenius form, then

$$\ln k_r' = \ln k_b + \tfrac{1}{2}\ln k_a - \tfrac{1}{2}\ln k_b$$

and $E_a' = E_2 + \frac{1}{2}E_1 - \frac{1}{2}E_4 = 140.6\ \text{kJ mol}^{-1}$

Hence $E_2 - \frac{1}{2}E_4 = E_a' - \frac{1}{2}E_1 = (140.6 - 80.4)\ \text{kJ mol}^{-1} = 60.2\ \text{kJ mol}^{-1}$

E_4 is expected to be small since reaction (4) is a radical-radical combination, so we set $E_4 \approx 0$; then $E_2 \approx \boxed{60\ \text{kJ mol}^{-1}}$

F17.3 The rates of the individual steps are

$$A \rightarrow B \qquad \frac{d[B]}{dt} = I_a$$

$$B \rightarrow A \qquad \frac{d[B]}{dt} = -k_r[B]^2$$

In the photostationary state, $I_a - k_r[B]^2 = 0$. Hence,

$$[B] = \boxed{\left(\frac{I_a}{k_r}\right)^{1/2}}$$

This concentration can differ significantly from an equilibrium distribution because changing the illumination may change the rate of the forward reaction without affecting the reverse reaction. Contrast this situation to the corresponding equilibrium expression, in which $[B]_{eq}$ depends on a ratio of rate constants for the forward and reverse reactions. In the equilibrium case, the rates of forward and reverse reactions cannot be changed **independently**.

Topic 87 Collision theory

Discussion question

D87.1 Collision theory expresses a rate of reaction as a fraction of the rate of collision, on the assumption that reaction happens only between colliding molecules, and then only if the collision has enough energy and the proper orientation. So the rate of reaction is directly proportional to the rate of collision, and the expression of this rate comes directly from kinetic-molecular theory. The fraction of collisions energetic enough for reaction also come from kinetic-molecular theory via the Boltzmann distribution of energy.

Exercises

E87.1(a) The collision frequency is (eqn 78.10a)

$$z = \sigma v_{\text{rel}} \mathcal{N}$$

$$\text{where } v_{\text{rel}} = \left(\frac{16kT}{\pi m}\right)^{1/2} \text{ [78.9b]}, \quad \sigma = \pi d^2 \text{ [87.3b]} = 4\pi R^2, \quad \text{and } \mathcal{N} = \frac{p}{kT}$$

$$\text{Therefore, } z = \sigma \mathcal{N}\left(\frac{16kT}{\pi m}\right)^{1/2} = 16pR^2\left(\frac{\pi}{mkT}\right)^{1/2}$$

$$= 16 \times (120 \times 10^3 \text{ Pa}) \times (190 \times 10^{-12} \text{ m})^2$$

$$\times \left(\frac{\pi}{17.03\, m_u \times 1.661 \times 10^{-27} \text{ kg } m_u^{-1} \times 1.381 \times 10^{-23} \text{ J K}^{-1} \times 303 \text{ K}}\right)^{1/2}$$

$$= \boxed{1.13 \times 10^{10} \text{ s}^{-1}}$$

The collision density is [*Justification* 87.1]

$$Z = \frac{z\mathcal{N}_A}{2} = \frac{z}{2}\left(\frac{p}{kT}\right) = \frac{1.13 \times 10^{10} \text{ s}^{-1}}{2}\left(\frac{120 \times 10^3 \text{ Pa}}{1.381 \times 10^{-23} \text{ J K}^{-1} \times 303 \text{ K}}\right)$$

$$= \boxed{1.62 \times 10^{35} \text{ s}^{-1} \text{ m}^{-3}}$$

For the percentage increase at constant volume, note that $\mathcal{N}$ is constant at constant volume, so the only constant-volume temperature dependence on z (and on Z) is in the speed factor.

$$z \propto T^{1/2} \quad \text{so} \quad \frac{1}{z}\left(\frac{\partial z}{\partial T}\right)_V = \frac{1}{2T} \quad \text{and} \quad \frac{1}{Z}\left(\frac{\partial Z}{\partial T}\right)_V = \frac{1}{2T}$$

Therefore $$\frac{\delta z}{z} = \frac{\delta T}{Z} \approx \frac{\delta T}{2T} = \frac{1}{2}\left(\frac{10\ \text{K}}{303\ \text{K}}\right) = 0.017$$

so both z and Z increase by about $\boxed{1.7 \text{ per cent}}$.

E87.2(a) The fraction of collisions having at least E_a along the line of flight may be inferred by dividing out of the collision-theory rate constant (eqn 87.9) those factors that can be identified as belonging to the steric factor or collision rate: $f = e^{-E_a/RT}$

(a) (i) $$\frac{E_a}{RT} = \frac{20\times10^3\ \text{J mol}^{-1}}{(8.3145\ \text{J K}^{-1}\ \text{mol}^{-1})\times(350\ \text{K})} = 6.9 \quad \text{so} \quad f = e^{-6.9} = \boxed{1.04\times10^{-3}}$$

(ii) $$\frac{E_a}{RT} = \frac{20\times10^3\ \text{J mol}^{-1}}{(8.3145\ \text{J K}^{-1}\ \text{mol}^{-1})\times(900\ \text{K})} = 2.67 \quad \text{so} \quad f = e^{-2.67} = \boxed{0.069}$$

(b) (i) $$\frac{E_a}{RT} = \frac{100\times10^3\ \text{J mol}^{-1}}{(8.3145\ \text{J K}^{-1}\ \text{mol}^{-1})\times(350\ \text{K})} = 34.4 \quad \text{so} \quad f = e^{-34.4} = \boxed{1.19\times10^{-15}}$$

(ii) $$\frac{E_a}{RT} = \frac{100\times10^3\ \text{J mol}^{-1}}{(8.3145\ \text{J K}^{-1}\ \text{mol}^{-1})\times(900\ \text{K})} = 13.4 \quad \text{so} \quad f = e^{-13.4} = \boxed{1.57\times10^{-6}}$$

E87.3(a) A straightforward approach would be to compute $f = e^{-E_a/RT}$ at the new temperature and compare it to that at the old temperature. An approximate approach would be to note that f changes from $f_0 = e^{-E_a/RT}$ to $\exp\left(\frac{-E_a}{RT(1+x)}\right)$, where x is the fractional increase in the temperature. If x is small, the exponent changes from $-E_a/RT$ to approximately $-E_a(1-x)/RT$ and f changes from f_0 to

$$f \approx e^{-E_a(1-x)/RT} = e^{-E_a/RT}\left(e^{-E_a/RT}\right)^{-x} = f_0 f_0^{-x}$$

Thus the new fraction is the old one times a factor of f_0^{-x}. The increase in f expressed as a percentage is

$$\frac{f - f_0}{f_0}\times100\% = \frac{f_0 f_0^{-x} - f_0}{f_0}\times100\% = (f_0^{-x} - 1)\times100\%$$

(a) (i) $f_0^{-x} = (1.04\times10^{-3})^{-10/350} = 1.22$ and the percentage change is $\boxed{22\%}$

(ii) $f_0^{-x} = (0.069)^{-10/900} = 1.03$ and the percentage change is $\boxed{3\%}$

(b) (i) $f_0^{-x} = (1.19\times10^{-15})^{-10/350} = 2.7$ and the percentage change is $\boxed{170\%}$

(ii) $f_0^{-x} = (1.57\times10^{-6})^{-10/900} = 1.16$ and the percentage change is $\boxed{16\%}$

E87.4(a) $k_r = P\sigma\left(\frac{8kT}{\pi\mu}\right)^{1/2} N_A e^{-E_a/RT}$ [87.9]

$$k_r = 0.36\times(10^{-9}\,\text{m})^2\times\left(\frac{8\times(1.381\times10^{-23}\,\text{J K}^{-1})\times(650\,\text{K})}{\pi\times(3.32\times10^{-27}\,\text{kg})}\right)^{1/2}\times(6.022\times10^{23}\,\text{mol}^{-1})$$

$$\times\exp\left(\frac{-171\times10^{3}\,\text{J}}{(8.3145\,\text{J K}^{-1}\,\text{mol}^{-1})\times(650\,\text{K})}\right)$$

$$=\boxed{1.03\times10^{-5}\,\text{m}^3\,\text{mol}^{-1}\,\text{s}^{-1}}=\boxed{1.03\times10^{-2}\,\text{dm}^3\,\text{mol}^{-1}\,\text{s}^{-1}}$$

E87.5(a) The steric factor, P, is

$$P=\frac{\sigma^*}{\sigma}\ \text{[Topic 87.3]}$$

The mean collision cross-section is $\sigma=\pi d^2$ with $d=(d_A+d_B)/2$

Get the diameters from the collision cross-sections:

$$d_A=(\sigma_A/\pi)^{1/2} \quad \text{and} \quad d_B=(\sigma_B/\pi)^{1/2},$$

$$\text{so } \sigma=\frac{\pi}{4}\left\{\left(\frac{\sigma_A}{\pi}\right)^{1/2}+\left(\frac{\sigma_B}{\pi}\right)^{1/2}\right\}^2=\frac{(\sigma_A^{1/2}+\sigma_B^{1/2})^2}{4}=\frac{\{(0.95\,\text{nm}^2)^{1/2}+(0.65\,\text{nm}^2)^{1/2}\}^2}{4}$$

$$=\boxed{0.79\,\text{nm}^2}$$

$$\text{Therefore, } P=\frac{9.2\times10^{-22}\,\text{m}^2}{0.79\times(10^{-9}\,\text{m})^2}=\boxed{1.16\times10^{-3}}$$

Problems

P87.1 Comparing the Arrhenius equation (85.4) to eqn 87.9, with $\sigma^* = P\sigma$ yields

$$A=N_A\sigma^*\left(\frac{8kT}{\pi\mu}\right)^{1/2}$$

where $\mu = m(CH_3)/2 = (15.03\,m_u/2)\times 1.6605\times10^{-27}\,\text{kg}\,m_u^{-1} = 1.248\times10^{-26}\,\text{kg}$

$$\text{So}\quad A=\sigma^*\times(6.022\times10^{23}\,\text{mol}^{-1})\times\left(\frac{(8)\times(1.381\times10^{-23}\,\text{J K}^{-1})\times(298\,\text{K})}{(\pi)\times(1.248\times10^{-26}\,\text{kg})}\right)^{1/2}$$

$$=(5.52\times10^{26})\times(\sigma^*)\,\text{mol}^{-1}\,\text{m s}^{-1}.$$

$$\text{(a) } \sigma^*=\frac{2.4\times10^{10}\,\text{mol}^{-1}\,\text{dm}^3\,\text{s}^{-1}}{5.52\times10^{26}\,\text{mol}^{-1}\,\text{m s}^{-1}}=\frac{2.4\times10^{7}\,\text{mol}^{-1}\,\text{m}^3\,\text{s}^{-1}}{5.52\times10^{26}\,\text{mol}^{-1}\,\text{m s}^{-1}}=\boxed{4.3\overline{5}\times10^{-20}\,\text{m}^2}$$

(b) Take $\sigma\approx\pi d^2$ and estimate d as twice the bond length; therefore

$$\sigma=(\pi)\times(154\times2\times10^{-12}\,\text{m})^2=3.0\times10^{-19}\,\text{m}^2$$

$$\text{Hence}\quad P=\frac{\sigma^*}{\sigma}=\frac{4.3\overline{5}\times10^{-20}}{3.0\times10^{-19}}=\boxed{0.15}$$

P87.3 For radical recombination it has been found experimentally that $E_a \approx 0$. The maximum rate of recombination is obtained when $P=1$ (or more), and then (see Problem 87.1)

$$k_r = A = \sigma^* N_A\left(\frac{8kT}{\pi\mu}\right)^{1/2} = 4\sigma^* N_A\left(\frac{kT}{\pi m}\right)^{1/2} \quad [\mu = \tfrac{1}{2}m]$$

$$\sigma^* \approx \pi d^2 = \pi\times(308\times10^{-12}\text{ m})^2 = 3.0\times10^{-19}\text{ m}^2$$

Hence
$$k_r = (4)\times(3.0\times10^{-19}\text{ m}^2)\times(6.022\times10^{23}\text{ mol}^{-1})$$
$$\times\left(\frac{(1.381\times10^{-23}\text{ J K}^{-1})\times(298\text{ K})}{(\pi)\times(15.03\ m_u)\times(1.6605\times10^{-27}\text{ kg}/m_u)}\right)^{1/2}$$
$$= 1.7\times10^{8}\text{ m}^3\text{ mol}^{-1}\text{ s}^{-1} = \boxed{1.7\times10^{11}\text{ mol}^{-1}\text{ dm}^3\text{ s}^{-1}}$$

The rate constant is for the rate law

$$v = k_r[CH_3]^2$$

Therefore $$-\frac{d[CH_3]}{dt} = 2k_r[CH_3]^2$$

and its solution is $$\frac{1}{[CH_3]} - \frac{1}{[CH_3]_0} = 2k_r t$$

For 90% recombination, $[CH_3] = 0.10\times[CH_3]_0$, which occurs when

$$2k_r t = \frac{9}{[CH_3]_0} \quad \text{or} \quad t = \frac{9}{2k_r[CH_3]_0}$$

The mole fraction of CH_3 radicals when 10 mol % of ethane is dissociated is

$$\frac{(2)\times(0.10)}{1+0.10} = 0.18$$

The initial partial pressure of CH_3 radicals is thus

$$p_0 = 0.18\,p = 1.8\times10^4\text{ Pa}$$

and $$[CH_3]_0 = \frac{1.8\times10^4\text{ Pa}}{RT}$$

Therefore $$t = \frac{9RT}{(2k_r)\times(1.8\times10^4\text{ Pa})} = \frac{(9)\times(8.3145\text{ J K}^{-1}\text{ mol}^{-1})\times(298\text{ K})}{(1.7\times10^8\text{ m}^3\text{ mol}^{-1}\text{ s}^{-1})\times(3.6\times10^4\text{ Pa})}$$
$$= \boxed{3.6\text{ ns}}.$$

P87.5 Linear regression analysis of ln(rate constant) against $1/T$ yields the following results:

$$\ln(k_r/22.4\text{ dm}^3\text{ mol}^{-1}\text{ min}^{-1}) = C + B/T$$

where $C = 34.36$ (standard deviation 0.36),

$B = -23\,227$ K (standard deviation 252 K),

and $R = 0.99976$ (indicating a good fit).

$$\ln(k_r'/22.4\ \text{dm}^3\ \text{mol}^{-1}\ \text{min}^{-1}) = C' + B'/T$$

where $C' = 28.30$ (standard deviation = 0.84),

$B' = -21\,065$ K (standard deviation = 582 K),

and $R = 0.99848$ (indicating a good fit).

The regression parameters can be used in the calculation of the pre-exponential factor (A) and the activation energy (E_a) using $\ln k_r = \ln A - E_a/RT$.

$$\ln A = C + \ln(22.4) = 37.47$$

so $$A = 1.87\times10^{16}\,\text{dm}^3\ \text{mol}^{-1}\ \text{min}^{-1} = \boxed{3.12\times10^{14}\,\text{dm}^3\ \text{mol}^{-1}\ \text{s}^{-1}}$$

$$E_a = -RB = -(8.3145\ \text{J K}^{-1}\ \text{mol}^{-1})\times(-23\,227\ \text{K})\times\left(\frac{1\ \text{kJ}}{10^3\ \text{J}}\right) = \boxed{193\ \text{kJ mol}^{-1}}$$

$$\ln A' = C' + \ln(22.4) = 31.41$$

so $$A' = 4.37\times10^{13}\ \text{dm}^3\ \text{mol}^{-1}\ \text{min}^{-1} = \boxed{7.29\times10^{11}\ \text{dm}^3\ \text{mol}^{-1}\ \text{s}^{-1}}$$

$$E_a' = -RB' = -(8.3145\ \text{J K}^{-1}\text{mol}^{-1})\times(-21\,065\ \text{K})\times\left(\frac{1\ \text{kJ}}{10^3\ \text{J}}\right) = \boxed{175\ \text{kJ mol}^{-1}}$$

To summarize:

	$A/(\text{dm}^3\ \text{mol}^{-1}\ \text{s}^{-1})$	$E_a/(\text{kJ mol}^{-1})$
forward reaction	3.12×10^{14}	193
reverse reaction	7.29×10^{11}	175

Both sets of data, k and k', fit the Arrhenius equation very well. To the extent that the Arrhenius equation (85.4) is consistent with collision theory (whose rate constant is given by eqn 87.9), these data are also consistent with collision theory. The numerical values for A' and A may be compared to the results of *Brief illustration* 87.3. The pre-exponential factor for the reverse reaction is comparable to the one estimated in that problem based on collision density; however, the pre-factor for the forward reaction appears to be much larger than collision density. Because they imply an excessively high collision density, these data are not really compatible with collision theory.

Topic 88 Diffusion-controlled reactions

Discussion question

D88.1 A reaction in solution can be regarded as the outcome of two stages: one is the encounter of two reactant species; this is followed by their reaction in the second stage, if they acquire their activation energy. If the rate-determining step is the former, then the reaction is said to be diffusion controlled. If the rate-determining step is the latter, then the reaction is activation controlled. For a reaction of the form $A + B \rightarrow P$ that obeys the second-order rate law $v = k_r[A][B]$, in the diffusion-controlled regime,

$$k_r = k_d = 4\pi R^* D N_A$$

where D is the sum of the diffusion coefficients of the two reactant species and R^* is the distance at which reaction occurs. A further approximation is that each molecule obeys the Stokes–Einstein relation and Stokes' law, and then

$$k_d \approx \frac{8RT}{3\eta} \quad [88.5]$$

where η is the viscosity of the medium. The result suggests that k_d is independent of the radii of the reactants. It also suggests that the rate constant depends only weakly on temperature, so the activation energy (eqn 85.3) is small.

Exercises

E88.1(a) The rate constant for a diffusion-controlled bimolecular reaction is

$$k_d = 4\pi R^* D N_A \quad [88.3]$$

where $D = D_A + D_B = 2\times(6\times10^{-9}\,\mathrm{m^2\,s^{-1}}) = 1.2\times10^{-8}\,\mathrm{m^2\,s^{-1}}$

$$k_d = 4\pi\times(0.5\times10^{-9}\,\mathrm{m})\times(1.2\times10^{-8}\,\mathrm{m^2\,s^{-1}})\times(6.022\times10^{23}\,\mathrm{mol^{-1}})$$

$$k_d = \boxed{4.\overline{5}\times10^{7}\,\mathrm{m^3\,mol^{-1}\,s^{-1}}} = \boxed{4.\overline{5}\times10^{10}\,\mathrm{dm^3\,mol^{-1}\,s^{-1}}}$$

E88.2(a) The rate constant for a diffusion-controlled bimolecular reaction is

$$k_d = \frac{8RT}{3\eta}\ [88.5] = \frac{8\times(8.3145\ \mathrm{J\,K^{-1}\,mol^{-1}})\times(298\ \mathrm{K})}{3\eta} = \frac{6.61\times10^{3}\ \mathrm{J\,mol^{-1}}}{\eta}$$

(a) For water, $\eta = 1.00\times10^{-3}\ \mathrm{kg\,m^{-1}\,s^{-1}}$

$$k_d = \frac{6.61\times10^{3}\ \mathrm{J\,mol^{-1}}}{1.00\times10^{-3}\ \mathrm{kg\,m^{-1}\,s^{-1}}} = \boxed{6.61\times10^{6}\ \mathrm{m^3\,mol^{-1}\,s^{-1}}}$$

$$= \boxed{6.61\times10^{9}\ \mathrm{dm^3\,mol^{-1}\,s^{-1}}}$$

(b) For pentane, $\eta = 2.2\times10^{-4}\ \mathrm{kg\,m^{-1}\,s^{-1}}$

$$k_d = \frac{6.61\times10^{3}\ \mathrm{J\,mol^{-1}}}{2.2\times10^{-4}\ \mathrm{kg\,m^{-1}\,s^{-1}}} = \boxed{3.0\times10^{7}\ \mathrm{m^3\,mol^{-1}\,s^{-1}}}$$

$$= \boxed{3.0\times10^{10}\ \mathrm{dm^3\,mol^{-1}\,s^{-1}}}$$

E88.3(a) The rate constant for a diffusion-controlled bimolecular reaction is (eqn 88.5)

$$k_d = \frac{8RT}{3\eta} = \frac{8\times(8.3145\ \mathrm{J\,K^{-1}\,mol^{-1}})\times(320\mathrm{K})}{3\times(0.89\times10^{-3}\ \mathrm{kg\,m^{-1}\,s^{-1}})}$$

$$= \boxed{8.0\times10^{6}\ \mathrm{m^3\,mol^{-1}\,s^{-1}}} = \boxed{8.0\times10^{9}\ \mathrm{dm^3\,mol^{-1}\,s^{-1}}}$$

Since this reaction is elementary bimolecular it is second-order; hence

$$t_{1/2} = \frac{1}{2k_d[\mathrm{A}]_0}\quad [\text{Table 83.3, with } k_r = 2k_d \text{ because 2 atoms are consumed}]$$

so $$t_{1/2} = \frac{1}{2\times(8.0\times10^{9}\ \mathrm{dm^3\,mol^{-1}\,s^{-1}})\times(1.5\times10^{-3}\ \mathrm{mol\,dm^{-3}})} = \boxed{4.2\times10^{-8}\ \mathrm{s}}$$

E88.4(a) Since the reaction is diffusion controlled, the rate-limiting step is bimolecular and therefore second-order; hence

$$\frac{\mathrm{d[P]}}{\mathrm{d}t} = k_d[\mathrm{A}][\mathrm{B}]$$

where

$$k_d = 4\pi R^* D N_A\,[88.3] = 4\pi N_A R^*(D_A + D_B)$$

$$= 4\pi N_A \times (R_A + R_B)\times\frac{kT}{6\pi\eta}\left(\frac{1}{R_A}+\frac{1}{R_B}\right)[80.19] = \frac{2RT}{3\eta}(R_A+R_B)\times\left(\frac{1}{R_A}+\frac{1}{R_B}\right)$$

$$k_d = \frac{2\times(8.3145\ \mathrm{J\,K^{-1}\,mol^{-1}})\times(313\mathrm{K})}{3\times(2.93\times10^{-3}\ \mathrm{kg\,m^{-1}\,s^{-1}})}\times(655+1\,820)\times\left(\frac{1}{655}+\frac{1}{1\,820}\right)$$

$$= 3.04\times10^{6}\ \mathrm{m^3\,mol^{-1}\,s^{-1}} = 3.04\times10^{9}\ \mathrm{dm^3\,mol^{-1}\,s^{-1}}.$$

Therefore, the initial rate is

$$\frac{d[P]}{dt} = (3.04\times10^9\ dm^3\ mol^{-1}\ s^{-1})\times(0.170\,mol\ dm^{-3})\times(0.350\,mol\ dm^{-3})$$

$$= \boxed{1.81\times10^8\ mol\ dm^{-3}\ s^{-1}}$$

Comment. If the approximation of eqn 88.5 is used, $k_d = 2.37\times10^9\ dm^3\ mol^{-1}\ s^{-1}$. In this case the approximation results in a difference of about 25% compared to the expression used above.

Problems

P88.1 We are to show that eqn 88.9

$$[J]^* = [J]e^{-k_r t}$$

is a solution of eqn 88.8,

$$\frac{\partial[J]^*}{\partial t} = D\frac{\partial^2[J]^*}{\partial x^2} - k_r[J]^*$$

provided that [J] is a solution of

$$\frac{\partial[J]}{\partial t} = D\frac{\partial^2[J]}{\partial x^2}$$

Evaluate the derivatives of [J]*:

$$\frac{\partial[J]^*}{\partial t} = \frac{\partial[J]}{\partial t}e^{-k_r t} - k_r[J]e^{-k_r t} \quad \text{and} \quad \frac{\partial^2[J]^*}{\partial x^2} = \frac{\partial^2[J]}{\partial x^2}e^{-k_r t}$$

Use the fact that [J] is a solution in the absence of reaction

$$D\frac{\partial^2[J]^*}{\partial x^2} = D\frac{\partial^2[J]}{\partial x^2}e^{-k_r t} = \frac{\partial[J]}{\partial t}e^{-k_r t} = \frac{\partial[J]^*}{\partial t} + k_r[J]e^{-k_r t}$$

which gives us back eqn 88.8, as required.

P88.3 (a) The rate constant of a diffusion-limited reaction is

$$k_d = \frac{8RT}{3\eta}[88.5] = \frac{8\times(8.3145\ J\ K^{-1}\ mol^{-1})\times(298\ K)}{3\times(1.06\times10^{-3}\ kg\ m^{-1}\ s^{-1})}$$

$$= \boxed{6.23\times10^6\ m^3\ mol^{-1}\ s^{-1}} = \boxed{6.23\times10^9\ dm^3\ mol^{-1}\ s^{-1}}$$

(b) The rate constant is related to the diffusion constants and reaction distance by

$$k_d = 4\pi R^* D N_A \; [88.3]$$

$$\text{so} \quad R^* = \frac{k_d}{4\pi D N_A} = \frac{(2.77\times10^9\ \text{dm}^3\ \text{mol}^{-1}\ \text{s}^{-1})\times(10^{-3}\ \text{m}^3\ \text{dm}^{-3})}{4\pi\times(1\times10^{-9}\ \text{m}^2\ \text{s}^{-1})\times(6.022\times10^{23}\ \text{mol}^{-1})}$$

$$= \boxed{4\times10^{-10}\ \text{m}} = \boxed{0.4\ \text{nm}}$$

Topic 89 **Transition-state theory**

Discussion question

D89.1 The Eyring equation (eqn 89.10) results from activated complex theory which is an attempt to account for the rate constants of bimolecular reactions of the form $A+B \rightleftharpoons C^{\ddagger} \rightarrow P$ in terms of the formation of an activated complex. In the formulation of the theory, it is assumed that the activated complex and the reactants are in equilibrium, and the concentration of activated complex is calculated in terms of an equilibrium constant, which in turn is calculated from the partition functions of the reactants and a postulated form of the activated complex. It is further supposed that one normal mode of the activated complex, the one corresponding to displacement along the reaction coordinate, has a very low force constant. Displacement along this mode leads to products, provided that the complex enters a certain configuration of its atoms, known as the transition state. The derivation of the equilibrium constant from the partition functions leads to eqn 89.9 and in turn to eqn 89.10, the Eyring equation.

Exercises

E89.1(a) The enthalpy of activation for a bimolecular solution reaction is [*Brief illustration* 89.4]

$$\Delta^{\ddagger}H = E_a - RT = (8\,681\ \mathrm{K} - 303\ \mathrm{K}) \times 8.3145\ \mathrm{J\ mol^{-1}\ K^{-1}} = \boxed{+69.7\ \mathrm{kJ\ mol^{-1}}}$$

$$k_r = Be^{\Delta^{\ddagger}S/R}e^{-\Delta^{\ddagger}H/RT}, \quad B = \left(\frac{kT}{h}\right)\times\left(\frac{RT}{p^{\ominus}}\right) = \frac{kRT^2}{hp^{\ominus}}$$

$$= Be^{\Delta^{\ddagger}S/R}e^{-E_a/RT}e = Ae^{-E_a/RT}$$

Therefore, $A = eBe^{\Delta^{\ddagger}S/R}$, implying that $\Delta^{\ddagger}S = R\left(\ln\frac{A}{B} - 1\right)$

$$B = \frac{(1.381\times10^{-23}\ \mathrm{J\ K^{-1}})\times(8.3145\ \mathrm{J\ K^{-1}\ mol^{-1}})\times(303\ \mathrm{K})^2}{6.626\times10^{-34}\ \mathrm{J\,s}\times10^5\ \mathrm{Pa}}$$

$$= 1.59\times10^{11}\ \mathrm{m^3\ mol^{-1}\ s^{-1}} = 1.59\times10^{14}\ \mathrm{dm^3\ mol^{-1}\ s^{-1}}$$

and hence $\Delta^{\ddagger}S = R\left[\ln\left(\frac{2.05\times10^{13}\ \mathrm{dm^3\ mol^{-1}\ s^{-1}}}{1.59\times10^{14}\ \mathrm{dm^3\ mol^{-1}\ s^{-1}}}\right) - 1\right]$

$$= 8.3145\ \mathrm{J\ K^{-1}\ mol^{-1}} \times (-3.05) = \boxed{-25\ \mathrm{J\ K^{-1}\ mol^{-1}}}$$

E89.2(a) The enthalpy of activation for a bimolecular solution reaction is [*Brief illustration* 89.4]

$$\Delta^{\ddagger}H = E_a - RT = 8.3145\,\mathrm{J\,K^{-1}\,mol^{-1}} \times (9\,134\,\mathrm{K} - 303\,\mathrm{K}) = \boxed{+73.4\,\mathrm{kJ\,mol^{-1}}}$$

The entropy of activation is [Exercise 89.2(a)]

$$\Delta^{\ddagger}S = R\left(\ln\frac{A}{B} - 1\right)$$

$$\text{with } B = \frac{kRT^2}{hp^{\ominus}} = 1.59\times10^{14}\,\mathrm{dm^3\,mol^{-1}\,s^{-1}}$$

$$\text{Therefore, } \Delta^{\ddagger}S = 8.3145\,\mathrm{J\,K^{-1}\,mol^{-1}} \times \left[\ln\left(\frac{7.78\times10^{14}}{1.59\times10^{14}}\right) - 1\right] = +4.9\,\mathrm{J\,K^{-1}\,mol^{-1}}$$

$$\text{Hence, } \Delta^{\ddagger}G = \Delta^{\ddagger}H - T\Delta^{\ddagger}S = \{73.4 - (303)\times(4.9\times10^{-3})\}\,\mathrm{kJ\,mol^{-1}} = \boxed{+71.9\,\mathrm{kJ\,mol^{-1}}}$$

E89.3(a) Use eqn 89.15(a) to relate a bimolecular gas-phase rate constant to activation energy and entropy:

$$k_r = e^2 B e^{\Delta^{\ddagger}S/R} e^{-E_a/RT}$$

$$\text{where } B = \left(\frac{kT}{h}\right)\times\left(\frac{RT}{p^{\ominus}}\right)\ [89.14] = \frac{(1.381\times10^{-23}\,\mathrm{J\,K^{-1}})\times(338\,\mathrm{K})^2\times(8.3145\,\mathrm{J\,mol^{-1}\,K^{-1}})}{(6.626\times10^{-34}\,\mathrm{J\,s})\times(10^5\,\mathrm{Pa})}$$

$$= 1.98\times10^{11}\,\mathrm{m^3\,mol^{-1}\,s^{-1}}$$

Solve for the entropy of activation:

$$\Delta^{\ddagger}S = R\left(\ln\frac{k_r}{B} - 2\right) + \frac{E_a}{T}$$

The derivations in Topic 89.2(a) are based on a k_r that contains concentration units, whereas the rate constant given here has pressure units. So k_r in the above expressions is not the rate constant given in the exercise, but

$$k_r = 7.84\times10^{-3}\,\mathrm{kPa^{-1}\,s^{-1}} \times RT$$

$$k_r = 7.84\times10^{-3}\times(103\,\mathrm{Pa})^{-1}\,\mathrm{s^{-1}} \times 8.3145\,\mathrm{J\,K^{-1}\,mol^{-1}} \times 338\,\mathrm{K} = 0.0220\,\mathrm{m^3\,mol^{-1}\,s^{-1}}$$

$$\text{Hence } \Delta^{\ddagger}S = 8.3145\,\mathrm{J\,K^{-1}\,mol^{-1}} \times \left(\ln\frac{0.0220\,\mathrm{m^3\,mol^{-1}\,s^{-1}}}{1.98\times10^{11}\,\mathrm{m^3\,mol^{-1}\,s^{-1}}} - 2\right) + \frac{58.6\times10^3\,\mathrm{J\,mol^{-1}}}{338\,\mathrm{K}}$$

$$= \boxed{-91\,\mathrm{J\,K^{-1}\,mol^{-1}}}$$

E89.4(a) For a bimolecular gas-phase reaction [Exercise 89.3(a)],

$$\Delta^{\ddagger}S = R\left(\ln\frac{k_r}{B} - 2\right) + \frac{E_a}{T} = R\left(\ln\frac{A}{B} - \frac{E_a}{RT} - 2\right) + \frac{E_a}{T} = R\left(\ln\frac{A}{B} - 2\right)$$

$$\text{where } B = \frac{kRT^2}{hp^{\ominus}}$$

For two structureless particles, the rate constant is the same as that of collision theory

$$k_r = N_A\sigma^*\left(\frac{8kT}{\pi\mu}\right)^{1/2} e^{-\Delta E_0/RT}\ \text{[Example 89.1]}$$

The activation energy is [85.3]

$$E_a = RT^2\frac{d\ln k_r}{dT} = RT^2\frac{d}{dT}\left(\ln N_A\sigma^* + \frac{1}{2}\ln\frac{8k}{\pi\mu} + \frac{1}{2}\ln T - \frac{\Delta E_0}{RT}\right)$$

$$= RT^2\left(\frac{1}{2T} + \frac{\Delta E_0}{RT^2}\right) = \Delta E_0 + \frac{RT}{2},$$

so the prefactor is

$$A = k_r e^{E_a/RT} = N_A\sigma^*\left(\frac{8kT}{\pi\mu}\right)^{1/2} e^{-\Delta E_0/RT}\left(e^{\Delta E_0/RT}e^{1/2}\right) = N_A\sigma^*\left(\frac{8kT}{\pi\mu}\right)^{1/2} e^{1/2}$$

$$\text{Hence } \Delta^{\ddagger}S = R\left\{\ln N_A\sigma^*\left(\frac{8kT}{\pi\mu}\right)^{1/2} + \frac{1}{2} - \ln\frac{kRT^2}{p^{\ominus}h} - 2\right\} = R\left\{\ln\frac{\sigma^* p^{\ominus}h}{(kT)^{3/2}}\left(\frac{8}{\pi\mu}\right)^{1/2} - \frac{3}{2}\right\}$$

For identical particles,

$$\mu = m/2 = (65\ m_u)(1.661\times10^{-27}\ \text{kg}\ m_u^{-1})/2 = 5.4\times10^{-26}\ \text{kg},$$

and hence

$$\Delta^{\ddagger}S = 8.3145\ \text{J K}^{-1}\ \text{mol}^{-1}$$

$$\times\left\{\ln\frac{0.35\times(10^{-9}\ \text{m})^2\times10^5\ \text{Pa}\times6.626\times10^{-34}\ \text{J s}}{(1.381\times10^{-23}\ \text{J K}^{-1}\times300\ \text{K})^{3/2}}\left(\frac{8}{\pi\times5.4\times10^{-26}\ \text{kg}}\right)^{1/2} - \frac{3}{2}\right\}$$

$$= \boxed{-74\ \text{J K}^{-1}\ \text{mol}^{-1}}$$

E89.5(a) At low pressure, the reaction can be assumed to be bimolecular. (The rate constant is certainly second-order. In addition, see Topic 91.)

(a) $\Delta^{\ddagger}S = R\left(\ln\frac{A}{B} - 2\right)$ [Exercise 89.4(a)]

$$\text{where } B = \frac{kRT^2}{hp^{\ominus}}\ \text{[Exercise 89.4(a)]} = \frac{1.381\times10^{-23}\ \text{J K}^{-1}\times8.3145\ \text{J K}^{-1}\ \text{mol}^{-1}\times(298\ \text{K})^2}{6.626\times10^{-34}\ \text{J s}\times10^5\ \text{Pa}}$$

$$= 1.54\times10^{11}\ \text{m}^3\ \text{mol}^{-1}\ \text{s}^{-1} = 1.54\times10^{14}\ \text{dm}^3\ \text{mol}^{-1}\ \text{s}^{-1}$$

$$\text{Hence } \Delta^{\ddagger}S = 8.3145\ \text{J K}^{-1}\ \text{mol}^{-1}\times\left(\ln\frac{4.6\times10^{12}\ \text{dm}^3\ \text{mol}^{-1}\ \text{s}^{-1}}{1.54\times10^{14}\ \text{dm}^3\ \text{mol}^{-1}\ \text{s}^{-1}} - 2\right)$$

$$= \boxed{-46\ \text{J K}^{-1}\ \text{mol}^{-1}}$$

(b) The enthalpy of activation for a bimolecular gas-phase reaction is [Topic 89.2(a)]

$$\Delta^{\ddagger}H = E_a - 2RT = 10.0\ \text{kJ mol}^{-1} - 2\times8.3145\ \text{J mol}^{-1}\ \text{K}^{-1}\times298\ \text{K} = \boxed{+5.0\ \text{kJ mol}^{-1}}$$

(c) The Gibbs energy of activation at 298 K is

$$\Delta^{\ddagger}G = \Delta^{\ddagger}H - T\Delta^{\ddagger}S = 5.0\ \text{kJ mol}^{-1} - (298\ \text{K})\times(-46\times10^{-3}\ \text{kJ K}^{-1}\ \text{mol}^{-1})$$

$$\Delta^{\ddagger}G = \boxed{+18.7\ \text{kJ mol}^{-1}}$$

E89.6(a) Use eqn 89.18 to examine the effect of ionic strength on a rate constant:

$$\log k_r = \log k_r^{\circ} + 2A|z_Az_B|I^{1/2}$$

$$\text{Hence } \log k_r^{\circ} = \log k_r - 2A|z_Az_B|I^{1/2} = \log 12.2 - 2\times0.509\times|1\times(-1)|\times(0.0525)^{1/2} = 0.85$$

$$\text{and } k_r^{\circ} = \boxed{7.1\ \text{dm}^6\ \text{mol}^{-2}\ \text{min}^{-1}}$$

Problems

P89.1 Draw up the following table for an Arrhenius plot:

θ / °C	−24.82	−20.73	−17.02	−13.00	−8.95
T / K	248.33	252.42	256.13	260.15	264.20
10^3 K / T	4.027	3.962	3.904	3.844	3.785
$10^4\,k_r$ / s^{-1}	1.22	2.31	4.39	8.50	14.3
ln (k_r / s^{-1})	−9.01	−8.37	−7.73	−7.07	−6.55

The points are plotted in Figure 89.1.

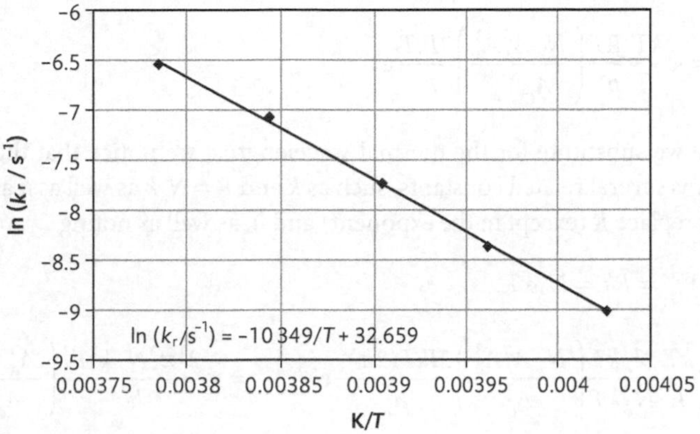

Fig 89.1

A least squares fit of the data yields the intercept +32.7 at $1/T = 0$, which implies that $\ln\left(\dfrac{A}{\mathrm{s}^{-1}}\right) = 32.7$, and hence that $A = 1.53\times10^{14}\ \mathrm{s}^{-1}$. The slope is $-1.035\times10^4\ \mathrm{K} = E_a/R$, and hence $\boxed{E_a = 86.0\ \mathrm{kJ\ mol^{-1}}}$

In solution $\Delta^{\ddagger}H = E_a - RT$ [Topic 89.2(a)], so at −20°C

$$\Delta^{\ddagger}H = 86.0\ \mathrm{kJ\ mol^{-1}} - (8.3145\ \mathrm{J\ mol^{-1}\ K^{-1}})\times(253\ \mathrm{K}) = \boxed{+83.9\ \mathrm{kJ\ mol^{-1}}}$$

We assume that the reaction is first-order for which, by analogy to Topic 89.1(d)

$$K^{\ddagger} = \frac{kT}{h\nu}\bar{K}^{\ddagger} \quad \text{and} \quad k_r = k^{\ddagger}K^{\ddagger} = \nu\times\frac{kT}{h\nu}\times\bar{K}^{\ddagger}$$

with $\Delta^{\ddagger}G = -RT\ln\bar{K}^{\ddagger}$

$$\text{Therefore, } k_r = A\mathrm{e}^{-E_a/RT} = \frac{kT}{h}\mathrm{e}^{-\Delta^{\ddagger}G/RT} = \frac{kT}{h}\mathrm{e}^{\Delta^{\ddagger}S/R}\,\mathrm{e}^{-\Delta^{\ddagger}H/RT}$$

We can identify $\Delta^{\ddagger}S$ by writing

$$k_r = \frac{kT}{h}\mathrm{e}^{\Delta^{\ddagger}S/R}\,\mathrm{e}^{-E_a/RT}\,\mathrm{e} = A\mathrm{e}^{-E_a/RT}$$

and hence obtain

$$\Delta^{\ddagger}S = R\left[\ln\left(\frac{hA}{kT}\right)-1\right]$$

$$= 8.3145\ \mathrm{J\,K^{-1}\,mol^{-1}}\times\left[\ln\left(\frac{(6.626\times10^{-34}\ \mathrm{J\,s})\times(1.53\times10^{14}\ \mathrm{s^{-1}})}{(1.381\times10^{-23}\ \mathrm{J\,K^{-1}})\times(253\ \mathrm{K})}\right)-1\right]$$

$$= \boxed{+19.6\ \mathrm{J\,K^{-1}\,mol^{-1}}}.$$

Therefore, $\Delta^{\ddagger}G = \Delta^{\ddagger}H - T\Delta^{\ddagger}S = 83.9\ \mathrm{kJ\,mol^{-1}} - 253\ \mathrm{K}\times19.6\ \mathrm{J\,K^{-1}\,mol^{-1}} =$ $\boxed{+79.0\ \mathrm{kJ\,mol^{-1}}}$

P89.3 In effect, we are asked to simplify the expression in the middle of Example 89.1:

$$k_r = \kappa\frac{kT}{h}\frac{RT}{p^{\ominus}}\left(\frac{N_A\Lambda_A^3\Lambda_B^3}{\Lambda_{C^{\ddagger}}^3V_m^{\ominus}}\right)\frac{2IkT}{\hbar^2}e^{-\Delta E_0/RT}$$

Before we substitute for the thermal wavelengths, we notice that the expression contains several related constants, such as k and $R = N_Ak$ as well as h and $\hbar = h/2\pi$. Let us replace R (except in the exponent) and $\hbar$, as well as noting

$$p^{\ominus}V_m^{\ominus} = RT = N_AkT$$

so $$k_r = \kappa\frac{kT}{h}\frac{N_AkT}{N_AkT}\left(\frac{N_A\Lambda_A^3\Lambda_B^3}{\Lambda_{C^{\ddagger}}^3}\right)\frac{2IkT(2\pi)^2}{h^2}e^{-\Delta E_0/RT} = \kappa\frac{2^3\pi^2N_Ak^2T^2}{h^3}\left(\frac{\Lambda_A^3\Lambda_B^3}{\Lambda_{C^{\ddagger}}^3}\right)Ie^{-\Delta E_0/RT}$$

The thermal wavelengths are

$$\Lambda_J = \frac{h}{(2\pi m_JkT)^{1/2}} = \frac{h}{(2\pi kT)^{1/2}m_J^{1/2}}$$

where we separate the one piece of the expression (the mass) that differs for different species. Thus the factor that involves the thermal wavelengths can be written as

$$\frac{\Lambda_A^3\Lambda_B^3}{\Lambda_{C^{\ddagger}}^3} = \frac{h^3(m_A+m_B)^{3/2}}{(2\pi kT)^{3/2}m_A^{3/2}m_B^{3/2}} = \frac{h^3}{(2\pi kT)^{3/2}}\left(\frac{m_A+m_B}{m_Am_B}\right)^{3/2} = \frac{h^3}{(2\pi kT)^{3/2}\mu^{3/2}}$$

Seeing the reduced mass μ reminds us to substitute for the moment of inertia $I = \mu r^2$. So

$$k_r = \kappa\frac{2^3\pi^2N_Ak^2T^2}{h^3}\left(\frac{h^3}{(2\pi kT)^{3/2}\mu^{3/2}}\right)\mu r^2e^{-\Delta E_0/RT} = \kappa r^2\frac{2^{3/2}\pi^{1/2}N_Ak^{1/2}T^{1/2}}{\mu^{1/2}}e^{-\Delta E_0/RT}$$

Identifying the reactive cross section $\sigma^* = \kappa\pi r^2$ and collecting the remaining terms gives the expression at the end of the Example:

$$k_r = \sigma^*N_A\left(\frac{8kT}{\pi\mu}\right)^{1/2}e^{-\Delta E_0/RT}$$

P89.5 According to the Debye-Hückel limiting law, the logarithms of ionic activity coefficients, γ, are proportional to $I^{1/2}$ (where I is ionic strength). The ionic strength in a 2:1 electrolyte solution is three times the molal concentration. If the limiting law holds, then a plot of log k_r versus $I^{1/2}$ should give a straight line whose y-intercept is log $k_r°$ and whose slope is $2Az_Az_B$, where z_A and z_B are charge numbers of the component ions of the activated complex (eqn 89.18).

The line based on the limiting law appears curved. The zero-ionic-strength rate constant based on it is

$$k_r^\circ = 10^{-0.690}\ \mathrm{dm^{3/2}\ mol^{-1/2}\ s^{-1}} = 0.204\ \mathrm{dm^{3/2}\ mol^{-1/2}\ s^{-1}}$$

The slope is positive, so the complex must overcome repulsive interactions. The product of charges, however, works out to be 0.5, not easily interpretable in terms of charge numbers.

The extended Debye–Hückel law (eqn 72.29 with $C = 0$) has log γ proportional to $\left(\dfrac{I^{1/2}}{1+BI^{1/2}}\right)$, so it requires plotting log k_r versus $\left(\dfrac{I^{1/2}}{1+BI^{1/2}}\right)$, and it also has a slope of $2Az_Az_B$ and a y-intercept of log $k_r°$. B is an empirical parameter, though, and we don't know its value. (In fact, the extended law reduces to the limiting law with $B = 0$.) But it is easy to try a few values, and the resulting curves are plotted in Fig. 89.2.

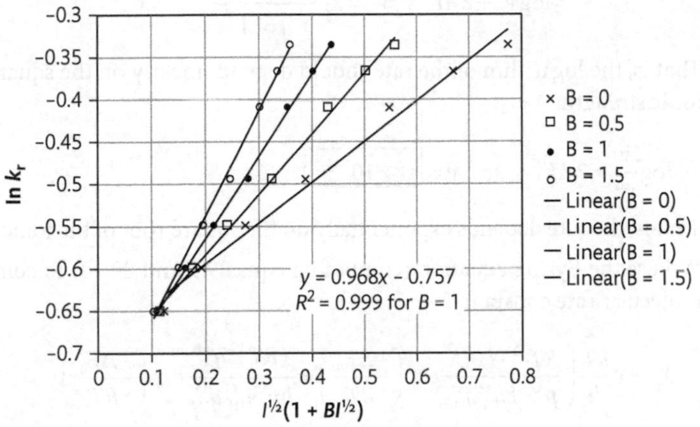

Fig 89.2

The line based on the extended law with $B = 1$ appears straighter and has a better correlation coefficient than the limiting law; it also has a better correlation coefficient than $B = 0.5$ or $B = 1.5$ curves. The zero-ionic-strength rate constant based on $B = 1$ is

$$k_r^\circ = 10^{-0.757}\ \mathrm{dm^{3/2}\ mol^{-1/2}\ s^{-1}} = 0.175\ \mathrm{dm^{3/2}\ mol^{-1/2}\ s^{-1}}$$

The product of charges works out to be 0.9, nearly 1, interpretable in terms of a complex of [two univalent ions of the same sign].

The transformed data are in the following table.

$[Na_2SO_4]$ / (mol kg^{-1})	0.2	0.15	0.1	0.05	0.025	0.0125	0.005
k_r / (dm$^{3/2}$ mol$^{-1/2}$ s^{-1})	0.462	0.430	0.390	0.321	0.283	0.252	0.224
$I^{1/2}$	0.775	0.671	0.548	0.387	0.274	0.194	0.122
$I^{1/2}/(1+I^{1/2})$	0.436	0.401	0.354	0.279	0.215	0.162	0.109
$\log k_r$	−0.335	−0.367	−0.409	−0.493	−0.548	−0.599	−0.650

P89.7

$$K_a = \frac{[H^+][A^-]}{[HA]\gamma_{HA}}\gamma_\pm^2 \approx \frac{[H^+][A^-]\gamma_\pm^2}{[HA]}$$

Therefore, $[H^+] = \dfrac{[HA]K_a}{[A^-]\gamma_\pm^2}$

and $$\log[H^+] = \log K_a + \log\frac{[HA]}{[A^-]} - 2\log\gamma_\pm = \boxed{\log K_a + \log\frac{[HA]}{[A^-]} + 2AI^{1/2}}$$

Write $v = k_r[H^+][B]$.

Then $$\log v = \log(k_r[B]) + \log[H^+]$$
$$= \log(k_r[B]) + \log\frac{[HA]}{[A^-]} + 2AI^{1/2} + \log K_a$$
$$= \log v^\circ + 2AI^{1/2}, \quad v^\circ = k_r\frac{[B][HA]K_a}{[A^-]}.$$

That is, the logarithm of the rate should depend linearly on the square root of the ionic strength.

$$\log\frac{v}{v^\circ} = 2AI^{1/2} \quad \text{so} \quad v = \boxed{v^\circ \times 10^{2AI^{1/2}}}$$

That is, the rate depends exponentially on the square root of the ionic strength.

P89.9 We use the Eyring equation (combining eqns 89.9 and 89.10) to compute the bimolecular rate constant

$$k_r = \kappa\frac{kT}{h}\left(\frac{RT}{p^\ominus}\right)\frac{N_A\bar{q}^\ominus_{C^\ddagger}}{q^\ominus_H q^\ominus_{D_2}}\exp\left(\frac{-\Delta E_0}{RT}\right) \approx \frac{(RT)^2\bar{q}^\ominus_{C^\ddagger}}{hp^\ominus q^\ominus_H q^\ominus_{D_2}}\exp\left(\frac{-\Delta E_0}{RT}\right)$$

We are to consider a variety of activated complexes, but the reactants, (H and D_2) and their partition functions do not change. Consider them first. The partition function of H is solely translational:

$$q^\ominus_H = \frac{RT}{p^\ominus\Lambda_H^3} \text{ and } \Lambda_H = \left(\frac{h^2}{2\pi kTm_H}\right)^{1/2} \text{ so } q^\ominus_H = \frac{RT(2\pi kTm_H)^{3/2}}{p^\ominus h^3}$$

We have neglected the spin degeneracy of H, which will cancel the spin degeneracy of the activated complex. The partition function of D_2 has a rotational term as well.

$$q^\ominus_{D_2} = \frac{RT}{p^\ominus\Lambda_{D_2}^3}\times\frac{kT}{\sigma hc\tilde{B}_{D_2}} = \frac{RkT^2(2\pi kTm_{D_2})^{3/2}}{2p^\ominus h^4 c\tilde{B}_{D_2}}$$

We have neglected the vibrational partition function of D_2, which is very close to unity at the temperature in question. The symmetry number σ is 2 for a homonuclear diatomic, and the rotational constant is 30.44 cm^{-1}. Now, the partition function of the activated complex will have a translational piece that is the same regardless of the model:

$$\bar{q}^{\ominus}_{C^\ddagger} = q^{T\ominus}_{C^\ddagger} \times q^{R}_{C^\ddagger} \times \bar{q}^{V}_{C^\ddagger}$$

$$\text{where} \quad q^{T\ominus}_{C^\ddagger} = \frac{RT(2\pi kTm_{HD_2})^{3/2}}{p^{\ominus}h^3}$$

Let us aggregate the model-independent factors into a single term, F where

$$F = \frac{(RT)^2\bar{q}^{T\ominus}_{C^\ddagger}}{hp^{\ominus}q^{\ominus}_{H}q^{\ominus}_{D_2}}\exp\left(\frac{-\Delta E_0}{RT}\right) = \frac{2h^3c\tilde{B}_{D_2}m^{3/2}_{HD_2}}{kT(2\pi m_H m_{D_2}kT)^{3/2}}\exp\left(\frac{-\Delta E_0}{RT}\right)$$

$$= h^3c\tilde{B}_{D_2}\left(\frac{5^3}{2m^3_H(4)^3p^3T^3k^5}\right)^{1/2}\exp\left(\frac{-\Delta E_0}{RT}\right) = 2.71\times 10^4\ \text{dm}^3\ \text{mol}^{-1}\ \text{s}^{-1}$$

where we have taken $m_{HD_2} = 5m_H$ and $m_{D_2} = 4m_H$.

Now $k_r = F \times q^{R}_{C^\ddagger} \times \bar{q}^{V}_{C^\ddagger}$

The number of vibrational modes in the activated complex is $3\times3-6=3$ for a non-linear complex, one more for a linear complex; however, in either case, one mode is the reaction coordinate, and is removed from the partition function. Therefore, assuming all real vibrations to have the same wavenumber $\tilde{\nu}$

$$\bar{q}^{V}_{C^\ddagger} = q^{2}_{mode}(\text{non-linear}) \text{ or } q^{3}_{mode}(\text{linear})$$

$$\text{where} \quad q_{mode} = \left[1-\exp\left(\frac{-hc\tilde{\nu}}{kT}\right)\right]^{-1} = 1.028$$

if the vibrational wavenumbers are 1 000 cm^{-1}. The rotational partition function is

$$q^{R}_{C^\ddagger} = \frac{kT}{\sigma hc\tilde{B}}(\text{linear}) \text{ or } \frac{1}{\sigma}\left(\frac{kT}{hc}\right)^{3/2}\left(\frac{\pi}{\tilde{A}\tilde{B}\tilde{C}}\right)^{1/2}(\text{non-linear})$$

where the rotational constants are related to moments of inertia by

$$\tilde{B} = \frac{\hbar}{4\pi cI} \text{ where } I = \sum mr^2$$

and r is the distance from an atom to a rotational axis.

(a) The first model for the activated complex is triangular, with two equal sides of

$$s = 1.30\times 74\ \text{pm} = 96\ \text{pm}$$

and a base of

$$b = 1.20\times 74\ \text{pm} = 89\ \text{pm}$$

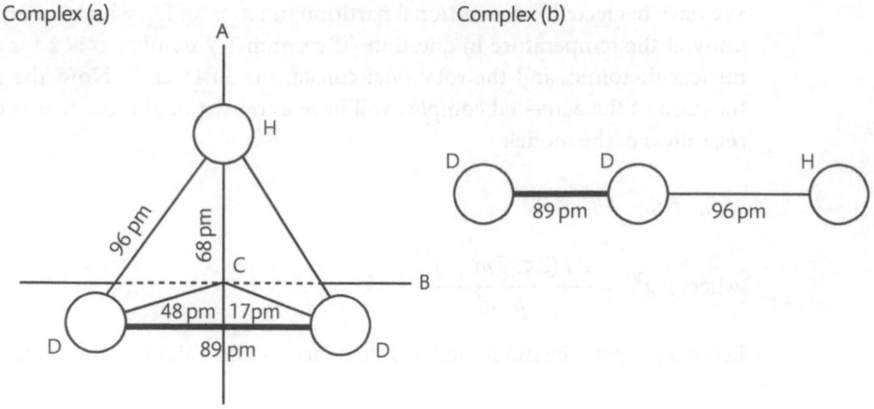

The moment of inertia about the axis of the altitude of the triangle (z axis) is

$$I_1 = 2m_D(b/2)^2 = m_H b^2 \text{ so } \tilde{A} = \frac{\hbar}{4\pi c m_H b^2} = 21.2 \text{ cm}^{-1}$$

To find the other moments of inertia, we need to find the center of mass. Clearly it is in the plane of the molecule and on the z axis; the center of mass is the position z at which

$$\sum_i m_i(z_i - z) = 0 = 2(2m_H)(0-z) + m_H(H-z)$$

where H is the height of the triangle,

$$H = [s^2 - (b/2)^2]^{1/2} = 85 \text{ pm}$$

so the center of mass is $z = H/5$.

The moment of inertia about the axis in the plane of the triangle perpendicular to the altitude is

$$I_2 = 2(2m_H)(H/5)^2 + m_H(4H/5)^2 = (4m_H/5)H^2$$

$$\text{so} \quad \tilde{B} = \frac{\hbar}{4\pi c(4m_H/5)H^2} = 28.3 \text{ cm}^{-1}$$

The distance from the center of mass to the D atoms is

$$r_D = [(H/5)^2 + (b/2)^2]^{1/2} = 48 \text{ pm}$$

and the moment of inertia about the axis perpendicular to the plane of the triangle is

$$I_3 = 2(2m_H)r_D^2 + m_H(4H/5)^2 = 2(2m_H)[(H/5)^2 + (b/2)^2] + m_H(4H/5)^2$$
$$= (4m_H/5)(s^2 + b^2).$$

so

$$\tilde{C} = \frac{\hbar}{4\pi c(4m_{\rm H}/5)(s^2+b^2)} = 12.2\ {\rm cm^{-1}}$$

The rotational partition function is

$$q_{\rm C^\ddagger}^{\rm R} = \frac{1}{\sigma}\left(\frac{kT}{hc}\right)^{3/2}\left(\frac{\pi}{\tilde{A}\tilde{B}\tilde{C}}\right)^{1/2} = 47.7$$

(The symmetry number σ is 2 for this model.) The vibrational partition function is

$$\bar{q}_{\rm C^\ddagger}^{\rm V} = q_{\rm mode}^2 = 1.057$$

So the rate constant is:

$$k_{\rm r} = F \times q_{\rm C^\ddagger}^{\rm R} \times \bar{q}_{\rm C^\ddagger}^{\rm V} = \boxed{1.37\times10^6\ {\rm dm^3\ mol^{-1}\ s^{-1}}}$$

(b) To compute the moment of inertia, we need the center of mass. Let the terminal D atom be at $x = 0$, the central D atom at $x = b$, and the H atom at $x = b + s$. The center of mass is the position X at which

$$\sum_i m_i(x_i - X) = 0 = 2m_{\rm H}(0-X) + 2m_{\rm H}(b-X) + m_{\rm H}(s+b-X)$$

$$5X = 3b + s \text{ so } X = (3b+s)/5$$

The moment of inertia is

$$I = \sum_i m_i(x_i - X)^2 = 2m_{\rm H}X^2 + 2m_{\rm H}(b-X)^2 + m_{\rm H}(s+b-X)^2$$

$$= 3.97\times10^{-47}\ {\rm m\ kg^2}$$

and $\tilde{B} = \dfrac{\hbar}{4\pi cI} = 7.06\ {\rm cm^{-1}}$

The rotational partition function is

$$q_{\rm C^\ddagger}^{\rm R} = \frac{kT}{\sigma hc\tilde{B}} = 39.4$$

(The symmetry number σ is 1 for this model.) The vibrational partition function is

$$\bar{q}_{\rm C^\ddagger}^{\rm V} = q_{\rm mode}^3 = 1.09$$

So the rate constant is

$$k_{\rm r} = F \times q_{\rm C^\ddagger}^{\rm R} \times \bar{q}_{\rm C^\ddagger}^{\rm V} = \boxed{1.16\times10^6\ {\rm dm^3\ mol^{-1}\ s^{-1}}}$$

(c) Both models are already pretty good, coming within a factor of 3 to 4 of the experimental result, and neither model has much room for improvement.

Consider how to try to change either model to reduce the rate constant toward the experimental value. The factor F is model-independent. The factor $\bar{q}^{V}_{C^‡}$ is nearly at its minimum possible value, 1, so stiffening the vibrational modes will have almost no effect. Only the factor $q^{R}_{C^‡}$ is amenable to lowering, and even that not by much. It would be decreased if the rotational constants were increased, which means decreasing the moments of inertia and the bond lengths. Reducing the lengths s and b in the models to the equilibrium bond length of H_2 would only drop k_r to 6.5×10^5 (model a) or 6.9×10^5 (model b) $dm^3\ mol^{-1}\ s^{-1}$, even with a stiffening of vibrations. Reducing the HD distance in model (a) to 80% of the H_2 bond length does produce a rate constant of $4.2 \times 10^5\ dm^3\ mol^{-1}\ s^{-1}$ (assuming stiff vibrations of 2 000 cm^{-1}); such a model is not intermediate in structure between reactants and products, though. It appears that the rate constant is rather insensitive to the geometry of the complex.

P89.11 The diffusion process described is unimolecular, hence first-order, and therefore analogous but not identical to the second-order case developed in this Topic. We may write

$$[A^‡] = K^‡[A] \quad \text{[analogous to 89.2]}$$

and $$-\frac{d[A]}{dt} = k^‡[A^‡] = \kappa\nu^‡[A^‡] \approx \nu^‡[A^‡] = \kappa\nu^‡K^‡[A] = k_r[A] \quad [89.3\text{-}89.5]$$

Thus $$k_r \approx \nu^‡K^‡ = \nu^‡\left(\frac{kT}{h\nu^‡}\right)\times\left(\frac{\bar{q}^‡}{q}\right)e^{-\Delta E_0/RT}$$

where $\bar{q}^‡$ and q are the (vibrational) partition functions at the top (missing one mode) and foot of the well, respectively. Let the y-direction be the direction of diffusion. Hence, for the activated atom the vibrational mode in this direction is lost, and

$\bar{q}^‡ = q_x^{‡V} q_z^{‡V}$ for the activated atom, and

$q = q_x^V q_y^V q_z^V$ for an atom at the bottom of a well.

For classical vibration, $$q^V \approx \frac{kT}{hc\tilde{\nu}}[52.16] = \frac{kT}{h\nu}$$

Hence $$k_r = \frac{kT}{h}\left(\frac{(kT/h\nu^‡)^2}{(kT/h\nu)^3}\right)e^{-\Delta E_0/RT} = \boxed{\frac{\nu^3}{(\nu^‡)^2}e^{-\Delta E_0/RT}} \approx \frac{\nu^3}{(\nu^‡)^2}e^{-E_a/RT}$$

(a) If $\nu^‡ = \nu$, then $k_r \approx \nu e^{-E_a/RT} = 10^{11}\ s^{-1} \times e^{-60000/(8.3145\times500)} = 5.4\times10^4\ s^{-1}$

$$D = \frac{\lambda^2}{2\tau}[81.16] \approx \frac{1}{2}\lambda^2 k_r \left[\tau = \frac{1}{k_r}\text{(period for vibration with enough energy)}\right]$$

$$= \frac{1}{2}\times(316\times10^{-12}\ m)^2\times5.4\times10^4\ s^{-1} = \boxed{2.7\times10^{-15}\ m^2\ s^{-1}}$$

(b) If $\nu^‡ = \nu/2$, then $k_r \approx 4\nu e^{-E_a/RT} = 2.2\times10^5\ s^{-1}$

$$D = 4\times(2.7\times10^{-15}\ m^2\ s^{-1}) = \boxed{1.1\times10^{-14}\ m^2\ s^{-1}}$$

Topic 90 **The dynamics of molecular collisions**

Discussion questions

D90.1 *Infrared chemiluminescence.* Chemical reactions may yield products in excited states. The emission of radiation as the molecules decay to lower energy states is called chemiluminescence. If the emission is from vibrationally excited states, then it is infrared chemiluminescence. The vibrationally excited product molecule in the example of Fig. 90.11 in the text is CO. By studying the intensities of the infrared emission spectrum, the populations of the vibrational states in the product CO may be determined and this information allows us to determine the relative rates of formation of CO in these excited states.

Laser-induced fluorescence (LIF) uses a laser to excite product molecules from a particular rotational-vibrational state and then "counts" those molecules by detecting fluorescence from the excited molecules. By tuning the exciting laser to different rotational-vibrational states, one can gather information about the distribution of product molecules among different states.

Multi-photon ionization (MPI). Multi-photon absorption is the absorption of two or more photons by the molecule in its transition to a higher electronic state. The frequencies of the photons satisfy the condition

$$\Delta E = h\nu_1 + h\nu_2 + \cdots$$

which is similar to the frequency condition for one-photon absorption. However, multi-photon selection rules are different from one-photon selection rules. Therefore, multi-photon processes allow examination of energy states that otherwise could not be reached. In multi-photon ionization, the second or third photon takes the molecule into the energy continuum above its highest lying energy state. This technique is especially useful for the study of weakly fluorescing molecules.

Resonant multi-photon ionization (REMPI). This is a variant of MPI described above, in which one or more photons promote a molecule to an electronically excited state and then additional photons generate ions from the excited state. The power of this method in the study of chemical reactions is its selectivity. In a chemically reacting system, individual reactants and products can be chosen by tuning the frequency of the laser generating the radiation to the electronic absorption band of specific molecules.

Reaction product imaging. In this technique, product ions are accelerated by an electric field toward a phosphorescent screen and the light emitted from the screen is imaged by a charge-coupled device. The significance of this experiment to the study of chemical reactions is that it allows for a detailed analysis of the angular distribution of products.

D90.3 The Rb atom must hit the I side of CH_3I in order to produce $RbI + CH_3$. The orientation of CH_3I can be controlled by exciting rotations about the CI axis with linearly polarized light; the optimal orientation aims the I side of CH_3I at the direction of approach of the beam of Rb atoms. Two possible alignments of the reactant beams are shown in Fig. 90.1. In the top depiction, the beams are antiparallel, thereby maximizing the likelihood of collision and the volume within which collision can occur (but also putting each beam source in the path of the other beam). In the lower depiction, the beam paths are at right angles, thereby minimizing the region in which the beams collide, but facilitating the study of that well-defined collision volume by a "probe" laser at right angles to both beams.

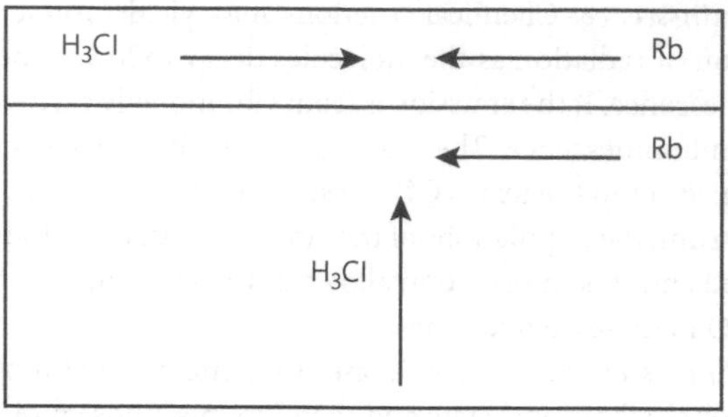

Fig 90.1

D90.5 Molecular beams may be used to prepare molecules in specific rotational and vibrational states and then to examine the results of collisions between such precisely prepared species. Topic 90.1(a) describes how molecular beams are prepared such that the molecules in them have a very narrow range of velocities and therefore relatively few collisions to redistribute their energies. Molecules in such beams can be prepared in specific vibrational states, for example, by having lasers excite vibrations. Crossing two molecular beams allows collisions to be staged between two sets of precisely characterized molecules. Detectors can then be used to study the results of those collisions, recording what molecules in what state are scattered where as a result.

Exercises

E90.1(a) Refer to Fig. 90.19 of the main text, which shows an attractive potential energy surface as well as trajectories of both a successful reaction and an unsuccessful one. The trajectories begin in the lower right, representing reactants. The successful trajectory passes through the transition state (marked by a circle with the symbol ‡ near it). That trajectory is fairly straight from the lower right through the transition state, indicating little or no vibrational excitation in the reactant. Therefore most of its energy is in translation. Since it has enough total energy to reach the transition state, we can say that the reactant is high in translational energy and low in vibrational energy. That successful trajectory moves from side to side along the valley representing products, so the product is high in vibrational energy and relatively lower in translational energy. The unsuccessful trajectory, by contrast, has a reactant high in vibrational energy; it moves from side to side in the reactant valley without reaching the transition state.

E90.2(a) The numerator of eqn 90.6 is

$$\int_0^{\infty} \overline{P}(E)\mathrm{e}^{-E/RT}\,\mathrm{d}E = \overline{P}\int_0^{\infty} \mathrm{e}^{-E/RT}\,\mathrm{d}E = -\frac{\overline{P}}{RT}\left(\mathrm{e}^{-E/RT}\right)\Big|_{E=0}^{E=-\infty} = \frac{\overline{P}}{RT}$$

Thus, if the cumulative reaction probability were independent of energy (could be taken outside the integral), then the numerator would decrease with increasing temperature and so, then would the rate constant. (In fact, the rate constant would decrease with increasing temperature even faster, because the denominator of eqn 90.6 would increase with increasing temperature.)

Problems

P90.1 The change in intensity of the beam, $\mathrm{d}I$, is proportional to the number of scatterers per unit volume, $\mathcal{N}$, the intensity of the beam, I, and the path length $\mathrm{d}L$. The constant of proportionality must be the collision cross-section σ, the "target area" of each scatterer. Thus, $\sigma \mathrm{d}L$ is the volume of scatterers to be encountered within the beam, and

$$\mathrm{d}I = -\sigma\mathcal{N}I\mathrm{d}L \quad or \quad \mathrm{d}\ln I = -\sigma\mathcal{N}\mathrm{d}L$$

If the incident intensity (at $L = 0$) is I_0 and the emergent intensity is I, we can write

$$\ln\frac{I}{I_0} = -\sigma\mathcal{N}L \quad \text{or} \quad \boxed{I = I_0\mathrm{e}^{-\sigma\mathcal{N}L}}$$

P90.3 Refer to Fig. 90.2.

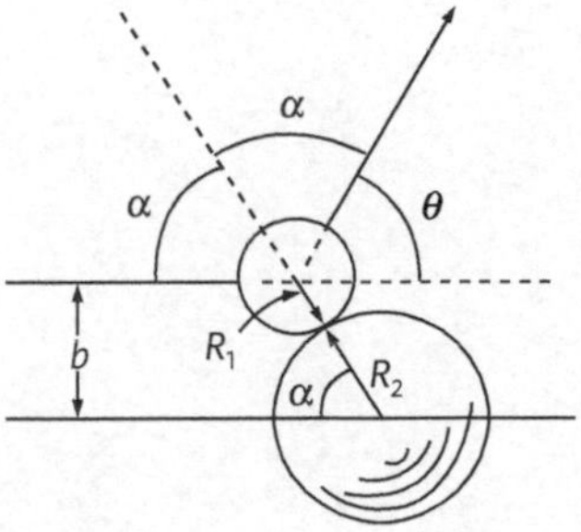

Fig 90.2

The scattering angle is $\theta = \pi - 2\alpha$ if specular reflection occurs in the collision (angle of impact equal to angle of departure from the surface). For $b \le R_1 + R_2$,

$$\sin\alpha = \frac{b}{R_1 + R_2}$$

$$\theta = \begin{cases} \pi - 2\arcsin\left(\dfrac{b}{R_1 + R_2}\right) & b \le R_1 + R_2 \\ 0 & b > R_1 + R_2 \end{cases}$$

The function is plotted in Fig. 90.3.

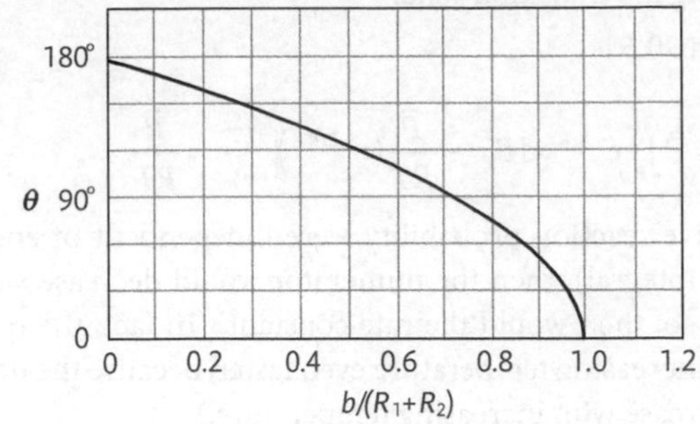

Fig 90.3

Focus 18: Integrated activities

F18.1 The standard molar partition function is the molecular partition function at standard pressure. See the Checklist of equations from Topic 52 for the partition function factors.

$$\frac{q_{\mathrm{m}}^{\ominus\mathrm{T}}}{N_{\mathrm{A}}}=\frac{V_{\mathrm{m}}^{\ominus}}{N_{\mathrm{A}}\Lambda^{3}}=\frac{RT}{N_{\mathrm{A}}p^{\ominus}\Lambda^{3}}=\frac{kT}{p^{\ominus}}\frac{(2\pi mkT)^{3/2}}{h^{3}}=\frac{(kT)^{5/2}}{p^{\ominus}h^{3}}\left(\frac{2\pi M}{N_{\mathrm{A}}}\right)^{3/2}$$

For $T\approx 300$ K, $M\approx 50$ g mol^{-1},

$$\frac{q_{\mathrm{m}}^{\ominus\mathrm{T}}}{N_{\mathrm{A}}}=\frac{\{(1.381\times10^{-23}\ \mathrm{J\,K^{-1}})(300\ \mathrm{K})\}^{5/2}}{(10^{5}\ \mathrm{Pa})(6.626\times10^{-34}\ \mathrm{J\,s})^{3}}\left(\frac{2\pi(50\times10^{-3}\ \mathrm{kg\,mol^{-1}})}{6.022\times10^{23}\ \mathrm{mol^{-1}}}\right)^{3/2}\approx\boxed{1.4\times10^{7}}$$

$$q^{\mathrm{R}}(\text{nonlinear})=\frac{1}{\sigma}\left(\frac{kT}{hc}\right)^{3/2}\left(\frac{\pi}{\tilde{A}\tilde{B}\tilde{C}}\right)^{1/2}$$

For a non-linear molecule, $T\approx 300$ K, $\tilde{A}\approx\tilde{B}\approx\tilde{C}\approx 2$ cm^{-1}, $\sigma\approx 2$,

$$q^{\mathrm{R}}=\frac{1}{2}\left(\frac{(1.381\times10^{-23}\ \mathrm{J\,K^{-1}})(300\ \mathrm{K})}{(6.626\times10^{-34}\ \mathrm{J\,s})(2.998\times10^{10}\ \mathrm{cm\,s^{-1}})}\right)^{3/2}\left(\frac{\pi}{(2\ \mathrm{cm^{-1}})^{3}}\right)^{1/2}=\boxed{900}$$

$$q^{\mathrm{R}}(\text{linear})=\frac{kT}{\sigma hc\tilde{B}}$$

For a linear molecule, $T\approx 300$ K, $\tilde{B}\approx 1$ cm^{-1}, $\sigma\approx 1$,

$$q^{\mathrm{R}}(\text{linear})=\frac{(1.381\times10^{-23}\ \mathrm{J\,K^{-1}})(300\ \mathrm{K})}{(1)(6.626\times10^{-34}\ \mathrm{J\,s})(2.998\times10^{10}\ \mathrm{cm\,s^{-1}})(1\ \mathrm{cm^{-1}})}\approx\boxed{200}$$

Energies of most excited vibrational states in small molecules and of most excited electronic states are high enough to make $q^{\mathrm{V}}\approx q^{\mathrm{E}}\approx\boxed{1}$

$$k_{\mathrm{r}}=\frac{\kappa kT}{h}\bar{K}_{\mathrm{c}}^{\ddagger}\,[89.10]=\left(\frac{\kappa kT}{h}\right)\times\left(\frac{RT}{p^{\ominus}}\right)\times\left(\frac{N_{\mathrm{A}}\bar{q}_{\mathrm{C}^{\ddagger}}^{\ominus}}{q_{\mathrm{A}}^{\ominus}q_{\mathrm{B}}^{\ominus}}\right)\mathrm{e}^{-\Delta E_{0}/RT}\,[89.9]=A\mathrm{e}^{-E_{\mathrm{a}}/RT}$$

If A and B are structureless molecules, then we use estimates from above to evaluate

$$\frac{q_{\mathrm{A}}^{\ominus}}{N_{\mathrm{A}}}=\frac{q_{\mathrm{A}}^{\ominus\mathrm{T}}}{N_{\mathrm{A}}}\approx1.4\times10^{7}\approx\frac{q_{\mathrm{B}}^{\ominus}}{N_{\mathrm{A}}}=\frac{q_{\mathrm{B}}^{\ominus\mathrm{T}}}{N_{\mathrm{A}}}$$

$$\frac{\bar{q}_{\mathrm{C}^{\ddagger}}^{\ominus}}{N_{\mathrm{A}}}=\frac{q_{\mathrm{C}^{\ddagger}}^{\ominus\mathrm{T}}q^{\mathrm{R}}(\text{linear})}{N_{\mathrm{A}}}\approx(2^{3/2})\times(1.4\times10^{7})\times(200)=8\times10^{9}$$

(The factor of $2^{3/2}$ comes from $m_C = m_A + m_B \approx 2m_A$ and $q^T \propto m^{3/2}$.)

$$\frac{RT}{p^{\ominus}} = \frac{(8.3145\ \mathrm{J\,K^{-1}\,mol^{-1}})\times(300\ \mathrm{K})}{10^5\ \mathrm{Pa}} = 2.5\times10^{-2}\ \mathrm{m^3\,mol^{-1}}$$

and $$\frac{\kappa kT}{h} \approx \frac{kT}{h} = \frac{(1.381\times10^{-23}\ \mathrm{J\,K^{-1}})\times(300\ \mathrm{K})}{6.626\times10^{-34}\ \mathrm{J\,s}} = 6.25\times10^{12}\ \mathrm{s^{-1}}$$

Strictly speaking, E_a is not exactly the same as ΔE_0, but they are approximately equal—close enough for the purposes of estimating the order of magnitude of the pre-exponential factor. So once we identify $E_a \approx \Delta E_0$, we identify the pre-exponential factor with everything *other than* the exponential term. Therefore, the pre-exponential factor

$$A \approx \frac{(6.25\times10^{12}\ \mathrm{s^{-1}})\times(2.5\times10^{-2}\ \mathrm{m^3\,mol^{-1}})\times(8\times10^9)}{(1.4\times10^7)^2}$$
$$\approx 6.3\times10^6\ \mathrm{m^3\,mol^{-1}\,s^{-1}} = \boxed{6.3\times10^9\ \mathrm{dm^3\,mol^{-1}\,s^{-1}}}.$$

According to collision theory [87.8]

$$A = P\sigma\left(\frac{8kT}{\pi\mu}\right)^{1/2} N_A$$

Take $\sigma \approx 0.5\ \mathrm{nm^2} = 5\times10^{-19}\ \mathrm{m^2}$ as a typical value for small molecules:

$$A = P\times5\times10^{-19}\ \mathrm{m^2}\times\left(\frac{8\times(1.381\times10^{-23}\ \mathrm{J\,K^{-1}})\times(300\ \mathrm{K})}{\pi\times(25\ m_u)\times(1.661\times10^{-27}\ \mathrm{kg}\ m_u^{-1})}\right)^{1/2}\times6.022\times10^{23}\ \mathrm{mol^{-1}}$$
$$= 1.5\times10^8\ \mathrm{m^3\,mol^{-1}\,s^{-1}}\times P.$$

The values are quite consistent, for they imply $P \approx 0.04$, which is certainly a plausible value.

If A and B are non-linear triatomics, then

$$\frac{q_A^{\ominus}}{N_A} \approx (1.4\times10^7)\times(900) = 1.3\times10^{10} \approx \frac{q_B^{\ominus}}{N_A}$$

$$\frac{q_{C^‡}^{\ominus}}{N_A} \approx (2^{3/2})\times(1.4\times10^7)\times(900) = 3.6\times10^{10}$$

and $$A \approx \frac{(6.25\times10^{12}\,\mathrm{s^{-1}})\times(2.5\times10^{-2}\ \mathrm{m^3\,mol^{-1}})\times(3.6\times10^{10})}{(1.3\times10^{10})^2}$$
$$\approx 33\ \mathrm{m^3\,mol^{-1}\,s^{-1}} = \boxed{3.3\times10^4\ \mathrm{dm^3\,mol^{-1}\,s^{-1}}}.$$

Comparison to the expression from collision theory implies $\boxed{P = 2\times10^{-7}}$

Topic 91 Unimolecular reactions

Discussion question

D91.1 The expression $k_r = k_a k_b[A]/(k_b + k'_a[A])$ for the effective rate constant of a unimolecular reaction $A \rightarrow P$ is based on the validity of the assumption of the existence of the pre-equilibrium $A + A \underset{k'_a}{\overset{k_a}{\rightleftharpoons}} A^* + A$. This can be a good assumption if both k_a and k'_a are much larger than k_b. The expression for the effective rate-constant, k_r, can be rearranged to

$$\frac{1}{k_r} = \frac{k'_a}{k_a k_b} + \frac{1}{k_a[A]}$$

Hence, a test of the theory is to plot $1/k_r$ against $1/[A]$ and to expect a straight line. Another test is based on the prediction from the Lindemann–Hinshelwood mechanism that as the concentration (and therefore the partial pressure) of A is reduced, the reaction should switch to overall second order kinetics. Whereas the mechanism agrees in general with the switch in order of unimolecular reactions, it does not agree in detail. A typical graph of $1/k_r$ against $1/[A]$ has a pronounced curvature, corresponding to a larger value of k_r (a smaller value of $1/k_r$) at high pressures (low $1/[A]$) than would be expected by extrapolation of the reasonably linear low pressure (high $1/[A]$) data.

Exercises

E91.1(a) $$\frac{1}{k_r} = \frac{k'_a}{k_a k_b} + \frac{1}{k_a p_A} \quad \text{[analogous to 91.8]}$$

Therefore, for two different pressures we have

$$\frac{1}{k_r(p_1)} - \frac{1}{k_r(p_2)} = \frac{1}{k_a}\left(\frac{1}{p_1} - \frac{1}{p_2}\right)$$

$$\text{so}\quad k_a = \frac{\dfrac{1}{p_1} - \dfrac{1}{p_2}}{\dfrac{1}{k_r(p_1)} - \dfrac{1}{k_r(p_2)}} = \frac{\dfrac{1}{12\ \text{Pa}} - \dfrac{1}{1.30\times10^{3}\ \text{Pa}}}{\dfrac{1}{2.10\times10^{-5}\ \text{s}^{-1}} - \dfrac{1}{2.50\times10^{-4}\ \text{s}^{-1}}} = \boxed{1.9\times10^{-6}\ \text{Pa}^{-1}\ \text{s}^{-1}}$$

or $1.9\ \text{MPa}^{-1}\ \text{s}^{-1}$

E91.2(a) According to RRK theory, the steric P-factor is given by eqn 91.9a

$$P=\left(1-\frac{E^*}{E}\right)^{s-1}$$

where s is the number of vibrational modes in the reacting molecule. For a non-linear molecule composed of N atoms, the number of modes is [Topic 44.1]

$$s=3N-6=3\times5-6=9.$$

Rearranging eqn 91.9a yields

$$\frac{E^*}{E}=1-P^{\frac{1}{s-1}}=1-(3.0\times10^{-5})^{\frac{1}{8}}=\boxed{0.73}$$

E91.3(a) According to RRK theory, the steric P-factor is given by eqn 91.9a

$$P=\left(1-\frac{E^*}{E}\right)^{s-1}=\left(1-\frac{200\text{ kJ mol}^{-1}}{250\text{ kJ mol}^{-1}}\right)^{10-1}=\boxed{5.1\times10^{-7}}$$

Problem

P91.1 $\dfrac{1}{k_r}=\dfrac{k_a'}{k_ak_b}+\dfrac{1}{k_ap}$ [analogous to 91.8]

We expect a straight line when $\dfrac{1}{k_r}$ is plotted against $\dfrac{1}{p}$. We draw up the following table:

p/Torr	84.1	11.0	2.89	0.569	0.120	0.067
$1/(p/\text{Torr})$	0.012	0.091	0.346	1.76	8.33	14.9
$10^4\,k_r/\text{s}^{-1}$	2.98	2.23	1.54	0.857	0.392	0.303
$10^{-4}/(k_r/\text{s}^{-1})$	0.336	0.448	0.629	1.17	2.55	3.30

These points are plotted in Fig. 91.1. There are marked deviations at low pressures, indicating that the Lindemann theory is deficient in that region.

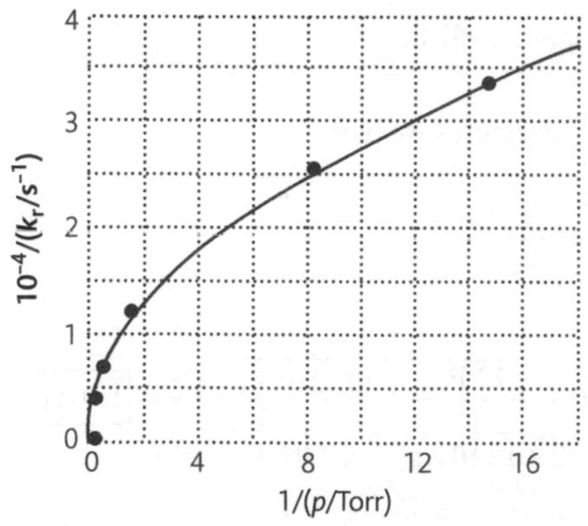

Fig 91.1

Topic 92 Enzymes

Discussion questions

92.1 The **Michaelis–Menten mechanism** of enzyme activity models the enzyme with one active site, weakly and reversibly, binding a substrate in homogeneous solution. It is a three-step mechanism. The first and second steps are the reversible formation of the enzyme–substrate complex (ES). The third step is the decay of the complex into the product. The steady-state approximation is applied to the concentration of the intermediate (ES) and its use simplifies the derivation of the final rate expression. However, the justification for the use of the approximation with this mechanism is suspect. Both rate constants for the reversible step may not be as large (in comparison to the rate constant for the decay to products) as they need to be for the approximation to be valid. The mechanism clearly indicates that the simplest form of the rate law, $v = v_{\max} = k_{\mathrm{b}}[\mathrm{E}]_0$, occurs when $[\mathrm{S}]_0 \gg K_{\mathrm{M}}$. In addition, the general form of the rate law does seem to match the principal experimental features of enzyme catalyzed reactions. It provides a mechanistic understanding of both the turnover number and catalytic efficiency. The model may be expanded to include multisubstrate reactions and inhibition.

92.3 Text Fig. 92.6 summarizes the important characteristics of the three major modes of enzyme inhibition: competitive inhibition, uncompetitive inhibition, and non-competitive inhibition. Mathematical models for inhibition, which are the analogues of the Michaelis–Menten and Lineweaver–Burk equations (92.3a and b), are presented in eqns 92.7 and 92.8.

$$\frac{1}{v} = \frac{\alpha'}{v_{\max}} + \left(\frac{\alpha K_{\mathrm{M}}}{v_{\max}}\right)\frac{1}{[\mathrm{S}]_0} \quad [92.8]$$

where $\alpha = 1+[\mathrm{I}]/K_{\mathrm{I}}, \alpha' = 1+[\mathrm{I}]/K_{\mathrm{I}}', K_{\mathrm{I}} = [\mathrm{E}][\mathrm{I}]/[\mathrm{EI}]$, and $K_{\mathrm{I}}' = [\mathrm{ES}][\mathrm{I}]/[\mathrm{ESI}]$

In **competitive inhibition** the inhibitor binds only to the active site of the enzyme and thereby inhibits the attachment of the substrate. This condition corresponds to $\alpha > 1$ and $\alpha' = 1$ (because ESI does not form). The slope of the Lineweaver–Burk plot increases by a factor of α relative to the slope for data on the uninhibited enzyme ($\alpha = \alpha' = 1$). The *y*-intercept does not change as a result of competitive inhibition.

In **uncompetitive inhibition** the inhibitor binds to a site of the enzyme that is removed from the active site, but only if the substrate is already present. The inhibition occurs because ESI reduces the concentration of ES, the active type of the complex. In this case $\alpha = 1$ (because EI does not form) and $\alpha' > 1$. The y-intercept of the Lineweaver–Burk plot increases by a factor of α' relative to the y-intercept for data on the uninhibited enzyme, but the slope does not change.

In **non-competitive inhibition** (also called **mixed inhibition**) the inhibitor binds to a site other than the active site, and its presence reduces the ability of the substrate to bind to the active site. Inhibition occurs at both the E and ES sites. This condition corresponds to $\alpha > 1$ and $\alpha' > 1$. Both the slope and y-intercept of the Lineweaver–Burk plot increase upon addition of the inhibitor. Fig. 92.6(c) shows the special case of $K_I = K_I'$ and $\alpha = \alpha'$, which results in intersection of the lines at the x-axis.

In all cases, the efficiency of the inhibitor may be obtained by determining K_M and v_{max} from a control experiment with uninhibited enzyme and then repeating the experiment with a known concentration of inhibitor. From the slope and y-intercept of the Lineweaver–Burk plot for the inhibited enzyme (eqn 92.8), the mode of inhibition, the values of α or α', and the values of K_I, or K_I' may be obtained.

Exercises

92.1(a) The fast, reversible step suggests the pre-equilibrium approximation:

$$K = \frac{[BH^+][A^-]}{[AH][B]} \quad \text{and} \quad [A^-] = \frac{K[AH][B]}{[BH^+]}$$

Thus, the rate of product formation is

$$\frac{d[P]}{dt} = k_b[AH][A^-] = \boxed{\frac{k_b K[AH]^2[B]}{[BH^+]}}$$

The application of the steady-state approximation to $[A^-]$ gives a similar, but significantly different, result:

$$\frac{d[A^-]}{dt} = k_a[AH][B] - k_a'[A^-][BH^+] - k_b[A^-][AH] = 0$$

Therefore, $[A^-] = \dfrac{k_a[AH][B]}{k_a'[BH^+] + k_b[AH]}$

and the rate of formation of product is

$$\frac{d[P]}{dt} = k_b[AH][A^-] = \frac{k_a k_b[AH]^2[B]}{k_a'[BH^+] + k_b[AH]}$$

92.2(a) Since $v = \dfrac{v_{max}}{1 + K_M/[S]_0}$ [92.3a],

$$\begin{aligned} v_{max} &= (1 + K_M/[S]_0)v \\ &= (1 + 0.046/0.105) \times (1.04 \text{ mmol dm}^{-3} \text{ s}^{-1}) \\ &= \boxed{1.50 \text{ mmol dm}^{-3} \text{ s}^{-1}} \end{aligned}$$

92.3(a) $$\upsilon=\frac{\upsilon_{\max}}{\alpha'+\alpha K_M/[S]_0} \quad [92.7]$$

In the absence of inhibition $\alpha=1$ and $\alpha'=1$. However, in **competitive inhibition** the inhibitor binds only to the active site of the enzyme and thereby inhibits the attachment of the substrate. This condition corresponds to $\alpha>1$ and $\alpha'=1$ (because ESI does not form). Thus,

$$\frac{\upsilon_{\text{no inhibition}}}{\upsilon_{\text{competitive inhibition}}}=\frac{1+\alpha K_M/[S]_0}{1+K_M/[S]_0}=2$$

$$\alpha=\frac{2(1+K_M/[S]_0)-1}{K_M/[S]_0}$$
$$=\frac{2(1+3.0/0.10)-1}{3.0/0.10}=2.0\overline{3}$$
$$[I]=(\alpha-1)K_I$$
$$=1.0\overline{3}\times(2.0\times10^{-5}\text{ mol dm}^{-3})=\boxed{2.0\times10^{-5}\text{ mol dm}^{-3}}$$

Problems

92.1 Assuming a rapid pre-equilibrium of E, S, and ES implies that

$$K=\frac{k_a}{k_a'}=\frac{[ES]}{[E][S]} \quad \text{and} \quad [ES]=K[E][S]$$

But the law of mass balance demands that $[E]=[E]_0-[ES]$ so $[ES]=K([E]_0-[ES])[S]$ and, solving for [ES], we find that

$$[ES]=\frac{[E]_0}{1+\dfrac{1}{K[S]_0}}$$

where the free substrate concentration has been replaced by $[S]_0$ because the substrate is typically in large excess relative to the enzyme. Now substitute the latter expression into the Michaelis-Menten rate law.

$$\upsilon=k_b[ES]=\frac{k_b[E]_0}{1+\dfrac{1}{K[S]_0}} \quad \text{where } v_{\max}=k_b[E]_0$$

$$\boxed{\upsilon=\frac{\upsilon_{\max}}{1+\dfrac{1}{K[S]_0}} \quad \text{Rate law based on rapid pre-equilibrium approximation}}.$$

With $K_M=(k_a'+k_b)/k_a$ eqn 92.1 is

$$\upsilon=\frac{\upsilon_{\max}}{1+\dfrac{K_M}{[S]_0}} \quad \text{Rate law based on steady-state approximation}$$

Inspection reveals that the two approximations are identical when $K_M = 1/K$, which implies that

$$(k_a' + k_b)/k_a = k_a'/k_a \text{ or } \boxed{k_a' >> k_b}$$

92.3 Assume that the steady-state approximation is appropriate for both intermediates ([ES] and [ES′]).

For [ES]:

$$\frac{d[ES]}{dt} = k_a[E][S] - k_a'[ES] - k_b[ES] = 0 \quad \text{and} \quad [ES] = \left(\frac{k_a}{k_a' + k_b}\right)[E][S].$$

For [ES′]:

$$\frac{d[ES']}{dt} = k_b[ES] - k_c[ES'] = 0 \quad \text{and} \quad [ES'] = \left(\frac{k_b}{k_c}\right)[ES].$$

We now have two equations in the three unknowns [E], [ES], and [ES′]. A third is provided by the mass balance expression $[E]_0 = [E] + [ES] + [ES']$. These three equations may be solved to give expressions for each of the three unknowns in terms of the rate constants, $[E]_0$, and [S]. (For practical purposes the free substrate concentration is replaced by $[S]_0$ because the substrate is typically in large excess relative to the enzyme.) The expression found for [ES′] is

$$[ES'] = \frac{v_{max}/k_c}{1 + K_M/[S]_0} \quad \text{where} \quad v_{max} = \left(\frac{k_b k_c}{k_b + k_c}\right)[E]_0 \quad \text{and} \quad K_M = \frac{k_c(k_a' + k_b)}{k_a(k_b + k_c)}$$

Substitution into the rate expression for product formation yields the desired equation.

$$v = k_c[ES'] = \frac{v_{max}}{1 + K_M/[S]_0}$$

92.5 We draw up the table below, which includes data rows required for a Lineweaver-Burk plot ($1/v$ against $1/[S]_0$). The linear regression fit is summarized in the Fig. 92.1 plot.

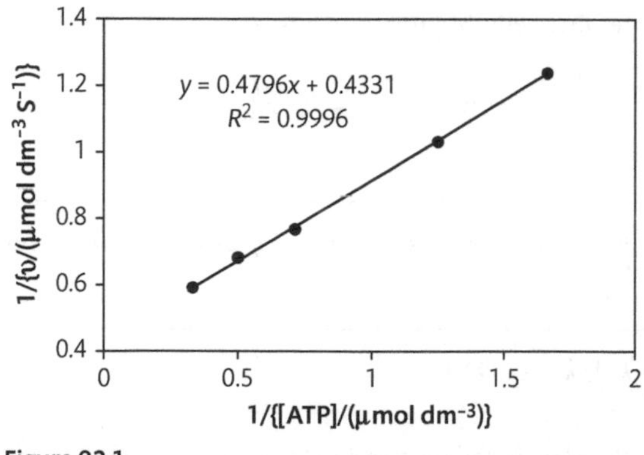

Figure 92.1

$$v_{max} = 1/\text{intercept [92.3b]} = 1/(0.433\ \mu\text{mol dm}^{-3}\ \text{s}^{-1}) = \boxed{2.31\ \mu\text{mol dm}^{-3}\ \text{s}^{-1}}$$

$$k_b = v_{max} / [\text{E}]_0\ [92.2b] = (2.31\ \mu\text{mol dm}^{-3}\ \text{s}^{-1}) / (0.020\ \mu\text{mol dm}^{-3}) = \boxed{115\ \text{s}^{-1}}$$

$$k_{cat} = k_b\ [92.4] = \boxed{115\ \text{s}^{-1}}$$

$$K_M = v_{max} \times \text{slope [92.3b]} = (2.31\ \mu\text{mol dm}^{-3}\ \text{s}^{-1})\times(0.480\ \text{s}) = \boxed{1.11\ \mu\text{mol dm}^{-3}}$$

$$\eta = k_{cat} / K_M\ [92.5] = (115\ \text{s}^{-1}) / (1.11\ \mu\text{mol dm}^{-3}) = \boxed{104\ \text{dm}^3\ \mu\text{mol}^{-1}\ \text{s}^{-1}}$$

92.7 (a) The dissociation equilibrium may be rearranged to give the following four relationships.

$$[\text{E}^-] = K_{\text{E,a}}[\text{EH}]/[\text{H}^+] \qquad [\text{EH}_2^+] = [\text{EH}][\text{H}^+]/K_{\text{E,b}}$$

$$[\text{ES}^-] = K_{\text{ES,a}}[\text{ESH}]/[\text{H}^+] \qquad [\text{ESH}_2] = [\text{ESH}][\text{H}^+]/K_{\text{ES,b}}$$

Mass balance provides an equation for [EH].

[ATP] / (μmol dm⁻³)	0.60	0.80	1.4	2.0	3.0
V/(μmol dm⁻³ s⁻¹)	0.81	0.97	1.30	1.47	1.69
1 / {[ATP]/(μmol dm⁻³)}	1.67	1.25	0.714	0.500	0.333
1 / {v/(μmol dm⁻³ s⁻¹)}	1.23	1.03	0.769	0.680	0.592

$$[\text{E}]_0 = [\text{E}^-]+[\text{EH}]+[\text{EH}_2^+]+[\text{ES}^-]+[\text{ESH}]+[\text{ESH}_2]$$

$$= \frac{K_{\text{E,a}}[\text{EH}]}{[\text{H}^+]}+[\text{EH}]+\frac{[\text{EH}][\text{H}^+]}{K_{\text{E,b}}}+\frac{K_{\text{ES,a}}[\text{ESH}]}{[\text{H}^+]}+[\text{ESH}]+\frac{[\text{ESH}][\text{H}^+]}{K_{\text{ES,b}}}$$

$$[\text{EH}] = \frac{[\text{E}]_0 - \left\{1+\frac{[\text{H}^+]}{K_{\text{ES,b}}}+\frac{K_{\text{ES,a}}}{[\text{H}^+]}\right\}[\text{ESH}]}{1+\frac{[\text{H}^+]}{K_{\text{E,b}}}+\frac{K_{\text{E,a}}}{[\text{H}^+]}}$$

$$= \frac{[\text{E}]_0 - c_1[\text{ESH}]}{c_2} \quad \text{where} \quad c_1 = 1+\frac{[\text{H}^+]}{K_{\text{ES,b}}}+\frac{K_{\text{ES,a}}}{[\text{H}^+]} \ \text{and}\ c_2 = 1+\frac{[\text{H}^+]}{K_{\text{E,b}}}+\frac{K_{\text{E,a}}}{[\text{H}^+]}$$

The steady-state approximation provides an equation for [ESH].

$$\frac{d[\text{ESH}]}{dt} = k_a[\text{EH}][\text{S}] - k_a'[\text{ESH}] - k_b[\text{ESH}] = 0$$

$$[\text{ESH}] = \frac{k_a}{k_a'+k_b}[\text{EH}][\text{S}] = K_M^{-1}[\text{EH}][\text{S}]$$

$$= K_M^{-1}[\text{S}]\left\{\frac{[\text{E}]_0 - c_1[\text{ESH}]}{c_2}\right\}$$

$$[\text{ESH}] = \frac{K_M^{-1}[\text{S}][\text{E}]_0 / c_2}{1+K_M^{-1}[\text{S}]c_1/c_2} = \frac{[\text{E}]_0/c_1}{1+K_M(c_2/c_1)/[\text{S}]}$$

The rate law becomes:

$$\upsilon = d[P]/dt = k_b[ESH] = \frac{k_b[E]_0/c_1}{1+K_M(c_2/c_1)/[S]} = \frac{\upsilon'_{max}}{1+K'_M/[S]}$$

where $\upsilon'_{max} = k_b[E]_0 \bigg/ \left\{1+\frac{[H^+]}{K_{ES,b}}+\frac{K_{ES,a}}{[H^+]}\right\} = \upsilon_{max} \bigg/ \left\{1+\frac{[H^+]}{K_{ES,b}}+\frac{K_{ES,a}}{[H^+]}\right\}$

and $K'_M = K_M \left\{1+\frac{[H^+]}{K_{E,b}}+\frac{K_{E,a}}{[H^+]}\right\} \bigg/ \left\{1+\frac{[H^+]}{K_{ES,b}}+\frac{K_{ES,a}}{[H^+]}\right\}$

(b) $\upsilon_{max} = 1.0\times10^{-6}$ mol dm^{-3} s^{-1}; $K_{ES,b} = 1.0\times10^{-6}$; $K_{ES,a} = 1.0\times10^{-8}$

Fig. 92.2 shows a plot of υ'_{max} against pH. The plot indicates a maximum value of υ'_{max} at pH = 7.0 for this set of equilibrium and kinetic constants. A formula for the pH of the maximum can be derived by finding the point at which $\frac{d\upsilon'_{max}}{d[H^+]} = 0$. This gives:

$$[H^+]_{max} = (K_{ES,a}\, K_{ES,b})^{1/2} = \sqrt{(1.0\times10^{-8}\ \text{mol dm}^{-3})(1.0\times10^{-6}\ \text{mol dm}^{-3})}$$
$$= 1.0\times10^{-7}\ \text{mol dm}^{-3}$$

which corresponds to $\boxed{\text{pH} = 7.0}$

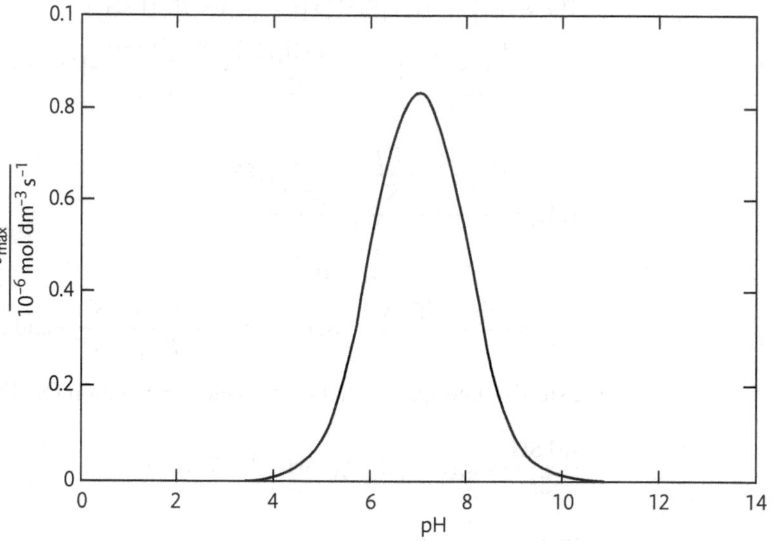

Figure 92.2

(c) $\upsilon_{max} = 1.0\times10^{-6}$ mol dm^{-3} s^{-1}; $K_{ES,b} = 1.0\times10^{-4}$; $K_{ES,a} = 1.0\times10^{-10}$

Fig. 92.3 shows a plot of υ'_{max} against pH. The plot once again indicates a maximum value of υ'_{max} at pH = 7.0 for this set of equilibrium and kinetic constants. However, the rate is high over a much larger pH range than appeared in part (b). This reflects the behaviour of the term $1+[H^+]/K_{ES,b}+K_{ES,a}/[H^+]$ in the denominator of the v'_{max} expression. When $K_{ES,b}$ is relatively large, large $[H^+]$ values (low pH) cause growth in the values of v'_{max}. However, when $K_{ES,a}$ is relatively small, very small $[H^+]$ values (high pH) cause a decline in the v'_{max} values.

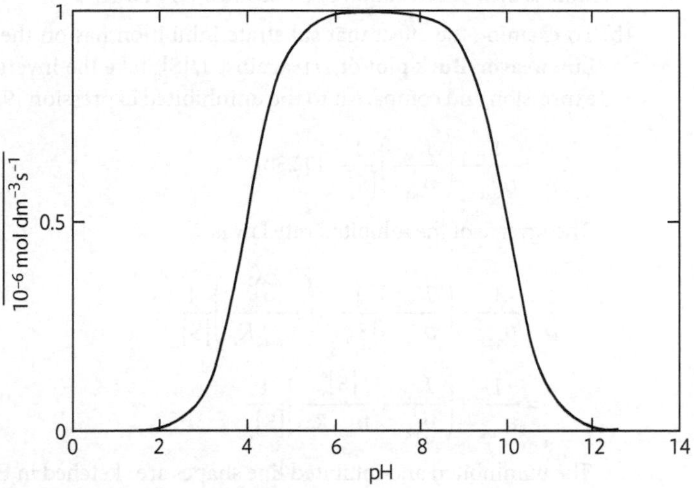

Figure 92.3

92.9 (a) We add to the Michaelis–Menten mechanism the inhibition by the substrate

$$SES \rightleftharpoons ES+S \qquad K_I=[ES][S]/[SES]$$

where the inhibited enzyme, SES, forms when S binds to ES and, thereby, prevents the formation of product. This inhibition might possibly occur when S is at a very high concentration. Enzyme mass balance is written in terms of [ES], K_I, K_M (= [E][S]/[ES]), and [S]. (For practical purposes the free substrate concentration is replaced by $[S]_0$ because the substrate is typically in large excess relative to the enzyme.)

$$\begin{aligned}[E]_0&=[E]+[ES]+[SES]\\&=\frac{K_M[ES]}{[S]}+[ES]+\frac{[ES][S]}{K_I}\\&=\left(1+\frac{K_M}{[S]}+\frac{[S]}{K_I}\right)[ES]\end{aligned}$$

Thus,

$$[ES]=\frac{[E]_0}{\left(1+\frac{K_M}{[S]}+\frac{[S]}{K_I}\right)}$$

and the expression for the rate of product formation becomes

$$v = k_b[\mathrm{ES}] = \frac{v_{\max}}{1 + \dfrac{K_M}{[\mathrm{S}]_0} + \dfrac{[\mathrm{S}]_0}{K_I}} \quad \text{where} \quad v_{\max} = k_b[\mathrm{E}]_0$$

The denominator term $[\mathrm{S}]_0/K_I$ reflects a reduced reaction rate caused by inhibition as the concentration of S becomes very large.

(b) To examine the effect that substrate inhibition has on the double reciprocal, **Lineweaver-Burk plot** of $1/v$ against $1/[\mathrm{S}]_0$ take the inverse of the above rate expression and compare it to the uninhibited expression [92.3b]:

$$\frac{1}{v} = \frac{1}{v_{\max}} + \left(\frac{K_M}{v_{\max}}\right)\frac{1}{[\mathrm{S}]_0} \quad [92.3b]$$

The inverse of the inhibited rate law is

$$\begin{aligned}\frac{1}{v} &= \frac{1}{v_{\max}} + \left(\frac{K_M}{v_{\max}}\right)\frac{1}{[\mathrm{S}]_0} + \left(\frac{[\mathrm{S}]_0^2}{v_{\max}K_I}\right)\frac{1}{[\mathrm{S}]_0} \\ &= \frac{1}{v_{\max}} + \left(\frac{K_M}{v_{\max}} + \frac{[\mathrm{S}]_0^2}{v_{\max}K_I}\right)\frac{1}{[\mathrm{S}]_0}\end{aligned}$$

The uninhibited and inhibited line shapes are sketched in Fig. 92.4.

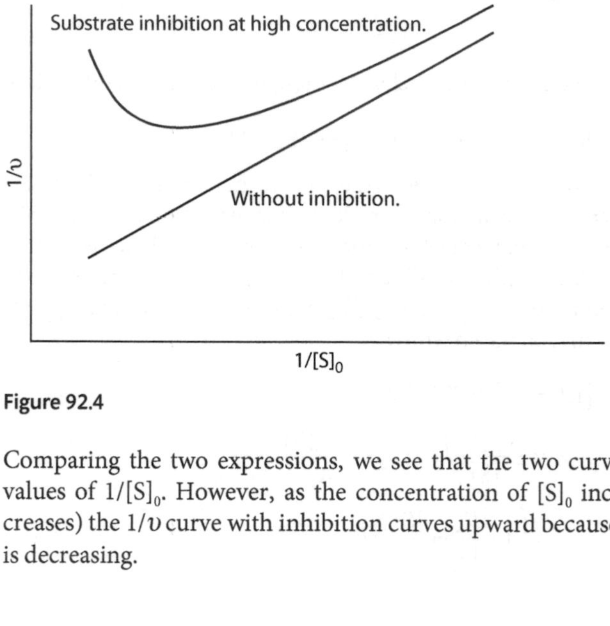

Figure 92.4

Comparing the two expressions, we see that the two curves match at high values of $1/[\mathrm{S}]_0$. However, as the concentration of $[\mathrm{S}]_0$ increases ($1/[\mathrm{S}]_0$ decreases) the $1/v$ curve with inhibition curves upward because the reaction rate is decreasing.

Topic 93 Photochemistry

Discussion question

D93.1 A primary quantum yield is associated with a primary photochemical event in an overall photochemical process. Primary photochemical events are those events that directly involve an excited state after absorbing light energy, events such as fluorescence, phosphorescence, internal conversion, intersystem crossing, or a photochemical reaction that involves the excited state directly. The primary quantum yield from a given primary process, then, is the ratio of the number of the product of that process to the number of photons absorbed (eqn 93.1a). Secondary processes are those that do not involve an excited state directly, but subsequent events. Examples include a chemical reaction involving a molecule that has received energy transferred from the absorbing molecule or a reaction involving a product of a primary photochemical reaction, which may involve secondary events as well. The overall quantum yield is the ratio of the amount of product (or products) of interest formed, whether by primary or secondary processes, to the amount of photons absorbed. Overall quantum yield of a given product can be determined by measuring the amount of the product(s) formed, the intensity of light employed, and the absorbance of the sample. Primary quantum yields of photochemical processes, such as fluorescence and phosphorescence, can be determined by techniques specific to the process. For example, fluorescence quantum yield can be measured by pulsed laser techniques, in which a short laser pulse is used to excite a molecule and then the fluorescence is detected as the excited molecules decay. In effect, the photons emitted by fluorescence are "counted" (integrated over time) and compared to the photons absorbed.

Exercises

E93.1(a) Number of photons absorbed = ϕ^{-1} × number of molecules that react [93.1a]. Therefore,

$$\text{Number absorbed} = \frac{(2.28\times10^{-3}\ \text{mol}/2)\times(6.022\times10^{23}\ \text{einstein}^{-1})}{2.1\times10^{2}\ \text{mol einstein}^{-1}} = \boxed{3.3\times10^{18}}$$

E93.2(a) The Stern-Volmer equation (eqn 93.8) relates the ratio of fluorescence quantum yields in the absence and presence of quenching

$$\frac{\phi_{\text{f},0}}{\phi_{\text{f}}} = 1+\tau_0 k_{\text{Q}}[\text{Q}] = \frac{I_{\text{f},0}}{I_{\text{f}}}$$

The last equality reflects the fact that fluorescence intensities are proportional to quantum yields. Solve this equation for [Q]:

$$[\mathrm{Q}] = \frac{(I_{\mathrm{f},0}/I_{\mathrm{f}})-1}{\tau_0 k_{\mathrm{Q}}} = \frac{2-1}{(6.0\times10^{-9}\ \mathrm{s})\times(3.0\times10^{8}\ \mathrm{dm^3\ mol^{-1}\ s^{-1}})} = \boxed{0.56\ \mathrm{mol\ dm^{-3}}}$$

Problems

P93.1 The quantum yield is defined as the amount of reacting molecules n_{A} divided by the amount of photons absorbed n_{abs}. The fraction of photons absorbed f_{abs} is one minus the fraction transmitted f_{trans}; and the amount of photons emitted n_{photon} can be inferred from the energy of the light source (power P times time t) and the energy per photon (hc/λ).

$$\phi = \frac{n_{\mathrm{A}}}{n_{\mathrm{abs}}} = \frac{n_{\mathrm{A}} hcN_{\mathrm{A}}}{(1-f_{\mathrm{trans}})\lambda Pt}$$

$$= \frac{(0.324\,\mathrm{mol})\times(6.626\times10^{-34}\ \mathrm{J\,s})\times(2.998\times10^{8}\ \mathrm{m\,s^{-1}})\times(6.022\times10^{23}\ \mathrm{mol^{-1}})}{(1-0.257)\times(320\times10^{-9}\ \mathrm{m})\times(87.5\,\mathrm{W})\times(28.0\,\mathrm{min})\times(60\,\mathrm{s\ min^{-1}})}$$

$$= \boxed{1.11}$$

P93.3 (a) The fluorescence intensity is proportional to the concentration of fluorescing species, so

$$\frac{I_{\mathrm{f}}}{I_0} = \frac{[\mathrm{S}]}{[\mathrm{S}]_0} = \mathrm{e}^{-t/\tau_0}\ [93.4] \quad \text{so} \quad \ln\left(\frac{I_{\mathrm{f}}}{I_0}\right) = -\frac{t}{\tau_0}$$

A plot of $\ln(I_{\mathrm{f}}/I_0)$ against t should be linear with a slope equal to $-1/\tau_0$ (i.e., $\tau_0 = -1/\text{slope}$) and an intercept equal to zero. See Fig. 93.1. The plot is linear, with slope $-0.150\ \mathrm{ns^{-1}}$, so

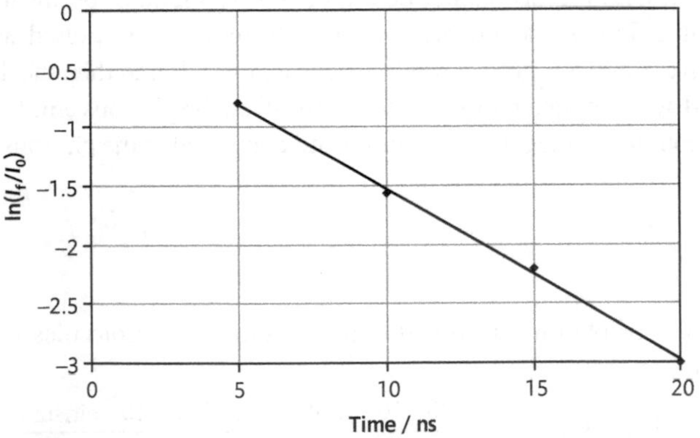

Fig 93.1

$$\tau_0 = -1/(-0.14\overline{5}\ \mathrm{ns^{-1}}) = \boxed{6.9\ \mathrm{ns}}$$

Alternatively, average the experimental values of $\frac{1}{t}\ln\left(\frac{I_{\mathrm{f}}}{I_0}\right)$ and check that the standard deviation is a small fraction of the average (it is). The average equals $-1/\tau_0$ (i.e., $\tau_0 = -1/\text{average}$).

(b) The quantum yield for fluorescence is related to the rate constants for the various decay mechanisms of the excited state by

$$\phi_f = \frac{k_f}{k_f + k_{ISC} + k_{IC}} \text{ [93.6]} = k_f \tau_0 \text{ [93.5]}$$

so $k_f = \phi_f / \tau_0 = 0.70/(6.9\text{ ns}) = \boxed{0.10\bar{1}\text{ ns}^{-1}}$

P93.5 Proceed as in Problem 93.3. In the absence of a quencher, a plot of ln I_f/I_0 against t should be linear with a slope equal to $-1/\tau_0$. The plot is in fact linear with a best-fit slope of $-1.00\bar{4}\ \mu s^{-1}$. (See Fig. 93.2)

$$\tau_0 = \frac{-1}{-1.00\bar{4}\ \mu s^{-1}} = 9.96\ \mu s$$

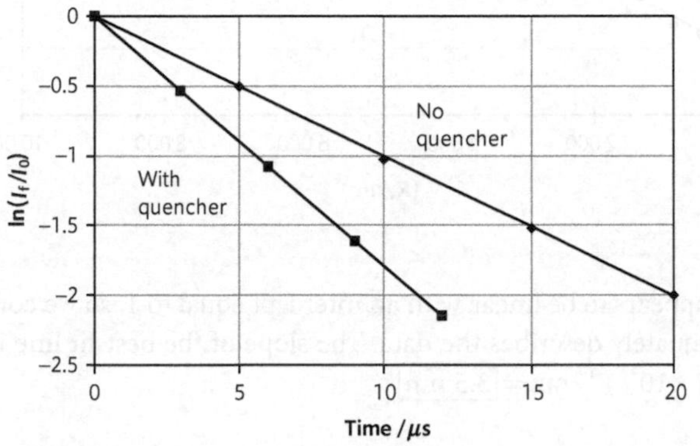

Fig 93.2

In the presence of a quencher, a graph of ln I_f/I_0 against t is still linear but with a slope equal to $-1/\tau$. This plot is found to be linear with a regression slope equal to $-1.78\bar{8}\ \mu s^{-1}$

$$\tau = \frac{-1}{-1.78\bar{8}\ \mu s^{-1}} = 5.59\ \mu s$$

The rate constant for quenching (i.e., for energy transfer to the quencher) can be obtained from

$$\frac{1}{\tau} = \frac{1}{\tau_0} + k_Q[\text{Q}] \text{ [Example 93.2]}$$

$$\text{Thus } k_Q = \frac{\tau^{-1} - \tau_0^{-1}}{[\text{N}_2]} = \frac{RT(\tau^{-1} - \tau_0^{-1})}{p_{\text{N}_2}}$$

$$= \frac{(0.08206\text{dm}^3\text{ atm K}^{-1}\text{ mol}^{-1})(300\text{K})(0.1788 - 0.1004)\times(10^{-6}\text{ s})^{-1}}{9.74\times10^{-4}\text{ atm}}$$

$$= \boxed{1.98\times10^9\text{ dm}^3\text{ mol}^{-1}\text{ s}^{-1}}$$

P93.7 $\eta_T = \dfrac{R_0^6}{R_0^6 + R^6}$ or $\dfrac{1}{\eta_T} = 1 + \left(\dfrac{R}{R_0}\right)^6$ [93.10]

A plot of η_T^{-1} vs. R^6 ought to be linear with a slope of $(R_0/\text{nm})^{-6}$.

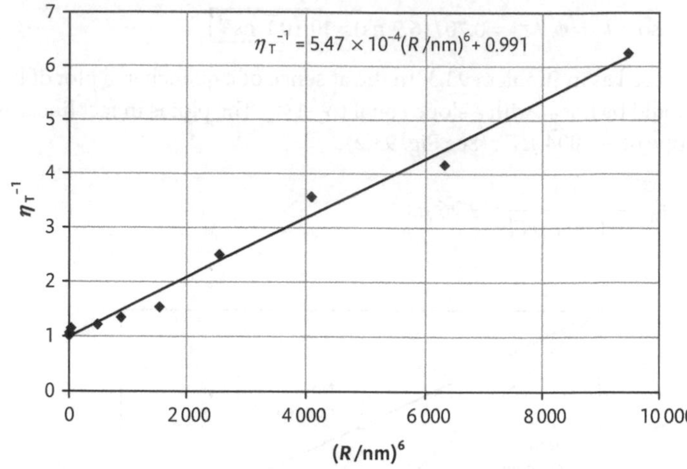

Fig 93.3

The plot appears to be linear with an intercept equal to 1, so we conclude that eqn 93.10 adequately describes the data. The slope of the best-fit line is $(R_0/\text{nm})^{-6}$, so $R_0 = (5.47 \times 10^{-4})^{-1/6}\,\text{nm} = \boxed{3.5\ \text{nm}}$

Topic 94 Electron transfer in homogeneous systems

Discussion questions

D94.1 An expression of the rate constant for electron transfer is given by eqn 94.6:

$$k_{et} = CH_{DA}(d)^2 e^{-\Delta^{\ddagger}G/RT}$$

This rate constant depends on the distance between donor and acceptor (d) through the function $H_{DA}(d)^2$, which is given by eqn 94.5; k_{et} decays exponentially with increasing d. The parameter β in eqns 94.5 and 94.9 determines how strongly k_{et} depends on distance. This parameter changes with the transfer medium. The standard Gibbs energy of the electron-transfer process ($\Delta_r G^{\ominus}$) affects the rate through the activation Gibbs energy and a quantity called the reorganization energy (ΔE_R), as shown in eqn 94.7. In systems where the reorganization energy is constant, the dependence of $\ln k_{et}$ on $\Delta_r G^{\ominus}$ is given by eqn 94.10; it is an inverted parabola in which the maximum rate occurs when $\Delta_r G^{\ominus} = \Delta E_R$. A more thorough discussion can be found in Topic 94.2.

D94.3 The inverted region is discussed just before *Brief illustration* 94.2 in the main text. The phenomenon in question refers to the fact that the rate constant for the reaction can decrease even as the reaction becomes more thermodynamically spontaneous (that is, as it becomes more exergonic). Equation 94.7 explains this phenomenon by relating the activation Gibbs energy ($\Delta^{\ddagger}G$), which affects reaction rate, and the standard reaction Gibbs energy ($\Delta_r G^{\ominus}$), which affects thermodynamic stability. Perhaps the best way to understand this relationship is with reference to Fig. 94.6 of the main text. The activation energy is determined by the crossing point between Gibbs energy curves for the product P and reactant R. One can see from the figure that as the parabola representing P is lowered, the energy of the crossing point decreases until the minimum in curve R is reached, and then the energy of the crossing point increases as the crossing point travels up the other side of curve R.

Exercises

E94.1(a) The rate constant for electron transfer is

$$k_{et} = C\{H_{DA}(d)\}^2 e^{-\Delta^{\ddagger}G/RT} \quad [94.6]$$

The reorganization energy, ΔE_R, appears in two of these factors:

$$\Delta^{\ddagger}G = \frac{(\Delta_r G^{\ominus} + \Delta E_R)^2}{4\Delta E_R} \text{ [94.7] and } C = \frac{1}{h}\left(\frac{\pi^3}{RT\Delta E_R}\right)^{1/2} \text{ [94.8]},$$

$$\text{so} \quad k_{et} = \frac{\{H_{DA}(d)\}^2}{h}\left(\frac{\pi^3}{RT\Delta E_R}\right)^{1/2} \exp\left(\frac{-(\Delta_r G^{\ominus} + \Delta E_R)^2}{4RT\Delta E_R}\right)$$

$$= \frac{\{H_{DA}(d)\}^2}{h}\left(\frac{\pi^3}{kT\Delta E_R}\right)^{1/2} \exp\left(\frac{-(\Delta_r G^{\ominus} + \Delta E_R)^2}{4kT\Delta E_R}\right)$$

depending on whether the energies are expressed in molar units or molecular units. The only unknown in this equation is ΔE_R. Isolating ΔE_R analytically is not possible; however, one can solve for it numerically using the root-finding command of a symbolic mathematics package, or graphically by plotting the right-hand side vs. the (constant) left-hand side and finding the value of ΔE_R at which the two lines cross. Before we put in numbers, we must make sure to use compatible units. We recognize that $H_{DA}(d)$ and $\Delta_r G^{\ominus}$ are both given in molecular units, but that the former is really a wavenumber rather than an energy. So we choose to express all energies per molecule in joules:

$$H_{DA}(d) = hc \times 0.04 \text{ cm}^{-1} = (6.626 \times 10^{-34} \text{ J s}) \times (2.998 \times 10^{10} \text{ cm s}^{-1}) \times (0.04 \text{ cm}^{-1})$$
$$H_{DA}(d) = 8 \times 10^{-25} \text{ J}$$

$$\frac{H_{DA}^2}{h}\left(\frac{\pi^3}{kT}\right)^{1/2} = \frac{(8\times10^{-25} \text{ J})^2}{6.626\times10^{-34} \text{ J s}}\left(\frac{\pi^3}{1.381\times10^{-23} \text{ J K}^{-1} \times 298 \text{ K}}\right)^{1/2} = 8\times10^{-5} \text{ J}^{0.5} \text{ s}^{-1}$$

$$\Delta_r G^{\ominus} = -0.185 \text{ eV} \times 1.602\times10^{-19} \text{ J eV}^{-1} = -2.96\times10^{-20} \text{ J}$$

and $4kT = 4 \times (1.381 \times 10^{-23} \text{ J K}^{-1}) \times (298 \text{ K}) = 1.65 \times 10^{-20} \text{ J}$

$$\text{Thus} \quad 37.5 = 8\times10^{-5}\left(\frac{\text{J}}{\Delta E_R}\right)^{1/2} \exp\left(\frac{-(-2.96\times10^{-20} \text{ J} + \Delta E_R)^2}{\Delta E_R \times 1.65\times10^{-20} \text{ J}}\right)$$

where $\Delta E_R = \boxed{4 \times 10^{-21} \text{ J}}$ or $\boxed{2 \text{ kJ mol}^{-1}}$.

E94.2(a) For the same donor and acceptor at different distances, eqn 94.9 applies:

$\ln k_{et} = -\beta d + \text{constant}$

The slope of a plot of k_{et} versus d is $-\beta$. The slope of a line defined by two points is:

$$\text{slope} = \frac{\Delta y}{\Delta x} = \frac{\ln k_{et,2} - \ln k_{et,1}}{d_2 - d_1} = -\beta = \frac{\ln 4.51\times10^4 - \ln 2.02\times10^5}{(1.23-1.11) \text{ nm}}$$

so $\beta = \boxed{12.\bar{5} \text{ nm}^{-1}}$

Problems

P94.1 First, assume that electron transfer is rate limiting, and not diffusion. In that case, the rate constant is (Topic 94.1)

$$k_r = Kk_{et} = K\kappa\nu^{\ddagger}e^{-\Delta^{\ddagger}G/RT} \text{ [94.4]}$$

(a) $\Delta^{\ddagger}G = \dfrac{(\Delta_r G^{\ominus} + \Delta E_R)^2}{4\Delta E_R}$ [94.7]

so $\Delta^{\ddagger}G_{DD} = \dfrac{(0+\Delta E_{R,DD})^2}{4\Delta E_{R,DD}} = \dfrac{\Delta E_{R,DD}}{4}$ and $\Delta^{\ddagger}G_{AA} = \dfrac{\Delta E_{R,AA}}{4}$

and $\Delta^{\ddagger}G_{DA} = \dfrac{(\Delta_r G^{\ominus} + \Delta E_{R,DA})^2}{4\Delta E_{R,DA}} = \dfrac{(\Delta_r G^{\ominus})^2 + 2\Delta_r G^{\ominus}\Delta E_{R,DA} + (\Delta E_{R,DA})^2}{4\Delta E_{R,DA}}$

(b) If $|\Delta_r G^{\ominus}| \ll \Delta E_{R,DA}$, then $\Delta^{\ddagger}G_{DA} \approx \dfrac{\Delta_r G^{\ominus}}{2} + \dfrac{\Delta E_{R,DA}}{4}$

Assume $\Delta E_{R,DA} = \dfrac{\Delta E_{R,AA} + \Delta E_{R,DD}}{2} = 2(\Delta^{\ddagger}G_{AA} + \Delta^{\ddagger}G_{DD})$

Hence $\Delta^{\ddagger}G_{DA} \approx \dfrac{\Delta_r G^{\ominus} + \Delta^{\ddagger}G_{AA} + \Delta^{\ddagger}G_{DD}}{2}$

(c) From the above expression based on eqn 94.4, the rate constants for the self-exchange reactions are

$$k_{AA} = K_{AA}\kappa\nu^{\ddagger}e^{-\Delta^{\ddagger}G_{AA}/RT} \quad \text{and} \quad k_{DD} = K_{DD}\kappa\nu^{\ddagger}e^{-\Delta^{\ddagger}G_{DD}/RT}$$

(d) Compare these results to the rate constant for the reaction of interest:

$$k_r = K_{DA}\kappa\nu^{\ddagger}e^{-\Delta^{\ddagger}G_{DA}/RT} \approx K_{DA}\kappa\nu^{\ddagger}e^{-\Delta_r G/2RT}e^{-\Delta^{\ddagger}G_{AA}/2RT}e^{-\Delta^{\ddagger}G_{DD}/2RT}$$

(e) Thus $k_r = (k_{AA}k_{DD})^{1/2}\dfrac{K_{DA}}{(K_{AA}K_{DD})^{1/2}}e^{-\Delta_r G/2RT}$

The constants K_{DA}, etc., are equilibrium constants for diffusive pairing, i.e., for steps like eqn 94.2a. It is reasonable to expect that $K_{DA} \approx (K_{AA}K_{DD})^{1/2}$, eliminating them from the expression. Finally, the equilibrium constant for the overall reaction is

$$e^{-\Delta_r G/RT} = K$$

so the exponential term in our expression is its square root. Therefore

$$k_r \approx (k_{AA}k_{DD}K)^{1/2}$$

P94.3 For a series of reactions with a fixed edge-to-edge distance and reorganization energy, the logarithm of the rate constant depends quadratically on the reaction free-energy:

$$\ln k_{et} = -\frac{(\Delta_r G^{\ominus})^2}{4kT\Delta E_R} - \frac{\Delta_r G^{\ominus}}{2kT} + \text{constant} \quad [94.10 \text{ in molecular units}]$$

Draw up the following table:

$\Delta_r G^\ominus$ / eV	$k_{et}/(10^6\ s^{-1})$	$\ln(k_{et}/s^{-1})$
−0.665	0.657	13.4
−0.705	1.52	14.2
−0.745	1.12	13.9
−0.975	8.99	16.0
−1.015	5.76	15.6
−1.055	10.1	16.1

and plot $\ln k_{et}$ vs. $\Delta_r G^\ominus$ (Fig. 94.1).

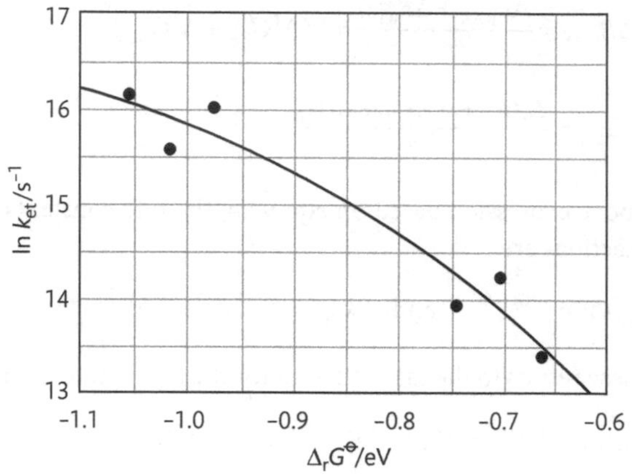

Fig 94.1

The least squares quadratic fit equation is

$$\ln k_{et}/\mathrm{s}^{-1} = 3.23 - 21.1(\Delta_r G^\ominus/\mathrm{eV}) - 8.48(\Delta_r G^\ominus/\mathrm{eV})^2, \quad r^2 = 0.938$$

The coefficient of the quadratic term is

$$-\frac{1}{4\lambda kT} = -\frac{8.48}{\mathrm{eV}^2}$$

so $$\Delta E_R = \frac{(\mathrm{eV})^2}{4(8.48)kT} = \frac{(1.602\times10^{-19}\ \mathrm{J\,eV^{-1}})(\mathrm{eV})^2}{2(8.48)(1.381\times10^{-23}\ \mathrm{J\,K^{-1}})(298\ \mathrm{K})} = \boxed{1.15\ \mathrm{eV}}$$

P94.5 The theoretical treatment of Topic 94 applies only at relatively high temperatures. At temperatures above 130 K, the reaction in question is observed to follow a temperature dependence consistent with eqn 94.6, namely increasing rate with increasing temperature. Below 130 K, the temperature dependent terms in eqn 94.6 are replaced by Frank–Condon factors; that is, temperature-dependent terms are replaced by temperature-independent wavefunction overlap integrals.

Focus 19: Integrated activities

F19.1 (a) $A + P \rightarrow P + P$ autocatalytic step, $v = k_r[A][P]$

Let $[A]=[A]_0 - x$ and $[P]=[P]_0 + x$

We substitute these definitions into the rate expression, simplify, and integrate.

$$v = -\frac{d[A]}{dt} = k_r[A][P]$$

$$-\frac{d([A]_0 - x)}{dt} = k_r([A]_0 - x)([P]_0 + x)$$

$$\frac{dx}{([A]_0 - x)([P]_0 + x)} = k_r\, dt$$

$$\frac{1}{[A]_0 + [P]_0}\left(\frac{1}{[A]_0 - x} + \frac{1}{[P]_0 + x}\right)dx = k_r\, dt$$

$$\frac{1}{[A]_0 + [P]_0}\int_0^x\left(\frac{1}{[A]_0 - x} + \frac{1}{[P]_0 + x}\right)dx = k_r\int_0^t dt$$

$$\frac{1}{[A]_0 + [P]_0}\left\{\ln\left(\frac{[A]_0}{[A]_0 - x}\right) + \ln\left(\frac{[P]_0 + x}{[P]_0}\right)\right\} = k_r t$$

$$\ln\left\{\left(\frac{[A]_0}{[P]_0}\right)\left(\frac{[P]_0 + x}{[A]_0 - x}\right)\right\} = k_r([A]_0 + [P]_0)t$$

$$\ln\left\{\left(\frac{[A]_0}{[P]_0}\right)\left(\frac{[P]}{[A]_0 + [P]_0 - [P]}\right)\right\} = k_r([A]_0 + [P]_0)t$$

$$\ln\left\{\left(\frac{1}{b}\right)\left(\frac{[P]}{[A]_0 + [P]_0 - [P]}\right)\right\} = at \quad \text{where } a = k_r([A]_0 + [P]_0) \text{ and } b = \frac{[P]_0}{[A]_0}$$

$$\frac{[P]}{[A]_0 + [P]_0 - [P]} = be^{at}$$

$$[P] = ([A]_0 + [P]_0)be^{at} - be^{at}[P]$$

$$(1 + be^{at})[P] = [P]_0\left(1 + \frac{[A]_0}{[P]_0}\right)be^{at} = [P]_0\left(1 + \frac{1}{b}\right)be^{at} = [P]_0(b+1)e^{at}$$

$$\boxed{\frac{[P]}{[P]_0} = (b+1)\frac{e^{at}}{1 + be^{at}}}$$

(b) See Fig. F19.1.

The growth to [P] reaches a maximum at very long times. As $t \to \infty$, the exponential term in the denominator of $[\mathrm{P}]/[\mathrm{P}]_0 = (b+1)\mathrm{e}^{at}/(1+b\mathrm{e}^{at})$ becomes so large that the denominator becomes $b\mathrm{e}^{at}$. Thus, $([\mathrm{P}]/[\mathrm{P}]_0)_{\max} = (b+1)\mathrm{e}^{at}/(b\mathrm{e}^{at}) = (b+1)/b$ where $b = [\mathrm{P}]_0/[\mathrm{A}]_0$ and this maximum occurs as $t \to \infty$.

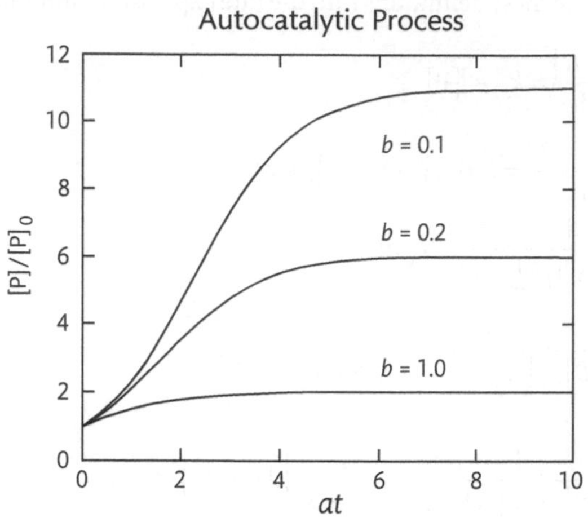

Figure F19.1

The autocatalytic curve $[\mathrm{P}]/[\mathrm{P}]_0 = (b+1)\mathrm{e}^{at}/(1+b\mathrm{e}^{at})$ has a shape that is very similar to that of the first-order process $[\mathrm{P}]/[\mathrm{A}]_0 = 1-\mathrm{e}^{-kt}$. However, $[\mathrm{P}]_{\max} = [\mathrm{A}]_0$ at $t \to \infty$ for the first-order process whereas $[\mathrm{P}]_{\max} = (1 + 1/b)[\mathrm{P}]_0$ for the autocatalytic mechanism. In a series of experiments at fixed $[\mathrm{A}]_0$ and assorted $[\mathrm{P}]_0$, only the autocatalytic mechanism will show variation in $[\mathrm{P}]_{\max}$. Another difference is that the autocatalytic curve is initially concave up, which gives an overall sigmoidal curve, whereas the first-order curve is concave down. See Fig. F19.2.

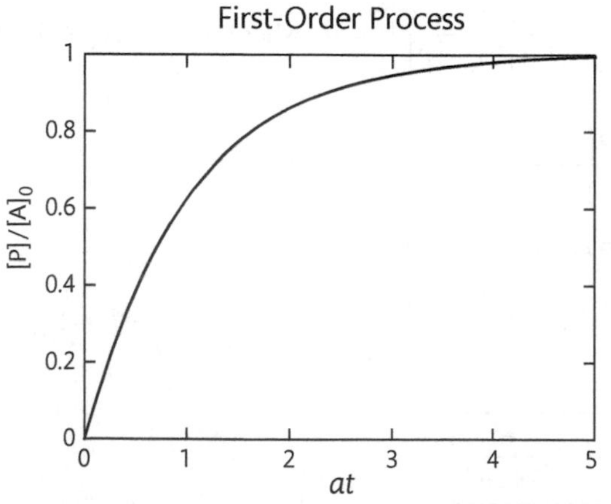

Figure F19.2

(c) Let $[\mathrm{P}]_{\upsilon_{\max}}$ be the concentration of P at which the reaction rate is a maximum and let $t_{\max}$ be the corresponding time.

$$\begin{aligned}\upsilon &= k_{\mathrm{r}}[\mathrm{A}][\mathrm{P}] = k([\mathrm{A}]_0 - x)([\mathrm{P}]_0 + x)\\ &= k_{\mathrm{r}}\{[\mathrm{A}]_0[\mathrm{P}]_0 + ([\mathrm{A}]_0 - [\mathrm{P}]_0)x - x^2\}\\ \frac{\mathrm{d}\upsilon}{\mathrm{d}t} &= k_{\mathrm{r}}([\mathrm{A}]_0 - [\mathrm{P}]_0 - 2x)\end{aligned}$$

The reaction rate is a maximum when $\mathrm{d}\upsilon/\mathrm{d}t = 0$. This occurs when

$$x = [\mathrm{P}]_{\upsilon_{\max}} - [\mathrm{P}]_0 = \frac{[\mathrm{A}]_0 - [\mathrm{P}]_0}{2} \quad \text{or} \quad \frac{[\mathrm{P}]_{\upsilon_{\max}}}{[\mathrm{P}]_0} = \frac{b+1}{2b}$$

Substitution into the final equation of part (a) gives:

$$\frac{[\mathrm{P}]_{\upsilon_{\max}}}{[\mathrm{P}]_0} = \frac{b+1}{2b} = (b+1)\frac{\mathrm{e}^{at_{\max}}}{1+b\mathrm{e}^{at_{\max}}}$$

Solving for $t_{\max}$:

$$\begin{aligned}1 + b\mathrm{e}^{at_{\max}} &= 2b\mathrm{e}^{at_{\max}}\\ \mathrm{e}^{at_{\max}} &= b^{-1}\\ at_{\max} &= \ln(b^{-1}) = -\ln(b)\\ t_{\max} &= \boxed{-\frac{1}{a}\ln(b)}\end{aligned}$$

(d) $$\frac{\mathrm{d}[\mathrm{P}]}{\mathrm{d}t} = k_{\mathrm{r}}[\mathrm{A}]^2[\mathrm{P}]$$

$$[\mathrm{A}] = A_0 - x, \quad [\mathrm{P}] = P_0 + x, \quad \frac{\mathrm{d}[\mathrm{P}]}{\mathrm{d}t} = \frac{\mathrm{d}x}{\mathrm{d}t} = k_{\mathrm{r}}(A_0 - x)^2(P_0 + x)$$

$$\int_0^x \frac{\mathrm{d}x}{(A_0 - x)^2(P_0 + x)} = k_{\mathrm{r}}t$$

Solve the integral by partial fractions.

$$\begin{aligned}\frac{1}{(A_0 - x)^2(P_0 + x)} &= \frac{\alpha}{(A_0 - x)^2} + \frac{\beta}{A_0 - x} + \frac{\gamma}{P_0 + x}\\ &= \frac{\alpha(P_0 + x) + \beta(A_0 - x)(P_0 + x) + \gamma(A_0 - x)^2}{(A_0 - x)^2(P_0 + x)}\end{aligned}$$

$$\left.\begin{aligned}P_0\alpha + A_0P_0\beta + A_0^2\gamma &= 1\\ \alpha + (A_0 - P_0)\beta - 2A_0\gamma &= 0\\ -\beta + \gamma &= 0\end{aligned}\right\}$$

This set of simultaneous equations solves to

$$\alpha = \frac{1}{A_0 + P_0}, \qquad \beta = \gamma = \frac{\alpha}{A_0 + P_0}$$

Therefore,

$$k_r t = \left(\frac{1}{A_0+P_0}\right)\int_0^x \left[\left(\frac{1}{A_0-x}\right)^2 + \left(\frac{1}{A_0+P_0}\right)\left(\frac{1}{A_0-x}+\frac{1}{P_0-x}\right)\right]dx$$

$$= \left(\frac{1}{A_0+P_0}\right)\left\{\left(\frac{1}{A_0-x}\right) - \left(\frac{1}{A_0}\right) + \left(\frac{1}{A_0+P_0}\right)\left[\ln\left(\frac{A_0}{A_0-x}\right) + \ln\left(\frac{P_0+x}{P_0}\right)\right]\right\}$$

$$\boxed{k_r t = \left(\frac{1}{A_0+P_0}\right)\left[\left(\frac{x}{A_0(A_0-x)}\right) + \left(\frac{1}{A_0+P_0}\right)\ln\left(\frac{A_0(P_0+x)}{(A_0-x)P_0}\right)\right]}$$

where $x = A_0 - [\mathrm{A}] = [\mathrm{P}] - P_0$

The maximum rate occurs at

$$\frac{d\upsilon_P}{dt} = 0, \qquad \upsilon_P = k_r[\mathrm{A}]^2[\mathrm{P}]$$

and hence at the solution of

$$2k_r\left(\frac{d[\mathrm{A}]}{dt}\right)[\mathrm{A}][\mathrm{P}] + k_r[\mathrm{A}]^2\frac{d[\mathrm{P}]}{dt} = 0$$

$$-2k_r[\mathrm{A}][\mathrm{P}]v_P + k_r[\mathrm{A}]^2 v_P = 0 \quad [\text{as } \upsilon_A = -\upsilon_P]$$

$$k_r[\mathrm{A}]([\mathrm{A}] - 2[\mathrm{p}])\upsilon_P = 0$$

The rate is a maximum when $[\mathrm{A}] = 2[\mathrm{P}]$, which occurs at

$$A_0 - x = 2P_0 + 2x \qquad \text{or} \quad x = \tfrac{1}{3}(A_0 - 2P_0)$$

Substituting this condition into the integrated rate law gives

$$\boxed{k_r t_{\max} = \left(\frac{1}{A_0+P_0}\right)^2\left[\left(\frac{A_0-2P_0}{2A_0}\right) + \ln\left(\frac{A_0}{2P_0}\right)\right]}$$

(e) $$\frac{d[\mathrm{P}]}{dt} = k_r[\mathrm{A}][\mathrm{P}]^2$$

$$\frac{dx}{dt} = k_r(A_0-x)(P_0+x)^2 \quad [x = [\mathrm{P}] - P_0]$$

$$k_r t = \int_0^x \frac{dx}{(A_0-x)(P_0+x)^2}$$

Integrate by partial fractions (as in Part (d)).

$$k_r t = \left(\frac{1}{A_0+P_0}\right)\int_0^x \left\{\left(\frac{1}{P_0+x}\right)^2 + \left(\frac{1}{A_0+P_0}\right)\left[\frac{1}{P_0+x}+\frac{1}{A_0-x}\right]\right\}\mathrm{d}x$$

$$= \left(\frac{1}{A_0+P_0}\right)\left\{\left(\frac{1}{P_0}-\frac{1}{P_0+x}\right)+\left(\frac{1}{A_0+P_0}\right)\left[\ln\left(\frac{P_0+x}{P_0}\right)+\ln\left(\frac{A_0}{A_0-x}\right)\right]\right\}$$

$$\boxed{k_r t = \left(\frac{1}{A_0+P_0}\right)\left[\left(\frac{x}{P_0(P_0+x)}\right)+\left(\frac{1}{A_0+P_0}\right)\ln\left(\frac{(P_0+x)A_0}{P_0(A_0-x)}\right)\right]}$$

where $x = [\mathrm{P}] - P_0 = A_0 - [\mathrm{A}]$

The rate is maximum when

$$\frac{\mathrm{d}\upsilon_\mathrm{P}}{\mathrm{d}t} = 2k_r[\mathrm{A}][\mathrm{P}]\left(\frac{\mathrm{d}[\mathrm{P}]}{\mathrm{d}t}\right) + k_r\left(\frac{\mathrm{d}[\mathrm{A}]}{\mathrm{d}t}\right)[\mathrm{P}]^2 = 2k_r[\mathrm{A}][\mathrm{P}]\upsilon_\mathrm{P} - k_r[\mathrm{P}]^2\upsilon_\mathrm{P}$$

$$= k_r[\mathrm{P}](2[\mathrm{A}]-[\mathrm{P}])\upsilon_\mathrm{P} = 0$$

That is, at $[\mathrm{P}] = 2[\mathrm{A}]$, which occurs at

$$x + P_0 = 2(A_0 - x) \quad \text{or} \quad x = \tfrac{1}{3}(2A_0 - P_0)$$

On substitution of this condition into the integrated rate law, we find

$$\boxed{k_r t_{\max} = \left(\frac{1}{A_0+P_0}\right)^2\left[\left(\frac{2A_0-P_0}{2P_0}\right)+\ln\left(\frac{2A_0}{P_0}\right)\right]}$$

F19.3 The time scales of atomic processes are rapid indeed: according to the following table, a nanosecond is an eternity. Note that the times given here are in some way typical values for times that may vary over two or three orders of magnitude. For example, vibrational wavenumbers can range from about 4 400 $\mathrm{cm^{-1}}$ (for H_2) to 100 $\mathrm{cm^{-1}}$ (for I_2) and even lower, with a corresponding range of associated times. Radiative decay rates of electronic states can vary even more widely: Times associated with phosphorescence can be in the millisecond and even second range. A large number of time scales for physical, chemical, and biological processes on the atomic and molecular scale are reported in Figure 2 of A.H. Zewail, Femtochemistry: atomic-scale dynamics of the chemical bond. *J. Phys. Chem. A* **104**, 5660 (2000).

Radiative decay of excited electronic states can range from about 10^{-9} s to 10^{-4} s–even longer for phosphorescence involving "forbidden" decay paths. Molecular rotational motion takes place on a scale of 10^{-12} s to 10^{-9} s. Molecular vibrations are faster still, about 10^{-14} s to 10^{-12} s. Proton transfer reactions occur on a timescale of about 10^{-10} s to 10^{-9} s, although protons can hop from molecule to molecule in water even more rapidly (1.5×10^{-12} s, Topic 80.2(b)). The mean time between collisions in liquids is similar to vibrational periods, around 10^{-13} s. One can estimate collision times in liquids very roughly by applying the expression for collisions in gases to liquid conditions.

process	t / ns	reference
radiative decay of electronic excited state	1×10^{1}	Topics 40 and 93
molecular rotational motion	3×10^{-2}	$B \approx 1\ \text{cm}^{-1}$
molecular vibrational motion	3×10^{-5}	$\tilde{\nu} \approx 1\,000\ \text{cm}^{-1}$
proton transfer	0.3	Zewail 2 000
collision frequency in liquids	4×10^{-4}	Topic 87‡

‡Use formula for gas collision frequency at 300 K, parameters for benzene from data section, and density of liquid benzene.

Reaction times in the biochemistry of vision includes the 200-fs photoisomerization of retinal from 11-*cis* to all-*trans* that gets the process started. Initial energy transfer (to a nearby pigment) has a time scale of around 10^{-13} s to 5×10^{-12} s, with longer-range transfer (to the reaction center) taking about 10^{-10} s. Electron transfer is also very fast (about 3 ps), with ultimate transfer (leading to oxidation of water and reduction of plastoquinone) taking from 10^{-10} s to 10^{-3} s.

Topic 95 **Solid surfaces**

Discussion questions

95.1 (a) A **terrace** is a flat layer of atoms on a surface. There can be more than one terrace on a surface, each at a different height. Steps are the joints between the terraces; the height of the step can be constant or variable.

(b) **Dislocations**, or discontinuities in the regularity of a crystal lattice, result in surface steps and terraces. Main types of dislocation include the **edge dislocation**, the **screw dislocation**, and the **mixed dislocation** that shows the characteristics of both the edge and screw dislocation. The edge dislocation can be envisioned by imagining small clumps of crystalline matter sticking together from either a melt or a solution. The lowest energy pattern sticks them together with valence requirements satisfied or atoms in a close-packed arrangement. This process is expected to form surface terraces because terraces yield the maximum possible number of nearest neighbors at a surface and the lowest possible surface energy. However, the very process of small clumps of matter rapidly sticking is very unlikely to always produce a perfect space-filled, crystalline structure. Crystal defects such as the half-plane of atoms shown in Fig. 95.1 may form near the surface of the growing crystal. This **edge dislocation** distorts adjacent planes into a high energy configuration that is inherently unstable but thermal agitations of the growth process cause the dislocation to propagate to the surface, thereby, forming a step (see Fig. 95.2).

A **screw dislocation** is shown in Fig. 95.3. Imagine a cut in the crystal, with the atoms to the left of the cut pushed up through a distance of one unit cell. The unit cells now form a continuous spiral around the end of the cut, which is called the **screw axis**. A path encircling the screw axis spirals up to the top of the crystal, and where the dislocation breaks through to the surface it takes the form of a spiral ramp.

The surface defect formed by a screw dislocation is a step, possibly with kinks, where growth can occur. The incoming particles lie in ranks on the ramp, and successive ranks reform the step at an angle to its initial position. As deposition continues the step rotates around the screw axis, and is not eliminated. Growth may therefore continue indefinitely. Several layers of deposition may occur, and the edges of the spirals might be cliffs several atoms high. Propagating spiral edges can also give rise to flat terraces. Terraces are formed if growth occurs simultaneously at neighboring left- and right-handed screw dislocations. Successive tables of atoms may form as counter-rotating defects collide on successive circuits, and the terraces formed may then fill up by further deposition at their edges to give flat crystal planes.

Edge Dislocation

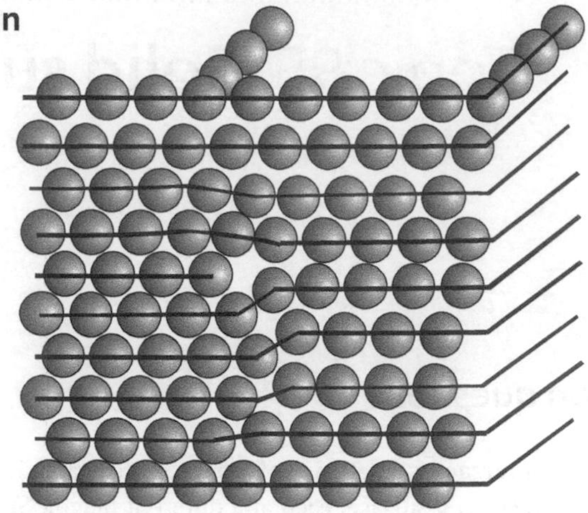

Figure 95.1

Step

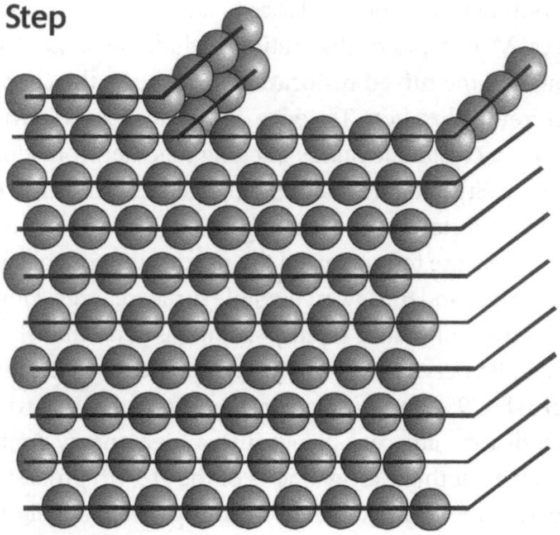

Figure 95.2

Screw Dislocation

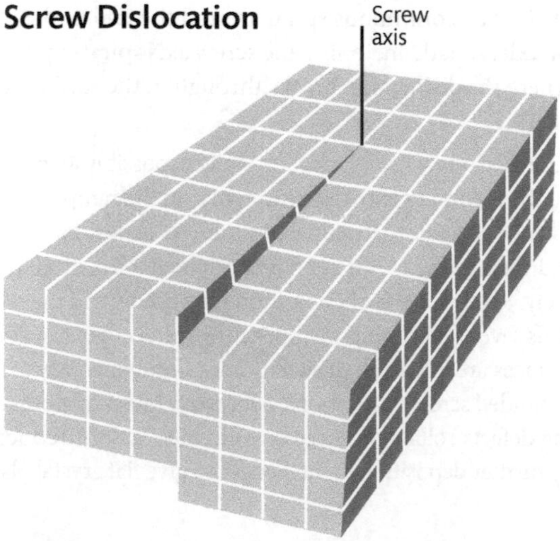

Fig 95.3

Exercises

95.1(a) The collision frequency, Z_W, of gas molecules with an ideally smooth surface area is given by eqn 95.1.

$$p = 0.10\ \mu\text{Torr} = 1.33\times10^{-5}\ \text{Pa}$$

$$Z_W = \frac{p}{(2\pi MkT/N_A)^{1/2}}\ [95.1;\ m = M/N_A]$$

$$= \frac{(1.33\times10^{-5}\ \text{Pa})\times(\text{m}/10^2\ \text{cm})^2}{\{2\pi\times(1.381\times10^{-23}\ \text{J K}^{-1})\times(298.15\ \text{K})\times(\text{kg mol}^{-1})/(6.022\times10^{23}\ \text{mol}^{-1})\}^{1/2}\{M/(\text{kg mol}^{-1})\}^{1/2}}$$

$$= \frac{6.42\times10^{12}}{\{M/(\text{kg mol}^{-1})\}^{1/2}}\ \text{cm}^{-2}\ \text{s}^{-1} \quad \text{at } 25°\ \text{C and } p = 0.10\ \mu\text{Torr}$$

(i) Hydrogen ($M = 0.002016\ \text{kg mol}^{-1}$), $Z_W = \boxed{1.4\times10^{14}\ \text{cm}^{-2}\ \text{s}^{-1}}$

(ii) Propane ($M = 0.04410\ \text{kg mol}^{-1}$), $Z_W = \boxed{3.1\times10^{13}\ \text{cm}^{-2}\ \text{s}^{-1}}$

95.2(a) $A = \pi d^2/4 = \pi(1.5\ \text{mm})^2/4 = 1.77\times10^{-6}\ \text{m}^2$

The collision frequency of the Ar gas molecules with surface area A equals $Z_W A$.

$$Z_W A = \frac{p}{(2\pi MkT/N_A)^{1/2}} A\ [95.1;\ m = M/N_A]$$

$$p = (Z_W A)\times(2\pi MkT/N_A)^{1/2}/A$$

$$= (4.5\times10^{20}\ \text{s}^{-1})$$

$$\times\{2\pi(39.95\times10^{-3}\ \text{kg mol}^{-1})\times(1.381\times10^{-23}\ \text{J K}^{-1})\times(425\ \text{K})/(6.022\times10^{23}\ \text{mol}^{-1})\}^{1/2}/(1.77\times10^{-6}\ \text{m}^2)$$

$$= 1.3\times10^4\ \text{Pa} = \boxed{0.13\ \text{bar}}$$

Problems

95.1 Fig 95.2(a) shows a dark univalent probe cation atop a two-dimensional square ionic lattice of grey univalent cations and white univalent anions. Let $d_0 = 200$ pm be the distance between nearest neighbors and let V_0 be the Coulombic interaction between nearest neighbors.

$$V_0 = -\frac{e^2}{4\pi\varepsilon_0 d_0} = -\frac{(1.602\times10^{-19}\ \text{C})^2}{(1.113\times10^{-10}\ \text{J}^{-1}\ \text{C}^2\ \text{m}^{-1})\times(200\times10^{-12}\ \text{m})} = -1.153\times10^{-18}\ \text{J}$$

The symmetry of the lattice w/r/t the probe cation consists of one region like that of Fig. 95.2(b) and two regions like that of Fig. 95.2(c) so we calculate the total Coulombic interaction of the probe with the lattice, $V_{\text{total Fig 95.2(a)}}$, by adding the interaction within Fig. 95.2(b) to twice the interaction within Fig. 95.2(c).

$$V_{\text{total Fig 95.2(a)}} = V_{\text{Fig 95.2(b)}} + 2V_{\text{Fig 95.2(c)}}$$

The calculation is pursued one column at a time and the column interactions with the probe cation are summed.

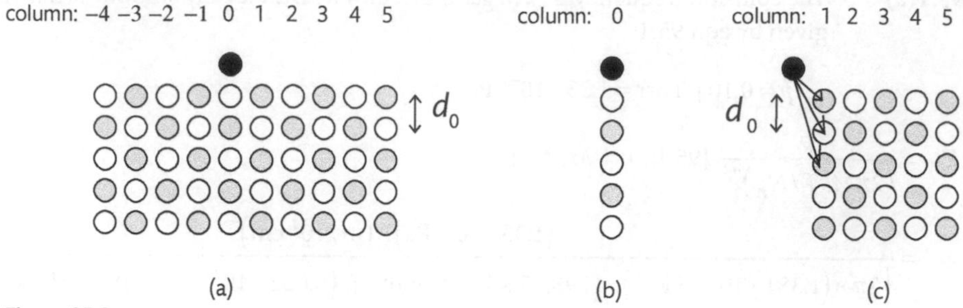

Figure 95.2

Probe-to-column 0 interaction:

$$V_{\text{Fig 95.2(b)}} = V_0 \times \left(1 - \frac{1}{2} + \frac{1}{3} - \frac{1}{4} + \frac{1}{5} \cdots\right) = V_0 \ln 2$$
$$= 0.6931\, V_0$$

Probe-to-column 1 interaction using the Pythagorean theorem for the probe-ion distance:

$$V_{\text{column 1}} = -V_0 \times \left(\frac{1}{2^{1/2}} - \frac{1}{5^{1/2}} + \frac{1}{10^{1/2}} - \frac{1}{17^{1/2}} \cdots\right) = -V_0 \sum_{n=1}^{\infty} \frac{(-1)^{n+1}}{\left(n^2+1^2\right)^{1/2}}$$

Similarly, the probe-to-column m interaction, using the Pythagorean theorem for the probe-ion distance, is

$$V_{\text{column } m} = -V_0 \sum_{n=1}^{\infty} \frac{(-1)^{n+m}}{\left(n^2+m^2\right)^{1/2}} \quad \text{for } 1 \le m < \infty$$

The total interaction for the region shown in Fig. 95.2(c) is the sum of the above expression over all m columns

$$V_{\text{Fig 95.2(c)}} = -V_0 \sum_{m=1}^{\infty} \sum_{n=1}^{\infty} \frac{(-1)^{n+m}}{\left(n^2+m^2\right)^{1/2}} \quad \text{(the sum is performed with a calculator or software)}$$
$$= -0.2893\, V_0$$

Thus, $V_{\text{total Fig 95.2(a)}} = (0.6931 - 2 \times 0.2893) V_0$

$$= 0.1145\, V_0 = (0.1145) \times \left(-1.1153 \times 10^{-18}\ \text{J}\right)$$
$$= -1.2770 \times 10^{-19}\ \text{J or } \boxed{-76.9\ \text{kJ mol}^{-1}}$$

Now consider the probe at the corner formed by a step and the terrace shown in Fig. 95.3. The symmetry of the lattice w/r/t the probe cation consists of two regions like that of Fig. 95.2(b) and three regions like that of Fig. 95.2(c) so we calculate the total Coulombic interaction of the probe with the lattice, $V_{\text{total Fig 95.3}}$, by adding twice the interaction within Fig. 95.2(b) to thrice the interaction within Fig. 95.2(c).

$$\begin{aligned}V_{\text{total Fig 95.3}} &= 2V_{\text{Fig 95.2(b)}} + 3V_{\text{Fig 95.2(c)}}\\ &= (2\times(0.6931) - 3\times(0.2893))V_0\\ &= 0.5183\,V_0 = 0.5183\times(-1.1153\times10^{-18}\text{ J})\\ &= -5.7806\times10^{-19}\text{ J or } \boxed{-348.1\text{ kJ mol}^{-1}}\end{aligned}$$

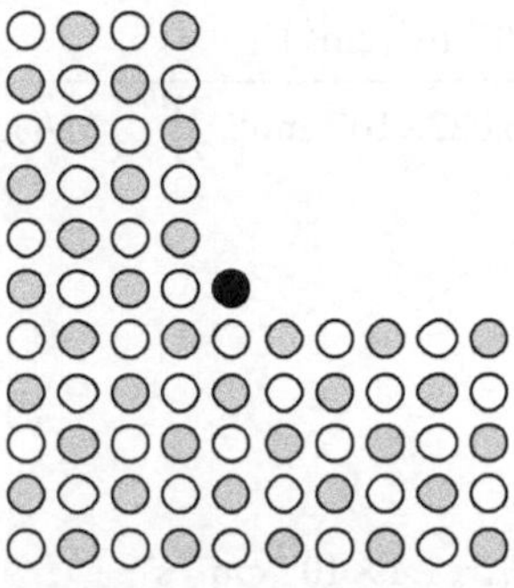

Fig 95.3

The potential energy of the probe cation is much lower at the corner formed by a step and a terrace than it is upon a simple terrace so the corner is the likely settling point.

95.3 Refer to Fig. 95.4.

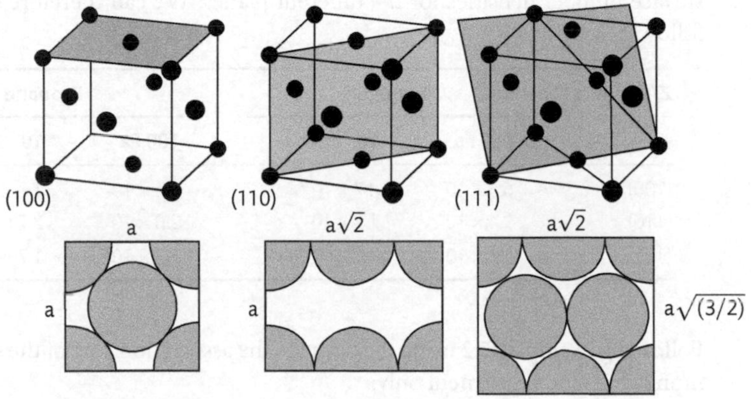

Fig 95.4

The (100) and (110) faces each expose two atoms, and the (111) face exposes four. The areas of the faces of each cell are (a) $(352\text{ pm})^2 = 1.24\times10^{-15}\text{ cm}^2$, (b) $\sqrt{2}\times(352\text{ pm})^2 = 1.75\times10^{-15}\text{ cm}^2$, and (c) $\sqrt{3}\times(352\text{ pm})^2 = 2.15\times10^{-15}\text{ cm}^2$. The numbers of atoms exposed per square centimetre, the surface number density, are therefore

(a) $\dfrac{2}{1.24\times10^{-15}\text{ cm}^2} = \boxed{1.61\times10^{15}\text{ cm}^{-2}}$

(b) $\dfrac{2}{1.75\times10^{-15}\,\text{cm}^2} = \boxed{1.14\times10^{15}\ \text{cm}^{-2}}$

(c) $\dfrac{4}{2.15\times10^{-15}\,\text{cm}^2} = \boxed{1.86\times10^{15}\ \text{cm}^{-2}}$

The collision frequency, Z_W, of gas molecules with a surface is given by eqn 18.15.

$$Z_W = \frac{p}{(2\pi MkT/N_A)^{1/2}} \quad [95.1;\ m = M/N_A]$$

$$= \frac{p\times\{(\text{kg m}^{-1}\,\text{s}^{-2})/\text{Pa}\}\times(10^{-4}\ \text{m}^2/\text{cm}^2)}{\{2\pi\times(1.381\times10^{-23}\ \text{J K}^{-1})\times(298.15\ \text{K})\times(\text{kg mol}^{-1})/(6.022\times10^{23}\ \text{mol}^{-1})\}^{1/2}\{M/(\text{kg mol}^{-1})\}^{1/2}}$$

$$= 4.825\times10^{17}\left(\frac{p/\text{Pa}}{\{M/(\text{kg mol}^{-1})\}^{1/2}}\right)\text{cm}^{-2}\ \text{s}^{-1} \quad \text{at } 25°\ \text{C}$$

(a) Hydrogen ($M = 0.002016$ kg mol^{-1})

(i) $p = 100$ Pa, $Z_W = 1.07\times10^{21}$ cm^{-2} s^{-1}

(ii) $p = 0.10$ μTorr $= 1.33\times10^{-5}$ Pa, $Z_W = 1.4\times10^{14}$ cm^{-2} s^{-1}

(b) Propane ($M = 0.04410$ kg mol^{-1})

(i) $p = 100$ Pa, $Z_W = 2.30\times10^{20}$ cm^{-2} s^{-1}

(ii) $p = 0.10$ μTorr $= 1.33\times10^{-5}$ Pa, $Z_W = 3.1\times10^{13}$ cm^{-2} s^{-1}

The frequency of collision per surface atom, Z, is calculated by dividing Z_W by the surface number densities for the different planes. We can therefore draw up the following table:

Z/(atom^{-1} s^{-1})	Hydrogen		Propane	
	100 Pa	10^{-7} Torr	100 Pa	10^{-7} Torr
(100)	6.6×10^5	8.7×10^{-2}	1.4×10^{-5}	1.9×10^{-2}
(110)	9.4×10^5	1.2×10^{-1}	2.0×10^5	2.7×10^{-2}
(111)	5.8×10^5	7.5×10^{-2}	1.2×10^5	1.7×10^{-2}

95.5 Following Example 95.2 using the term for the association part of the surface plasmon resonance experiment only:

$$\frac{dR}{dt} = k_{on}a_0(R_{eq} - R) \quad [95.4, \text{association only}]$$

$$\frac{dR}{R_{eq} - R} = k_{on}a_0 dt$$

$$\int_0^R \frac{dR}{R_{eq} - R} = \int_0^t k_{on}a_0\, dt = k_{on}a_0 t$$

$$-\ln(R_{\text{eq}} - R)|_0^R = k_{\text{on}}a_0t$$

$$-\ln\left(\frac{R_{\text{eq}} - R}{R_{\text{eq}}}\right) = k_{\text{on}}a_0t$$

$$\frac{R_{\text{eq}} - R}{R_{\text{eq}}} = e^{-k_{\text{on}}a_0t}$$

$$R = R_{\text{eq}}\left\{1 - e^{-k_{\text{on}}a_0t}\right\}$$

$$\boxed{R = R_{\text{eq}}\left\{1 - \mathrm{e}^{-k_{\text{r}}t}\right\} \text{ where } k_{\text{r}} = k_{\text{on}}a_0}$$

Following Example 95.2 using the term for the association part of the surface plasmon resonance experiment only:

$$\frac{\mathrm{d}R}{\mathrm{d}t} = -k_{\text{off}}R$$

$$\frac{\mathrm{d}R}{R} = -k_{\text{off}}\mathrm{d}t$$

$$\int_{R_{\text{eq}}}^{R}\frac{dR}{R} = -\int_0^t k_{\text{off}}\,\mathrm{d}t$$

$$\ln\left(\frac{R}{R_{\text{eq}}}\right) = -k_{\text{off}}t$$

$$\boxed{R = R_{\text{eq}}e^{-k_{\text{r}}t} \text{ where } k_{\text{r}} = k_{\text{off}}}\,.$$

Topic 96 **Adsorption and desorption**

Discussion question

96.1 The characteristic conditions of the **Langmuir isotherm** are:

1. Adsorption cannot proceed beyond monolayer coverage.
2. All sites are equivalent and the surface is uniform.
3. The ability of a molecule to adsorb at a given site is independent of the occupation of neighboring sites.

For the **BET isotherm** condition number 1 above is removed and the isotherm applies to multi-layer coverage.

In Example 96.1 it is shown that a gas exhibits the characteristics of a Langmuir adsorption isotherm when

$$\frac{p}{V}=\frac{p}{V_\infty}+\frac{1}{\alpha V_\infty}$$

where V is the volume of the adsorbate and V_∞ completes the monolayer coverage. Hence, a plot of p/V against p should give a straight line of slope $1/V_\infty$ and intercept $1/aV_\infty$.

In contrast the BET adsorption isotherm is followed when, as shown in Example 96.3,

$$\frac{z}{(1-z)V}=\frac{1}{cV_{\text{mon}}}+\frac{(c-1)z}{cV_{\text{mon}}}$$

where $z=p/p^*$ and p^* is the vapour pressure above a layer of adsorbate that is more than one molecule thick and which resembles a pure bulk liquid, V_{mon} is the volume corresponding to monolayer coverage, and c is a constant. Thus, the BET adsorption mechanism is indicated when a plot of $z/\{(1-z)V\}$ against z is linear.

Exercises

96.1(a) $$\theta=\frac{V}{V_\infty}=\frac{V}{V_{\text{mon}}}=\frac{\alpha p}{1+\alpha p}\ [96.2]$$

This rearranges to [Example 96.1]

$$\frac{p}{V}=\frac{p}{V_{\text{mon}}}+\frac{1}{\alpha V_{\text{mon}}}$$

Hence, $\dfrac{p_2}{V_2}-\dfrac{p_1}{V_1}=\dfrac{p_2}{V_{\text{mon}}}-\dfrac{p_1}{V_{\text{mon}}}$

Solving for V_{mon}

$$V_{\text{mon}}=\frac{p_2-p_1}{(p_2/V_2-p_1/V_1)}=\frac{(760-145.4)\,\text{Torr}}{(760/1.443-145.4/0.286)\ \text{Torr cm}^{-3}}=\boxed{33.6\ \text{cm}^3}$$

96.2(a) The enthalpy of adsorption is typical of $\boxed{\text{chemisorption}}$ (Table 95.2) for which $\tau_0\approx10^{-14}$ s because the adsorbate-substrate bond is stiff (see Brief illustration 96.2). The half-life for remaining on the surface is

$$t_{1/2}=\tau_0\mathrm{e}^{E_{\text{a,des}}/RT}\ [96.11]\approx(10^{-14}\ \text{s})\times(\mathrm{e}^{120\times10^3/(8.3145\times400)})\ [E_\text{d}\approx-\Delta_{\text{ad}}H]\approx\boxed{50\ \text{s}}$$

96.3(a) $\dfrac{m_1}{m_2}=\dfrac{\theta_1}{\theta_2}=\dfrac{p_1}{p_2}\times\dfrac{1+\alpha p_2}{1+\alpha p_1}$ [96.2 and $V\propto m/p$]

which solves to

$$\alpha=\frac{(m_1p_2/m_2p_1)-1}{p_2-(m_1p_2/m_2)}=\frac{(m_1/m_2)\times(p_2/p_1)-1}{1-(m_1/m_2)}\times\frac{1}{p_2}$$
$$=\frac{(0.44/0.19)\times(3.0/26.0)-1}{1-(0.44/0.19)}\times\frac{1}{3.0\ \text{kPa}}=0.19\ \text{kPa}^{-1}$$

Therefore,

$$\theta_1=\frac{(0.19\ \text{kPa}^{-1})\times(26.0\ \text{kPa})}{(1)+(0.19\ \text{kPa}^{-1})\times(26.0\ \text{kPa})}=\boxed{0.83}\quad[96.2]\quad\text{and}$$
$$\theta_2=\frac{(0.19)\times(3.0)}{(1)+(0.19)\times(3.0)}=\boxed{0.36}$$

96.4(a) $\theta=\dfrac{Kp}{1+Kp}$ [96.1, $\alpha=K=k_\text{a}/k_\text{d}$], which implies that $p=\left(\dfrac{\theta}{1-\theta}\right)\dfrac{1}{K}$.

(a) $p=(0.15/0.85)/0.75\ \text{kPa}^{-1}=\boxed{0.24\ \text{kPa}}$

(b) $p=(0.95/0.05)/0.75\ \text{kPa}^{-1}=\boxed{25\ \text{kPa}}$

96.5(a) $\left(\dfrac{\partial\ln(p/p^\ominus)}{\partial(1/T)}\right)_\theta=\dfrac{\Delta_{\text{ad}}H}{R}=-\dfrac{\Delta_{\text{des}}H}{R}$ [Example 96.2, $\Delta_{\text{ad}}H=-\Delta_{\text{des}}H$]

Assuming that $\Delta_{\text{des}}H$ is independent of temperature, integration and evaluation gives

$$\ln\frac{p_2}{p_1}=-\frac{\Delta_{\text{des}}H}{R}\left(\frac{1}{T_2}-\frac{1}{T_1}\right)=-\left(\frac{10.2\ \text{kJ mol}^{-1}}{8.3145\ \text{J K}^{-1}\ \text{mol}^{-1}}\right)\times\left(\frac{1}{313\ \text{K}}-\frac{1}{298\ \text{K}}\right)=0.197$$

which implies that $p_2=(12\ \text{kPa})\times(\mathrm{e}^{0.197})=\boxed{15\ \text{kPa}}$.

96.6(a) $\left(\dfrac{\partial\ln(p/p^\ominus)}{\partial(1/T)}\right)_\theta=\dfrac{\Delta_{\text{ad}}H^\ominus}{R}$ [Example 96.2]

Assuming that $\Delta_{ad}H^{\ominus}$ is independent of temperature, integration and evaluation gives

$$\ln\frac{p_2}{p_1}=\frac{\Delta_{ad}H^{\ominus}}{R}\left(\frac{1}{T_2}-\frac{1}{T_1}\right)$$

$$\Delta_{ad}H=R\left(\frac{1}{T_2}-\frac{1}{T_1}\right)^{-1}\ln\frac{p_2}{p_1}$$

$$=\left(8.3145\ \mathrm{J\,K^{-1}\ mol^{-1}}\right)\times\ln\left(\frac{3.2\times10^{3}\ \mathrm{kPa}}{490\ \mathrm{kPa}}\right)\times\left(\frac{1}{250\ \mathrm{K}}-\frac{1}{190\ \mathrm{K}}\right)^{-1}$$

$$=\boxed{-12.\bar{4}\ \mathrm{kJ\,mol^{-1}}}$$

96.7(a) The desorption time for a given volume is proportional to the half-life of the absorbed species and, consequently, the ratio of desorption times at two different temperatures is given by:

$$t(2)/t(1)=t_{1/2}(2)/t_{1/2}(1)=\mathrm{e}^{E_{a,des}/RT_2}/\mathrm{e}^{E_{a,des}/RT_1}\ [21.23]=\mathrm{e}^{E_{a,des}(1/T_2-1/T_1)/R}$$

Solving for the activation energy for desorption, $E_{a,des}$, gives:

$$E_{a,des}=R\ln\{t(2)/t(1)\}(1/T_2-1/T_1)^{-1}$$

$$=\left(8.3145\ \mathrm{J\,K^{-1}\ mol^{-1}}\right)\times\ln\left(\frac{2.0\ \mathrm{min}}{27\ \mathrm{min}}\right)\times\left(\frac{1}{1978\ \mathrm{K}}-\frac{1}{1856\ \mathrm{K}}\right)^{-1}$$

$$=\boxed{65\bar{1}\ \mathrm{kJ\,mol^{-1}}}$$

The desorption time, t, for the same volume at temperature t is given by:

$$t=t(1)\mathrm{e}^{E_{a,des}(1/T-1/T_1)/R}$$

$$=(27\ \mathrm{min})\exp\left\{\left(65\bar{1}\times10^{3}\ \mathrm{J\,mol^{-1}}\right)\times\left(\frac{1}{T}-\frac{1}{1\,856\ \mathrm{K}}\right)/\left(8.3145\ \mathrm{J\,K^{-1}\ mol^{-1}}\right)\right\}$$

$$=(27\ \mathrm{min})\exp\left\{(78.3)\times\left(\frac{1}{T/1\,000\ \mathrm{K}}-\frac{1}{1.856}\right)\right\}.$$

(a) At 298 K, $t=\boxed{1.6\times10^{97}\ \mathrm{min}}$, which is about forever.

(b) At 3000 K, $t=\boxed{2.8\times10^{-6}\ \mathrm{min}}$

96.8(a) The average time of molecular residence is proportional to the half-life of the absorbed species and, consequently, the ratio of average residence times at two different temperatures is given by:

$$t(2)/t(1)=t_{1/2}(2)/t_{1/2}(1)=\mathrm{e}^{E_{a,des}/RT_2}/\mathrm{e}^{E_{a,des}/RT_1}\ [96.11]=\mathrm{e}^{E_{a,des}(1/T_2-1/T_1)/R}$$

Solving for the activation energy for desorption, $E_{a,des}$, gives:

$$E_{a,des}=R\ln\{t(2)/t(1)\}(1/T_2-1/T_1)^{-1}$$

$$=\left(8.3145\ \mathrm{J\,K^{-1}\ mol^{-1}}\right)\times\ln\left(\frac{3.49\ \mathrm{s}}{0.36\ \mathrm{s}}\right)\times\left(\frac{1}{2362\ \mathrm{K}}-\frac{1}{2548\ \mathrm{K}}\right)^{-1}$$

$$=\boxed{61\bar{1}\ \mathrm{kJ\,mol^{-1}}}$$

96.9(a) At 400 K: $t_{1/2} = \tau_0 e^{E_{a,des}/RT}$ [96.11] $= (0.10\ \text{ps}) \times e^{0.301 E_{a,des}/\text{kJ mol}^{-1}}$

At 1 000 K: $t_{1/2} = \tau_0 e^{E_{a,des}/RT}$ [96.11] $= (0.10\ \text{ps}) \times e^{0.120 E_{a,des}/\text{kJ mol}^{-1}}$

(a) $E_{a,des} = 15\ \text{kJ mol}^{-1}$

$t_{1/2}(400\ \text{K}) = (0.10\ \text{ps}) \times e^{0.301\times 15} = \boxed{9.1\ \text{ps}}$,

$t_{1/2}(1\,000\ \text{K}) = (0.10\ \text{ps}) \times e^{0.120\times 15} = \boxed{0.60\ \text{ps}}$

(b) $E_{a,des} = 150\ \text{kJ mol}^{-1}$

$t_{1/2}(400\ \text{K}) = (0.10\ \text{ps}) \times e^{0.301\times 150} = \boxed{4.1\times 10^6\ \text{s}}$,

$t_{1/2}(1\,000\ \text{K}) = (0.10\ \text{ps}) \times e^{0.120\times 150} = \boxed{6.6\ \mu\text{s}}$

96.10(a) Rate of desorption $= k_r\theta = \dfrac{k_r\alpha p}{1+\alpha p}$ [96.1b, 96.2]

(a) On gold, $\theta \approx 1$, and $k_r\theta \approx$ constant, a $\boxed{\text{zeroth-order}}$ reaction.

(b) On platinum, $\theta \approx \alpha p$ (as $\alpha p \ll 1$), so the rate of desorption $= k_r Kp$ and the reaction is $\boxed{\text{first-order}}$.

Problems

96.1 We use Mathcad Prime 2 to study the functional dependences of this problem. Parameters within the worksheets are easily changed to view their effect.

(a) Inversion of eqn 96.2, $\dfrac{1}{\theta} = 1 + \dfrac{1}{\alpha p}$, show that for Langmuir adsorption isotherms without dissociation $1/\theta$ is linear in $1/p$ with an intercept of 1 and slope of $1/\alpha$.

$$\theta(p,\alpha) := \frac{\alpha\cdot\frac{p}{\text{bar}}}{1+\alpha\cdot\frac{p}{\text{bar}}} \qquad N := 200 \qquad i := 1..N \qquad p_i := \frac{i}{N}\cdot 10\cdot\text{bar}$$

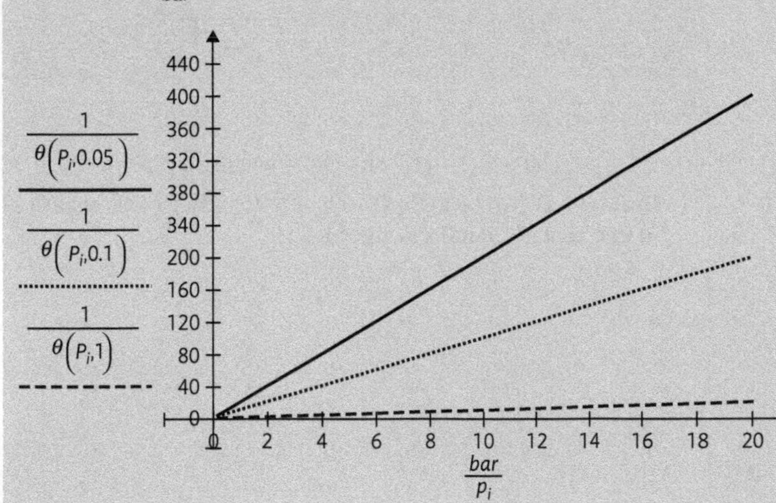

(b) Inversion of eqn 96.4, $\dfrac{1}{\theta} = 1 + \dfrac{1}{(\alpha p)^{1/2}}$, show that for Langmuir adsorption isotherms with dissociation $1/\theta$ is linear in $1/p^{1/2}$ with an intercept of 1 and slope of $1/\alpha^{1/2}$. Change the parameter "power" in the following worksheet so that it equals one, you'll see that the curves become non-linear. Thus, should an experimental data plot of $1/\theta_{\text{exp}}$ against $1/p_{\text{exp}}$ be non-linear while a data plot of $1/\theta_{\text{exp}}$ against $(1/p_{\text{exp}})^{1/2}$ is linear, we suspect dissociation.

$$\theta(p,\alpha) := \frac{\left(\alpha \cdot \frac{p}{\text{bar}}\right)^{\frac{1}{2}}}{1 + \left(\alpha \cdot \frac{p}{\text{bar}}\right)^{\frac{1}{2}}} \qquad N := 200 \qquad i := 1..N \qquad p_i := \frac{i}{N} \cdot 10 \cdot \text{bar}$$

$$\text{powers} := \frac{1}{2}$$

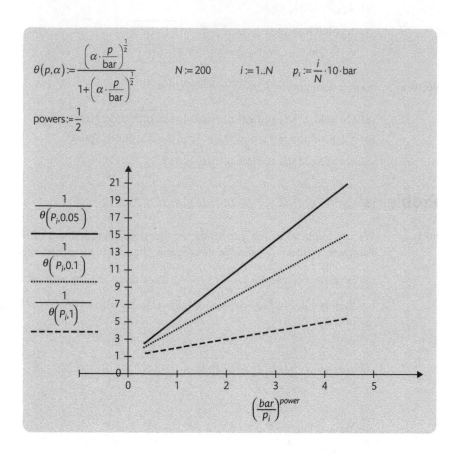

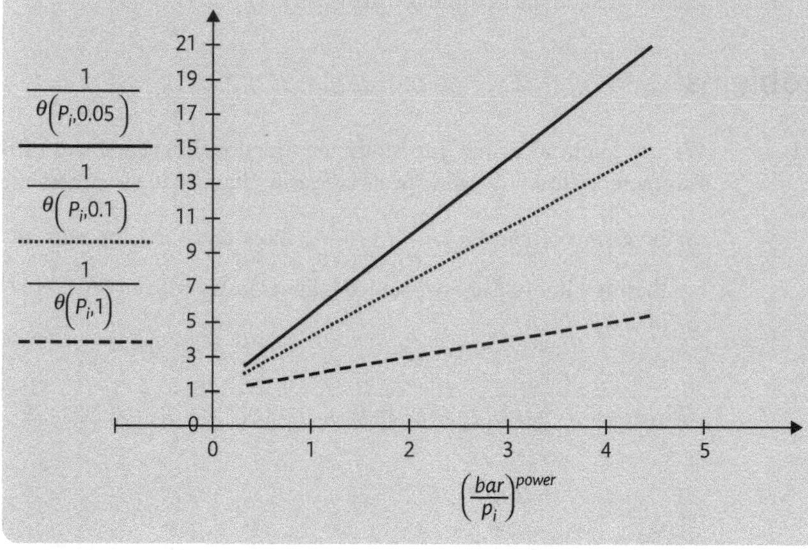

(c) $V/V_{\text{mon}} = cz/(1-z)\{1-(1-c)z\}$ [96.6] where $z = p/p^*$

Thus, $f \equiv zV_{\text{mon}}/(1-z)V = \{1-(1-c)z\}/c$ and a plot of f against z is linear with an intercept of $1/c$ and a slope of $1 - 1/c$.

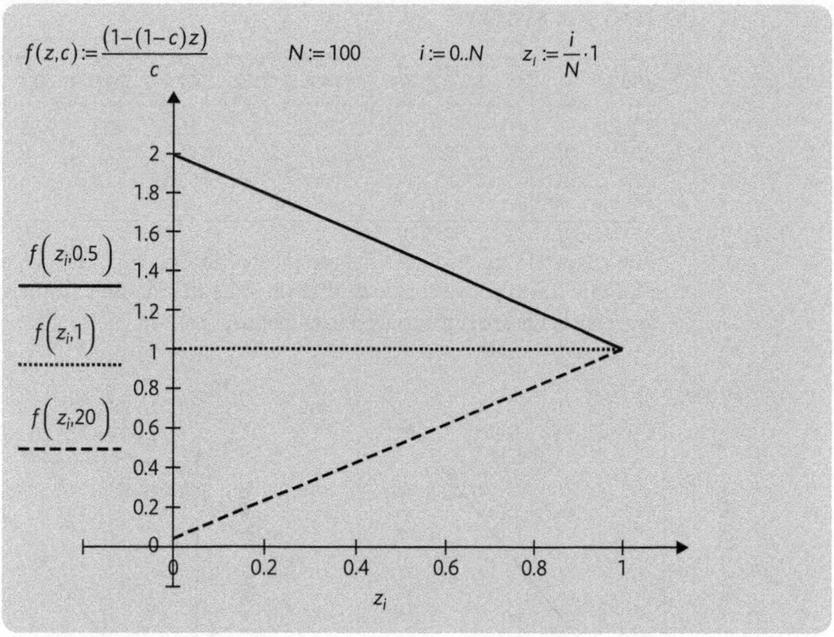

96.3 $\dfrac{V}{V_{\text{mon}}} = \dfrac{cz}{(1-z)\{1-(1-c)z\}}$ with $z = \dfrac{p}{p^*}$ [96.6]

This rearranges to

$$\frac{z}{(1-z)V} = \frac{1}{cV_{\text{mon}}} + \frac{(c-1)z}{cV_{\text{mon}}}$$

A plot of the left-hand side, $z/(1-z)V$, against z should result in a straight line if the data obeys the BET isotherm. Should it be linear, a linear regression fit of the plot yields values for the intercept and slope, which are related to c and V_{mon} by

$1/cV_{\text{mon}}$ = intercept and $(c-1)/cV_{\text{mon}}$ = slope.

Solving for c and V_{mon} yields

$c = 1 +$ slope/intercept and $V_{\text{mon}} = 1/(c \times \text{intercept})$.

We draw up the following tables:

(a) 0°C, $p^* = 429.6$ kPa

p/kPa	14.0	37.6	65.6	79.2	82.7	100.7	106.4
$10^3 z$	32.6	87.5	152.7	184.4	192.4	234.3	247.7
$\dfrac{10^3 z}{(1-z)(V/\text{cm}^3)}$	3.03	7.11	12.1	14.1	15.4	17.7	20.0

(b) 18°C, $p^* = 819.7$ kPa

p/kPa	5.3	8.4	14.4	29.2	62.1	74.0	80.1	102.0
$10^3 z$	6.5	10.2	17.6	35.6	75.8	90.3	97.8	124.4
$\dfrac{10^3 z}{(1-z)(V/\text{cm}^3)}$	0.70	1.06	1.74	3.27	6.35	7.58	8.08	10.1

The $z/(1-z)V$ against z points are plotted in Fig. 96.1. It is apparent that the plots are linear so we conclude that the data fits the BET isotherm. The linear regression fits are summarized in the figure.

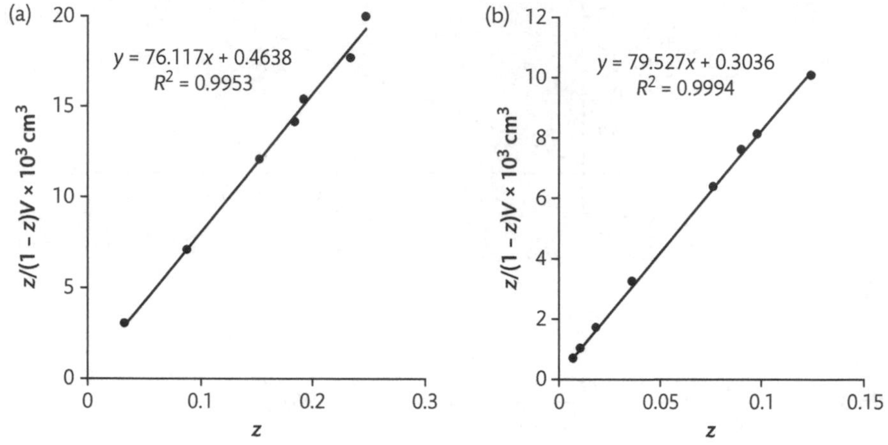

Figure 96.1

(a) intercept $= 0.4638 \times 10^{-3}\ \text{cm}^{-3}$ and slope $= 76.12 \times 10^{-3}\ \text{cm}^{-3}$

$c = 1 + \text{slope/intercept} = 1 + 76.12/0.4638 = \boxed{165}$

$V_{\text{mon}} = 1/(c \times \text{intercept}) = 1/(165 \times 0.4638 \times 10^{-3}\ \text{cm}^{-3}) = \boxed{13.1\ \text{cm}^3}$

(b) intercept $= 0.3036 \times 10^{-3}\ \text{cm}^{-3}$ and slope $= 79.53 \times 10^{-3}\ \text{cm}^{-3}$

$c = 1 + \text{slope/intercept} = 1 + 79.53/0.3036 = \boxed{263}$

$V_{\text{mon}} = 1/(c \times \text{intercept}) = 1/(263 \times 0.3036 \times 10^{-3}\ \text{cm}^{-3}) = \boxed{12.5\ \text{cm}^3}$

96.5 The Langmuir isotherm (eqn 96.2 with $\alpha = K$) is

$$\theta = \frac{Kp}{1+Kp} = \frac{n}{n_\infty} \text{ so } n(1+Kp) = n_\infty Kp \text{ and } \frac{p}{n} = \frac{p}{n_\infty} + \frac{1}{Kn_\infty}$$

So a plot of p/n against p should be a straight line with slope $1/n_\infty$ and y-intercept $1/Kn_\infty$. The transformed data and plot (Fig. 96.2) follow.

p/kPa	31.00	38.22	53.03	76.38	101.97	130.47	165.06	182.41	205.75	219.91
n/(mol kg^{-1})	1.00	1.17	1.54	2.04	2.49	2.90	3.22	3.30	3.35	3.36
$\dfrac{p/n}{\text{kPa mol}^{-1}\text{ kg}}$	31.00	32.67	34.44	37.44	40.95	44.99	51.26	55.28	61.42	65.45

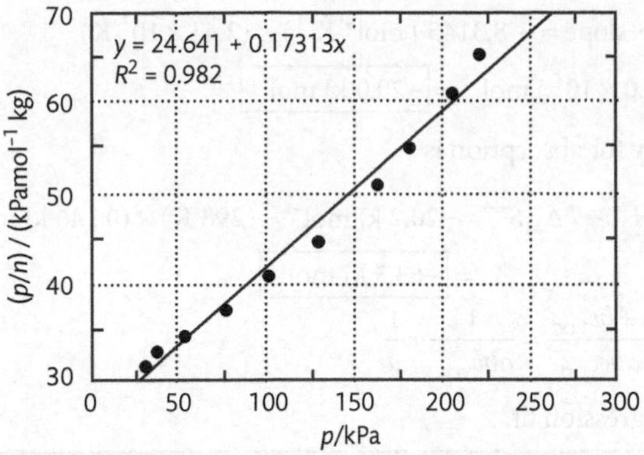

Figure 96.2

$$n_\infty = \frac{1}{0.17313\ \text{mol}^{-1}\ \text{kg}} = \boxed{5.78\ \text{mol kg}^{-1}}$$

The y-intercept is

$$b = \frac{1}{Kn_\infty} \text{ so } K = \frac{1}{bn_\infty} = \frac{1}{(24.641\ \text{kPa mol}^{-1}\ \text{kg}) \times (5.78\ \text{mol kg}^{-1})}$$

$$K = 7.02 \times 10^{-3}\ \text{kPa}^{-1} = \boxed{7.02\ \text{Pa}^{-1}}$$

96.7 Application of the van't Hoff equation (eqn 75.3) to adsorption equilibria yields

$$\left(\frac{\partial \ln K}{\partial T}\right)_\theta = \frac{\Delta_{\text{ad}} H^\ominus}{RT^2} \quad \text{or} \quad \left(\frac{\partial \ln K}{\partial (1/T)}\right)_\theta = \frac{-\Delta_{\text{ad}} H^\ominus}{R}$$

A plot (Fig. 96.3) of $\ln K$ against $1/T$ should be a straight line with slope $-\Delta_{\text{ad}} H^\ominus/R$. The transformed data and plot follow:

T/K	28.3	298	308	318
10^{-11} K	2.642	2.078	1.286	1.085
1 000 K/T	3.53	3.36	3.25	3.14
ln K	26.30	26.06	25.58	25.41

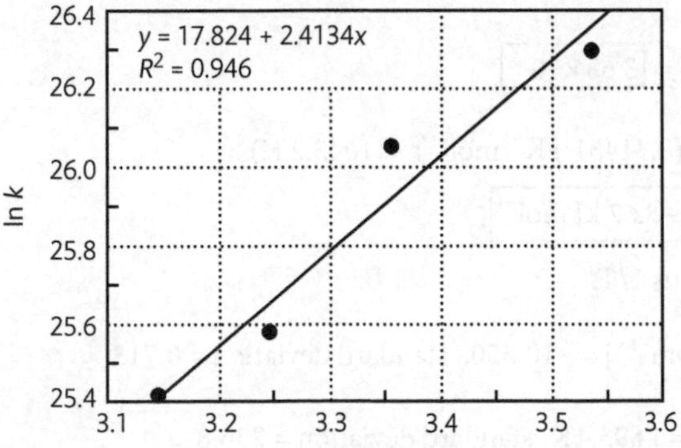

Figure 96.3

$$\Delta_{ad}H^{\ominus} = -R \times \text{slope} = -(8.3145\ \text{J mol}^{-1}\text{K}^{-1}) \times (2.41 \times 10^3\ \text{K})$$
$$= -20.0 \times 10^3\ \text{J mol}^{-1} = \boxed{-20.0\ \text{kJ mol}^{-1}}$$

The Gibbs energy for absorption is

$$\Delta_{ad}G^{\ominus} = \Delta_{ad}H^{\ominus} - T\Delta_{ad}S^{\ominus} = -20.2\ \text{kJ mol}^{-1} - (298\ \text{K}) \times (0.146\ \text{kJ mol}^{-1}\ \text{K}^{-1})$$
$$= \boxed{-63.5\ \text{kJ mol}^{-1}}$$

96.9 (a) $$\frac{1}{q_{\text{VOC,RH=0}}} = \frac{1+bc_{\text{VOC}}}{abc_{\text{VOC}}} = \frac{1}{abc_{\text{VOC}}} + \frac{1}{a}$$

Parameters of regression fit:

$\theta/°\text{C}$	$1/a$	$1/ab$	R	a	b/ppm^{-1}
33.6	9.07	709.8	0.9836	0.110	0.0128
41.5	10.14	890.4	0.9746	0.0986	0.0114
57.4	11.14	1 599	0.9943	0.0898	0.00697
76.4	13.58	2 063	0.9981	0.0736	0.00658
99	16.82	4 012	0.9916	0.0595	0.00419

The linear regression fit is generally good at all temperatures with

$\boxed{R \text{ values in the range 0.975 to 0.991}}$

(b) $$\ln a = \ln k_a - \frac{\Delta_{ad}H}{R}\frac{1}{T} \text{ and } \ln b = \ln k_b - \frac{\Delta_b H}{R}\frac{1}{T}$$

Linear regression analysis of $\ln a$ versus $1/T$ gives the intercept $\ln k_a$ and slope $-\Delta_{ad}H/R$ while a similar statement can be made for a $\ln b$ versus $1/T$ plot. The temperature must be in Kelvin.

For ln a versus $1/T$:

$$\ln k_a = -5.605, \quad \text{standard deviation} = 0.197$$

$$-\Delta_{ad}H/R = 1\,043.2\ \text{K}, \quad \text{standard deviation} = 65.4\ \text{K}$$

$$R = 0.9942 \quad [\text{good fit}]$$

$$k_a = e^{-5.605} = \boxed{3.68 \times 10^{-3}}$$

$$\Delta_{ad}H = -(8.31451\ \text{J K}^{-1}\text{mol}^{-1}) \times (1043.2\ \text{K})$$
$$= \boxed{-8.67\ \text{kJ mol}^{-1}}$$

For ln b versus $1/T$:

$$\ln\left(k_b/(\text{ppm}^{-1})\right) = -10.550, \text{ standard deviation} = 0.713$$

$$-\Delta_b H/R = 1\,895.4\ \text{K}, \text{ standard deviation} = 236.8$$

$$R = 0.9774 \quad \text{[good fit]}$$

$$k_b = e^{-10.550}\ \text{ppm}^{-1} = \boxed{2.62 \times 10^{-5}\ \text{ppm}^{-1}}$$

$$\Delta_b H = -(8.31451\ \text{J K}^{-1}\text{mol}^{-1}) \times (1\,895.4\ \text{K})$$

$$\boxed{\Delta_b H = -15.7\ \text{kJ mol}^{-1}}$$

(c) k_a may be interpreted to be the maximum adsorption capacity at an adsorption enthalpy of zero, while k_b is the maximum affinity in the case for which the adsorbant–surface bonding enthalpy is zero.

Topic 97 Heterogeneous catalysis

Discussion questions

97.1 In the Langmuir–Hinshelwood mechanism of surface-catalyzed reactions, the reaction takes place by encounters between molecular fragments and atoms already adsorbed on the surface. We therefore expect the rate law to be second-order in the extent of surface coverage:

$$A+B\rightarrow P \qquad \upsilon = k_r\theta_A\theta_B \quad [97.2]$$

Insertion of the appropriate isotherms for A and B then gives the reaction rate in terms of the partial pressures of the reactants. For example, if A and B follow Langmuir isotherms (eqn 96.2), and adsorb without dissociation, then it follows that the rate law is

$$\upsilon = \frac{k_r\alpha_A\alpha_B p_A p_B}{(1+\alpha_A p_A+\alpha_B p_B)^2} \quad [97.4]$$

The parameters α in the isotherms and the rate constant k_r are all temperature dependent, so the overall temperature dependence of the rate may be strongly non-Arrhenius (in the sense that the reaction rate is unlikely to be proportional to $\exp(-E_a/RT)$.

In the Eley-Rideal mechanism (ER mechanism) of a surface-catalysed reaction, a gas phase molecule collides with another molecule already adsorbed on the surface. The rate of formation of product is expected to be proportional to the partial pressure, p_B of the non-adsorbed gas B and the extent of surface coverage, θ_A, of the adsorbed gas A. It follows that the rate law should be

$$A+B\rightarrow P \qquad \upsilon = k_r p_B\theta_A \quad [97.5]$$

The rate constant, k, might be much larger than for the uncatalysed gas-phase reaction because the reaction on the surface has a low activation energy and the adsorption itself is often not activated.

If we know the adsorption isotherm for A, we can express the rate law in terms of its partial pressure, p_A. For example, if the adsorption of A follows a Langmuir isotherm in the pressure range of interest, then the rate law would be

$$\upsilon = \frac{k_r\alpha p_A p_B}{1+\alpha p_A} \quad [97.6]$$

If A were a diatomic molecule that adsorbed as atoms, we would substitute the isotherm given in eqn 96.4 instead.

According to eqn 97.6, when the partial pressure of A is high (in the sense $\alpha p_A \gg 1$), there is almost complete surface coverage, and the rate is equal to $k_r p_B$. Now the rate-determining step is the collision of B with the adsorbed fragments. When the pressure of A is low ($\beta p_A \ll 1$), perhaps because of its reaction, the rate is equal to $k_r \alpha p_A p_B$. Now the extent of surface coverage is important in the determination of the rate.

Exercises

97.1(a) Let us assume that the nitrogen molecules are close-packed, as shown in Fig. 97.1 as spheres, in the monolayer. Then, one molecule occupies the parallelogram area of $2\sqrt{3}\,r^2$ where r is the radius of the adsorbed molecule. Furthermore, let us assume that the collision cross-section of Table 78.1 ($\sigma = 0.43\ \text{nm}^2 = 4\pi r^2$) gives a reasonable estimate of r: $r = (\sigma/4\pi)^{1/2}$. With these assumptions the surface area occupied by one molecule is:

$$A_{\text{molecule}} = 2\sqrt{3}\,(\sigma/4\pi) = \sqrt{3}\,\sigma/2\pi$$
$$= \sqrt{3}\,(0.43\ \text{nm}^2)/2\pi = 0.12\ \text{nm}^2$$

In this model the surface area per gram of the catalyst equals $A_{\text{molecule}}N$ where N is the number of adsorbed molecules. N can be calculated with the 0° C data, a temperature that is so high compared to the boiling point of nitrogen that all molecules are likely to be desorbed from the surface as perfect gas.

$$N = \frac{pV}{kT} = \frac{(760\ \text{Torr})\times(133.3\ \text{Pa/Torr})\times(3.86\times10^{-6}\ \text{m}^3)}{(1.381\times10^{-23}\ \text{J K}^{-1})\times(273.15\ \text{K})} = 1.04\times10^{20}$$

$$A_{\text{molecule}}N = (0.12\times10^{-18}\ \text{m}^2)\times(1.04\times10^{20}) = \boxed{12\ \text{m}^2}$$

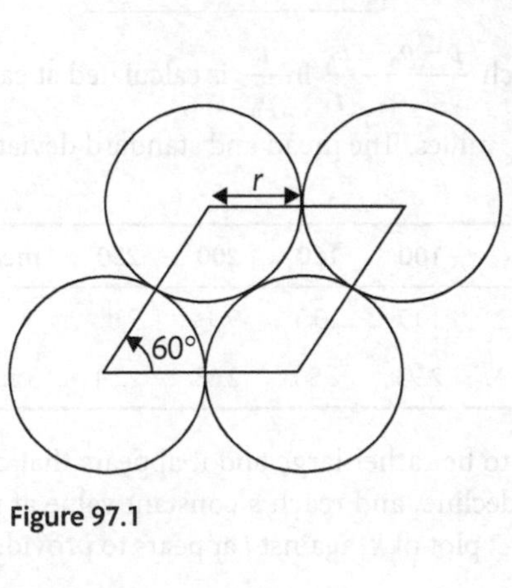

Figure 97.1

Problem

97.1 Proposed mechanism:

$$NH_3(g)+M_{surface} \xrightarrow{k_a} NH_3\cdot M_{surface} \rightarrow \xrightarrow{\text{rapid}} \rightarrow \tfrac{1}{2}N_2(g)$$
$$+\tfrac{3}{2}H_2\cdot M_{surface} \xrightleftharpoons{K=k_{a,H_2}/k_{d,H_2}} \tfrac{3}{2}H_2(g)+M_{surface}$$

This mechanism postulates that the decomposition rate is proportional both to p_{NH_3} and to the fraction of adsorption sites that remain unoccupied by the strongly adsorbed H_2, $1-\theta$. We assume that gaseous hydrogen and adsorbed hydrogen are in equilibrium at all $t \geq 30$ s. Then,

$$\theta=\frac{Kp_{H_2}}{1+Kp_{H_2}} \ [96.2, \alpha=K=k_{a,H_2}/k_{d,H_2}] \ \text{and} \ 1-\theta=\frac{1}{1+Kp_{H_2}}$$

For a strongly adsorbed hydrogen species, $Kp_{H_2} \gg 1$ and $1-\theta=1/Kp_{H_2}$. Since the reaction rate is proportional to the pressure of ammonia and the fraction of sites left uncovered by the strongly adsorbed hydrogen product, we can write

$$\frac{dp_{NH_3}}{dt}=-k_r p_{NH_3}(1-\theta)\approx \boxed{-\frac{k_r}{K}\frac{p_{NH_3}}{p_{H_2}}}$$

According to the reaction stoichiometry

$$p_{H_2}=\tfrac{3}{2}\{p_{0,NH_3}-p_{NH_3}\} \ [NH_3 \rightarrow \tfrac{1}{2}N_2+\tfrac{3}{2}H_2]$$

from which it follows that, with $p=p_{NH_3}$

$$\frac{-dp}{dt}=\frac{k_c p}{p_0-p} \quad \text{where} \quad k_c=\frac{2k_r}{3K}$$

This equation integrates as follows.

$$\int_{p_0}^{p}\left(1-\frac{p_0}{p}\right)dp=k_c\int_0^t dt \quad \text{so} \quad \boxed{k_c=\frac{p-p_0}{t}-\frac{p_0}{t}\ln\frac{p}{p_0}}$$

Thus, we prepare a table in which $\frac{p-p_0}{t}-\frac{p_0}{t}\ln\frac{p}{p_0}$ is calculated at each data pair and examine these calculated k_c values. The mean and standard deviation are also reported in the table.

t/s	0	30	60	100	160	200	250	mean	std. dev.
p_{NH_3}/kPa	13.3	11.7	11.2	10.7	10.3	9.9	9.6		
$k_c/10^{-3}$ kPa s^{-1}		3.49	3.09	2.93	2.50	2.63	2.54	3.02	0.67

The standard deviation is seen to be rather large and it appears that calculated k_c values are high at early times, decline, and reach a constant value at times larger than about 160 s. The Figure 97.2 plot of k_c against t appears to provide confirm of this trend so we accept the average of large time values as the best approximation of k_c:

$$\boxed{k_c = 2.5\times10^{-3}\ \text{kPa s}^{-1}}$$

The variation of k_c values may indicate that equilibrium between gaseous and adsorbed hydrogen is not achieved until the latter times.

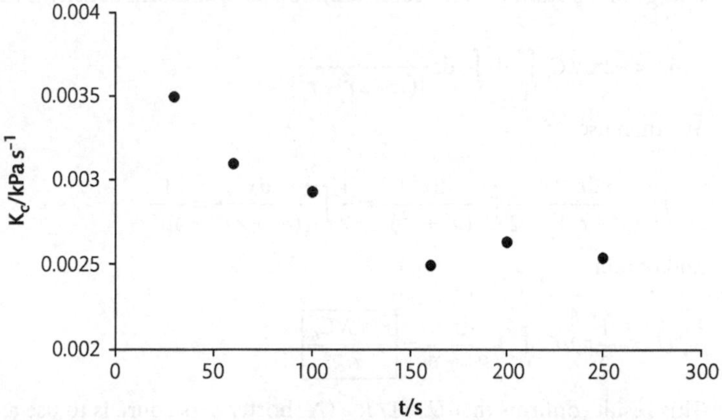

Figure 97.2

Focus 20: Integrated activities

F20.1 Refer to Fig. 97.3.

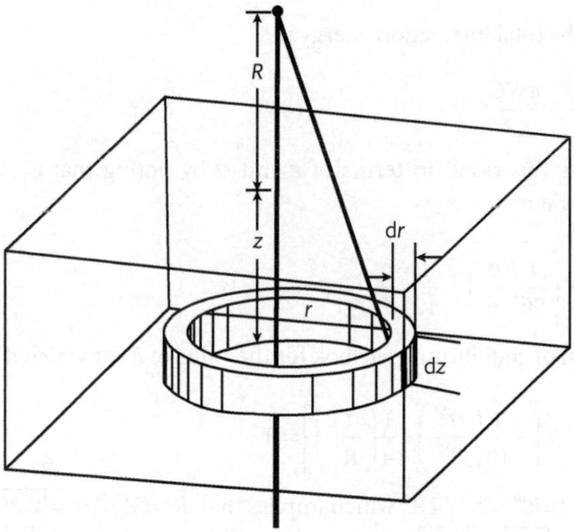

Figure 97.3

Let the number density of atoms in the solid be $\mathcal{N}$. Then the number in the annulus between r and $r + \mathrm{d}r$ and thickness $\mathrm{d}z$ at a depth z below the surface is $2\pi\mathcal{N}r\,\mathrm{d}r\,\mathrm{d}z$.

The interaction energy of these atoms and the single adsorbate atom at a height R above the surface is

$$dU = \frac{-2\pi\mathcal{N} r\,dr\,dz\,C_6}{\{(R+z)^2+r^2\}^3}$$

if the individual atoms interact as $-C_6/d^6$ with $d^2=(R+z)^2+r^2$. The total interaction energy of the atom with the semi-infinite slab of uniform density is therefore

$$U = -2\pi\mathcal{N}C_6\int_0^\infty dr\int_0^\infty dz\frac{r}{\{(R+z)^2+r^2\}^3}.$$

We then use

$$\int_0^\infty \frac{r\,dr}{(a^2+r^2)^3} = \frac{1}{2}\int_0^\infty \frac{d(r^2)}{(a^2+r^2)^3} = \frac{1}{2}\int_0^\infty \frac{dx}{(a^2+x)^3} = \frac{1}{4a^4}$$

and obtain

$$U = -\frac{1}{2}\pi\mathcal{N}C_6\int_0^\infty \frac{dz}{(R+z)^4} = \boxed{-\frac{\pi\mathcal{N}C_6}{6R^3}}.$$

This result confirms that $U \propto 1/R^3$. (A shorter procedure is to use a dimensional argument, but we need the explicit expression in the following.) When

$$V = 4\varepsilon\left[\left(\frac{\sigma}{R}\right)^{12} - \left(\frac{\sigma}{R}\right)^6\right] = \frac{C_{12}}{R^{12}} - \frac{C_6}{R^6}$$

we also need the contribution from C_{12}

$$U' = 2\pi\mathcal{N}C_{12}\int_0^\infty dr\int_0^\infty dz\frac{r}{\{(R+z)^2+r^2\}^6} = 2\pi\mathcal{N}C_{12}\times\frac{1}{10}\int_0^\infty \frac{dz}{(R+z)^{10}} = \frac{2\pi\mathcal{N}C_{12}}{90R^9}$$

and therefore the total interaction energy is

$$U = \frac{2\pi\mathcal{N}C_{12}}{90R^9} - \frac{\pi\mathcal{N}C_6}{6R^3}$$

We can express this result in terms of ε and σ by noting that $C_{12} = 4\varepsilon\sigma^{12}$ and $C_6 = 4\varepsilon\sigma^6$, for then

$$U = 8\pi\varepsilon\sigma^3\mathcal{N}\left[\frac{1}{90}\left(\frac{\sigma}{R}\right)^9 - \frac{1}{12}\left(\frac{\sigma}{R}\right)^3\right]$$

For the position of equilibrium, we look for the value of R for which $dU/dR = 0$.

$$\frac{dU}{dR} = 8\pi\varepsilon\sigma^3\mathcal{N}\left[-\frac{1}{10}\left(\frac{\sigma^9}{R^{10}}\right) + \frac{1}{4}\left(\frac{\sigma^3}{R^4}\right)\right] = 0$$

Therefore, $\sigma^9/10R^{10} = \sigma^3/4R^4$ which implies that $R = (2/5)^{1/6}\sigma = \boxed{0.858\,\sigma}$. For $\sigma = 342$ pm, $\boxed{R \approx 294\text{ pm}}$.

F20.3 The Coulombic force is

$$F = -\frac{dV}{dr} = -\frac{d}{dr}\left(\frac{Q_1Q_2}{4\pi\varepsilon_0 r}\right) = \frac{Q_1Q_2}{4\pi\varepsilon_0 r^2}$$

$$= \frac{\left(1.602\times10^{-19}\ \text{C}\right)^2}{\left(1.113\times10^{-10}\ \text{J}^{-1}\ \text{C}^2\ \text{m}^{-1}\right)\times\left(2.00\times10^{-9}\ \text{m}\right)^2}$$

$$= \boxed{5.77\times10^{-11}\ \text{N}}$$